INSTRUCTOR'S EDITION

FOURTH EDITION

Linear Algebra
and Its Applications

David C. Lay

University of Maryland—College Park

Addison-Wesley

Boston Columbus Indianapolis New York San Francisco Upper Saddle River
Amsterdam Cape Town Dubai London Madrid Milan Munich Paris Montréal Toronto
Delhi Mexico City São Paulo Sydney Hong Kong Seoul Singapore Taipei Tokyo

Editor-in-Chief: *Deirdre Lynch*
Senior Acquisitions Editor: *William Hoffmann*
Sponsoring Editor: *Caroline Celano*
Senior Content Editor: *Chere Bemelmans*
Editorial Assistant: *Brandon Rawnsley*
Senior Managing Editor: *Karen Wernholm*
Associate Managing Editor: *Tamela Ambush*
Digital Assets Manager: *Marianne Groth*
Supplements Production Coordinator: *Kerri McQueen*
Senior Media Producer: *Carl Cottrell*
QA Manager, Assessment Content: *Marty Wright*
Executive Marketing Manager: *Jeff Weidenaar*
Marketing Assistant: *Kendra Bassi*
Senior Author Support/Technology Specialist: *Joe Vetere*
Rights and Permissions Advisor: *Michael Joyce*
Image Manager: *Rachel Youdelman*
Senior Manufacturing Buyer: *Carol Melville*
Senior Media Buyer: *Ginny Michaud*
Design Manager: *Andrea Nix*
Senior Designer: *Beth Paquin*
Text Design: *Andrea Nix*
Production Coordination: *Tamela Ambush*
Composition: *Dennis Kletzing*
Illustrations: *Scientific Illustrators*

Cover Design: *Nancy Goulet, Studiowink*
Cover Image: *Shoula/Stone/Getty Images*

For permission to use copyrighted material, grateful acknowledgment is made to the copyright holders on page P1, which is hereby made part of this copyright page.

Many of the designations used by manufacturers and sellers to distinguish their products are claimed as trademarks. Where those designations appear in this book, and Pearson Education was aware of a trademark claim, the designations have been printed in initial caps or all caps.

Library of Congress Cataloging-in-Publication Data

Lay, David C.
    Linear algebra and its applications / David C. Lay. – 4th ed. update.
        p. cm.
    Includes index.
    ISBN-13: 978-0-321-38517-8   ISBN-13: 978-0-321-38518-5 (Instructor's Edition)
    ISBN-10: 0-321-38517-9   ISBN-10: 0-321-38518-7 (Instructor's Edition)
    1. Algebras, Linear–Textbooks. I. Title.
QA184.2.L39 2012
512′.5–dc22
                                    2010048460

Copyright © 2012, 2006, 1997, 1994 Pearson Education, Inc.

1 2 3 4 5 6 7 8 9 10—DOW—14 13 12 11 10

Addison-Wesley
is an imprint of

www.pearsonhighered.com

ISBN 13: 978-0-321-38517-8
ISBN 10: 0-321-38517-9

To my wife, Lillian, and our children, Christina, Deborah, and Melissa, whose support, encouragement, and faithful prayers made this book possible.

About the Author

David C. Lay holds a B.A. from Aurora University (Illinois), and an M.A. and Ph.D. from the University of California at Los Angeles. Lay has been an educator and research mathematician since 1966, mostly at the University of Maryland, College Park. He has also served as a visiting professor at the University of Amsterdam, the Free University in Amsterdam, and the University of Kaiserslautern, Germany. He has published more than 30 research articles on functional analysis and linear algebra.

As a founding member of the NSF-sponsored Linear Algebra Curriculum Study Group, Lay has been a leader in the current movement to modernize the linear algebra curriculum. Lay is also a co-author of several mathematics texts, including *Introduction to Functional Analysis* with Angus E. Taylor, *Calculus and Its Applications*, with L. J. Goldstein and D. I. Schneider, and *Linear Algebra Gems—Assets for Undergraduate Mathematics,* with D. Carlson, C. R. Johnson, and A. D. Porter.

Professor Lay has received four university awards for teaching excellence, including, in 1996, the title of Distinguished Scholar–Teacher of the University of Maryland. In 1994, he was given one of the Mathematical Association of America's Awards for Distinguished College or University Teaching of Mathematics. He has been elected by the university students to membership in Alpha Lambda Delta National Scholastic Honor Society and Golden Key National Honor Society. In 1989, Aurora University conferred on him the Outstanding Alumnus award. Lay is a member of the American Mathematical Society, the Canadian Mathematical Society, the International Linear Algebra Society, the Mathematical Association of America, Sigma Xi, and the Society for Industrial and Applied Mathematics. Since 1992, he has served several terms on the national board of the Association of Christians in the Mathematical Sciences.

Contents

Appendixes

Notes to the Instructor

These notes, which are not included in the student edition, offer some information that may help you organize your course. For instance, the notes show how to get to eigenvalues after five weeks, if you wish. Also described are new features of the *Fourth Edition*, including extensive web resources for your course, and class-tested material for a second course in linear algebra—available at *www.pearsonhighered.com/lay*.

INSTRUCTIONAL AIDS

The following resources are provided to make your linear algebra course more enjoyable for you and your students. The *Study Guide* certainly saved me time and made my teaching more effective.

PowerPoint Slides

A brisk pace at the beginning of the course helps to set the tone for the course. To get quickly through the first two sections in less than two lectures, consider using PowerPoint® slides. They permit you to focus on the process of row reduction rather than to write many numbers on the board. Students can receive a condensed version of the notes, with occasional blanks to fill in during the lecture. (Many students respond favorably to this gesture.) The PowerPoint slides are available for 25 core sections of the text. In addition, about 75 color figures from the text are available as PowerPoint slides. The PowerPoint slides are available for download at *www.pearsonhighered.com/irc*

Study Guide

The challenge of teaching linear algebra is heightened by the need to teach students how to learn mathematics. The *Study Guide* is designed to help you do this. It is available as a paperback (ISBN 0-321-38883-6). The *Guide* will reinforce many comments you are likely to make in class about the importance of reading the text carefully, working the exercises promptly, and the like. Appendixes in some sections explain mathematical concepts such as induction and logical implication.

The *Study Guide* can save you time in class. You can refer students to some solutions in the *Guide* instead of working all the exercises in class. The *Guide* also answers many questions that students might otherwise ask during office hours. For courses using technology, the *Guide* is a "lab manual" that gives keystroke instructions and command references for the most popular matrix programs.

The *Guide* has two unique pedagogical features:

1. The *Study Guide* supplies detailed solutions to every third odd-numbered problem (which includes most key exercises) and a solution to every odd-numbered writing problem for which the text's answer is only a *Hint*. A *Hint* shows students that they must work harder on a problem before reading the solution. (The *Study Guide* is separate from the text to prevent quick access to the solution.) Nearly 100 conceptual exercises have this design.

2. A series of subsections in the *Study Guide*, entitled "Mastering Linear Algebra Concepts," shows students how to make review sheets that identify connections between concepts, in a way that builds a robust mental "image" of each concept. Such mental images are a key to true comprehension and retention of ideas.

The *Fourth Edition* has an icon ⌐ SG ⌐ at appropriate points that directs the student to go to a "Mastering" subsection. An instructor can train students to make effective review sheets by requiring that each student submit review sheets, at least for the first two or three key concepts. In my courses, I glance at each sheet and add notes when key features are missing. I do not grade the sheets other than to record that an effort was made. If review sheets are not required, the students who need them most may never create them. In an anonymous survey, most of my students wrote that they really liked the review sheets and made additional sheets as the course progressed.

Technology Manuals

Instructor's *Technology Manuals* provide valuable help for faculty at all levels of technical expertise. They are all written by faculty specifically for this book. The writers share their experiences setting up the course, introducing students to technology, testing, and integrating technology projects that complement and extend the text's **[M]** exercises. Each manual also includes class-tested projects ready for duplication. The data for the projects can be accessed through *www.pearsonhighered.com/lay*.

The manuals are available to qualified instructors through the Pearson Instructor Resource Center (*www.pearsonhighered.com/irc*). MATLAB® (ISBN 0-321-53365-8), Maple™ (ISBN 0-321-75605-3), Mathematica® (ISBN 0-321-38885-2), and the TI-83+/86/89 (ISBN 0-321-38887-9).

Instructor's Solutions Manual

The Instructor's Solutions Manual manual (ISBN 0-321-38888-7) contains detailed solutions for all exercises, along with teaching notes for many sections.

Getting Started with Technology

If your course includes some work with MATLAB, Maple, Mathematica, or TI calculators, you can use one of the projects available at *www.pearsonhighered.com/lay* as the first introduction to the technology. The projects are referenced by the WEB icon on text page 90, but you can assign one whenever you wish to start the technology. Also, the *Study Guide* provides introductory material for first-time users.

MATLAB Projects

These exploratory projects invite students to discover basic mathematical and numerical issues in linear algebra. Written by Rick Smith, they were developed to accompany a computational linear algebra course at the University of Florida, which has used *Linear Algebra and Its Applications* for many years. The projects are referenced by an icon ⌐ **WEB** ⌐ at appropriate points in the text. About half of the projects explore fundamental concepts such as the column space, diagonalization, and orthogonal projections; several projects focus on numerical issues such as flops, iterative methods, and the SVD; and a few projects explore applications such as Lagrange interpolation and Markov chains. These projects are available at *www.pearsonhighered.com/lay*.

CHAPTERS ON THE WEB

Chapters 9 and 10, listed in the Table of Contents and posted online, are designed to be part of a second course in linear algebra that focuses on mathematics and applications of interest to students in mathematics, computer science, and engineering. These chapters offer opportunities for greater depth and breadth, in both the mathematics and the applications, than can be achieved in a single course.

For about five years, faculty at the University of Maryland and at several other schools have used my text nearly every semester (in the summer) as part of a second course—beginning with roughly three weeks review of basic linear algebra facts before starting the fresh material. The new material consists of Chapters 8, 9, and 10, which have been available online from Addison-Wesley. Chapters 8 and 9 were written and class-tested by my brother, Steven R. Lay, at Lee University (Cleveland, Tennessee). Chapter 10 was written and class tested by Thomas Polaski (Winthrop University, Rock Hill, South Carolina). I am indebted to both Steven and Tom for providing such interesting material.

For this edition, I have moved Chapter 8 to the main text, to give instructors more options for applications in the first course, and I recommend sections 1, 2, 3, and 6 of Chapter 8 as possibilities. This also gives instructors a chance to try out this new material before considering a second semester course. I would be delighted to receive comments and suggestions from anyone who reads or teaches this material. The two chapters are posted as PDF files on *www.pearsonhighered.com/irc*.

DESIGNING THE COURSE

The text is written for a one-semester course meeting three or four times per week. Many schools have adapted the text for use in a shorter course, and changes in this edition reflect suggestions from faculty teaching such courses. The topical organization table on the next page shows how the text can accommodate a variety of courses. The Core Topics listed there are likely to be included in most courses. Some of the Supplementary Topics will be considered as core topics by some instructors, but they are optional in the technical sense that subsequent sections make no essential use of them. All Applications are optional. A section-by-section commentary later in this note identifies connections between various sections. The balance between supplementary topics and applications will vary according to your own course objectives.

Extensive class testing for more than a decade has verified that each core section and supplementary section can usually be covered in one 50-minute period, with time for some discussion of homework. (I usually assign 15–20 of the odd-numbered problems, and students take from one to two hours on their homework.) To stay within this timeline, instructors can focus on the main points of each section and allow students to supplement lectures through careful reading of the text. (Many instructors have written that their students find the text easy to read.) Longer application sections have subsections that can be covered in one period or less. Some instructors plan an additional day per chapter to allow more in-depth coverage of selected topics.

Topical Organization Of *Linear Algebra and Its Applications*

	Core Topics		Supplementary Topics		Applications
1.	1.1	Systems of Linear Equations			
2.	1.2	Row Reduction and Echelon Forms			
3.	1.3	Vector Equations			
4.	1.4	The Matrix Equation $A\mathbf{x} = \mathbf{b}$			
5.	1.5	Solution Sets of Linear Systems		1.6	Applications of Linear Systems
6.	1.7	Linear Independence			
7.	1.8	Linear Transformations	1.9	The Matrix of a Linear Transformation	1.10	Linear Models in Business, Science, and Engineering
8.	2.1	Matrix Operations			
9.	2.2	Inverse of a Matrix			
10.	2.3	Characterizations of Invertible Matrices	2.4	Partitioned Matrices		
			2.5	Matrix Factorizations	2.6	The Leontief Input–Output Model
					2.7	Computer Graphics
			2.8	Subspaces of $\mathbb{R}^n$		
			2.9	Dimension and Rank (2.8 and 2.9 replace Chapter 4)		
			3.1	Introduction to Determinants		
			3.2	Properties of Determinants	3.3	Cramer's Rule, Volume, Linear Transformations
11.	4.1	Vector Spaces and Subspaces			
12.	4.2	Null Spaces, Column Spaces, and Linear Transformations			
13.	4.3	Linearly Independent Sets; Bases			
14.	4.4	Coordinate Systems			
15.	4.5	Dimension of a Vector Space			
16.	4.6	Rank	4.7	Change of Basis	4.8	Difference Equations
					4.9	Markov Chains
17.	5.1	Eigenvectors and Eigenvalues			
18.	5.2	The Characteristic Equation			
19.	5.3	Diagonalization	5.4	Eigenvectors and Linear Transformations		
			5.5	Complex Eigenvalues	5.6	Discrete Dynamical Systems
					5.7	Applications to Differential Equations
			5.8	Iterative Estimates for Eigenvalues		
20.	6.1	Inner Product, Length, and Orthogonality			
21.	6.2	Orthogonal Sets			
22.	6.3	Orthogonal Projections	6.4	Gram–Schmidt Process		
23.	6.5	Least-Squares Problems			6.6	Applications to Linear Models
			6.7	Inner Product Spaces	6.8	Applications of Inner Product Spaces
24.	7.1	Diagonalization of Symmetric Matrices			
25.	7.2	Quadratic Forms	7.3	Constrained Optimization		
			7.4	The Singular Value Decomposition	7.5	Applications to Image Processing and Statistics
			8.1	Affine Combinations		
			8.2	Affine Independence		
			8.3	Convex Combinations		
					8.4	Hyperplanes
					8.5	Polytopes
					8.6	Curves and Surfaces

SAMPLE SYLLABI

The first three syllabi are for semester-length courses, with at least 39 lecture periods of 50 minutes, including time for homework discussions and occasional quizzes, but not counting days for hour exams or the final examination. I have class-tested the text many times in each of these courses. The 39-day schedule is based on a 15-week semester, with three or four exams and two or three days' slack time, to allow for unplanned delays or digressions that often occur when a textbook is used for the first time. Extra review days, if any, should be planned within the 39-day schedule.

The fourth syllabus, for short courses, describes how to use new Sections 2.8 and 2.9 to cover six core topics in a week or less and get quickly to eigenvalues. The final syllabus shows how to incorporate the topics recommended by the Linear Algebra Curriculum Study Group into a semester course.

> **Course 1.** A 39-lecture class for freshman and sophomores, with a wide variety of majors.

Such a course could include all twenty-five core topics shown in the table on page xii and perhaps seven of the following supplementary topics:

1.9 The Matrix of a Linear Transformation

2.4 Partitioned Matrices

2.5 Matrix Factorizations

3.1 Introduction to Determinants

3.2 Properties of Determinants

4.7 Change of Basis

5.4 Eigenvectors and Linear Transformations

5.5 Complex Eigenvalues

6.4 Gram–Schmidt Process and QR Factorization

In addition to these 32 class days, you can probably fit in four or more applications and still have three review days. I recommend one application from each of Chapters 1, 2, 4, and 5. They are spread out to provide welcome relief from the theory. Early introduction of applications seems especially beneficial in a course for younger students.

Some courses, such as our course at the University of Maryland, are four credits instead of three. In this case, another five or more supplementary and application sections can be covered.

> **Course 2.** A 39-lecture course for juniors who have already seen determinants and had brief glimpses of matrix algebra in other courses and have had multivariable calculus.

This course can go a little faster through Sections 1.1–1.3, covering all three sections in two days, with students reading a lot on their own. (Later, the class will also speed through Section 6.1) Early applications are not as important for students who are already aware that linear algebra is useful. Also, Chapter 3 can be skipped entirely and the brief treatment in Section 5.2 used instead. As a result, the twenty-five core sections can be covered in 24 days, and perhaps nine of the supplementary sections can be covered, still leaving time for four or five days on the applications in Chapters

xiv   Notes to the Instructor

4–7. Section 5.7 (Applications to Differential Equations) was written with this upper-level course in mind. In a 14-week course, I usually get as far as the singular value decomposition (Section 7.4).

Course 3. A 39-lecture course for juniors and seniors in the social and management sciences who are aiming at graduate school and will likely take a course later in multivariate statistics.

This course covers all core sections. In Chapter 4, you should focus on $\mathbb{R}^n$, with some work on $\mathbb{P}_n$. For supplementary topics, you might select partitioned matrices (Section 2.4), inner product (Examples 1–6 of Section 6.7), constrained optimization (Section 7.3), and perhaps one or two other sections. That should leave room for applications from Sections 1.10 (Examples 1 and 3), 2.6, 4.9, 6.6, 6.8 (Examples 1 and 2), and 7.5, plus two days for review.

Course 4. A 27-lecture short course.

Short courses typically focus on linear algebra in $\mathbb{R}^n$, with perhaps only a brief mention of more general vector spaces. Such courses can take advantage of Sections 2.8 and 2.9. All vector space concepts needed for Chapters 5–7 can be covered in two to three days with the sections. The basic schedule is to cover the first ten core sections, insert Sections 2.8 and 2.9, and then cover core sections 5.1–5.3, 6.1, 6.2, 7.1, and 7.2. Total time: 19–22 days (not counting exams). There should be time to choose a few supplementary topics and one or two applications. (According to users, the application with the broadest appeal is Section 2.7, Computer Graphics.)

A single lecture on general vector spaces could be added at the end of the short course. If you wish to cover the six core sections of Chapter 4 instead of Sections 2.8 and 2.9, you will need about three extra days. Omit most of Section 4.4 (except perhaps Theorem 7), and treat Section 4.5 lightly.

Course 5. A course that covers the basic syllabus recommended by the Linear Algebra Curriculum Study Group (LACSG).

The LACSG "core" topics are all included in any course that covers all twenty-five topics listed in this text as core topics, plus the following sections:

  2.4 Partitioned Matrices

  2.5 Matrix Factorizations

  3.1 Introduction to Determinants

  3.2 Properties of Determinants (plus Theorem 7 and Example 1 from Section 3.3)

  6.4 The Gram–Schmidt Process

  6.6 Applications to Linear Models (Example 1 only)

This 31-lecture schedule is about three days longer than the LACSG time estimates, but the 31 days include time for linear transformations, general vector spaces, and positive definite matrices. Another five or six sections can be selected to fill out some of the supplementary topics and applications suggested by the LACSG. If you are using technology, Section 5.8 would be a natural choice, or you could spend extra time on special projects.

SECTION COMMENTARIES

The commentaries below point out connections between a given section and those that follow it. Obvious connections are usually not mentioned. Some subsections of core material are identified as optional when their omission will cause no serious problem later. The commentaries also list exercises that are particularly valuable pedagogically, anticipate future discussions, or treat an optional supplementary topic.

In each core section, I urge you to assign one of the sets of True/False questions. Students will probably spend from 5 to 15 minutes on the questions because they will have to read the text. That is the point of these exercises! Many instructors have used them enthusiastically for several years.

SECTION 1.1
Systems of Linear Equations

This section and the next form the foundation for the row reduction algorithm used throughout the text. The existence and uniqueness questions introduced here are of fundamental importance and appear at several points later in the text. The exercises focus only on the existence question. Systems with many solutions are treated in Section 1.2.

Key Exercises: 7, 19–22, 25. The heat flow problem in Exercise 33 is also discussed in exercises in Sections 1.10 and 2.5.

SECTION 1.2
Row Reduction and Echelon Forms

The pivot positions in a matrix will play an essential role in several theorems — on spanning, linear independence, invertibility, the determinant, and rank. Theorem 1 is fundamental. Example 3 is helpful but need not be presented in class. The parametric descriptions of solution sets are written in vector form in Section 1.5. The solution method is reviewed in Section 4.2 and is used throughout Chapters 5 and 7.

Key Exercises: 1–20, 23–28. Students should work at least four or five of Exercises 7–14 in preparation for Section 1.5. Interpolating polynomials in Exercises 33 and 34 will appear again in exercises at the end of Chapter 2.

SECTION 1.2
Row Reduction and Echelon Forms

The pivot positions in a matrix will play an essential role in several theorems — on spanning, linear independence, invertibility, the determinant, and rank. Theorem 1 is fundamental. Example 3 is helpful but need not be presented in class. The parametric descriptions of solution sets are written in vector form in Section 1.5. The solution method is reviewed in Section 4.2 and is used throughout Chapters 5 and 7.

Key Exercises: 1–20, 23–28. Students should work at least four or five of Exercises 7–14 in preparation for Section 1.5. Interpolating polynomials in Exercises 33 and 34 will appear again in exercises at the end of Chapter 2.

SECTION 1.3
Vector Equations

The traditional arrow representation of a vector is a handicap to understanding and visualizing Span $\{\mathbf{u}, \mathbf{v}\}$ as well as the set $\{\mathbf{p} + t\mathbf{v} : t \text{ in } \mathbb{R}\}$, discussed in Section 1.5. The set of free arrows (actually equivalence classes) in $\mathbb{R}^3$ is discussed in Section 4.1. Until then, a vector is a list of numbers, represented geometrically by a point. Sometimes an arrow is added to draw attention to the point.

Part or all of Section 6.1 could be inserted here, of course, along with more vector geometry, but the danger of doing so is that students will start to perceive the course as largely computational. I think that students' opinions of the course are set somewhere in the first two weeks, and they need to feel the conceptual emphasis early. Forcing students to deal with Span $\{\mathbf{v}_1, \ldots, \mathbf{v}_p\}$ has that effect. Example 7 is optional, but it is used in Example 6 of Section 1.8.

Key Exercises: 11–14, 17–22, 25, 26. A discussion of Exercise 25 will help students understand $[\,\mathbf{a}_1 \quad \mathbf{a}_2 \quad \mathbf{a}_3\,]$, $\{\mathbf{a}_1, \mathbf{a}_2, \mathbf{a}_3\}$, and Span $\{\mathbf{a}_1, \mathbf{a}_2, \mathbf{a}_3\}$.

SECTION 1.4
The Matrix Equation $A\mathbf{x} = \mathbf{b}$

First impressions are often lasting. That is why the connection between $A\mathbf{x}$ and a linear combination of vectors must be made immediately. The value of this definition is far greater than one might expect. I urge you to try it and to wait until later to give the modern definition of matrix multiplication, as in Section 2.1.

Nothing in the section is optional. Theorem 4 is extremely important. Students may have difficulty with its proof. (Example 3 will help.) The connection between Theorem 4(a) and an *existence* question could be made here, but I think it is better made when reviewing for an exam.

Key Exercises: 1–20, 27, 28, 31, 32. Students should be writing explanations as part of their answers.

Note: The term *rank* can be introduced here, if you wish, but that adds unnecessary terminology at this point. This chapter and the next will focus on the *locations* of pivot positions, which give more information than simply the number of pivot positions.

SECTION 1.5
Solution Sets of Linear Systems

The geometry helps students understand Span $\{\mathbf{u}, \mathbf{v}\}$, in preparation for later discussions of subspaces. The parametric vector form of a set will be used throughout the text to construct bases for null spaces and eigenspaces. Figure 6 will appear again in Section 4.8.

Key Exercises: 1–14, 29–34.

SECTION 1.6
Applications of Linear Systems

This optional section can provide a short homework assignment (on one or two of the applications) if your class needs a pause in the theoretical development of the chapter. You might omit the application to economics if you plan to cover Section 2.6.

SECTION 1.7
Linear Independence

The second half of Theorem 7 will be used several times later in the text. Exercise 30 could be stated as a theorem, in parallel with Theorem 4, but students *must* know the proof, to avoid confusing the two results. The *Study Guide* will help your students learn about linear independence.

Key Exercises: 9–20, 23–30.

Note: Now that your students know about spanning and linear independence, there may be a temptation to introduce vector space concepts such as *basis* and *subspace*. However, piling on too many concepts before the fundamentals are mastered is a mistake most texts make.

SECTION 1.8
Introduction to Linear Transformations

The connection with existence and uniqueness problems is made here and in Section 1.9. Introducing linear transformations here raises the level of the first part of the course and simplifies the work later in the course. Linear transformations appear in the following sections in the next three chapters:

2.1 The definition of AB is motivated by linear transformations.

2.3 Invertible transformations are discussed.

2.5 Matrix factorization is explained via linear transformations.

2.7 Linear transformations are at the heart of computer graphics.

3.3 The volume-change effect of a linear transformation is measured by a determinant.

4.2 General linear transformations are introduced without fanfare, because students are already familiar with concrete matrix transformations.

4.4 The coordinate mapping is a one-to-one linear mapping of V onto $\mathbb{R}^n$.

4.5 The coordinate mapping provides a transparent proof of the key theorem needed to discuss the dimension of a vector space.

4.8 This section is a showcase of the power of linear algebra, mainly because linear transformations can be used in the explanations.

Linear transformations appear frequently, too, in Chapters 5 and 7. Exercises 19 and 20 segue into Section 1.9. Exercise 27 is used in Section 2.7, Computer Graphics.

Key Exercises: 17–20, 25, 31. The *Study Guide* offers help for Exercise 31.

Note: If your students are using MATLAB or another matrix program, you might insert the definition of matrix multiplication from Section 2.1 after Section 1.8, and then assign a project that uses random matrices to explore properties of matrix multiplication. See Exercises 34–36 in Section 2.1. Meanwhile, in class, you can continue with your plans for Sections 1.9 and 1.10. When you get to Section 2.1, you won't have much to do.

SECTION 1.9
The Matrix of a Linear Transformation

This section is optional if you plan to treat linear transformations only lightly, but many instructors will want to cover at least Theorem 10 and a few geometric examples. The notions of *one-to-one* and *onto* appear in the Invertible Matrix Theorem (Section 2.3), in the discussion of the coordinate mapping (Sections 2.8 and 4.4), and in the proof of Theorem 17 in Section 4.8. Section 5.4 generalizes the construction in Theorem 10. The proof of Theorem 11 applies to any linear transformation, in case you want to discuss this later in Chapter 4.

Exercises 25–28 and 31–34 offer fairly easy writing practice. Exercises 31, 32, and 35 provide important links to earlier material.

SECTION 1.10
Linear Models in Business, Science, and Engineering

This section gives the students a break from the concentrated material in the previous sections. If only one application is covered, you should have plenty of time to get a good start on Section 2.1.

The nutrition problem is not mentioned later. The discussions of electrical circuits here and in Sections 2.5 and 5.7 are basically independent of one another. However, if you plan to treat Markov chains later in Section 4.9, then you should present the population movement examples. By considering population sizes here and population distributions in Section 4.9, you can sidestep the discussion of probabilities. You might also use the difference equations here to prepare for Chapter 5, although the introduction to that chapter will suffice.

SECTION 2.1
Matrix Operations

The definition here of a matrix product AB is not only well motivated, but it also gives the proper view of AB for nearly all matrix calculations. The dual fact about the rows of A and the rows of AB is seldom used, mainly because vectors here are usually written as columns. Although the section is long, it is easy and quite a bit can be assigned as outside reading.

Key Exercises: 13, 17–26. Exercises 23 and 26 are cited in the proof of Theorem 8 in Section 2.3. Exercises 27 and 28 are optional, but they are mentioned in Example 4 of Section 2.4. Outer products also appear in Exercises 31–34 of Section 4.6 and in the spectral decomposition of a symmetric matrix, in Section 7.1.

SECTION 2.2
The Inverse of a Matrix

The proof of Theorem 5 is important; students need to see that both uniqueness and existence must be proved. Elementary matrices are used in Section 2.5 (Matrix Factorizations) and in Section 3.2. The algorithm for finding A^{-1} is popular because it is so familiar and leads to easy exam questions. However, in most courses, time is better spent on the LU factorization in Section 2.5, for instance. (The LU factorization takes longer to present.)

Key Exercises: 11–24, 35. Exercise 12 is referenced in Section 2.3, after the proof of Theorem 8. Exercise 15 is useful and indicates how matrix products involving inverses are actually computed in practice. This exercise can be used in Section 4.7, to compute a change-of-coordinates matrix, and the exercise is mentioned in a numerical note in Section 5.4. Exercises 23 and 24 are cited in the proof of Theorem 8.

SECTION 2.3
Characterizations of Invertible Matrices

The Invertible Matrix Theorem ties together most of the concepts studied thus far. Additional statements are added to the theorem in Sections 2.9, 3.2, 4.6, 5.2, and 7.4. The subsection on invertible linear transformations is optional.

Key Exercises: 15–24.

SECTION 2.4
Partitioned Matrices

This section should be part of any modern course in linear algebra because partitioned matrix notation is widely used today in disciplines that employ linear algebra. I made it optional only because it is not yet a standard topic in many linear algebra courses. Block matrix multiplication is used to obtain the spectral decomposition of a symmetric matrix in Section 7.1 and the reduced singular value decomposition in Section 7.4.

Key Exercises: 13, 14, and 16. Exercises 1–10 provide excellent practice with matrix algebra and the Invertible Matrix Theorem. Exercises 19 and 20 are mentioned at the end of Section 2.5.

SECTION 2.5
Matrix Factorizations

Modern algorithms in numerical linear algebra are often described using matrix factorizations. For practical work, this section is more important than Sections 4.7 and 5.4, even though matrix factorizations can be explained nicely in terms of change of bases. See Exercises 22–26.

SECTION 2.6
The Leontief Input–Output Model

This section is independent of Section 1.10. The material here makes a good backdrop for the series expansion of $(I - C)^{-1}$, because this formula is actually used in some practical economic work. Exercise 8 gives an interpretation to entries of an inverse matrix that could be stated without the economic context.

SECTION 2.7
Applications to Computer Graphics

This section seems to have had universal appeal in the classes I have taught. A five- or ten-minute video demonstration of computer graphics will heighten student interest. The section is fun to teach, and it provides practice with composition of linear transformations. A common student mistake is to reverse the order of the matrices.

SECTION 2.8
Subspaces of $\mathbb{R}^2$

This section and the next extract everything you need from Chapter 4 to discuss the topics in Chapters 5–7 (except for the general inner product spaces in Sections 6.7 and 6.8). Omit Sections 2.8 and 2.9 if you plan to cover Chapter 4.

Theorem 12 and Example 6 are important for eigenvector calculations in Chapter 5. Examples 7 and 8 and Theorem 13 are used mainly for the Rank Theorem in

Section 2.9 and then for a discussion of the fundamental subspaces determined by a matrix, in Sections 6.1 and 7.4.

Key Exercises: 5–20, 23–26.

SECTION 2.9
Dimension and Rank

The concept of a coordinate vector is needed for Section 5.4. It also helps students to understand the change of variable $\mathbf{x} = P\mathbf{y}$ in Sections 5.6, 5.7, and 7.2. The notions of *dimension* and *rank* are important for later work in many sections. The Basis Theorem is crucial for the theory of homogeneous difference and differential equations (in Sections 5.6 and 5.7), and it is used in the Gram–Schmidt process (Section 6.4). Also, you will need the Basis Theorem if you plan to cover Section 4.8 later in the course.

Key Exercises: 9–16.

SECTION 3.1
Introduction to Determinants

Because Chapter 3 is mainly computational, it gives students time to absorb the theory from earlier chapters before they plunge into Chapter 4. The use of Theorem 1 in Section 3.1 enables you to get into the subject quickly. This is important, because a typical first course cannot afford to spend more than a week on determinants. Exercises 33–36 provide the first step in the inductive proof of Theorem 3, in the next section.

SECTION 3.2
Properties of Determinants

The development here is more efficient than some presentations, because the hard work is all packed into the proof of Theorem 3. The characterization of det A via pivots gives a concrete interpretation to the somewhat bizarre calculations with cofactors. Theorems 4 and 6 are important for later work.

SECTION 3.3
Cramer's Rule, Volume, and Linear Transformations

There are several independent topics here from which to choose. I tend to prefer the geometric interpretation of the determinant, because of its connection with calculus. But who can resist presenting the wonderful proof of Cramer's Rule? I learned it from Roger Horn and Charles Johnson. See page 21 of their book, *Matrix Analysis* (Cambridge University Press, Cambridge, 1986).

SECTION 4.1
Vector Spaces and Subspaces

The space $\mathbb{S}$ of signals is used in Section 4.8. The spaces of polynomials, $\mathbb{P}_n$, are used in many sections of Chapters 4 and 6. I designed this section to avoid the standard exercises in which a student is asked to check ten axioms on an array of sets (some of which may be artificially weird). In fact,

Theorem 1 provides the main homework tool in this section for showing that a set is a subspace. Students could be taught how to check the closure axioms, of course, but I think the time for that is better spent elsewhere in a first course.

Key Exercises: 1–18, 23, 24. The exercises here and in the next few sections emphasize $\mathbb{R}^n$, to give students time to absorb the abstract concepts.

SECTION 4.2
Null Spaces, Column Spaces, and Linear Transformations

This section provides a review of Chapter 1 using the new terminology. Linear transformations are introduced without fuss, because the students are already comfortable with the idea for $\mathbb{R}^n$. The comments for Section 1.8 list the sections in Chapter 4 where linear transformations are used.

Key Exercises: 3–6, 17–26. They are simple but helpful. The idea in Exercises 7–14 is to use Theorem 1 or Theorem 2 to show that a given set is a subspace.

SECTION 4.3
Linearly Independent Sets; Bases

The definition of *basis* is given initially for subspaces because this emphasizes that the basis elements must be in the subspace. Students often overlook this point when the definition is given only for a vector space. The subsection on bases for Nul A and Col A is essential for Sections 4.5 and 4.6.

Key Exercises: 21–25.

SECTION 4.4
Coordinate Systems

Section 4.7 depends heavily on this section, as does Section 5.4. It is possible to cover the $\mathbb{R}^n$ parts of the two later sections, however, if the first half of Section 4.4 (and perhaps Example 7) is covered. The linearity of the coordinate mapping is used in Section 5.4 to find the matrix of a transformation relative to two bases. The change-of-coordinates matrix appears in Theorem 8 and Exercise 27 of Section 5.4. The concept of an isomorphism is needed in the proof of Theorem 17 in Section 4.8.

Exercise 25 is used in Section 4.7 to show that the change-of-coordinates matrix is invertible.

SECTION 4.5
The Dimension of a Vector Space

Theorem 9 is true because a vector space isomorphic to $\mathbb{R}^n$ has the same algebraic properties as $\mathbb{R}^n$. A proof may not be needed to convince the class. If you skipped Theorem 8 in Section 4.4, you could give a proof such as the one on page 111 of *Introduction to Linear Algebra*, 2d ed., by Serge Lang (Springer-Verlag, New York, 1986). The Basis Theorem is used in Sections 4.8 and 6.4.

Exercise 32 is mentioned in the proof of Theorem 17 in Section 4.8.

SECTION 4.6
Rank

Figure 1, connected with Example 4, prepares the way for Theorem 3 in Section 6.1. Exercises 27–29 are optional, but they tie in neatly with Example 6 of Section 7.4.

SECTION 4.7
Change of Basis

The row reduction algorithm that produces $\underset{C \leftarrow B}{P}$ can also be deduced from Exercise 12 in Section 2.2, by row reducing $[P_C | P_B]$ to $[I | P_C^{-1} P_B]$. The change-of-coordinates matrix here will be interpreted in Section 5.4 as the matrix of the identity transformation relative to two bases.

SECTION 4.8
Applications to Difference Equations

This is an important section for engineering students and worth two days of class time, if you can arrange that. To spend only one day, you could cover up through Example 5, but assign Example 3 for reading (because it takes some time to present in class). Example 3 is the background for Exercise 26 in Section 6.5.

SECTION 4.9
Applications to Markov Chains

The migration matrix is examined again in Section 5.2, where an eigenvector decomposition shows explicitly why the sequence of state vectors $\mathbf{x}_k$ tends to a steady-state vector. The discussion in Section 5.2 does not depend on prior knowledge of Section 4.9.

SECTION 5.1
Eigenvectors and Eigenvalues

The subsection of eigenvectors and difference equations anticipates the discussion that follows in Section 5.2. Also, see Exercises 33 and 34. The **[M]** exercises do not involve the full power of an eigenvalue command because students need to make the connection between an eigenspace of A and a null space of $A - \lambda I$. After Section 5.3, students are encouraged to use their matrix programs to generate both eigenvalues and eigenvectors. (Supplementary Exercises 25 and 26 exhibit matrices that are not easily handled by current matrix programs.)

SECTION 5.2
The Characteristic Equation

Theorem 3 is review for students who have studied Chapter 3. For others, the theorem lists the facts they need to know about determinants. The idea of similarity, introduced here, is discussed again in Section 5.4, along with eight more exercises. The subsection on dynamical systems provides

a short introduction to the subject. The calculations here eliminate the need to supply details for the eigenvector decomposition (2) in Section 5.6.

Exercises 25 and 27 continue the development of difference equations. Exercises 9–14 can be omitted, unless you want your students to have some facility with such problems. If you covered partitioned matrices in Chapter 2, you might connect Exercise 14 in Section 2.4 with Supplementary Exercises 12–14 in Chapter 5.

SECTION 5.3
Diagonalization

This section is essential for the diagonalization of symmetric matrices, in Section 7.1. Theorem 7 is needed for Practice Problem 3 and Exercises 23–26, but is not used elsewhere. A number of exercises in later sections are stated for diagonalizable matrices, because the proofs are simpler for this class of matrices. (See Exercise 26 in Section 5.4, for example.)

Exercise 18 makes a good exam question, but students need to try such a problem before encountering it on an exam.

SECTION 5.4
Eigenvectors and Linear Transformations

Students should be encouraged to review Section 4.4 before the lecture here. To simplify this section, the focus is on the matrix for a linear transformation relative to a single basis, and no special notation is given to the matrix for a transformation that acts between two different spaces. If you wish to simplify the discussion further, you could derive the matrix formula (4) only for the case when $T : \mathbb{R}^n \rightarrow \mathbb{R}^n$, and then skip to Theorem 8.

SECTION 5.5
Complex Eigenvalues

Although some applications, such as in electrical engineering, naturally involve complex vector spaces, the majority of applications accessible to undergraduates require only real vector spaces. For that reason, I have chosen to discuss complex eigenvalues of only real matrices. A discussion of eigenvalues for complex matrices would take more time than most first courses have available. Moreover, in such discussion, the important case of a real matrix acting on $\mathbb{R}^n$ seldom receives the attention it deserves.

Even if there is no time for this section, students can be encouraged to look at the pictures! Figure 5 in Section 5.6 could be examined, too.

Exercises 23 and 24 provide the proof that a real symmetric matrix has only real eigenvalues. These exercises, together with the (real) Schur factorization described in Supplementary Exercise 16 in Chapter 6 and Exercise 32 in Section 7.1, give you pieces from which you can construct a proof of the spectral theorem for symmetric matrices, if you wish.

SECTION 5.6
Discrete Dynamical Systems

The results here are easier to describe than those for differential equations, and the discrete case deserves more attention in the undergraduate curriculum than it commonly now receives.

A brief treatment of this material could cover Example 1, the discussion following it, and some of Exercises 1–6. The last subsection on the spotted owls could be added to this, requiring only a little hand-waving if you have not discussed complex eigenvalues.

For my classes that follow Course Syllabus 2, I leave the ecology for outside reading and spend one day on the graphical description of trajectories, change of variable, and complex eigenvalues. Usually, two or three engineers in my class have already seen state-space methods, and their comments help to spark considerable enthusiasm for this material, particularly the change of variable that decouples a system.

SECTION 5.7
Applications to Differential Equations

This section presumes that students have seen the differential equation $y' = ky$. Sections 5.6 and 5.7 offer an opportunity to contrast discrete and continuous models, which is something engineering texts on signal processing, for example, often do. Students enjoy both sections, although they tend to confuse conditions such as $|\lambda| < 1$ and $\text{Re } \lambda < 0$ when both sections are covered.

SECTION 5.8
Iterative Estimates for Eigenvalues

Example 5 of Section 5.2 and the eigenvector decomposition in Section 5.6 both flow naturally into the power method. When discussing the power method, be careful to *avoid* the following incorrect statements: (1) for sufficiently large k, the vector $A^k\mathbf{x}$ is a good approximation to an eigenvector of A corresponding to the strictly dominant eigenvalue, λ_1; and (2) for large k, the vector $A(A^k\mathbf{x})$ is a good approximation to $\lambda_1(A^k\mathbf{x})$.

In Example 1, the points $A^k\mathbf{x}$ lie on a line parallel to the eigenspace for $\lambda = 2$. So statement (1) is false. In fact, $\mathbf{x} = (.1)\mathbf{v}_1 + \mathbf{v}_2$, where $\mathbf{v}_2$ is an eigenvector for $\lambda = 1$. Thus $A^k\mathbf{x} = 2^k(.1)\mathbf{v}_1 + \mathbf{v}_2$ and $A^{k+1}\mathbf{x} - 2A^k\mathbf{x} = \mathbf{v}_2$ for all k. Hence statement (2) is false. Also, see Exercise 21.

One or both statements essentially appear in a majority of the leading linear algebra texts that cover the power method. (However, their final algorithms for the power method itself are correct.) The statements are true when the second largest eigenvalue is less than 1 in magnitude. But,

in general, the sequence $\{A^k\mathbf{x}\}$ *must* be scaled to a sequence of unit vectors (in some convenient norm). The scaling is not simply a matter of keeping numbers from getting too large or too small.

SECTION 6.1
Inner Product, Length, and Orthogonality

The general material on orthogonal complements is essential for later work. Theorem 3 is an important general fact, but it is needed only for Supplementary Exercise 13 at the end of the chapter and in Section 7.4. The optional material on angles is not used later. Exercises 27–31 concern facts that are used later.

SECTION 6.2
Orthogonal Sets

The nonsquare matrices in Theorems 6 and 7 are needed for the QR factorization, in Section 6.4. It is important to emphasize that the term *orthogonal matrix* applies only to certain *square* matrices. The subsection on orthogonal projections not only sets the stage for the general case in Section 6.3, it also provides what is needed for the orthogonal diagonalization exercises in Section 7.1, because none of the eigenspaces there have dimension greater than 2. For this reason, the Gram–Schmidt process (Section 6.4) is not really needed in Chapter 7. Exercises 13 and 14 prepare for Section 6.3.

SECTION 6.3
Orthogonal Projections

Example 1 seems to help students understand Theorem 8. Theorem 8 is needed for the Gram–Schmidt process (but only for a subspace that itself has an orthogonal basis). Theorems 8 and 9 are needed for the discussions of least squares in Sections 6.5 and 6.6. Theorem 10 is used with the QR factorization to provide a good numerical method for solving least-squares problems, in Section 6.5.

Key Exercises: 19 and 20. They lead naturally into the Gram–Schmidt process.

SECTION 6.4
The Gram–Schmidt Process

The QR factorization encapsulates the essential outcome of the Gram–Schmidt process, just as the LU factorization describes the result of a row reduction process. For practical use of linear algebra, the factorizations are more important than the algorithms that produce them. In fact, the Gram–Schmidt process is *not* the appropriate way to compute the QR factorization. (See the Numerical Notes.) For that reason, one should consider deemphasizing the hand calculation of the Gram–Schmidt process, even though it provides easy exam questions.

The Gram–Schmidt process is used in Sections 6.7 and 6.8, in connection with various sets of orthogonal polynomials. The process is mentioned in Sections 7.1 and 7.4, but the one-dimensional projection constructed in Section 6.2 will suffice. The QR factorization is used in an optional subsection of Section 6.5, and it is needed in Supplementary Exercise 7 of Chapter 7 to produce the Cholesky factorization of a positive definite matrix.

SECTION 6.5
Least-Squares Problems

This is a core section, but it need not take a full day. Each example provides a stopping place. Theorem 13 and Example 1 are all that is needed for Section 6.6. Theorem 15, however, gives an illustration of why the QR factorization is important. Example 4 is related to Exercise 17 in Section 6.6.

SECTION 6.6
Applications to Linear Models

All science students will benefit from Example 1. The general linear model and the subsequent examples are aimed at students who may take a multivariate statistics course. In a general sophomore class, that may include more students than one might expect.

SECTION 6.7
Inner Product Spaces

The three types of inner products described here are matched by applications in Section 6.8. It is possible to spend just one day on selected portions of both sections. Example 1 matches the weighted least squares in Section 6.8. Examples 2–6 are applied to trend analysis in Section 6.8. This material is aimed at students who have not had much calculus or who intend to take more than one course in statistics.

For students who have seen some calculus, Example 7 is needed for the Fourier series in Section 6.8. Example 2 is used to motivate the inner product on $C[a, b]$. The Cauchy–Schwarz and triangle inequalities are not used later, but they should be part of the training of every mathematics student.

SECTION 6.8
Applications of Inner Product Spaces

The connections with Section 6.7 have already been identified. For my junior-level course (see Syllabus 2), I spend three days on the following topics: Theorems 13 and 15 in Section 6.5, plus Examples 1, 3, and 5; Example 1 in Section 6.6; Examples 2 and 7 in Section 6.7, with the motivation for the definite integral; and Fourier series in Section 6.8.

SECTION 7.1
Diagonalization of Symmetric Matrices

Theorems 1 and 2 and the calculations in Examples 2 and 3 are important for the sections that follow. Note that *symmetric matrix* means *real symmetric matrix*, because all matrices in the text have real entries, as mentioned at the beginning of Chapter 7.

Theorem 2 is easily proved for the 2×2 case:

$$\text{If } A = \begin{bmatrix} a & b \\ b & d \end{bmatrix},$$
$$\text{then } \lambda = \frac{1}{2}\left(a + d \pm \sqrt{(a-d)^2 + 4b^2}\right)$$

If $b = 0$, there is nothing to prove. Otherwise, there are two distinct eigenvalues, so A must be diagonalizable. In this case, an eigenvector for λ is $(d - \lambda, -b)$.

SECTION 7.2
Quadratic Forms

This section can provide a good conclusion to the course, because the mathematics here is widely used in applications. For instance, Exercises 23 and 24 can be used to develop the second derivative test for functions of two variables. However, if time permits, some interesting applications still lie ahead. Theorem 4 is used to prove Theorem 6 in Section 7.3, which in turn is used to develop the singular value decomposition.

SECTION 7.3
Constrained Optimization

Theorem 6 is the main result needed in the next two sections. Theorem 7 is mentioned in Example 2 of Section 7.4. Theorem 8 is needed at the very end of Section 7.5. The economic principles in Example 6 may be familiar to students who have had a course in macroeconomics.

SECTION 7.4
The Singular Value Decomposition

This section presents a modern topic of great importance in applications, particularly in computer calculations. Moreover, the singular value decomposition explains much about the structure of matrix transformations. The SVD does for an arbitrary matrix almost what an orthogonal diagonalization does for a symmetric matrix.

SECTION 7.5
Applications to Image Processing and Statistics

The application here has turned out to be of interest to a wide variety of students, including engineers. I cover this in Course Syllabus 3, described above, but in my other classes I only have time to mention the idea briefly.

SECTION 8.1
Affine Combinations

This section introduces special kinds of linear combinations used to describe geometric objects involving two or more variables, including affine sets and translations of subspaces. They set the stage for later applications to computer graphics and can be used in courses on linear programming.

SECTION 8.2
Affine Independence

Affine independence is a restricted type of linear dependence; it provides useful background for a course in computer graphics.

SECTION 8.3
Convex Combinations

Convex combinations are the basic objects studied in Section 8.5 (Polytopes), and they also play a role in computer graphics.

SECTION 8.4
Hyperplanes

This section generalizes the earlier material on lines and planes in $\mathbb{R}^3$ to higher dimensions, in which a hyperplane divides $\mathbb{R}^n$ into two regions.

SECTION 8.5
Polytopes

Polytopes are a class of compact convex sets used to study optimization problems in a variety of fields, including engineering design, linear programming, and business management.

SECTION 8.6
Curves and Surfaces

This section moves beyond lines and planes to curves and surfaces used in engineering and CAD (computer-aided design).

SECTION 9.1
Matrix Games (Online)

The two-person "zero-sum" games in this section introduce mathematical strategies for making rational decisions in a simplified but competitive environment. Optimal strategies here depend on geometric properties of convex sets. These results form the foundation for more complex decision problems in business and industrial environments, discussed in the remainder of the chapter.

SECTION 9.2
Linear Programming—Geometric Method (Online)

The simplified problems here at first involve only two variables. A solution is found by constructing a *feasible set* in the plane, with a boundary formed by three or more line segments. Each intersection of two lines identifies a point

that may be on the boundary of the feasible set. The solution of the problem is given by the coordinates of one of these boundary points.

When three variables are involved, the feasible set is in $\mathbb{R}^3$, bounded by planes that create "faces" on the sides of the feasible set. Each face is a polygonal region that has three or more boundary points, and the geometric situation becomes complicated. So this section only provides practice setting up such a linear programming problem in preparation for Section 9.3.

SECTION 9.3
Linear Programming—Simplex Method (Online)

Once the feasible set is in $\mathbb{R}^n$ for $n > 3$, an algebraic method of solution becomes essential. The feasible set is described by two inequalities, $A\mathbf{x} \leq \mathbf{b}$ and $\mathbf{x} \geq \mathbf{0}$. Suppose $\mathbf{x}_0$ is an extreme point on the feasible set, F. The next step is to examine all other extreme points of F that are connected by one edge to $\mathbf{x}_0$. Of these points, choose one that maximizes the value of the objective function and call it $\mathbf{x}_1$. Then repeat the process and obtain, if possible, another extreme point $\mathbf{x}_2$ at which the value of F increases. Continue the search until no further points are available. This is the Simplex Method, and Section 9.3 has the details that organize the search for the best possible extreme point.

SECTION 9.4
Duality (Online)

To each maximization problem in linear programming there corresponds a minimization problem, called the dual problem. Section 9.4 describes this dual problem and how to solve it. The section also shows how *any matrix game* can be solved using the primal and dual versions of a suitable linear programming problem.

SECTION 10.1
Introduction and Examples (Online)

This section reviews the terminology from Section 4.9 and introduces additional concepts needed to study Markov chains. New examples of Markov chains are developed for later use, including errors in a multi-stage process (such as signal transmission) and diffusion of gases between two compartments. Random walks on graphs are used as examples of unexpected behavior of Markov chains. Exercises include mathematical models for tennis and volleyball matches.

SECTION 10.2
The Steady-State Vector and Google's PageRank (Online)

The steady-state vector for a Markov chain, introduced in Section 4.9, is studied here in more detail. Theorem 1 describes a Markov chain that has a unique steady-state vector. Random surfing on the World-Wide Web is then modeled as a random walk on an immense directed graph. The surfing behavior is altered somewhat to allow Theorem 1 to apply to the Markov chain. The steady-state vector for the resulting Markov chain is essentially the vector that Google's PageRank algorithm uses to rank the popularity of web pages.

SECTION 10.3
Communication Classes (Online)

This section studies the ways in which Markov chains can fail to have a unique steady-state vector. This study involves partitioning the set of states of a Markov chain into subsets called communication classes. An important special case is an irreducible Markov chain, which has only one communication class. This section also introduces the mean return time for a state—a key idea that is the expected number of steps needed for the chain to return to its starting state.

SECTION 10.4
Classification of States and Periodicity (Online)

The states and communication classes of a Markov chain can be classified by whether the chain has a positive probability of never visiting that state (or class) after starting there. States and classes that have this property are called transient; those that do not are called recurrent. Each state or class of a Markov chain also has a period, which is a number that specifies how often the chain may return to that state or class. The recurrence or transience of the states in a Markov chain and their periods determine when the chain has a unique steady-state vector and how to interpret it when it does.

SECTION 10.5
The Fundamental Matrix (Online)

The fundamental matrix is derived from the transition ma-trix of a Markov chain that has both recurrent and transient classes. The entries in the fundamental matrix provide the expected number of visits to a given transient state before the chain is absorbed into a recurrent class. When the chain starts at a transient state, the sum of the entries in a column of the fundamental matrix gives the expected number of time steps until absorption. The fundamental matrix can be used to compute other important quantities, such as the expected transit time between any two states in an irreducible Markov chain.

SECTION 10.6
Markov Chains and Baseball Statistics (Online)

This case study shows how to model the offensive action in baseball as a Markov chain. The model uses 28 states that depend on which bases are occupied and how many outs have occurred; appropriate historical statistics are used to compute the transition probabilities. The results in Section 10.5 are then used to predict the number of earned runs a team will score in an inning and to compare the offensive abilities of different players. Exercises suggest how this model can be used to investigate baseball strategies, such as deciding whether to attempt a sacrifice or to attempt to steal a base.

Chapter 10 APPENDIX 1
Proof of Theorem 1 (Online)

This appendix provides a proof of Theorem 1 in Section 10.2.

Chapter 10 APPENDIX 2
Probability (Online)

This appendix provides the background in probability theory that is needed to develop a formal definition of a Markov chain. Proofs are provided for some results that are only stated or justified informally in Chapter 10.

I hope you enjoy using the text.

D.C.L.

Preface

The response of students and teachers to the first three editions of *Linear Algebra and Its Applications* has been most gratifying. This *Fourth Edition* provides substantial support both for teaching and for using technology in the course. As before, the text provides a modern elementary introduction to linear algebra and a broad selection of interesting applications. The material is accessible to students with the maturity that should come from successful completion of two semesters of college-level mathematics, usually calculus.

The main goal of the text is to help students master the basic concepts and skills they will use later in their careers. The topics here follow the recommendations of the Linear Algebra Curriculum Study Group, which were based on a careful investigation of the real needs of the students and a consensus among professionals in many disciplines that use linear algebra. Hopefully, this course will be one of the most useful and interesting mathematics classes taken by undergraduates.

WHAT'S NEW IN THIS EDITION

The main goal of this revision was to update the exercises and provide additional content, both in the book and online.

1. More than 25 percent of the exercises are new or updated, especially the computational exercises. The exercise sets remain one of the most important features of this book, and these new exercises follow the same high standard of the exercise sets of the past three editions. They are crafted in a way that retells the substance of each of the sections they follow, developing the students' confidence while challenging them to practice and generalize the new ideas they have just encountered.

2. Twenty-five percent of chapter openers are new. These introductory vignettes provide applications of linear algebra and the motivation for developing the mathematics that follows. The text returns to that application in a section toward the end of the chapter.

3. A New Chapter: Chapter 8, The Geometry of Vector Spaces, provides a fresh topic that my students have really enjoyed studying. Sections 1, 2, and 3 provide the basic geometric tools. Then Section 6 uses these ideas to study Bezier curves and surfaces, which are used in engineering and online computer graphics (in Adobe® Illustrator® and Macromedia® FreeHand®). These four sections can be covered in four or five 50-minute class periods.

    A second course in linear algebra applications typically begins with a substantial review of key ideas from the first course. If part of Chapter 8 is in the first course, the second course could include a brief review of sections 1 to 3 and then a focus on the geometry in sections 4 and 5. That would lead naturally into the online chapters 9 and 10, which have been used with Chapter 8 at a number of schools for the past five years.

4. The *Study Guide*, which has always been an integral part of the book, has been updated to cover the new Chapter 8. As with past editions, the *Study Guide* incorporates

detailed solutions to every third odd-numbered exercise as well as solutions to every odd-numbered writing exercise for which the text only provides a hint.

5. Two new chapters are now available online, and can be used in a second course:

   Chapter 9. Optimization
   Chapter 10. Finite-State Markov Chains

   An access code is required and is available to qualified adopters. For more information, visit *www.pearsonhighered.com/irc* or contact your Pearson representative.

6. PowerPoint® slides are now available for the 25 core sections of the text; also included are 75 figures from the text.

DISTINCTIVE FEATURES

Early Introduction of Key Concepts

Many fundamental ideas of linear algebra are introduced within the first seven lectures, in the concrete setting of $\mathbb{R}^n$, and then gradually examined from different points of view. Later generalizations of these concepts appear as natural extensions of familiar ideas, visualized through the geometric intuition developed in Chapter 1. A major achievement of this text is that the level of difficulty is fairly even throughout the course.

A Modern View of Matrix Multiplication

Good notation is crucial, and the text reflects the way scientists and engineers actually use linear algebra in practice. The definitions and proofs focus on the columns of a matrix rather than on the matrix entries. A central theme is to view a matrix–vector product $A\mathbf{x}$ as a linear combination of the columns of A. This modern approach simplifies many arguments, and it ties vector space ideas into the study of linear systems.

Linear Transformations

Linear transformations form a "thread" that is woven into the fabric of the text. Their use enhances the geometric flavor of the text. In Chapter 1, for instance, linear transformations provide a dynamic and graphical view of matrix–vector multiplication.

Eigenvalues and Dynamical Systems

Eigenvalues appear fairly early in the text, in Chapters 5 and 7. Because this material is spread over several weeks, students have more time than usual to absorb and review these critical concepts. Eigenvalues are motivated by and applied to discrete and continuous dynamical systems, which appear in Sections 1.10, 4.8, and 4.9, and in five sections of Chapter 5. Some courses reach Chapter 5 after about five weeks by covering Sections 2.8 and 2.9 instead of Chapter 4. These two optional sections present all the vector space concepts from Chapter 4 needed for Chapter 5.

Orthogonality and Least-Squares Problems

These topics receive a more comprehensive treatment than is commonly found in beginning texts. The Linear Algebra Curriculum Study Group has emphasized the need for a substantial unit on orthogonality and least-squares problems, because orthogonality plays such an important role in computer calculations and numerical linear algebra and because inconsistent linear systems arise so often in practical work.

PEDAGOGICAL FEATURES

Applications

A broad selection of applications illustrates the power of linear algebra to explain fundamental principles and simplify calculations in engineering, computer science, mathematics, physics, biology, economics, and statistics. Some applications appear in separate sections; others are treated in examples and exercises. In addition, each chapter opens with an introductory vignette that sets the stage for some application of linear algebra and provides a motivation for developing the mathematics that follows. Later, the text returns to that application in a section near the end of the chapter.

A Strong Geometric Emphasis

Every major concept in the course is given a geometric interpretation, because many students learn better when they can visualize an idea. There are substantially more drawings here than usual, and some of the figures have never before appeared in a linear algebra text.

Examples

This text devotes a larger proportion of its expository material to examples than do most linear algebra texts. There are more examples than an instructor would ordinarily present in class. But because the examples are written carefully, with lots of detail, students can read them on their own.

Theorems and Proofs

Important results are stated as theorems. Other useful facts are displayed in tinted boxes, for easy reference. Most of the theorems have formal proofs, written with the beginning student in mind. In a few cases, the essential calculations of a proof are exhibited in a carefully chosen example. Some routine verifications are saved for exercises, when they will benefit students.

Practice Problems

A few carefully selected Practice Problems appear just before each exercise set. Complete solutions follow the exercise set. These problems either focus on potential trouble spots in the exercise set or provide a "warm-up" for the exercises, and the solutions often contain helpful hints or warnings about the homework.

Exercises

The abundant supply of exercises ranges from routine computations to conceptual questions that require more thought. A good number of innovative questions pinpoint conceptual difficulties that I have found on student papers over the years. Each exercise set is carefully arranged in the same general order as the text; homework assignments are readily available when only part of a section is discussed. A notable feature of the exercises is their numerical simplicity. Problems "unfold" quickly, so students spend little time on numerical calculations. The exercises concentrate on teaching understanding rather than mechanical calculations. The exercises in the *Fourth Edition* maintain the integrity of the exercises from the third edition, while providing fresh problems for students and instructors.

Exercises marked with the symbol [M] are designed to be worked with the aid of a "Matrix program" (a computer program, such as MATLAB®, Maple™, Mathematica®, MathCad®, or Derive™, or a programmable calculator with matrix capabilities, such as those manufactured by Texas Instruments).

True/False Questions

To encourage students to read all of the text and to think critically, I have developed 300 simple true/false questions that appear in 33 sections of the text, just after the computational problems. They can be answered directly from the text, and they prepare students for the conceptual problems that follow. Students appreciate these questions—after they get used to the importance of reading the text carefully. Based on class testing and discussions with students, I decided not to put the answers in the text. (The *Study Guide* tells the students where to find the answers to the odd-numbered questions.) An additional 150 true/false questions (mostly at the ends of chapters) test understanding of the material. The text does provide simple T/F answers to most of these questions, but it omits the justifications for the answers (which usually require some thought).

Writing Exercises

An ability to write coherent mathematical statements in English is essential for all students of linear algebra, not just those who may go to graduate school in mathematics. The text includes many exercises for which a written justification is part of the answer. Conceptual exercises that require a short proof usually contain hints that help a student get started. For all odd-numbered writing exercises, either a solution is included at the back of the text or a hint is provided and the solution is given in the *Study Guide*, described below.

Computational Topics

The text stresses the impact of the computer on both the development and practice of linear algebra in science and engineering. Frequent Numerical Notes draw attention to issues in computing and distinguish between theoretical concepts, such as matrix inversion, and computer implementations, such as LU factorizations.

WEB SUPPORT

This Web site at *www.pearsonhighered.com/lay* contains support material for the textbook. For students, the Web site contains **review sheets** and **practice exams** (with solutions) that cover the main topics in the text. They come directly from courses I have taught in past years. Each review sheet identifies key definitions, theorems, and skills from a specified portion of the text.

Applications by Chapters

The Web site also contains seven Case Studies, which expand topics introduced at the beginning of each chapter, adding real-world data and opportunities for further exploration. In addition, more than 20 Application Projects either extend topics in the text or introduce new applications, such as cubic splines, airline flight routes, dominance matrices in sports competition, and error-correcting codes. Some mathematical applications are integration techniques, polynomial root location, conic sections, quadric surfaces, and extrema for functions of two variables. Numerical linear algebra topics, such as condition numbers, matrix factorizations, and the QR method for finding eigenvalues, are also included. Woven into each discussion are exercises that may involve large data sets (and thus require technology for their solution).

Getting Started with Technology

If your course includes some work with MATLAB, Maple, Mathematica, or TI calculators, you can read one of the projects on the Web site for an introduction to the technology. In addition, the *Study Guide* provides introductory material for first-time users.

Data Files

Hundreds of files contain data for about 900 numerical exercises in the text, Case Studies, and Application Projects. The data are available at *www.pearsonhighered.com/lay* in a variety of formats—for MATLAB, Maple, Mathematica, and the TI-83+/86/89 graphic calculators. By allowing students to access matrices and vectors for a particular problem with only a few keystrokes, the data files eliminate data entry errors and save time on homework.

MATLAB Projects

These exploratory projects invite students to discover basic mathematical and numerical issues in linear algebra. Written by Rick Smith, they were developed to accompany a computational linear algebra course at the University of Florida, which has used *Linear Algebra and Its Applications* for many years. The projects are referenced by an icon WEB at appropriate points in the text. About half of the projects explore fundamental concepts such as the column space, diagonalization, and orthogonal projections; several projects focus on numerical issues such as flops, iterative methods, and the SVD; and a few projects explore applications such as Lagrange interpolation and Markov chains.

SUPPLEMENTS

Study Guide

A printed version of the *Study Guide* is available at low cost. I wrote this *Guide* to be an integral part of the course. An icon SG in the text directs students to special subsections of the *Guide* that suggest how to master key concepts of the course. The *Guide* supplies a detailed solution to every third odd-numbered exercise, which allows students to check their work. A complete explanation is provided whenever an odd-numbered writing exercise has only a "Hint" in the answers. Frequent "Warnings" identify common errors and show how to prevent them. MATLAB boxes introduce commands as they are needed. Appendixes in the *Study Guide* provide comparable information about Maple, Mathematica, and TI graphing calculators (ISBN: 0-321-38883-6).

Instructor's Edition

For the convenience of instructors, this special edition includes brief answers to all exercises. A *Note to the Instructor* at the beginning of the text provides a commentary on the design and organization of the text, to help instructors plan their courses. It also describes other support available for instructors. (ISBN: 0-321-38518-7)

Instructor's Technology Manuals

Each manual provides detailed guidance for integrating a specific software package or graphic calculator throughout the course, written by faculty who have already used the technology with this text. The following manuals are available to qualified instructors through the Pearson Instructor Resource Center, *www.pearsonhighered.com/irc*: MATLAB (ISBN: 0-321-53365-8), Maple (ISBN: 0-321-75605-3), Mathematica (ISBN: 0-321-38885-2), and the TI-83+/86/89 (ISBN: 0-321-38887-9).

ACKNOWLEDGMENTS

I am indeed grateful to many groups of people who have helped me over the years with various aspects of this book.

I want to thank Israel Gohberg and Robert Ellis for more than fifteen years of research collaboration, which greatly shaped my view of linear algebra. And, it has been a privilege to be a member of the Linear Algebra Curriculum Study Group along with David Carlson, Charles Johnson, and Duane Porter. Their creative ideas about teaching linear algebra have influenced this text in significant ways.

I sincerely thank the following reviewers for their careful analyses and constructive suggestions:

Rafal Ablamowicz, *Tennessee Technological University*
Brian E. Blank, *Washington University in St. Louis*
Vahid Dabbaghian-Abdoly, *Simon Fraser University*
James L. Hartman, *The College of Wooster*
Richard P. Kubelka, *San Jose State University*
Martin Nikolov, *University of Connecticut*
Ilya M. Spitkovsky, *College of William & Mary*

John Alongi, *Northwestern University*
Steven Bellenot, *Florida State University*
Herman Gollwitzer, *Drexel University*
David R. Kincaid, *The University of Texas at Austin*
Douglas B. Meade, *University of South Carolina*
Tim Olson, *University of Florida*
Albert L. Vitter III, *Tulane University*

For this Fourth Edition, I thank my brother, Steven Lay, at Lee University, for his generous help and encouragement, and for his newly revised Chapters 8 and 9. I also thank Thomas Polaski, of Winthrop University, for his newly revised Chapter 10. For good advice and help with chapter introductory examples, I thank Raymond Rosentrater, of Westmont College. Another gifted professor, Judith McDonald, of Washington State University, developed many new exercises for the text. Her help and enthusiasm for the book was refreshing and inspiring.

I thank the technology experts who labored on the various supplements for the Fourth Edition, preparing the data, writing notes for the instructors, writing technology notes for the students in the *Study Guide*, and sharing their projects with us: Jeremy Case (MATLAB), Taylor University; Douglas Meade (Maple), University of South Carolina; Michael Miller (TI Calculator), Western Baptist College; and Marie Vanisko (Mathematica), Carroll College.

I thank Professor John Risley and graduate students David Aulicino, Sean Burke, and Hersh Goldberg for their technical expertise in helping develop online homework support for the text. I am grateful for the class testing of this online homework support by the following: Agnes Boskovitz, Malcolm Brooks, Elizabeth Ormerod, Alexander Isaev, and John Urbas at the Australian National University; John Scott and Leben Wee at Montgomery College, Maryland; and Xingru Zhang at SUNY University of Buffalo.

I appreciate the mathematical assistance provided by Blaise DeSesa, Jean Horn, Roger Lipsett, Paul Lorczak, Thomas Polaski, Sarah Streett, and Marie Vanisko, who checked the accuracy of calculations in the text.

Finally, I sincerely thank the staff at Addison-Wesley for all their help with the development and production of the Fourth Edition: Caroline Celano, sponsoring editor, Chere Bemelmans, senior content editor; Tamela Ambush, associate managing editor; Carl Cottrell, senior media producer; Jeff Weidenaar, executive marketing manager; Kendra Bassi, marketing assistant; and Andrea Nix, text design. Saved for last are the three good friends who have guided the development of the book nearly from the beginning—giving wise counsel and encouragement—Greg Tobin, publisher, Laurie Rosatone, former editor, and William Hoffman, current editor. Thank you all so much.

David C. Lay

A Note to Students

This course is potentially the most interesting and worthwhile undergraduate mathematics course you will complete. In fact, some students have written or spoken to me after graduation and said that they still use this text occasionally as a reference in their careers at major corporations and engineering graduate schools. The following remarks offer some practical advice and information to help you master the material and enjoy the course.

In linear algebra, the *concepts* are as important as the *computations*. The simple numerical exercises that begin each exercise set only help you check your understanding of basic procedures. Later in your career, computers will do the calculations, but you will have to choose the calculations, know how to interpret the results, and then explain the results to other people. For this reason, many exercises in the text ask you to explain or justify your calculations. A written explanation is often required as part of the answer. For odd-numbered exercises, you will find either the desired explanation or at least a good hint. You must avoid the temptation to look at such answers before you have tried to write out the solution yourself. Otherwise, you are likely to think you understand something when in fact you do not.

To master the concepts of linear algebra, you will have to read and reread the text carefully. New terms are in boldface type, sometimes enclosed in a definition box. A glossary of terms is included at the end of the text. Important facts are stated as theorems or are enclosed in tinted boxes, for easy reference. I encourage you to read the first five pages of the Preface to learn more about the structure of this text. This will give you a framework for understanding how the course may proceed.

In a practical sense, linear algebra is a language. You must learn this language the same way you would a foreign language—with daily work. Material presented in one section is not easily understood unless you have thoroughly studied the text and worked the exercises for the preceding sections. Keeping up with the course will save you lots of time and distress!

Numerical Notes

I hope you read the Numerical Notes in the text, even if you are not using a computer or graphic calculator with the text. In real life, most applications of linear algebra involve numerical computations that are subject to some numerical error, even though that error may be extremely small. The Numerical Notes will warn you of potential difficulties in using linear algebra later in your career, and if you study the notes now, you are more likely to remember them later.

If you enjoy reading the Numerical Notes, you may want to take a course later in numerical linear algebra. Because of the high demand for increased computing power, computer scientists and mathematicians work in numerical linear algebra to develop faster and more reliable algorithms for computations, and electrical engineers design faster and smaller computers to run the algorithms. This is an exciting field, and your first course in linear algebra will help you prepare for it.

Study Guide

To help you succeed in this course, I suggest that you purchase the *Study Guide* (*www.mypearsonstore.com*; 0-321-38883-6). Not only will it help you learn linear algebra, it also will show you how to study mathematics. At strategic points in your textbook, an icon [SG] will direct you to special subsections in the *Study Guide* entitled "Mastering Linear Algebra Concepts." There you will find suggestions for constructing effective review sheets of key concepts. The act of preparing the sheets is one of the secrets to success in the course, because you will construct *links between ideas*. These links are the "glue" that enables you to build a solid foundation for learning and *remembering* the main concepts in the course.

The *Study Guide* contains a detailed solution to every third odd-numbered exercise, plus solutions to all odd-numbered writing exercises for which only a hint is given in the Answers section of this book. The *Guide* is separate from the text because you must learn to write solutions by yourself, without much help. (I know from years of experience that easy access to solutions in the back of the text slows the mathematical development of most students.) The *Guide* also provides warnings of common errors and helpful hints that call attention to key exercises and potential exam questions.

If you have access to technology—MATLAB, Maple, Mathematica, or a TI graphing calculator—you can save many hours of homework time. The *Study Guide* is your "lab manual" that explains how to use each of these matrix utilities. It introduces new commands when they are needed. You can download from the website *www.pearsonhighered.com/lay* the data for more than 850 exercises in the text. (With a few keystrokes, you can display any numerical homework problem on your screen.) Special matrix commands will perform the computations for you!

What you do in your first few weeks of studying this course will set your pattern for the term and determine how well you finish the course. Please read "How to Study Linear Algebra" in the *Study Guide* as soon as possible. My students have found the strategies there very helpful, and I hope you will, too.

1

Linear Equations in Linear Algebra

INTRODUCTORY EXAMPLE

Linear Models in Economics and Engineering

It was late summer in 1949. Harvard Professor Wassily Leontief was carefully feeding the last of his punched cards into the university's Mark II computer. The cards contained economic information about the U.S. economy and represented a summary of more than 250,000 pieces of information produced by the U.S. Bureau of Labor Statistics after two years of intensive work. Leontief had divided the U.S. economy into 500 "sectors," such as the coal industry, the automotive industry, communications, and so on. For each sector, he had written a linear equation that described how the sector distributed its output to the other sectors of the economy. Because the Mark II, one of the largest computers of its day, could not handle the resulting system of 500 equations in 500 unknowns, Leontief had distilled the problem into a system of 42 equations in 42 unknowns.

Programming the Mark II computer for Leontief's 42 equations had required several months of effort, and he was anxious to see how long the computer would take to solve the problem. The Mark II hummed and blinked for 56 hours before finally producing a solution. We will discuss the nature of this solution in Sections 1.6 and 2.6.

Leontief, who was awarded the 1973 Nobel Prize in Economic Science, opened the door to a new era in mathematical modeling in economics. His efforts at Harvard in 1949 marked one of the first significant uses of computers to analyze what was then a large-scale mathematical model. Since that time, researchers in many other fields have employed computers to analyze mathematical models. Because of the massive amounts of data involved, the models are usually *linear*; that is, they are described by *systems of linear equations*.

The importance of linear algebra for applications has risen in direct proportion to the increase in computing power, with each new generation of hardware and software triggering a demand for even greater capabilities. Computer science is thus intricately linked with linear algebra through the explosive growth of parallel processing and large-scale computations.

Scientists and engineers now work on problems far more complex than even dreamed possible a few decades ago. Today, linear algebra has more potential value for students in many scientific and business fields than any other undergraduate mathematics subject! The material in this text provides the foundation for further work in many interesting areas. Here are a few possibilities; others will be described later.

- *Oil exploration.* When a ship searches for offshore oil deposits, its computers solve thousands of separate systems of linear equations *every day*. The seismic data for the equations are obtained from underwater shock waves created by explosions from air guns. The waves bounce off subsurface rocks and are measured by geophones attached to mile-long cables behind the ship.

- *Linear programming.* Many important management decisions today are made on the basis of linear programming models that utilize hundreds of variables. The airline industry, for instance,

employs linear programs that schedule flight crews, monitor the locations of aircraft, or plan the varied schedules of support services such as maintenance and terminal operations.

- *Electrical networks.* Engineers use simulation software to design electrical circuits and microchips involving millions of transistors. Such software relies on linear algebra techniques and systems of linear equations.

WEB

Systems of linear equations lie at the heart of linear algebra, and this chapter uses them to introduce some of the central concepts of linear algebra in a simple and concrete setting. Sections 1.1 and 1.2 present a systematic method for solving systems of linear equations. This algorithm will be used for computations throughout the text. Sections 1.3 and 1.4 show how a system of linear equations is equivalent to a *vector equation* and to a *matrix equation*. This equivalence will reduce problems involving linear combinations of vectors to questions about systems of linear equations. The fundamental concepts of spanning, linear independence, and linear transformations, studied in the second half of the chapter, will play an essential role throughout the text as we explore the beauty and power of linear algebra.

1.1 | SYSTEMS OF LINEAR EQUATIONS

A **linear equation** in the variables $x_1, \ldots, x_n$ is an equation that can be written in the form

$$a_1 x_1 + a_2 x_2 + \cdots + a_n x_n = b \tag{1}$$

where b and the **coefficients** $a_1, \ldots, a_n$ are real or complex numbers, usually known in advance. The subscript n may be any positive integer. In textbook examples and exercises, n is normally between 2 and 5. In real-life problems, n might be 50 or 5000, or even larger.

The equations

$$4x_1 - 5x_2 + 2 = x_1 \quad \text{and} \quad x_2 = 2(\sqrt{6} - x_1) + x_3$$

are both linear because they can be rearranged algebraically as in equation (1):

$$3x_1 - 5x_2 = -2 \quad \text{and} \quad 2x_1 + x_2 - x_3 = 2\sqrt{6}$$

The equations

$$4x_1 - 5x_2 = x_1 x_2 \quad \text{and} \quad x_2 = 2\sqrt{x_1} - 6$$

are not linear because of the presence of $x_1 x_2$ in the first equation and $\sqrt{x_1}$ in the second.

A **system of linear equations** (or a **linear system**) is a collection of one or more linear equations involving the same variables—say, $x_1, \ldots, x_n$. An example is

$$\begin{aligned} 2x_1 - x_2 + 1.5x_3 &= 8 \\ x_1 \qquad - 4x_3 &= -7 \end{aligned} \tag{2}$$

A **solution** of the system is a list $(s_1, s_2, \ldots, s_n)$ of numbers that makes each equation a true statement when the values $s_1, \ldots, s_n$ are substituted for $x_1, \ldots, x_n$, respectively. For instance, $(5, 6.5, 3)$ is a solution of system (2) because, when these values are substituted in (2) for x_1, x_2, x_3, respectively, the equations simplify to $8 = 8$ and $-7 = -7$.

The set of all possible solutions is called the **solution set** of the linear system. Two linear systems are called **equivalent** if they have the same solution set. That is, each solution of the first system is a solution of the second system, and each solution of the second system is a solution of the first.

Finding the solution set of a system of two linear equations in two variables is easy because it amounts to finding the intersection of two lines. A typical problem is

$$\begin{aligned} x_1 - 2x_2 &= -1 \\ -x_1 + 3x_2 &= 3 \end{aligned}$$

The graphs of these equations are lines, which we denote by ℓ_1 and ℓ_2. A pair of numbers (x_1, x_2) satisfies *both* equations in the system if and only if the point (x_1, x_2) lies on both ℓ_1 and ℓ_2. In the system above, the solution is the single point $(3, 2)$, as you can easily verify. See Fig. 1.

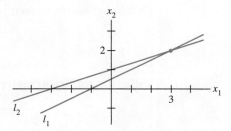

FIGURE 1 Exactly one solution.

Of course, two lines need not intersect in a single point—they could be parallel, or they could coincide and hence "intersect" at every point on the line. Figure 2 shows the graphs that correspond to the following systems:

(a) $\quad \begin{aligned} x_1 - 2x_2 &= -1 \\ -x_1 + 2x_2 &= 3 \end{aligned}$ 　　　(b) $\quad \begin{aligned} x_1 - 2x_2 &= -1 \\ -x_1 + 2x_2 &= 1 \end{aligned}$

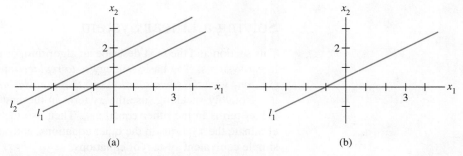

(a) 　　　　　　　　　　　　　　(b)

FIGURE 2 (a) No solution. (b) Infinitely many solutions.

Figures 1 and 2 illustrate the following general fact about linear systems, to be verified in Section 1.2.

A system of linear equations has

1. no solution, or
2. exactly one solution, or
3. infinitely many solutions.

A system of linear equations is said to be **consistent** if it has either one solution or infinitely many solutions; a system is **inconsistent** if it has no solution.

Matrix Notation

The essential information of a linear system can be recorded compactly in a rectangular array called a **matrix**. Given the system

$$
\begin{aligned}
x_1 - 2x_2 + x_3 &= 0 \\
2x_2 - 8x_3 &= 8 \\
-4x_1 + 5x_2 + 9x_3 &= -9
\end{aligned}
\tag{3}
$$

with the coefficients of each variable aligned in columns, the matrix

$$
\begin{bmatrix}
1 & -2 & 1 \\
0 & 2 & -8 \\
-4 & 5 & 9
\end{bmatrix}
$$

is called the **coefficient matrix** (or **matrix of coefficients**) of the system (3), and

$$
\begin{bmatrix}
1 & -2 & 1 & 0 \\
0 & 2 & -8 & 8 \\
-4 & 5 & 9 & -9
\end{bmatrix}
\tag{4}
$$

is called the **augmented matrix** of the system. (The second row here contains a zero because the second equation could be written as $0 \cdot x_1 + 2x_2 - 8x_3 = 8$.) An augmented matrix of a system consists of the coefficient matrix with an added column containing the constants from the right sides of the equations.

The **size** of a matrix tells how many rows and columns it has. The augmented matrix (4) above has 3 rows and 4 columns and is called a 3×4 (read "3 by 4") matrix. If m and n are positive integers, an **$m \times n$ matrix** is a rectangular array of numbers with m rows and n columns. (The number of rows always comes first.) Matrix notation will simplify the calculations in the examples that follow.

Solving a Linear System

This section and the next describe an algorithm, or a systematic procedure, for solving linear systems. The basic strategy is *to replace one system with an equivalent system (i.e., one with the same solution set) that is easier to solve*.

Roughly speaking, use the x_1 term in the first equation of a system to eliminate the x_1 terms in the other equations. Then use the x_2 term in the second equation to eliminate the x_2 terms in the other equations, and so on, until you finally obtain a very simple equivalent system of equations.

Three basic operations are used to simplify a linear system: Replace one equation by the sum of itself and a multiple of another equation, interchange two equations, and multiply all the terms in an equation by a nonzero constant. After the first example, you will see why these three operations do not change the solution set of the system.

EXAMPLE 1 Solve system (3).

SOLUTION The elimination procedure is shown here with and without matrix notation, and the results are placed side by side for comparison:

$$
\begin{aligned}
x_1 - 2x_2 + x_3 &= 0 \\
2x_2 - 8x_3 &= 8 \\
-4x_1 + 5x_2 + 9x_3 &= -9
\end{aligned}
\qquad
\begin{bmatrix}
1 & -2 & 1 & 0 \\
0 & 2 & -8 & 8 \\
-4 & 5 & 9 & -9
\end{bmatrix}
$$

Keep x_1 in the first equation and eliminate it from the other equations. To do so, add 4 times equation 1 to equation 3. After some practice, this type of calculation is usually performed mentally:

$$
\begin{array}{ll}
4 \cdot [\text{equation 1}]: & 4x_1 - 8x_2 + 4x_3 = 0 \\
+ [\text{equation 3}]: & -4x_1 + 5x_2 + 9x_3 = -9 \\
\hline
[\text{new equation 3}]: & -3x_2 + 13x_3 = -9
\end{array}
$$

The result of this calculation is written in place of the original third equation:

$$
\begin{aligned}
x_1 - 2x_2 + x_3 &= 0 \\
2x_2 - 8x_3 &= 8 \\
-3x_2 + 13x_3 &= -9
\end{aligned}
\qquad
\begin{bmatrix}
1 & -2 & 1 & 0 \\
0 & 2 & -8 & 8 \\
0 & -3 & 13 & -9
\end{bmatrix}
$$

Now, multiply equation 2 by $1/2$ in order to obtain 1 as the coefficient for x_2. (This calculation will simplify the arithmetic in the next step.)

$$
\begin{aligned}
x_1 - 2x_2 + x_3 &= 0 \\
x_2 - 4x_3 &= 4 \\
-3x_2 + 13x_3 &= -9
\end{aligned}
\qquad
\begin{bmatrix}
1 & -2 & 1 & 0 \\
0 & 1 & -4 & 4 \\
0 & -3 & 13 & -9
\end{bmatrix}
$$

Use the x_2 in equation 2 to eliminate the $-3x_2$ in equation 3. The "mental" computation is

$$
\begin{array}{ll}
3 \cdot [\text{equation 2}]: & 3x_2 - 12x_3 = 12 \\
+ [\text{equation 3}]: & -3x_2 + 13x_3 = -9 \\
\hline
[\text{new equation 3}]: & x_3 = 3
\end{array}
$$

The new system has a *triangular* form:[1]

$$
\begin{aligned}
x_1 - 2x_2 + x_3 &= 0 \\
x_2 - 4x_3 &= 4 \\
x_3 &= 3
\end{aligned}
\qquad
\begin{bmatrix}
1 & -2 & 1 & 0 \\
0 & 1 & -4 & 4 \\
0 & 0 & 1 & 3
\end{bmatrix}
$$

Eventually, you want to eliminate the $-2x_2$ term from equation 1, but it is more efficient to use the x_3 in equation 3 first, to eliminate the $-4x_3$ and $+x_3$ terms in equations 2 and 1. The two "mental" calculations are

$$
\begin{array}{ll}
4 \cdot [\text{eq. 3}]: & 4x_3 = 12 \\
+ [\text{eq. 2}]: & x_2 - 4x_3 = 4 \\
\hline
[\text{new eq. 2}]: & x_2 = 16
\end{array}
\qquad
\begin{array}{ll}
-1 \cdot [\text{eq. 3}]: & -x_3 = -3 \\
+ [\text{eq. 1}]: & x_1 - 2x_2 + x_3 = 0 \\
\hline
[\text{new eq. 1}]: & x_1 - 2x_2 = -3
\end{array}
$$

---

[1] The intuitive term *triangular* will be replaced by a precise term in the next section.

It is convenient to combine the results of these two operations:

$$\begin{aligned} x_1 - 2x_2 \quad\;\;\; &= -3 \\ x_2 \quad\;\; &= 16 \\ x_3 &= \;\; 3 \end{aligned} \qquad \begin{bmatrix} 1 & -2 & 0 & -3 \\ 0 & 1 & 0 & 16 \\ 0 & 0 & 1 & 3 \end{bmatrix}$$

Now, having cleaned out the column above the x_3 in equation 3, move back to the x_2 in equation 2 and use it to eliminate the $-2x_2$ above it. Because of the previous work with x_3, there is now no arithmetic involving x_3 terms. Add 2 times equation 2 to equation 1 and obtain the system:

$$\begin{cases} x_1 \quad\quad\;\;\;\, = 29 \\ \quad\; x_2 \quad\;\; = 16 \\ \quad\quad\;\; x_3 = \;\; 3 \end{cases} \qquad \begin{bmatrix} 1 & 0 & 0 & 29 \\ 0 & 1 & 0 & 16 \\ 0 & 0 & 1 & 3 \end{bmatrix}$$

The work is essentially done. It shows that the only solution of the original system is $(29, 16, 3)$. However, since there are so many calculations involved, it is a good practice to check the work. To verify that $(29, 16, 3)$ *is* a solution, substitute these values into the left side of the original system, and compute:

$$(29) - 2(16) + \;\; (3) = 29 - 32 + 3 = 0$$
$$2(16) - 8(3) = 32 - 24 = 8$$
$$-4(29) + 5(16) + 9(3) = -116 + 80 + 27 = -9$$

The results agree with the right side of the original system, so $(29, 16, 3)$ is a solution of the system. ∎

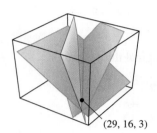

Each of the original equations determines a plane in three-dimensional space. The point $(29, 16, 3)$ lies in all three planes.

Example 1 illustrates how operations on equations in a linear system correspond to operations on the appropriate rows of the augmented matrix. The three basic operations listed earlier correspond to the following operations on the augmented matrix.

ELEMENTARY ROW OPERATIONS

1. (Replacement) Replace one row by the sum of itself and a multiple of another row.[2]

2. (Interchange) Interchange two rows.

3. (Scaling) Multiply all entries in a row by a nonzero constant.

Row operations can be applied to any matrix, not merely to one that arises as the augmented matrix of a linear system. Two matrices are called **row equivalent** if there is a sequence of elementary row operations that transforms one matrix into the other.

It is important to note that row operations are *reversible*. If two rows are interchanged, they can be returned to their original positions by another interchange. If a row is scaled by a nonzero constant c, then multiplying the new row by $1/c$ produces the original row. Finally, consider a replacement operation involving two rows—say, rows 1 and 2—and suppose that c times row 1 is added to row 2 to produce a new row 2. To "reverse" this operation, add $-c$ times row 1 to (new) row 2 and obtain the original row 2. See Exercises 29–32 at the end of this section.

---

[2] A common paraphrase of row replacement is "Add to one row a multiple of another row."

At the moment, we are interested in row operations on the augmented matrix of a system of linear equations. Suppose a system is changed to a new one via row operations. By considering each type of row operation, you can see that any solution of the original system remains a solution of the new system. Conversely, since the original system can be produced via row operations on the new system, each solution of the new system is also a solution of the original system. This discussion justifies the following statement.

> If the augmented matrices of two linear systems are row equivalent, then the two systems have the same solution set.

Though Example 1 is lengthy, you will find that after some practice, the calculations go quickly. Row operations in the text and exercises will usually be extremely easy to perform, allowing you to focus on the underlying concepts. Still, you must learn to perform row operations accurately because they will be used throughout the text.

The rest of this section shows how to use row operations to determine the size of a solution set, without completely solving the linear system.

Existence and Uniqueness Questions

Section 1.2 will show why a solution set for a linear system contains either no solutions, one solution, or infinitely many solutions. Answers to the following two questions will determine the nature of the solution set for a linear system.

To determine which possibility is true for a particular system, we ask two questions.

TWO FUNDAMENTAL QUESTIONS ABOUT A LINEAR SYSTEM

1. Is the system consistent; that is, does at least one solution *exist*?

2. If a solution exists, is it the *only* one; that is, is the solution *unique*?

These two questions will appear throughout the text, in many different guises. This section and the next will show how to answer these questions via row operations on the augmented matrix.

EXAMPLE 2   Determine if the following system is consistent:

$$
\begin{aligned}
x_1 - 2x_2 + x_3 &= 0 \\
2x_2 - 8x_3 &= 8 \\
-4x_1 + 5x_2 + 9x_3 &= -9
\end{aligned}
$$

SOLUTION   This is the system from Example 1. Suppose that we have performed the row operations necessary to obtain the triangular form

$$
\begin{aligned}
x_1 - 2x_2 + x_3 &= 0 \\
x_2 - 4x_3 &= 4 \\
x_3 &= 3
\end{aligned}
\qquad
\begin{bmatrix}
1 & -2 & 1 & 0 \\
0 & 1 & -4 & 4 \\
0 & 0 & 1 & 3
\end{bmatrix}
$$

At this point, we know x_3. Were we to substitute the value of x_3 into equation 2, we could compute x_2 and hence could determine x_1 from equation 1. So a solution exists; the system is consistent. (In fact, x_2 is uniquely determined by equation 2 since x_3 has only one possible value, and x_1 is therefore uniquely determined by equation 1. So the solution is unique.) ∎

EXAMPLE 3 Determine if the following system is consistent:

$$\begin{aligned} x_2 - 4x_3 &= 8 \\ 2x_1 - 3x_2 + 2x_3 &= 1 \\ 5x_1 - 8x_2 + 7x_3 &= 1 \end{aligned} \tag{5}$$

SOLUTION The augmented matrix is

$$\begin{bmatrix} 0 & 1 & -4 & 8 \\ 2 & -3 & 2 & 1 \\ 5 & -8 & 7 & 1 \end{bmatrix}$$

To obtain an x_1 in the first equation, interchange rows 1 and 2:

$$\begin{bmatrix} 2 & -3 & 2 & 1 \\ 0 & 1 & -4 & 8 \\ 5 & -8 & 7 & 1 \end{bmatrix}$$

To eliminate the $5x_1$ term in the third equation, add $-5/2$ times row 1 to row 3:

$$\begin{bmatrix} 2 & -3 & 2 & 1 \\ 0 & 1 & -4 & 8 \\ 0 & -1/2 & 2 & -3/2 \end{bmatrix} \tag{6}$$

Next, use the x_2 term in the second equation to eliminate the $-(1/2)x_2$ term from the third equation. Add $1/2$ times row 2 to row 3:

$$\begin{bmatrix} 2 & -3 & 2 & 1 \\ 0 & 1 & -4 & 8 \\ 0 & 0 & 0 & 5/2 \end{bmatrix} \tag{7}$$

The augmented matrix is now in triangular form. To interpret it correctly, go back to equation notation:

$$\begin{aligned} 2x_1 - 3x_2 + 2x_3 &= 1 \\ x_2 - 4x_3 &= 8 \\ 0 &= 5/2 \end{aligned} \tag{8}$$

The equation $0 = 5/2$ is a short form of $0x_1 + 0x_2 + 0x_3 = 5/2$. This system in triangular form obviously has a built-in contradiction. There are no values of x_1, x_2, x_3 that satisfy (8) because the equation $0 = 5/2$ is never true. Since (8) and (5) have the same solution set, the original system is inconsistent (i.e., has no solution). ∎

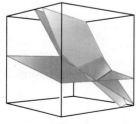

This system is inconsistent because there is no point that lies in all three planes.

Pay close attention to the augmented matrix in (7). Its last row is typical of an inconsistent system in triangular form.

> ┌─ NUMERICAL NOTE ──────────────────────────────────
>
> In real-world problems, systems of linear equations are solved by a computer. For a square coefficient matrix, computer programs nearly always use the elimination algorithm given here and in Section 1.2, modified slightly for improved accuracy.
>
> The vast majority of linear algebra problems in business and industry are solved with programs that use *floating point arithmetic*. Numbers are represented as decimals $\pm .d_1 \cdots d_p \times 10^r$, where r is an integer and the number p of digits to the right of the decimal point is usually between 8 and 16. Arithmetic with such numbers typically is inexact, because the result must be rounded (or truncated) to the number of digits stored. "Roundoff error" is also introduced when a number such as 1/3 is entered into the computer, since its decimal representation must be approximated by a finite number of digits. Fortunately, inaccuracies in floating point arithmetic seldom cause problems. The numerical notes in this book will occasionally warn of issues that you may need to consider later in your career.

PRACTICE PROBLEMS

Throughout the text, practice problems should be attempted before working the exercises. Solutions appear after each exercise set.

1. State in words the next elementary row operation that should be performed on the system in order to solve it. [More than one answer is possible in (a).]

    a.
    $$\begin{aligned} x_1 + 4x_2 - 2x_3 + 8x_4 &= 12 \\ x_2 - 7x_3 + 2x_4 &= -4 \\ 5x_3 - x_4 &= 7 \\ x_3 + 3x_4 &= -5 \end{aligned}$$

    b.
    $$\begin{aligned} x_1 - 3x_2 + 5x_3 - 2x_4 &= 0 \\ x_2 + 8x_3 \phantom{{}-2x_4} &= -4 \\ 2x_3 \phantom{{}-2x_4} &= 3 \\ x_4 &= 1 \end{aligned}$$

2. The augmented matrix of a linear system has been transformed by row operations into the form below. Determine if the system is consistent.
    $$\begin{bmatrix} 1 & 5 & 2 & -6 \\ 0 & 4 & -7 & 2 \\ 0 & 0 & 5 & 0 \end{bmatrix}$$

3. Is $(3, 4, -2)$ a solution of the following system?
    $$\begin{aligned} 5x_1 - x_2 + 2x_3 &= 7 \\ -2x_1 + 6x_2 + 9x_3 &= 0 \\ -7x_1 + 5x_2 - 3x_3 &= -7 \end{aligned}$$

4. For what values of h and k is the following system consistent?
    $$\begin{aligned} 2x_1 - x_2 &= h \\ -6x_1 + 3x_2 &= k \end{aligned}$$

1.1 EXERCISES

Solve each system in Exercises 1–4 by using elementary row operations on the equations or on the augmented matrix. Follow the systematic elimination procedure described in this section.

1.
$$x_1 + 5x_2 = 7$$
$$-2x_1 - 7x_2 = -5$$

2.
$$3x_1 + 6x_2 = -3$$
$$5x_1 + 7x_2 = 10$$

3. Find the point (x_1, x_2) that lies on the line $x_1 + 2x_2 = 4$ and on the line $x_1 - x_2 = 1$. See the figure.

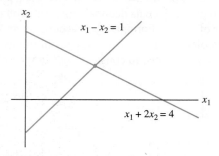

4. Find the point of intersection of the lines $x_1 + 2x_2 = -13$ and $3x_1 - 2x_2 = 1$

Consider each matrix in Exercises 5 and 6 as the augmented matrix of a linear system. State in words the next two elementary row operations that should be performed in the process of solving the system.

5.
$$\begin{bmatrix} 1 & -4 & -3 & 0 & 7 \\ 0 & 1 & 4 & 0 & 6 \\ 0 & 0 & 1 & 0 & 2 \\ 0 & 0 & 0 & 1 & -5 \end{bmatrix}$$

6.
$$\begin{bmatrix} 1 & -6 & 4 & 0 & -1 \\ 0 & 2 & -7 & 0 & 4 \\ 0 & 0 & 1 & 2 & -3 \\ 0 & 0 & 4 & 1 & 2 \end{bmatrix}$$

In Exercises 7–10, the augmented matrix of a linear system has been reduced by row operations to the form shown. In each case, continue the appropriate row operations and describe the solution set of the original system.

7.
$$\begin{bmatrix} 1 & 7 & 3 & -4 \\ 0 & 1 & -1 & 3 \\ 0 & 0 & 0 & 1 \\ 0 & 0 & 1 & -2 \end{bmatrix}$$

8.
$$\begin{bmatrix} 1 & -5 & 4 & 0 & 0 \\ 0 & 1 & 0 & 1 & 0 \\ 0 & 0 & 3 & 0 & 0 \\ 0 & 0 & 0 & 2 & 0 \end{bmatrix}$$

9.
$$\begin{bmatrix} 1 & -1 & 0 & 0 & -5 \\ 0 & 1 & -2 & 0 & -7 \\ 0 & 0 & 1 & -3 & 2 \\ 0 & 0 & 0 & 1 & 4 \end{bmatrix}$$

10.
$$\begin{bmatrix} 1 & 3 & 0 & -2 & -7 \\ 0 & 1 & 0 & 3 & 6 \\ 0 & 0 & 1 & 0 & 2 \\ 0 & 0 & 0 & 1 & -2 \end{bmatrix}$$

Solve the systems in Exercises 11–14.

11.
$$x_2 + 5x_3 = -4$$
$$x_1 + 4x_2 + 3x_3 = -2$$
$$2x_1 + 7x_2 + x_3 = -2$$

12.
$$x_1 - 5x_2 + 4x_3 = -3$$
$$2x_1 - 7x_2 + 3x_3 = -2$$
$$-2x_1 + x_2 + 7x_3 = -1$$

13.
$$x_1 - 3x_3 = 8$$
$$2x_1 + 2x_2 + 9x_3 = 7$$
$$x_2 + 5x_3 = -2$$

14.
$$2x_1 - 6x_3 = -8$$
$$x_2 + 2x_3 = 3$$
$$3x_1 + 6x_2 - 2x_3 = -4$$

Determine if the systems in Exercises 15 and 16 are consistent. Do not completely solve the systems.

15.
$$x_1 - 6x_2 = 5$$
$$x_2 - 4x_3 + x_4 = 0$$
$$-x_1 + 6x_2 + x_3 + 5x_4 = 3$$
$$-x_2 + 5x_3 + 4x_4 = 0$$

16.
$$2x_1 - 4x_4 = -10$$
$$3x_2 + 3x_3 = 0$$
$$x_3 + 4x_4 = -1$$
$$-3x_1 + 2x_2 + 3x_3 + x_4 = 5$$

17. Do the three lines $2x_1 + 3x_2 = -1$, $6x_1 + 5x_2 = 0$, and $2x_1 - 5x_2 = 7$ have a common point of intersection? Explain.

18. Do the three planes $2x_1 + 4x_2 + 4x_3 = 4$, $x_2 - 2x_3 = -2$, and $2x_1 + 3x_2 = 0$ have at least one common point of intersection? Explain.

In Exercises 19–22, determine the value(s) of h such that the matrix is the augmented matrix of a consistent linear system.

19.
$$\begin{bmatrix} 1 & h & 4 \\ 3 & 6 & 8 \end{bmatrix}$$

20.
$$\begin{bmatrix} 1 & h & -5 \\ 2 & -8 & 6 \end{bmatrix}$$

21.
$$\begin{bmatrix} 1 & 4 & -2 \\ 3 & h & -6 \end{bmatrix}$$

22.
$$\begin{bmatrix} -4 & 12 & h \\ 2 & -6 & -3 \end{bmatrix}$$

In Exercises 23 and 24, key statements from this section are either quoted directly, restated slightly (but still true), or altered in some way that makes them false in some cases. Mark each statement True or False, and *justify* your answer. (If true, give the approximate location where a similar statement appears, or refer to a definition or theorem. If false, give the location of a statement that has been quoted or used incorrectly, or cite an example that shows the statement is not true in all cases.) Similar true/false questions will appear in many sections of the text.

23. a. Every elementary row operation is reversible.

   b. A 5×6 matrix has six rows.

   c. The solution set of a linear system involving variables $x_1, \ldots, x_n$ is a list of numbers $(s_1, \ldots, s_n)$ that makes each equation in the system a true statement when the values $s_1, \ldots, s_n$ are substituted for $x_1, \ldots, x_n$, respectively.

   d. Two fundamental questions about a linear system involve existence and uniqueness.

24. a. Two matrices are row equivalent if they have the same number of rows.

   b. Elementary row operations on an augmented matrix never change the solution set of the associated linear system.

   c. Two equivalent linear systems can have different solution sets.

   d. A consistent system of linear equations has one or more solutions.

25. Find an equation involving g, h, and k that makes this augmented matrix correspond to a consistent system:
$$\begin{bmatrix} 1 & -4 & 7 & g \\ 0 & 3 & -5 & h \\ -2 & 5 & -9 & k \end{bmatrix}$$

26. Suppose the system below is consistent for all possible values of f and g. What can you say about the coefficients c and d? Justify your answer.
$$2x_1 + 4x_2 = f$$
$$cx_1 + dx_2 = g$$

27. Suppose a, b, c, and d are constants such that a is not zero and the system below is consistent for all possible values of f and g. What can you say about the numbers a, b, c, and d? Justify your answer.
$$ax_1 + bx_2 = f$$
$$cx_1 + dx_2 = g$$

28. Construct three different augmented matrices for linear systems whose solution set is $x_1 = 3$, $x_2 = -2$, $x_3 = -1$.

In Exercises 29–32, find the elementary row operation that transforms the first matrix into the second, and then find the reverse row operation that transforms the second matrix into the first.

29. $\begin{bmatrix} 0 & -2 & 5 \\ 1 & 3 & -5 \\ 3 & -1 & 6 \end{bmatrix}$, $\begin{bmatrix} 3 & -1 & 6 \\ 1 & 3 & -5 \\ 0 & -2 & 5 \end{bmatrix}$

30. $\begin{bmatrix} 1 & 3 & -4 \\ 0 & -2 & 6 \\ 0 & -5 & 10 \end{bmatrix}$, $\begin{bmatrix} 1 & 3 & -4 \\ 0 & -2 & 6 \\ 0 & 1 & -2 \end{bmatrix}$

31. $\begin{bmatrix} 1 & -2 & 1 & 0 \\ 0 & 5 & -2 & 8 \\ 4 & -1 & 3 & -6 \end{bmatrix}$, $\begin{bmatrix} 1 & -2 & 1 & 0 \\ 0 & 5 & -2 & 8 \\ 0 & 7 & -1 & -6 \end{bmatrix}$

32. $\begin{bmatrix} 1 & 2 & -5 & 0 \\ 0 & 1 & -3 & -2 \\ 0 & 4 & -12 & 7 \end{bmatrix}$, $\begin{bmatrix} 1 & 2 & -5 & 0 \\ 0 & 1 & -3 & -2 \\ 0 & 0 & 0 & 15 \end{bmatrix}$

An important concern in the study of heat transfer is to determine the steady-state temperature distribution of a thin plate when the temperature around the boundary is known. Assume the plate shown in the figure represents a cross section of a metal beam, with negligible heat flow in the direction perpendicular to the plate. Let $T_1, \ldots, T_4$ denote the temperatures at the four interior nodes of the mesh in the figure. The temperature at a node is approximately equal to the average of the four nearest nodes—to the left, above, to the right, and below.[3] For instance,
$$T_1 = (10 + 20 + T_2 + T_4)/4, \quad \text{or} \quad 4T_1 - T_2 - T_4 = 30$$

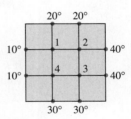

33. Write a system of four equations whose solution gives estimates for the temperatures $T_1, \ldots, T_4$.

34. Solve the system of equations from Exercise 33. [*Hint*: To speed up the calculations, interchange rows 1 and 4 before starting "replace" operations.]

---

[3] See Frank M. White, *Heat and Mass Transfer* (Reading, MA: Addison-Wesley Publishing, 1991), pp. 145–149.

SOLUTIONS TO PRACTICE PROBLEMS

1. a. For "hand computation," the best choice is to interchange equations 3 and 4. Another possibility is to multiply equation 3 by $1/5$. Or, replace equation 4 by its sum with $-1/5$ times row 3. (In any case, do not use the x_2 in equation 2 to eliminate the $4x_2$ in equation 1. Wait until a triangular form has been reached and the x_3 terms and x_4 terms have been eliminated from the first two equations.)

b. The system is in triangular form. Further simplification begins with the x_4 in the fourth equation. Use the x_4 to eliminate all x_4 terms above it. The appropriate step now is to add 2 times equation 4 to equation 1. (After that, move to equation 3, multiply it by $1/2$, and then use the equation to eliminate the x_3 terms above it.)

2. The system corresponding to the augmented matrix is

$$x_1 + 5x_2 + 2x_3 = -6$$
$$4x_2 - 7x_3 = 2$$
$$5x_3 = 0$$

The third equation makes $x_3 = 0$, which is certainly an allowable value for x_3. After eliminating the x_3 terms in equations 1 and 2, you could go on to solve for unique values for x_2 and x_1. Hence a solution exists, and it is unique. Contrast this situation with that in Example 3.

3. It is easy to check if a specific list of numbers is a solution. Set $x_1 = 3$, $x_2 = 4$, and $x_3 = -2$, and find that

$$5(3) - (4) + 2(-2) = 15 - 4 - 4 = 7$$
$$-2(3) + 6(4) + 9(-2) = -6 + 24 - 18 = 0$$
$$-7(3) + 5(4) - 3(-2) = -21 + 20 + 6 = 5$$

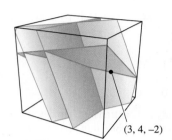

(3, 4, −2)

Since $(3, 4, -2)$ satisfies the first two equations, it is on the line of the intersection of the first two planes. Since $(3, 4, -2)$ does not satisfy all three equations, it does not lie on all three planes.

Although the first two equations are satisfied, the third is not, so $(3, 4, -2)$ is not a solution of the system. Notice the use of parentheses when making the substitutions. They are strongly recommended as a guard against arithmetic errors.

4. When the second equation is replaced by its sum with 3 times the first equation, the system becomes

$$2x_1 - x_2 = h$$
$$0 = k + 3h$$

If $k + 3h$ is nonzero, the system has no solution. The system is consistent for any values of h and k that make $k + 3h = 0$.

1.2 ROW REDUCTION AND ECHELON FORMS

This section refines the method of Section 1.1 into a row reduction algorithm that will enable us to analyze any system of linear equations.[1] By using only the first part of the algorithm, we will be able to answer the fundamental existence and uniqueness questions posed in Section 1.1.

The algorithm applies to any matrix, whether or not the matrix is viewed as an augmented matrix for a linear system. So the first part of this section concerns an arbitrary rectangular matrix and begins by introducing two important classes of matrices that include the "triangular" matrices of Section 1.1. In the definitions that follow, a *nonzero* row or column in a matrix means a row or column that contains at least one nonzero entry; a **leading entry** of a row refers to the leftmost nonzero entry (in a nonzero row).

---

[1] The algorithm here is a variant of what is commonly called *Gaussian elimination*. A similar elimination method for linear systems was used by Chinese mathematicians in about 250 B.C. The process was unknown in Western culture until the nineteenth century, when a famous German mathematician, Carl Friedrich Gauss, discovered it. A German engineer, Wilhelm Jordan, popularized the algorithm in an 1888 text on geodesy.

DEFINITION

A rectangular matrix is in **echelon form** (or **row echelon form**) if it has the following three properties:

1. All nonzero rows are above any rows of all zeros.
2. Each leading entry of a row is in a column to the right of the leading entry of the row above it.
3. All entries in a column below a leading entry are zeros.

If a matrix in echelon form satisfies the following additional conditions, then it is in **reduced echelon form** (or **reduced row echelon form**):

4. The leading entry in each nonzero row is 1.
5. Each leading 1 is the only nonzero entry in its column.

An **echelon matrix** (respectively, **reduced echelon matrix**) is one that is in echelon form (respectively, reduced echelon form). Property 2 says that the leading entries form an *echelon* ("steplike") pattern that moves down and to the right through the matrix. Property 3 is a simple consequence of property 2, but we include it for emphasis.

The "triangular" matrices of Section 1.1, such as

$$\begin{bmatrix} 2 & -3 & 2 & 1 \\ 0 & 1 & -4 & 8 \\ 0 & 0 & 0 & 5/2 \end{bmatrix} \quad \text{and} \quad \begin{bmatrix} 1 & 0 & 0 & 29 \\ 0 & 1 & 0 & 16 \\ 0 & 0 & 1 & 3 \end{bmatrix}$$

are in echelon form. In fact, the second matrix is in reduced echelon form. Here are additional examples.

EXAMPLE 1 The following matrices are in echelon form. The leading entries (■) may have any nonzero value; the starred entries (∗) may have any value (including zero).

The following matrices are in reduced echelon form because the leading entries are 1's, and there are 0's below *and above* each leading 1.

Any nonzero matrix may be **row reduced** (that is, transformed by elementary row operations) into more than one matrix in echelon form, using different sequences of row operations. However, the reduced echelon form one obtains from a matrix is unique. The following theorem is proved in Appendix A at the end of the text.

THEOREM 1

Uniqueness of the Reduced Echelon Form

Each matrix is row equivalent to one and only one reduced echelon matrix.

If a matrix A is row equivalent to an echelon matrix U, we call U **an echelon form** (or row echelon form) **of** A; if U is in reduced echelon form, we call U **the reduced echelon form of** A. [Most matrix programs and calculators with matrix capabilities use the abbreviation RREF for reduced (row) echelon form. Some use REF for (row) echelon form.]

Pivot Positions

When row operations on a matrix produce an echelon form, further row operations to obtain the reduced echelon form do not change the positions of the leading entries. Since the reduced echelon form is unique, *the leading entries are always in the same positions in any echelon form obtained from a given matrix.* These leading entries correspond to leading 1's in the reduced echelon form.

DEFINITION

A **pivot position** in a matrix A is a location in A that corresponds to a leading 1 in the reduced echelon form of A. A **pivot column** is a column of A that contains a pivot position.

In Example 1, the squares (■) identify the pivot positions. Many fundamental concepts in the first four chapters will be connected in one way or another with pivot positions in a matrix.

EXAMPLE 2 Row reduce the matrix A below to echelon form, and locate the pivot columns of A.

$$A = \begin{bmatrix} 0 & -3 & -6 & 4 & 9 \\ -1 & -2 & -1 & 3 & 1 \\ -2 & -3 & 0 & 3 & -1 \\ 1 & 4 & 5 & -9 & -7 \end{bmatrix}$$

SOLUTION Use the same basic strategy as in Section 1.1. The top of the leftmost nonzero column is the first pivot position. A nonzero entry, or *pivot*, must be placed in this position. A good choice is to interchange rows 1 and 4 (because the mental computations in the next step will not involve fractions).

┌─ Pivot

$$\begin{bmatrix} 1 & 4 & 5 & -9 & -7 \\ -1 & -2 & -1 & 3 & 1 \\ -2 & -3 & 0 & 3 & -1 \\ 0 & -3 & -6 & 4 & 9 \end{bmatrix}$$

└─ Pivot column

Create zeros below the pivot, 1, by adding multiples of the first row to the rows below, and obtain matrix (1) below. The pivot position in the second row must be as far left as possible—namely, in the second column. Choose the 2 in this position as the next pivot.

┌─ Pivot

$$\begin{bmatrix} 1 & 4 & 5 & -9 & -7 \\ 0 & 2 & 4 & -6 & -6 \\ 0 & 5 & 10 & -15 & -15 \\ 0 & -3 & -6 & 4 & 9 \end{bmatrix} \tag{1}$$

└─ Next pivot column

Add $-5/2$ times row 2 to row 3, and add $3/2$ times row 2 to row 4.

$$\begin{bmatrix} 1 & 4 & 5 & -9 & -7 \\ 0 & 2 & 4 & -6 & -6 \\ 0 & 0 & 0 & 0 & 0 \\ 0 & 0 & 0 & -5 & 0 \end{bmatrix} \qquad (2)$$

The matrix in (2) is different from any encountered in Section 1.1. There is no way to create a leading entry in column 3! (We can't use row 1 or 2 because doing so would destroy the echelon arrangement of the leading entries already produced.) However, if we interchange rows 3 and 4, we can produce a leading entry in column 4.

$$\begin{bmatrix} 1 & 4 & 5 & -9 & -7 \\ 0 & 2 & 4 & -6 & -6 \\ 0 & 0 & 0 & -5 & 0 \\ 0 & 0 & 0 & 0 & 0 \end{bmatrix} \qquad \text{General form:} \qquad \begin{bmatrix} \blacksquare & * & * & * & * \\ 0 & \blacksquare & * & * & * \\ 0 & 0 & 0 & \blacksquare & * \\ 0 & 0 & 0 & 0 & 0 \end{bmatrix}$$

Pivot columns

The matrix is in echelon form and thus reveals that columns 1, 2, and 4 of A are pivot columns.

Pivot positions

$$A = \begin{bmatrix} 0 & -3 & -6 & 4 & 9 \\ -1 & -2 & -1 & 3 & 1 \\ -2 & -3 & 0 & 3 & -1 \\ 1 & 4 & 5 & -9 & -7 \end{bmatrix} \qquad (3)$$

Pivot columns ∎

A **pivot**, as illustrated in Example 2, is a nonzero number in a pivot position that is used as needed to create zeros via row operations. The pivots in Example 2 were 1, 2, and -5. Notice that these numbers are not the same as the actual elements of A in the highlighted pivot positions shown in (3).

With Example 2 as a guide, we are ready to describe an efficient procedure for transforming a matrix into an echelon or reduced echelon matrix. Careful study and mastery of this procedure now will pay rich dividends later in the course.

The Row Reduction Algorithm

The algorithm that follows consists of four steps, and it produces a matrix in echelon form. A fifth step produces a matrix in reduced echelon form. We illustrate the algorithm by an example.

EXAMPLE 3   Apply elementary row operations to transform the following matrix first into echelon form and then into reduced echelon form:

$$\begin{bmatrix} 0 & 3 & -6 & 6 & 4 & -5 \\ 3 & -7 & 8 & -5 & 8 & 9 \\ 3 & -9 & 12 & -9 & 6 & 15 \end{bmatrix}$$

SOLUTION

STEP 1

Begin with the leftmost nonzero column. This is a pivot column. The pivot position is at the top.

$$\begin{bmatrix} 0 & 3 & -6 & 6 & 4 & -5 \\ 3 & -7 & 8 & -5 & 8 & 9 \\ 3 & -9 & 12 & -9 & 6 & 15 \end{bmatrix}$$

↑—— Pivot column

STEP 2

Select a nonzero entry in the pivot column as a pivot. If necessary, interchange rows to move this entry into the pivot position.

Interchange rows 1 and 3. (We could have interchanged rows 1 and 2 instead.)

— Pivot

$$\begin{bmatrix} 3 & -9 & 12 & -9 & 6 & 15 \\ 3 & -7 & 8 & -5 & 8 & 9 \\ 0 & 3 & -6 & 6 & 4 & -5 \end{bmatrix}$$

STEP 3

Use row replacement operations to create zeros in all positions below the pivot.

As a preliminary step, we could divide the top row by the pivot, 3. But with two 3's in column 1, it is just as easy to add -1 times row 1 to row 2.

— Pivot

$$\begin{bmatrix} 3 & -9 & 12 & -9 & 6 & 15 \\ 0 & 2 & -4 & 4 & 2 & -6 \\ 0 & 3 & -6 & 6 & 4 & -5 \end{bmatrix}$$

STEP 4

Cover (or ignore) the row containing the pivot position and cover all rows, if any, above it. Apply steps 1–3 to the submatrix that remains. Repeat the process until there are no more nonzero rows to modify.

With row 1 covered, step 1 shows that column 2 is the next pivot column; for step 2, select as a pivot the "top" entry in that column.

— Pivot

$$\begin{bmatrix} 3 & -9 & 12 & -9 & 6 & 15 \\ 0 & 2 & -4 & 4 & 2 & -6 \\ 0 & 3 & -6 & 6 & 4 & -5 \end{bmatrix}$$

↑—— New pivot column

For step 3, we could insert an optional step of dividing the "top" row of the submatrix by the pivot, 2. Instead, we add $-3/2$ times the "top" row to the row below. This produces

$$\begin{bmatrix} 3 & -9 & 12 & -9 & 6 & 15 \\ 0 & 2 & -4 & 4 & 2 & -6 \\ 0 & 0 & 0 & 0 & 1 & 4 \end{bmatrix}$$

When we cover the row containing the second pivot position for step 4, we are left with a new submatrix having only one row:

$$\begin{bmatrix} 3 & -9 & 12 & -9 & 6 & 15 \\ 0 & 2 & -4 & 4 & 2 & -6 \\ 0 & 0 & 0 & 0 & 1 & 4 \end{bmatrix}$$

             Pivot

Steps 1–3 require no work for this submatrix, and we have reached an echelon form of the full matrix. If we want the reduced echelon form, we perform one more step.

STEP 5

Beginning with the rightmost pivot and working upward and to the left, create zeros above each pivot. If a pivot is not 1, make it 1 by a scaling operation.

The rightmost pivot is in row 3. Create zeros above it, adding suitable multiples of row 3 to rows 2 and 1.

$$\begin{bmatrix} 3 & -9 & 12 & -9 & 0 & -9 \\ 0 & 2 & -4 & 4 & 0 & -14 \\ 0 & 0 & 0 & 0 & 1 & 4 \end{bmatrix} \quad \begin{array}{l} \leftarrow \text{Row } 1 + (-6) \cdot \text{row } 3 \\ \leftarrow \text{Row } 2 + (-2) \cdot \text{row } 3 \end{array}$$

The next pivot is in row 2. Scale this row, dividing by the pivot.

$$\begin{bmatrix} 3 & -9 & 12 & -9 & 0 & -9 \\ 0 & 1 & -2 & 2 & 0 & -7 \\ 0 & 0 & 0 & 0 & 1 & 4 \end{bmatrix} \quad \leftarrow \text{Row scaled by } \tfrac{1}{2}$$

Create a zero in column 2 by adding 9 times row 2 to row 1.

$$\begin{bmatrix} 3 & 0 & -6 & 9 & 0 & -72 \\ 0 & 1 & -2 & 2 & 0 & -7 \\ 0 & 0 & 0 & 0 & 1 & 4 \end{bmatrix} \quad \leftarrow \text{Row } 1 + (9) \cdot \text{row } 2$$

Finally, scale row 1, dividing by the pivot, 3.

$$\begin{bmatrix} 1 & 0 & -2 & 3 & 0 & -24 \\ 0 & 1 & -2 & 2 & 0 & -7 \\ 0 & 0 & 0 & 0 & 1 & 4 \end{bmatrix} \quad \leftarrow \text{Row scaled by } \tfrac{1}{3}$$

This is the reduced echelon form of the original matrix.   ■

  The combination of steps 1–4 is called the **forward phase** of the row reduction algorithm. Step 5, which produces the unique reduced echelon form, is called the **backward phase**.

NUMERICAL NOTE

In step 2 above, a computer program usually selects as a pivot the entry in a column having the largest absolute value. This strategy, called **partial pivoting**, is used because it reduces roundoff errors in the calculations.

Solutions of Linear Systems

The row reduction algorithm leads directly to an explicit description of the solution set of a linear system when the algorithm is applied to the augmented matrix of the system.

Suppose, for example, that the augmented matrix of a linear system has been changed into the equivalent *reduced* echelon form

$$\begin{bmatrix} 1 & 0 & -5 & 1 \\ 0 & 1 & 1 & 4 \\ 0 & 0 & 0 & 0 \end{bmatrix}$$

There are three variables because the augmented matrix has four columns. The associated system of equations is

$$
\begin{aligned}
x_1 \quad\;\; - 5x_3 &= 1 \\
x_2 + \;\; x_3 &= 4 \\
0 &= 0
\end{aligned}
\tag{4}
$$

The variables x_1 and x_2 corresponding to pivot columns in the matrix are called **basic variables**.[2] The other variable, x_3, is called a **free variable**.

Whenever a system is consistent, as in (4), the solution set can be described explicitly by solving the *reduced* system of equations for the basic variables in terms of the free variables. This operation is possible because the reduced echelon form places each basic variable in one and only one equation. In (4), solve the first equation for x_1 and the second for x_2. (Ignore the third equation; it offers no restriction on the variables.)

$$
\begin{cases}
x_1 = 1 + 5x_3 \\
x_2 = 4 - x_3 \\
x_3 \text{ is free}
\end{cases}
\tag{5}
$$

The statement "x_3 is free" means that you are free to choose any value for x_3. Once that is done, the formulas in (5) determine the values for x_1 and x_2. For instance, when $x_3 = 0$, the solution is $(1, 4, 0)$; when $x_3 = 1$, the solution is $(6, 3, 1)$. *Each different choice of x_3 determines a (different) solution of the system, and every solution of the system is determined by a choice of x_3.*

EXAMPLE 4 Find the general solution of the linear system whose augmented matrix has been reduced to

$$\begin{bmatrix} 1 & 6 & 2 & -5 & -2 & -4 \\ 0 & 0 & 2 & -8 & -1 & 3 \\ 0 & 0 & 0 & 0 & 1 & 7 \end{bmatrix}$$

SOLUTION The matrix is in echelon form, but we want the reduced echelon form before solving for the basic variables. The row reduction is completed next. The symbol $\sim$ before a matrix indicates that the matrix is row equivalent to the preceding matrix.

$$\begin{bmatrix} 1 & 6 & 2 & -5 & -2 & -4 \\ 0 & 0 & 2 & -8 & -1 & 3 \\ 0 & 0 & 0 & 0 & 1 & 7 \end{bmatrix} \sim \begin{bmatrix} 1 & 6 & 2 & -5 & 0 & 10 \\ 0 & 0 & 2 & -8 & 0 & 10 \\ 0 & 0 & 0 & 0 & 1 & 7 \end{bmatrix}$$

$$\sim \begin{bmatrix} 1 & 6 & 2 & -5 & 0 & 10 \\ 0 & 0 & 1 & -4 & 0 & 5 \\ 0 & 0 & 0 & 0 & 1 & 7 \end{bmatrix} \sim \begin{bmatrix} 1 & 6 & 0 & 3 & 0 & 0 \\ 0 & 0 & 1 & -4 & 0 & 5 \\ 0 & 0 & 0 & 0 & 1 & 7 \end{bmatrix}$$

---

[2] Some texts use the term *leading variables* because they correspond to the columns containing leading entries.

There are five variables because the augmented matrix has six columns. The associated system now is

$$\begin{array}{rcl} x_1 + 6x_2 \quad\quad + 3x_4 \quad\quad &=& 0 \\ x_3 - 4x_4 \quad &=& 5 \\ x_5 &=& 7 \end{array} \tag{6}$$

The pivot columns of the matrix are 1, 3, and 5, so the basic variables are x_1, x_3, and x_5. The remaining variables, x_2 and x_4, must be free. Solve for the basic variables to obtain the general solution:

$$\begin{cases} x_1 = -6x_2 - 3x_4 \\ x_2 \text{ is free} \\ x_3 = 5 + 4x_4 \\ x_4 \text{ is free} \\ x_5 = 7 \end{cases} \tag{7}$$

Note that the value of x_5 is already fixed by the third equation in system (6). ∎

Parametric Descriptions of Solution Sets

The descriptions in (5) and (7) are *parametric descriptions* of solution sets in which the free variables act as parameters. *Solving a system* amounts to finding a parametric description of the solution set or determining that the solution set is empty.

Whenever a system is consistent and has free variables, the solution set has many parametric descriptions. For instance, in system (4), we may add 5 times equation 2 to equation 1 and obtain the equivalent system

$$\begin{array}{rcl} x_1 + 5x_2 \quad\quad &=& 21 \\ x_2 + x_3 &=& 4 \end{array}$$

We could treat x_2 as a parameter and solve for x_1 and x_3 in terms of x_2, and we would have an accurate description of the solution set. However, to be consistent, we make the (arbitrary) convention of always using the free variables as the parameters for describing a solution set. (The answer section at the end of the text also reflects this convention.)

Whenever a system is inconsistent, the solution set is empty, even when the system has free variables. In this case, the solution set has *no* parametric representation.

Back-Substitution

Consider the following system, whose augmented matrix is in echelon form but is *not* in reduced echelon form:

$$\begin{array}{rcl} x_1 - 7x_2 + 2x_3 - 5x_4 + 8x_5 &=& 10 \\ x_2 - 3x_3 + 3x_4 + x_5 &=& -5 \\ x_4 - x_5 &=& 4 \end{array}$$

A computer program would solve this system by back-substitution, rather than by computing the reduced echelon form. That is, the program would solve equation 3 for x_4 in terms of x_5 and substitute the expression for x_4 into equation 2, solve equation 2 for x_2, and then substitute the expressions for x_2 and x_4 into equation 1 and solve for x_1.

Our matrix format for the backward phase of row reduction, which produces the reduced echelon form, has the same number of arithmetic operations as back-substitution. But the discipline of the matrix format substantially reduces the likelihood of errors during hand computations. The best strategy is to use only the *reduced* echelon form to solve a system! The *Study Guide* that accompanies this text offers several helpful suggestions for performing row operations accurately and rapidly.

NUMERICAL NOTE

In general, the forward phase of row reduction takes much longer than the backward phase. An algorithm for solving a system is usually measured in flops (or floating point operations). A **flop** is one arithmetic operation ($+, -, *, /$) on two real floating point numbers.[3] For an $n \times (n + 1)$ matrix, the reduction to echelon form can take $2n^3/3 + n^2/2 - 7n/6$ flops (which is approximately $2n^3/3$ flops when n is moderately large—say, $n \geq 30$). In contrast, further reduction to reduced echelon form needs at most n^2 flops.

Existence and Uniqueness Questions

Although a nonreduced echelon form is a poor tool for solving a system, this form is just the right device for answering two fundamental questions posed in Section 1.1.

EXAMPLE 5 Determine the existence and uniqueness of the solutions to the system

$$
\begin{aligned}
3x_2 - 6x_3 + 6x_4 + 4x_5 &= -5 \\
3x_1 - 7x_2 + 8x_3 - 5x_4 + 8x_5 &= 9 \\
3x_1 - 9x_2 + 12x_3 - 9x_4 + 6x_5 &= 15
\end{aligned}
$$

SOLUTION The augmented matrix of this system was row reduced in Example 3 to

$$
\begin{bmatrix}
3 & -9 & 12 & -9 & 6 & 15 \\
0 & 2 & -4 & 4 & 2 & -6 \\
0 & 0 & 0 & 0 & 1 & 4
\end{bmatrix}
\tag{8}
$$

The basic variables are x_1, x_2, and x_5; the free variables are x_3 and x_4. There is no equation such as $0 = 1$ that would indicate an inconsistent system, so we could use back-substitution to find a solution. But the *existence* of a solution is already clear in (8). Also, the solution is *not unique* because there are free variables. Each different choice of x_3 and x_4 determines a different solution. Thus the system has infinitely many solutions. ■

When a system is in echelon form and contains no equation of the form $0 = b$, with b nonzero, every nonzero equation contains a basic variable with a nonzero coefficient. Either the basic variables are completely determined (with no free variables) or at least one of the basic variables may be expressed in terms of one or more free variables. In the former case, there is a unique solution; in the latter case, there are infinitely many solutions (one for each choice of values for the free variables).

These remarks justify the following theorem.

---

[3] Traditionally, a *flop* was only a multiplication or division, because addition and subtraction took much less time and could be ignored. The definition of *flop* given here is preferred now, as a result of advances in computer architecture. See Golub and Van Loan, *Matrix Computations*, 2nd ed. (Baltimore: The Johns Hopkins Press, 1989), pp. 19–20.

THEOREM 2

Existence and Uniqueness Theorem

A linear system is consistent if and only if the rightmost column of the augmented matrix is *not* a pivot column—that is, if and only if an echelon form of the augmented matrix has *no* row of the form

$$[0 \ \cdots \ 0 \ b] \qquad \text{with } b \text{ nonzero}$$

If a linear system is consistent, then the solution set contains either (i) a unique solution, when there are no free variables, or (ii) infinitely many solutions, when there is at least one free variable.

The following procedure outlines how to find and describe all solutions of a linear system.

USING ROW REDUCTION TO SOLVE A LINEAR SYSTEM

1. Write the augmented matrix of the system.
2. Use the row reduction algorithm to obtain an equivalent augmented matrix in echelon form. Decide whether the system is consistent. If there is no solution, stop; otherwise, go to the next step.
3. Continue row reduction to obtain the reduced echelon form.
4. Write the system of equations corresponding to the matrix obtained in step 3.
5. Rewrite each nonzero equation from step 4 so that its one basic variable is expressed in terms of any free variables appearing in the equation.

PRACTICE PROBLEMS

1. Find the general solution of the linear system whose augmented matrix is

$$\begin{bmatrix} 1 & -3 & -5 & 0 \\ 0 & 1 & 1 & 3 \end{bmatrix}$$

2. Find the general solution of the system

$$\begin{aligned} x_1 - 2x_2 - \quad x_3 + 3x_4 &= 0 \\ -2x_1 + 4x_2 + 5x_3 - 5x_4 &= 3 \\ 3x_1 - 6x_2 - 6x_3 + 8x_4 &= 2 \end{aligned}$$

1.2 EXERCISES

In Exercises 1 and 2, determine which matrices are in reduced echelon form and which others are only in echelon form.

1. a. $\begin{bmatrix} 1 & 0 & 0 & 0 \\ 0 & 1 & 0 & 0 \\ 0 & 0 & 1 & 1 \end{bmatrix}$
b. $\begin{bmatrix} 1 & 0 & 1 & 0 \\ 0 & 1 & 1 & 0 \\ 0 & 0 & 0 & 1 \end{bmatrix}$
c. $\begin{bmatrix} 1 & 0 & 0 & 0 \\ 0 & 1 & 1 & 0 \\ 0 & 0 & 0 & 0 \\ 0 & 0 & 0 & 1 \end{bmatrix}$
d. $\begin{bmatrix} 1 & 1 & 0 & 1 & 1 \\ 0 & 2 & 0 & 2 & 2 \\ 0 & 0 & 0 & 3 & 3 \\ 0 & 0 & 0 & 0 & 4 \end{bmatrix}$

2. a. $\begin{bmatrix} 1 & 0 & 1 & 1 \\ 0 & 1 & 1 & 1 \\ 0 & 0 & 0 & 0 \end{bmatrix}$ **b.** $\begin{bmatrix} 1 & 0 & 0 & 0 \\ 0 & 2 & 0 & 0 \\ 0 & 0 & 1 & 1 \end{bmatrix}$

c. $\begin{bmatrix} 0 & 0 & 0 & 0 \\ 1 & 2 & 0 & 0 \\ 0 & 0 & 1 & 0 \\ 0 & 0 & 0 & 1 \end{bmatrix}$

d. $\begin{bmatrix} 0 & 1 & 1 & 1 & 1 \\ 0 & 0 & 1 & 1 & 1 \\ 0 & 0 & 0 & 0 & 1 \\ 0 & 0 & 0 & 0 & 0 \end{bmatrix}$

Row reduce the matrices in Exercises 3 and 4 to reduced echelon form. Circle the pivot positions in the final matrix and in the original matrix, and list the pivot columns.

3. $\begin{bmatrix} 1 & 2 & 4 & 8 \\ 2 & 4 & 6 & 8 \\ 3 & 6 & 9 & 12 \end{bmatrix}$ **4.** $\begin{bmatrix} 1 & 2 & 4 & 5 \\ 2 & 4 & 5 & 4 \\ 4 & 5 & 4 & 2 \end{bmatrix}$

5. Describe the possible echelon forms of a nonzero 2×2 matrix. Use the symbols ■, *, and 0, as in the first part of Example 1.

6. Repeat Exercise 5 for a nonzero 3×2 matrix.

Find the general solutions of the systems whose augmented matrices are given in Exercises 7–14.

7. $\begin{bmatrix} 1 & 3 & 4 & 7 \\ 3 & 9 & 7 & 6 \end{bmatrix}$ **8.** $\begin{bmatrix} 1 & -3 & 0 & -5 \\ -3 & 7 & 0 & 9 \end{bmatrix}$

9. $\begin{bmatrix} 0 & 1 & -2 & 3 \\ 1 & -3 & 4 & -6 \end{bmatrix}$ **10.** $\begin{bmatrix} 1 & -2 & -1 & 4 \\ -2 & 4 & -5 & 6 \end{bmatrix}$

11. $\begin{bmatrix} 3 & -2 & 4 & 0 \\ 9 & -6 & 12 & 0 \\ 6 & -4 & 8 & 0 \end{bmatrix}$ **12.** $\begin{bmatrix} 1 & 0 & -9 & 0 & 4 \\ 0 & 1 & 3 & 0 & -1 \\ 0 & 0 & 0 & 1 & -7 \\ 0 & 0 & 0 & 0 & 1 \end{bmatrix}$

13. $\begin{bmatrix} 1 & -3 & 0 & -1 & 0 & -2 \\ 0 & 1 & 0 & 0 & -4 & 1 \\ 0 & 0 & 0 & 1 & 9 & 4 \\ 0 & 0 & 0 & 0 & 0 & 0 \end{bmatrix}$

14. $\begin{bmatrix} 1 & 0 & -5 & 0 & -8 & 3 \\ 0 & 1 & 4 & -1 & 0 & 6 \\ 0 & 0 & 0 & 0 & 1 & 0 \\ 0 & 0 & 0 & 0 & 0 & 0 \end{bmatrix}$

Exercises 15 and 16 use the notation of Example 1 for matrices in echelon form. Suppose each matrix represents the augmented matrix for a system of linear equations. In each case, determine if the system is consistent. If the system is consistent, determine if the solution is unique.

15. a. $\begin{bmatrix} ■ & * & * & * \\ 0 & ■ & * & * \\ 0 & 0 & 0 & 0 \end{bmatrix}$

b. $\begin{bmatrix} 0 & ■ & * & * & * \\ 0 & 0 & ■ & * & * \\ 0 & 0 & 0 & ■ & 0 \end{bmatrix}$

16. a. $\begin{bmatrix} ■ & * & * \\ 0 & ■ & * \\ 0 & 0 & ■ \end{bmatrix}$

b. $\begin{bmatrix} ■ & * & * & * & * \\ 0 & 0 & ■ & * & * \\ 0 & 0 & 0 & ■ & * \end{bmatrix}$

In Exercises 17 and 18, determine the value(s) of h such that the matrix is the augmented matrix of a consistent linear system.

17. $\begin{bmatrix} 1 & -1 & 4 \\ -2 & 3 & h \end{bmatrix}$ **18.** $\begin{bmatrix} 1 & -3 & 1 \\ h & 6 & -2 \end{bmatrix}$

In Exercises 19 and 20, choose h and k such that the system has (a) no solution, (b) a unique solution, and (c) many solutions. Give separate answers for each part.

19. $\begin{aligned} x_1 + hx_2 &= 2 \\ 4x_1 + 8x_2 &= k \end{aligned}$ **20.** $\begin{aligned} x_1 - 3x_2 &= 1 \\ 2x_1 + hx_2 &= k \end{aligned}$

In Exercises 21 and 22, mark each statement True or False. Justify each answer.[4]

21. a. In some cases, a matrix may be row reduced to more than one matrix in reduced echelon form, using different sequences of row operations.

b. The row reduction algorithm applies only to augmented matrices for a linear system.

c. A basic variable in a linear system is a variable that corresponds to a pivot column in the coefficient matrix.

d. Finding a parametric description of the solution set of a linear system is the same as *solving* the system.

e. If one row in an echelon form of an augmented matrix is $[0 \ 0 \ 0 \ 5 \ 0]$, then the associated linear system is inconsistent.

22. a. The reduced echelon form of a matrix is unique.

b. If every column of an augmented matrix contains a pivot, then the corresponding system is consistent.

c. The pivot positions in a matrix depend on whether row interchanges are used in the row reduction process.

d. A general solution of a system is an explicit description of all solutions of the system.

e. Whenever a system has free variables, the solution set contains many solutions.

23. Suppose the coefficient matrix of a linear system of four equations in four variables has a pivot in each column. Explain why the system has a unique solution.

24. Suppose a system of linear equations has a 3×5 *augmented* matrix whose fifth column is not a pivot column. Is the system consistent? Why (or why not)?

---

[4] True/false questions of this type will appear in many sections. Methods for justifying your answers were described before Exercises 23 and 24 in Section 1.1.

25. Suppose the coefficient matrix of a system of linear equations has a pivot position in every row. Explain why the system is consistent.

26. Suppose a 3×5 *coefficient* matrix for a system has three pivot columns. Is the system consistent? Why or why not?

27. Restate the last sentence in Theorem 2 using the concept of pivot columns: "If a linear system is consistent, then the solution is unique if and only if _____."

28. What would you have to know about the pivot columns in an augmented matrix in order to know that the linear system is consistent and has a unique solution?

29. A system of linear equations with fewer equations than unknowns is sometimes called an *underdetermined system*. Can such a system have a unique solution? Explain.

30. Give an example of an inconsistent underdetermined system of two equations in three unknowns.

31. A system of linear equations with more equations than unknowns is sometimes called an *overdetermined system*. Can such a system be consistent? Illustrate your answer with a specific system of three equations in two unknowns.

32. Suppose an $n \times (n + 1)$ matrix is row reduced to reduced echelon form. Approximately what fraction of the total number of operations (flops) is involved in the backward phase of the reduction when $n = 20$? when $n = 200$?

Suppose experimental data are represented by a set of points in the plane. An **interpolating polynomial** for the data is a polynomial whose graph passes through every point. In scientific work, such a polynomial can be used, for example, to estimate values between the known data points. Another use is to create curves for graphical images on a computer screen. One method for finding an interpolating polynomial is to solve a system of linear equations.

WEB

33. Find the interpolating polynomial $p(t) = a_0 + a_1 t + a_2 t^2$ for the data $(1, 6)$, $(2, 15)$, $(3, 28)$. That is, find a_0, a_1, and a_2 such that

$$a_0 + a_1(1) + a_2(1)^2 = 6$$
$$a_0 + a_1(2) + a_2(2)^2 = 15$$
$$a_0 + a_1(3) + a_2(3)^2 = 28$$

34. [M] In a wind tunnel experiment, the force on a projectile due to air resistance was measured at different velocities:

Velocity (100 ft/sec)	0	2	4	6	8	10
Force (100 lb)	0	2.90	14.8	39.6	74.3	119

Find an interpolating polynomial for these data and estimate the force on the projectile when the projectile is traveling at 750 ft/sec. Use $p(t) = a_0 + a_1 t + a_2 t^2 + a_3 t^3 + a_4 t^4 + a_5 t^5$. What happens if you try to use a polynomial of degree less than 5? (Try a cubic polynomial, for instance.)[5]

_____
[5] Exercises marked with the symbol [M] are designed to be worked with the aid of a "Matrix program" (a computer program, such as MATLAB®, Maple™, Mathematica®, MathCad®, or Derive™, or a programmable calculator with matrix capabilities, such as those manufactured by Texas Instruments or Hewlett-Packard).

SOLUTIONS TO PRACTICE PROBLEMS

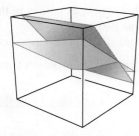

The general solution of the system of equations is the line of intersection of the two planes.

1. The reduced echelon form of the augmented matrix and the corresponding system are

$$\begin{bmatrix} 1 & 0 & -2 & 9 \\ 0 & 1 & 1 & 3 \end{bmatrix} \quad \text{and} \quad \begin{array}{r} x_1 \quad\ - 2x_3 = 9 \\ x_2 + \ x_3 = 3 \end{array}$$

The basic variables are x_1 and x_2, and the general solution is

$$\begin{cases} x_1 = 9 + 2x_3 \\ x_2 = 3 - x_3 \\ x_3 \text{ is free} \end{cases}$$

Note: It is essential that the general solution describe each variable, with any parameters clearly identified. The following statement does *not* describe the solution:

$$\begin{cases} x_1 = 9 + 2x_3 \\ x_2 = 3 - x_3 \\ x_3 = 3 - x_2 \quad \text{Incorrect solution} \end{cases}$$

This description implies that x_2 and x_3 are *both* free, which certainly is not the case.

2. Row reduce the system's augmented matrix:

$$\begin{bmatrix} 1 & -2 & -1 & 3 & 0 \\ -2 & 4 & 5 & -5 & 3 \\ 3 & -6 & -6 & 8 & 2 \end{bmatrix} \sim \begin{bmatrix} 1 & -2 & -1 & 3 & 0 \\ 0 & 0 & 3 & 1 & 3 \\ 0 & 0 & -3 & -1 & 2 \end{bmatrix}$$

$$\sim \begin{bmatrix} 1 & -2 & -1 & 3 & 0 \\ 0 & 0 & 3 & 1 & 3 \\ 0 & 0 & 0 & 0 & 5 \end{bmatrix}$$

This echelon matrix shows that the system is *inconsistent*, because its rightmost column is a pivot column; the third row corresponds to the equation $0 = 5$. There is no need to perform any more row operations. Note that the presence of the free variables in this problem is irrelevant because the system is inconsistent.

1.3 | VECTOR EQUATIONS

Important properties of linear systems can be described with the concept and notation of vectors. This section connects equations involving vectors to ordinary systems of equations. The term *vector* appears in a variety of mathematical and physical contexts, which we will discuss in Chapter 4, "Vector Spaces." Until then, *vector* will mean an *ordered list of numbers*. This simple idea enables us to get to interesting and important applications as quickly as possible.

Vectors in $\mathbb{R}^2$

A matrix with only one column is called a **column vector**, or simply a **vector**. Examples of vectors with two entries are

$$\mathbf{u} = \begin{bmatrix} 3 \\ -1 \end{bmatrix}, \qquad \mathbf{v} = \begin{bmatrix} .2 \\ .3 \end{bmatrix}, \qquad \mathbf{w} = \begin{bmatrix} w_1 \\ w_2 \end{bmatrix}$$

where w_1 and w_2 are any real numbers. The set of all vectors with two entries is denoted by $\mathbb{R}^2$ (read "r-two"). The $\mathbb{R}$ stands for the real numbers that appear as entries in the vectors, and the exponent 2 indicates that each vector contains two entries.[1]

Two vectors in $\mathbb{R}^2$ are **equal** if and only if their corresponding entries are equal. Thus $\begin{bmatrix} 4 \\ 7 \end{bmatrix}$ and $\begin{bmatrix} 7 \\ 4 \end{bmatrix}$ are *not* equal, because vectors in $\mathbb{R}^2$ are *ordered pairs* of real numbers.

Given two vectors $\mathbf{u}$ and $\mathbf{v}$ in $\mathbb{R}^2$, their **sum** is the vector $\mathbf{u} + \mathbf{v}$ obtained by adding corresponding entries of $\mathbf{u}$ and $\mathbf{v}$. For example,

$$\begin{bmatrix} 1 \\ -2 \end{bmatrix} + \begin{bmatrix} 2 \\ 5 \end{bmatrix} = \begin{bmatrix} 1+2 \\ -2+5 \end{bmatrix} = \begin{bmatrix} 3 \\ 3 \end{bmatrix}$$

Given a vector $\mathbf{u}$ and a real number c, the **scalar multiple** of $\mathbf{u}$ by c is the vector $c\mathbf{u}$ obtained by multiplying each entry in $\mathbf{u}$ by c. For instance,

$$\text{if} \quad \mathbf{u} = \begin{bmatrix} 3 \\ -1 \end{bmatrix} \quad \text{and} \quad c = 5, \quad \text{then} \quad c\mathbf{u} = 5\begin{bmatrix} 3 \\ -1 \end{bmatrix} = \begin{bmatrix} 15 \\ -5 \end{bmatrix}$$

---

[1] Most of the text concerns vectors and matrices that have only real entries. However, all definitions and theorems in Chapters 1–5, and in most of the rest of the text, remain valid if the entries are complex numbers. Complex vectors and matrices arise naturally, for example, in electrical engineering and physics.

The number c in $c\mathbf{u}$ is called a **scalar**; it is written in lightface type to distinguish it from the boldface vector $\mathbf{u}$.

The operations of scalar multiplication and vector addition can be combined, as in the following example.

EXAMPLE 1 Given $\mathbf{u} = \begin{bmatrix} 1 \\ -2 \end{bmatrix}$ and $\mathbf{v} = \begin{bmatrix} 2 \\ -5 \end{bmatrix}$, find $4\mathbf{u}$, $(-3)\mathbf{v}$, and $4\mathbf{u} + (-3)\mathbf{v}$.

SOLUTION

$$4\mathbf{u} = \begin{bmatrix} 4 \\ -8 \end{bmatrix}, \qquad (-3)\mathbf{v} = \begin{bmatrix} -6 \\ 15 \end{bmatrix}$$

and

$$4\mathbf{u} + (-3)\mathbf{v} = \begin{bmatrix} 4 \\ -8 \end{bmatrix} + \begin{bmatrix} -6 \\ 15 \end{bmatrix} = \begin{bmatrix} -2 \\ 7 \end{bmatrix}$$ ∎

Sometimes, for convenience (and also to save space), this text may write a column vector such as $\begin{bmatrix} 3 \\ -1 \end{bmatrix}$ in the form $(3, -1)$. In this case, the parentheses and the comma distinguish the vector $(3, -1)$ from the 1×2 row matrix $\begin{bmatrix} 3 & -1 \end{bmatrix}$, written with brackets and no comma. Thus

$$\begin{bmatrix} 3 \\ -1 \end{bmatrix} \neq \begin{bmatrix} 3 & -1 \end{bmatrix}$$

because the matrices have different shapes, even though they have the same entries.

Geometric Descriptions of $\mathbb{R}^2$

Consider a rectangular coordinate system in the plane. Because each point in the plane is determined by an ordered pair of numbers, *we can identify a geometric point* (a, b) *with the column vector* $\begin{bmatrix} a \\ b \end{bmatrix}$. So we may regard $\mathbb{R}^2$ as the set of all points in the plane. See Fig. 1.

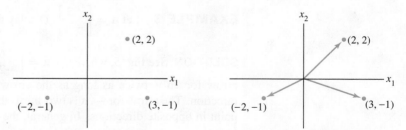

FIGURE 1 Vectors as points.　　　　**FIGURE 2** Vectors with arrows.

The geometric visualization of a vector such as $\begin{bmatrix} 3 \\ -1 \end{bmatrix}$ is often aided by including an arrow (directed line segment) from the origin $(0, 0)$ to the point $(3, -1)$, as in Fig. 2. In this case, the individual points along the arrow itself have no special significance.[2]

The sum of two vectors has a useful geometric representation. The following rule can be verified by analytic geometry.

---

[2] In physics, arrows can represent forces and usually are free to move about in space. This interpretation of vectors will be discussed in Section 4.1.

Parallelogram Rule for Addition

If **u** and **v** in $\mathbb{R}^2$ are represented as points in the plane, then **u** + **v** corresponds to the fourth vertex of the parallelogram whose other vertices are **u**, **0**, and **v**. See Fig. 3.

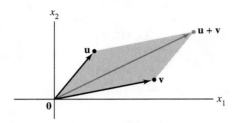

FIGURE 3 The parallelogram rule.

EXAMPLE 2 The vectors $\mathbf{u} = \begin{bmatrix} 2 \\ 2 \end{bmatrix}$, $\mathbf{v} = \begin{bmatrix} -6 \\ 1 \end{bmatrix}$, and $\mathbf{u} + \mathbf{v} = \begin{bmatrix} -4 \\ 3 \end{bmatrix}$ are displayed in Fig. 4. ■

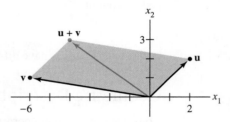

FIGURE 4

The next example illustrates the fact that the set of all scalar multiples of one fixed nonzero vector is a line through the origin, $(0, 0)$.

EXAMPLE 3 Let $\mathbf{u} = \begin{bmatrix} 3 \\ -1 \end{bmatrix}$. Display the vectors **u**, 2**u**, and $-\frac{2}{3}\mathbf{u}$ on a graph.

SOLUTION See Fig. 5, where **u**, $2\mathbf{u} = \begin{bmatrix} 6 \\ -2 \end{bmatrix}$, and $-\frac{2}{3}\mathbf{u} = \begin{bmatrix} -2 \\ 2/3 \end{bmatrix}$ are displayed. The arrow for 2**u** is twice as long as the arrow for **u**, and the arrows point in the same direction. The arrow for $-\frac{2}{3}\mathbf{u}$ is two-thirds the length of the arrow for **u**, and the arrows point in opposite directions. In general, the length of the arrow for $c\mathbf{u}$ is $|c|$ times the

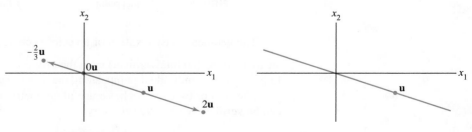

Typical multiples of **u**   The set of all multiples of **u**

FIGURE 5

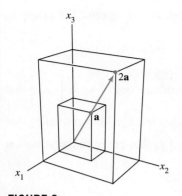

FIGURE 6
Scalar multiples .

length of the arrow for **u**. [Recall that the length of the line segment from $(0, 0)$ to (a, b) is $\sqrt{a^2 + b^2}$. We shall discuss this further in Chapter 6.] ∎

Vectors in $\mathbb{R}^3$

Vectors in $\mathbb{R}^3$ are 3×1 column matrices with three entries. They are represented geometrically by points in a three-dimensional coordinate space, with arrows from the origin sometimes included for visual clarity. The vectors $\mathbf{a} = \begin{bmatrix} 2 \\ 3 \\ 4 \end{bmatrix}$ and $2\mathbf{a}$ are displayed in Fig. 6.

Vectors in $\mathbb{R}^n$

If n is a positive integer, $\mathbb{R}^n$ (read "r-n") denotes the collection of all lists (or *ordered n-tuples*) of n real numbers, usually written as $n \times 1$ column matrices, such as

$$\mathbf{u} = \begin{bmatrix} u_1 \\ u_2 \\ \vdots \\ u_n \end{bmatrix}$$

The vector whose entries are all zero is called the **zero vector** and is denoted by **0**. (The number of entries in **0** will be clear from the context.)

Equality of vectors in $\mathbb{R}^n$ and the operations of scalar multiplication and vector addition in $\mathbb{R}^n$ are defined entry by entry just as in $\mathbb{R}^2$. These operations on vectors have the following properties, which can be verified directly from the corresponding properties for real numbers. See Practice Problem 1 and Exercises 33 and 34 at the end of this section.

Algebraic Properties of $\mathbb{R}^n$

For all $\mathbf{u}, \mathbf{v}, \mathbf{w}$ in $\mathbb{R}^n$ and all scalars c and d:

(i) $\mathbf{u} + \mathbf{v} = \mathbf{v} + \mathbf{u}$

(ii) $(\mathbf{u} + \mathbf{v}) + \mathbf{w} = \mathbf{u} + (\mathbf{v} + \mathbf{w})$

(iii) $\mathbf{u} + \mathbf{0} = \mathbf{0} + \mathbf{u} = \mathbf{u}$

(iv) $\mathbf{u} + (-\mathbf{u}) = -\mathbf{u} + \mathbf{u} = \mathbf{0}$, where $-\mathbf{u}$ denotes $(-1)\mathbf{u}$

(v) $c(\mathbf{u} + \mathbf{v}) = c\mathbf{u} + c\mathbf{v}$

(vi) $(c + d)\mathbf{u} = c\mathbf{u} + d\mathbf{u}$

(vii) $c(d\mathbf{u}) = (cd)(\mathbf{u})$

(viii) $1\mathbf{u} = \mathbf{u}$

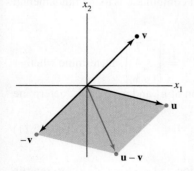

FIGURE 7
Vector subtraction.

For simplicity of notation, a vector such as $\mathbf{u} + (-1)\mathbf{v}$ is often written as $\mathbf{u} - \mathbf{v}$. Figure 7 shows $\mathbf{u} - \mathbf{v}$ as the sum of $\mathbf{u}$ and $-\mathbf{v}$.

Linear Combinations

Given vectors $\mathbf{v}_1, \mathbf{v}_2, \ldots, \mathbf{v}_p$ in $\mathbb{R}^n$ and given scalars $c_1, c_2, \ldots, c_p$, the vector $\mathbf{y}$ defined by

$$\mathbf{y} = c_1\mathbf{v}_1 + \cdots + c_p\mathbf{v}_p$$

is called a **linear combination** of $\mathbf{v}_1, \ldots, \mathbf{v}_p$ with **weights** $c_1, \ldots, c_p$. Property (ii) above permits us to omit parentheses when forming such a linear combination. The weights in a linear combination can be any real numbers, including zero. For example, some linear combinations of vectors $\mathbf{v}_1$ and $\mathbf{v}_2$ are

$$\sqrt{3}\,\mathbf{v}_1 + \mathbf{v}_2, \quad \tfrac{1}{2}\mathbf{v}_1 \ (= \tfrac{1}{2}\mathbf{v}_1 + 0\mathbf{v}_2), \quad \text{and} \quad \mathbf{0} \ (= 0\mathbf{v}_1 + 0\mathbf{v}_2)$$

EXAMPLE 4   Figure 8 identifies selected linear combinations of $\mathbf{v}_1 = \begin{bmatrix} -1 \\ 1 \end{bmatrix}$ and $\mathbf{v}_2 = \begin{bmatrix} 2 \\ 1 \end{bmatrix}$. (Note that sets of parallel grid lines are drawn through integer multiples of $\mathbf{v}_1$ and $\mathbf{v}_2$.) Estimate the linear combinations of $\mathbf{v}_1$ and $\mathbf{v}_2$ that generate the vectors $\mathbf{u}$ and $\mathbf{w}$.

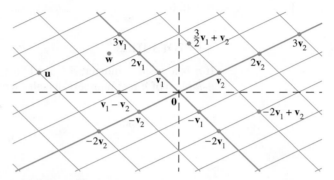

FIGURE 8 Linear combinations of $\mathbf{v}_1$ and $\mathbf{v}_2$.

SOLUTION The parallelogram rule shows that $\mathbf{u}$ is the sum of $3\mathbf{v}_1$ and $-2\mathbf{v}_2$; that is,

$$\mathbf{u} = 3\mathbf{v}_1 - 2\mathbf{v}_2$$

This expression for $\mathbf{u}$ can be interpreted as instructions for traveling from the origin to $\mathbf{u}$ along two straight paths. First, travel 3 units in the $\mathbf{v}_1$ direction to $3\mathbf{v}_1$, and then travel -2 units in the $\mathbf{v}_2$ direction (parallel to the line through $\mathbf{v}_2$ and $\mathbf{0}$). Next, although the vector $\mathbf{w}$ is not on a grid line, $\mathbf{w}$ appears to be about halfway between two pairs of grid lines, at the vertex of a parallelogram determined by $(5/2)\mathbf{v}_1$ and $(-1/2)\mathbf{v}_2$. (See Fig. 9.) Thus a reasonable estimate for $\mathbf{w}$ is

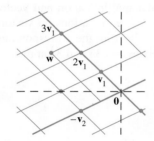

FIGURE 9

$$\mathbf{w} = \tfrac{5}{2}\mathbf{v}_1 - \tfrac{1}{2}\mathbf{v}_2 \qquad \blacksquare$$

The next example connects a problem about linear combinations to the fundamental existence question studied in Sections 1.1 and 1.2.

EXAMPLE 5   Let $\mathbf{a}_1 = \begin{bmatrix} 1 \\ -2 \\ -5 \end{bmatrix}$, $\mathbf{a}_2 = \begin{bmatrix} 2 \\ 5 \\ 6 \end{bmatrix}$, and $\mathbf{b} = \begin{bmatrix} 7 \\ 4 \\ -3 \end{bmatrix}$. Determine whether $\mathbf{b}$ can be generated (or written) as a linear combination of $\mathbf{a}_1$ and $\mathbf{a}_2$. That is, determine whether weights x_1 and x_2 exist such that

$$x_1\mathbf{a}_1 + x_2\mathbf{a}_2 = \mathbf{b} \qquad (1)$$

If vector equation (1) has a solution, find it.

SOLUTION Use the definitions of scalar multiplication and vector addition to rewrite the vector equation

$$x_1 \underset{\mathbf{a}_1}{\begin{bmatrix} 1 \\ -2 \\ -5 \end{bmatrix}} + x_2 \underset{\mathbf{a}_2}{\begin{bmatrix} 2 \\ 5 \\ 6 \end{bmatrix}} = \underset{\mathbf{b}}{\begin{bmatrix} 7 \\ 4 \\ -3 \end{bmatrix}}$$

which is the same as

$$\begin{bmatrix} x_1 \\ -2x_1 \\ -5x_1 \end{bmatrix} + \begin{bmatrix} 2x_2 \\ 5x_2 \\ 6x_2 \end{bmatrix} = \begin{bmatrix} 7 \\ 4 \\ -3 \end{bmatrix}$$

and

$$\begin{bmatrix} x_1 + 2x_2 \\ -2x_1 + 5x_2 \\ -5x_1 + 6x_2 \end{bmatrix} = \begin{bmatrix} 7 \\ 4 \\ -3 \end{bmatrix} \tag{2}$$

The vectors on the left and right sides of (2) are equal if and only if their corresponding entries are both equal. That is, x_1 and x_2 make the vector equation (1) true if and only if x_1 and x_2 satisfy the system

$$\begin{aligned} x_1 + 2x_2 &= 7 \\ -2x_1 + 5x_2 &= 4 \\ -5x_1 + 6x_2 &= -3 \end{aligned} \tag{3}$$

To solve this system, row reduce the augmented matrix of the system as follows:[3]

$$\begin{bmatrix} 1 & 2 & 7 \\ -2 & 5 & 4 \\ -5 & 6 & -3 \end{bmatrix} \sim \begin{bmatrix} 1 & 2 & 7 \\ 0 & 9 & 18 \\ 0 & 16 & 32 \end{bmatrix} \sim \begin{bmatrix} 1 & 2 & 7 \\ 0 & 1 & 2 \\ 0 & 16 & 32 \end{bmatrix} \sim \begin{bmatrix} 1 & 0 & 3 \\ 0 & 1 & 2 \\ 0 & 0 & 0 \end{bmatrix}$$

The solution of (3) is $x_1 = 3$ and $x_2 = 2$. Hence **b** is a linear combination of $\mathbf{a}_1$ and $\mathbf{a}_2$, with weights $x_1 = 3$ and $x_2 = 2$. That is,

$$3 \begin{bmatrix} 1 \\ -2 \\ -5 \end{bmatrix} + 2 \begin{bmatrix} 2 \\ 5 \\ 6 \end{bmatrix} = \begin{bmatrix} 7 \\ 4 \\ -3 \end{bmatrix} \qquad \blacksquare$$

Observe in Example 5 that the original vectors $\mathbf{a}_1$, $\mathbf{a}_2$, and **b** are the columns of the augmented matrix that we row reduced:

$$\begin{bmatrix} 1 & 2 & 7 \\ -2 & 5 & 4 \\ -5 & 6 & -3 \end{bmatrix}$$
$$\begin{array}{ccc} \uparrow & \uparrow & \uparrow \\ \mathbf{a}_1 & \mathbf{a}_2 & \mathbf{b} \end{array}$$

For brevity, write this matrix in a way that identifies its columns—namely,

$$\begin{bmatrix} \mathbf{a}_1 & \mathbf{a}_2 & \mathbf{b} \end{bmatrix} \tag{4}$$

It is clear how to write this augmented matrix immediately from vector equation (1), without going through the intermediate steps of Example 5. Take the vectors in the order in which they appear in (1) and put them into the columns of a matrix as in (4).

The discussion above is easily modified to establish the following fundamental fact.

> A vector equation
> $$x_1 \mathbf{a}_1 + x_2 \mathbf{a}_2 + \cdots + x_n \mathbf{a}_n = \mathbf{b}$$
> has the same solution set as the linear system whose augmented matrix is
> $$\begin{bmatrix} \mathbf{a}_1 & \mathbf{a}_2 & \cdots & \mathbf{a}_n & \mathbf{b} \end{bmatrix} \tag{5}$$
> In particular, **b** can be generated by a linear combination of $\mathbf{a}_1, \ldots, \mathbf{a}_n$ if and only if there exists a solution to the linear system corresponding to the matrix (5).

---

[3] The symbol $\sim$ between matrices denotes row equivalence (Section 1.2).

One of the key ideas in linear algebra is to study the set of all vectors that can be generated or written as a linear combination of a fixed set $\{\mathbf{v}_1, \ldots, \mathbf{v}_p\}$ of vectors.

DEFINITION

> If $\mathbf{v}_1, \ldots, \mathbf{v}_p$ are in $\mathbb{R}^n$, then the set of all linear combinations of $\mathbf{v}_1, \ldots, \mathbf{v}_p$ is denoted by Span$\{\mathbf{v}_1, \ldots, \mathbf{v}_p\}$ and is called the **subset of $\mathbb{R}^n$ spanned** (or **generated**) by $\mathbf{v}_1, \ldots, \mathbf{v}_p$. That is, Span$\{\mathbf{v}_1, \ldots, \mathbf{v}_p\}$ is the collection of all vectors that can be written in the form
>
> $$c_1\mathbf{v}_1 + c_2\mathbf{v}_2 + \cdots + c_p\mathbf{v}_p$$
>
> with $c_1, \ldots, c_p$ scalars.

Asking whether a vector $\mathbf{b}$ is in Span$\{\mathbf{v}_1, \ldots, \mathbf{v}_p\}$ amounts to asking whether the vector equation

$$x_1\mathbf{v}_1 + x_2\mathbf{v}_2 + \cdots + x_p\mathbf{v}_p = \mathbf{b}$$

has a solution, or, equivalently, asking whether the linear system with augmented matrix $[\, \mathbf{v}_1 \;\; \cdots \;\; \mathbf{v}_p \;\; \mathbf{b} \,]$ has a solution.

Note that Span$\{\mathbf{v}_1, \ldots, \mathbf{v}_p\}$ contains every scalar multiple of $\mathbf{v}_1$ (for example), since $c\mathbf{v}_1 = c\mathbf{v}_1 + 0\mathbf{v}_2 + \cdots + 0\mathbf{v}_p$. In particular, the zero vector must be in Span$\{\mathbf{v}_1, \ldots, \mathbf{v}_p\}$.

A Geometric Description of Span$\{\mathbf{v}\}$ and Span$\{\mathbf{u}, \mathbf{v}\}$

Let $\mathbf{v}$ be a nonzero vector in $\mathbb{R}^3$. Then Span$\{\mathbf{v}\}$ is the set of all scalar multiples of $\mathbf{v}$, which is the set of points on the line in $\mathbb{R}^3$ through $\mathbf{v}$ and $\mathbf{0}$. See Fig. 10.

If $\mathbf{u}$ and $\mathbf{v}$ are nonzero vectors in $\mathbb{R}^3$, with $\mathbf{v}$ not a multiple of $\mathbf{u}$, then Span$\{\mathbf{u}, \mathbf{v}\}$ is the plane in $\mathbb{R}^3$ that contains $\mathbf{u}$, $\mathbf{v}$, and $\mathbf{0}$. In particular, Span$\{\mathbf{u}, \mathbf{v}\}$ contains the line in $\mathbb{R}^3$ through $\mathbf{u}$ and $\mathbf{0}$ and the line through $\mathbf{v}$ and $\mathbf{0}$. See Fig. 11.

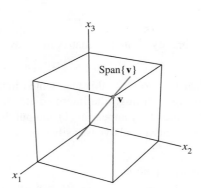

FIGURE 10 Span$\{\mathbf{v}\}$ as a line through the origin.

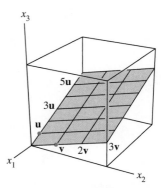

FIGURE 11 Span$\{\mathbf{u}, \mathbf{v}\}$ as a plane through the origin.

EXAMPLE 6 Let $\mathbf{a}_1 = \begin{bmatrix} 1 \\ -2 \\ 3 \end{bmatrix}$, $\mathbf{a}_2 = \begin{bmatrix} 5 \\ -13 \\ -3 \end{bmatrix}$, and $\mathbf{b} = \begin{bmatrix} -3 \\ 8 \\ 1 \end{bmatrix}$. Then Span$\{\mathbf{a}_1, \mathbf{a}_2\}$ is a plane through the origin in $\mathbb{R}^3$. Is $\mathbf{b}$ in that plane?

SOLUTION Does the equation $x_1\mathbf{a}_1 + x_2\mathbf{a}_2 = \mathbf{b}$ have a solution? To answer this, row reduce the augmented matrix $[\,\mathbf{a}_1\quad\mathbf{a}_2\quad\mathbf{b}\,]$:

$$\begin{bmatrix} 1 & 5 & -3 \\ -2 & -13 & 8 \\ 3 & -3 & 1 \end{bmatrix} \sim \begin{bmatrix} 1 & 5 & -3 \\ 0 & -3 & 2 \\ 0 & -18 & 10 \end{bmatrix} \sim \begin{bmatrix} 1 & 5 & -3 \\ 0 & -3 & 2 \\ 0 & 0 & -2 \end{bmatrix}$$

The third equation is $0 = -2$, which shows that the system has no solution. The vector equation $x_1\mathbf{a}_1 + x_2\mathbf{a}_2 = \mathbf{b}$ has no solution, and so $\mathbf{b}$ is *not* in Span $\{\mathbf{a}_1, \mathbf{a}_2\}$. ∎

Linear Combinations in Applications

The final example shows how scalar multiples and linear combinations can arise when a quantity such as "cost" is broken down into several categories. The basic principle for the example concerns the cost of producing several units of an item when the cost per unit is known:

$$\begin{Bmatrix} \text{number} \\ \text{of units} \end{Bmatrix} \cdot \begin{Bmatrix} \text{cost} \\ \text{per unit} \end{Bmatrix} = \begin{Bmatrix} \text{total} \\ \text{cost} \end{Bmatrix}$$

EXAMPLE 7 A company manufactures two products. For $1.00 worth of product B, the company spends $.45 on materials, $.25 on labor, and $.15 on overhead. For $1.00 worth of product C, the company spends $.40 on materials, $.30 on labor, and $.15 on overhead. Let

$$\mathbf{b} = \begin{bmatrix} .45 \\ .25 \\ .15 \end{bmatrix} \quad \text{and} \quad \mathbf{c} = \begin{bmatrix} .40 \\ .30 \\ .15 \end{bmatrix}$$

Then $\mathbf{b}$ and $\mathbf{c}$ represent the "costs per dollar of income" for the two products.

a. What economic interpretation can be given to the vector $100\mathbf{b}$?

b. Suppose the company wishes to manufacture x_1 dollars worth of product B and x_2 dollars worth of product C. Give a vector that describes the various costs the company will have (for materials, labor, and overhead).

SOLUTION

a. Compute

$$100\mathbf{b} = 100 \begin{bmatrix} .45 \\ .25 \\ .15 \end{bmatrix} = \begin{bmatrix} 45 \\ 25 \\ 15 \end{bmatrix}$$

The vector $100\mathbf{b}$ lists the various costs for producing $100 worth of product B—namely, $45 for materials, $25 for labor, and $15 for overhead.

b. The costs of manufacturing x_1 dollars worth of B are given by the vector $x_1\mathbf{b}$, and the costs of manufacturing x_2 dollars worth of C are given by $x_2\mathbf{c}$. Hence the total costs for both products are given by the vector $x_1\mathbf{b} + x_2\mathbf{c}$. ∎

PRACTICE PROBLEMS

1. Prove that $\mathbf{u} + \mathbf{v} = \mathbf{v} + \mathbf{u}$ for any $\mathbf{u}$ and $\mathbf{v}$ in $\mathbb{R}^n$.

2. For what value(s) of h will $\mathbf{y}$ be in Span$\{\mathbf{v}_1, \mathbf{v}_2, \mathbf{v}_3\}$ if

$$\mathbf{v}_1 = \begin{bmatrix} 1 \\ -1 \\ -2 \end{bmatrix}, \quad \mathbf{v}_2 = \begin{bmatrix} 5 \\ -4 \\ -7 \end{bmatrix}, \quad \mathbf{v}_3 = \begin{bmatrix} -3 \\ 1 \\ 0 \end{bmatrix}, \quad \text{and} \quad \mathbf{y} = \begin{bmatrix} -4 \\ 3 \\ h \end{bmatrix}$$

1.3 EXERCISES

In Exercises 1 and 2, compute $\mathbf{u} + \mathbf{v}$ and $\mathbf{u} - 2\mathbf{v}$.

1. $\mathbf{u} = \begin{bmatrix} -1 \\ 2 \end{bmatrix}$, $\mathbf{v} = \begin{bmatrix} -3 \\ -1 \end{bmatrix}$ **2.** $\mathbf{u} = \begin{bmatrix} 3 \\ 2 \end{bmatrix}$, $\mathbf{v} = \begin{bmatrix} 2 \\ -1 \end{bmatrix}$

In Exercises 3 and 4, display the following vectors using arrows on an xy-graph: $\mathbf{u}, \mathbf{v}, -\mathbf{v}, -2\mathbf{v}, \mathbf{u} + \mathbf{v}, \mathbf{u} - \mathbf{v}$, and $\mathbf{u} - 2\mathbf{v}$. Notice that $\mathbf{u} - \mathbf{v}$ is the vertex of a parallelogram whose other vertices are $\mathbf{u}, \mathbf{0}$, and $-\mathbf{v}$.

3. $\mathbf{u}$ and $\mathbf{v}$ as in Exercise 1   **4.** $\mathbf{u}$ and $\mathbf{v}$ as in Exercise 2

In Exercises 5 and 6, write a system of equations that is equivalent to the given vector equation.

5. $x_1 \begin{bmatrix} 3 \\ -2 \\ 8 \end{bmatrix} + x_2 \begin{bmatrix} 5 \\ 0 \\ -9 \end{bmatrix} = \begin{bmatrix} 2 \\ -3 \\ 8 \end{bmatrix}$

6. $x_1 \begin{bmatrix} 3 \\ -2 \end{bmatrix} + x_2 \begin{bmatrix} 7 \\ 3 \end{bmatrix} + x_3 \begin{bmatrix} -2 \\ 1 \end{bmatrix} = \begin{bmatrix} 0 \\ 0 \end{bmatrix}$

Use the accompanying figure to write each vector listed in Exercises 7 and 8 as a linear combination of $\mathbf{u}$ and $\mathbf{v}$. Is every vector in $\mathbb{R}^2$ a linear combination of $\mathbf{u}$ and $\mathbf{v}$?

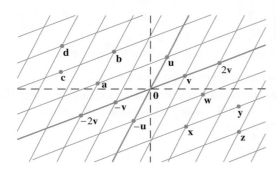

7. Vectors $\mathbf{a}, \mathbf{b}, \mathbf{c}$, and $\mathbf{d}$

8. Vectors $\mathbf{w}, \mathbf{x}, \mathbf{y}$, and $\mathbf{z}$

In Exercises 9 and 10, write a vector equation that is equivalent to the given system of equations.

9.
$$\begin{aligned} x_2 + 5x_3 &= 0 \\ 4x_1 + 6x_2 - x_3 &= 0 \\ -x_1 + 3x_2 - 8x_3 &= 0 \end{aligned}$$

10.
$$\begin{aligned} 3x_1 - 2x_2 + 4x_3 &= 3 \\ -2x_1 - 7x_2 + 5x_3 &= 1 \\ 5x_1 + 4x_2 - 3x_3 &= 2 \end{aligned}$$

In Exercises 11 and 12, determine if $\mathbf{b}$ is a linear combination of $\mathbf{a}_1, \mathbf{a}_2$, and $\mathbf{a}_3$.

11. $\mathbf{a}_1 = \begin{bmatrix} 1 \\ -2 \\ 0 \end{bmatrix}$, $\mathbf{a}_2 = \begin{bmatrix} 0 \\ 1 \\ 2 \end{bmatrix}$, $\mathbf{a}_3 = \begin{bmatrix} 5 \\ -6 \\ 8 \end{bmatrix}$, $\mathbf{b} = \begin{bmatrix} 2 \\ -1 \\ 6 \end{bmatrix}$

12. $\mathbf{a}_1 = \begin{bmatrix} 1 \\ 0 \\ 1 \end{bmatrix}$, $\mathbf{a}_2 = \begin{bmatrix} -2 \\ 3 \\ -2 \end{bmatrix}$, $\mathbf{a}_3 = \begin{bmatrix} -6 \\ 7 \\ 5 \end{bmatrix}$, $\mathbf{b} = \begin{bmatrix} 11 \\ -5 \\ 9 \end{bmatrix}$

In Exercises 13 and 14, determine if $\mathbf{b}$ is a linear combination of the vectors formed from the columns of the matrix A.

13. $A = \begin{bmatrix} 1 & -4 & 2 \\ 0 & 3 & 5 \\ -2 & 8 & -4 \end{bmatrix}$, $\mathbf{b} = \begin{bmatrix} 3 \\ -7 \\ -3 \end{bmatrix}$

14. $A = \begin{bmatrix} 1 & 0 & 5 \\ -2 & 1 & -6 \\ 0 & 2 & 8 \end{bmatrix}$, $\mathbf{b} = \begin{bmatrix} 2 \\ -1 \\ 6 \end{bmatrix}$

15. Let $\mathbf{a}_1 = \begin{bmatrix} 1 \\ 3 \\ -1 \end{bmatrix}$, $\mathbf{a}_2 = \begin{bmatrix} -5 \\ -8 \\ 2 \end{bmatrix}$, and $\mathbf{b} = \begin{bmatrix} 3 \\ -5 \\ h \end{bmatrix}$. For what value(s) of h is $\mathbf{b}$ in the plane spanned by $\mathbf{a}_1$ and $\mathbf{a}_2$?

16. Let $\mathbf{v}_1 = \begin{bmatrix} 1 \\ 0 \\ -2 \end{bmatrix}$, $\mathbf{v}_2 = \begin{bmatrix} -2 \\ 1 \\ 7 \end{bmatrix}$, and $\mathbf{y} = \begin{bmatrix} h \\ -3 \\ -5 \end{bmatrix}$. For what value(s) of h is $\mathbf{y}$ in the plane generated by $\mathbf{v}_1$ and $\mathbf{v}_2$?

In Exercises 17 and 18, list five vectors in Span $\{\mathbf{v}_1, \mathbf{v}_2\}$. For each vector, show the weights on $\mathbf{v}_1$ and $\mathbf{v}_2$ used to generate the vector and list the three entries of the vector. Do not make a sketch.

17. $\mathbf{v}_1 = \begin{bmatrix} 3 \\ 1 \\ 2 \end{bmatrix}$, $\mathbf{v}_2 = \begin{bmatrix} -4 \\ 0 \\ 1 \end{bmatrix}$

18. $\mathbf{v}_1 = \begin{bmatrix} 1 \\ 1 \\ -2 \end{bmatrix}$, $\mathbf{v}_2 = \begin{bmatrix} -2 \\ 3 \\ 0 \end{bmatrix}$

19. Give a geometric description of Span $\{\mathbf{v}_1, \mathbf{v}_2\}$ for the vectors $\mathbf{v}_1 = \begin{bmatrix} 8 \\ 2 \\ -6 \end{bmatrix}$ and $\mathbf{v}_2 = \begin{bmatrix} 12 \\ 3 \\ -9 \end{bmatrix}$.

20. Give a geometric description of Span $\{\mathbf{v}_1, \mathbf{v}_2\}$ for the vectors in Exercise 18.

21. Let $\mathbf{u} = \begin{bmatrix} 2 \\ -1 \end{bmatrix}$ and $\mathbf{v} = \begin{bmatrix} 2 \\ 1 \end{bmatrix}$. Show that $\begin{bmatrix} h \\ k \end{bmatrix}$ is in Span $\{\mathbf{u}, \mathbf{v}\}$ for all h and k.

22. Construct a 3×3 matrix A, with nonzero entries, and a vector $\mathbf{b}$ in $\mathbb{R}^3$ such that $\mathbf{b}$ is *not* in the set spanned by the columns of A.

In Exercises 23 and 24, mark each statement True or False. Justify each answer.

23. a. Another notation for the vector $\begin{bmatrix} -4 \\ 3 \end{bmatrix}$ is $[\,-4 \quad 3\,]$.

   b. The points in the plane corresponding to $\begin{bmatrix} -2 \\ 5 \end{bmatrix}$ and $\begin{bmatrix} -5 \\ 2 \end{bmatrix}$ lie on a line through the origin.

   c. An example of a linear combination of vectors $\mathbf{v}_1$ and $\mathbf{v}_2$ is the vector $\frac{1}{2}\mathbf{v}_1$.

d. The solution set of the linear system whose augmented matrix is $[\, \mathbf{a}_1 \quad \mathbf{a}_2 \quad \mathbf{a}_3 \quad \mathbf{b} \,]$ is the same as the solution set of the equation $x_1\mathbf{a}_1 + x_2\mathbf{a}_2 + x_3\mathbf{a}_3 = \mathbf{b}$.

e. The set Span $\{\mathbf{u}, \mathbf{v}\}$ is always visualized as a plane through the origin.

24. a. When $\mathbf{u}$ and $\mathbf{v}$ are nonzero vectors, Span $\{\mathbf{u}, \mathbf{v}\}$ contains only the line through $\mathbf{u}$ and the origin, and the line through $\mathbf{v}$ and the origin.

b. Any list of five real numbers is a vector in $\mathbb{R}^5$.

c. Asking whether the linear system corresponding to an augmented matrix $[\, \mathbf{a}_1 \quad \mathbf{a}_2 \quad \mathbf{a}_3 \quad \mathbf{b} \,]$ has a solution amounts to asking whether $\mathbf{b}$ is in Span $\{\mathbf{a}_1, \mathbf{a}_2, \mathbf{a}_3\}$.

d. The vector $\mathbf{v}$ results when a vector $\mathbf{u} - \mathbf{v}$ is added to the vector $\mathbf{v}$.

e. The weights $c_1, \ldots, c_p$ in a linear combination $c_1\mathbf{v}_1 + \cdots + c_p\mathbf{v}_p$ cannot all be zero.

25. Let $A = \begin{bmatrix} 1 & 0 & -4 \\ 0 & 3 & -2 \\ -2 & 6 & 3 \end{bmatrix}$ and $\mathbf{b} = \begin{bmatrix} 4 \\ 1 \\ -4 \end{bmatrix}$. Denote the columns of A by $\mathbf{a}_1, \mathbf{a}_2, \mathbf{a}_3$, and let $W = $ Span $\{\mathbf{a}_1, \mathbf{a}_2, \mathbf{a}_3\}$.

a. Is $\mathbf{b}$ in $\{\mathbf{a}_1, \mathbf{a}_2, \mathbf{a}_3\}$? How many vectors are in $\{\mathbf{a}_1, \mathbf{a}_2, \mathbf{a}_3\}$?

b. Is $\mathbf{b}$ in W? How many vectors are in W?

c. Show that $\mathbf{a}_1$ is in W. [*Hint:* Row operations are unnecessary.]

26. Let $A = \begin{bmatrix} 2 & 0 & 6 \\ -1 & 8 & 5 \\ 1 & -2 & 1 \end{bmatrix}$, let $\mathbf{b} = \begin{bmatrix} 10 \\ 3 \\ 7 \end{bmatrix}$, and let W be the set of all linear combinations of the columns of A.

a. Is $\mathbf{b}$ in W?

b. Show that the second column of A is in W.

27. A mining company has two mines. One day's operation at mine #1 produces ore that contains 30 metric tons of copper and 600 kilograms of silver, while one day's operation at mine #2 produces ore that contains 40 metric tons of copper and 380 kilograms of silver. Let $\mathbf{v}_1 = \begin{bmatrix} 30 \\ 600 \end{bmatrix}$ and $\mathbf{v}_2 = \begin{bmatrix} 40 \\ 380 \end{bmatrix}$. Then $\mathbf{v}_1$ and $\mathbf{v}_2$ represent the "output per day" of mine #1 and mine #2, respectively.

a. What physical interpretation can be given to the vector $5\mathbf{v}_1$?

b. Suppose the company operates mine #1 for x_1 days and mine #2 for x_2 days. Write a vector equation whose solution gives the number of days each mine should operate in order to produce 240 tons of copper and 2824 kilograms of silver. Do not solve the equation.

c. [**M**] Solve the equation in (b).

28. A steam plant burns two types of coal: anthracite (A) and bituminous (B). For each ton of A burned, the plant produces 27.6 million Btu of heat, 3100 grams (g) of sulfur dioxide, and 250 g of particulate matter (solid-particle pollutants). For each ton of B burned, the plant produces 30.2 million Btu, 6400 g of sulfur dioxide, and 360 g of particulate matter.

a. How much heat does the steam plant produce when it burns x_1 tons of A and x_2 tons of B?

b. Suppose the output of the steam plant is described by a vector that lists the amounts of heat, sulfur dioxide, and particulate matter. Express this output as a linear combination of two vectors, assuming that the plant burns x_1 tons of A and x_2 tons of B.

c. [**M**] Over a certain time period, the steam plant produced 162 million Btu of heat, 23,610 g of sulfur dioxide, and 1623 g of particulate matter. Determine how many tons of each type of coal the steam plant must have burned. Include a vector equation as part of your solution.

29. Let $\mathbf{v}_1, \ldots, \mathbf{v}_k$ be points in $\mathbb{R}^3$ and suppose that for $j = 1, \ldots, k$ an object with mass m_j is located at point $\mathbf{v}_j$. Physicists call such objects *point masses*. The total mass of the system of point masses is

$$m = m_1 + \cdots + m_k$$

The *center of gravity* (or *center of mass*) of the system is

$$\overline{\mathbf{v}} = \frac{1}{m}[m_1\mathbf{v}_1 + \cdots + m_k\mathbf{v}_k]$$

Compute the center of gravity of the system consisting of the following point masses (see the figure):

Point	Mass
$\mathbf{v}_1 = (2, -2, 4)$	4 g
$\mathbf{v}_2 = (-4, 2, 3)$	2 g
$\mathbf{v}_3 = (4, 0, -2)$	3 g
$\mathbf{v}_4 = (1, -6, 0)$	5 g

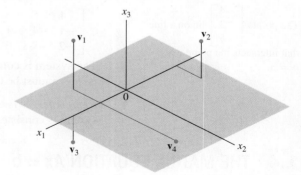

30. Let $\mathbf{v}$ be the center of mass of a system of point masses located at $\mathbf{v}_1, \ldots, \mathbf{v}_k$ as in Exercise 29. Is $\mathbf{v}$ in Span $\{\mathbf{v}_1, \ldots, \mathbf{v}_k\}$? Explain.

31. A thin triangular plate of uniform density and thickness has vertices at $\mathbf{v}_1 = (0, 1)$, $\mathbf{v}_2 = (8, 1)$, and $\mathbf{v}_3 = (2, 4)$, as in the figure below, and the mass of the plate is 3 g.

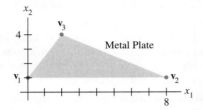

a. Find the (x, y)-coordinates of the center of mass of the plate. This "balance point" of the plate coincides with the center of mass of a system consisting of three 1-gram point masses located at the vertices of the plate.

b. Determine how to distribute an additional mass of 6 g at the three vertices of the plate to move the balance point of the plate to $(2, 2)$. [*Hint:* Let w_1, w_2, and w_3 denote the masses added at the three vertices, so that $w_1 + w_2 + w_3 = 6$.]

32. Consider the vectors $\mathbf{v}_1$, $\mathbf{v}_2$, $\mathbf{v}_3$, and $\mathbf{b}$ in $\mathbb{R}^2$, shown in the figure. Does the equation $x_1\mathbf{v}_1 + x_2\mathbf{v}_2 + x_3\mathbf{v}_3 = \mathbf{b}$ have a solution? Is the solution unique? Use the figure to explain your answers.

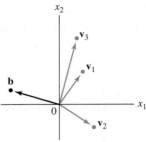

33. Use the vectors $\mathbf{u} = (u_1, \ldots, u_n)$, $\mathbf{v} = (v_1, \ldots, v_n)$, and $\mathbf{w} = (w_1, \ldots, w_n)$ to verify the following algebraic properties of $\mathbb{R}^n$.

a. $(\mathbf{u} + \mathbf{v}) + \mathbf{w} = \mathbf{u} + (\mathbf{v} + \mathbf{w})$

b. $c(\mathbf{u} + \mathbf{v}) = c\mathbf{u} + c\mathbf{v}$ for each scalar c

34. Use the vector $\mathbf{u} = (u_1, \ldots, u_n)$ to verify the following algebraic properties of $\mathbb{R}^n$.

a. $\mathbf{u} + (-\mathbf{u}) = (-\mathbf{u}) + \mathbf{u} = \mathbf{0}$

b. $c(d\mathbf{u}) = (cd)\mathbf{u}$ for all scalars c and d

SOLUTIONS TO PRACTICE PROBLEMS

1. Take arbitrary vectors $\mathbf{u} = (u_1, \ldots, u_n)$ and $\mathbf{v} = (v_1, \ldots, v_n)$ in $\mathbb{R}^n$, and compute

$$\mathbf{u} + \mathbf{v} = (u_1 + v_1, \ldots, u_n + v_n) \qquad \text{Definition of vector addition}$$
$$= (v_1 + u_1, \ldots, v_n + u_n) \qquad \text{Commutativity of addition in } \mathbb{R}$$
$$= \mathbf{v} + \mathbf{u} \qquad \text{Definition of vector addition}$$

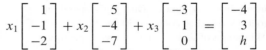

2. The vector $\mathbf{y}$ belongs to Span $\{\mathbf{v}_1, \mathbf{v}_2, \mathbf{v}_3\}$ if and only if there exist scalars x_1, x_2, x_3 such that

$$x_1 \begin{bmatrix} 1 \\ -1 \\ -2 \end{bmatrix} + x_2 \begin{bmatrix} 5 \\ -4 \\ -7 \end{bmatrix} + x_3 \begin{bmatrix} -3 \\ 1 \\ 0 \end{bmatrix} = \begin{bmatrix} -4 \\ 3 \\ h \end{bmatrix}$$

This vector equation is equivalent to a system of three linear equations in three unknowns. If you row reduce the augmented matrix for this system, you find that

$$\begin{bmatrix} 1 & 5 & -3 & -4 \\ -1 & -4 & 1 & 3 \\ -2 & -7 & 0 & h \end{bmatrix} \sim \begin{bmatrix} 1 & 5 & -3 & -4 \\ 0 & 1 & -2 & -1 \\ 0 & 3 & -6 & h-8 \end{bmatrix} \sim \begin{bmatrix} 1 & 5 & -3 & -4 \\ 0 & 1 & -2 & -1 \\ 0 & 0 & 0 & h-5 \end{bmatrix}$$

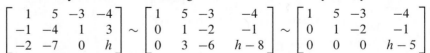

The system is consistent if and only if there is no pivot in the fourth column. That is, $h - 5$ must be 0. So $\mathbf{y}$ is in Span $\{\mathbf{v}_1, \mathbf{v}_2, \mathbf{v}_3\}$ if and only if $h = 5$.

Remember: The presence of a free variable in a system does not guarantee that the system is consistent.

The points $\begin{bmatrix} -4 \\ 3 \\ h \end{bmatrix}$ lie on a line that intersects the plane when $h = 5$.

1.4 THE MATRIX EQUATION $A\mathbf{x} = \mathbf{b}$

A fundamental idea in linear algebra is to view a linear combination of vectors as the product of a matrix and a vector. The following definition permits us to rephrase some of the concepts of Section 1.3 in new ways.

DEFINITION

> If A is an $m \times n$ matrix, with columns $\mathbf{a}_1, \ldots, \mathbf{a}_n$, and if $\mathbf{x}$ is in $\mathbb{R}^n$, then the **product of A and $\mathbf{x}$**, denoted by $A\mathbf{x}$, is **the linear combination of the columns of A using the corresponding entries in $\mathbf{x}$ as weights**; that is,
>
> $$A\mathbf{x} = \begin{bmatrix} \mathbf{a}_1 & \mathbf{a}_2 & \cdots & \mathbf{a}_n \end{bmatrix} \begin{bmatrix} x_1 \\ \vdots \\ x_n \end{bmatrix} = x_1\mathbf{a}_1 + x_2\mathbf{a}_2 + \cdots + x_n\mathbf{a}_n$$

Note that $A\mathbf{x}$ is defined only if the number of columns of A equals the number of entries in $\mathbf{x}$.

EXAMPLE 1

a. $\begin{bmatrix} 1 & 2 & -1 \\ 0 & -5 & 3 \end{bmatrix} \begin{bmatrix} 4 \\ 3 \\ 7 \end{bmatrix} = 4\begin{bmatrix} 1 \\ 0 \end{bmatrix} + 3\begin{bmatrix} 2 \\ -5 \end{bmatrix} + 7\begin{bmatrix} -1 \\ 3 \end{bmatrix}$

$\qquad = \begin{bmatrix} 4 \\ 0 \end{bmatrix} + \begin{bmatrix} 6 \\ -15 \end{bmatrix} + \begin{bmatrix} -7 \\ 21 \end{bmatrix} = \begin{bmatrix} 3 \\ 6 \end{bmatrix}$

b. $\begin{bmatrix} 2 & -3 \\ 8 & 0 \\ -5 & 2 \end{bmatrix} \begin{bmatrix} 4 \\ 7 \end{bmatrix} = 4\begin{bmatrix} 2 \\ 8 \\ -5 \end{bmatrix} + 7\begin{bmatrix} -3 \\ 0 \\ 2 \end{bmatrix} = \begin{bmatrix} 8 \\ 32 \\ -20 \end{bmatrix} + \begin{bmatrix} -21 \\ 0 \\ 14 \end{bmatrix} = \begin{bmatrix} -13 \\ 32 \\ -6 \end{bmatrix}$ ∎

EXAMPLE 2   For $\mathbf{v}_1, \mathbf{v}_2, \mathbf{v}_3$ in $\mathbb{R}^m$, write the linear combination $3\mathbf{v}_1 - 5\mathbf{v}_2 + 7\mathbf{v}_3$ as a matrix times a vector.

SOLUTION   Place $\mathbf{v}_1, \mathbf{v}_2, \mathbf{v}_3$ into the columns of a matrix A and place the weights $3, -5$, and 7 into a vector $\mathbf{x}$. That is,

$$3\mathbf{v}_1 - 5\mathbf{v}_2 + 7\mathbf{v}_3 = \begin{bmatrix} \mathbf{v}_1 & \mathbf{v}_2 & \mathbf{v}_3 \end{bmatrix} \begin{bmatrix} 3 \\ -5 \\ 7 \end{bmatrix} = A\mathbf{x}$$ ∎

Section 1.3 showed how to write a system of linear equations as a vector equation involving a linear combination of vectors. For example, the system

$$\begin{aligned} x_1 + 2x_2 - x_3 &= 4 \\ -5x_2 + 3x_3 &= 1 \end{aligned} \tag{1}$$

is equivalent to

$$x_1\begin{bmatrix} 1 \\ 0 \end{bmatrix} + x_2\begin{bmatrix} 2 \\ -5 \end{bmatrix} + x_3\begin{bmatrix} -1 \\ 3 \end{bmatrix} = \begin{bmatrix} 4 \\ 1 \end{bmatrix} \tag{2}$$

As in Example 2, the linear combination on the left side is a matrix times a vector, so that (2) becomes

$$\begin{bmatrix} 1 & 2 & -1 \\ 0 & -5 & 3 \end{bmatrix} \begin{bmatrix} x_1 \\ x_2 \\ x_3 \end{bmatrix} = \begin{bmatrix} 4 \\ 1 \end{bmatrix} \tag{3}$$

Equation (3) has the form $A\mathbf{x} = \mathbf{b}$. Such an equation is called a **matrix equation**, to distinguish it from a vector equation such as is shown in (2).

Notice how the matrix in (3) is just the matrix of coefficients of the system (1). Similar calculations show that any system of linear equations, or any vector equation such as (2), can be written as an equivalent matrix equation in the form $A\mathbf{x} = \mathbf{b}$. This simple observation will be used repeatedly throughout the text.

Here is the formal result.

THEOREM 3

If A is an $m \times n$ matrix, with columns $\mathbf{a}_1, \ldots, \mathbf{a}_n$, and if $\mathbf{b}$ is in $\mathbb{R}^m$, the matrix equation

$$A\mathbf{x} = \mathbf{b} \tag{4}$$

has the same solution set as the vector equation

$$x_1\mathbf{a}_1 + x_2\mathbf{a}_2 + \cdots + x_n\mathbf{a}_n = \mathbf{b} \tag{5}$$

which, in turn, has the same solution set as the system of linear equations whose augmented matrix is

$$\begin{bmatrix} \mathbf{a}_1 & \mathbf{a}_2 & \cdots & \mathbf{a}_n & \mathbf{b} \end{bmatrix} \tag{6}$$

Theorem 3 provides a powerful tool for gaining insight into problems in linear algebra, because a system of linear equations may now be viewed in three different but equivalent ways: as a matrix equation, as a vector equation, or as a system of linear equations. Whenever you construct a mathematical model of a problem in real life, you are free to choose whichever viewpoint is most natural. Then you may switch from one formulation of a problem to another whenever it is convenient. In any case, the matrix equation (4), the vector equation (5), and the system of equations are all solved in the same way—by row reducing the augmented matrix (6). Other methods of solution will be discussed later.

Existence of Solutions

The definition of $A\mathbf{x}$ leads directly to the following useful fact.

The equation $A\mathbf{x} = \mathbf{b}$ has a solution if and only if $\mathbf{b}$ is a linear combination of the columns of A.

Section 1.3 considered the existence question, "Is $\mathbf{b}$ in Span $\{\mathbf{a}_1, \ldots, \mathbf{a}_n\}$?" Equivalently, "Is $A\mathbf{x} = \mathbf{b}$ consistent?" A harder existence problem is to determine whether the equation $A\mathbf{x} = \mathbf{b}$ is consistent *for all* possible $\mathbf{b}$.

EXAMPLE 3 Let $A = \begin{bmatrix} 1 & 3 & 4 \\ -4 & 2 & -6 \\ -3 & -2 & -7 \end{bmatrix}$ and $\mathbf{b} = \begin{bmatrix} b_1 \\ b_2 \\ b_3 \end{bmatrix}$. Is the equation $A\mathbf{x} = \mathbf{b}$ consistent for all possible b_1, b_2, b_3?

SOLUTION Row reduce the augmented matrix for $A\mathbf{x} = \mathbf{b}$:

$$\begin{bmatrix} 1 & 3 & 4 & b_1 \\ -4 & 2 & -6 & b_2 \\ -3 & -2 & -7 & b_3 \end{bmatrix} \sim \begin{bmatrix} 1 & 3 & 4 & b_1 \\ 0 & 14 & 10 & b_2 + 4b_1 \\ 0 & 7 & 5 & b_3 + 3b_1 \end{bmatrix}$$

$$\sim \begin{bmatrix} 1 & 3 & 4 & b_1 \\ 0 & 14 & 10 & b_2 + 4b_1 \\ 0 & 0 & 0 & b_3 + 3b_1 - \frac{1}{2}(b_2 + 4b_1) \end{bmatrix}$$

The third entry in column 4 equals $b_1 - \frac{1}{2}b_2 + b_3$. The equation $A\mathbf{x} = \mathbf{b}$ is *not* consistent for every $\mathbf{b}$ because some choices of $\mathbf{b}$ can make $b_1 - \frac{1}{2}b_2 + b_3$ nonzero. ∎

The reduced matrix in Example 3 provides a description of all $\mathbf{b}$ for which the equation $A\mathbf{x} = \mathbf{b}$ *is* consistent: The entries in $\mathbf{b}$ must satisfy

$$b_1 - \tfrac{1}{2}b_2 + b_3 = 0$$

This is the equation of a plane through the origin in $\mathbb{R}^3$. The plane is the set of all linear combinations of the three columns of A. See Fig. 1.

The equation $A\mathbf{x} = \mathbf{b}$ in Example 3 fails to be consistent for all $\mathbf{b}$ because the echelon form of A has a row of zeros. If A had a pivot in all three rows, we would not care about the calculations in the augmented column because in this case an echelon form of the augmented matrix could not have a row such as $\begin{bmatrix} 0 & 0 & 0 & 1 \end{bmatrix}$.

In the next theorem, the sentence "The columns of A span $\mathbb{R}^m$" means that *every* $\mathbf{b}$ in $\mathbb{R}^m$ is a linear combination of the columns of A. In general, a set of vectors $\{\mathbf{v}_1, \ldots, \mathbf{v}_p\}$ in $\mathbb{R}^m$ **spans** (or **generates**) $\mathbb{R}^m$ if every vector in $\mathbb{R}^m$ is a linear combination of $\mathbf{v}_1, \ldots, \mathbf{v}_p$—that is, if Span $\{\mathbf{v}_1, \ldots, \mathbf{v}_p\} = \mathbb{R}^m$.

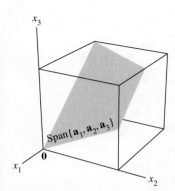

FIGURE 1

The columns of $A = \begin{bmatrix} \mathbf{a}_1 & \mathbf{a}_2 & \mathbf{a}_3 \end{bmatrix}$ span a plane through $\mathbf{0}$.

THEOREM 4

Let A be an $m \times n$ matrix. Then the following statements are logically equivalent. That is, for a particular A, either they are all true statements or they are all false.

a. For each $\mathbf{b}$ in $\mathbb{R}^m$, the equation $A\mathbf{x} = \mathbf{b}$ has a solution.

b. Each $\mathbf{b}$ in $\mathbb{R}^m$ is a linear combination of the columns of A.

c. The columns of A span $\mathbb{R}^m$.

d. A has a pivot position in every row.

Theorem 4 is one of the most useful theorems in this chapter. Statements (a), (b), and (c) are equivalent because of the definition of $A\mathbf{x}$ and what it means for a set of vectors to span $\mathbb{R}^m$. The discussion after Example 3 suggests why (a) and (d) are equivalent; a proof is given at the end of the section. The exercises will provide examples of how Theorem 4 is used.

Warning: Theorem 4 is about a *coefficient matrix*, not an augmented matrix. If an augmented matrix $\begin{bmatrix} A & \mathbf{b} \end{bmatrix}$ has a pivot position in every row, then the equation $A\mathbf{x} = \mathbf{b}$ may or may not be consistent.

Computation of $A\mathbf{x}$

The calculations in Example 1 were based on the definition of the product of a matrix A and a vector $\mathbf{x}$. The following simple example will lead to a more efficient method for calculating the entries in $A\mathbf{x}$ when working problems by hand.

EXAMPLE 4 Compute $A\mathbf{x}$, where $A = \begin{bmatrix} 2 & 3 & 4 \\ -1 & 5 & -3 \\ 6 & -2 & 8 \end{bmatrix}$ and $\mathbf{x} = \begin{bmatrix} x_1 \\ x_2 \\ x_3 \end{bmatrix}$.

SOLUTION From the definition,

$$\begin{bmatrix} 2 & 3 & 4 \\ -1 & 5 & -3 \\ 6 & -2 & 8 \end{bmatrix} \begin{bmatrix} x_1 \\ x_2 \\ x_3 \end{bmatrix} = x_1 \begin{bmatrix} 2 \\ -1 \\ 6 \end{bmatrix} + x_2 \begin{bmatrix} 3 \\ 5 \\ -2 \end{bmatrix} + x_3 \begin{bmatrix} 4 \\ -3 \\ 8 \end{bmatrix}$$

$$= \begin{bmatrix} 2x_1 \\ -x_1 \\ 6x_1 \end{bmatrix} + \begin{bmatrix} 3x_2 \\ 5x_2 \\ -2x_2 \end{bmatrix} + \begin{bmatrix} 4x_3 \\ -3x_3 \\ 8x_3 \end{bmatrix} \tag{7}$$

$$= \begin{bmatrix} 2x_1 + 3x_2 + 4x_3 \\ -x_1 + 5x_2 - 3x_3 \\ 6x_1 - 2x_2 + 8x_3 \end{bmatrix}$$

The first entry in the product $A\mathbf{x}$ is a sum of products (sometimes called a *dot product*), using the first row of A and the entries in $\mathbf{x}$. That is,

$$\begin{bmatrix} 2 & 3 & 4 \end{bmatrix} \begin{bmatrix} x_1 \\ x_2 \\ x_3 \end{bmatrix} = \begin{bmatrix} 2x_1 + 3x_2 + 4x_3 \end{bmatrix}$$

This matrix shows how to compute the first entry in $A\mathbf{x}$ directly, without writing down all the calculations shown in (7). Similarly, the second entry in $A\mathbf{x}$ can be calculated at once by multiplying the entries in the second row of A by the corresponding entries in $\mathbf{x}$ and then summing the resulting products:

$$\begin{bmatrix} -1 & 5 & -3 \end{bmatrix} \begin{bmatrix} x_1 \\ x_2 \\ x_3 \end{bmatrix} = \begin{bmatrix} -x_1 + 5x_2 - 3x_3 \end{bmatrix}$$

Likewise, the third entry in $A\mathbf{x}$ can be calculated from the third row of A and the entries in $\mathbf{x}$. ∎

> **Row–Vector Rule for Computing $A\mathbf{x}$**
>
> If the product $A\mathbf{x}$ is defined, then the ith entry in $A\mathbf{x}$ is the sum of the products of corresponding entries from row i of A and from the vector $\mathbf{x}$.

EXAMPLE 5

a. $\begin{bmatrix} 1 & 2 & -1 \\ 0 & -5 & 3 \end{bmatrix} \begin{bmatrix} 4 \\ 3 \\ 7 \end{bmatrix} = \begin{bmatrix} 1 \cdot 4 + 2 \cdot 3 + (-1) \cdot 7 \\ 0 \cdot 4 + (-5) \cdot 3 + 3 \cdot 7 \end{bmatrix} = \begin{bmatrix} 3 \\ 6 \end{bmatrix}$

b. $\begin{bmatrix} 2 & -3 \\ 8 & 0 \\ -5 & 2 \end{bmatrix} \begin{bmatrix} 4 \\ 7 \end{bmatrix} = \begin{bmatrix} 2 \cdot 4 + (-3) \cdot 7 \\ 8 \cdot 4 + 0 \cdot 7 \\ (-5) \cdot 4 + 2 \cdot 7 \end{bmatrix} = \begin{bmatrix} -13 \\ 32 \\ -6 \end{bmatrix}$

c. $\begin{bmatrix} 1 & 0 & 0 \\ 0 & 1 & 0 \\ 0 & 0 & 1 \end{bmatrix} \begin{bmatrix} r \\ s \\ t \end{bmatrix} = \begin{bmatrix} 1 \cdot r + 0 \cdot s + 0 \cdot t \\ 0 \cdot r + 1 \cdot s + 0 \cdot t \\ 0 \cdot r + 0 \cdot s + 1 \cdot t \end{bmatrix} = \begin{bmatrix} r \\ s \\ t \end{bmatrix}$ ∎

By definition, the matrix in Example 5(c) with 1's on the diagonal and 0's elsewhere is called an **identity matrix** and is denoted by I. The calculation in part (c) shows that $I\mathbf{x} = \mathbf{x}$ for every $\mathbf{x}$ in $\mathbb{R}^3$. There is an analogous $n \times n$ identity matrix, sometimes written as I_n. As in part (c), $I_n\mathbf{x} = \mathbf{x}$ for every $\mathbf{x}$ in $\mathbb{R}^n$.

Properties of the Matrix–Vector Product $A\mathbf{x}$

The facts in the next theorem are important and will be used throughout the text. The proof relies on the definition of $A\mathbf{x}$ and the algebraic properties of $\mathbb{R}^n$.

THEOREM 5

> If A is an $m \times n$ matrix, $\mathbf{u}$ and $\mathbf{v}$ are vectors in $\mathbb{R}^n$, and c is a scalar, then:
>
> a. $A(\mathbf{u} + \mathbf{v}) = A\mathbf{u} + A\mathbf{v}$;
>
> b. $A(c\mathbf{u}) = c(A\mathbf{u})$.

PROOF For simplicity, take $n = 3$, $A = [\,\mathbf{a}_1 \quad \mathbf{a}_2 \quad \mathbf{a}_3\,]$, and $\mathbf{u}, \mathbf{v}$ in $\mathbb{R}^3$. (The proof of the general case is similar.) For $i = 1, 2, 3$, let u_i and v_i be the ith entries in $\mathbf{u}$ and $\mathbf{v}$, respectively. To prove statement (a), compute $A(\mathbf{u} + \mathbf{v})$ as a linear combination of the columns of A using the entries in $\mathbf{u} + \mathbf{v}$ as weights.

$$A(\mathbf{u} + \mathbf{v}) = [\,\mathbf{a}_1 \quad \mathbf{a}_2 \quad \mathbf{a}_3\,] \begin{bmatrix} u_1 + v_1 \\ u_2 + v_2 \\ u_3 + v_3 \end{bmatrix}$$

$$= (u_1 + v_1)\mathbf{a}_1 + (u_2 + v_2)\mathbf{a}_2 + (u_3 + v_3)\mathbf{a}_3 \qquad \text{— Entries in } \mathbf{u} + \mathbf{v}$$

$$\text{— Columns of } A$$

$$= (u_1\mathbf{a}_1 + u_2\mathbf{a}_2 + u_3\mathbf{a}_3) + (v_1\mathbf{a}_1 + v_2\mathbf{a}_2 + v_3\mathbf{a}_3)$$

$$= A\mathbf{u} + A\mathbf{v}$$

To prove statement (b), compute $A(c\mathbf{u})$ as a linear combination of the columns of A using the entries in $c\mathbf{u}$ as weights.

$$A(c\mathbf{u}) = [\,\mathbf{a}_1 \quad \mathbf{a}_2 \quad \mathbf{a}_3\,] \begin{bmatrix} cu_1 \\ cu_2 \\ cu_3 \end{bmatrix} = (cu_1)\mathbf{a}_1 + (cu_2)\mathbf{a}_2 + (cu_3)\mathbf{a}_3$$

$$= c(u_1\mathbf{a}_1) + c(u_2\mathbf{a}_2) + c(u_3\mathbf{a}_3)$$

$$= c(u_1\mathbf{a}_1 + u_2\mathbf{a}_2 + u_3\mathbf{a}_3)$$

$$= c(A\mathbf{u}) \qquad \blacksquare$$

NUMERICAL NOTE

To optimize a computer algorithm to compute $A\mathbf{x}$, the sequence of calculations should involve data stored in contiguous memory locations. The most widely used professional algorithms for matrix computations are written in Fortran, a language that stores a matrix as a set of columns. Such algorithms compute $A\mathbf{x}$ as a linear combination of the columns of A. In contrast, if a program is written in the popular language C, which stores matrices by rows, $A\mathbf{x}$ should be computed via the alternative rule that uses the rows of A.

PROOF OF THEOREM 4 As was pointed out after Theorem 4, statements (a), (b), and (c) are logically equivalent. So, it suffices to show (for an arbitrary matrix A) that (a) and (d) are either both true or both false. That will tie all four statements together.

Let U be an echelon form of A. Given $\mathbf{b}$ in $\mathbb{R}^m$, we can row reduce the augmented matrix $[\,A \quad \mathbf{b}\,]$ to an augmented matrix $[\,U \quad \mathbf{d}\,]$ for some $\mathbf{d}$ in $\mathbb{R}^m$:

$$[\,A \quad \mathbf{b}\,] \sim \cdots \sim [\,U \quad \mathbf{d}\,]$$

If statement (d) is true, then each row of U contains a pivot position and there can be no pivot in the augmented column. So $A\mathbf{x} = \mathbf{b}$ has a solution for any $\mathbf{b}$, and (a) is true. If (d) is false, the last row of U is all zeros. Let $\mathbf{d}$ be any vector with a 1 in its last entry. Then $\begin{bmatrix} U & \mathbf{d} \end{bmatrix}$ represents an *inconsistent* system. Since row operations are reversible, $\begin{bmatrix} U & \mathbf{d} \end{bmatrix}$ can be transformed into the form $\begin{bmatrix} A & \mathbf{b} \end{bmatrix}$. The new system $A\mathbf{x} = \mathbf{b}$ is also inconsistent, and (a) is false. ∎

PRACTICE PROBLEMS

1. Let $A = \begin{bmatrix} 1 & 5 & -2 & 0 \\ -3 & 1 & 9 & -5 \\ 4 & -8 & -1 & 7 \end{bmatrix}$, $\mathbf{p} = \begin{bmatrix} 3 \\ -2 \\ 0 \\ -4 \end{bmatrix}$, and $\mathbf{b} = \begin{bmatrix} -7 \\ 9 \\ 0 \end{bmatrix}$. It can be shown that $\mathbf{p}$ is a solution of $A\mathbf{x} = \mathbf{b}$. Use this fact to exhibit $\mathbf{b}$ as a specific linear combination of the columns of A.

2. Let $A = \begin{bmatrix} 2 & 5 \\ 3 & 1 \end{bmatrix}$, $\mathbf{u} = \begin{bmatrix} 4 \\ -1 \end{bmatrix}$, and $\mathbf{v} = \begin{bmatrix} -3 \\ 5 \end{bmatrix}$. Verify Theorem 5(a) in this case by computing $A(\mathbf{u} + \mathbf{v})$ and $A\mathbf{u} + A\mathbf{v}$.

1.4 EXERCISES

Compute the products in Exercises 1–4 using (a) the definition, as in Example 1, and (b) the row–vector rule for computing $A\mathbf{x}$. If a product is undefined, explain why.

1. $\begin{bmatrix} -4 & 2 \\ 1 & 6 \\ 0 & 1 \end{bmatrix} \begin{bmatrix} 3 \\ -2 \\ 7 \end{bmatrix}$ 2. $\begin{bmatrix} 1 \\ 3 \\ 1 \end{bmatrix} \begin{bmatrix} 5 \\ -1 \end{bmatrix}$

3. $\begin{bmatrix} 1 & 2 \\ -3 & 1 \\ 1 & 6 \end{bmatrix} \begin{bmatrix} -2 \\ 3 \end{bmatrix}$ 4. $\begin{bmatrix} 1 & 3 & -4 \\ 3 & 2 & 1 \end{bmatrix} \begin{bmatrix} 1 \\ 2 \\ 1 \end{bmatrix}$

In Exercises 5–8, use the definition of $A\mathbf{x}$ to write the matrix equation as a vector equation, or vice versa.

5. $\begin{bmatrix} 1 & 2 & -3 & 1 \\ -2 & -3 & 1 & -1 \end{bmatrix} \begin{bmatrix} 2 \\ -1 \\ 1 \\ -1 \end{bmatrix} = \begin{bmatrix} -4 \\ 1 \end{bmatrix}$

6. $\begin{bmatrix} 2 & -3 \\ 3 & 2 \\ 8 & -5 \\ -2 & 1 \end{bmatrix} \begin{bmatrix} -3 \\ 5 \end{bmatrix} = \begin{bmatrix} -21 \\ 1 \\ -49 \\ 11 \end{bmatrix}$

7. $x_1 \begin{bmatrix} 4 \\ -1 \\ 7 \\ -4 \end{bmatrix} + x_2 \begin{bmatrix} -5 \\ 3 \\ -5 \\ 1 \end{bmatrix} + x_3 \begin{bmatrix} 7 \\ -8 \\ 0 \\ 2 \end{bmatrix} = \begin{bmatrix} 6 \\ -8 \\ 0 \\ -7 \end{bmatrix}$

8. $z_1 \begin{bmatrix} 2 \\ -4 \end{bmatrix} + z_2 \begin{bmatrix} -1 \\ 5 \end{bmatrix} + z_3 \begin{bmatrix} -4 \\ 3 \end{bmatrix} + z_4 \begin{bmatrix} 0 \\ 2 \end{bmatrix} = \begin{bmatrix} 5 \\ 12 \end{bmatrix}$

In Exercises 9 and 10, write the system first as a vector equation and then as a matrix equation.

9. $\begin{aligned} 5x_1 + x_2 - 3x_3 &= 8 \\ 2x_2 + 4x_3 &= 0 \end{aligned}$

10. $\begin{aligned} 4x_1 - x_2 &= 8 \\ 5x_1 + 3x_2 &= 2 \\ 3x_1 - x_2 &= 1 \end{aligned}$

Given A and $\mathbf{b}$ in Exercises 11 and 12, write the augmented matrix for the linear system that corresponds to the matrix equation $A\mathbf{x} = \mathbf{b}$. Then solve the system and write the solution as a vector.

11. $A = \begin{bmatrix} 1 & 3 & -4 \\ 1 & 5 & 2 \\ -3 & -7 & 6 \end{bmatrix}$, $\mathbf{b} = \begin{bmatrix} -2 \\ 4 \\ 12 \end{bmatrix}$

12. $A = \begin{bmatrix} 1 & 2 & -1 \\ -3 & -4 & 2 \\ 5 & 2 & 3 \end{bmatrix}$, $\mathbf{b} = \begin{bmatrix} 1 \\ 2 \\ -3 \end{bmatrix}$

13. Let $\mathbf{u} = \begin{bmatrix} 0 \\ 4 \\ 4 \end{bmatrix}$ and $A = \begin{bmatrix} 3 & -5 \\ -2 & 6 \\ 1 & 1 \end{bmatrix}$. Is $\mathbf{u}$ in the plane in $\mathbb{R}^3$ spanned by the columns of A? (See the figure.) Why or why not?

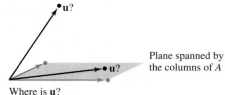

Where is $\mathbf{u}$?

14. Let $\mathbf{u} = \begin{bmatrix} 4 \\ -1 \\ 4 \end{bmatrix}$ and $A = \begin{bmatrix} 2 & 5 & -1 \\ 0 & 1 & -1 \\ 1 & 2 & 0 \end{bmatrix}$. Is $\mathbf{u}$ in the subset of $\mathbb{R}^3$ spanned by the columns of A? Why or why not?

15. Let $A = \begin{bmatrix} 3 & -1 \\ -9 & 3 \end{bmatrix}$ and $\mathbf{b} = \begin{bmatrix} b_1 \\ b_2 \end{bmatrix}$. Show that the equation $A\mathbf{x} = \mathbf{b}$ does not have a solution for all possible $\mathbf{b}$, and describe the set of all $\mathbf{b}$ for which $A\mathbf{x} = \mathbf{b}$ *does* have a solution.

16. Repeat the requests from Exercise 15 with

$$A = \begin{bmatrix} 1 & -2 & -1 \\ -2 & 2 & 0 \\ 4 & -1 & 3 \end{bmatrix}, \quad \text{and} \quad \mathbf{b} = \begin{bmatrix} b_1 \\ b_2 \\ b_3 \end{bmatrix}.$$

Exercises 17–20 refer to the matrices A and B below. Make appropriate calculations that justify your answers and mention an appropriate theorem.

$$A = \begin{bmatrix} 1 & 3 & 0 & 3 \\ -1 & -1 & -1 & 1 \\ 0 & -4 & 2 & -8 \\ 2 & 0 & 3 & -1 \end{bmatrix} \quad B = \begin{bmatrix} 1 & 4 & 1 & 2 \\ 0 & 1 & 3 & -4 \\ 0 & 2 & 6 & 7 \\ 2 & 9 & 5 & -7 \end{bmatrix}$$

17. How many rows of A contain a pivot position? Does the equation $A\mathbf{x} = \mathbf{b}$ have a solution for each $\mathbf{b}$ in $\mathbb{R}^4$?

18. Can every vector in $\mathbb{R}^4$ be written as a linear combination of the columns of the matrix B above? Do the columns of B span $\mathbb{R}^3$?

19. Can each vector in $\mathbb{R}^4$ be written as a linear combination of the columns of the matrix A above? Do the columns of A span $\mathbb{R}^4$?

20. Do the columns of B span $\mathbb{R}^4$? Does the equation $B\mathbf{x} = \mathbf{y}$ have a solution for each $\mathbf{y}$ in $\mathbb{R}^4$?

21. Let $\mathbf{v}_1 = \begin{bmatrix} 1 \\ 0 \\ -1 \\ 0 \end{bmatrix}$, $\mathbf{v}_2 = \begin{bmatrix} 0 \\ -1 \\ 0 \\ 1 \end{bmatrix}$, $\mathbf{v}_3 = \begin{bmatrix} 1 \\ 0 \\ 0 \\ -1 \end{bmatrix}$. Does $\{\mathbf{v}_1, \mathbf{v}_2, \mathbf{v}_3\}$ span $\mathbb{R}^4$? Why or why not?

22. Let $\mathbf{v}_1 = \begin{bmatrix} 0 \\ 0 \\ -3 \end{bmatrix}$, $\mathbf{v}_2 = \begin{bmatrix} 0 \\ -3 \\ 9 \end{bmatrix}$, $\mathbf{v}_3 = \begin{bmatrix} 4 \\ -2 \\ -6 \end{bmatrix}$. Does $\{\mathbf{v}_1, \mathbf{v}_2, \mathbf{v}_3\}$ span $\mathbb{R}^3$? Why or why not?

In Exercises 23 and 24, mark each statement True or False. Justify each answer.

23. a. The equation $A\mathbf{x} = \mathbf{b}$ is referred to as a *vector equation*.

   b. A vector $\mathbf{b}$ is a linear combination of the columns of a matrix A if and only if the equation $A\mathbf{x} = \mathbf{b}$ has at least one solution.

   c. The equation $A\mathbf{x} = \mathbf{b}$ is consistent if the augmented matrix $[\,A \quad \mathbf{b}\,]$ has a pivot position in every row.

   d. The first entry in the product $A\mathbf{x}$ is a sum of products.

   e. If the columns of an $m \times n$ matrix A span $\mathbb{R}^m$, then the equation $A\mathbf{x} = \mathbf{b}$ is consistent for each $\mathbf{b}$ in $\mathbb{R}^m$.

   f. If A is an $m \times n$ matrix and if the equation $A\mathbf{x} = \mathbf{b}$ is inconsistent for some $\mathbf{b}$ in $\mathbb{R}^m$, then A cannot have a pivot position in every row.

24. a. Every matrix equation $A\mathbf{x} = \mathbf{b}$ corresponds to a vector equation with the same solution set.

   b. If the equation $A\mathbf{x} = \mathbf{b}$ is consistent, then $\mathbf{b}$ is in the set spanned by the columns of A.

   c. Any linear combination of vectors can always be written in the form $A\mathbf{x}$ for a suitable matrix A and vector $\mathbf{x}$.

   d. If the coefficient matrix A has a pivot position in every row, then the equation $A\mathbf{x} = \mathbf{b}$ is inconsistent.

   e. The solution set of a linear system whose augmented matrix is $[\,\mathbf{a}_1 \quad \mathbf{a}_2 \quad \mathbf{a}_3 \quad \mathbf{b}\,]$ is the same as the solution set of $A\mathbf{x} = \mathbf{b}$, if $A = [\,\mathbf{a}_1 \quad \mathbf{a}_2 \quad \mathbf{a}_3\,]$.

   f. If A is an $m \times n$ matrix whose columns do not span $\mathbb{R}^m$, then the equation $A\mathbf{x} = \mathbf{b}$ is consistent for every $\mathbf{b}$ in $\mathbb{R}^m$.

25. Note that $\begin{bmatrix} 4 & -3 & 1 \\ 5 & -2 & 5 \\ -6 & 2 & -3 \end{bmatrix} \begin{bmatrix} -3 \\ -1 \\ 2 \end{bmatrix} = \begin{bmatrix} -7 \\ -3 \\ 10 \end{bmatrix}$. Use this fact (and no row operations) to find scalars c_1, c_2, c_3 such that $\begin{bmatrix} -7 \\ -3 \\ 10 \end{bmatrix} = c_1 \begin{bmatrix} 4 \\ 5 \\ -6 \end{bmatrix} + c_2 \begin{bmatrix} -3 \\ -2 \\ 2 \end{bmatrix} + c_3 \begin{bmatrix} 1 \\ 5 \\ -3 \end{bmatrix}$.

26. Let $\mathbf{u} = \begin{bmatrix} 7 \\ 2 \\ 5 \end{bmatrix}$, $\mathbf{v} = \begin{bmatrix} 3 \\ 1 \\ 3 \end{bmatrix}$, and $\mathbf{w} = \begin{bmatrix} 5 \\ 1 \\ 1 \end{bmatrix}$. It can be shown that $2\mathbf{u} - 3\mathbf{v} - \mathbf{w} = \mathbf{0}$. Use this fact (and no row operations) to find x_1 and x_2 that satisfy the equation $\begin{bmatrix} 7 & 3 \\ 2 & 1 \\ 5 & 3 \end{bmatrix} \begin{bmatrix} x_1 \\ x_2 \end{bmatrix} = \begin{bmatrix} 5 \\ 1 \\ 1 \end{bmatrix}$.

27. Rewrite the (numerical) matrix equation below in symbolic form as a vector equation, using symbols $\mathbf{v}_1, \mathbf{v}_2, \ldots$ for the vectors and $c_1, c_2, \ldots$ for scalars. Define what each symbol represents, using the data given in the matrix equation.

$$\begin{bmatrix} -3 & 5 & -4 & 9 & 7 \\ 5 & 8 & 1 & -2 & -4 \end{bmatrix} \begin{bmatrix} -3 \\ 1 \\ 2 \\ -1 \\ 2 \end{bmatrix} = \begin{bmatrix} 11 \\ -11 \end{bmatrix}$$

28. Let $\mathbf{q}_1, \mathbf{q}_2, \mathbf{q}_3$, and $\mathbf{v}$ represent vectors in $\mathbb{R}^5$, and let x_1, x_2, and x_3 denote scalars. Write the following vector equation as a matrix equation. Identify any symbols you choose to use.

$$x_1\mathbf{q}_1 + x_2\mathbf{q}_2 + x_3\mathbf{q}_3 = \mathbf{v}$$

29. Construct a 3×3 matrix, not in echelon form, whose columns span $\mathbb{R}^3$. Show that the matrix you construct has the desired property.

30. Construct a 3×3 matrix, not in echelon form, whose columns do *not* span $\mathbb{R}^3$. Show that the matrix you construct has the desired property.

31. Let A be a 3×2 matrix. Explain why the equation $A\mathbf{x} = \mathbf{b}$ cannot be consistent for all $\mathbf{b}$ in $\mathbb{R}^3$. Generalize your argument to the case of an arbitrary A with more rows than columns.

32. Could a set of three vectors in $\mathbb{R}^4$ span all of $\mathbb{R}^4$? Explain. What about n vectors in $\mathbb{R}^m$ when n is less than m?

33. Suppose A is a 4×3 matrix and $\mathbf{b}$ is a vector in $\mathbb{R}^4$ with the property that $A\mathbf{x} = \mathbf{b}$ has a unique solution. What can you say about the reduced echelon form of A? Justify your answer.

34. Let A be a 3×4 matrix, let $\mathbf{v}_1$ and $\mathbf{v}_2$ be vectors in $\mathbb{R}^3$, and let $\mathbf{w} = \mathbf{v}_1 + \mathbf{v}_2$. Suppose $\mathbf{v}_1 = A\mathbf{u}_1$ and $\mathbf{v}_2 = A\mathbf{u}_2$ for some vectors $\mathbf{u}_1$ and $\mathbf{u}_2$ in $\mathbb{R}^4$. What fact allows you to conclude that the system $A\mathbf{x} = \mathbf{w}$ is consistent? (*Note:* $\mathbf{u}_1$ and $\mathbf{u}_2$ denote vectors, not scalar entries in vectors.)

35. Let A be a 5×3 matrix, let $\mathbf{y}$ be a vector in $\mathbb{R}^3$, and let $\mathbf{z}$ be a vector in $\mathbb{R}^5$. Suppose $A\mathbf{y} = \mathbf{z}$. What fact allows you to conclude that the system $A\mathbf{x} = 5\mathbf{z}$ is consistent?

36. Suppose A is a 4×4 matrix and $\mathbf{b}$ is a vector in $\mathbb{R}^4$ with the property that $A\mathbf{x} = \mathbf{b}$ has a unique solution. Explain why the columns of A must span $\mathbb{R}^4$.

[M] In Exercises 37–40, determine if the columns of the matrix span $\mathbb{R}^4$.

37. $\begin{bmatrix} 7 & 2 & -5 & 8 \\ -5 & -3 & 4 & -9 \\ 6 & 10 & -2 & 7 \\ -7 & 9 & 2 & 15 \end{bmatrix}$ **38.** $\begin{bmatrix} 4 & -5 & -1 & 8 \\ 3 & -7 & -4 & 2 \\ 5 & -6 & -1 & 4 \\ 9 & 1 & 10 & 7 \end{bmatrix}$

39. $\begin{bmatrix} 10 & -7 & 1 & 4 & 6 \\ -8 & 4 & -6 & -10 & -3 \\ -7 & 11 & -5 & -1 & -8 \\ 3 & -1 & 10 & 12 & 12 \end{bmatrix}$

40. $\begin{bmatrix} 5 & 11 & -6 & -7 & 12 \\ -7 & -3 & -4 & 6 & -9 \\ 11 & 5 & 6 & -9 & -3 \\ -3 & 4 & -7 & 2 & 7 \end{bmatrix}$

41. **[M]** Find a column of the matrix in Exercise 39 that can be deleted and yet have the remaining matrix columns still span $\mathbb{R}^4$.

42. **[M]** Find a column of the matrix in Exercise 40 that can be deleted and yet have the remaining matrix columns still span $\mathbb{R}^4$. Can you delete more than one column?

SG Mastering Linear Algebra Concepts: Span 1–18

WEB

| SOLUTIONS TO PRACTICE PROBLEMS

1. The matrix equation

$$\begin{bmatrix} 1 & 5 & -2 & 0 \\ -3 & 1 & 9 & -5 \\ 4 & -8 & -1 & 7 \end{bmatrix} \begin{bmatrix} 3 \\ -2 \\ 0 \\ -4 \end{bmatrix} = \begin{bmatrix} -7 \\ 9 \\ 0 \end{bmatrix}$$

is equivalent to the vector equation

$$3 \begin{bmatrix} 1 \\ -3 \\ 4 \end{bmatrix} - 2 \begin{bmatrix} 5 \\ 1 \\ -8 \end{bmatrix} + 0 \begin{bmatrix} -2 \\ 9 \\ -1 \end{bmatrix} - 4 \begin{bmatrix} 0 \\ -5 \\ 7 \end{bmatrix} = \begin{bmatrix} -7 \\ 9 \\ 0 \end{bmatrix}$$

which expresses $\mathbf{b}$ as a linear combination of the columns of A.

2. $\mathbf{u} + \mathbf{v} = \begin{bmatrix} 4 \\ -1 \end{bmatrix} + \begin{bmatrix} -3 \\ 5 \end{bmatrix} = \begin{bmatrix} 1 \\ 4 \end{bmatrix}$

$A(\mathbf{u} + \mathbf{v}) = \begin{bmatrix} 2 & 5 \\ 3 & 1 \end{bmatrix} \begin{bmatrix} 1 \\ 4 \end{bmatrix} = \begin{bmatrix} 2 + 20 \\ 3 + 4 \end{bmatrix} = \begin{bmatrix} 22 \\ 7 \end{bmatrix}$

$A\mathbf{u} + A\mathbf{v} = \begin{bmatrix} 2 & 5 \\ 3 & 1 \end{bmatrix} \begin{bmatrix} 4 \\ -1 \end{bmatrix} + \begin{bmatrix} 2 & 5 \\ 3 & 1 \end{bmatrix} \begin{bmatrix} -3 \\ 5 \end{bmatrix}$

$= \begin{bmatrix} 3 \\ 11 \end{bmatrix} + \begin{bmatrix} 19 \\ -4 \end{bmatrix} = \begin{bmatrix} 22 \\ 7 \end{bmatrix}$

1.5 │ SOLUTION SETS OF LINEAR SYSTEMS

Solution sets of linear systems are important objects of study in linear algebra. They will appear later in several different contexts. This section uses vector notation to give explicit and geometric descriptions of such solution sets.

Homogeneous Linear Systems

A system of linear equations is said to be **homogeneous** if it can be written in the form $A\mathbf{x} = \mathbf{0}$, where A is an $m \times n$ matrix and $\mathbf{0}$ is the zero vector in $\mathbb{R}^m$. Such a system $A\mathbf{x} = \mathbf{0}$ *always* has at least one solution, namely, $\mathbf{x} = \mathbf{0}$ (the zero vector in $\mathbb{R}^n$). This zero solution is usually called the **trivial solution**. For a given equation $A\mathbf{x} = \mathbf{0}$, the important question is whether there exists a **nontrivial solution**, that is, a nonzero vector $\mathbf{x}$ that satisfies $A\mathbf{x} = \mathbf{0}$. The Existence and Uniqueness Theorem in Section 1.2 (Theorem 2) leads immediately to the following fact.

> The homogeneous equation $A\mathbf{x} = \mathbf{0}$ has a nontrivial solution if and only if the equation has at least one free variable.

EXAMPLE 1 Determine if the following homogeneous system has a nontrivial solution. Then describe the solution set.

$$3x_1 + 5x_2 - 4x_3 = 0$$
$$-3x_1 - 2x_2 + 4x_3 = 0$$
$$6x_1 + x_2 - 8x_3 = 0$$

SOLUTION Let A be the matrix of coefficients of the system and row reduce the augmented matrix $[\,A \quad \mathbf{0}\,]$ to echelon form:

$$\begin{bmatrix} 3 & 5 & -4 & 0 \\ -3 & -2 & 4 & 0 \\ 6 & 1 & -8 & 0 \end{bmatrix} \sim \begin{bmatrix} 3 & 5 & -4 & 0 \\ 0 & 3 & 0 & 0 \\ 0 & -9 & 0 & 0 \end{bmatrix} \sim \begin{bmatrix} 3 & 5 & -4 & 0 \\ 0 & 3 & 0 & 0 \\ 0 & 0 & 0 & 0 \end{bmatrix}$$

Since x_3 is a free variable, $A\mathbf{x} = \mathbf{0}$ has nontrivial solutions (one for each choice of x_3). To describe the solution set, continue the row reduction of $[\,A \quad \mathbf{0}\,]$ to *reduced* echelon form:

$$\begin{bmatrix} 1 & 0 & -\frac{4}{3} & 0 \\ 0 & 1 & 0 & 0 \\ 0 & 0 & 0 & 0 \end{bmatrix} \qquad \begin{aligned} x_1 \quad &- \tfrac{4}{3}x_3 = 0 \\ x_2 \quad &= 0 \\ 0 \, &= 0 \end{aligned}$$

Solve for the basic variables x_1 and x_2 and obtain $x_1 = \frac{4}{3}x_3$, $x_2 = 0$, with x_3 free. As a vector, the general solution of $A\mathbf{x} = \mathbf{0}$ has the form

$$\mathbf{x} = \begin{bmatrix} x_1 \\ x_2 \\ x_3 \end{bmatrix} = \begin{bmatrix} \frac{4}{3}x_3 \\ 0 \\ x_3 \end{bmatrix} = x_3 \begin{bmatrix} \frac{4}{3} \\ 0 \\ 1 \end{bmatrix} = x_3 \mathbf{v}, \quad \text{where } \mathbf{v} = \begin{bmatrix} \frac{4}{3} \\ 0 \\ 1 \end{bmatrix}$$

Here x_3 is factored out of the expression for the general solution vector. This shows that every solution of $A\mathbf{x} = \mathbf{0}$ in this case is a scalar multiple of $\mathbf{v}$. The trivial solution is obtained by choosing $x_3 = 0$. Geometrically, the solution set is a line through $\mathbf{0}$ in $\mathbb{R}^3$. See Fig. 1.

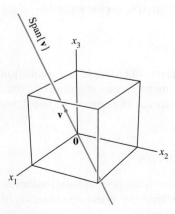

FIGURE 1

Notice that a nontrivial solution **x** can have some zero entries so long as not all of its entries are zero.

EXAMPLE 2 A single linear equation can be treated as a very simple system of equations. Describe all solutions of the homogeneous "system"

$$10x_1 - 3x_2 - 2x_3 = 0 \tag{1}$$

SOLUTION There is no need for matrix notation. Solve for the basic variable x_1 in terms of the free variables. The general solution is $x_1 = .3x_2 + .2x_3$, with x_2 and x_3 free. As a vector, the general solution is

$$\mathbf{x} = \begin{bmatrix} x_1 \\ x_2 \\ x_3 \end{bmatrix} = \begin{bmatrix} .3x_2 + .2x_3 \\ x_2 \\ x_3 \end{bmatrix} = \begin{bmatrix} .3x_2 \\ x_2 \\ 0 \end{bmatrix} + \begin{bmatrix} .2x_3 \\ 0 \\ x_3 \end{bmatrix}$$

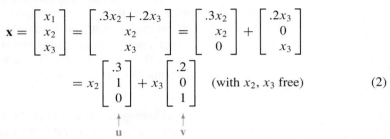

$$= x_2 \underset{\underset{\mathbf{u}}{\uparrow}}{\begin{bmatrix} .3 \\ 1 \\ 0 \end{bmatrix}} + x_3 \underset{\underset{\mathbf{v}}{\uparrow}}{\begin{bmatrix} .2 \\ 0 \\ 1 \end{bmatrix}} \quad \text{(with } x_2, x_3 \text{ free)} \tag{2}$$

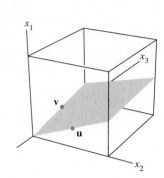

FIGURE 2

This calculation shows that every solution of (1) is a linear combination of the vectors **u** and **v**, shown in (2). That is, the solution set is Span $\{\mathbf{u}, \mathbf{v}\}$. Since neither **u** nor **v** is a scalar multiple of the other, the solution set is a plane through the origin. See Fig. 2. ■

Examples 1 and 2, along with the exercises, illustrate the fact that the solution set of a homogeneous equation $A\mathbf{x} = \mathbf{0}$ can always be expressed explicitly as Span $\{\mathbf{v}_1, \ldots, \mathbf{v}_p\}$ for suitable vectors $\mathbf{v}_1, \ldots, \mathbf{v}_p$. If the only solution is the zero vector, then the solution set is Span $\{\mathbf{0}\}$. If the equation $A\mathbf{x} = \mathbf{0}$ has only one free variable, the solution set is a line through the origin, as in Fig. 1. A plane through the origin, as in Fig. 2, provides a good mental image for the solution set of $A\mathbf{x} = \mathbf{0}$ when there are two or more free variables. Note, however, that a similar figure can be used to visualize Span $\{\mathbf{u}, \mathbf{v}\}$ even when **u** and **v** do not arise as solutions of $A\mathbf{x} = \mathbf{0}$. See Fig. 11 in Section 1.3.

Parametric Vector Form

The original equation (1) for the plane in Example 2 is an *implicit* description of the plane. Solving this equation amounts to finding an *explicit* description of the plane as the set spanned by **u** and **v**. Equation (2) is called a **parametric vector equation** of the plane. Sometimes such an equation is written as

$$\mathbf{x} = s\mathbf{u} + t\mathbf{v} \quad (s, t \text{ in } \mathbb{R})$$

to emphasize that the parameters vary over all real numbers. In Example 1, the equation $\mathbf{x} = x_3\mathbf{v}$ (with x_3 free), or $\mathbf{x} = t\mathbf{v}$ (with t in $\mathbb{R}$), is a parametric vector equation of a line. Whenever a solution set is described explicitly with vectors as in Examples 1 and 2, we say that the solution is in **parametric vector form**.

Solutions of Nonhomogeneous Systems

When a nonhomogeneous linear system has many solutions, the general solution can be written in parametric vector form as one vector plus an arbitrary linear combination of vectors that satisfy the corresponding homogeneous system.

EXAMPLE 3 Describe all solutions of $A\mathbf{x} = \mathbf{b}$, where

$$A = \begin{bmatrix} 3 & 5 & -4 \\ -3 & -2 & 4 \\ 6 & 1 & -8 \end{bmatrix} \quad \text{and} \quad \mathbf{b} = \begin{bmatrix} 7 \\ -1 \\ -4 \end{bmatrix}$$

SOLUTION Here A is the matrix of coefficients from Example 1. Row operations on $[\, A \quad \mathbf{b} \,]$ produce

$$\begin{bmatrix} 3 & 5 & -4 & 7 \\ -3 & -2 & 4 & -1 \\ 6 & 1 & -8 & -4 \end{bmatrix} \sim \begin{bmatrix} 1 & 0 & -\frac{4}{3} & -1 \\ 0 & 1 & 0 & 2 \\ 0 & 0 & 0 & 0 \end{bmatrix}, \qquad \begin{array}{rcl} x_1 & -\frac{4}{3}x_3 = & -1 \\ x_2 & = & 2 \\ 0 & = & 0 \end{array}$$

Thus $x_1 = -1 + \frac{4}{3}x_3$, $x_2 = 2$, and x_3 is free. As a vector, the general solution of $A\mathbf{x} = \mathbf{b}$ has the form

$$\mathbf{x} = \begin{bmatrix} x_1 \\ x_2 \\ x_3 \end{bmatrix} = \begin{bmatrix} -1 + \frac{4}{3}x_3 \\ 2 \\ x_3 \end{bmatrix} = \begin{bmatrix} -1 \\ 2 \\ 0 \end{bmatrix} + \begin{bmatrix} \frac{4}{3}x_3 \\ 0 \\ x_3 \end{bmatrix} = \underset{\mathbf{p}}{\underbrace{\begin{bmatrix} -1 \\ 2 \\ 0 \end{bmatrix}}} + x_3 \underset{\mathbf{v}}{\underbrace{\begin{bmatrix} \frac{4}{3} \\ 0 \\ 1 \end{bmatrix}}}$$

The equation $\mathbf{x} = \mathbf{p} + x_3\mathbf{v}$, or, writing t as a general parameter,

$$\mathbf{x} = \mathbf{p} + t\mathbf{v} \quad (t \text{ in } \mathbb{R}) \tag{3}$$

describes the solution set of $A\mathbf{x} = \mathbf{b}$ in parametric vector form. Recall from Example 1 that the solution set of $A\mathbf{x} = \mathbf{0}$ has the parametric vector equation

$$\mathbf{x} = t\mathbf{v} \quad (t \text{ in } \mathbb{R}) \tag{4}$$

[with the same $\mathbf{v}$ that appears in (3)]. Thus the solutions of $A\mathbf{x} = \mathbf{b}$ are obtained by adding the vector $\mathbf{p}$ to the solutions of $A\mathbf{x} = \mathbf{0}$. The vector $\mathbf{p}$ itself is just one particular solution of $A\mathbf{x} = \mathbf{b}$ [corresponding to $t = 0$ in (3)]. ∎

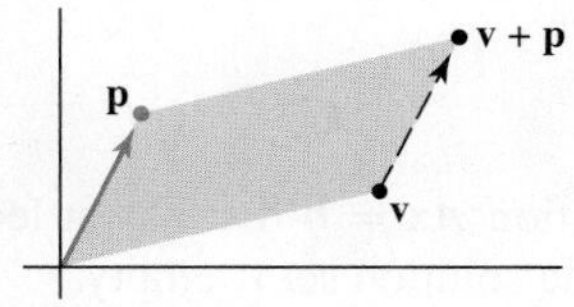

To describe the solution set of $A\mathbf{x} = \mathbf{b}$ geometrically, we can think of vector addition as a *translation*. Given $\mathbf{v}$ and $\mathbf{p}$ in $\mathbb{R}^2$ or $\mathbb{R}^3$, the effect of adding $\mathbf{p}$ to $\mathbf{v}$ is to *move* $\mathbf{v}$ in a direction parallel to the line through $\mathbf{p}$ and $\mathbf{0}$. We say that $\mathbf{v}$ is **translated by $\mathbf{p}$** to $\mathbf{v} + \mathbf{p}$. See Fig. 3. If each point on a line L in $\mathbb{R}^2$ or $\mathbb{R}^3$ is translated by a vector $\mathbf{p}$, the result is a line parallel to L. See Fig. 4.

Suppose L is the line through $\mathbf{0}$ and $\mathbf{v}$, described by equation (4). Adding $\mathbf{p}$ to each point on L produces the translated line described by equation (3). Note that $\mathbf{p}$ is on the line in equation (3). We call (3) **the equation of the line through $\mathbf{p}$ parallel to $\mathbf{v}$**. Thus *the solution set of $A\mathbf{x} = \mathbf{b}$ is a line through $\mathbf{p}$ parallel to the solution set of $A\mathbf{x} = \mathbf{0}$.* Figure 5 illustrates this case.

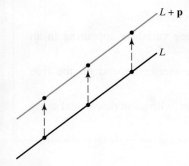

FIGURE 3
Adding $\mathbf{p}$ to $\mathbf{v}$ translates $\mathbf{v}$ to $\mathbf{v} + \mathbf{p}$.

FIGURE 4
Translated line.

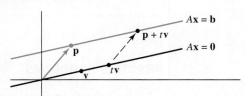

FIGURE 5 Parallel solution sets of $A\mathbf{x} = \mathbf{b}$ and $A\mathbf{x} = \mathbf{0}$.

The relation between the solution sets of $A\mathbf{x} = \mathbf{b}$ and $A\mathbf{x} = \mathbf{0}$ shown in Fig. 5 generalizes to any *consistent* equation $A\mathbf{x} = \mathbf{b}$, although the solution set will be larger than a line when there are several free variables. The following theorem gives the precise statement. See Exercise 25 for a proof.

THEOREM 6

> Suppose the equation $A\mathbf{x} = \mathbf{b}$ is consistent for some given $\mathbf{b}$, and let $\mathbf{p}$ be a solution. Then the solution set of $A\mathbf{x} = \mathbf{b}$ is the set of all vectors of the form $\mathbf{w} = \mathbf{p} + \mathbf{v}_h$, where $\mathbf{v}_h$ is any solution of the homogeneous equation $A\mathbf{x} = \mathbf{0}$.

Theorem 6 says that if $A\mathbf{x} = \mathbf{b}$ has a solution, then the solution set is obtained by translating the solution set of $A\mathbf{x} = \mathbf{0}$, using any particular solution $\mathbf{p}$ of $A\mathbf{x} = \mathbf{b}$ for the translation. Figure 6 illustrates the case in which there are two free variables. Even when $n > 3$, our mental image of the solution set of a consistent system $A\mathbf{x} = \mathbf{b}$ (with $\mathbf{b} \neq \mathbf{0}$) is either a single nonzero point or a line or plane not passing through the origin.

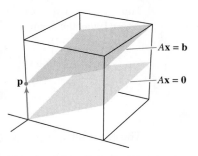

FIGURE 6 Parallel solution sets of $A\mathbf{x} = \mathbf{b}$ and $A\mathbf{x} = \mathbf{0}$.

Warning: Theorem 6 and Fig. 6 apply only to an equation $A\mathbf{x} = \mathbf{b}$ that has at least one nonzero solution $\mathbf{p}$. When $A\mathbf{x} = \mathbf{b}$ has no solution, the solution set is empty.

The following algorithm outlines the calculations shown in Examples 1, 2, and 3.

WRITING A SOLUTION SET (OF A CONSISTENT SYSTEM) IN PARAMETRIC VECTOR FORM

1. Row reduce the augmented matrix to reduced echelon form.
2. Express each basic variable in terms of any free variables appearing in an equation.
3. Write a typical solution $\mathbf{x}$ as a vector whose entries depend on the free variables, if any.
4. Decompose $\mathbf{x}$ into a linear combination of vectors (with numeric entries) using the free variables as parameters.

PRACTICE PROBLEMS

1. Each of the following equations determines a plane in $\mathbb{R}^3$. Do the two planes intersect? If so, describe their intersection.

$$x_1 + 4x_2 - 5x_3 = 0$$
$$2x_1 - x_2 + 8x_3 = 9$$

2. Write the general solution of $10x_1 - 3x_2 - 2x_3 = 7$ in parametric vector form, and relate the solution set to the one found in Example 2.

1.5 EXERCISES

In Exercises 1–4, determine if the system has a nontrivial solution. Try to use as few row operations as possible.

1.
$$2x_1 - 5x_2 + 8x_3 = 0$$
$$-2x_1 - 7x_2 + x_3 = 0$$
$$4x_1 + 2x_2 + 7x_3 = 0$$

2.
$$x_1 - 2x_2 + 3x_3 = 0$$
$$-2x_1 - 3x_2 - 4x_3 = 0$$
$$2x_1 - 4x_2 + 9x_3 = 0$$

3.
$$-3x_1 + 4x_2 - 8x_3 = 0$$
$$-2x_1 + 5x_2 + 4x_3 = 0$$

4.
$$5x_1 - 3x_2 + 2x_3 = 0$$
$$-3x_1 - 4x_2 + 2x_3 = 0$$

In Exercises 5 and 6, follow the method of Examples 1 and 2 to write the solution set of the given homogeneous system in parametric vector form.

5.
$$2x_1 + 2x_2 + 4x_3 = 0$$
$$-4x_1 - 4x_2 - 8x_3 = 0$$
$$- 3x_2 - 3x_3 = 0$$

6.
$$x_1 + 2x_2 - 3x_3 = 0$$
$$2x_1 + x_2 - 3x_3 = 0$$
$$-1x_1 + x_2 = 0$$

In Exercises 7–12, describe all solutions of $A\mathbf{x} = \mathbf{0}$ in parametric vector form, where A is row equivalent to the given matrix.

7. $\begin{bmatrix} 1 & 3 & -3 & 7 \\ 0 & 1 & -4 & 5 \end{bmatrix}$

8. $\begin{bmatrix} 1 & -3 & -8 & 5 \\ 0 & 1 & 2 & -4 \end{bmatrix}$

9. $\begin{bmatrix} 3 & -6 & 6 \\ -2 & 4 & -2 \end{bmatrix}$

10. $\begin{bmatrix} -1 & -4 & 0 & -4 \\ 2 & -8 & 0 & 8 \end{bmatrix}$

11. $\begin{bmatrix} 1 & -4 & -2 & 0 & 3 & -5 \\ 0 & 0 & 1 & 0 & 0 & -1 \\ 0 & 0 & 0 & 0 & 1 & -4 \\ 0 & 0 & 0 & 0 & 0 & 0 \end{bmatrix}$

12. $\begin{bmatrix} 1 & -2 & 3 & -6 & 5 & 0 \\ 0 & 0 & 0 & 1 & 4 & -6 \\ 0 & 0 & 0 & 0 & 0 & 1 \\ 0 & 0 & 0 & 0 & 0 & 0 \end{bmatrix}$

13. Suppose the solution set of a certain system of linear equations can be described as $x_1 = 5 + 4x_3$, $x_2 = -2 - 7x_3$, with x_3 free. Use vectors to describe this set as a line in $\mathbb{R}^3$.

14. Suppose the solution set of a certain system of linear equations can be described as $x_1 = 5x_4$, $x_2 = 3 - 2x_4$, $x_3 = 2 + 5x_4$, with x_4 free. Use vectors to describe this set as a "line" in $\mathbb{R}^4$.

15. Describe and compare the solution sets of $x_1 + 5x_2 - 3x_3 = 0$ and $x_1 + 5x_2 - 3x_3 = -2$.

16. Describe and compare the solution sets of $x_1 - 2x_2 + 3x_3 = 0$ and $x_1 - 2x_2 + 3x_3 = 4$.

17. Follow the method of Example 3 to describe the solutions of the following system in parametric vector form. Also, give a geometric description of the solution set and compare it to that in Exercise 5.

$$2x_1 + 2x_2 + 4x_3 = 8$$
$$-4x_1 - 4x_2 - 8x_3 = -16$$
$$- 3x_2 - 3x_3 = 12$$

18. As in Exercise 17, describe the solutions of the following system in parametric vector form, and provide a geometric comparison with the solution set in Exercise 6.

$$x_1 + 2x_2 - 3x_3 = 5$$
$$2x_1 + x_2 - 3x_3 = 13$$
$$-x_1 + x_2 = -8$$

In Exercises 19 and 20, find the parametric equation of the line through **a** parallel to **b**.

19. $\mathbf{a} = \begin{bmatrix} -2 \\ 0 \end{bmatrix}, \mathbf{b} = \begin{bmatrix} -5 \\ 3 \end{bmatrix}$

20. $\mathbf{a} = \begin{bmatrix} 3 \\ -2 \end{bmatrix}, \mathbf{b} = \begin{bmatrix} -7 \\ 6 \end{bmatrix}$

In Exercises 21 and 22, find a parametric equation of the line M through **p** and **q**. [*Hint*: M is parallel to the vector $\mathbf{q} - \mathbf{p}$. See the figure below.]

21. $\mathbf{p} = \begin{bmatrix} 3 \\ -3 \end{bmatrix}, \mathbf{q} = \begin{bmatrix} 4 \\ 1 \end{bmatrix}$

22. $\mathbf{p} = \begin{bmatrix} -3 \\ 2 \end{bmatrix}, \mathbf{q} = \begin{bmatrix} 0 \\ -3 \end{bmatrix}$

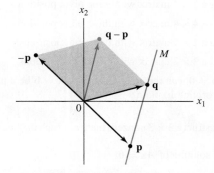

In Exercises 23 and 24, mark each statement True or False. Justify each answer.

23. a. A homogeneous equation is always consistent.

b. The equation $A\mathbf{x} = \mathbf{0}$ gives an explicit description of its solution set.

c. The homogeneous equation $A\mathbf{x} = \mathbf{0}$ has the trivial solution if and only if the equation has at least one free variable.

d. The equation $\mathbf{x} = \mathbf{p} + t\mathbf{v}$ describes a line through $\mathbf{v}$ parallel to $\mathbf{p}$.

e. The solution set of $A\mathbf{x} = \mathbf{b}$ is the set of all vectors of the form $\mathbf{w} = \mathbf{p} + \mathbf{v}_h$, where $\mathbf{v}_h$ is any solution of the equation $A\mathbf{x} = \mathbf{0}$.

24. a. A homogeneous system of equations can be inconsistent.

b. If $\mathbf{x}$ is a nontrivial solution of $A\mathbf{x} = \mathbf{0}$, then every entry in $\mathbf{x}$ is nonzero.

c. The effect of adding $\mathbf{p}$ to a vector is to move the vector in a direction parallel to $\mathbf{p}$.

d. The equation $A\mathbf{x} = \mathbf{b}$ is homogeneous if the zero vector is a solution.

e. If $A\mathbf{x} = \mathbf{b}$ is consistent, then the solution set of $A\mathbf{x} = \mathbf{b}$ is obtained by translating the solution set of $A\mathbf{x} = \mathbf{0}$.

25. Prove Theorem 6:

   a. Suppose $\mathbf{p}$ is a solution of $A\mathbf{x} = \mathbf{b}$, so that $A\mathbf{p} = \mathbf{b}$. Let $\mathbf{v}_h$ be any solution of the homogeneous equation $A\mathbf{x} = \mathbf{0}$, and let $\mathbf{w} = \mathbf{p} + \mathbf{v}_h$. Show that $\mathbf{w}$ is a solution of $A\mathbf{x} = \mathbf{b}$.

   b. Let $\mathbf{w}$ be any solution of $A\mathbf{x} = \mathbf{b}$, and define $\mathbf{v}_h = \mathbf{w} - \mathbf{p}$. Show that $\mathbf{v}_h$ is a solution of $A\mathbf{x} = \mathbf{0}$. This shows that every solution of $A\mathbf{x} = \mathbf{b}$ has the form $\mathbf{w} = \mathbf{p} + \mathbf{v}_h$, with $\mathbf{p}$ a particular solution of $A\mathbf{x} = \mathbf{b}$ and $\mathbf{v}_h$ a solution of $A\mathbf{x} = \mathbf{0}$.

26. Suppose A is the 3×3 *zero* matrix (with all zero entries). Describe the solution set of the equation $A\mathbf{x} = \mathbf{0}$.

27. Suppose $A\mathbf{x} = \mathbf{b}$ has a solution. Explain why the solution is unique precisely when $A\mathbf{x} = \mathbf{0}$ has only the trivial solution.

In Exercises 28–31, (a) does the equation $A\mathbf{x} = \mathbf{0}$ have a nontrivial solution and (b) does the equation $A\mathbf{x} = \mathbf{b}$ have at least one solution for every possible $\mathbf{b}$?

28. A is a 3×3 matrix with three pivot positions.

29. A is a 4×4 matrix with three pivot positions.

30. A is a 2×5 matrix with two pivot positions.

31. A is a 3×2 matrix with two pivot positions.

32. If $\mathbf{b} \neq \mathbf{0}$, can the solution set of $A\mathbf{x} = \mathbf{b}$ be a plane through the origin? Explain.

33. Construct a 3×3 nonzero matrix A such that the vector $\begin{bmatrix} 1 \\ 1 \\ 1 \end{bmatrix}$ is a solution of $A\mathbf{x} = \mathbf{0}$.

34. Construct a 3×3 nonzero matrix A such that the vector $\begin{bmatrix} 2 \\ -1 \\ 1 \end{bmatrix}$ is a solution of $A\mathbf{x} = \mathbf{0}$.

35. Given $A = \begin{bmatrix} -1 & -3 \\ 7 & 21 \\ -2 & -6 \end{bmatrix}$, find one nontrivial solution of $A\mathbf{x} = \mathbf{0}$ by inspection. [*Hint*: Think of the equation $A\mathbf{x} = \mathbf{0}$ written as a vector equation.]

36. Given $A = \begin{bmatrix} 3 & -2 \\ -6 & 4 \\ 12 & -8 \end{bmatrix}$, find one nontrivial solution of $A\mathbf{x} = \mathbf{0}$ by inspection.

37. Construct a 2×2 matrix A such that the solution set of the equation $A\mathbf{x} = \mathbf{0}$ is the line in $\mathbb{R}^2$ through $(4, 1)$ and the origin. Then, find a vector $\mathbf{b}$ in $\mathbb{R}^2$ such that the solution set of $A\mathbf{x} = \mathbf{b}$ is *not* a line in $\mathbb{R}^2$ parallel to the solution set of $A\mathbf{x} = \mathbf{0}$. Why does this *not* contradict Theorem 6?

38. Let A be an $m \times n$ matrix and let $\mathbf{w}$ be a vector in $\mathbb{R}^n$ that satisfies the equation $A\mathbf{x} = \mathbf{0}$. Show that for any scalar c, the vector $c\mathbf{w}$ also satisfies $A\mathbf{x} = \mathbf{0}$. [That is, show that $A(c\mathbf{w}) = \mathbf{0}$.]

39. Let A be an $m \times n$ matrix, and let $\mathbf{v}$ and $\mathbf{w}$ be vectors in $\mathbb{R}^n$ with the property that $A\mathbf{v} = \mathbf{0}$ and $A\mathbf{w} = \mathbf{0}$. Explain why $A(\mathbf{v} + \mathbf{w})$ must be the zero vector. Then explain why $A(c\mathbf{v} + d\mathbf{w}) = \mathbf{0}$ for each pair of scalars c and d.

40. Suppose A is a 3×3 matrix and $\mathbf{b}$ is a vector in $\mathbb{R}^3$ such that the equation $A\mathbf{x} = \mathbf{b}$ does *not* have a solution. Does there exist a vector $\mathbf{y}$ in $\mathbb{R}^3$ such that the equation $A\mathbf{x} = \mathbf{y}$ has a unique solution? Discuss.

SOLUTIONS TO PRACTICE PROBLEMS

1. Row reduce the augmented matrix:

$$\begin{bmatrix} 1 & 4 & -5 & 0 \\ 2 & -1 & 8 & 9 \end{bmatrix} \sim \begin{bmatrix} 1 & 4 & -5 & 0 \\ 0 & -9 & 18 & 9 \end{bmatrix} \sim \begin{bmatrix} 1 & 0 & 3 & 4 \\ 0 & 1 & -2 & -1 \end{bmatrix}$$

$$\begin{aligned} x_1 \quad + 3x_3 &= 4 \\ x_2 - 2x_3 &= -1 \end{aligned}$$

Thus $x_1 = 4 - 3x_3$, $x_2 = -1 + 2x_3$, with x_3 free. The general solution in parametric vector form is

$$\begin{bmatrix} x_1 \\ x_2 \\ x_3 \end{bmatrix} = \begin{bmatrix} 4 - 3x_3 \\ -1 + 2x_3 \\ x_3 \end{bmatrix} = \underset{\mathbf{p}}{\begin{bmatrix} 4 \\ -1 \\ 0 \end{bmatrix}} + x_3 \underset{\mathbf{v}}{\begin{bmatrix} -3 \\ 2 \\ 1 \end{bmatrix}}$$

The intersection of the two planes is the line through $\mathbf{p}$ in the direction of $\mathbf{v}$.

2. The augmented matrix $\begin{bmatrix} 10 & -3 & -2 & 7 \end{bmatrix}$ is row equivalent to $\begin{bmatrix} 1 & -.3 & -.2 & .7 \end{bmatrix}$, and the general solution is $x_1 = .7 + .3x_2 + .2x_3$, with x_2 and x_3 free. That is,

$$\mathbf{x} = \begin{bmatrix} x_1 \\ x_2 \\ x_3 \end{bmatrix} = \begin{bmatrix} .7 + .3x_2 + .2x_3 \\ x_2 \\ x_3 \end{bmatrix} = \begin{bmatrix} .7 \\ 0 \\ 0 \end{bmatrix} + x_2 \begin{bmatrix} .3 \\ 1 \\ 0 \end{bmatrix} + x_3 \begin{bmatrix} .2 \\ 0 \\ 1 \end{bmatrix}$$
$$= \quad \mathbf{p} \quad + \quad x_2\mathbf{u} \quad + \quad x_3\mathbf{v}$$

The solution set of the nonhomogeneous equation $A\mathbf{x} = \mathbf{b}$ is the translated plane $\mathbf{p} + \text{Span}\{\mathbf{u}, \mathbf{v}\}$, which passes through $\mathbf{p}$ and is parallel to the solution set of the homogeneous equation in Example 2.

1.6 | APPLICATIONS OF LINEAR SYSTEMS

You might expect that a real-life problem involving linear algebra would have only one solution, or perhaps no solution. The purpose of this section is to show how linear systems with many solutions can arise naturally. The applications here come from economics, chemistry, and network flow.

A Homogeneous System in Economics

WEB

The system of 500 equations in 500 variables, mentioned in this chapter's introduction, is now known as a Leontief "input–output" (or "production") model.[1] Section 2.6 will examine this model in more detail, when more theory and better notation are available. For now, we look at a simpler "exchange model," also due to Leontief.

Suppose a nation's economy is divided into many sectors, such as various manufacturing, communication, entertainment, and service industries. Suppose that for each sector we know its total output for one year and we know exactly how this output is divided or "exchanged" among the other sectors of the economy. Let the total dollar value of a sector's output be called the **price** of that output. Leontief proved the following result.

> There exist *equilibrium prices* that can be assigned to the total outputs of the various sectors in such a way that the income of each sector exactly balances its expenses.

The following example shows how to find the equilibrium prices.

EXAMPLE 1 Suppose an economy consists of the Coal, Electric (power), and Steel sectors, and the output of each sector is distributed among the various sectors as shown in Table 1 on page 50, where the entries in a column represent the fractional parts of a sector's total output.

The second column of Table 1, for instance, says that the total output of the Electric sector is divided as follows: 40% to Coal, 50% to Steel, and the remaining 10% to Electric. (Electric treats this 10% as an expense it incurs in order to operate its business.) Since all output must be taken into account, the decimal fractions in each column must sum to 1.

---

[1] See Wassily W. Leontief, "Input–Output Economics," *Scientific American*, October 1951, pp. 15–21.

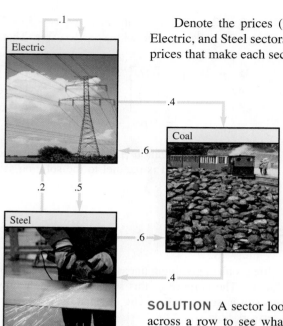

Denote the prices (i.e., dollar values) of the total annual outputs of the Coal, Electric, and Steel sectors by p_C, p_E, and p_S, respectively. If possible, find equilibrium prices that make each sector's income match its expenditures.

TABLE 1   A Simple Economy

Distribution of Output from:			
Coal	**Electric**	**Steel**	**Purchased by:**
.0	.4	.6	Coal
.6	.1	.2	Electric
.4	.5	.2	Steel

SOLUTION A sector looks down a column to see where its output goes, and it looks across a row to see what it needs as inputs. For instance, the first row of Table 1 says that Coal receives (and pays for) 40% of the Electric output and 60% of the Steel output. Since the respective values of the total outputs are p_E and p_S, Coal must spend $.4p_E$ dollars for its share of Electric's output and $.6p_S$ for its share of Steel's output. Thus Coal's total expenses are $.4p_E + .6p_S$. To make Coal's income, p_C, equal to its expenses, we want

$$p_C = .4p_E + .6p_S \tag{1}$$

The second row of the exchange table shows that the Electric sector spends $.6p_C$ for coal, $.1p_E$ for electricity, and $.2p_S$ for steel. Hence the income/expense requirement for Electric is

$$p_E = .6p_C + .1p_E + .2p_S \tag{2}$$

Finally, the third row of the exchange table leads to the final requirement:

$$p_S = .4p_C + .5p_E + .2p_S \tag{3}$$

To solve the system of equations (1), (2), and (3), move all the unknowns to the left sides of the equations and combine like terms. [For instance, on the left side of (2), write $p_E - .1p_E$ as $.9p_E$.]

$$p_C - .4p_E - .6p_S = 0$$
$$-.6p_C + .9p_E - .2p_S = 0$$
$$-.4p_C - .5p_E + .8p_S = 0$$

Row reduction is next. For simplicity here, decimals are rounded to two places.

$$\begin{bmatrix} 1 & -.4 & -.6 & 0 \\ -.6 & .9 & -.2 & 0 \\ -.4 & -.5 & .8 & 0 \end{bmatrix} \sim \begin{bmatrix} 1 & -.4 & -.6 & 0 \\ 0 & .66 & -.56 & 0 \\ 0 & -.66 & .56 & 0 \end{bmatrix} \sim \begin{bmatrix} 1 & -.4 & -.6 & 0 \\ 0 & .66 & -.56 & 0 \\ 0 & 0 & 0 & 0 \end{bmatrix}$$

$$\sim \begin{bmatrix} 1 & -.4 & -.6 & 0 \\ 0 & 1 & -.85 & 0 \\ 0 & 0 & 0 & 0 \end{bmatrix} \sim \begin{bmatrix} 1 & 0 & -.94 & 0 \\ 0 & 1 & -.85 & 0 \\ 0 & 0 & 0 & 0 \end{bmatrix}$$

The general solution is $p_C = .94p_S$, $p_E = .85p_S$, and p_S is free. The equilibrium price vector for the economy has the form

$$\mathbf{p} = \begin{bmatrix} p_C \\ p_E \\ p_S \end{bmatrix} = \begin{bmatrix} .94p_S \\ .85p_S \\ p_S \end{bmatrix} = p_S \begin{bmatrix} .94 \\ .85 \\ 1 \end{bmatrix}$$

Any (nonnegative) choice for p_S results in a choice of equilibrium prices. For instance, if we take p_S to be 100 (or $100 million), then $p_C = 94$ and $p_E = 85$. The incomes and expenditures of each sector will be equal if the output of Coal is priced at $94 million, that of Electric at $85 million, and that of Steel at $100 million. ∎

Balancing Chemical Equations

Chemical equations describe the quantities of substances consumed and produced by chemical reactions. For instance, when propane gas burns, the propane (C_3H_8) combines with oxygen (O_2) to form carbon dioxide (CO_2) and water (H_2O), according to an equation of the form

$$(x_1)C_3H_8 + (x_2)O_2 \rightarrow (x_3)CO_2 + (x_4)H_2O \tag{4}$$

To "balance" this equation, a chemist must find whole numbers $x_1, \ldots, x_4$ such that the total numbers of carbon (C), hydrogen (H), and oxygen (O) atoms on the left match the corresponding numbers of atoms on the right (because atoms are neither destroyed nor created in the reaction).

A systematic method for balancing chemical equations is to set up a vector equation that describes the numbers of atoms of each type present in a reaction. Since equation (4) involves three types of atoms (carbon, hydrogen, and oxygen), construct a vector in $\mathbb{R}^3$ for each reactant and product in (4) that lists the numbers of "atoms per molecule," as follows:

$$C_3H_8: \begin{bmatrix} 3 \\ 8 \\ 0 \end{bmatrix}, \quad O_2: \begin{bmatrix} 0 \\ 0 \\ 2 \end{bmatrix}, \quad CO_2: \begin{bmatrix} 1 \\ 0 \\ 2 \end{bmatrix}, \quad H_2O: \begin{bmatrix} 0 \\ 2 \\ 1 \end{bmatrix} \begin{matrix} \leftarrow \text{Carbon} \\ \leftarrow \text{Hydrogen} \\ \leftarrow \text{Oxygen} \end{matrix}$$

To balance equation (4), the coefficients $x_1, \ldots, x_4$ must satisfy

$$x_1 \begin{bmatrix} 3 \\ 8 \\ 0 \end{bmatrix} + x_2 \begin{bmatrix} 0 \\ 0 \\ 2 \end{bmatrix} = x_3 \begin{bmatrix} 1 \\ 0 \\ 2 \end{bmatrix} + x_4 \begin{bmatrix} 0 \\ 2 \\ 1 \end{bmatrix}$$

To solve, move all the terms to the left (changing the signs in the third and fourth vectors):

$$x_1 \begin{bmatrix} 3 \\ 8 \\ 0 \end{bmatrix} + x_2 \begin{bmatrix} 0 \\ 0 \\ 2 \end{bmatrix} + x_3 \begin{bmatrix} -1 \\ 0 \\ -2 \end{bmatrix} + x_4 \begin{bmatrix} 0 \\ -2 \\ -1 \end{bmatrix} = \begin{bmatrix} 0 \\ 0 \\ 0 \end{bmatrix}$$

Row reduction of the augmented matrix for this equation leads to the general solution

$$x_1 = \tfrac{1}{4}x_4, \quad x_2 = \tfrac{5}{4}x_4, \quad x_3 = \tfrac{3}{4}x_4, \quad \text{with } x_4 \text{ free}$$

Since the coefficients in a chemical equation must be integers, take $x_4 = 4$, in which case $x_1 = 1$, $x_2 = 5$, and $x_3 = 3$. The balanced equation is

$$C_3H_8 + 5O_2 \rightarrow 3CO_2 + 4H_2O$$

The equation would also be balanced if, for example, each coefficient were doubled. For most purposes, however, chemists prefer to use a balanced equation whose coefficients are the smallest possible whole numbers.

Network Flow

Systems of linear equations arise naturally when scientists, engineers, or economists study the flow of some quantity through a network. For instance, urban planners and traffic engineers monitor the pattern of traffic flow in a grid of city streets. Electrical engineers calculate current flow through electrical circuits. And economists analyze the distribution of products from manufacturers to consumers through a network of wholesalers and retailers. For many networks, the systems of equations involve hundreds or even thousands of variables and equations.

A *network* consists of a set of points called *junctions*, or *nodes*, with lines or arcs called *branches* connecting some or all of the junctions. The direction of flow in each branch is indicated, and the flow amount (or rate) is either shown or is denoted by a variable.

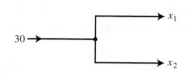

The basic assumption of network flow is that the total flow into the network equals the total flow out of the network and that the total flow into a junction equals the total flow out of the junction. For example, Fig. 1 shows 30 units flowing into a junction through one branch, with x_1 and x_2 denoting the flows out of the junction through other branches. Since the flow is "conserved" at each junction, we must have $x_1 + x_2 = 30$. In a similar fashion, the flow at each junction is described by a linear equation. The problem of network analysis is to determine the flow in each branch when partial information (such as the flow into and out of the network) is known.

FIGURE 1
A junction, or node.

EXAMPLE 2 The network in Fig. 2 shows the traffic flow (in vehicles per hour) over several one-way streets in downtown Baltimore during a typical early afternoon. Determine the general flow pattern for the network.

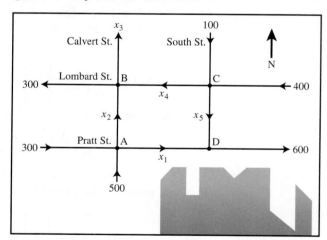

FIGURE 2 Baltimore streets.

SOLUTION Write equations that describe the flow, and then find the general solution of the system. Label the street intersections (junctions) and the unknown flows in the branches, as shown in Fig. 2. At each intersection, set the flow in equal to the flow out.

Intersection	Flow in		Flow out
A	$300 + 500$	$=$	$x_1 + x_2$
B	$x_2 + x_4$	$=$	$300 + x_3$
C	$100 + 400$	$=$	$x_4 + x_5$
D	$x_1 + x_5$	$=$	600

Also, the total flow into the network $(500 + 300 + 100 + 400)$ equals the total flow out of the network $(300 + x_3 + 600)$, which simplifies to $x_3 = 400$. Combine this equation with a rearrangement of the first four equations to obtain the following system of equations:

$$
\begin{aligned}
x_1 + x_2 & & & & &= 800 \\
x_2 - x_3 + x_4 & & & &= 300 \\
& x_4 + x_5 &= 500 \\
x_1 & & + x_5 &= 600 \\
x_3 & & &= 400
\end{aligned}
$$

Row reduction of the associated augmented matrix leads to

$$
\begin{aligned}
x_1 & & + x_5 &= 600 \\
x_2 & & - x_5 &= 200 \\
x_3 & &= 400 \\
& x_4 + x_5 &= 500
\end{aligned}
$$

The general flow pattern for the network is described by

$$
\begin{cases}
x_1 = 600 - x_5 \\
x_2 = 200 + x_5 \\
x_3 = 400 \\
x_4 = 500 - x_5 \\
x_5 \text{ is free}
\end{cases}
$$

A negative flow in a network branch corresponds to flow in the direction opposite to that shown on the model. Since the streets in this problem are one-way, none of the variables here can be negative. This fact leads to certain limitations on the possible values of the variables. For instance, $x_5 \leq 500$ because x_4 cannot be negative. Other constraints on the variables are considered in Practice Problem 2. ■

PRACTICE PROBLEMS

1. Suppose an economy has three sectors: Agriculture, Mining, and Manufacturing. Agriculture sells 5% of its output to Mining and 30% to Manufacturing, and retains the rest. Mining sells 20% of its output to Agriculture and 70% to Manufacturing, and retains the rest. Manufacturing sells 20% of its output to Agriculture and 30% to Mining, and retains the rest. Determine the exchange table for this economy, where the columns describe how the output of each sector is exchanged among the three sectors.

2. Consider the network flow studied in Example 2. Determine the possible range of values of x_1 and x_2. [*Hint:* The example showed that $x_5 \leq 500$. What does this imply about x_1 and x_2? Also, use the fact that $x_5 \geq 0$.]

1.6 EXERCISES

1. Suppose an economy has only two sectors: Goods and Services. Each year, Goods sells 80% of its output to Services and keeps the rest, while Services sells 70% of its output to Goods and retains the rest. Find equilibrium prices for the annual outputs of the Goods and Services sectors that make each sector's income match its expenditures.

2. Find another set of equilibrium prices for the economy in Example 1. Suppose the same economy used Japanese yen instead of dollars to measure the values of the various sectors' outputs. Would this change the problem in any way? Discuss.

3. Consider an economy with three sectors: Fuels and Power, Manufacturing, and Services. Fuels and Power sells 80% of its output to Manufacturing, 10% to Services, and retains the rest. Manufacturing sells 10% of its output to Fuels and Power, 80% to Services, and retains the rest. Services sells 20% to Fuels and Power, 40% to Manufacturing, and retains the rest.

   a. Construct the exchange table for this economy.

   b. Develop a system of equations that leads to prices at which each sector's income matches its expenses. Then write the augmented matrix that can be row reduced to find these prices.

   c. [M] Find a set of equilibrium prices when the price for the Services output is 100 units.

4. Suppose an economy has four sectors: Mining, Lumber, Energy, and Transportation. Mining sells 10% of its output to Lumber, 60% to Energy, and retains the rest. Lumber sells 15% of its output to Mining, 50% to Energy, 20% to Transportation, and retains the rest. Energy sells 20% of its output to Mining, 15% to Lumber, 20% to Transportation, and retains the rest. Transportation sells 20% of its output to Mining, 10% to Lumber, 50% to Energy, and retains the rest.

   a. Construct the exchange table for this economy.

   b. [M] Find a set of equilibrium prices for the economy.

5. An economy has four sectors: Agriculture, Manufacturing, Services, and Transportation. Agriculture sells 20% of its output to Manufacturing, 30% to Services, 30% to Transportation, and retains the rest. Manufacturing sells 35% of its output to Agriculture, 35% to Services, 20% to Transportation, and retains the rest. Services sells 10% of its output to Agriculture, 20% to Manufacturing, 20% to Transportation,

and retains the rest. Transportation sells 20% of its output to Agriculture, 30% to Manufacturing, 20% to Services, and retains the rest.

   a. Construct the exchange table for this economy.

   b. [M] Find a set of equilibrium prices for the economy if the value of Transportation is $10.00 per unit.

   c. The Services sector launches a successful "eat farm fresh" campaign, and increases its share of the output from the Agricultural sector to 40%, whereas the share of Agricultural production going to Manufacturing falls to 10%. Construct the exchange table for this new economy.

   d. [M] Find a set of equilibrium prices for this new economy if the value of Transportation is still $10.00 per unit. What effect has the "eat farm fresh" campaign had on the equilibrium prices for the sectors in this economy?

Balance the chemical equations in Exercises 6–11 using the vector equation approach discussed in this section.

6. Aluminum oxide and carbon react to create elemental aluminum and carbon dioxide:

$$Al_2O_3 + C \rightarrow Al + CO_2$$

[For each compound, construct a vector that lists the numbers of atoms of aluminum, oxygen, and carbon.]

7. Alka-Seltzer contains sodium bicarbonate ($NaHCO_3$) and citric acid ($H_3C_6H_5O_7$). When a tablet is dissolved in water, the following reaction produces sodium citrate, water, and carbon dioxide (gas):

$$NaHCO_3 + H_3C_6H_5O_7 \rightarrow Na_3C_6H_5O_7 + H_2O + CO_2$$

8. Limestone, $CaCO_3$, neutralizes the acid, H_3O, in acid rain by the following unbalanced equation:

$$H_3O + CaCO_3 \rightarrow H_2O + Ca + CO_2$$

9. Boron sulfide reacts violently with water to form boric acid and hydrogen sulfide gas (the smell of rotten eggs). The unbalanced equation is

$$B_2S_3 + H_2O \rightarrow H_3BO_3 + H_2S$$

10. [M] If possible, use exact arithmetic or a rational format for calculations in balancing the following chemical reaction:

$$PbN_6 + CrMn_2O_8 \rightarrow Pb_3O_4 + Cr_2O_3 + MnO_2 + NO$$

11. [M] The chemical reaction below can be used in some industrial processes, such as the production of arsene (AsH_3). Use exact arithmetic or a rational format for calculations to balance this equation.

$$MnS + As_2Cr_{10}O_{35} + H_2SO_4$$
$$\rightarrow HMnO_4 + AsH_3 + CrS_3O_{12} + H_2O$$

12. Find the general flow pattern of the network shown in the figure. Assuming that the flows are all nonnegative, what is the smallest possible value for x_4?

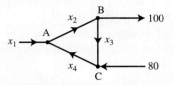

13. a. Find the general flow pattern of the network shown in the figure.

  b. Assuming that the flow must be in the directions indicated, find the minimum flows in the branches denoted by x_2, x_3, x_4, and x_5.

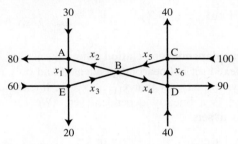

14. a. Find the general traffic pattern of the freeway network shown in the figure. (Flow rates are in cars/minute.)

  b. Describe the general traffic pattern when the road whose flow is x_5 is closed.

  c. When $x_5 = 0$, what is the minimum value of x_4?

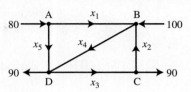

15. Intersections in England are often constructed as one-way "roundabouts," such as the one shown in the figure. Assume that traffic must travel in the directions shown. Find the general solution of the network flow. Find the smallest possible value for x_6.

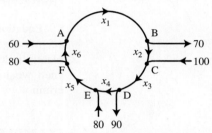

SOLUTIONS TO PRACTICE PROBLEMS

1. Write the percentages as decimals. Since all output must be taken into account, each column must sum to 1. This fact helps to fill in any missing entries.

Distribution of Output from:			
Agriculture	**Mining**	**Manufacturing**	**Purchased by:**
.65	.20	.20	Agriculture
.05	.10	.30	Mining
.30	.70	.50	Manufacturing

2. Since $x_5 \leq 500$, the equations D and A for x_1 and x_2 imply that $x_1 \geq 100$ and $x_2 \leq 700$. The fact that $x_5 \geq 0$ implies that $x_1 \leq 600$ and $x_2 \geq 200$. So, $100 \leq x_1 \leq 600$, and $200 \leq x_2 \leq 700$.

1.7 | LINEAR INDEPENDENCE

The homogeneous equations in Section 1.5 can be studied from a different perspective by writing them as vector equations. In this way, the focus shifts from the unknown solutions of $A\mathbf{x} = \mathbf{0}$ to the vectors that appear in the vector equations.

For instance, consider the equation

$$x_1 \begin{bmatrix} 1 \\ 2 \\ 3 \end{bmatrix} + x_2 \begin{bmatrix} 4 \\ 5 \\ 6 \end{bmatrix} + x_3 \begin{bmatrix} 2 \\ 1 \\ 0 \end{bmatrix} = \begin{bmatrix} 0 \\ 0 \\ 0 \end{bmatrix} \qquad (1)$$

This equation has a trivial solution, of course, where $x_1 = x_2 = x_3 = 0$. As in Section 1.5, the main issue is whether the trivial solution is the *only one*.

DEFINITION

> An indexed set of vectors $\{\mathbf{v}_1, \ldots, \mathbf{v}_p\}$ in $\mathbb{R}^n$ is said to be **linearly independent** if the vector equation
>
> $$x_1\mathbf{v}_1 + x_2\mathbf{v}_2 + \cdots + x_p\mathbf{v}_p = \mathbf{0}$$
>
> has only the trivial solution. The set $\{\mathbf{v}_1, \ldots, \mathbf{v}_p\}$ is said to be **linearly dependent** if there exist weights $c_1, \ldots, c_p$, not all zero, such that
>
> $$c_1\mathbf{v}_1 + c_2\mathbf{v}_2 + \cdots + c_p\mathbf{v}_p = \mathbf{0} \qquad (2)$$

Equation (2) is called a **linear dependence relation** among $\mathbf{v}_1, \ldots, \mathbf{v}_p$ when the weights are not all zero. An indexed set is linearly dependent if and only if it is not linearly independent. For brevity, we may say that $\mathbf{v}_1, \ldots, \mathbf{v}_p$ are linearly dependent when we mean that $\{\mathbf{v}_1, \ldots, \mathbf{v}_p\}$ is a linearly dependent set. We use analogous terminology for linearly independent sets.

EXAMPLE 1 Let $\mathbf{v}_1 = \begin{bmatrix} 1 \\ 2 \\ 3 \end{bmatrix}$, $\mathbf{v}_2 = \begin{bmatrix} 4 \\ 5 \\ 6 \end{bmatrix}$, and $\mathbf{v}_3 = \begin{bmatrix} 2 \\ 1 \\ 0 \end{bmatrix}$.

a. Determine if the set $\{\mathbf{v}_1, \mathbf{v}_2, \mathbf{v}_3\}$ is linearly independent.

b. If possible, find a linear dependence relation among $\mathbf{v}_1$, $\mathbf{v}_2$, and $\mathbf{v}_3$.

SOLUTION

a. We must determine if there is a nontrivial solution of equation (1) above. Row operations on the associated augmented matrix show that

$$\begin{bmatrix} 1 & 4 & 2 & 0 \\ 2 & 5 & 1 & 0 \\ 3 & 6 & 0 & 0 \end{bmatrix} \sim \begin{bmatrix} 1 & 4 & 2 & 0 \\ 0 & -3 & -3 & 0 \\ 0 & 0 & 0 & 0 \end{bmatrix}$$

Clearly, x_1 and x_2 are basic variables, and x_3 is free. Each nonzero value of x_3 determines a nontrivial solution of (1). Hence $\mathbf{v}_1, \mathbf{v}_2, \mathbf{v}_3$ are linearly dependent (and not linearly independent).

b. To find a linear dependence relation among $\mathbf{v}_1$, $\mathbf{v}_2$, and $\mathbf{v}_3$, completely row reduce the augmented matrix and write the new system:

$$\begin{bmatrix} 1 & 0 & -2 & 0 \\ 0 & 1 & 1 & 0 \\ 0 & 0 & 0 & 0 \end{bmatrix} \qquad \begin{aligned} x_1 \quad\;\; - 2x_3 &= 0 \\ x_2 + \;\; x_3 &= 0 \\ 0 &= 0 \end{aligned}$$

Thus $x_1 = 2x_3$, $x_2 = -x_3$, and x_3 is free. Choose any nonzero value for x_3—say, $x_3 = 5$. Then $x_1 = 10$ and $x_2 = -5$. Substitute these values into equation (1) and obtain

$$10\mathbf{v}_1 - 5\mathbf{v}_2 + 5\mathbf{v}_3 = \mathbf{0}$$

This is one (out of infinitely many) possible linear dependence relations among $\mathbf{v}_1$, $\mathbf{v}_2$, and $\mathbf{v}_3$. ∎

Linear Independence of Matrix Columns

Suppose that we begin with a matrix $A = [\,\mathbf{a}_1 \;\; \cdots \;\; \mathbf{a}_n\,]$ instead of a set of vectors. The matrix equation $A\mathbf{x} = \mathbf{0}$ can be written as

$$x_1\mathbf{a}_1 + x_2\mathbf{a}_2 + \cdots + x_n\mathbf{a}_n = \mathbf{0}$$

Each linear dependence relation among the columns of A corresponds to a nontrivial solution of $A\mathbf{x} = \mathbf{0}$. Thus we have the following important fact.

> The columns of a matrix A are linearly independent if and only if the equation $A\mathbf{x} = \mathbf{0}$ has *only* the trivial solution. $\qquad\qquad$ (3)

EXAMPLE 2 Determine if the columns of the matrix $A = \begin{bmatrix} 0 & 1 & 4 \\ 1 & 2 & -1 \\ 5 & 8 & 0 \end{bmatrix}$ are linearly independent.

SOLUTION To study $A\mathbf{x} = \mathbf{0}$, row reduce the augmented matrix:

$$\begin{bmatrix} 0 & 1 & 4 & 0 \\ 1 & 2 & -1 & 0 \\ 5 & 8 & 0 & 0 \end{bmatrix} \sim \begin{bmatrix} 1 & 2 & -1 & 0 \\ 0 & 1 & 4 & 0 \\ 0 & -2 & 5 & 0 \end{bmatrix} \sim \begin{bmatrix} 1 & 2 & -1 & 0 \\ 0 & 1 & 4 & 0 \\ 0 & 0 & 13 & 0 \end{bmatrix}$$

At this point, it is clear that there are three basic variables and no free variables. So the equation $A\mathbf{x} = \mathbf{0}$ has only the trivial solution, and the columns of A are linearly independent. ∎

Sets of One or Two Vectors

A set containing only one vector—say, $\mathbf{v}$—is linearly independent if and only if $\mathbf{v}$ is not the zero vector. This is because the vector equation $x_1\mathbf{v} = \mathbf{0}$ has only the trivial solution when $\mathbf{v} \neq \mathbf{0}$. The zero vector is linearly dependent because $x_1\mathbf{0} = \mathbf{0}$ has many nontrivial solutions.

The next example will explain the nature of a linearly dependent set of two vectors.

EXAMPLE 3 Determine if the following sets of vectors are linearly independent.

a. $\mathbf{v}_1 = \begin{bmatrix} 3 \\ 1 \end{bmatrix}, \mathbf{v}_2 = \begin{bmatrix} 6 \\ 2 \end{bmatrix}$ $\qquad\qquad$ b. $\mathbf{v}_1 = \begin{bmatrix} 3 \\ 2 \end{bmatrix}, \mathbf{v}_2 = \begin{bmatrix} 6 \\ 2 \end{bmatrix}$

SOLUTION

a. Notice that $\mathbf{v}_2$ is a multiple of $\mathbf{v}_1$, namely, $\mathbf{v}_2 = 2\mathbf{v}_1$. Hence $-2\mathbf{v}_1 + \mathbf{v}_2 = \mathbf{0}$, which shows that $\{\mathbf{v}_1, \mathbf{v}_2\}$ is linearly dependent.

b. The vectors $\mathbf{v}_1$ and $\mathbf{v}_2$ are certainly *not* multiples of one another. Could they be linearly dependent? Suppose c and d satisfy

$$c\mathbf{v}_1 + d\mathbf{v}_2 = \mathbf{0}$$

If $c \neq 0$, then we can solve for $\mathbf{v}_1$ in terms of $\mathbf{v}_2$, namely, $\mathbf{v}_1 = (-d/c)\mathbf{v}_2$. This result is impossible because $\mathbf{v}_1$ is *not* a multiple of $\mathbf{v}_2$. So c must be zero. Similarly, d must also be zero. Thus $\{\mathbf{v}_1, \mathbf{v}_2\}$ is a linearly independent set. ∎

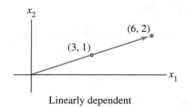

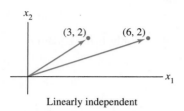

FIGURE 1

The arguments in Example 3 show that you can always decide *by inspection* when a set of two vectors is linearly dependent. Row operations are unnecessary. Simply check whether at least one of the vectors is a scalar times the other. (The test applies only to sets of *two* vectors.)

A set of two vectors $\{\mathbf{v}_1, \mathbf{v}_2\}$ is linearly dependent if at least one of the vectors is a multiple of the other. The set is linearly independent if and only if neither of the vectors is a multiple of the other.

In geometric terms, two vectors are linearly dependent if and only if they lie on the same line through the origin. Figure 1 shows the vectors from Example 3.

Sets of Two or More Vectors

The proof of the next theorem is similar to the solution of Example 3. Details are given at the end of this section.

THEOREM 7

Characterization of Linearly Dependent Sets

An indexed set $S = \{\mathbf{v}_1, \ldots, \mathbf{v}_p\}$ of two or more vectors is linearly dependent if and only if at least one of the vectors in S is a linear combination of the others. In fact, if S is linearly dependent and $\mathbf{v}_1 \neq \mathbf{0}$, then some $\mathbf{v}_j$ (with $j > 1$) is a linear combination of the preceding vectors, $\mathbf{v}_1, \ldots, \mathbf{v}_{j-1}$.

Warning: Theorem 7 does *not* say that *every* vector in a linearly dependent set is a linear combination of the preceding vectors. A vector in a linearly dependent set may fail to be a linear combination of the other vectors. See Practice Problem 3.

EXAMPLE 4   Let $\mathbf{u} = \begin{bmatrix} 3 \\ 1 \\ 0 \end{bmatrix}$ and $\mathbf{v} = \begin{bmatrix} 1 \\ 6 \\ 0 \end{bmatrix}$. Describe the set spanned by $\mathbf{u}$ and $\mathbf{v}$, and explain why a vector $\mathbf{w}$ is in Span $\{\mathbf{u}, \mathbf{v}\}$ if and only if $\{\mathbf{u}, \mathbf{v}, \mathbf{w}\}$ is linearly dependent.

SOLUTION   The vectors $\mathbf{u}$ and $\mathbf{v}$ are linearly independent because neither vector is a multiple of the other, and so they span a plane in $\mathbb{R}^3$. (See Section 1.3.) In fact, Span $\{\mathbf{u}, \mathbf{v}\}$ is the $x_1 x_2$-plane (with $x_3 = 0$). If $\mathbf{w}$ is a linear combination of $\mathbf{u}$ and $\mathbf{v}$, then $\{\mathbf{u}, \mathbf{v}, \mathbf{w}\}$ is linearly dependent, by Theorem 7. Conversely, suppose that $\{\mathbf{u}, \mathbf{v}, \mathbf{w}\}$ is linearly dependent. By Theorem 7, some vector in $\{\mathbf{u}, \mathbf{v}, \mathbf{w}\}$ is a linear combination of the preceding vectors (since $\mathbf{u} \neq \mathbf{0}$). That vector must be $\mathbf{w}$, since $\mathbf{v}$ is not a multiple of $\mathbf{u}$. So $\mathbf{w}$ is in Span $\{\mathbf{u}, \mathbf{v}\}$. See Fig. 2. ∎

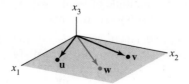

Linearly dependent,
w in Span{**u**, **v**}

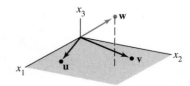

Linearly independent,
w *not* in Span{**u**, **v**}

FIGURE 2 Linear dependence in $\mathbb{R}^3$.

Example 4 generalizes to any set $\{\mathbf{u}, \mathbf{v}, \mathbf{w}\}$ in $\mathbb{R}^3$ with $\mathbf{u}$ and $\mathbf{v}$ linearly independent. The set $\{\mathbf{u}, \mathbf{v}, \mathbf{w}\}$ will be linearly dependent if and only if $\mathbf{w}$ is in the plane spanned by $\mathbf{u}$ and $\mathbf{v}$.

The next two theorems describe special cases in which the linear dependence of a set is automatic. Moreover, Theorem 8 will be a key result for work in later chapters.

THEOREM 8

If a set contains more vectors than there are entries in each vector, then the set is linearly dependent. That is, any set $\{\mathbf{v}_1, \ldots, \mathbf{v}_p\}$ in $\mathbb{R}^n$ is linearly dependent if $p > n$.

FIGURE 3

If $p > n$, the columns are linearly dependent.

PROOF Let $A = [\mathbf{v}_1 \ \cdots \ \mathbf{v}_p]$. Then A is $n \times p$, and the equation $A\mathbf{x} = \mathbf{0}$ corresponds to a system of n equations in p unknowns. If $p > n$, there are more variables than equations, so there must be a free variable. Hence $A\mathbf{x} = \mathbf{0}$ has a nontrivial solution, and the columns of A are linearly dependent. See Fig. 3 for a matrix version of this theorem. ∎

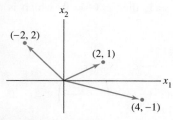

FIGURE 4

A linearly dependent set in $\mathbb{R}^2$.

Warning: Theorem 8 says nothing about the case in which the number of vectors in the set does *not* exceed the number of entries in each vector.

EXAMPLE 5 The vectors $\begin{bmatrix} 2 \\ 1 \end{bmatrix}, \begin{bmatrix} 4 \\ -1 \end{bmatrix}, \begin{bmatrix} -2 \\ 2 \end{bmatrix}$ are linearly dependent by Theorem 8, because there are three vectors in the set and there are only two entries in each vector. Notice, however, that none of the vectors is a multiple of one of the other vectors. See Fig. 4. ∎

THEOREM 9

If a set $S = \{\mathbf{v}_1, \ldots, \mathbf{v}_p\}$ in $\mathbb{R}^n$ contains the zero vector, then the set is linearly dependent.

PROOF By renumbering the vectors, we may suppose $\mathbf{v}_1 = \mathbf{0}$. Then the equation $1\mathbf{v}_1 + 0\mathbf{v}_2 + \cdots + 0\mathbf{v}_p = \mathbf{0}$ shows that S is linearly dependent. ∎

EXAMPLE 6 Determine by inspection if the given set is linearly dependent.

a. $\begin{bmatrix} 1 \\ 7 \\ 6 \end{bmatrix}, \begin{bmatrix} 2 \\ 0 \\ 9 \end{bmatrix}, \begin{bmatrix} 3 \\ 1 \\ 5 \end{bmatrix}, \begin{bmatrix} 4 \\ 1 \\ 8 \end{bmatrix}$ b. $\begin{bmatrix} 2 \\ 3 \\ 5 \end{bmatrix}, \begin{bmatrix} 0 \\ 0 \\ 0 \end{bmatrix}, \begin{bmatrix} 1 \\ 1 \\ 8 \end{bmatrix}$ c. $\begin{bmatrix} -2 \\ 4 \\ 6 \\ 10 \end{bmatrix}, \begin{bmatrix} 3 \\ -6 \\ -9 \\ 15 \end{bmatrix}$

SOLUTION

a. The set contains four vectors, each of which has only three entries. So the set is linearly dependent by Theorem 8.

b. Theorem 8 does not apply here because the number of vectors does not exceed the number of entries in each vector. Since the zero vector is in the set, the set is linearly dependent by Theorem 9.

c. Compare the corresponding entries of the two vectors. The second vector seems to be $-3/2$ times the first vector. This relation holds for the first three pairs of entries, but fails for the fourth pair. Thus neither of the vectors is a multiple of the other, and hence they are linearly independent. ∎

SG Mastering: Linear
Independence 1-31

In general, you should read a section thoroughly *several* times to absorb an important concept such as linear independence. The notes in the *Study Guide* for this section will help you learn to form mental images of key ideas in linear algebra. For instance, the following proof is worth reading carefully because it shows how the definition of linear independence can be *used*.

PROOF OF THEOREM 7 (Characterization of Linearly Dependent Sets)

If some $\mathbf{v}_j$ in S equals a linear combination of the other vectors, then $\mathbf{v}_j$ can be subtracted from both sides of the equation, producing a linear dependence relation with a nonzero weight (-1) on $\mathbf{v}_j$. [For instance, if $\mathbf{v}_1 = c_2\mathbf{v}_2 + c_3\mathbf{v}_3$, then $\mathbf{0} = (-1)\mathbf{v}_1 + c_2\mathbf{v}_2 + c_3\mathbf{v}_3 + 0\mathbf{v}_4 + \cdots + 0\mathbf{v}_p$.] Thus S is linearly dependent.

Conversely, suppose S is linearly dependent. If $\mathbf{v}_1$ is zero, then it is a (trivial) linear combination of the other vectors in S. Otherwise, $\mathbf{v}_1 \neq \mathbf{0}$, and there exist weights $c_1, \ldots, c_p$, not all zero, such that

$$c_1\mathbf{v}_1 + c_2\mathbf{v}_2 + \cdots + c_p\mathbf{v}_p = \mathbf{0}$$

Let j be the largest subscript for which $c_j \neq 0$. If $j = 1$, then $c_1\mathbf{v}_1 = \mathbf{0}$, which is impossible because $\mathbf{v}_1 \neq \mathbf{0}$. So $j > 1$, and

$$c_1\mathbf{v}_1 + \cdots + c_j\mathbf{v}_j + 0\mathbf{v}_{j+1} + \cdots + 0\mathbf{v}_p = \mathbf{0}$$

$$c_j\mathbf{v}_j = -c_1\mathbf{v}_1 - \cdots - c_{j-1}\mathbf{v}_{j-1}$$

$$\mathbf{v}_j = \left(-\frac{c_1}{c_j}\right)\mathbf{v}_1 + \cdots + \left(-\frac{c_{j-1}}{c_j}\right)\mathbf{v}_{j-1} \quad \blacksquare$$

PRACTICE PROBLEMS

Let $\mathbf{u} = \begin{bmatrix} 3 \\ 2 \\ -4 \end{bmatrix}$, $\mathbf{v} = \begin{bmatrix} -6 \\ 1 \\ 7 \end{bmatrix}$, $\mathbf{w} = \begin{bmatrix} 0 \\ -5 \\ 2 \end{bmatrix}$, and $\mathbf{z} = \begin{bmatrix} 3 \\ 7 \\ -5 \end{bmatrix}$.

1. Are the sets $\{\mathbf{u}, \mathbf{v}\}$, $\{\mathbf{u}, \mathbf{w}\}$, $\{\mathbf{u}, \mathbf{z}\}$, $\{\mathbf{v}, \mathbf{w}\}$, $\{\mathbf{v}, \mathbf{z}\}$, and $\{\mathbf{w}, \mathbf{z}\}$ each linearly independent? Why or why not?

2. Does the answer to Problem 1 imply that $\{\mathbf{u}, \mathbf{v}, \mathbf{w}, \mathbf{z}\}$ is linearly independent?

3. To determine if $\{\mathbf{u}, \mathbf{v}, \mathbf{w}, \mathbf{z}\}$ is linearly dependent, is it wise to check if, say, $\mathbf{w}$ is a linear combination of $\mathbf{u}$, $\mathbf{v}$, and $\mathbf{z}$?

4. Is $\{\mathbf{u}, \mathbf{v}, \mathbf{w}, \mathbf{z}\}$ linearly dependent?

1.7 EXERCISES

In Exercises 1–4, determine if the vectors are linearly independent. Justify each answer.

1. $\begin{bmatrix} 5 \\ 0 \\ 0 \end{bmatrix}$, $\begin{bmatrix} 7 \\ 2 \\ -6 \end{bmatrix}$, $\begin{bmatrix} 9 \\ 4 \\ -8 \end{bmatrix}$

2. $\begin{bmatrix} 0 \\ 2 \\ 3 \end{bmatrix}$, $\begin{bmatrix} 0 \\ 0 \\ -8 \end{bmatrix}$, $\begin{bmatrix} -1 \\ 3 \\ 1 \end{bmatrix}$

3. $\begin{bmatrix} 2 \\ -3 \end{bmatrix}$, $\begin{bmatrix} -4 \\ 6 \end{bmatrix}$

4. $\begin{bmatrix} -1 \\ 3 \end{bmatrix}$, $\begin{bmatrix} -3 \\ -9 \end{bmatrix}$

In Exercises 5–8, determine if the columns of the matrix form a linearly independent set. Justify each answer.

5. $\begin{bmatrix} 0 & -3 & 9 \\ 2 & 1 & -7 \\ -1 & 4 & -5 \\ 1 & -4 & -2 \end{bmatrix}$

6. $\begin{bmatrix} -4 & -3 & 0 \\ 0 & -1 & 5 \\ 1 & 1 & -5 \\ 2 & 1 & -10 \end{bmatrix}$

7. $\begin{bmatrix} 1 & 4 & -3 & 0 \\ -2 & -7 & 5 & 1 \\ -4 & -5 & 7 & 5 \end{bmatrix}$

8. $\begin{bmatrix} 1 & -2 & 3 & 2 \\ -2 & 4 & -6 & 2 \\ 0 & 1 & -1 & 3 \end{bmatrix}$

In Exercises 9 and 10, (a) for what values of h is $\mathbf{v}_3$ in Span $\{\mathbf{v}_1, \mathbf{v}_2\}$, and (b) for what values of h is $\{\mathbf{v}_1, \mathbf{v}_2, \mathbf{v}_3\}$ linearly *dependent*? Justify each answer.

9. $\mathbf{v}_1 = \begin{bmatrix} 1 \\ -3 \\ 2 \end{bmatrix}$, $\mathbf{v}_2 = \begin{bmatrix} -3 \\ 9 \\ -6 \end{bmatrix}$, $\mathbf{v}_3 = \begin{bmatrix} 5 \\ -7 \\ h \end{bmatrix}$

10. $\mathbf{v}_1 = \begin{bmatrix} 1 \\ -3 \\ -5 \end{bmatrix}$, $\mathbf{v}_2 = \begin{bmatrix} -3 \\ 9 \\ 15 \end{bmatrix}$, $\mathbf{v}_3 = \begin{bmatrix} 2 \\ -5 \\ h \end{bmatrix}$

In Exercises 11–14, find the value(s) of h for which the vectors are linearly *dependent*. Justify each answer.

11. $\begin{bmatrix} 2 \\ -2 \\ 4 \end{bmatrix}$, $\begin{bmatrix} 4 \\ -6 \\ 7 \end{bmatrix}$, $\begin{bmatrix} -2 \\ 2 \\ h \end{bmatrix}$ **12.** $\begin{bmatrix} 3 \\ -6 \\ 1 \end{bmatrix}$, $\begin{bmatrix} -6 \\ 4 \\ -3 \end{bmatrix}$, $\begin{bmatrix} 9 \\ h \\ 3 \end{bmatrix}$

13. $\begin{bmatrix} 1 \\ 5 \\ -3 \end{bmatrix}$, $\begin{bmatrix} -2 \\ -9 \\ 6 \end{bmatrix}$, $\begin{bmatrix} 3 \\ h \\ -9 \end{bmatrix}$ **14.** $\begin{bmatrix} 1 \\ -2 \\ -4 \end{bmatrix}$, $\begin{bmatrix} -3 \\ 7 \\ 6 \end{bmatrix}$, $\begin{bmatrix} 2 \\ 1 \\ h \end{bmatrix}$

Determine by inspection whether the vectors in Exercises 15–20 are linearly *independent*. Justify each answer.

15. $\begin{bmatrix} 5 \\ 1 \end{bmatrix}$, $\begin{bmatrix} 2 \\ 8 \end{bmatrix}$, $\begin{bmatrix} 1 \\ 3 \end{bmatrix}$, $\begin{bmatrix} -1 \\ 7 \end{bmatrix}$ **16.** $\begin{bmatrix} 2 \\ -4 \\ 8 \end{bmatrix}$, $\begin{bmatrix} -3 \\ 6 \\ -12 \end{bmatrix}$

17. $\begin{bmatrix} 5 \\ -3 \\ -1 \end{bmatrix}$, $\begin{bmatrix} 0 \\ 0 \\ 0 \end{bmatrix}$, $\begin{bmatrix} -7 \\ 2 \\ 4 \end{bmatrix}$ **18.** $\begin{bmatrix} 3 \\ 4 \end{bmatrix}$, $\begin{bmatrix} -1 \\ 5 \end{bmatrix}$, $\begin{bmatrix} 3 \\ 5 \end{bmatrix}$, $\begin{bmatrix} 7 \\ 1 \end{bmatrix}$

19. $\begin{bmatrix} -8 \\ 12 \\ -4 \end{bmatrix}$, $\begin{bmatrix} 2 \\ -3 \\ -1 \end{bmatrix}$ **20.** $\begin{bmatrix} 1 \\ 4 \\ -7 \end{bmatrix}$, $\begin{bmatrix} -2 \\ 5 \\ 3 \end{bmatrix}$, $\begin{bmatrix} 0 \\ 0 \\ 0 \end{bmatrix}$

In Exercises 21 and 22, mark each statement True or False. Justify each answer on the basis of a careful reading of the text.

21. a. The columns of a matrix A are linearly independent if the equation $A\mathbf{x} = \mathbf{0}$ has the trivial solution.

b. If S is a linearly dependent set, then each vector is a linear combination of the other vectors in S.

c. The columns of any 4×5 matrix are linearly dependent.

d. If $\mathbf{x}$ and $\mathbf{y}$ are linearly independent, and if $\{\mathbf{x}, \mathbf{y}, \mathbf{z}\}$ is linearly dependent, then $\mathbf{z}$ is in Span $\{\mathbf{x}, \mathbf{y}\}$.

22. a. If $\mathbf{u}$ and $\mathbf{v}$ are linearly independent, and if $\mathbf{w}$ is in Span $\{\mathbf{u}, \mathbf{v}\}$, then $\{\mathbf{u}, \mathbf{v}, \mathbf{w}\}$ is linearly dependent.

b. If three vectors in $\mathbb{R}^3$ lie in the same plane in $\mathbb{R}^3$, then they are linearly dependent.

c. If a set contains fewer vectors than there are entries in the vectors, then the set is linearly independent.

d. If a set in $\mathbb{R}^n$ is linearly dependent, then the set contains more than n vectors.

In Exercises 23–26, describe the possible echelon forms of the matrix. Use the notation of Example 1 in Section 1.2.

23. A is a 2×2 matrix with linearly dependent columns.

24. A is a 3×3 matrix with linearly independent columns.

25. A is a 4×2 matrix, $A = [\mathbf{a}_1 \quad \mathbf{a}_2]$, and $\mathbf{a}_2$ is not a multiple of $\mathbf{a}_1$.

26. A is a 4×3 matrix, $A = [\mathbf{a}_1 \quad \mathbf{a}_2 \quad \mathbf{a}_3]$, such that $\{\mathbf{a}_1, \mathbf{a}_2\}$ is linearly independent and $\mathbf{a}_3$ is not in Span $\{\mathbf{a}_1, \mathbf{a}_2\}$.

27. How many pivot columns must a 6×4 matrix have if its columns are linearly independent? Why?

28. How many pivot columns must a 4×6 matrix have if its columns span $\mathbb{R}^4$? Why?

29. Construct 3×2 matrices A and B such that $A\mathbf{x} = \mathbf{0}$ has a nontrivial solution, but $B\mathbf{x} = \mathbf{0}$ has only the trivial solution.

30. a. Fill in the blank in the following statement: "If A is an $m \times n$ matrix, then the columns of A are linearly independent if and only if A has _____ pivot columns."

b. Explain why the statement in (a) is true.

Exercises 31 and 32 should be solved *without performing row operations*. [*Hint:* Write $A\mathbf{x} = \mathbf{0}$ as a vector equation.]

31. Given $A = \begin{bmatrix} 2 & 3 & 5 \\ -5 & 1 & -4 \\ -3 & -1 & -4 \\ 1 & 0 & 1 \end{bmatrix}$, observe that the third column is the sum of the first two columns. Find a nontrivial solution of $A\mathbf{x} = \mathbf{0}$.

32. Given $A = \begin{bmatrix} 4 & 3 & -5 \\ -2 & -2 & 4 \\ -2 & -3 & 7 \end{bmatrix}$, observe that the first column minus three times the second column equals the third column. Find a nontrivial solution of $A\mathbf{x} = \mathbf{0}$.

Each statement in Exercises 33–38 is either true (in all cases) or false (for at least one example). If false, construct a specific example to show that the statement is not always true. Such an example is called a *counterexample* to the statement. If a statement is true, give a justification. (One specific example cannot explain why a statement is always true. You will have to do more work here than in Exercises 21 and 22.)

33. If $\mathbf{v}_1, \ldots, \mathbf{v}_4$ are in $\mathbb{R}^4$ and $\mathbf{v}_3 = 2\mathbf{v}_1 + \mathbf{v}_2$, then $\{\mathbf{v}_1, \mathbf{v}_2, \mathbf{v}_3, \mathbf{v}_4\}$ is linearly dependent.

34. If $\mathbf{v}_1$ and $\mathbf{v}_2$ are in $\mathbb{R}^4$ and $\mathbf{v}_2$ is not a scalar multiple of $\mathbf{v}_1$, then $\{\mathbf{v}_1, \mathbf{v}_2\}$ is linearly independent.

35. If $\mathbf{v}_1, \ldots, \mathbf{v}_5$ are in $\mathbb{R}^5$ and $\mathbf{v}_3 = \mathbf{0}$, then $\{\mathbf{v}_1, \mathbf{v}_2, \mathbf{v}_3, \mathbf{v}_4, \mathbf{v}_5\}$ is linearly dependent.

36. If $\mathbf{v}_1, \mathbf{v}_2, \mathbf{v}_3$ are in $\mathbb{R}^3$ and $\mathbf{v}_3$ is *not* a linear combination of $\mathbf{v}_1, \mathbf{v}_2$, then $\{\mathbf{v}_1, \mathbf{v}_2, \mathbf{v}_3\}$ is linearly independent.

37. If $\mathbf{v}_1, \ldots, \mathbf{v}_4$ are in $\mathbb{R}^4$ and $\{\mathbf{v}_1, \mathbf{v}_2, \mathbf{v}_3\}$ is linearly dependent, then $\{\mathbf{v}_1, \mathbf{v}_2, \mathbf{v}_3, \mathbf{v}_4\}$ is also linearly dependent.

38. If $\{\mathbf{v}_1, \ldots, \mathbf{v}_4\}$ is a linearly independent set of vectors in $\mathbb{R}^4$, then $\{\mathbf{v}_1, \mathbf{v}_2, \mathbf{v}_3\}$ is also linearly independent. [*Hint:* Think about $x_1\mathbf{v}_1 + x_2\mathbf{v}_2 + x_3\mathbf{v}_3 + 0 \cdot \mathbf{v}_4 = \mathbf{0}$.]

39. Suppose A is an $m \times n$ matrix with the property that for all $\mathbf{b}$ in $\mathbb{R}^m$ the equation $A\mathbf{x} = \mathbf{b}$ has at most one solution. Use the definition of linear independence to explain why the columns of A must be linearly independent.

40. Suppose an $m \times n$ matrix A has n pivot columns. Explain why for each **b** in $\mathbb{R}^m$ the equation $A\mathbf{x} = \mathbf{b}$ has at most one solution. [*Hint:* Explain why $A\mathbf{x} = \mathbf{b}$ cannot have infinitely many solutions.]

[**M**] In Exercises 41 and 42, use as many columns of A as possible to construct a matrix B with the property that the equation $B\mathbf{x} = \mathbf{0}$ has only the trivial solution. Solve $B\mathbf{x} = \mathbf{0}$ to verify your work.

41. $A = \begin{bmatrix} 3 & -4 & 10 & 7 & -4 \\ -5 & -3 & -7 & -11 & 15 \\ 4 & 3 & 5 & 2 & 1 \\ 8 & -7 & 23 & 4 & 15 \end{bmatrix}$

42. $A = \begin{bmatrix} 12 & 10 & -6 & 8 & 4 & -14 \\ -7 & -6 & 4 & -5 & -7 & 9 \\ 9 & 9 & -9 & 9 & 9 & -18 \\ -4 & -3 & -1 & 0 & -8 & 1 \\ 8 & 7 & -5 & 6 & 1 & -11 \end{bmatrix}$

43. [**M**] With A and B as in Exercise 41, select a column **v** of A that was not used in the construction of B and determine if **v** is in the set spanned by the columns of B. (Describe your calculations.)

44. [**M**] Repeat Exercise 43 with the matrices A and B from Exercise 42. Then give an explanation for what you discover, assuming that B was constructed as specified.

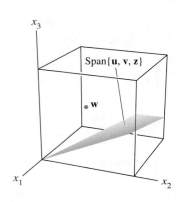

SOLUTIONS TO PRACTICE PROBLEMS

1. Yes. In each case, neither vector is a multiple of the other. Thus each set is linearly independent.

2. No. The observation in Practice Problem 1, by itself, says nothing about the linear independence of $\{\mathbf{u}, \mathbf{v}, \mathbf{w}, \mathbf{z}\}$.

3. No. When testing for linear independence, it is usually a poor idea to check if one selected vector is a linear combination of the others. It may happen that the selected vector is not a linear combination of the others and yet the whole set of vectors is linearly dependent. In this practice problem, **w** is not a linear combination of **u**, **v**, and **z**.

4. Yes, by Theorem 8. There are more vectors (four) than entries (three) in them.

1.8 INTRODUCTION TO LINEAR TRANSFORMATIONS

The difference between a matrix equation $A\mathbf{x} = \mathbf{b}$ and the associated vector equation $x_1\mathbf{a}_1 + \cdots + x_n\mathbf{a}_n = \mathbf{b}$ is merely a matter of notation. However, a matrix equation $A\mathbf{x} = \mathbf{b}$ can arise in linear algebra (and in applications such as computer graphics and signal processing) in a way that is not directly connected with linear combinations of vectors. This happens when we think of the matrix A as an object that "acts" on a vector **x** by multiplication to produce a new vector called $A\mathbf{x}$.

For instance, the equations

$$\underset{A}{\begin{bmatrix} 4 & -3 & 1 & 3 \\ 2 & 0 & 5 & 1 \end{bmatrix}} \underset{\mathbf{x}}{\begin{bmatrix} 1 \\ 1 \\ 1 \\ 1 \end{bmatrix}} = \underset{\mathbf{b}}{\begin{bmatrix} 5 \\ 8 \end{bmatrix}} \quad \text{and} \quad \underset{A}{\begin{bmatrix} 4 & -3 & 1 & 3 \\ 2 & 0 & 5 & 1 \end{bmatrix}} \underset{\mathbf{u}}{\begin{bmatrix} 1 \\ 4 \\ -1 \\ 3 \end{bmatrix}} = \underset{\mathbf{0}}{\begin{bmatrix} 0 \\ 0 \end{bmatrix}}$$

say that multiplication by A transforms **x** into **b** and transforms **u** into the zero vector. See Fig. 1.

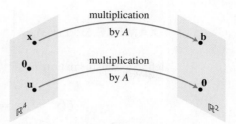

FIGURE 1 Transforming vectors via matrix multiplication.

From this new point of view, solving the equation $A\mathbf{x} = \mathbf{b}$ amounts to finding all vectors $\mathbf{x}$ in $\mathbb{R}^4$ that are transformed into the vector $\mathbf{b}$ in $\mathbb{R}^2$ under the "action" of multiplication by A.

The correspondence from $\mathbf{x}$ to $A\mathbf{x}$ is a *function* from one set of vectors to another. This concept generalizes the common notion of a function as a rule that transforms one real number into another.

A **transformation** (or **function** or **mapping**) T from $\mathbb{R}^n$ to $\mathbb{R}^m$ is a rule that assigns to each vector $\mathbf{x}$ in $\mathbb{R}^n$ a vector $T(\mathbf{x})$ in $\mathbb{R}^m$. The set $\mathbb{R}^n$ is called the **domain** of T, and $\mathbb{R}^m$ is called the **codomain** of T. The notation $T : \mathbb{R}^n \to \mathbb{R}^m$ indicates that the domain of T is $\mathbb{R}^n$ and the codomain is $\mathbb{R}^m$. For $\mathbf{x}$ in $\mathbb{R}^n$, the vector $T(\mathbf{x})$ in $\mathbb{R}^m$ is called the **image** of $\mathbf{x}$ (under the action of T). The set of all images $T(\mathbf{x})$ is called the **range** of T. See Fig. 2.

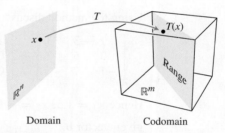

FIGURE 2 Domain, codomain, and range of $T : \mathbb{R}^n \to \mathbb{R}^m$.

The new terminology in this section is important because a dynamic view of matrix–vector multiplication is the key to understanding several ideas in linear algebra and to building mathematical models of physical systems that evolve over time. Such *dynamical systems* will be discussed in Sections 1.10, 4.8, and 4.9 and throughout Chapter 5.

Matrix Transformations

The rest of this section focuses on mappings associated with matrix multiplication. For each $\mathbf{x}$ in $\mathbb{R}^n$, $T(\mathbf{x})$ is computed as $A\mathbf{x}$, where A is an $m \times n$ matrix. For simplicity, we sometimes denote such a *matrix transformation* by $\mathbf{x} \mapsto A\mathbf{x}$. Observe that the domain of T is $\mathbb{R}^n$ when A has n columns and the codomain of T is $\mathbb{R}^m$ when each column of A has m entries. The range of T is the set of all linear combinations of the columns of A, because each image $T(\mathbf{x})$ is of the form $A\mathbf{x}$.

EXAMPLE 1   Let $A = \begin{bmatrix} 1 & -3 \\ 3 & 5 \\ -1 & 7 \end{bmatrix}$, $\mathbf{u} = \begin{bmatrix} 2 \\ -1 \end{bmatrix}$, $\mathbf{b} = \begin{bmatrix} 3 \\ 2 \\ -5 \end{bmatrix}$, $\mathbf{c} = \begin{bmatrix} 3 \\ 2 \\ 5 \end{bmatrix}$, and

define a transformation $T : \mathbb{R}^2 \to \mathbb{R}^3$ by $T(\mathbf{x}) = A\mathbf{x}$, so that

$$T(\mathbf{x}) = A\mathbf{x} = \begin{bmatrix} 1 & -3 \\ 3 & 5 \\ -1 & 7 \end{bmatrix} \begin{bmatrix} x_1 \\ x_2 \end{bmatrix} = \begin{bmatrix} x_1 - 3x_2 \\ 3x_1 + 5x_2 \\ -x_1 + 7x_2 \end{bmatrix}$$

a. Find $T(\mathbf{u})$, the image of $\mathbf{u}$ under the transformation T.

b. Find an $\mathbf{x}$ in $\mathbb{R}^2$ whose image under T is $\mathbf{b}$.

c. Is there more than one $\mathbf{x}$ whose image under T is $\mathbf{b}$?

d. Determine if $\mathbf{c}$ is in the range of the transformation T.

SOLUTION

a. Compute

$$T(\mathbf{u}) = A\mathbf{u} = \begin{bmatrix} 1 & -3 \\ 3 & 5 \\ -1 & 7 \end{bmatrix} \begin{bmatrix} 2 \\ -1 \end{bmatrix} = \begin{bmatrix} 5 \\ 1 \\ -9 \end{bmatrix}$$

b. Solve $T(\mathbf{x}) = \mathbf{b}$ for $\mathbf{x}$. That is, solve $A\mathbf{x} = \mathbf{b}$, or

$$\begin{bmatrix} 1 & -3 \\ 3 & 5 \\ -1 & 7 \end{bmatrix} \begin{bmatrix} x_1 \\ x_2 \end{bmatrix} = \begin{bmatrix} 3 \\ 2 \\ -5 \end{bmatrix} \tag{1}$$

Using the method discussed in Section 1.4, row reduce the augmented matrix:

$$\begin{bmatrix} 1 & -3 & 3 \\ 3 & 5 & 2 \\ -1 & 7 & -5 \end{bmatrix} \sim \begin{bmatrix} 1 & -3 & 3 \\ 0 & 14 & -7 \\ 0 & 4 & -2 \end{bmatrix} \sim \begin{bmatrix} 1 & -3 & 3 \\ 0 & 1 & -.5 \\ 0 & 0 & 0 \end{bmatrix} \sim \begin{bmatrix} 1 & 0 & 1.5 \\ 0 & 1 & -.5 \\ 0 & 0 & 0 \end{bmatrix} \tag{2}$$

Hence $x_1 = 1.5$, $x_2 = -.5$, and $\mathbf{x} = \begin{bmatrix} 1.5 \\ -.5 \end{bmatrix}$. The image of this $\mathbf{x}$ under T is the given vector $\mathbf{b}$.

c. Any $\mathbf{x}$ whose image under T is $\mathbf{b}$ must satisfy equation (1). From (2), it is clear that equation (1) has a unique solution. So there is exactly one $\mathbf{x}$ whose image is $\mathbf{b}$.

d. The vector $\mathbf{c}$ is in the range of T if $\mathbf{c}$ is the image of some $\mathbf{x}$ in $\mathbb{R}^2$, that is, if $\mathbf{c} = T(\mathbf{x})$ for some $\mathbf{x}$. This is just another way of asking if the system $A\mathbf{x} = \mathbf{c}$ is consistent. To find the answer, row reduce the augmented matrix:

$$\begin{bmatrix} 1 & -3 & 3 \\ 3 & 5 & 2 \\ -1 & 7 & 5 \end{bmatrix} \sim \begin{bmatrix} 1 & -3 & 3 \\ 0 & 14 & -7 \\ 0 & 4 & 8 \end{bmatrix} \sim \begin{bmatrix} 1 & -3 & 3 \\ 0 & 1 & 2 \\ 0 & 14 & -7 \end{bmatrix} \sim \begin{bmatrix} 1 & -3 & 3 \\ 0 & 1 & 2 \\ 0 & 0 & -35 \end{bmatrix}$$

The third equation, $0 = -35$, shows that the system is inconsistent. So $\mathbf{c}$ is *not* in the range of T.   ∎

The question in Example 1(c) is a *uniqueness* problem for a system of linear equations, translated here into the language of matrix transformations: Is $\mathbf{b}$ the image of a *unique* $\mathbf{x}$ in $\mathbb{R}^n$? Similarly, Example 1(d) is an *existence* problem: Does there *exist* an $\mathbf{x}$ whose image is $\mathbf{c}$?

The next two matrix transformations can be viewed geometrically. They reinforce the dynamic view of a matrix as something that transforms vectors into other vectors. Section 2.7 contains other interesting examples connected with computer graphics.

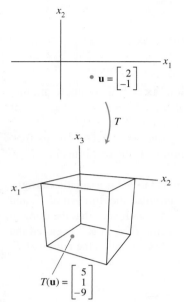

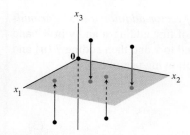

FIGURE 3
A projection transformation.

EXAMPLE 2 If $A = \begin{bmatrix} 1 & 0 & 0 \\ 0 & 1 & 0 \\ 0 & 0 & 0 \end{bmatrix}$, then the transformation $\mathbf{x} \mapsto A\mathbf{x}$ *projects* points in $\mathbb{R}^3$ onto the $x_1 x_2$-plane because

$$\begin{bmatrix} x_1 \\ x_2 \\ x_3 \end{bmatrix} \mapsto \begin{bmatrix} 1 & 0 & 0 \\ 0 & 1 & 0 \\ 0 & 0 & 0 \end{bmatrix} \begin{bmatrix} x_1 \\ x_2 \\ x_3 \end{bmatrix} = \begin{bmatrix} x_1 \\ x_2 \\ 0 \end{bmatrix}$$

See Fig. 3. ∎

EXAMPLE 3 Let $A = \begin{bmatrix} 1 & 3 \\ 0 & 1 \end{bmatrix}$. The transformation $T : \mathbb{R}^2 \to \mathbb{R}^2$ defined by $T(\mathbf{x}) = A\mathbf{x}$ is called a **shear transformation**. It can be shown that if T acts on each point in the 2×2 square shown in Fig. 4, then the set of images forms the shaded parallelogram. The key idea is to show that T maps line segments onto line segments (as shown in Exercise 27) and then to check that the corners of the square map onto the vertices of the parallelogram. For instance, the image of the point $\mathbf{u} = \begin{bmatrix} 0 \\ 2 \end{bmatrix}$ is

$$T(\mathbf{u}) = \begin{bmatrix} 1 & 3 \\ 0 & 1 \end{bmatrix}\begin{bmatrix} 0 \\ 2 \end{bmatrix} = \begin{bmatrix} 6 \\ 2 \end{bmatrix}, \text{ and the image of } \begin{bmatrix} 2 \\ 2 \end{bmatrix} \text{ is } \begin{bmatrix} 1 & 3 \\ 0 & 1 \end{bmatrix}\begin{bmatrix} 2 \\ 2 \end{bmatrix} = \begin{bmatrix} 8 \\ 2 \end{bmatrix}. \; T$$

deforms the square as if the top of the square were pushed to the right while the base is held fixed. Shear transformations appear in physics, geology, and crystallography. ∎

sheep

sheared sheep

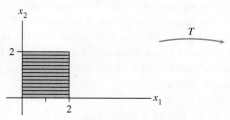

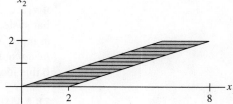

FIGURE 4 A shear transformation.

Linear Transformations

Theorem 5 in Section 1.4 shows that if A is $m \times n$, then the transformation $\mathbf{x} \mapsto A\mathbf{x}$ has the properties

$$A(\mathbf{u} + \mathbf{v}) = A\mathbf{u} + A\mathbf{v} \quad \text{and} \quad A(c\mathbf{u}) = cA\mathbf{u}$$

for all $\mathbf{u}, \mathbf{v}$ in $\mathbb{R}^n$ and all scalars c. These properties, written in function notation, identify the most important class of transformations in linear algebra.

DEFINITION

A transformation (or mapping) T is **linear** if:

(i) $T(\mathbf{u} + \mathbf{v}) = T(\mathbf{u}) + T(\mathbf{v})$ for all $\mathbf{u}, \mathbf{v}$ in the domain of T;

(ii) $T(c\mathbf{u}) = cT(\mathbf{u})$ for all scalars c and all $\mathbf{u}$ in the domain of T.

Every matrix transformation is a linear transformation. Important examples of linear transformations that are not matrix transformations will be discussed in Chapters 4 and 5.

Linear transformations *preserve the operations of vector addition and scalar multiplication.* Property (i) says that the result $T(\mathbf{u} + \mathbf{v})$ of first adding $\mathbf{u}$ and $\mathbf{v}$ in $\mathbb{R}^n$ and then applying T is the same as first applying T to $\mathbf{u}$ and to $\mathbf{v}$ and then adding $T(\mathbf{u})$ and $T(\mathbf{v})$ in $\mathbb{R}^m$. These two properties lead easily to the following useful facts.

If T is a linear transformation, then

$$T(\mathbf{0}) = \mathbf{0} \tag{3}$$

and

$$T(c\mathbf{u} + d\mathbf{v}) = cT(\mathbf{u}) + dT(\mathbf{v}) \tag{4}$$

for all vectors $\mathbf{u}, \mathbf{v}$ in the domain of T and all scalars c, d.

Property (3) follows from condition (ii) in the definition, because $T(\mathbf{0}) = T(0\mathbf{u}) = 0T(\mathbf{u}) = \mathbf{0}$. Property (4) requires both (i) and (ii):

$$T(c\mathbf{u} + d\mathbf{v}) = T(c\mathbf{u}) + T(d\mathbf{v}) = cT(\mathbf{u}) + dT(\mathbf{v})$$

Observe that *if a transformation satisfies* (4) *for all* $\mathbf{u}, \mathbf{v}$ *and* c, d, *it must be linear.* (Set $c = d = 1$ for preservation of addition, and set $d = 0$ for preservation of scalar multiplication.) Repeated application of (4) produces a useful generalization:

$$T(c_1\mathbf{v}_1 + \cdots + c_p\mathbf{v}_p) = c_1T(\mathbf{v}_1) + \cdots + c_pT(\mathbf{v}_p) \tag{5}$$

In engineering and physics, (5) is referred to as a *superposition principle.* Think of $\mathbf{v}_1, \ldots, \mathbf{v}_p$ as signals that go into a system and $T(\mathbf{v}_1), \ldots, T(\mathbf{v}_p)$ as the responses of that system to the signals. The system satisfies the superposition principle if whenever an input is expressed as a linear combination of such signals, the system's response is *the same* linear combination of the responses to the individual signals. We will return to this idea in Chapter 4.

EXAMPLE 4 Given a scalar r, define $T : \mathbb{R}^2 \to \mathbb{R}^2$ by $T(\mathbf{x}) = r\mathbf{x}$. T is called a **contraction** when $0 \le r \le 1$ and a **dilation** when $r > 1$. Let $r = 3$, and show that T is a linear transformation.

SOLUTION Let $\mathbf{u}, \mathbf{v}$ be in $\mathbb{R}^2$ and let c, d be scalars. Then

$$
\begin{aligned}
T(c\mathbf{u} + d\mathbf{v}) &= 3(c\mathbf{u} + d\mathbf{v}) && \text{Definition of } T \\
&= 3c\mathbf{u} + 3d\mathbf{v} && \left.\begin{aligned}\\\end{aligned}\right\}\text{Vector arithmetic} \\
&= c(3\mathbf{u}) + d(3\mathbf{v}) && \\
&= cT(\mathbf{u}) + dT(\mathbf{v}) &&
\end{aligned}
$$

Thus T is a linear transformation because it satisfies (4). See Fig. 5. ∎

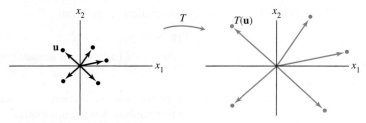

FIGURE 5 A dilation transformation.

EXAMPLE 5 Define a linear transformation $T : \mathbb{R}^2 \to \mathbb{R}^2$ by

$$T(\mathbf{x}) = \begin{bmatrix} 0 & -1 \\ 1 & 0 \end{bmatrix} \begin{bmatrix} x_1 \\ x_2 \end{bmatrix} = \begin{bmatrix} -x_2 \\ x_1 \end{bmatrix}$$

Find the images under T of $\mathbf{u} = \begin{bmatrix} 4 \\ 1 \end{bmatrix}$, $\mathbf{v} = \begin{bmatrix} 2 \\ 3 \end{bmatrix}$, and $\mathbf{u} + \mathbf{v} = \begin{bmatrix} 6 \\ 4 \end{bmatrix}$.

SOLUTION

$$T(\mathbf{u}) = \begin{bmatrix} 0 & -1 \\ 1 & 0 \end{bmatrix} \begin{bmatrix} 4 \\ 1 \end{bmatrix} = \begin{bmatrix} -1 \\ 4 \end{bmatrix}, \qquad T(\mathbf{v}) = \begin{bmatrix} 0 & -1 \\ 1 & 0 \end{bmatrix} \begin{bmatrix} 2 \\ 3 \end{bmatrix} = \begin{bmatrix} -3 \\ 2 \end{bmatrix},$$

$$T(\mathbf{u} + \mathbf{v}) = \begin{bmatrix} 0 & -1 \\ 1 & 0 \end{bmatrix} \begin{bmatrix} 6 \\ 4 \end{bmatrix} = \begin{bmatrix} -4 \\ 6 \end{bmatrix}$$

Note that $T(\mathbf{u} + \mathbf{v})$ is obviously equal to $T(\mathbf{u}) + T(\mathbf{v})$. It appears from Fig. 6 that T rotates $\mathbf{u}$, $\mathbf{v}$, and $\mathbf{u} + \mathbf{v}$ counterclockwise about the origin through 90°. In fact, T transforms the entire parallelogram determined by $\mathbf{u}$ and $\mathbf{v}$ into the one determined by $T(\mathbf{u})$ and $T(\mathbf{v})$. (See Exercise 28.) ∎

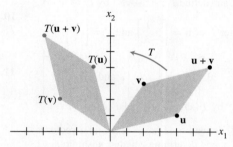

FIGURE 6 A rotation transformation.

The final example is not geometrical; instead, it shows how a linear mapping can transform one type of data into another.

EXAMPLE 6 A company manufactures two products, B and C. Using data from Example 7 in Section 1.3, we construct a "unit cost" matrix, $U = [\,\mathbf{b} \quad \mathbf{c}\,]$, whose columns describe the "costs per dollar of output" for the products:

$$\begin{array}{cc} & \text{Product} \\ & \begin{array}{cc} \text{B} & \text{C} \end{array} \end{array}$$

$$U = \begin{bmatrix} .45 & .40 \\ .25 & .35 \\ .15 & .15 \end{bmatrix} \begin{array}{l} \text{Materials} \\ \text{Labor} \\ \text{Overhead} \end{array}$$

Let $\mathbf{x} = (x_1, x_2)$ be a "production" vector, corresponding to x_1 dollars of product B and x_2 dollars of product C, and define $T : \mathbb{R}^2 \to \mathbb{R}^3$ by

$$T(\mathbf{x}) = U\mathbf{x} = x_1 \begin{bmatrix} .45 \\ .25 \\ .15 \end{bmatrix} + x_2 \begin{bmatrix} .40 \\ .30 \\ .15 \end{bmatrix} = \begin{bmatrix} \text{Total cost of materials} \\ \text{Total cost of labor} \\ \text{Total cost of overhead} \end{bmatrix}$$

The mapping T transforms a list of production quantities (measured in dollars) into a list of total costs. The linearity of this mapping is reflected in two ways:

1. If production is increased by a factor of, say, 4, from $\mathbf{x}$ to $4\mathbf{x}$, then the costs will increase by the same factor, from $T(\mathbf{x})$ to $4T(\mathbf{x})$.

2. If **x** and **y** are production vectors, then the total cost vector associated with the combined production **x** + **y** is precisely the sum of the cost vectors $T(\mathbf{x})$ and $T(\mathbf{y})$. ∎

PRACTICE PROBLEMS

1. Suppose $T : \mathbb{R}^5 \to \mathbb{R}^2$ and $T(\mathbf{x}) = A\mathbf{x}$ for some matrix A and for each **x** in $\mathbb{R}^5$. How many rows and columns does A have?

2. Let $A = \begin{bmatrix} 1 & 0 \\ 0 & -1 \end{bmatrix}$. Give a geometric description of the transformation $\mathbf{x} \mapsto A\mathbf{x}$.

3. The line segment from **0** to a vector **u** is the set of points of the form $t\mathbf{u}$, where $0 \leq t \leq 1$. Show that a linear transformation T maps this segment into the segment between **0** and $T(\mathbf{u})$.

1.8 EXERCISES

1. Let $A = \begin{bmatrix} 2 & 0 \\ 0 & 2 \end{bmatrix}$, and define $T : \mathbb{R}^2 \to \mathbb{R}^2$ by $T(\mathbf{x}) = A\mathbf{x}$.

Find the images under T of $\mathbf{u} = \begin{bmatrix} 1 \\ -3 \end{bmatrix}$ and $\mathbf{v} = \begin{bmatrix} a \\ b \end{bmatrix}$.

2. Let $A = \begin{bmatrix} \frac{1}{3} & 0 & 0 \\ 0 & \frac{1}{3} & 0 \\ 0 & 0 & \frac{1}{3} \end{bmatrix}$, $\mathbf{u} = \begin{bmatrix} 3 \\ 6 \\ -9 \end{bmatrix}$, and $\mathbf{v} = \begin{bmatrix} a \\ b \\ c \end{bmatrix}$.

Define $T : \mathbb{R}^3 \to \mathbb{R}^3$ by $T(\mathbf{x}) = A\mathbf{x}$. Find $T(\mathbf{u})$ and $T(\mathbf{v})$.

In Exercises 3–6, with T defined by $T(\mathbf{x}) = A\mathbf{x}$, find a vector **x** whose image under T is **b**, and determine whether **x** is unique.

3. $A = \begin{bmatrix} 1 & 0 & -3 \\ -3 & 1 & 6 \\ 2 & -2 & -1 \end{bmatrix}$, $\mathbf{b} = \begin{bmatrix} -2 \\ 3 \\ -1 \end{bmatrix}$

4. $A = \begin{bmatrix} 1 & -2 & 3 \\ 0 & 1 & -3 \\ 2 & -5 & 6 \end{bmatrix}$, $\mathbf{b} = \begin{bmatrix} -6 \\ -4 \\ -5 \end{bmatrix}$

5. $A = \begin{bmatrix} 1 & -5 & -7 \\ -3 & 7 & 5 \end{bmatrix}$, $\mathbf{b} = \begin{bmatrix} -2 \\ -2 \end{bmatrix}$

6. $A = \begin{bmatrix} 1 & -3 & 2 \\ 3 & -8 & 8 \\ 0 & 1 & 2 \\ 1 & 0 & 8 \end{bmatrix}$, $\mathbf{b} = \begin{bmatrix} 1 \\ 6 \\ 3 \\ 10 \end{bmatrix}$

7. Let A be a 6×5 matrix. What must a and b be in order to define $T : \mathbb{R}^a \to \mathbb{R}^b$ by $T(\mathbf{x}) = A\mathbf{x}$?

8. How many rows and columns must a matrix A have in order to define a mapping from $\mathbb{R}^5$ into $\mathbb{R}^7$ by the rule $T(\mathbf{x}) = A\mathbf{x}$?

For Exercises 9 and 10, find all **x** in $\mathbb{R}^4$ that are mapped into the zero vector by the transformation $\mathbf{x} \mapsto A\mathbf{x}$ for the given matrix A.

9. $A = \begin{bmatrix} 1 & -3 & 5 & -5 \\ 0 & 1 & -3 & 5 \\ 2 & -4 & 4 & -4 \end{bmatrix}$

10. $A = \begin{bmatrix} 3 & 2 & 10 & -6 \\ 1 & 0 & 2 & -4 \\ 0 & 1 & 2 & 3 \\ 1 & 4 & 10 & 8 \end{bmatrix}$

11. Let $\mathbf{b} = \begin{bmatrix} -1 \\ 1 \\ 0 \end{bmatrix}$, and let A be the matrix in Exercise 9. Is **b** in the range of the linear transformation $\mathbf{x} \mapsto A\mathbf{x}$? Why or why not?

12. Let $\mathbf{b} = \begin{bmatrix} -1 \\ 3 \\ -1 \\ 4 \end{bmatrix}$, and let A be the matrix in Exercise 10. Is **b** in the range of the linear transformation $\mathbf{x} \mapsto A\mathbf{x}$? Why or why not?

In Exercises 13–16, use a rectangular coordinate system to plot $\mathbf{u} = \begin{bmatrix} 5 \\ 2 \end{bmatrix}$, $\mathbf{v} = \begin{bmatrix} -2 \\ 4 \end{bmatrix}$, and their images under the given transformation T. (Make a separate and reasonably large sketch for each exercise.) Describe geometrically what T does to each vector **x** in $\mathbb{R}^2$.

13. $T(\mathbf{x}) = \begin{bmatrix} -1 & 0 \\ 0 & -1 \end{bmatrix} \begin{bmatrix} x_1 \\ x_2 \end{bmatrix}$

14. $T(\mathbf{x}) = \begin{bmatrix} 2 & 0 \\ 0 & 2 \end{bmatrix} \begin{bmatrix} x_1 \\ x_2 \end{bmatrix}$

15. $T(\mathbf{x}) = \begin{bmatrix} 0 & 1 \\ 1 & 0 \end{bmatrix} \begin{bmatrix} x_1 \\ x_2 \end{bmatrix}$

16. $T(\mathbf{x}) = \begin{bmatrix} 0 & 0 \\ 0 & 2 \end{bmatrix} \begin{bmatrix} x_1 \\ x_2 \end{bmatrix}$

17. Let $T : \mathbb{R}^2 \to \mathbb{R}^2$ be a linear transformation that maps $\mathbf{u} = \begin{bmatrix} 3 \\ 4 \end{bmatrix}$ into $\begin{bmatrix} 4 \\ 1 \end{bmatrix}$ and maps $\mathbf{v} = \begin{bmatrix} 3 \\ 3 \end{bmatrix}$ into $\begin{bmatrix} -1 \\ 3 \end{bmatrix}$. Use the fact that T is linear to find the images under T of $2\mathbf{u}$, $3\mathbf{v}$, and $2\mathbf{u} + 3\mathbf{v}$.

18. The figure shows vectors $\mathbf{u}$, $\mathbf{v}$, and $\mathbf{w}$, along with the images $T(\mathbf{u})$ and $T(\mathbf{v})$ under the action of a linear transformation $T : \mathbb{R}^2 \to \mathbb{R}^2$. Copy this figure carefully, and draw the image $T(\mathbf{w})$ as accurately as possible. [*Hint:* First, write $\mathbf{w}$ as a linear combination of $\mathbf{u}$ and $\mathbf{v}$.]

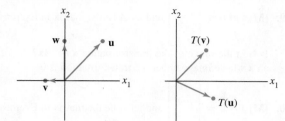

19. Let $\mathbf{e}_1 = \begin{bmatrix} 1 \\ 0 \end{bmatrix}$, $\mathbf{e}_2 = \begin{bmatrix} 0 \\ 1 \end{bmatrix}$, $\mathbf{y}_1 = \begin{bmatrix} 2 \\ 5 \end{bmatrix}$, and $\mathbf{y}_2 = \begin{bmatrix} -1 \\ 6 \end{bmatrix}$, and let $T : \mathbb{R}^2 \to \mathbb{R}^2$ be a linear transformation that maps $\mathbf{e}_1$ into $\mathbf{y}_1$ and maps $\mathbf{e}_2$ into $\mathbf{y}_2$. Find the images of $\begin{bmatrix} 5 \\ -3 \end{bmatrix}$ and $\begin{bmatrix} x_1 \\ x_2 \end{bmatrix}$.

20. Let $\mathbf{x} = \begin{bmatrix} x_1 \\ x_2 \end{bmatrix}$, $\mathbf{v}_1 = \begin{bmatrix} -3 \\ 5 \end{bmatrix}$, and $\mathbf{v}_2 = \begin{bmatrix} 7 \\ -2 \end{bmatrix}$, and let $T : \mathbb{R}^2 \to \mathbb{R}^2$ be a linear transformation that maps $\mathbf{x}$ into $x_1 \mathbf{v}_1 + x_2 \mathbf{v}_2$. Find a matrix A such that $T(\mathbf{x})$ is $A\mathbf{x}$ for each $\mathbf{x}$.

In Exercises 21 and 22, mark each statement True or False. Justify each answer.

21. a. A linear transformation is a special type of function.

b. If A is a 3×5 matrix and T is a transformation defined by $T(\mathbf{x}) = A\mathbf{x}$, then the domain of T is $\mathbb{R}^3$.

c. If A is an $m \times n$ matrix, then the range of the transformation $\mathbf{x} \mapsto A\mathbf{x}$ is $\mathbb{R}^m$.

d. Every linear transformation is a matrix transformation.

e. A transformation T is linear if and only if

$$T(c_1 \mathbf{v}_1 + c_2 \mathbf{v}_2) = c_1 T(\mathbf{v}_1) + c_2 T(\mathbf{v}_2)$$

for all $\mathbf{v}_1$ and $\mathbf{v}_2$ in the domain of T and for all scalars c_1 and c_2.

22. a. The range of the transformation $\mathbf{x} \mapsto A\mathbf{x}$ is the set of all linear combinations of the columns of A.

b. Every matrix transformation is a linear transformation.

c. If $T : \mathbb{R}^n \to \mathbb{R}^m$ is a linear transformation and if $\mathbf{c}$ is in $\mathbb{R}^m$, then a uniqueness question is "Is $\mathbf{c}$ in the range of T?"

d. A linear transformation preserves the operations of vector addition and scalar multiplication.

e. A linear transformation $T : \mathbb{R}^n \to \mathbb{R}^m$ always maps the origin of $\mathbb{R}^n$ to the origin of $\mathbb{R}^m$.

23. Define $f : \mathbb{R} \to \mathbb{R}$ by $f(x) = mx + b$.

a. Show that f is a linear transformation when $b = 0$.

b. Find a property of a linear transformation that is violated when $b \neq 0$.

c. Why is f called a linear function?

24. An *affine transformation* $T : \mathbb{R}^n \to \mathbb{R}^m$ has the form $T(\mathbf{x}) = A\mathbf{x} + \mathbf{b}$, with A an $m \times n$ matrix and $\mathbf{b}$ in $\mathbb{R}^m$. Show that T is *not* a linear transformation when $\mathbf{b} \neq \mathbf{0}$. (Affine transformations are important in computer graphics.)

25. Given $\mathbf{v} \neq \mathbf{0}$ and $\mathbf{p}$ in $\mathbb{R}^n$, the line through $\mathbf{p}$ in the direction of $\mathbf{v}$ has the parametric equation $\mathbf{x} = \mathbf{p} + t\mathbf{v}$. Show that a linear transformation $T : \mathbb{R}^n \to \mathbb{R}^n$ maps this line onto another line or onto a single point (a *degenerate line*).

26. a. Show that the line through vectors $\mathbf{p}$ and $\mathbf{q}$ in $\mathbb{R}^n$ may be written in the parametric form $\mathbf{x} = (1 - t)\mathbf{p} + t\mathbf{q}$. (Refer to the figure with Exercises 21 and 22 in Section 1.5.)

b. The line segment from $\mathbf{p}$ to $\mathbf{q}$ is the set of points of the form $(1 - t)\mathbf{p} + t\mathbf{q}$ for $0 \leq t \leq 1$ (as shown in the figure below). Show that a linear transformation T maps this line segment onto a line segment or onto a single point.

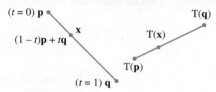

27. Let $\mathbf{u}$ and $\mathbf{v}$ be linearly independent vectors in $\mathbb{R}^3$, and let P be the plane through $\mathbf{u}$, $\mathbf{v}$, and $\mathbf{0}$. The parametric equation of P is $\mathbf{x} = s\mathbf{u} + t\mathbf{v}$ (with s, t in $\mathbb{R}$). Show that a linear transformation $T : \mathbb{R}^3 \to \mathbb{R}^3$ maps P onto a plane through $\mathbf{0}$, or onto a line through $\mathbf{0}$, or onto just the origin in $\mathbb{R}^3$. What must be true about $T(\mathbf{u})$ and $T(\mathbf{v})$ in order for the image of the plane P to be a plane?

28. Let $\mathbf{u}$ and $\mathbf{v}$ be vectors in $\mathbb{R}^n$. It can be shown that the set P of all points in the parallelogram determined by $\mathbf{u}$ and $\mathbf{v}$ has the form $a\mathbf{u} + b\mathbf{v}$, for $0 \leq a \leq 1, 0 \leq b \leq 1$. Let $T : \mathbb{R}^n \to \mathbb{R}^m$ be a linear transformation. Explain why the image of a point in P under the transformation T lies in the parallelogram determined by $T(\mathbf{u})$ and $T(\mathbf{v})$.

29. Let $T : \mathbb{R}^2 \to \mathbb{R}^2$ be the linear transformation that reflects each point through the x_2-axis. Make two sketches similar to Fig. 6 that illustrate properties (i) and (ii) of a linear transformation.

30. Suppose vectors $\mathbf{v}_1, \ldots, \mathbf{v}_p$ span $\mathbb{R}^n$, and let $T : \mathbb{R}^n \to \mathbb{R}^n$ be a linear transformation. Suppose $T(\mathbf{v}_i) = \mathbf{0}$ for $i = 1, \ldots, p$. Show that T is the zero transformation. That is, show that if $\mathbf{x}$ is any vector in $\mathbb{R}^n$, then $T(\mathbf{x}) = \mathbf{0}$.

31. Let $T : \mathbb{R}^n \to \mathbb{R}^m$ be a linear transformation, and let $\{\mathbf{v}_1, \mathbf{v}_2, \mathbf{v}_3\}$ be a linearly dependent set in $\mathbb{R}^n$. Explain why the set $\{T(\mathbf{v}_1), T(\mathbf{v}_2), T(\mathbf{v}_3)\}$ is linearly dependent.

In Exercises 32–36, column vectors are written as rows, such as $\mathbf{x} = (x_1, x_2)$, and $T(\mathbf{x})$ is written as $T(x_1, x_2)$.

32. Show that the transformation T defined by $T(x_1, x_2) = (x_1 - 2|x_2|, x_1 - 4x_2)$ is not linear.

33. Show that the transformation T defined by $T(x_1, x_2) = (x_1 - 2x_2, x_1 - 3, 2x_1 - 5x_2)$ is not linear.

34. Let $T : \mathbb{R}^3 \to \mathbb{R}^3$ be the transformation that reflects each vector $\mathbf{x} = (x_1, x_2, x_3)$ through the plane $x_3 = 0$ onto $T(\mathbf{x}) = (x_1, x_2, -x_3)$. Show that T is a linear transformation. [See Example 4 for ideas.]

35. Let $T : \mathbb{R}^3 \to \mathbb{R}^3$ be the transformation that projects each vector $\mathbf{x} = (x_1, x_2, x_3)$ onto the plane $x_2 = 0$, so $T(\mathbf{x}) = (x_1, 0, x_3)$. Show that T is a linear transformation.

36. Let $T : \mathbb{R}^n \to \mathbb{R}^m$ be a linear transformation. Suppose $\{\mathbf{u}, \mathbf{v}\}$ is a linearly independent set, but $\{T(\mathbf{u}), T(\mathbf{v})\}$ is a linearly dependent set. Show that $T(\mathbf{x}) = \mathbf{0}$ has a nontrivial solution. [*Hint:* Use the fact that $c_1 T(\mathbf{u}) + c_2 T(\mathbf{v}) = \mathbf{0}$ for some weights c_1 and c_2, not both zero.]

[M] In Exercises 37 and 38, the given matrix determines a linear transformation T. Find all $\mathbf{x}$ such that $T(\mathbf{x}) = \mathbf{0}$.

37. $\begin{bmatrix} 2 & 3 & 5 & -5 \\ -7 & 7 & 0 & 0 \\ -3 & 4 & 1 & 3 \\ -9 & 3 & -6 & -4 \end{bmatrix}$
38. $\begin{bmatrix} 3 & 4 & -7 & 0 \\ 5 & -8 & 7 & 4 \\ 6 & -8 & 6 & 4 \\ 9 & -7 & -2 & 0 \end{bmatrix}$

39. [M] Let $\mathbf{b} = \begin{bmatrix} 8 \\ 7 \\ 5 \\ -3 \end{bmatrix}$ and let A be the matrix in Exercise 37.

Is $\mathbf{b}$ in the range of the transformation $\mathbf{x} \mapsto A\mathbf{x}$? If so, find an $\mathbf{x}$ whose image under the transformation is $\mathbf{b}$.

40. [M] Let $\mathbf{b} = \begin{bmatrix} 4 \\ -4 \\ -4 \\ -7 \end{bmatrix}$ and let A be the matrix in Exercise 38.

Is $\mathbf{b}$ in the range of the transformation $\mathbf{x} \mapsto A\mathbf{x}$? If so, find an $\mathbf{x}$ whose image under the transformation is $\mathbf{b}$.

> **SG**   Mastering: Linear Transformations 1–34

SOLUTIONS TO PRACTICE PROBLEMS

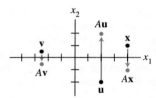

The transformation $\mathbf{x} \mapsto A\mathbf{x}$.

1. A must have five columns for $A\mathbf{x}$ to be defined. A must have two rows for the codomain of T to be $\mathbb{R}^2$.

2. Plot some random points (vectors) on graph paper to see what happens. A point such as $(4, 1)$ maps into $(4, -1)$. The transformation $\mathbf{x} \mapsto A\mathbf{x}$ reflects points through the x-axis (or x_1-axis).

3. Let $\mathbf{x} = t\mathbf{u}$ for some t such that $0 \le t \le 1$. Since T is linear, $T(t\mathbf{u}) = t\,T(\mathbf{u})$, which is a point on the line segment between $\mathbf{0}$ and $T(\mathbf{u})$.

1.9 │ THE MATRIX OF A LINEAR TRANSFORMATION

Whenever a linear transformation T arises geometrically or is described in words, we usually want a "formula" for $T(\mathbf{x})$. The discussion that follows shows that every linear transformation from $\mathbb{R}^n$ to $\mathbb{R}^m$ is actually a matrix transformation $\mathbf{x} \mapsto A\mathbf{x}$ and that important properties of T are intimately related to familiar properties of A. The key to finding A is to observe that T is completely determined by what it does to the columns of the $n \times n$ identity matrix I_n.

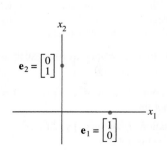

EXAMPLE 1   The columns of $I_2 = \begin{bmatrix} 1 & 0 \\ 0 & 1 \end{bmatrix}$ are $\mathbf{e}_1 = \begin{bmatrix} 1 \\ 0 \end{bmatrix}$ and $\mathbf{e}_2 = \begin{bmatrix} 0 \\ 1 \end{bmatrix}$. Suppose T is a linear transformation from $\mathbb{R}^2$ into $\mathbb{R}^3$ such that

$$T(\mathbf{e}_1) = \begin{bmatrix} 5 \\ -7 \\ 2 \end{bmatrix} \quad \text{and} \quad T(\mathbf{e}_2) = \begin{bmatrix} -3 \\ 8 \\ 0 \end{bmatrix}$$

With no additional information, find a formula for the image of an arbitrary $\mathbf{x}$ in $\mathbb{R}^2$.

SOLUTION Write

$$\mathbf{x} = \begin{bmatrix} x_1 \\ x_2 \end{bmatrix} = x_1 \begin{bmatrix} 1 \\ 0 \end{bmatrix} + x_2 \begin{bmatrix} 0 \\ 1 \end{bmatrix} = x_1\mathbf{e}_1 + x_2\mathbf{e}_2 \tag{1}$$

Since T is a *linear* transformation,

$$T(\mathbf{x}) = x_1 T(\mathbf{e}_1) + x_2 T(\mathbf{e}_2) \tag{2}$$

$$= x_1 \begin{bmatrix} 5 \\ -7 \\ 2 \end{bmatrix} + x_2 \begin{bmatrix} -3 \\ 8 \\ 0 \end{bmatrix} = \begin{bmatrix} 5x_1 - 3x_2 \\ -7x_1 + 8x_2 \\ 2x_1 + 0 \end{bmatrix} \qquad \blacksquare$$

The step from equation (1) to equation (2) explains why knowledge of $T(\mathbf{e}_1)$ and $T(\mathbf{e}_2)$ is sufficient to determine $T(\mathbf{x})$ for any $\mathbf{x}$. Moreover, since (2) expresses $T(\mathbf{x})$ as a linear combination of vectors, we can put these vectors into the columns of a matrix A and write (2) as

$$T(\mathbf{x}) = \begin{bmatrix} T(\mathbf{e}_1) & T(\mathbf{e}_2) \end{bmatrix} \begin{bmatrix} x_1 \\ x_2 \end{bmatrix} = A\mathbf{x}$$

THEOREM 10

Let $T : \mathbb{R}^n \to \mathbb{R}^m$ be a linear transformation. Then there exists a unique matrix A such that

$$T(\mathbf{x}) = A\mathbf{x} \quad \text{for all } \mathbf{x} \text{ in } \mathbb{R}^n$$

In fact, A is the $m \times n$ matrix whose jth column is the vector $T(\mathbf{e}_j)$, where $\mathbf{e}_j$ is the jth column of the identity matrix in $\mathbb{R}^n$:

$$A = \begin{bmatrix} T(\mathbf{e}_1) & \cdots & T(\mathbf{e}_n) \end{bmatrix} \tag{3}$$

PROOF Write $\mathbf{x} = I_n\mathbf{x} = \begin{bmatrix} \mathbf{e}_1 & \cdots & \mathbf{e}_n \end{bmatrix}\mathbf{x} = x_1\mathbf{e}_1 + \cdots + x_n\mathbf{e}_n$, and use the linearity of T to compute

$$T(\mathbf{x}) = T(x_1\mathbf{e}_1 + \cdots + x_n\mathbf{e}_n) = x_1 T(\mathbf{e}_1) + \cdots + x_n T(\mathbf{e}_n)$$

$$= \begin{bmatrix} T(\mathbf{e}_1) & \cdots & T(\mathbf{e}_n) \end{bmatrix} \begin{bmatrix} x_1 \\ \vdots \\ x_n \end{bmatrix} = A\mathbf{x}$$

The uniqueness of A is treated in Exercise 33. $\qquad \blacksquare$

The matrix A in (3) is called the **standard matrix for the linear transformation** T.

We know now that every linear transformation from $\mathbb{R}^n$ to $\mathbb{R}^m$ can be viewed as a matrix transformation, and vice versa. The term *linear transformation* focuses on a property of a mapping, while *matrix transformation* describes how such a mapping is implemented, as Examples 2 and 3 illustrate.

EXAMPLE 2 Find the standard matrix A for the dilation transformation $T(\mathbf{x}) = 3\mathbf{x}$, for $\mathbf{x}$ in $\mathbb{R}^2$.

SOLUTION Write

$$T(\mathbf{e}_1) = 3\mathbf{e}_1 = \begin{bmatrix} 3 \\ 0 \end{bmatrix} \quad \text{and} \quad T(\mathbf{e}_2) = 3\mathbf{e}_2 = \begin{bmatrix} 0 \\ 3 \end{bmatrix}$$

$$A = \begin{bmatrix} 3 & 0 \\ 0 & 3 \end{bmatrix}$$ ∎

EXAMPLE 3 Let $T : \mathbb{R}^2 \rightarrow \mathbb{R}^2$ be the transformation that rotates each point in $\mathbb{R}^2$ about the origin through an angle φ, with counterclockwise rotation for a positive angle. We could show geometrically that such a transformation is linear. (See Fig. 6 in Section 1.8.) Find the standard matrix A of this transformation.

SOLUTION $\begin{bmatrix} 1 \\ 0 \end{bmatrix}$ rotates into $\begin{bmatrix} \cos \varphi \\ \sin \varphi \end{bmatrix}$, and $\begin{bmatrix} 0 \\ 1 \end{bmatrix}$ rotates into $\begin{bmatrix} -\sin \varphi \\ \cos \varphi \end{bmatrix}$. See Fig. 1. By Theorem 10,

$$A = \begin{bmatrix} \cos \varphi & -\sin \varphi \\ \sin \varphi & \cos \varphi \end{bmatrix}$$

Example 5 in Section 1.8 is a special case of this transformation, with $\varphi = \pi/2$. ∎

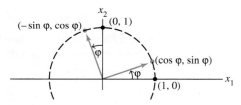

FIGURE 1 A rotation transformation.

Geometric Linear Transformations of $\mathbb{R}^2$

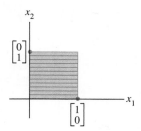

FIGURE 2

The unit square.

Examples 2 and 3 illustrate linear transformations that are described geometrically. Tables 1–4 illustrate other common geometric linear transformations of the plane. Because the transformations are linear, they are determined completely by what they do to the columns of I_2. Instead of showing only the images of $\mathbf{e}_1$ and $\mathbf{e}_2$, the tables show what a transformation does to the unit square (Fig. 2).

Other transformations can be constructed from those listed in Tables 1–4 by applying one transformation after another. For instance, a horizontal shear could be followed by a reflection in the x_2-axis. Section 2.1 will show that such a *composition* of linear transformations is linear. (Also, see Exercise 34.)

Existence and Uniqueness Questions

The concept of a linear transformation provides a new way to understand the existence and uniqueness questions asked earlier. The two definitions following Tables 1–4 give the appropriate terminology for transformations.

FIGURE 2

The unit square.

TABLE 1 **Reflections**

Transformation	Image of the Unit Square	Standard Matrix
Reflection through the x_1-axis		$\begin{bmatrix} 1 & 0 \\ 0 & -1 \end{bmatrix}$
Reflection through the x_2-axis		$\begin{bmatrix} -1 & 0 \\ 0 & 1 \end{bmatrix}$
Reflection through the line $x_2 = x_1$		$\begin{bmatrix} 0 & 1 \\ 1 & 0 \end{bmatrix}$
Reflection through the line $x_2 = -x_1$		$\begin{bmatrix} 0 & -1 \\ -1 & 0 \end{bmatrix}$
Reflection through the origin		$\begin{bmatrix} -1 & 0 \\ 0 & -1 \end{bmatrix}$

TABLE 2 Contractions and Expansions

Transformation	Image of the Unit Square	Standard Matrix
Horizontal contraction and expansion		$\begin{bmatrix} k & 0 \\ 0 & 1 \end{bmatrix}$
Vertical contraction and expansion		$\begin{bmatrix} 1 & 0 \\ 0 & k \end{bmatrix}$

TABLE 3 Shears

Transformation	Image of the Unit Square	Standard Matrix
Horizontal shear		$\begin{bmatrix} 1 & k \\ 0 & 1 \end{bmatrix}$
Vertical shear		$\begin{bmatrix} 1 & 0 \\ k & 1 \end{bmatrix}$

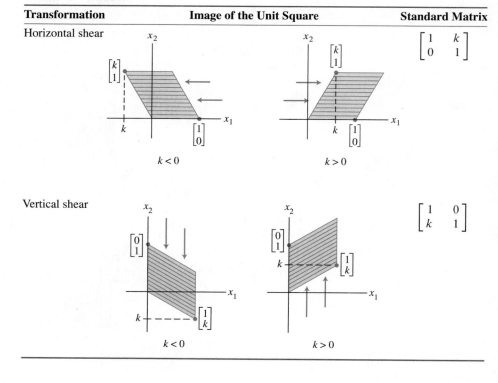

TABLE 4 Projections

Transformation	Image of the Unit Square	Standard Matrix
Projection onto the x_1-axis		$\begin{bmatrix} 1 & 0 \\ 0 & 0 \end{bmatrix}$
Projection onto the x_2-axis		$\begin{bmatrix} 0 & 0 \\ 0 & 1 \end{bmatrix}$

A mapping $T : \mathbb{R}^n \rightarrow \mathbb{R}^m$ is said to be **onto** $\mathbb{R}^m$ if each **b** in $\mathbb{R}^m$ is the image of *at least one* **x** in $\mathbb{R}^n$.

Equivalently, T is onto $\mathbb{R}^m$ when the range of T is all of the codomain $\mathbb{R}^m$. That is, T maps $\mathbb{R}^n$ onto $\mathbb{R}^m$ if, for each **b** in the codomain $\mathbb{R}^m$, there exists at least one solution of $T(\mathbf{x}) = \mathbf{b}$. "Does T map $\mathbb{R}^n$ onto $\mathbb{R}^m$?" is an existence question. The mapping T is *not* onto when there is some **b** in $\mathbb{R}^m$ for which the equation $T(\mathbf{x}) = \mathbf{b}$ has no solution. See Fig. 3.

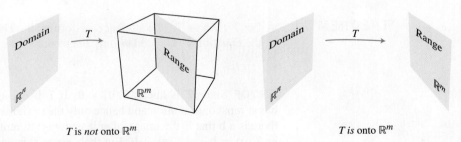

FIGURE 3 Is the range of T all of $\mathbb{R}^m$?

A mapping $T : \mathbb{R}^n \rightarrow \mathbb{R}^m$ is said to be **one-to-one** if each **b** in $\mathbb{R}^m$ is the image of *at most one* **x** in $\mathbb{R}^n$.

Equivalently, T is one-to-one if, for each $\mathbf{b}$ in $\mathbb{R}^m$, the equation $T(\mathbf{x}) = \mathbf{b}$ has either a unique solution or none at all. "Is T one-to-one?" is a uniqueness question. The mapping T is *not* one-to-one when some $\mathbf{b}$ in $\mathbb{R}^m$ is the image of more than one vector in $\mathbb{R}^n$. If there is no such $\mathbf{b}$, then T is one-to-one. See Fig. 4.

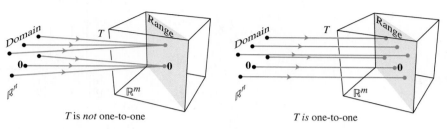

T is *not* one-to-one T *is* one-to-one

FIGURE 4 Is every $\mathbf{b}$ the image of at most one vector?

SG | Mastering: Existence and Uniqueness 1–39

The projection transformations shown in Table 4 are *not* one-to-one and do *not* map $\mathbb{R}^2$ onto $\mathbb{R}^2$. The transformations in Tables 1, 2, and 3 are one-to-one *and* do map $\mathbb{R}^2$ onto $\mathbb{R}^2$. Other possibilities are shown in the two examples below.

Example 4 and the theorems that follow show how the function properties of being one-to-one and mapping onto are related to important concepts studied earlier in this chapter.

EXAMPLE 4 Let T be the linear transformation whose standard matrix is

$$A = \begin{bmatrix} 1 & -4 & 8 & 1 \\ 0 & 2 & -1 & 3 \\ 0 & 0 & 0 & 5 \end{bmatrix}$$

Does T map $\mathbb{R}^4$ onto $\mathbb{R}^3$? Is T a one-to-one mapping?

SOLUTION Since A happens to be in echelon form, we can see at once that A has a pivot position in each row. By Theorem 4 in Section 1.4, for each $\mathbf{b}$ in $\mathbb{R}^3$, the equation $A\mathbf{x} = \mathbf{b}$ is consistent. In other words, the linear transformation T maps $\mathbb{R}^4$ (its domain) onto $\mathbb{R}^3$. However, since the equation $A\mathbf{x} = \mathbf{b}$ has a free variable (because there are four variables and only three basic variables), each $\mathbf{b}$ is the image of more than one $\mathbf{x}$. That is, T is *not* one-to-one. ■

THEOREM 11

Let $T : \mathbb{R}^n \rightarrow \mathbb{R}^m$ be a linear transformation. Then T is one-to-one if and only if the equation $T(\mathbf{x}) = \mathbf{0}$ has only the trivial solution.

PROOF Since T is linear, $T(\mathbf{0}) = \mathbf{0}$. If T is one-to-one, then the equation $T(\mathbf{x}) = \mathbf{0}$ has at most one solution and hence only the trivial solution. If T is not one-to-one, then there is a $\mathbf{b}$ that is the image of at least two different vectors in $\mathbb{R}^n$ — say, $\mathbf{u}$ and $\mathbf{v}$. That is, $T(\mathbf{u}) = \mathbf{b}$ and $T(\mathbf{v}) = \mathbf{b}$. But then, since T is linear,

$$T(\mathbf{u} - \mathbf{v}) = T(\mathbf{u}) - T(\mathbf{v}) = \mathbf{b} - \mathbf{b} = \mathbf{0}$$

The vector $\mathbf{u} - \mathbf{v}$ is not zero, since $\mathbf{u} \neq \mathbf{v}$. Hence the equation $T(\mathbf{x}) = \mathbf{0}$ has more than one solution. So, either the two conditions in the theorem are both true or they are both false. ■

THEOREM 12

Let $T : \mathbb{R}^n \rightarrow \mathbb{R}^m$ be a linear transformation and let A be the standard matrix for T. Then:

a. T maps $\mathbb{R}^n$ onto $\mathbb{R}^m$ if and only if the columns of A span $\mathbb{R}^m$;

b. T is one-to-one if and only if the columns of A are linearly independent.

PROOF

a. By Theorem 4 in Section 1.4, the columns of A span $\mathbb{R}^m$ if and only if for each $\mathbf{b}$ in $\mathbb{R}^m$ the equation $A\mathbf{x} = \mathbf{b}$ is consistent—in other words, if and only if for every $\mathbf{b}$, the equation $T(\mathbf{x}) = \mathbf{b}$ has at least one solution. This is true if and only if T maps $\mathbb{R}^n$ onto $\mathbb{R}^m$.

b. The equations $T(\mathbf{x}) = \mathbf{0}$ and $A\mathbf{x} = \mathbf{0}$ are the same except for notation. So, by Theorem 11, T is one-to-one if and only if $A\mathbf{x} = \mathbf{0}$ has only the trivial solution. This happens if and only if the columns of A are linearly independent, as was already noted in the boxed statement (3) in Section 1.7.   ∎

Statement (a) in Theorem 12 is equivalent to the statement "T maps $\mathbb{R}^n$ onto $\mathbb{R}^m$ if and only if every vector in $\mathbb{R}^m$ is a linear combination of the columns of A." See Theorem 4 in Section 1.4.

In the next example and in some exercises that follow, column vectors are written in rows, such as $\mathbf{x} = (x_1, x_2)$, and $T(\mathbf{x})$ is written as $T(x_1, x_2)$ instead of the more formal $T((x_1, x_2))$.

EXAMPLE 5   Let $T(x_1, x_2) = (3x_1 + x_2, 5x_1 + 7x_2, x_1 + 3x_2)$. Show that T is a one-to-one linear transformation. Does T map $\mathbb{R}^2$ onto $\mathbb{R}^3$?

SOLUTION   When $\mathbf{x}$ and $T(\mathbf{x})$ are written as column vectors, you can determine the standard matrix of T by inspection, visualizing the row–vector computation of each entry in $A\mathbf{x}$.

$$T(\mathbf{x}) = \begin{bmatrix} 3x_1 + x_2 \\ 5x_1 + 7x_2 \\ x_1 + 3x_2 \end{bmatrix} = \underbrace{\begin{bmatrix} ? & ? \\ ? & ? \\ ? & ? \end{bmatrix}}_{A} \begin{bmatrix} x_1 \\ x_2 \end{bmatrix} = \begin{bmatrix} 3 & 1 \\ 5 & 7 \\ 1 & 3 \end{bmatrix} \begin{bmatrix} x_1 \\ x_2 \end{bmatrix} \tag{4}$$

So T is indeed a linear transformation, with its standard matrix A shown in (4). The columns of A are linearly independent because they are not multiples. By Theorem 12(b), T is one-to-one. To decide if T is onto $\mathbb{R}^3$, examine the span of the columns of A. Since A is 3×2, the columns of A span $\mathbb{R}^3$ if and only if A has 3 pivot positions, by Theorem 4. This is impossible, since A has only 2 columns. So the columns of A do not span $\mathbb{R}^3$, and the associated linear transformation is not onto $\mathbb{R}^3$.   ∎

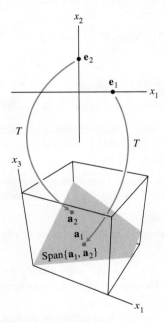

The transformation T is not onto $\mathbb{R}^3$.

PRACTICE PROBLEM

Let $T : \mathbb{R}^2 \rightarrow \mathbb{R}^2$ be the transformation that first performs a horizontal shear that maps $\mathbf{e}_2$ into $\mathbf{e}_2 - .5\mathbf{e}_1$ (but leaves $\mathbf{e}_1$ unchanged) and then reflects the result through the x_2-axis. Assuming that T is linear, find its standard matrix. [*Hint:* Determine the final location of the images of $\mathbf{e}_1$ and $\mathbf{e}_2$.]

1.9 EXERCISES

In Exercises 1–10, assume that T is a linear transformation. Find the standard matrix of T.

1. $T : \mathbb{R}^2 \to \mathbb{R}^4$, $T(\mathbf{e}_1) = (3, 1, 3, 1)$, and $T(\mathbf{e}_2) = (-5, 2, 0, 0)$, where $\mathbf{e}_1 = (1, 0)$ and $\mathbf{e}_2 = (0, 1)$.

2. $T : \mathbb{R}^3 \to \mathbb{R}^2$, $\quad T(\mathbf{e}_1) = (1, 4)$, $\quad T(\mathbf{e}_2) = (-2, 9)$, $\quad$ and $T(\mathbf{e}_3) = (3, -8)$, where $\mathbf{e}_1$, $\mathbf{e}_2$, and $\mathbf{e}_3$ are the columns of the 3×3 identity matrix.

3. $T : \mathbb{R}^2 \to \mathbb{R}^2$ is a vertical shear transformation that maps $\mathbf{e}_1$ into $\mathbf{e}_1 - 3\mathbf{e}_2$, but leaves $\mathbf{e}_2$ unchanged.

4. $T : \mathbb{R}^2 \to \mathbb{R}^2$ is a horizontal shear transformation that leaves $\mathbf{e}_1$ unchanged and maps $\mathbf{e}_2$ into $\mathbf{e}_2 + 2\mathbf{e}_1$.

5. $T : \mathbb{R}^2 \to \mathbb{R}^2$ rotates points (about the origin) through $\pi/2$ radians (counterclockwise).

6. $T : \mathbb{R}^2 \to \mathbb{R}^2$ rotates points (about the origin) through $-3\pi/2$ radians (clockwise).

7. $T : \mathbb{R}^2 \to \mathbb{R}^2$ first rotates points through $-3\pi/4$ radians (clockwise) and then reflects points through the horizontal x_1-axis. [*Hint:* $T(\mathbf{e}_1) = (-1/\sqrt{2}, 1/\sqrt{2})$.]

8. $T : \mathbb{R}^2 \to \mathbb{R}^2$ first performs a horizontal shear that transforms $\mathbf{e}_2$ into $\mathbf{e}_2 + 2\mathbf{e}_1$ (leaving $\mathbf{e}_1$ unchanged) and then reflects points through the line $x_2 = -x_1$.

9. $T : \mathbb{R}^2 \to \mathbb{R}^2$ first reflects points through the horizontal x_1-axis and then rotates points $-\pi/2$ radians.

10. $T : \mathbb{R}^2 \to \mathbb{R}^2$ first reflects points through the horizontal x_1-axis and then reflects points through the line $x_2 = x_1$.

11. A linear transformation $T : \mathbb{R}^2 \to \mathbb{R}^2$ first reflects points through the x_1-axis and then reflects points through the x_2-axis. Show that T can also be described as a linear transformation that rotates points about the origin. What is the angle of that rotation?

12. Show that the transformation in Exercise 10 is merely a rotation about the origin. What is the angle of the rotation?

13. Let $T : \mathbb{R}^2 \to \mathbb{R}^2$ be the linear transformation such that $T(\mathbf{e}_1)$ and $T(\mathbf{e}_2)$ are the vectors shown in the figure. Using the figure, sketch the vector $T(2, 1)$.

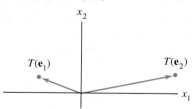

14. Let $T : \mathbb{R}^2 \to \mathbb{R}^2$ be a linear transformation with standard matrix $A = [\,\mathbf{a}_1 \quad \mathbf{a}_2\,]$, where $\mathbf{a}_1$ and $\mathbf{a}_2$ are shown in the figure at the top of column 2. Using the figure, draw the image of $\begin{bmatrix} 1 \\ -2 \end{bmatrix}$ under the transformation T.

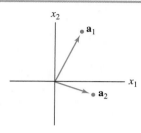

In Exercises 15 and 16, fill in the missing entries of the matrix, assuming that the equation holds for all values of the variables.

15. $\begin{bmatrix} ? & ? & ? \\ ? & ? & ? \\ ? & ? & ? \end{bmatrix} \begin{bmatrix} x_1 \\ x_2 \\ x_3 \end{bmatrix} = \begin{bmatrix} 2x_1 - 4x_2 \\ x_1 - x_3 \\ -x_2 + 3x_3 \end{bmatrix}$

16. $\begin{bmatrix} ? & ? \\ ? & ? \\ ? & ? \end{bmatrix} \begin{bmatrix} x_1 \\ x_2 \end{bmatrix} = \begin{bmatrix} 3x_1 - 2x_2 \\ x_1 + 4x_2 \\ x_2 \end{bmatrix}$

In Exercises 17–20, show that T is a linear transformation by finding a matrix that implements the mapping. Note that $x_1, x_2, \ldots$ are not vectors but are entries in vectors.

17. $T(x_1, x_2, x_3, x_4) = (x_1 + 2x_2, 0, 2x_2 + x_4, x_2 - x_4)$

18. $T(x_1, x_2) = (x_1 + 4x_2, 0, x_1 - 3x_2, x_1)$

19. $T(x_1, x_2, x_3) = (x_1 - 5x_2 + 4x_3, x_2 - 6x_3)$

20. $T(x_1, x_2, x_3, x_4) = 3x_1 + 4x_3 - 2x_4$ (Notice: $T : \mathbb{R}^4 \to \mathbb{R}$)

21. Let $T : \mathbb{R}^2 \to \mathbb{R}^2$ be a linear transformation such that $T(x_1, x_2) = (x_1 + x_2, 4x_1 + 5x_2)$. Find $\mathbf{x}$ such that $T(\mathbf{x}) = (3, 8)$.

22. Let $T : \mathbb{R}^2 \to \mathbb{R}^3$ be a linear transformation with $T(x_1, x_2) = (2x_1 - x_2, -3x_1 + x_2, 2x_1 - 3x_2)$. Find $\mathbf{x}$ such that $T(\mathbf{x}) = (0, -1, -4)$.

In Exercises 23 and 24, mark each statement True or False. Justify each answer.

23. a. A linear transformation $T : \mathbb{R}^n \to \mathbb{R}^m$ is completely determined by its effect on the columns of the $n \times n$ identity matrix.

b. If $T : \mathbb{R}^2 \to \mathbb{R}^2$ rotates vectors about the origin through an angle φ, then T is a linear transformation.

c. When two linear transformations are performed one after another, the combined effect may not always be a linear transformation.

d. A mapping $T : \mathbb{R}^n \to \mathbb{R}^m$ is onto $\mathbb{R}^m$ if every vector $\mathbf{x}$ in $\mathbb{R}^n$ maps onto some vector in $\mathbb{R}^m$.

e. If A is a 3×2 matrix, then the transformation $\mathbf{x} \mapsto A\mathbf{x}$ cannot be one-to-one.

24. a. If A is a 4×3 matrix, then the transformation $\mathbf{x} \mapsto A\mathbf{x}$ maps $\mathbb{R}^3$ onto $\mathbb{R}^4$.

b. Every linear transformation from $\mathbb{R}^n$ to $\mathbb{R}^m$ is a matrix transformation.

c. The columns of the standard matrix for a linear transformation from $\mathbb{R}^n$ to $\mathbb{R}^m$ are the images of the columns of the $n \times n$ identity matrix under T.

d. A mapping $T : \mathbb{R}^n \to \mathbb{R}^m$ is one-to-one if each vector in $\mathbb{R}^n$ maps onto a unique vector in $\mathbb{R}^m$.

e. The standard matrix of a horizontal shear transformation from $\mathbb{R}^2$ to $\mathbb{R}^2$ has the form $\begin{bmatrix} a & 0 \\ 0 & d \end{bmatrix}$, where a and d are ± 1.

In Exercises 25–28, determine if the specified linear transformation is (a) one-to-one and (b) onto. Justify each answer.

25. The transformation in Exercise 17

26. The transformation in Exercise 2

27. The transformation in Exercise 19

28. The transformation in Exercise 14

In Exercises 29 and 30, describe the possible echelon forms of the standard matrix for a linear transformation T. Use the notation of Example 1 in Section 1.2.

29. $T : \mathbb{R}^3 \to \mathbb{R}^4$ is one-to-one.   **30.** $T : \mathbb{R}^4 \to \mathbb{R}^3$ is onto.

31. Let $T : \mathbb{R}^n \to \mathbb{R}^m$ be a linear transformation, with A its standard matrix. Complete the following statement to make it true: "T is one-to-one if and only if A has ____ pivot columns." Explain why the statement is true. [*Hint:* Look in the exercises for Section 1.7.]

32. Let $T : \mathbb{R}^n \to \mathbb{R}^m$ be a linear transformation, with A its standard matrix. Complete the following statement to make it true: "T maps $\mathbb{R}^n$ onto $\mathbb{R}^m$ if and only if A has ____ pivot columns." Find some theorems that explain why the statement is true.

33. Verify the uniqueness of A in Theorem 10. Let $T : \mathbb{R}^n \to \mathbb{R}^m$ be a linear transformation such that $T(\mathbf{x}) = B\mathbf{x}$ for some $m \times n$ matrix B. Show that if A is the standard matrix for T, then $A = B$. [*Hint:* Show that A and B have the same columns.]

34. Let $S : \mathbb{R}^p \to \mathbb{R}^n$ and $T : \mathbb{R}^n \to \mathbb{R}^m$ be linear transformations. Show that the mapping $\mathbf{x} \mapsto T(S(\mathbf{x}))$ is a linear transformation (from $\mathbb{R}^p$ to $\mathbb{R}^m$). [*Hint:* Compute $T(S(c\mathbf{u} + d\mathbf{v}))$ for $\mathbf{u}, \mathbf{v}$ in $\mathbb{R}^p$ and scalars c and d. Justify each step of the computation, and explain why this computation gives the desired conclusion.]

35. If a linear transformation $T : \mathbb{R}^n \to \mathbb{R}^m$ maps $\mathbb{R}^n$ *onto* $\mathbb{R}^m$, can you give a relation between m and n? If T is one-to-one, what can you say about m and n?

36. Why is the question "Is the linear transformation T onto?" an existence question?

[M] In Exercises 37–40, let T be the linear transformation whose standard matrix is given. In Exercises 37 and 38, decide if T is a one-to-one mapping. In Exercises 39 and 40, decide if T maps $\mathbb{R}^5$ onto $\mathbb{R}^5$. Justify your answers.

37. $\begin{bmatrix} -5 & 6 & -5 & -6 \\ 8 & 3 & -3 & 8 \\ 2 & 9 & 5 & -12 \\ -3 & 2 & 7 & -12 \end{bmatrix}$ **38.** $\begin{bmatrix} 7 & 5 & 9 & -9 \\ 5 & 6 & 4 & -4 \\ 4 & 8 & 0 & 7 \\ -6 & -6 & 6 & 5 \end{bmatrix}$

39. $\begin{bmatrix} 4 & -7 & 3 & 7 & 5 \\ 6 & -8 & 5 & 12 & -8 \\ -7 & 10 & -8 & -9 & 14 \\ 3 & -5 & 4 & 2 & -6 \\ -5 & 6 & -6 & -7 & 3 \end{bmatrix}$

40. $\begin{bmatrix} 9 & 43 & 5 & 6 & -1 \\ 14 & 15 & -7 & -5 & 4 \\ -8 & -6 & 12 & -5 & -9 \\ -5 & -6 & -4 & 9 & 8 \\ 13 & 14 & 15 & 3 & 11 \end{bmatrix}$

SOLUTION TO PRACTICE PROBLEM

WEB

Follow what happens to $\mathbf{e}_1$ and $\mathbf{e}_2$. See Fig. 5. First, $\mathbf{e}_1$ is unaffected by the shear and then is reflected into $-\mathbf{e}_1$. So $T(\mathbf{e}_1) = -\mathbf{e}_1$. Second, $\mathbf{e}_2$ goes to $\mathbf{e}_2 - .5\mathbf{e}_1$ by the shear

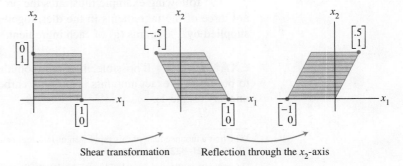

Shear transformation   Reflection through the x_2-axis

FIGURE 5 The composition of two transformations.

transformation. Since reflection through the x_2-axis changes $\mathbf{e}_1$ into $-\mathbf{e}_1$ and leaves $\mathbf{e}_2$ unchanged, the vector $\mathbf{e}_2 - .5\mathbf{e}_1$ goes to $\mathbf{e}_2 + .5\mathbf{e}_1$. So $T(\mathbf{e}_2) = \mathbf{e}_2 + .5\mathbf{e}_1$. Thus the standard matrix of T is

$$\begin{bmatrix} T(\mathbf{e}_1) & T(\mathbf{e}_2) \end{bmatrix} = \begin{bmatrix} -\mathbf{e}_1 & \mathbf{e}_2 + .5\mathbf{e}_1 \end{bmatrix} = \begin{bmatrix} -1 & .5 \\ 0 & 1 \end{bmatrix}$$

1.10 | LINEAR MODELS IN BUSINESS, SCIENCE, AND ENGINEERING

The mathematical models in this section are all *linear*; that is, each describes a problem by means of a linear equation, usually in vector or matrix form. The first model concerns nutrition but actually is representative of a general technique in linear programming problems. The second model comes from electrical engineering. The third model introduces the concept of a *linear difference equation*, a powerful mathematical tool for studying dynamic processes in a wide variety of fields such as engineering, ecology, economics, telecommunications, and the management sciences. Linear models are important because natural phenomena are often linear or nearly linear when the variables involved are held within reasonable bounds. Also, linear models are more easily adapted for computer calculation than are complex nonlinear models.

As you read about each model, pay attention to how its linearity reflects some property of the system being modeled.

Constructing a Nutritious Weight-Loss Diet

WEB

The formula for the Cambridge Diet, a popular diet in the 1980s, was based on years of research. A team of scientists headed by Dr. Alan H. Howard developed this diet at Cambridge University after more than eight years of clinical work with obese patients.[1] The very low-calorie powdered formula diet combines a precise balance of carbohydrate, high-quality protein, and fat, together with vitamins, minerals, trace elements, and electrolytes. Millions of persons have used the diet to achieve rapid and substantial weight loss.

To achieve the desired amounts and proportions of nutrients, Dr. Howard had to incorporate a large variety of foodstuffs in the diet. Each foodstuff supplied several of the required ingredients, but not in the correct proportions. For instance, nonfat milk was a major source of protein but contained too much calcium. So soy flour was used for part of the protein because soy flour contains little calcium. However, soy flour contains proportionally too much fat, so whey was added since it supplies less fat in relation to calcium. Unfortunately, whey contains too much carbohydrate. . . .

The following example illustrates the problem on a small scale. Listed in Table 1 are three of the ingredients in the diet, together with the amounts of certain nutrients supplied by 100 grams (g) of each ingredient.[2]

EXAMPLE 1 If possible, find some combination of nonfat milk, soy flour, and whey to provide the exact amounts of protein, carbohydrate, and fat supplied by the diet in one day (Table 1).

---

[1] The first announcement of this rapid weight-loss regimen was given in the *International Journal of Obesity* (1978) **2**, 321–332.

[2] Ingredients in the diet as of 1984; nutrient data for ingredients adapted from USDA Agricultural Handbooks No. 8-1 and 8-6, 1976.

TABLE 1

Amounts (g) Supplied per 100 g of Ingredient			Amounts (g) Supplied by Cambridge Diet in One Day
Nutrient	**Nonfat milk**	**Soy flour**	**Whey**	
Protein	36	51	13	33
Carbohydrate	52	34	74	45
Fat	0	7	1.1	3

SOLUTION Let x_1, x_2, and x_3, respectively, denote the number of units (100 g) of these foodstuffs. One approach to the problem is to derive equations for each nutrient separately. For instance, the product

$$\left\{ \begin{matrix} x_1 \text{ units of} \\ \text{nonfat milk} \end{matrix} \right\} \cdot \left\{ \begin{matrix} \text{protein per unit} \\ \text{of nonfat milk} \end{matrix} \right\}$$

gives the amount of protein supplied by x_1 units of nonfat milk. To this amount, we would then add similar products for soy flour and whey and set the resulting sum equal to the amount of protein we need. Analogous calculations would have to be made for each nutrient.

A more efficient method, and one that is conceptually simpler, is to consider a "nutrient vector" for each foodstuff and build just one vector equation. The amount of nutrients supplied by x_1 units of nonfat milk is the scalar multiple

$$\overset{\text{Scalar}}{\left\{ \begin{matrix} x_1 \text{ units of} \\ \text{nonfat milk} \end{matrix} \right\}} \cdot \overset{\text{Vector}}{\left\{ \begin{matrix} \text{nutrients per unit} \\ \text{of nonfat milk} \end{matrix} \right\}} = x_1 \mathbf{a}_1 \qquad (1)$$

where $\mathbf{a}_1$ is the first column in Table 1. Let $\mathbf{a}_2$ and $\mathbf{a}_3$ be the corresponding vectors for soy flour and whey, respectively, and let $\mathbf{b}$ be the vector that lists the total nutrients required (the last column of the table). Then $x_2 \mathbf{a}_2$ and $x_3 \mathbf{a}_3$ give the nutrients supplied by x_2 units of soy flour and x_3 units of whey, respectively. So the relevant equation is

$$x_1 \mathbf{a}_1 + x_2 \mathbf{a}_2 + x_3 \mathbf{a}_3 = \mathbf{b} \qquad (2)$$

Row reduction of the augmented matrix for the corresponding system of equations shows that

$$\begin{bmatrix} 36 & 51 & 13 & 33 \\ 52 & 34 & 74 & 45 \\ 0 & 7 & 1.1 & 3 \end{bmatrix} \sim \cdots \sim \begin{bmatrix} 1 & 0 & 0 & .277 \\ 0 & 1 & 0 & .392 \\ 0 & 0 & 1 & .233 \end{bmatrix}$$

To three significant digits, the diet requires .277 units of nonfat milk, .392 units of soy flour, and .233 units of whey in order to provide the desired amounts of protein, carbohydrate, and fat. ∎

It is important that the values of x_1, x_2, and x_3 found above are nonnegative. This is necessary for the solution to be physically feasible. (How could you use $-.233$ units of whey, for instance?) With a large number of nutrient requirements, it may be necessary to use a larger number of foodstuffs in order to produce a system of equations with a "nonnegative" solution. Thus many, many different combinations of foodstuffs may need to be examined in order to find a system of equations with such a solution. In fact, the manufacturer of the Cambridge Diet was able to supply 31 nutrients in precise amounts using only 33 ingredients.

The diet construction problem leads to the *linear* equation (2) because the amount of nutrients supplied by each foodstuff can be written as a scalar multiple of a vector, as in (1). That is, the nutrients supplied by a foodstuff are *proportional* to the amount of the foodstuff added to the diet mixture. Also, each nutrient in the mixture is the *sum* of the amounts from the various foodstuffs.

Problems of formulating specialized diets for humans and livestock occur frequently. Usually they are treated by linear programming techniques. Our method of constructing vector equations often simplifies the task of formulating such problems.

Linear Equations and Electrical Networks

WEB

Current flow in a simple electrical network can be described by a system of linear equations. A voltage source such as a battery forces a current of electrons to flow through the network. When the current passes through a resistor (such as a lightbulb or motor), some of the voltage is "used up"; by Ohm's law, this "voltage drop" across a resistor is given by

$$V = RI$$

where the voltage V is measured in *volts*, the resistance R in *ohms* (denoted by Ω), and the current flow I in *amperes* (*amps*, for short).

The network in Fig. 1 contains three closed loops. The currents flowing in loops 1, 2, and 3 are denoted by I_1, I_2, and I_3, respectively. The designated directions of such *loop currents* are arbitrary. If a current turns out to be negative, then the actual direction of current flow is opposite to that chosen in the figure. If the current direction shown is away from the positive (longer) side of a battery (⊣⊢) around to the negative (shorter) side, the voltage is positive; otherwise, the voltage is negative.

Current flow in a loop is governed by the following rule.

KIRCHHOFF'S VOLTAGE LAW

The algebraic sum of the RI voltage drops in one direction around a loop equals the algebraic sum of the voltage sources in the same direction around the loop.

EXAMPLE 2   Determine the loop currents in the network in Fig. 1.

SOLUTION For loop 1, the current I_1 flows through three resistors, and the sum of the RI voltage drops is

$$4I_1 + 4I_1 + 3I_1 = (4 + 4 + 3)I_1 = 11I_1$$

Current from loop 2 also flows in part of loop 1, through the short *branch* between A and B. The associated RI drop there is $3I_2$ volts. However, the current direction for the branch AB in loop 1 is opposite to that chosen for the flow in loop 2, so the algebraic sum of all RI drops for loop 1 is $11I_1 - 3I_2$. Since the voltage in loop 1 is $+30$ volts, Kirchhoff's voltage law implies that

$$11I_1 - 3I_2 = 30$$

The equation for loop 2 is

$$-3I_1 + 6I_2 - I_3 = 5$$

The term $-3I_1$ comes from the flow of the loop-1 current through the branch AB (with a negative voltage drop because the current flow there is opposite to the flow in loop 2). The term $6I_2$ is the sum of all resistances in loop 2, multiplied by the loop current. The term $-I_3 = -1 \cdot I_3$ comes from the loop-3 current flowing through the 1-ohm resistor in branch CD, in the direction opposite to the flow in loop 2. The loop-3 equation is

$$-I_2 + 3I_3 = -25$$

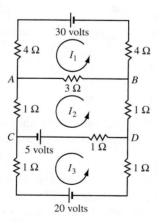

FIGURE 1

Note that the 5-volt battery in branch CD is counted as part of both loop 2 and loop 3, but it is -5 volts for loop 3 because of the direction chosen for the current in loop 3. The 20-volt battery is negative for the same reason.

The loop currents are found by solving the system

$$
\begin{aligned}
11I_1 - 3I_2 \quad &= \quad 30 \\
-3I_1 + 6I_2 - I_3 &= \quad 5 \\
- I_2 + 3I_3 &= -25
\end{aligned}
\tag{3}
$$

Row operations on the augmented matrix lead to the solution: $I_1 = 3$ amps, $I_2 = 1$ amp, and $I_3 = -8$ amps. The negative value of I_3 indicates that the actual current in loop 3 flows in the direction opposite to that shown in Fig. 1. ∎

It is instructive to look at system (3) as a vector equation:

$$
I_1 \begin{bmatrix} 11 \\ -3 \\ 0 \end{bmatrix} + I_2 \begin{bmatrix} -3 \\ 6 \\ -1 \end{bmatrix} + I_3 \begin{bmatrix} 0 \\ -1 \\ 3 \end{bmatrix} = \begin{bmatrix} 30 \\ 5 \\ -25 \end{bmatrix}
\tag{4}
$$
$$
\quad\ \uparrow\qquad\quad\ \uparrow\qquad\quad\ \uparrow\qquad\quad\ \uparrow
$$
$$
\quad\ \mathbf{r}_1\qquad\quad\ \mathbf{r}_2\qquad\quad\ \mathbf{r}_3\qquad\quad\ \mathbf{v}
$$

The first entry of each vector concerns the first loop, and similarly for the second and third entries. The first resistor vector $\mathbf{r}_1$ lists the resistance in the various loops through which current I_1 flows. A resistance is written negatively when I_1 flows against the flow direction in another loop. Examine Fig. 1 and see how to compute the entries in $\mathbf{r}_1$; then do the same for $\mathbf{r}_2$ and $\mathbf{r}_3$. The matrix form of equation (4),

$$
R\mathbf{i} = \mathbf{v}, \quad \text{where} \quad R = [\,\mathbf{r}_1 \quad \mathbf{r}_2 \quad \mathbf{r}_3\,] \quad \text{and} \quad \mathbf{i} = \begin{bmatrix} I_1 \\ I_2 \\ I_3 \end{bmatrix}
$$

provides a matrix version of Ohm's law. If all loop currents are chosen in the same direction (say, counterclockwise), then all entries off the main diagonal of R will be negative.

The matrix equation $R\mathbf{i} = \mathbf{v}$ makes the linearity of this model easy to see at a glance. For instance, if the voltage vector is doubled, then the current vector must double. Also, a *superposition principle* holds. That is, the solution of equation (4) is the sum of the solutions of the equations

$$
R\mathbf{i} = \begin{bmatrix} 30 \\ 0 \\ 0 \end{bmatrix}, \quad R\mathbf{i} = \begin{bmatrix} 0 \\ 5 \\ 0 \end{bmatrix}, \quad \text{and} \quad R\mathbf{i} = \begin{bmatrix} 0 \\ 0 \\ -25 \end{bmatrix}
$$

Each equation here corresponds to the circuit with only one voltage source (the other sources being replaced by wires that close each loop). The model for current flow is *linear* precisely because Ohm's law and Kirchhoff's law are linear: The voltage drop across a resistor is *proportional* to the current flowing through it (Ohm), and the *sum* of the voltage drops in a loop equals the sum of the voltage sources in the loop (Kirchhoff).

Loop currents in a network can be used to determine the current in any branch of the network. If only one loop current passes through a branch, such as from B to D in Fig. 1, the branch current equals the loop current. If more than one loop current passes through a branch, such as from A to B, the branch current is the algebraic sum of the loop currents in the branch (*Kirchhoff's current law*). For instance, the current in branch AB is $I_1 - I_2 = 3 - 1 = 2$ amps, in the direction of I_1. The current in branch CD is $I_2 - I_3 = 9$ amps.

Difference Equations

In many fields such as ecology, economics, and engineering, a need arises to model mathematically a dynamic system that changes over time. Several features of the system are each measured at discrete time intervals, producing a sequence of vectors $\mathbf{x}_0, \mathbf{x}_1, \mathbf{x}_2, \ldots$. The entries in $\mathbf{x}_k$ provide information about the *state* of the system at the time of the kth measurement.

If there is a matrix A such that $\mathbf{x}_1 = A\mathbf{x}_0$, $\mathbf{x}_2 = A\mathbf{x}_1$, and, in general,

$$\mathbf{x}_{k+1} = A\mathbf{x}_k \quad \text{for } k = 0, 1, 2, \ldots \tag{5}$$

then (5) is called a **linear difference equation** (or **recurrence relation**). Given such an equation, one can compute $\mathbf{x}_1$, $\mathbf{x}_2$, and so on, provided $\mathbf{x}_0$ is known. Sections 4.8 and 4.9, and several sections in Chapter 5, will develop formulas for $\mathbf{x}_k$ and describe what can happen to $\mathbf{x}_k$ as k increases indefinitely. The discussion below illustrates how a difference equation might arise.

A subject of interest to demographers is the movement of populations or groups of people from one region to another. The simple model here considers the changes in the population of a certain city and its surrounding suburbs over a period of years.

Fix an initial year—say, 2000—and denote the populations of the city and suburbs that year by r_0 and s_0, respectively. Let $\mathbf{x}_0$ be the population vector

$$\mathbf{x}_0 = \begin{bmatrix} r_0 \\ s_0 \end{bmatrix} \quad \begin{array}{l} \text{City population, 2000} \\ \text{Suburban population, 2000} \end{array}$$

For 2001 and subsequent years, denote the populations of the city and suburbs by the vectors

$$\mathbf{x}_1 = \begin{bmatrix} r_1 \\ s_1 \end{bmatrix}, \qquad \mathbf{x}_2 = \begin{bmatrix} r_2 \\ s_2 \end{bmatrix}, \qquad \mathbf{x}_3 = \begin{bmatrix} r_3 \\ s_3 \end{bmatrix}, \ldots$$

Our goal is to describe mathematically how these vectors might be related.

Suppose demographic studies show that each year about 5% of the city's population moves to the suburbs (and 95% remains in the city), while 3% of the suburban population moves to the city (and 97% remains in the suburbs). See Fig. 2.

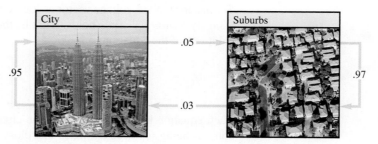

FIGURE 2 Annual percentage migration between city and suburbs.

After 1 year, the original r_0 persons in the city are now distributed between city and suburbs as

$$\begin{bmatrix} .95r_0 \\ .05r_0 \end{bmatrix} = r_0 \begin{bmatrix} .95 \\ .05 \end{bmatrix} \quad \begin{array}{l} \text{Remain in city} \\ \text{Move to suburbs} \end{array} \tag{6}$$

The s_0 persons in the suburbs in 2000 are distributed 1 year later as

$$s_0 \begin{bmatrix} .03 \\ .97 \end{bmatrix} \quad \begin{array}{l} \text{Move to city} \\ \text{Remain in suburbs} \end{array} \tag{7}$$

The vectors in (6) and (7) account for all of the population in 2001.[3] Thus

$$\begin{bmatrix} r_1 \\ s_1 \end{bmatrix} = r_0 \begin{bmatrix} .95 \\ .05 \end{bmatrix} + s_0 \begin{bmatrix} .03 \\ .97 \end{bmatrix} = \begin{bmatrix} .95 & .03 \\ .05 & .97 \end{bmatrix} \begin{bmatrix} r_0 \\ s_0 \end{bmatrix}$$

That is,

$$\mathbf{x}_1 = M\mathbf{x}_0 \tag{8}$$

where M is the **migration matrix** determined by the following table:

$$\begin{array}{cc} & \text{From:} \\ & \text{City} \quad \text{Suburbs} \quad \text{To:} \\ \begin{bmatrix} .95 & .03 \\ .05 & .97 \end{bmatrix} & \begin{array}{l} \text{City} \\ \text{Suburbs} \end{array} \end{array}$$

Equation (8) describes how the population changes from 2000 to 2001. If the migration percentages remain constant, then the change from 2001 to 2002 is given by

$$\mathbf{x}_2 = M\mathbf{x}_1$$

and similarly for 2002 to 2003 and subsequent years. In general,

$$\mathbf{x}_{k+1} = M\mathbf{x}_k \quad \text{for } k = 0, 1, 2, \dots \tag{9}$$

The sequence of vectors $\{\mathbf{x}_0, \mathbf{x}_1, \mathbf{x}_2, \dots\}$ describes the population of the city/suburban region over a period of years.

EXAMPLE 3 Compute the population of the region just described for the years 2001 and 2002, given that the population in 2000 was 600,000 in the city and 400,000 in the suburbs.

SOLUTION The initial population in 2000 is $\mathbf{x}_0 = \begin{bmatrix} 600,000 \\ 400,000 \end{bmatrix}$. For 2001,

$$\mathbf{x}_1 = \begin{bmatrix} .95 & .03 \\ .05 & .97 \end{bmatrix} \begin{bmatrix} 600,000 \\ 400,000 \end{bmatrix} = \begin{bmatrix} 582,000 \\ 418,000 \end{bmatrix}$$

For 2002,

$$\mathbf{x}_2 = M\mathbf{x}_1 = \begin{bmatrix} .95 & .03 \\ .05 & .97 \end{bmatrix} \begin{bmatrix} 582,000 \\ 418,000 \end{bmatrix} = \begin{bmatrix} 565,440 \\ 434,560 \end{bmatrix} \quad \blacksquare$$

The model for population movement in (9) is *linear* because the correspondence $\mathbf{x}_k \mapsto \mathbf{x}_{k+1}$ is a linear transformation. The linearity depends on two facts: the number of people who chose to move from one area to another is *proportional* to the number of people in that area, as shown in (6) and (7), and the cumulative effect of these choices is found by *adding* the movement of people from the different areas.

PRACTICE PROBLEM

Find a matrix A and vectors $\mathbf{x}$ and $\mathbf{b}$ such that the problem in Example 1 amounts to solving the equation $A\mathbf{x} = \mathbf{b}$.

---

[3] For simplicity, we ignore other influences on the population such as births, deaths, and migration into and out of the city/suburban region.

1.10 EXERCISES

1. The container of a breakfast cereal usually lists the number of calories and the amounts of protein, carbohydrate, and fat contained in one serving of the cereal. The amounts for two common cereals are given below. Suppose a mixture of these two cereals is to be prepared that contains exactly 295 calories, 9 g of protein, 48 g of carbohydrate, and 8 g of fat.

   a. Set up a vector equation for this problem. Include a statement of what the variables in your equation represent.

   b. Write an equivalent matrix equation, and then determine if the desired mixture of the two cereals can be prepared.

Nutrition Information per Serving

Nutrient	General Mills Cheerios®	Quaker® 100% Natural Cereal
Calories	110	130
Protein (g)	4	3
Carbohydrate (g)	20	18
Fat (g)	2	5

2. One serving of Shredded Wheat supplies 160 calories, 5 g of protein, 6 g of fiber, and 1 g of fat. One serving of Crispix® supplies 110 calories, 2 g of protein, .1 g of fiber, and .4 g of fat.

   a. Set up a matrix B and a vector $\mathbf{u}$ such that $B\mathbf{u}$ gives the amounts of calories, protein, fiber, and fat contained in a mixture of three servings of Shredded Wheat and two servings of Crispix.

   b. [M] Suppose that you want a cereal with more fiber than Crispix but fewer calories than Shredded Wheat. Is it possible for a mixture of the two cereals to supply 130 calories, 3.20 g of protein, 2.46 g of fiber, and .64 g of fat? If so, what is the mixture?

3. After taking a nutrition class, a big Annie's® Mac and Cheese fan decides to improve the levels of protein and fiber in her favorite lunch by adding broccoli and canned chicken. The nutritional information for the foods referred to in this exercise are given in the table below.

Nutrition Information per Serving

Nutrient	Mac and Cheese	Broccoli	Chicken	Shells
Calories	270	51	70	260
Protein (g)	10	5.4	15	9
Fiber (g)	2	5.2	0	5

   a. [M] If she wants to limit her lunch to 400 calories but get 30 g of protein and 10 g of fiber, what proportions of servings of Mac and Cheese, broccoli, and chicken should she use?

   b. [M] She found that there was too much broccoli in the proportions from part (a), so she decided to switch from

classical Mac and Cheese to Annie's® Whole Wheat Shells and White Cheddar. What proportions of servings of each food should she use to meet the same goals as in part (a)?

4. The Cambridge Diet supplies .8 g of calcium per day, in addition to the nutrients listed in the Table 1 for Example 1. The amounts of calcium per unit (100 g) supplied by the three ingredients in the Cambridge Diet are as follows: 1.26 g from nonfat milk, .19 g from soy flour, and .8 g from whey. Another ingredient in the diet mixture is isolated soy protein, which provides the following nutrients in each unit: 80 g of protein, 0 g of carbohydrate, 3.4 g of fat, and .18 g of calcium.

   a. Set up a matrix equation whose solution determines the amounts of nonfat milk, soy flour, whey, and isolated soy protein necessary to supply the precise amounts of protein, carbohydrate, fat, and calcium in the Cambridge Diet. State what the variables in the equation represent.

   b. [M] Solve the equation in (a) and discuss your answer.

In Exercises 5–8, write a matrix equation that determines the loop currents. [M] If MATLAB or another matrix program is available, solve the system for the loop currents.

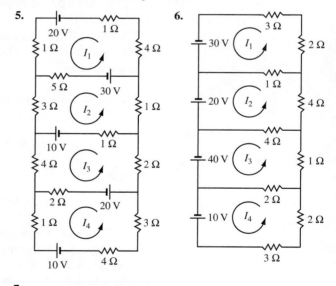

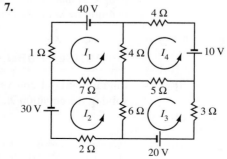

8.

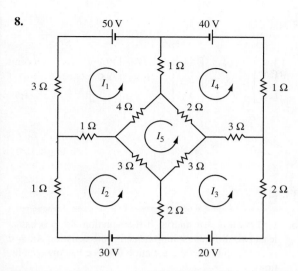

9. In a certain region, about 7% of a city's population moves to the surrounding suburbs each year, and about 5% of the suburban population moves into the city. In 2010, there were 800,000 residents in the city and 500,000 in the suburbs. Set up a difference equation that describes this situation, where x_0 is the initial population in 2010. Then estimate the populations in the city and in the suburbs two years later, in 2012. (Ignore other factors that might influence the population sizes.)

10. In a certain region, about 6% of a city's population moves to the surrounding suburbs each year, and about 4% of the suburban population moves into the city. In 2010, there were 10,000,000 residents in the city and 800,000 in the suburbs. Set up a difference equation that describes this situation, where x_0 is the initial population in 2010. Then estimate the populations in the city and in the suburbs two years later, in 2012.

11. In 1994, the population of California was 31,524,000, and the population living in the United States but *outside* California was 228,680,000. During the year, it is estimated that 516,100 persons moved from California to elsewhere in the United States, while 381,262 persons moved to California from elsewhere in the United States.[4]

a. Set up the migration matrix for this situation, using five decimal places for the migration rates into and out of California. Let your work show how you produced the migration matrix.

b. [M] Compute the projected populations in the year 2000 for California and elsewhere in the United States, assuming that the migration rates did not change during the 6-year period. (These calculations do not take into account births, deaths, or the substantial migration of persons into California and elsewhere in the United States from other countries.)

---

[4] Migration data supplied by the Demographic Research Unit of the California State Department of Finance.

12. [M] Budget® Rent A Car in Wichita, Kansas has a fleet of about 500 cars, at three locations. A car rented at one location may be returned to any of the three locations. The various fractions of cars returned to the three locations are shown in the matrix below. Suppose that on Monday there are 295 cars at the airport (or rented from there), 55 cars at the east side office, and 150 cars at the west side office. What will be the approximate distribution of cars on Wednesday?

Cars Rented From:

Airport	East	West	Returned To:
.97	.05	.10	Airport
.00	.90	.05	East
.03	.05	.85	West

13. [M] Let M and x_0 be as in Example 3.

a. Compute the population vectors x_k for $k = 1, \ldots, 20$. Discuss what you find.

b. Repeat part (a) with an initial population of 350,000 in the city and 650,000 in the suburbs. What do you find?

14. [M] Study how changes in boundary temperatures on a steel plate affect the temperatures at interior points on the plate.

a. Begin by estimating the temperatures T_1, T_2, T_3, T_4 at each of the sets of four points on the steel plate shown in the figure. In each case, the value of T_k is approximated by the average of the temperatures at the four closest points. See Exercises 33 and 34 in Section 1.1, where the values (in degrees) turn out to be $(20, 27.5, 30, 22.5)$. How is this list of values related to your results for the points in set (a) and set (b)?

b. Without making any computations, guess the interior temperatures in (a) when the boundary temperatures are all multiplied by 3. Check your guess.

c. Finally, make a general conjecture about the correspondence from the list of eight boundary temperatures to the list of four interior temperatures.

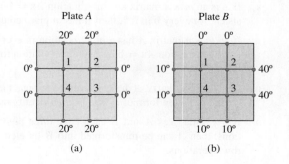

(a)   (b)

SOLUTION TO PRACTICE PROBLEM

$$A = \begin{bmatrix} 36 & 51 & 13 \\ 52 & 34 & 74 \\ 0 & 7 & 1.1 \end{bmatrix}, \quad \mathbf{x} = \begin{bmatrix} x_1 \\ x_2 \\ x_3 \end{bmatrix}, \quad \mathbf{b} = \begin{bmatrix} 33 \\ 45 \\ 3 \end{bmatrix}$$

CHAPTER 1 SUPPLEMENTARY EXERCISES

1. Mark each statement True or False. Justify each answer. (If true, cite appropriate facts or theorems. If false, explain why or give a counterexample that shows why the statement is not true in every case.

a. Every matrix is row equivalent to a unique matrix in echelon form.

b. Any system of n linear equations in n variables has at most n solutions.

c. If a system of linear equations has two different solutions, it must have infinitely many solutions.

d. If a system of linear equations has no free variables, then it has a unique solution.

e. If an augmented matrix $[\,A \quad \mathbf{b}\,]$ is transformed into $[\,C \quad \mathbf{d}\,]$ by elementary row operations, then the equations $A\mathbf{x} = \mathbf{b}$ and $C\mathbf{x} = \mathbf{d}$ have exactly the same solution sets.

f. If a system $A\mathbf{x} = \mathbf{b}$ has more than one solution, then so does the system $A\mathbf{x} = \mathbf{0}$.

g. If A is an $m \times n$ matrix and the equation $A\mathbf{x} = \mathbf{b}$ is consistent for some $\mathbf{b}$, then the columns of A span $\mathbb{R}^m$.

h. If an augmented matrix $[\,A \quad \mathbf{b}\,]$ can be transformed by elementary row operations into reduced echelon form, then the equation $A\mathbf{x} = \mathbf{b}$ is consistent.

i. If matrices A and B are row equivalent, they have the same reduced echelon form.

j. The equation $A\mathbf{x} = \mathbf{0}$ has the trivial solution if and only if there are no free variables.

k. If A is an $m \times n$ matrix and the equation $A\mathbf{x} = \mathbf{b}$ is consistent for every $\mathbf{b}$ in $\mathbb{R}^m$, then A has m pivot columns.

l. If an $m \times n$ matrix A has a pivot position in every row, then the equation $A\mathbf{x} = \mathbf{b}$ has a unique solution for each $\mathbf{b}$ in $\mathbb{R}^m$.

m. If an $n \times n$ matrix A has n pivot positions, then the reduced echelon form of A is the $n \times n$ identity matrix.

n. If 3×3 matrices A and B each have three pivot positions, then A can be transformed into B by elementary row operations.

o. If A is an $m \times n$ matrix, if the equation $A\mathbf{x} = \mathbf{b}$ has at least two different solutions, and if the equation $A\mathbf{x} = \mathbf{c}$ is consistent, then the equation $A\mathbf{x} = \mathbf{c}$ has many solutions.

p. If A and B are row equivalent $m \times n$ matrices and if the columns of A span $\mathbb{R}^m$, then so do the columns of B.

q. If none of the vectors in the set $S = \{\mathbf{v}_1, \mathbf{v}_2, \mathbf{v}_3\}$ in $\mathbb{R}^3$ is a multiple of one of the other vectors, then S is linearly independent.

r. If $\{\mathbf{u}, \mathbf{v}, \mathbf{w}\}$ is linearly independent, then $\mathbf{u}$, $\mathbf{v}$, and $\mathbf{w}$ are not in $\mathbb{R}^2$.

s. In some cases, it is possible for four vectors to span $\mathbb{R}^5$.

t. If $\mathbf{u}$ and $\mathbf{v}$ are in $\mathbb{R}^m$, then $-\mathbf{u}$ is in Span$\{\mathbf{u}, \mathbf{v}\}$.

u. If $\mathbf{u}$, $\mathbf{v}$, and $\mathbf{w}$ are nonzero vectors in $\mathbb{R}^2$, then $\mathbf{w}$ is a linear combination of $\mathbf{u}$ and $\mathbf{v}$.

v. If $\mathbf{w}$ is a linear combination of $\mathbf{u}$ and $\mathbf{v}$ in $\mathbb{R}^n$, then $\mathbf{u}$ is a linear combination of $\mathbf{v}$ and $\mathbf{w}$.

w. Suppose that $\mathbf{v}_1, \mathbf{v}_2$, and $\mathbf{v}_3$ are in $\mathbb{R}^5$, $\mathbf{v}_2$ is not a multiple of $\mathbf{v}_1$, and $\mathbf{v}_3$ is not a linear combination of $\mathbf{v}_1$ and $\mathbf{v}_2$. Then $\{\mathbf{v}_1, \mathbf{v}_2, \mathbf{v}_3\}$ is linearly independent.

x. A linear transformation is a function.

y. If A is a 6×5 matrix, the linear transformation $\mathbf{x} \mapsto A\mathbf{x}$ cannot map $\mathbb{R}^5$ onto $\mathbb{R}^6$.

z. If A is an $m \times n$ matrix with m pivot columns, then the linear transformation $\mathbf{x} \mapsto A\mathbf{x}$ is a one-to-one mapping.

2. Let a and b represent real numbers. Describe the possible solution sets of the (linear) equation $ax = b$. [*Hint:* The number of solutions depends upon a and b.]

3. The solutions (x, y, z) of a single linear equation

$$ax + by + cz = d$$

form a plane in $\mathbb{R}^3$ when a, b, and c are not all zero. Construct sets of three linear equations whose graphs (a) intersect in a single line, (b) intersect in a single point, and (c) have no points in common. Typical graphs are illustrated in the figure.

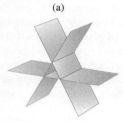

Three planes intersecting
in a line

(a)

Three planes intersecting
in a point

(b)

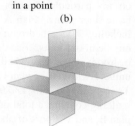

Three planes with no
intersection

(c)

Three planes with no
intersection

(c')

4. Suppose the coefficient matrix of a linear system of three equations in three variables has a pivot position in each column. Explain why the system has a unique solution.

5. Determine h and k such that the solution set of the system (i) is empty, (ii) contains a unique solution, and (iii) contains infinitely many solutions.

a. $x_1 + 3x_2 = k$
    $4x_1 + hx_2 = 8$

b. $-2x_1 + hx_2 = 1$
    $6x_1 + kx_2 = -2$

6. Consider the problem of determining whether the following system of equations is consistent:

$$4x_1 - 2x_2 + 7x_3 = -5$$
$$8x_1 - 3x_2 + 10x_3 = -3$$

a. Define appropriate vectors, and restate the problem in terms of linear combinations. Then solve that problem.

b. Define an appropriate matrix, and restate the problem using the phrase "columns of A."

c. Define an appropriate linear transformation T using the matrix in (b), and restate the problem in terms of T.

7. Consider the problem of determining whether the following system of equations is consistent for all b_1, b_2, b_3:

$$2x_1 - 4x_2 - 2x_3 = b_1$$
$$-5x_1 + x_2 + x_3 = b_2$$
$$7x_1 - 5x_2 - 3x_3 = b_3$$

a. Define appropriate vectors, and restate the problem in terms of Span $\{\mathbf{v}_1, \mathbf{v}_2, \mathbf{v}_3\}$. Then solve that problem.

b. Define an appropriate matrix, and restate the problem using the phrase "columns of A."

c. Define an appropriate linear transformation T using the matrix in (b), and restate the problem in terms of T.

8. Describe the possible echelon forms of the matrix A. Use the notation of Example 1 in Section 1.2.

a. A is a 2×3 matrix whose columns span $\mathbb{R}^2$.

b. A is a 3×3 matrix whose columns span $\mathbb{R}^3$.

9. Write the vector $\begin{bmatrix} 5 \\ 6 \end{bmatrix}$ as the sum of two vectors, one on the line $\{(x, y) : y = 2x\}$ and one on the line $\{(x, y) : y = x/2\}$.

10. Let $\mathbf{a}_1, \mathbf{a}_2$, and $\mathbf{b}$ be the vectors in $\mathbb{R}^2$ shown in the figure, and let $A = [\mathbf{a}_1 \quad \mathbf{a}_2]$. Does the equation $A\mathbf{x} = \mathbf{b}$ have a solution? If so, is the solution unique? Explain.

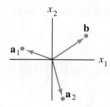

11. Construct a 2×3 matrix A, not in echelon form, such that the solution of $A\mathbf{x} = \mathbf{0}$ is a line in $\mathbb{R}^3$.

12. Construct a 2×3 matrix A, not in echelon form, such that the solution of $A\mathbf{x} = \mathbf{0}$ is a plane in $\mathbb{R}^3$.

13. Write the *reduced* echelon form of a 3×3 matrix A such that the first two columns of A are pivot columns and

$$A \begin{bmatrix} 3 \\ -2 \\ 1 \end{bmatrix} = \begin{bmatrix} 0 \\ 0 \\ 0 \end{bmatrix}.$$

14. Determine the value(s) of a such that $\left\{ \begin{bmatrix} 1 \\ a \end{bmatrix}, \begin{bmatrix} a \\ a+2 \end{bmatrix} \right\}$ is linearly independent.

15. In (a) and (b), suppose the vectors are linearly independent. What can you say about the numbers $a, \ldots, f$? Justify your answers. [*Hint:* Use a theorem for (b).]

a. $\begin{bmatrix} a \\ 0 \\ 0 \end{bmatrix}, \begin{bmatrix} b \\ c \\ 0 \end{bmatrix}, \begin{bmatrix} d \\ e \\ f \end{bmatrix}$

b. $\begin{bmatrix} a \\ 1 \\ 0 \\ 0 \end{bmatrix}, \begin{bmatrix} b \\ c \\ 1 \\ 0 \end{bmatrix}, \begin{bmatrix} d \\ e \\ f \\ 1 \end{bmatrix}$

16. Use Theorem 7 in Section 1.7 to explain why the columns of the matrix A are linearly independent.

$$A = \begin{bmatrix} 1 & 0 & 0 & 0 \\ 2 & 5 & 0 & 0 \\ 3 & 6 & 8 & 0 \\ 4 & 7 & 9 & 10 \end{bmatrix}$$

17. Explain why a set $\{\mathbf{v}_1, \mathbf{v}_2, \mathbf{v}_3, \mathbf{v}_4\}$ in $\mathbb{R}^5$ must be linearly independent when $\{\mathbf{v}_1, \mathbf{v}_2, \mathbf{v}_3\}$ is linearly independent and $\mathbf{v}_4$ is *not* in Span $\{\mathbf{v}_1, \mathbf{v}_2, \mathbf{v}_3\}$.

18. Suppose $\{\mathbf{v}_1, \mathbf{v}_2\}$ is a linearly independent set in $\mathbb{R}^n$. Show that $\{\mathbf{v}_1, \mathbf{v}_1 + \mathbf{v}_2\}$ is also linearly independent.

19. Suppose $\mathbf{v}_1, \mathbf{v}_2, \mathbf{v}_3$ are distinct points on one line in $\mathbb{R}^3$. The line need not pass through the origin. Show that $\{\mathbf{v}_1, \mathbf{v}_2, \mathbf{v}_3\}$ is linearly dependent.

20. Let $T : \mathbb{R}^n \to \mathbb{R}^m$ be a linear transformation, and suppose $T(\mathbf{u}) = \mathbf{v}$. Show that $T(-\mathbf{u}) = -\mathbf{v}$.

21. Let $T : \mathbb{R}^3 \to \mathbb{R}^3$ be the linear transformation that reflects each vector through the plane $x_2 = 0$. That is, $T(x_1, x_2, x_3) = (x_1, -x_2, x_3)$. Find the standard matrix of T.

22. Let A be a 3×3 matrix with the property that the linear transformation $\mathbf{x} \mapsto A\mathbf{x}$ maps $\mathbb{R}^3$ onto $\mathbb{R}^3$. Explain why the transformation must be one-to-one.

23. A *Givens rotation* is a linear transformation from $\mathbb{R}^n$ to $\mathbb{R}^n$ used in computer programs to create a zero entry in a vector (usually a column of a matrix). The standard matrix of a Givens rotation in $\mathbb{R}^2$ has the form

$$\begin{bmatrix} a & -b \\ b & a \end{bmatrix}, \qquad a^2 + b^2 = 1$$

Find a and b such that $\begin{bmatrix} 4 \\ 3 \end{bmatrix}$ is rotated into $\begin{bmatrix} 5 \\ 0 \end{bmatrix}$.

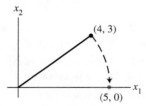

x_2

$(4, 3)$

$(5, 0)$

x_1

A Givens rotation in $\mathbb{R}^2$.

WEB

24. The following equation describes a Givens rotation in $\mathbb{R}^3$. Find a and b.

$$\begin{bmatrix} a & 0 & -b \\ 0 & 1 & 0 \\ b & 0 & a \end{bmatrix} \begin{bmatrix} 2 \\ 3 \\ 4 \end{bmatrix} = \begin{bmatrix} 2\sqrt{5} \\ 3 \\ 0 \end{bmatrix}, \qquad a^2 + b^2 = 1$$

25. A large apartment building is to be built using modular construction techniques. The arrangement of apartments on any particular floor is to be chosen from one of three basic floor plans. Plan A has 18 apartments on one floor, including 3 three-bedroom units, 7 two-bedroom units, and 8 one-bedroom units. Each floor of plan B includes 4 three-bedroom units, 4 two-bedroom units, and 8 one-bedroom units. Each floor of plan C includes 5 three-bedroom units, 3 two-bedroom units, and 9 one-bedroom units. Suppose the building contains a total of x_1 floors of plan A, x_2 floors of plan B, and x_3 floors of plan C.

a. What interpretation can be given to the vector $x_1 \begin{bmatrix} 3 \\ 7 \\ 8 \end{bmatrix}$?

b. Write a formal linear combination of vectors that expresses the total numbers of three-, two-, and one-bedroom apartments contained in the building.

c. **[M]** Is it possible to design the building with exactly 66 three-bedroom units, 74 two-bedroom units, and 136 one-bedroom units? If so, is there more than one way to do it? Explain your answer.

2 Matrix Algebra

Computer Models in Aircraft Design

To design the next generation of commercial and military aircraft, engineers at Boeing's Phantom Works use 3D modeling and computational fluid dynamics (CFD). They study the airflow around a virtual airplane to answer important design questions before physical models are created. This has drastically reduced design cycle times and cost—and linear algebra plays a crucial role in the process.

The virtual airplane begins as a mathematical "wire-frame" model that exists only in computer memory and on graphics display terminals. (A model of a Boeing 777 is shown.) This mathematical model organizes and influences each step of the design and manufacture of the airplane—both the exterior and interior. The CFD analysis concerns the exterior surface.

Although the finished skin of a plane may seem smooth, the geometry of the surface is complicated. In addition to wings and a fuselage, an aircraft has nacelles, stabilizers, slats, flaps, and ailerons. The way air flows around these structures determines how the plane moves through the sky. Equations that describe the airflow are complicated, and they must account for engine intake, engine exhaust, and the wakes left by the wings of the plane. To study the airflow, engineers need a highly refined description of the plane's surface.

A computer creates a model of the surface by first superimposing a three-dimensional grid of "boxes" on the original wire-frame model. Boxes in this grid lie either completely inside or completely outside the plane, or they intersect the surface of the plane. The computer selects the boxes that intersect the surface and subdivides them, retaining only the smaller boxes that still intersect the surface. The subdividing process is repeated until the grid is extremely fine. A typical grid can include over 400,000 boxes.

The process for finding the airflow around the plane involves repeatedly solving a system of linear equations $A\mathbf{x} = \mathbf{b}$ that may involve up to 2 million equations and variables. The vector $\mathbf{b}$ changes each time, based on data from the grid and solutions of previous equations. Using the fastest computers available commercially, a Phantom Works team can spend from a few hours to several days setting up and solving a single airflow problem. After the team analyzes the solution, they may make small changes to the airplane surface and begin the whole process again. Thousands of CFD runs may be required.

This chapter presents two important concepts that assist in the solution of such massive systems of equations:

- *Partitioned matrices:* A typical CFD system of equations has a "sparse" coefficient matrix with mostly zero entries. Grouping the variables correctly leads to a partitioned matrix with many zero blocks. Section 2.4 introduces such matrices and describes some of their applications.

- *Matrix factorizations:* Even when written with partitioned matrices, the system of equations is complicated. To further simplify the computations, the CFD software at Boeing uses what is called an LU factorization of the coefficient matrix. Section 2.5 discusses LU and other useful matrix factorizations. Further details about factorizations appear at several points later in the text.

To analyze a solution of an airflow system, engineers want to visualize the airflow over the surface of the plane. They use computer graphics, and linear algebra provides the engine for the graphics. The wire-frame model of the plane's surface is stored as data in many matrices. Once the image has been rendered on a computer screen, engineers can change its scale, zoom in or out of small regions, and rotate the image to see parts that may be hidden from view. Each of these operations is accomplished by appropriate

Modern CFD has revolutionized wing design. The Boeing Blended Wing Body is in design for the year 2020 or sooner.

matrix multiplications. Section 2.7 explains the basic ideas.

WEB

Our ability to analyze and solve equations will be greatly enhanced when we can perform algebraic operations with matrices. Furthermore, the definitions and theorems in this chapter provide some basic tools for handling the many applications of linear algebra that involve two or more matrices. For square matrices, the Invertible Matrix Theorem in Section 2.3 ties together most of the concepts treated earlier in the text. Sections 2.4 and 2.5 examine partitioned matrices and matrix factorizations, which appear in most modern uses of linear algebra. Sections 2.6 and 2.7 describe two interesting applications of matrix algebra, to economics and to computer graphics.

2.1 | MATRIX OPERATIONS

If A is an $m \times n$ matrix—that is, a matrix with m rows and n columns—then the scalar entry in the ith row and jth column of A is denoted by a_{ij} and is called the (i, j)-entry of A. See Fig. 1. For instance, the $(3, 2)$-entry is the number a_{32} in the third row, second column. Each column of A is a list of m real numbers, which identifies a vector in $\mathbb{R}^m$. Often, these columns are denoted by $\mathbf{a}_1, \ldots, \mathbf{a}_n$, and the matrix A is written as

$$A = \begin{bmatrix} \mathbf{a}_1 & \mathbf{a}_2 & \cdots & \mathbf{a}_n \end{bmatrix}$$

Observe that the number a_{ij} is the ith entry (from the top) of the jth column vector $\mathbf{a}_j$.

The **diagonal entries** in an $m \times n$ matrix $A = [a_{ij}]$ are $a_{11}, a_{22}, a_{33}, \ldots$, and they form the **main diagonal** of A. A **diagonal matrix** is a square $n \times n$ matrix whose nondiagonal entries are zero. An example is the $n \times n$ identity matrix, I_n. An $m \times n$ matrix whose entries are all zero is a **zero matrix** and is written as 0. The size of a zero matrix is usually clear from the context.

$$\begin{array}{c} \text{Column} \\ j \end{array}$$

$$\text{Row } i \begin{bmatrix} a_{11} & \cdots & a_{1j} & \cdots & a_{1n} \\ \vdots & & \vdots & & \vdots \\ a_{i1} & \cdots & a_{ij} & \cdots & a_{in} \\ \vdots & & \vdots & & \vdots \\ a_{m1} & \cdots & a_{mj} & \cdots & a_{mn} \end{bmatrix} = A$$

$$\qquad\qquad \uparrow \qquad\quad \uparrow \qquad\quad \uparrow$$
$$\qquad\qquad \mathbf{a}_1 \qquad\quad \mathbf{a}_j \qquad\quad \mathbf{a}_n$$

FIGURE 1 Matrix notation.

Sums and Scalar Multiples

The arithmetic for vectors described earlier has a natural extension to matrices. We say that two matrices are **equal** if they have the same size (i.e., the same number of rows and the same number of columns) and if their corresponding columns are equal, which amounts to saying that their corresponding entries are equal. If A and B are $m \times n$ matrices, then the **sum** $A + B$ is the $m \times n$ matrix whose columns are the sums of the corresponding columns in A and B. Since vector addition of the columns is done entrywise, each entry in $A + B$ is the sum of the corresponding entries in A and B. The sum $A + B$ is defined only when A and B are the same size.

EXAMPLE 1 Let

$$A = \begin{bmatrix} 4 & 0 & 5 \\ -1 & 3 & 2 \end{bmatrix}, \qquad B = \begin{bmatrix} 1 & 1 & 1 \\ 3 & 5 & 7 \end{bmatrix}, \qquad C = \begin{bmatrix} 2 & -3 \\ 0 & 1 \end{bmatrix}$$

Then

$$A + B = \begin{bmatrix} 5 & 1 & 6 \\ 2 & 8 & 9 \end{bmatrix}$$

but $A + C$ is not defined because A and C have different sizes. ∎

If r is a scalar and A is a matrix, then the **scalar multiple** rA is the matrix whose columns are r times the corresponding columns in A. As with vectors, $-A$ stands for $(-1)A$, and $A - B$ is the same as $A + (-1)B$.

EXAMPLE 2 If A and B are the matrices in Example 1, then

$$2B = 2\begin{bmatrix} 1 & 1 & 1 \\ 3 & 5 & 7 \end{bmatrix} = \begin{bmatrix} 2 & 2 & 2 \\ 6 & 10 & 14 \end{bmatrix}$$

$$A - 2B = \begin{bmatrix} 4 & 0 & 5 \\ -1 & 3 & 2 \end{bmatrix} - \begin{bmatrix} 2 & 2 & 2 \\ 6 & 10 & 14 \end{bmatrix} = \begin{bmatrix} 2 & -2 & 3 \\ -7 & -7 & -12 \end{bmatrix}$$ ∎

It was unnecessary in Example 2 to compute $A - 2B$ as $A + (-1)2B$ because the usual rules of algebra apply to sums and scalar multiples of matrices, as the following theorem shows.

THEOREM 1

Let A, B, and C be matrices of the same size, and let r and s be scalars.

a. $A + B = B + A$ d. $r(A + B) = rA + rB$

b. $(A + B) + C = A + (B + C)$ e. $(r + s)A = rA + sA$

c. $A + 0 = A$ f. $r(sA) = (rs)A$

Each equality in Theorem 1 is verified by showing that the matrix on the left side has the same size as the matrix on the right and that corresponding columns are equal. Size is no problem because A, B, and C are equal in size. The equality of columns follows immediately from analogous properties of vectors. For instance, if the jth columns of A, B, and C are $\mathbf{a}_j$, $\mathbf{b}_j$, and $\mathbf{c}_j$, respectively, then the jth columns of $(A + B) + C$ and $A + (B + C)$ are

$$(\mathbf{a}_j + \mathbf{b}_j) + \mathbf{c}_j \quad \text{and} \quad \mathbf{a}_j + (\mathbf{b}_j + \mathbf{c}_j)$$

respectively. Since these two vector sums are equal for each j, property (b) is verified.

Because of the associative property of addition, we can simply write $A + B + C$ for the sum, which can be computed either as $(A + B) + C$ or as $A + (B + C)$. The same applies to sums of four or more matrices.

Matrix Multiplication

When a matrix B multiplies a vector $\mathbf{x}$, it transforms $\mathbf{x}$ into the vector $B\mathbf{x}$. If this vector is then multiplied in turn by a matrix A, the resulting vector is $A(B\mathbf{x})$. See Fig. 2.

FIGURE 2 Multiplication by B and then A.

Thus $A(B\mathbf{x})$ is produced from $\mathbf{x}$ by a *composition* of mappings—the linear transformations studied in Section 1.8. Our goal is to represent this composite mapping as multiplication by a single matrix, denoted by AB, so that

$$A(B\mathbf{x}) = (AB)\mathbf{x} \tag{1}$$

See Fig. 3.

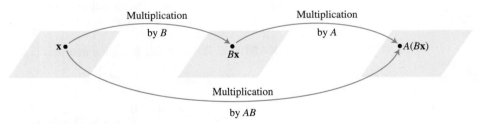

FIGURE 3 Multiplication by AB.

If A is $m \times n$, B is $n \times p$, and $\mathbf{x}$ is in $\mathbb{R}^p$, denote the columns of B by $\mathbf{b}_1, \ldots, \mathbf{b}_p$ and the entries in $\mathbf{x}$ by $x_1, \ldots, x_p$. Then

$$B\mathbf{x} = x_1\mathbf{b}_1 + \cdots + x_p\mathbf{b}_p$$

By the linearity of multiplication by A,

$$A(B\mathbf{x}) = A(x_1\mathbf{b}_1) + \cdots + A(x_p\mathbf{b}_p)$$
$$= x_1 A\mathbf{b}_1 + \cdots + x_p A\mathbf{b}_p$$

The vector $A(B\mathbf{x})$ is a linear combination of the vectors $A\mathbf{b}_1, \ldots, A\mathbf{b}_p$, using the entries in $\mathbf{x}$ as weights. In matrix notation, this linear combination is written as

$$A(B\mathbf{x}) = [\, A\mathbf{b}_1 \quad A\mathbf{b}_2 \quad \cdots \quad A\mathbf{b}_p \,]\mathbf{x}$$

Thus multiplication by $[\, A\mathbf{b}_1 \quad A\mathbf{b}_2 \quad \cdots \quad A\mathbf{b}_p \,]$ transforms $\mathbf{x}$ into $A(B\mathbf{x})$. We have found the matrix we sought!

DEFINITION

> If A is an $m \times n$ matrix, and if B is an $n \times p$ matrix with columns $\mathbf{b}_1, \ldots, \mathbf{b}_p$, then the product AB is the $m \times p$ matrix whose columns are $A\mathbf{b}_1, \ldots, A\mathbf{b}_p$. That is,
>
> $$AB = A[\, \mathbf{b}_1 \quad \mathbf{b}_2 \quad \cdots \quad \mathbf{b}_p \,] = [\, A\mathbf{b}_1 \quad A\mathbf{b}_2 \quad \cdots \quad A\mathbf{b}_p \,]$$

This definition makes equation (1) true for all $\mathbf{x}$ in $\mathbb{R}^p$. Equation (1) proves that the composite mapping in Fig. 3 is a linear transformation and that its standard matrix is AB. *Multiplication of matrices corresponds to composition of linear transformations.*

EXAMPLE 3   Compute AB, where $A = \begin{bmatrix} 2 & 3 \\ 1 & -5 \end{bmatrix}$ and $B = \begin{bmatrix} 4 & 3 & 6 \\ 1 & -2 & 3 \end{bmatrix}$.

SOLUTION   Write $B = [\, \mathbf{b}_1 \quad \mathbf{b}_2 \quad \mathbf{b}_3 \,]$, and compute:

$$A\mathbf{b}_1 = \begin{bmatrix} 2 & 3 \\ 1 & -5 \end{bmatrix}\begin{bmatrix} 4 \\ 1 \end{bmatrix}, \quad A\mathbf{b}_2 = \begin{bmatrix} 2 & 3 \\ 1 & -5 \end{bmatrix}\begin{bmatrix} 3 \\ -2 \end{bmatrix}, \quad A\mathbf{b}_3 = \begin{bmatrix} 2 & 3 \\ 1 & -5 \end{bmatrix}\begin{bmatrix} 6 \\ 3 \end{bmatrix}$$

$$= \begin{bmatrix} 11 \\ -1 \end{bmatrix} \qquad\qquad = \begin{bmatrix} 0 \\ 13 \end{bmatrix} \qquad\qquad = \begin{bmatrix} 21 \\ -9 \end{bmatrix}$$

Then

$$AB = A[\, \mathbf{b}_1 \quad \mathbf{b}_2 \quad \mathbf{b}_3 \,] = \begin{bmatrix} 11 & 0 & 21 \\ -1 & 13 & -9 \end{bmatrix}$$

$$\qquad\qquad\qquad A\mathbf{b}_1 \quad A\mathbf{b}_2 \quad A\mathbf{b}_3$$

Notice that since the first column of AB is $A\mathbf{b}_1$, this column is a linear combination of the columns of A using the entries in $\mathbf{b}_1$ as weights. A similar statement is true for each column of AB.

> Each column of AB is a linear combination of the columns of A using weights from the corresponding column of B.

Obviously, the number of columns of A must match the number of rows in B in order for a linear combination such as $A\mathbf{b}_1$ to be defined. Also, the definition of AB shows that *AB has the same number of rows as A and the same number of columns as B.*

EXAMPLE 4   If A is a 3×5 matrix and B is a 5×2 matrix, what are the sizes of AB and BA, if they are defined?

SOLUTION Since A has 5 columns and B has 5 rows, the product AB is defined and is a 3×2 matrix:

$$
\overset{A}{\begin{bmatrix} * & * & * & * & * \\ * & * & * & * & * \\ * & * & * & * & * \end{bmatrix}}
\overset{B}{\begin{bmatrix} * & * \\ * & * \\ * & * \\ * & * \\ * & * \end{bmatrix}}
=
\overset{AB}{\begin{bmatrix} * & * \\ * & * \\ * & * \end{bmatrix}}
$$

$$3 \times 5 \qquad\qquad 5 \times 2 \qquad\qquad 3 \times 2$$

Match

Size of AB

The product BA is *not* defined because the 2 columns of B do not match the 3 rows of A. ∎

The definition of AB is important for theoretical work and applications, but the following rule provides a more efficient method for calculating the individual entries in AB when working small problems by hand.

ROW–COLUMN RULE FOR COMPUTING *AB*

If the product AB is defined, then the entry in row i and column j of AB is the sum of the products of corresponding entries from row i of A and column j of B. If $(AB)_{ij}$ denotes the (i, j)-entry in AB, and if A is an $m \times n$ matrix, then

$$(AB)_{ij} = a_{i1}b_{1j} + a_{i2}b_{2j} + \cdots + a_{in}b_{nj}$$

To verify this rule, let $B = [\, \mathbf{b}_1 \ \cdots \ \mathbf{b}_p \,]$. Column j of AB is $A\mathbf{b}_j$, and we can compute $A\mathbf{b}_j$ by the row–vector rule for computing $A\mathbf{x}$ from Section 1.4. The ith entry in $A\mathbf{b}_j$ is the sum of the products of corresponding entries from row i of A and the vector $\mathbf{b}_j$, which is precisely the computation described in the rule for computing the (i, j)-entry of AB.

EXAMPLE 5 Use the row–column rule to compute two of the entries in AB for the matrices in Example 3. An inspection of the numbers involved will make it clear how the two methods for calculating AB produce the same matrix.

SOLUTION To find the entry in row 1 and column 3 of AB, consider row 1 of A and column 3 of B. Multiply corresponding entries and add the results, as shown below:

$$
AB = \begin{bmatrix} 2 & 3 \\ 1 & -5 \end{bmatrix}\begin{bmatrix} 4 & 3 & 6 \\ 1 & -2 & 3 \end{bmatrix} = \begin{bmatrix} \square & \square & 2(6)+3(3) \\ \square & \square & \square \end{bmatrix} = \begin{bmatrix} \square & \square & 21 \\ \square & \square & \square \end{bmatrix}
$$

For the entry in row 2 and column 2 of AB, use row 2 of A and column 2 of B:

$$
\begin{bmatrix} 2 & 3 \\ 1 & -5 \end{bmatrix}\begin{bmatrix} 4 & 3 & 6 \\ 1 & -2 & 3 \end{bmatrix} = \begin{bmatrix} \square & \square & 21 \\ \square & 1(3)+-5(-2) & \square \end{bmatrix} = \begin{bmatrix} \square & \square & 21 \\ \square & 13 & \square \end{bmatrix}
$$

∎

EXAMPLE 6   Find the entries in the second row of AB, where

$$A = \begin{bmatrix} 2 & -5 & 0 \\ -1 & 3 & -4 \\ 6 & -8 & -7 \\ -3 & 0 & 9 \end{bmatrix}, \qquad B = \begin{bmatrix} 4 & -6 \\ 7 & 1 \\ 3 & 2 \end{bmatrix}$$

SOLUTION   By the row–column rule, the entries of the second row of AB come from row 2 of A (and the columns of B):

$$\rightarrow \begin{bmatrix} 2 & -5 & 0 \\ -1 & 3 & -4 \\ 6 & -8 & -7 \\ -3 & 0 & 9 \end{bmatrix} \begin{bmatrix} 4 & -6 \\ 7 & 1 \\ 3 & 2 \end{bmatrix}$$

$$= \begin{bmatrix} \square & \square \\ -4+21-12 & 6+3-8 \\ \square & \square \\ \square & \square \end{bmatrix} = \begin{bmatrix} \square & \square \\ 5 & 1 \\ \square & \square \\ \square & \square \end{bmatrix} \qquad \blacksquare$$

Notice that since Example 6 requested only the second row of AB, we could have written just the second row of A to the left of B and computed

$$\begin{bmatrix} -1 & 3 & -4 \end{bmatrix} \begin{bmatrix} 4 & -6 \\ 7 & 1 \\ 3 & 2 \end{bmatrix} = \begin{bmatrix} 5 & 1 \end{bmatrix}$$

This observation about rows of AB is true in general and follows from the row–column rule. Let $\text{row}_i(A)$ denote the ith row of a matrix A. Then

$$\text{row}_i(AB) = \text{row}_i(A) \cdot B \tag{2}$$

Properties of Matrix Multiplication

The following theorem lists the standard properties of matrix multiplication. Recall that I_m represents the $m \times m$ identity matrix and $I_m\mathbf{x} = \mathbf{x}$ for all $\mathbf{x}$ in $\mathbb{R}^m$.

THEOREM 2

Let A be an $m \times n$ matrix, and let B and C have sizes for which the indicated sums and products are defined.

a. $A(BC) = (AB)C$ (associative law of multiplication)
b. $A(B + C) = AB + AC$ (left distributive law)
c. $(B + C)A = BA + CA$ (right distributive law)
d. $r(AB) = (rA)B = A(rB)$
    for any scalar r
e. $I_m A = A = AI_n$ (identity for matrix multiplication)

PROOF   Properties (b)–(e) are considered in the exercises. Property (a) follows from the fact that matrix multiplication corresponds to composition of linear transformations (which are functions), and it is known (or easy to check) that the composition of functions is associative. Here is another proof of (a) that rests on the "column definition" of the product of two matrices. Let

$$C = \begin{bmatrix} \mathbf{c}_1 & \cdots & \mathbf{c}_p \end{bmatrix}$$

By the definition of matrix multiplication,

$$BC = [\, B\mathbf{c}_1 \quad \cdots \quad B\mathbf{c}_p \,]$$
$$A(BC) = [\, A(B\mathbf{c}_1) \quad \cdots \quad A(B\mathbf{c}_p) \,]$$

Recall from equation (1) that the definition of AB makes $A(B\mathbf{x}) = (AB)\mathbf{x}$ for all $\mathbf{x}$, so

$$A(BC) = [\, (AB)\mathbf{c}_1 \quad \cdots \quad (AB)\mathbf{c}_p \,] = (AB)C \qquad \blacksquare$$

The associative and distributive laws in Theorems 1 and 2 say essentially that pairs of parentheses in matrix expressions can be inserted and deleted in the same way as in the algebra of real numbers. In particular, we can write ABC for the product, which can be computed either as $A(BC)$ or as $(AB)C$.[1] Similarly, a product $ABCD$ of four matrices can be computed as $A(BCD)$ or $(ABC)D$ or $A(BC)D$, and so on. It does not matter how we group the matrices when computing the product, so long as the left-to-right order of the matrices is preserved.

The left-to-right order in products is critical because AB and BA are usually not the same. This is not surprising, because the columns of AB are linear combinations of the columns of A, whereas the columns of BA are constructed from the columns of B. The position of the factors in the product AB is emphasized by saying that A is *right-multiplied* by B or that B is *left-multiplied* by A. If $AB = BA$, we say that A and B **commute** with one another.

EXAMPLE 7 Let $A = \begin{bmatrix} 5 & 1 \\ 3 & -2 \end{bmatrix}$ and $B = \begin{bmatrix} 2 & 0 \\ 4 & 3 \end{bmatrix}$. Show that these matrices do not commute. That is, verify that $AB \neq BA$.

SOLUTION

$$AB = \begin{bmatrix} 5 & 1 \\ 3 & -2 \end{bmatrix}\begin{bmatrix} 2 & 0 \\ 4 & 3 \end{bmatrix} = \begin{bmatrix} 14 & 3 \\ -2 & -6 \end{bmatrix}$$

$$BA = \begin{bmatrix} 2 & 0 \\ 4 & 3 \end{bmatrix}\begin{bmatrix} 5 & 1 \\ 3 & -2 \end{bmatrix} = \begin{bmatrix} 10 & 2 \\ 29 & -2 \end{bmatrix} \qquad \blacksquare$$

Example 7 illustrates the first of the following list of important differences between matrix algebra and the ordinary algebra of real numbers. See Exercises 9–12 for examples of these situations.

WARNINGS:

1. In general, $AB \neq BA$.

2. The cancellation laws do *not* hold for matrix multiplication. That is, if $AB = AC$, then it is *not* true in general that $B = C$. (See Exercise 10.)

3. If a product AB is the zero matrix, you *cannot* conclude in general that either $A = 0$ or $B = 0$. (See Exercise 12.)

Powers of a Matrix

WEB

If A is an $n \times n$ matrix and if k is a positive integer, then A^k denotes the product of k

---

[1] When B is square and C has fewer columns than A has rows, it is more efficient to compute $A(BC)$ than $(AB)C$.

copies of A:

$$A^k = \underbrace{A \cdots A}_{k}$$

If A is nonzero and if $\mathbf{x}$ is in $\mathbb{R}^n$, then $A^k\mathbf{x}$ is the result of left-multiplying $\mathbf{x}$ by A repeatedly k times. If $k = 0$, then $A^0\mathbf{x}$ should be $\mathbf{x}$ itself. Thus A^0 is interpreted as the identity matrix. Matrix powers are useful in both theory and applications (Sections 2.6, 4.9, and later in the text).

The Transpose of a Matrix

Given an $m \times n$ matrix A, the **transpose** of A is the $n \times m$ matrix, denoted by A^T, whose columns are formed from the corresponding rows of A.

EXAMPLE 8 Let

$$A = \begin{bmatrix} a & b \\ c & d \end{bmatrix}, \quad B = \begin{bmatrix} -5 & 2 \\ 1 & -3 \\ 0 & 4 \end{bmatrix}, \quad C = \begin{bmatrix} 1 & 1 & 1 & 1 \\ -3 & 5 & -2 & 7 \end{bmatrix}$$

Then

$$A^T = \begin{bmatrix} a & c \\ b & d \end{bmatrix}, \quad B^T = \begin{bmatrix} -5 & 1 & 0 \\ 2 & -3 & 4 \end{bmatrix}, \quad C^T = \begin{bmatrix} 1 & -3 \\ 1 & 5 \\ 1 & -2 \\ 1 & 7 \end{bmatrix} \quad \blacksquare$$

THEOREM 3

Let A and B denote matrices whose sizes are appropriate for the following sums and products.

a. $(A^T)^T = A$

b. $(A + B)^T = A^T + B^T$

c. For any scalar r, $(rA)^T = rA^T$

d. $(AB)^T = B^T A^T$

Proofs of (a)–(c) are straightforward and are omitted. For (d), see Exercise 33. Usually, $(AB)^T$ is not equal to $A^T B^T$, even when A and B have sizes such that the product $A^T B^T$ is defined.

The generalization of Theorem 3(d) to products of more than two factors can be stated in words as follows:

The transpose of a product of matrices equals the product of their transposes in the *reverse* order.

The exercises contain numerical examples that illustrate properties of transposes.

NUMERICAL NOTES

1. The fastest way to obtain AB on a computer depends on the way in which the computer stores matrices in its memory. The standard high-performance algorithms, such as in LAPACK, calculate AB by columns, as in our definition of the product. (A version of LAPACK written in C++ calculates AB by rows.)

2. The definition of AB lends itself well to parallel processing on a computer. The columns of B are assigned individually or in groups to different processors, which independently and hence simultaneously compute the corresponding columns of AB.

PRACTICE PROBLEMS

1. Since vectors in $\mathbb{R}^n$ may be regarded as $n \times 1$ matrices, the properties of transposes in Theorem 3 apply to vectors, too. Let

$$A = \begin{bmatrix} 1 & -3 \\ -2 & 4 \end{bmatrix} \quad \text{and} \quad \mathbf{x} = \begin{bmatrix} 5 \\ 3 \end{bmatrix}$$

Compute $(A\mathbf{x})^T$, $\mathbf{x}^T A^T$, $\mathbf{x}\mathbf{x}^T$, and $\mathbf{x}^T\mathbf{x}$. Is $A^T\mathbf{x}^T$ defined?

2. Let A be a 4×4 matrix and let $\mathbf{x}$ be a vector in $\mathbb{R}^4$. What is the fastest way to compute $A^2\mathbf{x}$? Count the multiplications.

2.1 EXERCISES

In Exercises 1 and 2, compute each matrix sum or product if it is defined. If an expression is undefined, explain why. Let

$$A = \begin{bmatrix} 2 & 0 & -1 \\ 4 & -5 & 2 \end{bmatrix}, \quad B = \begin{bmatrix} 7 & -5 & 1 \\ 1 & -4 & -3 \end{bmatrix},$$

$$C = \begin{bmatrix} 1 & 2 \\ -2 & 1 \end{bmatrix}, \quad D = \begin{bmatrix} 3 & 5 \\ -1 & 4 \end{bmatrix}, \quad E = \begin{bmatrix} -5 \\ 3 \end{bmatrix}$$

1. $-2A$, $B - 2A$, AC, CD

2. $A + 3B$, $2C - 3E$, DB, EC

In the rest of this exercise set and in those to follow, assume that each matrix expression is defined. That is, the sizes of the matrices (and vectors) involved "match" appropriately.

3. Let $A = \begin{bmatrix} 2 & -5 \\ 3 & -2 \end{bmatrix}$. Compute $3I_2 - A$ and $(3I_2)A$.

4. Compute $A - 5I_3$ and $(5I_3)A$, where

$$A = \begin{bmatrix} 5 & -1 & 3 \\ -4 & 3 & -6 \\ -3 & 1 & 2 \end{bmatrix}.$$

In Exercises 5 and 6, compute the product AB in two ways: (a) by the definition, where $A\mathbf{b}_1$ and $A\mathbf{b}_2$ are computed separately, and (b) by the row–column rule for computing AB.

5. $A = \begin{bmatrix} -1 & 3 \\ 2 & 4 \\ 5 & -3 \end{bmatrix}$, $B = \begin{bmatrix} 4 & -2 \\ -2 & 3 \end{bmatrix}$

6. $A = \begin{bmatrix} 4 & -3 \\ -3 & 5 \\ 0 & 1 \end{bmatrix}$, $B = \begin{bmatrix} 1 & 4 \\ 3 & -2 \end{bmatrix}$

7. If a matrix A is 5×3 and the product AB is 5×7, what is the size of B?

8. How many rows does B have if BC is a 5×4 matrix?

9. Let $A = \begin{bmatrix} 2 & 3 \\ -1 & 1 \end{bmatrix}$ and $B = \begin{bmatrix} 1 & 9 \\ -3 & k \end{bmatrix}$. What value(s) of k, if any, will make $AB = BA$?

10. Let $A = \begin{bmatrix} 3 & -6 \\ -1 & 2 \end{bmatrix}$, $B = \begin{bmatrix} -1 & 1 \\ 3 & 4 \end{bmatrix}$, and $C = \begin{bmatrix} -3 & -5 \\ 2 & 1 \end{bmatrix}$. Verify that $AB = AC$ and yet $B \neq C$.

11. Let $A = \begin{bmatrix} 1 & 2 & 3 \\ 2 & 4 & 5 \\ 3 & 5 & 6 \end{bmatrix}$ and $D = \begin{bmatrix} 5 & 0 & 0 \\ 0 & 3 & 0 \\ 0 & 0 & 2 \end{bmatrix}$. Compute AD and DA. Explain how the columns or rows of A change when A is multiplied by D on the right or on the left. Find a 3×3 matrix B, not the identity matrix or the zero matrix, such that $AB = BA$.

12. Let $A = \begin{bmatrix} 3 & -6 \\ -2 & 4 \end{bmatrix}$. Construct a 2×2 matrix B such that AB is the zero matrix. Use two different nonzero columns for B.

13. Let $\mathbf{r}_1, \ldots, \mathbf{r}_p$ be vectors in $\mathbb{R}^n$, and let Q be an $m \times n$ matrix. Write the matrix $[\, Q\mathbf{r}_1 \; \cdots \; Q\mathbf{r}_p \,]$ as a *product* of two matrices (neither of which is an identity matrix).

14. Let U be the 3×2 cost matrix described in Example 6 in Section 1.8. The first column of U lists the costs per dollar of output for manufacturing product B, and the second column lists the costs per dollar of output for product C. (The costs are categorized as materials, labor, and overhead.) Let $\mathbf{q}_1$ be a vector in $\mathbb{R}^2$ that lists the output (measured in dollars) of products B and C manufactured during the first quarter of the year, and let $\mathbf{q}_2$, $\mathbf{q}_3$, and $\mathbf{q}_4$ be the analogous vectors that list the amounts of products B and C manufactured in the second, third, and fourth quarters, respectively. Give an economic description of the data in the matrix UQ, where $Q = [\, \mathbf{q}_1 \; \mathbf{q}_2 \; \mathbf{q}_3 \; \mathbf{q}_4 \,]$.

Exercises 15 and 16 concern arbitrary matrices A, B, and C for which the indicated sums and products are defined. Mark each statement True or False. Justify each answer.

15. a. If A and B are 2×2 matrices with columns $\mathbf{a}_1$, $\mathbf{a}_2$, and $\mathbf{b}_1$, $\mathbf{b}_2$, respectively, then $AB = [\, \mathbf{a}_1\mathbf{b}_1 \; \mathbf{a}_2\mathbf{b}_2 \,]$.

   b. Each column of AB is a linear combination of the columns of B using weights from the corresponding column of A.

   c. $AB + AC = A(B + C)$

   d. $A^T + B^T = (A + B)^T$

   e. The transpose of a product of matrices equals the product of their transposes in the same order.

16. a. The first row of AB is the first row of A multiplied on the right by B.

   b. If A and B are 3×3 matrices and $B = [\, \mathbf{b}_1 \; \mathbf{b}_2 \; \mathbf{b}_3 \,]$, then $AB = [\, A\mathbf{b}_1 + A\mathbf{b}_2 + A\mathbf{b}_3 \,]$.

   c. If A is an $n \times n$ matrix, then $(A^2)^T = (A^T)^2$

   d. $(ABC)^T = C^T A^T B^T$

   e. The transpose of a sum of matrices equals the sum of their transposes.

17. If $A = \begin{bmatrix} 1 & -3 \\ -3 & 5 \end{bmatrix}$ and $AB = \begin{bmatrix} -3 & -11 \\ 1 & 17 \end{bmatrix}$, determine the first and second columns of B.

18. Suppose the third column of B is all zeros. What can be said about the third column of AB?

19. Suppose the third column of B is the sum of the first two columns. What can be said about the third column of AB? Why?

20. Suppose the first two columns, $\mathbf{b}_1$ and $\mathbf{b}_2$, of B are equal. What can be said about the columns of AB? Why?

21. Suppose the last column of AB is entirely zeros but B itself has no column of zeros. What can be said about the columns of A?

22. Show that if the columns of B are linearly dependent, then so are the columns of AB.

23. Suppose $CA = I_n$ (the $n \times n$ identity matrix). Show that the equation $A\mathbf{x} = \mathbf{0}$ has only the trivial solution. Explain why A cannot have more columns than rows.

24. Suppose A is a $3 \times n$ matrix whose columns span $\mathbb{R}^3$. Explain how to construct an $n \times 3$ matrix D such that $AD = I_3$.

25. Suppose A is an $m \times n$ matrix and there exist $n \times m$ matrices C and D such that $CA = I_n$ and $AD = I_m$. Prove that $m = n$ and $C = D$. [*Hint:* Think about the product CAD.]

26. Suppose $AD = I_m$ (the $m \times m$ identity matrix). Show that for any $\mathbf{b}$ in $\mathbb{R}^m$, the equation $A\mathbf{x} = \mathbf{b}$ has a solution. [*Hint:* Think about the equation $AD\mathbf{b} = \mathbf{b}$.] Explain why A cannot have more rows than columns.

In Exercises 27 and 28, view vectors in $\mathbb{R}^n$ as $n \times 1$ matrices. For $\mathbf{u}$ and $\mathbf{v}$ in $\mathbb{R}^n$, the matrix product $\mathbf{u}^T\mathbf{v}$ is a 1×1 matrix, called the **scalar product**, or **inner product**, of $\mathbf{u}$ and $\mathbf{v}$. It is usually written as a single real number without brackets. The matrix product $\mathbf{u}\mathbf{v}^T$ is an $n \times n$ matrix, called the **outer product** of $\mathbf{u}$ and $\mathbf{v}$. The products $\mathbf{u}^T\mathbf{v}$ and $\mathbf{u}\mathbf{v}^T$ will appear later in the text.

27. Let $\mathbf{u} = \begin{bmatrix} -3 \\ 2 \\ -5 \end{bmatrix}$ and $\mathbf{v} = \begin{bmatrix} a \\ b \\ c \end{bmatrix}$. Compute $\mathbf{u}^T\mathbf{v}$, $\mathbf{v}^T\mathbf{u}$, $\mathbf{u}\mathbf{v}^T$, and $\mathbf{v}\mathbf{u}^T$.

28. If $\mathbf{u}$ and $\mathbf{v}$ are in $\mathbb{R}^n$, how are $\mathbf{u}^T\mathbf{v}$ and $\mathbf{v}^T\mathbf{u}$ related? How are $\mathbf{u}\mathbf{v}^T$ and $\mathbf{v}\mathbf{u}^T$ related?

29. Prove Theorem 2(b) and 2(c). Use the row–column rule. The (i, j)-entry in $A(B + C)$ can be written as

$$a_{i1}(b_{1j} + c_{1j}) + \cdots + a_{in}(b_{nj} + c_{nj})$$

or

$$\sum_{k=1}^{n} a_{ik}(b_{kj} + c_{kj})$$

30. Prove Theorem 2(d). [*Hint:* The (i, j)-entry in $(rA)B$ is $(ra_{i1})b_{1j} + \cdots + (ra_{in})b_{nj}$.]

31. Show that $I_m A = A$ where A is an $m \times n$ matrix. Assume $I_m\mathbf{x} = \mathbf{x}$ for all $\mathbf{x}$ in $\mathbb{R}^m$.

32. Show that $AI_n = A$ when A is an $m \times n$ matrix. [*Hint:* Use the (column) definition of AI_n.]

33. Prove Theorem 3(d). [*Hint:* Consider the jth row of $(AB)^T$.]

34. Give a formula for $(AB\mathbf{x})^T$, where $\mathbf{x}$ is a vector and A and B are matrices of appropriate sizes.

35. [M] Read the documentation for your matrix program, and write the commands that will produce the following matrices (without keying in each entry of the matrix).

   a. A 4×5 matrix of zeros

   b. A 5×3 matrix of ones

   c. The 5×5 identity matrix

   d. A 4×4 diagonal matrix, with diagonal entries 3, 4, 2, 5

A useful way to test new ideas in matrix algebra, or to make conjectures, is to make calculations with matrices selected at random. Checking a property for a few matrices does not prove that the property holds in general, but it makes the property more believable. Also, if the property is actually false, making a few calculations may help to discover this.

36. **[M]** Write the command(s) that will create a 5×6 matrix with random entries. In what range of numbers do the entries lie? Tell how to create a 4×4 matrix with random integer entries between -9 and 9. [*Hint:* If x is a random number such that $0 < x < 1$, then $-9.5 < 19(x - .5) < 9.5$.]

37. **[M]** Construct random 4×4 matrices A and B to test whether $AB = BA$. The best way to do this is to compute $AB - BA$ and check whether this difference is the zero matrix. Then test $AB - BA$ for three more pairs of random 4×4 matrices. Report your conclusions.

38. **[M]** Construct a random 5×5 matrix A and test whether $(A + I)(A - I) = A^2 - I$. The best way to do this is to compute $(A + I)(A - I) - (A^2 - I)$ and verify that this difference is the zero matrix. Do this for three random matrices. Then test $(A + B)(A - B) = A^2 - B^2$ the same way for three pairs of random 4×4 matrices. Report your conclusions.

39. **[M]** Use at least three pairs of random 4×4 matrices A and B to test the equalities $(A + B)^T = A^T + B^T$ and $(AB)^T = B^T A^T$, as well as $(AB)^T = A^T B^T$. (See Exercise 37.) Report your conclusions. [*Note:* Most matrix programs use A' for A^T.]

40. **[M]** Let

$$S = \begin{bmatrix} 0 & 1 & 0 & 0 & 0 \\ 0 & 0 & 1 & 0 & 0 \\ 0 & 0 & 0 & 1 & 0 \\ 0 & 0 & 0 & 0 & 1 \\ 0 & 0 & 0 & 0 & 0 \end{bmatrix}$$

Compute S^k for $k = 2, \ldots, 6$.

41. **[M]** Describe in words what happens when A^5, A^{10}, A^{20}, and A^{30} are computed for

$$A = \begin{bmatrix} 1/4 & 1/2 & 1/4 \\ 1/2 & 1/3 & 1/6 \\ 1/4 & 1/6 & 7/12 \end{bmatrix}$$

SOLUTIONS TO PRACTICE PROBLEMS

1. $A\mathbf{x} = \begin{bmatrix} 1 & -3 \\ -2 & 4 \end{bmatrix}\begin{bmatrix} 5 \\ 3 \end{bmatrix} = \begin{bmatrix} -4 \\ 2 \end{bmatrix}$. So $(A\mathbf{x})^T = \begin{bmatrix} -4 & 2 \end{bmatrix}$. Also,

$$\mathbf{x}^T A^T = \begin{bmatrix} 5 & 3 \end{bmatrix}\begin{bmatrix} 1 & -2 \\ -3 & 4 \end{bmatrix} = \begin{bmatrix} -4 & 2 \end{bmatrix}.$$

The quantities $(A\mathbf{x})^T$ and $\mathbf{x}^T A^T$ are equal, by Theorem 3(d). Next,

$$\mathbf{x}\mathbf{x}^T = \begin{bmatrix} 5 \\ 3 \end{bmatrix}\begin{bmatrix} 5 & 3 \end{bmatrix} = \begin{bmatrix} 25 & 15 \\ 15 & 9 \end{bmatrix}$$

$$\mathbf{x}^T\mathbf{x} = \begin{bmatrix} 5 & 3 \end{bmatrix}\begin{bmatrix} 5 \\ 3 \end{bmatrix} = \begin{bmatrix} 25 + 9 \end{bmatrix} = 34$$

A 1×1 matrix such as $\mathbf{x}^T\mathbf{x}$ is usually written without the brackets. Finally, $A^T\mathbf{x}^T$ is not defined, because $\mathbf{x}^T$ does not have two rows to match the two columns of A^T.

2. The fastest way to compute $A^2\mathbf{x}$ is to compute $A(A\mathbf{x})$. The product $A\mathbf{x}$ requires 16 multiplications, 4 for each entry, and $A(A\mathbf{x})$ requires 16 more. In contrast, the product A^2 requires 64 multiplications, 4 for each of the 16 entries in A^2. After that, $A^2\mathbf{x}$ takes 16 more multiplications, for a total of 80.

2.2 | THE INVERSE OF A MATRIX

Matrix algebra provides tools for manipulating matrix equations and creating various useful formulas in ways similar to doing ordinary algebra with real numbers. This section investigates the matrix analogue of the reciprocal, or multiplicative inverse, of a nonzero number.

Recall that the multiplicative inverse of a number such as 5 is 1/5 or 5^{-1}. This inverse satisfies the equations

$$5^{-1} \cdot 5 = 1 \quad \text{and} \quad 5 \cdot 5^{-1} = 1$$

The matrix generalization requires *both* equations and avoids the slanted-line notation (for division) because matrix multiplication is not commutative. Furthermore, a full generalization is possible only if the matrices involved are square.[1]

An $n \times n$ matrix A is said to be **invertible** if there is an $n \times n$ matrix C such that

$$CA = I \quad \text{and} \quad AC = I$$

where $I = I_n$, the $n \times n$ identity matrix. In this case, C is an **inverse** of A. In fact, C is uniquely determined by A, because if B were another inverse of A, then $B = BI = B(AC) = (BA)C = IC = C$. This unique inverse is denoted by A^{-1}, so that

$$A^{-1}A = I \quad \text{and} \quad AA^{-1} = I$$

A matrix that is *not* invertible is sometimes called a **singular matrix**, and an invertible matrix is called a **nonsingular matrix**.

EXAMPLE 1   If $A = \begin{bmatrix} 2 & 5 \\ -3 & -7 \end{bmatrix}$ and $C = \begin{bmatrix} -7 & -5 \\ 3 & 2 \end{bmatrix}$, then

$$AC = \begin{bmatrix} 2 & 5 \\ -3 & -7 \end{bmatrix}\begin{bmatrix} -7 & -5 \\ 3 & 2 \end{bmatrix} = \begin{bmatrix} 1 & 0 \\ 0 & 1 \end{bmatrix} \quad \text{and}$$

$$CA = \begin{bmatrix} -7 & -5 \\ 3 & 2 \end{bmatrix}\begin{bmatrix} 2 & 5 \\ -3 & -7 \end{bmatrix} = \begin{bmatrix} 1 & 0 \\ 0 & 1 \end{bmatrix}$$

Thus $C = A^{-1}$.   ∎

Here is a simple formula for the inverse of a 2×2 matrix, along with a test to tell if the inverse exists.

THEOREM 4   Let $A = \begin{bmatrix} a & b \\ c & d \end{bmatrix}$. If $ad - bc \neq 0$, then A is invertible and

$$A^{-1} = \frac{1}{ad - bc}\begin{bmatrix} d & -b \\ -c & a \end{bmatrix}$$

If $ad - bc = 0$, then A is not invertible.

The simple proof of Theorem 4 is outlined in Exercises 25 and 26. The quantity $ad - bc$ is called the **determinant** of A, and we write

$$\det A = ad - bc$$

Theorem 4 says that a 2×2 matrix A is invertible if and only if $\det A \neq 0$.

---

[1] One could say that an $m \times n$ matrix A is invertible if there exist $n \times m$ matrices C and D such that $CA = I_n$ and $AD = I_m$. However, these equations imply that A is square and $C = D$. Thus A is invertible as defined above. See Exercises 23–25 in Section 2.1.

EXAMPLE 2   Find the inverse of $A = \begin{bmatrix} 3 & 4 \\ 5 & 6 \end{bmatrix}$.

SOLUTION   Since $\det A = 3(6) - 4(5) = -2 \neq 0$, A is invertible, and

$$A^{-1} = \frac{1}{-2}\begin{bmatrix} 6 & -4 \\ -5 & 3 \end{bmatrix} = \begin{bmatrix} 6/(-2) & -4/(-2) \\ -5/(-2) & 3/(-2) \end{bmatrix} = \begin{bmatrix} -3 & 2 \\ 5/2 & -3/2 \end{bmatrix} \qquad \blacksquare$$

Invertible matrices are indispensable in linear algebra—mainly for algebraic calculations and formula derivations, as in the next theorem. There are also occasions when an inverse matrix provides insight into a mathematical model of a real-life situation, as in Example 3, below.

THEOREM 5

If A is an invertible $n \times n$ matrix, then for each $\mathbf{b}$ in $\mathbb{R}^n$, the equation $A\mathbf{x} = \mathbf{b}$ has the unique solution $\mathbf{x} = A^{-1}\mathbf{b}$.

PROOF   Take any $\mathbf{b}$ in $\mathbb{R}^n$. A solution exists because if $A^{-1}\mathbf{b}$ is substituted for $\mathbf{x}$, then $A\mathbf{x} = A(A^{-1}\mathbf{b}) = (AA^{-1})\mathbf{b} = I\mathbf{b} = \mathbf{b}$. So $A^{-1}\mathbf{b}$ is a solution. To prove that the solution is unique, show that if $\mathbf{u}$ is any solution, then $\mathbf{u}$, in fact, must be $A^{-1}\mathbf{b}$. Indeed, if $A\mathbf{u} = \mathbf{b}$, we can multiply both sides by A^{-1} and obtain

$$A^{-1}A\mathbf{u} = A^{-1}\mathbf{b}, \quad I\mathbf{u} = A^{-1}\mathbf{b}, \quad \text{and} \quad \mathbf{u} = A^{-1}\mathbf{b} \qquad \blacksquare$$

EXAMPLE 3   A horizontal elastic beam is supported at each end and is subjected to forces at points 1, 2, 3, as shown in Fig. 1. Let $\mathbf{f}$ in $\mathbb{R}^3$ list the forces at these points, and let $\mathbf{y}$ in $\mathbb{R}^3$ list the amounts of deflection (that is, movement) of the beam at the three points. Using Hooke's law from physics, it can be shown that

$$\mathbf{y} = D\mathbf{f}$$

where D is a *flexibility matrix*. Its inverse is called the *stiffness matrix*. Describe the physical significance of the columns of D and D^{-1}.

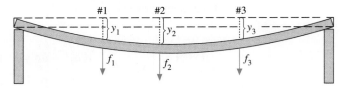

FIGURE 1   Deflection of an elastic beam.

SOLUTION   Write $I_3 = [\,\mathbf{e}_1 \quad \mathbf{e}_2 \quad \mathbf{e}_3\,]$ and observe that

$$D = DI_3 = [\,D\mathbf{e}_1 \quad D\mathbf{e}_2 \quad D\mathbf{e}_3\,]$$

Interpret the vector $\mathbf{e}_1 = (1,0,0)$ as a unit force applied downward at point 1 on the beam (with zero force at the other two points). Then $D\mathbf{e}_1$, the first column of D, lists the beam deflections due to a unit force at point 1. Similar descriptions apply to the second and third columns of D.

To study the stiffness matrix D^{-1}, observe that the equation $\mathbf{f} = D^{-1}\mathbf{y}$ computes a force vector $\mathbf{f}$ when a deflection vector $\mathbf{y}$ is given. Write

$$D^{-1} = D^{-1}I_3 = [\,D^{-1}\mathbf{e}_1 \quad D^{-1}\mathbf{e}_2 \quad D^{-1}\mathbf{e}_3\,]$$

Now interpret $\mathbf{e}_1$ as a deflection vector. Then $D^{-1}\mathbf{e}_1$ lists the forces that create the deflection. That is, the first column of D^{-1} lists the forces that must be applied at the three points to produce a unit deflection at point 1 and zero deflections at the other points. Similarly, columns 2 and 3 of D^{-1} list the forces required to produce unit deflections at points 2 and 3, respectively. In each column, one or two of the forces must be negative (point upward) to produce a unit deflection at the desired point and zero deflections at the other two points. If the flexibility is measured, for example, in inches of deflection per pound of load, then the stiffness matrix entries are given in pounds of load per inch of deflection. ■

The formula in Theorem 5 is seldom used to solve an equation $A\mathbf{x} = \mathbf{b}$ numerically because row reduction of $[\,A \quad \mathbf{b}\,]$ is nearly always faster. (Row reduction is usually more accurate, too, when computations involve rounding off numbers.) One possible exception is the 2×2 case. In this case, mental computations to solve $A\mathbf{x} = \mathbf{b}$ are sometimes easier using the formula for A^{-1}, as in the next example.

EXAMPLE 4   Use the inverse of the matrix A in Example 2 to solve the system

$$3x_1 + 4x_2 = 3$$
$$5x_1 + 6x_2 = 7$$

SOLUTION   This system is equivalent to $A\mathbf{x} = \mathbf{b}$, so

$$\mathbf{x} = A^{-1}\mathbf{b} = \begin{bmatrix} -3 & 2 \\ 5/2 & -3/2 \end{bmatrix} \begin{bmatrix} 3 \\ 7 \end{bmatrix} = \begin{bmatrix} 5 \\ -3 \end{bmatrix}$$ ■

The next theorem provides three useful facts about invertible matrices.

THEOREM 6

a. If A is an invertible matrix, then A^{-1} is invertible and

$$(A^{-1})^{-1} = A$$

b. If A and B are $n \times n$ invertible matrices, then so is AB, and the inverse of AB is the product of the inverses of A and B in the reverse order. That is,

$$(AB)^{-1} = B^{-1}A^{-1}$$

c. If A is an invertible matrix, then so is A^T, and the inverse of A^T is the transpose of A^{-1}. That is,

$$(A^T)^{-1} = (A^{-1})^T$$

PROOF   To verify statement (a), find a matrix C such that

$$A^{-1}C = I \quad \text{and} \quad CA^{-1} = I$$

In fact, these equations are satisfied with A in place of C. Hence A^{-1} is invertible, and A is its inverse. Next, to prove statement (b), compute:

$$(AB)(B^{-1}A^{-1}) = A(BB^{-1})A^{-1} = AIA^{-1} = AA^{-1} = I$$

A similar calculation shows that $(B^{-1}A^{-1})(AB) = I$. For statement (c), use Theorem 3(d), read from right to left, $(A^{-1})^T A^T = (AA^{-1})^T = I^T = I$. Similarly, $A^T (A^{-1})^T = I^T = I$. Hence A^T is invertible, and its inverse is $(A^{-1})^T$. ■

The following generalization of Theorem 6(b) is needed later.

The product of $n \times n$ invertible matrices is invertible, and the inverse is the product of their inverses in the reverse order.

There is an important connection between invertible matrices and row operations that leads to a method for computing inverses. As we shall see, an invertible matrix A is row equivalent to an identity matrix, and we can find A^{-1} by *watching the row reduction of A to I*.

Elementary Matrices

An **elementary matrix** is one that is obtained by performing a single elementary row operation on an identity matrix. The next example illustrates the three kinds of elementary matrices.

EXAMPLE 5 Let

$$E_1 = \begin{bmatrix} 1 & 0 & 0 \\ 0 & 1 & 0 \\ -4 & 0 & 1 \end{bmatrix}, \quad E_2 = \begin{bmatrix} 0 & 1 & 0 \\ 1 & 0 & 0 \\ 0 & 0 & 1 \end{bmatrix}, \quad E_3 = \begin{bmatrix} 1 & 0 & 0 \\ 0 & 1 & 0 \\ 0 & 0 & 5 \end{bmatrix},$$

$$A = \begin{bmatrix} a & b & c \\ d & e & f \\ g & h & i \end{bmatrix}$$

Compute $E_1 A$, $E_2 A$, and $E_3 A$, and describe how these products can be obtained by elementary row operations on A.

SOLUTION Verify that

$$E_1 A = \begin{bmatrix} a & b & c \\ d & e & f \\ g-4a & h-4b & i-4c \end{bmatrix}, \quad E_2 A = \begin{bmatrix} d & e & f \\ a & b & c \\ g & h & i \end{bmatrix},$$

$$E_3 A = \begin{bmatrix} a & b & c \\ d & e & f \\ 5g & 5h & 5i \end{bmatrix}.$$

Addition of -4 times row 1 of A to row 3 produces $E_1 A$. (This is a row replacement operation.) An interchange of rows 1 and 2 of A produces $E_2 A$, and multiplication of row 3 of A by 5 produces $E_3 A$. ∎

Left-multiplication (that is, multiplication on the left) by E_1 in Example 5 has the same effect on any $3 \times n$ matrix. It adds -4 times row 1 to row 3. In particular, since $E_1 \cdot I = E_1$, we see that E_1 *itself* is produced by this same row operation on the identity. Thus Example 5 illustrates the following general fact about elementary matrices. See Exercises 27 and 28.

> If an elementary row operation is performed on an $m \times n$ matrix A, the resulting matrix can be written as EA, where the $m \times m$ matrix E is created by performing the same row operation on I_m.

Since row operations are reversible, as shown in Section 1.1, elementary matrices are invertible, for if E is produced by a row operation on I, then there is another row operation of the same type that changes E back into I. Hence there is an elementary matrix F such that $FE = I$. Since E and F correspond to reverse operations, $EF = I$, too.

Each elementary matrix E is invertible. The inverse of E is the elementary matrix of the same type that transforms E back into I.

EXAMPLE 6   Find the inverse of $E_1 = \begin{bmatrix} 1 & 0 & 0 \\ 0 & 1 & 0 \\ -4 & 0 & 1 \end{bmatrix}$.

SOLUTION   To transform E_1 into I, add $+4$ times row 1 to row 3. The elementary matrix that does this is

$$E_1^{-1} = \begin{bmatrix} 1 & 0 & 0 \\ 0 & 1 & 0 \\ +4 & 0 & 1 \end{bmatrix}$$ ■

The following theorem provides the best way to "visualize" an invertible matrix, and the theorem leads immediately to a method for finding the inverse of a matrix.

THEOREM 7   An $n \times n$ matrix A is invertible if and only if A is row equivalent to I_n, and in this case, any sequence of elementary row operations that reduces A to I_n also transforms I_n into A^{-1}.

PROOF   Suppose that A is invertible. Then, since the equation $A\mathbf{x} = \mathbf{b}$ has a solution for each $\mathbf{b}$ (Theorem 5), A has a pivot position in every row (Theorem 4 in Section 1.4). Because A is square, the n pivot positions must be on the diagonal, which implies that the reduced echelon form of A is I_n. That is, $A \sim I_n$.

Now suppose, conversely, that $A \sim I_n$. Then, since each step of the row reduction of A corresponds to left-multiplication by an elementary matrix, there exist elementary matrices $E_1, \ldots, E_p$ such that

$$A \sim E_1 A \sim E_2(E_1 A) \sim \cdots \sim E_p(E_{p-1} \cdots E_1 A) = I_n$$

That is,

$$E_p \cdots E_1 A = I_n \tag{1}$$

Since the product $E_p \cdots E_1$ of invertible matrices is invertible, (1) leads to

$$(E_p \cdots E_1)^{-1}(E_p \cdots E_1)A = (E_p \cdots E_1)^{-1} I_n$$
$$A = (E_p \cdots E_1)^{-1}$$

Thus A is invertible, as it is the inverse of an invertible matrix (Theorem 6). Also,

$$A^{-1} = [(E_p \cdots E_1)^{-1}]^{-1} = E_p \cdots E_1$$

Then $A^{-1} = E_p \cdots E_1 \cdot I_n$, which says that A^{-1} results from applying $E_1, \ldots, E_p$ successively to I_n. This is the same sequence in (1) that reduced A to I_n. ■

An Algorithm for Finding A^{-1}

If we place A and I side-by-side to form an augmented matrix $[\,A \quad I\,]$, then row operations on this matrix produce identical operations on A and on I. By Theorem 7, either there are row operations that transform A to I_n and I_n to A^{-1} or else A is not invertible.

> **ALGORITHM FOR FINDING A^{-1}**
>
> Row reduce the augmented matrix $[\,A \quad I\,]$. If A is row equivalent to I, then $[\,A \quad I\,]$ is row equivalent to $[\,I \quad A^{-1}\,]$. Otherwise, A does not have an inverse.

EXAMPLE 7 Find the inverse of the matrix $A = \begin{bmatrix} 0 & 1 & 2 \\ 1 & 0 & 3 \\ 4 & -3 & 8 \end{bmatrix}$, if it exists.

SOLUTION

$$[\,A \quad I\,] = \begin{bmatrix} 0 & 1 & 2 & 1 & 0 & 0 \\ 1 & 0 & 3 & 0 & 1 & 0 \\ 4 & -3 & 8 & 0 & 0 & 1 \end{bmatrix} \sim \begin{bmatrix} 1 & 0 & 3 & 0 & 1 & 0 \\ 0 & 1 & 2 & 1 & 0 & 0 \\ 4 & -3 & 8 & 0 & 0 & 1 \end{bmatrix}$$

$$\sim \begin{bmatrix} 1 & 0 & 3 & 0 & 1 & 0 \\ 0 & 1 & 2 & 1 & 0 & 0 \\ 0 & -3 & -4 & 0 & -4 & 1 \end{bmatrix} \sim \begin{bmatrix} 1 & 0 & 3 & 0 & 1 & 0 \\ 0 & 1 & 2 & 1 & 0 & 0 \\ 0 & 0 & 2 & 3 & -4 & 1 \end{bmatrix}$$

$$\sim \begin{bmatrix} 1 & 0 & 3 & 0 & 1 & 0 \\ 0 & 1 & 2 & 1 & 0 & 0 \\ 0 & 0 & 1 & 3/2 & -2 & 1/2 \end{bmatrix}$$

$$\sim \begin{bmatrix} 1 & 0 & 0 & -9/2 & 7 & -3/2 \\ 0 & 1 & 0 & -2 & 4 & -1 \\ 0 & 0 & 1 & 3/2 & -2 & 1/2 \end{bmatrix}$$

Theorem 7 shows, since $A \sim I$, that A is invertible, and

$$A^{-1} = \begin{bmatrix} -9/2 & 7 & -3/2 \\ -2 & 4 & -1 \\ 3/2 & -2 & 1/2 \end{bmatrix}$$

It is a good idea to check the final answer:

$$AA^{-1} = \begin{bmatrix} 0 & 1 & 2 \\ 1 & 0 & 3 \\ 4 & -3 & 8 \end{bmatrix} \begin{bmatrix} -9/2 & 7 & -3/2 \\ -2 & 4 & -1 \\ 3/2 & -2 & 1/2 \end{bmatrix} = \begin{bmatrix} 1 & 0 & 0 \\ 0 & 1 & 0 \\ 0 & 0 & 1 \end{bmatrix}$$

It is not necessary to check that $A^{-1}A = I$ since A is invertible. ∎

Another View of Matrix Inversion

Denote the columns of I_n by $\mathbf{e}_1, \ldots, \mathbf{e}_n$. Then row reduction of $[\,A \quad I\,]$ to $[\,I \quad A^{-1}\,]$ can be viewed as the simultaneous solution of the n systems

$$A\mathbf{x} = \mathbf{e}_1, \quad A\mathbf{x} = \mathbf{e}_2, \quad \ldots, \quad A\mathbf{x} = \mathbf{e}_n \tag{2}$$

where the "augmented columns" of these systems have all been placed next to A to form $[\,A \quad \mathbf{e}_1 \quad \mathbf{e}_2 \quad \cdots \quad \mathbf{e}_n\,] = [\,A \quad I\,]$. The equation $AA^{-1} = I$ and the definition of matrix multiplication show that the columns of A^{-1} are precisely the solutions of the systems in (2). This observation is useful because some applied problems may require finding only one or two columns of A^{-1}. In this case, only the corresponding systems in (2) need be solved.

┌─ NUMERICAL NOTE ───

In practical work, A^{-1} is seldom computed, unless the entries of A^{-1} are needed.
Computing both A^{-1} and $A^{-1}\mathbf{b}$ takes about three times as many arithmetic
operations as solving $A\mathbf{x} = \mathbf{b}$ by row reduction, and row reduction may be more
accurate.

WEB

PRACTICE PROBLEMS

1. Use determinants to determine which of the following matrices are invertible.

   a. $\begin{bmatrix} 3 & -9 \\ 2 & 6 \end{bmatrix}$
   b. $\begin{bmatrix} 4 & -9 \\ 0 & 5 \end{bmatrix}$
   c. $\begin{bmatrix} 6 & -9 \\ -4 & 6 \end{bmatrix}$

2. Find the inverse of the matrix $A = \begin{bmatrix} 1 & -2 & -1 \\ -1 & 5 & 6 \\ 5 & -4 & 5 \end{bmatrix}$, if it exists.

2.2 EXERCISES

Find the inverses of the matrices in Exercises 1–4.

1. $\begin{bmatrix} 8 & 6 \\ 5 & 4 \end{bmatrix}$

2. $\begin{bmatrix} 3 & 2 \\ 8 & 5 \end{bmatrix}$

3. $\begin{bmatrix} 7 & 3 \\ -6 & -3 \end{bmatrix}$

4. $\begin{bmatrix} 2 & -4 \\ 4 & -6 \end{bmatrix}$

5. Use the inverse found in Exercise 1 to solve the system
$$8x_1 + 6x_2 = 2$$
$$5x_1 + 4x_2 = -1$$

6. Use the inverse found in Exercise 3 to solve the system
$$7x_1 + 3x_2 = -9$$
$$-6x_1 - 3x_2 = 4$$

7. Let $A = \begin{bmatrix} 1 & 2 \\ 5 & 12 \end{bmatrix}$, $\mathbf{b}_1 = \begin{bmatrix} -1 \\ 3 \end{bmatrix}$, $\mathbf{b}_2 = \begin{bmatrix} 1 \\ -5 \end{bmatrix}$, $\mathbf{b}_3 = \begin{bmatrix} 2 \\ 6 \end{bmatrix}$,
and $\mathbf{b}_4 = \begin{bmatrix} 3 \\ 5 \end{bmatrix}$.

   a. Find A^{-1}, and use it to solve the four equations
   $$A\mathbf{x} = \mathbf{b}_1, \quad A\mathbf{x} = \mathbf{b}_2, \quad A\mathbf{x} = \mathbf{b}_3, \quad A\mathbf{x} = \mathbf{b}_4$$

   b. The four equations in part (a) can be solved by the *same* set of row operations, since the coefficient matrix is the same in each case. Solve the four equations in part (a) by row reducing the augmented matrix $[\,A \quad \mathbf{b}_1 \quad \mathbf{b}_2 \quad \mathbf{b}_3 \quad \mathbf{b}_4\,]$.

8. Suppose P is invertible and $A = PBP^{-1}$. Solve for B in terms of A.

In Exercises 9 and 10, mark each statement True or False. Justify each answer.

9. a. In order for a matrix B to be the inverse of A, the equations $AB = I$ and $BA = I$ must both be true.

   b. If A and B are $n \times n$ and invertible, then $A^{-1}B^{-1}$ is the inverse of AB.

   c. If $A = \begin{bmatrix} a & b \\ c & d \end{bmatrix}$ and $ab - cd \neq 0$, then A is invertible.

   d. If A is an invertible $n \times n$ matrix, then the equation $A\mathbf{x} = \mathbf{b}$ is consistent for *each* $\mathbf{b}$ in $\mathbb{R}^n$.

   e. Each elementary matrix is invertible.

10. a. If A is invertible, then elementary row operations that reduce A to the identity I_n also reduce A^{-1} to I_n.

    b. If A is invertible, then the inverse of A^{-1} is A itself.

    c. A product of invertible $n \times n$ matrices is invertible, and the inverse of the product is the product of their inverses in the same order.

    d. If A is an $n \times n$ matrix and $A\mathbf{x} = \mathbf{e}_j$ is consistent for every $j \in \{1, 2, \ldots, n\}$, then A is invertible. Note: $\mathbf{e}_1, \ldots, \mathbf{e}_n$ represent the columns of the identity matrix.

    e. If A can be row reduced to the identity matrix, then A must be invertible.

11. Let A be an invertible $n \times n$ matrix, and let B be an $n \times p$ matrix. Show that the equation $AX = B$ has a unique solution $A^{-1}B$.

12. Use matrix algebra to show that if A is invertible and D satisfies $AD = I$, then $D = A^{-1}$.

13. Suppose $AB = AC$, where B and C are $n \times p$ matrices and A is invertible. Show that $B = C$. Is this true, in general, when A is not invertible?

14. Suppose $(B - C)D = 0$, where B and C are $m \times n$ matrices and D is invertible. Show that $B = C$.

15. Let A be an invertible $n \times n$ matrix, and let B be an $n \times p$ matrix. Explain why $A^{-1}B$ can be computed by row reduction:

If $[\,A \quad B\,] \sim \cdots \sim [\,I \quad X\,]$, then $X = A^{-1}B$.

If A is larger than 2×2, then row reduction of $[\,A \quad B\,]$ is much faster than computing both A^{-1} and $A^{-1}B$.

16. Suppose A and B are $n \times n$ matrices, B is invertible, and AB is invertible. Show that A is invertible. [*Hint:* Let $C = AB$, and solve this equation for A.]

17. Suppose A, B, and C are invertible $n \times n$ matrices. Show that ABC is also invertible by producing a matrix D such that $(ABC)D = I$ and $D(ABC) = I$.

18. Solve the equation $AB = BC$ for A, assuming that A, B, and C are square and B is invertible.

19. If A, B, and C are $n \times n$ invertible matrices, does the equation $C^{-1}(A + X)B^{-1} = I_n$ have a solution, X? If so, find it.

20. Suppose A, B, and X are $n \times n$ matrices with A, X, and $A - AX$ invertible, and suppose

$$(A - AX)^{-1} = X^{-1}B \tag{3}$$

a. Explain why B is invertible.

b. Solve equation (3) for X. If a matrix needs to be inverted, explain why that matrix is invertible.

21. Explain why the columns of an $n \times n$ matrix A are linearly independent when A is invertible.

22. Explain why the columns of an $n \times n$ matrix A span $\mathbb{R}^n$ when A is invertible. [*Hint:* Review Theorem 4 in Section 1.4.]

23. Suppose A is $n \times n$ and the equation $A\mathbf{x} = \mathbf{0}$ has only the trivial solution. Explain why A has n pivot columns and A is row equivalent to I_n. By Theorem 7, this shows that A must be invertible. (This exercise and Exercise 24 will be cited in Section 2.3.)

24. Suppose A is $n \times n$ and the equation $A\mathbf{x} = \mathbf{b}$ has a solution for each $\mathbf{b}$ in $\mathbb{R}^n$. Explain why A must be invertible. [*Hint:* Is A row equivalent to I_n?]

Exercises 25 and 26 prove Theorem 4 for $A = \begin{bmatrix} a & b \\ c & d \end{bmatrix}$.

25. Show that if $ad - bc = 0$, then the equation $A\mathbf{x} = \mathbf{0}$ has more than one solution. Why does this imply that A is not invertible? [*Hint:* First, consider $a = b = 0$. Then, if a and b are not both zero, consider the vector $\mathbf{x} = \begin{bmatrix} -b \\ a \end{bmatrix}$.]

26. Show that if $ad - bc \neq 0$, the formula for A^{-1} works.

Exercises 27 and 28 prove special cases of the facts about elementary matrices stated in the box following Example 5. Here A is a 3×3 matrix and $I = I_3$. (A general proof would require slightly more notation.)

27. Let A be a 3×3 matrix.

a. Use equation (2) from Section 2.1 to show that $\text{row}_i(A) = \text{row}_i(I) \cdot A$, for $i = 1, 2, 3$.

b. Show that if rows 1 and 2 of A are interchanged, then the result may be written as EA, where E is an elementary matrix formed by interchanging rows 1 and 2 of I.

c. Show that if row 3 of A is multiplied by 5, then the result may be written as EA, where E is formed by multiplying row 3 of I by 5.

28. Suppose row 2 of A is replaced by $\text{row}_2(A) - 3 \cdot \text{row}_1(A)$. Show that the result is EA, where E is formed from I by replacing $\text{row}_2(I)$ by $\text{row}_2(I) - 3 \cdot \text{row}_1(A)$.

Find the inverses of the matrices in Exercises 29–32, if they exist. Use the algorithm introduced in this section.

29. $\begin{bmatrix} 1 & -3 \\ 4 & -9 \end{bmatrix}$ **30.** $\begin{bmatrix} 3 & 6 \\ 4 & 7 \end{bmatrix}$

31. $\begin{bmatrix} 1 & 0 & -2 \\ -3 & 1 & 4 \\ 2 & -3 & 4 \end{bmatrix}$ **32.** $\begin{bmatrix} 1 & 2 & -1 \\ -4 & -7 & 3 \\ -2 & -6 & 4 \end{bmatrix}$

33. Use the algorithm from this section to find the inverses of

$$\begin{bmatrix} 1 & 0 & 0 \\ 1 & 1 & 0 \\ 1 & 1 & 1 \end{bmatrix} \quad \text{and} \quad \begin{bmatrix} 1 & 0 & 0 & 0 \\ 1 & 1 & 0 & 0 \\ 1 & 1 & 1 & 0 \\ 1 & 1 & 1 & 1 \end{bmatrix}.$$

Let A be the corresponding $n \times n$ matrix, and let B be its inverse. Guess the form of B, and then show that $AB = I$.

34. Repeat the strategy of Exercise 33 to guess the inverse B of

$$A = \begin{bmatrix} 1 & 0 & 0 & \cdots & 0 \\ 2 & 2 & 0 & & 0 \\ 3 & 3 & 3 & & 0 \\ \vdots & & & \ddots & \vdots \\ n & n & n & \cdots & n \end{bmatrix}.$$

Show that $AB = I$.

35. Let $A = \begin{bmatrix} -1 & -7 & -3 \\ 2 & 15 & 6 \\ 1 & 3 & 2 \end{bmatrix}$. Find the third column of A^{-1} without computing the other columns.

36. [M] Let $A = \begin{bmatrix} -25 & -9 & -27 \\ 536 & 185 & 537 \\ 154 & 52 & 143 \end{bmatrix}$. Find the second and third columns of A^{-1} without computing the first column.

37. Let $A = \begin{bmatrix} 1 & 2 \\ 1 & 3 \\ 1 & 5 \end{bmatrix}$. Construct a 2×3 matrix C (by trial and error) using only 1, -1, and 0 as entries, such that $CA = I_2$. Compute AC and note that $AC \neq I_3$.

38. Let $A = \begin{bmatrix} 1 & -1 & 1 & 0 \\ 0 & 1 & -1 & 1 \end{bmatrix}$. Construct a 4×2 matrix D using only 1 and 0 as entries, such that $AD = I_2$. Is it possible that $CA = I_4$ for some 4×2 matrix C? Why or why not?

39. [M] Let

$$D = \begin{bmatrix} .011 & .003 & .001 \\ .003 & .009 & .003 \\ .001 & .003 & .011 \end{bmatrix}$$

be a flexibility matrix, with flexibility measured in inches per pound. Suppose that forces of 40, 50, and 30 lb are applied at points 1, 2, and 3, respectively, in Fig. 1 of Example 3. Find the corresponding deflections.

40. [M] Compute the stiffness matrix D^{-1} for D in Exercise 39. List the forces needed to produce a deflection of .04 in. at point 3, with zero deflections at the other points.

41. [M] Let

$$D = \begin{bmatrix} .0130 & .0050 & .0020 & .0010 \\ .0050 & .0100 & .0040 & .0020 \\ .0020 & .0040 & .0100 & .0050 \\ .0010 & .0020 & .0050 & .0130 \end{bmatrix}$$

be a flexibility matrix for an elastic beam such as the one in Example 3, with four points at which force is applied. Units are centimeters per newton of force. Measurements at the four points show deflections of .07, .12, .16, and .12 cm. Determine the forces at the four points.

42. [M] With D as in Exercise 41, determine the forces that produce a deflection of .22 cm at the second point on the beam, with zero deflections at the other three points. How is the answer related to the entries in D^{-1}? [*Hint:* First answer the question when the deflection is 1 cm at the second point.]

SOLUTIONS TO PRACTICE PROBLEMS

1. a. $\det \begin{bmatrix} 3 & -9 \\ 2 & 6 \end{bmatrix} = 3 \cdot 6 - (-9) \cdot 2 = 18 + 18 = 36$. The determinant is nonzero, so the matrix is invertible.

b. $\det \begin{bmatrix} 4 & -9 \\ 0 & 5 \end{bmatrix} = 4 \cdot 5 - (-9) \cdot 0 = 20 \neq 0$. The matrix is invertible.

c. $\det \begin{bmatrix} 6 & -9 \\ -4 & 6 \end{bmatrix} = 6 \cdot 6 - (-9)(-4) = 36 - 36 = 0$. The matrix is not invertible.

2. $\begin{bmatrix} A & I \end{bmatrix} \sim \begin{bmatrix} 1 & -2 & -1 & 1 & 0 & 0 \\ -1 & 5 & 6 & 0 & 1 & 0 \\ 5 & -4 & 5 & 0 & 0 & 1 \end{bmatrix}$

$\sim \begin{bmatrix} 1 & -2 & -1 & 1 & 0 & 0 \\ 0 & 3 & 5 & 1 & 1 & 0 \\ 0 & 6 & 10 & -5 & 0 & 1 \end{bmatrix}$

$\sim \begin{bmatrix} 1 & -2 & -1 & 1 & 0 & 0 \\ 0 & 3 & 5 & 1 & 1 & 0 \\ 0 & 0 & 0 & -7 & -2 & 1 \end{bmatrix}$

So $\begin{bmatrix} A & I \end{bmatrix}$ is row equivalent to a matrix of the form $\begin{bmatrix} B & D \end{bmatrix}$, where B is square and has a row of zeros. Further row operations will not transform B into I, so we stop. A does not have an inverse.

2.3 | CHARACTERIZATIONS OF INVERTIBLE MATRICES

This section provides a review of most of the concepts introduced in Chapter 1, in relation to systems of n linear equations in n unknowns and to *square* matrices. The main result is Theorem 8.

THEOREM 8

The Invertible Matrix Theorem

Let A be a square $n \times n$ matrix. Then the following statements are equivalent. That is, for a given A, the statements are either all true or all false.

a. A is an invertible matrix.

b. A is row equivalent to the $n \times n$ identity matrix.

c. A has n pivot positions.

d. The equation $A\mathbf{x} = \mathbf{0}$ has only the trivial solution.

e. The columns of A form a linearly independent set.

f. The linear transformation $\mathbf{x} \mapsto A\mathbf{x}$ is one-to-one.

g. The equation $A\mathbf{x} = \mathbf{b}$ has at least one solution for each $\mathbf{b}$ in $\mathbb{R}^n$.

h. The columns of A span $\mathbb{R}^n$.

i. The linear transformation $\mathbf{x} \mapsto A\mathbf{x}$ maps $\mathbb{R}^n$ onto $\mathbb{R}^n$.

j. There is an $n \times n$ matrix C such that $CA = I$.

k. There is an $n \times n$ matrix D such that $AD = I$.

l. A^T is an invertible matrix.

FIGURE 1

First, we need some notation. If the truth of statement (a) always implies that statement (j) is true, we say that (a) *implies* (j) and write (a) $\Rightarrow$ (j). The proof will establish the "circle" of implications shown in Fig. 1. If any one of these five statements is true, then so are the others. Finally, the proof will link the remaining statements of the theorem to the statements in this circle.

PROOF If statement (a) is true, then A^{-1} works for C in (j), so (a) $\Rightarrow$ (j). Next, (j) $\Rightarrow$ (d) by Exercise 23 in Section 2.1. (Turn back and read the exercise.) Also, (d) $\Rightarrow$ (c) by Exercise 23 in Section 2.2. If A is square and has n pivot positions, then the pivots must lie on the main diagonal, in which case the reduced echelon form of A is I_n. Thus (c) $\Rightarrow$ (b). Also, (b) $\Rightarrow$ (a) by Theorem 7 in Section 2.2. This completes the circle in Fig. 1.

Next, (a) $\Rightarrow$ (k) because A^{-1} works for D. Also, (k) $\Rightarrow$ (g) by Exercise 26 in Section 2.1, and (g) $\Rightarrow$ (a) by Exercise 24 in Section 2.2. So (k) and (g) are linked to the circle. Further, (g), (h), and (i) are equivalent for any matrix, by Theorem 4 in Section 1.4 and Theorem 12(a) in Section 1.9. Thus, (h) and (i) are linked through (g) to the circle.

Since (d) is linked to the circle, so are (e) and (f), because (d), (e), and (f) are all equivalent for *any* matrix A. (See Section 1.7 and Theorem 12(b) in Section 1.9.) Finally, (a) $\Rightarrow$ (l) by Theorem 6(c) in Section 2.2, and (l) $\Rightarrow$ (a) by the same theorem with A and A^T interchanged. This completes the proof. ∎

Because of Theorem 5 in Section 2.2, statement (g) in Theorem 8 could also be written as "The equation $A\mathbf{x} = \mathbf{b}$ has a *unique* solution for each $\mathbf{b}$ in $\mathbb{R}^n$." This statement certainly implies (b) and hence implies that A is invertible.

The next fact follows from Theorem 8 and Exercise 12 in Section 2.2.

Let A and B be square matrices. If $AB = I$, then A and B are both invertible, with $B = A^{-1}$ and $A = B^{-1}$.

The Invertible Matrix Theorem divides the set of all $n \times n$ matrices into two disjoint classes: the invertible (nonsingular) matrices, and the noninvertible (singular) matrices. Each statement in the theorem describes a property of every $n \times n$ invertible matrix. The *negation* of a statement in the theorem describes a property of every $n \times n$ singular matrix. For instance, an $n \times n$ singular matrix is *not* row equivalent to I_n, does *not* have n pivot positions, and has linearly *dependent* columns. Negations of other statements are considered in the exercises.

EXAMPLE 1   Use the Invertible Matrix Theorem to decide if A is invertible:

$$A = \begin{bmatrix} 1 & 0 & -2 \\ 3 & 1 & -2 \\ -5 & -1 & 9 \end{bmatrix}$$

SOLUTION

$$A \sim \begin{bmatrix} 1 & 0 & -2 \\ 0 & 1 & 4 \\ 0 & -1 & -1 \end{bmatrix} \sim \begin{bmatrix} 1 & 0 & -2 \\ 0 & 1 & 4 \\ 0 & 0 & 3 \end{bmatrix}$$

So A has three pivot positions and hence is invertible, by the Invertible Matrix Theorem, statement (c).   ∎

SG   Expanded Table for the IMT 2-10

The power of the Invertible Matrix Theorem lies in the connections it provides among so many important concepts, such as linear independence of columns of a matrix A and the existence of solutions to equations of the form $A\mathbf{x} = \mathbf{b}$. It should be emphasized, however, that the Invertible Matrix Theorem *applies only to square matrices*. For example, if the columns of a 4×3 matrix are linearly independent, we cannot use the Invertible Matrix Theorem to conclude anything about the existence or nonexistence of solutions to equations of the form $A\mathbf{x} = \mathbf{b}$.

Invertible Linear Transformations

Recall from Section 2.1 that matrix multiplication corresponds to composition of linear transformations. When a matrix A is invertible, the equation $A^{-1}A\mathbf{x} = \mathbf{x}$ can be viewed as a statement about linear transformations. See Fig. 2.

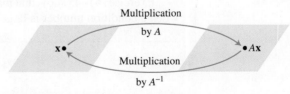

FIGURE 2 A^{-1} transforms $A\mathbf{x}$ back to $\mathbf{x}$.

A linear transformation $T : \mathbb{R}^n \to \mathbb{R}^n$ is said to be **invertible** if there exists a function $S : \mathbb{R}^n \to \mathbb{R}^n$ such that

$$S(T(\mathbf{x})) = \mathbf{x} \quad \text{for all } \mathbf{x} \text{ in } \mathbb{R}^n \tag{1}$$

$$T(S(\mathbf{x})) = \mathbf{x} \quad \text{for all } \mathbf{x} \text{ in } \mathbb{R}^n \tag{2}$$

The next theorem shows that if such an S exists, it is unique and must be a linear transformation. We call S the **inverse** of T and write it as T^{-1}.

THEOREM 9

Let $T : \mathbb{R}^n \to \mathbb{R}^n$ be a linear transformation and let A be the standard matrix for T. Then T is invertible if and only if A is an invertible matrix. In that case, the linear transformation S given by $S(\mathbf{x}) = A^{-1}\mathbf{x}$ is the unique function satisfying equations (1) and (2).

PROOF Suppose that T is invertible. Then (2) shows that T is onto $\mathbb{R}^n$, for if $\mathbf{b}$ is in $\mathbb{R}^n$ and $\mathbf{x} = S(\mathbf{b})$, then $T(\mathbf{x}) = T(S(\mathbf{b})) = \mathbf{b}$, so each $\mathbf{b}$ is in the range of T. Thus A is invertible, by the Invertible Matrix Theorem, statement (i).

Conversely, suppose that A is invertible, and let $S(\mathbf{x}) = A^{-1}\mathbf{x}$. Then, S is a linear transformation, and S obviously satisfies (1) and (2). For instance,

$$S(T(\mathbf{x})) = S(A\mathbf{x}) = A^{-1}(A\mathbf{x}) = \mathbf{x}$$

Thus T is invertible. The proof that S is unique is outlined in Exercise 38. ∎

EXAMPLE 2 What can you say about a one-to-one linear transformation T from $\mathbb{R}^n$ into $\mathbb{R}^n$?

SOLUTION The columns of the standard matrix A of T are linearly independent (by Theorem 12 in Section 1.9). So A is invertible, by the Invertible Matrix Theorem, and T maps $\mathbb{R}^n$ *onto* $\mathbb{R}^n$. Also, T is invertible, by Theorem 9. ∎

NUMERICAL NOTES

In practical work, you might occasionally encounter a "nearly singular" or **ill-conditioned** matrix—an invertible matrix that can become singular if some of its entries are changed ever so slightly. In this case, row reduction may produce fewer than n pivot positions, as a result of roundoff error. Also, roundoff error can sometimes make a singular matrix appear to be invertible.

WEB

Some matrix programs will compute a **condition number** for a square matrix. The larger the condition number, the closer the matrix is to being singular. The condition number of the identity matrix is 1. A singular matrix has an infinite condition number. In extreme cases, a matrix program may not be able to distinguish between a singular matrix and an ill-conditioned matrix.

Exercises 41–45 show that matrix computations can produce substantial error when a condition number is large.

PRACTICE PROBLEMS

1. Determine if $A = \begin{bmatrix} 2 & 3 & 4 \\ 2 & 3 & 4 \\ 2 & 3 & 4 \end{bmatrix}$ is invertible.

2. Suppose that for a certain $n \times n$ matrix A, statement (g) of the Invertible Matrix Theorem is *not* true. What can you say about equations of the form $A\mathbf{x} = \mathbf{b}$?

3. Suppose that A and B are $n \times n$ matrices and the equation $AB\mathbf{x} = \mathbf{0}$ has a nontrivial solution. What can you say about the matrix AB?

2.3 EXERCISES

Unless otherwise specified, assume that all matrices in these exercises are $n \times n$. Determine which of the matrices in Exercises 1–10 are invertible. Use as few calculations as possible. Justify your answers.

1. $\begin{bmatrix} 5 & 7 \\ -3 & -6 \end{bmatrix}$

2. $\begin{bmatrix} -4 & 2 \\ 6 & -3 \end{bmatrix}$

3. $\begin{bmatrix} 3 & 0 & 0 \\ -3 & -4 & 0 \\ 8 & 5 & -3 \end{bmatrix}$

4. $\begin{bmatrix} -5 & 1 & 4 \\ 0 & 0 & 0 \\ 1 & 4 & 9 \end{bmatrix}$

5. $\begin{bmatrix} 3 & 0 & -3 \\ 2 & 0 & 4 \\ -4 & 0 & 7 \end{bmatrix}$

6. $\begin{bmatrix} 1 & -3 & -6 \\ 0 & 4 & 3 \\ -3 & 6 & 0 \end{bmatrix}$

7. $\begin{bmatrix} -1 & -3 & 0 & 1 \\ 3 & 5 & 8 & -3 \\ -2 & -6 & 3 & 2 \\ 0 & -1 & 2 & 1 \end{bmatrix}$

8. $\begin{bmatrix} 3 & 4 & 7 & 4 \\ 0 & 1 & 4 & 6 \\ 0 & 0 & 2 & 8 \\ 0 & 0 & 0 & 1 \end{bmatrix}$

9. [M] $\begin{bmatrix} 4 & 0 & -3 & -7 \\ -6 & 9 & 9 & 9 \\ 7 & -5 & 10 & 19 \\ -1 & 2 & 4 & -1 \end{bmatrix}$

10. [M] $\begin{bmatrix} 5 & 3 & 1 & 7 & 9 \\ 6 & 4 & 2 & 8 & -8 \\ 7 & 5 & 3 & 10 & 9 \\ 9 & 6 & 4 & -9 & -5 \\ 8 & 5 & 2 & 11 & 4 \end{bmatrix}$

In Exercises 11 and 12, the matrices are all $n \times n$. Each part of the exercises is an *implication* of the form "If ⟨ statement 1 ⟩, then ⟨ statement 2 ⟩." Mark an implication as True if the truth of ⟨ statement 2 ⟩ *always* follows whenever ⟨ statement 1 ⟩ happens to be true. An implication is False if there is an instance in which ⟨ statement 2 ⟩ is false but ⟨ statement 1 ⟩ is true. Justify each answer.

11. a. If the equation $A\mathbf{x} = \mathbf{0}$ has only the trivial solution, then A is row equivalent to the $n \times n$ identity matrix.

   b. If the columns of A span $\mathbb{R}^n$, then the columns are linearly independent.

   c. If A is an $n \times n$ matrix, then the equation $A\mathbf{x} = \mathbf{b}$ has at least one solution for each $\mathbf{b}$ in $\mathbb{R}^n$.

   d. If the equation $A\mathbf{x} = \mathbf{0}$ has a nontrivial solution, then A has fewer than n pivot positions.

   e. If A^T is not invertible, then A is not invertible.

12. a. If there is an $n \times n$ matrix D such that $AD = I$, then $DA = I$.

   b. If the linear transformation $\mathbf{x} \mapsto A\mathbf{x}$ maps $\mathbb{R}^n$ into $\mathbb{R}^n$, then the row reduced echelon form of A is I.

   c. If the columns of A are linearly independent, then the columns of A span $\mathbb{R}^n$.

   d. If the equation $A\mathbf{x} = \mathbf{b}$ has at least one solution for each $\mathbf{b}$ in $\mathbb{R}^n$, then the transformation $\mathbf{x} \mapsto A\mathbf{x}$ is not one-to-one.

   e. If there is a $\mathbf{b}$ in $\mathbb{R}^n$ such that the equation $A\mathbf{x} = \mathbf{b}$ is consistent, then the solution is unique.

13. An $m \times n$ **upper triangular matrix** is one whose entries *below* the main diagonal are 0's (as in Exercise 8). When is a square upper triangular matrix invertible? Justify your answer.

14. An $m \times n$ **lower triangular matrix** is one whose entries *above* the main diagonal are 0's (as in Exercise 3). When is a square lower triangular matrix invertible? Justify your answer.

15. Is it possible for a 4×4 matrix to be invertible when its columns do not span $\mathbb{R}^4$? Why or why not?

16. If an $n \times n$ matrix A is invertible, then the columns of A^T are linearly independent. Explain why.

17. Can a square matrix with two identical columns be invertible? Why or why not?

18. Can a square matrix with two identical rows be invertible? Why or why not?

19. If the columns of a 7×7 matrix D are linearly independent, what can be said about the solutions of $D\mathbf{x} = \mathbf{b}$? Why?

20. If A is a 5×5 matrix and the equation $A\mathbf{x} = \mathbf{b}$ is consistent for every $\mathbf{b}$ in $\mathbb{R}^5$, is it possible that for some $\mathbf{b}$, the equation $A\mathbf{x} = \mathbf{b}$ has more than one solution? Why or why not?

21. If the equation $C\mathbf{u} = \mathbf{v}$ has more than one solution for some $\mathbf{v}$ in $\mathbb{R}^n$, can the columns of the $n \times n$ matrix C span $\mathbb{R}^n$? Why or why not?

22. If $n \times n$ matrices E and F have the property that $EF = I$, then E and F commute. Explain why.

23. Assume that F is an $n \times n$ matrix. If the equation $F\mathbf{x} = \mathbf{y}$ is inconsistent for some $\mathbf{y}$ in $\mathbb{R}^n$, what can you say about the equation $F\mathbf{x} = \mathbf{0}$? Why?

24. If an $n \times n$ matrix G cannot be row reduced to I_n, what can you say about the columns of G? Why?

25. Verify the boxed statement preceding Example 1.

26. Explain why the columns of A^2 span $\mathbb{R}^n$ whenever the columns of an $n \times n$ matrix A are linearly independent.

27. Let A and B be $n \times n$ matrices. Show that if AB is invertible, so is A. You cannot use Theorem 6(b), because you cannot *assume* that A and B are invertible. [*Hint:* There is a matrix W such that $ABW = I$. Why?]

28. Let A and B be $n \times n$ matrices. Show that if AB is invertible, so is B.

29. If A is an $n \times n$ matrix and the transformation $\mathbf{x} \mapsto A\mathbf{x}$ is one-to-one, what else can you say about this transformation? Justify your answer.

30. If A is an $n \times n$ matrix and the equation $Ax = b$ has more than one solution for some b, then the transformation $x \mapsto Ax$ is not one-to-one. What else can you say about this transformation? Justify your answer.

31. Suppose A is an $n \times n$ matrix with the property that the equation $Ax = b$ has at least one solution for each b in $\mathbb{R}^n$. Without using Theorems 5 or 8, explain why each equation $Ax = b$ has in fact exactly one solution.

32. Suppose A is an $n \times n$ matrix with the property that the equation $Ax = 0$ has only the trivial solution. Without using the Invertible Matrix Theorem, explain directly why the equation $Ax = b$ must have a solution for each b in $\mathbb{R}^n$.

In Exercises 33 and 34, T is a linear transformation from $\mathbb{R}^2$ into $\mathbb{R}^2$. Show that T is invertible and find a formula for T^{-1}.

33. $T(x_1, x_2) = (-5x_1 + 9x_2, 4x_1 - 7x_2)$

34. $T(x_1, x_2) = (2x_1 - 8x_2, -2x_1 + 7x_2)$

35. Let $T : \mathbb{R}^n \to \mathbb{R}^n$ be an invertible linear transformation. Explain why T is both one-to-one and onto $\mathbb{R}^n$. Use equations (1) and (2). Then give a second explanation using one or more theorems.

36. Suppose a linear transformation $T : \mathbb{R}^n \to \mathbb{R}^n$ has the property that $T(u) = T(v)$ for some pair of distinct vectors u and v in $\mathbb{R}^n$. Can T map $\mathbb{R}^n$ onto $\mathbb{R}^n$? Why or why not?

37. Suppose T and U are linear transformations from $\mathbb{R}^n$ to $\mathbb{R}^n$ such that $T(U(x)) = x$ for all x in $\mathbb{R}^n$. Is it true that $U(T(x)) = x$ for all x in $\mathbb{R}^n$? Why or why not?

38. Let $T : \mathbb{R}^n \to \mathbb{R}^n$ be an invertible linear transformation, and let S and U be functions from $\mathbb{R}^n$ into $\mathbb{R}^n$ such that $S(T(x)) = x$ and $U(T(x)) = x$ for all x in $\mathbb{R}^n$. Show that $U(v) = S(v)$ for all v in $\mathbb{R}^n$. This will show that T has a unique inverse, as asserted in Theorem 9. [*Hint:* Given any v in $\mathbb{R}^n$, we can write $v = T(x)$ for some x. Why? Compute $S(v)$ and $U(v)$.]

39. Let T be a linear transformation that maps $\mathbb{R}^n$ onto $\mathbb{R}^n$. Show that T^{-1} exists and maps $\mathbb{R}^n$ onto $\mathbb{R}^n$. Is T^{-1} also one-to-one?

40. Suppose T and S satisfy the invertibility equations (1) and (2), where T is a linear transformation. Show directly that S is a linear transformation. [*Hint:* Given u, v in $\mathbb{R}^n$, let $x = S(u), y = S(v)$. Then $T(x) = u, T(y) = v$. Why? Apply S to both sides of the equation $T(x) + T(y) = T(x + y)$. Also, consider $T(cx) = cT(x)$.]

41. **[M]** Suppose an experiment leads to the following system of equations:

$$\begin{aligned} 4.5x_1 + 3.1x_2 &= 19.249 \\ 1.6x_1 + 1.1x_2 &= 6.843 \end{aligned} \tag{3}$$

a. Solve system (3), and then solve system (4), below, in which the data on the right have been rounded to two decimal places. In each case, find the *exact* solution.

$$\begin{aligned} 4.5x_1 + 3.1x_2 &= 19.25 \\ 1.6x_1 + 1.1x_2 &= 6.84 \end{aligned} \tag{4}$$

b. The entries in system (4) differ from those in system (3) by less than .05%. Find the percentage error when using the solution of (4) as an approximation for the solution of (3).

c. Use a matrix program to produce the condition number of the coefficient matrix in (3).

Exercises 42–44 show how to use the condition number of a matrix A to estimate the accuracy of a computed solution of $Ax = b$. If the entries of A and b are accurate to about r significant digits and if the condition number of A is approximately 10^k (with k a positive integer), then the computed solution of $Ax = b$ should usually be accurate to at least $r - k$ significant digits.

42. **[M]** Let A be the matrix in Exercise 9. Find the condition number of A. Construct a random vector x in $\mathbb{R}^4$ and compute $b = Ax$. Then use a matrix program to compute the solution x_1 of $Ax = b$. To how many digits do x and x_1 agree? Find out the number of digits the matrix program stores accurately, and report how many digits of accuracy are lost when x_1 is used in place of the exact solution x.

43. **[M]** Repeat Exercise 42 for the matrix in Exercise 10.

44. **[M]** Solve an equation $Ax = b$ for a suitable b to find the last column of the inverse of the *fifth-order Hilbert matrix*

$$A = \begin{bmatrix} 1 & 1/2 & 1/3 & 1/4 & 1/5 \\ 1/2 & 1/3 & 1/4 & 1/5 & 1/6 \\ 1/3 & 1/4 & 1/5 & 1/6 & 1/7 \\ 1/4 & 1/5 & 1/6 & 1/7 & 1/8 \\ 1/5 & 1/6 & 1/7 & 1/8 & 1/9 \end{bmatrix}$$

How many digits in each entry of x do you expect to be correct? Explain. [*Note:* The exact solution is $(630, -12600, 56700, -88200, 44100)$.]

45. **[M]** Some matrix programs, such as MATLAB, have a command to create Hilbert matrices of various sizes. If possible, use an inverse command to compute the inverse of a twelfth-order or larger Hilbert matrix, A. Compute AA^{-1}. Report what you find.

| **SG** | Mastering: Reviewing and Reflecting 2–13 |

SOLUTIONS TO PRACTICE PROBLEMS

1. The columns of A are obviously linearly dependent because columns 2 and 3 are multiples of column 1. Hence A cannot be invertible, by the Invertible Matrix Theorem.

2. If statement (g) is *not* true, then the equation $A\mathbf{x} = \mathbf{b}$ is inconsistent for at least one $\mathbf{b}$ in $\mathbb{R}^n$.

3. Apply the Invertible Matrix Theorem to the matrix AB in place of A. Then statement (d) becomes: $AB\mathbf{x} = \mathbf{0}$ has only the trivial solution. This is not true. So AB is not invertible.

2.4 | PARTITIONED MATRICES

A key feature of our work with matrices has been the ability to regard a matrix A as a list of column vectors rather than just a rectangular array of numbers. This point of view has been so useful that we wish to consider other **partitions** of A, indicated by horizontal and vertical dividing rules, as in Example 1 below. Partitioned matrices appear in most modern applications of linear algebra because the notation highlights essential structures in matrix analysis, as in the chapter introductory example on aircraft design. This section provides an opportunity to review matrix algebra and use the Invertible Matrix Theorem.

EXAMPLE 1   The matrix

$$A = \begin{bmatrix} 3 & 0 & -1 & 5 & 9 & -2 \\ -5 & 2 & 4 & 0 & -3 & 1 \\ -8 & -6 & 3 & 1 & 7 & -4 \end{bmatrix}$$

can also be written as the 2×3 **partitioned** (or **block**) **matrix**

$$A = \begin{bmatrix} A_{11} & A_{12} & A_{13} \\ A_{21} & A_{22} & A_{23} \end{bmatrix}$$

whose entries are the *blocks* (or *submatrices*)

$$A_{11} = \begin{bmatrix} 3 & 0 & -1 \\ -5 & 2 & 4 \end{bmatrix}, \quad A_{12} = \begin{bmatrix} 5 & 9 \\ 0 & -3 \end{bmatrix}, \quad A_{13} = \begin{bmatrix} -2 \\ 1 \end{bmatrix}$$

$$A_{21} = \begin{bmatrix} -8 & -6 & 3 \end{bmatrix}, \quad A_{22} = \begin{bmatrix} 1 & 7 \end{bmatrix}, \quad A_{23} = \begin{bmatrix} -4 \end{bmatrix} \quad \blacksquare$$

EXAMPLE 2   When a matrix A appears in a mathematical model of a physical system such as an electrical network, a transportation system, or a large corporation, it may be natural to regard A as a partitioned matrix. For instance, if a microcomputer circuit board consists mainly of three VLSI (very large-scale integrated) microchips, then the matrix for the circuit board might have the general form

$$A = \begin{bmatrix} A_{11} & A_{12} & A_{13} \\ A_{21} & A_{22} & A_{23} \\ A_{31} & A_{32} & A_{33} \end{bmatrix}$$

The submatrices on the "diagonal" of A—namely, A_{11}, A_{22}, and A_{33}—concern the three VLSI chips, while the other submatrices depend on the interconnections among those microchips. $\blacksquare$

Addition and Scalar Multiplication

If matrices A and B are the same size and are partitioned in exactly the same way, then it is natural to make the same partition of the ordinary matrix sum $A + B$. In this case, each block of $A + B$ is the (matrix) sum of the corresponding blocks of A and B. Multiplication of a partitioned matrix by a scalar is also computed block by block.

Multiplication of Partitioned Matrices

Partitioned matrices can be multiplied by the usual row–column rule as if the block entries were scalars, provided that for a product AB, the column partition of A matches the row partition of B.

EXAMPLE 3 Let

$$
A = \begin{bmatrix} 2 & -3 & 1 & 0 & -4 \\ 1 & 5 & -2 & 3 & -1 \\ 0 & -4 & -2 & 7 & -1 \end{bmatrix} = \begin{bmatrix} A_{11} & A_{12} \\ A_{21} & A_{22} \end{bmatrix}, \quad
B = \begin{bmatrix} 6 & 4 \\ -2 & 1 \\ -3 & 7 \\ -1 & 3 \\ 5 & 2 \end{bmatrix} = \begin{bmatrix} B_1 \\ B_2 \end{bmatrix}
$$

The 5 columns of A are partitioned into a set of 3 columns and then a set of 2 columns. The 5 rows of B are partitioned in the same way—into a set of 3 rows and then a set of 2 rows. We say that the partitions of A and B are **conformable** for **block multiplication**. It can be shown that the ordinary product AB can be written as

$$
AB = \begin{bmatrix} A_{11} & A_{12} \\ A_{21} & A_{22} \end{bmatrix} \begin{bmatrix} B_1 \\ B_2 \end{bmatrix} = \begin{bmatrix} A_{11}B_1 + A_{12}B_2 \\ A_{21}B_1 + A_{22}B_2 \end{bmatrix} = \begin{bmatrix} -5 & 4 \\ -6 & 2 \\ 2 & 1 \end{bmatrix}
$$

It is important for each smaller product in the expression for AB to be written with the submatrix from A on the left, since matrix multiplication is not commutative. For instance,

$$
A_{11}B_1 = \begin{bmatrix} 2 & -3 & 1 \\ 1 & 5 & -2 \end{bmatrix} \begin{bmatrix} 6 & 4 \\ -2 & 1 \\ -3 & 7 \end{bmatrix} = \begin{bmatrix} 15 & 12 \\ 2 & -5 \end{bmatrix}
$$

$$
A_{12}B_2 = \begin{bmatrix} 0 & -4 \\ 3 & -1 \end{bmatrix} \begin{bmatrix} -1 & 3 \\ 5 & 2 \end{bmatrix} = \begin{bmatrix} -20 & -8 \\ -8 & 7 \end{bmatrix}
$$

Hence the top block in AB is

$$
A_{11}B_1 + A_{12}B_2 = \begin{bmatrix} 15 & 12 \\ 2 & -5 \end{bmatrix} + \begin{bmatrix} -20 & -8 \\ -8 & 7 \end{bmatrix} = \begin{bmatrix} -5 & 4 \\ -6 & 2 \end{bmatrix} \qquad \blacksquare
$$

The row–column rule for multiplication of block matrices provides the most general way to regard the product of two matrices. Each of the following views of a product has already been described using simple partitions of matrices: (1) the definition of $A\mathbf{x}$ using the columns of A, (2) the column definition of AB, (3) the row–column rule for computing AB, and (4) the rows of AB as products of the rows of A and the matrix B. A fifth view of AB, again using partitions, follows in Theorem 10 below.

The calculations in the next example prepare the way for Theorem 10. Here $\mathrm{col}_k(A)$ is the kth column of A, and $\mathrm{row}_k(B)$ is the kth row of B.

EXAMPLE 4 Let $A = \begin{bmatrix} -3 & 1 & 2 \\ 1 & -4 & 5 \end{bmatrix}$ and $B = \begin{bmatrix} a & b \\ c & d \\ e & f \end{bmatrix}$. Verify that

$$AB = \text{col}_1(A)\,\text{row}_1(B) + \text{col}_2(A)\,\text{row}_2(B) + \text{col}_3(A)\,\text{row}_3(B)$$

SOLUTION Each term above is an *outer product*. (See Exercises 27 and 28 in Section 2.1.) By the row–column rule for computing a matrix product,

$$\text{col}_1(A)\,\text{row}_1(B) = \begin{bmatrix} -3 \\ 1 \end{bmatrix}\begin{bmatrix} a & b \end{bmatrix} = \begin{bmatrix} -3a & -3b \\ a & b \end{bmatrix}$$

$$\text{col}_2(A)\,\text{row}_2(B) = \begin{bmatrix} 1 \\ -4 \end{bmatrix}\begin{bmatrix} c & d \end{bmatrix} = \begin{bmatrix} c & d \\ -4c & -4d \end{bmatrix}$$

$$\text{col}_3(A)\,\text{row}_3(B) = \begin{bmatrix} 2 \\ 5 \end{bmatrix}\begin{bmatrix} e & f \end{bmatrix} = \begin{bmatrix} 2e & 2f \\ 5e & 5f \end{bmatrix}$$

Thus

$$\sum_{k=1}^{3} \text{col}_k(A)\,\text{row}_k(B) = \begin{bmatrix} -3a + c + 2e & -3b + d + 2f \\ a - 4c + 5e & b - 4d + 5f \end{bmatrix}$$

This matrix is obviously AB. Notice that the $(1, 1)$-entry in AB is the sum of the $(1, 1)$-entries in the three outer products, the $(1, 2)$-entry in AB is the sum of the $(1, 2)$-entries in the three outer products, and so on. ∎

THEOREM 10

Column–Row Expansion of *AB*

If A is $m \times n$ and B is $n \times p$, then

$$AB = \begin{bmatrix} \text{col}_1(A) & \text{col}_2(A) & \cdots & \text{col}_n(A) \end{bmatrix}\begin{bmatrix} \text{row}_1(B) \\ \text{row}_2(B) \\ \vdots \\ \text{row}_n(B) \end{bmatrix} \tag{1}$$

$$= \text{col}_1(A)\,\text{row}_1(B) + \cdots + \text{col}_n(A)\,\text{row}_n(B)$$

PROOF For each row index i and column index j, the (i, j)-entry in $\text{col}_k(A)\,\text{row}_k(B)$ is the product of a_{ik} from $\text{col}_k(A)$ and b_{kj} from $\text{row}_k(B)$. Hence the (i, j)-entry in the sum shown in equation (1) is

$$\underset{(k=1)}{a_{i1}b_{1j}} + \underset{(k=2)}{a_{i2}b_{2j}} + \cdots + \underset{(k=n)}{a_{in}b_{nj}}$$

This sum is also the (i, j)-entry in AB, by the row–column rule. ∎

Inverses of Partitioned Matrices

The next example illustrates calculations involving inverses and partitioned matrices.

EXAMPLE 5 A matrix of the form

$$A = \begin{bmatrix} A_{11} & A_{12} \\ 0 & A_{22} \end{bmatrix}$$

is said to be *block upper triangular*. Assume that A_{11} is $p \times p$, A_{22} is $q \times q$, and A is invertible. Find a formula for A^{-1}.

SOLUTION Denote A^{-1} by B and partition B so that

$$\begin{bmatrix} A_{11} & A_{12} \\ 0 & A_{22} \end{bmatrix} \begin{bmatrix} B_{11} & B_{12} \\ B_{21} & B_{22} \end{bmatrix} = \begin{bmatrix} I_p & 0 \\ 0 & I_q \end{bmatrix} \tag{2}$$

This matrix equation provides four equations that will lead to the unknown blocks $B_{11}, \ldots, B_{22}$. Compute the product on the left side of equation (2), and equate each entry with the corresponding block in the identity matrix on the right. That is, set

$$A_{11}B_{11} + A_{12}B_{21} = I_p \tag{3}$$
$$A_{11}B_{12} + A_{12}B_{22} = 0 \tag{4}$$
$$A_{22}B_{21} = 0 \tag{5}$$
$$A_{22}B_{22} = I_q \tag{6}$$

By itself, equation (6) does not show that A_{22} is invertible. However, since A_{22} is square, the Invertible Matrix Theorem and (6) together show that A_{22} is invertible and $B_{22} = A_{22}^{-1}$. Next, left-multiply both sides of (5) by A_{22}^{-1} and obtain

$$B_{21} = A_{22}^{-1}0 = 0$$

so that (3) simplifies to

$$A_{11}B_{11} + 0 = I_p$$

Since A_{11} is square, this shows that A_{11} is invertible and $B_{11} = A_{11}^{-1}$. Finally, use these results with (4) to find that

$$A_{11}B_{12} = -A_{12}B_{22} = -A_{12}A_{22}^{-1} \quad \text{and} \quad B_{12} = -A_{11}^{-1}A_{12}A_{22}^{-1}$$

Thus

$$A^{-1} = \begin{bmatrix} A_{11} & A_{12} \\ 0 & A_{22} \end{bmatrix}^{-1} = \begin{bmatrix} A_{11}^{-1} & -A_{11}^{-1}A_{12}A_{22}^{-1} \\ 0 & A_{22}^{-1} \end{bmatrix} \quad \blacksquare$$

A **block diagonal matrix** is a partitioned matrix with zero blocks off the main diagonal (of blocks). Such a matrix is invertible if and only if each block on the diagonal is invertible. See Exercises 13 and 14.

NUMERICAL NOTES

1. When matrices are too large to fit in a computer's high-speed memory, partitioning permits the computer to work with only two or three submatrices at a time. For instance, one linear programming research team simplified a problem by partitioning the matrix into 837 rows and 51 columns. The problem's solution took about 4 minutes on a Cray supercomputer.[1]

2. Some high-speed computers, particularly those with vector pipeline architecture, perform matrix calculations more efficiently when the algorithms use partitioned matrices.[2]

3. Professional software for high-performance numerical linear algebra, such as LAPACK, makes intensive use of partitioned matrix calculations.

---

[1] The solution time doesn't sound too impressive until you learn that each of the 51 block columns contained about 250,000 individual columns. The original problem had 837 equations and over 12,750,000 variables! Nearly 100 million of the more than 10 billion entries in the matrix were nonzero. See Robert E. Bixby et al., "Very Large-Scale Linear Programming: A Case Study in Combining Interior Point and Simplex Methods," *Operations Research*, 40, no. 5 (1992): 885–897.

[2] The importance of block matrix algorithms for computer calculations is described in *Matrix Computations*, 3rd ed., by Gene H. Golub and Charles F. van Loan (Baltimore: Johns Hopkins University Press, 1996).

The exercises that follow give practice with matrix algebra and illustrate typical calculations found in applications.

PRACTICE PROBLEMS

1. Show that $\begin{bmatrix} I & 0 \\ A & I \end{bmatrix}$ is invertible and find its inverse.

2. Compute $X^T X$, where X is partitioned as $\begin{bmatrix} X_1 & X_2 \end{bmatrix}$.

2.4 EXERCISES

In Exercises 1–9, assume that the matrices are partitioned conformably for block multiplication. Compute the products shown in Exercises 1–4.

1. $\begin{bmatrix} I & 0 \\ E & I \end{bmatrix} \begin{bmatrix} A & B \\ C & D \end{bmatrix}$
     2. $\begin{bmatrix} E & 0 \\ 0 & F \end{bmatrix} \begin{bmatrix} P & Q \\ R & S \end{bmatrix}$

3. $\begin{bmatrix} 0 & I \\ I & 0 \end{bmatrix} \begin{bmatrix} A & B \\ C & D \end{bmatrix}$
     4. $\begin{bmatrix} I & 0 \\ -E & I \end{bmatrix} \begin{bmatrix} W & X \\ Y & Z \end{bmatrix}$

In Exercises 5–8, find formulas for X, Y, and Z in terms of A, B, and C, and justify your calculations. In some cases, you may need to make assumptions about the size of a matrix in order to produce a formula. [*Hint:* Compute the product on the left, and set it equal to the right side.]

5. $\begin{bmatrix} A & B \\ C & 0 \end{bmatrix} \begin{bmatrix} I & 0 \\ X & Y \end{bmatrix} = \begin{bmatrix} 0 & I \\ Z & 0 \end{bmatrix}$

6. $\begin{bmatrix} X & 0 \\ Y & Z \end{bmatrix} \begin{bmatrix} A & 0 \\ B & C \end{bmatrix} = \begin{bmatrix} I & 0 \\ 0 & I \end{bmatrix}$

7. $\begin{bmatrix} X & 0 & 0 \\ Y & 0 & I \end{bmatrix} \begin{bmatrix} A & Z \\ 0 & 0 \\ B & I \end{bmatrix} = \begin{bmatrix} I & 0 \\ 0 & I \end{bmatrix}$

8. $\begin{bmatrix} A & B \\ 0 & I \end{bmatrix} \begin{bmatrix} X & Y & Z \\ 0 & 0 & I \end{bmatrix} = \begin{bmatrix} I & 0 & 0 \\ 0 & 0 & I \end{bmatrix}$

9. Suppose B_{11} is an invertible matrix. Find matrices A_{21} and A_{31} (in terms of the blocks of B) such that the product below has the form indicated. Also, compute C_{22} (in terms of the blocks of B). [*Hint:* Compute the product on the left, and set it equal to the right side.]

$$\begin{bmatrix} I & 0 & 0 \\ A_{21} & I & 0 \\ A_{31} & 0 & I \end{bmatrix} \begin{bmatrix} B_{11} & B_{12} \\ B_{21} & B_{22} \\ B_{31} & B_{32} \end{bmatrix} = \begin{bmatrix} C_{11} & C_{12} \\ 0 & C_{22} \\ 0 & C_{32} \end{bmatrix}$$

10. The inverse of

$$\begin{bmatrix} I & 0 & 0 \\ A & I & 0 \\ B & D & I \end{bmatrix} \text{ is } \begin{bmatrix} I & 0 & 0 \\ P & I & 0 \\ Q & R & I \end{bmatrix}.$$

Find P, Q, and R.

In Exercises 11 and 12, mark each statement True or False. Justify each answer.

11. a. If $A = \begin{bmatrix} A_1 & A_2 \end{bmatrix}$ and $B = \begin{bmatrix} B_1 & B_2 \end{bmatrix}$, with A_1 and A_2 the same sizes as B_1 and B_2, respectively, then $A + B = \begin{bmatrix} A_1 + B_1 & A_2 + B_2 \end{bmatrix}$.

     b. If $A = \begin{bmatrix} A_{11} & A_{12} \\ A_{21} & A_{22} \end{bmatrix}$ and $B = \begin{bmatrix} B_1 \\ B_2 \end{bmatrix}$, then the partitions of A and B are conformable for block multiplication.

12. a. If A_1, A_2, B_1, and B_2 are $n \times n$ matrices, $A = \begin{bmatrix} A_1 \\ A_2 \end{bmatrix}$, and $B = \begin{bmatrix} B_1 & B_2 \end{bmatrix}$, then the product BA is defined, but AB is not.

     b. If $A = \begin{bmatrix} P & Q \\ R & S \end{bmatrix}$, then the transpose of A is

$$A^T = \begin{bmatrix} P^T & Q^T \\ R^T & S^T \end{bmatrix}.$$

13. Let $A = \begin{bmatrix} B & 0 \\ 0 & C \end{bmatrix}$, where B and C are square. Show that A is invertible if and only if both B and C are invertible.

14. Show that the block upper triangular matrix A in Example 5 is invertible if and only if both A_{11} and A_{22} are invertible. [*Hint:* If A_{11} and A_{22} are invertible, the formula for A^{-1} given in Example 5 actually works as the inverse of A.] This fact about A is an important part of several computer algorithms that estimate eigenvalues of matrices. Eigenvalues are discussed in Chapter 5.

15. When a deep space probe is launched, corrections may be necessary to place the probe on a precisely calculated trajectory. Radio telemetry provides a stream of vectors, $\mathbf{x}_1, \dots, \mathbf{x}_k$, giving information at different times about how the probe's position compares with its planned trajectory. Let X_k be the matrix $[\mathbf{x}_1 \; \cdots \; \mathbf{x}_k]$. The matrix $G_k = X_k X_k^T$ is computed as the radar data are analyzed. When $\mathbf{x}_{k+1}$ arrives, a new G_{k+1} must be computed. Since the data vectors arrive at high speed, the computational burden could be severe. But partitioned matrix multiplication helps tremendously. Compute the column–row expansions of G_k and G_{k+1}, and describe what must be computed in order to *update* G_k to form G_{k+1}.

The probe Galileo was launched October 18, 1989, and arrived near Jupiter in early December 1995.

16. Let $A = \begin{bmatrix} A_{11} & A_{12} \\ A_{21} & A_{22} \end{bmatrix}$. If A_{11} is invertible, then the matrix $S = A_{22} - A_{21}A_{11}^{-1}A_{12}$ is called the **Schur complement** of A_{11}. Likewise, if A_{22} is invertible, the matrix $A_{11} - A_{12}A_{22}^{-1}A_{21}$ is called the Schur complement of A_{22}. Suppose A_{11} is invertible. Find X and Y such that

$$\begin{bmatrix} A_{11} & A_{12} \\ A_{21} & A_{22} \end{bmatrix} = \begin{bmatrix} I & 0 \\ X & I \end{bmatrix}\begin{bmatrix} A_{11} & 0 \\ 0 & S \end{bmatrix}\begin{bmatrix} I & Y \\ 0 & I \end{bmatrix} \qquad (7)$$

17. Suppose the block matrix A on the left side of (7) is invertible and A_{11} is invertible. Show that the Schur complement S of A_{11} is invertible. [*Hint:* The outside factors on the right side of (7) are always invertible. Verify this.] When A and A_{11} are both invertible, (7) leads to a formula for A^{-1}, using S^{-1}, A_{11}^{-1}, and the other entries in A.

18. Let X be an $m \times n$ data matrix such that $X^T X$ is invertible, and let $M = I_m - X(X^T X)^{-1}X^T$. Add a column $\mathbf{x}_0$ to the data and form

$$W = \begin{bmatrix} X & \mathbf{x}_0 \end{bmatrix}$$

Compute $W^T W$. The $(1, 1)$-entry is $X^T X$. Show that the Schur complement (Exercise 16) of $X^T X$ can be written in the form $\mathbf{x}_0^T M \mathbf{x}_0$. It can be shown that the quantity $(\mathbf{x}_0^T M \mathbf{x}_0)^{-1}$ is the $(2, 2)$-entry in $(W^T W)^{-1}$. This entry has a useful statistical interpretation, under appropriate hypotheses.

In the study of engineering control of physical systems, a standard set of differential equations is transformed by Laplace transforms into the following system of linear equations:

$$\begin{bmatrix} A - sI_n & B \\ C & I_m \end{bmatrix}\begin{bmatrix} \mathbf{x} \\ \mathbf{u} \end{bmatrix} = \begin{bmatrix} \mathbf{0} \\ \mathbf{y} \end{bmatrix} \qquad (8)$$

where A is $n \times n$, B is $n \times m$, C is $m \times n$, and s is a variable. The vector $\mathbf{u}$ in $\mathbb{R}^m$ is the "input" to the system, $\mathbf{y}$ in $\mathbb{R}^m$ is the "output," and $\mathbf{x}$ in $\mathbb{R}^n$ is the "state" vector. (Actually, the vectors $\mathbf{x}$, $\mathbf{u}$, and $\mathbf{y}$ are functions of s, but this does not affect the algebraic calculations in Exercises 19 and 20.)

19. Assume $A - sI_n$ is invertible and view (8) as a system of two matrix equations. Solve the top equation for $\mathbf{x}$ and substitute into the bottom equation. The result is an equation of the form $W(s)\mathbf{u} = \mathbf{y}$, where $W(s)$ is a matrix that depends on s. $W(s)$ is called the *transfer function* of the system because it transforms the input $\mathbf{u}$ into the output $\mathbf{y}$. Find $W(s)$ and describe how it is related to the partitioned *system matrix* on the left side of (8). See Exercise 16.

20. Suppose the transfer function $W(s)$ in Exercise 19 is invertible for some s. It can be shown that the inverse transfer function $W(s)^{-1}$, which transforms outputs into inputs, is the Schur complement of $A - BC - sI_n$ for the matrix below. Find this Schur complement. See Exercise 16.

$$\begin{bmatrix} A - BC - sI_n & B \\ -C & I_m \end{bmatrix}$$

21. a. Verify that $A^2 = I$ when $A = \begin{bmatrix} 1 & 0 \\ 2 & -1 \end{bmatrix}$.

b. Use partitioned matrices to show that $M^2 = I$ when

$$M = \begin{bmatrix} 1 & 0 & 0 & 0 \\ 2 & -1 & 0 & 0 \\ 1 & 0 & -1 & 0 \\ 0 & 1 & -2 & 1 \end{bmatrix}$$

22. Generalize the idea of Exercise 21 by constructing a 6×6 matrix $M = \begin{bmatrix} A & 0 & 0 \\ 0 & B & 0 \\ C & 0 & D \end{bmatrix}$ such that $M^2 = I$. Make C a nonzero 2×2 matrix. Show that your construction works.

23. Use partitioned matrices to prove by induction that the product of two lower triangular matrices is also lower triangular. [*Hint:* A $(k + 1) \times (k + 1)$ matrix A_1 can be written in the form below, where a is a scalar, $\mathbf{v}$ is in $\mathbb{R}^k$, and A is a $k \times k$ lower triangular matrix. See the *Study Guide* for help with induction.]

$$A_1 = \begin{bmatrix} a & \mathbf{0}^T \\ \mathbf{v} & A \end{bmatrix}$$

| **SG** | The Principle of Induction 2–19 |

24. Use partitioned matrices to prove by induction that for $n = 2, 3, \ldots,$ the $n \times n$ matrix A shown below is invertible and B is its inverse.

$$A = \begin{bmatrix} 1 & 0 & 0 & \cdots & 0 \\ 1 & 1 & 0 & & 0 \\ 1 & 1 & 1 & & 0 \\ \vdots & & & \ddots & \\ 1 & 1 & 1 & \cdots & 1 \end{bmatrix},$$

$$B = \begin{bmatrix} 1 & 0 & 0 & \cdots & 0 \\ -1 & 1 & 0 & & 0 \\ 0 & -1 & 1 & & 0 \\ \vdots & & \ddots & \ddots & \\ 0 & & \cdots & -1 & 1 \end{bmatrix}$$

For the induction step, assume A and B are $(k + 1) \times (k + 1)$ matrices, and partition A and B in a form similar to that displayed in Exercise 23.

25. Without using row reduction, find the inverse of

$$A = \begin{bmatrix} 1 & 2 & 0 & 0 & 0 \\ 3 & 5 & 0 & 0 & 0 \\ 0 & 0 & 2 & 0 & 0 \\ 0 & 0 & 0 & 7 & 8 \\ 0 & 0 & 0 & 5 & 6 \end{bmatrix}$$

26. [M] For block operations, it may be necessary to access or enter submatrices of a large matrix. Describe the functions or commands of a matrix program that accomplish the following tasks. Suppose A is a 20×30 matrix.

a. Display the submatrix of A from rows 5 to 10 and columns 15 to 20.

b. Insert a 5×10 matrix B into A, beginning at row 5 and column 10.

c. Create a 50×50 matrix of the form $C = \begin{bmatrix} A & 0 \\ 0 & A^T \end{bmatrix}$.
[*Note:* It may not be necessary to specify the zero blocks in C.]

27. [M] Suppose memory or size restrictions prevent a matrix program from working with matrices having more than 32 rows and 32 columns, and suppose some project involves 50×50 matrices A and B. Describe the commands or operations of the matrix program that accomplish the following tasks.

a. Compute $A + B$.

b. Compute AB.

c. Solve $A\mathbf{x} = \mathbf{b}$ for some vector $\mathbf{b}$ in $\mathbb{R}^{50}$, assuming that A can be partitioned into a 2×2 block matrix $[A_{ij}]$, with A_{11} an invertible 20×20 matrix, A_{22} an invertible 30×30 matrix, and A_{12} a zero matrix. [*Hint:* Describe appropriate smaller systems to solve, without using any matrix inverses.]

SOLUTIONS TO PRACTICE PROBLEMS

1. If $\begin{bmatrix} I & 0 \\ A & I \end{bmatrix}$ is invertible, its inverse has the form $\begin{bmatrix} W & X \\ Y & Z \end{bmatrix}$. Verify that

$$\begin{bmatrix} I & 0 \\ A & I \end{bmatrix} \begin{bmatrix} W & X \\ Y & Z \end{bmatrix} = \begin{bmatrix} W & X \\ AW + Y & AX + Z \end{bmatrix}$$

So W, X, Y, Z must satisfy $W = I$, $X = 0$, $AW + Y = 0$, and $AX + Z = I$. It follows that $Y = -A$ and $Z = I$. Hence

$$\begin{bmatrix} I & 0 \\ A & I \end{bmatrix} \begin{bmatrix} I & 0 \\ -A & I \end{bmatrix} = \begin{bmatrix} I & 0 \\ 0 & I \end{bmatrix}$$

The product in the reverse order is also the identity, so the block matrix is invertible, and its inverse is $\begin{bmatrix} I & 0 \\ -A & I \end{bmatrix}$. (You could also appeal to the Invertible Matrix Theorem.)

2. $X^T X = \begin{bmatrix} X_1^T \\ X_2^T \end{bmatrix} \begin{bmatrix} X_1 & X_2 \end{bmatrix} = \begin{bmatrix} X_1^T X_1 & X_1^T X_2 \\ X_2^T X_1 & X_2^T X_2 \end{bmatrix}$. The partitions of X^T and X are automatically conformable for block multiplication because the columns of X^T are the rows of X. This partition of $X^T X$ is used in several computer algorithms for matrix computations.

2.5 | MATRIX FACTORIZATIONS

A *factorization* of a matrix A is an equation that expresses A as a product of two or more matrices. Whereas matrix multiplication involves a *synthesis* of data (combining the effects of two or more linear transformations into a single matrix), matrix factorization is an *analysis* of data. In the language of computer science, the expression of A as a product amounts to a *preprocessing* of the data in A, organizing that data into two or more parts whose structures are more useful in some way, perhaps more accessible for computation.

Matrix factorizations and, later, factorizations of linear transformations will appear at a number of key points throughout the text. This section focuses on a factorization that lies at the heart of several important computer programs widely used in applications, such as the airflow problem described in the chapter introduction. Several other factorizations, to be studied later, are introduced in the exercises.

The LU Factorization

The LU factorization, described below, is motivated by the fairly common industrial and business problem of solving a sequence of equations, all with the same coefficient matrix:

$$A\mathbf{x} = \mathbf{b}_1, \quad A\mathbf{x} = \mathbf{b}_2, \quad \ldots, \quad A\mathbf{x} = \mathbf{b}_p \tag{1}$$

See Exercise 32, for example. Also see Section 5.8, where the inverse power method is used to estimate eigenvalues of a matrix by solving equations like those in sequence (1), one at a time.

When A is invertible, one could compute A^{-1} and then compute $A^{-1}\mathbf{b}_1$, $A^{-1}\mathbf{b}_2$, and so on. However, it is more efficient to solve the first equation in sequence (1) by row reduction and obtain an LU factorization of A at the same time. Thereafter, the remaining equations in sequence (1) are solved with the LU factorization.

At first, assume that A is an $m \times n$ matrix that can be row reduced to echelon form, *without row interchanges*. (Later, we will treat the general case.) Then A can be written in the form $A = LU$, where L is an $m \times m$ lower triangular matrix with 1's on the diagonal and U is an $m \times n$ echelon form of A. For instance, see Fig. 1. Such a factorization is called an **LU factorization** of A. The matrix L is invertible and is called a *unit* lower triangular matrix.

$$A = \begin{bmatrix} 1 & 0 & 0 & 0 \\ * & 1 & 0 & 0 \\ * & * & 1 & 0 \\ * & * & * & 1 \end{bmatrix} \begin{bmatrix} \blacksquare & * & * & * & * \\ 0 & \blacksquare & * & * & * \\ 0 & 0 & 0 & \blacksquare & * \\ 0 & 0 & 0 & 0 & 0 \end{bmatrix}$$

$$\qquad\qquad L \qquad\qquad\qquad U$$

FIGURE 1 An LU factorization.

Before studying how to construct L and U, we should look at why they are so useful. When $A = LU$, the equation $A\mathbf{x} = \mathbf{b}$ can be written as $L(U\mathbf{x}) = \mathbf{b}$. Writing $\mathbf{y}$ for $U\mathbf{x}$, we can find $\mathbf{x}$ by solving the *pair* of equations

$$\boxed{\begin{aligned} L\mathbf{y} &= \mathbf{b} \\ U\mathbf{x} &= \mathbf{y} \end{aligned}} \tag{2}$$

First solve $L\mathbf{y} = \mathbf{b}$ for $\mathbf{y}$, and then solve $U\mathbf{x} = \mathbf{y}$ for $\mathbf{x}$. See Fig. 2. Each equation is easy to solve because L and U are triangular.

EXAMPLE 1 It can be verified that

$$A = \begin{bmatrix} 3 & -7 & -2 & 2 \\ -3 & 5 & 1 & 0 \\ 6 & -4 & 0 & -5 \\ -9 & 5 & -5 & 12 \end{bmatrix} = \begin{bmatrix} 1 & 0 & 0 & 0 \\ -1 & 1 & 0 & 0 \\ 2 & -5 & 1 & 0 \\ -3 & 8 & 3 & 1 \end{bmatrix} \begin{bmatrix} 3 & -7 & -2 & 2 \\ 0 & -2 & -1 & 2 \\ 0 & 0 & -1 & 1 \\ 0 & 0 & 0 & -1 \end{bmatrix} = LU$$

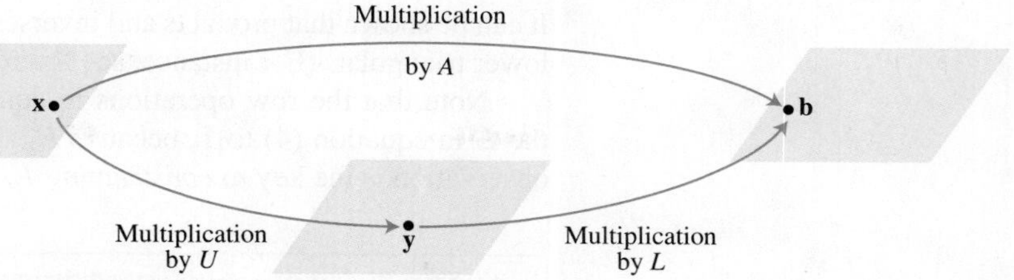

Multiplication
by A

$\mathbf{x}$ ● → ● $\mathbf{b}$

Multiplication
by U

$\mathbf{y}$

Multiplication
by L

FIGURE 2 Factorization of the mapping $\mathbf{x} \mapsto A\mathbf{x}$.

Use this LU factorization of A to solve $A\mathbf{x} = \mathbf{b}$, where $\mathbf{b} = \begin{bmatrix} -9 \\ 5 \\ 7 \\ 11 \end{bmatrix}$.

SOLUTION The solution of $L\mathbf{y} = \mathbf{b}$ needs only 6 multiplications and 6 additions, because the arithmetic takes place only in column 5. (The zeros below each pivot in L are created automatically by the choice of row operations.)

$$
\begin{bmatrix} L & \mathbf{b} \end{bmatrix} = \begin{bmatrix} 1 & 0 & 0 & 0 & -9 \\ -1 & 1 & 0 & 0 & 5 \\ 2 & -5 & 1 & 0 & 7 \\ -3 & 8 & 3 & 1 & 11 \end{bmatrix} \sim \begin{bmatrix} 1 & 0 & 0 & 0 & -9 \\ 0 & 1 & 0 & 0 & -4 \\ 0 & 0 & 1 & 0 & 5 \\ 0 & 0 & 0 & 1 & 1 \end{bmatrix} = \begin{bmatrix} I & \mathbf{y} \end{bmatrix}
$$

Then, for $U\mathbf{x} = \mathbf{y}$, the "backward" phase of row reduction requires 4 divisions, 6 multiplications, and 6 additions. (For instance, creating the zeros in column 4 of $\begin{bmatrix} U & \mathbf{y} \end{bmatrix}$ requires 1 division in row 4 and 3 multiplication–addition pairs to add multiples of row 4 to the rows above.)

$$
\begin{bmatrix} U & \mathbf{y} \end{bmatrix} = \begin{bmatrix} 3 & -7 & -2 & 2 & -9 \\ 0 & -2 & -1 & 2 & -4 \\ 0 & 0 & -1 & 1 & 5 \\ 0 & 0 & 0 & -1 & 1 \end{bmatrix} \sim \begin{bmatrix} 1 & 0 & 0 & 0 & 3 \\ 0 & 1 & 0 & 0 & 4 \\ 0 & 0 & 1 & 0 & -6 \\ 0 & 0 & 0 & 1 & -1 \end{bmatrix}, \quad \mathbf{x} = \begin{bmatrix} 3 \\ 4 \\ -6 \\ -1 \end{bmatrix}
$$

To find $\mathbf{x}$ requires 28 arithmetic operations, or "flops" (floating point operations), excluding the cost of finding L and U. In contrast, row reduction of $\begin{bmatrix} A & \mathbf{b} \end{bmatrix}$ to $\begin{bmatrix} I & \mathbf{x} \end{bmatrix}$ takes 62 operations. ∎

The computational efficiency of the LU factorization depends on knowing L and U. The next algorithm shows that the row reduction of A to an echelon form U amounts to an LU factorization because it produces L with essentially no extra work. After the first row reduction, L and U are available for solving additional equations whose coefficient matrix is A.

An LU Factorization Algorithm

Suppose A can be reduced to an echelon form U using only row replacements that add a multiple of one row to another row *below it*. In this case, there exist unit lower triangular elementary matrices $E_1, \ldots, E_p$ such that

$$
E_p \cdots E_1 A = U \tag{3}
$$

Then

$$
A = (E_p \cdots E_1)^{-1} U = LU
$$

where

$$
L = (E_p \cdots E_1)^{-1} \tag{4}
$$

It can be shown that products and inverses of unit lower triangular matrices are also unit lower triangular. (For instance, see Exercise 19.) Thus L is unit lower triangular.

Note that the row operations in equation (3), which reduce A to U, also reduce the L in equation (4) to I, because $E_p \cdots E_1 L = (E_p \cdots E_1)(E_p \cdots E_1)^{-1} = I$. This observation is the key to *constructing L*.

ALGORITHM FOR AN LU FACTORIZATION

1. Reduce A to an echelon form U by a sequence of row replacement operations, if possible.

2. Place entries in L such that the *same sequence of row operations* reduces L to I.

Step 1 is not always possible, but when it is, the argument above shows that an LU factorization exists. Example 2 will show how to implement step 2. By construction, L will satisfy

$$(E_p \cdots E_1)L = I$$

using the same $E_1, \ldots, E_p$ as in equation (3). Thus L will be invertible, by the Invertible Matrix Theorem, with $(E_p \cdots E_1) = L^{-1}$. From (3), $L^{-1}A = U$, and $A = LU$. So step 2 will produce an acceptable L.

EXAMPLE 2 Find an LU factorization of

$$A = \begin{bmatrix} 2 & 4 & -1 & 5 & -2 \\ -4 & -5 & 3 & -8 & 1 \\ 2 & -5 & -4 & 1 & 8 \\ -6 & 0 & 7 & -3 & 1 \end{bmatrix}$$

SOLUTION Since A has four rows, L should be 4×4. The first column of L is the first column of A divided by the top pivot entry:

$$L = \begin{bmatrix} 1 & 0 & 0 & 0 \\ -2 & 1 & 0 & 0 \\ 1 & & 1 & 0 \\ -3 & & & 1 \end{bmatrix}$$

Compare the first columns of A and L. *The row operations that create zeros in the first column of A will also create zeros in the first column of L.* To make this same correspondence of row operations on A hold for the rest of L, watch a row reduction of A to an echelon form U. That is, *highlight the entries* in each matrix that are used to determine the sequence of row operations that transform A into U. [See the highlighted entries in equation (5).]

$$A = \begin{bmatrix} 2 & 4 & -1 & 5 & -2 \\ -4 & -5 & 3 & -8 & 1 \\ 2 & -5 & -4 & 1 & 8 \\ -6 & 0 & 7 & -3 & 1 \end{bmatrix} \sim \begin{bmatrix} 2 & 4 & -1 & 5 & -2 \\ 0 & 3 & 1 & 2 & -3 \\ 0 & -9 & -3 & -4 & 10 \\ 0 & 12 & 4 & 12 & -5 \end{bmatrix} = A_1 \qquad (5)$$

$$\sim A_2 = \begin{bmatrix} 2 & 4 & -1 & 5 & -2 \\ 0 & 3 & 1 & 2 & -3 \\ 0 & 0 & 0 & 2 & 1 \\ 0 & 0 & 0 & 4 & 7 \end{bmatrix} \sim \begin{bmatrix} 2 & 4 & -1 & 5 & -2 \\ 0 & 3 & 1 & 2 & -3 \\ 0 & 0 & 0 & 2 & 1 \\ 0 & 0 & 0 & 0 & 5 \end{bmatrix} = U$$

The highlighted entries above determine the row reduction of A to U. At each pivot column, divide the highlighted entries by the pivot and place the result into L:

$$\begin{bmatrix} 2 \\ -4 \\ 2 \\ -6 \end{bmatrix} \begin{bmatrix} 3 \\ -9 \\ 12 \end{bmatrix} \begin{bmatrix} 2 \\ 4 \end{bmatrix} \begin{bmatrix} 5 \end{bmatrix}$$

$$\begin{matrix} \div 2 & \div 3 & \div 2 & \div 5 \\ \downarrow & \downarrow & \downarrow & \downarrow \end{matrix}$$

$$\begin{bmatrix} 1 & & & \\ -2 & 1 & & \\ 1 & -3 & 1 & \\ -3 & 4 & 2 & 1 \end{bmatrix}, \quad \text{and} \quad L = \begin{bmatrix} 1 & 0 & 0 & 0 \\ -2 & 1 & 0 & 0 \\ 1 & -3 & 1 & 0 \\ -3 & 4 & 2 & 1 \end{bmatrix}$$

An easy calculation verifies that this L and U satisfy $LU = A$. ∎

In practical work, row interchanges are nearly always needed, because partial pivoting is used for high accuracy. (Recall that this procedure selects, among the possible choices for a pivot, an entry in the column having the largest absolute value.) To handle row interchanges, the LU factorization above can be modified easily to produce an L that is *permuted lower triangular*, in the sense that a rearrangement (called a permutation) of the rows of L can make L (unit) lower triangular. The resulting *permuted LU factorization* solves $A\mathbf{x} = \mathbf{b}$ in the same way as before, except that the reduction of $[\, L \quad \mathbf{b}\,]$ to $[\, I \quad \mathbf{y}\,]$ follows the order of the pivots in L from left to right, starting with the pivot in the first column. A reference to an "LU factorization" usually includes the possibility that L might be permuted lower triangular. For details, see the *Study Guide*.

> **SG** Permuted LU Factorizations 2-23

─── NUMERICAL NOTES ───

The following operation counts apply to an $n \times n$ dense matrix A (with most entries nonzero) for n moderately large, say, $n \geq 30$.[1]

1. Computing an LU factorization of A takes about $2n^3/3$ flops (about the same as row reducing $[\, A \quad \mathbf{b}\,]$), whereas finding A^{-1} requires about $2n^3$ flops.

2. Solving $L\mathbf{y} = \mathbf{b}$ and $U\mathbf{x} = \mathbf{y}$ requires about $2n^2$ flops, because any $n \times n$ triangular system can be solved in about n^2 flops.

3. Multiplication of $\mathbf{b}$ by A^{-1} also requires about $2n^2$ flops, but the result may not be as accurate as that obtained from L and U (because of roundoff error when computing both A^{-1} and $A^{-1}\mathbf{b}$).

4. If A is sparse (with mostly zero entries), then L and U may be sparse, too, whereas A^{-1} is likely to be dense. In this case, a solution of $A\mathbf{x} = \mathbf{b}$ with an LU factorization is *much* faster than using A^{-1}. See Exercise 31.

> **WEB**

A Matrix Factorization in Electrical Engineering

Matrix factorization is intimately related to the problem of constructing an electrical network with specified properties. The following discussion gives just a glimpse of the connection between factorization and circuit design.

---

[1] See Section 3.8 in *Applied Linear Algebra*, 3rd ed., by Ben Noble and James W. Daniel (Englewood Cliffs, NJ: Prentice-Hall, 1988). Recall that for our purposes, a *flop* is $+$, $-$, $\times$, or $\div$.

Suppose the box in Fig. 3 represents some sort of electric circuit, with an input and output. Record the input voltage and current by $\begin{bmatrix} v_1 \\ i_1 \end{bmatrix}$ (with voltage v in volts and current i in amps), and record the output voltage and current by $\begin{bmatrix} v_2 \\ i_2 \end{bmatrix}$. Frequently, the transformation $\begin{bmatrix} v_1 \\ i_1 \end{bmatrix} \mapsto \begin{bmatrix} v_2 \\ i_2 \end{bmatrix}$ is linear. That is, there is a matrix A, called the *transfer matrix*, such that

$$\begin{bmatrix} v_2 \\ i_2 \end{bmatrix} = A \begin{bmatrix} v_1 \\ i_1 \end{bmatrix}$$

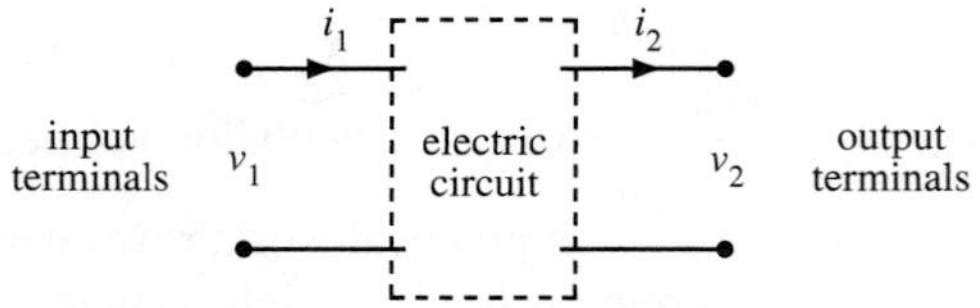

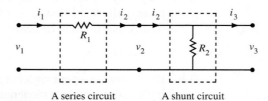

FIGURE 3 A circuit with input and output terminals.

Figure 4 shows a *ladder network*, where two circuits (there could be more) are connected in series, so that the output of one circuit becomes the input of the next circuit. The left circuit in Fig. 4 is called a *series circuit*, with resistance R_1 (in ohms).

A series circuit   A shunt circuit

FIGURE 4 A ladder network.

The right circuit in Fig. 4 is a *shunt circuit*, with resistance R_2. Using Ohm's law and Kirchhoff's laws, one can show that the transfer matrices of the series and shunt circuits, respectively, are

$$\begin{bmatrix} 1 & -R_1 \\ 0 & 1 \end{bmatrix} \quad \text{and} \quad \begin{bmatrix} 1 & 0 \\ -1/R_2 & 1 \end{bmatrix}$$

Transfer matrix   Transfer matrix
of series circuit   of shunt circuit

EXAMPLE 3

a. Compute the transfer matrix of the ladder network in Fig. 4.

b. Design a ladder network whose transfer matrix is $\begin{bmatrix} 1 & -8 \\ -.5 & 5 \end{bmatrix}$.

SOLUTION

a. Let A_1 and A_2 be the transfer matrices of the series and shunt circuits, respectively. Then an input vector $\mathbf{x}$ is transformed first into $A_1\mathbf{x}$ and then into $A_2(A_1\mathbf{x})$. The series connection of the circuits corresponds to composition of linear transformations, and the transfer matrix of the ladder network is (note the order)

$$A_2 A_1 = \begin{bmatrix} 1 & 0 \\ -1/R_2 & 1 \end{bmatrix} \begin{bmatrix} 1 & -R_1 \\ 0 & 1 \end{bmatrix} = \begin{bmatrix} 1 & -R_1 \\ -1/R_2 & 1 + R_1/R_2 \end{bmatrix} \qquad (6)$$

b. To factor the matrix $\begin{bmatrix} 1 & -8 \\ -.5 & 5 \end{bmatrix}$ into the product of transfer matrices, as in equation (6), look for R_1 and R_2 in Fig. 4 to satisfy

$$\begin{bmatrix} 1 & -R_1 \\ -1/R_2 & 1 + R_1/R_2 \end{bmatrix} = \begin{bmatrix} 1 & -8 \\ -.5 & 5 \end{bmatrix}$$

From the $(1, 2)$-entries, $R_1 = 8$ ohms, and from the $(2, 1)$-entries, $1/R_2 = .5$ ohm and $R_2 = 1/.5 = 2$ ohms. With these values, the network in Fig. 4 has the desired transfer matrix. ∎

A network transfer matrix summarizes the input–output behavior (the design specifications) of the network without reference to the interior circuits. To physically build a network with specified properties, an engineer first determines if such a network can be constructed (or *realized*). Then the engineer tries to factor the transfer matrix into matrices corresponding to smaller circuits that perhaps are already manufactured and ready for assembly. In the common case of alternating current, the entries in the transfer matrix are usually rational complex-valued functions. (See Exercises 19 and 20 in Section 2.4 and Example 2 in Section 3.3.) A standard problem is to find a *minimal realization* that uses the smallest number of electrical components.

PRACTICE PROBLEM

Find an LU factorization of $A = \begin{bmatrix} 2 & -4 & -2 & 3 \\ 6 & -9 & -5 & 8 \\ 2 & -7 & -3 & 9 \\ 4 & -2 & -2 & -1 \\ -6 & 3 & 3 & 4 \end{bmatrix}$. [*Note*: It will turn out that A has only three pivot columns, so the method of Example 2 will produce only the first three columns of L. The remaining two columns of L come from I_5.]

2.5 EXERCISES

In Exercises 1–6, solve the equation $A\mathbf{x} = \mathbf{b}$ by using the LU factorization given for A. In Exercises 1 and 2, also solve $A\mathbf{x} = \mathbf{b}$ by ordinary row reduction.

1. $A = \begin{bmatrix} 3 & -7 & -2 \\ -3 & 5 & 1 \\ 6 & -4 & 0 \end{bmatrix}$, $\mathbf{b} = \begin{bmatrix} -7 \\ 5 \\ 2 \end{bmatrix}$

$A = \begin{bmatrix} 1 & 0 & 0 \\ -1 & 1 & 0 \\ 2 & -5 & 1 \end{bmatrix}\begin{bmatrix} 3 & -7 & -2 \\ 0 & -2 & -1 \\ 0 & 0 & -1 \end{bmatrix}$

2. $A = \begin{bmatrix} 2 & -6 & 4 \\ -4 & 8 & 0 \\ 0 & -4 & 6 \end{bmatrix}$, $\mathbf{b} = \begin{bmatrix} 2 \\ -4 \\ 6 \end{bmatrix}$

$A = \begin{bmatrix} 1 & 0 & 0 \\ -2 & 1 & 0 \\ 0 & 1 & 1 \end{bmatrix}\begin{bmatrix} 2 & -6 & 4 \\ 0 & -4 & 8 \\ 0 & 0 & -2 \end{bmatrix}$

3. $A = \begin{bmatrix} 2 & -4 & 2 \\ -4 & 5 & 2 \\ 6 & -9 & 1 \end{bmatrix}$, $\mathbf{b} = \begin{bmatrix} 6 \\ 0 \\ 6 \end{bmatrix}$

$A = \begin{bmatrix} 1 & 0 & 0 \\ -2 & 1 & 0 \\ 3 & -1 & 1 \end{bmatrix}\begin{bmatrix} 2 & -4 & 2 \\ 0 & -3 & 6 \\ 0 & 0 & 1 \end{bmatrix}$

4. $A = \begin{bmatrix} 1 & -1 & 2 \\ 1 & -3 & 1 \\ 3 & 7 & 5 \end{bmatrix}$, $\mathbf{b} = \begin{bmatrix} 0 \\ -5 \\ 7 \end{bmatrix}$

$A = \begin{bmatrix} 1 & 0 & 0 \\ 1 & 1 & 0 \\ 3 & -5 & 1 \end{bmatrix}\begin{bmatrix} 1 & -1 & 2 \\ 0 & -2 & -1 \\ 0 & 0 & -6 \end{bmatrix}$

5. $A = \begin{bmatrix} 1 & -2 & -2 & -3 \\ 3 & -9 & 0 & -9 \\ -1 & 2 & 4 & 7 \\ -3 & -6 & 26 & 2 \end{bmatrix}$, $\mathbf{b} = \begin{bmatrix} 1 \\ 6 \\ 0 \\ 3 \end{bmatrix}$

$A = \begin{bmatrix} 1 & 0 & 0 & 0 \\ 3 & 1 & 0 & 0 \\ -1 & 0 & 1 & 0 \\ -3 & 4 & -2 & 1 \end{bmatrix}\begin{bmatrix} 1 & -2 & -2 & -3 \\ 0 & -3 & 6 & 0 \\ 0 & 0 & 2 & 4 \\ 0 & 0 & 0 & 1 \end{bmatrix}$

6. $A = \begin{bmatrix} 1 & 3 & 2 & 0 \\ -2 & -3 & -4 & 12 \\ 3 & 0 & 4 & -36 \\ -5 & -3 & -8 & 49 \end{bmatrix}$, $\mathbf{b} = \begin{bmatrix} 1 \\ -2 \\ -1 \\ 2 \end{bmatrix}$

$$A = \begin{bmatrix} 1 & 0 & 0 & 0 \\ -2 & 1 & 0 & 0 \\ 3 & -3 & 1 & 0 \\ -5 & 4 & -1 & 1 \end{bmatrix} \begin{bmatrix} 1 & 3 & 2 & 0 \\ 0 & 3 & 0 & 12 \\ 0 & 0 & -2 & 0 \\ 0 & 0 & 0 & 1 \end{bmatrix}$$

Find an LU factorization of the matrices in Exercises 7–16 (with L unit lower triangular). Note that MATLAB will usually produce a permuted LU factorization because it uses partial pivoting for numerical accuracy.

7. $\begin{bmatrix} 2 & 5 \\ -3 & -4 \end{bmatrix}$

8. $\begin{bmatrix} 6 & 4 \\ 12 & 5 \end{bmatrix}$

9. $\begin{bmatrix} 3 & 1 & 2 \\ -9 & 0 & -4 \\ 9 & 9 & 14 \end{bmatrix}$

10. $\begin{bmatrix} -5 & 0 & 4 \\ 10 & 2 & -5 \\ 10 & 10 & 16 \end{bmatrix}$

11. $\begin{bmatrix} 3 & 7 & 2 \\ 6 & 19 & 4 \\ -3 & -2 & 3 \end{bmatrix}$

12. $\begin{bmatrix} 2 & 3 & 2 \\ 4 & 13 & 9 \\ -6 & 5 & 4 \end{bmatrix}$

13. $\begin{bmatrix} 1 & 3 & -5 & -3 \\ -1 & -5 & 8 & 4 \\ 4 & 2 & -5 & -7 \\ -2 & -4 & 7 & 5 \end{bmatrix}$

14. $\begin{bmatrix} 1 & 3 & 1 & 5 \\ 5 & 20 & 6 & 31 \\ -2 & -1 & -1 & -4 \\ -1 & 7 & 1 & 7 \end{bmatrix}$

15. $\begin{bmatrix} 2 & 0 & 5 & 2 \\ -6 & 3 & -13 & -3 \\ 4 & 9 & 16 & 17 \end{bmatrix}$

16. $\begin{bmatrix} 2 & -3 & 4 \\ -4 & 8 & -7 \\ 6 & -5 & 14 \\ -6 & 9 & -12 \\ 8 & -6 & 19 \end{bmatrix}$

17. When A is invertible, MATLAB finds A^{-1} by factoring $A = LU$ (where L may be permuted lower triangular), inverting L and U, and then computing $U^{-1}L^{-1}$. Use this method to compute the inverse of A in Exercise 2. (Apply the algorithm in Section 2.2 to L and to U.)

18. Find A^{-1} as in Exercise 17, using A from Exercise 3.

19. Let A be a lower triangular $n \times n$ matrix with nonzero entries on the diagonal. Show that A is invertible and A^{-1} is lower triangular. [*Hint:* Explain why A can be changed into I using only row replacements and scaling. (Where are the pivots?) Also, explain why the row operations that reduce A to I change I into a lower triangular matrix.]

20. Let $A = LU$ be an LU factorization. Explain why A can be row reduced to U using only replacement operations. (This fact is the converse of what was proved in the text.)

21. Suppose $A = BC$, where B is invertible. Show that any sequence of row operations that reduces B to I also reduces A to C. The converse is not true, since the zero matrix may be factored as $0 = B \cdot 0$.

Exercises 22–26 provide a glimpse of some widely used matrix factorizations, some of which are discussed later in the text.

22. (*Reduced LU Factorization*) With A as in the Practice Problem, find a 5×3 matrix B and a 3×4 matrix C such that $A = BC$. Generalize this idea to the case where A is $m \times n$, $A = LU$, and U has only three nonzero rows.

23. (*Rank Factorization*) Suppose an $m \times n$ matrix A admits a factorization $A = CD$ where C is $m \times 4$ and D is $4 \times n$.

a. Show that A is the sum of four outer products. (See Section 2.4.)

b. Let $m = 400$ and $n = 100$. Explain why a computer programmer might prefer to store the data from A in the form of two matrices C and D.

24. (*QR Factorization*) Suppose $A = QR$, where Q and R are $n \times n$, R is invertible and upper triangular, and Q has the property that $Q^T Q = I$. Show that for each $\mathbf{b}$ in $\mathbb{R}^n$, the equation $A\mathbf{x} = \mathbf{b}$ has a unique solution. What computations with Q and R will produce the solution?

WEB

25. (*Singular Value Decomposition*) Suppose $A = UDV^T$, where U and V are $n \times n$ matrices with the property that $U^T U = I$ and $V^T V = I$, and where D is a diagonal matrix with positive numbers $\sigma_1, \ldots, \sigma_n$ on the diagonal. Show that A is invertible, and find a formula for A^{-1}.

26. (*Spectral Factorization*) Suppose a 3×3 matrix A admits a factorization as $A = PDP^{-1}$, where P is some invertible 3×3 matrix and D is the diagonal matrix

$$D = \begin{bmatrix} 2 & 0 & 0 \\ 0 & 3 & 0 \\ 0 & 0 & 1 \end{bmatrix}$$

Show that this factorization is useful when computing high powers of A. Find fairly simple formulas for A^2, A^3, and A^k (k a positive integer), using P and the entries in D.

27. Design two different ladder networks that each output 9 volts and 4 amps when the input is 12 volts and 6 amps.

28. Show that if three shunt circuits (with resistances R_1, R_2, R_3) are connected in series, the resulting network has the same transfer matrix as a single shunt circuit. Find a formula for the resistance in that circuit.

29. a. Compute the transfer matrix of the network in the figure below.

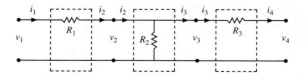

b. Let $A = \begin{bmatrix} 3 & -12 \\ -1/3 & 5/3 \end{bmatrix}$. Design a ladder network whose transfer matrix is A by finding a suitable matrix factorization of A.

30. Find a different factorization of the transfer matrix A in Exercise 29, and thereby design a different ladder network whose transfer matrix is A.

31. [M] Consider the heat plate in the following figure (refer to Exercise 33 in Section 1.1).

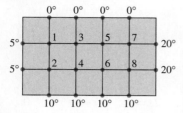

The solution to the steady-state heat flow problem for this plate is approximated by the solution to the equation $A\mathbf{x} = \mathbf{b}$, where $\mathbf{b} = (5, 15, 0, 10, 0, 10, 20, 30)$ and

$$A = \begin{bmatrix} 4 & -1 & -1 & & & & & \\ -1 & 4 & 0 & -1 & & & & \\ -1 & 0 & 4 & -1 & -1 & & & \\ & -1 & -1 & 4 & 0 & -1 & & \\ & & -1 & 0 & 4 & -1 & -1 & \\ & & & -1 & -1 & 4 & 0 & -1 \\ & & & & -1 & 0 & 4 & -1 \\ & & & & & -1 & -1 & 4 \end{bmatrix}$$

WEB

The missing entries in A are zeros. The nonzero entries of A lie within a band along the main diagonal. Such *band matrices* occur in a variety of applications and often are extremely large (with thousands of rows and columns but relatively narrow bands).

a. Use the method in Example 2 to construct an LU factorization of A, and note that both factors are band matrices (with two nonzero diagonals below or above the main diagonal). Compute $LU - A$ to check your work.

b. Use the LU factorization to solve $A\mathbf{x} = \mathbf{b}$.

c. Obtain A^{-1} and note that A^{-1} is a dense matrix with no band structure. When A is large, L and U can be stored in much less space than A^{-1}. This fact is another reason for preferring the LU factorization of A to A^{-1} itself.

32. [M] The band matrix A shown below can be used to estimate the unsteady conduction of heat in a rod when the temperatures at points $p_1, \ldots, p_4$ on the rod change with time.[2]

The constant C in the matrix depends on the physical nature of the rod, the distance Δx between the points on the rod, and the length of time Δt between successive temperature measurements. Suppose that for $k = 0, 1, 2, \ldots$, a vector $\mathbf{t}_k$ in $\mathbb{R}^4$ lists the temperatures at time $k\Delta t$. If the two ends of the rod are maintained at $0°$, then the temperature vectors satisfy the equation $A\mathbf{t}_{k+1} = \mathbf{t}_k$ $(k = 0, 1, \ldots)$, where

$$A = \begin{bmatrix} (1 + 2C) & -C & & \\ -C & (1 + 2C) & -C & \\ & -C & (1 + 2C) & -C \\ & & -C & (1 + 2C) \end{bmatrix}$$

a. Find the LU factorization of A when $C = 1$. A matrix such as A with three nonzero diagonals is called a *tridiagonal matrix*. The L and U factors are *bidiagonal matrices*.

b. Suppose $C = 1$ and $\mathbf{t}_0 = (10, 15, 15, 10)^T$. Use the LU factorization of A to find the temperature distributions $\mathbf{t}_1$, $\mathbf{t}_2$, $\mathbf{t}_3$, and $\mathbf{t}_4$.

---

[2] See Biswa N. Datta, *Numerical Linear Algebra and Applications* (Pacific Grove, CA: Brooks/Cole, 1994), pp. 200–201.

SOLUTION TO PRACTICE PROBLEM

$$A = \begin{bmatrix} 2 & -4 & -2 & 3 \\ 6 & -9 & -5 & 8 \\ 2 & -7 & -3 & 9 \\ 4 & -2 & -2 & -1 \\ -6 & 3 & 3 & 4 \end{bmatrix} \sim \begin{bmatrix} 2 & -4 & -2 & 3 \\ 0 & 3 & 1 & -1 \\ 0 & -3 & -1 & 6 \\ 0 & 6 & 2 & -7 \\ 0 & -9 & -3 & 13 \end{bmatrix}$$

$$\sim \begin{bmatrix} 2 & -4 & -2 & 3 \\ 0 & 3 & 1 & -1 \\ 0 & 0 & 0 & 5 \\ 0 & 0 & 0 & -5 \\ 0 & 0 & 0 & 10 \end{bmatrix} \sim \begin{bmatrix} 2 & -4 & -2 & 3 \\ 0 & 3 & 1 & -1 \\ 0 & 0 & 0 & 5 \\ 0 & 0 & 0 & 0 \\ 0 & 0 & 0 & 0 \end{bmatrix} = U$$

Divide the entries in each highlighted column by the pivot at the top. The resulting columns form the first three columns in the lower half of L. This suffices to make row reduction of L to I correspond to reduction of A to U. Use the last two columns of I_5

to make L unit lower triangular.

$$
\begin{bmatrix} 2 \\ 6 \\ 2 \\ 4 \\ -6 \end{bmatrix}
\begin{bmatrix} \\ 3 \\ -3 \\ 6 \\ -9 \end{bmatrix}
\begin{bmatrix} \\ \\ 5 \\ -5 \\ 10 \end{bmatrix}
$$

$$
\begin{array}{ccc}
\div 2 & \div 3 & \div 5 \\
\downarrow & \downarrow & \downarrow
\end{array}
$$

$$
\begin{bmatrix} 1 & & & \\ 3 & 1 & & \\ 1 & -1 & 1 & \cdots \\ 2 & 2 & -1 & \\ -3 & -3 & 2 & \end{bmatrix},
\quad
L = \begin{bmatrix} 1 & 0 & 0 & 0 & 0 \\ 3 & 1 & 0 & 0 & 0 \\ 1 & -1 & 1 & 0 & 0 \\ 2 & 2 & -1 & 1 & 0 \\ -3 & -3 & 2 & 0 & 1 \end{bmatrix}
$$

2.6 | THE LEONTIEF INPUT–OUTPUT MODEL

WEB

Linear algebra played an essential role in the Nobel prize–winning work of Wassily Leontief, as mentioned at the beginning of Chapter 1. The economic model described in this section is the basis for more elaborate models used in many parts of the world.

Suppose a nation's economy is divided into n sectors that produce goods or services, and let $\mathbf{x}$ be a **production vector** in $\mathbb{R}^n$ that lists the output of each sector for one year. Also, suppose another part of the economy (called the *open sector*) does not produce goods or services but only consumes them, and let $\mathbf{d}$ be a **final demand vector** (or **bill of final demands**) that lists the values of the goods and services demanded from the various sectors by the nonproductive part of the economy. The vector $\mathbf{d}$ can represent consumer demand, government consumption, surplus production, exports, or other external demands.

As the various sectors produce goods to meet consumer demand, the producers themselves create additional **intermediate demand** for goods they need as inputs for their own production. The interrelations between the sectors are very complex, and the connection between the final demand and the production is unclear. Leontief asked if there is a production level $\mathbf{x}$ such that the amounts produced (or "supplied") will exactly balance the total demand for that production, so that

$$
\left\{ \begin{array}{c} \text{amount} \\ \text{produced} \\ \mathbf{x} \end{array} \right\} = \left\{ \begin{array}{c} \text{intermediate} \\ \text{demand} \end{array} \right\} + \left\{ \begin{array}{c} \text{final} \\ \text{demand} \\ \mathbf{d} \end{array} \right\} \tag{1}
$$

The basic assumption of Leontief's input–output model is that for each sector, there is a **unit consumption vector** in $\mathbb{R}^n$ that lists the inputs needed *per unit of output* of the sector. All input and output units are measured in millions of dollars, rather than in quantities such as tons or bushels. (Prices of goods and services are held constant.)

As a simple example, suppose the economy consists of three sectors—manufacturing, agriculture, and services—with unit consumption vectors $\mathbf{c}_1$, $\mathbf{c}_2$, and $\mathbf{c}_3$, as shown in the table that follows.

Purchased from:	Inputs Consumed per Unit of Output		
	Manufacturing	**Agriculture**	**Services**
Manufacturing	.50	.40	.20
Agriculture	.20	.30	.10
Services	.10	.10	.30
	↑	↑	↑
	$\mathbf{c}_1$	$\mathbf{c}_2$	$\mathbf{c}_3$

EXAMPLE 1 What amounts will be consumed by the manufacturing sector if it decides to produce 100 units?

SOLUTION Compute

$$100\mathbf{c}_1 = 100\begin{bmatrix} .50 \\ .20 \\ .10 \end{bmatrix} = \begin{bmatrix} 50 \\ 20 \\ 10 \end{bmatrix}$$

To produce 100 units, manufacturing will order (i.e., "demand") and consume 50 units from other parts of the manufacturing sector, 20 units from agriculture, and 10 units from services. ∎

If manufacturing decides to produce x_1 units of output, then $x_1\mathbf{c}_1$ represents the *intermediate demands* of manufacturing, because the amounts in $x_1\mathbf{c}_1$ will be consumed in the process of creating the x_1 units of output. Likewise, if x_2 and x_3 denote the planned outputs of the agriculture and services sectors, $x_2\mathbf{c}_2$ and $x_3\mathbf{c}_3$ list their corresponding intermediate demands. The total intermediate demand from all three sectors is given by

$$\{\text{intermediate demand}\} = x_1\mathbf{c}_1 + x_2\mathbf{c}_2 + x_3\mathbf{c}_3$$
$$= C\mathbf{x} \tag{2}$$

where C is the **consumption matrix** $[\,\mathbf{c}_1 \quad \mathbf{c}_2 \quad \mathbf{c}_3\,]$, namely,

$$C = \begin{bmatrix} .50 & .40 & .20 \\ .20 & .30 & .10 \\ .10 & .10 & .30 \end{bmatrix} \tag{3}$$

Equations (1) and (2) yield Leontief's model.

THE LEONTIEF INPUT–OUTPUT MODEL, OR PRODUCTION EQUATION

$$\underset{\substack{\text{Amount} \\ \text{produced}}}{\mathbf{x}} \quad = \quad \underset{\substack{\text{Intermediate} \\ \text{demand}}}{C\mathbf{x}} \quad + \quad \underset{\substack{\text{Final} \\ \text{demand}}}{\mathbf{d}} \tag{4}$$

Equation (4) may also be written as $I\mathbf{x} - C\mathbf{x} = \mathbf{d}$, or

$$(I - C)\mathbf{x} = \mathbf{d} \tag{5}$$

EXAMPLE 2 Consider the economy whose consumption matrix is given by (3). Suppose the final demand is 50 units for manufacturing, 30 units for agriculture, and 20 units for services. Find the production level $\mathbf{x}$ that will satisfy this demand.

SOLUTION The coefficient matrix in (5) is

$$I - C = \begin{bmatrix} 1 & 0 & 0 \\ 0 & 1 & 0 \\ 0 & 0 & 1 \end{bmatrix} - \begin{bmatrix} .5 & .4 & .2 \\ .2 & .3 & .1 \\ .1 & .1 & .3 \end{bmatrix} = \begin{bmatrix} .5 & -.4 & -.2 \\ -.2 & .7 & -.1 \\ -.1 & -.1 & .7 \end{bmatrix}$$

To solve (5), row reduce the augmented matrix

$$\begin{bmatrix} .5 & -.4 & -.2 & 50 \\ -.2 & .7 & -.1 & 30 \\ -.1 & -.1 & .7 & 20 \end{bmatrix} \sim \begin{bmatrix} 5 & -4 & -2 & 500 \\ -2 & 7 & -1 & 300 \\ -1 & -1 & 7 & 200 \end{bmatrix} \sim \cdots \sim \begin{bmatrix} 1 & 0 & 0 & 226 \\ 0 & 1 & 0 & 119 \\ 0 & 0 & 1 & 78 \end{bmatrix}$$

The last column is rounded to the nearest whole unit. Manufacturing must produce approximately 226 units, agriculture 119 units, and services only 78 units. ∎

If the matrix $I - C$ is invertible, then we can apply Theorem 5 in Section 2.2, with A replaced by $(I - C)$, and from the equation $(I - C)\mathbf{x} = \mathbf{d}$ obtain $\mathbf{x} = (I - C)^{-1}\mathbf{d}$. The theorem below shows that in most practical cases, $I - C$ *is* invertible and the production vector $\mathbf{x}$ is economically feasible, in the sense that the entries in $\mathbf{x}$ are nonnegative.

In the theorem, the term **column sum** denotes the sum of the entries in a column of a matrix. Under ordinary circumstances, the column sums of a consumption matrix are less than 1 because a sector should require less than one unit's worth of inputs to produce one unit of output.

THEOREM 11

Let C be the consumption matrix for an economy, and let $\mathbf{d}$ be the final demand. If C and $\mathbf{d}$ have nonnegative entries and if each column sum of C is less than 1, then $(I - C)^{-1}$ exists and the production vector

$$\mathbf{x} = (I - C)^{-1}\mathbf{d}$$

has nonnegative entries and is the unique solution of

$$\mathbf{x} = C\mathbf{x} + \mathbf{d}$$

The following discussion will suggest why the theorem is true and will lead to a new way to compute $(I - C)^{-1}$.

A Formula for $(I - C)^{-1}$

Imagine that the demand represented by $\mathbf{d}$ is presented to the various industries at the beginning of the year, and the industries respond by setting their production levels at $\mathbf{x} = \mathbf{d}$, which will exactly meet the final demand. As the industries prepare to produce $\mathbf{d}$, they send out orders for their raw materials and other inputs. This creates an intermediate demand of $C\mathbf{d}$ for inputs.

To meet the additional demand of $C\mathbf{d}$, the industries will need as additional inputs the amounts in $C(C\mathbf{d}) = C^2\mathbf{d}$. Of course, this creates a second round of intermediate demand, and when the industries decide to produce even more to meet this new demand, they create a third round of demand, namely, $C(C^2\mathbf{d}) = C^3\mathbf{d}$. And so it goes.

Theoretically, this process could continue indefinitely, although in real life it would not take place in such a rigid sequence of events. We can diagram this hypothetical situation as follows:

	Demand That Must Be Met	Inputs Needed to Meet This Demand
Final demand	$\mathbf{d}$	$C\mathbf{d}$
Intermediate demand		
1st round	$C\mathbf{d}$	$C(C\mathbf{d}) = C^2\mathbf{d}$
2nd round	$C^2\mathbf{d}$	$C(C^2\mathbf{d}) = C^3\mathbf{d}$
3rd round	$C^3\mathbf{d}$	$C(C^3\mathbf{d}) = C^4\mathbf{d}$
$\vdots$		$\vdots$

The production level $\mathbf{x}$ that will meet all of this demand is

$$\mathbf{x} = \mathbf{d} + C\mathbf{d} + C^2\mathbf{d} + C^3\mathbf{d} + \cdots$$
$$= (I + C + C^2 + C^3 + \cdots)\mathbf{d} \tag{6}$$

To make sense of equation (6), consider the following algebraic identity:

$$(I - C)(I + C + C^2 + \cdots + C^m) = I - C^{m+1} \tag{7}$$

It can be shown that if the column sums in C are all strictly less than 1, then $I - C$ is invertible, C^m approaches the zero matrix as m gets arbitrarily large, and $I - C^{m+1} \rightarrow I$. (This fact is analogous to the fact that if a positive number t is less than 1, then $t^m \rightarrow 0$ as m increases.) Using equation (7), write

$$(I - C)^{-1} \approx I + C + C^2 + C^3 + \cdots + C^m$$
$$\text{when the column sums of } C \text{ are less than 1.} \tag{8}$$

The approximation in (8) means that the right side can be made as close to $(I - C)^{-1}$ as desired by taking m sufficiently large.

In actual input–output models, powers of the consumption matrix approach the zero matrix rather quickly. So (8) really provides a practical way to compute $(I - C)^{-1}$. Likewise, for any $\mathbf{d}$, the vectors $C^m\mathbf{d}$ approach the zero vector quickly, and (6) is a practical way to solve $(I - C)\mathbf{x} = \mathbf{d}$. If the entries in C and $\mathbf{d}$ are nonnegative, then (6) shows that the entries in $\mathbf{x}$ are nonnegative, too.

The Economic Importance of Entries in $(I - C)^{-1}$

The entries in $(I - C)^{-1}$ are significant because they can be used to predict how the production $\mathbf{x}$ will have to change when the final demand $\mathbf{d}$ changes. In fact, the entries in column j of $(I - C)^{-1}$ are the *increased* amounts the various sectors will have to produce in order to satisfy *an increase of 1 unit* in the final demand for output from sector j. See Exercise 8.

┌─ NUMERICAL NOTE ───

In any applied problem (not just in economics), an equation $A\mathbf{x} = \mathbf{b}$ can always be written as $(I - C)\mathbf{x} = \mathbf{b}$, with $C = I - A$. If the system is large and *sparse* (with mostly zero entries), it can happen that the column sums of the absolute values in C are less than 1. In this case, $C^m \rightarrow 0$. If C^m approaches zero quickly enough, (6) and (8) will provide practical formulas for solving $A\mathbf{x} = \mathbf{b}$ and finding A^{-1}.

| PRACTICE PROBLEM

Suppose an economy has two sectors: goods and services. One unit of output from goods requires inputs of .2 unit from goods and .5 unit from services. One unit of output from services requires inputs of .4 unit from goods and .3 unit from services. There is a final demand of 20 units of goods and 30 units of services. Set up the Leontief input–output model for this situation.

2.6 EXERCISES

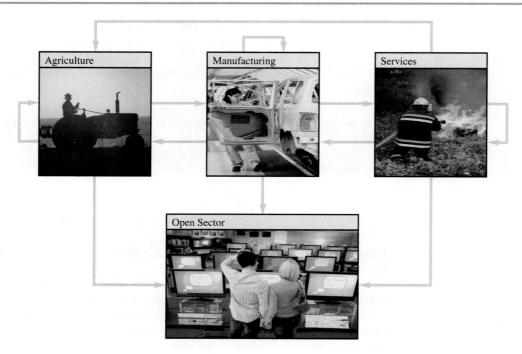

Exercises 1–4 refer to an economy that is divided into three sectors—manufacturing, agriculture, and services. For each unit of output, manufacturing requires .10 unit from other companies in that sector, .30 unit from agriculture, and .30 unit from services. For each unit of output, agriculture uses .20 unit of its own output, .60 unit from manufacturing, and .10 unit from services. For each unit of output, the services sector consumes .10 unit from services, .60 unit from manufacturing, but no agricultural products.

1. Construct the consumption matrix for this economy, and determine what intermediate demands are created if agriculture plans to produce 100 units.

2. Determine the production levels needed to satisfy a final demand of 20 units for agriculture, with no final demand for the other sectors. (Do not compute an inverse matrix.)

3. Determine the production levels needed to satisfy a final demand of 20 units for manufacturing, with no final demand for the other sectors. (Do not compute an inverse matrix.)

4. Determine the production levels needed to satisfy a final demand of 20 units for manufacturing, 20 units for agriculture, and 0 units for services.

5. Consider the production model $\mathbf{x} = C\mathbf{x} + \mathbf{d}$ for an economy with two sectors, where

$$C = \begin{bmatrix} .0 & .5 \\ .6 & .2 \end{bmatrix}, \qquad \mathbf{d} = \begin{bmatrix} 50 \\ 30 \end{bmatrix}$$

Use an inverse matrix to determine the production level necessary to satisfy the final demand.

6. Repeat Exercise 5 with $C = \begin{bmatrix} .2 & .5 \\ .6 & .1 \end{bmatrix}$ and $\mathbf{d} = \begin{bmatrix} 16 \\ 12 \end{bmatrix}$.

7. Let C and $\mathbf{d}$ be as in Exercise 5.

   a. Determine the production level necessary to satisfy a final demand for 1 unit of output from sector 1.

   b. Use an inverse matrix to determine the production level necessary to satisfy a final demand of $\begin{bmatrix} 51 \\ 30 \end{bmatrix}$.

c. Use the fact that $\begin{bmatrix} 51 \\ 30 \end{bmatrix} = \begin{bmatrix} 50 \\ 30 \end{bmatrix} + \begin{bmatrix} 1 \\ 0 \end{bmatrix}$ to explain how and why the answers to parts (a) and (b) and to Exercise 5 are related.

8. Let C be an $n \times n$ consumption matrix whose column sums are less than 1. Let $\mathbf{x}$ be the production vector that satisfies a final demand $\mathbf{d}$, and let $\Delta \mathbf{x}$ be a production vector that satisfies a different final demand $\Delta \mathbf{d}$.

   a. Show that if the final demand changes from $\mathbf{d}$ to $\mathbf{d} + \Delta \mathbf{d}$, then the new production level must be $\mathbf{x} + \Delta \mathbf{x}$. Thus $\Delta \mathbf{x}$ gives the amounts by which production must *change* in order to accommodate the *change* $\Delta \mathbf{d}$ in demand.

   b. Let $\Delta \mathbf{d}$ be the vector in $\mathbb{R}^n$ with 1 as the first entry and 0's elsewhere. Explain why the corresponding production $\Delta \mathbf{x}$ is the first column of $(I - C)^{-1}$. This shows that the first column of $(I - C)^{-1}$ gives the amounts the various sectors must produce to satisfy an increase of 1 unit in the final demand for output from sector 1.

9. Solve the Leontief production equation for an economy with three sectors, given that

$$C = \begin{bmatrix} .2 & .2 & .0 \\ .3 & .1 & .3 \\ .1 & .0 & .2 \end{bmatrix} \quad \text{and} \quad \mathbf{d} = \begin{bmatrix} 40 \\ 60 \\ 80 \end{bmatrix}$$

10. The consumption matrix C for the U.S. economy in 1972 has the property that *every entry* in the matrix $(I - C)^{-1}$ is nonzero (and positive).[1] What does that say about the effect of raising the demand for the output of just one sector of the economy?

11. The Leontief production equation, $\mathbf{x} = C\mathbf{x} + \mathbf{d}$, is usually accompanied by a dual **price equation**,

$$\mathbf{p} = C^T \mathbf{p} + \mathbf{v}$$

   where $\mathbf{p}$ is a **price vector** whose entries list the price per unit for each sector's output, and $\mathbf{v}$ is a **value added vector** whose entries list the value added per unit of output. (Value added includes wages, profit, depreciation, etc.) An important fact in economics is that the gross domestic product (GDP) can be expressed in two ways:

$$\{\text{gross domestic product}\} = \mathbf{p}^T \mathbf{d} = \mathbf{v}^T \mathbf{x}$$

   Verify the second equality. [*Hint*: Compute $\mathbf{p}^T \mathbf{x}$ in two ways.]

12. Let C be a consumption matrix such that $C^m \to 0$ as $m \to \infty$, and for $m = 1, 2, \ldots$, let $D_m = I + C + \cdots + C^m$. Find a difference equation that relates D_m and D_{m+1} and thereby obtain an iterative procedure for computing formula (8) for $(I - C)^{-1}$.

13. [M] The consumption matrix C below is based on input–output data for the U.S. economy in 1958, with data for 81 sectors grouped into 7 larger sectors: (1) nonmetal household and personal products, (2) final metal products (such as motor vehicles), (3) basic metal products and mining, (4) basic nonmetal products and agriculture, (5) energy, (6) services, and (7) entertainment and miscellaneous products.[2] Find the production levels needed to satisfy the final demand $\mathbf{d}$. (Units are in millions of dollars.)

$$\begin{bmatrix} .1588 & .0064 & .0025 & .0304 & .0014 & .0083 & .1594 \\ .0057 & .2645 & .0436 & .0099 & .0083 & .0201 & .3413 \\ .0264 & .1506 & .3557 & .0139 & .0142 & .0070 & .0236 \\ .3299 & .0565 & .0495 & .3636 & .0204 & .0483 & .0649 \\ .0089 & .0081 & .0333 & .0295 & .3412 & .0237 & .0020 \\ .1190 & .0901 & .0996 & .1260 & .1722 & .2368 & .3369 \\ .0063 & .0126 & .0196 & .0098 & .0064 & .0132 & .0012 \end{bmatrix},$$

$$\mathbf{d} = \begin{bmatrix} 74{,}000 \\ 56{,}000 \\ 10{,}500 \\ 25{,}000 \\ 17{,}500 \\ 196{,}000 \\ 5{,}000 \end{bmatrix}$$

14. [M] The demand vector in Exercise 13 is reasonable for 1958 data, but Leontief's discussion of the economy in the reference cited there used a demand vector closer to 1964 data:

$$\mathbf{d} = (99640, 75548, 14444, 33501, 23527, 263985, 6526)$$

   Find the production levels needed to satisfy this demand.

15. [M] Use equation (6) to solve the problem in Exercise 13. Set $\mathbf{x}^{(0)} = \mathbf{d}$, and for $k = 1, 2, \ldots$, compute $\mathbf{x}^{(k)} = \mathbf{d} + C\mathbf{x}^{(k-1)}$. How many steps are needed to obtain the answer in Exercise 13 to four significant figures?

---

[1] Wassily W. Leontief, "The World Economy of the Year 2000," *Scientific American*, September 1980, pp. 206–231.

[2] Wassily W. Leontief, "The Structure of the U.S. Economy," *Scientific American*, April 1965, pp. 30–32.

SOLUTION TO PRACTICE PROBLEM

The following data are given:

Purchased from:	Inputs Needed per Unit of Output		External Demand
	Goods	Services	
Goods	.2	.4	20
Services	.5	.3	30

The Leontief input–output model is $\mathbf{x} = C\mathbf{x} + \mathbf{d}$, where

$$C = \begin{bmatrix} .2 & .4 \\ .5 & .3 \end{bmatrix}, \qquad \mathbf{d} = \begin{bmatrix} 20 \\ 30 \end{bmatrix}$$

2.7 | APPLICATIONS TO COMPUTER GRAPHICS

Computer graphics are images displayed or animated on a computer screen. Applications of computer graphics are widespread and growing rapidly. For instance, computer-aided design (CAD) is an integral part of many engineering processes, such as the aircraft design process described in the chapter introduction. The entertainment industry has made the most spectacular use of computer graphics—from the special effects in *The Matrix* to PlayStation 2 and the Xbox.

Most interactive computer software for business and industry makes use of computer graphics in the screen displays and for other functions, such as graphical display of data, desktop publishing, and slide production for commercial and educational presentations. Consequently, anyone studying a computer language invariably spends time learning how to use at least two-dimensional (2D) graphics.

This section examines some of the basic mathematics used to manipulate and display graphical images such as a wire-frame model of an airplane. Such an image (or picture) consists of a number of points, connecting lines or curves, and information about how to fill in closed regions bounded by the lines and curves. Often, curved lines are approximated by short straight-line segments, and a figure is defined mathematically by a list of points.

Among the simplest 2D graphics symbols are letters used for labels on the screen. Some letters are stored as wire-frame objects; others that have curved portions are stored with additional mathematical formulas for the curves.

EXAMPLE 1   The capital letter N in Fig. 1 is determined by eight points, or *vertices*. The coordinates of the points can be stored in a data matrix, D.

Vertex:

$$\begin{array}{c} x\text{-coordinate} \\ y\text{-coordinate} \end{array} \begin{bmatrix} 0 & .5 & .5 & 6 & 6 & 5.5 & 5.5 & 0 \\ 0 & 0 & 6.42 & 0 & 8 & 8 & 1.58 & 8 \end{bmatrix} = D$$

(columns numbered 1 2 3 4 5 6 7 8)

In addition to D, it is necessary to specify which vertices are connected by lines, but we omit this detail. ∎

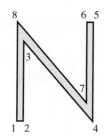

FIGURE 1
Regular N.

The main reason graphical objects are described by collections of straight-line segments is that the standard transformations in computer graphics map line segments onto other line segments. (For instance, see Exercise 26 in Section 1.8.) Once the vertices that describe an object have been transformed, their images can be connected with the appropriate straight lines to produce the complete image of the original object.

EXAMPLE 2 Given $A = \begin{bmatrix} 1 & .25 \\ 0 & 1 \end{bmatrix}$, describe the effect of the shear transformation $\mathbf{x} \mapsto A\mathbf{x}$ on the letter N in Example 1.

SOLUTION By definition of matrix multiplication, the columns of the product AD contain the images of the vertices of the letter N.

$$AD = \begin{array}{cccccccc} 1 & 2 & 3 & 4 & 5 & 6 & 7 & 8 \end{array}$$
$$AD = \begin{bmatrix} 0 & .5 & 2.105 & 6 & 8 & 7.5 & 5.895 & 2 \\ 0 & 0 & 6.420 & 0 & 8 & 8 & 1.580 & 8 \end{bmatrix}$$

The transformed vertices are plotted in Fig. 2, along with connecting line segments that correspond to those in the original figure. ∎

The italic N in Fig. 2 looks a bit too wide. To compensate, shrink the width by a scale transformation that affects the x-coordinates of the points.

EXAMPLE 3 Compute the matrix of the transformation that performs a shear transformation, as in Example 2, and then scales all x-coordinates by a factor of .75.

SOLUTION The matrix that multiplies the x-coordinate of a point by .75 is

$$S = \begin{bmatrix} .75 & 0 \\ 0 & 1 \end{bmatrix}$$

So the matrix of the composite transformation is

$$SA = \begin{bmatrix} .75 & 0 \\ 0 & 1 \end{bmatrix} \begin{bmatrix} 1 & .25 \\ 0 & 1 \end{bmatrix}$$
$$= \begin{bmatrix} .75 & .1875 \\ 0 & 1 \end{bmatrix}$$

The result of this composite transformation is shown in Fig. 3. ∎

The mathematics of computer graphics is intimately connected with matrix multiplication. Unfortunately, translating an object on a screen does not correspond directly to matrix multiplication because translation is not a linear transformation. The standard way to avoid this difficulty is to introduce what are called *homogeneous coordinates*.

Homogeneous Coordinates

Each point (x, y) in $\mathbb{R}^2$ can be identified with the point $(x, y, 1)$ on the plane in $\mathbb{R}^3$ that lies one unit above the xy-plane. We say that (x, y) has *homogeneous coordinates* $(x, y, 1)$. For instance, the point $(0, 0)$ has homogeneous coordinates $(0, 0, 1)$. Homogeneous coordinates for points are not added or multiplied by scalars, but they can be transformed via multiplication by 3×3 matrices.

EXAMPLE 4 A translation of the form $(x, y) \mapsto (x + h, y + k)$ is written in homogeneous coordinates as $(x, y, 1) \mapsto (x + h, y + k, 1)$. This transformation can be computed via matrix multiplication:

$$\begin{bmatrix} 1 & 0 & h \\ 0 & 1 & k \\ 0 & 0 & 1 \end{bmatrix} \begin{bmatrix} x \\ y \\ 1 \end{bmatrix} = \begin{bmatrix} x + h \\ y + k \\ 1 \end{bmatrix}$$
∎

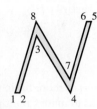

FIGURE 2
Slanted N.

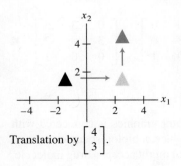

FIGURE 3
Composite transformation of N.

Translation by $\begin{bmatrix} 4 \\ 3 \end{bmatrix}$.

EXAMPLE 5 Any linear transformation on $\mathbb{R}^2$ is represented with respect to homogeneous coordinates by a partitioned matrix of the form $\begin{bmatrix} A & 0 \\ 0 & 1 \end{bmatrix}$, where A is a 2×2 matrix. Typical examples are

$$\begin{bmatrix} \cos\varphi & -\sin\varphi & 0 \\ \sin\varphi & \cos\varphi & 0 \\ 0 & 0 & 1 \end{bmatrix}, \qquad \begin{bmatrix} 0 & 1 & 0 \\ 1 & 0 & 0 \\ 0 & 0 & 1 \end{bmatrix}, \qquad \begin{bmatrix} s & 0 & 0 \\ 0 & t & 0 \\ 0 & 0 & 1 \end{bmatrix}$$

Counterclockwise rotation about the origin, angle φ Reflection through $y = x$ Scale x by s and y by t ∎

Composite Transformations

The movement of a figure on a computer screen often requires two or more basic transformations. The composition of such transformations corresponds to matrix multiplication when homogeneous coordinates are used.

EXAMPLE 6 Find the 3×3 matrix that corresponds to the composite transformation of a scaling by .3, a rotation of $90°$ about the origin, and finally a translation that adds $(-.5, 2)$ to each point of a figure.

SOLUTION If $\varphi = \pi/2$, then $\sin\varphi = 1$ and $\cos\varphi = 0$. From Examples 4 and 5, we have

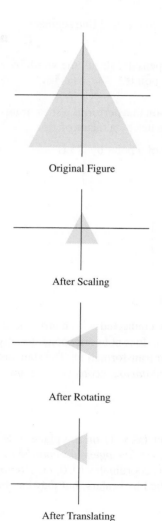

Original Figure

After Scaling

After Rotating

After Translating

$$\begin{bmatrix} x \\ y \\ 1 \end{bmatrix} \xrightarrow{\text{Scale}} \begin{bmatrix} .3 & 0 & 0 \\ 0 & .3 & 0 \\ 0 & 0 & 1 \end{bmatrix} \begin{bmatrix} x \\ y \\ 1 \end{bmatrix}$$

$$\xrightarrow{\text{Rotate}} \begin{bmatrix} 0 & -1 & 0 \\ 1 & 0 & 0 \\ 0 & 0 & 1 \end{bmatrix} \begin{bmatrix} .3 & 0 & 0 \\ 0 & .3 & 0 \\ 0 & 0 & 1 \end{bmatrix} \begin{bmatrix} x \\ y \\ 1 \end{bmatrix}$$

$$\xrightarrow{\text{Translate}} \begin{bmatrix} 1 & 0 & -.5 \\ 0 & 1 & 2 \\ 0 & 0 & 1 \end{bmatrix} \begin{bmatrix} 0 & -1 & 0 \\ 1 & 0 & 0 \\ 0 & 0 & 1 \end{bmatrix} \begin{bmatrix} .3 & 0 & 0 \\ 0 & .3 & 0 \\ 0 & 0 & 1 \end{bmatrix} \begin{bmatrix} x \\ y \\ 1 \end{bmatrix}$$

The matrix for the composite transformation is

$$\begin{bmatrix} 1 & 0 & -.5 \\ 0 & 1 & 2 \\ 0 & 0 & 1 \end{bmatrix} \begin{bmatrix} 0 & -1 & 0 \\ 1 & 0 & 0 \\ 0 & 0 & 1 \end{bmatrix} \begin{bmatrix} .3 & 0 & 0 \\ 0 & .3 & 0 \\ 0 & 0 & 1 \end{bmatrix}$$

$$= \begin{bmatrix} 0 & -1 & -.5 \\ 1 & 0 & 2 \\ 0 & 0 & 1 \end{bmatrix} \begin{bmatrix} .3 & 0 & 0 \\ 0 & .3 & 0 \\ 0 & 0 & 1 \end{bmatrix} = \begin{bmatrix} 0 & -.3 & -.5 \\ .3 & 0 & 2 \\ 0 & 0 & 1 \end{bmatrix}$$ ∎

3D Computer Graphics

Some of the newest and most exciting work in computer graphics is connected with molecular modeling. With 3D (three-dimensional) graphics, a biologist can examine a simulated protein molecule and search for active sites that might accept a drug molecule. The biologist can rotate and translate an experimental drug and attempt to attach it to the protein. This ability to *visualize* potential chemical reactions is vital to modern drug and cancer research. In fact, advances in drug design depend to some extent upon progress in the ability of computer graphics to construct realistic simulations of molecules and their interactions.[1]

Current research in molecular modeling is focused on *virtual reality*, an environment in which a researcher can see and *feel* the drug molecule slide into the protein. In Fig. 4, such tactile feedback is provided by a force-displaying remote manipulator.

FIGURE 4 Molecular modeling in virtual reality. (Computer Science Department, University of North Carolina at Chapel Hill. Photo by Bo Strain.)

Another design for virtual reality involves a helmet and glove that detect head, hand, and finger movements. The helmet contains two tiny computer screens, one for each eye. Making this virtual environment more realistic is a challenge to engineers, scientists, and mathematicians. The mathematics we examine here barely opens the door to this interesting field of research.

Homogeneous 3D Coordinates

By analogy with the 2D case, we say that $(x, y, z, 1)$ are homogeneous coordinates for the point (x, y, z) in $\mathbb{R}^3$. In general, (X, Y, Z, H) are **homogeneous coordinates** for (x, y, z) if $H \neq 0$ and

$$x = \frac{X}{H}, \qquad y = \frac{Y}{H}, \quad \text{and} \quad z = \frac{Z}{H} \tag{1}$$

Each nonzero scalar multiple of $(x, y, z, 1)$ gives a set of homogeneous coordinates for (x, y, z). For instance, both $(10, -6, 14, 2)$ and $(-15, 9, -21, -3)$ are homogeneous coordinates for $(5, -3, 7)$.

The next example illustrates the transformations used in molecular modeling to move a drug into a protein molecule.

EXAMPLE 7   Give 4×4 matrices for the following transformations:

a. Rotation about the y-axis through an angle of $30°$. (By convention, a positive angle is the counterclockwise direction when looking toward the origin from the positive half of the axis of rotation—in this case, the y-axis.)

---

[1] Robert Pool, "Computing in Science," *Science* **256**, 3 April 1992, p. 45.

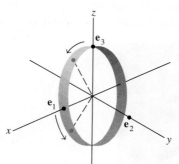

FIGURE 5

b. Translation by the vector $\mathbf{p} = (-6, 4, 5)$.

SOLUTION

a. First, construct the 3×3 matrix for the rotation. The vector $\mathbf{e}_1$ rotates down toward the negative z-axis, stopping at $(\cos 30°, 0, -\sin 30°) = (\sqrt{3}/2, 0, -.5)$. The vector $\mathbf{e}_2$ on the y-axis does not move, but $\mathbf{e}_3$ on the z-axis rotates down toward the positive x-axis, stopping at $(\sin 30°, 0, \cos 30°) = (.5, 0, \sqrt{3}/2)$. See Fig. 5. From Section 1.9, the standard matrix for this rotation is

$$\begin{bmatrix} \sqrt{3}/2 & 0 & .5 \\ 0 & 1 & 0 \\ -.5 & 0 & \sqrt{3}/2 \end{bmatrix}$$

So the rotation matrix for homogeneous coordinates is

$$A = \begin{bmatrix} \sqrt{3}/2 & 0 & .5 & 0 \\ 0 & 1 & 0 & 0 \\ -.5 & 0 & \sqrt{3}/2 & 0 \\ 0 & 0 & 0 & 1 \end{bmatrix}$$

b. We want $(x, y, z, 1)$ to map to $(x - 6, y + 4, z + 5, 1)$. The matrix that does this is

$$\begin{bmatrix} 1 & 0 & 0 & -6 \\ 0 & 1 & 0 & 4 \\ 0 & 0 & 1 & 5 \\ 0 & 0 & 0 & 1 \end{bmatrix} \qquad \blacksquare$$

Perspective Projections

A three-dimensional object is represented on the two-dimensional computer screen by projecting the object onto a *viewing plane*. (We ignore other important steps, such as selecting the portion of the viewing plane to display on the screen.) For simplicity, let the xy-plane represent the computer screen, and imagine that the eye of a viewer is along the positive z-axis, at a point $(0, 0, d)$. A *perspective projection* maps each point (x, y, z) onto an image point $(x^*, y^*, 0)$ so that the two points and the eye position, called the *center of projection*, are on a line. See Fig. 6(a).

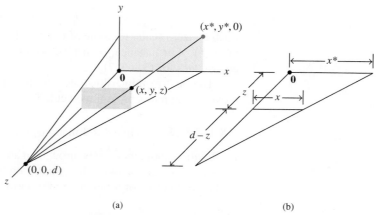

(a) (b)

FIGURE 6 Perspective projection of (x, y, z) onto $(x^*, y^*, 0)$.

The triangle in the xz-plane in Fig. 6(a) is redrawn in part (b) showing the lengths of line segments. Similar triangles show that

$$\frac{x^*}{d} = \frac{x}{d-z} \quad \text{and} \quad x^* = \frac{dx}{d-z} = \frac{x}{1-z/d}$$

Similarly,

$$y^* = \frac{y}{1-z/d}$$

Using homogeneous coordinates, we can represent the perspective projection by a matrix, say, P. We want $(x, y, z, 1)$ to map into $\left(\dfrac{x}{1-z/d}, \dfrac{y}{1-z/d}, 0, 1\right)$. Scaling these coordinates by $1 - z/d$, we can also use $(x, y, 0, 1 - z/d)$ as homogeneous coordinates for the image. Now it is easy to display P. In fact,

$$P\begin{bmatrix} x \\ y \\ z \\ 1 \end{bmatrix} = \begin{bmatrix} 1 & 0 & 0 & 0 \\ 0 & 1 & 0 & 0 \\ 0 & 0 & 0 & 0 \\ 0 & 0 & -1/d & 1 \end{bmatrix}\begin{bmatrix} x \\ y \\ z \\ 1 \end{bmatrix} = \begin{bmatrix} x \\ y \\ 0 \\ 1 - z/d \end{bmatrix}$$

EXAMPLE 8   Let S be the box with vertices $(3, 1, 5)$, $(5, 1, 5)$, $(5, 0, 5)$, $(3, 0, 5)$, $(3, 1, 4)$, $(5, 1, 4)$, $(5, 0, 4)$, and $(3, 0, 4)$. Find the image of S under the perspective projection with center of projection at $(0, 0, 10)$.

SOLUTION   Let P be the projection matrix, and let D be the data matrix for S using homogeneous coordinates. The data matrix for the image of S is

Vertex:

$$PD = \begin{bmatrix} 1 & 0 & 0 & 0 \\ 0 & 1 & 0 & 0 \\ 0 & 0 & 0 & 0 \\ 0 & 0 & -1/10 & 1 \end{bmatrix}\begin{array}{c}\begin{array}{cccccccc} 1 & 2 & 3 & 4 & 5 & 6 & 7 & 8 \end{array}\\\begin{bmatrix} 3 & 5 & 5 & 3 & 3 & 5 & 5 & 3 \\ 1 & 1 & 0 & 0 & 1 & 1 & 0 & 0 \\ 5 & 5 & 5 & 5 & 4 & 4 & 4 & 4 \\ 1 & 1 & 1 & 1 & 1 & 1 & 1 & 1 \end{bmatrix}\end{array}$$

$$= \begin{bmatrix} 3 & 5 & 5 & 3 & 3 & 5 & 5 & 3 \\ 1 & 1 & 0 & 0 & 1 & 1 & 0 & 0 \\ 0 & 0 & 0 & 0 & 0 & 0 & 0 & 0 \\ .5 & .5 & .5 & .5 & .6 & .6 & .6 & .6 \end{bmatrix}$$

To obtain $\mathbb{R}^3$ coordinates, use equation (1) before Example 7, and divide the top three entries in each column by the corresponding entry in the fourth row:

Vertex:

$$\begin{array}{c}\begin{array}{cccccccc} 1 & 2 & 3 & 4 & 5 & 6 & 7 & 8 \end{array}\\\begin{bmatrix} 6 & 10 & 10 & 6 & 5 & 8.3 & 8.3 & 5 \\ 2 & 2 & 0 & 0 & 1.7 & 1.7 & 0 & 0 \\ 0 & 0 & 0 & 0 & 0 & 0 & 0 & 0 \end{bmatrix}\end{array}$$

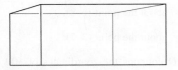

S under the perspective transformation.

This text's web site has some interesting applications of computer graphics, including a further discussion of perspective projections. One of the computer projects on the web site involves simple animation.

WEB

NUMERICAL NOTE

Continuous movement of graphical 3D objects requires intensive computation with 4×4 matrices, particularly when the surfaces are *rendered* to appear realistic, with texture and appropriate lighting. *High-end computer graphics boards* have 4×4 matrix operations and graphics algorithms embedded in their microchips and circuitry. Such boards can perform the billions of matrix multiplications per second needed for realistic color animation in 3D gaming programs.[2]

Further Reading

James D. Foley, Andries van Dam, Steven K. Feiner, and John F. Hughes, *Computer Graphics: Principles and Practice*, 3rd ed. (Boston, MA: Addison-Wesley, 2002), Chapters 5 and 6.

PRACTICE PROBLEM

Rotation of a figure about a point **p** in $\mathbb{R}^2$ is accomplished by first translating the figure by $-\mathbf{p}$, rotating about the origin, and then translating back by **p**. See Fig. 7. Construct the 3×3 matrix that rotates points $-30°$ about the point $(-2, 6)$, using homogeneous coordinates.

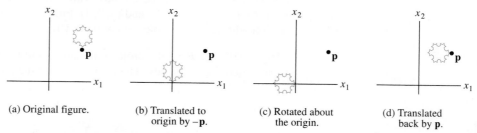

(a) Original figure.   (b) Translated to origin by $-\mathbf{p}$.   (c) Rotated about the origin.   (d) Translated back by **p**.

FIGURE 7 Rotation of figure about point **p**.

2.7 EXERCISES

1. What 3×3 matrix will have the same effect on homogeneous coordinates for $\mathbb{R}^2$ that the shear matrix A has in Example 2?

2. Use matrix multiplication to find the image of the triangle with data matrix $D = \begin{bmatrix} 4 & 2 & 5 \\ 0 & 2 & 3 \end{bmatrix}$ under the transformation that reflects points through the y-axis. Sketch both the original triangle and its image.

In Exercises 3–8, find the 3×3 matrices that produce the described composite 2D transformations, using homogeneous coordinates.

3. Translate by $(2, 1)$, and then rotate $90°$ about the origin.

4. Translate by $(-1, 4)$, and then scale the x-coordinate by 1/2 and the y-coordinate by 3/2.

5. Reflect points through the x-axis, and then rotate $45°$ about the origin.

6. Rotate points $45°$ about the origin, then reflect through the x-axis.

7. Rotate points through $60°$ about the point $(6, 8)$.

8. Rotate points through $45°$ about the point $(3, 7)$.

9. A 2×100 data matrix D contains the coordinates of 100 points. Compute the number of multiplications required to transform these points using two arbitrary 2×2 matrices A and B. Consider the two possibilities $A(BD)$ and $(AB)D$. Discuss the implications of your results for computer graphics calculations.

---

[2] See Jan Ozer, "High-Performance Graphics Boards," *PC Magazine* **19**, 1 September 2000, pp. 187–200. Also, "The Ultimate Upgrade Guide: Moving On Up," *PC Magazine* **21**, 29 January 2002, pp. 82–91.

10. Consider the following geometric 2D transformations: D, a dilation (in which x-coordinates and y-coordinates are scaled by the same factor); R, a rotation; and T, a translation. Does D *commute* with R? That is, is $D(R(\mathbf{x})) = R(D(\mathbf{x}))$ for all $\mathbf{x}$ in $\mathbb{R}^2$? Does D commute with T? Does R commute with T?

11. A rotation on a computer screen is sometimes implemented as the product of two shear-and-scale transformations, which can speed up calculations that determine how a graphic image actually appears in terms of screen pixels. (The screen consists of rows and columns of small dots, called *pixels*.) The first transformation A_1 shears vertically and then compresses each column of pixels; the second transformation A_2 shears horizontally and then stretches each row of pixels. Let

$$A_1 = \begin{bmatrix} 1 & 0 & 0 \\ \sin\varphi & \cos\varphi & 0 \\ 0 & 0 & 1 \end{bmatrix},$$

$$A_2 = \begin{bmatrix} \sec\varphi & -\tan\varphi & 0 \\ 0 & 1 & 0 \\ 0 & 0 & 1 \end{bmatrix},$$

and show that the composition of the two transformations is a rotation in $\mathbb{R}^2$.

12. A rotation in $\mathbb{R}^2$ usually requires four multiplications. Compute the product below, and show that the matrix for a rotation can be factored into three shear transformations (each of which requires only one multiplication).

$$\begin{bmatrix} 1 & -\tan\varphi/2 & 0 \\ 0 & 1 & 0 \\ 0 & 0 & 1 \end{bmatrix} \begin{bmatrix} 1 & 0 & 0 \\ \sin\varphi & 1 & 0 \\ 0 & 0 & 1 \end{bmatrix}$$
$$\begin{bmatrix} 1 & -\tan\varphi/2 & 0 \\ 0 & 1 & 0 \\ 0 & 0 & 1 \end{bmatrix}$$

13. The usual transformations on homogeneous coordinates for 2D computer graphics involve 3×3 matrices of the form $\begin{bmatrix} A & \mathbf{p} \\ \mathbf{0}^T & 1 \end{bmatrix}$ where A is a 2×2 matrix and $\mathbf{p}$ is in $\mathbb{R}^2$. Show that such a transformation amounts to a linear transformation on $\mathbb{R}^2$ followed by a translation. [*Hint:* Find an appropriate matrix factorization involving partitioned matrices.]

14. Show that the transformation in Exercise 7 is equivalent to a rotation about the origin followed by a translation by $\mathbf{p}$. Find $\mathbf{p}$.

15. What vector in $\mathbb{R}^3$ has homogeneous coordinates $(\frac{1}{2}, -\frac{1}{4}, -\frac{1}{8}, \frac{1}{24})$?

16. Are $(1, -2, -3, 4)$ and $(10, -20, -30, 40)$ homogeneous coordinates for the same point in $\mathbb{R}^3$? Why or why not?

17. Give the 4×4 matrix that rotates points in $\mathbb{R}^3$ about the x-axis through an angle of $60°$. (See the figure.)

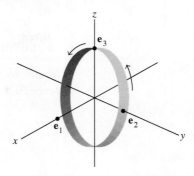

18. Give the 4×4 matrix that rotates points in $\mathbb{R}^3$ about the z-axis through an angle of $-30°$, and then translates by $\mathbf{p} = (5, -2, 1)$.

19. Let S be the triangle with vertices $(4.2, 1.2, 4)$, $(6, 4, 2)$, and $(2, 2, 6)$. Find the image of S under the perspective projection with center of projection at $(0, 0, 10)$.

20. Let S be the triangle with vertices $(7, 3, -5)$, $(12, 8, 2)$, and $(1, 2, 1)$. Find the image of S under the perspective projection with center of projection at $(0, 0, 10)$.

Exercises 21 and 22 concern the way in which color is specified for display in computer graphics. A color on a computer screen is encoded by three numbers (R, G, B) that list the amounts of energy an electron gun must transmit to red, green, and blue phosphor dots on the computer screen. (A fourth number specifies the luminance or intensity of the color.)

21. [M] The actual color a viewer sees on a screen is influenced by the specific type and amount of phosphors on the screen. So each computer screen manufacturer must convert between the (R, G, B) data and an international CIE standard for color, which uses three primary colors, called X, Y, and Z. A typical conversion for short-persistence phosphors is

$$\begin{bmatrix} .61 & .29 & .150 \\ .35 & .59 & .063 \\ .04 & .12 & .787 \end{bmatrix} \begin{bmatrix} R \\ G \\ B \end{bmatrix} = \begin{bmatrix} X \\ Y \\ Z \end{bmatrix}$$

A computer program will send a stream of color information to the screen, using standard CIE data (X, Y, Z). Find the equation that converts these data to the (R, G, B) data needed for the screen's electron gun.

22. [M] The signal broadcast by commercial television describes each color by a vector (Y, I, Q). If the screen is black and white, only the Y-coordinate is used. (This gives a better monochrome picture than using CIE data for colors.) The correspondence between YIQ and a "standard" RGB color is given by

$$\begin{bmatrix} Y \\ I \\ Q \end{bmatrix} = \begin{bmatrix} .299 & .587 & .114 \\ .596 & -.275 & -.321 \\ .212 & -.528 & .311 \end{bmatrix} \begin{bmatrix} R \\ G \\ B \end{bmatrix}$$

(A screen manufacturer would change the matrix entries to work for its *RGB* screens.) Find the equation that converts the *YIQ* data transmitted by the television station to the *RGB* data needed for the television screen.

SOLUTION TO PRACTICE PROBLEM

Assemble the matrices right-to-left for the three operations. Using $\mathbf{p} = (-2, 6)$, $\cos(-30°) = \sqrt{3}/2$, and $\sin(-30°) = -.5$, we have

$$
\underset{\substack{\text{Translate} \\ \text{back by } p}}{\begin{bmatrix} 1 & 0 & -2 \\ 0 & 1 & 6 \\ 0 & 0 & 1 \end{bmatrix}}
\underset{\substack{\text{Rotate around} \\ \text{the origin}}}{\begin{bmatrix} \sqrt{3}/2 & 1/2 & 0 \\ -1/2 & \sqrt{3}/2 & 0 \\ 0 & 0 & 1 \end{bmatrix}}
\underset{\substack{\text{Translate} \\ \text{by } -p}}{\begin{bmatrix} 1 & 0 & 2 \\ 0 & 1 & -6 \\ 0 & 0 & 1 \end{bmatrix}}
$$

$$
= \begin{bmatrix} \sqrt{3}/2 & 1/2 & \sqrt{3} - 5 \\ -1/2 & \sqrt{3}/2 & -3\sqrt{3} + 5 \\ 0 & 0 & 1 \end{bmatrix}
$$

2.8 | SUBSPACES OF $\mathbb{R}^n$

This section focuses on important sets of vectors in $\mathbb{R}^n$ called *subspaces*. Often subspaces arise in connection with some matrix A, and they provide useful information about the equation $A\mathbf{x} = \mathbf{b}$. The concepts and terminology in this section will be used repeatedly throughout the rest of the book.[1]

DEFINITION

A **subspace** of $\mathbb{R}^n$ is any set H in $\mathbb{R}^n$ that has three properties:

a. The zero vector is in H.

b. For each $\mathbf{u}$ and $\mathbf{v}$ in H, the sum $\mathbf{u} + \mathbf{v}$ is in H.

c. For each $\mathbf{u}$ in H and each scalar c, the vector $c\mathbf{u}$ is in H.

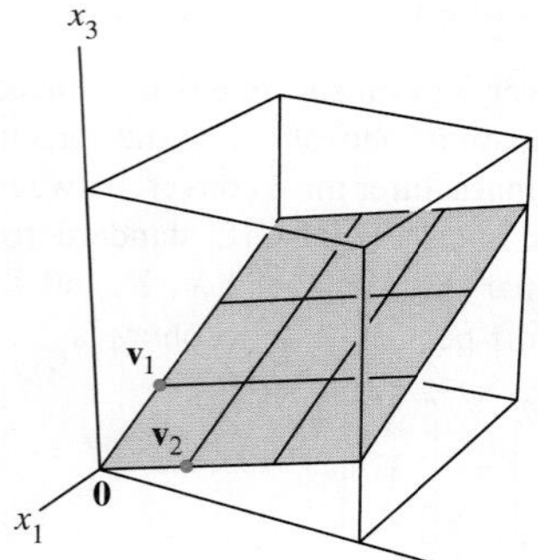

FIGURE 1

Span $\{\mathbf{v}_1, \mathbf{v}_2\}$ as a plane through the origin.

In words, a subspace is *closed* under addition and scalar multiplication. As you will see in the next few examples, most sets of vectors discussed in Chapter 1 are subspaces. For instance, a plane through the origin is the standard way to visualize the subspace in Example 1. See Fig. 1.

EXAMPLE 1   If $\mathbf{v}_1$ and $\mathbf{v}_2$ are in $\mathbb{R}^n$ and $H = \text{Span}\{\mathbf{v}_1, \mathbf{v}_2\}$, then H is a subspace of $\mathbb{R}^n$. To verify this statement, note that the zero vector is in H (because $0\mathbf{v}_1 + 0\mathbf{v}_2$ is a linear combination of $\mathbf{v}_1$ and $\mathbf{v}_2$). Now take two arbitrary vectors in H, say,

$$\mathbf{u} = s_1\mathbf{v}_1 + s_2\mathbf{v}_2 \quad \text{and} \quad \mathbf{v} = t_1\mathbf{v}_1 + t_2\mathbf{v}_2$$

Then

$$\mathbf{u} + \mathbf{v} = (s_1 + t_1)\mathbf{v}_1 + (s_2 + t_2)\mathbf{v}_2$$

which shows that $\mathbf{u} + \mathbf{v}$ is a linear combination of $\mathbf{v}_1$ and $\mathbf{v}_2$ and hence is in H. Also, for any scalar c, the vector $c\mathbf{u}$ is in H, because $c\mathbf{u} = c(s_1\mathbf{v}_1 + s_2\mathbf{v}_2) = (cs_1)\mathbf{v}_1 + (cs_2)\mathbf{v}_2$. ∎

---

[1] Sections 2.8 and 2.9 are included here to permit readers to postpone the study of most or all of the next two chapters and to skip directly to Chapter 5, if so desired. *Omit* these two sections if you plan to work through Chapter 4 before beginning Chapter 5.

If $\mathbf{v}_1$ is not zero and if $\mathbf{v}_2$ is a multiple of $\mathbf{v}_1$, then $\mathbf{v}_1$ and $\mathbf{v}_2$ simply span a *line* through the origin. So a line through the origin is another example of a subspace.

EXAMPLE 2 A line L *not* through the origin is *not* a subspace, because it does not contain the origin, as required. Also, Fig. 2 shows that L is not closed under addition or scalar multiplication. ∎

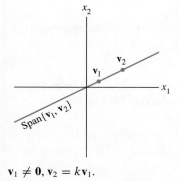

$\mathbf{v}_1 \neq \mathbf{0}, \mathbf{v}_2 = k\mathbf{v}_1.$

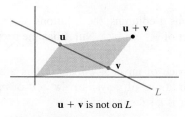

$\mathbf{u} + \mathbf{v}$ is not on L $2\mathbf{w}$ is not on L

FIGURE 2

EXAMPLE 3 For $\mathbf{v}_1, \ldots, \mathbf{v}_p$ in $\mathbb{R}^n$, the set of all linear combinations of $\mathbf{v}_1, \ldots, \mathbf{v}_p$ is a subspace of $\mathbb{R}^n$. The verification of this statement is similar to the argument given in Example 1. We shall now refer to Span $\{\mathbf{v}_1, \ldots, \mathbf{v}_p\}$ as **the subspace spanned** (or **generated**) by $\mathbf{v}_1, \ldots, \mathbf{v}_p$. ∎

Note that $\mathbb{R}^n$ is a subspace of itself because it has the three properties required for a subspace. Another special subspace is the set consisting of only the zero vector in $\mathbb{R}^n$. This set, called the **zero subspace**, also satisfies the conditions for a subspace.

Column Space and Null Space of a Matrix

Subspaces of $\mathbb{R}^n$ usually occur in applications and theory in one of two ways. In both cases, the subspace can be related to a matrix.

DEFINITION

> The **column space** of a matrix A is the set Col A of all linear combinations of the columns of A.

If $A = [\, \mathbf{a}_1 \ \cdots \ \mathbf{a}_n \,]$, with the columns in $\mathbb{R}^m$, then Col A is the same as Span $\{\mathbf{a}_1, \ldots, \mathbf{a}_n\}$. Example 4 shows that the **column space of an $m \times n$ matrix is a subspace of $\mathbb{R}^m$**. Note that Col A equals $\mathbb{R}^m$ only when the columns of A span $\mathbb{R}^m$. Otherwise, Col A is only part of $\mathbb{R}^m$.

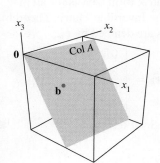

EXAMPLE 4 Let $A = \begin{bmatrix} 1 & -3 & -4 \\ -4 & 6 & -2 \\ -3 & 7 & 6 \end{bmatrix}$ and $\mathbf{b} = \begin{bmatrix} 3 \\ 3 \\ -4 \end{bmatrix}$. Determine whether $\mathbf{b}$ is in the column space of A.

SOLUTION The vector $\mathbf{b}$ is a linear combination of the columns of A if and only if $\mathbf{b}$ can be written as $A\mathbf{x}$ for some $\mathbf{x}$, that is, if and only if the equation $A\mathbf{x} = \mathbf{b}$ has a solution. Row reducing the augmented matrix $[\, A \ \ \mathbf{b} \,]$,

$$\begin{bmatrix} 1 & -3 & -4 & 3 \\ -4 & 6 & -2 & 3 \\ -3 & 7 & 6 & -4 \end{bmatrix} \sim \begin{bmatrix} 1 & -3 & -4 & 3 \\ 0 & -6 & -18 & 15 \\ 0 & -2 & -6 & 5 \end{bmatrix} \sim \begin{bmatrix} 1 & -3 & -4 & 3 \\ 0 & -6 & -18 & 15 \\ 0 & 0 & 0 & 0 \end{bmatrix}$$

we conclude that $A\mathbf{x} = \mathbf{b}$ is consistent and $\mathbf{b}$ is in Col A. ∎

The solution of Example 4 shows that when a system of linear equations is written in the form $A\mathbf{x} = \mathbf{b}$, the column space of A is the set of all $\mathbf{b}$ for which the system has a solution.

DEFINITION

> The **null space** of a matrix A is the set Nul A of all solutions of the homogeneous equation $A\mathbf{x} = \mathbf{0}$.

When A has n columns, the solutions of $A\mathbf{x} = \mathbf{0}$ belong to $\mathbb{R}^n$, and the null space of A is a subset of $\mathbb{R}^n$. In fact, Nul A has the properties of a *subspace* of $\mathbb{R}^n$.

THEOREM 12

> The null space of an $m \times n$ matrix A is a subspace of $\mathbb{R}^n$. Equivalently, the set of all solutions of a system $A\mathbf{x} = \mathbf{0}$ of m homogeneous linear equations in n unknowns is a subspace of $\mathbb{R}^n$.

PROOF The zero vector is in Nul A (because $A\mathbf{0} = \mathbf{0}$). To show that Nul A satisfies the other two properties required for a subspace, take any $\mathbf{u}$ and $\mathbf{v}$ in Nul A. That is, suppose $A\mathbf{u} = \mathbf{0}$ and $A\mathbf{v} = \mathbf{0}$. Then, by a property of matrix multiplication,

$$A(\mathbf{u} + \mathbf{v}) = A\mathbf{u} + A\mathbf{v} = \mathbf{0} + \mathbf{0} = \mathbf{0}$$

Thus $\mathbf{u} + \mathbf{v}$ satisfies $A\mathbf{x} = \mathbf{0}$, and so $\mathbf{u} + \mathbf{v}$ is in Nul A. Also, for any scalar c, $A(c\mathbf{u}) = c(A\mathbf{u}) = c(\mathbf{0}) = \mathbf{0}$, which shows that $c\mathbf{u}$ is in Nul A. ∎

To test whether a given vector $\mathbf{v}$ is in Nul A, just compute $A\mathbf{v}$ to see whether $A\mathbf{v}$ is the zero vector. Because Nul A is described by a condition that must be checked for each vector, we say that the null space is defined *implicitly*. In contrast, the column space is defined *explicitly*, because vectors in Col A can be constructed (by linear combinations) from the columns of A. To create an explicit description of Nul A, solve the equation $A\mathbf{x} = \mathbf{0}$ and write the solution in parametric vector form. (See Example 6, below.)[2]

Basis for a Subspace

Because a subspace typically contains an infinite number of vectors, some problems involving a subspace are handled best by working with a small finite set of vectors that span the subspace. The smaller the set, the better. It can be shown that the smallest possible spanning set must be linearly independent.

DEFINITION

> A **basis** for a subspace H of $\mathbb{R}^n$ is a linearly independent set in H that spans H.

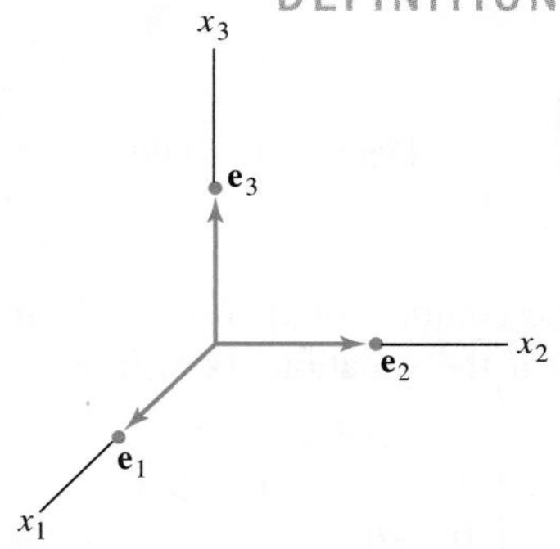

EXAMPLE 5 The columns of an invertible $n \times n$ matrix form a basis for all of $\mathbb{R}^n$ because they are linearly independent and span $\mathbb{R}^n$, by the Invertible Matrix Theorem. One such matrix is the $n \times n$ identity matrix. Its columns are denoted by $\mathbf{e}_1, \ldots, \mathbf{e}_n$:

$$\mathbf{e}_1 = \begin{bmatrix} 1 \\ 0 \\ \vdots \\ 0 \end{bmatrix}, \quad \mathbf{e}_2 = \begin{bmatrix} 0 \\ 1 \\ \vdots \\ 0 \end{bmatrix}, \quad \ldots, \quad \mathbf{e}_n = \begin{bmatrix} 0 \\ \vdots \\ 0 \\ 1 \end{bmatrix}$$

The set $\{\mathbf{e}_1, \ldots, \mathbf{e}_n\}$ is called the **standard basis** for $\mathbb{R}^n$. See Fig. 3. ∎

FIGURE 3

The standard basis for $\mathbb{R}^3$.

---

[2] The contrast between Nul A and Col A is discussed further in Section 4.2.

The next example shows that the standard procedure for writing the solution set of $A\mathbf{x} = \mathbf{0}$ in parametric vector form actually identifies a basis for Nul A. This fact will be used throughout Chapter 5.

EXAMPLE 6   Find a basis for the null space of the matrix

$$A = \begin{bmatrix} -3 & 6 & -1 & 1 & -7 \\ 1 & -2 & 2 & 3 & -1 \\ 2 & -4 & 5 & 8 & -4 \end{bmatrix}$$

SOLUTION   First, write the solution of $A\mathbf{x} = \mathbf{0}$ in parametric vector form:

$$\begin{bmatrix} A & \mathbf{0} \end{bmatrix} \sim \begin{bmatrix} 1 & -2 & 0 & -1 & 3 & 0 \\ 0 & 0 & 1 & 2 & -2 & 0 \\ 0 & 0 & 0 & 0 & 0 & 0 \end{bmatrix}, \quad \begin{array}{r} x_1 - 2x_2 \quad - x_4 + 3x_5 = 0 \\ x_3 + 2x_4 - 2x_5 = 0 \\ 0 = 0 \end{array}$$

The general solution is $x_1 = 2x_2 + x_4 - 3x_5$, $x_3 = -2x_4 + 2x_5$, with x_2, x_4, and x_5 free.

$$\begin{bmatrix} x_1 \\ x_2 \\ x_3 \\ x_4 \\ x_5 \end{bmatrix} = \begin{bmatrix} 2x_2 + x_4 - 3x_5 \\ x_2 \\ -2x_4 + 2x_5 \\ x_4 \\ x_5 \end{bmatrix} = x_2 \underbrace{\begin{bmatrix} 2 \\ 1 \\ 0 \\ 0 \\ 0 \end{bmatrix}}_{\mathbf{u}} + x_4 \underbrace{\begin{bmatrix} 1 \\ 0 \\ -2 \\ 1 \\ 0 \end{bmatrix}}_{\mathbf{v}} + x_5 \underbrace{\begin{bmatrix} -3 \\ 0 \\ 2 \\ 0 \\ 1 \end{bmatrix}}_{\mathbf{w}}$$

$$= x_2\mathbf{u} + x_4\mathbf{v} + x_5\mathbf{w} \tag{1}$$

Equation (1) shows that Nul A coincides with the set of all linear combinations of $\mathbf{u}$, $\mathbf{v}$, and $\mathbf{w}$. That is, $\{\mathbf{u}, \mathbf{v}, \mathbf{w}\}$ generates Nul A. In fact, this construction of $\mathbf{u}$, $\mathbf{v}$, and $\mathbf{w}$ automatically makes them linearly independent, because equation (1) shows that $\mathbf{0} = x_2\mathbf{u} + x_4\mathbf{v} + x_5\mathbf{w}$ only if the weights x_2, x_4, and x_5 are all zero. (Examine entries 2, 4, and 5 in the vector $x_2\mathbf{u} + x_4\mathbf{v} + x_5\mathbf{w}$.) So $\{\mathbf{u}, \mathbf{v}, \mathbf{w}\}$ is a *basis* for Nul A.   ∎

Finding a basis for the column space of a matrix is actually less work than finding a basis for the null space. However, the method requires some explanation. Let's begin with a simple case.

EXAMPLE 7   Find a basis for the column space of the matrix

$$B = \begin{bmatrix} 1 & 0 & -3 & 5 & 0 \\ 0 & 1 & 2 & -1 & 0 \\ 0 & 0 & 0 & 0 & 1 \\ 0 & 0 & 0 & 0 & 0 \end{bmatrix}$$

SOLUTION   Denote the columns of B by $\mathbf{b}_1, \ldots, \mathbf{b}_5$ and note that $\mathbf{b}_3 = -3\mathbf{b}_1 + 2\mathbf{b}_2$ and $\mathbf{b}_4 = 5\mathbf{b}_1 - \mathbf{b}_2$. The fact that $\mathbf{b}_3$ and $\mathbf{b}_4$ are combinations of the pivot columns means that any combination of $\mathbf{b}_1, \ldots, \mathbf{b}_5$ is actually just a combination of $\mathbf{b}_1$, $\mathbf{b}_2$, and $\mathbf{b}_5$. Indeed, if $\mathbf{v}$ is any vector in Col B, say,

$$\mathbf{v} = c_1\mathbf{b}_1 + c_2\mathbf{b}_2 + c_3\mathbf{b}_3 + c_4\mathbf{b}_4 + c_5\mathbf{b}_5$$

then, substituting for $\mathbf{b}_3$ and $\mathbf{b}_4$, we can write $\mathbf{v}$ in the form

$$\mathbf{v} = c_1\mathbf{b}_1 + c_2\mathbf{b}_2 + c_3(-3\mathbf{b}_1 + 2\mathbf{b}_2) + c_4(5\mathbf{b}_1 - \mathbf{b}_2) + c_5\mathbf{b}_5$$

which is a linear combination of $\mathbf{b}_1$, $\mathbf{b}_2$, and $\mathbf{b}_5$. So $\{\mathbf{b}_1, \mathbf{b}_2, \mathbf{b}_5\}$ spans Col B. Also, $\mathbf{b}_1$, $\mathbf{b}_2$, and $\mathbf{b}_5$ are linearly independent, because they are columns from an identity matrix. So the pivot columns of B form a basis for Col B.   ∎

The matrix B in Example 7 is in reduced echelon form. To handle a general matrix A, recall that linear dependence relations among the columns of A can be expressed in the form $A\mathbf{x} = \mathbf{0}$ for some $\mathbf{x}$. (If some columns are not involved in a particular dependence relation, then the corresponding entries in $\mathbf{x}$ are zero.) When A is row reduced to echelon form B, the columns are drastically changed, but the equations $A\mathbf{x} = \mathbf{0}$ and $B\mathbf{x} = \mathbf{0}$ have the same set of solutions. That is, the columns of A have *exactly the same linear dependence relationships* as the columns of B.

EXAMPLE 8 It can be verified that the matrix

$$A = [\,\mathbf{a}_1 \quad \mathbf{a}_2 \quad \cdots \quad \mathbf{a}_5\,] = \begin{bmatrix} 1 & 3 & 3 & 2 & -9 \\ -2 & -2 & 2 & -8 & 2 \\ 2 & 3 & 0 & 7 & 1 \\ 3 & 4 & -1 & 11 & -8 \end{bmatrix}$$

is row equivalent to the matrix B in Example 7. Find a basis for Col A.

SOLUTION From Example 7, the pivot columns of A are columns 1, 2, and 5. Also, $\mathbf{b}_3 = -3\mathbf{b}_1 + 2\mathbf{b}_2$ and $\mathbf{b}_4 = 5\mathbf{b}_1 - \mathbf{b}_2$. Since row operations do not affect linear dependence relations among the columns of the matrix, we should have

$$\mathbf{a}_3 = -3\mathbf{a}_1 + 2\mathbf{a}_2 \quad \text{and} \quad \mathbf{a}_4 = 5\mathbf{a}_1 - \mathbf{a}_2$$

Check that this is true! By the argument in Example 7, $\mathbf{a}_3$ and $\mathbf{a}_4$ are not needed to generate the column space of A. Also, $\{\mathbf{a}_1, \mathbf{a}_2, \mathbf{a}_5\}$ must be linearly independent, because any dependence relation among $\mathbf{a}_1, \mathbf{a}_2$, and $\mathbf{a}_5$ would imply the same dependence relation among $\mathbf{b}_1, \mathbf{b}_2$, and $\mathbf{b}_5$. Since $\{\mathbf{b}_1, \mathbf{b}_2, \mathbf{b}_5\}$ is linearly independent, $\{\mathbf{a}_1, \mathbf{a}_2, \mathbf{a}_5\}$ is also linearly independent and hence is a basis for Col A. ∎

The argument in Example 8 can be adapted to prove the following theorem.

THEOREM 13

> The pivot columns of a matrix A form a basis for the column space of A.

Warning: Be careful to use *pivot columns of A itself* for the basis of Col A. The columns of an echelon form B are often not in the column space of A. (For instance, in Examples 7 and 8, the columns of B all have zeros in their last entries and cannot generate the columns of A.)

PRACTICE PROBLEMS

1. Let $A = \begin{bmatrix} 1 & -1 & 5 \\ 2 & 0 & 7 \\ -3 & -5 & -3 \end{bmatrix}$ and $\mathbf{u} = \begin{bmatrix} -7 \\ 3 \\ 2 \end{bmatrix}$. Is $\mathbf{u}$ in Nul A? Is $\mathbf{u}$ in Col A? Justify each answer.

2. Given $A = \begin{bmatrix} 0 & 1 & 0 \\ 0 & 0 & 1 \\ 0 & 0 & 0 \end{bmatrix}$, find a vector in Nul A and a vector in Col A.

3. Suppose an $n \times n$ matrix A is invertible. What can you say about Col A? About Nul A?

SG | Mastering: Subspace, Col A, Nul A, Basis 2–37

2.8 EXERCISES

Exercises 1–4 display sets in $\mathbb{R}^2$. Assume the sets include the bounding lines. In each case, give a specific reason why the set H is *not* a subspace of $\mathbb{R}^2$. (For instance, find two vectors in H whose sum is *not* in H, or find a vector in H with a scalar multiple that is not in H. Draw a picture.)

1.

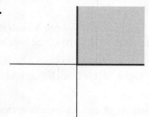

2.

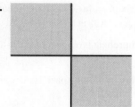

3.

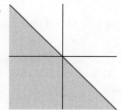

4.

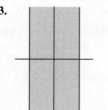

5. Let $\mathbf{v}_1 = \begin{bmatrix} 1 \\ 3 \\ -4 \end{bmatrix}$, $\mathbf{v}_2 = \begin{bmatrix} -2 \\ -3 \\ 7 \end{bmatrix}$, and $\mathbf{w} = \begin{bmatrix} -3 \\ -3 \\ 10 \end{bmatrix}$. Determine if $\mathbf{w}$ is in the subspace of $\mathbb{R}^3$ generated by $\mathbf{v}_1$ and $\mathbf{v}_2$.

6. Let $\mathbf{v}_1 = \begin{bmatrix} 1 \\ -3 \\ 2 \\ 3 \end{bmatrix}$, $\mathbf{v}_2 = \begin{bmatrix} 4 \\ -4 \\ 5 \\ 7 \end{bmatrix}$, $\mathbf{v}_3 = \begin{bmatrix} 5 \\ -3 \\ 6 \\ 5 \end{bmatrix}$, and $\mathbf{u} = \begin{bmatrix} -1 \\ -7 \\ -1 \\ 2 \end{bmatrix}$. Determine if $\mathbf{u}$ is in the subspace of $\mathbb{R}^4$ generated by $\{\mathbf{v}_1, \mathbf{v}_2, \mathbf{v}_3\}$.

7. Let
$$\mathbf{v}_1 = \begin{bmatrix} 2 \\ -8 \\ 6 \end{bmatrix}, \quad \mathbf{v}_2 = \begin{bmatrix} -3 \\ 8 \\ -7 \end{bmatrix}, \quad \mathbf{v}_3 = \begin{bmatrix} -4 \\ 6 \\ -7 \end{bmatrix},$$
$$\mathbf{p} = \begin{bmatrix} 6 \\ -10 \\ 11 \end{bmatrix}, \quad \text{and} \quad A = \begin{bmatrix} \mathbf{v}_1 & \mathbf{v}_2 & \mathbf{v}_3 \end{bmatrix}.$$

a. How many vectors are in $\{\mathbf{v}_1, \mathbf{v}_2, \mathbf{v}_3\}$?

b. How many vectors are in Col A?

c. Is $\mathbf{p}$ in Col A? Why or why not?

8. Let
$$\mathbf{v}_1 = \begin{bmatrix} -2 \\ 0 \\ 6 \end{bmatrix}, \quad \mathbf{v}_2 = \begin{bmatrix} -2 \\ 3 \\ 3 \end{bmatrix}, \quad \mathbf{v}_3 = \begin{bmatrix} 0 \\ -5 \\ 5 \end{bmatrix},$$
and $\mathbf{p} = \begin{bmatrix} -6 \\ 1 \\ 17 \end{bmatrix}$. Determine if $\mathbf{p}$ is in Col A, where $A = \begin{bmatrix} \mathbf{v}_1 & \mathbf{v}_2 & \mathbf{v}_3 \end{bmatrix}$.

9. With A and $\mathbf{p}$ as in Exercise 7, determine if $\mathbf{p}$ is in Nul A.

10. With $\mathbf{u} = \begin{bmatrix} -5 \\ 5 \\ 3 \end{bmatrix}$ and A as in Exercise 8, determine if $\mathbf{u}$ is in Nul A.

In Exercises 11 and 12, give integers p and q such that Nul A is a subspace of $\mathbb{R}^p$ and Col A is a subspace of $\mathbb{R}^q$.

11. $A = \begin{bmatrix} 3 & 2 & 1 & -5 \\ -9 & -4 & 1 & 7 \\ 9 & 2 & -5 & 1 \end{bmatrix}$

12. $A = \begin{bmatrix} 1 & 2 & 3 \\ 4 & 5 & 7 \\ -5 & -1 & 0 \\ 2 & 7 & 11 \\ 3 & 3 & 4 \end{bmatrix}$

13. For A as in Exercise 11, find a nonzero vector in Nul A and a nonzero vector in Col A.

14. For A as in Exercise 12, find a nonzero vector in Nul A and a nonzero vector in Col A.

Determine which sets in Exercises 15–20 are bases for $\mathbb{R}^2$ or $\mathbb{R}^3$. Justify each answer.

15. $\begin{bmatrix} 4 \\ -2 \end{bmatrix}, \begin{bmatrix} 16 \\ -3 \end{bmatrix}$ **16.** $\begin{bmatrix} -2 \\ 5 \end{bmatrix}, \begin{bmatrix} 4 \\ -10 \end{bmatrix}$

17. $\begin{bmatrix} 0 \\ 0 \\ -2 \end{bmatrix}, \begin{bmatrix} 5 \\ 0 \\ 4 \end{bmatrix}, \begin{bmatrix} 6 \\ 3 \\ 2 \end{bmatrix}$ **18.** $\begin{bmatrix} 1 \\ 1 \\ -3 \end{bmatrix}, \begin{bmatrix} 3 \\ -1 \\ 2 \end{bmatrix}, \begin{bmatrix} 5 \\ 1 \\ -4 \end{bmatrix}$

19. $\begin{bmatrix} 3 \\ -8 \\ 1 \end{bmatrix}, \begin{bmatrix} 6 \\ 2 \\ -5 \end{bmatrix}$ **20.** $\begin{bmatrix} 1 \\ -6 \\ -7 \end{bmatrix}, \begin{bmatrix} 3 \\ -6 \\ 7 \end{bmatrix}, \begin{bmatrix} -3 \\ 7 \\ 5 \end{bmatrix}, \begin{bmatrix} 0 \\ 7 \\ 9 \end{bmatrix}$

In Exercises 21 and 22, mark each statement True or False. Justify each answer.

21. a. A subspace of $\mathbb{R}^n$ is any set H such that (i) the zero vector is in H, (ii) $\mathbf{u}$, $\mathbf{v}$, and $\mathbf{u} + \mathbf{v}$ are in H, and (iii) c is a scalar and $c\mathbf{u}$ is in H.

   b. If $\mathbf{v}_1, \ldots, \mathbf{v}_p$ are in $\mathbb{R}^n$, then Span $\{\mathbf{v}_1, \ldots, \mathbf{v}_p\}$ is the same as the column space of the matrix $\begin{bmatrix} \mathbf{v}_1 & \cdots & \mathbf{v}_p \end{bmatrix}$.

   c. The set of all solutions of a system of m homogeneous equations in n unknowns is a subspace of $\mathbb{R}^m$.

   d. The columns of an invertible $n \times n$ matrix form a basis for $\mathbb{R}^n$.

   e. Row operations do not affect linear dependence relations among the columns of a matrix.

22. a. A subset H of $\mathbb{R}^n$ is a subspace if the zero vector is in H.

   b. If B is an echelon form of a matrix A, then the pivot columns of B form a basis for Col A.

   c. Given vectors $\mathbf{v}_1, \ldots, \mathbf{v}_p$ in $\mathbb{R}^n$, the set of all linear combinations of these vectors is a subspace of $\mathbb{R}^n$.

   d. Let H be a subspace of $\mathbb{R}^n$. If $\mathbf{x}$ is in H, and $\mathbf{y}$ is in $\mathbb{R}^n$, then $\mathbf{x} + \mathbf{y}$ is in H.

   e. The column space of a matrix A is the set of solutions of $A\mathbf{x} = \mathbf{b}$.

Exercises 23–26 display a matrix A and an echelon form of A. Find a basis for Col A and a basis for Nul A.

23. $A = \begin{bmatrix} 4 & 5 & 9 & -2 \\ 6 & 5 & 1 & 12 \\ 3 & 4 & 8 & -3 \end{bmatrix} \sim \begin{bmatrix} 1 & 2 & 6 & -5 \\ 0 & 1 & 5 & -6 \\ 0 & 0 & 0 & 0 \end{bmatrix}$

24. $A = \begin{bmatrix} 3 & -6 & 9 & 0 \\ 2 & -4 & 7 & 2 \\ 3 & -6 & 6 & -6 \end{bmatrix} \sim \begin{bmatrix} 1 & -2 & 5 & 4 \\ 0 & 0 & 3 & 6 \\ 0 & 0 & 0 & 0 \end{bmatrix}$

25. $A = \begin{bmatrix} 1 & 4 & 8 & -3 & -7 \\ -1 & 2 & 7 & 3 & 4 \\ -2 & 2 & 9 & 5 & 5 \\ 3 & 6 & 9 & -5 & -2 \end{bmatrix}$

$\sim \begin{bmatrix} 1 & 4 & 8 & 0 & 5 \\ 0 & 2 & 5 & 0 & -1 \\ 0 & 0 & 0 & 1 & 4 \\ 0 & 0 & 0 & 0 & 0 \end{bmatrix}$

26. $A = \begin{bmatrix} 3 & -1 & -3 & -1 & 8 \\ 3 & 1 & 3 & 0 & 2 \\ 0 & 3 & 9 & -1 & -4 \\ 6 & 3 & 9 & -2 & 6 \end{bmatrix}$

$\sim \begin{bmatrix} 3 & -1 & -3 & 0 & 6 \\ 0 & 2 & 6 & 0 & -4 \\ 0 & 0 & 0 & -1 & 2 \\ 0 & 0 & 0 & 0 & 0 \end{bmatrix}$

27. Construct a 3×3 matrix A and a nonzero vector $\mathbf{b}$ such that $\mathbf{b}$ is in Col A, but $\mathbf{b}$ is not the same as any one of the columns of A.

28. Construct a 3×3 matrix A and a vector $\mathbf{b}$ such that $\mathbf{b}$ is *not* in Col A.

29. Construct a nonzero 3×3 matrix A and a nonzero vector $\mathbf{b}$ such that $\mathbf{b}$ is in Nul A.

30. Suppose the columns of a matrix $A = [\mathbf{a}_1 \cdots \mathbf{a}_p]$ are linearly independent. Explain why $\{\mathbf{a}_1, \ldots, \mathbf{a}_p\}$ is a basis for Col A.

In Exercises 31–36, respond as comprehensively as possible, and justify your answer.

31. Suppose F is a 5×5 matrix whose column space is not equal to $\mathbb{R}^5$. What can be said about Nul F?

32. If B is a 7×7 matrix and Col $B = \mathbb{R}^7$, what can be said about solutions of equations of the form $B\mathbf{x} = \mathbf{b}$ for $\mathbf{b}$ in $\mathbb{R}^7$?

33. If C is a 6×6 matrix and Nul C is the zero subspace, what can be said about solutions of equations of the form $C\mathbf{x} = \mathbf{b}$ for $\mathbf{b}$ in $\mathbb{R}^6$?

34. What can be said about the shape of an $m \times n$ matrix A when the columns of A form a basis for $\mathbb{R}^m$?

35. If B is a 5×5 matrix and Nul B is *not* the zero subspace, what can be said about Col B?

36. What can be said about Nul C when C is a 6×4 matrix with linearly independent columns?

[M] In Exercises 37 and 38, construct bases for the column space and the null space of the given matrix A. Justify your work.

37. $A = \begin{bmatrix} 3 & -5 & 0 & -1 & 3 \\ -7 & 9 & -4 & 9 & -11 \\ -5 & 7 & -2 & 5 & -7 \\ 3 & -7 & -3 & 4 & 0 \end{bmatrix}$

38. $A = \begin{bmatrix} 5 & 3 & 2 & -6 & -8 \\ 4 & 1 & 3 & -8 & -7 \\ 5 & 1 & 4 & 5 & 19 \\ -7 & -5 & -2 & 8 & 5 \end{bmatrix}$

| WEB | Column Space and Null Space |

| WEB | A Basis for Col A |

| **SOLUTIONS TO PRACTICE PROBLEMS**

1. To determine whether **u** is in Nul A, simply compute

$$A\mathbf{u} = \begin{bmatrix} 1 & -1 & 5 \\ 2 & 0 & 7 \\ -3 & -5 & -3 \end{bmatrix} \begin{bmatrix} -7 \\ 3 \\ 2 \end{bmatrix} = \begin{bmatrix} 0 \\ 0 \\ 0 \end{bmatrix}$$

The result shows that **u** is in Nul A. Deciding whether **u** is in Col A requires more work. Reduce the augmented matrix $[A \quad \mathbf{u}]$ to echelon form to determine whether the equation $A\mathbf{x} = \mathbf{u}$ is consistent:

$$\begin{bmatrix} 1 & -1 & 5 & -7 \\ 2 & 0 & 7 & 3 \\ -3 & -5 & -3 & 2 \end{bmatrix} \sim \begin{bmatrix} 1 & -1 & 5 & -7 \\ 0 & 2 & -3 & 17 \\ 0 & -8 & 12 & -19 \end{bmatrix} \sim \begin{bmatrix} 1 & -1 & 5 & -7 \\ 0 & 2 & -3 & 17 \\ 0 & 0 & 0 & 49 \end{bmatrix}$$

The equation $A\mathbf{x} = \mathbf{u}$ has no solution, so **u** is not in Col A.

2. In contrast to Practice Problem 1, finding a vector in Nul A requires more work than testing whether a specified vector is in Nul A. However, since A is already in reduced echelon form, the equation $A\mathbf{x} = \mathbf{0}$ shows that if $\mathbf{x} = (x_1, x_2, x_3)$, then $x_2 = 0$, $x_3 = 0$, and x_1 is a free variable. Thus, a basis for Nul A is $\mathbf{v} = (1, 0, 0)$. Finding just one vector in Col A is trivial, since each column of A is in Col A. In this particular case, the same vector **v** is in both Nul A *and* Col A. For most $n \times n$ matrices, the zero vector of $\mathbb{R}^n$ is the only vector in both Nul A and Col A.

3. If A is invertible, then the columns of A span $\mathbb{R}^n$, by the Invertible Matrix Theorem. By definition, the columns of any matrix always span the column space, so in this case Col A is all of $\mathbb{R}^n$. In symbols, Col $A = \mathbb{R}^n$. Also, since A is invertible, the equation $A\mathbf{x} = \mathbf{0}$ has only the trivial solution. This means that Nul A is the zero subspace. In symbols, Nul $A = \{\mathbf{0}\}$.

2.9   DIMENSION AND RANK

This section continues the discussion of subspaces and bases for subspaces, beginning with the concept of a coordinate system. The definition and example below should make a useful new term, *dimension,* seem quite natural, at least for subspaces of $\mathbb{R}^3$.

Coordinate Systems

The main reason for selecting a basis for a subspace H, instead of merely a spanning set, is that each vector in H can be written in only one way as a linear combination of the basis vectors. To see why, suppose $\mathcal{B} = \{\mathbf{b}_1, \ldots, \mathbf{b}_p\}$ is a basis for H, and suppose a vector **x** in H can be generated in two ways, say,

$$\mathbf{x} = c_1\mathbf{b}_1 + \cdots + c_p\mathbf{b}_p \quad \text{and} \quad \mathbf{x} = d_1\mathbf{b}_1 + \cdots + d_p\mathbf{b}_p \tag{1}$$

Then, subtracting gives

$$\mathbf{0} = \mathbf{x} - \mathbf{x} = (c_1 - d_1)\mathbf{b}_1 + \cdots + (c_p - d_p)\mathbf{b}_p \tag{2}$$

Since $\mathcal{B}$ is linearly independent, the weights in (2) must all be zero. That is, $c_j = d_j$ for $1 \le j \le p$, which shows that the two representations in (1) are actually the same.

DEFINITION

Suppose the set $\mathcal{B} = \{\mathbf{b}_1, \ldots, \mathbf{b}_p\}$ is a basis for a subspace H. For each $\mathbf{x}$ in H, the **coordinates of x relative to the basis** $\mathcal{B}$ are the weights $c_1, \ldots, c_p$ such that $\mathbf{x} = c_1\mathbf{b}_1 + \cdots + c_p\mathbf{b}_p$, and the vector in $\mathbb{R}^p$

$$[\mathbf{x}]_{\mathcal{B}} = \begin{bmatrix} c_1 \\ \vdots \\ c_p \end{bmatrix}$$

is called the **coordinate vector of x (relative to** $\mathcal{B}$) or the $\mathcal{B}$-**coordinate vector of x.**[1]

EXAMPLE 1 Let $\mathbf{v}_1 = \begin{bmatrix} 3 \\ 6 \\ 2 \end{bmatrix}$, $\mathbf{v}_2 = \begin{bmatrix} -1 \\ 0 \\ 1 \end{bmatrix}$, $\mathbf{x} = \begin{bmatrix} 3 \\ 12 \\ 7 \end{bmatrix}$, and $\mathcal{B} = \{\mathbf{v}_1, \mathbf{v}_2\}$. Then $\mathcal{B}$ is a basis for $H = \text{Span}\{\mathbf{v}_1, \mathbf{v}_2\}$ because $\mathbf{v}_1$ and $\mathbf{v}_2$ are linearly independent. Determine if $\mathbf{x}$ is in H, and if it is, find the coordinate vector of $\mathbf{x}$ relative to $\mathcal{B}$.

SOLUTION If $\mathbf{x}$ is in H, then the following vector equation is consistent:

$$c_1\begin{bmatrix} 3 \\ 6 \\ 2 \end{bmatrix} + c_2\begin{bmatrix} -1 \\ 0 \\ 1 \end{bmatrix} = \begin{bmatrix} 3 \\ 12 \\ 7 \end{bmatrix}$$

The scalars c_1 and c_2, if they exist, are the $\mathcal{B}$-coordinates of $\mathbf{x}$. Row operations show that

$$\begin{bmatrix} 3 & -1 & 3 \\ 6 & 0 & 12 \\ 2 & 1 & 7 \end{bmatrix} \sim \begin{bmatrix} 1 & 0 & 2 \\ 0 & 1 & 3 \\ 0 & 0 & 0 \end{bmatrix}$$

Thus $c_1 = 2$, $c_2 = 3$, and $[\mathbf{x}]_{\mathcal{B}} = \begin{bmatrix} 2 \\ 3 \end{bmatrix}$. The basis $\mathcal{B}$ determines a "coordinate system" on H, which can be visualized by the grid shown in Fig. 1. ∎

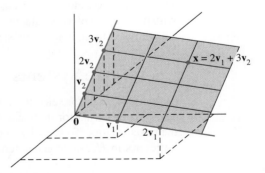

FIGURE 1 A coordinate system on a plane H in $\mathbb{R}^3$.

---

[1] It is important that the elements of $\mathcal{B}$ are numbered because the entries in $[\mathbf{x}]_{\mathcal{B}}$ depend on the order of the vectors in $\mathcal{B}$.

Notice that although points in H are also in $\mathbb{R}^3$, they are completely determined by their coordinate vectors, which belong to $\mathbb{R}^2$. The grid on the plane in Fig. 1 makes H "look" like $\mathbb{R}^2$. The correspondence $\mathbf{x} \mapsto [\mathbf{x}]_{\mathcal{B}}$ is a one-to-one correspondence between H and $\mathbb{R}^2$ that preserves linear combinations. We call such a correspondence an *isomorphism*, and we say that H is *isomorphic* to $\mathbb{R}^2$.

In general, if $\mathcal{B} = \{\mathbf{b}_1, \ldots, \mathbf{b}_p\}$ is a basis for H, then the mapping $\mathbf{x} \mapsto [\mathbf{x}]_{\mathcal{B}}$ is a one-to-one correspondence that makes H look and act the same as $\mathbb{R}^p$ (even though the vectors in H themselves may have more than p entries). (Section 4.4 has more details.)

The Dimension of a Subspace

It can be shown that if a subspace H has a basis of p vectors, then every basis of H must consist of exactly p vectors. (See Exercises 27 and 28.) Thus the following definition makes sense.

DEFINITION

> The **dimension** of a nonzero subspace H, denoted by $\dim H$, is the number of vectors in any basis for H. The dimension of the zero subspace $\{\mathbf{0}\}$ is defined to be zero.[2]

The space $\mathbb{R}^n$ has dimension n. Every basis for $\mathbb{R}^n$ consists of n vectors. A plane through $\mathbf{0}$ in $\mathbb{R}^3$ is two-dimensional, and a line through $\mathbf{0}$ is one-dimensional.

EXAMPLE 2   Recall that the null space of the matrix A in Example 6 in Section 2.8 had a basis of 3 vectors. So the dimension of Nul A in this case is 3. Observe how each basis vector corresponds to a free variable in the equation $A\mathbf{x} = \mathbf{0}$. Our construction always produces a basis in this way. So, to find the dimension of Nul A, simply identify and count the number of free variables in $A\mathbf{x} = \mathbf{0}$.   ∎

DEFINITION

> The **rank** of a matrix A, denoted by rank A, is the dimension of the column space of A.

Since the pivot columns of A form a basis for Col A, the rank of A is just the number of pivot columns in A.

EXAMPLE 3   Determine the rank of the matrix

$$A = \begin{bmatrix} 2 & 5 & -3 & -4 & 8 \\ 4 & 7 & -4 & -3 & 9 \\ 6 & 9 & -5 & 2 & 4 \\ 0 & -9 & 6 & 5 & -6 \end{bmatrix}$$

SOLUTION Reduce A to echelon form:

$$A \sim \begin{bmatrix} 2 & 5 & -3 & -4 & 8 \\ 0 & -3 & 2 & 5 & -7 \\ 0 & -6 & 4 & 14 & -20 \\ 0 & -9 & 6 & 5 & -6 \end{bmatrix} \sim \cdots \sim \begin{bmatrix} 2 & 5 & -3 & -4 & 8 \\ 0 & -3 & 2 & 5 & -7 \\ 0 & 0 & 0 & 4 & -6 \\ 0 & 0 & 0 & 0 & 0 \end{bmatrix}$$

Pivot columns

The matrix A has 3 pivot columns, so rank $A = 3$.   ∎

---

[2] The zero subspace has *no* basis (because the zero vector by itself forms a linearly dependent set).

The row reduction in Example 3 reveals that there are two free variables in $A\mathbf{x} = \mathbf{0}$, because two of the five columns of A are *not* pivot columns. (The nonpivot columns correspond to the free variables in $A\mathbf{x} = \mathbf{0}$.) Since the number of pivot columns plus the number of nonpivot columns is exactly the number of columns, the dimensions of Col A and Nul A have the following useful connection. (See the Rank Theorem in Section 4.6 for additional details.)

THEOREM 14

The Rank Theorem

If a matrix A has n columns, then rank $A + \dim \text{Nul } A = n$.

The following theorem is important for applications and will be needed in Chapters 5 and 6. The theorem (proved in Section 4.5) is certainly plausible, if you think of a p-dimensional subspace as isomorphic to $\mathbb{R}^p$. The Invertible Matrix Theorem shows that p vectors in $\mathbb{R}^p$ are linearly independent if and only if they also span $\mathbb{R}^p$.

THEOREM 15

The Basis Theorem

Let H be a p-dimensional subspace of $\mathbb{R}^n$. Any linearly independent set of exactly p elements in H is automatically a basis for H. Also, any set of p elements of H that spans H is automatically a basis for H.

Rank and the Invertible Matrix Theorem

The various vector space concepts associated with a matrix provide several more statements for the Invertible Matrix Theorem. They are presented below to follow the statements in the original theorem in Section 2.3.

THEOREM

The Invertible Matrix Theorem (continued)

Let A be an $n \times n$ matrix. Then the following statements are each equivalent to the statement that A is an invertible matrix.

m. The columns of A form a basis of $\mathbb{R}^n$.

n. Col $A = \mathbb{R}^n$

o. $\dim \text{Col } A = n$

p. rank $A = n$

q. Nul $A = \{\mathbf{0}\}$

r. $\dim \text{Nul } A = 0$

PROOF Statement (m) is logically equivalent to statements (e) and (h) regarding linear independence and spanning. The other five statements are linked to the earlier ones of the theorem by the following chain of almost trivial implications:

$$(g) \Rightarrow (n) \Rightarrow (o) \Rightarrow (p) \Rightarrow (r) \Rightarrow (q) \Rightarrow (d)$$

Statement (g), which says that the equation $A\mathbf{x} = \mathbf{b}$ has at least one solution for each $\mathbf{b}$ in $\mathbb{R}^n$, implies statement (n), because Col A is precisely the set of all $\mathbf{b}$ such that the equation $A\mathbf{x} = \mathbf{b}$ is consistent. The implications (n) $\Rightarrow$ (o) $\Rightarrow$ (p) follow from the definitions of *dimension* and *rank*. If the rank of A is n, the number of columns of A, then $\dim \text{Nul } A = 0$, by the Rank Theorem, and so Nul $A = \{\mathbf{0}\}$. Thus (p) $\Rightarrow$ (r) $\Rightarrow$ (q).

Also, statement (q) implies that the equation $A\mathbf{x} = \mathbf{0}$ has only the trivial solution, which is statement (d). Since statements (d) and (g) are already known to be equivalent to the statement that A is invertible, the proof is complete. ∎

NUMERICAL NOTES

Many algorithms discussed in this text are useful for understanding concepts and making simple computations by hand. However, the algorithms are often unsuitable for large-scale problems in real life.

Rank determination is a good example. It would seem easy to reduce a matrix to echelon form and count the pivots. But unless exact arithmetic is performed on a matrix whose entries are specified exactly, row operations can change the apparent rank of a matrix. For instance, if the value of x in the matrix $\begin{bmatrix} 5 & 7 \\ 5 & x \end{bmatrix}$ is not stored exactly as 7 in a computer, then the rank may be 1 or 2, depending on whether the computer treats $x - 7$ as zero.

In practical applications, the effective rank of a matrix A is often determined from the singular value decomposition of A, to be discussed in Section 7.4.

WEB

PRACTICE PROBLEMS

1. Determine the dimension of the subspace H of $\mathbb{R}^3$ spanned by the vectors $\mathbf{v}_1$, $\mathbf{v}_2$, and $\mathbf{v}_3$. (First, find a basis for H.)

$$\mathbf{v}_1 = \begin{bmatrix} 2 \\ -8 \\ 6 \end{bmatrix}, \quad \mathbf{v}_2 = \begin{bmatrix} 3 \\ -7 \\ -1 \end{bmatrix}, \quad \mathbf{v}_3 = \begin{bmatrix} -1 \\ 6 \\ -7 \end{bmatrix}$$

2. Consider the basis

$$\mathcal{B} = \left\{ \begin{bmatrix} 1 \\ .2 \end{bmatrix}, \begin{bmatrix} .2 \\ 1 \end{bmatrix} \right\}$$

for $\mathbb{R}^2$. If $[\mathbf{x}]_{\mathcal{B}} = \begin{bmatrix} 3 \\ 2 \end{bmatrix}$, what is $\mathbf{x}$?

3. Could $\mathbb{R}^3$ possibly contain a four-dimensional subspace? Explain.

2.9 EXERCISES

In Exercises 1 and 2, find the vector $\mathbf{x}$ determined by the given coordinate vector $[\mathbf{x}]_{\mathcal{B}}$ and the given basis $\mathcal{B}$. Illustrate your answer with a figure, as in the solution of Practice Problem 2.

1. $\mathcal{B} = \left\{ \begin{bmatrix} 1 \\ 1 \end{bmatrix}, \begin{bmatrix} 2 \\ -1 \end{bmatrix} \right\}$, $[\mathbf{x}]_{\mathcal{B}} = \begin{bmatrix} 3 \\ 2 \end{bmatrix}$

2. $\mathcal{B} = \left\{ \begin{bmatrix} -3 \\ 1 \end{bmatrix}, \begin{bmatrix} 3 \\ 2 \end{bmatrix} \right\}$, $[\mathbf{x}]_{\mathcal{B}} = \begin{bmatrix} -1 \\ 2 \end{bmatrix}$

In Exercises 3–6, the vector $\mathbf{x}$ is in a subspace H with a basis $\mathcal{B} = \{\mathbf{b}_1, \mathbf{b}_2\}$. Find the $\mathcal{B}$-coordinate vector of $\mathbf{x}$.

3. $\mathbf{b}_1 = \begin{bmatrix} 2 \\ -3 \end{bmatrix}$, $\mathbf{b}_2 = \begin{bmatrix} -1 \\ 5 \end{bmatrix}$, $\mathbf{x} = \begin{bmatrix} 0 \\ 7 \end{bmatrix}$

4. $\mathbf{b}_1 = \begin{bmatrix} 1 \\ -5 \end{bmatrix}$, $\mathbf{b}_2 = \begin{bmatrix} -2 \\ 3 \end{bmatrix}$, $\mathbf{x} = \begin{bmatrix} 1 \\ 9 \end{bmatrix}$

5. $\mathbf{b}_1 = \begin{bmatrix} 1 \\ 4 \\ -3 \end{bmatrix}$, $\mathbf{b}_2 = \begin{bmatrix} -2 \\ -7 \\ 5 \end{bmatrix}$, $\mathbf{x} = \begin{bmatrix} 2 \\ 9 \\ -7 \end{bmatrix}$

6. $\mathbf{b}_1 = \begin{bmatrix} -3 \\ 2 \\ -4 \end{bmatrix}$, $\mathbf{b}_2 = \begin{bmatrix} 7 \\ -3 \\ 5 \end{bmatrix}$, $\mathbf{x} = \begin{bmatrix} 5 \\ 0 \\ -2 \end{bmatrix}$

7. Let $\mathbf{b}_1 = \begin{bmatrix} 3 \\ 0 \end{bmatrix}$, $\mathbf{b}_2 = \begin{bmatrix} -1 \\ 2 \end{bmatrix}$, $\mathbf{w} = \begin{bmatrix} 7 \\ -2 \end{bmatrix}$, $\mathbf{x} = \begin{bmatrix} 4 \\ 1 \end{bmatrix}$, and $\mathcal{B} = \{\mathbf{b}_1, \mathbf{b}_2\}$. Use the figure to estimate $[\mathbf{w}]_{\mathcal{B}}$ and $[\mathbf{x}]_{\mathcal{B}}$. Confirm your estimate of $[\mathbf{x}]_{\mathcal{B}}$ by using it and $\{\mathbf{b}_1, \mathbf{b}_2\}$ to compute $\mathbf{x}$.

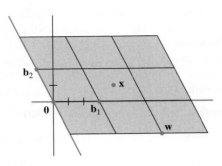

8. Let $\mathbf{b}_1 = \begin{bmatrix} 0 \\ 2 \end{bmatrix}$, $\mathbf{b}_2 = \begin{bmatrix} 2 \\ 1 \end{bmatrix}$, $\mathbf{x} = \begin{bmatrix} -2 \\ 3 \end{bmatrix}$, $\mathbf{y} = \begin{bmatrix} 2 \\ 4 \end{bmatrix}$, $\mathbf{z} = \begin{bmatrix} -1 \\ -2.5 \end{bmatrix}$, and $\mathcal{B} = \{\mathbf{b}_1, \mathbf{b}_2\}$. Use the figure to estimate $[\mathbf{x}]_{\mathcal{B}}$, $[\mathbf{y}]_{\mathcal{B}}$, and $[\mathbf{z}]_{\mathcal{B}}$. Confirm your estimates of $[\mathbf{y}]_{\mathcal{B}}$ and $[\mathbf{z}]_{\mathcal{B}}$ by using them and $\{\mathbf{b}_1, \mathbf{b}_2\}$ to compute $\mathbf{y}$ and $\mathbf{z}$.

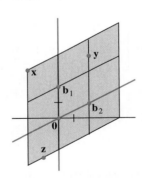

Exercises 9–12 display a matrix A and an echelon form of A. Find bases for Col A and Nul A, and then state the dimensions of these subspaces.

9. $A = \begin{bmatrix} 1 & 3 & 2 & -6 \\ 3 & 9 & 1 & 5 \\ 2 & 6 & -1 & 9 \\ 5 & 15 & 0 & 14 \end{bmatrix} \sim \begin{bmatrix} 1 & 3 & 3 & 2 \\ 0 & 0 & 5 & -7 \\ 0 & 0 & 0 & 5 \\ 0 & 0 & 0 & 0 \end{bmatrix}$

10. $A = \begin{bmatrix} 1 & -2 & -1 & 5 & 4 \\ 2 & -1 & 1 & 5 & 6 \\ -2 & 0 & -2 & 1 & -6 \\ 3 & 1 & 4 & 1 & 5 \end{bmatrix}$

$\sim \begin{bmatrix} 1 & -2 & -1 & 2 & 0 \\ 0 & 1 & 1 & 0 & 3 \\ 0 & 0 & 0 & 1 & 0 \\ 0 & 0 & 0 & 0 & 1 \end{bmatrix}$

11. $A = \begin{bmatrix} 2 & 4 & -5 & 2 & -3 \\ 3 & 6 & -8 & 3 & -5 \\ 0 & 0 & 9 & 0 & 9 \\ -3 & -6 & -7 & -3 & -10 \end{bmatrix}$

$\sim \begin{bmatrix} 1 & 2 & -5 & 1 & -4 \\ 0 & 0 & 5 & 0 & 5 \\ 0 & 0 & 0 & 0 & 0 \\ 0 & 0 & 0 & 0 & 0 \end{bmatrix}$

12. $A = \begin{bmatrix} 1 & 2 & -4 & 4 & 6 \\ 5 & 1 & -9 & 2 & 10 \\ 4 & 6 & -9 & 12 & 15 \\ 3 & 4 & -5 & 8 & 9 \end{bmatrix}$

$\sim \begin{bmatrix} 1 & 2 & 8 & 4 & -6 \\ 0 & 2 & 3 & 4 & -1 \\ 0 & 0 & 5 & 0 & -5 \\ 0 & 0 & 0 & 0 & 0 \end{bmatrix}$

In Exercises 13 and 14, find a basis for the subspace spanned by the given vectors. What is the dimension of the subspace?

13. $\begin{bmatrix} 1 \\ -3 \\ 2 \\ -4 \end{bmatrix}$, $\begin{bmatrix} -3 \\ 9 \\ -6 \\ 12 \end{bmatrix}$, $\begin{bmatrix} 2 \\ -1 \\ 4 \\ 2 \end{bmatrix}$, $\begin{bmatrix} -4 \\ 5 \\ -3 \\ 7 \end{bmatrix}$

14. $\begin{bmatrix} 1 \\ -1 \\ -2 \\ 3 \end{bmatrix}$, $\begin{bmatrix} 2 \\ -3 \\ -1 \\ 4 \end{bmatrix}$, $\begin{bmatrix} 0 \\ -1 \\ 3 \\ -2 \end{bmatrix}$, $\begin{bmatrix} -1 \\ 4 \\ -7 \\ 7 \end{bmatrix}$, $\begin{bmatrix} 3 \\ -7 \\ 6 \\ -9 \end{bmatrix}$

15. Suppose a 4×6 matrix A has four pivot columns. Is Col $A = \mathbb{R}^4$? Is Nul $A = \mathbb{R}^2$? Explain your answers.

16. Suppose a 4×7 matrix A has three pivot columns. Is Col $A = \mathbb{R}^3$? What is the dimension of Nul A? Explain your answers.

In Exercises 17 and 18, mark each statement True or False. Justify each answer. Here A is an $m \times n$ matrix.

17. a. If $\mathcal{B} = \{\mathbf{v}_1, \ldots, \mathbf{v}_p\}$ is a basis for a subspace H and if $\mathbf{x} = c_1\mathbf{v}_1 + \cdots + c_p\mathbf{v}_p$, then $c_1, \ldots, c_p$ are the coordinates of $\mathbf{x}$ relative to the basis $\mathcal{B}$.

b. Each line in $\mathbb{R}^n$ is a one-dimensional subspace of $\mathbb{R}^n$.

c. The dimension of Col A is the number of pivot columns in A.

d. The dimensions of Col A and Nul A add up to the number of columns in A.

e. If a set of p vectors spans a p-dimensional subspace H of $\mathbb{R}^n$, then these vectors form a basis for H.

18. a. If $\mathcal{B}$ is a basis for a subspace H, then each vector in H can be written in only one way as a linear combination of the vectors in $\mathcal{B}$.

b. The dimension of Nul A is the number of variables in the equation $A\mathbf{x} = \mathbf{0}$.

c. The dimension of the column space of A is rank A.

d. If $\mathcal{B} = \{\mathbf{v}_1, \ldots, \mathbf{v}_p\}$ is a basis for a subspace H of $\mathbb{R}^n$, then the correspondence $\mathbf{x} \mapsto [\mathbf{x}]_\mathcal{B}$ makes H look and act the same as $\mathbb{R}^p$.

e. If H is a p-dimensional subspace of $\mathbb{R}^n$, then a linearly independent set of p vectors in H is a basis for H.

In Exercises 19–24, justify each answer or construction.

19. If the subspace of all solutions of $A\mathbf{x} = \mathbf{0}$ has a basis consisting of three vectors and if A is a 5×7 matrix, what is the rank of A?

20. What is the rank of a 6×8 matrix whose null space is three-dimensional?

21. If the rank of a 9×8 matrix A is 7, what is the dimension of the solution space of $A\mathbf{x} = \mathbf{0}$?

22. Show that a set $\{\mathbf{v}_1, \ldots, \mathbf{v}_5\}$ in $\mathbb{R}^n$ is linearly dependent if dim Span $\{\mathbf{v}_1, \ldots, \mathbf{v}_5\} = 4$.

23. If possible, construct a 3×5 matrix A such that dim Nul $A = 3$ and dim Col $A = 2$.

24. Construct a 3×4 matrix with rank 1.

25. Let A be an $n \times p$ matrix whose column space is p-dimensional. Explain why the columns of A must be linearly independent.

26. Suppose columns 1, 3, 4, 5, and 7 of a matrix A are linearly independent (but are not necessarily pivot columns) and the rank of A is 5. Explain why the five columns mentioned must be a basis for the column space of A.

27. Suppose vectors $\mathbf{b}_1, \ldots, \mathbf{b}_p$ span a subspace W, and let $\{\mathbf{a}_1, \ldots, \mathbf{a}_q\}$ be any set in W containing more than p vectors. Fill in the details of the following argument to show that $\{\mathbf{a}_1, \ldots, \mathbf{a}_q\}$ must be linearly dependent. First, let $B = [\mathbf{b}_1 \ \cdots \ \mathbf{b}_p]$ and $A = [\mathbf{a}_1 \ \cdots \ \mathbf{a}_q]$.

a. Explain why for each vector $\mathbf{a}_j$, there exists a vector $\mathbf{c}_j$ in $\mathbb{R}^p$ such that $\mathbf{a}_j = B\mathbf{c}_j$.

b. Let $C = [\mathbf{c}_1 \ \cdots \ \mathbf{c}_q]$. Explain why there is a nonzero vector $\mathbf{u}$ such that $C\mathbf{u} = \mathbf{0}$.

c. Use B and C to show that $A\mathbf{u} = \mathbf{0}$. This shows that the columns of A are linearly dependent.

28. Use Exercise 27 to show that if $\mathcal{A}$ and $\mathcal{B}$ are bases for a subspace W of $\mathbb{R}^n$, then $\mathcal{A}$ cannot contain more vectors than $\mathcal{B}$, and, conversely, $\mathcal{B}$ cannot contain more vectors than $\mathcal{A}$.

29. [M] Let $H = \text{Span} \{\mathbf{v}_1, \mathbf{v}_2\}$ and $\mathcal{B} = \{\mathbf{v}_1, \mathbf{v}_2\}$. Show that $\mathbf{x}$ is in H, and find the $\mathcal{B}$-coordinate vector of $\mathbf{x}$, when

$$\mathbf{v}_1 = \begin{bmatrix} 15 \\ -5 \\ 12 \\ 7 \end{bmatrix}, \quad \mathbf{v}_2 = \begin{bmatrix} 14 \\ -10 \\ 13 \\ 17 \end{bmatrix}, \quad \mathbf{x} = \begin{bmatrix} 16 \\ 0 \\ 11 \\ -3 \end{bmatrix}$$

30. [M] Let $H = \text{Span} \{\mathbf{v}_1, \mathbf{v}_2, \mathbf{v}_3\}$ and $\mathcal{B} = \{\mathbf{v}_1, \mathbf{v}_2, \mathbf{v}_3\}$. Show that $\mathcal{B}$ is a basis for H and $\mathbf{x}$ is in H, and find the $\mathcal{B}$-coordinate vector of $\mathbf{x}$, when

$$\mathbf{v}_1 = \begin{bmatrix} -6 \\ 3 \\ -9 \\ 4 \end{bmatrix}, \mathbf{v}_2 = \begin{bmatrix} 8 \\ 0 \\ 7 \\ -3 \end{bmatrix}, \mathbf{v}_3 = \begin{bmatrix} -9 \\ 4 \\ -8 \\ 3 \end{bmatrix}, \mathbf{x} = \begin{bmatrix} 11 \\ -2 \\ 17 \\ -8 \end{bmatrix}$$

> **SG**   Mastering: Dimension and Rank 2–41

SOLUTIONS TO PRACTICE PROBLEMS

1. Construct $A = [\mathbf{v}_1 \ \ \mathbf{v}_2 \ \ \mathbf{v}_3]$ so that the subspace spanned by $\mathbf{v}_1, \mathbf{v}_2, \mathbf{v}_3$ is the column space of A. A basis for this space is provided by the pivot columns of A.

$$A = \begin{bmatrix} 2 & 3 & -1 \\ -8 & -7 & 6 \\ 6 & -1 & -7 \end{bmatrix} \sim \begin{bmatrix} 2 & 3 & -1 \\ 0 & 5 & 2 \\ 0 & -10 & -4 \end{bmatrix} \sim \begin{bmatrix} 2 & 3 & -1 \\ 0 & 5 & 2 \\ 0 & 0 & 0 \end{bmatrix}$$

The first two columns of A are pivot columns and form a basis for H. Thus dim $H = 2$.

2. If $[\mathbf{x}]_\mathcal{B} = \begin{bmatrix} 3 \\ 2 \end{bmatrix}$, then $\mathbf{x}$ is formed from a linear combination of the basis vectors using weights 3 and 2:

$$\mathbf{x} = 3\mathbf{b}_1 + 2\mathbf{b}_2 = 3 \begin{bmatrix} 1 \\ .2 \end{bmatrix} + 2 \begin{bmatrix} .2 \\ 1 \end{bmatrix} = \begin{bmatrix} 3.4 \\ 2.6 \end{bmatrix}$$

The basis $\{\mathbf{b}_1, \mathbf{b}_2\}$ determines a *coordinate system* for $\mathbb{R}^2$, illustrated by the grid in the figure. Note how $\mathbf{x}$ is 3 units in the $\mathbf{b}_1$-direction and 2 units in the $\mathbf{b}_2$-direction.

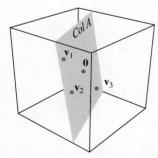

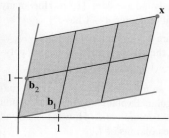

3. A four-dimensional subspace would contain a basis of four linearly independent vectors. This is impossible inside $\mathbb{R}^3$. Since any linearly independent set in $\mathbb{R}^3$ has no more than three vectors, any subspace of $\mathbb{R}^3$ has dimension no more than 3. The space $\mathbb{R}^3$ itself is the only three-dimensional subspace of $\mathbb{R}^3$. Other subspaces of $\mathbb{R}^3$ have dimension 2, 1, or 0.

CHAPTER 2 SUPPLEMENTARY EXERCISES

1. Assume that the matrices mentioned in the statements below have appropriate sizes. Mark each statement True or False. Justify each answer.

   a. If A and B are $m \times n$, then both AB^T and A^TB are defined.

   b. If $AB = C$ and C has 2 columns, then A has 2 columns.

   c. Left-multiplying a matrix B by a diagonal matrix A, with nonzero entries on the diagonal, scales the rows of B.

   d. If $BC = BD$, then $C = D$.

   e. If $AC = 0$, then either $A = 0$ or $C = 0$.

   f. If A and B are $n \times n$, then $(A + B)(A - B) = A^2 - B^2$.

   g. An elementary $n \times n$ matrix has either n or $n + 1$ nonzero entries.

   h. The transpose of an elementary matrix is an elementary matrix.

   i. An elementary matrix must be square.

   j. Every square matrix is a product of elementary matrices.

   k. If A is a 3×3 matrix with three pivot positions, there exist elementary matrices $E_1, \ldots, E_p$ such that $E_p \cdots E_1 A = I$.

   l. If $AB = I$, then A is invertible.

   m. If A and B are square and invertible, then AB is invertible, and $(AB)^{-1} = A^{-1}B^{-1}$.

   n. If $AB = BA$ and if A is invertible, then $A^{-1}B = BA^{-1}$.

   o. If A is invertible and if $r \neq 0$, then $(rA)^{-1} = rA^{-1}$.

   p. If A is a 3×3 matrix and the equation $A\mathbf{x} = \begin{bmatrix} 1 \\ 0 \\ 0 \end{bmatrix}$ has a unique solution, then A is invertible.

2. Find the matrix C whose inverse is $C^{-1} = \begin{bmatrix} 4 & 5 \\ 6 & 7 \end{bmatrix}$.

3. Let $A = \begin{bmatrix} 0 & 0 & 0 \\ 1 & 0 & 0 \\ 0 & 1 & 0 \end{bmatrix}$. Show that $A^3 = 0$. Use matrix algebra to compute the product $(I - A)(I + A + A^2)$.

4. Suppose $A^n = 0$ for some $n > 1$. Find an inverse for $I - A$.

5. Suppose an $n \times n$ matrix A satisfies the equation $A^2 - 2A + I = 0$. Show that $A^3 = 3A - 2I$ and $A^4 = 4A - 3I$.

6. Let $A = \begin{bmatrix} 1 & 0 \\ 0 & -1 \end{bmatrix}$, $B = \begin{bmatrix} 0 & 1 \\ 1 & 0 \end{bmatrix}$. These are *Pauli spin* matrices used in the study of electron spin in quantum mechanics. Show that $A^2 = I$, $B^2 = I$, and $AB = -BA$. Matrices such that $AB = -BA$ are said to *anticommute*.

7. Let $A = \begin{bmatrix} 1 & 3 & 8 \\ 2 & 4 & 11 \\ 1 & 2 & 5 \end{bmatrix}$ and $B = \begin{bmatrix} -3 & 5 \\ 1 & 5 \\ 3 & 4 \end{bmatrix}$. Compute $A^{-1}B$ without computing A^{-1}. [*Hint*: $A^{-1}B$ is the solution of the equation $AX = B$.]

8. Find a matrix A such that the transformation $\mathbf{x} \mapsto A\mathbf{x}$ maps $\begin{bmatrix} 1 \\ 3 \end{bmatrix}$ and $\begin{bmatrix} 2 \\ 7 \end{bmatrix}$ into $\begin{bmatrix} 1 \\ 1 \end{bmatrix}$ and $\begin{bmatrix} 3 \\ 1 \end{bmatrix}$, respectively. [*Hint*: Write a matrix equation involving A, and solve for A.]

9. Suppose $AB = \begin{bmatrix} 5 & 4 \\ -2 & 3 \end{bmatrix}$ and $B = \begin{bmatrix} 7 & 3 \\ 2 & 1 \end{bmatrix}$. Find A.

10. Suppose A is invertible. Explain why A^TA is also invertible. Then show that $A^{-1} = (A^TA)^{-1}A^T$.

11. Let $x_1, \ldots, x_n$ be fixed numbers. The matrix below, called a *Vandermonde matrix*, occurs in applications such as signal processing, error-correcting codes, and polynomial interpolation.

$$V = \begin{bmatrix} 1 & x_1 & x_1^2 & \cdots & x_1^{n-1} \\ 1 & x_2 & x_2^2 & \cdots & x_2^{n-1} \\ \vdots & \vdots & \vdots & & \vdots \\ 1 & x_n & x_n^2 & \cdots & x_n^{n-1} \end{bmatrix}$$

Given $\mathbf{y} = (y_1, \ldots, y_n)$ in $\mathbb{R}^n$, suppose $\mathbf{c} = (c_0, \ldots, c_{n-1})$ in $\mathbb{R}^n$ satisfies $V\mathbf{c} = \mathbf{y}$, and define the polynomial

$$p(t) = c_0 + c_1 t + c_2 t^2 + \cdots + c_{n-1}t^{n-1}$$

   a. Show that $p(x_1) = y_1, \ldots, p(x_n) = y_n$. We call $p(t)$ an *interpolating polynomial for the points* $(x_1, y_1), \ldots, (x_n, y_n)$ because the graph of $p(t)$ passes through the points.

   b. Suppose $x_1, \ldots, x_n$ are distinct numbers. Show that the columns of V are linearly independent. [*Hint*: How many zeros can a polynomial of degree $n - 1$ have?]

   c. Prove: "If $x_1, \ldots, x_n$ are distinct numbers, and $y_1, \ldots, y_n$ are arbitrary numbers, then there is an interpolating polynomial of degree $\leq n - 1$ for $(x_1, y_1), \ldots, (x_n, y_n)$."

12. Let $A = LU$, where L is an invertible lower triangular matrix and U is upper triangular. Explain why the first column of A is a multiple of the first column of L. How is the second column of A related to the columns of L?

13. Given $\mathbf{u}$ in $\mathbb{R}^n$ with $\mathbf{u}^T\mathbf{u} = 1$, let $P = \mathbf{u}\mathbf{u}^T$ (an outer product) and $Q = I - 2P$. Justify statements (a), (b), and (c).

    a. $P^2 = P$ b. $P^T = P$ c. $Q^2 = I$

    The transformation $\mathbf{x} \mapsto P\mathbf{x}$ is called a *projection*, and $\mathbf{x} \mapsto Q\mathbf{x}$ is called a *Householder reflection*. Such reflections are used in computer programs to create multiple zeros in a vector (usually a column of a matrix).

14. Let $\mathbf{u} = \begin{bmatrix} 0 \\ 0 \\ 1 \end{bmatrix}$ and $\mathbf{x} = \begin{bmatrix} 1 \\ 5 \\ 3 \end{bmatrix}$. Determine P and Q as in Exercise 13, and compute $P\mathbf{x}$ and $Q\mathbf{x}$. The figure shows that $Q\mathbf{x}$ is the reflection of $\mathbf{x}$ through the x_1x_2-plane.

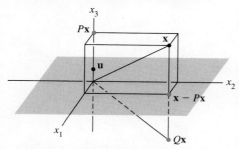

A Householder reflection through the plane $x_3 = 0$.

15. Suppose $C = E_3 E_2 E_1 B$, where E_1, E_2, and E_3 are elementary matrices. Explain why C is row equivalent to B.

16. Let A be an $n \times n$ singular matrix. Describe how to construct an $n \times n$ nonzero matrix B such that $AB = 0$.

17. Let A be a 6×4 matrix and B a 4×6 matrix. Show that the 6×6 matrix AB cannot be invertible.

18. Suppose A is a 5×3 matrix and there exists a 3×5 matrix C such that $CA = I_3$. Suppose further that for some given $\mathbf{b}$ in $\mathbb{R}^5$, the equation $A\mathbf{x} = \mathbf{b}$ has at least one solution. Show that this solution is unique.

19. [M] Certain dynamical systems can be studied by examining powers of a matrix, such as those below. Determine what happens to A^k and B^k as k increases (for example, try $k = 2, \ldots, 16$). Try to identify what is special about A and B. Investigate large powers of other matrices of this type, and make a conjecture about such matrices.

$$A = \begin{bmatrix} .4 & .2 & .3 \\ .3 & .6 & .3 \\ .3 & .2 & .4 \end{bmatrix}, \quad B = \begin{bmatrix} 0 & .2 & .3 \\ .1 & .6 & .3 \\ .9 & .2 & .4 \end{bmatrix}$$

20. [M] Let A_n be the $n \times n$ matrix with 0's on the main diagonal and 1's elsewhere. Compute A_n^{-1} for $n = 4, 5,$ and 6, and make a conjecture about the general form of A_n^{-1} for larger values of n.

3 Determinants

WEB

Random Paths and Distortion

In his autobiographical book "Surely You're Joking, Mr. Feynman," the Nobel Prize–winning physicist Richard Feynman tells of observing ants in his Princeton graduate school apartment. He studied the ants' behavior by providing paper ferries to sugar suspended on a string where the ants would not accidentally find it. When an ant would step onto a paper ferry, Feynman would transport the ant to the food and then back. After the ants learned to use the ferry, he relocated the return landing. The colony soon confused the outbound and return ferry landings, indicating that their "learning" consisted of creating and following trails. Feynman confirmed this conjecture by laying glass slides on the floor. Once the ants established trails on the glass slides, he rearranged the slides and therefore the trails on them. The ants followed the repositioned trails and Feynman could direct the ants where he wished.

Suppose Feynman had decided to conduct additional investigations using a globe built of wire mesh on which an ant must follow individual wires and choose between going left and right at each intersection. If several ants and an equal number of food sources are placed on the globe, how likely is it that each ant would find its own food source rather than encountering another ant's trail and following it to a shared resource?[1]

In order to record the actual routes of the ants and to communicate the results to others, it is convenient to use a rectangular map of the globe. There are many ways to create such maps. One simple way is to use the longitude and latitude on the globe as x and y coordinates on the map. As is the case with all maps, the result is not a faithful representation of the globe. Features near the "equator" look much the same on the globe and the map, but regions near the "poles" of the globe are distorted. Images of polar regions are much larger than the images of similar sized regions near the equator. To fit in with its surroundings on the map, the image of an ant near one of the poles should be larger than one near the equator. How much larger?

Surprisingly, both the ant-path and the area distortion problems are best answered through the use of the determinant, the subject of this chapter. Indeed, the determinant has so many uses that a summary of the applications known in the early 1900's filled a four volume treatise by Thomas Muir. With changes in emphasis and the greatly increased sizes of the matrices used in modern applications, many uses that were important then are no longer critical today. Nevertheless, the determinant still plays an important role.

---

[1] The solution to the ant-path problem (and two other applications) can be found in a June 2005, *Mathematical Monthly* article by Arthur Benjamin and Naomi Cameron.

Beyond introducing the determinant in Section 3.1, this chapter presents two important ideas. Section 3.2 derives an invertibility criterion for a square matrix that plays a pivotal role in Chapter 5. Section 3.3 shows how the determinant measures the amount by which a linear transformation changes the area of a figure. When applied locally, this technique answers the question of a map's expansion rate near the poles. This idea plays a critical role in multivariable calculus in the form of the Jacobian.

3.1 INTRODUCTION TO DETERMINANTS

Recall from Section 2.2 that a 2×2 matrix is invertible if and only if its determinant is nonzero. To extend this useful fact to larger matrices, we need a definition for the determinant of an $n \times n$ matrix. We can discover the definition for the 3×3 case by watching what happens when an invertible 3×3 matrix A is row reduced.

Consider $A = [a_{ij}]$ with $a_{11} \neq 0$. If we multiply the second and third rows of A by a_{11} and then subtract appropriate multiples of the first row from the other two rows, we find that A is row equivalent to the following two matrices:

$$\begin{bmatrix} a_{11} & a_{12} & a_{13} \\ a_{11}a_{21} & a_{11}a_{22} & a_{11}a_{23} \\ a_{11}a_{31} & a_{11}a_{32} & a_{11}a_{33} \end{bmatrix} \sim \begin{bmatrix} a_{11} & a_{12} & a_{13} \\ 0 & a_{11}a_{22} - a_{12}a_{21} & a_{11}a_{23} - a_{13}a_{21} \\ 0 & a_{11}a_{32} - a_{12}a_{31} & a_{11}a_{33} - a_{13}a_{31} \end{bmatrix} \quad (1)$$

Since A is invertible, either the $(2, 2)$-entry or the $(3, 2)$-entry on the right in (1) is nonzero. Let us suppose that the $(2, 2)$-entry is nonzero. (Otherwise, we can make a row interchange before proceeding.) Multiply row 3 by $a_{11}a_{22} - a_{12}a_{21}$, and then to the new row 3 add $-(a_{11}a_{32} - a_{12}a_{31})$ times row 2. This will show that

$$A \sim \begin{bmatrix} a_{11} & a_{12} & a_{13} \\ 0 & a_{11}a_{22} - a_{12}a_{21} & a_{11}a_{23} - a_{13}a_{21} \\ 0 & 0 & a_{11}\Delta \end{bmatrix}$$

where

$$\Delta = a_{11}a_{22}a_{33} + a_{12}a_{23}a_{31} + a_{13}a_{21}a_{32} - a_{11}a_{23}a_{32} - a_{12}a_{21}a_{33} - a_{13}a_{22}a_{31} \quad (2)$$

Since A is invertible, Δ must be nonzero. The converse is true, too, as we will see in Section 3.2. We call Δ in (2) the **determinant** of the 3×3 matrix A.

Recall that the determinant of a 2×2 matrix, $A = [a_{ij}]$, is the number

$$\det A = a_{11}a_{22} - a_{12}a_{21}$$

For a 1×1 matrix—say, $A = [a_{11}]$—we define $\det A = a_{11}$. To generalize the definition of the determinant to larger matrices, we'll use 2×2 determinants to rewrite the 3×3 determinant Δ described above. Since the terms in Δ can be grouped as $(a_{11}a_{22}a_{33} - a_{11}a_{23}a_{32}) - (a_{12}a_{21}a_{33} - a_{12}a_{23}a_{31}) + (a_{13}a_{21}a_{32} - a_{13}a_{22}a_{31})$,

$$\Delta = a_{11} \cdot \det \begin{bmatrix} a_{22} & a_{23} \\ a_{32} & a_{33} \end{bmatrix} - a_{12} \cdot \det \begin{bmatrix} a_{21} & a_{23} \\ a_{31} & a_{33} \end{bmatrix} + a_{13} \cdot \det \begin{bmatrix} a_{21} & a_{22} \\ a_{31} & a_{32} \end{bmatrix}$$

For brevity, write

$$\Delta = a_{11} \cdot \det A_{11} - a_{12} \cdot \det A_{12} + a_{13} \cdot \det A_{13} \quad (3)$$

where A_{11}, A_{12}, and A_{13} are obtained from A by deleting the first row and one of the three columns. For any square matrix A, let A_{ij} denote the submatrix formed by deleting the ith row and jth column of A. For instance, if

$$A = \begin{bmatrix} 1 & -2 & 5 & 0 \\ 2 & 0 & 4 & -1 \\ 3 & 1 & 0 & 7 \\ 0 & 4 & -2 & 0 \end{bmatrix}$$

then A_{32} is obtained by crossing out row 3 and column 2,

$$\begin{bmatrix} 1 & -2 & 5 & 0 \\ 2 & 0 & 4 & -1 \\ 3 & 1 & 0 & 7 \\ 0 & 4 & -2 & 0 \end{bmatrix}$$

so that

$$A_{32} = \begin{bmatrix} 1 & 5 & 0 \\ 2 & 4 & -1 \\ 0 & -2 & 0 \end{bmatrix}$$

We can now give a *recursive* definition of a determinant. When $n = 3$, det A is defined using determinants of the 2×2 submatrices A_{1j}, as in (3) above. When $n = 4$, det A uses determinants of the 3×3 submatrices A_{1j}. In general, an $n \times n$ determinant is defined by determinants of $(n-1) \times (n-1)$ submatrices.

DEFINITION

For $n \geq 2$, the **determinant** of an $n \times n$ matrix $A = [a_{ij}]$ is the sum of n terms of the form $\pm a_{1j}$ det A_{1j}, with plus and minus signs alternating, where the entries $a_{11}, a_{12}, \ldots, a_{1n}$ are from the first row of A. In symbols,

$$\det A = a_{11} \det A_{11} - a_{12} \det A_{12} + \cdots + (-1)^{1+n} a_{1n} \det A_{1n}$$

$$= \sum_{j=1}^{n} (-1)^{1+j} a_{1j} \det A_{1j}$$

EXAMPLE 1 Compute the determinant of

$$A = \begin{bmatrix} 1 & 5 & 0 \\ 2 & 4 & -1 \\ 0 & -2 & 0 \end{bmatrix}$$

SOLUTION Compute det $A = a_{11} \det A_{11} - a_{12} \det A_{12} + a_{13} \det A_{13}$:

$$\det A = 1 \cdot \det \begin{bmatrix} 4 & -1 \\ -2 & 0 \end{bmatrix} - 5 \cdot \det \begin{bmatrix} 2 & -1 \\ 0 & 0 \end{bmatrix} + 0 \cdot \det \begin{bmatrix} 2 & 4 \\ 0 & -2 \end{bmatrix}$$

$$= 1(0 - 2) - 5(0 - 0) + 0(-4 - 0) = -2 \qquad \blacksquare$$

Another common notation for the determinant of a matrix uses a pair of vertical lines in place of brackets. Thus the calculation in Example 1 can be written as

$$\det A = 1 \begin{vmatrix} 4 & -1 \\ -2 & 0 \end{vmatrix} - 5 \begin{vmatrix} 2 & -1 \\ 0 & 0 \end{vmatrix} + 0 \begin{vmatrix} 2 & 4 \\ 0 & -2 \end{vmatrix} = \cdots = -2$$

To state the next theorem, it is convenient to write the definition of det A in a slightly different form. Given $A = [a_{ij}]$, the (i, j)**-cofactor** of A is the number C_{ij} given by

$$C_{ij} = (-1)^{i+j} \det A_{ij} \qquad (4)$$

Then

$$\det A = a_{11} C_{11} + a_{12} C_{12} + \cdots + a_{1n} C_{1n}$$

This formula is called a **cofactor expansion across the first row** of A. We omit the proof of the following fundamental theorem to avoid a lengthy digression.

THEOREM 1

The determinant of an $n \times n$ matrix A can be computed by a cofactor expansion across any row or down any column. The expansion across the ith row using the cofactors in (4) is

$$\det A = a_{i1}C_{i1} + a_{i2}C_{i2} + \cdots + a_{in}C_{in}$$

The cofactor expansion down the jth column is

$$\det A = a_{1j}C_{1j} + a_{2j}C_{2j} + \cdots + a_{nj}C_{nj}$$

The plus or minus sign in the (i, j)-cofactor depends on the position of a_{ij} in the matrix, regardless of the sign of a_{ij} itself. The factor $(-1)^{i+j}$ determines the following checkerboard pattern of signs:

$$\begin{bmatrix} + & - & + & \cdots \\ - & + & - & \\ + & - & + & \\ \vdots & & & \ddots \end{bmatrix}$$

EXAMPLE 2   Use a cofactor expansion across the third row to compute $\det A$, where

$$A = \begin{bmatrix} 1 & 5 & 0 \\ 2 & 4 & -1 \\ 0 & -2 & 0 \end{bmatrix}$$

SOLUTION Compute

$$\det A = a_{31}C_{31} + a_{32}C_{32} + a_{33}C_{33}$$
$$= (-1)^{3+1}a_{31} \det A_{31} + (-1)^{3+2}a_{32} \det A_{32} + (-1)^{3+3}a_{33} \det A_{33}$$
$$= 0 \begin{vmatrix} 5 & 0 \\ 4 & -1 \end{vmatrix} - (-2) \begin{vmatrix} 1 & 0 \\ 2 & -1 \end{vmatrix} + 0 \begin{vmatrix} 1 & 5 \\ 2 & 4 \end{vmatrix}$$
$$= 0 + 2(-1) + 0 = -2 \qquad \blacksquare$$

Theorem 1 is helpful for computing the determinant of a matrix that contains many zeros. For example, if a row is mostly zeros, then the cofactor expansion across that row has many terms that are zero, and the cofactors in those terms need not be calculated. The same approach works with a column that contains many zeros.

EXAMPLE 3   Compute $\det A$, where

$$A = \begin{bmatrix} 3 & -7 & 8 & 9 & -6 \\ 0 & 2 & -5 & 7 & 3 \\ 0 & 0 & 1 & 5 & 0 \\ 0 & 0 & 2 & 4 & -1 \\ 0 & 0 & 0 & -2 & 0 \end{bmatrix}$$

SOLUTION The cofactor expansion down the first column of A has all terms equal to zero except the first. Thus

$$\det A = 3 \cdot \begin{vmatrix} 2 & -5 & 7 & 3 \\ 0 & 1 & 5 & 0 \\ 0 & 2 & 4 & -1 \\ 0 & 0 & -2 & 0 \end{vmatrix} - 0 \cdot C_{21} + 0 \cdot C_{31} - 0 \cdot C_{41} + 0 \cdot C_{51}$$

Henceforth we will omit the zero terms in the cofactor expansion. Next, expand this 4×4 determinant down the first column, in order to take advantage of the zeros there. We have

$$\det A = 3 \cdot 2 \cdot \begin{vmatrix} 1 & 5 & 0 \\ 2 & 4 & -1 \\ 0 & -2 & 0 \end{vmatrix}$$

This 3×3 determinant was computed in Example 1 and found to equal -2. Hence $\det A = 3 \cdot 2 \cdot (-2) = -12$. ∎

The matrix in Example 3 was nearly triangular. The method in that example is easily adapted to prove the following theorem.

THEOREM 2   If A is a triangular matrix, then $\det A$ is the product of the entries on the main diagonal of A.

The strategy in Example 3 of looking for zeros works extremely well when an entire row or column consists of zeros. In such a case, the cofactor expansion along such a row or column is a sum of zeros! So the determinant is zero. Unfortunately, most cofactor expansions are not so quickly evaluated.

NUMERICAL NOTE

By today's standards, a 25×25 matrix is small. Yet it would be impossible to calculate a 25×25 determinant by cofactor expansion. In general, a cofactor expansion requires over $n!$ multiplications, and $25!$ is approximately 1.5×10^{25}.

If a computer performs one trillion multiplications per second, it would have to run for over 500,000 years to compute a 25×25 determinant by this method. Fortunately, there are faster methods, as we'll soon discover.

Exercises 19–38 explore important properties of determinants, mostly for the 2×2 case. The results from Exercises 33–36 will be used in the next section to derive the analogous properties for $n \times n$ matrices.

PRACTICE PROBLEM

Compute $\begin{vmatrix} 5 & -7 & 2 & 2 \\ 0 & 3 & 0 & -4 \\ -5 & -8 & 0 & 3 \\ 0 & 5 & 0 & -6 \end{vmatrix}$.

3.1 EXERCISES

Compute the determinants in Exercises 1–8 using a cofactor expansion across the first row. In Exercises 1–4, also compute the determinant by a cofactor expansion down the second column.

1. $\begin{vmatrix} 3 & 0 & 4 \\ 2 & 3 & 2 \\ 0 & 5 & -1 \end{vmatrix}$

2. $\begin{vmatrix} 0 & 5 & 1 \\ 4 & -3 & 0 \\ 2 & 4 & 1 \end{vmatrix}$

3. $\begin{vmatrix} 2 & -4 & 3 \\ 3 & 1 & 2 \\ 1 & 4 & -1 \end{vmatrix}$

4. $\begin{vmatrix} 1 & 3 & 5 \\ 2 & 1 & 1 \\ 3 & 4 & 2 \end{vmatrix}$

5. $\begin{vmatrix} 2 & 3 & -4 \\ 4 & 0 & 5 \\ 5 & 1 & 6 \end{vmatrix}$

6. $\begin{vmatrix} 5 & -2 & 4 \\ 0 & 3 & -5 \\ 2 & -4 & 7 \end{vmatrix}$

7. $\begin{vmatrix} 4 & 3 & 0 \\ 6 & 5 & 2 \\ 9 & 7 & 3 \end{vmatrix}$
   8. $\begin{vmatrix} 8 & 1 & 6 \\ 4 & 0 & 3 \\ 3 & -2 & 5 \end{vmatrix}$

Compute the determinants in Exercises 9–14 by cofactor expansions. At each step, choose a row or column that involves the least amount of computation.

9. $\begin{vmatrix} 6 & 0 & 0 & 5 \\ 1 & 7 & 2 & -5 \\ 2 & 0 & 0 & 0 \\ 8 & 3 & 1 & 8 \end{vmatrix}$
   10. $\begin{vmatrix} 1 & -2 & 5 & 2 \\ 0 & 0 & 3 & 0 \\ 2 & -6 & -7 & 5 \\ 5 & 0 & 4 & 4 \end{vmatrix}$

11. $\begin{vmatrix} 3 & 5 & -8 & 4 \\ 0 & -2 & 3 & -7 \\ 0 & 0 & 1 & 5 \\ 0 & 0 & 0 & 2 \end{vmatrix}$
   12. $\begin{vmatrix} 4 & 0 & 0 & 0 \\ 7 & -1 & 0 & 0 \\ 2 & 6 & 3 & 0 \\ 5 & -8 & 4 & -3 \end{vmatrix}$

13. $\begin{vmatrix} 4 & 0 & -7 & 3 & -5 \\ 0 & 0 & 2 & 0 & 0 \\ 7 & 3 & -6 & 4 & -8 \\ 5 & 0 & 5 & 2 & -3 \\ 0 & 0 & 9 & -1 & 2 \end{vmatrix}$

14. $\begin{vmatrix} 6 & 3 & 2 & 4 & 0 \\ 9 & 0 & -4 & 1 & 0 \\ 8 & -5 & 6 & 7 & 1 \\ 3 & 0 & 0 & 0 & 0 \\ 4 & 2 & 3 & 2 & 0 \end{vmatrix}$

The expansion of a 3×3 determinant can be remembered by the following device. Write a second copy of the first two columns to the right of the matrix, and compute the determinant by multiplying entries on six diagonals:

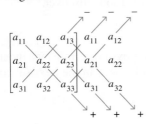

Add the downward diagonal products and subtract the upward products. Use this method to compute the determinants in Exercises 15–18. **Warning:** *This trick does not generalize in any reasonable way to 4×4 or larger matrices.*

15. $\begin{vmatrix} 3 & 0 & 4 \\ 2 & 3 & 2 \\ 0 & 5 & -1 \end{vmatrix}$
   16. $\begin{vmatrix} 0 & 5 & 1 \\ 4 & -3 & 0 \\ 2 & 4 & 1 \end{vmatrix}$

17. $\begin{vmatrix} 2 & -4 & 3 \\ 3 & 1 & 2 \\ 1 & 4 & -1 \end{vmatrix}$
   18. $\begin{vmatrix} 1 & 3 & 5 \\ 2 & 1 & 1 \\ 3 & 4 & 2 \end{vmatrix}$

In Exercises 19–24, explore the effect of an elementary row operation on the determinant of a matrix. In each case, state the row operation and describe how it affects the determinant.

19. $\begin{bmatrix} a & b \\ c & d \end{bmatrix}, \begin{bmatrix} c & d \\ a & b \end{bmatrix}$
   20. $\begin{bmatrix} a & b \\ c & d \end{bmatrix}, \begin{bmatrix} a & b \\ kc & kd \end{bmatrix}$

21. $\begin{bmatrix} 3 & 4 \\ 5 & 6 \end{bmatrix}, \begin{bmatrix} 3 & 4 \\ 5+3k & 6+4k \end{bmatrix}$

22. $\begin{bmatrix} a & b \\ c & d \end{bmatrix}, \begin{bmatrix} a+kc & b+kd \\ c & d \end{bmatrix}$

23. $\begin{bmatrix} 1 & 1 & 1 \\ -3 & 8 & -4 \\ 2 & -3 & 2 \end{bmatrix}, \begin{bmatrix} k & k & k \\ -3 & 8 & -4 \\ 2 & -3 & 2 \end{bmatrix}$

24. $\begin{bmatrix} a & b & c \\ 3 & 2 & 2 \\ 6 & 5 & 6 \end{bmatrix}, \begin{bmatrix} 3 & 2 & 2 \\ a & b & c \\ 6 & 5 & 6 \end{bmatrix}$

Compute the determinants of the elementary matrices given in Exercises 25–30. (See Section 2.2.)

25. $\begin{bmatrix} 1 & 0 & 0 \\ 0 & 1 & 0 \\ 0 & k & 1 \end{bmatrix}$
   26. $\begin{bmatrix} 1 & 0 & 0 \\ 0 & 1 & 0 \\ k & 0 & 1 \end{bmatrix}$

27. $\begin{bmatrix} k & 0 & 0 \\ 0 & 1 & 0 \\ 0 & 0 & 1 \end{bmatrix}$
   28. $\begin{bmatrix} 1 & 0 & 0 \\ 0 & k & 0 \\ 0 & 0 & 1 \end{bmatrix}$

29. $\begin{bmatrix} 0 & 1 & 0 \\ 1 & 0 & 0 \\ 0 & 0 & 1 \end{bmatrix}$
   30. $\begin{bmatrix} 0 & 0 & 1 \\ 0 & 1 & 0 \\ 1 & 0 & 0 \end{bmatrix}$

Use Exercises 25–28 to answer the questions in Exercises 31 and 32. Give reasons for your answers.

31. What is the determinant of an elementary row replacement matrix?

32. What is the determinant of an elementary scaling matrix with k on the diagonal?

In Exercises 33–36, verify that $\det EA = (\det E)(\det A)$, where E is the elementary matrix shown and $A = \begin{bmatrix} a & b \\ c & d \end{bmatrix}$.

33. $\begin{bmatrix} 0 & 1 \\ 1 & 0 \end{bmatrix}$
   34. $\begin{bmatrix} 1 & 0 \\ 0 & k \end{bmatrix}$

35. $\begin{bmatrix} 1 & k \\ 0 & 1 \end{bmatrix}$
   36. $\begin{bmatrix} 1 & 0 \\ k & 1 \end{bmatrix}$

37. Let $A = \begin{bmatrix} 3 & 1 \\ 4 & 2 \end{bmatrix}$. Write $5A$. Is $\det 5A = 5 \det A$?

38. Let $A = \begin{bmatrix} a & b \\ c & d \end{bmatrix}$ and let k be a scalar. Find a formula that relates $\det kA$ to k and $\det A$.

In Exercises 39 and 40, A is an $n \times n$ matrix. Mark each statement True or False. Justify each answer.

39. a. An $n \times n$ determinant is defined by determinants of $(n-1) \times (n-1)$ submatrices.

   b. The (i, j)-cofactor of a matrix A is the matrix A_{ij} obtained by deleting from A its ith row and jth column.

40. a. The cofactor expansion of $\det A$ down a column is the negative of the cofactor expansion along a row.

b. The determinant of a triangular matrix is the sum of the entries on the main diagonal.

41. Let $\mathbf{u} = \begin{bmatrix} 3 \\ 0 \end{bmatrix}$ and $\mathbf{v} = \begin{bmatrix} 1 \\ 2 \end{bmatrix}$. Compute the area of the parallelogram determined by $\mathbf{u}$, $\mathbf{v}$, $\mathbf{u} + \mathbf{v}$, and $\mathbf{0}$, and compute the determinant of $[\,\mathbf{u}\ \ \mathbf{v}\,]$. How do they compare? Replace the first entry of $\mathbf{v}$ by an arbitrary number x, and repeat the problem. Draw a picture and explain what you find.

42. Let $u = \begin{bmatrix} a \\ b \end{bmatrix}$ and $\mathbf{v} = \begin{bmatrix} c \\ 0 \end{bmatrix}$, where a, b, c are positive (for simplicity). Compute the area of the parallelogram determined by $\mathbf{u}$, $\mathbf{v}$, $\mathbf{u} + \mathbf{v}$, and $\mathbf{0}$, and compute the determinants of the matrices $[\,\mathbf{u}\ \ \mathbf{v}\,]$ and $[\,\mathbf{v}\ \ \mathbf{u}\,]$. Draw a picture and explain what you find.

43. [M] Is it true that $\det(A + B) = \det A + \det B$? To find out, generate random 5×5 matrices A and B, and compute $\det(A + B) - \det A - \det B$. (Refer to Exercise 37 in Section 2.1.) Repeat the calculations for three other pairs of $n \times n$ matrices, for various values of n. Report your results.

44. [M] Is it true that $\det AB = (\det A)(\det B)$? Experiment with four pairs of random matrices as in Exercise 43, and make a conjecture.

45. [M] Construct a random 4×4 matrix A with integer entries between -9 and 9, and compare $\det A$ with $\det A^T$, $\det(-A)$, $\det(2A)$, and $\det(10A)$. Repeat with two other random 4×4 integer matrices, and make conjectures about how these determinants are related. (Refer to Exercise 36 in Section 2.1.) Then check your conjectures with several random 5×5 and 6×6 integer matrices. Modify your conjectures, if necessary, and report your results.

46. [M] How is $\det A^{-1}$ related to $\det A$? Experiment with random $n \times n$ integer matrices for $n = 4, 5$, and 6, and make a conjecture. *Note:* In the unlikely event that you encounter a matrix with a zero determinant, reduce it to echelon form and discuss what you find.

| **SOLUTION TO PRACTICE PROBLEM**

Take advantage of the zeros. Begin with a cofactor expansion down the third column to obtain a 3×3 matrix, which may be evaluated by an expansion down its first column.

$$\begin{vmatrix} 5 & -7 & 2 & 2 \\ 0 & 3 & 0 & -4 \\ -5 & -8 & 0 & 3 \\ 0 & 5 & 0 & -6 \end{vmatrix} = (-1)^{1+3}2 \begin{vmatrix} 0 & 3 & -4 \\ -5 & -8 & 3 \\ 0 & 5 & -6 \end{vmatrix}$$

$$= 2 \cdot (-1)^{2+1}(-5) \begin{vmatrix} 3 & -4 \\ 5 & -6 \end{vmatrix} = 20$$

The $(-1)^{2+1}$ in the next-to-last calculation came from the $(2, 1)$-position of the -5 in the 3×3 determinant.

3.2 │ PROPERTIES OF DETERMINANTS

The secret of determinants lies in how they change when row operations are performed. The following theorem generalizes the results of Exercises 19–24 in Section 3.1. The proof is at the end of this section.

THEOREM 3

Row Operations

Let A be a square matrix.

a. If a multiple of one row of A is added to another row to produce a matrix B, then $\det B = \det A$.

b. If two rows of A are interchanged to produce B, then $\det B = -\det A$.

c. If one row of A is multiplied by k to produce B, then $\det B = k \cdot \det A$.

The following examples show how to use Theorem 3 to find determinants efficiently.

EXAMPLE 1 Compute det A, where $A = \begin{bmatrix} 1 & -4 & 2 \\ -2 & 8 & -9 \\ -1 & 7 & 0 \end{bmatrix}$.

SOLUTION The strategy is to reduce A to echelon form and then to use the fact that the determinant of a triangular matrix is the product of the diagonal entries. The first two row replacements in column 1 do not change the determinant:

$$\det A = \begin{vmatrix} 1 & -4 & 2 \\ -2 & 8 & -9 \\ -1 & 7 & 0 \end{vmatrix} = \begin{vmatrix} 1 & -4 & 2 \\ 0 & 0 & -5 \\ -1 & 7 & 0 \end{vmatrix} = \begin{vmatrix} 1 & -4 & 2 \\ 0 & 0 & -5 \\ 0 & 3 & 2 \end{vmatrix}$$

An interchange of rows 2 and 3 reverses the sign of the determinant, so

$$\det A = - \begin{vmatrix} 1 & -4 & 2 \\ 0 & 3 & 2 \\ 0 & 0 & -5 \end{vmatrix} = -(1)(3)(-5) = 15 \qquad \blacksquare$$

A common use of Theorem 3(c) in hand calculations is to *factor out a common multiple of one row* of a matrix. For instance,

$$\begin{vmatrix} * & * & * \\ 5k & -2k & 3k \\ * & * & * \end{vmatrix} = k \begin{vmatrix} * & * & * \\ 5 & -2 & 3 \\ * & * & * \end{vmatrix}$$

where the starred entries are unchanged. We use this step in the next example.

EXAMPLE 2 Compute det A, where $A = \begin{bmatrix} 2 & -8 & 6 & 8 \\ 3 & -9 & 5 & 10 \\ -3 & 0 & 1 & -2 \\ 1 & -4 & 0 & 6 \end{bmatrix}$.

SOLUTION To simplify the arithmetic, we want a 1 in the upper-left corner. We could interchange rows 1 and 4. Instead, we factor out 2 from the top row, and then proceed with row replacements in the first column:

$$\det A = 2 \begin{vmatrix} 1 & -4 & 3 & 4 \\ 3 & -9 & 5 & 10 \\ -3 & 0 & 1 & -2 \\ 1 & -4 & 0 & 6 \end{vmatrix} = 2 \begin{vmatrix} 1 & -4 & 3 & 4 \\ 0 & 3 & -4 & -2 \\ 0 & -12 & 10 & 10 \\ 0 & 0 & -3 & 2 \end{vmatrix}$$

Next, we could factor out another 2 from row 3 or use the 3 in the second column as a pivot. We choose the latter operation, adding 4 times row 2 to row 3:

$$\det A = 2 \begin{vmatrix} 1 & -4 & 3 & 4 \\ 0 & 3 & -4 & -2 \\ 0 & 0 & -6 & 2 \\ 0 & 0 & -3 & 2 \end{vmatrix}$$

Finally, adding $-1/2$ times row 3 to row 4, and computing the "triangular" determinant, we find that

$$\det A = 2 \begin{vmatrix} 1 & -4 & 3 & 4 \\ 0 & 3 & -4 & -2 \\ 0 & 0 & -6 & 2 \\ 0 & 0 & 0 & 1 \end{vmatrix} = 2 \cdot (1)(3)(-6)(1) = -36 \qquad \blacksquare$$

$$U = \begin{bmatrix} \blacksquare & * & * & * \\ 0 & \blacksquare & * & * \\ 0 & 0 & \blacksquare & * \\ 0 & 0 & 0 & \blacksquare \end{bmatrix}$$

det $U \neq 0$

$$U = \begin{bmatrix} \blacksquare & * & * & * \\ 0 & \blacksquare & * & * \\ 0 & 0 & 0 & \blacksquare \\ 0 & 0 & 0 & 0 \end{bmatrix}$$

det $U = 0$

FIGURE 1

Typical echelon forms of square matrices.

Suppose a square matrix A has been reduced to an echelon form U by row replacements and row interchanges. (This is always possible. See the row reduction algorithm in Section 1.2.) If there are r interchanges, then Theorem 3 shows that

$$\det A = (-1)^r \det U$$

Since U is in echelon form, it is triangular, and so $\det U$ is the product of the diagonal entries $u_{11}, \ldots, u_{nn}$. If A is invertible, the entries u_{ii} are all pivots (because $A \sim I_n$ and the u_{ii} have not been scaled to 1's). Otherwise, at least u_{nn} is zero, and the product $u_{11} \cdots u_{nn}$ is zero. See Fig. 1. Thus

$$\det A = \begin{cases} (-1)^r \cdot \begin{pmatrix} \text{product of} \\ \text{pivots in } U \end{pmatrix} & \text{when } A \text{ is invertible} \\ 0 & \text{when } A \text{ is not invertible} \end{cases} \tag{1}$$

It is interesting to note that although the echelon form U described above is not unique (because it is not completely row reduced), and the pivots are not unique, the *product* of the pivots *is* unique, except for a possible minus sign.

Formula (1) not only gives a concrete interpretation of $\det A$ but also proves the main theorem of this section:

THEOREM 4   A square matrix A is invertible if and only if $\det A \neq 0$.

Theorem 4 adds the statement "$\det A \neq 0$" to the Invertible Matrix Theorem. A useful corollary is that $\det A = 0$ when the columns of A are linearly dependent. Also, $\det A = 0$ when the *rows* of A are linearly dependent. (Rows of A are columns of A^T, and linearly dependent columns of A^T make A^T singular. When A^T is singular, so is A, by the Invertible Matrix Theorem.) In practice, linear dependence is obvious when two columns or two rows are the same or a column or a row is zero.

EXAMPLE 3   Compute $\det A$, where $A = \begin{bmatrix} 3 & -1 & 2 & -5 \\ 0 & 5 & -3 & -6 \\ -6 & 7 & -7 & 4 \\ -5 & -8 & 0 & 9 \end{bmatrix}$.

SOLUTION Add 2 times row 1 to row 3 to obtain

$$\det A = \det \begin{bmatrix} 3 & -1 & 2 & -5 \\ 0 & 5 & -3 & -6 \\ 0 & 5 & -3 & -6 \\ -5 & -8 & 0 & 9 \end{bmatrix} = 0$$

because the second and third rows of the second matrix are equal. ∎

NUMERICAL NOTES

1. Most computer programs that compute $\det A$ for a general matrix A use the method of formula (1) above.

2. It can be shown that evaluation of an $n \times n$ determinant using row operations requires about $2n^3/3$ arithmetic operations. Any modern microcomputer can calculate a 25×25 determinant in a fraction of a second, since only about 10,000 operations are required.

WEB

Computers can also handle large "sparse" matrices, with special routines that take advantage of the presence of many zeros. Of course, zero entries can speed hand computations, too. The calculations in the next example combine the power of row operations with the strategy from Section 3.1 of using zero entries in cofactor expansions.

EXAMPLE 4 Compute det A, where $A = \begin{bmatrix} 0 & 1 & 2 & -1 \\ 2 & 5 & -7 & 3 \\ 0 & 3 & 6 & 2 \\ -2 & -5 & 4 & -2 \end{bmatrix}$.

SOLUTION A good way to begin is to use the 2 in column 1 as a pivot, eliminating the -2 below it. Then use a cofactor expansion to reduce the size of the determinant, followed by another row replacement operation. Thus

$$\det A = \begin{vmatrix} 0 & 1 & 2 & -1 \\ 2 & 5 & -7 & 3 \\ 0 & 3 & 6 & 2 \\ 0 & 0 & -3 & 1 \end{vmatrix} = -2 \begin{vmatrix} 1 & 2 & -1 \\ 3 & 6 & 2 \\ 0 & -3 & 1 \end{vmatrix} = -2 \begin{vmatrix} 1 & 2 & -1 \\ 0 & 0 & 5 \\ 0 & -3 & 1 \end{vmatrix}$$

An interchange of rows 2 and 3 would produce a "triangular determinant." Another approach is to make a cofactor expansion down the first column:

$$\det A = (-2)(1) \begin{vmatrix} 0 & 5 \\ -3 & 1 \end{vmatrix} = -2 \cdot (15) = -30 \qquad \blacksquare$$

Column Operations

We can perform operations on the columns of a matrix in a way that is analogous to the row operations we have considered. The next theorem shows that column operations have the same effects on determinants as row operations.

THEOREM 5 If A is an $n \times n$ matrix, then det $A^T =$ det A.

PROOF The theorem is obvious for $n = 1$. Suppose the theorem is true for $k \times k$ determinants and let $n = k + 1$. Then the cofactor of a_{1j} in A equals the cofactor of a_{j1} in A^T, because the cofactors involve $k \times k$ determinants. Hence the cofactor expansion of det A along the first *row* equals the cofactor expansion of det A^T down the first *column*. That is, A and A^T have equal determinants. Thus the theorem is true for $n = 1$, and the truth of the theorem for one value of n implies its truth for the next value of n. By the principle of induction, the theorem is true for all $n \geq 1$. $\blacksquare$

Because of Theorem 5, each statement in Theorem 3 is true when the word *row* is replaced everywhere by *column*. To verify this property, one merely applies the original Theorem 3 to A^T. A row operation on A^T amounts to a column operation on A.

Column operations are useful for both theoretical purposes and hand computations. However, for simplicity we'll perform only row operations in numerical calculations.

Determinants and Matrix Products

The proof of the following useful theorem is at the end of the section. Applications are in the exercises.

THEOREM 6   Multiplicative Property

If A and B are $n \times n$ matrices, then det $AB = (\det A)(\det B)$.

EXAMPLE 5   Verify Theorem 6 for $A = \begin{bmatrix} 6 & 1 \\ 3 & 2 \end{bmatrix}$ and $B = \begin{bmatrix} 4 & 3 \\ 1 & 2 \end{bmatrix}$.

SOLUTION

$$AB = \begin{bmatrix} 6 & 1 \\ 3 & 2 \end{bmatrix}\begin{bmatrix} 4 & 3 \\ 1 & 2 \end{bmatrix} = \begin{bmatrix} 25 & 20 \\ 14 & 13 \end{bmatrix}$$

and

$$\det AB = 25 \cdot 13 - 20 \cdot 14 = 325 - 280 = 45$$

Since det $A = 9$ and det $B = 5$,

$$(\det A)(\det B) = 9 \cdot 5 = 45 = \det AB \qquad \blacksquare$$

Warning: A common misconception is that Theorem 6 has an analogue for *sums* of matrices. However, $\det(A + B)$ is *not* equal to det $A +$ det B, in general.

A Linearity Property of the Determinant Function

For an $n \times n$ matrix A, we can consider det A as a function of the n column vectors in A. We will show that if all columns except one are held fixed, then det A is a *linear function* of that one (vector) variable.

Suppose that the jth column of A is allowed to vary, and write

$$A = \begin{bmatrix} \mathbf{a}_1 & \cdots & \mathbf{a}_{j-1} & \mathbf{x} & \mathbf{a}_{j+1} & \cdots & \mathbf{a}_n \end{bmatrix}$$

Define a transformation T from $\mathbb{R}^n$ to $\mathbb{R}$ by

$$T(\mathbf{x}) = \det \begin{bmatrix} \mathbf{a}_1 & \cdots & \mathbf{a}_{j-1} & \mathbf{x} & \mathbf{a}_{j+1} & \cdots & \mathbf{a}_n \end{bmatrix}$$

Then,

$$T(c\mathbf{x}) = cT(\mathbf{x}) \quad \text{for all scalars } c \text{ and all } \mathbf{x} \text{ in } \mathbb{R}^n \qquad (2)$$

$$T(\mathbf{u} + \mathbf{v}) = T(\mathbf{u}) + T(\mathbf{v}) \quad \text{for all } \mathbf{u}, \mathbf{v} \text{ in } \mathbb{R}^n \qquad (3)$$

Property (2) is Theorem 3(c) applied to the columns of A. A proof of property (3) follows from a cofactor expansion of det A down the jth column. (See Exercise 43.) This (multi-) linearity property of the determinant turns out to have many useful consequences that are studied in more advanced courses.

Proofs of Theorems 3 and 6

It is convenient to prove Theorem 3 when it is stated in terms of the elementary matrices discussed in Section 2.2. We call an elementary matrix E a *row replacement* (*matrix*) if E is obtained from the identity I by adding a multiple of one row to another row; E is an *interchange* if E is obtained by interchanging two rows of I; and E is *a scale by r* if E is obtained by multiplying a row of I by a nonzero scalar r. With this terminology, Theorem 3 can be reformulated as follows:

If A is an $n \times n$ matrix and E is an $n \times n$ elementary matrix, then

$$\det EA = (\det E)(\det A)$$

where

$$\det E = \begin{cases} 1 & \text{if } E \text{ is a row replacement} \\ -1 & \text{if } E \text{ is an interchange} \\ r & \text{if } E \text{ is a scale by } r \end{cases}$$

PROOF OF THEOREM 3 The proof is by induction on the size of A. The case of a 2×2 matrix was verified in Exercises 33–36 of Section 3.1. Suppose the theorem has been verified for determinants of $k \times k$ matrices with $k \geq 2$, let $n = k + 1$, and let A be $n \times n$. The action of E on A involves either two rows or only one row. So we can expand $\det EA$ across a row that is unchanged by the action of E, say, row i. Let A_{ij} (respectively, B_{ij}) be the matrix obtained by deleting row i and column j from A (respectively, EA). Then the rows of B_{ij} are obtained from the rows of A_{ij} by the same type of elementary row operation that E performs on A. Since these submatrices are only $k \times k$, the induction assumption implies that

$$\det B_{ij} = \alpha \cdot \det A_{ij}$$

where $\alpha = 1, -1$, or r, depending on the nature of E. The cofactor expansion across row i is

$$\det EA = a_{i1}(-1)^{i+1} \det B_{i1} + \cdots + a_{in}(-1)^{i+n} \det B_{in}$$
$$= \alpha a_{i1}(-1)^{i+1} \det A_{i1} + \cdots + \alpha a_{in}(-1)^{i+n} \det A_{in}$$
$$= \alpha \cdot \det A$$

In particular, taking $A = I_n$, we see that $\det E = 1, -1$, or r, depending on the nature of E. Thus the theorem is true for $n = 2$, and the truth of the theorem for one value of n implies its truth for the next value of n. By the principle of induction, the theorem must be true for $n \geq 2$. The theorem is trivially true for $n = 1$. ∎

PROOF OF THEOREM 6 If A is not invertible, then neither is AB, by Exercise 27 in Section 2.3. In this case, $\det AB = (\det A)(\det B)$, because both sides are zero, by Theorem 4. If A is invertible, then A and the identity matrix I_n are row equivalent by the Invertible Matrix Theorem. So there exist elementary matrices $E_1, \ldots, E_p$ such that

$$A = E_p E_{p-1} \cdots E_1 \cdot I_n = E_p E_{p-1} \cdots E_1$$

For brevity, write $|A|$ for $\det A$. Then repeated application of Theorem 3, as rephrased above, shows that

$$|AB| = |E_p \cdots E_1 B| = |E_p||E_{p-1} \cdots E_1 B| = \cdots$$
$$= |E_p| \cdots |E_1||B| = \cdots = |E_p \cdots E_1||B|$$
$$= |A||B|$$

∎

PRACTICE PROBLEMS

1. Compute $\begin{vmatrix} 1 & -3 & 1 & -2 \\ 2 & -5 & -1 & -2 \\ 0 & -4 & 5 & 1 \\ -3 & 10 & -6 & 8 \end{vmatrix}$ in as few steps as possible.

2. Use a determinant to decide if $\mathbf{v}_1, \mathbf{v}_2, \mathbf{v}_3$ are linearly independent, when

$$\mathbf{v}_1 = \begin{bmatrix} 5 \\ -7 \\ 9 \end{bmatrix}, \qquad \mathbf{v}_2 = \begin{bmatrix} -3 \\ 3 \\ -5 \end{bmatrix}, \qquad \mathbf{v}_3 = \begin{bmatrix} 2 \\ -7 \\ 5 \end{bmatrix}$$

3.2 EXERCISES

Each equation in Exercises 1–4 illustrates a property of determinants. State the property.

1. $\begin{vmatrix} 0 & 5 & -2 \\ 1 & -3 & 6 \\ 4 & -1 & 8 \end{vmatrix} = - \begin{vmatrix} 1 & -3 & 6 \\ 0 & 5 & -2 \\ 4 & -1 & 8 \end{vmatrix}$

2. $\begin{vmatrix} 2 & -6 & 4 \\ 3 & 5 & -2 \\ 1 & 6 & 3 \end{vmatrix} = 2 \begin{vmatrix} 1 & -3 & 2 \\ 3 & 5 & -2 \\ 1 & 6 & 3 \end{vmatrix}$

3. $\begin{vmatrix} 1 & 3 & -4 \\ 2 & 0 & -3 \\ 5 & -4 & 7 \end{vmatrix} = \begin{vmatrix} 1 & 3 & -4 \\ 0 & -6 & 5 \\ 5 & -4 & 7 \end{vmatrix}$

4. $\begin{vmatrix} 1 & 2 & 3 \\ 0 & 5 & -4 \\ 3 & 7 & 4 \end{vmatrix} = \begin{vmatrix} 1 & 2 & 3 \\ 0 & 5 & -4 \\ 0 & 1 & -5 \end{vmatrix}$

Find the determinants in Exercises 5–10 by row reduction to echelon form.

5. $\begin{vmatrix} 1 & 5 & -6 \\ -1 & -4 & 4 \\ -2 & -7 & 9 \end{vmatrix}$

6. $\begin{vmatrix} 1 & 5 & -3 \\ 3 & -3 & 3 \\ 2 & 13 & -7 \end{vmatrix}$

7. $\begin{vmatrix} 1 & 3 & 0 & 2 \\ -2 & -5 & 7 & 4 \\ 3 & 5 & 2 & 1 \\ 1 & -1 & 2 & -3 \end{vmatrix}$

8. $\begin{vmatrix} 1 & 3 & 3 & -4 \\ 0 & 1 & 2 & -5 \\ 2 & 5 & 4 & -3 \\ -3 & -7 & -5 & 2 \end{vmatrix}$

9. $\begin{vmatrix} 1 & -1 & -3 & 0 \\ 0 & 1 & 5 & 4 \\ -1 & 2 & 8 & 5 \\ 3 & -1 & -2 & 3 \end{vmatrix}$

10. $\begin{vmatrix} 1 & 3 & -1 & 0 & -2 \\ 0 & 2 & -4 & -1 & -6 \\ -2 & -6 & 2 & 3 & 9 \\ 3 & 7 & -3 & 8 & -7 \\ 3 & 5 & 5 & 2 & 7 \end{vmatrix}$

Combine the methods of row reduction and cofactor expansion to compute the determinants in Exercises 11–14.

11. $\begin{vmatrix} 2 & 5 & -3 & -1 \\ 3 & 0 & 1 & -3 \\ -6 & 0 & -4 & 9 \\ 4 & 10 & -4 & -1 \end{vmatrix}$

12. $\begin{vmatrix} -1 & 2 & 3 & 0 \\ 3 & 4 & 3 & 0 \\ 5 & 4 & 6 & 6 \\ 4 & 2 & 4 & 3 \end{vmatrix}$

13. $\begin{vmatrix} 2 & 5 & 4 & 1 \\ 4 & 7 & 6 & 2 \\ 6 & -2 & -4 & 0 \\ -6 & 7 & 7 & 0 \end{vmatrix}$

14. $\begin{vmatrix} -3 & -2 & 1 & -4 \\ 1 & 3 & 0 & -3 \\ -3 & 4 & -2 & 8 \\ 3 & -4 & 0 & 4 \end{vmatrix}$

Find the determinants in Exercises 15–20, where
$\begin{vmatrix} a & b & c \\ d & e & f \\ g & h & i \end{vmatrix} = 7.$

15. $\begin{vmatrix} a & b & c \\ d & e & f \\ 5g & 5h & 5i \end{vmatrix}$

16. $\begin{vmatrix} a & b & c \\ 3d & 3e & 3f \\ g & h & i \end{vmatrix}$

17. $\begin{vmatrix} a & b & c \\ g & h & i \\ d & e & f \end{vmatrix}$

18. $\begin{vmatrix} g & h & i \\ a & b & c \\ d & e & f \end{vmatrix}$

19. $\begin{vmatrix} a & b & c \\ 2d+a & 2e+b & 2f+c \\ g & h & i \end{vmatrix}$

20. $\begin{vmatrix} a+d & b+e & c+f \\ d & e & f \\ g & h & i \end{vmatrix}$

In Exercises 21–23, use determinants to find out if the matrix is invertible.

21. $\begin{bmatrix} 2 & 3 & 0 \\ 1 & 3 & 4 \\ 1 & 2 & 1 \end{bmatrix}$

22. $\begin{bmatrix} 5 & 0 & -1 \\ 1 & -3 & -2 \\ 0 & 5 & 3 \end{bmatrix}$

23. $\begin{bmatrix} 2 & 0 & 0 & 8 \\ 1 & -7 & -5 & 0 \\ 3 & 8 & 6 & 0 \\ 0 & 7 & 5 & 4 \end{bmatrix}$

In Exercises 24–26, use determinants to decide if the set of vectors is linearly independent.

24. $\begin{bmatrix} 4 \\ 6 \\ -7 \end{bmatrix}, \begin{bmatrix} -7 \\ 0 \\ 2 \end{bmatrix}, \begin{bmatrix} -3 \\ -5 \\ 6 \end{bmatrix}$

25. $\begin{bmatrix} 7 \\ -4 \\ -6 \end{bmatrix}, \begin{bmatrix} -8 \\ 5 \\ 7 \end{bmatrix}, \begin{bmatrix} 7 \\ 0 \\ -5 \end{bmatrix}$

26. $\begin{bmatrix} 3 \\ 5 \\ -6 \\ 4 \end{bmatrix}, \begin{bmatrix} 2 \\ -6 \\ 0 \\ 7 \end{bmatrix}, \begin{bmatrix} -2 \\ -1 \\ 3 \\ 0 \end{bmatrix}, \begin{bmatrix} 0 \\ 0 \\ 0 \\ -3 \end{bmatrix}$

In Exercises 27 and 28, A and B are $n \times n$ matrices. Mark each statement True or False. Justify each answer.

27. a. A row replacement operation does not affect the determinant of a matrix.

b. The determinant of A is the product of the pivots in any echelon form U of A, multiplied by $(-1)^r$, where r is the number of row interchanges made during row reduction from A to U.

c. If the columns of A are linearly dependent, then $\det A = 0$.

d. $\det(A + B) = \det A + \det B$.

28. a. If two row interchanges are made in succession, then the new determinant equals the old determinant.

b. The determinant of A is the product of the diagonal entries in A.

c. If $\det A$ is zero, then two rows or two columns are the same, or a row or a column is zero.

d. $\det A^T = (-1)\det A$.

29. Compute $\det B^5$, where $B = \begin{bmatrix} 1 & 0 & 1 \\ 1 & 1 & 2 \\ 1 & 2 & 1 \end{bmatrix}$.

30. Use Theorem 3 (but not Theorem 4) to show that if two rows of a square matrix A are equal, then $\det A = 0$. The same is true for two columns. Why?

In Exercises 31–36, mention an appropriate theorem in your explanation.

31. Show that if A is invertible, then $\det A^{-1} = \dfrac{1}{\det A}$.

32. Find a formula for $\det(rA)$ when A is an $n \times n$ matrix.

33. Let A and B be square matrices. Show that even though AB and BA may not be equal, it is always true that $\det AB = \det BA$.

34. Let A and P be square matrices, with P invertible. Show that $\det(PAP^{-1}) = \det A$.

35. Let U be a square matrix such that $U^T U = I$. Show that $\det U = \pm 1$.

36. Suppose that A is a square matrix such that $\det A^4 = 0$. Explain why A cannot be invertible.

Verify that $\det AB = (\det A)(\det B)$ for the matrices in Exercises 37 and 38. (Do not use Theorem 6.)

37. $A = \begin{bmatrix} 3 & 0 \\ 6 & 1 \end{bmatrix}$, $B = \begin{bmatrix} 2 & 0 \\ 5 & 4 \end{bmatrix}$

38. $A = \begin{bmatrix} 3 & 6 \\ -1 & -2 \end{bmatrix}$, $B = \begin{bmatrix} 4 & 2 \\ -1 & -1 \end{bmatrix}$

39. Let A and B be 3×3 matrices, with $\det A = 4$ and $\det B = -3$. Use properties of determinants (in the text and in the exercises above) to compute:

a. $\det AB$ b. $\det 5A$ c. $\det B^T$

d. $\det A^{-1}$ e. $\det A^3$

40. Let A and B be 4×4 matrices, with $\det A = -1$ and $\det B = 2$. Compute:

a. $\det AB$ b. $\det B^5$ c. $\det 2A$

d. $\det A^T A$ e. $\det B^{-1}AB$

41. Verify that $\det A = \det B + \det C$, where

$$A = \begin{bmatrix} a+e & b+f \\ c & d \end{bmatrix},\ B = \begin{bmatrix} a & b \\ c & d \end{bmatrix},\ C = \begin{bmatrix} e & f \\ c & d \end{bmatrix}$$

42. Let $A = \begin{bmatrix} 1 & 0 \\ 0 & 1 \end{bmatrix}$ and $B = \begin{bmatrix} a & b \\ c & d \end{bmatrix}$. Show that $\det(A + B) = \det A + \det B$ if and only if $a + d = 0$.

43. Verify that $\det A = \det B + \det C$, where

$$A = \begin{bmatrix} a_{11} & a_{12} & u_1 + v_1 \\ a_{21} & a_{22} & u_2 + v_2 \\ a_{31} & a_{32} & u_3 + v_3 \end{bmatrix},$$

$$B = \begin{bmatrix} a_{11} & a_{12} & u_1 \\ a_{21} & a_{22} & u_2 \\ a_{31} & a_{32} & u_3 \end{bmatrix},\ C = \begin{bmatrix} a_{11} & a_{12} & v_1 \\ a_{21} & a_{22} & v_2 \\ a_{31} & a_{32} & v_3 \end{bmatrix}$$

Note, however, that A is *not* the same as $B + C$.

44. Right-multiplication by an elementary matrix E affects the *columns* of A in the same way that left-multiplication affects the *rows*. Use Theorems 5 and 3 and the obvious fact that E^T is another elementary matrix to show that

$$\det AE = (\det E)(\det A)$$

Do not use Theorem 6.

45. [M] Compute $\det A^T A$ and $\det AA^T$ for several random 4×5 matrices and several random 5×6 matrices. What can you say about $A^T A$ and AA^T when A has more columns than rows?

46. [M] If $\det A$ is close to zero, is the matrix A nearly singular? Experiment with the nearly singular 4×4 matrix A in Exercise 9 of Section 2.3. Compute the determinants of A, $10A$, and $0.1A$. In contrast, compute the condition numbers of these matrices. Repeat these calculations when A is the 4×4 identity matrix. Discuss your results.

SOLUTIONS TO PRACTICE PROBLEMS

1. Perform row replacements to create zeros in the first column and then create a row of zeros.

$$\begin{vmatrix} 1 & -3 & 1 & -2 \\ 2 & -5 & -1 & -2 \\ 0 & -4 & 5 & 1 \\ -3 & 10 & -6 & 8 \end{vmatrix} = \begin{vmatrix} 1 & -3 & 1 & -2 \\ 0 & 1 & -3 & 2 \\ 0 & -4 & 5 & 1 \\ 0 & 1 & -3 & 2 \end{vmatrix} = \begin{vmatrix} 1 & -3 & 1 & -2 \\ 0 & 1 & -3 & 2 \\ 0 & -4 & 5 & 1 \\ 0 & 0 & 0 & 0 \end{vmatrix} = 0$$

2. $\det [\, \mathbf{v}_1 \quad \mathbf{v}_2 \quad \mathbf{v}_3 \,] = \begin{vmatrix} 5 & -3 & 2 \\ -7 & 3 & -7 \\ 9 & -5 & 5 \end{vmatrix} = \begin{vmatrix} 5 & -3 & 2 \\ -2 & 0 & -5 \\ 9 & -5 & 5 \end{vmatrix}$ Row 1 added to row 2

$$= -(-3)\begin{vmatrix} -2 & -5 \\ 9 & 5 \end{vmatrix} - (-5)\begin{vmatrix} 5 & 2 \\ -2 & -5 \end{vmatrix} \quad \text{Cofactors of column 2}$$

$$= 3 \cdot (35) + 5 \cdot (-21) = 0$$

By Theorem 4, the matrix $[\, \mathbf{v}_1 \quad \mathbf{v}_2 \quad \mathbf{v}_3 \,]$ is not invertible. The columns are linearly dependent, by the Invertible Matrix Theorem.

3.3 | CRAMER'S RULE, VOLUME, AND LINEAR TRANSFORMATIONS

This section applies the theory of the preceding sections to obtain important theoretical formulas and a geometric interpretation of the determinant.

Cramer's Rule

Cramer's rule is needed in a variety of theoretical calculations. For instance, it can be used to study how the solution of $A\mathbf{x} = \mathbf{b}$ is affected by changes in the entries of $\mathbf{b}$. However, the formula is inefficient for hand calculations, except for 2×2 or perhaps 3×3 matrices.

For any $n \times n$ matrix A and any $\mathbf{b}$ in $\mathbb{R}^n$, let $A_i(\mathbf{b})$ be the matrix obtained from A by replacing column i by the vector $\mathbf{b}$.

$$A_i(\mathbf{b}) = [\mathbf{a}_1 \quad \cdots \quad \underset{\underset{\text{col } i}{\uparrow}}{\mathbf{b}} \quad \cdots \quad \mathbf{a}_n]$$

THEOREM 7

> **Cramer's Rule**
>
> Let A be an invertible $n \times n$ matrix. For any $\mathbf{b}$ in $\mathbb{R}^n$, the unique solution $\mathbf{x}$ of $A\mathbf{x} = \mathbf{b}$ has entries given by
>
> $$x_i = \frac{\det A_i(\mathbf{b})}{\det A}, \qquad i = 1, 2, \ldots, n \tag{1}$$

PROOF Denote the columns of A by $\mathbf{a}_1, \ldots, \mathbf{a}_n$ and the columns of the $n \times n$ identity matrix I by $\mathbf{e}_1, \ldots, \mathbf{e}_n$. If $A\mathbf{x} = \mathbf{b}$, the definition of matrix multiplication shows that

$$A \cdot I_i(\mathbf{x}) = A [\, \mathbf{e}_1 \quad \cdots \quad \mathbf{x} \quad \cdots \quad \mathbf{e}_n \,] = [\, A\mathbf{e}_1 \quad \cdots \quad A\mathbf{x} \quad \cdots \quad A\mathbf{e}_n \,]$$

$$= [\, \mathbf{a}_1 \quad \cdots \quad \mathbf{b} \quad \cdots \quad \mathbf{a}_n \,] = A_i(\mathbf{b})$$

By the multiplicative property of determinants,

$$(\det A)(\det I_i(\mathbf{x})) = \det A_i(\mathbf{b})$$

The second determinant on the left is simply x_i. (Make a cofactor expansion along the ith row.) Hence $(\det A) \cdot x_i = \det A_i(\mathbf{b})$. This proves (1) because A is invertible and $\det A \neq 0$. ∎

EXAMPLE 1 Use Cramer's rule to solve the system

$$3x_1 - 2x_2 = 6$$
$$-5x_1 + 4x_2 = 8$$

SOLUTION View the system as $A\mathbf{x} = \mathbf{b}$. Using the notation introduced above,

$$A = \begin{bmatrix} 3 & -2 \\ -5 & 4 \end{bmatrix}, \qquad A_1(\mathbf{b}) = \begin{bmatrix} 6 & -2 \\ 8 & 4 \end{bmatrix}, \qquad A_2(\mathbf{b}) = \begin{bmatrix} 3 & 6 \\ -5 & 8 \end{bmatrix}$$

Since $\det A = 2$, the system has a unique solution. By Cramer's rule,

$$x_1 = \frac{\det A_1(\mathbf{b})}{\det A} = \frac{24 + 16}{2} = 20$$

$$x_2 = \frac{\det A_2(\mathbf{b})}{\det A} = \frac{24 + 30}{2} = 27 \qquad \blacksquare$$

Application to Engineering

A number of important engineering problems, particularly in electrical engineering and control theory, can be analyzed by *Laplace transforms*. This approach converts an appropriate system of linear differential equations into a system of linear algebraic equations whose coefficients involve a parameter. The next example illustrates the type of algebraic system that may arise.

EXAMPLE 2 Consider the following system in which s is an unspecified parameter. Determine the values of s for which the system has a unique solution, and use Cramer's rule to describe the solution.

$$3sx_1 - 2x_2 = 4$$
$$-6x_1 + sx_2 = 1$$

SOLUTION View the system as $A\mathbf{x} = \mathbf{b}$. Then

$$A = \begin{bmatrix} 3s & -2 \\ -6 & s \end{bmatrix}, \qquad A_1(\mathbf{b}) = \begin{bmatrix} 4 & -2 \\ 1 & s \end{bmatrix}, \qquad A_2(\mathbf{b}) = \begin{bmatrix} 3s & 4 \\ -6 & 1 \end{bmatrix}$$

Since

$$\det A = 3s^2 - 12 = 3(s + 2)(s - 2)$$

the system has a unique solution precisely when $s \neq \pm 2$. For such an s, the solution is (x_1, x_2), where

$$x_1 = \frac{\det A_1(\mathbf{b})}{\det A} = \frac{4s + 2}{3(s + 2)(s - 2)}$$

$$x_2 = \frac{\det A_2(\mathbf{b})}{\det A} = \frac{3s + 24}{3(s + 2)(s - 2)} = \frac{s + 8}{(s + 2)(s - 2)} \qquad \blacksquare$$

A Formula for A^{-1}

Cramer's rule leads easily to a general formula for the inverse of an $n \times n$ matrix A. The jth column of A^{-1} is a vector $\mathbf{x}$ that satisfies

$$A\mathbf{x} = \mathbf{e}_j$$

where $\mathbf{e}_j$ is the jth column of the identity matrix, and the ith entry of $\mathbf{x}$ is the (i, j)-entry of A^{-1}. By Cramer's rule,

$$\{(i, j)\text{-entry of } A^{-1}\} = x_i = \frac{\det A_i(\mathbf{e}_j)}{\det A} \qquad (2)$$

Recall that A_{ji} denotes the submatrix of A formed by deleting row j and column i. A cofactor expansion down column i of $A_i(\mathbf{e}_j)$ shows that

$$\det A_i(\mathbf{e}_j) = (-1)^{i+j} \det A_{ji} = C_{ji} \tag{3}$$

where C_{ji} is a cofactor of A. By (2), the (i, j)-entry of A^{-1} is the cofactor C_{ji} divided by $\det A$. [Note that the subscripts on C_{ji} are the reverse of (i, j).] Thus

$$A^{-1} = \frac{1}{\det A} \begin{bmatrix} C_{11} & C_{21} & \cdots & C_{n1} \\ C_{12} & C_{22} & \cdots & C_{n2} \\ \vdots & \vdots & & \vdots \\ C_{1n} & C_{2n} & \cdots & C_{nn} \end{bmatrix} \tag{4}$$

The matrix of cofactors on the right side of (4) is called the **adjugate** (or **classical adjoint**) of A, denoted by adj A. (The term *adjoint* also has another meaning in advanced texts on linear transformations.) The next theorem simply restates (4).

THEOREM 8

An Inverse Formula

Let A be an invertible $n \times n$ matrix. Then

$$A^{-1} = \frac{1}{\det A} \text{ adj } A$$

EXAMPLE 3 Find the inverse of the matrix $A = \begin{bmatrix} 2 & 1 & 3 \\ 1 & -1 & 1 \\ 1 & 4 & -2 \end{bmatrix}$.

SOLUTION The nine cofactors are

$$C_{11} = + \begin{vmatrix} -1 & 1 \\ 4 & -2 \end{vmatrix} = -2, \quad C_{12} = - \begin{vmatrix} 1 & 1 \\ 1 & -2 \end{vmatrix} = 3, \quad C_{13} = + \begin{vmatrix} 1 & -1 \\ 1 & 4 \end{vmatrix} = 5$$

$$C_{21} = - \begin{vmatrix} 1 & 3 \\ 4 & -2 \end{vmatrix} = 14, \quad C_{22} = + \begin{vmatrix} 2 & 3 \\ 1 & -2 \end{vmatrix} = -7, \quad C_{23} = - \begin{vmatrix} 2 & 1 \\ 1 & 4 \end{vmatrix} = -7$$

$$C_{31} = + \begin{vmatrix} 1 & 3 \\ -1 & 1 \end{vmatrix} = 4, \quad C_{32} = - \begin{vmatrix} 2 & 3 \\ 1 & 1 \end{vmatrix} = 1, \quad C_{33} = + \begin{vmatrix} 2 & 1 \\ 1 & -1 \end{vmatrix} = -3$$

The adjugate matrix is the *transpose* of the matrix of cofactors. [For instance, C_{12} goes in the $(2, 1)$ position.] Thus

$$\text{adj } A = \begin{bmatrix} -2 & 14 & 4 \\ 3 & -7 & 1 \\ 5 & -7 & -3 \end{bmatrix}$$

We could compute $\det A$ directly, but the following computation provides a check on the calculations above *and* produces $\det A$:

$$(\text{adj } A) \cdot A = \begin{bmatrix} -2 & 14 & 4 \\ 3 & -7 & 1 \\ 5 & -7 & -3 \end{bmatrix} \begin{bmatrix} 2 & 1 & 3 \\ 1 & -1 & 1 \\ 1 & 4 & -2 \end{bmatrix} = \begin{bmatrix} 14 & 0 & 0 \\ 0 & 14 & 0 \\ 0 & 0 & 14 \end{bmatrix} = 14I$$

Since $(\text{adj } A)A = 14I$, Theorem 8 shows that $\det A = 14$ and

$$A^{-1} = \frac{1}{14} \begin{bmatrix} -2 & 14 & 4 \\ 3 & -7 & 1 \\ 5 & -7 & -3 \end{bmatrix} = \begin{bmatrix} -1/7 & 1 & 2/7 \\ 3/14 & -1/2 & 1/14 \\ 5/14 & -1/2 & -3/14 \end{bmatrix} \quad \blacksquare$$

NUMERICAL NOTES

Theorem 8 is useful mainly for theoretical calculations. The formula for A^{-1} permits one to deduce properties of the inverse without actually calculating it. Except for special cases, the algorithm in Section 2.2 gives a much better way to compute A^{-1}, if the inverse is really needed.

Cramer's rule is also a theoretical tool. It can be used to study how sensitive the solution of $A\mathbf{x} = \mathbf{b}$ is to changes in an entry in $\mathbf{b}$ or in A (perhaps due to experimental error when acquiring the entries for $\mathbf{b}$ or A). When A is a 3×3 matrix with *complex* entries, Cramer's rule is sometimes selected for hand computation because row reduction of $[\,A \quad \mathbf{b}\,]$ with complex arithmetic can be messy, and the determinants are fairly easy to compute. For a larger $n \times n$ matrix (real or complex), Cramer's rule is hopelessly inefficient. Computing just *one* determinant takes about as much work as solving $A\mathbf{x} = \mathbf{b}$ by row reduction.

Determinants as Area or Volume

In the next application, we verify the geometric interpretation of determinants described in the chapter introduction. Although a general discussion of length and distance in $\mathbb{R}^n$ will not be given until Chapter 6, we assume here that the usual Euclidean concepts of length, area, and volume are already understood for $\mathbb{R}^2$ and $\mathbb{R}^3$.

THEOREM 9

If A is a 2×2 matrix, the area of the parallelogram determined by the columns of A is $|\det A|$. If A is a 3×3 matrix, the volume of the parallelepiped determined by the columns of A is $|\det A|$.

SG   A Geometric Proof 3-12

PROOF   The theorem is obviously true for any 2×2 diagonal matrix:

$$\left| \det \begin{bmatrix} a & 0 \\ 0 & d \end{bmatrix} \right| = |ad| = \left\{ \begin{array}{l} \text{area of} \\ \text{rectangle} \end{array} \right\}$$

See Fig. 1. It will suffice to show that any 2×2 matrix $A = [\,\mathbf{a}_1 \quad \mathbf{a}_2\,]$ can be transformed into a diagonal matrix in a way that changes neither the area of the associated parallelogram nor $|\det A|$. From Section 3.2, we know that the absolute value of the determinant is unchanged when two columns are interchanged or a multiple of one column is added to another. And it is easy to see that such operations suffice to transform A into a diagonal matrix. Column interchanges do not change the parallelogram at all. So it suffices to prove the following simple geometric observation that applies to vectors in $\mathbb{R}^2$ or $\mathbb{R}^3$:

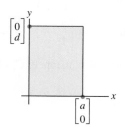

FIGURE 1

Area $= |ad|$.

Let $\mathbf{a}_1$ and $\mathbf{a}_2$ be nonzero vectors. Then for any scalar c, the area of the parallelogram determined by $\mathbf{a}_1$ and $\mathbf{a}_2$ equals the area of the parallelogram determined by $\mathbf{a}_1$ and $\mathbf{a}_2 + c\mathbf{a}_1$.

To prove this statement, we may assume that $\mathbf{a}_2$ is not a multiple of $\mathbf{a}_1$, for otherwise the two parallelograms would be degenerate and have zero area. If L is the line through $\mathbf{0}$ and $\mathbf{a}_1$, then $\mathbf{a}_2 + L$ is the line through $\mathbf{a}_2$ parallel to L, and $\mathbf{a}_2 + c\mathbf{a}_1$ is on this line. See Fig. 2. The points $\mathbf{a}_2$ and $\mathbf{a}_2 + c\mathbf{a}_1$ have the same perpendicular distance to L. Hence the two parallelograms in Fig. 2 have the same area, since they share the base from $\mathbf{0}$ to $\mathbf{a}_1$. This completes the proof for $\mathbb{R}^2$.

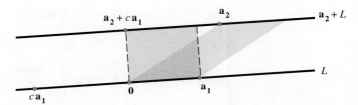

FIGURE 2 Two parallelograms of equal area.

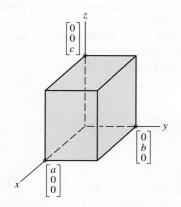

FIGURE 3
Volume $= |abc|$.

The proof for $\mathbb{R}^3$ is similar. The theorem is obviously true for a 3×3 diagonal matrix. See Fig. 3. And any 3×3 matrix A can be transformed into a diagonal matrix using column operations that do not change $|\det A|$. (Think about doing row operations on A^T.) So it suffices to show that these operations do not affect the volume of the parallelepiped determined by the columns of A.

A parallelepiped is shown in Fig. 4 as a shaded box with two sloping sides. Its volume is the area of the base in the plane Span $\{\mathbf{a}_1, \mathbf{a}_3\}$ times the altitude of $\mathbf{a}_2$ above Span $\{\mathbf{a}_1, \mathbf{a}_3\}$. Any vector $\mathbf{a}_2 + c\mathbf{a}_1$ has the same altitude because $\mathbf{a}_2 + c\mathbf{a}_1$ lies in the plane $\mathbf{a}_2 + $ Span $\{\mathbf{a}_1, \mathbf{a}_3\}$, which is parallel to Span $\{\mathbf{a}_1, \mathbf{a}_3\}$. Hence the volume of the parallelepiped is unchanged when $[\,\mathbf{a}_1 \;\; \mathbf{a}_2 \;\; \mathbf{a}_3\,]$ is changed to $[\,\mathbf{a}_1 \;\; \mathbf{a}_2 + c\mathbf{a}_1 \;\; \mathbf{a}_3\,]$. Thus a column replacement operation does not affect the volume of the parallelepiped. Since column interchanges have no effect on the volume, the proof is complete. ∎

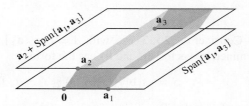

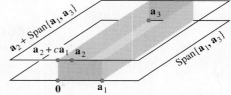

FIGURE 4 Two parallelepipeds of equal volume.

EXAMPLE 4   Calculate the area of the parallelogram determined by the points $(-2, -2)$, $(0, 3)$, $(4, -1)$, and $(6, 4)$. See Fig. 5(a).

SOLUTION First translate the parallelogram to one having the origin as a vertex. For example, subtract the vertex $(-2, -2)$ from each of the four vertices. The new parallelogram has the same area, and its vertices are $(0, 0)$, $(2, 5)$, $(6, 1)$, and $(8, 6)$. See

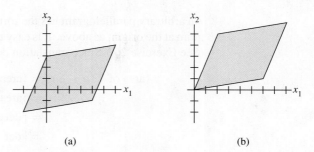

(a)   (b)

FIGURE 5 Translating a parallelogram does not change its area.

Fig. 5(b). This parallelogram is determined by the columns of

$$A = \begin{bmatrix} 2 & 6 \\ 5 & 1 \end{bmatrix}$$

Since $|\det A| = |-28|$, the area of the parallelogram is 28. ∎

Linear Transformations

Determinants can be used to describe an important geometric property of linear transformations in the plane and in $\mathbb{R}^3$. If T is a linear transformation and S is a set in the domain of T, let $T(S)$ denote the set of images of points in S. We are interested in how the area (or volume) of $T(S)$ compares with the area (or volume) of the original set S. For convenience, when S is a region bounded by a parallelogram, we also refer to S as a parallelogram.

THEOREM 10

Let $T : \mathbb{R}^2 \to \mathbb{R}^2$ be the linear transformation determined by a 2×2 matrix A. If S is a parallelogram in $\mathbb{R}^2$, then

$$\{\text{area of } T(S)\} = |\det A| \cdot \{\text{area of } S\} \tag{5}$$

If T is determined by a 3×3 matrix A, and if S is a parallelepiped in $\mathbb{R}^3$, then

$$\{\text{volume of } T(S)\} = |\det A| \cdot \{\text{volume of } S\} \tag{6}$$

PROOF Consider the 2×2 case, with $A = [\, \mathbf{a}_1 \ \ \mathbf{a}_2 \,]$. A parallelogram at the origin in $\mathbb{R}^2$ determined by vectors $\mathbf{b}_1$ and $\mathbf{b}_2$ has the form

$$S = \{s_1\mathbf{b}_1 + s_2\mathbf{b}_2 : 0 \le s_1 \le 1, \ 0 \le s_2 \le 1\}$$

The image of S under T consists of points of the form

$$T(s_1\mathbf{b}_1 + s_2\mathbf{b}_2) = s_1 T(\mathbf{b}_1) + s_2 T(\mathbf{b}_2)$$
$$= s_1 A\mathbf{b}_1 + s_2 A\mathbf{b}_2$$

where $0 \le s_1 \le 1$, $0 \le s_2 \le 1$. It follows that $T(S)$ is the parallelogram determined by the columns of the matrix $[\, A\mathbf{b}_1 \ \ A\mathbf{b}_2 \,]$. This matrix can be written as AB, where $B = [\, \mathbf{b}_1 \ \ \mathbf{b}_2 \,]$. By Theorem 9 and the product theorem for determinants,

$$\{\text{area of } T(S)\} = |\det AB| = |\det A| \cdot |\det B|$$
$$= |\det A| \cdot \{\text{area of } S\} \tag{7}$$

An arbitrary parallelogram has the form $\mathbf{p} + S$, where $\mathbf{p}$ is a vector and S is a parallelogram at the origin, as above. It is easy to see that T transforms $\mathbf{p} + S$ into $T(\mathbf{p}) + T(S)$. (See Exercise 26.) Since translation does not affect the area of a set,

$$\{\text{area of } T(\mathbf{p} + S)\} = \{\text{area of } T(\mathbf{p}) + T(S)\}$$
$$= \{\text{area of } T(S)\} \qquad \text{Translation}$$
$$= |\det A| \cdot \{\text{area of } S\} \qquad \text{By equation (7)}$$
$$= |\det A| \cdot \{\text{area of } \mathbf{p} + S\} \qquad \text{Translation}$$

This shows that (5) holds for all parallelograms in $\mathbb{R}^2$. The proof of (6) for the 3×3 case is analogous. ∎

When we attempt to generalize Theorem 10 to a region in $\mathbb{R}^2$ or $\mathbb{R}^3$ that is not bounded by straight lines or planes, we must face the problem of how to define and compute its area or volume. This is a question studied in calculus, and we shall only outline the basic idea for $\mathbb{R}^2$. If R is a planar region that has a finite area, then R can be approximated by a grid of small squares that lie inside R. By making the squares sufficiently small, the area of R may be approximated as closely as desired by the sum of the areas of the small squares. See Fig. 6.

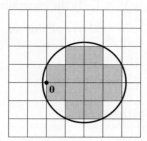 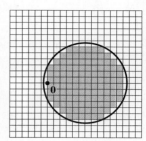

FIGURE 6 Approximating a planar region by a union of squares. The approximation improves as the grid becomes finer.

If T is a linear transformation associated with a 2×2 matrix A, then the image of a planar region R under T is approximated by the images of the small squares inside R. The proof of Theorem 10 shows that each such image is a parallelogram whose area is $|\det A|$ times the area of the square. If R' is the union of the squares inside R, then the area of $T(R')$ is $|\det A|$ times the area of R'. See Fig. 7. Also, the area of $T(R')$ is close to the area of $T(R)$. An argument involving a limiting process may be given to justify the following generalization of Theorem 10.

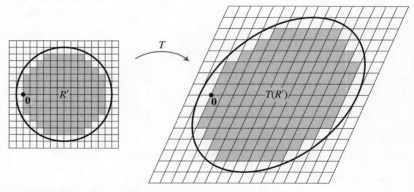

FIGURE 7 Approximating $T(R)$ by a union of parallelograms.

The conclusions of Theorem 10 hold whenever S is a region in $\mathbb{R}^2$ with finite area or a region in $\mathbb{R}^3$ with finite volume.

EXAMPLE 5   Let a and b be positive numbers. Find the area of the region E bounded by the ellipse whose equation is

$$\frac{x_1^2}{a^2} + \frac{x_2^2}{b^2} = 1$$

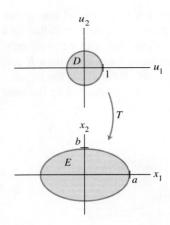

SOLUTION We claim that E is the image of the unit disk D under the linear transformation T determined by the matrix $A = \begin{bmatrix} a & 0 \\ 0 & b \end{bmatrix}$, because if $\mathbf{u} = \begin{bmatrix} u_1 \\ u_2 \end{bmatrix}$, $\mathbf{x} = \begin{bmatrix} x_1 \\ x_2 \end{bmatrix}$, and $\mathbf{x} = A\mathbf{u}$, then

$$u_1 = \frac{x_1}{a} \quad \text{and} \quad u_2 = \frac{x_2}{b}$$

It follows that $\mathbf{u}$ is in the unit disk, with $u_1^2 + u_2^2 \le 1$, if and only if $\mathbf{x}$ is in E, with $(x_1/a)^2 + (x_2/b)^2 \le 1$. By the generalization of Theorem 10,

$$\begin{aligned} \{\text{area of ellipse}\} &= \{\text{area of } T(D)\} \\ &= |\det A| \cdot \{\text{area of } D\} \\ &= ab \cdot \pi(1)^2 = \pi ab \end{aligned} \qquad \blacksquare$$

PRACTICE PROBLEM

Let S be the parallelogram determined by the vectors $\mathbf{b}_1 = \begin{bmatrix} 1 \\ 3 \end{bmatrix}$ and $\mathbf{b}_2 = \begin{bmatrix} 5 \\ 1 \end{bmatrix}$, and let $A = \begin{bmatrix} 1 & -.1 \\ 0 & 2 \end{bmatrix}$. Compute the area of the image of S under the mapping $\mathbf{x} \mapsto A\mathbf{x}$.

3.3 EXERCISES

Use Cramer's rule to compute the solutions of the systems in Exercises 1–6.

1. $5x_1 + 7x_2 = 3$
   $2x_1 + 4x_2 = 1$

2. $4x_1 + x_2 = 6$
   $5x_1 + 2x_2 = 7$

3. $3x_1 - 2x_2 = 7$
   $-5x_1 + 6x_2 = -5$

4. $-5x_1 + 3x_2 = 9$
   $3x_1 - x_2 = -5$

5. $2x_1 + x_2 \quad = 7$
   $-3x_1 \quad + x_3 = -8$
   $x_2 + 2x_3 = -3$

6. $2x_1 + x_2 + x_3 = 4$
   $-x_1 + \quad 2x_3 = 2$
   $3x_1 + x_2 + 3x_3 = -2$

In Exercises 7–10, determine the values of the parameter s for which the system has a unique solution, and describe the solution.

7. $6sx_1 + 4x_2 = 5$
   $9x_1 + 2sx_2 = -2$

8. $3sx_1 - 5x_2 = 3$
   $9x_1 + 5sx_2 = 2$

9. $sx_1 - 2sx_2 = -1$
   $3x_1 + 6sx_2 = 4$

10. $2sx_1 + x_2 = 1$
   $3sx_1 + 6sx_2 = 2$

In Exercises 11–16, compute the adjugate of the given matrix, and then use Theorem 8 to give the inverse of the matrix.

11. $\begin{bmatrix} 0 & -2 & -1 \\ 3 & 0 & 0 \\ -1 & 1 & 1 \end{bmatrix}$

12. $\begin{bmatrix} 1 & 1 & 3 \\ 2 & -2 & 1 \\ 0 & 1 & 0 \end{bmatrix}$

13. $\begin{bmatrix} 3 & 5 & 4 \\ 1 & 0 & 1 \\ 2 & 1 & 1 \end{bmatrix}$

14. $\begin{bmatrix} 3 & 6 & 7 \\ 0 & 2 & 1 \\ 2 & 3 & 4 \end{bmatrix}$

15. $\begin{bmatrix} 3 & 0 & 0 \\ -1 & 1 & 0 \\ -2 & 3 & 2 \end{bmatrix}$

16. $\begin{bmatrix} 1 & 2 & 4 \\ 0 & -3 & 1 \\ 0 & 0 & 3 \end{bmatrix}$

17. Show that if A is 2×2, then Theorem 8 gives the same formula for A^{-1} as that given by Theorem 4 in Section 2.2.

18. Suppose that all the entries in A are integers and $\det A = 1$. Explain why all the entries in A^{-1} are integers.

In Exercises 19–22, find the area of the parallelogram whose vertices are listed.

19. $(0,0), (5,2), (6,4), (11,6)$

20. $(0,0), (-1,3), (4,-5), (3,-2)$

21. $(-1,0), (0,5), (1,-4), (2,1)$

22. $(0,-2), (6,-1), (-3,1), (3,2)$

23. Find the volume of the parallelepiped with one vertex at the origin and adjacent vertices at $(1,0,-2)$, $(1,2,4)$, and $(7,1,0)$.

24. Find the volume of the parallelepiped with one vertex at the origin and adjacent vertices at $(1,4,0)$, $(-2,-5,2)$, and $(-1,2,-1)$.

25. Use the concept of volume to explain why the determinant of a 3×3 matrix A is zero if and only if A is not invertible. Do not appeal to Theorem 4 in Section 3.2. [*Hint:* Think about the columns of A.]

26. Let $T : \mathbb{R}^m \to \mathbb{R}^n$ be a linear transformation, and let $\mathbf{p}$ be a vector and S a set in $\mathbb{R}^m$. Show that the image of $\mathbf{p} + S$ under T is the translated set $T(\mathbf{p}) + T(S)$ in $\mathbb{R}^n$.

27. Let S be the parallelogram determined by the vectors $\mathbf{b}_1 = \begin{bmatrix} -2 \\ 3 \end{bmatrix}$ and $\mathbf{b}_2 = \begin{bmatrix} -2 \\ 5 \end{bmatrix}$, and let $A = \begin{bmatrix} 6 & -2 \\ -3 & 2 \end{bmatrix}$. Compute the area of the image of S under the mapping $\mathbf{x} \mapsto A\mathbf{x}$.

28. Repeat Exercise 27 with $\mathbf{b}_1 = \begin{bmatrix} 4 \\ -7 \end{bmatrix}$, $\mathbf{b}_2 = \begin{bmatrix} 0 \\ 1 \end{bmatrix}$, and $A = \begin{bmatrix} 7 & 2 \\ 1 & 1 \end{bmatrix}$.

29. Find a formula for the area of the triangle whose vertices are $\mathbf{0}$, $\mathbf{v}_1$, and $\mathbf{v}_2$ in $\mathbb{R}^2$.

30. Let R be the triangle with vertices at (x_1, y_1), (x_2, y_2), and (x_3, y_3). Show that

$$\{\text{area of triangle}\} = \frac{1}{2} \det \begin{bmatrix} x_1 & y_1 & 1 \\ x_2 & y_2 & 1 \\ x_3 & y_3 & 1 \end{bmatrix}$$

[*Hint:* Translate R to the origin by subtracting one of the vertices, and use Exercise 29.]

31. Let $T : \mathbb{R}^3 \to \mathbb{R}^3$ be the linear transformation determined by the matrix $A = \begin{bmatrix} a & 0 & 0 \\ 0 & b & 0 \\ 0 & 0 & c \end{bmatrix}$, where a, b, and c are positive numbers. Let S be the unit ball, whose bounding surface has the equation $x_1^2 + x_2^2 + x_3^2 = 1$.

  a. Show that $T(S)$ is bounded by the ellipsoid with the equation $\dfrac{x_1^2}{a^2} + \dfrac{x_2^2}{b^2} + \dfrac{x_3^2}{c^2} = 1$.

  b. Use the fact that the volume of the unit ball is $4\pi/3$ to determine the volume of the region bounded by the ellipsoid in part (a).

32. Let S be the tetrahedron in $\mathbb{R}^3$ with vertices at the vectors $\mathbf{0}$, $\mathbf{e}_1$, $\mathbf{e}_2$, and $\mathbf{e}_3$, and let S' be the tetrahedron with vertices at vectors $\mathbf{0}$, $\mathbf{v}_1$, $\mathbf{v}_2$, and $\mathbf{v}_3$. See the figure.

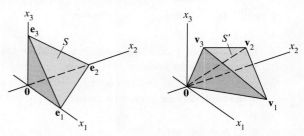

  a. Describe a linear transformation that maps S onto S'.

  b. Find a formula for the volume of the tetrahedron S' using the fact that

$$\{\text{volume of } S\} = (1/3)\{\text{area of base}\} \cdot \{\text{height}\}$$

33. [M] Test the inverse formula of Theorem 8 for a random 4×4 matrix A. Use your matrix program to compute the cofactors of the 3×3 submatrices, construct the adjugate, and set $B = (\text{adj } A)/(\det A)$. Then compute $B - \text{inv}(A)$, where $\text{inv}(A)$ is the inverse of A as computed by the matrix program. Use floating point arithmetic with the maximum possible number of decimal places. Report your results.

34. [M] Test Cramer's rule for a random 4×4 matrix A and a random 4×1 vector $\mathbf{b}$. Compute each entry in the solution of $A\mathbf{x} = \mathbf{b}$, and compare these entries with the entries in $A^{-1}\mathbf{b}$. Write the command (or keystrokes) for your matrix program that uses Cramer's rule to produce the second entry of $\mathbf{x}$.

35. [M] If your version of MATLAB has the `flops` command, use it to count the number of floating point operations to compute A^{-1} for a random 30×30 matrix. Compare this number with the number of flops needed to form $(\text{adj } A)/(\det A)$.

SOLUTION TO PRACTICE PROBLEM

The area of S is $\left| \det \begin{bmatrix} 1 & 5 \\ 3 & 1 \end{bmatrix} \right| = 14$, and $\det A = 2$. By Theorem 10, the area of the image of S under the mapping $\mathbf{x} \mapsto A\mathbf{x}$ is

$$|\det A| \cdot \{\text{area of } S\} = 2 \cdot 14 = 28$$

CHAPTER 3 SUPPLEMENTARY EXERCISES

1. Mark each statement True or False. Justify each answer. Assume that all matrices here are square.

  a. If A is a 2×2 matrix with a zero determinant, then one column of A is a multiple of the other.

  b. If two rows of a 3×3 matrix A are the same, then $\det A = 0$.

  c. If A is a 3×3 matrix, then $\det 5A = 5 \det A$.

  d. If A and B are $n \times n$ matrices, with $\det A = 2$ and $\det B = 3$, then $\det(A + B) = 5$.

  e. If A is $n \times n$ and $\det A = 2$, then $\det A^3 = 6$.

  f. If B is produced by interchanging two rows of A, then $\det B = \det A$.

  g. If B is produced by multiplying row 3 of A by 5, then $\det B = 5 \cdot \det A$.

h. If B is formed by adding to one row of A a linear combination of the other rows, then det $B = $ det A.

i. det $A^T = -$ det A.

j. det$(-A) = -$ det A.

k. det $A^TA \geq 0$.

l. Any system of n linear equations in n variables can be solved by Cramer's rule.

m. If $\mathbf{u}$ and $\mathbf{v}$ are in $\mathbb{R}^2$ and det $[\,\mathbf{u} \quad \mathbf{v}\,] = 10$, then the area of the triangle in the plane with vertices at $\mathbf{0}, \mathbf{u}$, and $\mathbf{v}$ is 10.

n. If $A^3 = 0$, then det $A = 0$.

o. If A is invertible, then det $A^{-1} = $ det A.

p. If A is invertible, then $(\det A)(\det A^{-1}) = 1$.

Use row operations to show that the determinants in Exercises 2–4 are all zero.

2. $\begin{vmatrix} 12 & 13 & 14 \\ 15 & 16 & 17 \\ 18 & 19 & 20 \end{vmatrix}$

3. $\begin{vmatrix} 1 & a & b+c \\ 1 & b & a+c \\ 1 & c & a+b \end{vmatrix}$

4. $\begin{vmatrix} a & b & c \\ a+x & b+x & c+x \\ a+y & b+y & c+y \end{vmatrix}$

Compute the determinants in Exercises 5 and 6.

5. $\begin{vmatrix} 9 & 1 & 9 & 9 & 9 \\ 9 & 0 & 9 & 9 & 2 \\ 4 & 0 & 0 & 5 & 0 \\ 9 & 0 & 3 & 9 & 0 \\ 6 & 0 & 0 & 7 & 0 \end{vmatrix}$

6. $\begin{vmatrix} 4 & 8 & 8 & 8 & 5 \\ 0 & 1 & 0 & 0 & 0 \\ 6 & 8 & 8 & 8 & 7 \\ 0 & 8 & 8 & 3 & 0 \\ 0 & 8 & 2 & 0 & 0 \end{vmatrix}$

7. Show that the equation of the line in $\mathbb{R}^2$ through distinct points (x_1, y_1) and (x_2, y_2) can be written as

$$\det \begin{bmatrix} 1 & x & y \\ 1 & x_1 & y_1 \\ 1 & x_2 & y_2 \end{bmatrix} = 0$$

8. Find a 3×3 determinant equation similar to that in Exercise 7 that describes the equation of the line through (x_1, y_1) with slope m.

Exercises 9 and 10 concern determinants of the following *Vandermonde matrices*.

$$T = \begin{bmatrix} 1 & a & a^2 \\ 1 & b & b^2 \\ 1 & c & c^2 \end{bmatrix}, \quad V(t) = \begin{bmatrix} 1 & t & t^2 & t^3 \\ 1 & x_1 & x_1^2 & x_1^3 \\ 1 & x_2 & x_2^2 & x_2^3 \\ 1 & x_3 & x_3^2 & x_3^3 \end{bmatrix}$$

9. Use row operations to show that

$$\det T = (b-a)(c-a)(c-b)$$

10. Let $f(t) = \det V$, with x_1, x_2, x_3 all distinct. Explain why $f(t)$ is a cubic polynomial, show that the coefficient of t^3 is nonzero, and find three points on the graph of f.

11. Determine the area of the parallelogram determined by the points $(1, 4)$, $(-1, 5)$, $(3, 9)$, and $(5, 8)$. How can you tell that the quadrilateral determined by the points is actually a parallelogram?

12. Use the concept of area of a parallelogram to write a statement about a 2×2 matrix A that is true if and only if A is invertible.

13. Show that if A is invertible, then adj A is invertible, and

$$(\text{adj } A)^{-1} = \frac{1}{\det A} A$$

[*Hint:* Given matrices B and C, what calculation(s) would show that C is the inverse of B?]

14. Let $A, B, C, D,$ and I be $n \times n$ matrices. Use the definition or properties of a determinant to justify the following formulas. Part (c) is useful in applications of eigenvalues (Chapter 5).

a. $\det \begin{bmatrix} A & 0 \\ 0 & I \end{bmatrix} = \det A$ b. $\det \begin{bmatrix} I & 0 \\ C & D \end{bmatrix} = \det D$

c. $\det \begin{bmatrix} A & 0 \\ C & D \end{bmatrix} = (\det A)(\det D) = \det \begin{bmatrix} A & B \\ 0 & D \end{bmatrix}$

15. Let $A, B, C,$ and D be $n \times n$ matrices with A invertible.

a. Find matrices X and Y to produce the block LU factorization

$$\begin{bmatrix} A & B \\ C & D \end{bmatrix} = \begin{bmatrix} I & 0 \\ X & I \end{bmatrix} \begin{bmatrix} A & B \\ 0 & Y \end{bmatrix}$$

and then show that

$$\det \begin{bmatrix} A & B \\ C & D \end{bmatrix} = (\det A) \cdot \det(D - CA^{-1}B)$$

b. Show that if $AC = CA$, then

$$\det \begin{bmatrix} A & B \\ C & D \end{bmatrix} = \det(AD - CB)$$

16. Let J be the $n \times n$ matrix of all 1's, and consider $A = (a-b)I + bJ$; that is,

$$A = \begin{bmatrix} a & b & b & \cdots & b \\ b & a & b & \cdots & b \\ b & b & a & \cdots & b \\ \vdots & \vdots & \vdots & \ddots & \vdots \\ b & b & b & \cdots & a \end{bmatrix}$$

Confirm that $\det A = (a-b)^{n-1}[a + (n-1)b]$ as follows:

a. Subtract row 2 from row 1, row 3 from row 2, and so on, and explain why this does not change the determinant of the matrix.

b. With the resulting matrix from part (a), add column 1 to column 2, then add this new column 2 to column 3, and so on, and explain why this does not change the determinant.

c. Find the determinant of the resulting matrix from (b).

17. Let A be the original matrix given in Exercise 16, and let

$$B = \begin{bmatrix} a-b & b & b & \cdots & b \\ 0 & a & b & \cdots & b \\ 0 & b & a & \cdots & b \\ \vdots & \vdots & \vdots & \ddots & \vdots \\ 0 & b & b & \cdots & a \end{bmatrix},$$

$$C = \begin{bmatrix} b & b & b & \cdots & b \\ b & a & b & \cdots & b \\ b & b & a & \cdots & b \\ \vdots & \vdots & \vdots & \ddots & \vdots \\ b & b & b & \cdots & a \end{bmatrix}$$

Notice that A, B, and C are nearly the same except that the first column of A equals the sum of the first columns of B and C. A *linearity property* of the determinant function, discussed in Section 3.2, says that $\det A = \det B + \det C$. Use this fact to prove the formula in Exercise 16 by induction on the size of matrix A.

18. [M] Apply the result of Exercise 16 to find the determinants of the following matrices, and confirm your answers using a matrix program.

$$\begin{bmatrix} 3 & 8 & 8 & 8 \\ 8 & 3 & 8 & 8 \\ 8 & 8 & 3 & 8 \\ 8 & 8 & 8 & 3 \end{bmatrix} \qquad \begin{bmatrix} 8 & 3 & 3 & 3 & 3 \\ 3 & 8 & 3 & 3 & 3 \\ 3 & 3 & 8 & 3 & 3 \\ 3 & 3 & 3 & 8 & 3 \\ 3 & 3 & 3 & 3 & 8 \end{bmatrix}$$

19. [M] Use a matrix program to compute the determinants of the following matrices.

$$\begin{bmatrix} 1 & 1 & 1 \\ 1 & 2 & 2 \\ 1 & 2 & 3 \end{bmatrix} \qquad \begin{bmatrix} 1 & 1 & 1 & 1 \\ 1 & 2 & 2 & 2 \\ 1 & 2 & 3 & 3 \\ 1 & 2 & 3 & 4 \end{bmatrix}$$

$$\begin{bmatrix} 1 & 1 & 1 & 1 & 1 \\ 1 & 2 & 2 & 2 & 2 \\ 1 & 2 & 3 & 3 & 3 \\ 1 & 2 & 3 & 4 & 4 \\ 1 & 2 & 3 & 4 & 5 \end{bmatrix}$$

Use the results to guess the determinant of the matrix below, and confirm your guess by using row operations to evaluate that determinant.

$$\begin{bmatrix} 1 & 1 & 1 & \cdots & 1 \\ 1 & 2 & 2 & \cdots & 2 \\ 1 & 2 & 3 & \cdots & 3 \\ \vdots & \vdots & \vdots & \ddots & \vdots \\ 1 & 2 & 3 & \cdots & n \end{bmatrix}$$

20. [M] Use the method of Exercise 19 to guess the determinant of

$$\begin{bmatrix} 1 & 1 & 1 & \cdots & 1 \\ 1 & 3 & 3 & \cdots & 3 \\ 1 & 3 & 6 & \cdots & 6 \\ \vdots & \vdots & \vdots & \ddots & \vdots \\ 1 & 3 & 6 & \cdots & 3(n-1) \end{bmatrix}$$

Justify your conjecture. [*Hint:* Use Exercise 14(c) and the result of Exercise 19.]

4

Vector Spaces

Space Flight and Control Systems

Twelve stories high and weighing 75 tons, *Columbia* rose majestically off the launching pad on a cool Palm Sunday morning in April 1981. A product of ten years' intensive research and development, the first U.S. space shuttle was a triumph of control systems engineering design, involving many branches of engineering—aeronautical, chemical, electrical, hydraulic, and mechanical.

The space shuttle's control systems are absolutely critical for flight. Because the shuttle is an unstable airframe, it requires constant computer monitoring during atmospheric flight. The flight control system sends a stream of commands to aerodynamic control surfaces and 44 small thruster jets. Figure 1 shows a typical closed-loop feedback system that controls the pitch of the shuttle

during flight. (The pitch is the elevation angle of the nose cone.) The junction symbols ($\otimes$) show where signals from various sensors are added to the computer signals flowing along the top of the figure.

Mathematically, the input and output signals to an engineering system are functions. It is important in applications that these functions can be added, as in Fig. 1, and multiplied by scalars. These two operations on functions have algebraic properties that are completely analogous to the operations of adding vectors in $\mathbb{R}^n$ and multiplying a vector by a scalar, as we shall see in Sections 4.1 and 4.8. For this reason, the set of all possible inputs (functions) is called a *vector space*. The mathematical foundation for systems engineering rests

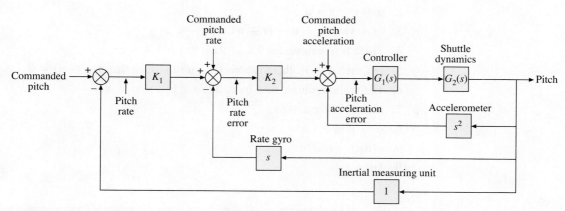

FIGURE 1 Pitch control system for the space shuttle. (*Source:* Adapted from *Space Shuttle GN&C Operations Manual*, Rockwell International, ©1988.)

on vector spaces of functions, and Chapter 4 extends the theory of vectors in $\mathbb{R}^n$ to include such functions. Later on, you will see how other vector spaces arise in engineering, physics, and statistics.

WEB

The mathematical seeds planted in Chapters 1 and 2 germinate and begin to blossom in this chapter. The beauty and power of linear algebra will be seen more clearly when you view $\mathbb{R}^n$ as only one of a variety of vector spaces that arise naturally in applied problems. Actually, a study of vector spaces is not much different from a study of $\mathbb{R}^n$ itself, because you can use your geometric experience with $\mathbb{R}^2$ and $\mathbb{R}^3$ to visualize many general concepts.

Beginning with basic definitions in Section 4.1, the general vector space framework develops gradually throughout the chapter. A goal of Sections 4.3–4.5 is to demonstrate how closely other vector spaces resemble $\mathbb{R}^n$. Section 4.6 on rank is one of the high points of the chapter, using vector space terminology to tie together important facts about rectangular matrices. Section 4.8 applies the theory of the chapter to discrete signals and difference equations used in digital control systems such as in the space shuttle. Markov chains, in Section 4.9, provide a change of pace from the more theoretical sections of the chapter and make good examples for concepts to be introduced in Chapter 5.

4.1 | VECTOR SPACES AND SUBSPACES

Much of the theory in Chapters 1 and 2 rested on certain simple and obvious algebraic properties of $\mathbb{R}^n$, listed in Section 1.3. In fact, many other mathematical systems have the same properties. The specific properties of interest are listed in the following definition.

DEFINITION

A **vector space** is a nonempty set V of objects, called *vectors*, on which are defined two operations, called *addition* and *multiplication by scalars* (real numbers), subject to the ten axioms (or rules) listed below.[1] The axioms must hold for all vectors $\mathbf{u}$, $\mathbf{v}$, and $\mathbf{w}$ in V and for all scalars c and d.

1. The sum of $\mathbf{u}$ and $\mathbf{v}$, denoted by $\mathbf{u} + \mathbf{v}$, is in V.
2. $\mathbf{u} + \mathbf{v} = \mathbf{v} + \mathbf{u}$.
3. $(\mathbf{u} + \mathbf{v}) + \mathbf{w} = \mathbf{u} + (\mathbf{v} + \mathbf{w})$.
4. There is a **zero** vector $\mathbf{0}$ in V such that $\mathbf{u} + \mathbf{0} = \mathbf{u}$.
5. For each $\mathbf{u}$ in V, there is a vector $-\mathbf{u}$ in V such that $\mathbf{u} + (-\mathbf{u}) = \mathbf{0}$.
6. The scalar multiple of $\mathbf{u}$ by c, denoted by $c\mathbf{u}$, is in V.
7. $c(\mathbf{u} + \mathbf{v}) = c\mathbf{u} + c\mathbf{v}$.
8. $(c + d)\mathbf{u} = c\mathbf{u} + d\mathbf{u}$.
9. $c(d\mathbf{u}) = (cd)\mathbf{u}$.
10. $1\mathbf{u} = \mathbf{u}$.

---

[1] Technically, V is a *real vector space*. All of the theory in this chapter also holds for a *complex vector space* in which the scalars are complex numbers. We will look at this briefly in Chapter 5. Until then, all scalars are assumed to be real.

Using only these axioms, one can show that the zero vector in Axiom 4 is unique, and the vector $-\mathbf{u}$, called the **negative** of $\mathbf{u}$, in Axiom 5 is unique for each $\mathbf{u}$ in V. See Exercises 25 and 26. Proofs of the following simple facts are also outlined in the exercises:

For each $\mathbf{u}$ in V and scalar c,

$$0\mathbf{u} = \mathbf{0} \tag{1}$$
$$c\mathbf{0} = \mathbf{0} \tag{2}$$
$$-\mathbf{u} = (-1)\mathbf{u} \tag{3}$$

EXAMPLE 1 The spaces $\mathbb{R}^n$, where $n \geq 1$, are the premier examples of vector spaces. The geometric intuition developed for $\mathbb{R}^3$ will help you understand and visualize many concepts throughout the chapter. ∎

EXAMPLE 2 Let V be the set of all arrows (directed line segments) in three-dimensional space, with two arrows regarded as equal if they have the same length and point in the same direction. Define addition by the parallelogram rule (from Section 1.3), and for each $\mathbf{v}$ in V, define $c\mathbf{v}$ to be the arrow whose length is $|c|$ times the length of $\mathbf{v}$, pointing in the same direction as $\mathbf{v}$ if $c \geq 0$ and otherwise pointing in the opposite direction. (See Fig. 1.) Show that V is a vector space. This space is a common model in physical problems for various forces.

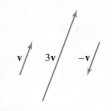

FIGURE 1

SOLUTION The definition of V is geometric, using concepts of length and direction. No xyz-coordinate system is involved. An arrow of zero length is a single point and represents the zero vector. The negative of $\mathbf{v}$ is $(-1)\mathbf{v}$. So Axioms 1, 4, 5, 6, and 10 are evident. The rest are verified by geometry. For instance, see Figs. 2 and 3. ∎

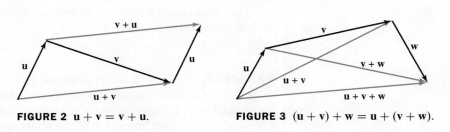

FIGURE 2 $\mathbf{u} + \mathbf{v} = \mathbf{v} + \mathbf{u}$.

FIGURE 3 $(\mathbf{u} + \mathbf{v}) + \mathbf{w} = \mathbf{u} + (\mathbf{v} + \mathbf{w})$.

EXAMPLE 3 Let $\mathbb{S}$ be the space of all doubly infinite sequences of numbers (usually written in a row rather than a column):

$$\{y_k\} = (\ldots, y_{-2}, y_{-1}, y_0, y_1, y_2, \ldots)$$

If $\{z_k\}$ is another element of $\mathbb{S}$, then the sum $\{y_k\} + \{z_k\}$ is the sequence $\{y_k + z_k\}$ formed by adding corresponding terms of $\{y_k\}$ and $\{z_k\}$. The scalar multiple $c\{y_k\}$ is the sequence $\{cy_k\}$. The vector space axioms are verified in the same way as for $\mathbb{R}^n$.

Elements of $\mathbb{S}$ arise in engineering, for example, whenever a signal is measured (or sampled) at discrete times. A signal might be electrical, mechanical, optical, and so on. The major control systems for the space shuttle, mentioned in the chapter introduction, use discrete (or digital) signals. For convenience, we will call $\mathbb{S}$ the space of (discrete-time) **signals**. A signal may be visualized by a graph as in Fig. 4. ∎

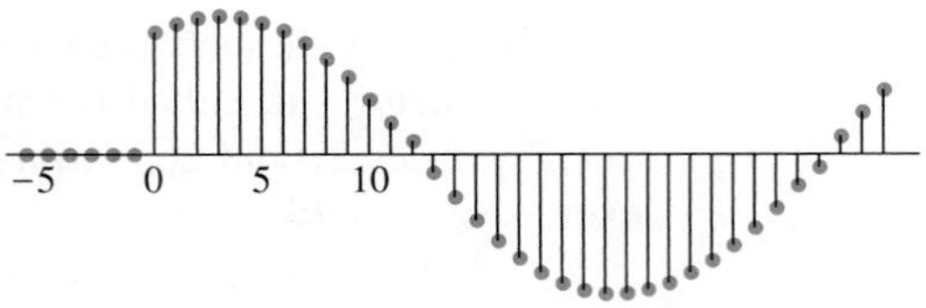

FIGURE 4 A discrete-time signal.

EXAMPLE 4   For $n \geq 0$, the set $\mathbb{P}_n$ of polynomials of degree at most n consists of all polynomials of the form

$$\mathbf{p}(t) = a_0 + a_1 t + a_2 t^2 + \cdots + a_n t^n \tag{4}$$

where the coefficients $a_0, \ldots, a_n$ and the variable t are real numbers. The *degree* of $\mathbf{p}$ is the highest power of t in (4) whose coefficient is not zero. If $\mathbf{p}(t) = a_0 \neq 0$, the degree of $\mathbf{p}$ is zero. If all the coefficients are zero, $\mathbf{p}$ is called the *zero polynomial*. The zero polynomial is included in $\mathbb{P}_n$ even though its degree, for technical reasons, is not defined.

If $\mathbf{p}$ is given by (4) and if $\mathbf{q}(t) = b_0 + b_1 t + \cdots + b_n t^n$, then the sum $\mathbf{p} + \mathbf{q}$ is defined by

$$(\mathbf{p} + \mathbf{q})(t) = \mathbf{p}(t) + \mathbf{q}(t)$$
$$= (a_0 + b_0) + (a_1 + b_1)t + \cdots + (a_n + b_n)t^n$$

The scalar multiple $c\mathbf{p}$ is the polynomial defined by

$$(c\mathbf{p})(t) = c\mathbf{p}(t) = ca_0 + (ca_1)t + \cdots + (ca_n)t^n$$

These definitions satisfy Axioms 1 and 6 because $\mathbf{p} + \mathbf{q}$ and $c\mathbf{p}$ are polynomials of degree less than or equal to n. Axioms 2, 3, and 7–10 follow from properties of the real numbers. Clearly, the zero polynomial acts as the zero vector in Axiom 4. Finally, $(-1)\mathbf{p}$ acts as the negative of $\mathbf{p}$, so Axiom 5 is satisfied. Thus $\mathbb{P}_n$ is a vector space.

The vector spaces $\mathbb{P}_n$ for various n are used, for instance, in statistical trend analysis of data, discussed in Section 6.8.   ∎

EXAMPLE 5   Let V be the set of all real-valued functions defined on a set $\mathbb{D}$. (Typically, $\mathbb{D}$ is the set of real numbers or some interval on the real line.) Functions are added in the usual way: $\mathbf{f} + \mathbf{g}$ is the function whose value at t in the domain $\mathbb{D}$ is $\mathbf{f}(t) + \mathbf{g}(t)$. Likewise, for a scalar c and an $\mathbf{f}$ in V, the scalar multiple $c\mathbf{f}$ is the function whose value at t is $c\mathbf{f}(t)$. For instance, if $\mathbb{D} = \mathbb{R}$, $\mathbf{f}(t) = 1 + \sin 2t$, and $\mathbf{g}(t) = 2 + .5t$, then

$$(\mathbf{f} + \mathbf{g})(t) = 3 + \sin 2t + .5t \quad \text{and} \quad (2\mathbf{g})(t) = 4 + t$$

Two functions in V are equal if and only if their values are equal for every t in $\mathbb{D}$. Hence the zero vector in V is the function that is identically zero, $\mathbf{f}(t) = 0$ for all t, and the negative of $\mathbf{f}$ is $(-1)\mathbf{f}$. Axioms 1 and 6 are obviously true, and the other axioms follow from properties of the real numbers, so V is a vector space.   ∎

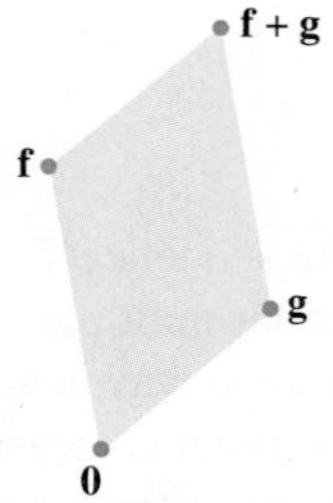

FIGURE 5

The sum of two vectors (functions).

It is important to think of each function in the vector space V of Example 5 as a single object, as just one "point" or vector in the vector space. The sum of two vectors $\mathbf{f}$ and $\mathbf{g}$ (functions in V, or elements of *any* vector space) can be visualized as in Fig. 5, because this can help you carry over to a general vector space the geometric intuition you have developed while working with the vector space $\mathbb{R}^n$. See the *Study Guide* for help as you learn to adopt this more general point of view.

Subspaces

In many problems, a vector space consists of an appropriate subset of vectors from some larger vector space. In this case, only three of the ten vector space axioms need to be checked; the rest are automatically satisfied.

DEFINITION

> A **subspace** of a vector space V is a subset H of V that has three properties:
>
> a. The zero vector of V is in H.[2]
> b. H is closed under vector addition. That is, for each $\mathbf{u}$ and $\mathbf{v}$ in H, the sum $\mathbf{u} + \mathbf{v}$ is in H.
> c. H is closed under multiplication by scalars. That is, for each $\mathbf{u}$ in H and each scalar c, the vector $c\mathbf{u}$ is in H.

Properties (a), (b), and (c) guarantee that a subspace H of V is itself a *vector space*, under the vector space operations already defined in V. To verify this, note that properties (a), (b), and (c) are Axioms 1, 4, and 6. Axioms 2, 3, and 7–10 are automatically true in H because they apply to all elements of V, including those in H. Axiom 5 is also true in H, because if $\mathbf{u}$ is in H, then $(-1)\mathbf{u}$ is in H by property (c), and we know from equation (3) on page 191 that $(-1)\mathbf{u}$ is the vector $-\mathbf{u}$ in Axiom 5.

So every subspace is a vector space. Conversely, every vector space is a subspace (of itself and possibly of other larger spaces). The term *subspace* is used when at least two vector spaces are in mind, with one inside the other, and the phrase *subspace of V* identifies V as the larger space. (See Fig. 6.)

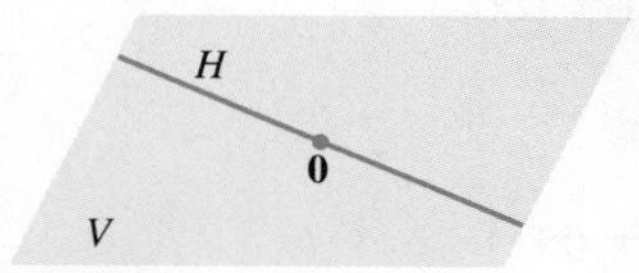

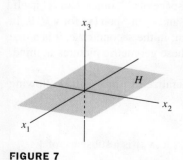

FIGURE 6

A subspace of V.

EXAMPLE 6 The set consisting of only the zero vector in a vector space V is a subspace of V, called the **zero subspace** and written as $\{\mathbf{0}\}$. ∎

EXAMPLE 7 Let $\mathbb{P}$ be the set of all polynomials with real coefficients, with operations in $\mathbb{P}$ defined as for functions. Then $\mathbb{P}$ is a subspace of the space of all real-valued functions defined on $\mathbb{R}$. Also, for each $n \geq 0$, $\mathbb{P}_n$ is a subspace of $\mathbb{P}$, because $\mathbb{P}_n$ is a subset of $\mathbb{P}$ that contains the zero polynomial, the sum of two polynomials in $\mathbb{P}_n$ is also in $\mathbb{P}_n$, and a scalar multiple of a polynomial in $\mathbb{P}_n$ is also in $\mathbb{P}_n$. ∎

EXAMPLE 8 The vector space $\mathbb{R}^2$ is *not* a subspace of $\mathbb{R}^3$ because $\mathbb{R}^2$ is not even a subset of $\mathbb{R}^3$. (The vectors in $\mathbb{R}^3$ all have three entries, whereas the vectors in $\mathbb{R}^2$ have only two.) The set

$$H = \left\{ \begin{bmatrix} s \\ t \\ 0 \end{bmatrix} : s \text{ and } t \text{ are real} \right\}$$

FIGURE 7

The $x_1 x_2$-plane as a subspace of $\mathbb{R}^3$.

is a subset of $\mathbb{R}^3$ that "looks" and "acts" like $\mathbb{R}^2$, although it is logically distinct from $\mathbb{R}^2$. See Fig. 7. Show that H is a subspace of $\mathbb{R}^3$.

SOLUTION The zero vector is in H, and H is closed under vector addition and scalar multiplication because these operations on vectors in H always produce vectors whose third entries are zero (and so belong to H). Thus H is a subspace of $\mathbb{R}^3$. ∎

---

[2] Some texts replace property (a) in this definition by the assumption that H is nonempty. Then (a) could be deduced from (c) and the fact that $0\mathbf{u} = \mathbf{0}$. But the best way to test for a subspace is to look first for the zero vector. If $\mathbf{0}$ is in H, then properties (b) and (c) must be checked. If $\mathbf{0}$ is *not* in H, then H cannot be a subspace and the other properties need not be checked.

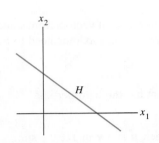

FIGURE 8

A line that is not a vector space.

EXAMPLE 9   A plane in $\mathbb{R}^3$ *not* through the origin is not a subspace of $\mathbb{R}^3$, because the plane does not contain the zero vector of $\mathbb{R}^3$. Similarly, a line in $\mathbb{R}^2$ *not* through the origin, such as in Fig. 8, is *not* a subspace of $\mathbb{R}^2$. ∎

A Subspace Spanned by a Set

The next example illustrates one of the most common ways of describing a subspace. As in Chapter 1, the term **linear combination** refers to any sum of scalar multiples of vectors, and Span $\{\mathbf{v}_1, \ldots, \mathbf{v}_p\}$ denotes the set of all vectors that can be written as linear combinations of $\mathbf{v}_1, \ldots, \mathbf{v}_p$.

EXAMPLE 10   Given $\mathbf{v}_1$ and $\mathbf{v}_2$ in a vector space V, let $H = \text{Span}\,\{\mathbf{v}_1, \mathbf{v}_2\}$. Show that H is a subspace of V.

SOLUTION The zero vector is in H, since $\mathbf{0} = 0\mathbf{v}_1 + 0\mathbf{v}_2$. To show that H is closed under vector addition, take two arbitrary vectors in H, say,

$$\mathbf{u} = s_1\mathbf{v}_1 + s_2\mathbf{v}_2 \quad \text{and} \quad \mathbf{w} = t_1\mathbf{v}_1 + t_2\mathbf{v}_2$$

By Axioms 2, 3, and 8 for the vector space V,

$$\mathbf{u} + \mathbf{w} = (s_1\mathbf{v}_1 + s_2\mathbf{v}_2) + (t_1\mathbf{v}_1 + t_2\mathbf{v}_2)$$
$$= (s_1 + t_1)\mathbf{v}_1 + (s_2 + t_2)\mathbf{v}_2$$

So $\mathbf{u} + \mathbf{w}$ is in H. Furthermore, if c is any scalar, then by Axioms 7 and 9,

$$c\mathbf{u} = c(s_1\mathbf{v}_1 + s_2\mathbf{v}_2) = (cs_1)\mathbf{v}_1 + (cs_2)\mathbf{v}_2$$

which shows that $c\mathbf{u}$ is in H and H is closed under scalar multiplication. Thus H is a subspace of V. ∎

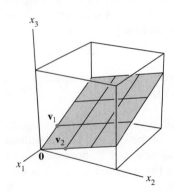

FIGURE 9

An example of a subspace.

In Section 4.5, you will see that every nonzero subspace of $\mathbb{R}^3$, other than $\mathbb{R}^3$ itself, is either Span $\{\mathbf{v}_1, \mathbf{v}_2\}$ for some linearly independent $\mathbf{v}_1$ and $\mathbf{v}_2$ or Span $\{\mathbf{v}\}$ for $\mathbf{v} \neq \mathbf{0}$. In the first case, the subspace is a plane through the origin; in the second case, it is a line through the origin. (See Fig. 9.) It is helpful to keep these geometric pictures in mind, even for an abstract vector space.

The argument in Example 10 can easily be generalized to prove the following theorem.

THEOREM 1   If $\mathbf{v}_1, \ldots, \mathbf{v}_p$ are in a vector space V, then Span $\{\mathbf{v}_1, \ldots, \mathbf{v}_p\}$ is a subspace of V.

We call Span $\{\mathbf{v}_1, \ldots, \mathbf{v}_p\}$ **the subspace spanned** (or **generated**) by $\{\mathbf{v}_1, \ldots, \mathbf{v}_p\}$. Given any subspace H of V, a **spanning** (or **generating**) **set** for H is a set $\{\mathbf{v}_1, \ldots, \mathbf{v}_p\}$ in H such that $H = \text{Span}\,\{\mathbf{v}_1, \ldots, \mathbf{v}_p\}$.

The next example shows how to use Theorem 1.

EXAMPLE 11   Let H be the set of all vectors of the form $(a - 3b, b - a, a, b)$, where a and b are arbitrary scalars. That is, let $H = \{(a - 3b, b - a, a, b) : a \text{ and } b \text{ in } \mathbb{R}\}$. Show that H is a subspace of $\mathbb{R}^4$.

SOLUTION Write the vectors in H as column vectors. Then an arbitrary vector in H has the form

$$\begin{bmatrix} a - 3b \\ b - a \\ a \\ b \end{bmatrix} = a \underset{\underset{\mathbf{v}_1}{\uparrow}}{\begin{bmatrix} 1 \\ -1 \\ 1 \\ 0 \end{bmatrix}} + b \underset{\underset{\mathbf{v}_2}{\uparrow}}{\begin{bmatrix} -3 \\ 1 \\ 0 \\ 1 \end{bmatrix}}$$

This calculation shows that $H = \text{Span}\{\mathbf{v}_1, \mathbf{v}_2\}$, where $\mathbf{v}_1$ and $\mathbf{v}_2$ are the vectors indicated above. Thus H is a subspace of $\mathbb{R}^4$ by Theorem 1. ∎

Example 11 illustrates a useful technique of expressing a subspace H as the set of linear combinations of some small collection of vectors. If $H = \text{Span}\{\mathbf{v}_1, \ldots, \mathbf{v}_p\}$, we can think of the vectors $\mathbf{v}_1, \ldots, \mathbf{v}_p$ in the spanning set as "handles" that allow us to hold on to the subspace H. Calculations with the infinitely many vectors in H are often reduced to operations with the finite number of vectors in the spanning set.

EXAMPLE 12 For what value(s) of h will $\mathbf{y}$ be in the subspace of $\mathbb{R}^3$ spanned by $\mathbf{v}_1, \mathbf{v}_2, \mathbf{v}_3$, if

$$\mathbf{v}_1 = \begin{bmatrix} 1 \\ -1 \\ -2 \end{bmatrix}, \qquad \mathbf{v}_2 = \begin{bmatrix} 5 \\ -4 \\ -7 \end{bmatrix}, \qquad \mathbf{v}_3 = \begin{bmatrix} -3 \\ 1 \\ 0 \end{bmatrix}, \quad \text{and} \quad \mathbf{y} = \begin{bmatrix} -4 \\ 3 \\ h \end{bmatrix}$$

SOLUTION This question is Practice Problem 2 in Section 1.3, written here with the term *subspace* rather than Span $\{\mathbf{v}_1, \mathbf{v}_2, \mathbf{v}_3\}$. The solution there shows that $\mathbf{y}$ is in Span $\{\mathbf{v}_1, \mathbf{v}_2, \mathbf{v}_3\}$ if and only if $h = 5$. That solution is worth reviewing now, along with Exercises 11–16 and 19–21 in Section 1.3. ∎

Although many vector spaces in this chapter will be subspaces of $\mathbb{R}^n$, it is important to keep in mind that the abstract theory applies to other vector spaces as well. Vector spaces of functions arise in many applications, and they will receive more attention later.

PRACTICE PROBLEMS

1. Show that the set H of all points in $\mathbb{R}^2$ of the form $(3s, 2 + 5s)$ is not a vector space, by showing that it is not closed under scalar multiplication. (Find a specific vector $\mathbf{u}$ in H and a scalar c such that $c\mathbf{u}$ is not in H.)

2. Let $W = \text{Span}\{\mathbf{v}_1, \ldots, \mathbf{v}_p\}$, where $\mathbf{v}_1, \ldots, \mathbf{v}_p$ are in a vector space V. Show that $\mathbf{v}_k$ is in W for $1 \le k \le p$. [*Hint:* First write an equation that shows that $\mathbf{v}_1$ is in W. Then adjust your notation for the general case.]

WEB

4.1 EXERCISES

1. Let V be the first quadrant in the xy-plane; that is, let

$$V = \left\{ \begin{bmatrix} x \\ y \end{bmatrix} : x \ge 0, y \ge 0 \right\}$$

a. If $\mathbf{u}$ and $\mathbf{v}$ are in V, is $\mathbf{u} + \mathbf{v}$ in V? Why?

b. Find a specific vector $\mathbf{u}$ in V and a specific scalar c such

that $c\mathbf{u}$ is *not* in V. (This is enough to show that V is *not* a vector space.)

2. Let W be the union of the first and third quadrants in the xy-plane. That is, let $W = \left\{ \begin{bmatrix} x \\ y \end{bmatrix} : xy \ge 0 \right\}$.

a. If $\mathbf{u}$ is in W and c is any scalar, is $c\mathbf{u}$ in W? Why?

b. Find specific vectors **u** and **v** in W such that **u** + **v** is not in W. This is enough to show that W is *not* a vector space.

3. Let H be the set of points inside and on the unit circle in the xy-plane. That is, let $H = \left\{ \begin{bmatrix} x \\ y \end{bmatrix} : x^2 + y^2 \le 1 \right\}$. Find a specific example—two vectors or a vector and a scalar—to show that H is not a subspace of $\mathbb{R}^2$.

4. Construct a geometric figure that illustrates why a line in $\mathbb{R}^2$ *not* through the origin is not closed under vector addition.

In Exercises 5–8, determine if the given set is a subspace of $\mathbb{P}_n$ for an appropriate value of n. Justify your answers.

5. All polynomials of the form $\mathbf{p}(t) = at^2$, where a is in $\mathbb{R}$.

6. All polynomials of the form $\mathbf{p}(t) = a + t^2$, where a is in $\mathbb{R}$.

7. All polynomials of degree at most 3, with integers as coefficients.

8. All polynomials in $\mathbb{P}_n$ such that $\mathbf{p}(0) = 0$.

9. Let H be the set of all vectors of the form $\begin{bmatrix} -2t \\ 5t \\ 3t \end{bmatrix}$. Find a vector **v** in $\mathbb{R}^3$ such that $H = \text{Span}\{\mathbf{v}\}$. Why does this show that H is a subspace of $\mathbb{R}^3$?

10. Let H be the set of all vectors of the form $\begin{bmatrix} 3t \\ 0 \\ -7t \end{bmatrix}$, where t is any real number. Show that H is a subspace of $\mathbb{R}^3$. (Use the method of Exercise 9.)

11. Let W be the set of all vectors of the form $\begin{bmatrix} 2b + 3c \\ -b \\ 2c \end{bmatrix}$, where b and c are arbitrary. Find vectors **u** and **v** such that $W = \text{Span}\{\mathbf{u}, \mathbf{v}\}$. Why does this show that W is a subspace of $\mathbb{R}^3$?

12. Let W be the set of all vectors of the form $\begin{bmatrix} 2s + 4t \\ 2s \\ 2s - 3t \\ 5t \end{bmatrix}$. Show that W is a subspace of $\mathbb{R}^4$. (Use the method of Exercise 11.)

13. Let $\mathbf{v}_1 = \begin{bmatrix} 1 \\ 0 \\ -1 \end{bmatrix}$, $\mathbf{v}_2 = \begin{bmatrix} 2 \\ 1 \\ 3 \end{bmatrix}$, $\mathbf{v}_3 = \begin{bmatrix} 4 \\ 2 \\ 6 \end{bmatrix}$, and $\mathbf{w} = \begin{bmatrix} 3 \\ 1 \\ 2 \end{bmatrix}$.

a. Is **w** in $\{\mathbf{v}_1, \mathbf{v}_2, \mathbf{v}_3\}$? How many vectors are in $\{\mathbf{v}_1, \mathbf{v}_2, \mathbf{v}_3\}$?

b. How many vectors are in $\text{Span}\{\mathbf{v}_1, \mathbf{v}_2, \mathbf{v}_3\}$?

c. Is **w** in the subspace spanned by $\{\mathbf{v}_1, \mathbf{v}_2, \mathbf{v}_3\}$? Why?

14. Let $\mathbf{v}_1, \mathbf{v}_2, \mathbf{v}_3$ be as in Exercise 13, and let $\mathbf{w} = \begin{bmatrix} 1 \\ 3 \\ 14 \end{bmatrix}$. Is **w** in the subspace spanned by $\{\mathbf{v}_1, \mathbf{v}_2, \mathbf{v}_3\}$? Why?

In Exercises 15–18, let W be the set of all vectors of the form shown, where a, b, and c represent arbitrary real numbers. In each case, either find a set S of vectors that spans W or give an example to show that W is *not* a vector space.

15. $\begin{bmatrix} 2a + 3b \\ -1 \\ 2a - 5b \end{bmatrix}$

16. $\begin{bmatrix} 1 \\ 3a - 5b \\ 3b + 2a \end{bmatrix}$

17. $\begin{bmatrix} 2a - b \\ 3b - c \\ 3c - a \\ 3b \end{bmatrix}$

18. $\begin{bmatrix} 4a + 3b \\ 0 \\ a + 3b + c \\ 3b - 2c \end{bmatrix}$

19. If a mass m is placed at the end of a spring, and if the mass is pulled downward and released, the mass–spring system will begin to oscillate. The displacement y of the mass from its resting position is given by a function of the form

$$y(t) = c_1 \cos \omega t + c_2 \sin \omega t \tag{5}$$

where ω is a constant that depends on the spring and the mass. (See the figure below.) Show that the set of all functions described in (5) (with ω fixed and c_1, c_2 arbitrary) is a vector space.

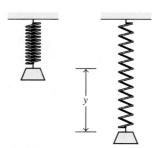

20. The set of all continuous real-valued functions defined on a closed interval $[a, b]$ in $\mathbb{R}$ is denoted by $C[a, b]$. This set is a subspace of the vector space of all real-valued functions defined on $[a, b]$.

a. What facts about continuous functions should be proved in order to demonstrate that $C[a, b]$ is indeed a subspace as claimed? (These facts are usually discussed in a calculus class.)

b. Show that $\{\mathbf{f} \text{ in } C[a, b] : \mathbf{f}(a) = \mathbf{f}(b)\}$ is a subspace of $C[a, b]$.

For fixed positive integers m and n, the set $M_{m \times n}$ of all $m \times n$ matrices is a vector space, under the usual operations of addition of matrices and multiplication by real scalars.

21. Determine if the set H of all matrices of the form $\begin{bmatrix} a & b \\ 0 & d \end{bmatrix}$ is a subspace of $M_{2 \times 2}$.

22. Let F be a fixed 3×2 matrix, and let H be the set of all matrices A in $M_{2 \times 4}$ with the property that $FA = 0$ (the zero matrix in $M_{3 \times 4}$). Determine if H is a subspace of $M_{2 \times 4}$.

In Exercises 23 and 24, mark each statement True or False. Justify each answer.

23. a. If $\mathbf{f}$ is a function in the vector space V of all real-valued functions on $\mathbb{R}$ and if $\mathbf{f}(t) = 0$ for some t, then $\mathbf{f}$ is the zero vector in V.

    b. A vector is an arrow in three-dimensional space.

    c. A subset H of a vector space V is a subspace of V if the zero vector is in H.

    d. A subspace is also a vector space.

    e. Analog signals are used in the major control systems for the space shuttle, mentioned in the introduction to the chapter.

24. a. A vector is any element of a vector space.

    b. If $\mathbf{u}$ is a vector in a vector space V, then $(-1)\mathbf{u}$ is the same as the negative of $\mathbf{u}$.

    c. A vector space is also a subspace.

    d. $\mathbb{R}^2$ is a subspace of $\mathbb{R}^3$.

    e. A subset H of a vector space V is a subspace of V if the following conditions are satisfied: (i) the zero vector of V is in H, (ii) $\mathbf{u}$, $\mathbf{v}$, and $\mathbf{u} + \mathbf{v}$ are in H, and (iii) c is a scalar and $c\mathbf{u}$ is in H.

Exercises 25–29 show how the axioms for a vector space V can be used to prove the elementary properties described after the definition of a vector space. Fill in the blanks with the appropriate axiom numbers. Because of Axiom 2, Axioms 4 and 5 imply, respectively, that $\mathbf{0} + \mathbf{u} = \mathbf{u}$ and $-\mathbf{u} + \mathbf{u} = \mathbf{0}$ for all $\mathbf{u}$.

25. Complete the following proof that the zero vector is unique. Suppose that $\mathbf{w}$ in V has the property that $\mathbf{u} + \mathbf{w} = \mathbf{w} + \mathbf{u} = \mathbf{u}$ for all $\mathbf{u}$ in V. In particular, $\mathbf{0} + \mathbf{w} = \mathbf{0}$. But $\mathbf{0} + \mathbf{w} = \mathbf{w}$, by Axiom _____. Hence $\mathbf{w} = \mathbf{0} + \mathbf{w} = \mathbf{0}$.

26. Complete the following proof that $-\mathbf{u}$ is the *unique vector* in V such that $\mathbf{u} + (-\mathbf{u}) = \mathbf{0}$. Suppose that $\mathbf{w}$ satisfies $\mathbf{u} + \mathbf{w} = \mathbf{0}$. Adding $-\mathbf{u}$ to both sides, we have

$$(-\mathbf{u}) + [\mathbf{u} + \mathbf{w}] = (-\mathbf{u}) + \mathbf{0}$$
$$[(-\mathbf{u}) + \mathbf{u}] + \mathbf{w} = (-\mathbf{u}) + \mathbf{0} \qquad \text{by Axiom _____ (a)}$$
$$\mathbf{0} + \mathbf{w} = (-\mathbf{u}) + \mathbf{0} \qquad \text{by Axiom _____ (b)}$$
$$\mathbf{w} = -\mathbf{u} \qquad \text{by Axiom _____ (c)}$$

27. Fill in the missing axiom numbers in the following proof that $0\mathbf{u} = \mathbf{0}$ for every $\mathbf{u}$ in V.

$$0\mathbf{u} = (0 + 0)\mathbf{u} = 0\mathbf{u} + 0\mathbf{u} \qquad \text{by Axiom _____ (a)}$$

Add the negative of $0\mathbf{u}$ to both sides:

$$0\mathbf{u} + (-0\mathbf{u}) = [0\mathbf{u} + 0\mathbf{u}] + (-0\mathbf{u})$$
$$0\mathbf{u} + (-0\mathbf{u}) = 0\mathbf{u} + [0\mathbf{u} + (-0\mathbf{u})] \qquad \text{by Axiom _____ (b)}$$
$$\mathbf{0} = 0\mathbf{u} + \mathbf{0} \qquad \text{by Axiom _____ (c)}$$
$$\mathbf{0} = 0\mathbf{u} \qquad \text{by Axiom _____ (d)}$$

28. Fill in the missing axiom numbers in the following proof that

$c\mathbf{0} = \mathbf{0}$ for every scalar c.

$$c\mathbf{0} = c(\mathbf{0} + \mathbf{0}) \qquad \text{by Axiom _____ (a)}$$
$$= c\mathbf{0} + c\mathbf{0} \qquad \text{by Axiom _____ (b)}$$

Add the negative of $c\mathbf{0}$ to both sides:

$$c\mathbf{0} + (-c\mathbf{0}) = [c\mathbf{0} + c\mathbf{0}] + (-c\mathbf{0})$$
$$c\mathbf{0} + (-c\mathbf{0}) = c\mathbf{0} + [c\mathbf{0} + (-c\mathbf{0})] \qquad \text{by Axiom _____ (c)}$$
$$\mathbf{0} = c\mathbf{0} + \mathbf{0} \qquad \text{by Axiom _____ (d)}$$
$$\mathbf{0} = c\mathbf{0} \qquad \text{by Axiom _____ (e)}$$

29. Prove that $(-1)\mathbf{u} = -\mathbf{u}$. [*Hint:* Show that $\mathbf{u} + (-1)\mathbf{u} = \mathbf{0}$. Use some axioms and the results of Exercises 27 and 26.]

30. Suppose $c\mathbf{u} = \mathbf{0}$ for some nonzero scalar c. Show that $\mathbf{u} = \mathbf{0}$. Mention the axioms or properties you use.

31. Let $\mathbf{u}$ and $\mathbf{v}$ be vectors in a vector space V, and let H be any subspace of V that contains both $\mathbf{u}$ and $\mathbf{v}$. Explain why H also contains Span $\{\mathbf{u}, \mathbf{v}\}$. This shows that Span $\{\mathbf{u}, \mathbf{v}\}$ is the smallest subspace of V that contains both $\mathbf{u}$ and $\mathbf{v}$.

32. Let H and K be subspaces of a vector space V. The **intersection** of H and K, written as $H \cap K$, is the set of $\mathbf{v}$ in V that belong to both H and K. Show that $H \cap K$ is a subspace of V. (See the figure.) Give an example in $\mathbb{R}^2$ to show that the union of two subspaces is not, in general, a subspace.

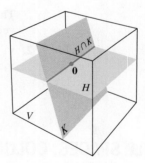

33. Given subspaces H and K of a vector space V, the **sum** of H and K, written as $H + K$, is the set of all vectors in V that can be written as the sum of two vectors, one in H and the other in K; that is,

$$H + K = \{\mathbf{w} : \mathbf{w} = \mathbf{u} + \mathbf{v} \text{ for some } \mathbf{u} \text{ in } H$$
$$\text{and some } \mathbf{v} \text{ in } K\}$$

    a. Show that $H + K$ is a subspace of V.

    b. Show that H is a subspace of $H + K$ and K is a subspace of $H + K$.

34. Suppose $\mathbf{u}_1, \ldots, \mathbf{u}_p$ and $\mathbf{v}_1, \ldots, \mathbf{v}_q$ are vectors in a vector space V, and let

$$H = \text{Span}\{\mathbf{u}_1, \ldots, \mathbf{u}_p\} \text{ and } K = \text{Span}\{\mathbf{v}_1, \ldots, \mathbf{v}_q\}$$

Show that $H + K = \text{Span}\{\mathbf{u}_1, \ldots, \mathbf{u}_p, \mathbf{v}_1, \ldots, \mathbf{v}_q\}$.

35. [M] Show that **w** is in the subspace of $\mathbb{R}^4$ spanned by $\mathbf{v}_1, \mathbf{v}_2, \mathbf{v}_3$, where

$$\mathbf{w} = \begin{bmatrix} 9 \\ -4 \\ -4 \\ 7 \end{bmatrix}, \mathbf{v}_1 = \begin{bmatrix} 8 \\ -4 \\ -3 \\ 9 \end{bmatrix}, \mathbf{v}_2 = \begin{bmatrix} -4 \\ 3 \\ -2 \\ -8 \end{bmatrix}, \mathbf{v}_3 = \begin{bmatrix} -7 \\ 6 \\ -5 \\ -18 \end{bmatrix}$$

36. [M] Determine if **y** is in the subspace of $\mathbb{R}^4$ spanned by the columns of A, where

$$\mathbf{y} = \begin{bmatrix} -4 \\ -8 \\ 6 \\ -5 \end{bmatrix}, \quad A = \begin{bmatrix} 3 & -5 & -9 \\ 8 & 7 & -6 \\ -5 & -8 & 3 \\ 2 & -2 & -9 \end{bmatrix}$$

37. [M] The vector space $H = \text{Span}\{1, \cos^2 t, \cos^4 t, \cos^6 t\}$ contains at least two interesting functions that will be used in a later exercise:

$$\mathbf{f}(t) = 1 - 8\cos^2 t + 8\cos^4 t$$
$$\mathbf{g}(t) = -1 + 18\cos^2 t - 48\cos^4 t + 32\cos^6 t$$

Study the graph of **f** for $0 \le t \le 2\pi$, and guess a simple formula for $\mathbf{f}(t)$. Verify your conjecture by graphing the difference between $1 + \mathbf{f}(t)$ and your formula for $\mathbf{f}(t)$. (Hopefully, you will see the constant function 1.) Repeat for **g**.

38. [M] Repeat Exercise 37 for the functions

$$\mathbf{f}(t) = 3\sin t - 4\sin^3 t$$
$$\mathbf{g}(t) = 1 - 8\sin^2 t + 8\sin^4 t$$
$$\mathbf{h}(t) = 5\sin t - 20\sin^3 t + 16\sin^5 t$$

in the vector space $\text{Span}\{1, \sin t, \sin^2 t, \ldots, \sin^5 t\}$.

SOLUTIONS TO PRACTICE PROBLEMS

1. Take any **u** in H—say, $\mathbf{u} = \begin{bmatrix} 3 \\ 7 \end{bmatrix}$—and take any $c \ne 1$—say, $c = 2$. Then $c\mathbf{u} = \begin{bmatrix} 6 \\ 14 \end{bmatrix}$. If this is in H, then there is some s such that

$$\begin{bmatrix} 3s \\ 2 + 5s \end{bmatrix} = \begin{bmatrix} 6 \\ 14 \end{bmatrix}$$

That is, $s = 2$ and $s = 12/5$, which is impossible. So $2\mathbf{u}$ is not in H and H is not a vector space.

2. $\mathbf{v}_1 = 1\mathbf{v}_1 + 0\mathbf{v}_2 + \cdots + 0\mathbf{v}_p$. This expresses $\mathbf{v}_1$ as a linear combination of $\mathbf{v}_1, \ldots, \mathbf{v}_p$, so $\mathbf{v}_1$ is in W. In general, $\mathbf{v}_k$ is in W because

$$\mathbf{v}_k = 0\mathbf{v}_1 + \cdots + 0\mathbf{v}_{k-1} + 1\mathbf{v}_k + 0\mathbf{v}_{k+1} + \cdots + 0\mathbf{v}_p$$

4.2 | NULL SPACES, COLUMN SPACES, AND LINEAR TRANSFORMATIONS

In applications of linear algebra, subspaces of $\mathbb{R}^n$ usually arise in one of two ways: (1) as the set of all solutions to a system of homogeneous linear equations or (2) as the set of all linear combinations of certain specified vectors. In this section, we compare and contrast these two descriptions of subspaces, allowing us to practice using the concept of a subspace. Actually, as you will soon discover, we have been working with subspaces ever since Section 1.3. The main new feature here is the terminology. The section concludes with a discussion of the kernel and range of a linear transformation.

The Null Space of a Matrix

Consider the following system of homogeneous equations:

$$\begin{aligned} x_1 - 3x_2 - 2x_3 &= 0 \\ -5x_1 + 9x_2 + x_3 &= 0 \end{aligned} \tag{1}$$

In matrix form, this system is written as $A\mathbf{x} = \mathbf{0}$, where

$$A = \begin{bmatrix} 1 & -3 & -2 \\ -5 & 9 & 1 \end{bmatrix} \tag{2}$$

Recall that the set of all $\mathbf{x}$ that satisfy (1) is called the **solution set** of the system (1). Often it is convenient to relate this set directly to the matrix A and the equation $A\mathbf{x} = \mathbf{0}$. We call the set of $\mathbf{x}$ that satisfy $A\mathbf{x} = \mathbf{0}$ the **null space** of the matrix A.

DEFINITION

> The **null space** of an $m \times n$ matrix A, written as Nul A, is the set of all solutions of the homogeneous equation $A\mathbf{x} = \mathbf{0}$. In set notation,
>
> $$\text{Nul } A = \{\mathbf{x} : \mathbf{x} \text{ is in } \mathbb{R}^n \text{ and } A\mathbf{x} = \mathbf{0}\}$$

A more dynamic description of Nul A is the set of all $\mathbf{x}$ in $\mathbb{R}^n$ that are mapped into the zero vector of $\mathbb{R}^m$ via the linear transformation $\mathbf{x} \mapsto A\mathbf{x}$. See Fig. 1.

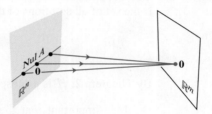

FIGURE 1

EXAMPLE 1 Let A be the matrix in (2) above, and let $\mathbf{u} = \begin{bmatrix} 5 \\ 3 \\ -2 \end{bmatrix}$. Determine if $\mathbf{u}$ belongs to the null space of A.

SOLUTION To test if $\mathbf{u}$ satisfies $A\mathbf{u} = \mathbf{0}$, simply compute

$$A\mathbf{u} = \begin{bmatrix} 1 & -3 & -2 \\ -5 & 9 & 1 \end{bmatrix} \begin{bmatrix} 5 \\ 3 \\ -2 \end{bmatrix} = \begin{bmatrix} 5 - 9 + 4 \\ -25 + 27 - 2 \end{bmatrix} = \begin{bmatrix} 0 \\ 0 \end{bmatrix}$$

Thus $\mathbf{u}$ is in Nul A. ∎

The term *space* in *null space* is appropriate because the null space of a matrix is a vector space, as shown in the next theorem.

THEOREM 2

> The null space of an $m \times n$ matrix A is a subspace of $\mathbb{R}^n$. Equivalently, the set of all solutions to a system $A\mathbf{x} = \mathbf{0}$ of m homogeneous linear equations in n unknowns is a subspace of $\mathbb{R}^n$.

PROOF Certainly Nul A is a subset of $\mathbb{R}^n$ because A has n columns. We must show that Nul A satisfies the three properties of a subspace. Of course, $\mathbf{0}$ is in Nul A. Next, let $\mathbf{u}$ and $\mathbf{v}$ represent any two vectors in Nul A. Then

$$A\mathbf{u} = \mathbf{0} \quad \text{and} \quad A\mathbf{v} = \mathbf{0}$$

To show that $\mathbf{u} + \mathbf{v}$ is in Nul A, we must show that $A(\mathbf{u} + \mathbf{v}) = \mathbf{0}$. Using a property of matrix multiplication, compute

$$A(\mathbf{u} + \mathbf{v}) = A\mathbf{u} + A\mathbf{v} = \mathbf{0} + \mathbf{0} = \mathbf{0}$$

Thus $\mathbf{u} + \mathbf{v}$ is in Nul A, and Nul A is closed under vector addition. Finally, if c is any scalar, then

$$A(c\mathbf{u}) = c(A\mathbf{u}) = c(\mathbf{0}) = \mathbf{0}$$

which shows that $c\mathbf{u}$ is in Nul A. Thus Nul A is a subspace of $\mathbb{R}^n$. ∎

EXAMPLE 2 Let H be the set of all vectors in $\mathbb{R}^4$ whose coordinates a, b, c, d satisfy the equations $a - 2b + 5c = d$ and $c - a = b$. Show that H is a subspace of $\mathbb{R}^4$.

SOLUTION Rearrange the equations that describe the elements of H, and note that H is the set of all solutions of the following system of homogeneous linear equations:

$$a - 2b + 5c - d = 0$$
$$-a - b + c \quad\quad = 0$$

By Theorem 2, H is a subspace of $\mathbb{R}^4$. ∎

It is important that the linear equations defining the set H are homogeneous. Otherwise, the set of solutions will definitely *not* be a subspace (because the zero vector is not a solution of a nonhomogeneous system). Also, in some cases, the set of solutions could be empty.

An Explicit Description of Nul A

There is no obvious relation between vectors in Nul A and the entries in A. We say that Nul A is defined *implicitly*, because it is defined by a condition that must be checked. No explicit list or description of the elements in Nul A is given. However, *solving* the equation $A\mathbf{x} = \mathbf{0}$ amounts to producing an *explicit* description of Nul A. The next example reviews the procedure from Section 1.5.

EXAMPLE 3 Find a spanning set for the null space of the matrix

$$A = \begin{bmatrix} -3 & 6 & -1 & 1 & -7 \\ 1 & -2 & 2 & 3 & -1 \\ 2 & -4 & 5 & 8 & -4 \end{bmatrix}$$

SOLUTION The first step is to find the general solution of $A\mathbf{x} = \mathbf{0}$ in terms of free variables. Row reduce the augmented matrix $[\, A \quad \mathbf{0} \,]$ to *reduced* echelon form in order to write the basic variables in terms of the free variables:

$$\begin{bmatrix} 1 & -2 & 0 & -1 & 3 & 0 \\ 0 & 0 & 1 & 2 & -2 & 0 \\ 0 & 0 & 0 & 0 & 0 & 0 \end{bmatrix},$$

$$x_1 - 2x_2 \quad - x_4 + 3x_5 = 0$$
$$x_3 + 2x_4 - 2x_5 = 0$$
$$0 = 0$$

The general solution is $x_1 = 2x_2 + x_4 - 3x_5$, $x_3 = -2x_4 + 2x_5$, with x_2, x_4, and x_5 free. Next, decompose the vector giving the general solution into a linear combination of vectors where *the weights are the free variables*. That is,

$$\begin{bmatrix} x_1 \\ x_2 \\ x_3 \\ x_4 \\ x_5 \end{bmatrix} = \begin{bmatrix} 2x_2 + x_4 - 3x_5 \\ x_2 \\ -2x_4 + 2x_5 \\ x_4 \\ x_5 \end{bmatrix} = x_2 \underset{\mathbf{u}}{\begin{bmatrix} 2 \\ 1 \\ 0 \\ 0 \\ 0 \end{bmatrix}} + x_4 \underset{\mathbf{v}}{\begin{bmatrix} 1 \\ 0 \\ -2 \\ 1 \\ 0 \end{bmatrix}} + x_5 \underset{\mathbf{w}}{\begin{bmatrix} -3 \\ 0 \\ 2 \\ 0 \\ 1 \end{bmatrix}}$$

$$= x_2 \mathbf{u} + x_4 \mathbf{v} + x_5 \mathbf{w} \tag{3}$$

Every linear combination of $\mathbf{u}$, $\mathbf{v}$, and $\mathbf{w}$ is an element of Nul A. Thus $\{\mathbf{u}, \mathbf{v}, \mathbf{w}\}$ is a spanning set for Nul A.   ∎

Two points should be made about the solution of Example 3 that apply to all problems of this type where Nul A contains nonzero vectors. We will use these facts later.

1. The spanning set produced by the method in Example 3 is automatically linearly independent because the free variables are the weights on the spanning vectors. For instance, look at the 2nd, 4th, and 5th entries in the solution vector in (3) and note that $x_2 \mathbf{u} + x_4 \mathbf{v} + x_5 \mathbf{w}$ can be $\mathbf{0}$ only if the weights x_2, x_4, and x_5 are all zero.

2. When Nul A contains nonzero vectors, the number of vectors in the spanning set for Nul A equals the number of free variables in the equation $A\mathbf{x} = \mathbf{0}$.

The Column Space of a Matrix

Another important subspace associated with a matrix is its column space. Unlike the null space, the column space is defined explicitly via linear combinations.

DEFINITION

> The **column space** of an $m \times n$ matrix A, written as Col A, is the set of all linear combinations of the columns of A. If $A = [\mathbf{a}_1 \cdots \mathbf{a}_n]$, then
> $$\text{Col } A = \text{Span}\{\mathbf{a}_1, \dots, \mathbf{a}_n\}$$

Since Span $\{\mathbf{a}_1, \dots, \mathbf{a}_n\}$ is a subspace, by Theorem 1, the next theorem follows from the definition of Col A and the fact that the columns of A are in $\mathbb{R}^m$.

THEOREM 3

> The column space of an $m \times n$ matrix A is a subspace of $\mathbb{R}^m$.

Note that a typical vector in Col A can be written as $A\mathbf{x}$ for some $\mathbf{x}$ because the notation $A\mathbf{x}$ stands for a linear combination of the columns of A. That is,

$$\boxed{\text{Col } A = \{\mathbf{b} : \mathbf{b} = A\mathbf{x} \text{ for some } \mathbf{x} \text{ in } \mathbb{R}^n\}}$$

The notation $A\mathbf{x}$ for vectors in Col A also shows that Col A is the *range* of the linear transformation $\mathbf{x} \mapsto A\mathbf{x}$. We will return to this point of view at the end of the section.

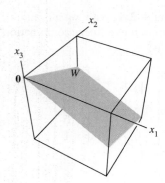

EXAMPLE 4 Find a matrix A such that $W = \text{Col } A$.

$$W = \left\{ \begin{bmatrix} 6a - b \\ a + b \\ -7a \end{bmatrix} : a, b \text{ in } \mathbb{R} \right\}$$

SOLUTION First, write W as a set of linear combinations.

$$W = \left\{ a \begin{bmatrix} 6 \\ 1 \\ -7 \end{bmatrix} + b \begin{bmatrix} -1 \\ 1 \\ 0 \end{bmatrix} : a, b \text{ in } \mathbb{R} \right\} = \text{Span} \left\{ \begin{bmatrix} 6 \\ 1 \\ -7 \end{bmatrix}, \begin{bmatrix} -1 \\ 1 \\ 0 \end{bmatrix} \right\}$$

Second, use the vectors in the spanning set as the columns of A. Let $A = \begin{bmatrix} 6 & -1 \\ 1 & 1 \\ -7 & 0 \end{bmatrix}$.

Then $W = \text{Col } A$, as desired. ∎

Recall from Theorem 4 in Section 1.4 that the columns of A span $\mathbb{R}^m$ if and only if the equation $A\mathbf{x} = \mathbf{b}$ has a solution for each $\mathbf{b}$. We can restate this fact as follows:

> The column space of an $m \times n$ matrix A is all of $\mathbb{R}^m$ if and only if the equation $A\mathbf{x} = \mathbf{b}$ has a solution for each $\mathbf{b}$ in $\mathbb{R}^m$.

The Contrast Between Nul A and Col A

It is natural to wonder how the null space and column space of a matrix are related. In fact, the two spaces are quite dissimilar, as Examples 5–7 will show. Nevertheless, a surprising connection between the null space and column space will emerge in Section 4.6, after more theory is available.

EXAMPLE 5 Let

$$A = \begin{bmatrix} 2 & 4 & -2 & 1 \\ -2 & -5 & 7 & 3 \\ 3 & 7 & -8 & 6 \end{bmatrix}$$

a. If the column space of A is a subspace of $\mathbb{R}^k$, what is k?
b. If the null space of A is a subspace of $\mathbb{R}^k$, what is k?

SOLUTION

a. The columns of A each have three entries, so Col A is a subspace of $\mathbb{R}^k$, where $k = 3$.

b. A vector $\mathbf{x}$ such that $A\mathbf{x}$ is defined must have four entries, so Nul A is a subspace of $\mathbb{R}^k$, where $k = 4$. ∎

When a matrix is not square, as in Example 5, the vectors in Nul A and Col A live in entirely different "universes." For example, no linear combination of vectors in $\mathbb{R}^3$ can produce a vector in $\mathbb{R}^4$. When A is square, Nul A and Col A do have the zero vector in common, and in special cases it is possible that some nonzero vectors belong to both Nul A and Col A.

EXAMPLE 6 With A as in Example 5, find a nonzero vector in Col A and a nonzero vector in Nul A.

SOLUTION It is easy to find a vector in Col A. Any column of A will do, say, $\begin{bmatrix} 2 \\ -2 \\ 3 \end{bmatrix}$.

To find a nonzero vector in Nul A, row reduce the augmented matrix $[\, A \quad \mathbf{0} \,]$ and obtain

$$[\, A \quad \mathbf{0} \,] \sim \begin{bmatrix} 1 & 0 & 9 & 0 & 0 \\ 0 & 1 & -5 & 0 & 0 \\ 0 & 0 & 0 & 1 & 0 \end{bmatrix}$$

Thus, if $\mathbf{x}$ satisfies $A\mathbf{x} = \mathbf{0}$, then $x_1 = -9x_3$, $x_2 = 5x_3$, $x_4 = 0$, and x_3 is free. Assigning a nonzero value to x_3—say, $x_3 = 1$—we obtain a vector in Nul A, namely, $\mathbf{x} = (-9, 5, 1, 0)$. $\blacksquare$

EXAMPLE 7 With A as in Example 5, let $\mathbf{u} = \begin{bmatrix} 3 \\ -2 \\ -1 \\ 0 \end{bmatrix}$ and $\mathbf{v} = \begin{bmatrix} 3 \\ -1 \\ 3 \end{bmatrix}$.

a. Determine if $\mathbf{u}$ is in Nul A. Could $\mathbf{u}$ be in Col A?
b. Determine if $\mathbf{v}$ is in Col A. Could $\mathbf{v}$ be in Nul A?

SOLUTION

a. An explicit description of Nul A is not needed here. Simply compute the product $A\mathbf{u}$.

$$A\mathbf{u} = \begin{bmatrix} 2 & 4 & -2 & 1 \\ -2 & -5 & 7 & 3 \\ 3 & 7 & -8 & 6 \end{bmatrix} \begin{bmatrix} 3 \\ -2 \\ -1 \\ 0 \end{bmatrix} = \begin{bmatrix} 0 \\ -3 \\ 3 \end{bmatrix} \neq \begin{bmatrix} 0 \\ 0 \\ 0 \end{bmatrix}$$

Obviously, $\mathbf{u}$ is *not* a solution of $A\mathbf{x} = \mathbf{0}$, so $\mathbf{u}$ is not in Nul A. Also, with four entries, $\mathbf{u}$ could not possibly be in Col A, since Col A is a subspace of $\mathbb{R}^3$.

b. Reduce $[\, A \quad \mathbf{v} \,]$ to an echelon form.

$$[\, A \quad \mathbf{v} \,] = \begin{bmatrix} 2 & 4 & -2 & 1 & 3 \\ -2 & -5 & 7 & 3 & -1 \\ 3 & 7 & -8 & 6 & 3 \end{bmatrix} \sim \begin{bmatrix} 2 & 4 & -2 & 1 & 3 \\ 0 & 1 & -5 & -4 & -2 \\ 0 & 0 & 0 & 17 & 1 \end{bmatrix}$$

At this point, it is clear that the equation $A\mathbf{x} = \mathbf{v}$ is consistent, so $\mathbf{v}$ is in Col A. With only three entries, $\mathbf{v}$ could not possibly be in Nul A, since Nul A is a subspace of $\mathbb{R}^4$. $\blacksquare$

The table on page 204 summarizes what we have learned about Nul A and Col A. Item 8 is a restatement of Theorems 11 and 12(a) in Section 1.9.

Kernel and Range of a Linear Transformation

Subspaces of vector spaces other than $\mathbb{R}^n$ are often described in terms of a linear transformation instead of a matrix. To make this precise, we generalize the definition given in Section 1.8.

Contrast Between Nul A and Col A for an $m \times n$ Matrix A

Nul A	Col A
1. Nul A is a subspace of $\mathbb{R}^n$.	1. Col A is a subspace of $\mathbb{R}^m$.
2. Nul A is implicitly defined; that is, you are given only a condition ($A\mathbf{x} = \mathbf{0}$) that vectors in Nul A must satisfy.	2. Col A is explicitly defined; that is, you are told how to build vectors in Col A.
3. It takes time to find vectors in Nul A. Row operations on $[\,A \quad \mathbf{0}\,]$ are required.	3. It is easy to find vectors in Col A. The columns of A are displayed; others are formed from them.
4. There is no obvious relation between Nul A and the entries in A.	4. There is an obvious relation between Col A and the entries in A, since each column of A is in Col A.
5. A typical vector $\mathbf{v}$ in Nul A has the property that $A\mathbf{v} = \mathbf{0}$.	5. A typical vector $\mathbf{v}$ in Col A has the property that the equation $A\mathbf{x} = \mathbf{v}$ is consistent.
6. Given a specific vector $\mathbf{v}$, it is easy to tell if $\mathbf{v}$ is in Nul A. Just compute $A\mathbf{v}$.	6. Given a specific vector $\mathbf{v}$, it may take time to tell if $\mathbf{v}$ is in Col A. Row operations on $[\,A \quad \mathbf{v}\,]$ are required.
7. Nul $A = \{\mathbf{0}\}$ if and only if the equation $A\mathbf{x} = \mathbf{0}$ has only the trivial solution.	7. Col $A = \mathbb{R}^m$ if and only if the equation $A\mathbf{x} = \mathbf{b}$ has a solution for every $\mathbf{b}$ in $\mathbb{R}^m$.
8. Nul $A = \{\mathbf{0}\}$ if and only if the linear transformation $\mathbf{x} \mapsto A\mathbf{x}$ is one-to-one.	8. Col $A = \mathbb{R}^m$ if and only if the linear transformation $\mathbf{x} \mapsto A\mathbf{x}$ maps $\mathbb{R}^n$ *onto* $\mathbb{R}^m$.

DEFINITION

A **linear transformation** T from a vector space V into a vector space W is a rule that assigns to each vector $\mathbf{x}$ in V a unique vector $T(\mathbf{x})$ in W, such that

(i) $T(\mathbf{u} + \mathbf{v}) = T(\mathbf{u}) + T(\mathbf{v})$ for all $\mathbf{u}$, $\mathbf{v}$ in V, and

(ii) $T(c\mathbf{u}) = cT(\mathbf{u})$ for all $\mathbf{u}$ in V and all scalars c.

The **kernel** (or **null space**) of such a T is the set of all $\mathbf{u}$ in V such that $T(\mathbf{u}) = \mathbf{0}$ (the zero vector in W). The **range** of T is the set of all vectors in W of the form $T(\mathbf{x})$ for some $\mathbf{x}$ in V. If T happens to arise as a matrix transformation—say, $T(\mathbf{x}) = A\mathbf{x}$ for some matrix A—then the kernel and the range of T are just the null space and the column space of A, as defined earlier.

It is not difficult to show that the kernel of T is a subspace of V. The proof is essentially the same as the one for Theorem 2. Also, the range of T is a subspace of W. See Fig. 2 and Exercise 30.

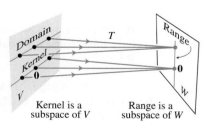

FIGURE 2 Subspaces associated with a linear transformation.

In applications, a subspace usually arises as either the kernel or the range of an appropriate linear transformation. For instance, the set of all solutions of a homogeneous linear differential equation turns out to be the kernel of a linear transformation.

Typically, such a linear transformation is described in terms of one or more derivatives of a function. To explain this in any detail would take us too far afield at this point. So we consider only two examples. The first explains why the operation of differentiation is a linear transformation.

EXAMPLE 8 (Calculus required) Let V be the vector space of all real-valued functions f defined on an interval $[a, b]$ with the property that they are differentiable and their derivatives are continuous functions on $[a, b]$. Let W be the vector space $C[a, b]$ of all continuous functions on $[a, b]$, and let $D : V \to W$ be the transformation that changes f in V into its derivative f'. In calculus, two simple differentiation rules are

$$D(f + g) = D(f) + D(g) \quad \text{and} \quad D(cf) = cD(f)$$

That is, D is a linear transformation. It can be shown that the kernel of D is the set of constant functions on $[a, b]$ and the range of D is the set W of all continuous functions on $[a, b]$. ∎

EXAMPLE 9 (Calculus required) The differential equation

$$y'' + \omega^2 y = 0 \tag{4}$$

where ω is a constant, is used to describe a variety of physical systems, such as the vibration of a weighted spring, the movement of a pendulum, and the voltage in an inductance-capacitance electrical circuit. The set of solutions of (4) is precisely the kernel of the linear transformation that maps a function $y = f(t)$ into the function $f''(t) + \omega^2 f(t)$. Finding an explicit description of this vector space is a problem in differential equations. The solution set turns out to be the space described in Exercise 19 in Section 4.1. ∎

PRACTICE PROBLEMS

1. Let $W = \left\{ \begin{bmatrix} a \\ b \\ c \end{bmatrix} : a - 3b - c = 0 \right\}$. Show in two different ways that W is a subspace of $\mathbb{R}^3$. (Use two theorems.)

2. Let $A = \begin{bmatrix} 7 & -3 & 5 \\ -4 & 1 & -5 \\ -5 & 2 & -4 \end{bmatrix}$, $\mathbf{v} = \begin{bmatrix} 2 \\ 1 \\ -1 \end{bmatrix}$, and $\mathbf{w} = \begin{bmatrix} 7 \\ 6 \\ -3 \end{bmatrix}$. Suppose you know that the equations $A\mathbf{x} = \mathbf{v}$ and $A\mathbf{x} = \mathbf{w}$ are both consistent. What can you say about the equation $A\mathbf{x} = \mathbf{v} + \mathbf{w}$?

4.2 EXERCISES

1. Determine if $\mathbf{w} = \begin{bmatrix} 1 \\ 3 \\ -4 \end{bmatrix}$ is in Nul A, where

$A = \begin{bmatrix} 3 & -5 & -3 \\ 6 & -2 & 0 \\ -8 & 4 & 1 \end{bmatrix}$.

2. Determine if $\mathbf{w} = \begin{bmatrix} 1 \\ -1 \\ 1 \end{bmatrix}$ is in Nul A, where

$A = \begin{bmatrix} 2 & 6 & 4 \\ -3 & 2 & 5 \\ -5 & -4 & 1 \end{bmatrix}$.

In Exercises 3–6, find an explicit description of Nul A, by listing vectors that span the null space.

3. $A = \begin{bmatrix} 1 & 2 & 4 & 0 \\ 0 & 1 & 3 & -2 \end{bmatrix}$

4. $A = \begin{bmatrix} 1 & -3 & 2 & 0 \\ 0 & 0 & 3 & 0 \end{bmatrix}$

5. $A = \begin{bmatrix} 1 & -4 & 0 & 2 & 0 \\ 0 & 0 & 1 & -5 & 0 \\ 0 & 0 & 0 & 0 & 2 \end{bmatrix}$

6. $A = \begin{bmatrix} 1 & 3 & -4 & -3 & 1 \\ 0 & 1 & -3 & 1 & 0 \\ 0 & 0 & 0 & 0 & 0 \end{bmatrix}$

In Exercises 7–14, either use an appropriate theorem to show that the given set, W, is a vector space, or find a specific example to the contrary.

7. $\left\{ \begin{bmatrix} a \\ b \\ c \end{bmatrix} : a + b + c = 2 \right\}$

8. $\left\{ \begin{bmatrix} r \\ s \\ t \end{bmatrix} : 3r - 2 = 3s + t \right\}$

9. $\left\{ \begin{bmatrix} p \\ q \\ r \\ s \end{bmatrix} : \begin{array}{l} p - 3q = 4s \\ 2p = s + 5r \end{array} \right\}$

10. $\left\{ \begin{bmatrix} a \\ b \\ c \\ d \end{bmatrix} : \begin{array}{l} 3a + b = c \\ a + b + 2c = 2d \end{array} \right\}$

11. $\left\{ \begin{bmatrix} s - 2t \\ 3 + 3s \\ 3s + t \\ 2s \end{bmatrix} : s, t \text{ real} \right\}$

12. $\left\{ \begin{bmatrix} 3p - 5q \\ 4q \\ p \\ q + 1 \end{bmatrix} : p, q \text{ real} \right\}$

13. $\left\{ \begin{bmatrix} c - 6d \\ d \\ c \end{bmatrix} : c, d \text{ real} \right\}$

14. $\left\{ \begin{bmatrix} -s + 3t \\ s - 2t \\ 5s - t \end{bmatrix} : s, t \text{ real} \right\}$

In Exercises 15 and 16, find A such that the given set is Col A.

15. $\left\{ \begin{bmatrix} 2s + t \\ r - s + 2t \\ 3r + s \\ 2r - s - t \end{bmatrix} : r, s, t \text{ real} \right\}$

16. $\left\{ \begin{bmatrix} b - c \\ 2b + 3d \\ b + 3c - 3d \\ c + d \end{bmatrix} : b, c, d \text{ real} \right\}$

For the matrices in Exercises 17–20, (a) find k such that Nul A is a subspace of $\mathbb{R}^k$, and (b) find k such that Col A is a subspace of $\mathbb{R}^k$.

17. $A = \begin{bmatrix} 6 & -4 \\ -3 & 2 \\ -9 & 6 \\ 9 & -6 \end{bmatrix}$

18. $A = \begin{bmatrix} 5 & -2 & 3 \\ -1 & 0 & -1 \\ 0 & -2 & -2 \\ -5 & 7 & 2 \end{bmatrix}$

19. $A = \begin{bmatrix} 4 & 5 & -2 & 6 & 0 \\ 1 & 1 & 0 & 1 & 0 \end{bmatrix}$

20. $A = \begin{bmatrix} 1 & -3 & 2 & 0 & -5 \end{bmatrix}$

21. With A as in Exercise 17, find a nonzero vector in Nul A and a nonzero vector in Col A.

22. With A as in Exercise 18, find a nonzero vector in Nul A and a nonzero vector in Col A.

23. Let $A = \begin{bmatrix} -2 & 4 \\ -1 & 2 \end{bmatrix}$ and $\mathbf{w} = \begin{bmatrix} 2 \\ 1 \end{bmatrix}$. Determine if $\mathbf{w}$ is in Col A. Is $\mathbf{w}$ in Nul A?

24. Let $A = \begin{bmatrix} 10 & -8 & -2 & -2 \\ 0 & 2 & 2 & -2 \\ 1 & -1 & 6 & 0 \\ 1 & 1 & 0 & -2 \end{bmatrix}$ and $\mathbf{w} = \begin{bmatrix} 2 \\ 2 \\ 0 \\ 2 \end{bmatrix}$. Determine if $\mathbf{w}$ is in Col A. Is $\mathbf{w}$ in Nul A?

In Exercises 25 and 26, A denotes an $m \times n$ matrix. Mark each statement True or False. Justify each answer.

25. a. The null space of A is the solution set of the equation $A\mathbf{x} = \mathbf{0}$.

  b. The null space of an $m \times n$ matrix is in $\mathbb{R}^m$.

  c. The column space of A is the range of the mapping $\mathbf{x} \mapsto A\mathbf{x}$.

  d. If the equation $A\mathbf{x} = \mathbf{b}$ is consistent, then Col A is $\mathbb{R}^m$.

  e. The kernel of a linear transformation is a vector space.

  f. Col A is the set of all vectors that can be written as $A\mathbf{x}$ for some $\mathbf{x}$.

26. a. A null space is a vector space.

  b. The column space of an $m \times n$ matrix is in $\mathbb{R}^m$.

  c. Col A is the set of all solutions of $A\mathbf{x} = \mathbf{b}$.

  d. Nul A is the kernel of the mapping $\mathbf{x} \mapsto A\mathbf{x}$.

  e. The range of a linear transformation is a vector space.

  f. The set of all solutions of a homogeneous linear differential equation is the kernel of a linear transformation.

27. It can be shown that a solution of the system below is $x_1 = 3$, $x_2 = 2$, and $x_3 = -1$. Use this fact and the theory from this section to explain why another solution is $x_1 = 30$, $x_2 = 20$, and $x_3 = -10$. (Observe how the solutions are related, but make no other calculations.)

$$\begin{array}{r} x_1 - 3x_2 - 3x_3 = 0 \\ -2x_1 + 4x_2 + 2x_3 = 0 \\ -x_1 + 5x_2 + 7x_3 = 0 \end{array}$$

28. Consider the following two systems of equations:

$$\begin{array}{rr} 5x_1 + x_2 - 3x_3 = 0 & \quad 5x_1 + x_2 - 3x_3 = 0 \\ -9x_1 + 2x_2 + 5x_3 = 1 & \quad -9x_1 + 2x_2 + 5x_3 = 5 \\ 4x_1 + x_2 - 6x_3 = 9 & \quad 4x_1 + x_2 - 6x_3 = 45 \end{array}$$

It can be shown that the first system has a solution. Use this fact and the theory from this section to explain why the second system must also have a solution. (Make no row operations.)

29. Prove Theorem 3 as follows: Given an $m \times n$ matrix A, an element in Col A has the form $A\mathbf{x}$ for some $\mathbf{x}$ in $\mathbb{R}^n$. Let $A\mathbf{x}$ and $A\mathbf{w}$ represent any two vectors in Col A.

a. Explain why the zero vector is in Col A.

b. Show that the vector $A\mathbf{x} + A\mathbf{w}$ is in Col A.

c. Given a scalar c, show that $c(A\mathbf{x})$ is in Col A.

30. Let $T : V \to W$ be a linear transformation from a vector space V into a vector space W. Prove that the range of T is a subspace of W. [*Hint:* Typical elements of the range have the form $T(\mathbf{x})$ and $T(\mathbf{w})$ for some $\mathbf{x}, \mathbf{w}$ in V.]

31. Define $T : \mathbb{P}_2 \to \mathbb{R}^2$ by $T(\mathbf{p}) = \begin{bmatrix} \mathbf{p}(0) \\ \mathbf{p}(1) \end{bmatrix}$. For instance, if

$\mathbf{p}(t) = 3 + 5t + 7t^2$, then $T(\mathbf{p}) = \begin{bmatrix} 3 \\ 15 \end{bmatrix}$.

a. Show that T is a linear transformation. [*Hint:* For arbitrary polynomials $\mathbf{p}, \mathbf{q}$ in $\mathbb{P}_2$, compute $T(\mathbf{p} + \mathbf{q})$ and $T(c\mathbf{p})$.]

b. Find a polynomial $\mathbf{p}$ in $\mathbb{P}_2$ that spans the kernel of T, and describe the range of T.

32. Define a linear transformation $T : \mathbb{P}_2 \to \mathbb{R}^2$ by $T(\mathbf{p}) = \begin{bmatrix} \mathbf{p}(0) \\ \mathbf{p}(0) \end{bmatrix}$. Find polynomials $\mathbf{p}_1$ and $\mathbf{p}_2$ in $\mathbb{P}_2$ that span the kernel of T, and describe the range of T.

33. Let $M_{2\times2}$ be the vector space of all 2×2 matrices, and define $T : M_{2\times2} \to M_{2\times2}$ by $T(A) = A + A^T$, where

$A = \begin{bmatrix} a & b \\ c & d \end{bmatrix}$.

a. Show that T is a linear transformation.

b. Let B be any element of $M_{2\times2}$ such that $B^T = B$. Find an A in $M_{2\times2}$ such that $T(A) = B$.

c. Show that the range of T is the set of B in $M_{2\times2}$ with the property that $B^T = B$.

d. Describe the kernel of T.

34. (*Calculus required*) Define $T : C[0, 1] \to C[0, 1]$ as follows: For $\mathbf{f}$ in $C[0, 1]$, let $T(\mathbf{f})$ be the antiderivative $\mathbf{F}$ of $\mathbf{f}$ such that $\mathbf{F}(0) = 0$. Show that T is a linear transformation, and describe the kernel of T. (See the notation in Exercise 20 of Section 4.1.)

35. Let V and W be vector spaces, and let $T : V \to W$ be a linear transformation. Given a subspace U of V, let $T(U)$ denote the set of all images of the form $T(\mathbf{x})$, where $\mathbf{x}$ is in U. Show that $T(U)$ is a subspace of W.

36. Given $T : V \to W$ as in Exercise 35, and given a subspace Z of W, let U be the set of all $\mathbf{x}$ in V such that $T(\mathbf{x})$ is in Z. Show that U is a subspace of V.

37. [M] Determine whether $\mathbf{w}$ is in the column space of A, the null space of A, or both, where

$$\mathbf{w} = \begin{bmatrix} 1 \\ 1 \\ -1 \\ -3 \end{bmatrix}, \quad A = \begin{bmatrix} 7 & 6 & -4 & 1 \\ -5 & -1 & 0 & -2 \\ 9 & -11 & 7 & -3 \\ 19 & -9 & 7 & 1 \end{bmatrix}$$

38. [M] Determine whether $\mathbf{w}$ is in the column space of A, the null space of A, or both, where

$$\mathbf{w} = \begin{bmatrix} 1 \\ 2 \\ 1 \\ 0 \end{bmatrix}, \quad A = \begin{bmatrix} -8 & 5 & -2 & 0 \\ -5 & 2 & 1 & -2 \\ 10 & -8 & 6 & -3 \\ 3 & -2 & 1 & 0 \end{bmatrix}$$

39. [M] Let $\mathbf{a}_1, \ldots, \mathbf{a}_5$ denote the columns of the matrix A, where

$$A = \begin{bmatrix} 5 & 1 & 2 & 2 & 0 \\ 3 & 3 & 2 & -1 & -12 \\ 8 & 4 & 4 & -5 & 12 \\ 2 & 1 & 1 & 0 & -2 \end{bmatrix}, \quad B = [\,\mathbf{a}_1 \quad \mathbf{a}_2 \quad \mathbf{a}_4\,]$$

a. Explain why $\mathbf{a}_3$ and $\mathbf{a}_5$ are in the column space of B.

b. Find a set of vectors that spans Nul A.

c. Let $T : \mathbb{R}^5 \to \mathbb{R}^4$ be defined by $T(\mathbf{x}) = A\mathbf{x}$. Explain why T is neither one-to-one nor onto.

40. [M] Let $H = \text{Span}\{\mathbf{v}_1, \mathbf{v}_2\}$ and $K = \text{Span}\{\mathbf{v}_3, \mathbf{v}_4\}$, where

$$\mathbf{v}_1 = \begin{bmatrix} 5 \\ 3 \\ 8 \end{bmatrix}, \mathbf{v}_2 = \begin{bmatrix} 1 \\ 3 \\ 4 \end{bmatrix}, \mathbf{v}_3 = \begin{bmatrix} 2 \\ -1 \\ 5 \end{bmatrix}, \mathbf{v}_4 = \begin{bmatrix} 0 \\ -12 \\ -28 \end{bmatrix}.$$

Then H and K are subspaces of $\mathbb{R}^3$. In fact, H and K are planes in $\mathbb{R}^3$ through the origin, and they intersect in a line through $\mathbf{0}$. Find a nonzero vector $\mathbf{w}$ that generates that line. [*Hint:* $\mathbf{w}$ can be written as $c_1\mathbf{v}_1 + c_2\mathbf{v}_2$ and also as $c_3\mathbf{v}_3 + c_4\mathbf{v}_4$. To build $\mathbf{w}$, solve the equation $c_1\mathbf{v}_1 + c_2\mathbf{v}_2 = c_3\mathbf{v}_3 + c_4\mathbf{v}_4$ for the unknown c_j's.]

SG | Mastering: Vector Space, Subspace, Col A, and Nul A 4–6

SOLUTIONS TO PRACTICE PROBLEMS

1. *First method:* W is a subspace of $\mathbb{R}^3$ by Theorem 2 because W is the set of all solutions to a system of homogeneous linear equations (where the system has only one equation). Equivalently, W is the null space of the 1×3 matrix $A = [\,1 \quad -3 \quad -1\,]$.

Second method: Solve the equation $a - 3b - c = 0$ for the leading variable a in terms of the free variables b and c. Any solution has the form $\begin{bmatrix} 3b + c \\ b \\ c \end{bmatrix}$, where b and c are arbitrary, and

$$\begin{bmatrix} 3b + c \\ b \\ c \end{bmatrix} = b \underset{\underset{\mathbf{v}_1}{\uparrow}}{\begin{bmatrix} 3 \\ 1 \\ 0 \end{bmatrix}} + c \underset{\underset{\mathbf{v}_2}{\uparrow}}{\begin{bmatrix} 1 \\ 0 \\ 1 \end{bmatrix}}$$

This calculation shows that $W = \text{Span}\{\mathbf{v}_1, \mathbf{v}_2\}$. Thus W is a subspace of $\mathbb{R}^3$ by Theorem 1. We could also solve the equation $a - 3b - c = 0$ for b or c and get alternative descriptions of W as a set of linear combinations of two vectors.

2. Both $\mathbf{v}$ and $\mathbf{w}$ are in Col A. Since Col A is a vector space, $\mathbf{v} + \mathbf{w}$ must be in Col A. That is, the equation $A\mathbf{x} = \mathbf{v} + \mathbf{w}$ is consistent.

4.3 | LINEARLY INDEPENDENT SETS; BASES

In this section we identify and study the subsets that span a vector space V or a subspace H as "efficiently" as possible. The key idea is that of linear independence, defined as in $\mathbb{R}^n$.

An indexed set of vectors $\{\mathbf{v}_1, \ldots, \mathbf{v}_p\}$ in V is said to be **linearly independent** if the vector equation

$$c_1 \mathbf{v}_1 + c_2 \mathbf{v}_2 + \cdots + c_p \mathbf{v}_p = \mathbf{0} \tag{1}$$

has *only* the trivial solution, $c_1 = 0, \ldots, c_p = 0$.[1]

The set $\{\mathbf{v}_1, \ldots, \mathbf{v}_p\}$ is said to be **linearly dependent** if (1) has a nontrivial solution, that is, if there are some weights, $c_1, \ldots, c_p$, *not all zero*, such that (1) holds. In such a case, (1) is called a **linear dependence relation** among $\mathbf{v}_1, \ldots, \mathbf{v}_p$.

Just as in $\mathbb{R}^n$, a set containing a single vector $\mathbf{v}$ is linearly independent if and only if $\mathbf{v} \neq \mathbf{0}$. Also, a set of two vectors is linearly dependent if and only if one of the vectors is a multiple of the other. And any set containing the zero vector is linearly dependent. The following theorem has the same proof as Theorem 7 in Section 1.7.

THEOREM 4

An indexed set $\{\mathbf{v}_1, \ldots, \mathbf{v}_p\}$ of two or more vectors, with $\mathbf{v}_1 \neq \mathbf{0}$, is linearly dependent if and only if some $\mathbf{v}_j$ (with $j > 1$) is a linear combination of the preceding vectors, $\mathbf{v}_1, \ldots, \mathbf{v}_{j-1}$.

The main difference between linear dependence in $\mathbb{R}^n$ and in a general vector space is that when the vectors are not n-tuples, the homogeneous equation (1) usually cannot be written as a system of n linear equations. That is, the vectors cannot be made into the columns of a matrix A in order to study the equation $A\mathbf{x} = \mathbf{0}$. We must rely instead on the definition of linear dependence and on Theorem 4.

EXAMPLE 1 Let $\mathbf{p}_1(t) = 1$, $\mathbf{p}_2(t) = t$, and $\mathbf{p}_3(t) = 4 - t$. Then $\{\mathbf{p}_1, \mathbf{p}_2, \mathbf{p}_3\}$ is linearly dependent in $\mathbb{P}$ because $\mathbf{p}_3 = 4\mathbf{p}_1 - \mathbf{p}_2$. ∎

---

[1] It is convenient to use $c_1, \ldots, c_p$ in (1) for the scalars instead of $x_1, \ldots, x_p$, as we did in Chapter 1.

EXAMPLE 2   The set $\{\sin t, \cos t\}$ is linearly independent in $C[0, 1]$, the space of all continuous functions on $0 \le t \le 1$, because $\sin t$ and $\cos t$ are not multiples of one another *as vectors in* $C[0, 1]$. That is, there is no scalar c such that $\cos t = c \cdot \sin t$ for all t in $[0, 1]$. (Look at the graphs of $\sin t$ and $\cos t$.) However, $\{\sin t \cos t, \sin 2t\}$ is linearly dependent because of the identity: $\sin 2t = 2 \sin t \cos t$, for all t. ∎

DEFINITION

Let H be a subspace of a vector space V.   An indexed set of vectors $\mathcal{B} = \{\mathbf{b}_1, \dots, \mathbf{b}_p\}$ in V is a **basis** for H if

(i) $\mathcal{B}$ is a linearly independent set, and

(ii) the subspace spanned by $\mathcal{B}$ coincides with H; that is,

$$H = \text{Span}\{\mathbf{b}_1, \dots, \mathbf{b}_p\}$$

The definition of a basis applies to the case when $H = V$, because any vector space is a subspace of itself. Thus a basis of V is a linearly independent set that spans V. Observe that when $H \ne V$, condition (ii) includes the requirement that each of the vectors $\mathbf{b}_1, \dots, \mathbf{b}_p$ must belong to H, because Span $\{\mathbf{b}_1, \dots, \mathbf{b}_p\}$ contains $\mathbf{b}_1, \dots, \mathbf{b}_p$, as shown in Section 4.1.

EXAMPLE 3   Let A be an invertible $n \times n$ matrix—say, $A = [\,\mathbf{a}_1 \ \cdots \ \mathbf{a}_n\,]$. Then the columns of A form a basis for $\mathbb{R}^n$ because they are linearly independent and they span $\mathbb{R}^n$, by the Invertible Matrix Theorem. ∎

EXAMPLE 4   Let $\mathbf{e}_1, \dots, \mathbf{e}_n$ be the columns of the $n \times n$ identity matrix, I_n. That is,

$$\mathbf{e}_1 = \begin{bmatrix} 1 \\ 0 \\ \vdots \\ 0 \end{bmatrix}, \quad \mathbf{e}_2 = \begin{bmatrix} 0 \\ 1 \\ \vdots \\ 0 \end{bmatrix}, \quad \dots, \quad \mathbf{e}_n = \begin{bmatrix} 0 \\ \vdots \\ 0 \\ 1 \end{bmatrix}$$

The set $\{\mathbf{e}_1, \dots, \mathbf{e}_n\}$ is called the **standard basis** for $\mathbb{R}^n$ (Fig. 1). ∎

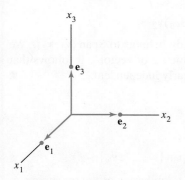

FIGURE 1
The standard basis for $\mathbb{R}^3$.

EXAMPLE 5   Let $\mathbf{v}_1 = \begin{bmatrix} 3 \\ 0 \\ -6 \end{bmatrix}$, $\mathbf{v}_2 = \begin{bmatrix} -4 \\ 1 \\ 7 \end{bmatrix}$, and $\mathbf{v}_3 = \begin{bmatrix} -2 \\ 1 \\ 5 \end{bmatrix}$.   Determine if $\{\mathbf{v}_1, \mathbf{v}_2, \mathbf{v}_3\}$ is a basis for $\mathbb{R}^3$.

SOLUTION   Since there are exactly three vectors here in $\mathbb{R}^3$, we can use any of several methods to determine if the matrix $A = [\,\mathbf{v}_1 \ \mathbf{v}_2 \ \mathbf{v}_3\,]$ is invertible. For instance, two row replacements reveal that A has three pivot positions. Thus A is invertible. As in Example 3, the columns of A form a basis for $\mathbb{R}^3$. ∎

EXAMPLE 6   Let $S = \{1, t, t^2, \dots, t^n\}$. Verify that S is a basis for $\mathbb{P}_n$. This basis is called the **standard basis** for $\mathbb{P}_n$.

SOLUTION   Certainly S spans $\mathbb{P}_n$. To show that S is linearly independent, suppose that $c_0, \dots, c_n$ satisfy

$$c_0 \cdot 1 + c_1 t + c_2 t^2 + \cdots + c_n t^n = \mathbf{0}(t) \qquad (2)$$

This equality means that the polynomial on the left has the same values as the zero polynomial on the right. A fundamental theorem in algebra says that the only polynomial in $\mathbb{P}_n$ with more than n zeros is the zero polynomial. That is, equation (2) holds for all t only if $c_0 = \cdots = c_n = 0$. This proves that S is linearly independent and hence is a basis for $\mathbb{P}_n$. See Fig. 2. ∎

Problems involving linear independence and spanning in $\mathbb{P}_n$ are handled best by a technique to be discussed in Section 4.4.

The Spanning Set Theorem

As we will see, a basis is an "efficient" spanning set that contains no unnecessary vectors. In fact, a basis can be constructed from a spanning set by discarding unneeded vectors.

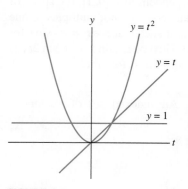

FIGURE 2

The standard basis for $\mathbb{P}_2$.

EXAMPLE 7   Let

$$\mathbf{v}_1 = \begin{bmatrix} 0 \\ 2 \\ -1 \end{bmatrix}, \quad \mathbf{v}_2 = \begin{bmatrix} 2 \\ 2 \\ 0 \end{bmatrix}, \quad \mathbf{v}_3 = \begin{bmatrix} 6 \\ 16 \\ -5 \end{bmatrix}, \quad \text{and} \quad H = \text{Span}\{\mathbf{v}_1, \mathbf{v}_2, \mathbf{v}_3\}.$$

Note that $\mathbf{v}_3 = 5\mathbf{v}_1 + 3\mathbf{v}_2$, and show that $\text{Span}\{\mathbf{v}_1, \mathbf{v}_2, \mathbf{v}_3\} = \text{Span}\{\mathbf{v}_1, \mathbf{v}_2\}$. Then find a basis for the subspace H.

SOLUTION   Every vector in $\text{Span}\{\mathbf{v}_1, \mathbf{v}_2\}$ belongs to H because

$$c_1\mathbf{v}_1 + c_2\mathbf{v}_2 = c_1\mathbf{v}_1 + c_2\mathbf{v}_2 + 0\mathbf{v}_3$$

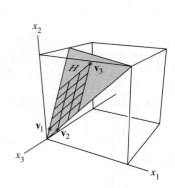

Now let $\mathbf{x}$ be any vector in H—say, $\mathbf{x} = c_1\mathbf{v}_1 + c_2\mathbf{v}_2 + c_3\mathbf{v}_3$. Since $\mathbf{v}_3 = 5\mathbf{v}_1 + 3\mathbf{v}_2$, we may substitute

$$\begin{aligned} \mathbf{x} &= c_1\mathbf{v}_1 + c_2\mathbf{v}_2 + c_3(5\mathbf{v}_1 + 3\mathbf{v}_2) \\ &= (c_1 + 5c_3)\mathbf{v}_1 + (c_2 + 3c_3)\mathbf{v}_2 \end{aligned}$$

Thus $\mathbf{x}$ is in $\text{Span}\{\mathbf{v}_1, \mathbf{v}_2\}$, so every vector in H already belongs to $\text{Span}\{\mathbf{v}_1, \mathbf{v}_2\}$. We conclude that H and $\text{Span}\{\mathbf{v}_1, \mathbf{v}_2\}$ are actually the same set of vectors. It follows that $\{\mathbf{v}_1, \mathbf{v}_2\}$ is a basis of H since $\{\mathbf{v}_1, \mathbf{v}_2\}$ is obviously linearly independent. ∎

The next theorem generalizes Example 7.

THEOREM 5

The Spanning Set Theorem

Let $S = \{\mathbf{v}_1, \ldots, \mathbf{v}_p\}$ be a set in V, and let $H = \text{Span}\{\mathbf{v}_1, \ldots, \mathbf{v}_p\}$.

a. If one of the vectors in S—say, $\mathbf{v}_k$—is a linear combination of the remaining vectors in S, then the set formed from S by removing $\mathbf{v}_k$ still spans H.

b. If $H \neq \{\mathbf{0}\}$, some subset of S is a basis for H.

PROOF

a. By rearranging the list of vectors in S, if necessary, we may suppose that $\mathbf{v}_p$ is a linear combination of $\mathbf{v}_1, \ldots, \mathbf{v}_{p-1}$—say,

$$\mathbf{v}_p = a_1\mathbf{v}_1 + \cdots + a_{p-1}\mathbf{v}_{p-1} \tag{3}$$

Given any $\mathbf{x}$ in H, we may write

$$\mathbf{x} = c_1\mathbf{v}_1 + \cdots + c_{p-1}\mathbf{v}_{p-1} + c_p\mathbf{v}_p \tag{4}$$

for suitable scalars $c_1, \ldots, c_p$. Substituting the expression for $\mathbf{v}_p$ from (3) into (4), it is easy to see that $\mathbf{x}$ is a linear combination of $\mathbf{v}_1, \ldots, \mathbf{v}_{p-1}$. Thus $\{\mathbf{v}_1, \ldots, \mathbf{v}_{p-1}\}$ spans H, because $\mathbf{x}$ was an arbitrary element of H.

b. If the original spanning set S is linearly independent, then it is already a basis for H. Otherwise, one of the vectors in S depends on the others and can be deleted, by part (a). So long as there are two or more vectors in the spanning set, we can repeat this process until the spanning set is linearly independent and hence is a basis for H. If the spanning set is eventually reduced to one vector, that vector will be nonzero (and hence linearly independent) because $H \neq \{\mathbf{0}\}$.   ■

Bases for Nul A and Col A

We already know how to find vectors that span the null space of a matrix A. The discussion in Section 4.2 pointed out that our method always produces a linearly independent set when Nul A contains nonzero vectors. So, in this case, that method produces a *basis* for Nul A.

The next two examples describe a simple algorithm for finding a basis for the column space.

EXAMPLE 8   Find a basis for Col B, where

$$B = \begin{bmatrix} \mathbf{b}_1 & \mathbf{b}_2 & \cdots & \mathbf{b}_5 \end{bmatrix} = \begin{bmatrix} 1 & 4 & 0 & 2 & 0 \\ 0 & 0 & 1 & -1 & 0 \\ 0 & 0 & 0 & 0 & 1 \\ 0 & 0 & 0 & 0 & 0 \end{bmatrix}$$

SOLUTION   Each nonpivot column of B is a linear combination of the pivot columns. In fact, $\mathbf{b}_2 = 4\mathbf{b}_1$ and $\mathbf{b}_4 = 2\mathbf{b}_1 - \mathbf{b}_3$. By the Spanning Set Theorem, we may discard $\mathbf{b}_2$ and $\mathbf{b}_4$, and $\{\mathbf{b}_1, \mathbf{b}_3, \mathbf{b}_5\}$ will still span Col B. Let

$$S = \{\mathbf{b}_1, \mathbf{b}_3, \mathbf{b}_5\} = \left\{ \begin{bmatrix} 1 \\ 0 \\ 0 \\ 0 \end{bmatrix}, \begin{bmatrix} 0 \\ 1 \\ 0 \\ 0 \end{bmatrix}, \begin{bmatrix} 0 \\ 0 \\ 1 \\ 0 \end{bmatrix} \right\}$$

Since $\mathbf{b}_1 \neq \mathbf{0}$ and no vector in S is a linear combination of the vectors that precede it, S is linearly independent (Theorem 4). Thus S is a basis for Col B.   ■

What about a matrix A that is *not* in reduced echelon form? Recall that any linear dependence relationship among the columns of A can be expressed in the form $A\mathbf{x} = \mathbf{0}$, where $\mathbf{x}$ is a column of weights. (If some columns are not involved in a particular dependence relation, then their weights are zero.) When A is row reduced to a matrix B, the columns of B are often totally different from the columns of A. However, the equations $A\mathbf{x} = \mathbf{0}$ and $B\mathbf{x} = \mathbf{0}$ have exactly the same set of solutions. If $A = \begin{bmatrix} \mathbf{a}_1 & \cdots & \mathbf{a}_n \end{bmatrix}$ and $B = \begin{bmatrix} \mathbf{b}_1 & \cdots & \mathbf{b}_n \end{bmatrix}$, then the vector equations

$$x_1\mathbf{a}_1 + \cdots + x_n\mathbf{a}_n = \mathbf{0} \quad \text{and} \quad x_1\mathbf{b}_1 + \cdots + x_n\mathbf{b}_n = \mathbf{0}$$

also have the same set of solutions. That is, the columns of A have *exactly the same linear dependence relationships* as the columns of B.

EXAMPLE 9   It can be shown that the matrix

$$A = \begin{bmatrix} \mathbf{a}_1 & \mathbf{a}_2 & \cdots & \mathbf{a}_5 \end{bmatrix} = \begin{bmatrix} 1 & 4 & 0 & 2 & -1 \\ 3 & 12 & 1 & 5 & 5 \\ 2 & 8 & 1 & 3 & 2 \\ 5 & 20 & 2 & 8 & 8 \end{bmatrix}$$

is row equivalent to the matrix B in Example 8. Find a basis for Col A.

SOLUTION In Example 8 we saw that

$$\mathbf{b}_2 = 4\mathbf{b}_1 \quad \text{and} \quad \mathbf{b}_4 = 2\mathbf{b}_1 - \mathbf{b}_3$$

so we can expect that

$$\mathbf{a}_2 = 4\mathbf{a}_1 \quad \text{and} \quad \mathbf{a}_4 = 2\mathbf{a}_1 - \mathbf{a}_3$$

Check that this is indeed the case! Thus we may discard $\mathbf{a}_2$ and $\mathbf{a}_4$ when selecting a minimal spanning set for Col A. In fact, $\{\mathbf{a}_1, \mathbf{a}_3, \mathbf{a}_5\}$ must be linearly independent because any linear dependence relationship among $\mathbf{a}_1$, $\mathbf{a}_3$, $\mathbf{a}_5$ would imply a linear dependence relationship among $\mathbf{b}_1$, $\mathbf{b}_3$, $\mathbf{b}_5$. But we know that $\{\mathbf{b}_1, \mathbf{b}_3, \mathbf{b}_5\}$ is a linearly independent set. Thus $\{\mathbf{a}_1, \mathbf{a}_3, \mathbf{a}_5\}$ is a basis for Col A. The columns we have used for this basis are the pivot columns of A. ∎

Examples 8 and 9 illustrate the following useful fact.

THEOREM 6   The pivot columns of a matrix A form a basis for Col A.

PROOF The general proof uses the arguments discussed above. Let B be the reduced echelon form of A. The set of pivot columns of B is linearly independent, for no vector in the set is a linear combination of the vectors that precede it. Since A is row equivalent to B, the pivot columns of A are linearly independent as well, because any linear dependence relation among the columns of A corresponds to a linear dependence relation among the columns of B. For this same reason, every nonpivot column of A is a linear combination of the pivot columns of A. Thus the nonpivot columns of A may be discarded from the spanning set for Col A, by the Spanning Set Theorem. This leaves the pivot columns of A as a basis for Col A. ∎

Warning: The pivot columns of a matrix A are evident when A has been reduced only to *echelon* form. But, be careful to use the *pivot columns of A itself* for the basis of Col A. Row operations can change the column space of a matrix. The columns of an echelon form B of A are often not in the column space of A. For instance, the columns of matrix B in Example 8 all have zeros in their last entries, so they cannot span the column space of matrix A in Example 9.

Two Views of a Basis

When the Spanning Set Theorem is used, the deletion of vectors from a spanning set must stop when the set becomes linearly independent. If an additional vector is deleted, it will not be a linear combination of the remaining vectors, and hence the smaller set will no longer span V. Thus a basis is a spanning set that is as small as possible.

A basis is also a linearly independent set that is as large as possible. If S is a basis for V, and if S is enlarged by one vector—say, $\mathbf{w}$—from V, then the new set cannot be linearly independent, because S spans V, and $\mathbf{w}$ is therefore a linear combination of the elements in S.

EXAMPLE 10 The following three sets in $\mathbb{R}^3$ show how a linearly independent set can be enlarged to a basis and how further enlargement destroys the linear independence of the set. Also, a spanning set can be shrunk to a basis, but further shrinking destroys the spanning property.

$$\left\{\begin{bmatrix} 1 \\ 0 \\ 0 \end{bmatrix}, \begin{bmatrix} 2 \\ 3 \\ 0 \end{bmatrix}\right\} \qquad \left\{\begin{bmatrix} 1 \\ 0 \\ 0 \end{bmatrix}, \begin{bmatrix} 2 \\ 3 \\ 0 \end{bmatrix}, \begin{bmatrix} 4 \\ 5 \\ 6 \end{bmatrix}\right\} \qquad \left\{\begin{bmatrix} 1 \\ 0 \\ 0 \end{bmatrix}, \begin{bmatrix} 2 \\ 3 \\ 0 \end{bmatrix}, \begin{bmatrix} 4 \\ 5 \\ 6 \end{bmatrix}, \begin{bmatrix} 7 \\ 8 \\ 9 \end{bmatrix}\right\}$$

Linearly independent   A basis   Spans $\mathbb{R}^3$ but is
but does not span $\mathbb{R}^3$ for $\mathbb{R}^3$ linearly dependent   ■

PRACTICE PROBLEMS

1. Let $\mathbf{v}_1 = \begin{bmatrix} 1 \\ -2 \\ 3 \end{bmatrix}$ and $\mathbf{v}_2 = \begin{bmatrix} -2 \\ 7 \\ -9 \end{bmatrix}$. Determine if $\{\mathbf{v}_1, \mathbf{v}_2\}$ is a basis for $\mathbb{R}^3$. Is $\{\mathbf{v}_1, \mathbf{v}_2\}$ a basis for $\mathbb{R}^2$?

2. Let $\mathbf{v}_1 = \begin{bmatrix} 1 \\ -3 \\ 4 \end{bmatrix}$, $\mathbf{v}_2 = \begin{bmatrix} 6 \\ 2 \\ -1 \end{bmatrix}$, $\mathbf{v}_3 = \begin{bmatrix} 2 \\ -2 \\ 3 \end{bmatrix}$, and $\mathbf{v}_4 = \begin{bmatrix} -4 \\ -8 \\ 9 \end{bmatrix}$. Find a basis for the subspace W spanned by $\{\mathbf{v}_1, \mathbf{v}_2, \mathbf{v}_3, \mathbf{v}_4\}$.

3. Let $\mathbf{v}_1 = \begin{bmatrix} 1 \\ 0 \\ 0 \end{bmatrix}$, $\mathbf{v}_2 = \begin{bmatrix} 0 \\ 1 \\ 0 \end{bmatrix}$, and $H = \left\{\begin{bmatrix} s \\ s \\ 0 \end{bmatrix} : s \text{ in } \mathbb{R}\right\}$. Then every vector in H is a linear combination of $\mathbf{v}_1$ and $\mathbf{v}_2$ because

$$\begin{bmatrix} s \\ s \\ 0 \end{bmatrix} = s \begin{bmatrix} 1 \\ 0 \\ 0 \end{bmatrix} + s \begin{bmatrix} 0 \\ 1 \\ 0 \end{bmatrix}$$

SG Mastering: Basis 4-9

Is $\{\mathbf{v}_1, \mathbf{v}_2\}$ a basis for H?

4.3 EXERCISES

Determine whether the sets in Exercises 1–8 are bases for $\mathbb{R}^3$. Of the sets that are *not* bases, determine which ones are linearly independent and which ones span $\mathbb{R}^3$. Justify your answers.

1. $\begin{bmatrix} 1 \\ 0 \\ 0 \end{bmatrix}, \begin{bmatrix} 1 \\ 1 \\ 0 \end{bmatrix}, \begin{bmatrix} 1 \\ 1 \\ 1 \end{bmatrix}$ 2. $\begin{bmatrix} 1 \\ 1 \\ 0 \end{bmatrix}, \begin{bmatrix} 0 \\ 0 \\ 0 \end{bmatrix}, \begin{bmatrix} 0 \\ 1 \\ 1 \end{bmatrix}$

3. $\begin{bmatrix} 1 \\ 0 \\ -3 \end{bmatrix}, \begin{bmatrix} 3 \\ 1 \\ -4 \end{bmatrix}, \begin{bmatrix} -2 \\ -1 \\ 1 \end{bmatrix}$ 4. $\begin{bmatrix} 2 \\ -1 \\ 1 \end{bmatrix}, \begin{bmatrix} 2 \\ -3 \\ 2 \end{bmatrix}, \begin{bmatrix} -8 \\ 5 \\ 4 \end{bmatrix}$

5. $\begin{bmatrix} 3 \\ -3 \\ 0 \end{bmatrix}, \begin{bmatrix} -3 \\ 7 \\ 0 \end{bmatrix}, \begin{bmatrix} 0 \\ 0 \\ 0 \end{bmatrix}, \begin{bmatrix} 0 \\ -3 \\ 5 \end{bmatrix}$ 6. $\begin{bmatrix} 1 \\ 2 \\ -4 \end{bmatrix}, \begin{bmatrix} -4 \\ 3 \\ 6 \end{bmatrix}$

7. $\begin{bmatrix} -2 \\ 3 \\ 0 \end{bmatrix}, \begin{bmatrix} 6 \\ -1 \\ 5 \end{bmatrix}$ 8. $\begin{bmatrix} 1 \\ -2 \\ 3 \end{bmatrix}, \begin{bmatrix} 0 \\ 3 \\ -1 \end{bmatrix}, \begin{bmatrix} 2 \\ -1 \\ 5 \end{bmatrix}, \begin{bmatrix} 0 \\ 0 \\ -1 \end{bmatrix}$

Find bases for the null spaces of the matrices given in Exercises 9 and 10. Refer to the remarks that follow Example 3 in Section 4.2.

9. $\begin{bmatrix} 1 & 0 & -2 & -2 \\ 0 & 1 & 1 & 4 \\ 3 & -1 & -7 & 3 \end{bmatrix}$ 10. $\begin{bmatrix} 1 & 1 & -2 & 1 & 5 \\ 0 & 1 & 0 & -1 & -2 \\ 0 & 0 & -8 & 0 & 16 \end{bmatrix}$

11. Find a basis for the set of vectors in $\mathbb{R}^3$ in the plane $x - 3y + 2z = 0$. [*Hint:* Think of the equation as a "system" of homogeneous equations.]

12. Find a basis for the set of vectors in $\mathbb{R}^2$ on the line $y = -3x$.

In Exercises 13 and 14, assume that A is row equivalent to B. Find bases for Nul A and Col A.

13. $A = \begin{bmatrix} -2 & 4 & -2 & -4 \\ 2 & -6 & -3 & 1 \\ -3 & 8 & 2 & -3 \end{bmatrix}$, $B = \begin{bmatrix} 1 & 0 & 6 & 5 \\ 0 & 2 & 5 & 3 \\ 0 & 0 & 0 & 0 \end{bmatrix}$

14. $A = \begin{bmatrix} 1 & 2 & 3 & -4 & 8 \\ 1 & 2 & 0 & 2 & 8 \\ 2 & 4 & -3 & 10 & 9 \\ 3 & 6 & 0 & 6 & 9 \end{bmatrix}$,

$B = \begin{bmatrix} 1 & 2 & 0 & 2 & 5 \\ 0 & 0 & 3 & -6 & 3 \\ 0 & 0 & 0 & 0 & -7 \\ 0 & 0 & 0 & 0 & 0 \end{bmatrix}$

In Exercises 15–18, find a basis for the space spanned by the given vectors, $\mathbf{v}_1, \ldots, \mathbf{v}_5$.

15. $\begin{bmatrix} 1 \\ 0 \\ -2 \\ 3 \end{bmatrix}, \begin{bmatrix} 0 \\ 1 \\ 2 \\ 3 \end{bmatrix}, \begin{bmatrix} 2 \\ -2 \\ -8 \\ 0 \end{bmatrix}, \begin{bmatrix} 2 \\ -1 \\ 10 \\ 3 \end{bmatrix}, \begin{bmatrix} 3 \\ -1 \\ -6 \\ 9 \end{bmatrix}$

16. $\begin{bmatrix} 1 \\ 0 \\ 0 \\ 1 \end{bmatrix}, \begin{bmatrix} -2 \\ 0 \\ 0 \\ 2 \end{bmatrix}, \begin{bmatrix} 3 \\ -1 \\ 1 \\ -1 \end{bmatrix}, \begin{bmatrix} 5 \\ -3 \\ 3 \\ -4 \end{bmatrix}, \begin{bmatrix} 2 \\ -1 \\ 1 \\ 0 \end{bmatrix}$

17. [M] $\begin{bmatrix} 2 \\ 0 \\ -4 \\ -6 \\ 0 \end{bmatrix}, \begin{bmatrix} 4 \\ 0 \\ 2 \\ -4 \\ 4 \end{bmatrix}, \begin{bmatrix} -2 \\ -4 \\ 0 \\ 1 \\ -7 \end{bmatrix}, \begin{bmatrix} 8 \\ 4 \\ 8 \\ -3 \\ 15 \end{bmatrix}, \begin{bmatrix} -8 \\ 4 \\ 0 \\ 0 \\ 1 \end{bmatrix}$

18. [M] $\begin{bmatrix} -3 \\ 2 \\ 6 \\ 0 \\ -7 \end{bmatrix}, \begin{bmatrix} 3 \\ 0 \\ -9 \\ 0 \\ 6 \end{bmatrix}, \begin{bmatrix} 0 \\ 2 \\ -4 \\ 0 \\ -1 \end{bmatrix}, \begin{bmatrix} 6 \\ -2 \\ -14 \\ 0 \\ 13 \end{bmatrix}, \begin{bmatrix} -6 \\ 3 \\ 0 \\ -1 \\ 0 \end{bmatrix}$

19. Let $\mathbf{v}_1 = \begin{bmatrix} 4 \\ -3 \\ 7 \end{bmatrix}, \mathbf{v}_2 = \begin{bmatrix} 1 \\ 9 \\ -2 \end{bmatrix}, \mathbf{v}_3 = \begin{bmatrix} 7 \\ 11 \\ 6 \end{bmatrix}$, and also let $H = \text{Span}\{\mathbf{v}_1, \mathbf{v}_2, \mathbf{v}_3\}$. It can be verified that $4\mathbf{v}_1 + 5\mathbf{v}_2 - 3\mathbf{v}_3 = \mathbf{0}$. Use this information to find a basis for H. There is more than one answer.

20. Let $\mathbf{v}_1 = \begin{bmatrix} 3 \\ 4 \\ -2 \\ -5 \end{bmatrix}, \mathbf{v}_2 = \begin{bmatrix} 4 \\ 3 \\ 2 \\ 4 \end{bmatrix}$, and $\mathbf{v}_3 = \begin{bmatrix} 2 \\ 5 \\ -6 \\ -14 \end{bmatrix}$. It can be verified that $2\mathbf{v}_1 - \mathbf{v}_2 - \mathbf{v}_3 = \mathbf{0}$. Use this information to find a basis for $H = \text{Span}\{\mathbf{v}_1, \mathbf{v}_2, \mathbf{v}_3\}$.

In Exercises 21 and 22, mark each statement True or False. Justify each answer.

21. a. A single vector by itself is linearly dependent.

  b. If $H = \text{Span}\{\mathbf{b}_1, \ldots, \mathbf{b}_p\}$, then $\{\mathbf{b}_1, \ldots, \mathbf{b}_p\}$ is a basis for H.

  c. The columns of an invertible $n \times n$ matrix form a basis for $\mathbb{R}^n$.

  d. A basis is a spanning set that is as large as possible.

  e. In some cases, the linear dependence relations among the columns of a matrix can be affected by certain elementary row operations on the matrix.

22. a. A linearly independent set in a subspace H is a basis for H.

  b. If a finite set S of nonzero vectors spans a vector space V, then some subset of S is a basis for V.

  c. A basis is a linearly independent set that is as large as possible.

  d. The standard method for producing a spanning set for Nul A, described in Section 4.2, sometimes fails to produce a basis for Nul A.

  e. If B is an echelon form of a matrix A, then the pivot columns of B form a basis for Col A.

23. Suppose $\mathbb{R}^4 = \text{Span}\{\mathbf{v}_1, \ldots, \mathbf{v}_4\}$. Explain why $\{\mathbf{v}_1, \ldots, \mathbf{v}_4\}$ is a basis for $\mathbb{R}^4$.

24. Let $\mathcal{B} = \{\mathbf{v}_1, \ldots, \mathbf{v}_n\}$ be a linearly independent set in $\mathbb{R}^n$. Explain why $\mathcal{B}$ must be a basis for $\mathbb{R}^n$.

25. Let $\mathbf{v}_1 = \begin{bmatrix} 1 \\ 0 \\ 1 \end{bmatrix}, \mathbf{v}_2 = \begin{bmatrix} 0 \\ 1 \\ 1 \end{bmatrix}, \mathbf{v}_3 = \begin{bmatrix} 0 \\ 1 \\ 0 \end{bmatrix}$, and let H be the set of vectors in $\mathbb{R}^3$ whose second and third entries are equal. Then every vector in H has a unique expansion as a linear combination of $\mathbf{v}_1, \mathbf{v}_2, \mathbf{v}_3$, because

$$\begin{bmatrix} s \\ t \\ t \end{bmatrix} = s \begin{bmatrix} 1 \\ 0 \\ 1 \end{bmatrix} + (t - s) \begin{bmatrix} 0 \\ 1 \\ 1 \end{bmatrix} + s \begin{bmatrix} 0 \\ 1 \\ 0 \end{bmatrix}$$

for any s and t. Is $\{\mathbf{v}_1, \mathbf{v}_2, \mathbf{v}_3\}$ a basis for H? Why or why not?

26. In the vector space of all real-valued functions, find a basis for the subspace spanned by $\{\sin t, \sin 2t, \sin t \cos t\}$.

27. Let V be the vector space of functions that describe the vibration of a mass–spring system. (Refer to Exercise 19 in Section 4.1.) Find a basis for V.

28. (*RLC circuit*) The circuit in the figure consists of a resistor (R ohms), an inductor (L henrys), a capacitor (C farads), and an initial voltage source. Let $b = R/(2L)$, and suppose R, L, and C have been selected so that b also equals $1/\sqrt{LC}$. (This is done, for instance, when the circuit is used in a voltmeter.) Let $v(t)$ be the voltage (in volts) at time t, measured across the capacitor. It can be shown that v is in the null space H of the linear transformation that maps $v(t)$ into $Lv''(t) + Rv'(t) + (1/C)v(t)$, and H consists of all functions of the form $v(t) = e^{-bt}(c_1 + c_2 t)$. Find a basis for H.

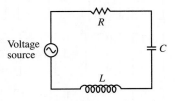

Exercises 29 and 30 show that every basis for $\mathbb{R}^n$ must contain exactly n vectors.

29. Let $S = \{\mathbf{v}_1, \ldots, \mathbf{v}_k\}$ be a set of k vectors in $\mathbb{R}^n$, with $k < n$. Use a theorem from Section 1.4 to explain why S cannot be a basis for $\mathbb{R}^n$.

30. Let $S = \{\mathbf{v}_1, \ldots, \mathbf{v}_k\}$ be a set of k vectors in $\mathbb{R}^n$, with $k > n$. Use a theorem from Chapter 1 to explain why S cannot be a basis for $\mathbb{R}^n$.

Exercises 31 and 32 reveal an important connection between linear independence and linear transformations and provide practice using the definition of linear dependence. Let V and W be vector spaces, let $T : V \rightarrow W$ be a linear transformation, and let $\{\mathbf{v}_1, \ldots, \mathbf{v}_p\}$ be a subset of V.

31. Show that if $\{\mathbf{v}_1, \ldots, \mathbf{v}_p\}$ is linearly dependent in V, then the set of images, $\{T(\mathbf{v}_1), \ldots, T(\mathbf{v}_p)\}$, is linearly dependent in W. This fact shows that if a linear transformation maps a set $\{\mathbf{v}_1, \ldots, \mathbf{v}_p\}$ onto a linearly *independent* set $\{T(\mathbf{v}_1), \ldots, T(\mathbf{v}_p)\}$, then the original set is linearly independent, too (because it cannot be linearly dependent).

32. Suppose that T is a one-to-one transformation, so that an equation $T(\mathbf{u}) = T(\mathbf{v})$ always implies $\mathbf{u} = \mathbf{v}$. Show that if the set of images $\{T(\mathbf{v}_1), \ldots, T(\mathbf{v}_p)\}$ is linearly dependent, then $\{\mathbf{v}_1, \ldots, \mathbf{v}_p\}$ is linearly dependent. This fact shows that *a one-to-one linear transformation maps a linearly independent set onto a linearly independent set* (because in this case the set of images cannot be linearly dependent).

33. Consider the polynomials $\mathbf{p}_1(t) = 1 + t^2$ and $\mathbf{p}_2(t) = 1 - t^2$. Is $\{\mathbf{p}_1, \mathbf{p}_2\}$ a linearly independent set in $\mathbb{P}_3$? Why or why not?

34. Consider the polynomials $\mathbf{p}_1(t) = 1 + t$, $\mathbf{p}_2(t) = 1 - t$, and $\mathbf{p}_3(t) = 2$ (for all t). By inspection, write a linear dependence relation among $\mathbf{p}_1$, $\mathbf{p}_2$, and $\mathbf{p}_3$. Then find a basis for Span $\{\mathbf{p}_1, \mathbf{p}_2, \mathbf{p}_3\}$.

35. Let V be a vector space that contains a linearly independent set $\{\mathbf{u}_1, \mathbf{u}_2, \mathbf{u}_3, \mathbf{u}_4\}$. Describe how to construct a set of vectors $\{\mathbf{v}_1, \mathbf{v}_2, \mathbf{v}_3, \mathbf{v}_4\}$ in V such that $\{\mathbf{v}_1, \mathbf{v}_3\}$ is a basis for Span $\{\mathbf{v}_1, \mathbf{v}_2, \mathbf{v}_3, \mathbf{v}_4\}$.

36. [M] Let $H = \text{Span}\,\{\mathbf{u}_1, \mathbf{u}_2, \mathbf{u}_3\}$ and $K = \text{Span}\,\{\mathbf{v}_1, \mathbf{v}_2, \mathbf{v}_3\}$, where

$$\mathbf{u}_1 = \begin{bmatrix} 1 \\ 2 \\ 0 \\ -1 \end{bmatrix}, \quad \mathbf{u}_2 = \begin{bmatrix} 0 \\ 2 \\ -1 \\ 1 \end{bmatrix}, \quad \mathbf{u}_3 = \begin{bmatrix} 3 \\ 4 \\ 1 \\ -4 \end{bmatrix},$$

$$\mathbf{v}_1 = \begin{bmatrix} -2 \\ -2 \\ -1 \\ 3 \end{bmatrix}, \quad \mathbf{v}_2 = \begin{bmatrix} 2 \\ 3 \\ 2 \\ -6 \end{bmatrix}, \quad \mathbf{v}_3 = \begin{bmatrix} -1 \\ 4 \\ 6 \\ -2 \end{bmatrix}.$$

Find bases for H, K, and $H + K$. (See Exercises 33 and 34 in Section 4.1.)

37. [M] Show that $\{t, \sin t, \cos 2t, \sin t \cos t\}$ is a linearly independent set of functions defined on $\mathbb{R}$. Start by assuming that

$$c_1 \cdot t + c_2 \cdot \sin t + c_3 \cdot \cos 2t + c_4 \cdot \sin t \cos t = 0 \qquad (5)$$

Equation (5) must hold for all real t, so choose several specific values of t (say, $t = 0, .1, .2$) until you get a system of enough equations to determine that all the c_j must be zero.

38. [M] Show that $\{1, \cos t, \cos^2 t, \ldots, \cos^6 t\}$ is a linearly independent set of functions defined on $\mathbb{R}$. Use the method of Exercise 37. (This result will be needed in Exercise 34 in Section 4.5.)

WEB

SOLUTIONS TO PRACTICE PROBLEMS

1. Let $A = [\,\mathbf{v}_1 \quad \mathbf{v}_2\,]$. Row operations show that

$$A = \begin{bmatrix} 1 & -2 \\ -2 & 7 \\ 3 & -9 \end{bmatrix} \sim \begin{bmatrix} 1 & -2 \\ 0 & 3 \\ 0 & 0 \end{bmatrix}$$

Not every row of A contains a pivot position. So the columns of A do not span $\mathbb{R}^3$, by Theorem 4 in Section 1.4. Hence $\{\mathbf{v}_1, \mathbf{v}_2\}$ is not a basis for $\mathbb{R}^3$. Since $\mathbf{v}_1$ and $\mathbf{v}_2$ are not in $\mathbb{R}^2$, they cannot possibly be a basis for $\mathbb{R}^2$. However, since $\mathbf{v}_1$ and $\mathbf{v}_2$ are obviously linearly independent, they are a basis for a subspace of $\mathbb{R}^3$, namely, Span $\{\mathbf{v}_1, \mathbf{v}_2\}$.

2. Set up a matrix A whose column space is the space spanned by $\{\mathbf{v}_1, \mathbf{v}_2, \mathbf{v}_3, \mathbf{v}_4\}$, and then row reduce A to find its pivot columns.

$$A = \begin{bmatrix} 1 & 6 & 2 & -4 \\ -3 & 2 & -2 & -8 \\ 4 & -1 & 3 & 9 \end{bmatrix} \sim \begin{bmatrix} 1 & 6 & 2 & -4 \\ 0 & 20 & 4 & -20 \\ 0 & -25 & -5 & 25 \end{bmatrix} \sim \begin{bmatrix} 1 & 6 & 2 & -4 \\ 0 & 5 & 1 & -5 \\ 0 & 0 & 0 & 0 \end{bmatrix}$$

The first two columns of A are the pivot columns and hence form a basis of Col $A = W$. Hence $\{\mathbf{v}_1, \mathbf{v}_2\}$ is a basis for W. Note that the reduced echelon form of A is not needed in order to locate the pivot columns.

3. Neither $\mathbf{v}_1$ nor $\mathbf{v}_2$ is in H, so $\{\mathbf{v}_1, \mathbf{v}_2\}$ cannot be a basis for H. In fact, $\{\mathbf{v}_1, \mathbf{v}_2\}$ is a basis for the *plane* of all vectors of the form $(c_1, c_2, 0)$, but H is only a *line*.

4.4 | COORDINATE SYSTEMS

An important reason for specifying a basis $\mathcal{B}$ for a vector space V is to impose a "coordinate system" on V. This section will show that if $\mathcal{B}$ contains n vectors, then the coordinate system will make V act like $\mathbb{R}^n$. If V is already $\mathbb{R}^n$ itself, then $\mathcal{B}$ will determine a coordinate system that gives a new "view" of V.

The existence of coordinate systems rests on the following fundamental result.

THEOREM 7

The Unique Representation Theorem

Let $\mathcal{B} = \{\mathbf{b}_1, \ldots, \mathbf{b}_n\}$ be a basis for a vector space V. Then for each $\mathbf{x}$ in V, there exists a unique set of scalars $c_1, \ldots, c_n$ such that

$$\mathbf{x} = c_1\mathbf{b}_1 + \cdots + c_n\mathbf{b}_n \qquad (1)$$

PROOF Since $\mathcal{B}$ spans V, there exist scalars such that (1) holds. Suppose $\mathbf{x}$ also has the representation

$$\mathbf{x} = d_1\mathbf{b}_1 + \cdots + d_n\mathbf{b}_n$$

for scalars $d_1, \ldots, d_n$. Then, subtracting, we have

$$\mathbf{0} = \mathbf{x} - \mathbf{x} = (c_1 - d_1)\mathbf{b}_1 + \cdots + (c_n - d_n)\mathbf{b}_n \qquad (2)$$

Since $\mathcal{B}$ is linearly independent, the weights in (2) must all be zero. That is, $c_j = d_j$ for $1 \le j \le n$. ∎

DEFINITION

Suppose $\mathcal{B} = \{\mathbf{b}_1, \ldots, \mathbf{b}_n\}$ is a basis for V and $\mathbf{x}$ is in V. The **coordinates of x relative to the basis** $\mathcal{B}$ (or the $\mathcal{B}$**-coordinates of x**) are the weights $c_1, \ldots, c_n$ such that $\mathbf{x} = c_1\mathbf{b}_1 + \cdots + c_n\mathbf{b}_n$.

If $c_1, \ldots, c_n$ are the $\mathcal{B}$-coordinates of $\mathbf{x}$, then the vector in $\mathbb{R}^n$

$$[\mathbf{x}]_{\mathcal{B}} = \begin{bmatrix} c_1 \\ \vdots \\ c_n \end{bmatrix}$$

is the **coordinate vector of x (relative to $\mathcal{B}$)**, or the $\mathcal{B}$**-coordinate vector of x**. The mapping $\mathbf{x} \mapsto [\mathbf{x}]_{\mathcal{B}}$ is the **coordinate mapping (determined by $\mathcal{B}$)**.[1]

---

[1] The concept of a coordinate mapping assumes that the basis $\mathcal{B}$ is an indexed set whose vectors are listed in some fixed preassigned order. This property makes the definition of $[\mathbf{x}]_{\mathcal{B}}$ unambiguous.

EXAMPLE 1 Consider a basis $\mathcal{B} = \{\mathbf{b}_1, \mathbf{b}_2\}$ for $\mathbb{R}^2$, where $\mathbf{b}_1 = \begin{bmatrix} 1 \\ 0 \end{bmatrix}$ and $\mathbf{b}_2 = \begin{bmatrix} 1 \\ 2 \end{bmatrix}$. Suppose an $\mathbf{x}$ in $\mathbb{R}^2$ has the coordinate vector $[\,\mathbf{x}\,]_{\mathcal{B}} = \begin{bmatrix} -2 \\ 3 \end{bmatrix}$. Find $\mathbf{x}$.

SOLUTION The $\mathcal{B}$-coordinates of $\mathbf{x}$ tell how to build $\mathbf{x}$ from the vectors in $\mathcal{B}$. That is,

$$\mathbf{x} = (-2)\mathbf{b}_1 + 3\mathbf{b}_2 = (-2)\begin{bmatrix} 1 \\ 0 \end{bmatrix} + 3\begin{bmatrix} 1 \\ 2 \end{bmatrix} = \begin{bmatrix} 1 \\ 6 \end{bmatrix} \quad\blacksquare$$

EXAMPLE 2 The entries in the vector $\mathbf{x} = \begin{bmatrix} 1 \\ 6 \end{bmatrix}$ are the coordinates of $\mathbf{x}$ relative to the *standard basis* $\mathcal{E} = \{\mathbf{e}_1, \mathbf{e}_2\}$, since

$$\begin{bmatrix} 1 \\ 6 \end{bmatrix} = 1 \cdot \begin{bmatrix} 1 \\ 0 \end{bmatrix} + 6 \cdot \begin{bmatrix} 0 \\ 1 \end{bmatrix} = 1 \cdot \mathbf{e}_1 + 6 \cdot \mathbf{e}_2$$

If $\mathcal{E} = \{\mathbf{e}_1, \mathbf{e}_2\}$, then $[\,\mathbf{x}\,]_{\mathcal{E}} = \mathbf{x}$. $\quad\blacksquare$

A Graphical Interpretation of Coordinates

A coordinate system on a set consists of a one-to-one mapping of the points in the set into $\mathbb{R}^n$. For example, ordinary graph paper provides a coordinate system for the plane when one selects perpendicular axes and a unit of measurement on each axis. Figure 1 shows the standard basis $\{\mathbf{e}_1, \mathbf{e}_2\}$, the vectors $\mathbf{b}_1 (= \mathbf{e}_1)$ and $\mathbf{b}_2$ from Example 1, and the vector $\mathbf{x} = \begin{bmatrix} 1 \\ 6 \end{bmatrix}$. The coordinates 1 and 6 give the location of $\mathbf{x}$ relative to the standard basis: 1 unit in the $\mathbf{e}_1$ direction and 6 units in the $\mathbf{e}_2$ direction.

Figure 2 shows the vectors $\mathbf{b}_1$, $\mathbf{b}_2$, and $\mathbf{x}$ from Fig. 1. (Geometrically, the three vectors lie on a vertical line in both figures.) However, the standard coordinate grid was erased and replaced by a grid especially adapted to the basis $\mathcal{B}$ in Example 1. The coordinate vector $[\,\mathbf{x}\,]_{\mathcal{B}} = \begin{bmatrix} -2 \\ 3 \end{bmatrix}$ gives the location of $\mathbf{x}$ on this new coordinate system: -2 units in the $\mathbf{b}_1$ direction and 3 units in the $\mathbf{b}_2$ direction.

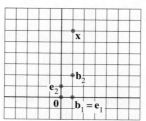

FIGURE 1 Standard graph paper.

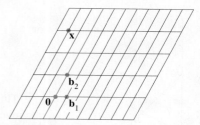

FIGURE 2 $\mathcal{B}$-graph paper.

EXAMPLE 3 In crystallography, the description of a crystal lattice is aided by choosing a basis $\{\mathbf{u}, \mathbf{v}, \mathbf{w}\}$ for $\mathbb{R}^3$ that corresponds to three adjacent edges of one "unit cell" of the crystal. An entire lattice is constructed by stacking together many copies of one cell. There are fourteen basic types of unit cells; three are displayed in Fig. 3.[2]

---

[2] Adapted from *The Science and Engineering of Materials*, 4th Ed., by Donald R. Askeland (Boston: Prindle, Weber & Schmidt, ©2002), p. 36.

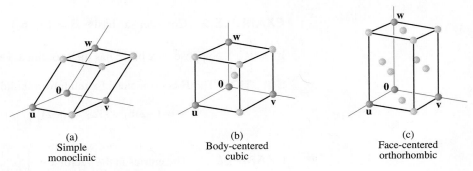

FIGURE 3 Examples of unit cells.

The coordinates of atoms within the crystal are given relative to the basis for the lattice. For instance,

$$\begin{bmatrix} 1/2 \\ 1/2 \\ 1 \end{bmatrix}$$

identifies the top face-centered atom in the cell in Fig. 3(c). ∎

Coordinates in $\mathbb{R}^n$

When a basis $\mathcal{B}$ for $\mathbb{R}^n$ is fixed, the $\mathcal{B}$-coordinate vector of a specified $\mathbf{x}$ is easily found, as in the next example.

EXAMPLE 4 Let $\mathbf{b}_1 = \begin{bmatrix} 2 \\ 1 \end{bmatrix}$, $\mathbf{b}_2 = \begin{bmatrix} -1 \\ 1 \end{bmatrix}$, $\mathbf{x} = \begin{bmatrix} 4 \\ 5 \end{bmatrix}$, and $\mathcal{B} = \{\mathbf{b}_1, \mathbf{b}_2\}$. Find the coordinate vector $[\,\mathbf{x}\,]_{\mathcal{B}}$ of $\mathbf{x}$ relative to $\mathcal{B}$.

SOLUTION The $\mathcal{B}$-coordinates c_1, c_2 of $\mathbf{x}$ satisfy

$$c_1 \underset{\mathbf{b}_1}{\begin{bmatrix} 2 \\ 1 \end{bmatrix}} + c_2 \underset{\mathbf{b}_2}{\begin{bmatrix} -1 \\ 1 \end{bmatrix}} = \underset{\mathbf{x}}{\begin{bmatrix} 4 \\ 5 \end{bmatrix}}$$

or

$$\underset{\mathbf{b}_1 \quad \mathbf{b}_2}{\begin{bmatrix} 2 & -1 \\ 1 & 1 \end{bmatrix}} \begin{bmatrix} c_1 \\ c_2 \end{bmatrix} = \underset{\mathbf{x}}{\begin{bmatrix} 4 \\ 5 \end{bmatrix}} \qquad (3)$$

This equation can be solved by row operations on an augmented matrix or by using the inverse of the matrix on the left. In any case, the solution is $c_1 = 3$, $c_2 = 2$. Thus $\mathbf{x} = 3\mathbf{b}_1 + 2\mathbf{b}_2$, and

$$[\,\mathbf{x}\,]_{\mathcal{B}} = \begin{bmatrix} c_1 \\ c_2 \end{bmatrix} = \begin{bmatrix} 3 \\ 2 \end{bmatrix}$$ ∎

See Fig. 4.

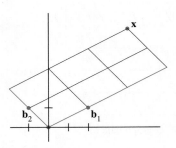

FIGURE 4

The $\mathcal{B}$-coordinate vector of $\mathbf{x}$ is $(3, 2)$.

The matrix in (3) changes the $\mathcal{B}$-coordinates of a vector $\mathbf{x}$ into the standard coordinates for $\mathbf{x}$. An analogous change of coordinates can be carried out in $\mathbb{R}^n$ for a basis $\mathcal{B} = \{\mathbf{b}_1, \ldots, \mathbf{b}_n\}$. Let

$$P_{\mathcal{B}} = [\,\mathbf{b}_1 \quad \mathbf{b}_2 \quad \cdots \quad \mathbf{b}_n\,]$$

Then the vector equation

$$\mathbf{x} = c_1\mathbf{b}_1 + c_2\mathbf{b}_2 + \cdots + c_n\mathbf{b}_n$$

is equivalent to

$$\boxed{\mathbf{x} = P_\mathcal{B}[\,\mathbf{x}\,]_\mathcal{B}} \qquad (4)$$

We call $P_\mathcal{B}$ the **change-of-coordinates matrix** from $\mathcal{B}$ to the standard basis in $\mathbb{R}^n$. Left-multiplication by $P_\mathcal{B}$ transforms the coordinate vector $[\,\mathbf{x}\,]_\mathcal{B}$ into $\mathbf{x}$. The change-of-coordinates equation (4) is important and will be needed at several points in Chapters 5 and 7.

Since the columns of $P_\mathcal{B}$ form a basis for $\mathbb{R}^n$, $P_\mathcal{B}$ is invertible (by the Invertible Matrix Theorem). Left-multiplication by $P_\mathcal{B}^{-1}$ converts $\mathbf{x}$ into its $\mathcal{B}$-coordinate vector:

$$P_\mathcal{B}^{-1}\mathbf{x} = [\,\mathbf{x}\,]_\mathcal{B}$$

The correspondence $\mathbf{x} \mapsto [\,\mathbf{x}\,]_\mathcal{B}$, produced here by $P_\mathcal{B}^{-1}$, is the coordinate mapping mentioned earlier. Since $P_\mathcal{B}^{-1}$ is an invertible matrix, the coordinate mapping is a one-to-one linear transformation from $\mathbb{R}^n$ onto $\mathbb{R}^n$, by the Invertible Matrix Theorem. (See also Theorem 12 in Section 1.9.) This property of the coordinate mapping is also true in a general vector space that has a basis, as we shall see.

The Coordinate Mapping

Choosing a basis $\mathcal{B} = \{\mathbf{b}_1, \ldots, \mathbf{b}_n\}$ for a vector space V introduces a coordinate system in V. The coordinate mapping $\mathbf{x} \mapsto [\,\mathbf{x}\,]_\mathcal{B}$ connects the possibly unfamiliar space V to the familiar space $\mathbb{R}^n$. See Fig. 5. Points in V can now be identified by their new "names."

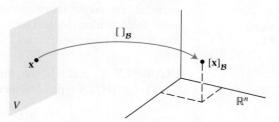

FIGURE 5 The coordinate mapping from V onto $\mathbb{R}^n$.

THEOREM 8

Let $\mathcal{B} = \{\mathbf{b}_1, \ldots, \mathbf{b}_n\}$ be a basis for a vector space V. Then the coordinate mapping $\mathbf{x} \mapsto [\,\mathbf{x}\,]_\mathcal{B}$ is a one-to-one linear transformation from V onto $\mathbb{R}^n$.

PROOF Take two typical vectors in V, say,

$$\mathbf{u} = c_1\mathbf{b}_1 + \cdots + c_n\mathbf{b}_n$$
$$\mathbf{w} = d_1\mathbf{b}_1 + \cdots + d_n\mathbf{b}_n$$

Then, using vector operations,

$$\mathbf{u} + \mathbf{w} = (c_1 + d_1)\mathbf{b}_1 + \cdots + (c_n + d_n)\mathbf{b}_n$$

It follows that

$$[\mathbf{u} + \mathbf{w}]_{\mathcal{B}} = \begin{bmatrix} c_1 + d_1 \\ \vdots \\ c_n + d_n \end{bmatrix} = \begin{bmatrix} c_1 \\ \vdots \\ c_n \end{bmatrix} + \begin{bmatrix} d_1 \\ \vdots \\ d_n \end{bmatrix} = [\mathbf{u}]_{\mathcal{B}} + [\mathbf{w}]_{\mathcal{B}}$$

So the coordinate mapping preserves addition. If r is any scalar, then

$$r\mathbf{u} = r(c_1\mathbf{b}_1 + \cdots + c_n\mathbf{b}_n) = (rc_1)\mathbf{b}_1 + \cdots + (rc_n)\mathbf{b}_n$$

So

$$[r\mathbf{u}]_{\mathcal{B}} = \begin{bmatrix} rc_1 \\ \vdots \\ rc_n \end{bmatrix} = r\begin{bmatrix} c_1 \\ \vdots \\ c_n \end{bmatrix} = r[\mathbf{u}]_{\mathcal{B}}$$

Thus the coordinate mapping also preserves scalar multiplication and hence is a linear transformation. See Exercises 23 and 24 for verification that the coordinate mapping is one-to-one and maps V onto $\mathbb{R}^n$. ∎

The linearity of the coordinate mapping extends to linear combinations, just as in Section 1.8. If $\mathbf{u}_1, \ldots, \mathbf{u}_p$ are in V and if $c_1, \ldots, c_p$ are scalars, then

$$[c_1\mathbf{u}_1 + \cdots + c_p\mathbf{u}_p]_{\mathcal{B}} = c_1[\mathbf{u}_1]_{\mathcal{B}} + \cdots + c_p[\mathbf{u}_p]_{\mathcal{B}} \tag{5}$$

In words, (5) says that the $\mathcal{B}$-coordinate vector of a linear combination of $\mathbf{u}_1, \ldots, \mathbf{u}_p$ is the *same* linear combination of their coordinate vectors.

The coordinate mapping in Theorem 8 is an important example of an *isomorphism* from V onto $\mathbb{R}^n$. In general, a one-to-one linear transformation from a vector space V onto a vector space W is called an **isomorphism** from V onto W (*iso* from the Greek for "the same," and *morph* from the Greek for "form" or "structure"). The notation and terminology for V and W may differ, but the two spaces are indistinguishable as vector spaces. *Every vector space calculation in V is accurately reproduced in W, and vice versa.* In particular, any real vector space with a basis of n vectors is indistinguishable from $\mathbb{R}^n$. See Exercises 25 and 26.

| SG | Isomorphic Vector
         Spaces 4–11

EXAMPLE 5   Let $\mathcal{B}$ be the standard basis of the space $\mathbb{P}_3$ of polynomials; that is, let $\mathcal{B} = \{1, t, t^2, t^3\}$. A typical element $\mathbf{p}$ of $\mathbb{P}_3$ has the form

$$\mathbf{p}(t) = a_0 + a_1 t + a_2 t^2 + a_3 t^3$$

Since $\mathbf{p}$ is already displayed as a linear combination of the standard basis vectors, we conclude that

$$[\mathbf{p}]_{\mathcal{B}} = \begin{bmatrix} a_0 \\ a_1 \\ a_2 \\ a_3 \end{bmatrix}$$

Thus the coordinate mapping $\mathbf{p} \mapsto [\mathbf{p}]_{\mathcal{B}}$ is an isomorphism from $\mathbb{P}_3$ onto $\mathbb{R}^4$. All vector space operations in $\mathbb{P}_3$ correspond to operations in $\mathbb{R}^4$. ∎

If we think of $\mathbb{P}_3$ and $\mathbb{R}^4$ as displays on two computer screens that are connected via the coordinate mapping, then every vector space operation in $\mathbb{P}_3$ on one screen is exactly duplicated by a corresponding vector operation in $\mathbb{R}^4$ on the other screen. The vectors on the $\mathbb{P}_3$ screen look different from those on the $\mathbb{R}^4$ screen, but they "act" as vectors in exactly the same way. See Fig. 6.

FIGURE 6 The space $\mathbb{P}_3$ is isomorphic to $\mathbb{R}^4$.

EXAMPLE 6   Use coordinate vectors to verify that the polynomials $1 + 2t^2$, $4 + t + 5t^2$, and $3 + 2t$ are linearly dependent in $\mathbb{P}_2$.

SOLUTION The coordinate mapping from Example 5 produces the coordinate vectors $(1, 0, 2)$, $(4, 1, 5)$, and $(3, 2, 0)$, respectively. Writing these vectors as the *columns* of a matrix A, we can determine their independence by row reducing the augmented matrix for $A\mathbf{x} = \mathbf{0}$:

$$\begin{bmatrix} 1 & 4 & 3 & 0 \\ 0 & 1 & 2 & 0 \\ 2 & 5 & 0 & 0 \end{bmatrix} \sim \begin{bmatrix} 1 & 4 & 3 & 0 \\ 0 & 1 & 2 & 0 \\ 0 & 0 & 0 & 0 \end{bmatrix}$$

The columns of A are linearly dependent, so the corresponding polynomials are linearly dependent. In fact, it is easy to check that column 3 of A is 2 times column 2 minus 5 times column 1. The corresponding relation for the polynomials is

$$3 + 2t = 2(4 + t + 5t^2) - 5(1 + 2t^2) \qquad ■$$

The final example concerns a plane in $\mathbb{R}^3$ that is isomorphic to $\mathbb{R}^2$.

EXAMPLE 7   Let

$$\mathbf{v}_1 = \begin{bmatrix} 3 \\ 6 \\ 2 \end{bmatrix}, \quad \mathbf{v}_2 = \begin{bmatrix} -1 \\ 0 \\ 1 \end{bmatrix}, \quad \mathbf{x} = \begin{bmatrix} 3 \\ 12 \\ 7 \end{bmatrix},$$

and $\mathcal{B} = \{\mathbf{v}_1, \mathbf{v}_2\}$. Then $\mathcal{B}$ is a basis for $H = \text{Span}\,\{\mathbf{v}_1, \mathbf{v}_2\}$. Determine if $\mathbf{x}$ is in H, and if it is, find the coordinate vector of $\mathbf{x}$ relative to $\mathcal{B}$.

SOLUTION If $\mathbf{x}$ is in H, then the following vector equation is consistent:

$$c_1 \begin{bmatrix} 3 \\ 6 \\ 2 \end{bmatrix} + c_2 \begin{bmatrix} -1 \\ 0 \\ 1 \end{bmatrix} = \begin{bmatrix} 3 \\ 12 \\ 7 \end{bmatrix}$$

The scalars c_1 and c_2, if they exist, are the $\mathcal{B}$-coordinates of $\mathbf{x}$. Using row operations, we obtain

$$\begin{bmatrix} 3 & -1 & 3 \\ 6 & 0 & 12 \\ 2 & 1 & 7 \end{bmatrix} \sim \begin{bmatrix} 1 & 0 & 2 \\ 0 & 1 & 3 \\ 0 & 0 & 0 \end{bmatrix}$$

Thus $c_1 = 2$, $c_2 = 3$, and $[\mathbf{x}]_\mathcal{B} = \begin{bmatrix} 2 \\ 3 \end{bmatrix}$. The coordinate system on H determined by $\mathcal{B}$ is shown in Fig. 7. ∎

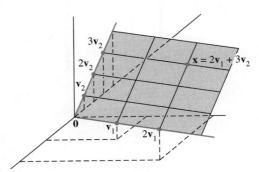

FIGURE 7 A coordinate system on a plane H in $\mathbb{R}^3$.

If a different basis for H were chosen, would the associated coordinate system also make H isomorphic to $\mathbb{R}^2$? Surely, this must be true. We shall prove it in the next section.

PRACTICE PROBLEMS

1. Let $\mathbf{b}_1 = \begin{bmatrix} 1 \\ 0 \\ 0 \end{bmatrix}$, $\mathbf{b}_2 = \begin{bmatrix} -3 \\ 4 \\ 0 \end{bmatrix}$, $\mathbf{b}_3 = \begin{bmatrix} 3 \\ -6 \\ 3 \end{bmatrix}$, and $\mathbf{x} = \begin{bmatrix} -8 \\ 2 \\ 3 \end{bmatrix}$.

   a. Show that the set $\mathcal{B} = \{\mathbf{b}_1, \mathbf{b}_2, \mathbf{b}_3\}$ is a basis of $\mathbb{R}^3$.

   b. Find the change-of-coordinates matrix from $\mathcal{B}$ to the standard basis.

   c. Write the equation that relates $\mathbf{x}$ in $\mathbb{R}^3$ to $[\mathbf{x}]_\mathcal{B}$.

   d. Find $[\mathbf{x}]_\mathcal{B}$, for the $\mathbf{x}$ given above.

2. The set $\mathcal{B} = \{1 + t, 1 + t^2, t + t^2\}$ is a basis for $\mathbb{P}_2$. Find the coordinate vector of $\mathbf{p}(t) = 6 + 3t - t^2$ relative to $\mathcal{B}$.

4.4 EXERCISES

In Exercises 1–4, find the vector $\mathbf{x}$ determined by the given coordinate vector $[\mathbf{x}]_\mathcal{B}$ and the given basis $\mathcal{B}$.

1. $\mathcal{B} = \left\{ \begin{bmatrix} 3 \\ -5 \end{bmatrix}, \begin{bmatrix} -4 \\ 6 \end{bmatrix} \right\}$, $[\mathbf{x}]_\mathcal{B} = \begin{bmatrix} 5 \\ 3 \end{bmatrix}$

2. $\mathcal{B} = \left\{ \begin{bmatrix} 3 \\ 2 \end{bmatrix}, \begin{bmatrix} -4 \\ 1 \end{bmatrix} \right\}$, $[\mathbf{x}]_\mathcal{B} = \begin{bmatrix} -2 \\ 5 \end{bmatrix}$

3. $\mathcal{B} = \left\{ \begin{bmatrix} 1 \\ -2 \\ 3 \end{bmatrix}, \begin{bmatrix} 5 \\ 0 \\ -2 \end{bmatrix}, \begin{bmatrix} 4 \\ -3 \\ 0 \end{bmatrix} \right\}$, $[\mathbf{x}]_\mathcal{B} = \begin{bmatrix} 1 \\ 0 \\ -2 \end{bmatrix}$

4. $\mathcal{B} = \left\{ \begin{bmatrix} -2 \\ 2 \\ 0 \end{bmatrix}, \begin{bmatrix} 3 \\ 0 \\ 2 \end{bmatrix}, \begin{bmatrix} 4 \\ -1 \\ 3 \end{bmatrix} \right\}$, $[\mathbf{x}]_\mathcal{B} = \begin{bmatrix} -3 \\ 2 \\ -1 \end{bmatrix}$

In Exercises 5–8, find the coordinate vector $[\mathbf{x}]_\mathcal{B}$ of $\mathbf{x}$ relative to the given basis $\mathcal{B} = \{\mathbf{b}_1, \ldots, \mathbf{b}_n\}$.

5. $\mathbf{b}_1 = \begin{bmatrix} 1 \\ -2 \end{bmatrix}$, $\mathbf{b}_2 = \begin{bmatrix} 3 \\ -5 \end{bmatrix}$, $\mathbf{x} = \begin{bmatrix} -1 \\ 1 \end{bmatrix}$

6. $\mathbf{b}_1 = \begin{bmatrix} 1 \\ -4 \end{bmatrix}$, $\mathbf{b}_2 = \begin{bmatrix} 2 \\ -3 \end{bmatrix}$, $\mathbf{x} = \begin{bmatrix} -1 \\ -6 \end{bmatrix}$

7. $\mathbf{b}_1 = \begin{bmatrix} 1 \\ -1 \\ -3 \end{bmatrix}$, $\mathbf{b}_2 = \begin{bmatrix} -3 \\ 4 \\ 9 \end{bmatrix}$, $\mathbf{b}_3 = \begin{bmatrix} 2 \\ -2 \\ 4 \end{bmatrix}$, $\mathbf{x} = \begin{bmatrix} 8 \\ -9 \\ 6 \end{bmatrix}$

8. $\mathbf{b}_1 = \begin{bmatrix} 1 \\ 1 \\ 3 \end{bmatrix}$, $\mathbf{b}_2 = \begin{bmatrix} 2 \\ 0 \\ 8 \end{bmatrix}$, $\mathbf{b}_3 = \begin{bmatrix} 1 \\ -1 \\ 3 \end{bmatrix}$, $\mathbf{x} = \begin{bmatrix} 0 \\ 0 \\ -2 \end{bmatrix}$

In Exercises 9 and 10, find the change-of-coordinates matrix from $\mathcal{B}$ to the standard basis in $\mathbb{R}^n$.

9. $\mathcal{B} = \left\{ \begin{bmatrix} 1 \\ -3 \end{bmatrix}, \begin{bmatrix} 2 \\ -5 \end{bmatrix} \right\}$

10. $\mathcal{B} = \left\{ \begin{bmatrix} 3 \\ 0 \\ 6 \end{bmatrix}, \begin{bmatrix} 2 \\ 2 \\ -4 \end{bmatrix}, \begin{bmatrix} 1 \\ -2 \\ 3 \end{bmatrix} \right\}$

In Exercises 11 and 12, use an inverse matrix to find $[\mathbf{x}]_{\mathcal{B}}$ for the given $\mathbf{x}$ and $\mathcal{B}$.

11. $\mathcal{B} = \left\{ \begin{bmatrix} 1 \\ -2 \end{bmatrix}, \begin{bmatrix} -3 \\ 5 \end{bmatrix} \right\}$, $\mathbf{x} = \begin{bmatrix} 2 \\ -5 \end{bmatrix}$

12. $\mathcal{B} = \left\{ \begin{bmatrix} 1 \\ -1 \end{bmatrix}, \begin{bmatrix} 2 \\ -1 \end{bmatrix} \right\}$, $\mathbf{x} = \begin{bmatrix} 2 \\ 3 \end{bmatrix}$

13. The set $\mathcal{B} = \{1 + t^2, t + t^2, 1 + 2t + t^2\}$ is a basis for $\mathbb{P}_2$. Find the coordinate vector of $\mathbf{p}(t) = 1 + 4t + 7t^2$ relative to $\mathcal{B}$.

14. The set $\mathcal{B} = \{1 - t^2, t - t^2, 2 - t + t^2\}$ is a basis for $\mathbb{P}_2$. Find the coordinate vector of $\mathbf{p}(t) = 1 + 3t - 6t^2$ relative to $\mathcal{B}$.

In Exercises 15 and 16, mark each statement True or False. Justify each answer. Unless stated otherwise, $\mathcal{B}$ is a basis for a vector space V.

15. a. If $\mathbf{x}$ is in V and if $\mathcal{B}$ contains n vectors, then the $\mathcal{B}$-coordinate vector of $\mathbf{x}$ is in $\mathbb{R}^n$.

   b. If $P_{\mathcal{B}}$ is the change-of-coordinates matrix, then $[\mathbf{x}]_{\mathcal{B}} = P_{\mathcal{B}}\mathbf{x}$, for $\mathbf{x}$ in V.

   c. The vector spaces $\mathbb{P}_3$ and $\mathbb{R}^3$ are isomorphic.

16. a. If $\mathcal{B}$ is the standard basis for $\mathbb{R}^n$, then the $\mathcal{B}$-coordinate vector of an $\mathbf{x}$ in $\mathbb{R}^n$ is $\mathbf{x}$ itself.

   b. The correspondence $[\mathbf{x}]_{\mathcal{B}} \mapsto \mathbf{x}$ is called the coordinate mapping.

   c. In some cases, a plane in $\mathbb{R}^3$ can be isomorphic to $\mathbb{R}^2$.

17. The vectors $\mathbf{v}_1 = \begin{bmatrix} 1 \\ -3 \end{bmatrix}$, $\mathbf{v}_2 = \begin{bmatrix} 2 \\ -8 \end{bmatrix}$, $\mathbf{v}_3 = \begin{bmatrix} -3 \\ 7 \end{bmatrix}$ span $\mathbb{R}^2$ but do not form a basis. Find two different ways to express $\begin{bmatrix} 1 \\ 1 \end{bmatrix}$ as a linear combination of $\mathbf{v}_1, \mathbf{v}_2, \mathbf{v}_3$.

18. Let $\mathcal{B} = \{\mathbf{b}_1, \ldots, \mathbf{b}_n\}$ be a basis for a vector space V. Explain why the $\mathcal{B}$-coordinate vectors of $\mathbf{b}_1, \ldots, \mathbf{b}_n$ are the columns $\mathbf{e}_1, \ldots, \mathbf{e}_n$ of the $n \times n$ identity matrix.

19. Let S be a finite set in a vector space V with the property that every $\mathbf{x}$ in V has a unique representation as a linear combination of elements of S. Show that S is a basis of V.

20. Suppose $\{\mathbf{v}_1, \ldots, \mathbf{v}_4\}$ is a linearly dependent spanning set for a vector space V. Show that each $\mathbf{w}$ in V can be expressed in more than one way as a linear combination of $\mathbf{v}_1, \ldots, \mathbf{v}_4$. [*Hint:* Let $\mathbf{w} = k_1\mathbf{v}_1 + \cdots + k_4\mathbf{v}_4$ be an arbitrary vector in V. Use the linear dependence of $\{\mathbf{v}_1, \ldots, \mathbf{v}_4\}$ to produce another representation of $\mathbf{w}$ as a linear combination of $\mathbf{v}_1, \ldots, \mathbf{v}_4$.]

21. Let $\mathcal{B} = \left\{ \begin{bmatrix} 1 \\ -4 \end{bmatrix}, \begin{bmatrix} -2 \\ 9 \end{bmatrix} \right\}$. Since the coordinate mapping determined by $\mathcal{B}$ is a linear transformation from $\mathbb{R}^2$ into $\mathbb{R}^2$, this mapping must be implemented by some 2×2 matrix A. Find it. [*Hint:* Multiplication by A should transform a vector $\mathbf{x}$ into its coordinate vector $[\mathbf{x}]_{\mathcal{B}}$.]

22. Let $\mathcal{B} = \{\mathbf{b}_1, \ldots, \mathbf{b}_n\}$ be a basis for $\mathbb{R}^n$. Produce a description of an $n \times n$ matrix A that implements the coordinate mapping $\mathbf{x} \mapsto [\mathbf{x}]_{\mathcal{B}}$. (See Exercise 21.)

Exercises 23–26 concern a vector space V, a basis $\mathcal{B} = \{\mathbf{b}_1, \ldots, \mathbf{b}_n\}$, and the coordinate mapping $\mathbf{x} \mapsto [\mathbf{x}]_{\mathcal{B}}$.

23. Show that the coordinate mapping is one-to-one. (*Hint:* Suppose $[\mathbf{u}]_{\mathcal{B}} = [\mathbf{w}]_{\mathcal{B}}$ for some $\mathbf{u}$ and $\mathbf{w}$ in V, and show that $\mathbf{u} = \mathbf{w}$.)

24. Show that the coordinate mapping is *onto* $\mathbb{R}^n$. That is, given any $\mathbf{y}$ in $\mathbb{R}^n$, with entries $y_1, \ldots, y_n$, produce $\mathbf{u}$ in V such that $[\mathbf{u}]_{\mathcal{B}} = \mathbf{y}$.

25. Show that a subset $\{\mathbf{u}_1, \ldots, \mathbf{u}_p\}$ in V is linearly independent if and only if the set of coordinate vectors $\{[\mathbf{u}_1]_{\mathcal{B}}, \ldots, [\mathbf{u}_p]_{\mathcal{B}}\}$ is linearly independent in $\mathbb{R}^n$. *Hint:* Since the coordinate mapping is one-to-one, the following equations have the same solutions, $c_1, \ldots, c_p$.

$$c_1\mathbf{u}_1 + \cdots + c_p\mathbf{u}_p = \mathbf{0} \qquad \text{The zero vector in } V$$
$$[c_1\mathbf{u}_1 + \cdots + c_p\mathbf{u}_p]_{\mathcal{B}} = [\mathbf{0}]_{\mathcal{B}} \qquad \text{The zero vector in } \mathbb{R}^n$$

26. Given vectors $\mathbf{u}_1, \ldots, \mathbf{u}_p$, and $\mathbf{w}$ in V, show that $\mathbf{w}$ is a linear combination of $\mathbf{u}_1, \ldots, \mathbf{u}_p$ if and only if $[\mathbf{w}]_{\mathcal{B}}$ is a linear combination of the coordinate vectors $[\mathbf{u}_1]_{\mathcal{B}}, \ldots, [\mathbf{u}_p]_{\mathcal{B}}$.

In Exercises 27–30, use coordinate vectors to test the linear independence of the sets of polynomials. Explain your work.

27. $1 + 2t^3, 2 + t - 3t^2, -t + 2t^2 - t^3$

28. $1 - 2t^2 - t^3, t + 2t^3, 1 + t - 2t^2$

29. $(1 - t)^2, t - 2t^2 + t^3, (1 - t)^3$

30. $(2 - t)^3, (3 - t)^2, 1 + 6t - 5t^2 + t^3$

31. Use coordinate vectors to test whether the following sets of polynomials span $\mathbb{P}_2$. Justify your conclusions.

   a. $1 - 3t + 5t^2, -3 + 5t - 7t^2, -4 + 5t - 6t^2, 1 - t^2$

   b. $5t + t^2, 1 - 8t - 2t^2, -3 + 4t + 2t^2, 2 - 3t$

32. Let $\mathbf{p}_1(t) = 1 + t^2, \mathbf{p}_2(t) = t - 3t^2, \mathbf{p}_3(t) = 1 + t - 3t^2$.

   a. Use coordinate vectors to show that these polynomials form a basis for $\mathbb{P}_2$.

   b. Consider the basis $\mathcal{B} = \{\mathbf{p}_1, \mathbf{p}_2, \mathbf{p}_3\}$ for $\mathbb{P}_2$. Find $\mathbf{q}$ in $\mathbb{P}_2$, given that $[\mathbf{q}]_{\mathcal{B}} = \begin{bmatrix} -1 \\ 1 \\ 2 \end{bmatrix}$.

In Exercises 33 and 34, determine whether the sets of polynomials form a basis for $\mathbb{P}_3$. Justify your conclusions.

33. [M] $3 + 7t, 5 + t - 2t^3, t - 2t^2, 1 + 16t - 6t^2 + 2t^3$

34. [M] $5 - 3t + 4t^2 + 2t^3, 9 + t + 8t^2 - 6t^3, 6 - 2t + 5t^2, t^3$

35. [M] Let $H = \text{Span}\{v_1, v_2\}$ and $\mathcal{B} = \{v_1, v_2\}$. Show that x is in H and find the $\mathcal{B}$-coordinate vector of x, for

$$v_1 = \begin{bmatrix} 11 \\ -5 \\ 10 \\ 7 \end{bmatrix}, v_2 = \begin{bmatrix} 14 \\ -8 \\ 13 \\ 10 \end{bmatrix}, x = \begin{bmatrix} 19 \\ -13 \\ 18 \\ 15 \end{bmatrix}$$

36. [M] Let $H = \text{Span}\{v_1, v_2, v_3\}$ and $\mathcal{B} = \{v_1, v_2, v_3\}$. Show that $\mathcal{B}$ is a basis for H and x is in H, and find the $\mathcal{B}$-coordinate vector of x, for

$$v_1 = \begin{bmatrix} -6 \\ 4 \\ -9 \\ 4 \end{bmatrix}, v_2 = \begin{bmatrix} 8 \\ -3 \\ 7 \\ -3 \end{bmatrix}, v_3 = \begin{bmatrix} -9 \\ 5 \\ -8 \\ 3 \end{bmatrix}, x = \begin{bmatrix} 4 \\ 7 \\ -8 \\ 3 \end{bmatrix}$$

[M] Exercises 37 and 38 concern the crystal lattice for titanium, which has the hexagonal structure shown on the left in the accompanying figure. The vectors $\begin{bmatrix} 2.6 \\ -1.5 \\ 0 \end{bmatrix}, \begin{bmatrix} 0 \\ 3 \\ 0 \end{bmatrix}, \begin{bmatrix} 0 \\ 0 \\ 4.8 \end{bmatrix}$ in $\mathbb{R}^3$ form a basis for the unit cell shown on the right. The numbers here are Ångstrom units (1 Å $= 10^{-8}$ cm). In alloys of titanium, some additional atoms may be in the unit cell at the *octahedral* and *tetrahedral* sites (so named because of the geometric objects formed by atoms at these locations).

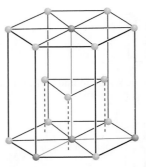 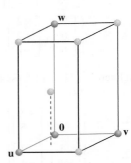

The hexagonal close-packed lattice and its unit cell.

37. One of the octahedral sites is $\begin{bmatrix} 1/2 \\ 1/4 \\ 1/6 \end{bmatrix}$, relative to the lattice basis. Determine the coordinates of this site relative to the standard basis of $\mathbb{R}^3$.

38. One of the tetrahedral sites is $\begin{bmatrix} 1/2 \\ 1/2 \\ 1/3 \end{bmatrix}$. Determine the coordinates of this site relative to the standard basis of $\mathbb{R}^3$.

SOLUTIONS TO PRACTICE PROBLEMS

1. a. It is evident that the matrix $P_\mathcal{B} = [\,b_1 \quad b_2 \quad b_3\,]$ is row-equivalent to the identity matrix. By the Invertible Matrix Theorem, $P_\mathcal{B}$ is invertible and its columns form a basis for $\mathbb{R}^3$.

   b. From part (a), the change-of-coordinates matrix is $P_\mathcal{B} = \begin{bmatrix} 1 & -3 & 3 \\ 0 & 4 & -6 \\ 0 & 0 & 3 \end{bmatrix}$.

   c. $x = P_\mathcal{B}[\,x\,]_\mathcal{B}$

   d. To solve the equation in (c), it is probably easier to row reduce an augmented matrix than to compute $P_\mathcal{B}^{-1}$:

$$\begin{matrix} \begin{bmatrix} 1 & -3 & 3 & -8 \\ 0 & 4 & -6 & 2 \\ 0 & 0 & 3 & 3 \end{bmatrix} & \sim & \begin{bmatrix} 1 & 0 & 0 & -5 \\ 0 & 1 & 0 & 2 \\ 0 & 0 & 1 & 1 \end{bmatrix} \\ & & \\ P_\mathcal{B} \quad\quad x & & I \quad\quad [\,x\,]_\mathcal{B} \end{matrix}$$

   Hence

$$[\,x_\mathcal{B}\,] = \begin{bmatrix} -5 \\ 2 \\ 1 \end{bmatrix}$$

2. The coordinates of $p(t) = 6 + 3t - t^2$ with respect to $\mathcal{B}$ satisfy

$$c_1(1 + t) + c_2(1 + t^2) + c_3(t + t^2) = 6 + 3t - t^2$$

Equating coefficients of like powers of t, we have

$$
\begin{aligned}
c_1 + c_2 \quad\quad &= \quad 6 \\
c_1 \quad\quad + c_3 &= \quad 3 \\
c_2 + c_3 &= -1
\end{aligned}
$$

Solving, we find that $c_1 = 5$, $c_2 = 1$, $c_3 = -2$, and $[\mathbf{p}]_{\mathcal{B}} = \begin{bmatrix} 5 \\ 1 \\ -2 \end{bmatrix}$.

4.5 | THE DIMENSION OF A VECTOR SPACE

Theorem 8 in Section 4.4 implies that a vector space V with a basis $\mathcal{B}$ containing n vectors is isomorphic to $\mathbb{R}^n$. This section shows that this number n is an intrinsic property (called the dimension) of the space V that does not depend on the particular choice of basis. The discussion of dimension will give additional insight into properties of bases.

The first theorem generalizes a well-known result about the vector space $\mathbb{R}^n$.

THEOREM 9

> If a vector space V has a basis $\mathcal{B} = \{\mathbf{b}_1, \ldots, \mathbf{b}_n\}$, then any set in V containing more than n vectors must be linearly dependent.

PROOF Let $\{\mathbf{u}_1, \ldots, \mathbf{u}_p\}$ be a set in V with more than n vectors. The coordinate vectors $[\mathbf{u}_1]_{\mathcal{B}}, \ldots, [\mathbf{u}_p]_{\mathcal{B}}$ form a linearly dependent set in $\mathbb{R}^n$, because there are more vectors (p) than entries (n) in each vector. So there exist scalars $c_1, \ldots, c_p$, not all zero, such that

$$
c_1[\mathbf{u}_1]_{\mathcal{B}} + \cdots + c_p[\mathbf{u}_p]_{\mathcal{B}} = \begin{bmatrix} 0 \\ \vdots \\ 0 \end{bmatrix} \qquad \text{The zero vector in } \mathbb{R}^n
$$

Since the coordinate mapping is a linear transformation,

$$
[c_1\mathbf{u}_1 + \cdots + c_p\mathbf{u}_p]_{\mathcal{B}} = \begin{bmatrix} 0 \\ \vdots \\ 0 \end{bmatrix}
$$

The zero vector on the right displays the n weights needed to build the vector $c_1\mathbf{u}_1 + \cdots + c_p\mathbf{u}_p$ from the basis vectors in $\mathcal{B}$. That is, $c_1\mathbf{u}_1 + \cdots + c_p\mathbf{u}_p = 0 \cdot \mathbf{b}_1 + \cdots + 0 \cdot \mathbf{b}_n = \mathbf{0}$. Since the c_i are not all zero, $\{\mathbf{u}_1, \ldots, \mathbf{u}_p\}$ is linearly dependent.[1] ∎

Theorem 9 implies that if a vector space V has a basis $\mathcal{B} = \{\mathbf{b}_1, \ldots, \mathbf{b}_n\}$, then each linearly independent set in V has no more than n vectors.

---

[1] Theorem 9 also applies to infinite sets in V. An infinite set is said to be linearly dependent if some finite subset is linearly dependent; otherwise, the set is linearly independent. If S is an infinite set in V, take any subset $\{\mathbf{u}_1, \ldots, \mathbf{u}_p\}$ of S, with $p > n$. The proof above shows that this subset is linearly dependent, and hence so is S.

THEOREM 10

If a vector space V has a basis of n vectors, then every basis of V must consist of exactly n vectors.

PROOF Let $\mathcal{B}_1$ be a basis of n vectors and $\mathcal{B}_2$ be any other basis (of V). Since $\mathcal{B}_1$ is a basis and $\mathcal{B}_2$ is linearly independent, $\mathcal{B}_2$ has no more than n vectors, by Theorem 9. Also, since $\mathcal{B}_2$ is a basis and $\mathcal{B}_1$ is linearly independent, $\mathcal{B}_2$ has at least n vectors. Thus $\mathcal{B}_2$ consists of exactly n vectors. ∎

If a nonzero vector space V is spanned by a finite set S, then a subset of S is a basis for V, by the Spanning Set Theorem. In this case, Theorem 10 ensures that the following definition makes sense.

DEFINITION

If V is spanned by a finite set, then V is said to be **finite-dimensional**, and the **dimension** of V, written as $\dim V$, is the number of vectors in a basis for V. The dimension of the zero vector space $\{\mathbf{0}\}$ is defined to be zero. If V is not spanned by a finite set, then V is said to be **infinite-dimensional**.

EXAMPLE 1 The standard basis for $\mathbb{R}^n$ contains n vectors, so $\dim \mathbb{R}^n = n$. The standard polynomial basis $\{1, t, t^2\}$ shows that $\dim \mathbb{P}_2 = 3$. In general, $\dim \mathbb{P}_n = n + 1$. The space $\mathbb{P}$ of all polynomials is infinite-dimensional (Exercise 27). ∎

EXAMPLE 2 Let $H = \text{Span}\{\mathbf{v}_1, \mathbf{v}_2\}$, where $\mathbf{v}_1 = \begin{bmatrix} 3 \\ 6 \\ 2 \end{bmatrix}$ and $\mathbf{v}_2 = \begin{bmatrix} -1 \\ 0 \\ 1 \end{bmatrix}$. Then H is the plane studied in Example 7 in Section 4.4. A basis for H is $\{\mathbf{v}_1, \mathbf{v}_2\}$, since $\mathbf{v}_1$ and $\mathbf{v}_2$ are not multiples and hence are linearly independent. Thus $\dim H = 2$. ∎

EXAMPLE 3 Find the dimension of the subspace

$$H = \left\{ \begin{bmatrix} a - 3b + 6c \\ 5a + 4d \\ b - 2c - d \\ 5d \end{bmatrix} : a, b, c, d \text{ in } \mathbb{R} \right\}$$

SOLUTION It is easy to see that H is the set of all linear combinations of the vectors

$$\mathbf{v}_1 = \begin{bmatrix} 1 \\ 5 \\ 0 \\ 0 \end{bmatrix}, \quad \mathbf{v}_2 = \begin{bmatrix} -3 \\ 0 \\ 1 \\ 0 \end{bmatrix}, \quad \mathbf{v}_3 = \begin{bmatrix} 6 \\ 0 \\ -2 \\ 0 \end{bmatrix}, \quad \mathbf{v}_4 = \begin{bmatrix} 0 \\ 4 \\ -1 \\ 5 \end{bmatrix}$$

Clearly, $\mathbf{v}_1 \neq \mathbf{0}$, $\mathbf{v}_2$ is not a multiple of $\mathbf{v}_1$, but $\mathbf{v}_3$ is a multiple of $\mathbf{v}_2$. By the Spanning Set Theorem, we may discard $\mathbf{v}_3$ and still have a set that spans H. Finally, $\mathbf{v}_4$ is not a linear combination of $\mathbf{v}_1$ and $\mathbf{v}_2$. So $\{\mathbf{v}_1, \mathbf{v}_2, \mathbf{v}_4\}$ is linearly independent (by Theorem 4 in Section 4.3) and hence is a basis for H. Thus $\dim H = 3$. ∎

EXAMPLE 4 The subspaces of $\mathbb{R}^3$ can be classified by dimension. See Fig. 1.

0-dimensional subspaces. Only the zero subspace.

1-dimensional subspaces. Any subspace spanned by a single nonzero vector. Such subspaces are lines through the origin.

2-dimensional subspaces. Any subspace spanned by two linearly independent vectors. Such subspaces are planes through the origin.

3-dimensional subspaces. Only $\mathbb{R}^3$ itself. Any three linearly independent vectors in $\mathbb{R}^3$ span all of $\mathbb{R}^3$, by the Invertible Matrix Theorem. ∎

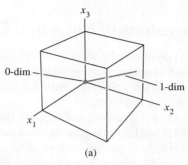 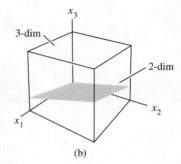

FIGURE 1 Sample subspaces of $\mathbb{R}^3$.

Subspaces of a Finite-Dimensional Space

The next theorem is a natural counterpart to the Spanning Set Theorem.

THEOREM 11

Let H be a subspace of a finite-dimensional vector space V. Any linearly independent set in H can be expanded, if necessary, to a basis for H. Also, H is finite-dimensional and

$$\dim H \leq \dim V$$

PROOF If $H = \{\mathbf{0}\}$, then certainly $\dim H = 0 \leq \dim V$. Otherwise, let $S = \{\mathbf{u}_1, \ldots, \mathbf{u}_k\}$ be any linearly independent set in H. If S spans H, then S is a basis for H. Otherwise, there is some $\mathbf{u}_{k+1}$ in H that is not in Span S. But then $\{\mathbf{u}_1, \ldots, \mathbf{u}_k, \mathbf{u}_{k+1}\}$ will be linearly independent, because no vector in the set can be a linear combination of vectors that precede it (by Theorem 4).

So long as the new set does not span H, we can continue this process of expanding S to a larger linearly independent set in H. But the number of vectors in a linearly independent expansion of S can never exceed the dimension of V, by Theorem 9. So eventually the expansion of S will span H and hence will be a basis for H, and $\dim H \leq \dim V$. ∎

When the dimension of a vector space or subspace is known, the search for a basis is simplified by the next theorem. It says that if a set has the right number of elements, then one has only to show either that the set is linearly independent or that it spans the space. The theorem is of critical importance in numerous applied problems (involving differential equations or difference equations, for example) where linear independence is much easier to verify than spanning.

THEOREM 12

The Basis Theorem

Let V be a p-dimensional vector space, $p \geq 1$. Any linearly independent set of exactly p elements in V is automatically a basis for V. Any set of exactly p elements that spans V is automatically a basis for V.

PROOF By Theorem 11, a linearly independent set S of p elements can be extended to a basis for V. But that basis must contain exactly p elements, since dim $V = p$. So S must already be a basis for V. Now suppose that S has p elements and spans V. Since V is nonzero, the Spanning Set Theorem implies that a subset S' of S is a basis of V. Since dim $V = p$, S' must contain p vectors. Hence $S = S'$. ∎

The Dimensions of Nul A and Col A

Since the pivot columns of a matrix A form a basis for Col A, we know the dimension of Col A as soon as we know the pivot columns. The dimension of Nul A might seem to require more work, since finding a basis for Nul A usually takes more time than a basis for Col A. But there is a shortcut!

Let A be an $m \times n$ matrix, and suppose the equation $A\mathbf{x} = \mathbf{0}$ has k free variables. From Section 4.2, we know that the standard method of finding a spanning set for Nul A will produce exactly k linearly independent vectors—say, $\mathbf{u}_1, \ldots, \mathbf{u}_k$—one for each free variable. So $\{\mathbf{u}_1, \ldots, \mathbf{u}_k\}$ is a basis for Nul A, and the number of free variables determines the size of the basis. Let us summarize these facts for future reference.

The dimension of Nul A is the number of free variables in the equation $A\mathbf{x} = \mathbf{0}$, and the dimension of Col A is the number of pivot columns in A.

EXAMPLE 5 Find the dimensions of the null space and the column space of

$$A = \begin{bmatrix} -3 & 6 & -1 & 1 & -7 \\ 1 & -2 & 2 & 3 & -1 \\ 2 & -4 & 5 & 8 & -4 \end{bmatrix}$$

SOLUTION Row reduce the augmented matrix $[A \ \mathbf{0}]$ to echelon form:

$$\begin{bmatrix} 1 & -2 & 2 & 3 & -1 & 0 \\ 0 & 0 & 1 & 2 & -2 & 0 \\ 0 & 0 & 0 & 0 & 0 & 0 \end{bmatrix}$$

There are three free variables—x_2, x_4, and x_5. Hence the dimension of Nul A is 3. Also, dim Col $A = 2$ because A has two pivot columns. ∎

PRACTICE PROBLEMS

Decide whether each statement is True or False, and give a reason for each answer. Here V is a nonzero finite-dimensional vector space.

1. If dim $V = p$ and if S is a linearly dependent subset of V, then S contains more than p vectors.

2. If S spans V and if T is a subset of V that contains more vectors than S, then T is linearly dependent.

4.5 EXERCISES

For each subspace in Exercises 1–8, (a) find a basis for the subspace, and (b) state the dimension.

1. $\left\{ \begin{bmatrix} s - 2t \\ s + t \\ 3t \end{bmatrix} : s, t \text{ in } \mathbb{R} \right\}$
2. $\left\{ \begin{bmatrix} 2a \\ -4b \\ -2a \end{bmatrix} : a, b \text{ in } \mathbb{R} \right\}$

3. $\left\{ \begin{bmatrix} 2c \\ a - b \\ b - 3c \\ a + 2b \end{bmatrix} : a, b, c \text{ in } \mathbb{R} \right\}$
4. $\left\{ \begin{bmatrix} p + 2q \\ -p \\ 3p - q \\ p + q \end{bmatrix} : p, q \text{ in } \mathbb{R} \right\}$

5. $\left\{ \begin{bmatrix} p - 2q \\ 2p + 5r \\ -2q + 2r \\ -3p + 6r \end{bmatrix} : p, q, r \text{ in } \mathbb{R} \right\}$

6. $\left\{ \begin{bmatrix} 3a - c \\ -b - 3c \\ -7a + 6b + 5c \\ -3a + c \end{bmatrix} : a, b, c \text{ in } \mathbb{R} \right\}$

7. $\{(a, b, c) : a - 3b + c = 0, b - 2c = 0, 2b - c = 0\}$

8. $\{(a, b, c, d) : a - 3b + c = 0\}$

9. Find the dimension of the subspace of all vectors in $\mathbb{R}^3$ whose first and third entries are equal.

10. Find the dimension of the subspace H of $\mathbb{R}^2$ spanned by $\begin{bmatrix} 1 \\ -5 \end{bmatrix}, \begin{bmatrix} -2 \\ 10 \end{bmatrix}, \begin{bmatrix} -3 \\ 15 \end{bmatrix}$.

In Exercises 11 and 12, find the dimension of the subspace spanned by the given vectors.

11. $\begin{bmatrix} 1 \\ 0 \\ 2 \end{bmatrix}, \begin{bmatrix} 3 \\ 1 \\ 1 \end{bmatrix}, \begin{bmatrix} -2 \\ -1 \\ 1 \end{bmatrix}, \begin{bmatrix} 5 \\ 2 \\ 2 \end{bmatrix}$

12. $\begin{bmatrix} 1 \\ -2 \\ 0 \end{bmatrix}, \begin{bmatrix} -3 \\ -6 \\ 0 \end{bmatrix}, \begin{bmatrix} -2 \\ 3 \\ 5 \end{bmatrix}, \begin{bmatrix} -3 \\ 5 \\ 5 \end{bmatrix}$

Determine the dimensions of Nul A and Col A for the matrices shown in Exercises 13–18.

13. $A = \begin{bmatrix} 1 & -6 & 9 & 0 & -2 \\ 0 & 1 & 2 & -4 & 5 \\ 0 & 0 & 0 & 5 & 1 \\ 0 & 0 & 0 & 0 & 0 \end{bmatrix}$

14. $A = \begin{bmatrix} 1 & 2 & -4 & 3 & -2 & 6 & 0 \\ 0 & 0 & 0 & 1 & 0 & -3 & 7 \\ 0 & 0 & 0 & 0 & 1 & 4 & -2 \\ 0 & 0 & 0 & 0 & 0 & 0 & 1 \end{bmatrix}$

15. $A = \begin{bmatrix} 1 & 2 & 3 & 0 & 0 \\ 0 & 0 & 1 & 0 & 1 \\ 0 & 0 & 0 & 1 & 0 \end{bmatrix}$
16. $A = \begin{bmatrix} 3 & 2 \\ -6 & 5 \end{bmatrix}$

17. $A = \begin{bmatrix} 1 & -1 & 0 \\ 0 & 1 & 3 \\ 0 & 0 & 1 \end{bmatrix}$
18. $A = \begin{bmatrix} 1 & 1 & -1 \\ 0 & 2 & 0 \\ 0 & 0 & 0 \end{bmatrix}$

In Exercises 19 and 20, V is a vector space. Mark each statement True or False. Justify each answer.

19. a. The number of pivot columns of a matrix equals the dimension of its column space.

   b. A plane in $\mathbb{R}^3$ is a two-dimensional subspace of $\mathbb{R}^3$.

   c. The dimension of the vector space $\mathbb{P}_4$ is 4.

   d. If dim $V = n$ and S is a linearly independent set in V, then S is a basis for V.

   e. If a set $\{\mathbf{v}_1, \ldots, \mathbf{v}_p\}$ spans a finite-dimensional vector space V and if T is a set of more than p vectors in V, then T is linearly dependent.

20. a. $\mathbb{R}^2$ is a two-dimensional subspace of $\mathbb{R}^3$.

   b. The number of variables in the equation $A\mathbf{x} = \mathbf{0}$ equals the dimension of Nul A.

   c. A vector space is infinite-dimensional if it is spanned by an infinite set.

   d. If dim $V = n$ and if S spans V, then S is a basis of V.

   e. The only three-dimensional subspace of $\mathbb{R}^3$ is $\mathbb{R}^3$ itself.

21. The first four Hermite polynomials are $1, 2t, -2 + 4t^2$, and $-12t + 8t^3$. These polynomials arise naturally in the study of certain important differential equations in mathematical physics.[2] Show that the first four Hermite polynomials form a basis of $\mathbb{P}_3$.

22. The first four Laguerre polynomials are $1, 1 - t, 2 - 4t + t^2$, and $6 - 18t + 9t^2 - t^3$. Show that these polynomials form a basis of $\mathbb{P}_3$.

23. Let $\mathcal{B}$ be the basis of $\mathbb{P}_3$ consisting of the Hermite polynomials in Exercise 21, and let $\mathbf{p}(t) = -1 + 8t^2 + 8t^3$. Find the coordinate vector of $\mathbf{p}$ relative to $\mathcal{B}$.

24. Let $\mathcal{B}$ be the basis of $\mathbb{P}_2$ consisting of the first three Laguerre polynomials listed in Exercise 22, and let $\mathbf{p}(t) = 5 + 5t - 2t^2$. Find the coordinate vector of $\mathbf{p}$ relative to $\mathcal{B}$.

25. Let S be a subset of an n-dimensional vector space V, and suppose S contains fewer than n vectors. Explain why S cannot span V.

26. Let H be an n-dimensional subspace of an n-dimensional vector space V. Show that $H = V$.

27. Explain why the space $\mathbb{P}$ of all polynomials is an infinite-dimensional space.

[2] See *Introduction to Functional Analysis*, 2d ed., by A. E. Taylor and David C. Lay (New York: John Wiley & Sons, 1980), pp. 92–93. Other sets of polynomials are discussed there, too.

28. Show that the space $C(\mathbb{R})$ of all continuous functions defined on the real line is an infinite-dimensional space.

In Exercises 29 and 30, V is a nonzero finite-dimensional vector space, and the vectors listed belong to V. Mark each statement True or False. Justify each answer. (These questions are more difficult than those in Exercises 19 and 20.)

29. a. If there exists a set $\{\mathbf{v}_1, \ldots, \mathbf{v}_p\}$ that spans V, then $\dim V \leq p$.

   b. If there exists a linearly independent set $\{\mathbf{v}_1, \ldots, \mathbf{v}_p\}$ in V, then $\dim V \geq p$.

   c. If $\dim V = p$, then there exists a spanning set of $p + 1$ vectors in V.

30. a. If there exists a linearly dependent set $\{\mathbf{v}_1, \ldots, \mathbf{v}_p\}$ in V, then $\dim V \leq p$.

   b. If every set of p elements in V fails to span V, then $\dim V > p$.

   c. If $p \geq 2$ and $\dim V = p$, then every set of $p - 1$ nonzero vectors is linearly independent.

Exercises 31 and 32 concern finite-dimensional vector spaces V and W and a linear transformation $T : V \rightarrow W$.

31. Let H be a nonzero subspace of V, and let $T(H)$ be the set of images of vectors in H. Then $T(H)$ is a subspace of W, by Exercise 35 in Section 4.2. Prove that $\dim T(H) \leq \dim H$.

32. Let H be a nonzero subspace of V, and suppose T is a one-to-one (linear) mapping of V into W. Prove that $\dim T(H) = \dim H$. If T happens to be a one-to-one mapping of V *onto* W, then $\dim V = \dim W$. Isomorphic finite-dimensional vector spaces have the same dimension.

33. [M] According to Theorem 11, a linearly independent set $\{\mathbf{v}_1, \ldots, \mathbf{v}_k\}$ in $\mathbb{R}^n$ can be expanded to a basis for $\mathbb{R}^n$. One way to do this is to create $A = [\,\mathbf{v}_1 \;\; \cdots \;\; \mathbf{v}_k \;\; \mathbf{e}_1 \;\; \cdots \;\; \mathbf{e}_n\,]$, with $\mathbf{e}_1, \ldots, \mathbf{e}_n$ the columns of the identity matrix; the pivot columns of A form a basis for $\mathbb{R}^n$.

   a. Use the method described to extend the following vectors to a basis for $\mathbb{R}^5$:

$$\mathbf{v}_1 = \begin{bmatrix} -9 \\ -7 \\ 8 \\ -5 \\ 7 \end{bmatrix}, \quad \mathbf{v}_2 = \begin{bmatrix} 9 \\ 4 \\ 1 \\ 6 \\ -7 \end{bmatrix}, \quad \mathbf{v}_3 = \begin{bmatrix} 6 \\ 7 \\ -8 \\ 5 \\ -7 \end{bmatrix}$$

   b. Explain why the method works in general: Why are the original vectors $\mathbf{v}_1, \ldots, \mathbf{v}_k$ included in the basis found for Col A? Why is Col $A = \mathbb{R}^n$?

34. [M] Let $\mathcal{B} = \{1, \cos t, \cos^2 t, \ldots, \cos^6 t\}$ and $\mathcal{C} = \{1, \cos t, \cos 2t, \ldots, \cos 6t\}$. Assume the following trigonometric identities (see Exercise 37 in Section 4.1).

$$\cos 2t = -1 + 2\cos^2 t$$
$$\cos 3t = -3\cos t + 4\cos^3 t$$
$$\cos 4t = 1 - 8\cos^2 t + 8\cos^4 t$$
$$\cos 5t = 5\cos t - 20\cos^3 t + 16\cos^5 t$$
$$\cos 6t = -1 + 18\cos^2 t - 48\cos^4 t + 32\cos^6 t$$

Let H be the subspace of functions spanned by the functions in $\mathcal{B}$. Then $\mathcal{B}$ is a basis for H, by Exercise 38 in Section 4.3.

   a. Write the $\mathcal{B}$-coordinate vectors of the vectors in $\mathcal{C}$, and use them to show that $\mathcal{C}$ is a linearly independent set in H.

   b. Explain why $\mathcal{C}$ is a basis for H.

SOLUTIONS TO PRACTICE PROBLEMS

1. False. Consider the set $\{\mathbf{0}\}$.

2. True. By the Spanning Set Theorem, S contains a basis for V; call that basis S'. Then T will contain more vectors than S'. By Theorem 9, T is linearly dependent.

4.6 | RANK

With the aid of vector space concepts, this section takes a look *inside* a matrix and reveals several interesting and useful relationships hidden in its rows and columns.

For instance, imagine placing 2000 random numbers into a 40×50 matrix A and then determining both the maximum number of linearly independent columns in A and the maximum number of linearly independent columns in A^T (rows in A). Remarkably, the two numbers are the same. As we'll soon see, their common value is the *rank* of the matrix. To explain why, we need to examine the subspace spanned by the rows of A.

The Row Space

If A is an $m \times n$ matrix, each row of A has n entries and thus can be identified with a vector in $\mathbb{R}^n$. The set of all linear combinations of the row vectors is called the **row space** of A and is denoted by Row A. Each row has n entries, so Row A is a subspace of $\mathbb{R}^n$. Since the rows of A are identified with the columns of A^T, we could also write Col A^T in place of Row A.

EXAMPLE 1   Let

$$A = \begin{bmatrix} -2 & -5 & 8 & 0 & -17 \\ 1 & 3 & -5 & 1 & 5 \\ 3 & 11 & -19 & 7 & 1 \\ 1 & 7 & -13 & 5 & -3 \end{bmatrix} \quad \text{and} \quad \begin{array}{l} \mathbf{r}_1 = (-2, -5, 8, 0, -17) \\ \mathbf{r}_2 = (1, 3, -5, 1, 5) \\ \mathbf{r}_3 = (3, 11, -19, 7, 1) \\ \mathbf{r}_4 = (1, 7, -13, 5, -3) \end{array}$$

The row space of A is the subspace of $\mathbb{R}^5$ spanned by $\{\mathbf{r}_1, \mathbf{r}_2, \mathbf{r}_3, \mathbf{r}_4\}$. That is, Row $A =$ Span $\{\mathbf{r}_1, \mathbf{r}_2, \mathbf{r}_3, \mathbf{r}_4\}$. It is natural to write row vectors horizontally; however, they may also be written as column vectors if that is more convenient.   ■

If we knew some linear dependence relations among the rows of matrix A in Example 1, we could use the Spanning Set Theorem to shrink the spanning set to a basis. Unfortunately, row operations on A will not give us that information, because row operations change the row-dependence relations. But row reducing A is certainly worthwhile, as the next theorem shows!

THEOREM 13

If two matrices A and B are row equivalent, then their row spaces are the same. If B is in echelon form, the nonzero rows of B form a basis for the row space of A as well as for that of B.

PROOF If B is obtained from A by row operations, the rows of B are linear combinations of the rows of A. It follows that any linear combination of the rows of B is automatically a linear combination of the rows of A. Thus the row space of B is contained in the row space of A. Since row operations are reversible, the same argument shows that the row space of A is a subset of the row space of B. So the two row spaces are the same. If B is in echelon form, its nonzero rows are linearly independent because no nonzero row is a linear combination of the nonzero rows below it. (Apply Theorem 4 to the nonzero rows of B in reverse order, with the first row last.) Thus the nonzero rows of B form a basis of the (common) row space of B and A.   ■

The main result of this section involves the three spaces: Row A, Col A, and Nul A. The following example prepares the way for this result and shows how *one* sequence of row operations on A leads to bases for all three spaces.

EXAMPLE 2   Find bases for the row space, the column space, and the null space of the matrix

$$A = \begin{bmatrix} -2 & -5 & 8 & 0 & -17 \\ 1 & 3 & -5 & 1 & 5 \\ 3 & 11 & -19 & 7 & 1 \\ 1 & 7 & -13 & 5 & -3 \end{bmatrix}$$

SOLUTION To find bases for the row space and the column space, row reduce A to an echelon form:

$$A \sim B = \begin{bmatrix} 1 & 3 & -5 & 1 & 5 \\ 0 & 1 & -2 & 2 & -7 \\ 0 & 0 & 0 & -4 & 20 \\ 0 & 0 & 0 & 0 & 0 \end{bmatrix}$$

By Theorem 13, the first three rows of B form a basis for the row space of A (as well as for the row space of B). Thus

Basis for Row A: $\{(1, 3, -5, 1, 5), (0, 1, -2, 2, -7), (0, 0, 0, -4, 20)\}$

For the column space, observe from B that the pivots are in columns 1, 2, and 4. Hence columns 1, 2, and 4 of A (not B) form a basis for Col A:

$$\text{Basis for Col } A: \left\{ \begin{bmatrix} -2 \\ 1 \\ 3 \\ 1 \end{bmatrix}, \begin{bmatrix} -5 \\ 3 \\ 11 \\ 7 \end{bmatrix}, \begin{bmatrix} 0 \\ 1 \\ 7 \\ 5 \end{bmatrix} \right\}$$

Notice that any echelon form of A provides (in its nonzero rows) a basis for Row A and also identifies the pivot columns of A for Col A. However, for Nul A, we need the *reduced echelon form*. Further row operations on B yield

$$A \sim B \sim C = \begin{bmatrix} 1 & 0 & 1 & 0 & 1 \\ 0 & 1 & -2 & 0 & 3 \\ 0 & 0 & 0 & 1 & -5 \\ 0 & 0 & 0 & 0 & 0 \end{bmatrix}$$

The equation $A\mathbf{x} = \mathbf{0}$ is equivalent to $C\mathbf{x} = \mathbf{0}$, that is,

$$\begin{aligned} x_1 + \quad\quad x_3 \quad\quad\; + \; x_5 &= 0 \\ x_2 - 2x_3 \quad\quad + \; 3x_5 &= 0 \\ x_4 - 5x_5 &= 0 \end{aligned}$$

So $x_1 = -x_3 - x_5$, $x_2 = 2x_3 - 3x_5$, $x_4 = 5x_5$, with x_3 and x_5 free variables. The usual calculations (discussed in Section 4.2) show that

$$\text{Basis for Nul } A: \left\{ \begin{bmatrix} -1 \\ 2 \\ 1 \\ 0 \\ 0 \end{bmatrix}, \begin{bmatrix} -1 \\ -3 \\ 0 \\ 5 \\ 1 \end{bmatrix} \right\}$$

Observe that, unlike the basis for Col A, the bases for Row A and Nul A have no simple connection with the entries in A itself.[1] ∎

---

[1] It is possible to find a basis for the row space Row A that uses rows of A. First form A^T, and then row reduce until the pivot columns of A^T are found. These pivot columns of A^T are rows of A, and they form a basis for the row space of A.

Warning: Although the first three rows of B in Example 2 are linearly independent, it is wrong to conclude that the first three rows of A are linearly independent. (In fact, the third row of A is 2 times the first row plus 7 times the second row.) Row operations may change the linear dependence relations among the *rows* of a matrix.

The Rank Theorem

WEB

The next theorem describes fundamental relations among the dimensions of Col A, Row A, and Nul A.

DEFINITION

The **rank** of A is the dimension of the column space of A.

Since Row A is the same as Col A^T, the dimension of the row space of A is the rank of A^T. The dimension of the null space is sometimes called the **nullity** of A, though we will not use this term.

An alert reader may have already discovered part or all of the next theorem while working the exercises in Section 4.5 or reading Example 2 above.

THEOREM 14

The Rank Theorem

The dimensions of the column space and the row space of an $m \times n$ matrix A are equal. This common dimension, the rank of A, also equals the number of pivot positions in A and satisfies the equation

$$\text{rank } A + \dim \text{Nul } A = n$$

PROOF   By Theorem 6 in Section 4.3, rank A is the number of pivot columns in A. Equivalently, rank A is the number of pivot positions in an echelon form B of A. Furthermore, since B has a nonzero row for each pivot, and since these rows form a basis for the row space of A, the rank of A is also the dimension of the row space.

From Section 4.5, the dimension of Nul A equals the number of free variables in the equation $A\mathbf{x} = \mathbf{0}$. Expressed another way, the dimension of Nul A is the number of columns of A that are *not* pivot columns. (It is the number of these columns, not the columns themselves, that is related to Nul A.) Obviously,

$$\left\{ \begin{array}{c} \text{number of} \\ \text{pivot columns} \end{array} \right\} + \left\{ \begin{array}{c} \text{number of} \\ \text{nonpivot columns} \end{array} \right\} = \left\{ \begin{array}{c} \text{number of} \\ \text{columns} \end{array} \right\}$$

This proves the theorem.   ∎

The ideas behind Theorem 14 are visible in the calculations in Example 2. The three pivot positions in the echelon form B determine the basic variables and identify the basis vectors for Col A and those for Row A.

EXAMPLE 3

a. If A is a 7×9 matrix with a two-dimensional null space, what is the rank of A?

b. Could a 6×9 matrix have a two-dimensional null space?

SOLUTION

a. Since A has 9 columns, $(\text{rank } A) + 2 = 9$, and hence rank $A = 7$.

b. No. If a 6×9 matrix, call it B, had a two-dimensional null space, it would have to have rank 7, by the Rank Theorem. But the columns of B are vectors in $\mathbb{R}^6$, and so the dimension of Col B cannot exceed 6; that is, rank B cannot exceed 6. ∎

The next example provides a nice way to visualize the subspaces we have been studying. In Chapter 6, we will learn that Row A and Nul A have only the zero vector in common and are actually "perpendicular" to each other. The same fact will apply to Row $A^T (= \text{Col } A)$ and Nul A^T. So Fig. 1, which accompanies Example 4, creates a good mental image for the general case. (The value of studying A^T along with A is demonstrated in Exercise 29.)

EXAMPLE 4   Let $A = \begin{bmatrix} 3 & 0 & -1 \\ 3 & 0 & -1 \\ 4 & 0 & 5 \end{bmatrix}$. It is readily checked that Nul A is the x_2-axis, Row A is the $x_1 x_3$-plane, Col A is the plane whose equation is $x_1 - x_2 = 0$, and Nul A^T is the set of all multiples of $(1, -1, 0)$. Figure 1 shows Nul A and Row A in the domain of the linear transformation $\mathbf{x} \mapsto A\mathbf{x}$; the range of this mapping, Col A, is shown in a separate copy of $\mathbb{R}^3$, along with Nul A^T. ∎

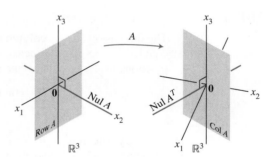

FIGURE 1 Subspaces determined by a matrix A.

Applications to Systems of Equations

The Rank Theorem is a powerful tool for processing information about systems of linear equations. The next example simulates the way a real-life problem using linear equations might be stated, without explicit mention of linear algebra terms such as matrix, subspace, and dimension.

EXAMPLE 5   A scientist has found two solutions to a homogeneous system of 40 equations in 42 variables. The two solutions are not multiples, and all other solutions can be constructed by adding together appropriate multiples of these two solutions. Can the scientist be *certain* that an associated nonhomogeneous system (with the same coefficients) has a solution?

SOLUTION Yes. Let A be the 40×42 coefficient matrix of the system. The given information implies that the two solutions are linearly independent and span Nul A. So dim Nul $A = 2$. By the Rank Theorem, dim Col $A = 42 - 2 = 40$. Since $\mathbb{R}^{40}$ is the only subspace of $\mathbb{R}^{40}$ whose dimension is 40, Col A must be all of $\mathbb{R}^{40}$. This means that every nonhomogeneous equation $A\mathbf{x} = \mathbf{b}$ has a solution. ∎

Rank and the Invertible Matrix Theorem

The various vector space concepts associated with a matrix provide several more statements for the Invertible Matrix Theorem. The new statements listed here follow those in the original Invertible Matrix Theorem in Section 2.3.

THEOREM

The Invertible Matrix Theorem (continued)

Let A be an $n \times n$ matrix. Then the following statements are each equivalent to the statement that A is an invertible matrix.

m. The columns of A form a basis of $\mathbb{R}^n$.

n. $\operatorname{Col} A = \mathbb{R}^n$

o. $\dim \operatorname{Col} A = n$

p. $\operatorname{rank} A = n$

q. $\operatorname{Nul} A = \{\mathbf{0}\}$

r. $\dim \operatorname{Nul} A = 0$

PROOF Statement (m) is logically equivalent to statements (e) and (h) regarding linear independence and spanning. The other five statements are linked to the earlier ones of the theorem by the following chain of almost trivial implications:

$$(g) \Rightarrow (n) \Rightarrow (o) \Rightarrow (p) \Rightarrow (r) \Rightarrow (q) \Rightarrow (d)$$

Statement (g), which says that the equation $A\mathbf{x} = \mathbf{b}$ has at least one solution for each $\mathbf{b}$ in $\mathbb{R}^n$, implies (n), because $\operatorname{Col} A$ is precisely the set of all $\mathbf{b}$ such that the equation $A\mathbf{x} = \mathbf{b}$ is consistent. The implications (n) $\Rightarrow$ (o) $\Rightarrow$ (p) follow from the definitions of dimension and rank. If the rank of A is n, the number of columns of A, then $\dim \operatorname{Nul} A = 0$, by the Rank Theorem, and so $\operatorname{Nul} A = \{\mathbf{0}\}$. Thus (p) $\Rightarrow$ (r) $\Rightarrow$ (q). Also, (q) implies that the equation $A\mathbf{x} = \mathbf{0}$ has only the trivial solution, which is statement (d). Since statements (d) and (g) are already known to be equivalent to the statement that A is invertible, the proof is complete. ∎

SG Expanded Table for the IMT 4-19

We have refrained from adding to the Invertible Matrix Theorem obvious statements about the row space of A, because the row space is the column space of A^T. Recall from statement (l) of the Invertible Matrix Theorem that A is invertible if and only if A^T is invertible. Hence every statement in the Invertible Matrix Theorem can also be stated for A^T. To do so would double the length of the theorem and produce a list of over 30 statements!

┌─ NUMERICAL NOTE ─────────────────────────────

Many algorithms discussed in this text are useful for understanding concepts and making simple computations by hand. However, the algorithms are often unsuitable for large-scale problems in real life.

Rank determination is a good example. It would seem easy to reduce a matrix to echelon form and count the pivots. But unless exact arithmetic is performed on a matrix whose entries are specified exactly, row operations can change the apparent rank of a matrix. For instance, if the value of x in the matrix $\begin{bmatrix} 5 & 7 \\ 5 & x \end{bmatrix}$ is not stored exactly as 7 in a computer, then the rank may be 1 or 2, depending on whether the computer treats $x - 7$ as zero.

In practical applications, the effective rank of a matrix A is often determined from the singular value decomposition of A, to be discussed in Section 7.4. This decomposition is also a reliable source of bases for Col A, Row A, Nul A, and Nul A^T.

WEB

PRACTICE PROBLEMS

The matrices below are row equivalent.

$$A = \begin{bmatrix} 2 & -1 & 1 & -6 & 8 \\ 1 & -2 & -4 & 3 & -2 \\ -7 & 8 & 10 & 3 & -10 \\ 4 & -5 & -7 & 0 & 4 \end{bmatrix}, \quad B = \begin{bmatrix} 1 & -2 & -4 & 3 & -2 \\ 0 & 3 & 9 & -12 & 12 \\ 0 & 0 & 0 & 0 & 0 \\ 0 & 0 & 0 & 0 & 0 \end{bmatrix}$$

1. Find rank A and dim Nul A.
2. Find bases for Col A and Row A.
3. What is the next step to perform to find a basis for Nul A?
4. How many pivot columns are in a row echelon form of A^T?

4.6 EXERCISES

In Exercises 1–4, assume that the matrix A is row equivalent to B. Without calculations, list rank A and dim Nul A. Then find bases for Col A, Row A, and Nul A.

1. $A = \begin{bmatrix} 1 & -4 & 9 & -7 \\ -1 & 2 & -4 & 1 \\ 5 & -6 & 10 & 7 \end{bmatrix}$,

$B = \begin{bmatrix} 1 & 0 & -1 & 5 \\ 0 & -2 & 5 & -6 \\ 0 & 0 & 0 & 0 \end{bmatrix}$

2. $A = \begin{bmatrix} 1 & 3 & 4 & -1 & 2 \\ 2 & 6 & 6 & 0 & -3 \\ 3 & 9 & 3 & 6 & -3 \\ 3 & 9 & 0 & 9 & 0 \end{bmatrix}$,

$B = \begin{bmatrix} 1 & 3 & 4 & -1 & 2 \\ 0 & 0 & 1 & -1 & 1 \\ 0 & 0 & 0 & 0 & -5 \\ 0 & 0 & 0 & 0 & 0 \end{bmatrix}$

3. $A = \begin{bmatrix} 2 & 6 & -6 & 6 & 3 & 6 \\ -2 & -3 & 6 & -3 & 0 & -6 \\ 4 & 9 & -12 & 9 & 3 & 12 \\ -2 & 3 & 6 & 3 & 3 & -6 \end{bmatrix}$,

$B = \begin{bmatrix} 2 & 6 & -6 & 6 & 3 & 6 \\ 0 & 3 & 0 & 3 & 3 & 0 \\ 0 & 0 & 0 & 0 & 3 & 0 \\ 0 & 0 & 0 & 0 & 0 & 0 \end{bmatrix}$

4. $A = \begin{bmatrix} 1 & 1 & -2 & 0 & 1 & -2 \\ 1 & 2 & -3 & 0 & -2 & -3 \\ 1 & -1 & 0 & 0 & 1 & 6 \\ 1 & -2 & 2 & 1 & -3 & 0 \\ 1 & -2 & 1 & 0 & 2 & -1 \end{bmatrix}$,

$B = \begin{bmatrix} 1 & 1 & -2 & 0 & 1 & -2 \\ 0 & 1 & -1 & 0 & -3 & -1 \\ 0 & 0 & 1 & 1 & -13 & -1 \\ 0 & 0 & 0 & 0 & 1 & -1 \\ 0 & 0 & 0 & 0 & 0 & 1 \end{bmatrix}$

5. If a 4×7 matrix A has rank 3, find dim Nul A, dim Row A, and rank A^T.

6. If a 7×5 matrix A has rank 2, find dim Nul A, dim Row A, and rank A^T.

7. Suppose a 4×7 matrix A has four pivot columns. Is Col $A = \mathbb{R}^4$? Is Nul $A = \mathbb{R}^3$? Explain your answers.

8. Suppose a 6×8 matrix A has four pivot columns. What is dim Nul A? Is Col $A = \mathbb{R}^4$? Why or why not?

9. If the null space of a 4×6 matrix A is 3-dimensional, what is the dimension of the column space of A? Is Col $A = \mathbb{R}^3$? Why or why not?

10. If the null space of an 8×7 matrix A is 5-dimensional, what is the dimension of the column space of A?

11. If the null space of an 8×5 matrix A is 3-dimensional, what is the dimension of the row space of A?

12. If the null space of a 5×4 matrix A is 2-dimensional, what is the dimension of the row space of A?

13. If A is a 7×5 matrix, what is the largest possible rank of A? If A is a 5×7 matrix, what is the largest possible rank of A? Explain your answers.

14. If A is a 5×4 matrix, what is the largest possible dimension of the row space of A? If A is a 4×5 matrix, what is the largest possible dimension of the row space of A? Explain.

15. If A is a 3×7 matrix, what is the smallest possible dimension of Nul A?

16. If A is a 7×5 matrix, what is the smallest possible dimension of Nul A?

In Exercises 17 and 18, A is an $m \times n$ matrix. Mark each statement True or False. Justify each answer.

17. a. The row space of A is the same as the column space of A^T.

    b. If B is any echelon form of A, and if B has three nonzero rows, then the first three rows of A form a basis for Row A.

    c. The dimensions of the row space and the column space of A are the same, even if A is not square.

    d. The sum of the dimensions of the row space and the null space of A equals the number of rows in A.

    e. On a computer, row operations can change the apparent rank of a matrix.

18. a. If B is any echelon form of A, then the pivot columns of B form a basis for the column space of A.

    b. Row operations preserve the linear dependence relations among the rows of A.

    c. The dimension of the null space of A is the number of columns of A that are *not* pivot columns.

    d. The row space of A^T is the same as the column space of A.

    e. If A and B are row equivalent, then their row spaces are the same.

19. Suppose the solutions of a homogeneous system of five linear equations in six unknowns are all multiples of one nonzero solution. Will the system necessarily have a solution for every possible choice of constants on the right sides of the equations? Explain.

20. Suppose a nonhomogeneous system of six linear equations in eight unknowns has a solution, with two free variables. Is it possible to change some constants on the equations' right sides to make the new system inconsistent? Explain.

21. Suppose a nonhomogeneous system of nine linear equations in ten unknowns has a solution for all possible constants on the right sides of the equations. Is it possible to find two nonzero solutions of the associated homogeneous system that are *not* multiples of each other? Discuss.

22. Is is possible that all solutions of a homogeneous system of ten linear equations in twelve variables are multiples of one fixed nonzero solution? Discuss.

23. A homogeneous system of twelve linear equations in eight unknowns has two fixed solutions that are not multiples of each other, and all other solutions are linear combinations of these two solutions. Can the set of all solutions be described with fewer than twelve homogeneous linear equations? If so, how many? Discuss.

24. Is it possible for a nonhomogeneous system of seven equations in six unknowns to have a unique solution for some right-hand side of constants? Is it possible for such a system to have a unique solution for every right-hand side? Explain.

25. A scientist solves a nonhomogeneous system of ten linear equations in twelve unknowns and finds that three of the unknowns are free variables. Can the scientist be certain that, if the right sides of the equations are changed, the new nonhomogeneous system will have a solution? Discuss.

26. In statistical theory, a common requirement is that a matrix be of *full rank*. That is, the rank should be as large as possible. Explain why an $m \times n$ matrix with more rows than columns has full rank if and only if its columns are linearly independent.

Exercises 27–29 concern an $m \times n$ matrix A and what are often called the *fundamental subspaces* determined by A.

27. Which of the subspaces Row A, Col A, Nul A, Row A^T, Col A^T, and Nul A^T are in $\mathbb{R}^m$ and which are in $\mathbb{R}^n$? How many distinct subspaces are in this list?

28. Justify the following equalities:

    a. dim Row A + dim Nul $A = n$ Number of columns of A

    b. dim Col A + dim Nul $A^T = m$ Number of rows of A

29. Use Exercise 28 to explain why the equation $A\mathbf{x} = \mathbf{b}$ has a solution for all $\mathbf{b}$ in $\mathbb{R}^m$ if and only if the equation $A^T\mathbf{x} = \mathbf{0}$ has only the trivial solution.

30. Suppose A is $m \times n$ and $\mathbf{b}$ is in $\mathbb{R}^m$. What has to be true about the two numbers rank $[\, A \quad \mathbf{b} \,]$ and rank A in order for the equation $A\mathbf{x} = \mathbf{b}$ to be consistent?

Rank 1 matrices are important in some computer algorithms and several theoretical contexts, including the singular value decomposition in Chapter 7. It can be shown that an $m \times n$ matrix A has rank 1 if and only if it is an outer product; that is, $A = \mathbf{u}\mathbf{v}^T$ for some $\mathbf{u}$ in $\mathbb{R}^m$ and $\mathbf{v}$ in $\mathbb{R}^n$. Exercises 31–33 suggest why this property is true.

31. Verify that rank $\mathbf{u}\mathbf{v}^T \leq 1$ if $\mathbf{u} = \begin{bmatrix} 2 \\ -3 \\ 5 \end{bmatrix}$ and $\mathbf{v} = \begin{bmatrix} a \\ b \\ c \end{bmatrix}$.

32. Let $\mathbf{u} = \begin{bmatrix} 1 \\ 2 \end{bmatrix}$. Find $\mathbf{v}$ in $\mathbb{R}^3$ such that $\begin{bmatrix} 1 & -3 & 4 \\ 2 & -6 & 8 \end{bmatrix} = \mathbf{u}\mathbf{v}^T$.

33. Let A be any 2×3 matrix such that rank $A = 1$, let $\mathbf{u}$ be the first column of A, and suppose $\mathbf{u} \neq \mathbf{0}$. Explain why there is a vector $\mathbf{v}$ in $\mathbb{R}^3$ such that $A = \mathbf{u}\mathbf{v}^T$. How could this construction be modified if the first column of A were zero?

34. Let A be an $m \times n$ matrix of rank $r > 0$ and let U be an echelon form of A. Explain why there exists an invertible matrix E such that $A = EU$, and use this factorization to write A as the sum of r rank 1 matrices. [*Hint:* See Theorem 10 in Section 2.4.]

35. [M] Let $A = \begin{bmatrix} 7 & -9 & -4 & 5 & 3 & -3 & -7 \\ -4 & 6 & 7 & -2 & -6 & -5 & 5 \\ 5 & -7 & -6 & 5 & -6 & 2 & 8 \\ -3 & 5 & 8 & -1 & -7 & -4 & 8 \\ 6 & -8 & -5 & 4 & 4 & 9 & 3 \end{bmatrix}$.

  a. Construct matrices C and N whose columns are bases for Col A and Nul A, respectively, and construct a matrix R whose *rows* form a basis for Row A.

  b. Construct a matrix M whose columns form a basis for Nul A^T, form the matrices $S = [\, R^T \quad N \,]$ and $T = [\, C \quad M \,]$, and explain why S and T should be square. Verify that both S and T are invertible.

36. [M] Repeat Exercise 35 for a random integer-valued 6×7 matrix A whose rank is at most 4. One way to make A is to create a random integer-valued 6×4 matrix J and a random integer-valued 4×7 matrix K, and set $A = JK$. (See Supplementary Exercise 12 at the end of the chapter; and see the *Study Guide* for matrix-generating programs.)

37. [M] Let A be the matrix in Exercise 35. Construct a matrix C whose columns are the pivot columns of A, and construct a matrix R whose rows are the nonzero rows of the reduced echelon form of A. Compute CR, and discuss what you see.

38. [M] Repeat Exercise 37 for three random integer-valued 5×7 matrices A whose ranks are 5, 4, and 3. Make a conjecture about how CR is related to A for any matrix A. Prove your conjecture.

SOLUTIONS TO PRACTICE PROBLEMS

1. A has two pivot columns, so rank $A = 2$. Since A has 5 columns altogether, dim Nul $A = 5 - 2 = 3$.

2. The pivot columns of A are the first two columns. So a basis for Col A is

$$\{\mathbf{a}_1, \mathbf{a}_2\} = \left\{ \begin{bmatrix} 2 \\ 1 \\ -7 \\ 4 \end{bmatrix}, \begin{bmatrix} -1 \\ -2 \\ 8 \\ -5 \end{bmatrix} \right\}$$

The nonzero rows of B form a basis for Row A, namely, $\{(1, -2, -4, 3, -2), (0, 3, 9, -12, 12)\}$. In this particular example, it happens that any two rows of A form a basis for the row space, because the row space is two-dimensional and none of the rows of A is a multiple of another row. In general, the nonzero rows of an echelon form of A should be used as a basis for Row A, not the rows of A itself.

3. For Nul A, the next step is to perform row operations on B to obtain the reduced echelon form of A.

4. Rank $A^T = $ rank A, by the Rank Theorem, because Col $A^T = $ Row A. So A^T has two pivot positions.

4.7 CHANGE OF BASIS

When a basis $\mathcal{B}$ is chosen for an n-dimensional vector space V, the associated coordinate mapping onto $\mathbb{R}^n$ provides a coordinate system for V. Each $\mathbf{x}$ in V is identified uniquely by its $\mathcal{B}$-coordinate vector $[\mathbf{x}]_{\mathcal{B}}$.[1]

In some applications, a problem is described initially using a basis $\mathcal{B}$, but the problem's solution is aided by changing $\mathcal{B}$ to a new basis $\mathcal{C}$. (Examples will be given in Chapters 5 and 7.) Each vector is assigned a new $\mathcal{C}$-coordinate vector. In this section, we study how $[\mathbf{x}]_{\mathcal{C}}$ and $[\mathbf{x}]_{\mathcal{B}}$ are related for each $\mathbf{x}$ in V.

To visualize the problem, consider the two coordinate systems in Fig. 1. In Fig. 1(a), $\mathbf{x} = 3\mathbf{b}_1 + \mathbf{b}_2$, while in Fig. 1(b), the same $\mathbf{x}$ is shown as $\mathbf{x} = 6\mathbf{c}_1 + 4\mathbf{c}_2$. That is,

$$[\mathbf{x}]_{\mathcal{B}} = \begin{bmatrix} 3 \\ 1 \end{bmatrix} \quad \text{and} \quad [\mathbf{x}]_{\mathcal{C}} = \begin{bmatrix} 6 \\ 4 \end{bmatrix}$$

Our problem is to find the connection between the two coordinate vectors. Example 1 shows how to do this, provided we know how $\mathbf{b}_1$ and $\mathbf{b}_2$ are formed from $\mathbf{c}_1$ and $\mathbf{c}_2$.

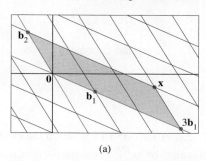
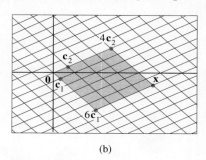

(a)   (b)

FIGURE 1 Two coordinate systems for the same vector space.

EXAMPLE 1 Consider two bases $\mathcal{B} = \{\mathbf{b}_1, \mathbf{b}_2\}$ and $\mathcal{C} = \{\mathbf{c}_1, \mathbf{c}_2\}$ for a vector space V, such that

$$\mathbf{b}_1 = 4\mathbf{c}_1 + \mathbf{c}_2 \quad \text{and} \quad \mathbf{b}_2 = -6\mathbf{c}_1 + \mathbf{c}_2 \tag{1}$$

Suppose

$$\mathbf{x} = 3\mathbf{b}_1 + \mathbf{b}_2 \tag{2}$$

That is, suppose $[\mathbf{x}]_{\mathcal{B}} = \begin{bmatrix} 3 \\ 1 \end{bmatrix}$. Find $[\mathbf{x}]_{\mathcal{C}}$.

SOLUTION Apply the coordinate mapping determined by $\mathcal{C}$ to $\mathbf{x}$ in (2). Since the coordinate mapping is a linear transformation,

$$[\mathbf{x}]_{\mathcal{C}} = [3\mathbf{b}_1 + \mathbf{b}_2]_{\mathcal{C}}$$
$$= 3[\mathbf{b}_1]_{\mathcal{C}} + [\mathbf{b}_2]_{\mathcal{C}}$$

We can write this vector equation as a matrix equation, using the vectors in the linear combination as the columns of a matrix:

$$[\mathbf{x}]_{\mathcal{C}} = \begin{bmatrix} [\mathbf{b}_1]_{\mathcal{C}} & [\mathbf{b}_2]_{\mathcal{C}} \end{bmatrix} \begin{bmatrix} 3 \\ 1 \end{bmatrix} \tag{3}$$

---

[1] Think of $[\mathbf{x}]_{\mathcal{B}}$ as a "name" for $\mathbf{x}$ that lists the weights used to build $\mathbf{x}$ as a linear combination of the basis vectors in $\mathcal{B}$.

This formula gives $[\mathbf{x}]_{\mathcal{C}}$, once we know the columns of the matrix. From (1),

$$[\mathbf{b}_1]_{\mathcal{C}} = \begin{bmatrix} 4 \\ 1 \end{bmatrix} \quad \text{and} \quad [\mathbf{b}_2]_{\mathcal{C}} = \begin{bmatrix} -6 \\ 1 \end{bmatrix}$$

Thus (3) provides the solution:

$$[\mathbf{x}]_{\mathcal{C}} = \begin{bmatrix} 4 & -6 \\ 1 & 1 \end{bmatrix} \begin{bmatrix} 3 \\ 1 \end{bmatrix} = \begin{bmatrix} 6 \\ 4 \end{bmatrix}$$

The $\mathcal{C}$-coordinates of $\mathbf{x}$ match those of the $\mathbf{x}$ in Fig. 1. ∎

The argument used to derive formula (3) can be generalized to yield the following result. (See Exercises 15 and 16.)

THEOREM 15

Let $\mathcal{B} = \{\mathbf{b}_1, \ldots, \mathbf{b}_n\}$ and $\mathcal{C} = \{\mathbf{c}_1, \ldots, \mathbf{c}_n\}$ be bases of a vector space V. Then there is a unique $n \times n$ matrix $\underset{\mathcal{C} \leftarrow \mathcal{B}}{P}$ such that

$$[\mathbf{x}]_{\mathcal{C}} = \underset{\mathcal{C} \leftarrow \mathcal{B}}{P}[\mathbf{x}]_{\mathcal{B}} \tag{4}$$

The columns of $\underset{\mathcal{C} \leftarrow \mathcal{B}}{P}$ are the $\mathcal{C}$-coordinate vectors of the vectors in the basis $\mathcal{B}$. That is,

$$\underset{\mathcal{C} \leftarrow \mathcal{B}}{P} = \begin{bmatrix} [\mathbf{b}_1]_{\mathcal{C}} & [\mathbf{b}_2]_{\mathcal{C}} & \cdots & [\mathbf{b}_n]_{\mathcal{C}} \end{bmatrix} \tag{5}$$

The matrix $\underset{\mathcal{C} \leftarrow \mathcal{B}}{P}$ in Theorem 15 is called the **change-of-coordinates matrix from** $\boldsymbol{\mathcal{B}}$ **to** $\boldsymbol{\mathcal{C}}$. Multiplication by $\underset{\mathcal{C} \leftarrow \mathcal{B}}{P}$ converts $\mathcal{B}$-coordinates into $\mathcal{C}$-coordinates.[2] Figure 2 illustrates the change-of-coordinates equation (4).

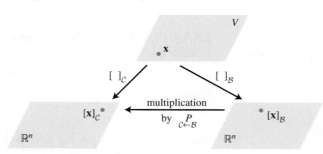

FIGURE 2 Two coordinate systems for V.

The columns of $\underset{\mathcal{C} \leftarrow \mathcal{B}}{P}$ are linearly independent because they are the coordinate vectors of the linearly independent set $\mathcal{B}$. (See Exercise 25 in Section 4.4.) Since $\underset{\mathcal{C} \leftarrow \mathcal{B}}{P}$ is square, it must be invertible, by the Invertible Matrix Theorem. Left-multiplying both sides of equation (4) by $(\underset{\mathcal{C} \leftarrow \mathcal{B}}{P})^{-1}$ yields

$$(\underset{\mathcal{C} \leftarrow \mathcal{B}}{P})^{-1}[\mathbf{x}]_{\mathcal{C}} = [\mathbf{x}]_{\mathcal{B}}$$

---

[2] To remember how to construct the matrix, think of $\underset{\mathcal{C} \leftarrow \mathcal{B}}{P}[\mathbf{x}]_{\mathcal{B}}$ as a linear combination of the columns of $\underset{\mathcal{C} \leftarrow \mathcal{B}}{P}$. The matrix-vector product is a $\mathcal{C}$-coordinate vector, so the columns of $\underset{\mathcal{C} \leftarrow \mathcal{B}}{P}$ should be $\mathcal{C}$-coordinate vectors, too.

Thus $(_{C \leftarrow B}^{\quad P})^{-1}$ is the matrix that converts C-coordinates into B-coordinates. That is,

$$(_{C \leftarrow B}^{\quad P})^{-1} = {}_{B \leftarrow C}^{\quad P} \tag{6}$$

Change of Basis in $\mathbb{R}^n$

If $B = \{\mathbf{b}_1, \ldots, \mathbf{b}_n\}$ and $\mathcal{E}$ is the *standard basis* $\{\mathbf{e}_1, \ldots, \mathbf{e}_n\}$ in $\mathbb{R}^n$, then $[\mathbf{b}_1]_\mathcal{E} = \mathbf{b}_1$, and likewise for the other vectors in B. In this case, $_{\mathcal{E} \leftarrow B}^{\quad P}$ is the same as the change-of-coordinates matrix P_B introduced in Section 4.4, namely,

$$P_B = [\,\mathbf{b}_1 \quad \mathbf{b}_2 \quad \cdots \quad \mathbf{b}_n\,]$$

To change coordinates between two nonstandard bases in $\mathbb{R}^n$, we need Theorem 15. The theorem shows that to solve the change-of-basis problem, we need the coordinate vectors of the old basis relative to the new basis.

EXAMPLE 2   Let $\mathbf{b}_1 = \begin{bmatrix} -9 \\ 1 \end{bmatrix}$, $\mathbf{b}_2 = \begin{bmatrix} -5 \\ -1 \end{bmatrix}$, $\mathbf{c}_1 = \begin{bmatrix} 1 \\ -4 \end{bmatrix}$, $\mathbf{c}_2 = \begin{bmatrix} 3 \\ -5 \end{bmatrix}$, and consider the bases for $\mathbb{R}^2$ given by $B = \{\mathbf{b}_1, \mathbf{b}_2\}$ and $C = \{\mathbf{c}_1, \mathbf{c}_2\}$. Find the change-of-coordinates matrix from B to C.

SOLUTION   The matrix $_{C \leftarrow B}^{\quad P}$ involves the C-coordinate vectors of $\mathbf{b}_1$ and $\mathbf{b}_2$. Let $[\,\mathbf{b}_1\,]_C = \begin{bmatrix} x_1 \\ x_2 \end{bmatrix}$ and $[\,\mathbf{b}_2\,]_C = \begin{bmatrix} y_1 \\ y_2 \end{bmatrix}$. Then, by definition,

$$[\,\mathbf{c}_1 \quad \mathbf{c}_2\,]\begin{bmatrix} x_1 \\ x_2 \end{bmatrix} = \mathbf{b}_1 \quad \text{and} \quad [\,\mathbf{c}_1 \quad \mathbf{c}_2\,]\begin{bmatrix} y_1 \\ y_2 \end{bmatrix} = \mathbf{b}_2$$

To solve both systems simultaneously, augment the coefficient matrix with $\mathbf{b}_1$ *and* $\mathbf{b}_2$, and row reduce:

$$[\,\mathbf{c}_1 \quad \mathbf{c}_2 \mid \mathbf{b}_1 \quad \mathbf{b}_2\,] = \begin{bmatrix} 1 & 3 & -9 & -5 \\ -4 & -5 & 1 & -1 \end{bmatrix} \sim \begin{bmatrix} 1 & 0 & 6 & 4 \\ 0 & 1 & -5 & -3 \end{bmatrix} \tag{7}$$

Thus

$$[\,\mathbf{b}_1\,]_C = \begin{bmatrix} 6 \\ -5 \end{bmatrix} \quad \text{and} \quad [\,\mathbf{b}_2\,]_C = \begin{bmatrix} 4 \\ -3 \end{bmatrix}$$

The desired change-of-coordinates matrix is therefore

$$_{C \leftarrow B}^{\quad P} = [\,[\,\mathbf{b}_1\,]_C \quad [\,\mathbf{b}_2\,]_C\,] = \begin{bmatrix} 6 & 4 \\ -5 & -3 \end{bmatrix}$$ ∎

Observe that the matrix $_{C \leftarrow B}^{\quad P}$ in Example 2 already appeared in (7). This is not surprising because the first column of $_{C \leftarrow B}^{\quad P}$ results from row reducing $[\,\mathbf{c}_1 \quad \mathbf{c}_2 \mid \mathbf{b}_1\,]$ to $[\,I \mid [\,\mathbf{b}_1\,]_C\,]$, and similarly for the second column of $_{C \leftarrow B}^{\quad P}$. Thus

$$[\,\mathbf{c}_1 \quad \mathbf{c}_2 \mid \mathbf{b}_1 \quad \mathbf{b}_2\,] \sim [\,I \mid {}_{C \leftarrow B}^{\quad P}\,]$$

An analogous procedure works for finding the change-of-coordinates matrix between any two bases in $\mathbb{R}^n$.

EXAMPLE 3   Let $\mathbf{b}_1 = \begin{bmatrix} 1 \\ -3 \end{bmatrix}$, $\mathbf{b}_2 = \begin{bmatrix} -2 \\ 4 \end{bmatrix}$, $\mathbf{c}_1 = \begin{bmatrix} -7 \\ 9 \end{bmatrix}$, $\mathbf{c}_2 = \begin{bmatrix} -5 \\ 7 \end{bmatrix}$, and consider the bases for $\mathbb{R}^2$ given by $\mathcal{B} = \{\mathbf{b}_1, \mathbf{b}_2\}$ and $\mathcal{C} = \{\mathbf{c}_1, \mathbf{c}_2\}$.

a. Find the change-of-coordinates matrix from $\mathcal{C}$ to $\mathcal{B}$.

b. Find the change-of-coordinates matrix from $\mathcal{B}$ to $\mathcal{C}$.

SOLUTION

a. Notice that $\underset{\mathcal{B}\leftarrow\mathcal{C}}{P}$ is needed rather than $\underset{\mathcal{C}\leftarrow\mathcal{B}}{P}$, and compute

$$\begin{bmatrix} \mathbf{b}_1 & \mathbf{b}_2 & \vdots & \mathbf{c}_1 & \mathbf{c}_2 \end{bmatrix} = \begin{bmatrix} 1 & -2 & \vdots & -7 & -5 \\ -3 & 4 & \vdots & 9 & 7 \end{bmatrix} \sim \begin{bmatrix} 1 & 0 & \vdots & 5 & 3 \\ 0 & 1 & \vdots & 6 & 4 \end{bmatrix}$$

So

$$\underset{\mathcal{B}\leftarrow\mathcal{C}}{P} = \begin{bmatrix} 5 & 3 \\ 6 & 4 \end{bmatrix}$$

b. By part (a) and property (6) above (with $\mathcal{B}$ and $\mathcal{C}$ interchanged),

$$\underset{\mathcal{C}\leftarrow\mathcal{B}}{P} = (\underset{\mathcal{B}\leftarrow\mathcal{C}}{P})^{-1} = \frac{1}{2}\begin{bmatrix} 4 & -3 \\ -6 & 5 \end{bmatrix} = \begin{bmatrix} 2 & -3/2 \\ -3 & 5/2 \end{bmatrix} \quad\blacksquare$$

Another description of the change-of-coordinates matrix $\underset{\mathcal{C}\leftarrow\mathcal{B}}{P}$ uses the change-of-coordinate matrices $P_{\mathcal{B}}$ and $P_{\mathcal{C}}$ that convert $\mathcal{B}$-coordinates and $\mathcal{C}$-coordinates, respectively, into standard coordinates. Recall that for each $\mathbf{x}$ in $\mathbb{R}^n$,

$$P_{\mathcal{B}}[\mathbf{x}]_{\mathcal{B}} = \mathbf{x}, \quad P_{\mathcal{C}}[\mathbf{x}]_{\mathcal{C}} = \mathbf{x}, \quad \text{and} \quad [\mathbf{x}]_{\mathcal{C}} = P_{\mathcal{C}}^{-1}\mathbf{x}$$

Thus

$$[\mathbf{x}]_{\mathcal{C}} = P_{\mathcal{C}}^{-1}\mathbf{x} = P_{\mathcal{C}}^{-1}P_{\mathcal{B}}[\mathbf{x}]_{\mathcal{B}}$$

In $\mathbb{R}^n$, the change-of-coordinates matrix $\underset{\mathcal{C}\leftarrow\mathcal{B}}{P}$ may be computed as $P_{\mathcal{C}}^{-1}P_{\mathcal{B}}$. Actually, for matrices larger than 2×2, an algorithm analogous to the one in Example 3 is faster than computing $P_{\mathcal{C}}^{-1}$ and then $P_{\mathcal{C}}^{-1}P_{\mathcal{B}}$. See Exercise 12 in Section 2.2.

PRACTICE PROBLEMS

1. Let $\mathcal{F} = \{\mathbf{f}_1, \mathbf{f}_2\}$ and $\mathcal{G} = \{\mathbf{g}_1, \mathbf{g}_2\}$ be bases for a vector space V, and let P be a matrix whose columns are $[\mathbf{f}_1]_{\mathcal{G}}$ and $[\mathbf{f}_2]_{\mathcal{G}}$. Which of the following equations is satisfied by P for all $\mathbf{v}$ in V?

(i) $[\mathbf{v}]_{\mathcal{F}} = P[\mathbf{v}]_{\mathcal{G}}$ (ii) $[\mathbf{v}]_{\mathcal{G}} = P[\mathbf{v}]_{\mathcal{F}}$

2. Let $\mathcal{B}$ and $\mathcal{C}$ be as in Example 1. Use the results of that example to find the change-of-coordinates matrix from $\mathcal{C}$ to $\mathcal{B}$.

4.7 EXERCISES

1. Let $\mathcal{B} = \{\mathbf{b}_1, \mathbf{b}_2\}$ and $\mathcal{C} = \{\mathbf{c}_1, \mathbf{c}_2\}$ be bases for a vector space V, and suppose $\mathbf{b}_1 = 6\mathbf{c}_1 - 2\mathbf{c}_2$ and $\mathbf{b}_2 = 9\mathbf{c}_1 - 4\mathbf{c}_2$.

a. Find the change-of-coordinates matrix from $\mathcal{B}$ to $\mathcal{C}$.

b. Find $[\mathbf{x}]_{\mathcal{C}}$ for $\mathbf{x} = -3\mathbf{b}_1 + 2\mathbf{b}_2$. Use part (a).

2. Let $\mathcal{B} = \{\mathbf{b}_1, \mathbf{b}_2\}$ and $\mathcal{C} = \{\mathbf{c}_1, \mathbf{c}_2\}$ be bases for a vector space V, and suppose $\mathbf{b}_1 = -2\mathbf{c}_1 + 4\mathbf{c}_2$ and $\mathbf{b}_2 = 3\mathbf{c}_1 - 6\mathbf{c}_2$.

a. Find the change-of-coordinates matrix from $\mathcal{B}$ to $\mathcal{C}$.

b. Find $[\mathbf{x}]_{\mathcal{C}}$ for $\mathbf{x} = 2\mathbf{b}_1 + 3\mathbf{b}_2$.

3. Let $\mathcal{U} = \{\mathbf{u}_1, \mathbf{u}_2\}$ and $\mathcal{W} = \{\mathbf{w}_1, \mathbf{w}_2\}$ be bases for V, and let P be a matrix whose columns are $[\,\mathbf{u}_1\,]_{\mathcal{W}}$ and $[\mathbf{u}_2]_{\mathcal{W}}$. Which of the following equations is satisfied by P for all $\mathbf{x}$ in V?

(i) $[\,\mathbf{x}\,]_{\mathcal{U}} = P[\,\mathbf{x}\,]_{\mathcal{W}}$ (ii) $[\,\mathbf{x}\,]_{\mathcal{W}} = P[\,\mathbf{x}\,]_{\mathcal{U}}$

4. Let $\mathcal{A} = \{\mathbf{a}_1, \mathbf{a}_2, \mathbf{a}_3\}$ and $\mathcal{D} = \{\mathbf{d}_1, \mathbf{d}_2, \mathbf{d}_3\}$ be bases for V, and let $P = [\,[\mathbf{d}_1]_{\mathcal{A}} \ \ [\mathbf{d}_2]_{\mathcal{A}} \ \ [\mathbf{d}_3]_{\mathcal{A}}\,]$. Which of the following equations is satisfied by P for all $\mathbf{x}$ in V?

(i) $[\,\mathbf{x}\,]_{\mathcal{A}} = P[\,\mathbf{x}\,]_{\mathcal{D}}$ (ii) $[\,\mathbf{x}\,]_{\mathcal{D}} = P[\,\mathbf{x}\,]_{\mathcal{A}}$

5. Let $\mathcal{A} = \{\mathbf{a}_1, \mathbf{a}_2, \mathbf{a}_3\}$ and $\mathcal{B} = \{\mathbf{b}_1, \mathbf{b}_2, \mathbf{b}_3\}$ be bases for a vector space V, and suppose $\mathbf{a}_1 = 4\mathbf{b}_1 - \mathbf{b}_2$, $\mathbf{a}_2 = -\mathbf{b}_1 + \mathbf{b}_2 + \mathbf{b}_3$, and $\mathbf{a}_3 = \mathbf{b}_2 - 2\mathbf{b}_3$.

a. Find the change-of-coordinates matrix from $\mathcal{A}$ to $\mathcal{B}$.

b. Find $[\,\mathbf{x}\,]_{\mathcal{B}}$ for $\mathbf{x} = 3\mathbf{a}_1 + 4\mathbf{a}_2 + \mathbf{a}_3$.

6. Let $\mathcal{D} = \{\mathbf{d}_1, \mathbf{d}_2, \mathbf{d}_3\}$ and $\mathcal{F} = \{\mathbf{f}_1, \mathbf{f}_2, \mathbf{f}_3\}$ be bases for a vector space V, and suppose $\mathbf{f}_1 = 2\mathbf{d}_1 - \mathbf{d}_2 + \mathbf{d}_3$, $\mathbf{f}_2 = 3\mathbf{d}_2 + \mathbf{d}_3$, and $\mathbf{f}_3 = -3\mathbf{d}_1 + 2\mathbf{d}_3$.

a. Find the change-of-coordinates matrix from $\mathcal{F}$ to $\mathcal{D}$.

b. Find $[\,\mathbf{x}\,]_{\mathcal{D}}$ for $\mathbf{x} = \mathbf{f}_1 - 2\mathbf{f}_2 + 2\mathbf{f}_3$.

In Exercises 7–10, let $\mathcal{B} = \{\mathbf{b}_1, \mathbf{b}_2\}$ and $\mathcal{C} = \{\mathbf{c}_1, \mathbf{c}_2\}$ be bases for $\mathbb{R}^2$. In each exercise, find the change-of-coordinates matrix from $\mathcal{B}$ to $\mathcal{C}$ and the change-of-coordinates matrix from $\mathcal{C}$ to $\mathcal{B}$.

7. $\mathbf{b}_1 = \begin{bmatrix} 7 \\ 5 \end{bmatrix}$, $\mathbf{b}_2 = \begin{bmatrix} -3 \\ -1 \end{bmatrix}$, $\mathbf{c}_1 = \begin{bmatrix} 1 \\ -5 \end{bmatrix}$, $\mathbf{c}_2 = \begin{bmatrix} -2 \\ 2 \end{bmatrix}$

8. $\mathbf{b}_1 = \begin{bmatrix} -1 \\ 8 \end{bmatrix}$, $\mathbf{b}_2 = \begin{bmatrix} 1 \\ -7 \end{bmatrix}$, $\mathbf{c}_1 = \begin{bmatrix} 1 \\ 2 \end{bmatrix}$, $\mathbf{c}_2 = \begin{bmatrix} 1 \\ 1 \end{bmatrix}$

9. $\mathbf{b}_1 = \begin{bmatrix} 4 \\ 4 \end{bmatrix}$, $\mathbf{b}_2 = \begin{bmatrix} 8 \\ 4 \end{bmatrix}$, $\mathbf{c}_1 = \begin{bmatrix} 2 \\ 2 \end{bmatrix}$, $\mathbf{c}_2 = \begin{bmatrix} -2 \\ 2 \end{bmatrix}$

10. $\mathbf{b}_1 = \begin{bmatrix} 6 \\ -12 \end{bmatrix}$, $\mathbf{b}_2 = \begin{bmatrix} 4 \\ 2 \end{bmatrix}$, $\mathbf{c}_1 = \begin{bmatrix} 4 \\ 2 \end{bmatrix}$, $\mathbf{c}_2 = \begin{bmatrix} 3 \\ 9 \end{bmatrix}$

In Exercises 11 and 12, $\mathcal{B}$ and $\mathcal{C}$ are bases for a vector space V. Mark each statement True or False. Justify each answer.

11. a. The columns of the change-of-coordinates matrix $\underset{\mathcal{C}\leftarrow\mathcal{B}}{P}$ are $\mathcal{B}$-coordinate vectors of the vectors in $\mathcal{C}$.

b. If $V = \mathbb{R}^n$ and $\mathcal{C}$ is the *standard* basis for V, then $\underset{\mathcal{C}\leftarrow\mathcal{B}}{P}$ is the same as the change-of-coordinates matrix $P_{\mathcal{B}}$ introduced in Section 4.4.

12. a. The columns of $\underset{\mathcal{C}\leftarrow\mathcal{B}}{P}$ are linearly independent.

b. If $V = \mathbb{R}^2$, $\mathcal{B} = \{\mathbf{b}_1, \mathbf{b}_2\}$, and $\mathcal{C} = \{\mathbf{c}_1, \mathbf{c}_2\}$, then row reduction of $[\,\mathbf{c}_1 \ \ \mathbf{c}_2 \ \ \mathbf{b}_1 \ \ \mathbf{b}_2\,]$ to $[\,I \ \ P\,]$ produces a matrix P that satisfies $[\,\mathbf{x}\,]_{\mathcal{B}} = P[\,\mathbf{x}\,]_{\mathcal{C}}$ for all $\mathbf{x}$ in V.

13. In $\mathbb{P}_2$, find the change-of-coordinates matrix from the basis $\mathcal{B} = \{1 - 2t + t^2, 3 - 5t + 4t^2, 2t + 3t^2\}$ to the standard basis $\mathcal{C} = \{1, t, t^2\}$. Then find the $\mathcal{B}$-coordinate vector for $-1 + 2t$.

14. In $\mathbb{P}_2$, find the change-of-coordinates matrix from the basis $\mathcal{B} = \{1 - 3t^2, 2 + t - 5t^2, 1 + 2t\}$ to the standard basis. Then write t^2 as a linear combination of the polynomials in $\mathcal{B}$.

Exercises 15 and 16 provide a proof of Theorem 15. Fill in a justification for each step.

15. Given $\mathbf{v}$ in V, there exist scalars $x_1, \ldots, x_n$, such that

$$\mathbf{v} = x_1\mathbf{b}_1 + x_2\mathbf{b}_2 + \cdots + x_n\mathbf{b}_n$$

because (a) _____. Apply the coordinate mapping determined by the basis $\mathcal{C}$, and obtain

$$[\mathbf{v}]_{\mathcal{C}} = x_1[\mathbf{b}_1]_{\mathcal{C}} + x_2[\mathbf{b}_2]_{\mathcal{C}} + \cdots + x_n[\mathbf{b}_n]_{\mathcal{C}}$$

because (b) _____. This equation may be written in the form

$$[\,\mathbf{v}\,]_{\mathcal{C}} = \begin{bmatrix} [\,\mathbf{b}_1\,]_{\mathcal{C}} & [\,\mathbf{b}_2\,]_{\mathcal{C}} & \cdots & [\,\mathbf{b}_n\,]_{\mathcal{C}} \end{bmatrix} \begin{bmatrix} x_1 \\ \vdots \\ x_n \end{bmatrix} \quad (8)$$

by the definition of (c) _____. This shows that the matrix $\underset{\mathcal{C}\leftarrow\mathcal{B}}{P}$ shown in (5) satisfies $[\mathbf{v}]_{\mathcal{C}} = \underset{\mathcal{C}\leftarrow\mathcal{B}}{P}[\mathbf{v}]_{\mathcal{B}}$ for each $\mathbf{v}$ in V, because the vector on the right side of (8) is (d) _____.

16. Suppose Q is any matrix such that

$$[\mathbf{v}]_{\mathcal{C}} = Q[\mathbf{v}]_{\mathcal{B}} \quad \text{for each } \mathbf{v} \text{ in } V \quad (9)$$

Set $\mathbf{v} = \mathbf{b}_1$ in (9). Then (9) shows that $[\mathbf{b}_1]_{\mathcal{C}}$ is the first column of Q because (a) _____. Similarly, for $k = 2, \ldots, n$, the kth column of Q is (b) _____ because (c) _____. This shows that the matrix $\underset{\mathcal{C}\leftarrow\mathcal{B}}{P}$ defined by (5) in Theorem 15 is the only matrix that satisfies condition (4).

17. [M] Let $\mathcal{B} = \{\mathbf{x}_0, \ldots, \mathbf{x}_6\}$ and $\mathcal{C} = \{\mathbf{y}_0, \ldots, \mathbf{y}_6\}$, where $\mathbf{x}_k$ is the function $\cos^k t$ and $\mathbf{y}_k$ is the function $\cos kt$. Exercise 34 in Section 4.5 showed that both $\mathcal{B}$ and $\mathcal{C}$ are bases for the vector space $H = \text{Span}\,\{\mathbf{x}_0, \ldots, \mathbf{x}_6\}$.

a. Set $P = [\,[\mathbf{y}_0]_{\mathcal{B}} \quad \cdots \quad [\mathbf{y}_6]_{\mathcal{B}}\,]$, and calculate P^{-1}.

b. Explain why the columns of P^{-1} are the $\mathcal{C}$-coordinate vectors of $\mathbf{x}_0, \ldots, \mathbf{x}_6$. Then use these coordinate vectors to write trigonometric identities that express powers of $\cos t$ in terms of the functions in $\mathcal{C}$.

See the *Study Guide*.

18. [M] (*Calculus required*)[3] Recall from calculus that integrals such as

$$\int (5\cos^3 t - 6\cos^4 t + 5\cos^5 t - 12\cos^6 t)\,dt \quad (10)$$

are tedious to compute. (The usual method is to apply integration by parts repeatedly and use the half-angle formula.) Use the matrix P or P^{-1} from Exercise 17 to transform (10); then compute the integral.

---

[3] The idea for Exercises 17 and 18 and five related exercises in earlier sections came from a paper by Jack W. Rogers, Jr., of Auburn University, presented at a meeting of the International Linear Algebra Society, August 1995. See "Applications of Linear Algebra in Calculus," *American Mathematical Monthly* **104** (1), 1997.

19. **[M]** Let

$$P = \begin{bmatrix} 1 & 2 & -1 \\ -3 & -5 & 0 \\ 4 & 6 & 1 \end{bmatrix},$$

$$\mathbf{v}_1 = \begin{bmatrix} -2 \\ 2 \\ 3 \end{bmatrix}, \ \mathbf{v}_2 = \begin{bmatrix} -8 \\ 5 \\ 2 \end{bmatrix}, \ \mathbf{v}_3 = \begin{bmatrix} -7 \\ 2 \\ 6 \end{bmatrix}$$

a. Find a basis $\{\mathbf{u}_1, \mathbf{u}_2, \mathbf{u}_3\}$ for $\mathbb{R}^3$ such that P is the change-of-coordinates matrix from $\{\mathbf{u}_1, \mathbf{u}_2, \mathbf{u}_3\}$ to the basis $\{\mathbf{v}_1, \mathbf{v}_2, \mathbf{v}_3\}$. [*Hint:* What do the columns of $\underset{\mathcal{C} \leftarrow \mathcal{B}}{P}$ represent?]

b. Find a basis $\{\mathbf{w}_1, \mathbf{w}_2, \mathbf{w}_3\}$ for $\mathbb{R}^3$ such that P is the change-of-coordinates matrix from $\{\mathbf{v}_1, \mathbf{v}_2, \mathbf{v}_3\}$ to $\{\mathbf{w}_1, \mathbf{w}_2, \mathbf{w}_3\}$.

20. Let $\mathcal{B} = \{\mathbf{b}_1, \mathbf{b}_2\}$, $\mathcal{C} = \{\mathbf{c}_1, \mathbf{c}_2\}$, and $\mathcal{D} = \{\mathbf{d}_1, \mathbf{d}_2\}$ be bases for a two-dimensional vector space.

a. Write an equation that relates the matrices $\underset{\mathcal{C} \leftarrow \mathcal{B}}{P}$, $\underset{\mathcal{D} \leftarrow \mathcal{C}}{P}$, and $\underset{\mathcal{D} \leftarrow \mathcal{B}}{P}$. Justify your result.

b. **[M]** Use a matrix program either to help you find the equation or to check the equation you write. Work with three bases for $\mathbb{R}^2$. (See Exercises 7–10.)

SOLUTIONS TO PRACTICE PROBLEMS

1. Since the columns of P are $\mathcal{G}$-coordinate vectors, a vector of the form $P\mathbf{x}$ must be a $\mathcal{G}$-coordinate vector. Thus P satisfies equation (ii).

2. The coordinate vectors found in Example 1 show that

$$\underset{\mathcal{C} \leftarrow \mathcal{B}}{P} = \begin{bmatrix} [\mathbf{b}_1]_{\mathcal{C}} & [\mathbf{b}_2]_{\mathcal{C}} \end{bmatrix} = \begin{bmatrix} 4 & -6 \\ 1 & 1 \end{bmatrix}$$

Hence

$$\underset{\mathcal{B} \leftarrow \mathcal{C}}{P} = (\underset{\mathcal{C} \leftarrow \mathcal{B}}{P})^{-1} = \frac{1}{10} \begin{bmatrix} 1 & 6 \\ -1 & 4 \end{bmatrix} = \begin{bmatrix} .1 & .6 \\ -.1 & .4 \end{bmatrix}$$

4.8 APPLICATIONS TO DIFFERENCE EQUATIONS

Now that powerful computers are widely available, more and more scientific and engineering problems are being treated in a way that uses discrete, or digital, data rather than continuous data. Difference equations are often the appropriate tool to analyze such data. Even when a differential equation is used to model a continuous process, a numerical solution is often produced from a related difference equation.

This section highlights some fundamental properties of linear difference equations that are best explained using linear algebra.

Discrete-Time Signals

The vector space $\mathbb{S}$ of discrete-time signals was introduced in Section 4.1. A **signal** in $\mathbb{S}$ is a function defined only on the integers and is visualized as a sequence of numbers, say, $\{y_k\}$. Figure 1 shows three typical signals whose general terms are $(.7)^k$, 1^k, and $(-1)^k$, respectively.

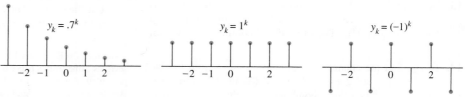

FIGURE 1 Three signals in $\mathbb{S}$.

Digital signals obviously arise in electrical and control systems engineering, but discrete-data sequences are also generated in biology, physics, economics, demography, and many other areas, wherever a process is measured, or *sampled*, at discrete time intervals. When a process begins at a specific time, it is sometimes convenient to write a signal as a sequence of the form $(y_0, y_1, y_2, \ldots)$. The terms y_k for $k < 0$ either are assumed to be zero or are simply omitted.

EXAMPLE 1   The crystal-clear sounds from a compact disc player are produced from music that has been sampled at the rate of 44,100 times per second. See Fig. 2. At each measurement, the amplitude of the music signal is recorded as a number, say, y_k. The original music is composed of many different sounds of varying frequencies, yet the sequence $\{y_k\}$ contains enough information to reproduce all the frequencies in the sound up to about 20,000 cycles per second, higher than the human ear can sense. ∎

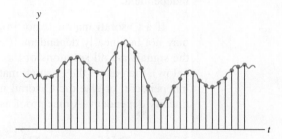

FIGURE 2 Sampled data from a music signal.

Linear Independence in the Space $\mathbb{S}$ of Signals

To simplify notation, we consider a set of only three signals in $\mathbb{S}$, say, $\{u_k\}$, $\{v_k\}$, and $\{w_k\}$. They are linearly independent precisely when the equation

$$c_1 u_k + c_2 v_k + c_3 w_k = 0 \quad \text{for all } k \tag{1}$$

implies that $c_1 = c_2 = c_3 = 0$. The phrase "for all k" means for all integers—positive, negative, and zero. One could also consider signals that start with $k = 0$, for example, in which case, "for all k" would mean for all integers $k \geq 0$.

Suppose c_1, c_2, c_3 satisfy (1). Then equation (1) holds for any three consecutive values of k, say, $k, k + 1$, and $k + 2$. Thus (1) implies that

$$c_1 u_{k+1} + c_2 v_{k+1} + c_3 w_{k+1} = 0 \quad \text{for all } k$$

and

$$c_1 u_{k+2} + c_2 v_{k+2} + c_3 w_{k+2} = 0 \quad \text{for all } k$$

Hence c_1, c_2, c_3 satisfy

$$\begin{bmatrix} u_k & v_k & w_k \\ u_{k+1} & v_{k+1} & w_{k+1} \\ u_{k+2} & v_{k+2} & w_{k+2} \end{bmatrix} \begin{bmatrix} c_1 \\ c_2 \\ c_3 \end{bmatrix} = \begin{bmatrix} 0 \\ 0 \\ 0 \end{bmatrix} \quad \text{for all } k \tag{2}$$

SG   The Casorati Test 4–30

The coefficient matrix in this system is called the **Casorati matrix** of the signals, and the determinant of the matrix is called the **Casoratian** of $\{u_k\}$, $\{v_k\}$, and $\{w_k\}$. If the Casorati matrix is invertible for at least one value of k, then (2) will imply that $c_1 = c_2 = c_3 = 0$, which will prove that the three signals are linearly independent.

EXAMPLE 2 Verify that 1^k, $(-2)^k$, and 3^k are linearly independent signals.

SOLUTION The Casorati matrix is

$$\begin{bmatrix} 1^k & (-2)^k & 3^k \\ 1^{k+1} & (-2)^{k+1} & 3^{k+1} \\ 1^{k+2} & (-2)^{k+2} & 3^{k+2} \end{bmatrix}$$

Row operations can show fairly easily that this matrix is always invertible. However, it is faster to substitute a value for k — say, $k = 0$ — and row reduce the numerical matrix:

$$\begin{bmatrix} 1 & 1 & 1 \\ 1 & -2 & 3 \\ 1 & 4 & 9 \end{bmatrix} \sim \begin{bmatrix} 1 & 1 & 1 \\ 0 & -3 & 2 \\ 0 & 3 & 8 \end{bmatrix} \sim \begin{bmatrix} 1 & 1 & 1 \\ 0 & -3 & 2 \\ 0 & 0 & 10 \end{bmatrix}$$

The Casorati matrix is invertible for $k = 0$. So 1^k, $(-2)^k$, and 3^k are linearly independent. ∎

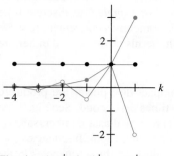

The signals 1^k, $(-2)^k$, and 3^k.

If a Casorati matrix is not invertible, the associated signals being tested may or may not be linearly dependent. (See Exercise 33.) However, it can be shown that if the signals are all solutions of the *same* homogeneous difference equation (described below), then either the Casorati matrix is invertible for all k and the signals are linearly independent, or else the Casorati matrix is not invertible for all k and the signals are linearly dependent. A nice proof using linear transformations is in the *Study Guide*.

Linear Difference Equations

Given scalars $a_0, \ldots, a_n$, with a_0 and a_n nonzero, and given a signal $\{z_k\}$, the equation

$$a_0 y_{k+n} + a_1 y_{k+n-1} + \cdots + a_{n-1} y_{k+1} + a_n y_k = z_k \quad \text{for all } k \tag{3}$$

is called a **linear difference equation** (or **linear recurrence relation**) **of order n**. For simplicity, a_0 is often taken equal to 1. If $\{z_k\}$ is the zero sequence, the equation is **homogeneous**; otherwise, the equation is **nonhomogeneous**.

EXAMPLE 3 In digital signal processing, a difference equation such as (3) describes a **linear filter**, and $a_0, \ldots, a_n$ are called the **filter coefficients**. If $\{y_k\}$ is treated as the input and $\{z_k\}$ as the output, then the solutions of the associated homogeneous equation are the signals that are filtered *out* and transformed into the zero signal. Let us feed two different signals into the filter

$$.35 y_{k+2} + .5 y_{k+1} + .35 y_k = z_k$$

Here $.35$ is an abbreviation for $\sqrt{2}/4$. The first signal is created by sampling the continuous signal $y = \cos(\pi t/4)$ at integer values of t, as in Fig. 3(a). The discrete signal is

$$\{y_k\} = \{\ldots, \cos(0), \cos(\pi/4), \cos(2\pi/4), \cos(3\pi/4), \ldots\}$$

For simplicity, write $\pm.7$ in place of $\pm\sqrt{2}/2$, so that

$$\{y_k\} = \{\ldots, 1, .7, 0, -.7, -1, -.7, 0, .7, 1, .7, 0, \ldots\}$$
$$\underset{k=0}{\uparrow}$$

Table 1 shows a calculation of the output sequence $\{z_k\}$, where $.35(.7)$ is an abbreviation for $(\sqrt{2}/4)(\sqrt{2}/2) = .25$. The output is $\{y_k\}$, shifted by one term.

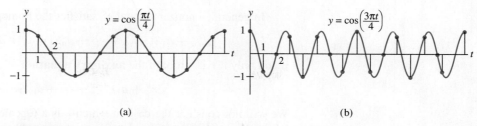

FIGURE 3 Discrete signals with different frequencies.

TABLE 1 Computing the Output of a Filter

k	y_k	y_{k+1}	y_{k+2}	$.35y_k$	$+\ .5y_{k+1}$	$+\ .35y_{k+2}$	$=$	z_k
0	1	.7	0	$.35(1)$	$+\ .5(.7)$	$+\ .35(0)$	$=$	.7
1	.7	0	$-.7$	$.35(.7)$	$+\ .5(0)$	$+\ .35(-.7)$	$=$	0
2	0	$-.7$	-1	$.35(0)$	$+\ .5(-.7)$	$+\ .35(-1)$	$=$	$-.7$
3	$-.7$	-1	$-.7$	$.35(-.7)$	$+\ .5(-1)$	$+\ .35(-.7)$	$=$	-1
4	-1	$-.7$	0	$.35(-1)$	$+\ .5(-.7)$	$+\ .35(0)$	$=$	$-.7$
5	$-.7$	0	.7	$.35(-.7)$	$+\ .5(0)$	$+\ .35(.7)$	$=$	0
$\vdots$	$\vdots$							$\vdots$

A different input signal is produced from the higher frequency signal $y = \cos(3\pi t/4)$, shown in Fig. 3(b). Sampling at the same rate as before produces a new input sequence:

$$\{w_k\} = \{\dots, 1,\ -.7,\ 0,\ .7,\ -1,\ .7,\ 0,\ -.7,\ 1,\ -.7,\ 0, \dots\}$$
$$\uparrow$$
$$k = 0$$

When $\{w_k\}$ is fed into the filter, the output is the zero sequence. The filter, called a *low-pass filter*, lets $\{y_k\}$ pass through, but stops the higher frequency $\{w_k\}$. ∎

In many applications, a sequence $\{z_k\}$ is specified for the right side of a difference equation (3), and a $\{y_k\}$ that satisfies (3) is called a **solution** of the equation. The next example shows how to find solutions for a homogeneous equation.

EXAMPLE 4 Solutions of a homogeneous difference equation often have the form $y_k = r^k$ for some r. Find some solutions of the equation

$$y_{k+3} - 2y_{k+2} - 5y_{k+1} + 6y_k = 0 \quad \text{for all } k \tag{4}$$

SOLUTION Substitute r^k for y_k in the equation and factor the left side:

$$r^{k+3} - 2r^{k+2} - 5r^{k+1} + 6r^k = 0 \tag{5}$$
$$r^k(r^3 - 2r^2 - 5r + 6) = 0$$
$$r^k(r - 1)(r + 2)(r - 3) = 0 \tag{6}$$

Since (5) is equivalent to (6), r^k satisfies the difference equation (4) if and only if r satisfies (6). Thus 1^k, $(-2)^k$, and 3^k are all solutions of (4). For instance, to verify that 3^k is a solution of (4), compute

$$3^{k+3} - 2 \cdot 3^{k+2} - 5 \cdot 3^{k+1} + 6 \cdot 3^k$$
$$= 3^k(27 - 18 - 15 + 6) = 0 \quad \text{for all } k$$

∎

In general, a nonzero signal r^k satisfies the homogeneous difference equation

$$y_{k+n} + a_1 y_{k+n-1} + \cdots + a_{n-1} y_{k+1} + a_n y_k = 0 \quad \text{for all } k$$

if and only if r is a root of the **auxiliary equation**

$$r^n + a_1 r^{n-1} + \cdots + a_{n-1} r + a_n = 0$$

We will not consider the case in which r is a repeated root of the auxiliary equation. When the auxiliary equation has a *complex root*, the difference equation has solutions of the form $s^k \cos k\omega$ and $s^k \sin k\omega$, for constants s and ω. This happened in Example 3.

Solution Sets of Linear Difference Equations

Given $a_1, \ldots, a_n$, consider the mapping $T : \mathbb{S} \to \mathbb{S}$ that transforms a signal $\{y_k\}$ into a signal $\{w_k\}$ given by

$$w_k = y_{k+n} + a_1 y_{k+n-1} + \cdots + a_{n-1} y_{k+1} + a_n y_k$$

It is readily checked that T is a *linear* transformation. This implies that the solution set of the homogeneous equation

$$y_{k+n} + a_1 y_{k+n-1} + \cdots + a_{n-1} y_{k+1} + a_n y_k = 0 \quad \text{for all } k$$

is the kernel of T (the set of signals that T maps into the zero signal), and hence the solution set is a *subspace* of $\mathbb{S}$. Any linear combination of solutions is again a solution.

The next theorem, a simple but basic result, will lead to more information about the solution sets of difference equations.

THEOREM 16

If $a_n \neq 0$ and if $\{z_k\}$ is given, the equation

$$y_{k+n} + a_1 y_{k+n-1} + \cdots + a_{n-1} y_{k+1} + a_n y_k = z_k \quad \text{for all } k \qquad (7)$$

has a unique solution whenever $y_0, \ldots, y_{n-1}$ are specified.

PROOF If $y_0, \ldots, y_{n-1}$ are specified, use (7) to *define*

$$y_n = z_0 - [a_1 y_{n-1} + \cdots + a_{n-1} y_1 + a_n y_0]$$

And now that $y_1, \ldots, y_n$ are specified, use (7) to define y_{n+1}. In general, use the recurrence relation

$$y_{n+k} = z_k - [a_1 y_{k+n-1} + \cdots + a_n y_k] \qquad (8)$$

to define y_{n+k} for $k \geq 0$. To define y_k for $k < 0$, use the recurrence relation

$$y_k = \frac{1}{a_n} z_k - \frac{1}{a_n} [y_{k+n} + a_1 y_{k+n-1} + \cdots + a_{n-1} y_{k+1}] \qquad (9)$$

This produces a signal that satisfies (7). Conversely, any signal that satisfies (7) for all k certainly satisfies (8) and (9), so the solution of (7) is unique. ∎

THEOREM 17

The set H of all solutions of the nth-order homogeneous linear difference equation

$$y_{k+n} + a_1 y_{k+n-1} + \cdots + a_{n-1} y_{k+1} + a_n y_k = 0 \quad \text{for all } k \qquad (10)$$

is an n-dimensional vector space.

PROOF As was pointed out earlier, H is a subspace of $\mathbb{S}$ because H is the kernel of a linear transformation. For $\{y_k\}$ in H, let $F\{y_k\}$ be the vector in $\mathbb{R}^n$ given by $(y_0, y_1, \ldots, y_{n-1})$. It is readily verified that $F : H \to \mathbb{R}^n$ is a linear transformation. Given any vector $(y_0, y_1, \ldots, y_{n-1})$ in $\mathbb{R}^n$, Theorem 16 says that there is a unique signal $\{y_k\}$ in H such that $F\{y_k\} = (y_0, y_1, \ldots, y_{n-1})$. This means that F is a one-to-one linear transformation of H onto $\mathbb{R}^n$; that is, F is an isomorphism. Thus $\dim H = \dim \mathbb{R}^n = n$. (See Exercise 32 in Section 4.5.) ∎

EXAMPLE 5 Find a basis for the set of all solutions to the difference equation

$$y_{k+3} - 2y_{k+2} - 5y_{k+1} + 6y_k = 0 \quad \text{for all } k$$

SOLUTION Our work in linear algebra really pays off now! We know from Examples 2 and 4 that 1^k, $(-2)^k$, and 3^k are linearly independent solutions. In general, it can be difficult to verify directly that a set of signals *spans* the solution space. But that is no problem here because of two key theorems—Theorem 17, which shows that the solution space is exactly three-dimensional, and the Basis Theorem in Section 4.5, which says that a linearly independent set of n vectors in an n-dimensional space is automatically a basis. So 1^k, $(-2)^k$, and 3^k form a basis for the solution space. ∎

The standard way to describe the "general solution" of the difference equation (10) is to exhibit a basis for the subspace of all solutions. Such a basis is usually called a **fundamental set of solutions** of (10). In practice, if you can find n linearly independent signals that satisfy (10), they will automatically span the n-dimensional solution space, as explained in Example 5.

Nonhomogeneous Equations

The general solution of the nonhomogeneous difference equation

$$y_{k+n} + a_1 y_{k+n-1} + \cdots + a_{n-1} y_{k+1} + a_n y_k = z_k \quad \text{for all } k \tag{11}$$

can be written as one particular solution of (11) plus an arbitrary linear combination of a fundamental set of solutions of the corresponding homogeneous equation (10). This fact is analogous to the result in Section 1.5 showing that the solution sets of $A\mathbf{x} = \mathbf{b}$ and $A\mathbf{x} = \mathbf{0}$ are parallel. Both results have the same explanation: The mapping $\mathbf{x} \mapsto A\mathbf{x}$ is linear, and the mapping that transforms the signal $\{y_k\}$ into the signal $\{z_k\}$ in (11) is linear. See Exercise 35.

EXAMPLE 6 Verify that the signal $y_k = k^2$ satisfies the difference equation

$$y_{k+2} - 4y_{k+1} + 3y_k = -4k \quad \text{for all } k \tag{12}$$

Then find a description of all solutions of this equation.

SOLUTION Substitute k^2 for y_k on the left side of (12):

$$(k + 2)^2 - 4(k + 1)^2 + 3k^2$$
$$= (k^2 + 4k + 4) - 4(k^2 + 2k + 1) + 3k^2$$
$$= -4k$$

So k^2 is indeed a solution of (12). The next step is to solve the homogeneous equation

$$y_{k+2} - 4y_{k+1} + 3y_k = 0 \tag{13}$$

The auxiliary equation is

$$r^2 - 4r + 3 = (r - 1)(r - 3) = 0$$

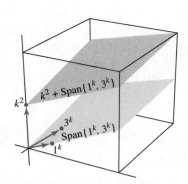

FIGURE 4

Solution sets of difference equations (12) and (13).

The roots are $r = 1, 3$. So two solutions of the homogeneous difference equation are 1^k and 3^k. They are obviously not multiples of each other, so they are linearly independent signals. By Theorem 17, the solution space is two-dimensional, so 3^k and 1^k form a basis for the set of solutions of equation (13). Translating that set by a particular solution of the nonhomogeneous equation (12), we obtain the general solution of (12):

$$k^2 + c_1 1^k + c_2 3^k, \quad \text{or} \quad k^2 + c_1 + c_2 3^k$$

Figure 4 gives a geometric visualization of the two solution sets. Each point in the figure corresponds to one signal in $\mathbb{S}$. ∎

Reduction to Systems of First-Order Equations

A modern way to study a homogeneous nth-order linear difference equation is to replace it by an equivalent system of first-order difference equations, written in the form

$$\mathbf{x}_{k+1} = A\mathbf{x}_k \quad \text{for all } k$$

where the vectors $\mathbf{x}_k$ are in $\mathbb{R}^n$ and A is an $n \times n$ matrix.

A simple example of such a (vector-valued) difference equation was already studied in Section 1.10. Further examples will be covered in Sections 4.9 and 5.6.

EXAMPLE 7   Write the following difference equation as a first-order system:

$$y_{k+3} - 2y_{k+2} - 5y_{k+1} + 6y_k = 0 \quad \text{for all } k$$

SOLUTION   For each k, set

$$\mathbf{x}_k = \begin{bmatrix} y_k \\ y_{k+1} \\ y_{k+2} \end{bmatrix}$$

The difference equation says that $y_{k+3} = -6y_k + 5y_{k+1} + 2y_{k+2}$, so

$$\mathbf{x}_{k+1} = \begin{bmatrix} y_{k+1} \\ y_{k+2} \\ y_{k+3} \end{bmatrix} = \begin{bmatrix} 0 & + & y_{k+1} & + 0 \\ 0 & + 0 & & + y_{k+2} \\ -6y_k & + 5y_{k+1} & + 2y_{k+2} \end{bmatrix} = \begin{bmatrix} 0 & 1 & 0 \\ 0 & 0 & 1 \\ -6 & 5 & 2 \end{bmatrix} \begin{bmatrix} y_k \\ y_{k+1} \\ y_{k+2} \end{bmatrix}$$

That is,

$$\mathbf{x}_{k+1} = A\mathbf{x}_k \quad \text{for all } k, \quad \text{where} \quad A = \begin{bmatrix} 0 & 1 & 0 \\ 0 & 0 & 1 \\ -6 & 5 & 2 \end{bmatrix} \quad ∎$$

In general, the equation

$$y_{k+n} + a_1 y_{k+n-1} + \cdots + a_{n-1} y_{k+1} + a_n y_k = 0 \quad \text{for all } k$$

can be rewritten as $\mathbf{x}_{k+1} = A\mathbf{x}_k$ for all k, where

$$\mathbf{x}_k = \begin{bmatrix} y_k \\ y_{k+1} \\ \vdots \\ y_{k+n-1} \end{bmatrix}, \quad A = \begin{bmatrix} 0 & 1 & 0 & \cdots & 0 \\ 0 & 0 & 1 & & 0 \\ \vdots & & & \ddots & \vdots \\ 0 & 0 & 0 & & 1 \\ -a_n & -a_{n-1} & -a_{n-2} & \cdots & -a_1 \end{bmatrix}$$

Further Reading

Hamming, R. W., *Digital Filters*, 3rd ed. (Englewood Cliffs, NJ: Prentice-Hall, 1989), pp. 1–37.

Kelly, W. G., and A. C. Peterson, *Difference Equations*, 2nd ed. (San Diego: Harcourt-Academic Press, 2001).

Mickens, R. E., *Difference Equations*, 2nd ed. (New York: Van Nostrand Reinhold, 1990), pp. 88–141.

Oppenheim, A. V., and A. S. Willsky, *Signals and Systems*, 2nd ed. (Upper Saddle River, NJ: Prentice-Hall, 1997), pp. 1–14, 21–30, 38–43.

PRACTICE PROBLEM

It can be shown that the signals 2^k, $3^k \sin \frac{k\pi}{2}$, and $3^k \cos \frac{k\pi}{2}$ are solutions of

$$y_{k+3} - 2y_{k+2} + 9y_{k+1} - 18y_k = 0$$

Show that these signals form a basis for the set of all solutions of the difference equation.

4.8 EXERCISES

Verify that the signals in Exercises 1 and 2 are solutions of the accompanying difference equation.

1. $2^k, (-4)^k$; $y_{k+2} + 2y_{k+1} - 8y_k = 0$

2. $5^k, (-5)^k$; $y_{k+2} - 25y_k = 0$

Show that the signals in Exercises 3–6 form a basis for the solution set of the accompanying difference equation.

3. The signals and equation in Exercise 1

4. The signals and equation in Exercise 2

5. $(-2)^k, k(-2)^k$; $y_{k+2} + 4y_{k+1} + 4y_k = 0$

6. $4^k \cos \left(\frac{k\pi}{2}\right), 4^k \sin \left(\frac{k\pi}{2}\right)$; $y_{k+2} + 16y_k = 0$

In Exercises 7–12, assume the signals listed are solutions of the given difference equation. Do the signals form a basis for the solution space of the equation? Justify your answers using appropriate theorems.

7. $1^k, 2^k, (-2)^k$; $y_{k+3} - y_{k+2} - 4y_{k+1} + 4y_k = 0$

8. $(-1)^k, 2^k, 3^k$; $y_{k+3} - 4y_{k+2} + 1y_{k+1} + 6y_k = 0$

9. $2^k, 5^k \cos \left(\frac{k\pi}{2}\right), 5^k \sin \left(\frac{k\pi}{2}\right)$;
$y_{k+3} - 2y_{k+2} + 25y_{k+1} - 50y_k = 0$

10. $(-2)^k, k(-2)^k, 3^k$; $y_{k+3} + y_{k+2} - 8y_{k+1} - 12y_k = 0$

11. $(-1)^k, 2^k$; $y_{k+3} - 3y_{k+2} + 4y_k = 0$

12. $3^k, (-2)^k$; $y_{k+4} - 13y_{k+2} + 36y_k = 0$

In Exercises 13–16, find a basis for the solution space of the difference equation. Prove that the solutions you find span the solution set.

13. $y_{k+2} - y_{k+1} + \frac{2}{9}y_k = 0$ 14. $y_{k+2} - 5y_{k+1} + 6y_k = 0$

15. $6y_{k+2} + y_{k+1} - 2y_k = 0$ 16. $y_{k+2} - 25y_k = 0$

Exercises 17 and 18 concern a simple model of the national economy described by the difference equation

$$Y_{k+2} - a(1+b)Y_{k+1} + abY_k = 1 \tag{14}$$

Here Y_k is the total national income during year k, a is a constant less than 1, called the *marginal propensity to consume*, and b is a positive *constant of adjustment* that describes how changes in consumer spending affect the annual rate of private investment.[1]

17. Find the general solution of equation (14) when $a = .9$ and $b = \frac{4}{9}$. What happens to Y_k as k increases? [*Hint:* First find a particular solution of the form $Y_k = T$, where T is a constant, called the equilibrium level of national income.]

18. Find the general solution of equation (14) when $a = .9$ and $b = .5$.

---

[1] For example, see *Discrete Dynamical Systems*, by James T. Sandefur (Oxford: Clarendon Press, 1990), pp. 267–276. The original *accelerator-multiplier model* is attributed to the economist P. A. Samuelson.

A lightweight cantilevered beam is supported at N points spaced 10 ft apart, and a weight of 500 lb is placed at the end of the beam, 10 ft from the first support, as in the figure. Let y_k be the bending moment at the kth support. Then $y_1 = 5000$ ft-lb. Suppose the beam is rigidly attached at the Nth support and the bending moment there is zero. In between, the moments satisfy the *three-moment equation*

$$y_{k+2} + 4y_{k+1} + y_k = 0 \quad \text{for } k = 1, 2, \ldots, N-2 \qquad (15)$$

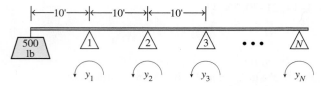

Bending moments on a cantilevered beam.

19. Find the general solution of difference equation (15). Justify your answer.

20. Find the particular solution of (15) that satisfies the *boundary conditions* $y_1 = 5000$ and $y_N = 0$. (The answer involves N.)

21. When a signal is produced from a sequence of measurements made on a process (a chemical reaction, a flow of heat through a tube, a moving robot arm, etc.), the signal usually contains random *noise* produced by measurement errors. A standard method of preprocessing the data to reduce the noise is to smooth or filter the data. One simple filter is a *moving average* that replaces each y_k by its average with the two adjacent values:

$$\tfrac{1}{3}y_{k+1} + \tfrac{1}{3}y_k + \tfrac{1}{3}y_{k-1} = z_k \quad \text{for } k = 1, 2, \ldots$$

Suppose a signal y_k, for $k = 0, \ldots, 14$, is

9, 5, 7, 3, 2, 4, 6, 5, 7, 6, 8, 10, 9, 5, 7

Use the filter to compute $z_1, \ldots, z_{13}$. Make a broken-line graph that superimposes the original signal and the smoothed signal.

22. Let $\{y_k\}$ be the sequence produced by sampling the continuous signal $2\cos\frac{\pi t}{4} + \cos\frac{3\pi t}{4}$ at $t = 0, 1, 2, \ldots$, as shown in the figure. The values of y_k, beginning with $k = 0$, are

3, .7, 0, −.7, −3, −.7, 0, .7, 3, .7, 0, ...

where .7 is an abbreviation for $\sqrt{2}/2$.

a. Compute the output signal $\{z_k\}$ when $\{y_k\}$ is fed into the filter in Example 3.

b. Explain how and why the output in part (a) is related to the calculations in Example 3.

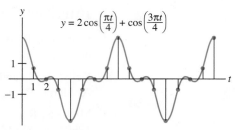

Sampled data from $2\cos\frac{\pi t}{4} + \cos\frac{3\pi t}{4}$.

Exercises 23 and 24 refer to a difference equation of the form $y_{k+1} - ay_k = b$, for suitable constants a and b.

23. A loan of $10,000 has an interest rate of 1% per month and a monthly payment of $450. The loan is made at month $k = 0$, and the first payment is made one month later, at $k = 1$. For $k = 0, 1, 2, \ldots$, let y_k be the unpaid balance of the loan just after the kth monthly payment. Thus

$$\underset{\substack{\text{New} \\ \text{balance}}}{y_1} = \underset{\substack{\text{Balance} \\ \text{due}}}{10,000} + \underset{\substack{\text{Interest} \\ \text{added}}}{(.01)10,000} - \underset{\text{Payment}}{450}$$

a. Write a difference equation satisfied by $\{y_k\}$.

b. [M] Create a table showing k and the balance y_k at month k. List the program or the keystrokes you used to create the table.

c. [M] What will k be when the last payment is made? How much will the last payment be? How much money did the borrower pay in total?

24. At time $k = 0$, an initial investment of $1000 is made into a savings account that pays 6% interest per year compounded monthly. (The interest rate per month is .005.) Each month after the initial investment, an additional $200 is added to the account. For $k = 0, 1, 2, \ldots$, let y_k be the amount in the account at time k, just after a deposit has been made.

a. Write a difference equation satisfied by $\{y_k\}$.

b. [M] Create a table showing k and the total amount in the savings account at month k, for $k = 0$ through 60. List your program or the keystrokes you used to create the table.

c. [M] How much will be in the account after two years (that is, 24 months), four years, and five years? How much of the five-year total is interest?

In Exercises 25–28, show that the given signal is a solution of the difference equation. Then find the general solution of that difference equation.

25. $y_k = k^2$; $y_{k+2} + 3y_{k+1} - 4y_k = 7 + 10k$

26. $y_k = 1 + k$; $y_{k+2} - 6y_{k+1} + 5y_k = -4$

27. $y_k = k - 2$; $y_{k+2} - 4y_k = 8 - 3k$

28. $y_k = 1 + 2k$; $y_{k+2} - 25y_k = -48k - 20$

Write the difference equations in Exercises 29 and 30 as first-order systems, $\mathbf{x}_{k+1} = A\mathbf{x}_k$, for all k.

29. $y_{k+4} + 3y_{k+3} - 8y_{k+2} + 6y_{k+1} - 2y_k = 0$

30. $y_{k+3} - 5y_{k+2} + 8y_k = 0$

31. Is the following difference equation of order 3? Explain.

$$y_{k+3} + 5y_{k+2} + 6y_{k+1} = 0$$

32. What is the order of the following difference equation? Explain your answer.

$$y_{k+3} + a_1 y_{k+2} + a_2 y_{k+1} + a_3 y_k = 0$$

33. Let $y_k = k^2$ and $z_k = 2k|k|$. Are the signals $\{y_k\}$ and $\{z_k\}$ linearly independent? Evaluate the associated Casorati matrix $C(k)$ for $k = 0$, $k = -1$, and $k = -2$, and discuss your results.

34. Let f, g, and h be linearly independent functions defined for all real numbers, and construct three signals by sampling the values of the functions at the integers:

$$u_k = f(k), \qquad v_k = g(k), \qquad w_k = h(k)$$

Must the signals be linearly independent in $\mathbb{S}$? Discuss.

35. Let a and b be nonzero numbers. Show that the mapping T defined by $T\{y_k\} = \{w_k\}$, where

$$w_k = y_{k+2} + ay_{k+1} + by_k$$

is a linear transformation from $\mathbb{S}$ into $\mathbb{S}$.

36. Let V be a vector space, and let $T : V \rightarrow V$ be a linear transformation. Given $\mathbf{z}$ in V, suppose $\mathbf{x}_p$ in V satisfies $T(\mathbf{x}_p) = \mathbf{z}$, and let $\mathbf{u}$ be any vector in the kernel of T. Show that $\mathbf{u} + \mathbf{x}_p$ satisfies the nonhomogeneous equation $T(\mathbf{x}) = \mathbf{z}$.

37. Let $\mathbb{S}_0$ be the vector space of all sequences of the form $(y_0, y_1, y_2, \ldots)$, and define linear transformations T and D from $\mathbb{S}_0$ into $\mathbb{S}_0$ by

$$T(y_0, y_1, y_2, \ldots) = (y_1, y_2, y_3, \ldots)$$
$$D(y_0, y_1, y_2, \ldots) = (0, y_0, y_1, y_2, \ldots)$$

Show that $TD = I$ (the identity transformation on $\mathbb{S}_0$) and yet $DT \neq I$.

| **SOLUTION TO PRACTICE PROBLEM**

Examine the Casorati matrix:

$$C(k) = \begin{bmatrix} 2^k & 3^k \sin \frac{k\pi}{2} & 3^k \cos \frac{k\pi}{2} \\ 2^{k+1} & 3^{k+1} \sin \frac{(k+1)\pi}{2} & 3^{k+1} \cos \frac{(k+1)\pi}{2} \\ 2^{k+2} & 3^{k+2} \sin \frac{(k+2)\pi}{2} & 3^{k+2} \cos \frac{(k+2)\pi}{2} \end{bmatrix}$$

Set $k = 0$ and row reduce the matrix to verify that it has three pivot positions and hence is invertible:

$$C(0) = \begin{bmatrix} 1 & 0 & 1 \\ 2 & 3 & 0 \\ 4 & 0 & -9 \end{bmatrix} \sim \begin{bmatrix} 1 & 0 & 1 \\ 0 & 3 & -2 \\ 0 & 0 & -13 \end{bmatrix}$$

The Casorati matrix is invertible at $k = 0$, so the signals are linearly independent. Since there are three signals, and the solution space H of the difference equation has dimension 3 (Theorem 17), the signals form a basis for H, by the Basis Theorem.

4.9 | APPLICATIONS TO MARKOV CHAINS

The Markov chains described in this section are used as mathematical models of a wide variety of situations in biology, business, chemistry, engineering, physics, and elsewhere. In each case, the model is used to describe an experiment or measurement that is performed many times in the same way, where the outcome of each trial of the experiment will be one of several specified possible outcomes, and where the outcome of one trial depends only on the immediately preceding trial.

For example, if the population of a city and its suburbs were measured each year, then a vector such as

$$\mathbf{x}_0 = \begin{bmatrix} .60 \\ .40 \end{bmatrix} \tag{1}$$

could indicate that 60% of the population lives in the city and 40% in the suburbs. The decimals in $\mathbf{x}_0$ add up to 1 because they account for the entire population of the region. Percentages are more convenient for our purposes here than population totals.

A vector with nonnegative entries that add up to 1 is called a **probability vector**. A **stochastic matrix** is a square matrix whose columns are probability vectors. A **Markov chain** is a sequence of probability vectors $\mathbf{x}_0, \mathbf{x}_1, \mathbf{x}_2, \ldots$, together with a stochastic matrix P, such that

$$\mathbf{x}_1 = P\mathbf{x}_0, \quad \mathbf{x}_2 = P\mathbf{x}_1, \quad \mathbf{x}_3 = P\mathbf{x}_2, \quad \ldots$$

Thus the Markov chain is described by the first-order difference equation

$$\mathbf{x}_{k+1} = P\mathbf{x}_k \quad \text{for } k = 0, 1, 2, \ldots$$

When a Markov chain of vectors in $\mathbb{R}^n$ describes a system or a sequence of experiments, the entries in $\mathbf{x}_k$ list, respectively, the probabilities that the system is in each of n possible states, or the probabilities that the outcome of the experiment is one of n possible outcomes. For this reason, $\mathbf{x}_k$ is often called a **state vector**.

EXAMPLE 1 Section 1.10 examined a model for population movement between a city and its suburbs. See Fig. 1. The annual migration between these two parts of the metropolitan region was governed by the *migration matrix M*:

$$\begin{array}{cc} & \text{From:} \\ & \begin{array}{cc} \text{City} & \text{Suburbs} \end{array} \quad \text{To:} \\ M = \begin{bmatrix} .95 & .03 \\ .05 & .97 \end{bmatrix} & \begin{array}{l} \text{City} \\ \text{Suburbs} \end{array} \end{array}$$

That is, each year 5% of the city population moves to the suburbs, and 3% of the suburban population moves to the city. The columns of M are probability vectors, so M is a stochastic matrix. Suppose the 2000 population of the region is 600,000 in the city and 400,000 in the suburbs. Then the initial distribution of the population in the region is given by $\mathbf{x}_0$ in (1) above. What is the distribution of the population in 2001? In 2002?

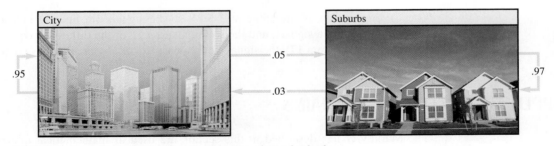

FIGURE 1 Annual percentage migration between city and suburbs.

SOLUTION In Example 3 of Section 1.10, we saw that after one year, the population vector $\begin{bmatrix} 600{,}000 \\ 400{,}000 \end{bmatrix}$ changed to

$$\begin{bmatrix} .95 & .03 \\ .05 & .97 \end{bmatrix} \begin{bmatrix} 600{,}000 \\ 400{,}000 \end{bmatrix} = \begin{bmatrix} 582{,}000 \\ 418{,}000 \end{bmatrix}$$

If we divide both sides of this equation by the total population of 1 million, and use the fact that $kM\mathbf{x} = M(k\mathbf{x})$, we find that

$$\begin{bmatrix} .95 & .03 \\ .05 & .97 \end{bmatrix} \begin{bmatrix} .600 \\ .400 \end{bmatrix} = \begin{bmatrix} .582 \\ .418 \end{bmatrix}$$

The vector $\mathbf{x}_1 = \begin{bmatrix} .582 \\ .418 \end{bmatrix}$ gives the population distribution in 2001. That is, 58.2% of the region lived in the city and 41.8% lived in the suburbs. Similarly, the population distribution in 2002 is described by a vector $\mathbf{x}_2$, where

$$\mathbf{x}_2 = M\mathbf{x}_1 = \begin{bmatrix} .95 & .03 \\ .05 & .97 \end{bmatrix} \begin{bmatrix} .582 \\ .418 \end{bmatrix} = \begin{bmatrix} .565 \\ .435 \end{bmatrix} \qquad ∎$$

EXAMPLE 2   Suppose the voting results of a congressional election at a certain voting precinct are represented by a vector $\mathbf{x}$ in $\mathbb{R}^3$:

$$\mathbf{x} = \begin{bmatrix} \% \text{ voting Democratic (D)} \\ \% \text{ voting Republican (R)} \\ \% \text{ voting Libertarian (L)} \end{bmatrix}$$

Suppose we record the outcome of the congressional election every two years by a vector of this type and the outcome of one election depends only on the results of the preceding election. Then the sequence of vectors that describe the votes every two years may be a Markov chain. As an example of a stochastic matrix P for this chain, we take

$$\begin{array}{c} \text{From:} \\ \begin{array}{ccc} \text{D} & \text{R} & \text{L} \end{array} \\ P = \begin{bmatrix} .70 & .10 & .30 \\ .20 & .80 & .30 \\ .10 & .10 & .40 \end{bmatrix} \begin{array}{l} \text{To:} \\ \text{D} \\ \text{R} \\ \text{L} \end{array} \end{array}$$

The entries in the first column, labeled D, describe what the persons voting Democratic in one election will do in the next election. Here we have supposed that 70% will vote D again in the next election, 20% will vote R, and 10% will vote L. Similar interpretations hold for the other columns of P. A diagram for this matrix is shown in Fig. 2.

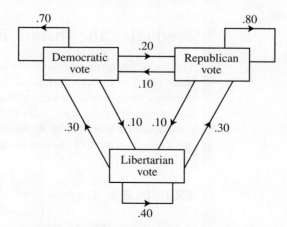

FIGURE 2 Voting changes from one election to the next.

If the "transition" percentages remain constant over many years from one election to the next, then the sequence of vectors that give the voting outcomes forms a Markov chain. Suppose the outcome of one election is given by

$$\mathbf{x}_0 = \begin{bmatrix} .55 \\ .40 \\ .05 \end{bmatrix}$$

Determine the likely outcome of the next election and the likely outcome of the election after that.

SOLUTION The outcome of the next election is described by the state vector $\mathbf{x}_1$ and that of the election after that by $\mathbf{x}_2$, where

$$\mathbf{x}_1 = P\mathbf{x}_0 = \begin{bmatrix} .70 & .10 & .30 \\ .20 & .80 & .30 \\ .10 & .10 & .40 \end{bmatrix} \begin{bmatrix} .55 \\ .40 \\ .05 \end{bmatrix} = \begin{bmatrix} .440 \\ .445 \\ .115 \end{bmatrix} \qquad \begin{array}{l} \text{44\% will vote D.} \\ \text{44.5\% will vote R.} \\ \text{11.5\% will vote L.} \end{array}$$

$$\mathbf{x}_2 = P\mathbf{x}_1 = \begin{bmatrix} .70 & .10 & .30 \\ .20 & .80 & .30 \\ .10 & .10 & .40 \end{bmatrix} \begin{bmatrix} .440 \\ .445 \\ .115 \end{bmatrix} = \begin{bmatrix} .3870 \\ .4785 \\ .1345 \end{bmatrix} \qquad \begin{array}{l} \text{38.7\% will vote D.} \\ \text{47.8\% will vote R.} \\ \text{13.5\% will vote L.} \end{array}$$

To understand why $\mathbf{x}_1$ does indeed give the outcome of the next election, suppose 1000 persons voted in the "first" election, with 550 voting D, 400 voting R, and 50 voting L. (See the percentages in $\mathbf{x}_0$.) In the next election, 70% of the 550 will vote D again, 10% of the 400 will switch from R to D, and 30% of the 50 will switch from L to D. Thus the total D vote will be

$$.70(550) + .10(400) + .30(50) = 385 + 40 + 15 = 440 \qquad (2)$$

Thus 44% of the vote next time will be for the D candidate. The calculation in (2) is essentially the same as that used to compute the first entry in $\mathbf{x}_1$. Analogous calculations could be made for the other entries in $\mathbf{x}_1$, for the entries in $\mathbf{x}_2$, and so on. ∎

Predicting the Distant Future

The most interesting aspect of Markov chains is the study of a chain's long-term behavior. For instance, what can be said in Example 2 about the voting after many elections have passed (assuming that the given stochastic matrix continues to describe the transition percentages from one election to the next)? Or, what happens to the population distribution in Example 1 "in the long run"? Before answering these questions, we turn to a numerical example.

EXAMPLE 3 Let $P = \begin{bmatrix} .5 & .2 & .3 \\ .3 & .8 & .3 \\ .2 & 0 & .4 \end{bmatrix}$ and $\mathbf{x}_0 = \begin{bmatrix} 1 \\ 0 \\ 0 \end{bmatrix}$. Consider a system whose state is described by the Markov chain $\mathbf{x}_{k+1} = P\mathbf{x}_k$, for $k = 0, 1, \ldots$ What happens to the system as time passes? Compute the state vectors $\mathbf{x}_1, \ldots, \mathbf{x}_{15}$ to find out.

SOLUTION

$$\mathbf{x}_1 = P\mathbf{x}_0 = \begin{bmatrix} .5 & .2 & .3 \\ .3 & .8 & .3 \\ .2 & 0 & .4 \end{bmatrix} \begin{bmatrix} 1 \\ 0 \\ 0 \end{bmatrix} = \begin{bmatrix} .5 \\ .3 \\ .2 \end{bmatrix}$$

$$\mathbf{x}_2 = P\mathbf{x}_1 = \begin{bmatrix} .5 & .2 & .3 \\ .3 & .8 & .3 \\ .2 & 0 & .4 \end{bmatrix} \begin{bmatrix} .5 \\ .3 \\ .2 \end{bmatrix} = \begin{bmatrix} .37 \\ .45 \\ .18 \end{bmatrix}$$

$$\mathbf{x}_3 = P\mathbf{x}_2 = \begin{bmatrix} .5 & .2 & .3 \\ .3 & .8 & .3 \\ .2 & 0 & .4 \end{bmatrix} \begin{bmatrix} .37 \\ .45 \\ .18 \end{bmatrix} = \begin{bmatrix} .329 \\ .525 \\ .146 \end{bmatrix}$$

The results of further calculations are shown below, with entries rounded to four or five significant figures.

$$\mathbf{x}_4 = \begin{bmatrix} .3133 \\ .5625 \\ .1242 \end{bmatrix}, \quad \mathbf{x}_5 = \begin{bmatrix} .3064 \\ .5813 \\ .1123 \end{bmatrix}, \quad \mathbf{x}_6 = \begin{bmatrix} .3032 \\ .5906 \\ .1062 \end{bmatrix}, \quad \mathbf{x}_7 = \begin{bmatrix} .3016 \\ .5953 \\ .1031 \end{bmatrix}$$

$$\mathbf{x}_8 = \begin{bmatrix} .3008 \\ .5977 \\ .1016 \end{bmatrix}, \quad \mathbf{x}_9 = \begin{bmatrix} .3004 \\ .5988 \\ .1008 \end{bmatrix}, \quad \mathbf{x}_{10} = \begin{bmatrix} .3002 \\ .5994 \\ .1004 \end{bmatrix}, \quad \mathbf{x}_{11} = \begin{bmatrix} .3001 \\ .5997 \\ .1002 \end{bmatrix}$$

$$\mathbf{x}_{12} = \begin{bmatrix} .30005 \\ .59985 \\ .10010 \end{bmatrix}, \quad \mathbf{x}_{13} = \begin{bmatrix} .30002 \\ .59993 \\ .10005 \end{bmatrix}, \quad \mathbf{x}_{14} = \begin{bmatrix} .30001 \\ .59996 \\ .10002 \end{bmatrix}, \quad \mathbf{x}_{15} = \begin{bmatrix} .30001 \\ .59998 \\ .10001 \end{bmatrix}$$

These vectors seem to be approaching $\mathbf{q} = \begin{bmatrix} .3 \\ .6 \\ .1 \end{bmatrix}$. The probabilities are hardly changing from one value of k to the next. Observe that the following calculation is exact (with no rounding error):

$$P\mathbf{q} = \begin{bmatrix} .5 & .2 & .3 \\ .3 & .8 & .3 \\ .2 & 0 & .4 \end{bmatrix} \begin{bmatrix} .3 \\ .6 \\ .1 \end{bmatrix} = \begin{bmatrix} .15 + .12 + .03 \\ .09 + .48 + .03 \\ .06 + 0 + .04 \end{bmatrix} = \begin{bmatrix} .30 \\ .60 \\ .10 \end{bmatrix} = \mathbf{q}$$

When the system is in state $\mathbf{q}$, there is no change in the system from one measurement to the next. ∎

Steady-State Vectors

If P is a stochastic matrix, then a **steady-state vector** (or **equilibrium vector**) for P is a probability vector $\mathbf{q}$ such that

$$P\mathbf{q} = \mathbf{q}$$

It can be shown that every stochastic matrix has a steady-state vector. In Example 3, $\mathbf{q}$ is a steady-state vector for P.

EXAMPLE 4 The probability vector $\mathbf{q} = \begin{bmatrix} .375 \\ .625 \end{bmatrix}$ is a steady-state vector for the population migration matrix M in Example 1, because

$$M\mathbf{q} = \begin{bmatrix} .95 & .03 \\ .05 & .97 \end{bmatrix} \begin{bmatrix} .375 \\ .625 \end{bmatrix} = \begin{bmatrix} .35625 + .01875 \\ .01875 + .60625 \end{bmatrix} = \begin{bmatrix} .375 \\ .625 \end{bmatrix} = \mathbf{q} \qquad ∎$$

If the total population of the metropolitan region in Example 1 is 1 million, then **q** from Example 4 would correspond to having 375,000 persons in the city and 625,000 in the suburbs. At the end of one year, the migration *out of* the city would be $(.05)(375,000) = 18,750$ persons, and the migration *into* the city from the suburbs would be $(.03)(625,000) = 18,750$ persons. As a result, the population in the city would remain the same. Similarly, the suburban population would be stable.

The next example shows how to *find* a steady-state vector.

EXAMPLE 5 Let $P = \begin{bmatrix} .6 & .3 \\ .4 & .7 \end{bmatrix}$. Find a steady-state vector for P.

SOLUTION First, solve the equation $P\mathbf{x} = \mathbf{x}$.

$$P\mathbf{x} - \mathbf{x} = \mathbf{0}$$
$$P\mathbf{x} - I\mathbf{x} = \mathbf{0} \qquad \text{Recall from Section 1.4 that } I\mathbf{x} = \mathbf{x}.$$
$$(P - I)\mathbf{x} = \mathbf{0}$$

For P as above,

$$P - I = \begin{bmatrix} .6 & .3 \\ .4 & .7 \end{bmatrix} - \begin{bmatrix} 1 & 0 \\ 0 & 1 \end{bmatrix} = \begin{bmatrix} -.4 & .3 \\ .4 & -.3 \end{bmatrix}$$

To find all solutions of $(P - I)\mathbf{x} = \mathbf{0}$, row reduce the augmented matrix:

$$\begin{bmatrix} -.4 & .3 & 0 \\ .4 & -.3 & 0 \end{bmatrix} \sim \begin{bmatrix} -.4 & .3 & 0 \\ 0 & 0 & 0 \end{bmatrix} \sim \begin{bmatrix} 1 & -3/4 & 0 \\ 0 & 0 & 0 \end{bmatrix}$$

Then $x_1 = \frac{3}{4}x_2$ and x_2 is free. The general solution is $x_2 \begin{bmatrix} 3/4 \\ 1 \end{bmatrix}$.

Next, choose a simple basis for the solution space. One obvious choice is $\begin{bmatrix} 3/4 \\ 1 \end{bmatrix}$ but a better choice with no fractions is $\mathbf{w} = \begin{bmatrix} 3 \\ 4 \end{bmatrix}$ (corresponding to $x_2 = 4$).

Finally, find a probability vector in the set of all solutions of $P\mathbf{x} = \mathbf{x}$. This process is easy, since every solution is a multiple of the solution $\mathbf{w}$ above. Divide $\mathbf{w}$ by the sum of its entries and obtain

$$\mathbf{q} = \begin{bmatrix} 3/7 \\ 4/7 \end{bmatrix}$$

As a check, compute

$$P\mathbf{q} = \begin{bmatrix} 6/10 & 3/10 \\ 4/10 & 7/10 \end{bmatrix} \begin{bmatrix} 3/7 \\ 4/7 \end{bmatrix} = \begin{bmatrix} 18/70 + 12/70 \\ 12/70 + 28/70 \end{bmatrix} = \begin{bmatrix} 30/70 \\ 40/70 \end{bmatrix} = \mathbf{q} \qquad \blacksquare$$

The next theorem shows that what happened in Example 3 is typical of many stochastic matrices. We say that a stochastic matrix is **regular** if some matrix power P^k contains only strictly positive entries. For P in Example 3,

$$P^2 = \begin{bmatrix} .37 & .26 & .33 \\ .45 & .70 & .45 \\ .18 & .04 & .22 \end{bmatrix}$$

Since every entry in P^2 is strictly positive, P is a regular stochastic matrix.

Also, we say that a sequence of vectors $\{\mathbf{x}_k : k = 1, 2, \ldots\}$ **converges** to a vector $\mathbf{q}$ as $k \to \infty$ if the entries in $\mathbf{x}_k$ can be made as close as desired to the corresponding entries in $\mathbf{q}$ by taking k sufficiently large.

THEOREM 18

> If P is an $n \times n$ regular stochastic matrix, then P has a unique steady-state vector $\mathbf{q}$. Further, if $\mathbf{x}_0$ is any initial state and $\mathbf{x}_{k+1} = P\mathbf{x}_k$ for $k = 0, 1, 2, \ldots$, then the Markov chain $\{\mathbf{x}_k\}$ converges to $\mathbf{q}$ as $k \to \infty$.

This theorem is proved in standard texts on Markov chains. The amazing part of the theorem is that the initial state has no effect on the long-term behavior of the Markov chain. You will see later (in Section 5.2) why this is true for several stochastic matrices studied here.

EXAMPLE 6   In Example 2, what percentage of the voters are likely to vote for the Republican candidate in some election many years from now, assuming that the election outcomes form a Markov chain?

SOLUTION For computations by hand, the *wrong* approach is to pick some initial vector $\mathbf{x}_0$ and compute $\mathbf{x}_1, \ldots, \mathbf{x}_k$ for some large value of k. You have no way of knowing how many vectors to compute, and you cannot be sure of the limiting values of the entries in $\mathbf{x}_k$.

The correct approach is to compute the steady-state vector and then appeal to Theorem 18. Given P as in Example 2, form $P - I$ by subtracting 1 from each diagonal entry in P. Then row reduce the augmented matrix:

$$[(P - I) \quad \mathbf{0}] = \begin{bmatrix} -.3 & .1 & .3 & 0 \\ .2 & -.2 & .3 & 0 \\ .1 & .1 & -.6 & 0 \end{bmatrix}$$

Recall from earlier work with decimals that the arithmetic is simplified by multiplying each row by 10.[1]

$$\begin{bmatrix} -3 & 1 & 3 & 0 \\ 2 & -2 & 3 & 0 \\ 1 & 1 & -6 & 0 \end{bmatrix} \sim \begin{bmatrix} 1 & 0 & -9/4 & 0 \\ 0 & 1 & -15/4 & 0 \\ 0 & 0 & 0 & 0 \end{bmatrix}$$

The general solution of $(P - I)\mathbf{x} = \mathbf{0}$ is $x_1 = \frac{9}{4}x_3$, $x_2 = \frac{15}{4}x_3$, and x_3 is free. Choosing $x_3 = 4$, we obtain a basis for the solution space whose entries are integers, and from this we easily find the steady-state vector whose entries sum to 1:

$$\mathbf{w} = \begin{bmatrix} 9 \\ 15 \\ 4 \end{bmatrix}, \quad \text{and} \quad \mathbf{q} = \begin{bmatrix} 9/28 \\ 15/28 \\ 4/28 \end{bmatrix} \approx \begin{bmatrix} .32 \\ .54 \\ .14 \end{bmatrix}$$

The entries in $\mathbf{q}$ describe the distribution of votes at an election to be held many years from now (assuming the stochastic matrix continues to describe the changes from one election to the next). Thus, eventually, about 54% of the vote will be for the Republican candidate. ∎

---

[1] *Warning:* Don't multiply only P by 10. Instead, multiply the augmented matrix for equation $(P - I)\mathbf{x} = \mathbf{0}$ by 10.

NUMERICAL NOTE

You may have noticed that if $\mathbf{x}_{k+1} = P\mathbf{x}_k$ for $k = 0, 1, \ldots$, then

$$\mathbf{x}_2 = P\mathbf{x}_1 = P(P\mathbf{x}_0) = P^2\mathbf{x}_0,$$

and, in general,

$$\mathbf{x}_k = P^k\mathbf{x}_0 \quad \text{for } k = 0, 1, \ldots$$

To compute a specific vector such as $\mathbf{x}_3$, fewer arithmetic operations are needed to compute $\mathbf{x}_1$, $\mathbf{x}_2$, and $\mathbf{x}_3$, rather than P^3 and $P^3\mathbf{x}_0$. However, if P is small—say, 30×30—the machine computation time is insignificant for both methods, and a command to compute $P^3\mathbf{x}_0$ might be preferred because it requires fewer human keystrokes.

PRACTICE PROBLEMS

1. Suppose the residents of a metropolitan region move according to the probabilities in the migration matrix M in Example 1 and a resident is chosen "at random." Then a state vector for a certain year may be interpreted as giving the probabilities that the person is a city resident or a suburban resident at that time.

   a. Suppose the person chosen is a city resident now, so that $\mathbf{x}_0 = \begin{bmatrix} 1 \\ 0 \end{bmatrix}$. What is the likelihood that the person will live in the suburbs next year?

   b. What is the likelihood that the person will be living in the suburbs in two years?

2. Let $P = \begin{bmatrix} .6 & .2 \\ .4 & .8 \end{bmatrix}$ and $\mathbf{q} = \begin{bmatrix} .3 \\ .7 \end{bmatrix}$. Is $\mathbf{q}$ a steady-state vector for P?

3. What percentage of the population in Example 1 will live in the suburbs after many years?

4.9 EXERCISES

1. A small remote village receives radio broadcasts from two radio stations, a news station and a music station. Of the listeners who are tuned to the news station, 70% will remain listening to the news after the station break that occurs each half hour, while 30% will switch to the music station at the station break. Of the listeners who are tuned to the music station, 60% will switch to the news station at the station break, while 40% will remain listening to the music. Suppose everyone is listening to the news at 8:15 A.M.

   a. Give the stochastic matrix that describes how the radio listeners tend to change stations at each station break. Label the rows and columns.

   b. Give the initial state vector.

   c. What percentage of the listeners will be listening to the music station at 9:25 A.M. (after the station breaks at 8:30 and 9:00 A.M.)?

2. A laboratory animal may eat any one of three foods each day. Laboratory records show that if the animal chooses one food on one trial, it will choose the same food on the next trial with a probability of 60%, and it will choose the other foods on the next trial with equal probabilities of 20%.

   a. What is the stochastic matrix for this situation?

   b. If the animal chooses food #1 on an initial trial, what is the probability that it will choose food #2 on the second trial after the initial trial?

3. On any given day, a student is either healthy or ill. Of the students who are healthy today, 95% will be healthy

tomorrow. Of the students who are ill today, 55% will still be ill tomorrow.

a. What is the stochastic matrix for this situation?

b. Suppose 20% of the students are ill on Monday. What fraction or percentage of the students are likely to be ill on Tuesday? On Wednesday?

c. If a student is healthy today, what is the probability that he or she will be healthy two days from now?

4. The weather in Columbus is either good, indifferent, or bad on any given day. If the weather is good today, there is a 40% chance it will be good tomorrow, a 30% chance it will be indifferent, and a 30% chance it will be bad. If the weather is indifferent today, there is a 50% chance it will be good tomorrow, and a 20% chance it will be indifferent. Finally, if the weather is bad today, there is a 30% chance it will be good tomorrow and a 40% chance it will be indifferent.

a. What is the stochastic matrix for this situation?

b. Suppose there is a 50% chance of good weather today and a 50% chance of indifferent weather. What are the chances of bad weather tomorrow?

c. Suppose the predicted weather for Monday is 60% indifferent weather and 40% bad weather. What are the chances for good weather on Wednesday?

In Exercises 5–8, find the steady-state vector.

5. $\begin{bmatrix} .1 & .5 \\ .9 & .5 \end{bmatrix}$

6. $\begin{bmatrix} .4 & .8 \\ .6 & .2 \end{bmatrix}$

7. $\begin{bmatrix} .7 & .1 & .1 \\ .2 & .8 & .2 \\ .1 & .1 & .7 \end{bmatrix}$

8. $\begin{bmatrix} .4 & .5 & .8 \\ 0 & .5 & .1 \\ .6 & 0 & .1 \end{bmatrix}$

9. Determine if $P = \begin{bmatrix} .2 & 1 \\ .8 & 0 \end{bmatrix}$ is a regular stochastic matrix.

10. Determine if $P = \begin{bmatrix} 1 & .3 \\ 0 & .7 \end{bmatrix}$ is a regular stochastic matrix.

11. a. Find the steady-state vector for the Markov chain in Exercise 1.

b. At some time late in the day, what fraction of the listeners will be listening to the news?

12. Refer to Exercise 2. Which food will the animal prefer after many trials?

13. a. Find the steady-state vector for the Markov chain in Exercise 3.

b. What is the probability that after many days a specific student is ill? Does it matter if that person is ill today?

14. Refer to Exercise 4. In the long run, how likely is it for the weather in Columbus to be good on a given day?

15. [M] The Demographic Research Unit of the California State Department of Finance supplied data for the following migration matrix, which describes the movement of the United States population during 1989. In 1989, about 11.7% of the total population lived in California. What percentage of the total population would eventually live in California if the listed migration probabilities were to remain constant over many years?

From:

CA	Rest of U.S.	To:
.9821	.0029	California
.0179	.9971	Rest of U.S.

16. [M] In Detroit, Hertz Rent A Car has a fleet of about 2000 cars. The pattern of rental and return locations is given by the fractions in the table below. On a typical day, about how many cars will be rented or ready to rent from the downtown location?

Cars Rented from:

City Airport	Downtown	Metro Airport	Returned to:
.90	.01	.09	City Airport
.01	.90	.01	Downtown
.09	.09	.90	Metro Airport

17. Let P be an $n \times n$ stochastic matrix. The following argument shows that the equation $P\mathbf{x} = \mathbf{x}$ has a nontrivial solution. (In fact, a steady-state solution exists with nonnegative entries. A proof is given in some advanced texts.) Justify each assertion below. (Mention a theorem when appropriate.)

a. If all the other rows of $P - I$ are added to the bottom row, the result is a row of zeros.

b. The rows of $P - I$ are linearly dependent.

c. The dimension of the row space of $P - I$ is less than n.

d. $P - I$ has a nontrivial null space.

18. Show that every 2×2 stochastic matrix has at least one steady-state vector. Any such matrix can be written in the form $P = \begin{bmatrix} 1 - \alpha & \beta \\ \alpha & 1 - \beta \end{bmatrix}$, where α and β are constants between 0 and 1. (There are two linearly independent steady-state vectors if $\alpha = \beta = 0$. Otherwise, there is only one.)

19. Let S be the $1 \times n$ row matrix with a 1 in each column,

$$S = [1 \quad 1 \quad \cdots \quad 1]$$

a. Explain why a vector $\mathbf{x}$ in $\mathbb{R}^n$ is a probability vector if and only if its entries are nonnegative and $S\mathbf{x} = 1$. (A 1×1 matrix such as the product $S\mathbf{x}$ is usually written without the matrix bracket symbols.)

b. Let P be an $n \times n$ stochastic matrix. Explain why $SP = S$.

c. Let P be an $n \times n$ stochastic matrix, and let $\mathbf{x}$ be a probability vector. Show that $P\mathbf{x}$ is also a probability vector.

20. Use Exercise 19 to show that if P is an $n \times n$ stochastic matrix, then so is P^2.

21. [M] Examine powers of a regular stochastic matrix.

a. Compute P^k for $k = 2, 3, 4, 5$, when

$$P = \begin{bmatrix} .3355 & .3682 & .3067 & .0389 \\ .2663 & .2723 & .3277 & .5451 \\ .1935 & .1502 & .1589 & .2395 \\ .2047 & .2093 & .2067 & .1765 \end{bmatrix}$$

Display calculations to four decimal places. What happens to the columns of P^k as k increases? Compute the steady-state vector for P.

b. Compute Q^k for $k = 10, 20, \ldots, 80$, when

$$Q = \begin{bmatrix} .97 & .05 & .10 \\ 0 & .90 & .05 \\ .03 & .05 & .85 \end{bmatrix}$$

(Stability for Q^k to four decimal places may require $k = 116$ or more.) Compute the steady-state vector for

Q. Conjecture what might be true for any regular stochastic matrix.

c. Use Theorem 18 to explain what you found in parts (a) and (b).

22. [M] Compare two methods for finding the steady-state vector $\mathbf{q}$ of a regular stochastic matrix P: (1) computing $\mathbf{q}$ as in Example 5, or (2) computing P^k for some large value of k and using one of the columns of P^k as an approximation for $\mathbf{q}$. [The *Study Guide* describes a program *nulbasis* that almost automates method (1).]

Experiment with the largest random stochastic matrices your matrix program will allow, and use $k = 100$ or some other large value. For each method, describe the time *you* need to enter the keystrokes and run your program. (Some versions of MATLAB have commands `flops` and `tic` `...toc` that record the number of floating point operations and the total elapsed time MATLAB uses.) Contrast the advantages of each method, and state which you prefer.

SOLUTIONS TO PRACTICE PROBLEMS

1. a. Since 5% of the city residents will move to the suburbs within one year, there is a 5% chance of choosing such a person. Without further knowledge about the person, we say that there is a 5% chance the person will move to the suburbs. This fact is contained in the second entry of the state vector $\mathbf{x}_1$, where

$$\mathbf{x}_1 = M\mathbf{x}_0 = \begin{bmatrix} .95 & .03 \\ .05 & .97 \end{bmatrix} \begin{bmatrix} 1 \\ 0 \end{bmatrix} = \begin{bmatrix} .95 \\ .05 \end{bmatrix}$$

b. The likelihood that the person will be living in the suburbs after two years is 9.6%, because

$$\mathbf{x}_2 = M\mathbf{x}_1 = \begin{bmatrix} .95 & .03 \\ .05 & .97 \end{bmatrix} \begin{bmatrix} .95 \\ .05 \end{bmatrix} = \begin{bmatrix} .904 \\ .096 \end{bmatrix}$$

2. The steady-state vector satisfies $P\mathbf{x} = \mathbf{x}$. Since

$$P\mathbf{q} = \begin{bmatrix} .6 & .2 \\ .4 & .8 \end{bmatrix} \begin{bmatrix} .3 \\ .7 \end{bmatrix} = \begin{bmatrix} .32 \\ .68 \end{bmatrix} \neq \mathbf{q}$$

we conclude that $\mathbf{q}$ is *not* the steady-state vector for P.

3. M in Example 1 is a regular stochastic matrix because its entries are all strictly positive. So we may use Theorem 18. We already know the steady-state vector from Example 4. Thus the population distribution vectors $\mathbf{x}_k$ converge to

$$\mathbf{q} = \begin{bmatrix} .375 \\ .625 \end{bmatrix}$$

WEB

Eventually 62.5% of the population will live in the suburbs.

CHAPTER 4 SUPPLEMENTARY EXERCISES

1. Mark each statement True or False. Justify each answer. (If true, cite appropriate facts or theorems. If false, explain why or give a counterexample that shows why the statement is not true in every case.) In parts (a)–(f), $\mathbf{v}_1, \ldots, \mathbf{v}_p$ are vectors in a nonzero finite-dimensional vector space V, and $S = \{\mathbf{v}_1, \ldots, \mathbf{v}_p\}$.

a. The set of all linear combinations of $\mathbf{v}_1, \ldots, \mathbf{v}_p$ is a vector space.

b. If $\{\mathbf{v}_1, \ldots, \mathbf{v}_{p-1}\}$ spans V, then S spans V.

c. If $\{\mathbf{v}_1, \ldots, \mathbf{v}_{p-1}\}$ is linearly independent, then so is S.

d. If S is linearly independent, then S is a basis for V.

e. If Span $S = V$, then some subset of S is a basis for V.

f. If dim $V = p$ and Span $S = V$, then S cannot be linearly dependent.

g. A plane in $\mathbb{R}^3$ is a two-dimensional subspace.

h. The nonpivot columns of a matrix are always linearly dependent.

i. Row operations on a matrix A can change the linear dependence relations among the rows of A.

j. Row operations on a matrix can change the null space.

k. The rank of a matrix equals the number of nonzero rows.

l. If an $m \times n$ matrix A is row equivalent to an echelon matrix U and if U has k nonzero rows, then the dimension of the solution space of $A\mathbf{x} = \mathbf{0}$ is $m - k$.

m. If B is obtained from a matrix A by several elementary row operations, then rank $B = $ rank A.

n. The nonzero rows of a matrix A form a basis for Row A.

o. If matrices A and B have the same reduced echelon form, then Row $A = $ Row B.

p. If H is a subspace of $\mathbb{R}^3$, then there is a 3×3 matrix A such that $H = $ Col A.

q. If A is $m \times n$ and rank $A = m$, then the linear transformation $\mathbf{x} \mapsto A\mathbf{x}$ is one-to-one.

r. If A is $m \times n$ and the linear transformation $\mathbf{x} \mapsto A\mathbf{x}$ is onto, then rank $A = m$.

s. A change-of-coordinates matrix is always invertible.

t. If $\mathcal{B} = \{\mathbf{b}_1, \ldots, \mathbf{b}_n\}$ and $\mathcal{C} = \{\mathbf{c}_1, \ldots, \mathbf{c}_n\}$ are bases for a vector space V, then the jth column of the change-of-coordinates matrix $\underset{C \leftarrow B}{P}$ is the coordinate vector $[\mathbf{c}_j]_\mathcal{B}$.

2. Find a basis for the set of all vectors of the form
$$\begin{bmatrix} a - 2b + 5c \\ 2a + 5b - 8c \\ -a - 4b + 7c \\ 3a + b + c \end{bmatrix}. \quad \text{(Be careful.)}$$

3. Let $\mathbf{u}_1 = \begin{bmatrix} -2 \\ 4 \\ -6 \end{bmatrix}$, $\mathbf{u}_2 = \begin{bmatrix} 1 \\ 2 \\ -5 \end{bmatrix}$, $\mathbf{b} = \begin{bmatrix} b_1 \\ b_2 \\ b_3 \end{bmatrix}$, and $W = \text{Span}\{\mathbf{u}_1, \mathbf{u}_2\}$. Find an *implicit* description of W; that is, find a set of one or more homogeneous equations that characterize the points of W. [*Hint:* When is $\mathbf{b}$ in W?]

4. Explain what is wrong with the following discussion: Let $\mathbf{f}(t) = 3 + t$ and $\mathbf{g}(t) = 3t + t^2$, and note that $\mathbf{g}(t) = t\mathbf{f}(t)$. Then $\{\mathbf{f}, \mathbf{g}\}$ is linearly dependent because $\mathbf{g}$ is a multiple of $\mathbf{f}$.

5. Consider the polynomials $\mathbf{p}_1(t) = 1 + t$, $\mathbf{p}_2(t) = 1 - t$, $\mathbf{p}_3(t) = 4$, $\mathbf{p}_4(t) = t + t^2$, and $\mathbf{p}_5(t) = 1 + 2t + t^2$, and let H be the subspace of $\mathbb{P}_5$ spanned by the set $S = \{\mathbf{p}_1, \mathbf{p}_2, \mathbf{p}_3, \mathbf{p}_4, \mathbf{p}_5\}$. Use the method described in the proof of the Spanning Set Theorem (Section 4.3) to produce a basis for H. (Explain how to select appropriate members of S.)

6. Suppose $\mathbf{p}_1, \mathbf{p}_2, \mathbf{p}_3, \mathbf{p}_4$ are specific polynomials that span a two-dimensional subspace H of $\mathbb{P}_5$. Describe how one can find a basis for H by examining the four polynomials and making almost no computations.

7. What would you have to know about the solution set of a homogeneous system of 18 linear equations in 20 variables in order to know that every associated nonhomogeneous equation has a solution? Discuss.

8. Let H be an n-dimensional subspace of an n-dimensional vector space V. Explain why $H = V$.

9. Let $T : \mathbb{R}^n \to \mathbb{R}^m$ be a linear transformation.

a. What is the dimension of the range of T if T is a one-to-one mapping? Explain.

b. What is the dimension of the kernel of T (see Section 4.2) if T maps $\mathbb{R}^n$ onto $\mathbb{R}^m$? Explain.

10. Let S be a maximal linearly independent subset of a vector space V. That is, S has the property that if a vector not in S is adjoined to S, then the new set will no longer be linearly independent. Prove that S must be a basis for V. [*Hint:* What if S were linearly independent but not a basis of V?]

11. Let S be a finite minimal spanning set of a vector space V. That is, S has the property that if a vector is removed from S, then the new set will no longer span V. Prove that S must be a basis for V.

Exercises 12–17 develop properties of rank that are sometimes needed in applications. Assume the matrix A is $m \times n$.

12. Show from parts (a) and (b) that rank AB cannot exceed the rank of A or the rank of B. (In general, the rank of a product of matrices cannot exceed the rank of any factor in the product.)

a. Show that if B is $n \times p$, then rank $AB \leq$ rank A. [*Hint:* Explain why every vector in the column space of AB is in the column space of A.]

b. Show that if B is $n \times p$, then rank $AB \leq$ rank B. [*Hint:* Use part (a) to study rank$(AB)^T$.]

13. Show that if P is an invertible $m \times m$ matrix, then rank $PA = $ rank A. [*Hint:* Apply Exercise 12 to PA and $P^{-1}(PA)$.]

14. Show that if Q is invertible, then rank $AQ = $ rank A. [*Hint:* Use Exercise 13 to study rank$(AQ)^T$.]

15. Let A be an $m \times n$ matrix, and let B be an $n \times p$ matrix such that $AB = 0$. Show that rank A + rank $B \leq n$. [*Hint:* One of the four subspaces Nul A, Col A, Nul B, and Col B is contained in one of the other three subspaces.]

16. If A is an $m \times n$ matrix of rank r, then a *rank factorization* of A is an equation of the form $A = CR$, where C is an $m \times r$ matrix of rank r and R is an $r \times n$ matrix of rank r.

Such a factorization always exists (Exercise 38 in Section 4.6). Given any two $m \times n$ matrices A and B, use rank factorizations of A and B to prove that

$$\text{rank}(A + B) \leq \text{rank } A + \text{rank } B$$

[*Hint:* Write $A + B$ as the product of two partitioned matrices.]

17. A **submatrix** of a matrix A is any matrix that results from deleting some (or no) rows and/or columns of A. It can be shown that A has rank r if and only if A contains an invertible $r \times r$ submatrix and no larger square submatrix is invertible. Demonstrate part of this statement by explaining (a) why an $m \times n$ matrix A of rank r has an $m \times r$ submatrix A_1 of rank r, and (b) why A_1 has an invertible $r \times r$ submatrix A_2.

The concept of rank plays an important role in the design of engineering control systems, such as the space shuttle system mentioned in this chapter's introductory example. A *state-space model* of a control system includes a difference equation of the form

$$\mathbf{x}_{k+1} = A\mathbf{x}_k + B\mathbf{u}_k \quad \text{for } k = 0, 1, \ldots \tag{1}$$

where A is $n \times n$, B is $n \times m$, $\{\mathbf{x}_k\}$ is a sequence of "state vectors" in $\mathbb{R}^n$ that describe the state of the system at discrete times, and $\{\mathbf{u}_k\}$ is a *control*, or *input*, sequence. The pair (A, B) is said to be **controllable** if

$$\text{rank } [\, B \quad AB \quad A^2B \quad \cdots \quad A^{n-1}B \,] = n \tag{2}$$

The matrix that appears in (2) is called the **controllability matrix** for the system. If (A, B) is controllable, then the system can be controlled, or driven from the state $\mathbf{0}$ to any specified state $\mathbf{v}$ (in $\mathbb{R}^n$) in at most n steps, simply by choosing an appropriate control sequence in $\mathbb{R}^m$. This fact is illustrated in Exercise 18 for $n = 4$

and $m = 2$. For a further discussion of controllability, see this text's web site (Case Study for Chapter 4).

WEB

18. Suppose A is a 4×4 matrix and B is a 4×2 matrix, and let $\mathbf{u}_0, \ldots, \mathbf{u}_3$ represent a sequence of input vectors in $\mathbb{R}^2$.

a. Set $\mathbf{x}_0 = \mathbf{0}$, compute $\mathbf{x}_1, \ldots, \mathbf{x}_4$ from equation (1), and write a formula for $\mathbf{x}_4$ involving the controllability matrix M appearing in equation (2). (*Note:* The matrix M is constructed as a partitioned matrix. Its overall size here is 4×8.)

b. Suppose (A, B) is controllable and $\mathbf{v}$ is any vector in $\mathbb{R}^4$. Explain why there exists a control sequence $\mathbf{u}_0, \ldots, \mathbf{u}_3$ in $\mathbb{R}^2$ such that $\mathbf{x}_4 = \mathbf{v}$.

Determine if the matrix pairs in Exercises 19–22 are controllable.

19. $A = \begin{bmatrix} .9 & 1 & 0 \\ 0 & -.9 & 0 \\ 0 & 0 & .5 \end{bmatrix}$, $B = \begin{bmatrix} 0 \\ 1 \\ 1 \end{bmatrix}$

20. $A = \begin{bmatrix} .8 & -.3 & 0 \\ .2 & .5 & 1 \\ 0 & 0 & -.5 \end{bmatrix}$, $B = \begin{bmatrix} 1 \\ 1 \\ 0 \end{bmatrix}$

21. [M] $A = \begin{bmatrix} 0 & 1 & 0 & 0 \\ 0 & 0 & 1 & 0 \\ 0 & 0 & 0 & 1 \\ -2 & -4.2 & -4.8 & -3.6 \end{bmatrix}$, $B = \begin{bmatrix} 1 \\ 0 \\ 0 \\ -1 \end{bmatrix}$

22. [M] $A = \begin{bmatrix} 0 & 1 & 0 & 0 \\ 0 & 0 & 1 & 0 \\ 0 & 0 & 0 & 1 \\ -1 & -13 & -12.2 & -1.5 \end{bmatrix}$, $B = \begin{bmatrix} 1 \\ 0 \\ 0 \\ -1 \end{bmatrix}$

5

Eigenvalues and Eigenvectors

Dynamical Systems and Spotted Owls

In 1990, the northern spotted owl became the center of a nationwide controversy over the use and misuse of the majestic forests in the Pacific Northwest. Environmentalists convinced the federal government that the owl was threatened with extinction if logging continued in the old-growth forests (with trees over 200 years old), where the owls prefer to live. The timber industry, anticipating the loss of 30,000 to 100,000 jobs as a result of new government restrictions on logging, argued that the owl should not be classified as a "threatened species" and cited a number of published scientific reports to support its case.[1]

Caught in the crossfire of the two lobbying groups, mathematical ecologists intensified their drive to understand the population dynamics of the spotted owl. The life cycle of a spotted owl divides naturally into three stages: juvenile (up to 1 year old), subadult (1 to 2 years), and adult (over 2 years). The owls mate for life during the subadult and adult stages, begin to breed as adults, and live for up to 20 years. Each owl pair requires about 1000 hectares (4 square miles) for its own home territory. A critical time in the life cycle is when the juveniles leave the nest. To survive and become a subadult, a juvenile must successfully find a new home range (and usually a mate).

A first step in studying the population dynamics is to model the population at yearly intervals, at times denoted by $k = 0, 1, 2, \ldots$. Usually, one assumes that there is a 1:1 ratio of males to females in each life stage and counts only the females. The population at year k can be described by a vector $\mathbf{x}_k = (j_k, s_k, a_k)$, where j_k, s_k, and a_k are the numbers of females in the juvenile, subadult, and adult stages, respectively.

Using actual field data from demographic studies, R. Lamberson and co-workers considered the following *stage-matrix model*:[2]

$$\begin{bmatrix} j_{k+1} \\ s_{k+1} \\ a_{k+1} \end{bmatrix} = \begin{bmatrix} 0 & 0 & .33 \\ .18 & 0 & 0 \\ 0 & .71 & .94 \end{bmatrix} \begin{bmatrix} j_k \\ s_k \\ a_k \end{bmatrix}$$

Here the number of new juvenile females in year $k + 1$ is .33 times the number of adult females in year k (based on the average birth rate per owl pair). Also, 18% of the juveniles survive to become subadults, and 71% of the subadults and 94% of the adults survive to be counted as adults.

The stage-matrix model is a difference equation of the form $\mathbf{x}_{k+1} = A\mathbf{x}_k$. Such an equation is often called a

---

[1] "The Great Spotted Owl War," *Reader's Digest*, November 1992, pp. 91–95.

[2] R. H. Lamberson, R. McKelvey, B. R. Noon, and C. Voss, "A Dynamic Analysis of the Viability of the Northern Spotted Owl in a Fragmented Forest Environment," *Conservation Biology* **6** (1992), 505–512. Also, a private communication from Professor Lamberson, 1993.

dynamical system (or a **discrete linear dynamical system**) because it describes the changes in a system as time passes.

The 18% juvenile survival rate in the Lamberson stage matrix is the entry affected most by the amount of old-growth forest available. Actually, 60% of the juveniles normally survive to leave the nest, but in the Willow Creek region of California studied by Lamberson and his colleagues, only 30% of the juveniles that left the nest were able to find new home ranges. The rest perished during the search process.

A significant reason for the failure of owls to find new home ranges is the increasing fragmentation of old-growth timber stands due to clear-cutting of scattered areas on the old-growth land. When an owl leaves the protective canopy of the forest and crosses a clear-cut area, the risk of attack by predators increases dramatically. Section 5.6 will show that the model described above predicts the eventual demise of the spotted owl, but that if 50% of the juveniles who survive to leave the nest also find new home ranges, then the owl population will thrive.

WEB

The goal of this chapter is to dissect the action of a linear transformation $\mathbf{x} \mapsto A\mathbf{x}$ into elements that are easily visualized. Except for a brief digression in Section 5.4, all matrices in the chapter are square. The main applications described here are to discrete dynamical systems, including the spotted owls discussed above. However, the basic concepts—eigenvectors and eigenvalues—are useful throughout pure and applied mathematics, and they appear in settings far more general than we consider here. Eigenvalues are also used to study differential equations and *continuous* dynamical systems, they provide critical information in engineering design, and they arise naturally in fields such as physics and chemistry.

5.1 | EIGENVECTORS AND EIGENVALUES

Although a transformation $\mathbf{x} \mapsto A\mathbf{x}$ may move vectors in a variety of directions, it often happens that there are special vectors on which the action of A is quite simple.

EXAMPLE 1   Let $A = \begin{bmatrix} 3 & -2 \\ 1 & 0 \end{bmatrix}$, $\mathbf{u} = \begin{bmatrix} -1 \\ 1 \end{bmatrix}$, and $\mathbf{v} = \begin{bmatrix} 2 \\ 1 \end{bmatrix}$. The images of $\mathbf{u}$ and $\mathbf{v}$ under multiplication by A are shown in Fig. 1. In fact, $A\mathbf{v}$ is just $2\mathbf{v}$. So A only "stretches," or dilates, $\mathbf{v}$. ∎

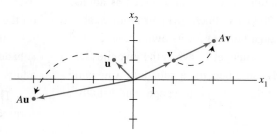

FIGURE 1 Effects of multiplication by A.

As another example, readers of Section 4.9 will recall that if A is a stochastic matrix, then the steady-state vector $\mathbf{q}$ for A satisfies the equation $A\mathbf{x} = \mathbf{x}$. That is, $A\mathbf{q} = 1 \cdot \mathbf{q}$.

This section studies equations such as

$$Ax = 2x \quad \text{or} \quad Ax = -4x$$

where special vectors are transformed by A into scalar multiples of themselves.

DEFINITION

An **eigenvector** of an $n \times n$ matrix A is a nonzero vector x such that $Ax = \lambda x$ for some scalar λ. A scalar λ is called an **eigenvalue** of A if there is a nontrivial solution x of $Ax = \lambda x$; such an x is called an *eigenvector corresponding to* λ.[1]

It is easy to determine if a given vector is an eigenvector of a matrix. It is also easy to decide if a specified scalar is an eigenvalue.

EXAMPLE 2   Let $A = \begin{bmatrix} 1 & 6 \\ 5 & 2 \end{bmatrix}$, $u = \begin{bmatrix} 6 \\ -5 \end{bmatrix}$, and $v = \begin{bmatrix} 3 \\ -2 \end{bmatrix}$. Are u and v eigenvectors of A?

SOLUTION

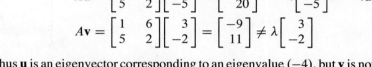

$$Au = \begin{bmatrix} 1 & 6 \\ 5 & 2 \end{bmatrix}\begin{bmatrix} 6 \\ -5 \end{bmatrix} = \begin{bmatrix} -24 \\ 20 \end{bmatrix} = -4\begin{bmatrix} 6 \\ -5 \end{bmatrix} = -4u$$

$$Av = \begin{bmatrix} 1 & 6 \\ 5 & 2 \end{bmatrix}\begin{bmatrix} 3 \\ -2 \end{bmatrix} = \begin{bmatrix} -9 \\ 11 \end{bmatrix} \neq \lambda\begin{bmatrix} 3 \\ -2 \end{bmatrix}$$

Thus u is an eigenvector corresponding to an eigenvalue (-4), but v is not an eigenvector of A, because Av is not a multiple of v.   ∎

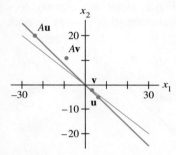

$Au = -4u$, but $Av \neq \lambda v$.

EXAMPLE 3   Show that 7 is an eigenvalue of matrix A in Example 2, and find the corresponding eigenvectors.

SOLUTION   The scalar 7 is an eigenvalue of A if and only if the equation

$$Ax = 7x \tag{1}$$

has a nontrivial solution. But (1) is equivalent to $Ax - 7x = 0$, or

$$(A - 7I)x = 0 \tag{2}$$

To solve this homogeneous equation, form the matrix

$$A - 7I = \begin{bmatrix} 1 & 6 \\ 5 & 2 \end{bmatrix} - \begin{bmatrix} 7 & 0 \\ 0 & 7 \end{bmatrix} = \begin{bmatrix} -6 & 6 \\ 5 & -5 \end{bmatrix}$$

The columns of $A - 7I$ are obviously linearly dependent, so (2) has nontrivial solutions. Thus 7 *is* an eigenvalue of A. To find the corresponding eigenvectors, use row operations:

$$\begin{bmatrix} -6 & 6 & 0 \\ 5 & -5 & 0 \end{bmatrix} \sim \begin{bmatrix} 1 & -1 & 0 \\ 0 & 0 & 0 \end{bmatrix}$$

The general solution has the form $x_2\begin{bmatrix} 1 \\ 1 \end{bmatrix}$. Each vector of this form with $x_2 \neq 0$ is an eigenvector corresponding to $\lambda = 7$.   ∎

---

[1] Note that an eigenvector must be *nonzero*, by definition, but an eigenvalue may be zero. The case in which the number 0 is an eigenvalue is discussed after Example 5.

Warning: Although row reduction was used in Example 3 to find eigen*vectors*, it cannot be used to find eigen*values*. An echelon form of a matrix A usually does *not* display the eigenvalues of A.

The equivalence of equations (1) and (2) obviously holds for any λ in place of $\lambda = 7$. Thus λ is an eigenvalue of an $n \times n$ matrix A if and only if the equation

$$(A - \lambda I)\mathbf{x} = \mathbf{0} \tag{3}$$

has a nontrivial solution. The set of *all* solutions of (3) is just the null space of the matrix $A - \lambda I$. So this set is a *subspace* of $\mathbb{R}^n$ and is called the **eigenspace** of A corresponding to λ. The eigenspace consists of the zero vector and all the eigenvectors corresponding to λ.

Example 3 shows that for matrix A in Example 2, the eigenspace corresponding to $\lambda = 7$ consists of *all* multiples of $(1, 1)$, which is the line through $(1, 1)$ and the origin. From Example 2, you can check that the eigenspace corresponding to $\lambda = -4$ is the line through $(6, -5)$. These eigenspaces are shown in Fig. 2, along with eigenvectors $(1, 1)$ and $(3/2, -5/4)$ and the geometric action of the transformation $\mathbf{x} \mapsto A\mathbf{x}$ on each eigenspace.

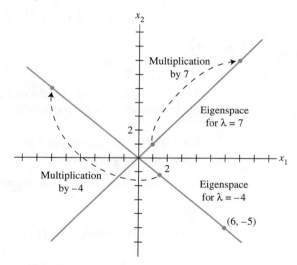

FIGURE 2 Eigenspaces for $\lambda = -4$ and $\lambda = 7$.

EXAMPLE 4 Let $A = \begin{bmatrix} 4 & -1 & 6 \\ 2 & 1 & 6 \\ 2 & -1 & 8 \end{bmatrix}$. An eigenvalue of A is 2. Find a basis for the corresponding eigenspace.

SOLUTION Form

$$A - 2I = \begin{bmatrix} 4 & -1 & 6 \\ 2 & 1 & 6 \\ 2 & -1 & 8 \end{bmatrix} - \begin{bmatrix} 2 & 0 & 0 \\ 0 & 2 & 0 \\ 0 & 0 & 2 \end{bmatrix} = \begin{bmatrix} 2 & -1 & 6 \\ 2 & -1 & 6 \\ 2 & -1 & 6 \end{bmatrix}$$

and row reduce the augmented matrix for $(A - 2I)\mathbf{x} = \mathbf{0}$:

$$\begin{bmatrix} 2 & -1 & 6 & 0 \\ 2 & -1 & 6 & 0 \\ 2 & -1 & 6 & 0 \end{bmatrix} \sim \begin{bmatrix} 2 & -1 & 6 & 0 \\ 0 & 0 & 0 & 0 \\ 0 & 0 & 0 & 0 \end{bmatrix}$$

At this point, it is clear that 2 is indeed an eigenvalue of A because the equation $(A - 2I)\mathbf{x} = \mathbf{0}$ has free variables. The general solution is

$$\begin{bmatrix} x_1 \\ x_2 \\ x_3 \end{bmatrix} = x_2 \begin{bmatrix} 1/2 \\ 1 \\ 0 \end{bmatrix} + x_3 \begin{bmatrix} -3 \\ 0 \\ 1 \end{bmatrix}, \quad x_2 \text{ and } x_3 \text{ free}$$

The eigenspace, shown in Fig. 3, is a two-dimensional subspace of $\mathbb{R}^3$. A basis is

$$\left\{ \begin{bmatrix} 1 \\ 2 \\ 0 \end{bmatrix}, \begin{bmatrix} -3 \\ 0 \\ 1 \end{bmatrix} \right\} \qquad \blacksquare$$

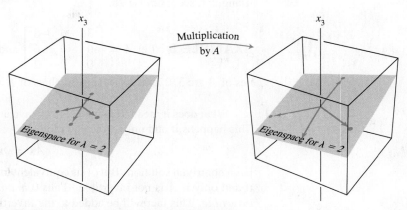

FIGURE 3 A acts as a dilation on the eigenspace.

> **NUMERICAL NOTE**
>
> Example 4 shows a good method for manual computation of eigenvectors in simple cases when an eigenvalue is known. Using a matrix program and row reduction to find an eigenspace (for a specified eigenvalue) usually works, too, but this is not entirely reliable. Roundoff error can lead occasionally to a reduced echelon form with the wrong number of pivots. The best computer programs compute approximations for eigenvalues and eigenvectors simultaneously, to any desired degree of accuracy, for matrices that are not too large. The size of matrices that can be analyzed increases each year as computing power and software improve.

The following theorem describes one of the few special cases in which eigenvalues can be found precisely. Calculation of eigenvalues will also be discussed in Section 5.2.

THEOREM 1   The eigenvalues of a triangular matrix are the entries on its main diagonal.

PROOF For simplicity, consider the 3×3 case. If A is upper triangular, then $A - \lambda I$

has the form

$$A - \lambda I = \begin{bmatrix} a_{11} & a_{12} & a_{13} \\ 0 & a_{22} & a_{23} \\ 0 & 0 & a_{33} \end{bmatrix} - \begin{bmatrix} \lambda & 0 & 0 \\ 0 & \lambda & 0 \\ 0 & 0 & \lambda \end{bmatrix}$$

$$= \begin{bmatrix} a_{11} - \lambda & a_{12} & a_{13} \\ 0 & a_{22} - \lambda & a_{23} \\ 0 & 0 & a_{33} - \lambda \end{bmatrix}$$

The scalar λ is an eigenvalue of A if and only if the equation $(A - \lambda I)\mathbf{x} = \mathbf{0}$ has a nontrivial solution, that is, if and only if the equation has a free variable. Because of the zero entries in $A - \lambda I$, it is easy to see that $(A - \lambda I)\mathbf{x} = \mathbf{0}$ has a free variable if and only if at least one of the entries on the diagonal of $A - \lambda I$ is zero. This happens if and only if λ equals one of the entries a_{11}, a_{22}, a_{33} in A. For the case in which A is lower triangular, see Exercise 28. ∎

EXAMPLE 5 Let $A = \begin{bmatrix} 3 & 6 & -8 \\ 0 & 0 & 6 \\ 0 & 0 & 2 \end{bmatrix}$ and $B = \begin{bmatrix} 4 & 0 & 0 \\ -2 & 1 & 0 \\ 5 & 3 & 4 \end{bmatrix}$. The eigenvalues of A are 3, 0, and 2. The eigenvalues of B are 4 and 1. ∎

What does it mean for a matrix A to have an eigenvalue of 0, such as in Example 5? This happens if and only if the equation

$$A\mathbf{x} = 0\mathbf{x} \tag{4}$$

has a nontrivial solution. But (4) is equivalent to $A\mathbf{x} = \mathbf{0}$, which has a nontrivial solution if and only if A is not invertible. Thus 0 *is an eigenvalue of A if and only if A is not invertible*. This fact will be added to the Invertible Matrix Theorem in Section 5.2.

The following important theorem will be needed later. Its proof illustrates a typical calculation with eigenvectors.

THEOREM 2 If $\mathbf{v}_1, \ldots, \mathbf{v}_r$ are eigenvectors that correspond to distinct eigenvalues $\lambda_1, \ldots, \lambda_r$ of an $n \times n$ matrix A, then the set $\{\mathbf{v}_1, \ldots, \mathbf{v}_r\}$ is linearly independent.

PROOF Suppose $\{\mathbf{v}_1, \ldots, \mathbf{v}_r\}$ is linearly dependent. Since $\mathbf{v}_1$ is nonzero, Theorem 7 in Section 1.7 says that one of the vectors in the set is a linear combination of the preceding vectors. Let p be the least index such that $\mathbf{v}_{p+1}$ is a linear combination of the preceding (linearly independent) vectors. Then there exist scalars $c_1, \ldots, c_p$ such that

$$c_1 \mathbf{v}_1 + \cdots + c_p \mathbf{v}_p = \mathbf{v}_{p+1} \tag{5}$$

Multiplying both sides of (5) by A and using the fact that $A\mathbf{v}_k = \lambda_k \mathbf{v}_k$ for each k, we obtain

$$c_1 A\mathbf{v}_1 + \cdots + c_p A\mathbf{v}_p = A\mathbf{v}_{p+1}$$
$$c_1 \lambda_1 \mathbf{v}_1 + \cdots + c_p \lambda_p \mathbf{v}_p = \lambda_{p+1} \mathbf{v}_{p+1} \tag{6}$$

Multiplying both sides of (5) by λ_{p+1} and subtracting the result from (6), we have

$$c_1(\lambda_1 - \lambda_{p+1})\mathbf{v}_1 + \cdots + c_p(\lambda_p - \lambda_{p+1})\mathbf{v}_p = \mathbf{0} \tag{7}$$

Since $\{\mathbf{v}_1, \ldots, \mathbf{v}_p\}$ is linearly independent, the weights in (7) are all zero. But none of the factors $\lambda_i - \lambda_{p+1}$ are zero, because the eigenvalues are distinct. Hence $c_i = 0$ for $i = 1, \ldots, p$. But then (5) says that $\mathbf{v}_{p+1} = \mathbf{0}$, which is impossible. Hence $\{\mathbf{v}_1, \ldots, \mathbf{v}_r\}$ cannot be linearly dependent and therefore must be linearly independent. ∎

Eigenvectors and Difference Equations

This section concludes by showing how to construct solutions of the first-order difference equation discussed in the chapter introductory example:

$$\mathbf{x}_{k+1} = A\mathbf{x}_k \quad (k = 0, 1, 2, \ldots) \tag{8}$$

If A is an $n \times n$ matrix, then (8) is a *recursive* description of a sequence $\{\mathbf{x}_k\}$ in $\mathbb{R}^n$. A **solution** of (8) is an explicit description of $\{\mathbf{x}_k\}$ whose formula for each $\mathbf{x}_k$ does not depend directly on A or on the preceding terms in the sequence other than the initial term $\mathbf{x}_0$.

The simplest way to build a solution of (8) is to take an eigenvector $\mathbf{x}_0$ and its corresponding eigenvalue λ and let

$$\mathbf{x}_k = \lambda^k \mathbf{x}_0 \quad (k = 1, 2, \ldots) \tag{9}$$

This sequence is a solution because

$$A\mathbf{x}_k = A(\lambda^k \mathbf{x}_0) = \lambda^k(A\mathbf{x}_0) = \lambda^k(\lambda \mathbf{x}_0) = \lambda^{k+1}\mathbf{x}_0 = \mathbf{x}_{k+1}$$

Linear combinations of solutions in the form of equation (9) are solutions, too! See Exercise 33.

PRACTICE PROBLEMS

1. Is 5 an eigenvalue of $A = \begin{bmatrix} 6 & -3 & 1 \\ 3 & 0 & 5 \\ 2 & 2 & 6 \end{bmatrix}$?

2. If $\mathbf{x}$ is an eigenvector of A corresponding to λ, what is $A^3\mathbf{x}$?

3. Suppose that $\mathbf{b}_1$ and $\mathbf{b}_2$ are eigenvectors corresponding to distinct eigenvalues λ_1 and λ_2, respectively, and suppose that $\mathbf{b}_3$ and $\mathbf{b}_4$ are linearly independent eigenvectors corresponding to a third distinct eigenvalue λ_3. Does it necessarily follow that $\{\mathbf{b}_1, \mathbf{b}_2, \mathbf{b}_3, \mathbf{b}_4\}$ is a linearly independent set? [*Hint*: Consider the equation $c_1\mathbf{b}_1 + c_2\mathbf{b}_2 + (c_3\mathbf{b}_3 + c_4\mathbf{b}_4) = \mathbf{0}$.]

5.1 EXERCISES

1. Is $\lambda = 2$ an eigenvalue of $\begin{bmatrix} 3 & 2 \\ 3 & 8 \end{bmatrix}$? Why or why not?

2. Is $\lambda = -3$ an eigenvalue of $\begin{bmatrix} -1 & 4 \\ 6 & 9 \end{bmatrix}$? Why or why not?

3. Is $\begin{bmatrix} 1 \\ 3 \end{bmatrix}$ an eigenvector of $\begin{bmatrix} 1 & -1 \\ 6 & -4 \end{bmatrix}$? If so, find the eigenvalue.

4. Is $\begin{bmatrix} -1 \\ 1 \end{bmatrix}$ an eigenvector of $\begin{bmatrix} 5 & 2 \\ 3 & 6 \end{bmatrix}$? If so, find the eigenvalue.

5. Is $\begin{bmatrix} 3 \\ -2 \\ 1 \end{bmatrix}$ an eigenvector of $\begin{bmatrix} -4 & 3 & 3 \\ 2 & -3 & -2 \\ -1 & 0 & -2 \end{bmatrix}$? If so, find the eigenvalue.

6. Is $\begin{bmatrix} 1 \\ -2 \\ 2 \end{bmatrix}$ an eigenvector of $\begin{bmatrix} 3 & 6 & 7 \\ 3 & 2 & 7 \\ 5 & 6 & 4 \end{bmatrix}$? If so, find the eigenvalue.

7. Is $\lambda = 4$ an eigenvalue of $\begin{bmatrix} 3 & 0 & -1 \\ 2 & 3 & 1 \\ -3 & 4 & 5 \end{bmatrix}$? If so, find one corresponding eigenvector.

8. Is $\lambda = 1$ an eigenvalue of $\begin{bmatrix} 4 & -2 & 3 \\ 0 & -1 & 3 \\ -1 & 2 & -2 \end{bmatrix}$? If so, find one corresponding eigenvector.

In Exercises 9–16, find a basis for the eigenspace corresponding to each listed eigenvalue.

9. $A = \begin{bmatrix} 3 & 0 \\ 2 & 1 \end{bmatrix}$, $\lambda = 1, 3$

10. $A = \begin{bmatrix} -4 & 2 \\ 3 & 1 \end{bmatrix}$, $\lambda = -5$

11. $A = \begin{bmatrix} 1 & -3 \\ -4 & 5 \end{bmatrix}$, $\lambda = -1, 7$

12. $A = \begin{bmatrix} 4 & 1 \\ 3 & 6 \end{bmatrix}$, $\lambda = 3, 7$

13. $A = \begin{bmatrix} 4 & 0 & 1 \\ -2 & 1 & 0 \\ -2 & 0 & 1 \end{bmatrix}$, $\lambda = 1, 2, 3$

14. $A = \begin{bmatrix} 4 & 0 & -1 \\ 3 & 0 & 3 \\ 2 & -2 & 5 \end{bmatrix}$, $\lambda = 3$

15. $A = \begin{bmatrix} -4 & 1 & 1 \\ 2 & -3 & 2 \\ 3 & 3 & -2 \end{bmatrix}$, $\lambda = -5$

16. $A = \begin{bmatrix} 5 & 0 & -1 & 0 \\ 1 & 3 & 0 & 0 \\ 2 & -1 & 3 & 0 \\ 4 & -2 & -2 & 4 \end{bmatrix}$, $\lambda = 4$

Find the eigenvalues of the matrices in Exercises 17 and 18.

17. $\begin{bmatrix} 0 & 0 & 0 \\ 0 & 3 & 4 \\ 0 & 0 & -2 \end{bmatrix}$ **18.** $\begin{bmatrix} 5 & 0 & 0 \\ 0 & 0 & 0 \\ -1 & 0 & 3 \end{bmatrix}$

19. For $A = \begin{bmatrix} 1 & 2 & 3 \\ 1 & 2 & 3 \\ 1 & 2 & 3 \end{bmatrix}$, find one eigenvalue, with no calculation. Justify your answer.

20. Without calculation, find one eigenvalue and two linearly independent eigenvectors of $A = \begin{bmatrix} 2 & 2 & 2 \\ 2 & 2 & 2 \\ 2 & 2 & 2 \end{bmatrix}$. Justify your answer.

In Exercises 21 and 22, A is an $n \times n$ matrix. Mark each statement True or False. Justify each answer

21. a. If $A\mathbf{x} = \lambda\mathbf{x}$ for some vector $\mathbf{x}$, then λ is an eigenvalue of A.

   b. A matrix A is not invertible if and only if 0 is an eigenvalue of A.

   c. A number c is an eigenvalue of A if and only if the equation $(A - cI)\mathbf{x} = \mathbf{0}$ has a nontrivial solution.

   d. Finding an eigenvector of A may be difficult, but checking whether a given vector is in fact an eigenvector is easy.

   e. To find the eigenvalues of A, reduce A to echelon form.

22. a. If $A\mathbf{x} = \lambda\mathbf{x}$ for some scalar λ, then $\mathbf{x}$ is an eigenvector of A.

   b. If $\mathbf{v}_1$ and $\mathbf{v}_2$ are linearly independent eigenvectors, then they correspond to distinct eigenvalues.

   c. A steady-state vector for a stochastic matrix is actually an eigenvector.

   d. The eigenvalues of a matrix are on its main diagonal.

   e. An eigenspace of A is a null space of a certain matrix.

23. Explain why a 2×2 matrix can have at most two distinct eigenvalues. Explain why an $n \times n$ matrix can have at most n distinct eigenvalues.

24. Construct an example of a 2×2 matrix with only one distinct eigenvalue.

25. Let λ be an eigenvalue of an invertible matrix A. Show that λ^{-1} is an eigenvalue of A^{-1}. [*Hint:* Suppose a nonzero $\mathbf{x}$ satisfies $A\mathbf{x} = \lambda\mathbf{x}$.]

26. Show that if A^2 is the zero matrix, then the only eigenvalue of A is 0.

27. Show that λ is an eigenvalue of A if and only if λ is an eigenvalue of A^T. [*Hint:* Find out how $A - \lambda I$ and $A^T - \lambda I$ are related.]

28. Use Exercise 27 to complete the proof of Theorem 1 for the case in which A is lower triangular.

29. Consider an $n \times n$ matrix A with the property that the row sums all equal the same number s. Show that s is an eigenvalue of A. [*Hint:* Find an eigenvector.]

30. Consider an $n \times n$ matrix A with the property that the column sums all equal the same number s. Show that s is an eigenvalue of A. [*Hint:* Use Exercises 27 and 29.]

In Exercises 31 and 32, let A be the matrix of the linear transformation T. Without writing A, find an eigenvalue of A and describe the eigenspace.

31. T is the transformation on $\mathbb{R}^2$ that reflects points across some line through the origin.

32. T is the transformation on $\mathbb{R}^3$ that rotates points about some line through the origin.

33. Let $\mathbf{u}$ and $\mathbf{v}$ be eigenvectors of a matrix A, with corresponding eigenvalues λ and μ, and let c_1 and c_2 be scalars. Define

$$\mathbf{x}_k = c_1 \lambda^k \mathbf{u} + c_2 \mu^k \mathbf{v} \quad (k = 0, 1, 2, \ldots)$$

   a. What is $\mathbf{x}_{k+1}$, by definition?

   b. Compute $A\mathbf{x}_k$ from the formula for $\mathbf{x}_k$, and show that $A\mathbf{x}_k = \mathbf{x}_{k+1}$. This calculation will prove that the sequence $\{\mathbf{x}_k\}$ defined above satisfies the difference equation $\mathbf{x}_{k+1} = A\mathbf{x}_k$ ($k = 0, 1, 2, \ldots$).

34. Describe how you might try to build a solution of a difference equation $\mathbf{x}_{k+1} = A\mathbf{x}_k$ ($k = 0, 1, 2, \ldots$) if you were given the initial $\mathbf{x}_0$ and this vector did not happen to be an eigenvector of A. [*Hint:* How might you relate $\mathbf{x}_0$ to eigenvectors of A?]

35. Let $\mathbf{u}$ and $\mathbf{v}$ be the vectors shown in the figure, and suppose $\mathbf{u}$ and $\mathbf{v}$ are eigenvectors of a 2×2 matrix A that correspond to eigenvalues 2 and 3, respectively. Let $T : \mathbb{R}^2 \to \mathbb{R}^2$ be the linear transformation given by $T(\mathbf{x}) = A\mathbf{x}$ for each $\mathbf{x}$ in $\mathbb{R}^2$, and let $\mathbf{w} = \mathbf{u} + \mathbf{v}$. Make a copy of the figure, and on the same coordinate system, carefully plot the vectors $T(\mathbf{u})$, $T(\mathbf{v})$, and $T(\mathbf{w})$.

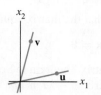

36. Repeat Exercise 35, assuming $\mathbf{u}$ and $\mathbf{v}$ are eigenvectors of A that correspond to eigenvalues -1 and 3, respectively.

[M] In Exercises 37–40, use a matrix program to find the eigenvalues of the matrix. Then use the method of Example 4 with a row reduction routine to produce a basis for each eigenspace.

37. $\begin{bmatrix} 12 & 1 & 4 \\ 2 & 11 & 4 \\ 1 & 3 & 7 \end{bmatrix}$

38. $\begin{bmatrix} 5 & -2 & 2 & -4 \\ 7 & -4 & 2 & -4 \\ 4 & -4 & 2 & 0 \\ 3 & -1 & 1 & -3 \end{bmatrix}$

39. $\begin{bmatrix} 12 & -90 & 30 & 30 & 30 \\ 8 & -49 & 15 & 15 & 15 \\ 16 & -52 & 12 & 0 & 20 \\ 0 & -30 & 10 & 22 & 10 \\ 8 & -41 & 15 & 15 & 7 \end{bmatrix}$

40. $\begin{bmatrix} -23 & 57 & -9 & -15 & -59 \\ -10 & 12 & -10 & 2 & -22 \\ 11 & 5 & -3 & -19 & -15 \\ -27 & 31 & -27 & 25 & -37 \\ -5 & -15 & -5 & 1 & 31 \end{bmatrix}$

SOLUTIONS TO PRACTICE PROBLEMS

1. The number 5 is an eigenvalue of A if and only if the equation $(A - 5I)\mathbf{x} = \mathbf{0}$ has a nontrivial solution. Form

$$A - 5I = \begin{bmatrix} 6 & -3 & 1 \\ 3 & 0 & 5 \\ 2 & 2 & 6 \end{bmatrix} - \begin{bmatrix} 5 & 0 & 0 \\ 0 & 5 & 0 \\ 0 & 0 & 5 \end{bmatrix} = \begin{bmatrix} 1 & -3 & 1 \\ 3 & -5 & 5 \\ 2 & 2 & 1 \end{bmatrix}$$

and row reduce the augmented matrix:

$$\begin{bmatrix} 1 & -3 & 1 & 0 \\ 3 & -5 & 5 & 0 \\ 2 & 2 & 1 & 0 \end{bmatrix} \sim \begin{bmatrix} 1 & -3 & 1 & 0 \\ 0 & 4 & 2 & 0 \\ 0 & 8 & -1 & 0 \end{bmatrix} \sim \begin{bmatrix} 1 & -3 & 1 & 0 \\ 0 & 4 & 2 & 0 \\ 0 & 0 & -5 & 0 \end{bmatrix}$$

At this point, it is clear that the homogeneous system has no free variables. Thus $A - 5I$ is an invertible matrix, which means that 5 is *not* an eigenvalue of A.

2. If $\mathbf{x}$ is an eigenvector of A corresponding to λ, then $A\mathbf{x} = \lambda\mathbf{x}$ and so

$$A^2\mathbf{x} = A(\lambda\mathbf{x}) = \lambda A\mathbf{x} = \lambda^2\mathbf{x}$$

Again, $A^3\mathbf{x} = A(A^2\mathbf{x}) = A(\lambda^2\mathbf{x}) = \lambda^2 A\mathbf{x} = \lambda^3\mathbf{x}$. The general pattern, $A^k\mathbf{x} = \lambda^k\mathbf{x}$, is proved by induction.

3. Yes. Suppose $c_1\mathbf{b}_1 + c_2\mathbf{b}_2 + c_3\mathbf{b}_3 + c_4\mathbf{b}_4 = \mathbf{0}$. Since any linear combination of eigenvectors from the same eigenvalue is again an eigenvector for that eigenvalue, $c_3\mathbf{b}_3 + c_4\mathbf{b}_4$ is an eigenvector for λ_3. By Theorem 2, the vectors $\mathbf{b}_1, \mathbf{b}_2$, and $c_3\mathbf{b}_3 + c_4\mathbf{b}_4$ are linearly independent, so

$$c_1\mathbf{b}_1 + c_2\mathbf{b}_2 + (c_3\mathbf{b}_3 + c_4\mathbf{b}_4) = \mathbf{0}$$

implies $c_1 = c_2 = 0$. But then, c_3 and c_4 must also be zero since $\mathbf{b}_3$ and $\mathbf{b}_4$ are linearly independent. Hence all the coefficients in the original equation must be zero, and the vectors $\mathbf{b}_1, \mathbf{b}_2, \mathbf{b}_3$, and $\mathbf{b}_4$ are linearly independent.

5.2 | THE CHARACTERISTIC EQUATION

Useful information about the eigenvalues of a square matrix A is encoded in a special scalar equation called the characteristic equation of A. A simple example will lead to the general case.

EXAMPLE 1   Find the eigenvalues of $A = \begin{bmatrix} 2 & 3 \\ 3 & -6 \end{bmatrix}$.

SOLUTION   We must find all scalars λ such that the matrix equation

$$(A - \lambda I)\mathbf{x} = \mathbf{0}$$

has a nontrivial solution. By the Invertible Matrix Theorem in Section 2.3, this problem is equivalent to finding all λ such that the matrix $A - \lambda I$ is *not* invertible, where

$$A - \lambda I = \begin{bmatrix} 2 & 3 \\ 3 & -6 \end{bmatrix} - \begin{bmatrix} \lambda & 0 \\ 0 & \lambda \end{bmatrix} = \begin{bmatrix} 2 - \lambda & 3 \\ 3 & -6 - \lambda \end{bmatrix}$$

By Theorem 4 in Section 2.2, this matrix fails to be invertible precisely when its determinant is zero. So the eigenvalues of A are the solutions of the equation

$$\det(A - \lambda I) = \det \begin{bmatrix} 2 - \lambda & 3 \\ 3 & -6 - \lambda \end{bmatrix} = 0$$

Recall that

$$\det \begin{bmatrix} a & b \\ c & d \end{bmatrix} = ad - bc$$

So

$$\begin{aligned} \det(A - \lambda I) &= (2 - \lambda)(-6 - \lambda) - (3)(3) \\ &= -12 + 6\lambda - 2\lambda + \lambda^2 - 9 \\ &= \lambda^2 + 4\lambda - 21 \\ &= (\lambda - 3)(\lambda + 7) \end{aligned}$$

If $\det(A - \lambda I) = 0$, then $\lambda = 3$ or $\lambda = -7$. So the eigenvalues of A are 3 and -7.   ∎

The determinant in Example 1 transformed the matrix equation $(A - \lambda I)\mathbf{x} = \mathbf{0}$, which involves *two* unknowns (λ and $\mathbf{x}$), into the scalar equation $\lambda^2 + 4\lambda - 21 = 0$, which involves only one unknown. The same idea works for $n \times n$ matrices. However, before turning to larger matrices, we summarize the properties of determinants needed to study eigenvalues.

Determinants

Let A be an $n \times n$ matrix, let U be any echelon form obtained from A by row replacements and row interchanges (without scaling), and let r be the number of such row interchanges. Then the **determinant** of A, written as $\det A$, is $(-1)^r$ times the product of the diagonal entries $u_{11}, \ldots, u_{nn}$ in U. If A is invertible, then $u_{11}, \ldots, u_{nn}$ are all *pivots* (because $A \sim I_n$ and the u_{ii} have not been scaled to 1's). Otherwise, at least u_{nn} is zero, and the product $u_{11} \cdots u_{nn}$ is zero. Thus[1]

$$\det A = \begin{cases} (-1)^r \cdot \begin{pmatrix} \text{product of} \\ \text{pivots in } U \end{pmatrix}, & \text{when } A \text{ is invertible} \\ 0, & \text{when } A \text{ is not invertible} \end{cases} \tag{1}$$

---

[1] Formula (1) was derived in Section 3.2. Readers who have not studied Chapter 3 may use this formula as the definition of $\det A$. It is a remarkable and nontrivial fact that any echelon form U obtained from A without scaling gives the same value for $\det A$.

EXAMPLE 2 Compute det A for $A = \begin{bmatrix} 1 & 5 & 0 \\ 2 & 4 & -1 \\ 0 & -2 & 0 \end{bmatrix}$.

SOLUTION The following row reduction uses one row interchange:

$$A \sim \begin{bmatrix} 1 & 5 & 0 \\ 0 & -6 & -1 \\ 0 & -2 & 0 \end{bmatrix} \sim \begin{bmatrix} 1 & 5 & 0 \\ 0 & -2 & 0 \\ 0 & -6 & -1 \end{bmatrix} \sim \begin{bmatrix} 1 & 5 & 0 \\ 0 & -2 & 0 \\ 0 & 0 & -1 \end{bmatrix} = U_1$$

So det A equals $(-1)^1(1)(-2)(-1) = -2$. The following alternative row reduction avoids the row interchange and produces a different echelon form. The last step adds $-1/3$ times row 2 to row 3:

$$A \sim \begin{bmatrix} 1 & 5 & 0 \\ 0 & -6 & -1 \\ 0 & -2 & 0 \end{bmatrix} \sim \begin{bmatrix} 1 & 5 & 0 \\ 0 & -6 & -1 \\ 0 & 0 & 1/3 \end{bmatrix} = U_2$$

This time det A is $(-1)^0(1)(-6)(1/3) = -2$, the same as before. ∎

Formula (1) for the determinant shows that A is invertible if and only if det A is nonzero. This fact, and the characterization of invertibility found in Section 5.1, can be added to the Invertible Matrix Theorem.

THEOREM

> **The Invertible Matrix Theorem (continued)**
>
> Let A be an $n \times n$ matrix. Then A is invertible if and only if:
>
> s. The number 0 is *not* an eigenvalue of A.
>
> t. The determinant of A is *not* zero.

When A is a 3×3 matrix, $|\det A|$ turns out to be the *volume* of the parallelepiped determined by the columns $\mathbf{a}_1$, $\mathbf{a}_2$, $\mathbf{a}_3$ of A, as in Fig. 1. (See Section 3.3 for details.) This volume is *nonzero* if and only if the vectors $\mathbf{a}_1$, $\mathbf{a}_2$, $\mathbf{a}_3$ are linearly independent, in which case the matrix A is invertible. (If the vectors are nonzero and linearly dependent, they lie in a plane or along a line.)

The next theorem lists facts needed from Sections 3.1 and 3.2. Part (a) is included here for convenient reference.

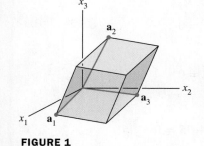

FIGURE 1

THEOREM 3

> **Properties of Determinants**
>
> Let A and B be $n \times n$ matrices.
>
> a. A is invertible if and only if det $A \neq 0$.
>
> b. $\det AB = (\det A)(\det B)$.
>
> c. $\det A^T = \det A$.
>
> d. If A is triangular, then det A is the product of the entries on the main diagonal of A.
>
> e. A row replacement operation on A does not change the determinant. A row interchange changes the sign of the determinant. A row scaling also scales the determinant by the same scalar factor.

The Characteristic Equation

Theorem 3(a) shows how to determine when a matrix of the form $A - \lambda I$ is *not* invertible. The scalar equation $\det(A - \lambda I) = 0$ is called the **characteristic equation** of A, and the argument in Example 1 justifies the following fact.

> A scalar λ is an eigenvalue of an $n \times n$ matrix A if and only if λ satisfies the characteristic equation
> $$\det(A - \lambda I) = 0$$

EXAMPLE 3 Find the characteristic equation of

$$A = \begin{bmatrix} 5 & -2 & 6 & -1 \\ 0 & 3 & -8 & 0 \\ 0 & 0 & 5 & 4 \\ 0 & 0 & 0 & 1 \end{bmatrix}$$

SOLUTION Form $A - \lambda I$, and use Theorem 3(d):

$$\det(A - \lambda I) = \det \begin{bmatrix} 5-\lambda & -2 & 6 & -1 \\ 0 & 3-\lambda & -8 & 0 \\ 0 & 0 & 5-\lambda & 4 \\ 0 & 0 & 0 & 1-\lambda \end{bmatrix}$$

$$= (5-\lambda)(3-\lambda)(5-\lambda)(1-\lambda)$$

The characteristic equation is

$$(5-\lambda)^2(3-\lambda)(1-\lambda) = 0$$

or

$$(\lambda - 5)^2(\lambda - 3)(\lambda - 1) = 0$$

Expanding the product, we can also write

$$\lambda^4 - 14\lambda^3 + 68\lambda^2 - 130\lambda + 75 = 0 \qquad \blacksquare$$

In Examples 1 and 3, $\det(A - \lambda I)$ is a polynomial in λ. It can be shown that if A is an $n \times n$ matrix, then $\det(A - \lambda I)$ is a polynomial of degree n called the **characteristic polynomial** of A.

The eigenvalue 5 in Example 3 is said to have *multiplicity* 2 because $(\lambda - 5)$ occurs two times as a factor of the characteristic polynomial. In general, the (**algebraic**) **multiplicity** of an eigenvalue λ is its multiplicity as a root of the characteristic equation.

EXAMPLE 4 The characteristic polynomial of a 6×6 matrix is $\lambda^6 - 4\lambda^5 - 12\lambda^4$. Find the eigenvalues and their multiplicities.

SOLUTION Factor the polynomial

$$\lambda^6 - 4\lambda^5 - 12\lambda^4 = \lambda^4(\lambda^2 - 4\lambda - 12) = \lambda^4(\lambda - 6)(\lambda + 2)$$

The eigenvalues are 0 (multiplicity 4), 6 (multiplicity 1), and -2 (multiplicity 1). $\qquad \blacksquare$

We could also list the eigenvalues in Example 4 as $0, 0, 0, 0, 6$, and -2, so that the eigenvalues are repeated according to their multiplicities.

Because the characteristic equation for an $n \times n$ matrix involves an nth-degree polynomial, the equation has exactly n roots, counting multiplicities, provided complex roots are allowed. Such complex roots, called *complex eigenvalues*, will be discussed in Section 5.5. Until then, we consider only real eigenvalues, and scalars will continue to be real numbers.

The characteristic equation is important for theoretical purposes. In practical work, however, eigenvalues of any matrix larger than 2×2 should be found by a computer, unless the matrix is triangular or has other special properties. Although a 3×3 characteristic polynomial is easy to compute by hand, factoring it can be difficult (unless the matrix is carefully chosen). See the Numerical Notes at the end of this section.

SG Factoring a Polynomial 5-8

Similarity

The next theorem illustrates one use of the characteristic polynomial, and it provides the foundation for several iterative methods that *approximate* eigenvalues. If A and B are $n \times n$ matrices, then A **is similar to** B if there is an invertible matrix P such that $P^{-1}AP = B$, or, equivalently, $A = PBP^{-1}$. Writing Q for P^{-1}, we have $Q^{-1}BQ = A$. So B is also similar to A, and we say simply that A and B **are similar**. Changing A into $P^{-1}AP$ is called a **similarity transformation**.

THEOREM 4

If $n \times n$ matrices A and B are similar, then they have the same characteristic polynomial and hence the same eigenvalues (with the same multiplicities).

PROOF If $B = P^{-1}AP$, then

$$B - \lambda I = P^{-1}AP - \lambda P^{-1}P = P^{-1}(AP - \lambda P) = P^{-1}(A - \lambda I)P$$

Using the multiplicative property (b) in Theorem 3, we compute

$$\det(B - \lambda I) = \det[P^{-1}(A - \lambda I)P]$$
$$= \det(P^{-1}) \cdot \det(A - \lambda I) \cdot \det(P) \qquad (2)$$

Since $\det(P^{-1}) \cdot \det(P) = \det(P^{-1}P) = \det I = 1$, we see from equation (2) that $\det(B - \lambda I) = \det(A - \lambda I)$. ∎

WARNINGS:
1. The matrices
$$\begin{bmatrix} 2 & 1 \\ 0 & 2 \end{bmatrix} \text{ and } \begin{bmatrix} 1 & 0 \\ 0 & 2 \end{bmatrix}$$
are not similar even though they have the same eigenvalues.
2. Similarity is not the same as row equivalence. (If A is row equivalent to B, then $B = EA$ for some invertible matrix E.) Row operations on a matrix usually change its eigenvalues.

Application to Dynamical Systems

Eigenvalues and eigenvectors hold the key to the discrete evolution of a dynamical system, as mentioned in the chapter introduction.

EXAMPLE 5 Let $A = \begin{bmatrix} .95 & .03 \\ .05 & .97 \end{bmatrix}$. Analyze the long-term behavior of the dynamical system defined by $\mathbf{x}_{k+1} = A\mathbf{x}_k$ ($k = 0, 1, 2, \ldots$), with $\mathbf{x}_0 = \begin{bmatrix} .6 \\ .4 \end{bmatrix}$.

SOLUTION The first step is to find the eigenvalues of A and a basis for each eigenspace. The characteristic equation for A is

$$0 = \det \begin{bmatrix} .95 - \lambda & .03 \\ .05 & .97 - \lambda \end{bmatrix} = (.95 - \lambda)(.97 - \lambda) - (.03)(.05)$$

$$= \lambda^2 - 1.92\lambda + .92$$

By the quadratic formula

$$\lambda = \frac{1.92 \pm \sqrt{(1.92)^2 - 4(.92)}}{2} = \frac{1.92 \pm \sqrt{.0064}}{2}$$

$$= \frac{1.92 \pm .08}{2} = 1 \quad \text{or} \quad .92$$

It is readily checked that eigenvectors corresponding to $\lambda = 1$ and $\lambda = .92$ are multiples of

$$\mathbf{v}_1 = \begin{bmatrix} 3 \\ 5 \end{bmatrix} \quad \text{and} \quad \mathbf{v}_2 = \begin{bmatrix} 1 \\ -1 \end{bmatrix}$$

respectively.

The next step is to write the given $\mathbf{x}_0$ in terms of $\mathbf{v}_1$ and $\mathbf{v}_2$. This can be done because $\{\mathbf{v}_1, \mathbf{v}_2\}$ is obviously a basis for $\mathbb{R}^2$. (Why?) So there exist weights c_1 and c_2 such that

$$\mathbf{x}_0 = c_1\mathbf{v}_1 + c_2\mathbf{v}_2 = [\,\mathbf{v}_1 \quad \mathbf{v}_2\,]\begin{bmatrix} c_1 \\ c_2 \end{bmatrix} \tag{3}$$

In fact,

$$\begin{bmatrix} c_1 \\ c_2 \end{bmatrix} = [\,\mathbf{v}_1 \quad \mathbf{v}_2\,]^{-1}\mathbf{x}_0 = \begin{bmatrix} 3 & 1 \\ 5 & -1 \end{bmatrix}^{-1}\begin{bmatrix} .60 \\ .40 \end{bmatrix}$$

$$= \frac{1}{-8}\begin{bmatrix} -1 & -1 \\ -5 & 3 \end{bmatrix}\begin{bmatrix} .60 \\ .40 \end{bmatrix} = \begin{bmatrix} .125 \\ .225 \end{bmatrix} \tag{4}$$

Because $\mathbf{v}_1$ and $\mathbf{v}_2$ in (3) are eigenvectors of A, with $A\mathbf{v}_1 = \mathbf{v}_1$ and $A\mathbf{v}_2 = .92\mathbf{v}_2$, we easily compute each $\mathbf{x}_k$:

$$\mathbf{x}_1 = A\mathbf{x}_0 = c_1 A\mathbf{v}_1 + c_2 A\mathbf{v}_2 \qquad \text{\small Using linearity of } \mathbf{x} \mapsto A\mathbf{x}$$

$$= c_1\mathbf{v}_1 + c_2(.92)\mathbf{v}_2 \qquad \text{\small } \mathbf{v}_1 \text{ and } \mathbf{v}_2 \text{ are eigenvectors.}$$

$$\mathbf{x}_2 = A\mathbf{x}_1 = c_1 A\mathbf{v}_1 + c_2(.92)A\mathbf{v}_2$$

$$= c_1\mathbf{v}_1 + c_2(.92)^2\mathbf{v}_2$$

and so on. In general,

$$\mathbf{x}_k = c_1\mathbf{v}_1 + c_2(.92)^k\mathbf{v}_2 \quad (k = 0, 1, 2, \ldots)$$

Using c_1 and c_2 from (4),

$$\mathbf{x}_k = .125\begin{bmatrix} 3 \\ 5 \end{bmatrix} + .225(.92)^k\begin{bmatrix} 1 \\ -1 \end{bmatrix} \quad (k = 0, 1, 2, \ldots) \tag{5}$$

This explicit formula for $\mathbf{x}_k$ gives the solution of the difference equation $\mathbf{x}_{k+1} = A\mathbf{x}_k$. As $k \to \infty$, $(.92)^k$ tends to zero and $\mathbf{x}_k$ tends to $\begin{bmatrix} .375 \\ .625 \end{bmatrix} = .125\mathbf{v}_1$. ∎

The calculations in Example 5 have an interesting application to a Markov chain discussed in Section 4.9. Those who read that section may recognize that matrix A in Example 5 above is the same as the migration matrix M in Section 4.9, $\mathbf{x}_0$ is the initial population distribution between city and suburbs, and $\mathbf{x}_k$ represents the population distribution after k years.

Theorem 18 in Section 4.9 stated that for a matrix such as A, the sequence $\mathbf{x}_k$ tends to a steady-state vector. Now we know *why* the $\mathbf{x}_k$ behave this way, at least for the migration matrix. The steady-state vector is $.125\mathbf{v}_1$, a multiple of the eigenvector $\mathbf{v}_1$, and formula (5) for $\mathbf{x}_k$ shows precisely why $\mathbf{x}_k \to .125\mathbf{v}_1$.

─ NUMERICAL NOTES ─

1. Computer software such as Mathematica and Maple can use symbolic calculations to find the characteristic polynomial of a moderate-sized matrix. But there is no formula or finite algorithm to solve the characteristic equation of a general $n \times n$ matrix for $n \geq 5$.

2. The best numerical methods for finding eigenvalues avoid the characteristic polynomial entirely. In fact, MATLAB finds the characteristic polynomial of a matrix A by first computing the eigenvalues $\lambda_1, \ldots, \lambda_n$ of A and then expanding the product $(\lambda - \lambda_1)(\lambda - \lambda_2) \cdots (\lambda - \lambda_n)$.

3. Several common algorithms for estimating the eigenvalues of a matrix A are based on Theorem 4. The powerful *QR algorithm* is discussed in the exercises. Another technique, called *Jacobi's method*, works when $A = A^T$ and computes a sequence of matrices of the form

$$A_1 = A \quad \text{and} \quad A_{k+1} = P_k^{-1} A_k P_k \quad (k = 1, 2, \ldots)$$

Each matrix in the sequence is similar to A and so has the same eigenvalues as A. The nondiagonal entries of A_{k+1} tend to zero as k increases, and the diagonal entries tend to approach the eigenvalues of A.

4. Other methods of estimating eigenvalues are discussed in Section 5.8.

PRACTICE PROBLEM

Find the characteristic equation and eigenvalues of $A = \begin{bmatrix} 1 & -4 \\ 4 & 2 \end{bmatrix}$.

5.2 EXERCISES

Find the characteristic polynomial and the real eigenvalues of the matrices in Exercises 1–8.

1. $\begin{bmatrix} 2 & 7 \\ 7 & 2 \end{bmatrix}$

2. $\begin{bmatrix} -4 & -1 \\ 6 & 1 \end{bmatrix}$

3. $\begin{bmatrix} -4 & 2 \\ 6 & 7 \end{bmatrix}$

4. $\begin{bmatrix} 8 & 2 \\ 3 & 3 \end{bmatrix}$

5. $\begin{bmatrix} 8 & 4 \\ 4 & 8 \end{bmatrix}$

6. $\begin{bmatrix} 9 & -2 \\ 2 & 5 \end{bmatrix}$

7. $\begin{bmatrix} 5 & 3 \\ -4 & 4 \end{bmatrix}$

8. $\begin{bmatrix} -4 & 3 \\ 2 & 1 \end{bmatrix}$

Exercises 9–14 require techniques from Section 3.1. Find the characteristic polynomial of each matrix, using either a cofactor expansion or the special formula for 3×3 determinants described prior to Exercises 15–18 in Section 3.1. [*Note:* Finding the characteristic polynomial of a 3×3 matrix is not easy to do with just row operations, because the variable λ is involved.]

9. $\begin{bmatrix} 4 & 0 & -1 \\ 0 & 4 & -1 \\ 1 & 0 & 2 \end{bmatrix}$ **10.** $\begin{bmatrix} 3 & 1 & 1 \\ 0 & 5 & 0 \\ -2 & 0 & 7 \end{bmatrix}$

11. $\begin{bmatrix} 3 & 0 & 0 \\ 2 & 1 & 4 \\ 1 & 0 & 4 \end{bmatrix}$ **12.** $\begin{bmatrix} -1 & 0 & 2 \\ 3 & 1 & 0 \\ 0 & 1 & 2 \end{bmatrix}$

13. $\begin{bmatrix} 6 & -2 & 0 \\ -2 & 9 & 0 \\ 5 & 8 & 3 \end{bmatrix}$ **14.** $\begin{bmatrix} 4 & 0 & -1 \\ -1 & 0 & 4 \\ 0 & 2 & 3 \end{bmatrix}$

For the matrices in Exercises 15–17, list the real eigenvalues, repeated according to their multiplicities.

15. $\begin{bmatrix} 5 & 5 & 0 & 2 \\ 0 & 2 & -3 & 6 \\ 0 & 0 & 3 & -2 \\ 0 & 0 & 0 & 5 \end{bmatrix}$ **16.** $\begin{bmatrix} 3 & 0 & 0 & 0 \\ 6 & 2 & 0 & 0 \\ 0 & 3 & 6 & 0 \\ 2 & 3 & 3 & -5 \end{bmatrix}$

17. $\begin{bmatrix} 3 & 0 & 0 & 0 & 0 \\ -5 & 1 & 0 & 0 & 0 \\ 3 & 8 & 0 & 0 & 0 \\ 0 & -7 & 2 & 1 & 0 \\ -4 & 1 & 9 & -2 & 3 \end{bmatrix}$

18. It can be shown that the algebraic multiplicity of an eigenvalue λ is always greater than or equal to the dimension of the eigenspace corresponding to λ. Find h in the matrix A below such that the eigenspace for $\lambda = 4$ is two-dimensional:

$$A = \begin{bmatrix} 4 & 2 & 3 & 3 \\ 0 & 2 & h & 3 \\ 0 & 0 & 4 & 14 \\ 0 & 0 & 0 & 2 \end{bmatrix}$$

19. Let A be an $n \times n$ matrix, and suppose A has n real eigenvalues, $\lambda_1, \ldots, \lambda_n$, repeated according to multiplicities, so that

$$\det (A - \lambda I) = (\lambda_1 - \lambda)(\lambda_2 - \lambda) \cdots (\lambda_n - \lambda)$$

Explain why $\det A$ is the product of the n eigenvalues of A. (This result is true for any square matrix when complex eigenvalues are considered.)

20. Use a property of determinants to show that A and A^T have the same characteristic polynomial.

In Exercises 21 and 22, A and B are $n \times n$ matrices. Mark each statement True or False. Justify each answer.

21. a. The determinant of A is the product of the diagonal entries in A.

b. An elementary row operation on A does not change the determinant.

c. $(\det A)(\det B) = \det AB$

d. If $\lambda + 5$ is a factor of the characteristic polynomial of A, then 5 is an eigenvalue of A.

22. a. If A is 3×3, with columns $\mathbf{a}_1, \mathbf{a}_2, \mathbf{a}_3$, then $\det A$ equals the volume of the parallelepiped determined by $\mathbf{a}_1, \mathbf{a}_2, \mathbf{a}_3$.

b. $\det A^T = (-1) \det A$.

c. The multiplicity of a root r of the characteristic equation of A is called the algebraic multiplicity of r as an eigenvalue of A.

d. A row replacement operation on A does not change the eigenvalues.

A widely used method for estimating eigenvalues of a general matrix A is the *QR algorithm*. Under suitable conditions, this algorithm produces a sequence of matrices, all similar to A, that become almost upper triangular, with diagonal entries that approach the eigenvalues of A. The main idea is to factor A (or another matrix similar to A) in the form $A = Q_1 R_1$, where $Q_1^T = Q_1^{-1}$ and R_1 is upper triangular. The factors are interchanged to form $A_1 = R_1 Q_1$, which is again factored as $A_1 = Q_2 R_2$; then to form $A_2 = R_2 Q_2$, and so on. The similarity of $A, A_1, \ldots$ follows from the more general result in Exercise 23.

23. Show that if $A = QR$ with Q invertible, then A is similar to $A_1 = RQ$.

24. Show that if A and B are similar, then $\det A = \det B$.

25. Let $A = \begin{bmatrix} .6 & .3 \\ .4 & .7 \end{bmatrix}$, $\mathbf{v}_1 = \begin{bmatrix} 3/7 \\ 4/7 \end{bmatrix}$, and $\mathbf{x}_0 = \begin{bmatrix} .5 \\ .5 \end{bmatrix}$. [*Note:* A is the stochastic matrix studied in Example 5 in Section 4.9.]

a. Find a basis for $\mathbb{R}^2$ consisting of $\mathbf{v}_1$ and another eigenvector $\mathbf{v}_2$ of A.

b. Verify that $\mathbf{x}_0$ may be written in the form $\mathbf{x}_0 = \mathbf{v}_1 + c\mathbf{v}_2$.

c. For $k = 1, 2, \ldots$, define $\mathbf{x}_k = A^k \mathbf{x}_0$. Compute $\mathbf{x}_1$ and $\mathbf{x}_2$, and write a formula for $\mathbf{x}_k$. Then show that $\mathbf{x}_k \to \mathbf{v}_1$ as k increases.

26. Let $A = \begin{bmatrix} a & b \\ c & d \end{bmatrix}$. Use formula (1) for a determinant (given before Example 2) to show that $\det A = ad - bc$. Consider two cases: $a \neq 0$ and $a = 0$.

27. Let $A = \begin{bmatrix} .5 & .2 & .3 \\ .3 & .8 & .3 \\ .2 & 0 & .4 \end{bmatrix}$, $\mathbf{v}_1 = \begin{bmatrix} .3 \\ .6 \\ .1 \end{bmatrix}$, $\mathbf{v}_2 = \begin{bmatrix} 1 \\ -3 \\ 2 \end{bmatrix}$,

$\mathbf{v}_3 = \begin{bmatrix} -1 \\ 0 \\ 1 \end{bmatrix}$, and $\mathbf{w} = \begin{bmatrix} 1 \\ 1 \\ 1 \end{bmatrix}$.

a. Show that $\mathbf{v}_1, \mathbf{v}_2, \mathbf{v}_3$ are eigenvectors of A. [*Note:* A is the stochastic matrix studied in Example 3 of Section 4.9.]

b. Let $\mathbf{x}_0$ be any vector in $\mathbb{R}^3$ with nonnegative entries whose sum is 1. (In Section 4.9, $\mathbf{x}_0$ was called a probability vector.) Explain why there are constants c_1, c_2, c_3 such that $\mathbf{x}_0 = c_1\mathbf{v}_1 + c_2\mathbf{v}_2 + c_3\mathbf{v}_3$. Compute $\mathbf{w}^T\mathbf{x}_0$, and deduce that $c_1 = 1$.

c. For $k = 1, 2, \ldots$, define $\mathbf{x}_k = A^k\mathbf{x}_0$, with $\mathbf{x}_0$ as in part (b). Show that $\mathbf{x}_k \to \mathbf{v}_1$ as k increases.

28. [M] Construct a random integer-valued 4×4 matrix A, and verify that A and A^T have the same characteristic polynomial (the same eigenvalues with the same multiplicities). Do A and A^T have the same eigenvectors? Make the same analysis of a 5×5 matrix. Report the matrices and your conclusions.

29. [M] Construct a random integer-valued 4×4 matrix A.

   a. Reduce A to echelon form U with no row scaling, and use U in formula (1) (before Example 2) to compute $\det A$. (If A happens to be singular, start over with a new random matrix.)

   b. Compute the eigenvalues of A and the product of these eigenvalues (as accurately as possible).

   c. List the matrix A, and, to four decimal places, list the pivots in U and the eigenvalues of A. Compute $\det A$ with your matrix program, and compare it with the products you found in (a) and (b).

30. [M] Let $A = \begin{bmatrix} -6 & 28 & 21 \\ 4 & -15 & -12 \\ -8 & a & 25 \end{bmatrix}$. For each value of a in the set $\{32, 31.9, 31.8, 32.1, 32.2\}$, compute the characteristic polynomial of A and the eigenvalues. In each case, create a graph of the characteristic polynomial $p(t) = \det(A - tI)$ for $0 \le t \le 3$. If possible, construct all graphs on one coordinate system. Describe how the graphs reveal the changes in the eigenvalues as a changes.

SOLUTION TO PRACTICE PROBLEM

The characteristic equation is

$$0 = \det(A - \lambda I) = \det \begin{bmatrix} 1 - \lambda & -4 \\ 4 & 2 - \lambda \end{bmatrix}$$

$$= (1 - \lambda)(2 - \lambda) - (-4)(4) = \lambda^2 - 3\lambda + 18$$

From the quadratic formula,

$$\lambda = \frac{3 \pm \sqrt{(-3)^2 - 4(18)}}{2} = \frac{3 \pm \sqrt{-63}}{2}$$

It is clear that the characteristic equation has no real solutions, so A has no real eigenvalues. The matrix A is acting on the real vector space $\mathbb{R}^2$, and there is no nonzero vector $\mathbf{v}$ in $\mathbb{R}^2$ such that $A\mathbf{v} = \lambda \mathbf{v}$ for some scalar λ.

5.3 DIAGONALIZATION

In many cases, the eigenvalue–eigenvector information contained within a matrix A can be displayed in a useful factorization of the form $A = PDP^{-1}$ where D is a diagonal matrix. In this section, the factorization enables us to compute A^k quickly for large values of k, a fundamental idea in several applications of linear algebra. Later, in Sections 5.6 and 5.7, the factorization will be used to analyze (and *decouple*) dynamical systems.

The following example illustrates that powers of a diagonal matrix are easy to compute.

EXAMPLE 1 If $D = \begin{bmatrix} 5 & 0 \\ 0 & 3 \end{bmatrix}$, then $D^2 = \begin{bmatrix} 5 & 0 \\ 0 & 3 \end{bmatrix} \begin{bmatrix} 5 & 0 \\ 0 & 3 \end{bmatrix} = \begin{bmatrix} 5^2 & 0 \\ 0 & 3^2 \end{bmatrix}$

and

$$D^3 = DD^2 = \begin{bmatrix} 5 & 0 \\ 0 & 3 \end{bmatrix} \begin{bmatrix} 5^2 & 0 \\ 0 & 3^2 \end{bmatrix} = \begin{bmatrix} 5^3 & 0 \\ 0 & 3^3 \end{bmatrix}$$

In general,

$$D^k = \begin{bmatrix} 5^k & 0 \\ 0 & 3^k \end{bmatrix} \quad \text{for } k \ge 1$$

∎

If $A = PDP^{-1}$ for some invertible P and diagonal D, then A^k is also easy to compute, as the next example shows.

EXAMPLE 2   Let $A = \begin{bmatrix} 7 & 2 \\ -4 & 1 \end{bmatrix}$. Find a formula for A^k, given that $A = PDP^{-1}$, where

$$P = \begin{bmatrix} 1 & 1 \\ -1 & -2 \end{bmatrix} \quad \text{and} \quad D = \begin{bmatrix} 5 & 0 \\ 0 & 3 \end{bmatrix}$$

SOLUTION   The standard formula for the inverse of a 2×2 matrix yields

$$P^{-1} = \begin{bmatrix} 2 & 1 \\ -1 & -1 \end{bmatrix}$$

Then, by associativity of matrix multiplication,

$$A^2 = (PDP^{-1})(PDP^{-1}) = PD \underbrace{(P^{-1}P)}_{I} DP^{-1} = PDDP^{-1}$$

$$= PD^2 P^{-1} = \begin{bmatrix} 1 & 1 \\ -1 & -2 \end{bmatrix} \begin{bmatrix} 5^2 & 0 \\ 0 & 3^2 \end{bmatrix} \begin{bmatrix} 2 & 1 \\ -1 & -1 \end{bmatrix}$$

Again,

$$A^3 = (PDP^{-1})A^2 = (PDP^{-1})\underbrace{}_{I}PD^2 P^{-1} = PDD^2 P^{-1} = PD^3 P^{-1}$$

In general, for $k \geq 1$,

$$A^k = PD^k P^{-1} = \begin{bmatrix} 1 & 1 \\ -1 & -2 \end{bmatrix} \begin{bmatrix} 5^k & 0 \\ 0 & 3^k \end{bmatrix} \begin{bmatrix} 2 & 1 \\ -1 & -1 \end{bmatrix}$$

$$= \begin{bmatrix} 2 \cdot 5^k - 3^k & 5^k - 3^k \\ 2 \cdot 3^k - 2 \cdot 5^k & 2 \cdot 3^k - 5^k \end{bmatrix} \qquad \blacksquare$$

A square matrix A is said to be **diagonalizable** if A is similar to a diagonal matrix, that is, if $A = PDP^{-1}$ for some invertible matrix P and some diagonal matrix D. The next theorem gives a characterization of diagonalizable matrices and tells how to construct a suitable factorization.

THEOREM 5

The Diagonalization Theorem

An $n \times n$ matrix A is diagonalizable if and only if A has n linearly independent eigenvectors.

In fact, $A = PDP^{-1}$, with D a diagonal matrix, if and only if the columns of P are n linearly independent eigenvectors of A. In this case, the diagonal entries of D are eigenvalues of A that correspond, respectively, to the eigenvectors in P.

In other words, A is diagonalizable if and only if there are enough eigenvectors to form a basis of $\mathbb{R}^n$. We call such a basis an **eigenvector basis** of $\mathbb{R}^n$.

PROOF   First, observe that if P is any $n \times n$ matrix with columns $\mathbf{v}_1, \ldots, \mathbf{v}_n$, and if D is any diagonal matrix with diagonal entries $\lambda_1, \ldots, \lambda_n$, then

$$AP = A[\, \mathbf{v}_1 \quad \mathbf{v}_2 \quad \cdots \quad \mathbf{v}_n \,] = [\, A\mathbf{v}_1 \quad A\mathbf{v}_2 \quad \cdots \quad A\mathbf{v}_n \,] \qquad (1)$$

while

$$PD = P \begin{bmatrix} \lambda_1 & 0 & \cdots & 0 \\ 0 & \lambda_2 & \cdots & 0 \\ \vdots & \vdots & & \vdots \\ 0 & 0 & \cdots & \lambda_n \end{bmatrix} = [\, \lambda_1 \mathbf{v}_1 \quad \lambda_2 \mathbf{v}_2 \quad \cdots \quad \lambda_n \mathbf{v}_n \,] \qquad (2)$$

Now suppose A is diagonalizable and $A = PDP^{-1}$. Then right-multiplying this relation by P, we have $AP = PD$. In this case, equations (1) and (2) imply that

$$[\, A\mathbf{v}_1 \quad A\mathbf{v}_2 \quad \cdots \quad A\mathbf{v}_n \,] = [\, \lambda_1\mathbf{v}_1 \quad \lambda_2\mathbf{v}_2 \quad \cdots \quad \lambda_n\mathbf{v}_n \,] \tag{3}$$

Equating columns, we find that

$$A\mathbf{v}_1 = \lambda_1\mathbf{v}_1, \quad A\mathbf{v}_2 = \lambda_2\mathbf{v}_2, \quad \ldots, \quad A\mathbf{v}_n = \lambda_n\mathbf{v}_n \tag{4}$$

Since P is invertible, its columns $\mathbf{v}_1, \ldots, \mathbf{v}_n$ must be linearly independent. Also, since these columns are nonzero, the equations in (4) show that $\lambda_1, \ldots, \lambda_n$ are eigenvalues and $\mathbf{v}_1, \ldots, \mathbf{v}_n$ are corresponding eigenvectors. This argument proves the "only if" parts of the first and second statements, along with the third statement, of the theorem.

Finally, given any n eigenvectors $\mathbf{v}_1, \ldots, \mathbf{v}_n$, use them to construct the columns of P and use corresponding eigenvalues $\lambda_1, \ldots, \lambda_n$ to construct D. By equations (1)–(3), $AP = PD$. This is true without any condition on the eigenvectors. If, in fact, the eigenvectors are linearly independent, then P is invertible (by the Invertible Matrix Theorem), and $AP = PD$ implies that $A = PDP^{-1}$. ∎

Diagonalizing Matrices

EXAMPLE 3 Diagonalize the following matrix, if possible.

$$A = \begin{bmatrix} 1 & 3 & 3 \\ -3 & -5 & -3 \\ 3 & 3 & 1 \end{bmatrix}$$

That is, find an invertible matrix P and a diagonal matrix D such that $A = PDP^{-1}$.

SOLUTION There are four steps to implement the description in Theorem 5.

*Step 1. **Find the eigenvalues of A.*** As mentioned in Section 5.2, the mechanics of this step are appropriate for a computer when the matrix is larger than 2×2. To avoid unnecessary distractions, the text will usually supply information needed for this step. In the present case, the characteristic equation turns out to involve a cubic polynomial that can be factored:

$$0 = \det(A - \lambda I) = -\lambda^3 - 3\lambda^2 + 4$$
$$= -(\lambda - 1)(\lambda + 2)^2$$

The eigenvalues are $\lambda = 1$ and $\lambda = -2$.

*Step 2. **Find three linearly independent eigenvectors of A.*** *Three* vectors are needed because A is a 3×3 matrix. This is the critical step. If it fails, then Theorem 5 says that A cannot be diagonalized. The method in Section 5.1 produces a basis for each eigenspace:

$$\text{Basis for } \lambda = 1: \quad \mathbf{v}_1 = \begin{bmatrix} 1 \\ -1 \\ 1 \end{bmatrix}$$

$$\text{Basis for } \lambda = -2: \quad \mathbf{v}_2 = \begin{bmatrix} -1 \\ 1 \\ 0 \end{bmatrix} \quad \text{and} \quad \mathbf{v}_3 = \begin{bmatrix} -1 \\ 0 \\ 1 \end{bmatrix}$$

You can check that $\{\mathbf{v}_1, \mathbf{v}_2, \mathbf{v}_3\}$ is a linearly independent set.

Step 3. ***Construct P from the vectors in step 2.*** The order of the vectors is unimportant. Using the order chosen in step 2, form

$$P = \begin{bmatrix} \mathbf{v}_1 & \mathbf{v}_2 & \mathbf{v}_3 \end{bmatrix} = \begin{bmatrix} 1 & -1 & -1 \\ -1 & 1 & 0 \\ 1 & 0 & 1 \end{bmatrix}$$

Step 4. ***Construct D from the corresponding eigenvalues.*** In this step, it is essential that the order of the eigenvalues matches the order chosen for the columns of P. Use the eigenvalue $\lambda = -2$ twice, once for each of the eigenvectors corresponding to $\lambda = -2$:

$$D = \begin{bmatrix} 1 & 0 & 0 \\ 0 & -2 & 0 \\ 0 & 0 & -2 \end{bmatrix}$$

It is a good idea to check that P and D really work. To avoid computing P^{-1}, simply verify that $AP = PD$. This is equivalent to $A = PDP^{-1}$ when P is invertible. (However, be sure that P is invertible!) Compute

$$AP = \begin{bmatrix} 1 & 3 & 3 \\ -3 & -5 & -3 \\ 3 & 3 & 1 \end{bmatrix} \begin{bmatrix} 1 & -1 & -1 \\ -1 & 1 & 0 \\ 1 & 0 & 1 \end{bmatrix} = \begin{bmatrix} 1 & 2 & 2 \\ -1 & -2 & 0 \\ 1 & 0 & -2 \end{bmatrix}$$

$$PD = \begin{bmatrix} 1 & -1 & -1 \\ -1 & 1 & 0 \\ 1 & 0 & 1 \end{bmatrix} \begin{bmatrix} 1 & 0 & 0 \\ 0 & -2 & 0 \\ 0 & 0 & -2 \end{bmatrix} = \begin{bmatrix} 1 & 2 & 2 \\ -1 & -2 & 0 \\ 1 & 0 & -2 \end{bmatrix} \qquad \blacksquare$$

EXAMPLE 4 Diagonalize the following matrix, if possible.

$$A = \begin{bmatrix} 2 & 4 & 3 \\ -4 & -6 & -3 \\ 3 & 3 & 1 \end{bmatrix}$$

SOLUTION The characteristic equation of A turns out to be exactly the same as that in Example 3:

$$0 = \det (A - \lambda I) = -\lambda^3 - 3\lambda^2 + 4 = -(\lambda - 1)(\lambda + 2)^2$$

The eigenvalues are $\lambda = 1$ and $\lambda = -2$. However, it is easy to verify that each eigenspace is only one-dimensional:

$$\text{Basis for } \lambda = 1: \qquad \mathbf{v}_1 = \begin{bmatrix} 1 \\ -1 \\ 1 \end{bmatrix}$$

$$\text{Basis for } \lambda = -2: \qquad \mathbf{v}_2 = \begin{bmatrix} -1 \\ 1 \\ 0 \end{bmatrix}$$

There are no other eigenvalues, and every eigenvector of A is a multiple of either $\mathbf{v}_1$ or $\mathbf{v}_2$. Hence it is impossible to construct a basis of $\mathbb{R}^3$ using eigenvectors of A. By Theorem 5, A is *not* diagonalizable. $\qquad \blacksquare$

The following theorem provides a *sufficient* condition for a matrix to be diagonalizable.

THEOREM 6   An $n \times n$ matrix with n distinct eigenvalues is diagonalizable.

PROOF Let $\mathbf{v}_1, \ldots, \mathbf{v}_n$ be eigenvectors corresponding to the n distinct eigenvalues of a matrix A. Then $\{\mathbf{v}_1, \ldots, \mathbf{v}_n\}$ is linearly independent, by Theorem 2 in Section 5.1. Hence A is diagonalizable, by Theorem 5. ∎

It is not *necessary* for an $n \times n$ matrix to have n distinct eigenvalues in order to be diagonalizable. The 3×3 matrix in Example 3 is diagonalizable even though it has only two distinct eigenvalues.

EXAMPLE 5 Determine if the following matrix is diagonalizable.

$$A = \begin{bmatrix} 5 & -8 & 1 \\ 0 & 0 & 7 \\ 0 & 0 & -2 \end{bmatrix}$$

SOLUTION This is easy! Since the matrix is triangular, its eigenvalues are obviously 5, 0, and -2. Since A is a 3×3 matrix with three distinct eigenvalues, A is diagonalizable. ∎

Matrices Whose Eigenvalues Are Not Distinct

If an $n \times n$ matrix A has n distinct eigenvalues, with corresponding eigenvectors $\mathbf{v}_1, \ldots, \mathbf{v}_n$, and if $P = [\,\mathbf{v}_1 \; \cdots \; \mathbf{v}_n\,]$, then P is automatically invertible because its columns are linearly independent, by Theorem 2. When A is diagonalizable but has fewer than n distinct eigenvalues, it is still possible to build P in a way that makes P automatically invertible, as the next theorem shows.[1]

THEOREM 7

Let A be an $n \times n$ matrix whose distinct eigenvalues are $\lambda_1, \ldots, \lambda_p$.

a. For $1 \leq k \leq p$, the dimension of the eigenspace for λ_k is less than or equal to the multiplicity of the eigenvalue λ_k.

b. The matrix A is diagonalizable if and only if the sum of the dimensions of the eigenspaces equals n, and this happens if and only if (i) the characteristic polynomial factors completely into linear factors and (ii) the dimension of the eigenspace for each λ_k equals the multiplicity of λ_k.

c. If A is diagonalizable and $\mathcal{B}_k$ is a basis for the eigenspace corresponding to λ_k for each k, then the total collection of vectors in the sets $\mathcal{B}_1, \ldots, \mathcal{B}_p$ forms an eigenvector basis for $\mathbb{R}^n$.

EXAMPLE 6 Diagonalize the following matrix, if possible.

$$A = \begin{bmatrix} 5 & 0 & 0 & 0 \\ 0 & 5 & 0 & 0 \\ 1 & 4 & -3 & 0 \\ -1 & -2 & 0 & -3 \end{bmatrix}$$

---

[1] The proof of Theorem 7 is somewhat lengthy but not difficult. For instance, see S. Friedberg, A. Insel, and L. Spence, *Linear Algebra*, 4th ed. (Englewood Cliffs, NJ: Prentice-Hall, 2002), Section 5.2.

SOLUTION Since A is a triangular matrix, the eigenvalues are 5 and -3, each with multiplicity 2. Using the method in Section 5.1, we find a basis for each eigenspace.

$$\text{Basis for } \lambda = 5: \quad \mathbf{v}_1 = \begin{bmatrix} -8 \\ 4 \\ 1 \\ 0 \end{bmatrix} \quad \text{and} \quad \mathbf{v}_2 = \begin{bmatrix} -16 \\ 4 \\ 0 \\ 1 \end{bmatrix}$$

$$\text{Basis for } \lambda = -3: \quad \mathbf{v}_3 = \begin{bmatrix} 0 \\ 0 \\ 1 \\ 0 \end{bmatrix} \quad \text{and} \quad \mathbf{v}_4 = \begin{bmatrix} 0 \\ 0 \\ 0 \\ 1 \end{bmatrix}$$

The set $\{\mathbf{v}_1, \ldots, \mathbf{v}_4\}$ is linearly independent, by Theorem 7. So the matrix $P = [\,\mathbf{v}_1 \quad \cdots \quad \mathbf{v}_4\,]$ is invertible, and $A = PDP^{-1}$, where

$$P = \begin{bmatrix} -8 & -16 & 0 & 0 \\ 4 & 4 & 0 & 0 \\ 1 & 0 & 1 & 0 \\ 0 & 1 & 0 & 1 \end{bmatrix} \quad \text{and} \quad D = \begin{bmatrix} 5 & 0 & 0 & 0 \\ 0 & 5 & 0 & 0 \\ 0 & 0 & -3 & 0 \\ 0 & 0 & 0 & -3 \end{bmatrix} \quad \blacksquare$$

PRACTICE PROBLEMS

1. Compute A^8, where $A = \begin{bmatrix} 4 & -3 \\ 2 & -1 \end{bmatrix}$.

2. Let $A = \begin{bmatrix} -3 & 12 \\ -2 & 7 \end{bmatrix}$, $\mathbf{v}_1 = \begin{bmatrix} 3 \\ 1 \end{bmatrix}$, and $\mathbf{v}_2 = \begin{bmatrix} 2 \\ 1 \end{bmatrix}$. Suppose you are told that $\mathbf{v}_1$ and $\mathbf{v}_2$ are eigenvectors of A. Use this information to diagonalize A.

3. Let A be a 4×4 matrix with eigenvalues 5, 3, and -2, and suppose you know that the eigenspace for $\lambda = 3$ is two-dimensional. Do you have enough information to determine if A is diagonalizable?

WEB

5.3 EXERCISES

In Exercises 1 and 2, let $A = PDP^{-1}$ and compute A^4.

1. $P = \begin{bmatrix} 5 & 7 \\ 2 & 3 \end{bmatrix}$, $D = \begin{bmatrix} 2 & 0 \\ 0 & 1 \end{bmatrix}$

2. $P = \begin{bmatrix} 1 & 2 \\ 2 & 3 \end{bmatrix}$, $D = \begin{bmatrix} 1 & 0 \\ 0 & 3 \end{bmatrix}$

In Exercises 3 and 4, use the factorization $A = PDP^{-1}$ to compute A^k, where k represents an arbitrary positive integer.

3. $\begin{bmatrix} a & 0 \\ 2(a-b) & b \end{bmatrix} = \begin{bmatrix} 1 & 0 \\ 2 & 1 \end{bmatrix} \begin{bmatrix} a & 0 \\ 0 & b \end{bmatrix} \begin{bmatrix} 1 & 0 \\ -2 & 1 \end{bmatrix}$

4. $\begin{bmatrix} 1 & -6 \\ 2 & -6 \end{bmatrix} = \begin{bmatrix} 3 & -2 \\ 2 & -1 \end{bmatrix} \begin{bmatrix} -3 & 0 \\ 0 & -2 \end{bmatrix} \begin{bmatrix} -1 & 2 \\ -2 & 3 \end{bmatrix}$

In Exercises 5 and 6, the matrix A is factored in the form PDP^{-1}. Use the Diagonalization Theorem to find the eigenvalues of A and a basis for each eigenspace.

5. $A = \begin{bmatrix} 2 & -1 & -1 \\ 1 & 4 & 1 \\ -1 & -1 & 2 \end{bmatrix}$

$= \begin{bmatrix} 1 & -1 & 0 \\ -1 & 1 & -1 \\ 0 & -1 & 1 \end{bmatrix} \begin{bmatrix} 3 & 0 & 0 \\ 0 & 2 & 0 \\ 0 & 0 & 3 \end{bmatrix} \begin{bmatrix} 0 & -1 & -1 \\ -1 & -1 & -1 \\ -1 & -1 & 0 \end{bmatrix}$

6. $A = \begin{bmatrix} 3 & 0 & 0 \\ -3 & 4 & 9 \\ 0 & 0 & 3 \end{bmatrix}$

$= \begin{bmatrix} 3 & 0 & -1 \\ 0 & 1 & -3 \\ 1 & 0 & 0 \end{bmatrix} \begin{bmatrix} 3 & 0 & 0 \\ 0 & 4 & 0 \\ 0 & 0 & 3 \end{bmatrix} \begin{bmatrix} 0 & 0 & 1 \\ -3 & 1 & 9 \\ -1 & 0 & 3 \end{bmatrix}$

Diagonalize the matrices in Exercises 7–20, if possible. The real eigenvalues for Exercises 11–16 and 18 are included below the matrix.

7. $\begin{bmatrix} 1 & 0 \\ 6 & -1 \end{bmatrix}$

8. $\begin{bmatrix} 3 & 2 \\ 0 & 3 \end{bmatrix}$

9. $\begin{bmatrix} 2 & -1 \\ 1 & 4 \end{bmatrix}$ **10.** $\begin{bmatrix} 1 & 3 \\ 4 & 2 \end{bmatrix}$

11. $\begin{bmatrix} 0 & 1 & 1 \\ 2 & 1 & 2 \\ 3 & 3 & 2 \end{bmatrix}$ **12.** $\begin{bmatrix} 3 & 1 & 1 \\ 1 & 3 & 1 \\ 1 & 1 & 3 \end{bmatrix}$

$\lambda = -1, 5$ $\lambda = 2, 5$

13. $\begin{bmatrix} 2 & 2 & -1 \\ 1 & 3 & -1 \\ -1 & -2 & 2 \end{bmatrix}$ **14.** $\begin{bmatrix} 2 & 0 & -2 \\ 1 & 3 & 2 \\ 0 & 0 & 3 \end{bmatrix}$

$\lambda = 1, 5$ $\lambda = 2, 3$

15. $\begin{bmatrix} 0 & -1 & -1 \\ 1 & 2 & 1 \\ -1 & -1 & 0 \end{bmatrix}$ **16.** $\begin{bmatrix} 1 & 2 & -3 \\ 2 & 5 & -2 \\ 1 & 3 & 1 \end{bmatrix}$

$\lambda = 0, 1$ $\lambda = 0$

17. $\begin{bmatrix} 2 & 0 & 0 \\ 2 & 2 & 0 \\ 2 & 2 & 2 \end{bmatrix}$ **18.** $\begin{bmatrix} 2 & -2 & -2 \\ 3 & -3 & -2 \\ 2 & -2 & -2 \end{bmatrix}$

$\lambda = -2, -1, 0$

19. $\begin{bmatrix} 5 & -3 & 0 & 9 \\ 0 & 3 & 1 & -2 \\ 0 & 0 & 2 & 0 \\ 0 & 0 & 0 & 2 \end{bmatrix}$ **20.** $\begin{bmatrix} 3 & 0 & 0 & 0 \\ 0 & 2 & 0 & 0 \\ 0 & 0 & 2 & 0 \\ 1 & 0 & 0 & 3 \end{bmatrix}$

In Exercises 21 and 22, A, B, P, and D are $n \times n$ matrices. Mark each statement True or False. Justify each answer. (Study Theorems 5 and 6 and the examples in this section carefully before you try these exercises.)

21. a. A is diagonalizable if $A = PDP^{-1}$ for some matrix D and some invertible matrix P.

b. If $\mathbb{R}^n$ has a basis of eigenvectors of A, then A is diagonalizable.

c. A is diagonalizable if and only if A has n eigenvalues, counting multiplicities.

d. If A is diagonalizable, then A is invertible.

22. a. A is diagonalizable if A has n eigenvectors.

b. If A is diagonalizable, then A has n distinct eigenvalues.

c. If $AP = PD$, with D diagonal, then the nonzero columns of P must be eigenvectors of A.

d. If A is invertible, then A is diagonalizable.

23. A is a 5×5 matrix with two eigenvalues. One eigenspace is three-dimensional, and the other eigenspace is two-dimensional. Is A diagonalizable? Why?

24. A is a 3×3 matrix with two eigenvalues. Each eigenspace is one-dimensional. Is A diagonalizable? Why?

25. A is a 4×4 matrix with three eigenvalues. One eigenspace is one-dimensional, and one of the other eigenspaces is two-dimensional. Is it possible that A is *not* diagonalizable? Justify your answer.

26. A is a 7×7 matrix with three eigenvalues. One eigenspace is two-dimensional, and one of the other eigenspaces is three-dimensional. Is it possible that A is *not* diagonalizable? Justify your answer.

27. Show that if A is both diagonalizable and invertible, then so is A^{-1}.

28. Show that if A has n linearly independent eigenvectors, then so does A^T. [*Hint:* Use the Diagonalization Theorem.]

29. A factorization $A = PDP^{-1}$ is not unique. Demonstrate this for the matrix A in Example 2. With $D_1 = \begin{bmatrix} 3 & 0 \\ 0 & 5 \end{bmatrix}$, use the information in Example 2 to find a matrix P_1 such that $A = P_1 D_1 P_1^{-1}$.

30. With A and D as in Example 2, find an invertible P_2 unequal to the P in Example 2, such that $A = P_2 D P_2^{-1}$.

31. Construct a nonzero 2×2 matrix that is invertible but not diagonalizable.

32. Construct a nondiagonal 2×2 matrix that is diagonalizable but not invertible.

[M] Diagonalize the matrices in Exercises 33–36. Use your matrix program's eigenvalue command to find the eigenvalues, and then compute bases for the eigenspaces as in Section 5.1.

33. $\begin{bmatrix} 9 & -4 & -2 & -4 \\ -56 & 32 & -28 & 44 \\ -14 & -14 & 6 & -14 \\ 42 & -33 & 21 & -45 \end{bmatrix}$

34. $\begin{bmatrix} 4 & -9 & -7 & 8 & 2 \\ -7 & -9 & 0 & 7 & 14 \\ 5 & 10 & 5 & -5 & -10 \\ -2 & 3 & 7 & 0 & 4 \\ -3 & -13 & -7 & 10 & 11 \end{bmatrix}$

35. $\begin{bmatrix} 13 & -12 & 9 & -15 & 9 \\ 6 & -5 & 9 & -15 & 9 \\ 6 & -12 & -5 & 6 & 9 \\ 6 & -12 & 9 & -8 & 9 \\ -6 & 12 & 12 & -6 & -2 \end{bmatrix}$

36. $\begin{bmatrix} 24 & -6 & 2 & 6 & 2 \\ 72 & 51 & 9 & -99 & 9 \\ 0 & -63 & 15 & 63 & 63 \\ 72 & 15 & 9 & -63 & 9 \\ 0 & 63 & 21 & -63 & -27 \end{bmatrix}$

| SOLUTIONS TO PRACTICE PROBLEMS

1. $\det(A - \lambda I) = \lambda^2 - 3\lambda + 2 = (\lambda - 2)(\lambda - 1)$. The eigenvalues are 2 and 1, and the corresponding eigenvectors are $\mathbf{v}_1 = \begin{bmatrix} 3 \\ 2 \end{bmatrix}$ and $\mathbf{v}_2 = \begin{bmatrix} 1 \\ 1 \end{bmatrix}$. Next, form

$$P = \begin{bmatrix} 3 & 1 \\ 2 & 1 \end{bmatrix}, \quad D = \begin{bmatrix} 2 & 0 \\ 0 & 1 \end{bmatrix}, \quad \text{and} \quad P^{-1} = \begin{bmatrix} 1 & -1 \\ -2 & 3 \end{bmatrix}$$

Since $A = PDP^{-1}$,

$$\begin{aligned}
A^8 = PD^8 P^{-1} &= \begin{bmatrix} 3 & 1 \\ 2 & 1 \end{bmatrix} \begin{bmatrix} 2^8 & 0 \\ 0 & 1^8 \end{bmatrix} \begin{bmatrix} 1 & -1 \\ -2 & 3 \end{bmatrix} \\
&= \begin{bmatrix} 3 & 1 \\ 2 & 1 \end{bmatrix} \begin{bmatrix} 256 & 0 \\ 0 & 1 \end{bmatrix} \begin{bmatrix} 1 & -1 \\ -2 & 3 \end{bmatrix} \\
&= \begin{bmatrix} 766 & -765 \\ 510 & -509 \end{bmatrix}
\end{aligned}$$

2. Compute $A\mathbf{v}_1 = \begin{bmatrix} -3 & 12 \\ -2 & 7 \end{bmatrix} \begin{bmatrix} 3 \\ 1 \end{bmatrix} = \begin{bmatrix} 3 \\ 1 \end{bmatrix} = 1 \cdot \mathbf{v}_1$, and

$$A\mathbf{v}_2 = \begin{bmatrix} -3 & 12 \\ -2 & 7 \end{bmatrix} \begin{bmatrix} 2 \\ 1 \end{bmatrix} = \begin{bmatrix} 6 \\ 3 \end{bmatrix} = 3 \cdot \mathbf{v}_2$$

So, $\mathbf{v}_1$ and $\mathbf{v}_2$ are eigenvectors for the eigenvalues 1 and 3, respectively. Thus

$$A = PDP^{-1}, \quad \text{where} \quad P = \begin{bmatrix} 3 & 2 \\ 1 & 1 \end{bmatrix} \quad \text{and} \quad D = \begin{bmatrix} 1 & 0 \\ 0 & 3 \end{bmatrix}$$

3. Yes, A is diagonalizable. There is a basis $\{\mathbf{v}_1, \mathbf{v}_2\}$ for the eigenspace corresponding to $\lambda = 3$. In addition, there will be at least one eigenvector for $\lambda = 5$ and one for $\lambda = -2$. Call them $\mathbf{v}_3$ and $\mathbf{v}_4$. Then $\{\mathbf{v}_1, \mathbf{v}_2, \mathbf{v}_3, \mathbf{v}_4\}$ is linearly independent by Theorem 2 and Practice Problem 3 in Section 5.1. There can be no additional eigenvectors that are linearly independent from $\mathbf{v}_1, \mathbf{v}_2, \mathbf{v}_3, \mathbf{v}_4$, because the vectors are all in $\mathbb{R}^4$. Hence the eigenspaces for $\lambda = 5$ and $\lambda = -2$ are both one-dimensional. It follows that A is diagonalizable by Theorem 7(b).

5.4 | EIGENVECTORS AND LINEAR TRANSFORMATIONS

The goal of this section is to understand the matrix factorization $A = PDP^{-1}$ as a statement about linear transformations. We shall see that the transformation $\mathbf{x} \mapsto A\mathbf{x}$ is essentially the same as the very simple mapping $\mathbf{u} \mapsto D\mathbf{u}$, when viewed from the proper perspective. A similar interpretation will apply to A and D even when D is not a diagonal matrix.

Recall from Section 1.9 that any linear transformation T from $\mathbb{R}^n$ to $\mathbb{R}^m$ can be implemented via left-multiplication by a matrix A, called the *standard matrix* of T. Now we need the same sort of representation for any linear transformation between two finite-dimensional vector spaces.

The Matrix of a Linear Transformation

Let V be an n-dimensional vector space, let W be an m-dimensional vector space, and let T be any linear transformation from V to W. To associate a matrix with T, choose (ordered) bases $\mathcal{B}$ and $\mathcal{C}$ for V and W, respectively.

Given any $\mathbf{x}$ in V, the coordinate vector $[\,\mathbf{x}\,]_{\mathcal{B}}$ is in $\mathbb{R}^n$ and the coordinate vector of its image, $[\,T(\mathbf{x})\,]_{\mathcal{C}}$, is in $\mathbb{R}^m$, as shown in Fig. 1.

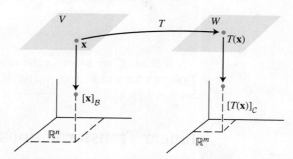

FIGURE 1   A linear transformation from V to W.

The connection between $[\,\mathbf{x}\,]_{\mathcal{B}}$ and $[\,T(\mathbf{x})\,]_{\mathcal{C}}$ is easy to find. Let $\{\mathbf{b}_1, \ldots, \mathbf{b}_n\}$ be the basis $\mathcal{B}$ for V. If $\mathbf{x} = r_1\mathbf{b}_1 + \cdots + r_n\mathbf{b}_n$, then

$$[\mathbf{x}]_{\mathcal{B}} = \begin{bmatrix} r_1 \\ \vdots \\ r_n \end{bmatrix}$$

and

$$T(\mathbf{x}) = T(r_1\mathbf{b}_1 + \cdots + r_n\mathbf{b}_n) = r_1 T(\mathbf{b}_1) + \cdots + r_n T(\mathbf{b}_n) \tag{1}$$

because T is linear. Now, since the coordinate mapping from W to $\mathbb{R}^m$ is linear (Theorem 8 in Section 4.4), equation (1) leads to

$$[\,T(\mathbf{x})\,]_{\mathcal{C}} = r_1[\,T(\mathbf{b}_1)\,]_{\mathcal{C}} + \cdots + r_n[\,T(\mathbf{b}_n)\,]_{\mathcal{C}} \tag{2}$$

Since $\mathcal{C}$-coordinate vectors are in $\mathbb{R}^m$, the vector equation (2) can be written as a matrix equation, namely,

$$[\,T(\mathbf{x})\,]_{\mathcal{C}} = M[\,\mathbf{x}\,]_{\mathcal{B}} \tag{3}$$

where

$$M = \big[\, [\,T(\mathbf{b}_1)\,]_{\mathcal{C}} \quad [\,T(\mathbf{b}_2)\,]_{\mathcal{C}} \quad \cdots \quad [\,T(\mathbf{b}_n)\,]_{\mathcal{C}} \,\big] \tag{4}$$

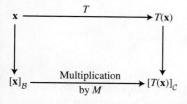

FIGURE 2

The matrix M is a matrix representation of T, called the **matrix for T relative to the bases $\mathcal{B}$ and $\mathcal{C}$.** See Fig. 2.

Equation (3) says that, so far as coordinate vectors are concerned, the action of T on $\mathbf{x}$ may be viewed as left-multiplication by M.

EXAMPLE 1   Suppose $\mathcal{B} = \{\mathbf{b}_1, \mathbf{b}_2\}$ is a basis for V and $\mathcal{C} = \{\mathbf{c}_1, \mathbf{c}_2, \mathbf{c}_3\}$ is a basis for W. Let $T : V \to W$ be a linear transformation with the property that

$$T(\mathbf{b}_1) = 3\mathbf{c}_1 - 2\mathbf{c}_2 + 5\mathbf{c}_3 \quad \text{and} \quad T(\mathbf{b}_2) = 4\mathbf{c}_1 + 7\mathbf{c}_2 - \mathbf{c}_3$$

Find the matrix M for T relative to $\mathcal{B}$ and $\mathcal{C}$.

SOLUTION The $\mathcal{C}$-coordinate vectors of the *images* of $\mathbf{b}_1$ and $\mathbf{b}_2$ are

$$[\, T(\mathbf{b}_1)\,]_\mathcal{C} = \begin{bmatrix} 3 \\ -2 \\ 5 \end{bmatrix} \quad \text{and} \quad [\, T(\mathbf{b}_2)\,]_\mathcal{C} = \begin{bmatrix} 4 \\ 7 \\ -1 \end{bmatrix}$$

Hence

$$M = \begin{bmatrix} 3 & 4 \\ -2 & 7 \\ 5 & -1 \end{bmatrix}$$

∎

If $\mathcal{B}$ and $\mathcal{C}$ are bases for the same space V and if T is the identity transformation $T(\mathbf{x}) = \mathbf{x}$ for $\mathbf{x}$ in V, then matrix M in (4) is just a change-of-coordinates matrix (see Section 4.7).

Linear Transformations from V into V

In the common case where W is the same as V and the basis $\mathcal{C}$ is the same as $\mathcal{B}$, the matrix M in (4) is called the **matrix for T relative to $\mathcal{B}$**, or simply the **$\mathcal{B}$-matrix for T**, and is denoted by $[\,T\,]_\mathcal{B}$. See Fig. 3.

The $\mathcal{B}$-matrix for $T : V \to V$ satisfies

$$[\, T(\mathbf{x})\,]_\mathcal{B} = [\,T\,]_\mathcal{B}[\,\mathbf{x}\,]_\mathcal{B}, \quad \text{for all } \mathbf{x} \text{ in } V \tag{5}$$

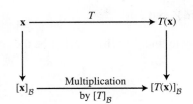

FIGURE 3

EXAMPLE 2 The mapping $T : \mathbb{P}_2 \to \mathbb{P}_2$ defined by

$$T(a_0 + a_1 t + a_2 t^2) = a_1 + 2a_2 t$$

is a linear transformation. (Calculus students will recognize T as the differentiation operator.)

a. Find the $\mathcal{B}$-matrix for T, when $\mathcal{B}$ is the basis $\{1, t, t^2\}$.

b. Verify that $[\, T(\mathbf{p})\,]_\mathcal{B} = [\,T\,]_\mathcal{B}[\,\mathbf{p}\,]_\mathcal{B}$ for each $\mathbf{p}$ in $\mathbb{P}_2$.

SOLUTION

a. Compute the images of the basis vectors:

$$\begin{aligned} T(1) &= 0 & &\text{The zero polynomial} \\ T(t) &= 1 & &\text{The polynomial whose value is always 1} \\ T(t^2) &= 2t \end{aligned}$$

Then write the $\mathcal{B}$-coordinate vectors of $T(1)$, $T(t)$, and $T(t^2)$ (which are found by inspection in this example) and place them together as the $\mathcal{B}$-matrix for T:

$$[\,T(1)\,]_\mathcal{B} = \begin{bmatrix} 0 \\ 0 \\ 0 \end{bmatrix}, \quad [\,T(t)\,]_\mathcal{B} = \begin{bmatrix} 1 \\ 0 \\ 0 \end{bmatrix}, \quad [\,T(t^2)\,]_\mathcal{B} = \begin{bmatrix} 0 \\ 2 \\ 0 \end{bmatrix}$$

$$[\,T\,]_\mathcal{B} = \begin{bmatrix} 0 & 1 & 0 \\ 0 & 0 & 2 \\ 0 & 0 & 0 \end{bmatrix}$$

b. For a general $\mathbf{p}(t) = a_0 + a_1 t + a_2 t^2$,

$$[\,T(\mathbf{p})\,]_{\mathcal{B}} = [\,a_1 + 2a_2 t\,]_{\mathcal{B}} = \begin{bmatrix} a_1 \\ 2a_2 \\ 0 \end{bmatrix}$$

$$= \begin{bmatrix} 0 & 1 & 0 \\ 0 & 0 & 2 \\ 0 & 0 & 0 \end{bmatrix} \begin{bmatrix} a_0 \\ a_1 \\ a_2 \end{bmatrix} = [\,T\,]_{\mathcal{B}}[\,\mathbf{p}\,]_{\mathcal{B}}$$

See Fig. 4.

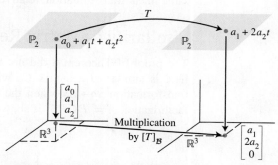

FIGURE 4 Matrix representation of a linear transformation.

Linear Transformations on $\mathbb{R}^n$

In an applied problem involving $\mathbb{R}^n$, a linear transformation T usually appears first as a matrix transformation, $\mathbf{x} \mapsto A\mathbf{x}$. If A is diagonalizable, then there is a basis $\mathcal{B}$ for $\mathbb{R}^n$ consisting of eigenvectors of A. Theorem 8 below shows that, in this case, the $\mathcal{B}$-matrix for T is diagonal. Diagonalizing A amounts to finding a diagonal matrix representation of $\mathbf{x} \mapsto A\mathbf{x}$.

THEOREM 8

Diagonal Matrix Representation

Suppose $A = PDP^{-1}$, where D is a diagonal $n \times n$ matrix. If $\mathcal{B}$ is the basis for $\mathbb{R}^n$ formed from the columns of P, then D is the $\mathcal{B}$-matrix for the transformation $\mathbf{x} \mapsto A\mathbf{x}$.

PROOF Denote the columns of P by $\mathbf{b}_1, \ldots, \mathbf{b}_n$, so that $\mathcal{B} = \{\mathbf{b}_1, \ldots, \mathbf{b}_n\}$ and $P = [\,\mathbf{b}_1 \ \cdots \ \mathbf{b}_n\,]$. In this case, P is the change-of-coordinates matrix $P_{\mathcal{B}}$ discussed in Section 4.4, where

$$P[\,\mathbf{x}\,]_{\mathcal{B}} = \mathbf{x} \quad \text{and} \quad [\,\mathbf{x}\,]_{\mathcal{B}} = P^{-1}\mathbf{x}$$

If $T(\mathbf{x}) = A\mathbf{x}$ for $\mathbf{x}$ in $\mathbb{R}^n$, then

$$\begin{aligned}
[\,T\,]_{\mathcal{B}} &= \left[\, [\,T(\mathbf{b}_1)\,]_{\mathcal{B}} \ \cdots \ [\,T(\mathbf{b}_n)\,]_{\mathcal{B}} \,\right] && \text{Definition of } [\,T\,]_{\mathcal{B}} \\
&= \left[\, [\,A\mathbf{b}_1\,]_{\mathcal{B}} \ \cdots \ [\,A\mathbf{b}_n\,]_{\mathcal{B}} \,\right] && \text{Since } T(\mathbf{x}) = A\mathbf{x} \\
&= [\, P^{-1}A\mathbf{b}_1 \ \cdots \ P^{-1}A\mathbf{b}_n \,] && \text{Change of coordinates} \\
&= P^{-1}A[\, \mathbf{b}_1 \ \cdots \ \mathbf{b}_n \,] && \text{Matrix multiplication} \\
&= P^{-1}AP && \text{(6)}
\end{aligned}$$

Since $A = PDP^{-1}$, we have $[\,T\,]_{\mathcal{B}} = P^{-1}AP = D$.

EXAMPLE 3   Define $T : \mathbb{R}^2 \to \mathbb{R}^2$ by $T(\mathbf{x}) = A\mathbf{x}$, where $A = \begin{bmatrix} 7 & 2 \\ -4 & 1 \end{bmatrix}$. Find a basis $\mathcal{B}$ for $\mathbb{R}^2$ with the property that the $\mathcal{B}$-matrix for T is a diagonal matrix.

SOLUTION   From Example 2 in Section 5.3, we know that $A = PDP^{-1}$, where

$$P = \begin{bmatrix} 1 & 1 \\ -1 & -2 \end{bmatrix} \quad \text{and} \quad D = \begin{bmatrix} 5 & 0 \\ 0 & 3 \end{bmatrix}$$

The columns of P, call them $\mathbf{b}_1$ and $\mathbf{b}_2$, are eigenvectors of A. By Theorem 8, D is the $\mathcal{B}$-matrix for T when $\mathcal{B} = \{\mathbf{b}_1, \mathbf{b}_2\}$. The mappings $\mathbf{x} \mapsto A\mathbf{x}$ and $\mathbf{u} \mapsto D\mathbf{u}$ describe the same linear transformation, relative to different bases. ∎

Similarity of Matrix Representations

The proof of Theorem 8 did not use the information that D was diagonal. Hence, if A is similar to a matrix C, with $A = PCP^{-1}$, then C is the $\mathcal{B}$-matrix for the transformation $\mathbf{x} \mapsto A\mathbf{x}$ when the basis $\mathcal{B}$ is formed from the columns of P. The factorization $A = PCP^{-1}$ is shown in Fig. 5.

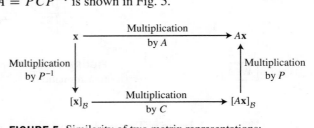

FIGURE 5 Similarity of two matrix representations: $A = PCP^{-1}$.

Conversely, if $T : \mathbb{R}^n \to \mathbb{R}^n$ is defined by $T(\mathbf{x}) = A\mathbf{x}$, and if $\mathcal{B}$ is any basis for $\mathbb{R}^n$, then the $\mathcal{B}$-matrix for T is similar to A. In fact, the calculations in the proof of Theorem 8 show that if P is the matrix whose columns come from the vectors in $\mathcal{B}$, then $[T]_\mathcal{B} = P^{-1}AP$. Thus, the set of all matrices similar to a matrix A coincides with the set of all matrix representations of the transformation $\mathbf{x} \mapsto A\mathbf{x}$.

EXAMPLE 4   Let $A = \begin{bmatrix} 4 & -9 \\ 4 & -8 \end{bmatrix}$, $\mathbf{b}_1 = \begin{bmatrix} 3 \\ 2 \end{bmatrix}$, and $\mathbf{b}_2 = \begin{bmatrix} 2 \\ 1 \end{bmatrix}$. The characteristic polynomial of A is $(\lambda + 2)^2$, but the eigenspace for the eigenvalue -2 is only one-dimensional; so A is not diagonalizable. However, the basis $\mathcal{B} = \{\mathbf{b}_1, \mathbf{b}_2\}$ has the property that the $\mathcal{B}$-matrix for the transformation $\mathbf{x} \mapsto A\mathbf{x}$ is a triangular matrix called the *Jordan form* of A.[1] Find this $\mathcal{B}$-matrix.

SOLUTION   If $P = [\,\mathbf{b}_1 \quad \mathbf{b}_2\,]$, then the $\mathcal{B}$-matrix is $P^{-1}AP$. Compute

$$AP = \begin{bmatrix} 4 & -9 \\ 4 & -8 \end{bmatrix}\begin{bmatrix} 3 & 2 \\ 2 & 1 \end{bmatrix} = \begin{bmatrix} -6 & -1 \\ -4 & 0 \end{bmatrix}$$

$$P^{-1}AP = \begin{bmatrix} -1 & 2 \\ 2 & -3 \end{bmatrix}\begin{bmatrix} -6 & -1 \\ -4 & 0 \end{bmatrix} = \begin{bmatrix} -2 & 1 \\ 0 & -2 \end{bmatrix}$$

Notice that the eigenvalue of A is on the diagonal. ∎

---

[1] Every square matrix A is similar to a matrix in Jordan form. The basis used to produce a Jordan form consists of eigenvectors and so-called "generalized eigenvectors" of A. See Chapter 9 of *Applied Linear Algebra*, 3rd ed. (Englewood Cliffs, NJ: Prentice-Hall, 1988), by B. Noble and J. W. Daniel.

┌─ NUMERICAL NOTE ──┐

An efficient way to compute a $\mathcal{B}$-matrix $P^{-1}AP$ is to compute AP and then to row reduce the augmented matrix $[\,P \quad AP\,]$ to $[\,I \quad P^{-1}AP\,]$. A separate computation of P^{-1} is unnecessary. See Exercise 15 in Section 2.2.

└──┘

PRACTICE PROBLEMS

1. Find $T(a_0 + a_1t + a_2t^2)$, if T is the linear transformation from $\mathbb{P}_2$ to $\mathbb{P}_2$ whose matrix relative to $\mathcal{B} = \{1, t, t^2\}$ is

$$[\,T\,]_{\mathcal{B}} = \begin{bmatrix} 3 & 4 & 0 \\ 0 & 5 & -1 \\ 1 & -2 & 7 \end{bmatrix}$$

2. Let A, B, and C be $n \times n$ matrices. The text has shown that if A is similar to B, then B is similar to A. This property, together with the statements below, shows that "similar to" is an *equivalence relation*. (Row equivalence is another example of an equivalence relation.) Verify parts (a) and (b).

   a. A is similar to A.

   b. If A is similar to B and B is similar to C, then A is similar to C.

5.4 EXERCISES

1. Let $\mathcal{B} = \{\mathbf{b}_1, \mathbf{b}_2, \mathbf{b}_3\}$ and $\mathcal{D} = \{\mathbf{d}_1, \mathbf{d}_2\}$ be bases for vector spaces V and W, respectively. Let $T : V \to W$ be a linear transformation with the property that

$$T(\mathbf{b}_1) = 3\mathbf{d}_1 - 5\mathbf{d}_2, \quad T(\mathbf{b}_2) = -\mathbf{d}_1 + 6\mathbf{d}_2, \quad T(\mathbf{b}_3) = 4\mathbf{d}_2$$

   Find the matrix for T relative to $\mathcal{B}$ and $\mathcal{D}$.

2. Let $\mathcal{D} = \{\mathbf{d}_1, \mathbf{d}_2\}$ and $\mathcal{B} = \{\mathbf{b}_1, \mathbf{b}_2\}$ be bases for vector spaces V and W, respectively. Let $T : V \to W$ be a linear transformation with the property that

$$T(\mathbf{d}_1) = 3\mathbf{b}_1 - 3\mathbf{b}_2, \quad T(\mathbf{d}_2) = -2\mathbf{b}_1 + 5\mathbf{b}_2$$

   Find the matrix for T relative to $\mathcal{D}$ and $\mathcal{B}$.

3. Let $\mathcal{E} = \{\mathbf{e}_1, \mathbf{e}_2, \mathbf{e}_3\}$ be the standard basis for $\mathbb{R}^3$, let $\mathcal{B} = \{\mathbf{b}_1, \mathbf{b}_2, \mathbf{b}_3\}$ be a basis for a vector space V, and let $T : \mathbb{R}^3 \to V$ be a linear transformation with the property that

$$T(x_1, x_2, x_3) = (2x_3 - x_2)\mathbf{b}_1 - (2x_2)\mathbf{b}_2 + (x_1 + 3x_3)\mathbf{b}_3$$

   a. Compute $T(\mathbf{e}_1)$, $T(\mathbf{e}_2)$, and $T(\mathbf{e}_3)$.

   b. Compute $[T(\mathbf{e}_1)]_{\mathcal{B}}$, $[T(\mathbf{e}_2)]_{\mathcal{B}}$, and $[T(\mathbf{e}_3)]_{\mathcal{B}}$.

   c. Find the matrix for T relative to $\mathcal{E}$ and $\mathcal{B}$.

4. Let $\mathcal{B} = \{\mathbf{b}_1, \mathbf{b}_2, \mathbf{b}_3\}$ be a basis for a vector space V and let $T : V \to \mathbb{R}^2$ be a linear transformation with the property that

$$T(x_1\mathbf{b}_1 + x_2\mathbf{b}_2 + x_3\mathbf{b}_3) = \begin{bmatrix} 2x_1 - 3x_2 + x_3 \\ -2x_1 + 5x_3 \end{bmatrix}$$

   Find the matrix for T relative to $\mathcal{B}$ and the standard basis for $\mathbb{R}^2$.

5. Let $T : \mathbb{P}_2 \to \mathbb{P}_3$ be the transformation that maps a polynomial $\mathbf{p}(t)$ into the polynomial $(t + 3)\mathbf{p}(t)$.

   a. Find the image of $\mathbf{p}(t) = 3 - 2t + t^2$.

   b. Show that T is a linear transformation.

   c. Find the matrix for T relative to the bases $\{1, t, t^2\}$ and $\{1, t, t^2, t^3\}$.

6. Let $T : \mathbb{P}_2 \to \mathbb{P}_4$ be the transformation that maps a polynomial $\mathbf{p}(t)$ into the polynomial $\mathbf{p}(t) + 2t^2\mathbf{p}(t)$.

   a. Find the image of $\mathbf{p}(t) = 3 - 2t + t^2$.

   b. Show that T is a linear transformation.

   c. Find the matrix for T relative to the bases $\{1, t, t^2\}$ and $\{1, t, t^2, t^3, t^4\}$.

7. Assume the mapping $T : \mathbb{P}_2 \to \mathbb{P}_2$ defined by

$$T(a_0 + a_1t + a_2t^2) = 3a_0 + (5a_0 - 2a_1)t + (4a_1 + a_2)t^2$$

   is linear. Find the matrix representation of T relative to the basis $\mathcal{B} = \{1, t, t^2\}$.

8. Let $\mathcal{B} = \{\mathbf{b}_1, \mathbf{b}_2, \mathbf{b}_3\}$ be a basis for a vector space V. Find $T(4\mathbf{b}_1 - 3\mathbf{b}_2)$ when T is a linear transformation from V to V whose matrix relative to $\mathcal{B}$ is

$$[\,T\,]_{\mathcal{B}} = \begin{bmatrix} 0 & 0 & 1 \\ 2 & 1 & -2 \\ 1 & 3 & 1 \end{bmatrix}$$

9. Define $T : \mathbb{P}_2 \to \mathbb{R}^3$ by $T(\mathbf{p}) = \begin{bmatrix} \mathbf{p}(-1) \\ \mathbf{p}(0) \\ \mathbf{p}(1) \end{bmatrix}$.

   a. Find the image under T of $\mathbf{p}(t) = 5 + 3t$.

   b. Show that T is a linear transformation.

   c. Find the matrix for T relative to the basis $\{1, t, t^2\}$ for $\mathbb{P}_2$ and the standard basis for $\mathbb{R}^3$.

10. Define $T : \mathbb{P}_3 \to \mathbb{R}^4$ by $T(\mathbf{p}) = \begin{bmatrix} \mathbf{p}(-2) \\ \mathbf{p}(3) \\ \mathbf{p}(1) \\ \mathbf{p}(0) \end{bmatrix}$.

   a. Show that T is a linear transformation.

   b. Find the matrix for T relative to the basis $\{1, t, t^2, t^3\}$ for $\mathbb{P}_3$ and the standard basis for $\mathbb{R}^4$.

In Exercises 11 and 12, find the $\mathcal{B}$-matrix for the transformation $\mathbf{x} \mapsto A\mathbf{x}$, where $\mathcal{B} = \{\mathbf{b}_1, \mathbf{b}_2\}$.

11. $A = \begin{bmatrix} -4 & -1 \\ 6 & 1 \end{bmatrix}$, $\mathbf{b}_1 = \begin{bmatrix} -1 \\ 2 \end{bmatrix}$, $\mathbf{b}_2 = \begin{bmatrix} -1 \\ 1 \end{bmatrix}$

12. $A = \begin{bmatrix} -6 & -2 \\ 4 & 0 \end{bmatrix}$, $\mathbf{b}_1 = \begin{bmatrix} 0 \\ 1 \end{bmatrix}$, $\mathbf{b}_2 = \begin{bmatrix} -1 \\ 2 \end{bmatrix}$

In Exercises 13–16, define $T : \mathbb{R}^2 \to \mathbb{R}^2$ by $T(\mathbf{x}) = A\mathbf{x}$. Find a basis $\mathcal{B}$ for $\mathbb{R}^2$ with the property that $[T]_{\mathcal{B}}$ is diagonal.

13. $A = \begin{bmatrix} 0 & 1 \\ -3 & 4 \end{bmatrix}$ **14.** $A = \begin{bmatrix} 2 & 3 \\ 3 & 2 \end{bmatrix}$

15. $A = \begin{bmatrix} 1 & 2 \\ 3 & -4 \end{bmatrix}$ **16.** $A = \begin{bmatrix} 4 & -2 \\ -1 & 5 \end{bmatrix}$

17. Let $A = \begin{bmatrix} 4 & 1 \\ -1 & 2 \end{bmatrix}$ and $\mathcal{B} = \{\mathbf{b}_1, \mathbf{b}_2\}$, for $\mathbf{b}_1 = \begin{bmatrix} 1 \\ -1 \end{bmatrix}$, $\mathbf{b}_2 = \begin{bmatrix} -1 \\ 2 \end{bmatrix}$. Define $T : \mathbb{R}^2 \to \mathbb{R}^2$ by $T(\mathbf{x}) = A\mathbf{x}$.

   a. Verify that $\mathbf{b}_1$ is an eigenvector of A but that A is not diagonalizable.

   b. Find the $\mathcal{B}$-matrix for T.

18. Define $T : \mathbb{R}^3 \to \mathbb{R}^3$ by $T(\mathbf{x}) = A\mathbf{x}$, where A is a 3×3 matrix with eigenvalues 5, 5, and -2. Does there exist a basis $\mathcal{B}$ for $\mathbb{R}^3$ such that the $\mathcal{B}$-matrix for T is a diagonal matrix? Discuss.

Verify the statements in Exercises 19–24. The matrices are square.

19. If A is invertible and similar to B, then B is invertible and A^{-1} is similar to B^{-1}. [*Hint:* $P^{-1}AP = B$ for some invertible P. Explain why B is invertible. Then find an invertible Q such that $Q^{-1}A^{-1}Q = B^{-1}$.]

20. If A is similar to B, then A^2 is similar to B^2.

21. If B is similar to A and C is similar to A, then B is similar to C.

22. If A is diagonalizable and B is similar to A, then B is also diagonalizable.

23. If $B = P^{-1}AP$ and $\mathbf{x}$ is an eigenvector of A corresponding to an eigenvalue λ, then $P^{-1}\mathbf{x}$ is an eigenvector of B corresponding also to λ.

24. If A and B are similar, then they have the same rank. [*Hint:* Refer to Supplementary Exercises 13 and 14 in Chapter 4.]

25. The *trace* of a square matrix A is the sum of the diagonal entries in A and is denoted by tr A. It can be verified that $\text{tr}(FG) = \text{tr}(GF)$ for any two $n \times n$ matrices F and G. Show that if A and B are similar, then tr $A = $ tr B.

26. It can be shown that the trace of a matrix A equals the sum of the eigenvalues of A. Verify this statement for the case when A is diagonalizable.

27. Let V be $\mathbb{R}^n$ with a basis $\mathcal{B} = \{\mathbf{b}_1, \ldots, \mathbf{b}_n\}$; let W be $\mathbb{R}^n$ with the standard basis, denoted here by $\mathcal{E}$; and consider the identity transformation $I : \mathbb{R}^n \to \mathbb{R}^n$, where $I(\mathbf{x}) = \mathbf{x}$. Find the matrix for I relative to $\mathcal{B}$ and $\mathcal{E}$. What was this matrix called in Section 4.4?

28. Let V be a vector space with a basis $\mathcal{B} = \{\mathbf{b}_1, \ldots, \mathbf{b}_n\}$, let W be the same space V with a basis $\mathcal{C} = \{\mathbf{c}_1, \ldots, \mathbf{c}_n\}$, and let I be the identity transformation $I : V \to W$. Find the matrix for I relative to $\mathcal{B}$ and $\mathcal{C}$. What was this matrix called in Section 4.7?

29. Let V be a vector space with a basis $\mathcal{B} = \{\mathbf{b}_1, \ldots, \mathbf{b}_n\}$. Find the $\mathcal{B}$-matrix for the identity transformation $I : V \to V$.

[M] In Exercises 30 and 31, find the $\mathcal{B}$-matrix for the transformation $\mathbf{x} \mapsto A\mathbf{x}$ where $\mathcal{B} = \{\mathbf{b}_1, \mathbf{b}_2, \mathbf{b}_3\}$.

30. $A = \begin{bmatrix} 6 & -2 & -2 \\ 3 & 1 & -2 \\ 2 & -2 & 2 \end{bmatrix}$,

   $\mathbf{b}_1 = \begin{bmatrix} 1 \\ 1 \\ 1 \end{bmatrix}$, $\mathbf{b}_2 = \begin{bmatrix} 2 \\ 1 \\ 3 \end{bmatrix}$, $\mathbf{b}_3 = \begin{bmatrix} -1 \\ -1 \\ 0 \end{bmatrix}$

31. $A = \begin{bmatrix} -7 & -48 & -16 \\ 1 & 14 & 6 \\ -3 & -45 & -19 \end{bmatrix}$,

   $\mathbf{b}_1 = \begin{bmatrix} -3 \\ 1 \\ -3 \end{bmatrix}$, $\mathbf{b}_2 = \begin{bmatrix} -2 \\ 1 \\ -3 \end{bmatrix}$, $\mathbf{b}_3 = \begin{bmatrix} 3 \\ -1 \\ 0 \end{bmatrix}$

32. [M] Let T be the transformation whose standard matrix is given below. Find a basis for $\mathbb{R}^4$ with the property that $[T]_{\mathcal{B}}$ is diagonal.

$$A = \begin{bmatrix} -6 & 4 & 0 & 9 \\ -3 & 0 & 1 & 6 \\ -1 & -2 & 1 & 0 \\ -4 & 4 & 0 & 7 \end{bmatrix}$$

SOLUTIONS TO PRACTICE PROBLEMS

1. Let $\mathbf{p}(t) = a_0 + a_1 t + a_2 t^2$ and compute

$$[T(\mathbf{p})]_\mathcal{B} = [T]_\mathcal{B}[\mathbf{p}]_\mathcal{B} = \begin{bmatrix} 3 & 4 & 0 \\ 0 & 5 & -1 \\ 1 & -2 & 7 \end{bmatrix} \begin{bmatrix} a_0 \\ a_1 \\ a_2 \end{bmatrix} = \begin{bmatrix} 3a_0 + 4a_1 \\ 5a_1 - a_2 \\ a_0 - 2a_1 + 7a_2 \end{bmatrix}$$

So $T(\mathbf{p}) = (3a_0 + 4a_1) + (5a_1 - a_2)t + (a_0 - 2a_1 + 7a_2)t^2$.

2. a. $A = (I)^{-1}AI$, so A is similar to A.

   b. By hypothesis, there exist invertible matrices P and Q with the property that $B = P^{-1}AP$ and $C = Q^{-1}BQ$. Substitute the formula for B into the formula for C, and use a fact about the inverse of a product:

   $$C = Q^{-1}BQ = Q^{-1}(P^{-1}AP)Q = (PQ)^{-1}A(PQ)$$

   This equation has the proper form to show that A is similar to C.

5.5 | COMPLEX EIGENVALUES

Since the characteristic equation of an $n \times n$ matrix involves a polynomial of degree n, the equation always has exactly n roots, counting multiplicities, *provided that possibly complex roots are included*. This section shows that if the characteristic equation of a real matrix A has some complex roots, then these roots provide critical information about A. The key is to let A act on the space $\mathbb{C}^n$ of n-tuples of complex numbers.[1]

Our interest in $\mathbb{C}^n$ does not arise from a desire to "generalize" the results of the earlier chapters, although that would in fact open up significant new applications of linear algebra.[2] Rather, this study of complex eigenvalues is essential in order to uncover "hidden" information about certain matrices with real entries that arise in a variety of real-life problems. Such problems include many real dynamical systems that involve periodic motion, vibration, or some type of rotation in space.

The matrix eigenvalue–eigenvector theory already developed for $\mathbb{R}^n$ applies equally well to $\mathbb{C}^n$. So a complex scalar λ satisfies $\det(A - \lambda I) = 0$ if and only if there is a nonzero vector $\mathbf{x}$ in $\mathbb{C}^n$ such that $A\mathbf{x} = \lambda\mathbf{x}$. We call λ a (**complex**) **eigenvalue** and $\mathbf{x}$ a (**complex**) **eigenvector** corresponding to λ.

EXAMPLE 1   If $A = \begin{bmatrix} 0 & -1 \\ 1 & 0 \end{bmatrix}$, then the linear transformation $\mathbf{x} \mapsto A\mathbf{x}$ on $\mathbb{R}^2$ rotates the plane counterclockwise through a quarter-turn. The action of A is periodic, since after four quarter-turns, a vector is back where it started. Obviously, no nonzero vector is mapped into a multiple of itself, so A has no eigenvectors in $\mathbb{R}^2$ and hence no real eigenvalues. In fact, the characteristic equation of A is

$$\lambda^2 + 1 = 0$$

---

[1] Refer to Appendix B for a brief discussion of complex numbers. Matrix algebra and concepts about real vector spaces carry over to the case with complex entries and scalars. In particular, $A(c\mathbf{x} + d\mathbf{y}) = cA\mathbf{x} + dA\mathbf{y}$, for A an $m \times n$ matrix with complex entries, $\mathbf{x}, \mathbf{y}$ in $\mathbb{C}^n$, and c, d in $\mathbb{C}$.

[2] A second course in linear algebra often discusses such topics. They are of particular importance in electrical engineering.

The only roots are complex: $\lambda = i$ and $\lambda = -i$. However, if we permit A to act on $\mathbb{C}^2$, then

$$\begin{bmatrix} 0 & -1 \\ 1 & 0 \end{bmatrix} \begin{bmatrix} 1 \\ -i \end{bmatrix} = \begin{bmatrix} i \\ 1 \end{bmatrix} = i \begin{bmatrix} 1 \\ -i \end{bmatrix}$$

$$\begin{bmatrix} 0 & -1 \\ 1 & 0 \end{bmatrix} \begin{bmatrix} 1 \\ i \end{bmatrix} = \begin{bmatrix} -i \\ 1 \end{bmatrix} = -i \begin{bmatrix} 1 \\ i \end{bmatrix}$$

Thus i and $-i$ are eigenvalues, with $\begin{bmatrix} 1 \\ -i \end{bmatrix}$ and $\begin{bmatrix} 1 \\ i \end{bmatrix}$ as corresponding eigenvectors. (A method for *finding* complex eigenvectors is discussed in Example 2.) ∎

The main focus of this section will be on the matrix in the next example.

EXAMPLE 2   Let $A = \begin{bmatrix} .5 & -.6 \\ .75 & 1.1 \end{bmatrix}$. Find the eigenvalues of A, and find a basis for each eigenspace.

SOLUTION The characteristic equation of A is

$$0 = \det \begin{bmatrix} .5 - \lambda & -.6 \\ .75 & 1.1 - \lambda \end{bmatrix} = (.5 - \lambda)(1.1 - \lambda) - (-.6)(.75)$$

$$= \lambda^2 - 1.6\lambda + 1$$

From the quadratic formula, $\lambda = \frac{1}{2}[1.6 \pm \sqrt{(-1.6)^2 - 4}] = .8 \pm .6i$. For the eigenvalue $\lambda = .8 - .6i$, construct

$$A - (.8 - .6i)I = \begin{bmatrix} .5 & -.6 \\ .75 & 1.1 \end{bmatrix} - \begin{bmatrix} .8 - .6i & 0 \\ 0 & .8 - .6i \end{bmatrix}$$

$$= \begin{bmatrix} -.3 + .6i & -.6 \\ .75 & .3 + .6i \end{bmatrix} \tag{1}$$

Row reduction of the usual augmented matrix is quite unpleasant by hand because of the complex arithmetic. However, here is a nice observation that really simplifies matters: Since $.8 - .6i$ is an eigenvalue, the system

$$\begin{aligned} (-.3 + .6i)x_1 - \qquad .6x_2 &= 0 \\ .75x_1 + (.3 + .6i)x_2 &= 0 \end{aligned} \tag{2}$$

has a nontrivial solution (with x_1 and x_2 possibly complex numbers). Therefore, *both equations in (2) determine the same relationship between x_1 and x_2, and either equation can be used to express one variable in terms of the other*.[3]

The second equation in (2) leads to

$$.75x_1 = (-.3 - .6i)x_2$$
$$x_1 = (-.4 - .8i)x_2$$

Choose $x_2 = 5$ to eliminate the decimals, and obtain $x_1 = -2 - 4i$. A basis for the eigenspace corresponding to $\lambda = .8 - .6i$ is

$$\mathbf{v}_1 = \begin{bmatrix} -2 - 4i \\ 5 \end{bmatrix}$$

---

[3] Another way to see this is to realize that the matrix in equation (1) is not invertible, so its rows are linearly dependent (as vectors in $\mathbb{C}^2$), and hence one row is a (complex) multiple of the other.

Analogous calculations for $\lambda = .8 + .6i$ produce the eigenvector

$$\mathbf{v}_2 = \begin{bmatrix} -2 + 4i \\ 5 \end{bmatrix}$$

As a check on the work, compute

$$A\mathbf{v}_2 = \begin{bmatrix} .5 & -.6 \\ .75 & 1.1 \end{bmatrix} \begin{bmatrix} -2 + 4i \\ 5 \end{bmatrix} = \begin{bmatrix} -4 + 2i \\ 4 + 3i \end{bmatrix} = (.8 + .6i)\mathbf{v}_2 \qquad \blacksquare$$

Surprisingly, the matrix A in Example 2 determines a transformation $\mathbf{x} \mapsto A\mathbf{x}$ that is essentially a rotation. This fact becomes evident when appropriate points are plotted.

EXAMPLE 3   One way to see how multiplication by the matrix A in Example 2 affects points is to plot an arbitrary initial point—say, $\mathbf{x}_0 = (2, 0)$—and then to plot successive images of this point under repeated multiplications by A. That is, plot

$$\mathbf{x}_1 = A\mathbf{x}_0 = \begin{bmatrix} .5 & -.6 \\ .75 & 1.1 \end{bmatrix} \begin{bmatrix} 2 \\ 0 \end{bmatrix} = \begin{bmatrix} 1.0 \\ 1.5 \end{bmatrix}$$

$$\mathbf{x}_2 = A\mathbf{x}_1 = \begin{bmatrix} .5 & -.6 \\ .75 & 1.1 \end{bmatrix} \begin{bmatrix} 1.0 \\ 1.5 \end{bmatrix} = \begin{bmatrix} -.4 \\ 2.4 \end{bmatrix}$$

$$\mathbf{x}_3 = A\mathbf{x}_2, \ldots$$

Figure 1 shows $\mathbf{x}_0, \ldots, \mathbf{x}_8$ as larger dots. The smaller dots are the locations of $\mathbf{x}_9, \ldots,$ $\mathbf{x}_{100}$. The sequence lies along an elliptical orbit. $\qquad \blacksquare$

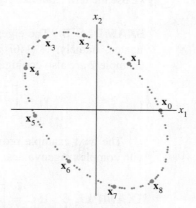

FIGURE 1 Iterates of a point $\mathbf{x}_0$ under the action of a matrix with a complex eigenvalue.

Of course, Fig. 1 does not explain *why* the rotation occurs. The secret to the rotation is hidden in the real and imaginary parts of a complex eigenvector.

Real and Imaginary Parts of Vectors

The complex conjugate of a complex vector $\mathbf{x}$ in $\mathbb{C}^n$ is the vector $\bar{\mathbf{x}}$ in $\mathbb{C}^n$ whose entries are the complex conjugates of the entries in $\mathbf{x}$. The **real** and **imaginary parts** of a complex vector $\mathbf{x}$ are the vectors $\text{Re}\,\mathbf{x}$ and $\text{Im}\,\mathbf{x}$ in $\mathbb{R}^n$ formed from the real and imaginary parts of the entries of $\mathbf{x}$.

EXAMPLE 4 If $\mathbf{x} = \begin{bmatrix} 3 - i \\ i \\ 2 + 5i \end{bmatrix} = \begin{bmatrix} 3 \\ 0 \\ 2 \end{bmatrix} + i \begin{bmatrix} -1 \\ 1 \\ 5 \end{bmatrix}$, then

$$\mathrm{Re}\,\mathbf{x} = \begin{bmatrix} 3 \\ 0 \\ 2 \end{bmatrix}, \quad \mathrm{Im}\,\mathbf{x} = \begin{bmatrix} -1 \\ 1 \\ 5 \end{bmatrix}, \quad \text{and} \quad \overline{\mathbf{x}} = \begin{bmatrix} 3 \\ 0 \\ 2 \end{bmatrix} - i \begin{bmatrix} -1 \\ 1 \\ 5 \end{bmatrix} = \begin{bmatrix} 3 + i \\ -i \\ 2 - 5i \end{bmatrix} \quad \blacksquare$$

If B is an $m \times n$ matrix with possibly complex entries, then $\overline{B}$ denotes the matrix whose entries are the complex conjugates of the entries in B. Properties of conjugates for complex numbers carry over to complex matrix algebra:

$$\overline{r\mathbf{x}} = \overline{r}\,\overline{\mathbf{x}}, \quad \overline{B\mathbf{x}} = \overline{B}\,\overline{\mathbf{x}}, \quad \overline{BC} = \overline{B}\,\overline{C}, \quad \text{and} \quad \overline{rB} = \overline{r}\,\overline{B}$$

Eigenvalues and Eigenvectors of a Real Matrix That Acts on $\mathbb{C}^n$

Let A be an $n \times n$ matrix whose entries are real. Then $\overline{A\mathbf{x}} = \overline{A}\,\overline{\mathbf{x}} = A\overline{\mathbf{x}}$. If λ is an eigenvalue of A and $\mathbf{x}$ is a corresponding eigenvector in $\mathbb{C}^n$, then

$$A\overline{\mathbf{x}} = \overline{A\mathbf{x}} = \overline{\lambda\mathbf{x}} = \overline{\lambda}\,\overline{\mathbf{x}}$$

Hence $\overline{\lambda}$ is also an eigenvalue of A, with $\overline{\mathbf{x}}$ a corresponding eigenvector. This shows that *when A is real, its complex eigenvalues occur in conjugate pairs.* (Here and elsewhere, we use the term *complex eigenvalue* to refer to an eigenvalue $\lambda = a + bi$, with $b \neq 0$.)

EXAMPLE 5 The eigenvalues of the real matrix in Example 2 are complex conjugates, namely, $.8 - .6i$ and $.8 + .6i$. The corresponding eigenvectors found in Example 2 are also conjugates:

$$\mathbf{v}_1 = \begin{bmatrix} -2 - 4i \\ 5 \end{bmatrix} \quad \text{and} \quad \mathbf{v}_2 = \begin{bmatrix} -2 + 4i \\ 5 \end{bmatrix} = \overline{\mathbf{v}}_1 \quad \blacksquare$$

The next example provides the basic "building block" for all real 2×2 matrices with complex eigenvalues.

EXAMPLE 6 If $C = \begin{bmatrix} a & -b \\ b & a \end{bmatrix}$, where a and b are real and not both zero, then the eigenvalues of C are $\lambda = a \pm bi$. (See the Practice Problem at the end of this section.) Also, if $r = |\lambda| = \sqrt{a^2 + b^2}$, then

$$C = r \begin{bmatrix} a/r & -b/r \\ b/r & a/r \end{bmatrix} = \begin{bmatrix} r & 0 \\ 0 & r \end{bmatrix} \begin{bmatrix} \cos\varphi & -\sin\varphi \\ \sin\varphi & \cos\varphi \end{bmatrix}$$

where φ is the angle between the positive x-axis and the ray from $(0, 0)$ through (a, b). See Fig. 2 and Appendix B. The angle φ is called the *argument* of $\lambda = a + bi$. Thus the transformation $\mathbf{x} \mapsto C\mathbf{x}$ may be viewed as the composition of a rotation through the angle φ and a scaling by $|\lambda|$ (see Fig. 3). $\quad \blacksquare$

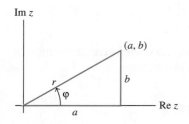

FIGURE 2

Finally, we are ready to uncover the rotation that is hidden within a real matrix having a complex eigenvalue.

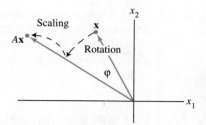

FIGURE 3 A rotation followed by a scaling.

EXAMPLE 7 Let $A = \begin{bmatrix} .5 & -.6 \\ .75 & 1.1 \end{bmatrix}$, $\lambda = .8 - .6i$, and $\mathbf{v}_1 = \begin{bmatrix} -2 - 4i \\ 5 \end{bmatrix}$, as in Example 2. Also, let P be the 2×2 real matrix

$$P = \begin{bmatrix} \operatorname{Re} \mathbf{v}_1 & \operatorname{Im} \mathbf{v}_1 \end{bmatrix} = \begin{bmatrix} -2 & -4 \\ 5 & 0 \end{bmatrix}$$

and let

$$C = P^{-1}AP = \frac{1}{20}\begin{bmatrix} 0 & 4 \\ -5 & -2 \end{bmatrix}\begin{bmatrix} .5 & -.6 \\ .75 & 1.1 \end{bmatrix}\begin{bmatrix} -2 & -4 \\ 5 & 0 \end{bmatrix} = \begin{bmatrix} .8 & -.6 \\ .6 & 8 \end{bmatrix}$$

By Example 6, C is a pure rotation because $|\lambda|^2 = (.8)^2 + (.6)^2 = 1$. From $C = P^{-1}AP$, we obtain

$$A = PCP^{-1} = P\begin{bmatrix} .8 & -.6 \\ .6 & 8 \end{bmatrix}P^{-1}$$

Here is the rotation "inside" A! The matrix P provides a change of variable, say, $\mathbf{x} = P\mathbf{u}$. The action of A amounts to a change of variable from $\mathbf{x}$ to $\mathbf{u}$, followed by a rotation, and then a return to the original variable. See Fig. 4. The rotation produces an ellipse, as in Fig. 1, instead of a circle, because the coordinate system determined by the columns of P is not rectangular and does not have equal unit lengths on the two axes. ■

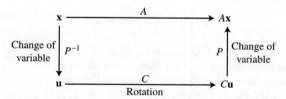

FIGURE 4 Rotation due to a complex eigenvalue.

The next theorem shows that the calculations in Example 7 can be carried out for any 2×2 real matrix A having a complex eigenvalue λ. The proof uses the fact that if the entries in A are real, then $A(\operatorname{Re}\mathbf{x}) = \operatorname{Re} A\mathbf{x}$ and $A(\operatorname{Im}\mathbf{x}) = \operatorname{Im} A\mathbf{x}$, and if $\mathbf{x}$ is an eigenvector for a complex eigenvalue, then $\operatorname{Re}\mathbf{x}$ and $\operatorname{Im}\mathbf{x}$ are linearly independent in $\mathbb{R}^2$. (See Exercises 25 and 26.) The details are omitted.

THEOREM 9

Let A be a real 2×2 matrix with a complex eigenvalue $\lambda = a - bi$ ($b \neq 0$) and an associated eigenvector $\mathbf{v}$ in $\mathbb{C}^2$. Then

$$A = PCP^{-1}, \quad \text{where} \quad P = \begin{bmatrix} \operatorname{Re}\mathbf{v} & \operatorname{Im}\mathbf{v} \end{bmatrix} \quad \text{and} \quad C = \begin{bmatrix} a & -b \\ b & a \end{bmatrix}$$

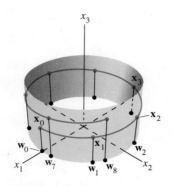

FIGURE 5

Iterates of two points under the action of a 3×3 matrix with a complex eigenvalue.

The phenomenon displayed in Example 7 persists in higher dimensions. For instance, if A is a 3×3 matrix with a complex eigenvalue, then there is a plane in $\mathbb{R}^3$ on which A acts as a rotation (possibly combined with scaling). Every vector in that plane is rotated into another point on the same plane. We say that the plane is **invariant** under A.

EXAMPLE 8 The matrix $A = \begin{bmatrix} .8 & -.6 & 0 \\ .6 & .8 & 0 \\ 0 & 0 & 1.07 \end{bmatrix}$ has eigenvalues $.8 \pm .6i$ and 1.07. Any vector $\mathbf{w}_0$ in the $x_1 x_2$-plane (with third coordinate 0) is rotated by A into another point in the plane. Any vector $\mathbf{x}_0$ not in the plane has its x_3-coordinate multiplied by 1.07. The iterates of the points $\mathbf{w}_0 = (2, 0, 0)$ and $\mathbf{x}_0 = (2, 0, 1)$ under multiplication by A are shown in Fig. 5. ∎

PRACTICE PROBLEM

Show that if a and b are real, then the eigenvalues of $A = \begin{bmatrix} a & -b \\ b & a \end{bmatrix}$ are $a \pm bi$, with corresponding eigenvectors $\begin{bmatrix} 1 \\ -i \end{bmatrix}$ and $\begin{bmatrix} 1 \\ i \end{bmatrix}$.

5.5 EXERCISES

Let each matrix in Exercises 1–6 act on $\mathbb{C}^2$. Find the eigenvalues and a basis for each eigenspace in $\mathbb{C}^2$.

1. $\begin{bmatrix} 1 & -2 \\ 1 & 3 \end{bmatrix}$ **2.** $\begin{bmatrix} 3 & -3 \\ 3 & 3 \end{bmatrix}$

3. $\begin{bmatrix} 5 & 1 \\ -8 & 1 \end{bmatrix}$ **4.** $\begin{bmatrix} 1 & -2 \\ 1 & 3 \end{bmatrix}$

5. $\begin{bmatrix} 3 & 1 \\ -2 & 5 \end{bmatrix}$ **6.** $\begin{bmatrix} 7 & -5 \\ 1 & 3 \end{bmatrix}$

In Exercises 7–12, use Example 6 to list the eigenvalues of A. In each case, the transformation $\mathbf{x} \mapsto A\mathbf{x}$ is the composition of a rotation and a scaling. Give the angle φ of the rotation, where $-\pi < \varphi \le \pi$, and give the scale factor r.

7. $\begin{bmatrix} \sqrt{3} & -1 \\ 1 & \sqrt{3} \end{bmatrix}$ **8.** $\begin{bmatrix} 3 & 3\sqrt{3} \\ -3\sqrt{3} & 3 \end{bmatrix}$

9. $\begin{bmatrix} 0 & 2 \\ -2 & 0 \end{bmatrix}$ **10.** $\begin{bmatrix} 0 & .5 \\ -.5 & 0 \end{bmatrix}$

11. $\begin{bmatrix} -\sqrt{3} & 1 \\ -1 & -\sqrt{3} \end{bmatrix}$ **12.** $\begin{bmatrix} 3 & -\sqrt{3} \\ \sqrt{3} & 3 \end{bmatrix}$

In Exercises 13–20, find an invertible matrix P and a matrix C of the form $\begin{bmatrix} a & -b \\ b & a \end{bmatrix}$ such that the given matrix has the form $A = PCP^{-1}$.

13. $\begin{bmatrix} 1 & -2 \\ 1 & 3 \end{bmatrix}$ **14.** $\begin{bmatrix} 3 & -3 \\ 1 & 1 \end{bmatrix}$

15. $\begin{bmatrix} 0 & 5 \\ -2 & 2 \end{bmatrix}$ **16.** $\begin{bmatrix} 4 & -2 \\ 1 & 6 \end{bmatrix}$

17. $\begin{bmatrix} -11 & -4 \\ 20 & 5 \end{bmatrix}$ **18.** $\begin{bmatrix} 3 & -5 \\ 2 & 5 \end{bmatrix}$

19. $\begin{bmatrix} 1.52 & -.7 \\ .56 & .4 \end{bmatrix}$ **20.** $\begin{bmatrix} -3 & -8 \\ 4 & 5 \end{bmatrix}$

21. In Example 2, solve the first equation in (2) for x_2 in terms of x_1, and from that produce the eigenvector $\mathbf{y} = \begin{bmatrix} 2 \\ -1 + 2i \end{bmatrix}$ for the matrix A. Show that this $\mathbf{y}$ is a (complex) multiple of the vector $\mathbf{v}_1$ used in Example 2.

22. Let A be a complex (or real) $n \times n$ matrix, and let $\mathbf{x}$ in $\mathbb{C}^n$ be an eigenvector corresponding to an eigenvalue λ in $\mathbb{C}$. Show that for each nonzero complex scalar μ, the vector $\mu\mathbf{x}$ is an eigenvector of A.

Chapter 7 will focus on matrices A with the property that $A^T = A$. Exercises 23 and 24 show that every eigenvalue of such a matrix is necessarily real.

23. Let A be an $n \times n$ real matrix with the property that $A^T = A$, let $\mathbf{x}$ be any vector in $\mathbb{C}^n$, and let $q = \overline{\mathbf{x}}^T A \mathbf{x}$. The equalities below show that q is a real number by verifying that $\overline{q} = q$. Give a reason for each step.

$$\overline{q} = \overline{\overline{\mathbf{x}}^T A \mathbf{x}} = \mathbf{x}^T \overline{A \mathbf{x}} = \mathbf{x}^T A \overline{\mathbf{x}} = (\mathbf{x}^T A \overline{\mathbf{x}})^T = \overline{\mathbf{x}}^T A^T \mathbf{x} = q$$
$$\quad\;\; \text{(a)} \qquad \text{(b)} \qquad \text{(c)} \qquad \text{(d)} \qquad \text{(e)}$$

24. Let A be an $n \times n$ real matrix with the property that $A^T = A$. Show that if $A\mathbf{x} = \lambda\mathbf{x}$ for some nonzero vector $\mathbf{x}$ in $\mathbb{C}^n$, then, in fact, λ is real and the real part of $\mathbf{x}$ is an eigenvector of A. [*Hint:* Compute $\overline{\mathbf{x}}^T A\mathbf{x}$, and use Exercise 23. Also, examine the real and imaginary parts of $A\mathbf{x}$.]

25. Let A be a real $n \times n$ matrix, and let $\mathbf{x}$ be a vector in $\mathbb{C}^n$. Show that $\mathrm{Re}(A\mathbf{x}) = A(\mathrm{Re}\,\mathbf{x})$ and $\mathrm{Im}(A\mathbf{x}) = A(\mathrm{Im}\,\mathbf{x})$.

26. Let A be a real 2×2 matrix with a complex eigenvalue $\lambda = a - bi$ ($b \neq 0$) and an associated eigenvector $\mathbf{v}$ in $\mathbb{C}^2$.

  a. Show that $A(\mathrm{Re}\,\mathbf{v}) = a\,\mathrm{Re}\,\mathbf{v} + b\,\mathrm{Im}\,\mathbf{v}$ and $A(\mathrm{Im}\,\mathbf{v}) = -b\,\mathrm{Re}\,\mathbf{v} + a\,\mathrm{Im}\,\mathbf{v}$. [*Hint:* Write $\mathbf{v} = \mathrm{Re}\,\mathbf{v} + i\,\mathrm{Im}\,\mathbf{v}$, and compute $A\mathbf{v}$.]

  b. Verify that if P and C are given as in Theorem 9, then $AP = PC$.

[M] In Exercises 27 and 28, find a factorization of the given matrix A in the form $A = PCP^{-1}$, where C is a block-diagonal matrix with 2×2 blocks of the form shown in Example 6. (For each conjugate pair of eigenvalues, use the real and imaginary parts of one eigenvector in $\mathbb{C}^4$ to create two columns of P.)

27. $A = \begin{bmatrix} 26 & 33 & 23 & 20 \\ -6 & -8 & -1 & -13 \\ -14 & -19 & -16 & 3 \\ -20 & -20 & -20 & -14 \end{bmatrix}$

28. $A = \begin{bmatrix} 7 & 11 & 20 & 17 \\ -20 & -40 & -86 & -74 \\ 0 & -5 & -10 & -10 \\ 10 & 28 & 60 & 53 \end{bmatrix}$

SOLUTION TO PRACTICE PROBLEM

Remember that it is easy to test whether a vector is an eigenvector. There is no need to examine the characteristic equation. Compute

$$A\mathbf{x} = \begin{bmatrix} a & -b \\ b & a \end{bmatrix}\begin{bmatrix} 1 \\ -i \end{bmatrix} = \begin{bmatrix} a + bi \\ b - ai \end{bmatrix} = (a + bi)\begin{bmatrix} 1 \\ -i \end{bmatrix}$$

Thus $\begin{bmatrix} 1 \\ -i \end{bmatrix}$ is an eigenvector corresponding to $\lambda = a + bi$. From the discussion in this section, $\begin{bmatrix} 1 \\ i \end{bmatrix}$ must be an eigenvector corresponding to $\overline{\lambda} = a - bi$.

5.6 DISCRETE DYNAMICAL SYSTEMS

Eigenvalues and eigenvectors provide the key to understanding the long-term behavior, or *evolution*, of a dynamical system described by a difference equation $\mathbf{x}_{k+1} = A\mathbf{x}_k$. Such an equation was used to model population movement in Section 1.10, various Markov chains in Section 4.9, and the spotted owl population in the introductory example for this chapter. The vectors $\mathbf{x}_k$ give information about the system as time (denoted by k) passes. In the spotted owl example, for instance, $\mathbf{x}_k$ listed the numbers of owls in three age classes at time k.

The applications in this section focus on ecological problems because they are easier to state and explain than, say, problems in physics or engineering. However, dynamical systems arise in many scientific fields. For instance, standard undergraduate courses in control systems discuss several aspects of dynamical systems. The modern *state-space* design method in such courses relies heavily on matrix algebra.[1] The *steady-state response* of a control system is the engineering equivalent of what we call here the "long-term behavior" of the dynamical system $\mathbf{x}_{k+1} = A\mathbf{x}_k$.

---

[1] See G. F. Franklin, J. D. Powell, and A. Emami-Naeimi, *Feedback Control of Dynamic Systems*, 5th ed. (Upper Saddle River, NJ: Prentice-Hall, 2006). This undergraduate text has a nice introduction to dynamic models (Chapter 2). State-space design is covered in Chapters 7 and 8.

Until Example 6, we assume that A is diagonalizable, with n linearly independent eigenvectors, $\mathbf{v}_1, \ldots, \mathbf{v}_n$, and corresponding eigenvalues, $\lambda_1, \ldots, \lambda_n$. For convenience, assume the eigenvectors are arranged so that $|\lambda_1| \geq |\lambda_2| \geq \cdots \geq |\lambda_n|$. Since $\{\mathbf{v}_1, \ldots, \mathbf{v}_n\}$ is a basis for $\mathbb{R}^n$, any initial vector $\mathbf{x}_0$ can be written uniquely as

$$\mathbf{x}_0 = c_1\mathbf{v}_1 + \cdots + c_n\mathbf{v}_n \tag{1}$$

This *eigenvector decomposition* of $\mathbf{x}_0$ determines what happens to the sequence $\{\mathbf{x}_k\}$. The next calculation generalizes the simple case examined in Example 5 of Section 5.2. Since the $\mathbf{v}_i$ are eigenvectors,

$$\begin{aligned}
\mathbf{x}_1 = A\mathbf{x}_0 &= c_1 A\mathbf{v}_1 + \cdots + c_n A\mathbf{v}_n \\
&= c_1\lambda_1\mathbf{v}_1 + \cdots + c_n\lambda_n\mathbf{v}_n
\end{aligned}$$

In general,

$$\mathbf{x}_k = c_1(\lambda_1)^k\mathbf{v}_1 + \cdots + c_n(\lambda_n)^k\mathbf{v}_n \qquad (k = 0, 1, 2, \ldots) \tag{2}$$

The examples that follow illustrate what can happen in (2) as $k \to \infty$.

A Predator–Prey System

Deep in the redwood forests of California, dusky-footed wood rats provide up to 80% of the diet for the spotted owl, the main predator of the wood rat. Example 1 uses a linear dynamical system to model the physical system of the owls and the rats. (Admittedly, the model is unrealistic in several respects, but it can provide a starting point for the study of more complicated nonlinear models used by environmental scientists.)

EXAMPLE 1 Denote the owl and wood rat populations at time k by $\mathbf{x}_k = \begin{bmatrix} O_k \\ R_k \end{bmatrix}$, where k is the time in months, O_k is the number of owls in the region studied, and R_k is the number of rats (measured in thousands). Suppose

$$\begin{aligned}
O_{k+1} &= (.5)O_k + (.4)R_k \\
R_{k+1} &= -p \cdot O_k + (1.1)R_k
\end{aligned} \tag{3}$$

where p is a positive parameter to be specified. The $(.5)O_k$ in the first equation says that with no wood rats for food, only half of the owls will survive each month, while the $(1.1)R_k$ in the second equation says that with no owls as predators, the rat population will grow by 10% per month. If rats are plentiful, the $(.4)R_k$ will tend to make the owl population rise, while the negative term $-p \cdot O_k$ measures the deaths of rats due to predation by owls. (In fact, $1000p$ is the average number of rats eaten by one owl in one month.) Determine the evolution of this system when the predation parameter p is .104.

SOLUTION When $p = .104$, the eigenvalues of the coefficient matrix A for the equations in (3) turn out to be $\lambda_1 = 1.02$ and $\lambda_2 = .58$. Corresponding eigenvectors are

$$\mathbf{v}_1 = \begin{bmatrix} 10 \\ 13 \end{bmatrix}, \qquad \mathbf{v}_2 = \begin{bmatrix} 5 \\ 1 \end{bmatrix}$$

An initial $\mathbf{x}_0$ can be written as $\mathbf{x}_0 = c_1\mathbf{v}_1 + c_2\mathbf{v}_2$. Then, for $k \geq 0$,

$$\begin{aligned}
\mathbf{x}_k &= c_1(1.02)^k\mathbf{v}_1 + c_2(.58)^k\mathbf{v}_2 \\
&= c_1(1.02)^k \begin{bmatrix} 10 \\ 13 \end{bmatrix} + c_2(.58)^k \begin{bmatrix} 5 \\ 1 \end{bmatrix}
\end{aligned}$$

As $k \to \infty$, $(.58)^k$ rapidly approaches zero. Assume $c_1 > 0$. Then, for all sufficiently large k, $\mathbf{x}_k$ is approximately the same as $c_1(1.02)^k \mathbf{v}_1$, and we write

$$\mathbf{x}_k \approx c_1(1.02)^k \begin{bmatrix} 10 \\ 13 \end{bmatrix} \tag{4}$$

The approximation in (4) improves as k increases, and so for large k,

$$\mathbf{x}_{k+1} \approx c_1(1.02)^{k+1} \begin{bmatrix} 10 \\ 13 \end{bmatrix} = (1.02)c_1(1.02)^k \begin{bmatrix} 10 \\ 13 \end{bmatrix} \approx 1.02\mathbf{x}_k \tag{5}$$

The approximation in (5) says that eventually both entries of $\mathbf{x}_k$ (the numbers of owls and rats) grow by a factor of almost 1.02 each month, a 2% monthly growth rate. By (4), $\mathbf{x}_k$ is approximately a multiple of (10, 13), so the entries in $\mathbf{x}_k$ are nearly in the same ratio as 10 to 13. That is, for every 10 owls there are about 13 thousand rats. ∎

Example 1 illustrates two general facts about a dynamical system $\mathbf{x}_{k+1} = A\mathbf{x}_k$ in which A is $n \times n$, its eigenvalues satisfy $|\lambda_1| \geq 1$ and $1 > |\lambda_j|$ for $j = 2, \ldots, n$, and $\mathbf{v}_1$ is an eigenvector corresponding to λ_1. If $\mathbf{x}_0$ is given by equation (1), with $c_1 \neq 0$, then for all sufficiently large k,

$$\mathbf{x}_{k+1} \approx \lambda_1 \mathbf{x}_k \tag{6}$$

and

$$\mathbf{x}_k \approx c_1(\lambda_1)^k \mathbf{v}_1 \tag{7}$$

The approximations in (6) and (7) can be made as close as desired by taking k sufficiently large. By (6), the $\mathbf{x}_k$ eventually grow almost by a factor of λ_1 each time, so λ_1 determines the eventual growth rate of the system. Also, by (7), the ratio of any two entries in $\mathbf{x}_k$ (for large k) is nearly the same as the ratio of the corresponding entries in $\mathbf{v}_1$. The case in which $\lambda_1 = 1$ is illustrated in Example 5 in Section 5.2.

Graphical Description of Solutions

When A is 2×2, algebraic calculations can be supplemented by a geometric description of a system's evolution. We can view the equation $\mathbf{x}_{k+1} = A\mathbf{x}_k$ as a description of what happens to an initial point $\mathbf{x}_0$ in $\mathbb{R}^2$ as it is transformed repeatedly by the mapping $\mathbf{x} \mapsto A\mathbf{x}$. The graph of $\mathbf{x}_0, \mathbf{x}_1, \ldots$ is called a **trajectory** of the dynamical system.

EXAMPLE 2 Plot several trajectories of the dynamical system $\mathbf{x}_{k+1} = A\mathbf{x}_k$, when

$$A = \begin{bmatrix} .80 & 0 \\ 0 & .64 \end{bmatrix}$$

SOLUTION The eigenvalues of A are .8 and .64, with eigenvectors $\mathbf{v}_1 = \begin{bmatrix} 1 \\ 0 \end{bmatrix}$ and $\mathbf{v}_2 = \begin{bmatrix} 0 \\ 1 \end{bmatrix}$. If $\mathbf{x}_0 = c_1\mathbf{v}_1 + c_2\mathbf{v}_2$, then

$$\mathbf{x}_k = c_1(.8)^k \begin{bmatrix} 1 \\ 0 \end{bmatrix} + c_2(.64)^k \begin{bmatrix} 0 \\ 1 \end{bmatrix}$$

Of course, $\mathbf{x}_k$ tends to $\mathbf{0}$ because $(.8)^k$ and $(.64)^k$ both approach 0 as $k \to \infty$. But *the way* $\mathbf{x}_k$ *goes toward* $\mathbf{0}$ is interesting. Figure 1 (on page 304) shows the first few terms of several trajectories that begin at points on the boundary of the box with corners at $(\pm 3, \pm 3)$. The points on each trajectory are connected by a thin curve, to make the trajectory easier to see. ∎

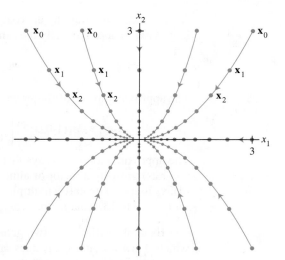

FIGURE 1 The origin as an attractor.

In Example 2, the origin is called an **attractor** of the dynamical system because all trajectories tend toward **0**. This occurs whenever both eigenvalues are less than 1 in magnitude. The direction of greatest attraction is along the line through **0** and the eigenvector $\mathbf{v}_2$ for the eigenvalue of smaller magnitude.

In the next example, both eigenvalues of A are larger than 1 in magnitude, and **0** is called a **repeller** of the dynamical system. All solutions of $\mathbf{x}_{k+1} = A\mathbf{x}_k$ except the (constant) zero solution are unbounded and tend away from the origin.[2]

EXAMPLE 3 Plot several typical solutions of the equation $\mathbf{x}_{k+1} = A\mathbf{x}_k$, where

$$A = \begin{bmatrix} 1.44 & 0 \\ 0 & 1.2 \end{bmatrix}$$

SOLUTION The eigenvalues of A are 1.44 and 1.2. If $\mathbf{x}_0 = \begin{bmatrix} c_1 \\ c_2 \end{bmatrix}$, then

$$\mathbf{x}_k = c_1 (1.44)^k \begin{bmatrix} 1 \\ 0 \end{bmatrix} + c_2 (1.2)^k \begin{bmatrix} 0 \\ 1 \end{bmatrix}$$

Both terms grow in size, but the first term grows faster. So the direction of greatest repulsion is the line through **0** and the eigenvector for the eigenvalue of larger magnitude. Figure 2 shows several trajectories that begin at points quite close to **0**. ∎

In the next example, **0** is called a **saddle point** because the origin attracts solutions from some directions and repels them in other directions. This occurs whenever one eigenvalue is greater than 1 in magnitude and the other is less than 1 in magnitude. The direction of greatest attraction is determined by an eigenvector for the eigenvalue of smaller magnitude. The direction of greatest repulsion is determined by an eigenvector for the eigenvalue of greater magnitude.

[2] The origin is the only possible attractor or repeller in a *linear* dynamical system, but there can be multiple attractors and repellers in a more general dynamical system for which the mapping $\mathbf{x}_k \mapsto \mathbf{x}_{k+1}$ is not linear. In such a system, attractors and repellers are defined in terms of the eigenvalues of a special matrix (with variable entries) called the *Jacobian matrix* of the system.

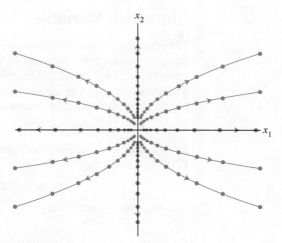

FIGURE 2 The origin as a repeller.

EXAMPLE 4 Plot several typical solutions of the equation $\mathbf{y}_{k+1} = D\mathbf{y}_k$, where

$$D = \begin{bmatrix} 2.0 & 0 \\ 0 & 0.5 \end{bmatrix}$$

(We write D and $\mathbf{y}$ here instead of A and $\mathbf{x}$ because this example will be used later.) Show that a solution $\{\mathbf{y}_k\}$ is unbounded if its initial point is not on the x_2-axis.

SOLUTION The eigenvalues of D are 2 and .5. If $\mathbf{y}_0 = \begin{bmatrix} c_1 \\ c_2 \end{bmatrix}$, then

$$\mathbf{y}_k = c_1 2^k \begin{bmatrix} 1 \\ 0 \end{bmatrix} + c_2 (.5)^k \begin{bmatrix} 0 \\ 1 \end{bmatrix} \tag{8}$$

If $\mathbf{y}_0$ is on the x_2-axis, then $c_1 = 0$ and $\mathbf{y}_k \to \mathbf{0}$ as $k \to \infty$. But if $\mathbf{y}_0$ is not on the x_2-axis, then the first term in the sum for $\mathbf{y}_k$ becomes arbitrarily large, and so $\{\mathbf{y}_k\}$ is unbounded. Figure 3 shows ten trajectories that begin near or on the x_2-axis. ∎

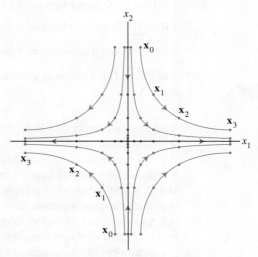

FIGURE 3 The origin as a saddle point.

Change of Variable

The preceding three examples involved diagonal matrices. To handle the nondiagonal case, we return for a moment to the $n \times n$ case in which eigenvectors of A form a basis $\{\mathbf{v}_1, \ldots, \mathbf{v}_n\}$ for $\mathbb{R}^n$. Let $P = [\,\mathbf{v}_1 \;\; \cdots \;\; \mathbf{v}_n\,]$, and let D be the diagonal matrix with the corresponding eigenvalues on the diagonal. Given a sequence $\{\mathbf{x}_k\}$ satisfying $\mathbf{x}_{k+1} = A\mathbf{x}_k$, define a new sequence $\{\mathbf{y}_k\}$ by

$$\mathbf{y}_k = P^{-1}\mathbf{x}_k, \quad \text{or equivalently,} \quad \mathbf{x}_k = P\mathbf{y}_k$$

Substituting these relations into the equation $\mathbf{x}_{k+1} = A\mathbf{x}_k$ and using the fact that $A = PDP^{-1}$, we find that

$$P\mathbf{y}_{k+1} = AP\mathbf{y}_k = (PDP^{-1})P\mathbf{y}_k = PD\mathbf{y}_k$$

Left-multiplying both sides by P^{-1}, we obtain

$$\mathbf{y}_{k+1} = D\mathbf{y}_k$$

If we write $\mathbf{y}_k$ as $\mathbf{y}(k)$ and denote the entries in $\mathbf{y}(k)$ by $y_1(k), \ldots, y_n(k)$, then

$$
\begin{bmatrix} y_1(k+1) \\ y_2(k+1) \\ \vdots \\ y_n(k+1) \end{bmatrix}
=
\begin{bmatrix} \lambda_1 & 0 & \cdots & 0 \\ 0 & \lambda_2 & & \vdots \\ \vdots & & \ddots & 0 \\ 0 & \cdots & 0 & \lambda_n \end{bmatrix}
\begin{bmatrix} y_1(k) \\ y_2(k) \\ \vdots \\ y_n(k) \end{bmatrix}
$$

The change of variable from $\mathbf{x}_k$ to $\mathbf{y}_k$ has *decoupled* the system of difference equations. The evolution of $y_1(k)$, for example, is unaffected by what happens to $y_2(k), \ldots, y_n(k)$, because $y_1(k+1) = \lambda_1 \cdot y_1(k)$ for each k.

The equation $\mathbf{x}_k = P\mathbf{y}_k$ says that $\mathbf{y}_k$ is the coordinate vector of $\mathbf{x}_k$ with respect to the eigenvector basis $\{\mathbf{v}_1, \ldots, \mathbf{v}_n\}$. We can decouple the system $\mathbf{x}_{k+1} = A\mathbf{x}_k$ by making calculations in the new eigenvector coordinate system. When $n = 2$, this amounts to using graph paper with axes in the directions of the two eigenvectors.

EXAMPLE 5   Show that the origin is a saddle point for solutions of $\mathbf{x}_{k+1} = A\mathbf{x}_k$, where

$$A = \begin{bmatrix} 1.25 & -.75 \\ -.75 & 1.25 \end{bmatrix}$$

Find the directions of greatest attraction and greatest repulsion.

SOLUTION   Using standard techniques, we find that A has eigenvalues 2 and .5, with corresponding eigenvectors $\mathbf{v}_1 = \begin{bmatrix} 1 \\ -1 \end{bmatrix}$ and $\mathbf{v}_2 = \begin{bmatrix} 1 \\ 1 \end{bmatrix}$, respectively. Since $|2| > 1$ and $|.5| < 1$, the origin is a saddle point of the dynamical system. If $\mathbf{x}_0 = c_1\mathbf{v}_1 + c_2\mathbf{v}_2$, then

$$\mathbf{x}_k = c_1 2^k \mathbf{v}_1 + c_2(.5)^k \mathbf{v}_2 \tag{9}$$

This equation looks just like equation (8) in Example 4, with $\mathbf{v}_1$ and $\mathbf{v}_2$ in place of the standard basis.

On graph paper, draw axes through $\mathbf{0}$ and the eigenvectors $\mathbf{v}_1$ and $\mathbf{v}_2$. See Fig. 4. Movement along these axes corresponds to movement along the standard axes in Fig. 3. In Fig. 4, the direction of greatest *repulsion* is the line through $\mathbf{0}$ and the eigenvector $\mathbf{v}_1$ whose eigenvalue is greater than 1 in magnitude. If $\mathbf{x}_0$ is on this line, the c_2 in (9) is zero and $\mathbf{x}_k$ moves quickly away from $\mathbf{0}$. The direction of greatest *attraction* is determined by the eigenvector $\mathbf{v}_2$ whose eigenvalue is less than 1 in magnitude.

A number of trajectories are shown in Fig. 4. When this graph is viewed in terms of the eigenvector axes, the picture "looks" essentially the same as the picture in Fig. 3. ∎

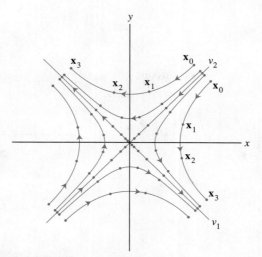

FIGURE 4 The origin as a saddle point.

Complex Eigenvalues

When a real 2×2 matrix A has complex eigenvalues, A is not diagonalizable (when acting on $\mathbb{R}^2$), but the dynamical system $\mathbf{x}_{k+1} = A\mathbf{x}_k$ is easy to describe. Example 3 of Section 5.5 illustrated the case in which the eigenvalues have absolute value 1. The iterates of a point $\mathbf{x}_0$ spiraled around the origin along an elliptical trajectory.

If A has two complex eigenvalues whose absolute value is greater than 1, then $\mathbf{0}$ is a repeller and iterates of $\mathbf{x}_0$ will spiral outward around the origin. If the absolute values of the complex eigenvalues are less than 1, then the origin is an attractor and the iterates of $\mathbf{x}_0$ spiral inward toward the origin, as in the following example.

EXAMPLE 6 It can be verified that the matrix

$$A = \begin{bmatrix} .8 & .5 \\ -.1 & 1.0 \end{bmatrix}$$

has eigenvalues $.9 \pm .2i$, with eigenvectors $\begin{bmatrix} 1 \mp 2i \\ 1 \end{bmatrix}$. Figure 5 (on page 308) shows three trajectories of the system $\mathbf{x}_{k+1} = A\mathbf{x}_k$, with initial vectors $\begin{bmatrix} 0 \\ 2.5 \end{bmatrix}$, $\begin{bmatrix} 3 \\ 0 \end{bmatrix}$, and $\begin{bmatrix} 0 \\ -2.5 \end{bmatrix}$. ∎

Survival of the Spotted Owls

Recall from this chapter's introductory example that the spotted owl population in the Willow Creek area of California was modeled by a dynamical system $\mathbf{x}_{k+1} = A\mathbf{x}_k$ in which the entries in $\mathbf{x}_k = (j_k, s_k, a_k)$ listed the numbers of females (at time k) in the juvenile, subadult, and adult life stages, respectively, and A is the stage-matrix

$$A = \begin{bmatrix} 0 & 0 & .33 \\ .18 & 0 & 0 \\ 0 & .71 & .94 \end{bmatrix} \tag{10}$$

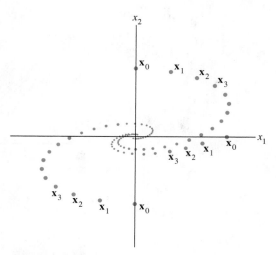

FIGURE 5 Rotation associated with complex eigenvalues.

MATLAB shows that the eigenvalues of A are approximately $\lambda_1 = .98$, $\lambda_2 = -.02 + .21i$, and $\lambda_3 = -.02 - .21i$. Observe that all three eigenvalues are less than 1 in magnitude, because $|\lambda_2|^2 = |\lambda_3|^2 = (-.02)^2 + (.21)^2 = .0445$.

For the moment, let A act on the complex vector space $\mathbb{C}^3$. Then, because A has three distinct eigenvalues, the three corresponding eigenvectors are linearly independent and form a basis for $\mathbb{C}^3$. Denote the eigenvectors by $\mathbf{v}_1$, $\mathbf{v}_2$, and $\mathbf{v}_3$. Then the general solution of $\mathbf{x}_{k+1} = A\mathbf{x}_k$ (using vectors in $\mathbb{C}^3$) has the form

$$\mathbf{x}_k = c_1(\lambda_1)^k \mathbf{v}_1 + c_2(\lambda_2)^k \mathbf{v}_2 + c_3(\lambda_3)^k \mathbf{v}_3 \tag{11}$$

If $\mathbf{x}_0$ is a real initial vector, then $\mathbf{x}_1 = A\mathbf{x}_0$ is real because A is real. Similarly, the equation $\mathbf{x}_{k+1} = A\mathbf{x}_k$ shows that each $\mathbf{x}_k$ on the left side of (11) is real, even though it is expressed as a sum of complex vectors. However, each term on the right side of (11) is approaching the zero vector, because the eigenvalues are all less than 1 in magnitude. Therefore the real sequence $\mathbf{x}_k$ approaches the zero vector, too. Sadly, this model predicts that the spotted owls will eventually all perish.

Is there hope for the spotted owl? Recall from the introductory example that the 18% entry in the matrix A in (10) comes from the fact that although 60% of the juvenile owls live long enough to leave the nest and search for new home territories, only 30% of that group survive the search and find new home ranges. Search survival is strongly influenced by the number of clear-cut areas in the forest, which make the search more difficult and dangerous.

Some owl populations live in areas with few or no clear-cut areas. It may be that a larger percentage of the juvenile owls there survive and find new home ranges. Of course, the problem of the spotted owl is more complex than we have described, but the final example provides a happy ending to the story.

EXAMPLE 7 Suppose the search survival rate of the juvenile owls is 50%, so the $(2, 1)$-entry in the stage-matrix A in (10) is .3 instead of .18. What does the stage-matrix model predict about this spotted owl population?

SOLUTION Now the eigenvalues of A turn out to be approximately $\lambda_1 = 1.01$, $\lambda_2 = -.03 + .26i$, and $\lambda_3 = -.03 - .26i$. An eigenvector for λ_1 is approximately $\mathbf{v}_1 = (10, 3, 31)$. Let $\mathbf{v}_2$ and $\mathbf{v}_3$ be (complex) eigenvectors for λ_2 and λ_3. In this case, equation

(11) becomes

$$\mathbf{x}_k = c_1(1.01)^k \mathbf{v}_1 + c_2(-.03 + .26i)^k \mathbf{v}_2 + c_3(-.03 - .26i)^k \mathbf{v}_3$$

As $k \to \infty$, the second two vectors tend to zero. So $\mathbf{x}_k$ becomes more and more like the (real) vector $c_1(1.01)^k \mathbf{v}_1$. The approximations in equations (6) and (7), following Example 1, apply here. Also, it can be shown that the constant c_1 in the initial decomposition of $\mathbf{x}_0$ is positive when the entries in $\mathbf{x}_0$ are nonnegative. Thus the owl population will grow slowly, with a long-term growth rate of 1.01. The eigenvector $\mathbf{v}_1$ describes the eventual distribution of the owls by life stages: for every 31 adults, there will be about 10 juveniles and 3 subadults. ∎

Further Reading

Franklin, G. F., J. D. Powell, and M. L. Workman. *Digital Control of Dynamic Systems*, 3rd ed. Reading, MA: Addison-Wesley, 1998.

Sandefur, James T. *Discrete Dynamical Systems—Theory and Applications*. Oxford: Oxford University Press, 1990.

Tuchinsky, Philip. *Management of a Buffalo Herd*, UMAP Module 207. Lexington, MA: COMAP, 1980.

PRACTICE PROBLEMS

1. The matrix A below has eigenvalues 1, $\frac{2}{3}$, and $\frac{1}{3}$, with corresponding eigenvectors $\mathbf{v}_1$, $\mathbf{v}_2$, and $\mathbf{v}_3$:

$$A = \frac{1}{9} \begin{bmatrix} 7 & -2 & 0 \\ -2 & 6 & 2 \\ 0 & 2 & 5 \end{bmatrix}, \quad \mathbf{v}_1 = \begin{bmatrix} -2 \\ 2 \\ 1 \end{bmatrix}, \quad \mathbf{v}_2 = \begin{bmatrix} 2 \\ 1 \\ 2 \end{bmatrix}, \quad \mathbf{v}_3 = \begin{bmatrix} 1 \\ 2 \\ -2 \end{bmatrix}$$

Find the general solution of the equation $\mathbf{x}_{k+1} = A\mathbf{x}_k$ if $\mathbf{x}_0 = \begin{bmatrix} 1 \\ 11 \\ -2 \end{bmatrix}$.

2. What happens to the sequence $\{\mathbf{x}_k\}$ in Practice Problem 1 as $k \to \infty$?

5.6 EXERCISES

1. Let A be a 2×2 matrix with eigenvalues 3 and $1/3$ and corresponding eigenvectors $\mathbf{v}_1 = \begin{bmatrix} 1 \\ 1 \end{bmatrix}$ and $\mathbf{v}_2 = \begin{bmatrix} -1 \\ 1 \end{bmatrix}$. Let $\{\mathbf{x}_k\}$ be a solution of the difference equation $\mathbf{x}_{k+1} = A\mathbf{x}_k$, $\mathbf{x}_0 = \begin{bmatrix} 9 \\ 1 \end{bmatrix}$.

   a. Compute $\mathbf{x}_1 = A\mathbf{x}_0$. [*Hint:* You do not need to know A itself.]

   b. Find a formula for $\mathbf{x}_k$ involving k and the eigenvectors $\mathbf{v}_1$ and $\mathbf{v}_2$.

2. Suppose the eigenvalues of a 3×3 matrix A are 3, $4/5$, and $3/5$, with corresponding eigenvectors $\begin{bmatrix} 1 \\ 0 \\ -3 \end{bmatrix}$, $\begin{bmatrix} 2 \\ 1 \\ -5 \end{bmatrix}$, and $\begin{bmatrix} -3 \\ -3 \\ 7 \end{bmatrix}$. Let $\mathbf{x}_0 = \begin{bmatrix} -2 \\ -5 \\ 3 \end{bmatrix}$. Find the solution of the equation $\mathbf{x}_{k+1} = A\mathbf{x}_k$ for the specified $\mathbf{x}_0$, and describe what happens as $k \to \infty$.

In Exercises 3–6, assume that any initial vector x_0 has an eigenvector decomposition such that the coefficient c_1 in equation (1) of this section is positive.[3]

3. Determine the evolution of the dynamical system in Example 1 when the predation parameter p is .2 in equation (3). (Give a formula for x_k.) Does the owl population grow or decline? What about the wood rat population?

4. Determine the evolution of the dynamical system in Example 1 when the predation parameter p is .125. (Give a formula for x_k.) As time passes, what happens to the sizes of the owl and wood rat populations? The system tends toward what is sometimes called an unstable equilibrium. What do you think might happen to the system if some aspect of the model (such as birth rates or the predation rate) were to change slightly?

5. In old-growth forests of Douglas fir, the spotted owl dines mainly on flying squirrels. Suppose the predator–prey matrix for these two populations is $A = \begin{bmatrix} .4 & .3 \\ -p & 1.2 \end{bmatrix}$. Show that if the predation parameter p is .325, both populations grow. Estimate the long-term growth rate and the eventual ratio of owls to flying squirrels.

6. Show that if the predation parameter p in Exercise 5 is .5, both the owls and the squirrels will eventually perish. Find a value of p for which populations of both owls and squirrels tend toward constant levels. What are the relative population sizes in this case?

7. Let A have the properties described in Exercise 1.
   a. Is the origin an attractor, a repeller, or a saddle point of the dynamical system $x_{k+1} = Ax_k$?
   b. Find the directions of greatest attraction and/or repulsion for this dynamical system.
   c. Make a graphical description of the system, showing the directions of greatest attraction or repulsion. Include a rough sketch of several typical trajectories (without computing specific points).

8. Determine the nature of the origin (attractor, repeller, or saddle point) for the dynamical system $x_{k+1} = Ax_k$ if A has the properties described in Exercise 2. Find the directions of greatest attraction or repulsion.

In Exercises 9–14, classify the origin as an attractor, repeller, or saddle point of the dynamical system $x_{k+1} = Ax_k$. Find the directions of greatest attraction and/or repulsion.

9. $A = \begin{bmatrix} 1.7 & -.3 \\ -1.2 & .8 \end{bmatrix}$

10. $A = \begin{bmatrix} .3 & .4 \\ -.3 & 1.1 \end{bmatrix}$

11. $A = \begin{bmatrix} .4 & .5 \\ -.4 & 1.3 \end{bmatrix}$

12. $A = \begin{bmatrix} .5 & .6 \\ -.3 & 1.4 \end{bmatrix}$

13. $A = \begin{bmatrix} .8 & .3 \\ -.4 & 1.5 \end{bmatrix}$

14. $A = \begin{bmatrix} 1.7 & .6 \\ -.4 & .7 \end{bmatrix}$

15. Let $A = \begin{bmatrix} .4 & 0 & .2 \\ .3 & .8 & .3 \\ .3 & .2 & .5 \end{bmatrix}$. The vector $v_1 = \begin{bmatrix} .1 \\ .6 \\ .3 \end{bmatrix}$ is an eigenvector for A, and two eigenvalues are .5 and .2. Construct the solution of the dynamical system $x_{k+1} = Ax_k$ that satisfies $x_0 = (0, .3, .7)$. What happens to x_k as $k \to \infty$?

16. [M] Produce the general solution of the dynamical system $x_{k+1} = Ax_k$ when A is the stochastic matrix for the Hertz Rent A Car model in Exercise 16 of Section 4.9.

17. Construct a stage-matrix model for an animal species that has two life stages: juvenile (up to 1 year old) and adult. Suppose the female adults give birth each year to an average of 1.6 female juveniles. Each year, 30% of the juveniles survive to become adults and 80% of the adults survive. For $k \geq 0$, let $x_k = (j_k, a_k)$, where the entries in x_k are the numbers of female juveniles and female adults in year k.
   a. Construct the stage-matrix A such that $x_{k+1} = Ax_k$ for $k \geq 0$.
   b. Show that the population is growing, compute the eventual growth rate of the population, and give the eventual ratio of juveniles to adults.
   c. [M] Suppose that initially there are 15 juveniles and 10 adults in the population. Produce four graphs that show how the population changes over eight years: (a) the number of juveniles, (b) the number of adults, (c) the total population, and (d) the ratio of juveniles to adults (each year). When does the ratio in (d) seem to stabilize? Include a listing of the program or keystrokes used to produce the graphs for (c) and (d).

18. A herd of American buffalo (bison) can be modeled by a stage matrix similar to that for the spotted owls. The females can be divided into calves (up to 1 year old), yearlings (1 to 2 years), and adults. Suppose an average of 42 female calves are born each year per 100 adult females. (Only adults produce offspring.) Each year, about 60% of the calves survive, 75% of the yearlings survive, and 95% of the adults survive. For $k \geq 0$, let $x_k = (c_k, y_k, a_k)$, where the entries in x_k are the numbers of females in each life stage at year k.
   a. Construct the stage-matrix A for the buffalo herd, such that $x_{k+1} = Ax_k$ for $k \geq 0$.
   b. [M] Show that the buffalo herd is growing, determine the expected growth rate after many years, and give the expected numbers of calves and yearlings present per 100 adults.

---

[3] One of the limitations of the model in Example 1 is that there always exist initial population vectors x_0 with positive entries such that the coefficient c_1 is negative. The approximation (7) is still valid, but the entries in x_k eventually become negative.

SOLUTIONS TO PRACTICE PROBLEMS

1. The first step is to write $\mathbf{x}_0$ as a linear combination of $\mathbf{v}_1$, $\mathbf{v}_2$, and $\mathbf{v}_3$. Row reduction of $[\,\mathbf{v}_1 \quad \mathbf{v}_2 \quad \mathbf{v}_3 \quad \mathbf{x}_0\,]$ produces the weights $c_1 = 2$, $c_2 = 1$, and $c_3 = 3$, so that

$$\mathbf{x}_0 = 2\mathbf{v}_1 + 1\mathbf{v}_2 + 3\mathbf{v}_3$$

Since the eigenvalues are 1, $\frac{2}{3}$, and $\frac{1}{3}$, the general solution is

$$\mathbf{x}_k = 2 \cdot 1^k \mathbf{v}_1 + 1 \cdot \left(\frac{2}{3}\right)^k \mathbf{v}_2 + 3 \cdot \left(\frac{1}{3}\right)^k \mathbf{v}_3$$

$$= 2\begin{bmatrix} -2 \\ 2 \\ 1 \end{bmatrix} + \left(\frac{2}{3}\right)^k \begin{bmatrix} 2 \\ 1 \\ 2 \end{bmatrix} + 3 \cdot \left(\frac{1}{3}\right)^k \begin{bmatrix} 1 \\ 2 \\ -2 \end{bmatrix} \qquad (12)$$

2. As $k \to \infty$, the second and third terms in (12) tend to the zero vector, and

$$\mathbf{x}_k = 2\mathbf{v}_1 + \left(\frac{2}{3}\right)^k \mathbf{v}_2 + 3\left(\frac{1}{3}\right)^k \mathbf{v}_3 \to 2\mathbf{v}_1 = \begin{bmatrix} -4 \\ 4 \\ 2 \end{bmatrix}$$

5.7 APPLICATIONS TO DIFFERENTIAL EQUATIONS

This section describes continuous analogues of the difference equations studied in Section 5.6. In many applied problems, several quantities are varying continuously in time, and they are related by a system of differential equations:

$$x_1' = a_{11}x_1 + \cdots + a_{1n}x_n$$
$$x_2' = a_{21}x_1 + \cdots + a_{2n}x_n$$
$$\vdots$$
$$x_n' = a_{n1}x_1 + \cdots + a_{nn}x_n$$

Here $x_1, \ldots, x_n$ are differentiable functions of t, with derivatives $x_1', \ldots, x_n'$, and the a_{ij} are constants. The crucial feature of this system is that it is *linear*. To see this, write the system as a matrix differential equation

$$\mathbf{x}'(t) = A\mathbf{x}(t) \qquad (1)$$

where

$$\mathbf{x}(t) = \begin{bmatrix} x_1(t) \\ \vdots \\ x_n(t) \end{bmatrix}, \quad \mathbf{x}'(t) = \begin{bmatrix} x_1'(t) \\ \vdots \\ x_n'(t) \end{bmatrix}, \quad \text{and} \quad A = \begin{bmatrix} a_{11} & \cdots & a_{1n} \\ \vdots & & \vdots \\ a_{n1} & \cdots & a_{nn} \end{bmatrix}$$

A **solution** of equation (1) is a vector-valued function that satisfies (1) for all t in some interval of real numbers, such as $t \geq 0$.

Equation (1) is *linear* because both differentiation of functions and multiplication of vectors by a matrix are linear transformations. Thus, if $\mathbf{u}$ and $\mathbf{v}$ are solutions of $\mathbf{x}' = A\mathbf{x}$, then $c\mathbf{u} + d\mathbf{v}$ is also a solution, because

$$(c\mathbf{u} + d\mathbf{v})' = c\mathbf{u}' + d\mathbf{v}'$$
$$= cA\mathbf{u} + dA\mathbf{v} = A(c\mathbf{u} + d\mathbf{v})$$

(Engineers call this property *superposition* of solutions.) Also, the identically zero function is a (trivial) solution of (1). In the terminology of Chapter 4, the set of all solutions of (1) is a *subspace* of the set of all continuous functions with values in $\mathbb{R}^n$.

Standard texts on differential equations show that there always exists what is called a **fundamental set of solutions** to (1). If A is $n \times n$, then there are n linearly independent functions in a fundamental set, and each solution of (1) is a unique linear combination of these n functions. That is, a fundamental set of solutions is a *basis* for the set of all solutions of (1), and the solution set is an n-dimensional vector space of functions. If a vector $\mathbf{x}_0$ is specified, then the **initial value problem** is to construct the (unique) function $\mathbf{x}$ such that $\mathbf{x}' = A\mathbf{x}$ and $\mathbf{x}(0) = \mathbf{x}_0$.

When A is a diagonal matrix, the solutions of (1) can be produced by elementary calculus. For instance, consider

$$\begin{bmatrix} x_1'(t) \\ x_2'(t) \end{bmatrix} = \begin{bmatrix} 3 & 0 \\ 0 & -5 \end{bmatrix} \begin{bmatrix} x_1(t) \\ x_2(t) \end{bmatrix} \tag{2}$$

that is,

$$\begin{aligned} x_1'(t) &= 3x_1(t) \\ x_2'(t) &= -5x_2(t) \end{aligned} \tag{3}$$

The system (2) is said to be *decoupled* because each derivative of a function depends only on the function itself, not on some combination or "coupling" of both $x_1(t)$ and $x_2(t)$. From calculus, the solutions of (3) are $x_1(t) = c_1 e^{3t}$ and $x_2(t) = c_2 e^{-5t}$, for any constants c_1 and c_2. Each solution of equation (2) can be written in the form

$$\begin{bmatrix} x_1(t) \\ x_2(t) \end{bmatrix} = \begin{bmatrix} c_1 e^{3t} \\ c_2 e^{-5t} \end{bmatrix} = c_1 \begin{bmatrix} 1 \\ 0 \end{bmatrix} e^{3t} + c_2 \begin{bmatrix} 0 \\ 1 \end{bmatrix} e^{-5t}$$

This example suggests that for the general equation $\mathbf{x}' = A\mathbf{x}$, a solution might be a linear combination of functions of the form

$$\mathbf{x}(t) = \mathbf{v} e^{\lambda t} \tag{4}$$

for some scalar λ and some fixed nonzero vector $\mathbf{v}$. [If $\mathbf{v} = \mathbf{0}$, the function $\mathbf{x}(t)$ is identically zero and hence satisfies $\mathbf{x}' = A\mathbf{x}$.] Observe that

$$\mathbf{x}'(t) = \lambda \mathbf{v} e^{\lambda t} \qquad \text{By calculus, since } \mathbf{v} \text{ is a constant vector}$$
$$A\mathbf{x}(t) = A\mathbf{v} e^{\lambda t} \qquad \text{Multiplying both sides of (4) by } A$$

Since $e^{\lambda t}$ is never zero, $\mathbf{x}'(t)$ will equal $A\mathbf{x}(t)$ if and only if $\lambda \mathbf{v} = A\mathbf{v}$, that is, if and only if λ is an eigenvalue of A and $\mathbf{v}$ is a corresponding eigenvector. Thus each eigenvalue–eigenvector pair provides a solution (4) of $\mathbf{x}' = A\mathbf{x}$. Such solutions are sometimes called *eigenfunctions* of the differential equation. Eigenfunctions provide the key to solving systems of differential equations.

EXAMPLE 1   The circuit in Fig. 1 can be described by the differential equation

$$\begin{bmatrix} x_1'(t) \\ x_2'(t) \end{bmatrix} = \begin{bmatrix} -(1/R_1 + 1/R_2)/C_1 & 1/(R_2 C_1) \\ 1/(R_2 C_2) & -1/(R_2 C_2) \end{bmatrix} \begin{bmatrix} x_1(t) \\ x_2(t) \end{bmatrix}$$

where $x_1(t)$ and $x_2(t)$ are the voltages across the two capacitors at time t. Suppose resistor R_1 is 1 ohm, R_2 is 2 ohms, capacitor C_1 is 1 farad, and C_2 is .5 farad, and suppose there is an initial charge of 5 volts on capacitor C_1 and 4 volts on capacitor C_2. Find formulas for $x_1(t)$ and $x_2(t)$ that describe how the voltages change over time.

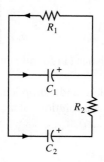

FIGURE 1

SOLUTION Let A denote the matrix displayed above, and let $\mathbf{x}(t) = \begin{bmatrix} x_1(t) \\ x_2(t) \end{bmatrix}$. For the data given, $A = \begin{bmatrix} -1.5 & .5 \\ 1 & -1 \end{bmatrix}$, and $\mathbf{x}(0) = \begin{bmatrix} 5 \\ 4 \end{bmatrix}$. The eigenvalues of A are $\lambda_1 = -.5$ and $\lambda_2 = -2$, with corresponding eigenvectors

$$\mathbf{v}_1 = \begin{bmatrix} 1 \\ 2 \end{bmatrix} \quad \text{and} \quad \mathbf{v}_2 = \begin{bmatrix} -1 \\ 1 \end{bmatrix}$$

The eigenfunctions $\mathbf{x}_1(t) = \mathbf{v}_1 e^{\lambda_1 t}$ and $\mathbf{x}_2(t) = \mathbf{v}_2 e^{\lambda_2 t}$ both satisfy $\mathbf{x}' = A\mathbf{x}$, and so does any linear combination of $\mathbf{x}_1$ and $\mathbf{x}_2$. Set

$$\mathbf{x}(t) = c_1 \mathbf{v}_1 e^{\lambda_1 t} + c_2 \mathbf{v}_2 e^{\lambda_2 t} = c_1 \begin{bmatrix} 1 \\ 2 \end{bmatrix} e^{-.5t} + c_2 \begin{bmatrix} -1 \\ 1 \end{bmatrix} e^{-2t}$$

and note that $\mathbf{x}(0) = c_1 \mathbf{v}_1 + c_2 \mathbf{v}_2$. Since $\mathbf{v}_1$ and $\mathbf{v}_2$ are obviously linearly independent and hence span $\mathbb{R}^2$, c_1 and c_2 can be found to make $\mathbf{x}(0)$ equal to $\mathbf{x}_0$. In fact, the equation

$$c_1 \underset{\mathbf{v}_1}{\underset{\uparrow}{\begin{bmatrix} 1 \\ 2 \end{bmatrix}}} + c_2 \underset{\mathbf{v}_2}{\underset{\uparrow}{\begin{bmatrix} -1 \\ 1 \end{bmatrix}}} = \underset{\mathbf{x}_0}{\underset{\uparrow}{\begin{bmatrix} 5 \\ 4 \end{bmatrix}}}$$

leads easily to $c_1 = 3$ and $c_2 = -2$. Thus the desired solution of the differential equation $\mathbf{x}' = A\mathbf{x}$ is

$$\mathbf{x}(t) = 3 \begin{bmatrix} 1 \\ 2 \end{bmatrix} e^{-.5t} - 2 \begin{bmatrix} -1 \\ 1 \end{bmatrix} e^{-2t}$$

or

$$\begin{bmatrix} x_1(t) \\ x_2(t) \end{bmatrix} = \begin{bmatrix} 3e^{-.5t} + 2e^{-2t} \\ 6e^{-.5t} - 2e^{-2t} \end{bmatrix}$$

Figure 2 shows the graph, or *trajectory*, of $\mathbf{x}(t)$, for $t \geq 0$, along with trajectories for some other initial points. The trajectories of the two eigenfunctions $\mathbf{x}_1$ and $\mathbf{x}_2$ lie in the eigenspaces of A.

The functions $\mathbf{x}_1$ and $\mathbf{x}_2$ both decay to zero as $t \to \infty$, but the values of $\mathbf{x}_2$ decay faster because its exponent is more negative. The entries in the corresponding eigenvector $\mathbf{v}_2$ show that the voltages across the capacitors will decay to zero as rapidly as possible if the initial voltages are equal in magnitude but opposite in sign. ∎

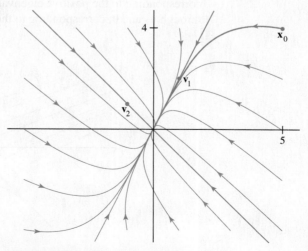

FIGURE 2 The origin as an attractor.

In Fig. 2, the origin is called an **attractor**, or **sink**, of the dynamical system because all trajectories are drawn into the origin. The direction of greatest attraction is along the trajectory of the eigenfunction $\mathbf{x}_2$ (along the line through $\mathbf{0}$ and $\mathbf{v}_2$) corresponding to the more negative eigenvalue, $\lambda = -2$. Trajectories that begin at points not on this line become asymptotic to the line through $\mathbf{0}$ and $\mathbf{v}_1$ because their components in the $\mathbf{v}_2$ direction decay so rapidly.

If the eigenvalues in Example 1 were positive instead of negative, the corresponding trajectories would be similar in shape, but the trajectories would be traversed *away* from the origin. In such a case, the origin is called a **repeller**, or **source**, of the dynamical system, and the direction of greatest repulsion is the line containing the trajectory of the eigenfunction corresponding to the more positive eigenvalue.

EXAMPLE 2 Suppose a particle is moving in a planar force field and its position vector $\mathbf{x}$ satisfies $\mathbf{x}' = A\mathbf{x}$ and $\mathbf{x}(0) = \mathbf{x}_0$, where

$$A = \begin{bmatrix} 4 & -5 \\ -2 & 1 \end{bmatrix}, \qquad \mathbf{x}_0 = \begin{bmatrix} 2.9 \\ 2.6 \end{bmatrix}$$

Solve this initial value problem for $t \geq 0$, and sketch the trajectory of the particle.

SOLUTION The eigenvalues of A turn out to be $\lambda_1 = 6$ and $\lambda_2 = -1$, with corresponding eigenvectors $\mathbf{v}_1 = (-5, 2)$ and $\mathbf{v}_2 = (1, 1)$. For any constants c_1 and c_2, the function

$$\mathbf{x}(t) = c_1 \mathbf{v}_1 e^{\lambda_1 t} + c_2 \mathbf{v}_2 e^{\lambda_2 t} = c_1 \begin{bmatrix} -5 \\ 2 \end{bmatrix} e^{6t} + c_2 \begin{bmatrix} 1 \\ 1 \end{bmatrix} e^{-t}$$

is a solution of $\mathbf{x}' = A\mathbf{x}$. We want c_1 and c_2 to satisfy $\mathbf{x}(0) = \mathbf{x}_0$, that is,

$$c_1 \begin{bmatrix} -5 \\ 2 \end{bmatrix} + c_2 \begin{bmatrix} 1 \\ 1 \end{bmatrix} = \begin{bmatrix} 2.9 \\ 2.6 \end{bmatrix} \quad \text{or} \quad \begin{bmatrix} -5 & 1 \\ 2 & 1 \end{bmatrix} \begin{bmatrix} c_1 \\ c_2 \end{bmatrix} = \begin{bmatrix} 2.9 \\ 2.6 \end{bmatrix}$$

Calculations show that $c_1 = -3/70$ and $c_2 = 188/70$, and so the desired function is

$$\mathbf{x}(t) = \frac{-3}{70} \begin{bmatrix} -5 \\ 2 \end{bmatrix} e^{6t} + \frac{188}{70} \begin{bmatrix} 1 \\ 1 \end{bmatrix} e^{-t}$$

Trajectories of $\mathbf{x}$ and other solutions are shown in Fig. 3. ∎

In Fig. 3, the origin is called a **saddle point** of the dynamical system because some trajectories approach the origin at first and then change direction and move away from the origin. A saddle point arises whenever the matrix A has both positive and negative eigenvalues. The direction of greatest repulsion is the line through $\mathbf{v}_1$ and $\mathbf{0}$, corresponding to the positive eigenvalue. The direction of greatest attraction is the line through $\mathbf{v}_2$ and $\mathbf{0}$, corresponding to the negative eigenvalue.

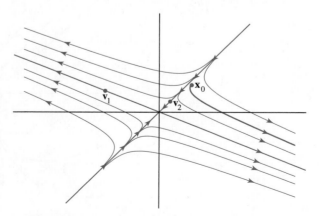

FIGURE 3 The origin as a saddle point.

Decoupling a Dynamical System

The following discussion shows that the method of Examples 1 and 2 produces a fundamental set of solutions for any dynamical system described by $\mathbf{x}' = A\mathbf{x}$ when A is $n \times n$ and has n linearly independent eigenvectors, that is, when A is diagonalizable. Suppose the eigenfunctions for A are

$$\mathbf{v}_1 e^{\lambda_1 t}, \quad \ldots, \mathbf{v}_n e^{\lambda_n t}$$

with $\mathbf{v}_1, \ldots, \mathbf{v}_n$ linearly independent eigenvectors. Let $P = [\,\mathbf{v}_1 \ \cdots \ \mathbf{v}_n\,]$, and let D be the diagonal matrix with entries $\lambda_1, \ldots, \lambda_n$, so that $A = PDP^{-1}$. Now make a *change of variable*, defining a new function $\mathbf{y}$ by

$$\mathbf{y}(t) = P^{-1}\mathbf{x}(t) \quad \text{or, equivalently,} \quad \mathbf{x}(t) = P\mathbf{y}(t)$$

The equation $\mathbf{x}(t) = P\mathbf{y}(t)$ says that $\mathbf{y}(t)$ is the coordinate vector of $\mathbf{x}(t)$ relative to the eigenvector basis. Substitution of $P\mathbf{y}$ for $\mathbf{x}$ in the equation $\mathbf{x}' = A\mathbf{x}$ gives

$$\frac{d}{dt}(P\mathbf{y}) = A(P\mathbf{y}) = (PDP^{-1})P\mathbf{y} = PD\mathbf{y} \tag{5}$$

Since P is a constant matrix, the left side of (5) is $P\mathbf{y}'$. Left-multiply both sides of (5) by P^{-1} and obtain $\mathbf{y}' = D\mathbf{y}$, or

$$\begin{bmatrix} y_1'(t) \\ y_2'(t) \\ \vdots \\ y_n'(t) \end{bmatrix} = \begin{bmatrix} \lambda_1 & 0 & \cdots & 0 \\ 0 & \lambda_2 & & \vdots \\ \vdots & & \ddots & 0 \\ 0 & \cdots & 0 & \lambda_n \end{bmatrix} \begin{bmatrix} y_1(t) \\ y_2(t) \\ \vdots \\ y_n(t) \end{bmatrix}$$

The change of variable from $\mathbf{x}$ to $\mathbf{y}$ has *decoupled* the system of differential equations, because the derivative of each scalar function y_k depends only on y_k. (Review the analogous change of variables in Section 5.6.) Since $y_1' = \lambda_1 y_1$, we have $y_1(t) = c_1 e^{\lambda_1 t}$, with similar formulas for $y_2, \ldots, y_n$. Thus

$$\mathbf{y}(t) = \begin{bmatrix} c_1 e^{\lambda_1 t} \\ \vdots \\ c_n e^{\lambda_n t} \end{bmatrix}, \quad \text{where} \quad \begin{bmatrix} c_1 \\ \vdots \\ c_n \end{bmatrix} = \mathbf{y}(0) = P^{-1}\mathbf{x}(0) = P^{-1}\mathbf{x}_0$$

To obtain the general solution $\mathbf{x}$ of the original system, compute

$$\mathbf{x}(t) = P\mathbf{y}(t) = [\,\mathbf{v}_1 \ \cdots \ \mathbf{v}_n\,]\mathbf{y}(t)$$
$$= c_1\mathbf{v}_1 e^{\lambda_1 t} + \cdots + c_n\mathbf{v}_n e^{\lambda_n t}$$

This is the eigenfunction expansion constructed as in Example 1.

Complex Eigenvalues

In the next example, a real matrix A has a pair of complex eigenvalues λ and $\overline{\lambda}$, with associated complex eigenvectors $\mathbf{v}$ and $\overline{\mathbf{v}}$. (Recall from Section 5.5 that for a real matrix, complex eigenvalues and associated eigenvectors come in conjugate pairs.) So two solutions of $\mathbf{x}' = A\mathbf{x}$ are

$$\mathbf{x}_1(t) = \mathbf{v}e^{\lambda t} \quad \text{and} \quad \mathbf{x}_2(t) = \overline{\mathbf{v}}e^{\overline{\lambda} t} \tag{6}$$

It can be shown that $\mathbf{x}_2(t) = \overline{\mathbf{x}_1(t)}$ by using a power series representation for the complex exponential function. Although the complex eigenfunctions $\mathbf{x}_1$ and $\mathbf{x}_2$ are convenient for some calculations (particularly in electrical engineering), real functions are more appropriate for many purposes. Fortunately, the real and imaginary parts of $\mathbf{x}_1$ are (real) solutions of $\mathbf{x}' = A\mathbf{x}$, because they are linear combinations of the solutions in (6):

$$\text{Re}(\mathbf{v}e^{\lambda t}) = \frac{1}{2}[\,\mathbf{x}_1(t) + \overline{\mathbf{x}_1(t)}\,], \qquad \text{Im}(\mathbf{v}e^{\lambda t}) = \frac{1}{2i}[\,\mathbf{x}_1(t) - \overline{\mathbf{x}_1(t)}\,]$$

To understand the nature of $\text{Re}(\mathbf{v}e^{\lambda t})$, recall from calculus that for any number x, the exponential function e^x can be computed from the power series:

$$e^x = 1 + x + \frac{1}{2!}x^2 + \cdots + \frac{1}{n!}x^n + \cdots$$

This series can be used to define $e^{\lambda t}$ when λ is complex:

$$e^{\lambda t} = 1 + (\lambda t) + \frac{1}{2!}(\lambda t)^2 + \cdots + \frac{1}{n!}(\lambda t)^n + \cdots$$

By writing $\lambda = a + bi$ (with a and b real), and using similar power series for the cosine and sine functions, one can show that

$$e^{(a+bi)t} = e^{at} \cdot e^{ibt} = e^{at}(\cos bt + i \sin bt) \tag{7}$$

Hence

$$\begin{aligned} \mathbf{v}e^{\lambda t} &= (\text{Re}\,\mathbf{v} + i\,\text{Im}\,\mathbf{v}) \cdot e^{at}(\cos bt + i \sin bt) \\ &= [\,(\text{Re}\,\mathbf{v})\cos bt - (\text{Im}\,\mathbf{v})\sin bt\,]e^{at} \\ &\quad + i[\,(\text{Re}\,\mathbf{v})\sin bt + (\text{Im}\,\mathbf{v})\cos bt\,]e^{at} \end{aligned}$$

So two real solutions of $\mathbf{x}' = A\mathbf{x}$ are

$$\begin{aligned} \mathbf{y}_1(t) &= \text{Re}\,\mathbf{x}_1(t) = [\,(\text{Re}\,\mathbf{v})\cos bt - (\text{Im}\,\mathbf{v})\sin bt\,]\,e^{at} \\ \mathbf{y}_2(t) &= \text{Im}\,\mathbf{x}_1(t) = [\,(\text{Re}\,\mathbf{v})\sin bt + (\text{Im}\,\mathbf{v})\cos bt\,]\,e^{at} \end{aligned}$$

It can be shown that $\mathbf{y}_1$ and $\mathbf{y}_2$ are linearly independent functions (when $b \neq 0$).[1]

EXAMPLE 3   The circuit in Fig. 4 can be described by the equation

$$\begin{bmatrix} i_L' \\ v_C' \end{bmatrix} = \begin{bmatrix} -R_2/L & -1/L \\ 1/C & -1/(R_1 C) \end{bmatrix} \begin{bmatrix} i_L \\ v_C \end{bmatrix}$$

where i_L is the current passing through the inductor L and v_C is the voltage drop across the capacitor C. Suppose R_1 is 5 ohms, R_2 is .8 ohm, C is .1 farad, and L is .4 henry. Find formulas for i_L and v_C, if the initial current through the inductor is 3 amperes and the initial voltage across the capacitor is 3 volts.

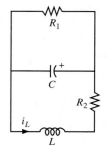

FIGURE 4

SOLUTION   For the data given, $A = \begin{bmatrix} -2 & -2.5 \\ 10 & -2 \end{bmatrix}$ and $\mathbf{x}_0 = \begin{bmatrix} 3 \\ 3 \end{bmatrix}$. The method discussed in Section 5.5 produces the eigenvalue $\lambda = -2 + 5i$ and the corresponding eigenvector $\mathbf{v}_1 = \begin{bmatrix} i \\ 2 \end{bmatrix}$. The complex solutions of $\mathbf{x}' = A\mathbf{x}$ are complex linear combinations of

$$\mathbf{x}_1(t) = \begin{bmatrix} i \\ 2 \end{bmatrix} e^{(-2+5i)t} \quad \text{and} \quad \mathbf{x}_2(t) = \begin{bmatrix} -i \\ 2 \end{bmatrix} e^{(-2-5i)t}$$

---

[1] Since $\mathbf{x}_2(t)$ is the complex conjugate of $\mathbf{x}_1(t)$, the real and imaginary parts of $\mathbf{x}_2(t)$ are $\mathbf{y}_1(t)$ and $-\mathbf{y}_2(t)$, respectively. Thus one can use either $\mathbf{x}_1(t)$ or $\mathbf{x}_2(t)$, but not both, to produce two real linearly independent solutions of $\mathbf{x}' = A\mathbf{x}$.

Next, use equation (7) to write

$$\mathbf{x}_1(t) = \begin{bmatrix} i \\ 2 \end{bmatrix} e^{-2t} (\cos 5t + i \sin 5t)$$

The real and imaginary parts of $\mathbf{x}_1$ provide real solutions:

$$\mathbf{y}_1(t) = \begin{bmatrix} -\sin 5t \\ 2\cos 5t \end{bmatrix} e^{-2t}, \qquad \mathbf{y}_2(t) = \begin{bmatrix} \cos 5t \\ 2\sin 5t \end{bmatrix} e^{-2t}$$

Since $\mathbf{y}_1$ and $\mathbf{y}_2$ are linearly independent functions, they form a basis for the two-dimensional real vector space of solutions of $\mathbf{x}' = A\mathbf{x}$. Thus the general solution is

$$\mathbf{x}(t) = c_1 \begin{bmatrix} -\sin 5t \\ 2\cos 5t \end{bmatrix} e^{-2t} + c_2 \begin{bmatrix} \cos 5t \\ 2\sin 5t \end{bmatrix} e^{-2t}$$

To satisfy $\mathbf{x}(0) = \begin{bmatrix} 3 \\ 3 \end{bmatrix}$, we need $c_1 \begin{bmatrix} 0 \\ 2 \end{bmatrix} + c_2 \begin{bmatrix} 1 \\ 0 \end{bmatrix} = \begin{bmatrix} 3 \\ 3 \end{bmatrix}$, which leads to $c_1 = 1.5$ and $c_2 = 3$. Thus

$$\mathbf{x}(t) = 1.5 \begin{bmatrix} -\sin 5t \\ 2\cos 5t \end{bmatrix} e^{-2t} + 3 \begin{bmatrix} \cos 5t \\ 2\sin 5t \end{bmatrix} e^{-2t}$$

or

$$\begin{bmatrix} i_L(t) \\ v_C(t) \end{bmatrix} = \begin{bmatrix} -1.5\sin 5t + 3\cos 5t \\ 3\cos 5t + 6\sin 5t \end{bmatrix} e^{-2t}$$

See Fig. 5. ∎

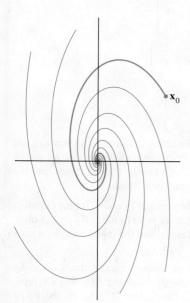

FIGURE 5
The origin as a spiral point.

In Fig. 5, the origin is called a **spiral point** of the dynamical system. The rotation is caused by the sine and cosine functions that arise from a complex eigenvalue. The trajectories spiral inward because the factor e^{-2t} tends to zero. Recall that -2 is the real part of the eigenvalue in Example 3. When A has a complex eigenvalue with positive real part, the trajectories spiral outward. If the real part of the eigenvalue is zero, the trajectories form ellipses around the origin.

PRACTICE PROBLEMS

A real 3×3 matrix A has eigenvalues $-.5$, $.2 + .3i$, and $.2 - .3i$, with corresponding eigenvectors

$$\mathbf{v}_1 = \begin{bmatrix} 1 \\ -2 \\ 1 \end{bmatrix}, \quad \mathbf{v}_2 = \begin{bmatrix} 1 + 2i \\ 4i \\ 2 \end{bmatrix}, \quad \text{and} \quad \mathbf{v}_3 = \begin{bmatrix} 1 - 2i \\ -4i \\ 2 \end{bmatrix}$$

1. Is A diagonalizable as $A = PDP^{-1}$, using complex matrices?
2. Write the general solution of $\mathbf{x}' = A\mathbf{x}$ using complex eigenfunctions, and then find the general real solution.
3. Describe the shapes of typical trajectories.

5.7 EXERCISES

1. A particle moving in a planar force field has a position vector $\mathbf{x}$ that satisfies $\mathbf{x}' = A\mathbf{x}$. The 2×2 matrix A has eigenvalues 4 and 2, with corresponding eigenvectors $\mathbf{v}_1 = \begin{bmatrix} -3 \\ 1 \end{bmatrix}$ and $\mathbf{v}_2 = \begin{bmatrix} -1 \\ 1 \end{bmatrix}$. Find the position of the particle at time t, assuming that $\mathbf{x}(0) = \begin{bmatrix} -6 \\ 1 \end{bmatrix}$.

2. Let A be a 2×2 matrix with eigenvalues -3 and -1 and corresponding eigenvectors $\mathbf{v}_1 = \begin{bmatrix} -1 \\ 1 \end{bmatrix}$ and $\mathbf{v}_2 = \begin{bmatrix} 1 \\ 1 \end{bmatrix}$. Let $\mathbf{x}(t)$ be the position of a particle at time t. Solve the initial value problem $\mathbf{x}' = A\mathbf{x}$, $\mathbf{x}(0) = \begin{bmatrix} 2 \\ 3 \end{bmatrix}$.

In Exercises 3–6, solve the initial value problem $\mathbf{x}'(t) = A\mathbf{x}(t)$ for $t \geq 0$, with $\mathbf{x}(0) = (3, 2)$. Classify the nature of the origin as an attractor, repeller, or saddle point of the dynamical system described by $\mathbf{x}' = A\mathbf{x}$. Find the directions of greatest attraction and/or repulsion. When the origin is a saddle point, sketch typical trajectories.

3. $A = \begin{bmatrix} 2 & 3 \\ -1 & -2 \end{bmatrix}$ **4.** $A = \begin{bmatrix} -2 & -5 \\ 1 & 4 \end{bmatrix}$

5. $A = \begin{bmatrix} 7 & -1 \\ 3 & 3 \end{bmatrix}$ **6.** $A = \begin{bmatrix} 1 & -2 \\ 3 & -4 \end{bmatrix}$

In Exercises 7 and 8, make a change of variable that decouples the equation $\mathbf{x}' = A\mathbf{x}$. Write the equation $\mathbf{x}(t) = P\mathbf{y}(t)$ and show the calculation that leads to the uncoupled system $\mathbf{y}' = D\mathbf{y}$, specifying P and D.

7. A as in Exercise 5 **8.** A as in Exercise 6

In Exercises 9–18, construct the general solution of $\mathbf{x}' = A\mathbf{x}$ involving complex eigenfunctions and then obtain the general real solution. Describe the shapes of typical trajectories.

9. $A = \begin{bmatrix} -3 & 2 \\ -1 & -1 \end{bmatrix}$ **10.** $A = \begin{bmatrix} 3 & 1 \\ -2 & 1 \end{bmatrix}$

11. $A = \begin{bmatrix} -3 & -9 \\ 2 & 3 \end{bmatrix}$ **12.** $A = \begin{bmatrix} -7 & 10 \\ -4 & 5 \end{bmatrix}$

13. $A = \begin{bmatrix} 4 & -3 \\ 6 & -2 \end{bmatrix}$ **14.** $A = \begin{bmatrix} -2 & 1 \\ -8 & 2 \end{bmatrix}$

15. [M] $A = \begin{bmatrix} -8 & -12 & -6 \\ 2 & 1 & 2 \\ 7 & 12 & 5 \end{bmatrix}$

16. [M] $A = \begin{bmatrix} -6 & -11 & 16 \\ 2 & 5 & -4 \\ -4 & -5 & 10 \end{bmatrix}$

17. [M] $A = \begin{bmatrix} 30 & 64 & 23 \\ -11 & -23 & -9 \\ 6 & 15 & 4 \end{bmatrix}$

18. [M] $A = \begin{bmatrix} 53 & -30 & -2 \\ 90 & -52 & -3 \\ 20 & -10 & 2 \end{bmatrix}$

19. [M] Find formulas for the voltages v_1 and v_2 (as functions of time t) for the circuit in Example 1, assuming that $R_1 = 1/5$ ohm, $R_2 = 1/3$ ohm, $C_1 = 4$ farads, $C_2 = 3$ farads, and the initial charge on each capacitor is 4 volts.

20. [M] Find formulas for the voltages v_1 and v_2 for the circuit in Example 1, assuming that $R_1 = 1/15$ ohm, $R_2 = 1/3$ ohm, $C_1 = 9$ farads, $C_2 = 2$ farads, and the initial charge on each capacitor is 3 volts.

21. [M] Find formulas for the current i_L and the voltage v_C for the circuit in Example 3, assuming that $R_1 = 1$ ohm, $R_2 = .125$ ohm, $C = .2$ farad, $L = .125$ henry, the initial current is 0 amp, and the initial voltage is 15 volts.

22. [M] The circuit in the figure is described by the equation

$$\begin{bmatrix} i_L' \\ v_C' \end{bmatrix} = \begin{bmatrix} 0 & 1/L \\ -1/C & -1/(RC) \end{bmatrix} \begin{bmatrix} i_L \\ v_C \end{bmatrix}$$

where i_L is the current through the inductor L and v_C is the voltage drop across the capacitor C. Find formulas for i_L and v_C when $R = .5$ ohm, $C = 2.5$ farads, $L = .5$ henry, the initial current is 0 amp, and the initial voltage is 12 volts.

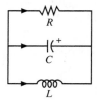

SOLUTIONS TO PRACTICE PROBLEMS

1. Yes, the 3×3 matrix is diagonalizable because it has three distinct eigenvalues. Theorem 2 in Section 5.1 and Theorem 5 in Section 5.3 are valid when complex scalars are used. (The proofs are essentially the same as for real scalars.)

2. The general solution has the form

$$\mathbf{x}(t) = c_1 \begin{bmatrix} 1 \\ -2 \\ 1 \end{bmatrix} e^{-.5t} + c_2 \begin{bmatrix} 1 + 2i \\ 4i \\ 2 \end{bmatrix} e^{(.2+.3i)t} + c_3 \begin{bmatrix} 1 - 2i \\ -4i \\ 2 \end{bmatrix} e^{(.2-.3i)t}$$

The scalars c_1, c_2, c_3 here can be any complex numbers. The first term in $\mathbf{x}(t)$ is real. Two more real solutions can be produced using the real and imaginary parts of the second term in $\mathbf{x}(t)$:

$$\begin{bmatrix} 1 + 2i \\ 4i \\ 2 \end{bmatrix} e^{.2t}(\cos .3t + i \sin .3t)$$

The general real solution has the following form, with *real* scalars c_1, c_2, c_3:

$$c_1 \begin{bmatrix} 1 \\ -2 \\ 1 \end{bmatrix} e^{-.5t} + c_2 \begin{bmatrix} \cos .3t - 2\sin .3t \\ -4\sin .3t \\ 2\cos .3t \end{bmatrix} e^{.2t} + c_3 \begin{bmatrix} \sin .3t + 2\cos .3t \\ 4\cos .3t \\ 2\sin .3t \end{bmatrix} e^{.2t}$$

3. Any solution with $c_2 = c_3 = 0$ is attracted to the origin because of the negative exponential factor. Other solutions have components that grow without bound, and the trajectories spiral outward.

   Be careful not to mistake this problem for one in Section 5.6. There the condition for attraction toward $\mathbf{0}$ was that an eigenvalue be less than 1 in magnitude, to make $|\lambda|^k \to 0$. Here the real part of the eigenvalue must be negative, to make $e^{\lambda t} \to 0$.

5.8   ITERATIVE ESTIMATES FOR EIGENVALUES

In scientific applications of linear algebra, eigenvalues are seldom known precisely. Fortunately, a close numerical approximation is usually quite satisfactory. In fact, some applications require only a rough approximation to the largest eigenvalue. The first algorithm described below can work well for this case. Also, it provides a foundation for a more powerful method that can give fast estimates for other eigenvalues as well.

The Power Method

The power method applies to an $n \times n$ matrix A with a **strictly dominant eigenvalue** λ_1, which means that λ_1 must be larger in absolute value than all the other eigenvalues. In this case, the power method produces a scalar sequence that approaches λ_1 and a vector sequence that approaches a corresponding eigenvector. The background for the method rests on the eigenvector decomposition used at the beginning of Section 5.6.

Assume for simplicity that A is diagonalizable and $\mathbb{R}^n$ has a basis of eigenvectors $\mathbf{v}_1, \dots, \mathbf{v}_n$, arranged so their corresponding eigenvalues $\lambda_1, \dots, \lambda_n$ decrease in size, with the strictly dominant eigenvalue first. That is,

$$|\lambda_1| > |\lambda_2| \geq |\lambda_3| \geq \cdots \geq |\lambda_n| \tag{1}$$

$\underset{\text{Strictly larger}}{\downarrow}$

As we saw in equation (2) of Section 5.6, if $\mathbf{x}$ in $\mathbb{R}^n$ is written as $\mathbf{x} = c_1\mathbf{v}_1 + \cdots + c_n\mathbf{v}_n$, then

$$A^k\mathbf{x} = c_1(\lambda_1)^k\mathbf{v}_1 + c_2(\lambda_2)^k\mathbf{v}_2 + \cdots + c_n(\lambda_n)^k\mathbf{v}_n \quad (k = 1, 2, \dots)$$

Assume $c_1 \neq 0$. Then, dividing by $(\lambda_1)^k$,

$$\frac{1}{(\lambda_1)^k} A^k\mathbf{x} = c_1\mathbf{v}_1 + c_2\left(\frac{\lambda_2}{\lambda_1}\right)^k \mathbf{v}_2 + \cdots + c_n\left(\frac{\lambda_n}{\lambda_1}\right)^k \mathbf{v}_n \quad (k = 1, 2, \dots) \tag{2}$$

From inequality (1), the fractions $\lambda_2/\lambda_1, \dots, \lambda_n/\lambda_1$ are all less than 1 in magnitude and so their powers go to zero. Hence

$$(\lambda_1)^{-k} A^k\mathbf{x} \to c_1\mathbf{v}_1 \quad \text{as } k \to \infty \tag{3}$$

Thus, for large k, a scalar multiple of $A^k \mathbf{x}$ determines almost the same *direction* as the eigenvector $c_1 \mathbf{v}_1$. Since positive scalar multiples do not change the direction of a vector, $A^k \mathbf{x}$ itself points almost in the same direction as $\mathbf{v}_1$ or $-\mathbf{v}_1$, provided $c_1 \neq 0$.

EXAMPLE 1   Let $A = \begin{bmatrix} 1.8 & .8 \\ .2 & 1.2 \end{bmatrix}$, $\mathbf{v}_1 = \begin{bmatrix} 4 \\ 1 \end{bmatrix}$, and $\mathbf{x} = \begin{bmatrix} -.5 \\ 1 \end{bmatrix}$. Then A has eigenvalues 2 and 1, and the eigenspace for $\lambda_1 = 2$ is the line through $\mathbf{0}$ and $\mathbf{v}_1$. For $k = 0, \ldots, 8$, compute $A^k \mathbf{x}$ and construct the line through $\mathbf{0}$ and $A^k \mathbf{x}$. What happens as k increases?

SOLUTION   The first three calculations are

$$A\mathbf{x} = \begin{bmatrix} 1.8 & .8 \\ .2 & 1.2 \end{bmatrix} \begin{bmatrix} -.5 \\ 1 \end{bmatrix} = \begin{bmatrix} -.1 \\ 1.1 \end{bmatrix}$$

$$A^2\mathbf{x} = A(A\mathbf{x}) = \begin{bmatrix} 1.8 & .8 \\ .2 & 1.2 \end{bmatrix} \begin{bmatrix} -.1 \\ 1.1 \end{bmatrix} = \begin{bmatrix} .7 \\ 1.3 \end{bmatrix}$$

$$A^3\mathbf{x} = A(A^2\mathbf{x}) = \begin{bmatrix} 1.8 & .8 \\ .2 & 1.2 \end{bmatrix} \begin{bmatrix} .7 \\ 1.3 \end{bmatrix} = \begin{bmatrix} 2.3 \\ 1.7 \end{bmatrix}$$

Analogous calculations complete Table 1.

TABLE 1   Iterates of a Vector

k	0	1	2	3	4	5	6	7	8
$A^k\mathbf{x}$	$\begin{bmatrix} -.5 \\ 1 \end{bmatrix}$	$\begin{bmatrix} -.1 \\ 1.1 \end{bmatrix}$	$\begin{bmatrix} .7 \\ 1.3 \end{bmatrix}$	$\begin{bmatrix} 2.3 \\ 1.7 \end{bmatrix}$	$\begin{bmatrix} 5.5 \\ 2.5 \end{bmatrix}$	$\begin{bmatrix} 11.9 \\ 4.1 \end{bmatrix}$	$\begin{bmatrix} 24.7 \\ 7.3 \end{bmatrix}$	$\begin{bmatrix} 50.3 \\ 13.7 \end{bmatrix}$	$\begin{bmatrix} 101.5 \\ 26.5 \end{bmatrix}$

The vectors $\mathbf{x}, A\mathbf{x}, \ldots, A^4\mathbf{x}$ are shown in Fig. 1. The other vectors are growing too long to display. However, line segments are drawn showing the directions of those vectors. In fact, the directions of the vectors are what we really want to see, not the vectors themselves. The lines seem to be approaching the line representing the eigenspace spanned by $\mathbf{v}_1$. More precisely, the angle between the line (subspace) determined by $A^k \mathbf{x}$ and the line (eigenspace) determined by $\mathbf{v}_1$ goes to zero as $k \to \infty$. ∎

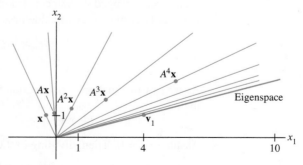

FIGURE 1   Directions determined by $\mathbf{x}, A\mathbf{x}, A^2\mathbf{x}, \ldots, A^7\mathbf{x}$.

The vectors $(\lambda_1)^{-k} A^k \mathbf{x}$ in (3) are scaled to make them converge to $c_1 \mathbf{v}_1$, provided $c_1 \neq 0$. We cannot scale $A^k \mathbf{x}$ in this way because we do not know λ_1. But we can scale each $A^k \mathbf{x}$ to make its largest entry a 1. It turns out that the resulting sequence $\{\mathbf{x}_k\}$ will converge to a multiple of $\mathbf{v}_1$ whose largest entry is 1. Figure 2 shows the scaled sequence for Example 1. The eigenvalue λ_1 can be estimated from the sequence $\{\mathbf{x}_k\}$, too. When $\mathbf{x}_k$ is close to an eigenvector for λ_1, the vector $A\mathbf{x}_k$ is close to $\lambda_1\mathbf{x}_k$, with each entry in $A\mathbf{x}_k$ approximately λ_1 times the corresponding entry in $\mathbf{x}_k$. Because the largest entry in $\mathbf{x}_k$ is 1, the largest entry in $A\mathbf{x}_k$ is close to λ_1. (Careful proofs of these statements are omitted.)

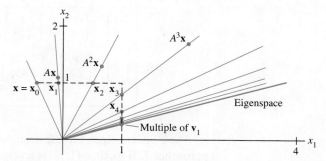

FIGURE 2 Scaled multiples of $\mathbf{x}$, $A\mathbf{x}$, $A^2\mathbf{x}$, ..., $A^7\mathbf{x}$.

THE POWER METHOD FOR ESTIMATING A STRICTLY DOMINANT EIGENVALUE

1. Select an initial vector $\mathbf{x}_0$ whose largest entry is 1.
2. For $k = 0, 1, \ldots,$
   a. Compute $A\mathbf{x}_k$.
   b. Let μ_k be an entry in $A\mathbf{x}_k$ whose absolute value is as large as possible.
   c. Compute $\mathbf{x}_{k+1} = (1/\mu_k)A\mathbf{x}_k$.
3. For almost all choices of $\mathbf{x}_0$, the sequence $\{\mu_k\}$ approaches the dominant eigenvalue, and the sequence $\{\mathbf{x}_k\}$ approaches a corresponding eigenvector.

EXAMPLE 2 Apply the power method to $A = \begin{bmatrix} 6 & 5 \\ 1 & 2 \end{bmatrix}$ with $\mathbf{x}_0 = \begin{bmatrix} 0 \\ 1 \end{bmatrix}$. Stop when $k = 5$, and estimate the dominant eigenvalue and a corresponding eigenvector of A.

SOLUTION Calculations in this example and the next were made with MATLAB, which computes with 16-digit accuracy, although we show only a few significant figures here. To begin, compute $A\mathbf{x}_0$ and identify the largest entry μ_0 in $A\mathbf{x}_0$:

$$A\mathbf{x}_0 = \begin{bmatrix} 6 & 5 \\ 1 & 2 \end{bmatrix}\begin{bmatrix} 0 \\ 1 \end{bmatrix} = \begin{bmatrix} 5 \\ 2 \end{bmatrix}, \quad \mu_0 = 5$$

Scale $A\mathbf{x}_0$ by $1/\mu_0$ to get $\mathbf{x}_1$, compute $A\mathbf{x}_1$, and identify the largest entry in $A\mathbf{x}_1$:

$$\mathbf{x}_1 = \frac{1}{\mu_0}A\mathbf{x}_0 = \frac{1}{5}\begin{bmatrix} 5 \\ 2 \end{bmatrix} = \begin{bmatrix} 1 \\ .4 \end{bmatrix}$$

$$A\mathbf{x}_1 = \begin{bmatrix} 6 & 5 \\ 1 & 2 \end{bmatrix}\begin{bmatrix} 1 \\ .4 \end{bmatrix} = \begin{bmatrix} 8 \\ 1.8 \end{bmatrix}, \quad \mu_1 = 8$$

Scale $A\mathbf{x}_1$ by $1/\mu_1$ to get $\mathbf{x}_2$, compute $A\mathbf{x}_2$, and identify the largest entry in $A\mathbf{x}_2$:

$$\mathbf{x}_2 = \frac{1}{\mu_1}A\mathbf{x}_1 = \frac{1}{8}\begin{bmatrix} 8 \\ 1.8 \end{bmatrix} = \begin{bmatrix} 1 \\ .225 \end{bmatrix}$$

$$A\mathbf{x}_2 = \begin{bmatrix} 6 & 5 \\ 1 & 2 \end{bmatrix}\begin{bmatrix} 1 \\ .225 \end{bmatrix} = \begin{bmatrix} 7.125 \\ 1.450 \end{bmatrix}, \quad \mu_2 = 7.125$$

Scale $A\mathbf{x}_2$ by $1/\mu_2$ to get $\mathbf{x}_3$, and so on. The results of MATLAB calculations for the first five iterations are arranged in Table 2.

TABLE 2   The Power Method for Example 2

k	0	1	2	3	4	5
$\mathbf{x}_k$	$\begin{bmatrix} 0 \\ 1 \end{bmatrix}$	$\begin{bmatrix} 1 \\ .4 \end{bmatrix}$	$\begin{bmatrix} 1 \\ .225 \end{bmatrix}$	$\begin{bmatrix} 1 \\ .2035 \end{bmatrix}$	$\begin{bmatrix} 1 \\ .2005 \end{bmatrix}$	$\begin{bmatrix} 1 \\ .20007 \end{bmatrix}$
$A\mathbf{x}_k$	$\begin{bmatrix} 5 \\ 2 \end{bmatrix}$	$\begin{bmatrix} 8 \\ 1.8 \end{bmatrix}$	$\begin{bmatrix} 7.125 \\ 1.450 \end{bmatrix}$	$\begin{bmatrix} 7.0175 \\ 1.4070 \end{bmatrix}$	$\begin{bmatrix} 7.0025 \\ 1.4010 \end{bmatrix}$	$\begin{bmatrix} 7.00036 \\ 1.40014 \end{bmatrix}$
μ_k	5	8	7.125	7.0175	7.0025	7.00036

The evidence from Table 2 strongly suggests that $\{\mathbf{x}_k\}$ approaches $(1, .2)$ and $\{\mu_k\}$ approaches 7. If so, then $(1, .2)$ is an eigenvector and 7 is the dominant eigenvalue. This is easily verified by computing

$$A \begin{bmatrix} 1 \\ .2 \end{bmatrix} = \begin{bmatrix} 6 & 5 \\ 1 & 2 \end{bmatrix} \begin{bmatrix} 1 \\ .2 \end{bmatrix} = \begin{bmatrix} 7 \\ 1.4 \end{bmatrix} = 7 \begin{bmatrix} 1 \\ .2 \end{bmatrix}$$
■

The sequence $\{\mu_k\}$ in Example 2 converged quickly to $\lambda_1 = 7$ because the second eigenvalue of A was much smaller. (In fact, $\lambda_2 = 1$.) In general, the rate of convergence depends on the ratio $|\lambda_2/\lambda_1|$, because the vector $c_2(\lambda_2/\lambda_1)^k \mathbf{v}_2$ in equation (2) is the main source of error when using a scaled version of $A^k\mathbf{x}$ as an estimate of $c_1\mathbf{v}_1$. (The other fractions λ_j/λ_1 are likely to be smaller.) If $|\lambda_2/\lambda_1|$ is close to 1, then $\{\mu_k\}$ and $\{\mathbf{x}_k\}$ can converge very slowly, and other approximation methods may be preferred.

With the power method, there is a slight chance that the chosen initial vector $\mathbf{x}$ will have no component in the $\mathbf{v}_1$ direction (when $c_1 = 0$). But computer rounding errors during the calculations of the $\mathbf{x}_k$ are likely to create a vector with at least a small component in the direction of $\mathbf{v}_1$. If that occurs, the $\mathbf{x}_k$ will start to converge to a multiple of $\mathbf{v}_1$.

The Inverse Power Method

This method provides an approximation for *any* eigenvalue, provided a good initial estimate α of the eigenvalue λ is known. In this case, we let $B = (A - \alpha I)^{-1}$ and apply the power method to B. It can be shown that if the eigenvalues of A are $\lambda_1, \ldots, \lambda_n$, then the eigenvalues of B are

$$\frac{1}{\lambda_1 - \alpha}, \quad \frac{1}{\lambda_2 - \alpha}, \quad \ldots, \quad \frac{1}{\lambda_n - \alpha}$$

and the corresponding eigenvectors are the same as those for A. (See Exercises 15 and 16.)

Suppose, for example, that α is closer to λ_2 than to the other eigenvalues of A. Then $1/(\lambda_2 - \alpha)$ will be a strictly dominant eigenvalue of B. If α is really close to λ_2, then $1/(\lambda_2 - \alpha)$ is *much* larger than the other eigenvalues of B, and the inverse power method produces a very rapid approximation to λ_2 for almost all choices of $\mathbf{x}_0$. The following algorithm gives the details.

THE INVERSE POWER METHOD FOR ESTIMATING AN EIGENVALUE λ OF A

1. Select an initial estimate α sufficiently close to λ.
2. Select an initial vector $\mathbf{x}_0$ whose largest entry is 1.
3. For $k = 0, 1, \ldots,$
   a. Solve $(A - \alpha I)\mathbf{y}_k = \mathbf{x}_k$ for $\mathbf{y}_k$.
   b. Let μ_k be an entry in $\mathbf{y}_k$ whose absolute value is as large as possible.
   c. Compute $\nu_k = \alpha + (1/\mu_k)$.
   d. Compute $\mathbf{x}_{k+1} = (1/\mu_k)\mathbf{y}_k$.
4. For almost all choices of $\mathbf{x}_0$, the sequence $\{\nu_k\}$ approaches the eigenvalue λ of A, and the sequence $\{\mathbf{x}_k\}$ approaches a corresponding eigenvector.

Notice that B, or rather $(A - \alpha I)^{-1}$, does not appear in the algorithm. Instead of computing $(A - \alpha I)^{-1}\mathbf{x}_k$ to get the next vector in the sequence, it is better to *solve* the equation $(A - \alpha I)\mathbf{y}_k = \mathbf{x}_k$ for $\mathbf{y}_k$ (and then scale $\mathbf{y}_k$ to produce $\mathbf{x}_{k+1}$). Since this equation for $\mathbf{y}_k$ must be solved for each k, an LU factorization of $A - \alpha I$ will speed up the process.

EXAMPLE 3 It is not uncommon in some applications to need to know the smallest eigenvalue of a matrix A and to have at hand rough estimates of the eigenvalues. Suppose 21, 3.3, and 1.9 are estimates for the eigenvalues of the matrix A below. Find the smallest eigenvalue, accurate to six decimal places.

$$A = \begin{bmatrix} 10 & -8 & -4 \\ -8 & 13 & 4 \\ -4 & 5 & 4 \end{bmatrix}$$

SOLUTION The two smallest eigenvalues seem close together, so we use the inverse power method for $A - 1.9I$. Results of a MATLAB calculation are shown in Table 3. Here $\mathbf{x}_0$ was chosen arbitrarily, $\mathbf{y}_k = (A - 1.9I)^{-1}\mathbf{x}_k$, μ_k is the largest entry in $\mathbf{y}_k$, $\nu_k = 1.9 + 1/\mu_k$, and $\mathbf{x}_{k+1} = (1/\mu_k)\mathbf{y}_k$. As it turns out, the initial eigenvalue estimate was fairly good, and the inverse power sequence converged quickly. The smallest eigenvalue is exactly 2. ∎

TABLE 3 The Inverse Power Method

k	0	1	2	3	4
$\mathbf{x}_k$	$\begin{bmatrix} 1 \\ 1 \\ 1 \end{bmatrix}$	$\begin{bmatrix} .5736 \\ .0646 \\ 1 \end{bmatrix}$	$\begin{bmatrix} .5054 \\ .0045 \\ 1 \end{bmatrix}$	$\begin{bmatrix} .5004 \\ .0003 \\ 1 \end{bmatrix}$	$\begin{bmatrix} .50003 \\ .00002 \\ 1 \end{bmatrix}$
$\mathbf{y}_k$	$\begin{bmatrix} 4.45 \\ .50 \\ 7.76 \end{bmatrix}$	$\begin{bmatrix} 5.0131 \\ .0442 \\ 9.9197 \end{bmatrix}$	$\begin{bmatrix} 5.0012 \\ .0031 \\ 9.9949 \end{bmatrix}$	$\begin{bmatrix} 5.0001 \\ .0002 \\ 9.9996 \end{bmatrix}$	$\begin{bmatrix} 5.000006 \\ .000015 \\ 9.999975 \end{bmatrix}$
μ_k	7.76	9.9197	9.9949	9.9996	9.999975
ν_k	2.03	2.0008	2.00005	2.000004	2.0000002

If an estimate for the smallest eigenvalue of a matrix is not available, one can simply take $\alpha = 0$ in the inverse power method. This choice of α works reasonably well if the smallest eigenvalue is much closer to zero than to the other eigenvalues.

The two algorithms presented in this section are practical tools for many simple situations, and they provide an introduction to the problem of eigenvalue estimation. A more robust and widely used iterative method is the QR algorithm. For instance, it is the heart of the MATLAB command `eig(A)`, which rapidly computes eigenvalues and eigenvectors of A. A brief description of the QR algorithm was given in the exercises for Section 5.2. Further details are presented in most modern numerical analysis texts.

| PRACTICE PROBLEM

How can you tell if a given vector $\mathbf{x}$ is a good approximation to an eigenvector of a matrix A? If it is, how would you estimate the corresponding eigenvalue? Experiment with

$$A = \begin{bmatrix} 5 & 8 & 4 \\ 8 & 3 & -1 \\ 4 & -1 & 2 \end{bmatrix} \quad \text{and} \quad \mathbf{x} = \begin{bmatrix} 1.0 \\ -4.3 \\ 8.1 \end{bmatrix}$$

5.8 EXERCISES

In Exercises 1–4, the matrix A is followed by a sequence $\{\mathbf{x}_k\}$ produced by the power method. Use these data to estimate the largest eigenvalue of A, and give a corresponding eigenvector.

1. $A = \begin{bmatrix} 4 & 3 \\ 1 & 2 \end{bmatrix}$;

$$\begin{bmatrix} 1 \\ 0 \end{bmatrix}, \begin{bmatrix} 1 \\ .25 \end{bmatrix}, \begin{bmatrix} 1 \\ .3158 \end{bmatrix}, \begin{bmatrix} 1 \\ .3298 \end{bmatrix}, \begin{bmatrix} 1 \\ .3326 \end{bmatrix}$$

2. $A = \begin{bmatrix} 1.8 & -.8 \\ -3.2 & 4.2 \end{bmatrix}$;

$$\begin{bmatrix} 1 \\ 0 \end{bmatrix}, \begin{bmatrix} -.5625 \\ 1 \end{bmatrix}, \begin{bmatrix} -.3021 \\ 1 \end{bmatrix}, \begin{bmatrix} -.2601 \\ 1 \end{bmatrix}, \begin{bmatrix} -.2520 \\ 1 \end{bmatrix}$$

3. $A = \begin{bmatrix} .5 & .2 \\ .4 & .7 \end{bmatrix}$;

$$\begin{bmatrix} 1 \\ 0 \end{bmatrix}, \begin{bmatrix} 1 \\ .8 \end{bmatrix}, \begin{bmatrix} .6875 \\ 1 \end{bmatrix}, \begin{bmatrix} .5577 \\ 1 \end{bmatrix}, \begin{bmatrix} .5188 \\ 1 \end{bmatrix}$$

4. $A = \begin{bmatrix} 4.1 & -6 \\ 3 & -4.4 \end{bmatrix}$;

$$\begin{bmatrix} 1 \\ 1 \end{bmatrix}, \begin{bmatrix} 1 \\ .7368 \end{bmatrix}, \begin{bmatrix} 1 \\ .7541 \end{bmatrix}, \begin{bmatrix} 1 \\ .7490 \end{bmatrix}, \begin{bmatrix} 1 \\ .7502 \end{bmatrix}$$

5. Let $A = \begin{bmatrix} 15 & 16 \\ -20 & -21 \end{bmatrix}$. The vectors $\mathbf{x}, \ldots, A^5\mathbf{x}$ are $\begin{bmatrix} 1 \\ 1 \end{bmatrix}$,

$$\begin{bmatrix} 31 \\ -41 \end{bmatrix}, \begin{bmatrix} -191 \\ 241 \end{bmatrix}, \begin{bmatrix} 991 \\ -1241 \end{bmatrix}, \begin{bmatrix} -4991 \\ 6241 \end{bmatrix}, \begin{bmatrix} 24991 \\ -31241 \end{bmatrix}.$$

Find a vector with a 1 in the second entry that is close to an eigenvector of A. Use four decimal places. Check your estimate, and give an estimate for the dominant eigenvalue of A.

6. Let $A = \begin{bmatrix} -2 & -3 \\ 6 & 7 \end{bmatrix}$. Repeat Exercise 5, using the following sequence $\mathbf{x}, A\mathbf{x}, \ldots, A^5\mathbf{x}$.

$$\begin{bmatrix} 1 \\ 1 \end{bmatrix}, \begin{bmatrix} -5 \\ 13 \end{bmatrix}, \begin{bmatrix} -29 \\ 61 \end{bmatrix}, \begin{bmatrix} -125 \\ 253 \end{bmatrix}, \begin{bmatrix} -509 \\ 1021 \end{bmatrix}, \begin{bmatrix} -2045 \\ 4093 \end{bmatrix}$$

[M] Exercises 7–12 require MATLAB or other computational aid. In Exercises 7 and 8, use the power method with the $\mathbf{x}_0$ given. List $\{\mathbf{x}_k\}$ and $\{\mu_k\}$ for $k = 1, \ldots, 5$. In Exercises 9 and 10, list μ_5 and μ_6.

7. $A = \begin{bmatrix} 6 & 7 \\ 8 & 5 \end{bmatrix}$, $\mathbf{x}_0 = \begin{bmatrix} 1 \\ 0 \end{bmatrix}$

8. $A = \begin{bmatrix} 2 & 1 \\ 4 & 5 \end{bmatrix}$, $\mathbf{x}_0 = \begin{bmatrix} 1 \\ 0 \end{bmatrix}$

9. $A = \begin{bmatrix} 8 & 0 & 12 \\ 1 & -2 & 1 \\ 0 & 3 & 0 \end{bmatrix}$, $\mathbf{x}_0 = \begin{bmatrix} 1 \\ 0 \\ 0 \end{bmatrix}$

10. $A = \begin{bmatrix} 1 & 2 & -2 \\ 1 & 1 & 9 \\ 0 & 1 & 9 \end{bmatrix}$, $\mathbf{x}_0 = \begin{bmatrix} 1 \\ 0 \\ 0 \end{bmatrix}$

Another estimate can be made for an eigenvalue when an approximate eigenvector is available. Observe that if $A\mathbf{x} = \lambda\mathbf{x}$, then $\mathbf{x}^T A\mathbf{x} = \mathbf{x}^T(\lambda\mathbf{x}) = \lambda(\mathbf{x}^T\mathbf{x})$, and the **Rayleigh quotient**

$$R(\mathbf{x}) = \frac{\mathbf{x}^T A\mathbf{x}}{\mathbf{x}^T\mathbf{x}}$$

equals λ. If $\mathbf{x}$ is close to an eigenvector for λ, then this quotient is close to λ. When A is a symmetric matrix ($A^T = A$), the Rayleigh quotient $R(\mathbf{x}_k) = (\mathbf{x}_k^T A\mathbf{x}_k)/(\mathbf{x}_k^T\mathbf{x}_k)$ will have roughly twice as many digits of accuracy as the scaling factor μ_k in the power method. Verify this increased accuracy in Exercises 11 and 12 by computing μ_k and $R(\mathbf{x}_k)$ for $k = 1, \ldots, 4$.

11. $A = \begin{bmatrix} 5 & 2 \\ 2 & 2 \end{bmatrix}$, $\mathbf{x}_0 = \begin{bmatrix} 1 \\ 0 \end{bmatrix}$

12. $A = \begin{bmatrix} -3 & 2 \\ 2 & 0 \end{bmatrix}$, $\mathbf{x}_0 = \begin{bmatrix} 1 \\ 0 \end{bmatrix}$

Exercises 13 and 14 apply to a 3×3 matrix A whose eigenvalues are estimated to be 4, -4, and 3.

13. If the eigenvalues close to 4 and -4 are known to have different absolute values, will the power method work? Is it likely to be useful?

14. Suppose the eigenvalues close to 4 and -4 are known to have exactly the same absolute value. Describe how one might obtain a sequence that estimates the eigenvalue close to 4.

15. Suppose $A\mathbf{x} = \lambda\mathbf{x}$ with $\mathbf{x} \neq \mathbf{0}$. Let α be a scalar different from the eigenvalues of A, and let $B = (A - \alpha I)^{-1}$. Subtract $\alpha\mathbf{x}$ from both sides of the equation $A\mathbf{x} = \lambda\mathbf{x}$, and use algebra to show that $1/(\lambda - \alpha)$ is an eigenvalue of B, with $\mathbf{x}$ a corresponding eigenvector.

16. Suppose μ is an eigenvalue of the B in Exercise 15, and that $\mathbf{x}$ is a corresponding eigenvector, so that $(A - \alpha I)^{-1}\mathbf{x} = \mu\mathbf{x}$. Use this equation to find an eigenvalue of A in terms of μ and α. [*Note:* $\mu \neq 0$ because B is invertible.]

17. [**M**] Use the inverse power method to estimate the middle eigenvalue of the A in Example 3, with accuracy to four decimal places. Set $\mathbf{x}_0 = (1, 0, 0)$.

18. [**M**] Let A be as in Exercise 9. Use the inverse power method with $\mathbf{x}_0 = (1, 0, 0)$ to estimate the eigenvalue of A near $\alpha = -1.4$, with an accuracy to four decimal places.

[**M**] In Exercises 19 and 20, find (a) the largest eigenvalue and (b) the eigenvalue closest to zero. In each case, set $\mathbf{x}_0 = (1, 0, 0, 0)$ and carry out approximations until the approximating sequence seems accurate to four decimal places. Include the approximate eigenvector.

19. $A = \begin{bmatrix} 10 & 7 & 8 & 7 \\ 7 & 5 & 6 & 5 \\ 8 & 6 & 10 & 9 \\ 7 & 5 & 9 & 10 \end{bmatrix}$

20. $A = \begin{bmatrix} 1 & 2 & 3 & 2 \\ 2 & 12 & 13 & 11 \\ -2 & 3 & 0 & 2 \\ 4 & 5 & 7 & 2 \end{bmatrix}$

21. A common misconception is that if A has a strictly dominant eigenvalue, then, for any sufficiently large value of k, the vector $A^k\mathbf{x}$ is approximately equal to an eigenvector of A. For the three matrices below, study what happens to $A^k\mathbf{x}$ when $\mathbf{x} = (.5, .5)$, and try to draw general conclusions (for a 2×2 matrix).

a. $A = \begin{bmatrix} .8 & 0 \\ 0 & .2 \end{bmatrix}$ b. $A = \begin{bmatrix} 1 & 0 \\ 0 & .8 \end{bmatrix}$

c. $A = \begin{bmatrix} 8 & 0 \\ 0 & 2 \end{bmatrix}$

| **SOLUTION TO PRACTICE PROBLEM** |

For the given A and $\mathbf{x}$,

$$Ax = \begin{bmatrix} 5 & 8 & 4 \\ 8 & 3 & -1 \\ 4 & -1 & 2 \end{bmatrix} \begin{bmatrix} 1.00 \\ -4.30 \\ 8.10 \end{bmatrix} = \begin{bmatrix} 3.00 \\ -13.00 \\ 24.50 \end{bmatrix}$$

If $A\mathbf{x}$ is nearly a multiple of $\mathbf{x}$, then the ratios of corresponding entries in the two vectors should be nearly constant. So compute:

{entry in $A\mathbf{x}$}	÷ {entry in $\mathbf{x}$}	= {ratio}
3.00	1.00	3.000
−13.00	−4.30	3.023
24.50	8.10	3.025

Each entry in $A\mathbf{x}$ is about 3 times the corresponding entry in $\mathbf{x}$, so $\mathbf{x}$ is close to an eigenvector. Any of the ratios above is an estimate for the eigenvalue. (To five decimal places, the eigenvalue is 3.02409.)

WEB

CHAPTER 5 SUPPLEMENTARY EXERCISES

Throughout these supplementary exercises, A and B represent square matrices of appropriate sizes.

1. Mark each statement as True or False. Justify each answer.

    a. If A is invertible and 1 is an eigenvalue for A, then 1 is also an eigenvalue of A^{-1}.

    b. If A is row equivalent to the identity matrix I, then A is diagonalizable.

    c. If A contains a row or column of zeros, then 0 is an eigenvalue of A.

    d. Each eigenvalue of A is also an eigenvalue of A^2.

    e. Each eigenvector of A is also an eigenvector of A^2.

    f. Each eigenvector of an invertible matrix A is also an eigenvector of A^{-1}.

    g. Eigenvalues must be nonzero scalars.

    h. Eigenvectors must be nonzero vectors.

    i. Two eigenvectors corresponding to the same eigenvalue are always linearly dependent.

    j. Similar matrices always have exactly the same eigenvalues.

    k. Similar matrices always have exactly the same eigenvectors.

    l. The sum of two eigenvectors of a matrix A is also an eigenvector of A.

    m. The eigenvalues of an upper triangular matrix A are exactly the nonzero entries on the diagonal of A.

    n. The matrices A and A^T have the same eigenvalues, counting multiplicities.

    o. If a 5×5 matrix A has fewer than 5 distinct eigenvalues, then A is not diagonalizable.

    p. There exists a 2×2 matrix that has no eigenvectors in $\mathbb{R}^2$.

    q. If A is diagonalizable, then the columns of A are linearly independent.

    r. A nonzero vector cannot correspond to two different eigenvalues of A.

    s. A (square) matrix A is invertible if and only if there is a coordinate system in which the transformation $\mathbf{x} \mapsto A\mathbf{x}$ is represented by a diagonal matrix.

    t. If each vector $\mathbf{e}_j$ in the standard basis for $\mathbb{R}^n$ is an eigenvector of A, then A is a diagonal matrix.

    u. If A is similar to a diagonalizable matrix B, then A is also diagonalizable.

    v. If A and B are invertible $n \times n$ matrices, then AB is similar to BA.

    w. An $n \times n$ matrix with n linearly independent eigenvectors is invertible.

    x. If A is an $n \times n$ diagonalizable matrix, then each vector in $\mathbb{R}^n$ can be written as a linear combination of eigenvectors of A.

2. Show that if $\mathbf{x}$ is an eigenvector of the matrix product AB and $B\mathbf{x} \neq \mathbf{0}$, then $B\mathbf{x}$ is an eigenvector of BA.

3. Suppose $\mathbf{x}$ is an eigenvector of A corresponding to an eigenvalue λ.

    a. Show that $\mathbf{x}$ is an eigenvector of $5I - A$. What is the corresponding eigenvalue?

    b. Show that $\mathbf{x}$ is an eigenvector of $5I - 3A + A^2$. What is the corresponding eigenvalue?

4. Use mathematical induction to show that if λ is an eigenvalue of an $n \times n$ matrix A, with $\mathbf{x}$ a corresponding eigenvector, then, for each positive integer m, λ^m is an eigenvalue of A^m, with $\mathbf{x}$ a corresponding eigenvector.

5. If $p(t) = c_0 + c_1 t + c_2 t^2 + \cdots + c_n t^n$, define $p(A)$ to be the matrix formed by replacing each power of t in $p(t)$ by the corresponding power of A (with $A^0 = I$). That is,

$$p(A) = c_0 I + c_1 A + c_2 A^2 + \cdots + c_n A^n$$

Show that if λ is an eigenvalue of A, then one eigenvalue of $p(A)$ is $p(\lambda)$.

6. Suppose $A = PDP^{-1}$, where P is 2×2 and $D = \begin{bmatrix} 2 & 0 \\ 0 & 7 \end{bmatrix}$.

    a. Let $B = 5I - 3A + A^2$. Show that B is diagonalizable by finding a suitable factorization of B.

    b. Given $p(t)$ and $p(A)$ as in Exercise 5, show that $p(A)$ is diagonalizable.

7. Suppose A is diagonalizable and $p(t)$ is the characteristic polynomial of A. Define $p(A)$ as in Exercise 5, and show that $p(A)$ is the zero matrix. This fact, which is also true for *any* square matrix, is called the *Cayley–Hamilton theorem*.

8. a. Let A be a diagonalizable $n \times n$ matrix. Show that if the multiplicity of an eigenvalue λ is n, then $A = \lambda I$.

    b. Use part (a) to show that the matrix $A = \begin{bmatrix} 3 & 1 \\ 0 & 3 \end{bmatrix}$ is not diagonalizable.

9. Show that $I - A$ is invertible when all the eigenvalues of A are less than 1 in magnitude. [*Hint:* What would be true if $I - A$ were not invertible?]

10. Show that if A is diagonalizable, with all eigenvalues less than 1 in magnitude, then A^k tends to the zero matrix as $k \to \infty$. [*Hint:* Consider $A^k \mathbf{x}$ where $\mathbf{x}$ represents any one of the columns of I.]

11. Let $\mathbf{u}$ be an eigenvector of A corresponding to an eigenvalue λ, and let H be the line in $\mathbb{R}^n$ through $\mathbf{u}$ and the origin.

    a. Explain why H is invariant under A in the sense that $A\mathbf{x}$ is in H whenever $\mathbf{x}$ is in H.

b. Let K be a one-dimensional subspace of $\mathbb{R}^n$ that is invariant under A. Explain why K contains an eigenvector of A.

12. Let $G = \begin{bmatrix} A & X \\ 0 & B \end{bmatrix}$. Use formula (1) for the determinant in Section 5.2 to explain why $\det G = (\det A)(\det B)$. From this, deduce that the characteristic polynomial of G is the product of the characteristic polynomials of A and B.

Use Exercise 12 to find the eigenvalues of the matrices in Exercises 13 and 14.

13. $A = \begin{bmatrix} 3 & -2 & 8 \\ 0 & 5 & -2 \\ 0 & -4 & 3 \end{bmatrix}$

14. $A = \begin{bmatrix} 1 & 5 & -6 & -7 \\ 2 & 4 & 5 & 2 \\ 0 & 0 & -7 & -4 \\ 0 & 0 & 3 & 1 \end{bmatrix}$

15. Let J be the $n \times n$ matrix of all 1's, and consider $A = (a - b)I + bJ$; that is,

$$A = \begin{bmatrix} a & b & b & \cdots & b \\ b & a & b & \cdots & b \\ b & b & a & \cdots & b \\ \vdots & \vdots & \vdots & \ddots & \vdots \\ b & b & b & \cdots & a \end{bmatrix}$$

Use the results of Exercise 16 in the Supplementary Exercises for Chapter 3 to show that the eigenvalues of A are $a - b$ and $a + (n - 1)b$. What are the multiplicities of these eigenvalues?

16. Apply the result of Exercise 15 to find the eigenvalues of the

matrices $\begin{bmatrix} 1 & 2 & 2 \\ 2 & 1 & 2 \\ 2 & 2 & 1 \end{bmatrix}$ and $\begin{bmatrix} 7 & 3 & 3 & 3 & 3 \\ 3 & 7 & 3 & 3 & 3 \\ 3 & 3 & 7 & 3 & 3 \\ 3 & 3 & 3 & 7 & 3 \\ 3 & 3 & 3 & 3 & 7 \end{bmatrix}$.

17. Let $A = \begin{bmatrix} a_{11} & a_{12} \\ a_{21} & a_{22} \end{bmatrix}$. Recall from Exercise 25 in Section 5.4 that $\operatorname{tr} A$ (the trace of A) is the sum of the diagonal entries in A. Show that the characteristic polynomial of A is

$$\lambda^2 - (\operatorname{tr} A)\lambda + \det A$$

Then show that the eigenvalues of a 2×2 matrix A are both real if and only if $\det A \leq \left(\dfrac{\operatorname{tr} A}{2}\right)^2$.

18. Let $A = \begin{bmatrix} .4 & -.3 \\ .4 & 1.2 \end{bmatrix}$. Explain why A^k approaches $\begin{bmatrix} -.5 & -.75 \\ 1.0 & 1.50 \end{bmatrix}$ as $k \to \infty$.

Exercises 19–23 concern the polynomial

$$p(t) = a_0 + a_1 t + \cdots + a_{n-1} t^{n-1} + t^n$$

and an $n \times n$ matrix C_p called the **companion matrix** of p:

$$C_p = \begin{bmatrix} 0 & 1 & 0 & \cdots & 0 \\ 0 & 0 & 1 & & 0 \\ \vdots & & & & \vdots \\ 0 & 0 & 0 & & 1 \\ -a_0 & -a_1 & -a_2 & \cdots & -a_{n-1} \end{bmatrix}$$

19. Write the companion matrix C_p for $p(t) = 6 - 5t + t^2$, and then find the characteristic polynomial of C_p.

20. Let $p(t) = (t - 2)(t - 3)(t - 4) = -24 + 26t - 9t^2 + t^3$. Write the companion matrix for $p(t)$, and use techniques from Chapter 3 to find its characteristic polynomial.

21. Use mathematical induction to prove that for $n \geq 2$,

$$\det(C_p - \lambda I) = (-1)^n(a_0 + a_1\lambda + \cdots + a_{n-1}\lambda^{n-1} + \lambda^n)$$
$$= (-1)^n p(\lambda)$$

[*Hint:* Expanding by cofactors down the first column, show that $\det(C_p - \lambda I)$ has the form $(-\lambda)B + (-1)^n a_0$, where B is a certain polynomial (by the induction assumption).]

22. Let $p(t) = a_0 + a_1 t + a_2 t^2 + t^3$, and let λ be a zero of p.

   a. Write the companion matrix for p.

   b. Explain why $\lambda^3 = -a_0 - a_1\lambda - a_2\lambda^2$, and show that $(1, \lambda, \lambda^2)$ is an eigenvector of the companion matrix for p.

23. Let p be the polynomial in Exercise 22, and suppose the equation $p(t) = 0$ has distinct roots $\lambda_1, \lambda_2, \lambda_3$. Let V be the Vandermonde matrix

$$V = \begin{bmatrix} 1 & 1 & 1 \\ \lambda_1 & \lambda_2 & \lambda_3 \\ \lambda_1^2 & \lambda_2^2 & \lambda_3^2 \end{bmatrix}$$

(The transpose of V was considered in Supplementary Exercise 11 in Chapter 2.) Use Exercise 22 and a theorem from this chapter to deduce that V is invertible (but do not compute V^{-1}). Then explain why $V^{-1}C_p V$ is a diagonal matrix.

24. [M] The MATLAB command `roots(p)` computes the roots of the polynomial equation $p(t) = 0$. Read a MATLAB manual, and then describe the basic idea behind the algorithm for the `roots` command.

25. [M] Use a matrix program to diagonalize

$$A = \begin{bmatrix} -3 & -2 & 0 \\ 14 & 7 & -1 \\ -6 & -3 & 1 \end{bmatrix}$$

if possible. Use the eigenvalue command to create the diagonal matrix D. If the program has a command that produces eigenvectors, use it to create an invertible matrix P. Then compute $AP - PD$ and PDP^{-1}. Discuss your results.

26. [M] Repeat Exercise 25 for $A = \begin{bmatrix} -8 & 5 & -2 & 0 \\ -5 & 2 & 1 & -2 \\ 10 & -8 & 6 & -3 \\ 3 & -2 & 1 & 0 \end{bmatrix}$.

6

Orthogonality and Least Squares

INTRODUCTORY EXAMPLE

The North American Datum and GPS Navigation

Imagine starting a massive project that you estimate will take ten years and require the efforts of scores of people to construct and solve a 1,800,000 by 900,000 system of linear equations. That is exactly what the National Geodetic Survey did in 1974, when it set out to update the North American Datum (NAD)—a network of 268,000 precisely located reference points that span the entire North American continent, together with Greenland, Hawaii, the Virgin Islands, Puerto Rico, and other Caribbean islands.

The recorded latitudes and longitudes in the NAD must be determined to within a few centimeters because they form the basis for all surveys, maps, legal property boundaries, and layouts of civil engineering projects such as highways and public utility lines. However, more than 200,000 new points had been added to the datum since the last adjustment in 1927, and errors had gradually accumulated over the years, due to imprecise measurements and shifts in the earth's crust. Data gathering for the NAD readjustment was completed in 1983.

The system of equations for the NAD had no solution in the ordinary sense, but rather had a *least-squares solution*, which assigned latitudes and longitudes to the reference points in a way that corresponded best to the 1.8 million observations. The least-squares solution was found in 1986 by solving a related system of so-called

normal equations, which involved 928,735 equations in 928,735 variables.[1]

More recently, knowledge of reference points on the ground has become crucial for accurately determining the locations of satellites in the satellite-based *Global Positioning System (GPS)*. A GPS satellite calculates its position relative to the earth by measuring the time it takes for signals to arrive from three ground transmitters. To do this, the satellites use precise atomic clocks that have been synchronized with ground stations (whose locations are known accurately because of the NAD).

The *Global Positioning System* is used both for determining the locations of new reference points on the ground and for finding a user's position on the ground relative to established maps. When a car driver (or a mountain climber) turns on a GPS receiver, the receiver measures the relative arrival times of signals from at least three satellites. This information, together with the transmitted data about the satellites' locations and message times, is used to adjust the GPS receiver's time and to determine its approximate location on the earth. Given information from a fourth satellite, the GPS receiver can even establish its approximate altitude.

---

[1] A mathematical discussion of the solution strategy (along with details of the entire NAD project) appears in *North American Datum of 1983*, Charles R. Schwarz (ed.), National Geodetic Survey, National Oceanic and Atmospheric Administration (NOAA) Professional Paper NOS 2, 1989.

Both the NAD and GPS problems are solved by finding a vector that "approximately satisfies" an inconsistent system of equations. A careful explanation of this apparent contradiction will require ideas developed in the first five sections of this chapter.

WEB

In order to find an approximate solution to an inconsistent system of equations that has no actual solution, a well-defined notion of nearness is needed. Section 6.1 introduces the concepts of distance and orthogonality in a vector space. Sections 6.2 and 6.3 show how orthogonality can be used to identify the point within a subspace W that is nearest to a point $\mathbf{y}$ lying outside of W. By taking W to be the column space of a matrix, Section 6.5 develops a method for producing approximate ("least-squares") solutions for inconsistent linear systems, such as the system solved for the NAD report.

Section 6.4 provides another opportunity to see orthogonal projections at work, creating a matrix factorization widely used in numerical linear algebra. The remaining sections examine some of the many least-squares problems that arise in applications, including those in vector spaces more general than $\mathbb{R}^n$.

6.1 INNER PRODUCT, LENGTH, AND ORTHOGONALITY

Geometric concepts of length, distance, and perpendicularity, which are well known for $\mathbb{R}^2$ and $\mathbb{R}^3$, are defined here for $\mathbb{R}^n$. These concepts provide powerful geometric tools for solving many applied problems, including the least-squares problems mentioned above. All three notions are defined in terms of the inner product of two vectors.

The Inner Product

If $\mathbf{u}$ and $\mathbf{v}$ are vectors in $\mathbb{R}^n$, then we regard $\mathbf{u}$ and $\mathbf{v}$ as $n \times 1$ matrices. The transpose $\mathbf{u}^T$ is a $1 \times n$ matrix, and the matrix product $\mathbf{u}^T\mathbf{v}$ is a 1×1 matrix, which we write as a single real number (a scalar) without brackets. The number $\mathbf{u}^T\mathbf{v}$ is called the **inner product** of $\mathbf{u}$ and $\mathbf{v}$, and often it is written as $\mathbf{u}\cdot\mathbf{v}$. This inner product, mentioned in the exercises for Section 2.1, is also referred to as a **dot product**. If

$$\mathbf{u} = \begin{bmatrix} u_1 \\ u_2 \\ \vdots \\ u_n \end{bmatrix} \quad \text{and} \quad \mathbf{v} = \begin{bmatrix} v_1 \\ v_2 \\ \vdots \\ v_n \end{bmatrix}$$

then the inner product of $\mathbf{u}$ and $\mathbf{v}$ is

$$\begin{bmatrix} u_1 & u_2 & \cdots & u_n \end{bmatrix} \begin{bmatrix} v_1 \\ v_2 \\ \vdots \\ v_n \end{bmatrix} = u_1 v_1 + u_2 v_2 + \cdots + u_n v_n$$

EXAMPLE 1 Compute $\mathbf{u} \cdot \mathbf{v}$ and $\mathbf{v} \cdot \mathbf{u}$ for $\mathbf{u} = \begin{bmatrix} 2 \\ -5 \\ -1 \end{bmatrix}$ and $\mathbf{v} = \begin{bmatrix} 3 \\ 2 \\ -3 \end{bmatrix}$.

SOLUTION

$$\mathbf{u} \cdot \mathbf{v} = \mathbf{u}^T \mathbf{v} = \begin{bmatrix} 2 & -5 & -1 \end{bmatrix} \begin{bmatrix} 3 \\ 2 \\ -3 \end{bmatrix} = (2)(3) + (-5)(2) + (-1)(-3) = -1$$

$$\mathbf{v} \cdot \mathbf{u} = \mathbf{v}^T \mathbf{u} = \begin{bmatrix} 3 & 2 & -3 \end{bmatrix} \begin{bmatrix} 2 \\ -5 \\ -1 \end{bmatrix} = (3)(2) + (2)(-5) + (-3)(-1) = -1 \quad \blacksquare$$

It is clear from the calculations in Example 1 why $\mathbf{u} \cdot \mathbf{v} = \mathbf{v} \cdot \mathbf{u}$. This commutativity of the inner product holds in general. The following properties of the inner product are easily deduced from properties of the transpose operation in Section 2.1. (See Exercises 21 and 22 at the end of this section.)

THEOREM 1

Let $\mathbf{u}$, $\mathbf{v}$, and $\mathbf{w}$ be vectors in $\mathbb{R}^n$, and let c be a scalar. Then

a. $\mathbf{u} \cdot \mathbf{v} = \mathbf{v} \cdot \mathbf{u}$

b. $(\mathbf{u} + \mathbf{v}) \cdot \mathbf{w} = \mathbf{u} \cdot \mathbf{w} + \mathbf{v} \cdot \mathbf{w}$

c. $(c\mathbf{u}) \cdot \mathbf{v} = c(\mathbf{u} \cdot \mathbf{v}) = \mathbf{u} \cdot (c\mathbf{v})$

d. $\mathbf{u} \cdot \mathbf{u} \geq 0$, and $\mathbf{u} \cdot \mathbf{u} = 0$ if and only if $\mathbf{u} = \mathbf{0}$

Properties (b) and (c) can be combined several times to produce the following useful rule:

$$(c_1 \mathbf{u}_1 + \cdots + c_p \mathbf{u}_p) \cdot \mathbf{w} = c_1 (\mathbf{u}_1 \cdot \mathbf{w}) + \cdots + c_p (\mathbf{u}_p \cdot \mathbf{w})$$

The Length of a Vector

If $\mathbf{v}$ is in $\mathbb{R}^n$, with entries $v_1, \ldots, v_n$, then the square root of $\mathbf{v} \cdot \mathbf{v}$ is defined because $\mathbf{v} \cdot \mathbf{v}$ is nonnegative.

DEFINITION

The **length** (or **norm**) of $\mathbf{v}$ is the nonnegative scalar $\|\mathbf{v}\|$ defined by

$$\|\mathbf{v}\| = \sqrt{\mathbf{v} \cdot \mathbf{v}} = \sqrt{v_1^2 + v_2^2 + \cdots + v_n^2}, \quad \text{and} \quad \|\mathbf{v}\|^2 = \mathbf{v} \cdot \mathbf{v}$$

Suppose $\mathbf{v}$ is in $\mathbb{R}^2$, say, $\mathbf{v} = \begin{bmatrix} a \\ b \end{bmatrix}$. If we identify $\mathbf{v}$ with a geometric point in the plane, as usual, then $\|\mathbf{v}\|$ coincides with the standard notion of the length of the line segment from the origin to $\mathbf{v}$. This follows from the Pythagorean Theorem applied to a triangle such as the one in Fig. 1.

A similar calculation with the diagonal of a rectangular box shows that the definition of length of a vector $\mathbf{v}$ in $\mathbb{R}^3$ coincides with the usual notion of length.

For any scalar c, the length of $c\mathbf{v}$ is $|c|$ times the length of $\mathbf{v}$. That is,

$$\|c\mathbf{v}\| = |c| \|\mathbf{v}\|$$

(To see this, compute $\|c\mathbf{v}\|^2 = (c\mathbf{v}) \cdot (c\mathbf{v}) = c^2 \mathbf{v} \cdot \mathbf{v} = c^2 \|\mathbf{v}\|^2$ and take square roots.)

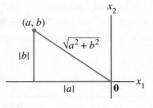

FIGURE 1
Interpretation of $\|\mathbf{v}\|$ as length.

A vector whose length is 1 is called a **unit vector**. If we *divide* a nonzero vector $\mathbf{v}$ by its length—that is, multiply by $1/\|\mathbf{v}\|$—we obtain a unit vector $\mathbf{u}$ because the length of $\mathbf{u}$ is $(1/\|\mathbf{v}\|)\|\mathbf{v}\|$. The process of creating $\mathbf{u}$ from $\mathbf{v}$ is sometimes called **normalizing** $\mathbf{v}$, and we say that $\mathbf{u}$ is *in the same direction* as $\mathbf{v}$.

Several examples that follow use the space-saving notation for (column) vectors.

EXAMPLE 2 Let $\mathbf{v} = (1, -2, 2, 0)$. Find a unit vector $\mathbf{u}$ in the same direction as $\mathbf{v}$.

SOLUTION First, compute the length of $\mathbf{v}$:

$$\|\mathbf{v}\|^2 = \mathbf{v}\cdot\mathbf{v} = (1)^2 + (-2)^2 + (2)^2 + (0)^2 = 9$$
$$\|\mathbf{v}\| = \sqrt{9} = 3$$

Then, multiply $\mathbf{v}$ by $1/\|\mathbf{v}\|$ to obtain

$$\mathbf{u} = \frac{1}{\|\mathbf{v}\|}\mathbf{v} = \frac{1}{3}\mathbf{v} = \frac{1}{3}\begin{bmatrix} 1 \\ -2 \\ 2 \\ 0 \end{bmatrix} = \begin{bmatrix} 1/3 \\ -2/3 \\ 2/3 \\ 0 \end{bmatrix}$$

To check that $\|\mathbf{u}\| = 1$, it suffices to show that $\|\mathbf{u}\|^2 = 1$.

$$\|\mathbf{u}\|^2 = \mathbf{u}\cdot\mathbf{u} = \left(\tfrac{1}{3}\right)^2 + \left(-\tfrac{2}{3}\right)^2 + \left(\tfrac{2}{3}\right)^2 + (0)^2$$
$$= \tfrac{1}{9} + \tfrac{4}{9} + \tfrac{4}{9} + 0 = 1 \qquad\blacksquare$$

EXAMPLE 3 Let W be the subspace of $\mathbb{R}^2$ spanned by $\mathbf{x} = (\tfrac{2}{3}, 1)$. Find a unit vector $\mathbf{z}$ that is a basis for W.

SOLUTION W consists of all multiples of $\mathbf{x}$, as in Fig. 2(a). Any nonzero vector in W is a basis for W. To simplify the calculation, "scale" $\mathbf{x}$ to eliminate fractions. That is, multiply $\mathbf{x}$ by 3 to get

$$\mathbf{y} = \begin{bmatrix} 2 \\ 3 \end{bmatrix}$$

Now compute $\|\mathbf{y}\|^2 = 2^2 + 3^2 = 13$, $\|\mathbf{y}\| = \sqrt{13}$, and normalize $\mathbf{y}$ to get

$$\mathbf{z} = \frac{1}{\sqrt{13}}\begin{bmatrix} 2 \\ 3 \end{bmatrix} = \begin{bmatrix} 2/\sqrt{13} \\ 3/\sqrt{13} \end{bmatrix}$$

See Fig. 2(b). Another unit vector is $(-2/\sqrt{13}, -3/\sqrt{13})$. $\qquad\blacksquare$

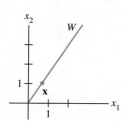

(a)

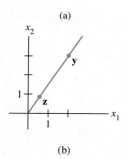

(b)

FIGURE 2
Normalizing a vector to produce a unit vector.

Distance in $\mathbb{R}^n$

We are ready now to describe how close one vector is to another. Recall that if a and b are real numbers, the distance on the number line between a and b is the number $|a - b|$. Two examples are shown in Fig. 3. This definition of distance in $\mathbb{R}$ has a direct analogue in $\mathbb{R}^n$.

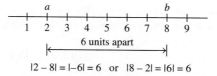

$|2 - 8| = |-6| = 6$ or $|8 - 2| = |6| = 6$

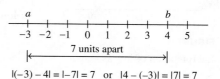

$|(-3) - 4| = |-7| = 7$ or $|4 - (-3)| = |7| = 7$

FIGURE 3 Distances in $\mathbb{R}$.

DEFINITION

For **u** and **v** in $\mathbb{R}^n$, the **distance between u and v**, written as $\text{dist}(\mathbf{u}, \mathbf{v})$, is the length of the vector $\mathbf{u} - \mathbf{v}$. That is,

$$\text{dist}(\mathbf{u}, \mathbf{v}) = \|\mathbf{u} - \mathbf{v}\|$$

In $\mathbb{R}^2$ and $\mathbb{R}^3$, this definition of distance coincides with the usual formulas for the Euclidean distance between two points, as the next two examples show.

EXAMPLE 4 Compute the distance between the vectors $\mathbf{u} = (7, 1)$ and $\mathbf{v} = (3, 2)$.

SOLUTION Calculate

$$\mathbf{u} - \mathbf{v} = \begin{bmatrix} 7 \\ 1 \end{bmatrix} - \begin{bmatrix} 3 \\ 2 \end{bmatrix} = \begin{bmatrix} 4 \\ -1 \end{bmatrix}$$

$$\|\mathbf{u} - \mathbf{v}\| = \sqrt{4^2 + (-1)^2} = \sqrt{17}$$

The vectors $\mathbf{u}$, $\mathbf{v}$, and $\mathbf{u} - \mathbf{v}$ are shown in Fig. 4. When the vector $\mathbf{u} - \mathbf{v}$ is added to $\mathbf{v}$, the result is $\mathbf{u}$. Notice that the parallelogram in Fig. 4 shows that the distance from $\mathbf{u}$ to $\mathbf{v}$ is the same as the distance from $\mathbf{u} - \mathbf{v}$ to $\mathbf{0}$. ■

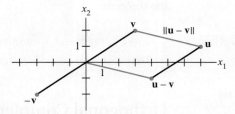

FIGURE 4 The distance between **u** and **v** is the length of **u** − **v**.

EXAMPLE 5 If $\mathbf{u} = (u_1, u_2, u_3)$ and $\mathbf{v} = (v_1, v_2, v_3)$, then

$$\text{dist}(\mathbf{u}, \mathbf{v}) = \|\mathbf{u} - \mathbf{v}\| = \sqrt{(\mathbf{u} - \mathbf{v}) \cdot (\mathbf{u} - \mathbf{v})}$$

$$= \sqrt{(u_1 - v_1)^2 + (u_2 - v_2)^2 + (u_3 - v_3)^2}$$ ■

Orthogonal Vectors

The rest of this chapter depends on the fact that the concept of perpendicular lines in ordinary Euclidean geometry has an analogue in $\mathbb{R}^n$.

Consider $\mathbb{R}^2$ or $\mathbb{R}^3$ and two lines through the origin determined by vectors $\mathbf{u}$ and $\mathbf{v}$. The two lines shown in Fig. 5 are geometrically perpendicular if and only if the distance from $\mathbf{u}$ to $\mathbf{v}$ is the same as the distance from $\mathbf{u}$ to $-\mathbf{v}$. This is the same as requiring the squares of the distances to be the same. Now

$$
\begin{aligned}
\left[\,\text{dist}(\mathbf{u}, -\mathbf{v})\,\right]^2 &= \|\mathbf{u} - (-\mathbf{v})\|^2 = \|\mathbf{u} + \mathbf{v}\|^2 \\
&= (\mathbf{u} + \mathbf{v}) \cdot (\mathbf{u} + \mathbf{v}) \\
&= \mathbf{u} \cdot (\mathbf{u} + \mathbf{v}) + \mathbf{v} \cdot (\mathbf{u} + \mathbf{v}) && \text{Theorem 1(b)} \\
&= \mathbf{u} \cdot \mathbf{u} + \mathbf{u} \cdot \mathbf{v} + \mathbf{v} \cdot \mathbf{u} + \mathbf{v} \cdot \mathbf{v} && \text{Theorem 1(a), (b)} \\
&= \|\mathbf{u}\|^2 + \|\mathbf{v}\|^2 + 2\mathbf{u} \cdot \mathbf{v} && \text{Theorem 1(a)} \qquad (1)
\end{aligned}
$$

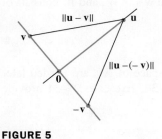

FIGURE 5

The same calculations with $\mathbf{v}$ and $-\mathbf{v}$ interchanged show that

$$[\text{dist}(\mathbf{u}, \mathbf{v})]^2 = \|\mathbf{u}\|^2 + \|-\mathbf{v}\|^2 + 2\mathbf{u}\cdot(-\mathbf{v})$$
$$= \|\mathbf{u}\|^2 + \|\mathbf{v}\|^2 - 2\mathbf{u}\cdot\mathbf{v}$$

The two squared distances are equal if and only if $2\mathbf{u}\cdot\mathbf{v} = -2\mathbf{u}\cdot\mathbf{v}$, which happens if and only if $\mathbf{u}\cdot\mathbf{v} = 0$.

This calculation shows that when vectors $\mathbf{u}$ and $\mathbf{v}$ are identified with geometric points, the corresponding lines through the points and the origin are perpendicular if and only if $\mathbf{u}\cdot\mathbf{v} = 0$. The following definition generalizes to $\mathbb{R}^n$ this notion of perpendicularity (or *orthogonality*, as it is commonly called in linear algebra).

DEFINITION

> Two vectors $\mathbf{u}$ and $\mathbf{v}$ in $\mathbb{R}^n$ are **orthogonal** (to each other) if $\mathbf{u}\cdot\mathbf{v} = 0$.

Observe that the zero vector is orthogonal to every vector in $\mathbb{R}^n$ because $\mathbf{0}^T\mathbf{v} = 0$ for all $\mathbf{v}$.

The next theorem provides a useful fact about orthogonal vectors. The proof follows immediately from the calculation in (1) above and the definition of orthogonality. The right triangle shown in Fig. 6 provides a visualization of the lengths that appear in the theorem.

THEOREM 2

> **The Pythagorean Theorem**
>
> Two vectors $\mathbf{u}$ and $\mathbf{v}$ are orthogonal if and only if $\|\mathbf{u} + \mathbf{v}\|^2 = \|\mathbf{u}\|^2 + \|\mathbf{v}\|^2$.

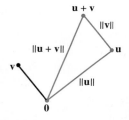

FIGURE 6

Orthogonal Complements

To provide practice using inner products, we introduce a concept here that will be of use in Section 6.3 and elsewhere in the chapter. If a vector $\mathbf{z}$ is orthogonal to every vector in a subspace W of $\mathbb{R}^n$, then $\mathbf{z}$ is said to be **orthogonal to** W. The set of all vectors $\mathbf{z}$ that are orthogonal to W is called the **orthogonal complement** of W and is denoted by $W^\perp$ (and read as "W perpendicular" or simply "W perp").

EXAMPLE 6 Let W be a plane through the origin in $\mathbb{R}^3$, and let L be the line through the origin and perpendicular to W. If $\mathbf{z}$ and $\mathbf{w}$ are nonzero, $\mathbf{z}$ is on L, and $\mathbf{w}$ is in W, then the line segment from $\mathbf{0}$ to $\mathbf{z}$ is perpendicular to the line segment from $\mathbf{0}$ to $\mathbf{w}$; that is, $\mathbf{z}\cdot\mathbf{w} = 0$. See Fig. 7. So each vector on L is orthogonal to every $\mathbf{w}$ in W. In fact, L consists of *all* vectors that are orthogonal to the $\mathbf{w}$'s in W, and W consists of all vectors orthogonal to the $\mathbf{z}$'s in L. That is,

$$L = W^\perp \quad \text{and} \quad W = L^\perp \qquad \blacksquare$$

FIGURE 7

A plane and line through $\mathbf{0}$ as orthogonal complements.

The following two facts about $W^\perp$, with W a subspace of $\mathbb{R}^n$, are needed later in the chapter. Proofs are suggested in Exercises 29 and 30. Exercises 27–31 provide excellent practice using properties of the inner product.

> 1. A vector $\mathbf{x}$ is in $W^\perp$ if and only if $\mathbf{x}$ is orthogonal to every vector in a set that spans W.
> 2. $W^\perp$ is a subspace of $\mathbb{R}^n$.

The next theorem and Exercise 31 verify the claims made in Section 4.6 concerning the subspaces shown in Fig. 8. (Also see Exercise 28 in Section 4.6.)

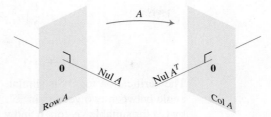

FIGURE 8 The fundamental subspaces determined by an $m \times n$ matrix A.

THEOREM 3

Let A be an $m \times n$ matrix. The orthogonal complement of the row space of A is the null space of A, and the orthogonal complement of the column space of A is the null space of A^T:

$$(\text{Row } A)^{\perp} = \text{Nul } A \quad \text{and} \quad (\text{Col } A)^{\perp} = \text{Nul } A^T$$

PROOF The row–column rule for computing $A\mathbf{x}$ shows that if $\mathbf{x}$ is in Nul A, then $\mathbf{x}$ is orthogonal to each row of A (with the rows treated as vectors in $\mathbb{R}^n$). Since the rows of A span the row space, $\mathbf{x}$ is orthogonal to Row A. Conversely, if $\mathbf{x}$ is orthogonal to Row A, then $\mathbf{x}$ is certainly orthogonal to each row of A, and hence $A\mathbf{x} = \mathbf{0}$. This proves the first statement of the theorem. Since this statement is true for any matrix, it is true for A^T. That is, the orthogonal complement of the row space of A^T is the null space of A^T. This proves the second statement, because Row $A^T = \text{Col } A$. ∎

Angles in $\mathbb{R}^2$ and $\mathbb{R}^3$ (Optional)

If $\mathbf{u}$ and $\mathbf{v}$ are nonzero vectors in either $\mathbb{R}^2$ or $\mathbb{R}^3$, then there is a nice connection between their inner product and the angle ϑ between the two line segments from the origin to the points identified with $\mathbf{u}$ and $\mathbf{v}$. The formula is

$$\mathbf{u} \cdot \mathbf{v} = \|\mathbf{u}\| \, \|\mathbf{v}\| \cos \vartheta \tag{2}$$

To verify this formula for vectors in $\mathbb{R}^2$, consider the triangle shown in Fig. 9, with sides of lengths $\|\mathbf{u}\|$, $\|\mathbf{v}\|$, and $\|\mathbf{u} - \mathbf{v}\|$. By the law of cosines,

$$\|\mathbf{u} - \mathbf{v}\|^2 = \|\mathbf{u}\|^2 + \|\mathbf{v}\|^2 - 2\|\mathbf{u}\| \, \|\mathbf{v}\| \cos \vartheta$$

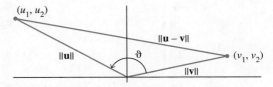

FIGURE 9 The angle between two vectors.

which can be rearranged to produce

$$\|\mathbf{u}\| \, \|\mathbf{v}\| \cos \vartheta = \frac{1}{2} \left[\|\mathbf{u}\|^2 + \|\mathbf{v}\|^2 - \|\mathbf{u} - \mathbf{v}\|^2 \right]$$

$$= \frac{1}{2} \left[u_1^2 + u_2^2 + v_1^2 + v_2^2 - (u_1 - v_1)^2 - (u_2 - v_2)^2 \right]$$

$$= u_1 v_1 + u_2 v_2$$

$$= \mathbf{u} \cdot \mathbf{v}$$

The verification for $\mathbb{R}^3$ is similar. When $n > 3$, formula (2) may be used to *define* the angle between two vectors in $\mathbb{R}^n$. In statistics, for instance, the value of $\cos \vartheta$ defined by (2) for suitable vectors $\mathbf{u}$ and $\mathbf{v}$ is what statisticians call a *correlation coefficient*.

PRACTICE PROBLEMS

Let $\mathbf{a} = \begin{bmatrix} -2 \\ 1 \end{bmatrix}$, $\mathbf{b} = \begin{bmatrix} -3 \\ 1 \end{bmatrix}$, $\mathbf{c} = \begin{bmatrix} 4/3 \\ -1 \\ 2/3 \end{bmatrix}$, and $\mathbf{d} = \begin{bmatrix} 5 \\ 6 \\ -1 \end{bmatrix}$.

1. Compute $\dfrac{\mathbf{a} \cdot \mathbf{b}}{\mathbf{a} \cdot \mathbf{a}}$ and $\left(\dfrac{\mathbf{a} \cdot \mathbf{b}}{\mathbf{a} \cdot \mathbf{a}} \right) \mathbf{a}$.

2. Find a unit vector $\mathbf{u}$ in the direction of $\mathbf{c}$.

3. Show that $\mathbf{d}$ is orthogonal to $\mathbf{c}$.

4. Use the results of Practice Problems 2 and 3 to explain why $\mathbf{d}$ must be orthogonal to the unit vector $\mathbf{u}$.

6.1 EXERCISES

Compute the quantities in Exercises 1–8 using the vectors

$$\mathbf{u} = \begin{bmatrix} -1 \\ 2 \end{bmatrix}, \quad \mathbf{v} = \begin{bmatrix} 4 \\ 6 \end{bmatrix}, \quad \mathbf{w} = \begin{bmatrix} 3 \\ -1 \\ -5 \end{bmatrix}, \quad \mathbf{x} = \begin{bmatrix} 6 \\ -2 \\ 3 \end{bmatrix}$$

1. $\mathbf{u} \cdot \mathbf{u}$, $\mathbf{v} \cdot \mathbf{u}$, and $\dfrac{\mathbf{v} \cdot \mathbf{u}}{\mathbf{u} \cdot \mathbf{u}}$

2. $\mathbf{w} \cdot \mathbf{w}$, $\mathbf{x} \cdot \mathbf{w}$, and $\dfrac{\mathbf{x} \cdot \mathbf{w}}{\mathbf{w} \cdot \mathbf{w}}$

3. $\dfrac{1}{\mathbf{w} \cdot \mathbf{w}} \mathbf{w}$

4. $\dfrac{1}{\mathbf{u} \cdot \mathbf{u}} \mathbf{u}$

5. $\left(\dfrac{\mathbf{u} \cdot \mathbf{v}}{\mathbf{v} \cdot \mathbf{v}} \right) \mathbf{v}$

6. $\left(\dfrac{\mathbf{x} \cdot \mathbf{w}}{\mathbf{x} \cdot \mathbf{x}} \right) \mathbf{x}$

7. $\|\mathbf{w}\|$

8. $\|\mathbf{x}\|$

In Exercises 9–12, find a unit vector in the direction of the given vector.

9. $\begin{bmatrix} -30 \\ 40 \end{bmatrix}$

10. $\begin{bmatrix} -6 \\ 4 \\ -3 \end{bmatrix}$

11. $\begin{bmatrix} 7/4 \\ 1/2 \\ 1 \end{bmatrix}$

12. $\begin{bmatrix} 8/3 \\ 2 \end{bmatrix}$

13. Find the distance between $\mathbf{x} = \begin{bmatrix} 10 \\ -3 \end{bmatrix}$ and $\mathbf{y} = \begin{bmatrix} -1 \\ -5 \end{bmatrix}$.

14. Find the distance between $\mathbf{u} = \begin{bmatrix} 0 \\ -5 \\ 2 \end{bmatrix}$ and $\mathbf{z} = \begin{bmatrix} -4 \\ -1 \\ 8 \end{bmatrix}$.

Determine which pairs of vectors in Exercises 15–18 are orthogonal.

15. $\mathbf{a} = \begin{bmatrix} 8 \\ -5 \end{bmatrix}$, $\mathbf{b} = \begin{bmatrix} -2 \\ -3 \end{bmatrix}$

16. $\mathbf{u} = \begin{bmatrix} 12 \\ 3 \\ -5 \end{bmatrix}$, $\mathbf{v} = \begin{bmatrix} 2 \\ -3 \\ 3 \end{bmatrix}$

17. $\mathbf{u} = \begin{bmatrix} 3 \\ 2 \\ -5 \\ 0 \end{bmatrix}$, $\mathbf{v} = \begin{bmatrix} -4 \\ 1 \\ -2 \\ 6 \end{bmatrix}$

18. $\mathbf{y} = \begin{bmatrix} -3 \\ 7 \\ 4 \\ 0 \end{bmatrix}$, $\mathbf{z} = \begin{bmatrix} 1 \\ -8 \\ 15 \\ -7 \end{bmatrix}$

In Exercises 19 and 20, all vectors are in $\mathbb{R}^n$. Mark each statement True or False. Justify each answer.

19. a. $\mathbf{v} \cdot \mathbf{v} = \|\mathbf{v}\|^2$.

   b. For any scalar c, $\mathbf{u} \cdot (c\mathbf{v}) = c(\mathbf{u} \cdot \mathbf{v})$.

   c. If the distance from $\mathbf{u}$ to $\mathbf{v}$ equals the distance from $\mathbf{u}$ to $-\mathbf{v}$, then $\mathbf{u}$ and $\mathbf{v}$ are orthogonal.

   d. For a square matrix A, vectors in Col A are orthogonal to vectors in Nul A.

   e. If vectors $\mathbf{v}_1, \ldots, \mathbf{v}_p$ span a subspace W and if $\mathbf{x}$ is orthogonal to each $\mathbf{v}_j$ for $j = 1, \ldots, p$, then $\mathbf{x}$ is in $W^{\perp}$.

20. a. $\mathbf{u}\cdot\mathbf{v} - \mathbf{v}\cdot\mathbf{u} = 0$.

b. For any scalar c, $\|c\mathbf{v}\| = c\|\mathbf{v}\|$.

c. If $\mathbf{x}$ is orthogonal to every vector in a subspace W, then $\mathbf{x}$ is in $W^\perp$.

d. If $\|\mathbf{u}\|^2 + \|\mathbf{v}\|^2 = \|\mathbf{u}+\mathbf{v}\|^2$, then $\mathbf{u}$ and $\mathbf{v}$ are orthogonal.

e. For an $m \times n$ matrix A, vectors in the null space of A are orthogonal to vectors in the row space of A.

21. Use the transpose definition of the inner product to verify parts (b) and (c) of Theorem 1. Mention the appropriate facts from Chapter 2.

22. Let $\mathbf{u} = (u_1, u_2, u_3)$.   Explain why $\mathbf{u}\cdot\mathbf{u} \geq 0$.   When is $\mathbf{u}\cdot\mathbf{u} = 0$?

23. Let $\mathbf{u} = \begin{bmatrix} 2 \\ -5 \\ -1 \end{bmatrix}$ and $\mathbf{v} = \begin{bmatrix} -7 \\ -4 \\ 6 \end{bmatrix}$. Compute and compare $\mathbf{u}\cdot\mathbf{v}$, $\|\mathbf{u}\|^2$, $\|\mathbf{v}\|^2$, and $\|\mathbf{u}+\mathbf{v}\|^2$. Do not use the Pythagorean Theorem.

24. Verify the *parallelogram law* for vectors $\mathbf{u}$ and $\mathbf{v}$ in $\mathbb{R}^n$:
$$\|\mathbf{u}+\mathbf{v}\|^2 + \|\mathbf{u}-\mathbf{v}\|^2 = 2\|\mathbf{u}\|^2 + 2\|\mathbf{v}\|^2$$

25. Let $\mathbf{v} = \begin{bmatrix} a \\ b \end{bmatrix}$. Describe the set H of vectors $\begin{bmatrix} x \\ y \end{bmatrix}$ that are orthogonal to $\mathbf{v}$. [*Hint:* Consider $\mathbf{v} = \mathbf{0}$ and $\mathbf{v} \neq \mathbf{0}$.]

26. Let $\mathbf{u} = \begin{bmatrix} 5 \\ -6 \\ 7 \end{bmatrix}$, and let W be the set of all $\mathbf{x}$ in $\mathbb{R}^3$ such that $\mathbf{u}\cdot\mathbf{x} = 0$. What theorem in Chapter 4 can be used to show that W is a subspace of $\mathbb{R}^3$? Describe W in geometric language.

27. Suppose a vector $\mathbf{y}$ is orthogonal to vectors $\mathbf{u}$ and $\mathbf{v}$. Show that $\mathbf{y}$ is orthogonal to the vector $\mathbf{u}+\mathbf{v}$.

28. Suppose $\mathbf{y}$ is orthogonal to $\mathbf{u}$ and $\mathbf{v}$. Show that $\mathbf{y}$ is orthogonal to every $\mathbf{w}$ in Span $\{\mathbf{u}, \mathbf{v}\}$. [*Hint:* An arbitrary $\mathbf{w}$ in Span $\{\mathbf{u}, \mathbf{v}\}$ has the form $\mathbf{w} = c_1\mathbf{u} + c_2\mathbf{v}$. Show that $\mathbf{y}$ is orthogonal to such a vector $\mathbf{w}$.]

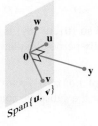

29. Let $W = \text{Span}\{\mathbf{v}_1, \ldots, \mathbf{v}_p\}$. Show that if $\mathbf{x}$ is orthogonal to each $\mathbf{v}_j$, for $1 \leq j \leq p$, then $\mathbf{x}$ is orthogonal to every vector in W.

30. Let W be a subspace of $\mathbb{R}^n$, and let $W^\perp$ be the set of all vectors orthogonal to W. Show that $W^\perp$ is a subspace of $\mathbb{R}^n$ using the following steps.

a. Take $\mathbf{z}$ in $W^\perp$, and let $\mathbf{u}$ represent any element of W. Then $\mathbf{z}\cdot\mathbf{u} = 0$. Take any scalar c and show that $c\mathbf{z}$ is orthogonal to $\mathbf{u}$. (Since $\mathbf{u}$ was an arbitrary element of W, this will show that $c\mathbf{z}$ is in $W^\perp$.)

b. Take $\mathbf{z}_1$ and $\mathbf{z}_2$ in $W^\perp$, and let $\mathbf{u}$ be any element of W. Show that $\mathbf{z}_1 + \mathbf{z}_2$ is orthogonal to $\mathbf{u}$. What can you conclude about $\mathbf{z}_1 + \mathbf{z}_2$? Why?

c. Finish the proof that $W^\perp$ is a subspace of $\mathbb{R}^n$.

31. Show that if $\mathbf{x}$ is in both W and $W^\perp$, then $\mathbf{x} = \mathbf{0}$.

32. [M] Construct a pair $\mathbf{u}$, $\mathbf{v}$ of random vectors in $\mathbb{R}^4$, and let
$$A = \begin{bmatrix} .5 & .5 & .5 & .5 \\ .5 & .5 & -.5 & -.5 \\ .5 & -.5 & .5 & -.5 \\ .5 & -.5 & -.5 & .5 \end{bmatrix}$$

a. Denote the columns of A by $\mathbf{a}_1, \ldots, \mathbf{a}_4$. Compute the length of each column, and compute $\mathbf{a}_1\cdot\mathbf{a}_2$, $\mathbf{a}_1\cdot\mathbf{a}_3$, $\mathbf{a}_1\cdot\mathbf{a}_4$, $\mathbf{a}_2\cdot\mathbf{a}_3$, $\mathbf{a}_2\cdot\mathbf{a}_4$, and $\mathbf{a}_3\cdot\mathbf{a}_4$.

b. Compute and compare the lengths of $\mathbf{u}$, $A\mathbf{u}$, $\mathbf{v}$, and $A\mathbf{v}$.

c. Use equation (2) in this section to compute the cosine of the angle between $\mathbf{u}$ and $\mathbf{v}$. Compare this with the cosine of the angle between $A\mathbf{u}$ and $A\mathbf{v}$.

d. Repeat parts (b) and (c) for two other pairs of random vectors. What do you conjecture about the effect of A on vectors?

33. [M] Generate random vectors $\mathbf{x}$, $\mathbf{y}$, and $\mathbf{v}$ in $\mathbb{R}^4$ with integer entries (and $\mathbf{v} \neq \mathbf{0}$), and compute the quantities
$$\left(\frac{\mathbf{x}\cdot\mathbf{v}}{\mathbf{v}\cdot\mathbf{v}}\right)\mathbf{v}, \quad \left(\frac{\mathbf{y}\cdot\mathbf{v}}{\mathbf{v}\cdot\mathbf{v}}\right)\mathbf{v}, \quad \frac{(\mathbf{x}+\mathbf{y})\cdot\mathbf{v}}{\mathbf{v}\cdot\mathbf{v}}\mathbf{v}, \quad \frac{(10\mathbf{x})\cdot\mathbf{v}}{\mathbf{v}\cdot\mathbf{v}}\mathbf{v}$$

Repeat the computations with new random vectors $\mathbf{x}$ and $\mathbf{y}$. What do you conjecture about the mapping $\mathbf{x} \mapsto T(\mathbf{x}) = \left(\frac{\mathbf{x}\cdot\mathbf{v}}{\mathbf{v}\cdot\mathbf{v}}\right)\mathbf{v}$ (for $\mathbf{v} \neq \mathbf{0}$)? Verify your conjecture algebraically.

34. [M] Let $A = \begin{bmatrix} -6 & 3 & -27 & -33 & -13 \\ 6 & -5 & 25 & 28 & 14 \\ 8 & -6 & 34 & 38 & 18 \\ 12 & -10 & 50 & 41 & 23 \\ 14 & -21 & 49 & 29 & 33 \end{bmatrix}$. Construct a matrix N whose columns form a basis for Nul A, and construct a matrix R whose *rows* form a basis for Row A (see Section 4.6 for details). Perform a matrix computation with N and R that illustrates a fact from Theorem 3.

SOLUTIONS TO PRACTICE PROBLEMS

1. $\mathbf{a} \cdot \mathbf{b} = 7$, $\mathbf{a} \cdot \mathbf{a} = 5$. Hence $\dfrac{\mathbf{a} \cdot \mathbf{b}}{\mathbf{a} \cdot \mathbf{a}} = \dfrac{7}{5}$, and $\left(\dfrac{\mathbf{a} \cdot \mathbf{b}}{\mathbf{a} \cdot \mathbf{a}}\right)\mathbf{a} = \dfrac{7}{5}\mathbf{a} = \begin{bmatrix} -14/5 \\ 7/5 \end{bmatrix}$.

2. Scale $\mathbf{c}$, multiplying by 3 to get $\mathbf{y} = \begin{bmatrix} 4 \\ -3 \\ 2 \end{bmatrix}$. Compute $\|\mathbf{y}\|^2 = 29$ and $\|\mathbf{y}\| = \sqrt{29}$.

   The unit vector in the direction of both $\mathbf{c}$ and $\mathbf{y}$ is $\mathbf{u} = \dfrac{1}{\|\mathbf{y}\|}\mathbf{y} = \begin{bmatrix} 4/\sqrt{29} \\ -3/\sqrt{29} \\ 2/\sqrt{29} \end{bmatrix}$.

3. $\mathbf{d}$ is orthogonal to $\mathbf{c}$, because

$$\mathbf{d} \cdot \mathbf{c} = \begin{bmatrix} 5 \\ 6 \\ -1 \end{bmatrix} \cdot \begin{bmatrix} 4/3 \\ -1 \\ 2/3 \end{bmatrix} = \frac{20}{3} - 6 - \frac{2}{3} = 0$$

4. $\mathbf{d}$ is orthogonal to $\mathbf{u}$ because $\mathbf{u}$ has the form $k\mathbf{c}$ for some k, and

$$\mathbf{d} \cdot \mathbf{u} = \mathbf{d} \cdot (k\mathbf{c}) = k(\mathbf{d} \cdot \mathbf{c}) = k(0) = 0$$

6.2 | ORTHOGONAL SETS

A set of vectors $\{\mathbf{u}_1, \ldots, \mathbf{u}_p\}$ in $\mathbb{R}^n$ is said to be an **orthogonal set** if each pair of distinct vectors from the set is orthogonal, that is, if $\mathbf{u}_i \cdot \mathbf{u}_j = 0$ whenever $i \neq j$.

EXAMPLE 1 Show that $\{\mathbf{u}_1, \mathbf{u}_2, \mathbf{u}_3\}$ is an orthogonal set, where

$$\mathbf{u}_1 = \begin{bmatrix} 3 \\ 1 \\ 1 \end{bmatrix}, \quad \mathbf{u}_2 = \begin{bmatrix} -1 \\ 2 \\ 1 \end{bmatrix}, \quad \mathbf{u}_3 = \begin{bmatrix} -1/2 \\ -2 \\ 7/2 \end{bmatrix}$$

SOLUTION Consider the three possible pairs of distinct vectors, namely, $\{\mathbf{u}_1, \mathbf{u}_2\}$, $\{\mathbf{u}_1, \mathbf{u}_3\}$, and $\{\mathbf{u}_2, \mathbf{u}_3\}$.

$$\mathbf{u}_1 \cdot \mathbf{u}_2 = 3(-1) + 1(2) + 1(1) = 0$$
$$\mathbf{u}_1 \cdot \mathbf{u}_3 = 3\left(-\tfrac{1}{2}\right) + 1(-2) + 1\left(\tfrac{7}{2}\right) = 0$$
$$\mathbf{u}_2 \cdot \mathbf{u}_3 = -1\left(-\tfrac{1}{2}\right) + 2(-2) + 1\left(\tfrac{7}{2}\right) = 0$$

Each pair of distinct vectors is orthogonal, and so $\{\mathbf{u}_1, \mathbf{u}_2, \mathbf{u}_3\}$ is an orthogonal set. See Fig. 1; the three line segments there are mutually perpendicular. ∎

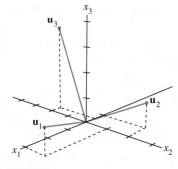

FIGURE 1

THEOREM 4

> If $S = \{\mathbf{u}_1, \ldots, \mathbf{u}_p\}$ is an orthogonal set of nonzero vectors in $\mathbb{R}^n$, then S is linearly independent and hence is a basis for the subspace spanned by S.

PROOF If $\mathbf{0} = c_1\mathbf{u}_1 + \cdots + c_p\mathbf{u}_p$ for some scalars $c_1, \ldots, c_p$, then

$$\begin{aligned}
0 = \mathbf{0} \cdot \mathbf{u}_1 &= (c_1\mathbf{u}_1 + c_2\mathbf{u}_2 + \cdots + c_p\mathbf{u}_p) \cdot \mathbf{u}_1 \\
&= (c_1\mathbf{u}_1) \cdot \mathbf{u}_1 + (c_2\mathbf{u}_2) \cdot \mathbf{u}_1 + \cdots + (c_p\mathbf{u}_p) \cdot \mathbf{u}_1 \\
&= c_1(\mathbf{u}_1 \cdot \mathbf{u}_1) + c_2(\mathbf{u}_2 \cdot \mathbf{u}_1) + \cdots + c_p(\mathbf{u}_p \cdot \mathbf{u}_1) \\
&= c_1(\mathbf{u}_1 \cdot \mathbf{u}_1)
\end{aligned}$$

because $\mathbf{u}_1$ is orthogonal to $\mathbf{u}_2, \ldots, \mathbf{u}_p$. Since $\mathbf{u}_1$ is nonzero, $\mathbf{u}_1 \cdot \mathbf{u}_1$ is not zero and so $c_1 = 0$. Similarly, $c_2, \ldots, c_p$ must be zero. Thus S is linearly independent. ∎

DEFINITION

An **orthogonal basis** for a subspace W of $\mathbb{R}^n$ is a basis for W that is also an orthogonal set.

The next theorem suggests why an orthogonal basis is much nicer than other bases. The weights in a linear combination can be computed easily.

THEOREM 5

Let $\{\mathbf{u}_1, \ldots, \mathbf{u}_p\}$ be an orthogonal basis for a subspace W of $\mathbb{R}^n$. For each $\mathbf{y}$ in W, the weights in the linear combination

$$\mathbf{y} = c_1\mathbf{u}_1 + \cdots + c_p\mathbf{u}_p$$

are given by

$$c_j = \frac{\mathbf{y} \cdot \mathbf{u}_j}{\mathbf{u}_j \cdot \mathbf{u}_j} \qquad (j = 1, \ldots, p)$$

PROOF As in the preceding proof, the orthogonality of $\{\mathbf{u}_1, \ldots, \mathbf{u}_p\}$ shows that

$$\mathbf{y} \cdot \mathbf{u}_1 = (c_1\mathbf{u}_1 + c_2\mathbf{u}_2 + \cdots + c_p\mathbf{u}_p) \cdot \mathbf{u}_1 = c_1(\mathbf{u}_1 \cdot \mathbf{u}_1)$$

Since $\mathbf{u}_1 \cdot \mathbf{u}_1$ is not zero, the equation above can be solved for c_1. To find c_j for $j = 2, \ldots, p$, compute $\mathbf{y} \cdot \mathbf{u}_j$ and solve for c_j. ∎

EXAMPLE 2   The set $S = \{\mathbf{u}_1, \mathbf{u}_2, \mathbf{u}_3\}$ in Example 1 is an orthogonal basis for $\mathbb{R}^3$. Express the vector $\mathbf{y} = \begin{bmatrix} 6 \\ 1 \\ -8 \end{bmatrix}$ as a linear combination of the vectors in S.

SOLUTION Compute

$$\mathbf{y} \cdot \mathbf{u}_1 = 11, \qquad \mathbf{y} \cdot \mathbf{u}_2 = -12, \qquad \mathbf{y} \cdot \mathbf{u}_3 = -33$$
$$\mathbf{u}_1 \cdot \mathbf{u}_1 = 11, \qquad \mathbf{u}_2 \cdot \mathbf{u}_2 = 6, \qquad \mathbf{u}_3 \cdot \mathbf{u}_3 = 33/2$$

By Theorem 5,

$$\mathbf{y} = \frac{\mathbf{y} \cdot \mathbf{u}_1}{\mathbf{u}_1 \cdot \mathbf{u}_1}\mathbf{u}_1 + \frac{\mathbf{y} \cdot \mathbf{u}_2}{\mathbf{u}_2 \cdot \mathbf{u}_2}\mathbf{u}_2 + \frac{\mathbf{y} \cdot \mathbf{u}_3}{\mathbf{u}_3 \cdot \mathbf{u}_3}\mathbf{u}_3$$

$$= \frac{11}{11}\mathbf{u}_1 + \frac{-12}{6}\mathbf{u}_2 + \frac{-33}{33/2}\mathbf{u}_3$$

$$= \mathbf{u}_1 - 2\mathbf{u}_2 - 2\mathbf{u}_3 \qquad\qquad ∎$$

Notice how easy it is to compute the weights needed to build $\mathbf{y}$ from an orthogonal basis. If the basis were not orthogonal, it would be necessary to solve a system of linear equations in order to find the weights, as in Chapter 1.

We turn next to a construction that will become a key step in many calculations involving orthogonality, and it will lead to a geometric interpretation of Theorem 5.

An Orthogonal Projection

Given a nonzero vector $\mathbf{u}$ in $\mathbb{R}^n$, consider the problem of decomposing a vector $\mathbf{y}$ in $\mathbb{R}^n$ into the sum of two vectors, one a multiple of $\mathbf{u}$ and the other orthogonal to $\mathbf{u}$. We wish to write

$$\mathbf{y} = \hat{\mathbf{y}} + \mathbf{z} \qquad\qquad (1)$$

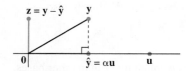

FIGURE 2
Finding α to make $\mathbf{y} - \hat{\mathbf{y}}$ orthogonal to $\mathbf{u}$.

where $\hat{\mathbf{y}} = \alpha\mathbf{u}$ for some scalar α and $\mathbf{z}$ is some vector orthogonal to $\mathbf{u}$. See Fig. 2. Given any scalar α, let $\mathbf{z} = \mathbf{y} - \alpha\mathbf{u}$, so that (1) is satisfied. Then $\mathbf{y} - \hat{\mathbf{y}}$ is orthogonal to $\mathbf{u}$ if and only if

$$0 = (\mathbf{y} - \alpha\mathbf{u})\cdot\mathbf{u} = \mathbf{y}\cdot\mathbf{u} - (\alpha\mathbf{u})\cdot\mathbf{u} = \mathbf{y}\cdot\mathbf{u} - \alpha(\mathbf{u}\cdot\mathbf{u})$$

That is, (1) is satisfied with $\mathbf{z}$ orthogonal to $\mathbf{u}$ if and only if $\alpha = \dfrac{\mathbf{y}\cdot\mathbf{u}}{\mathbf{u}\cdot\mathbf{u}}$ and $\hat{\mathbf{y}} = \dfrac{\mathbf{y}\cdot\mathbf{u}}{\mathbf{u}\cdot\mathbf{u}}\mathbf{u}$. The vector $\hat{\mathbf{y}}$ is called the **orthogonal projection of y onto u**, and the vector $\mathbf{z}$ is called the **component of y orthogonal to u**.

If c is any nonzero scalar and if $\mathbf{u}$ is replaced by $c\mathbf{u}$ in the definition of $\hat{\mathbf{y}}$, then the orthogonal projection of $\mathbf{y}$ onto $c\mathbf{u}$ is exactly the same as the orthogonal projection of $\mathbf{y}$ onto $\mathbf{u}$ (Exercise 31). Hence this projection is determined by the *subspace* L spanned by $\mathbf{u}$ (the line through $\mathbf{u}$ and $\mathbf{0}$). Sometimes $\hat{\mathbf{y}}$ is denoted by $\text{proj}_L \mathbf{y}$ and is called the **orthogonal projection of y onto L**. That is,

$$\hat{\mathbf{y}} = \text{proj}_L \mathbf{y} = \frac{\mathbf{y}\cdot\mathbf{u}}{\mathbf{u}\cdot\mathbf{u}}\mathbf{u} \tag{2}$$

EXAMPLE 3 Let $\mathbf{y} = \begin{bmatrix} 7 \\ 6 \end{bmatrix}$ and $\mathbf{u} = \begin{bmatrix} 4 \\ 2 \end{bmatrix}$. Find the orthogonal projection of $\mathbf{y}$ onto $\mathbf{u}$. Then write $\mathbf{y}$ as the sum of two orthogonal vectors, one in Span $\{\mathbf{u}\}$ and one orthogonal to $\mathbf{u}$.

SOLUTION Compute

$$\mathbf{y}\cdot\mathbf{u} = \begin{bmatrix} 7 \\ 6 \end{bmatrix}\cdot\begin{bmatrix} 4 \\ 2 \end{bmatrix} = 40$$

$$\mathbf{u}\cdot\mathbf{u} = \begin{bmatrix} 4 \\ 2 \end{bmatrix}\cdot\begin{bmatrix} 4 \\ 2 \end{bmatrix} = 20$$

The orthogonal projection of $\mathbf{y}$ onto $\mathbf{u}$ is

$$\hat{\mathbf{y}} = \frac{\mathbf{y}\cdot\mathbf{u}}{\mathbf{u}\cdot\mathbf{u}}\mathbf{u} = \frac{40}{20}\mathbf{u} = 2\begin{bmatrix} 4 \\ 2 \end{bmatrix} = \begin{bmatrix} 8 \\ 4 \end{bmatrix}$$

and the component of $\mathbf{y}$ orthogonal to $\mathbf{u}$ is

$$\mathbf{y} - \hat{\mathbf{y}} = \begin{bmatrix} 7 \\ 6 \end{bmatrix} - \begin{bmatrix} 8 \\ 4 \end{bmatrix} = \begin{bmatrix} -1 \\ 2 \end{bmatrix}$$

The sum of these two vectors is $\mathbf{y}$. That is,

$$\underset{\underset{\mathbf{y}}{\uparrow}}{\begin{bmatrix} 7 \\ 6 \end{bmatrix}} = \underset{\underset{\hat{\mathbf{y}}}{\uparrow}}{\begin{bmatrix} 8 \\ 4 \end{bmatrix}} + \underset{\underset{(\mathbf{y} - \hat{\mathbf{y}})}{\uparrow}}{\begin{bmatrix} -1 \\ 2 \end{bmatrix}}$$

This decomposition of $\mathbf{y}$ is illustrated in Fig. 3. *Note:* If the calculations above are correct, then $\{\hat{\mathbf{y}}, \mathbf{y} - \hat{\mathbf{y}}\}$ will be an orthogonal set. As a check, compute

$$\hat{\mathbf{y}}\cdot(\mathbf{y} - \hat{\mathbf{y}}) = \begin{bmatrix} 8 \\ 4 \end{bmatrix}\cdot\begin{bmatrix} -1 \\ 2 \end{bmatrix} = -8 + 8 = 0 \qquad \blacksquare$$

Since the line segment in Fig. 3 between $\mathbf{y}$ and $\hat{\mathbf{y}}$ is perpendicular to L, by construction of $\hat{\mathbf{y}}$, the point identified with $\hat{\mathbf{y}}$ is the closest point of L to $\mathbf{y}$. (This can be proved from geometry. We will assume this for $\mathbb{R}^2$ now and prove it for $\mathbb{R}^n$ in Section 6.3.)

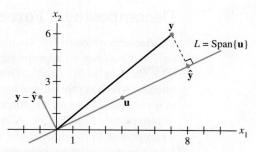

FIGURE 3 The orthogonal projection of **y** onto a line L through the origin.

EXAMPLE 4 Find the distance in Fig. 3 from **y** to L.

SOLUTION The distance from **y** to L is the length of the perpendicular line segment from **y** to the orthogonal projection $\hat{\mathbf{y}}$. This length equals the length of $\mathbf{y} - \hat{\mathbf{y}}$. Thus the distance is

$$\|\mathbf{y} - \hat{\mathbf{y}}\| = \sqrt{(-1)^2 + 2^2} = \sqrt{5}$$
∎

A Geometric Interpretation of Theorem 5

The formula for the orthogonal projection $\hat{\mathbf{y}}$ in (2) has the same appearance as each of the terms in Theorem 5. Thus Theorem 5 decomposes a vector **y** into a sum of orthogonal projections onto one-dimensional subspaces.

It is easy to visualize the case in which $W = \mathbb{R}^2 = \text{Span}\{\mathbf{u}_1, \mathbf{u}_2\}$, with $\mathbf{u}_1$ and $\mathbf{u}_2$ orthogonal. Any **y** in $\mathbb{R}^2$ can be written in the form

$$\mathbf{y} = \frac{\mathbf{y} \cdot \mathbf{u}_1}{\mathbf{u}_1 \cdot \mathbf{u}_1} \mathbf{u}_1 + \frac{\mathbf{y} \cdot \mathbf{u}_2}{\mathbf{u}_2 \cdot \mathbf{u}_2} \mathbf{u}_2 \tag{3}$$

The first term in (3) is the projection of **y** onto the subspace spanned by $\mathbf{u}_1$ (the line through $\mathbf{u}_1$ and the origin), and the second term is the projection of **y** onto the subspace spanned by $\mathbf{u}_2$. Thus (3) expresses **y** as the sum of its projections onto the (orthogonal) axes determined by $\mathbf{u}_1$ and $\mathbf{u}_2$. See Fig. 4.

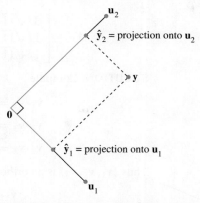

FIGURE 4 A vector decomposed into the sum of two projections.

Theorem 5 decomposes each **y** in $\text{Span}\{\mathbf{u}_1, \ldots, \mathbf{u}_p\}$ into the sum of p projections onto one-dimensional subspaces that are mutually orthogonal.

Decomposing a Force into Component Forces

The decomposition in Fig. 4 can occur in physics when some sort of force is applied to an object. Choosing an appropriate coordinate system allows the force to be represented by a vector $\mathbf{y}$ in $\mathbb{R}^2$ or $\mathbb{R}^3$. Often the problem involves some particular direction of interest, which is represented by another vector $\mathbf{u}$. For instance, if the object is moving in a straight line when the force is applied, the vector $\mathbf{u}$ might point in the direction of movement, as in Fig. 5. A key step in the problem is to decompose the force into a component in the direction of $\mathbf{u}$ and a component orthogonal to $\mathbf{u}$. The calculations would be analogous to those made in Example 3 above.

FIGURE 5

Orthonormal Sets

A set $\{\mathbf{u}_1, \ldots, \mathbf{u}_p\}$ is an **orthonormal set** if it is an orthogonal set of unit vectors. If W is the subspace spanned by such a set, then $\{\mathbf{u}_1, \ldots, \mathbf{u}_p\}$ is an **orthonormal basis** for W, since the set is automatically linearly independent, by Theorem 4.

The simplest example of an orthonormal set is the standard basis $\{\mathbf{e}_1, \ldots, \mathbf{e}_n\}$ for $\mathbb{R}^n$. Any nonempty subset of $\{\mathbf{e}_1, \ldots, \mathbf{e}_n\}$ is orthonormal, too. Here is a more complicated example.

EXAMPLE 5 Show that $\{\mathbf{v}_1, \mathbf{v}_2, \mathbf{v}_3\}$ is an orthonormal basis of $\mathbb{R}^3$, where

$$\mathbf{v}_1 = \begin{bmatrix} 3/\sqrt{11} \\ 1/\sqrt{11} \\ 1/\sqrt{11} \end{bmatrix}, \quad \mathbf{v}_2 = \begin{bmatrix} -1/\sqrt{6} \\ 2/\sqrt{6} \\ 1/\sqrt{6} \end{bmatrix}, \quad \mathbf{v}_3 = \begin{bmatrix} -1/\sqrt{66} \\ -4/\sqrt{66} \\ 7/\sqrt{66} \end{bmatrix}$$

SOLUTION Compute

$$\mathbf{v}_1 \cdot \mathbf{v}_2 = -3/\sqrt{66} + 2/\sqrt{66} + 1/\sqrt{66} = 0$$
$$\mathbf{v}_1 \cdot \mathbf{v}_3 = -3/\sqrt{726} - 4/\sqrt{726} + 7/\sqrt{726} = 0$$
$$\mathbf{v}_2 \cdot \mathbf{v}_3 = 1/\sqrt{396} - 8/\sqrt{396} + 7/\sqrt{396} = 0$$

Thus $\{\mathbf{v}_1, \mathbf{v}_2, \mathbf{v}_3\}$ is an orthogonal set. Also,

$$\mathbf{v}_1 \cdot \mathbf{v}_1 = 9/11 + 1/11 + 1/11 = 1$$
$$\mathbf{v}_2 \cdot \mathbf{v}_2 = 1/6 + 4/6 + 1/6 = 1$$
$$\mathbf{v}_3 \cdot \mathbf{v}_3 = 1/66 + 16/66 + 49/66 = 1$$

which shows that $\mathbf{v}_1$, $\mathbf{v}_2$, and $\mathbf{v}_3$ are unit vectors. Thus $\{\mathbf{v}_1, \mathbf{v}_2, \mathbf{v}_3\}$ is an orthonormal set. Since the set is linearly independent, its three vectors form a basis for $\mathbb{R}^3$. See Fig. 6. ∎

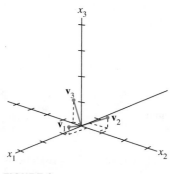

FIGURE 6

When the vectors in an orthogonal set of nonzero vectors are *normalized* to have unit length, the new vectors will still be orthogonal, and hence the new set will be an orthonormal set. See Exercise 32. It is easy to check that the vectors in Fig. 6 (Example 5) are simply the unit vectors in the directions of the vectors in Fig. 1 (Example 1).

Matrices whose columns form an orthonormal set are important in applications and in computer algorithms for matrix computations. Their main properties are given in Theorems 6 and 7.

THEOREM 6 An $m \times n$ matrix U has orthonormal columns if and only if $U^T U = I$.

PROOF To simplify notation, we suppose that U has only three columns, each a vector in $\mathbb{R}^m$. The proof of the general case is essentially the same. Let $U = [\, \mathbf{u}_1 \quad \mathbf{u}_2 \quad \mathbf{u}_3 \,]$ and compute

$$
U^T U = \begin{bmatrix} \mathbf{u}_1^T \\ \mathbf{u}_2^T \\ \mathbf{u}_3^T \end{bmatrix} [\, \mathbf{u}_1 \quad \mathbf{u}_2 \quad \mathbf{u}_3 \,] = \begin{bmatrix} \mathbf{u}_1^T \mathbf{u}_1 & \mathbf{u}_1^T \mathbf{u}_2 & \mathbf{u}_1^T \mathbf{u}_3 \\ \mathbf{u}_2^T \mathbf{u}_1 & \mathbf{u}_2^T \mathbf{u}_2 & \mathbf{u}_2^T \mathbf{u}_3 \\ \mathbf{u}_3^T \mathbf{u}_1 & \mathbf{u}_3^T \mathbf{u}_2 & \mathbf{u}_3^T \mathbf{u}_3 \end{bmatrix} \tag{4}
$$

The entries in the matrix at the right are inner products, using transpose notation. The columns of U are orthogonal if and only if

$$
\mathbf{u}_1^T \mathbf{u}_2 = \mathbf{u}_2^T \mathbf{u}_1 = 0, \quad \mathbf{u}_1^T \mathbf{u}_3 = \mathbf{u}_3^T \mathbf{u}_1 = 0, \quad \mathbf{u}_2^T \mathbf{u}_3 = \mathbf{u}_3^T \mathbf{u}_2 = 0 \tag{5}
$$

The columns of U all have unit length if and only if

$$
\mathbf{u}_1^T \mathbf{u}_1 = 1, \quad \mathbf{u}_2^T \mathbf{u}_2 = 1, \quad \mathbf{u}_3^T \mathbf{u}_3 = 1 \tag{6}
$$

The theorem follows immediately from (4)–(6). ∎

THEOREM 7 Let U be an $m \times n$ matrix with orthonormal columns, and let $\mathbf{x}$ and $\mathbf{y}$ be in $\mathbb{R}^n$. Then

a. $\|U\mathbf{x}\| = \|\mathbf{x}\|$

b. $(U\mathbf{x}) \cdot (U\mathbf{y}) = \mathbf{x} \cdot \mathbf{y}$

c. $(U\mathbf{x}) \cdot (U\mathbf{y}) = 0$ if and only if $\mathbf{x} \cdot \mathbf{y} = 0$

Properties (a) and (c) say that the linear mapping $\mathbf{x} \mapsto U\mathbf{x}$ preserves lengths and orthogonality. These properties are crucial for many computer algorithms. See Exercise 25 for the proof of Theorem 7.

EXAMPLE 6 Let $U = \begin{bmatrix} 1/\sqrt{2} & 2/3 \\ 1/\sqrt{2} & -2/3 \\ 0 & 1/3 \end{bmatrix}$ and $\mathbf{x} = \begin{bmatrix} \sqrt{2} \\ 3 \end{bmatrix}$. Notice that U has orthonormal columns and

$$
U^T U = \begin{bmatrix} 1/\sqrt{2} & 1/\sqrt{2} & 0 \\ 2/3 & -2/3 & 1/3 \end{bmatrix} \begin{bmatrix} 1/\sqrt{2} & 2/3 \\ 1/\sqrt{2} & -2/3 \\ 0 & 1/3 \end{bmatrix} = \begin{bmatrix} 1 & 0 \\ 0 & 1 \end{bmatrix}
$$

Verify that $\|U\mathbf{x}\| = \|\mathbf{x}\|$.

SOLUTION

$$U\mathbf{x} = \begin{bmatrix} 1/\sqrt{2} & 2/3 \\ 1/\sqrt{2} & -2/3 \\ 0 & 1/3 \end{bmatrix} \begin{bmatrix} \sqrt{2} \\ 3 \end{bmatrix} = \begin{bmatrix} 3 \\ -1 \\ 1 \end{bmatrix}$$

$$\|U\mathbf{x}\| = \sqrt{9+1+1} = \sqrt{11}$$

$$\|\mathbf{x}\| = \sqrt{2+9} = \sqrt{11} \qquad \blacksquare$$

Theorems 6 and 7 are particularly useful when applied to *square* matrices. An **orthogonal matrix** is a square invertible matrix U such that $U^{-1} = U^T$. By Theorem 6, such a matrix has orthonormal columns.[1] It is easy to see that any *square* matrix with orthonormal columns is an orthogonal matrix. Surprisingly, such a matrix must have orthonormal *rows*, too. See Exercises 27 and 28. Orthogonal matrices will appear frequently in Chapter 7.

EXAMPLE 7 The matrix

$$U = \begin{bmatrix} 3/\sqrt{11} & -1/\sqrt{6} & -1/\sqrt{66} \\ 1/\sqrt{11} & 2/\sqrt{6} & -4/\sqrt{66} \\ 1/\sqrt{11} & 1/\sqrt{6} & 7/\sqrt{66} \end{bmatrix}$$

is an orthogonal matrix because it is square and because its columns are orthonormal, by Example 5. Verify that the rows are orthonormal, too! $\qquad \blacksquare$

PRACTICE PROBLEMS

1. Let $\mathbf{u}_1 = \begin{bmatrix} -1/\sqrt{5} \\ 2/\sqrt{5} \end{bmatrix}$ and $\mathbf{u}_2 = \begin{bmatrix} 2/\sqrt{5} \\ 1/\sqrt{5} \end{bmatrix}$. Show that $\{\mathbf{u}_1, \mathbf{u}_2\}$ is an orthonormal basis for $\mathbb{R}^2$.

2. Let $\mathbf{y}$ and L be as in Example 3 and Fig. 3. Compute the orthogonal projection $\hat{\mathbf{y}}$ of $\mathbf{y}$ onto L using $\mathbf{u} = \begin{bmatrix} 2 \\ 1 \end{bmatrix}$ instead of the $\mathbf{u}$ in Example 3.

3. Let U and $\mathbf{x}$ be as in Example 6, and let $\mathbf{y} = \begin{bmatrix} -3\sqrt{2} \\ 6 \end{bmatrix}$. Verify that $U\mathbf{x} \cdot U\mathbf{y} = \mathbf{x} \cdot \mathbf{y}$.

6.2 EXERCISES

In Exercises 1–6, determine which sets of vectors are orthogonal.

1. $\begin{bmatrix} -1 \\ 4 \\ -3 \end{bmatrix}, \begin{bmatrix} 5 \\ 2 \\ 1 \end{bmatrix}, \begin{bmatrix} 3 \\ -4 \\ -7 \end{bmatrix}$ 2. $\begin{bmatrix} 1 \\ -2 \\ 1 \end{bmatrix}, \begin{bmatrix} 0 \\ 1 \\ 2 \end{bmatrix}, \begin{bmatrix} -5 \\ -2 \\ 1 \end{bmatrix}$

3. $\begin{bmatrix} 2 \\ -7 \\ -1 \end{bmatrix}, \begin{bmatrix} -6 \\ -3 \\ 9 \end{bmatrix}, \begin{bmatrix} 3 \\ 1 \\ -1 \end{bmatrix}$ 4. $\begin{bmatrix} 2 \\ -5 \\ -3 \end{bmatrix}, \begin{bmatrix} 0 \\ 0 \\ 0 \end{bmatrix}, \begin{bmatrix} 4 \\ -2 \\ 6 \end{bmatrix}$

5. $\begin{bmatrix} 3 \\ -2 \\ 1 \\ 3 \end{bmatrix}, \begin{bmatrix} -1 \\ 3 \\ -3 \\ 4 \end{bmatrix}, \begin{bmatrix} 3 \\ 8 \\ 7 \\ 0 \end{bmatrix}$ 6. $\begin{bmatrix} 5 \\ -4 \\ 0 \\ 3 \end{bmatrix}, \begin{bmatrix} -4 \\ 1 \\ -3 \\ 8 \end{bmatrix}, \begin{bmatrix} 3 \\ 3 \\ 5 \\ -1 \end{bmatrix}$

In Exercises 7–10, show that $\{\mathbf{u}_1, \mathbf{u}_2\}$ or $\{\mathbf{u}_1, \mathbf{u}_2, \mathbf{u}_3\}$ is an orthogonal basis for $\mathbb{R}^2$ or $\mathbb{R}^3$, respectively. Then express $\mathbf{x}$ as a linear combination of the $\mathbf{u}$'s.

7. $\mathbf{u}_1 = \begin{bmatrix} 2 \\ -3 \end{bmatrix}$, $\mathbf{u}_2 = \begin{bmatrix} 6 \\ 4 \end{bmatrix}$, and $\mathbf{x} = \begin{bmatrix} 9 \\ -7 \end{bmatrix}$

---

[1] A better name might be *orthonormal matrix*, and this term is found in some statistics texts. However, *orthogonal matrix* is the standard term in linear algebra.

8. $\mathbf{u}_1 = \begin{bmatrix} 3 \\ 1 \end{bmatrix}$, $\mathbf{u}_2 = \begin{bmatrix} -2 \\ 6 \end{bmatrix}$, and $\mathbf{x} = \begin{bmatrix} -6 \\ 3 \end{bmatrix}$

9. $\mathbf{u}_1 = \begin{bmatrix} 1 \\ 0 \\ 1 \end{bmatrix}$, $\mathbf{u}_2 = \begin{bmatrix} -1 \\ 4 \\ 1 \end{bmatrix}$, $\mathbf{u}_3 = \begin{bmatrix} 2 \\ 1 \\ -2 \end{bmatrix}$, and $\mathbf{x} = \begin{bmatrix} 8 \\ -4 \\ -3 \end{bmatrix}$

10. $\mathbf{u}_1 = \begin{bmatrix} 3 \\ -3 \\ 0 \end{bmatrix}$, $\mathbf{u}_2 = \begin{bmatrix} 2 \\ 2 \\ -1 \end{bmatrix}$, $\mathbf{u}_3 = \begin{bmatrix} 1 \\ 1 \\ 4 \end{bmatrix}$, and $\mathbf{x} = \begin{bmatrix} 5 \\ -3 \\ 1 \end{bmatrix}$

11. Compute the orthogonal projection of $\begin{bmatrix} 1 \\ 7 \end{bmatrix}$ onto the line through $\begin{bmatrix} -4 \\ 2 \end{bmatrix}$ and the origin.

12. Compute the orthogonal projection of $\begin{bmatrix} 1 \\ -1 \end{bmatrix}$ onto the line through $\begin{bmatrix} -1 \\ 3 \end{bmatrix}$ and the origin.

13. Let $\mathbf{y} = \begin{bmatrix} 2 \\ 3 \end{bmatrix}$ and $\mathbf{u} = \begin{bmatrix} 4 \\ -7 \end{bmatrix}$. Write $\mathbf{y}$ as the sum of two orthogonal vectors, one in Span $\{\mathbf{u}\}$ and one orthogonal to $\mathbf{u}$.

14. Let $\mathbf{y} = \begin{bmatrix} 2 \\ 6 \end{bmatrix}$ and $\mathbf{u} = \begin{bmatrix} 7 \\ 1 \end{bmatrix}$. Write $\mathbf{y}$ as the sum of a vector in Span $\{\mathbf{u}\}$ and a vector orthogonal to $\mathbf{u}$.

15. Let $\mathbf{y} = \begin{bmatrix} 3 \\ 1 \end{bmatrix}$ and $\mathbf{u} = \begin{bmatrix} 8 \\ 6 \end{bmatrix}$. Compute the distance from $\mathbf{y}$ to the line through $\mathbf{u}$ and the origin.

16. Let $\mathbf{y} = \begin{bmatrix} -3 \\ 9 \end{bmatrix}$ and $\mathbf{u} = \begin{bmatrix} 1 \\ 2 \end{bmatrix}$. Compute the distance from $\mathbf{y}$ to the line through $\mathbf{u}$ and the origin.

In Exercises 17–22, determine which sets of vectors are orthonormal. If a set is only orthogonal, normalize the vectors to produce an orthonormal set.

17. $\begin{bmatrix} 1/3 \\ 1/3 \\ 1/3 \end{bmatrix}$, $\begin{bmatrix} -1/2 \\ 0 \\ 1/2 \end{bmatrix}$ **18.** $\begin{bmatrix} 0 \\ 1 \\ 0 \end{bmatrix}$, $\begin{bmatrix} 0 \\ -1 \\ 0 \end{bmatrix}$

19. $\begin{bmatrix} -.6 \\ .8 \end{bmatrix}$, $\begin{bmatrix} .8 \\ .6 \end{bmatrix}$ **20.** $\begin{bmatrix} -2/3 \\ 1/3 \\ 2/3 \end{bmatrix}$, $\begin{bmatrix} 1/3 \\ 2/3 \\ 0 \end{bmatrix}$

21. $\begin{bmatrix} 1/\sqrt{10} \\ 3/\sqrt{20} \\ 3/\sqrt{20} \end{bmatrix}$, $\begin{bmatrix} 3/\sqrt{10} \\ -1/\sqrt{20} \\ -1/\sqrt{20} \end{bmatrix}$, $\begin{bmatrix} 0 \\ -1/\sqrt{2} \\ 1/\sqrt{2} \end{bmatrix}$

22. $\begin{bmatrix} 1/\sqrt{18} \\ 4/\sqrt{18} \\ 1/\sqrt{18} \end{bmatrix}$, $\begin{bmatrix} 1/\sqrt{2} \\ 0 \\ -1/\sqrt{2} \end{bmatrix}$, $\begin{bmatrix} -2/3 \\ 1/3 \\ -2/3 \end{bmatrix}$

In Exercises 23 and 24, all vectors are in $\mathbb{R}^n$. Mark each statement True or False. Justify each answer.

23. a. Not every linearly independent set in $\mathbb{R}^n$ is an orthogonal set.

   b. If $\mathbf{y}$ is a linear combination of nonzero vectors from an orthogonal set, then the weights in the linear combination can be computed without row operations on a matrix.

   c. If the vectors in an orthogonal set of nonzero vectors are normalized, then some of the new vectors may not be orthogonal.

   d. A matrix with orthonormal columns is an orthogonal matrix.

   e. If L is a line through $\mathbf{0}$ and if $\hat{\mathbf{y}}$ is the orthogonal projection of $\mathbf{y}$ onto L, then $\|\hat{\mathbf{y}}\|$ gives the distance from $\mathbf{y}$ to L.

24. a. Not every orthogonal set in $\mathbb{R}^n$ is linearly independent.

   b. If a set $S = \{\mathbf{u}_1, \ldots, \mathbf{u}_p\}$ has the property that $\mathbf{u}_i \cdot \mathbf{u}_j = 0$ whenever $i \neq j$, then S is an orthonormal set.

   c. If the columns of an $m \times n$ matrix A are orthonormal, then the linear mapping $\mathbf{x} \mapsto A\mathbf{x}$ preserves lengths.

   d. The orthogonal projection of $\mathbf{y}$ onto $\mathbf{v}$ is the same as the orthogonal projection of $\mathbf{y}$ onto $c\mathbf{v}$ whenever $c \neq 0$.

   e. An orthogonal matrix is invertible.

25. Prove Theorem 7. [*Hint:* For (a), compute $\|U\mathbf{x}\|^2$, or prove (b) first.]

26. Suppose W is a subspace of $\mathbb{R}^n$ spanned by n nonzero orthogonal vectors. Explain why $W = \mathbb{R}^n$.

27. Let U be a square matrix with orthonormal columns. Explain why U is invertible. (Mention the theorems you use.)

28. Let U be an $n \times n$ orthogonal matrix. Show that the rows of U form an orthonormal basis of $\mathbb{R}^n$.

29. Let U and V be $n \times n$ orthogonal matrices. Explain why UV is an orthogonal matrix. [That is, explain why UV is invertible and its inverse is $(UV)^T$.]

30. Let U be an orthogonal matrix, and construct V by interchanging some of the columns of U. Explain why V is an orthogonal matrix.

31. Show that the orthogonal projection of a vector $\mathbf{y}$ onto a line L through the origin in $\mathbb{R}^2$ does not depend on the choice of the nonzero $\mathbf{u}$ in L used in the formula for $\hat{\mathbf{y}}$. To do this, suppose $\mathbf{y}$ and $\mathbf{u}$ are given and $\hat{\mathbf{y}}$ has been computed by formula (2) in this section. Replace $\mathbf{u}$ in that formula by $c\mathbf{u}$, where c is an unspecified nonzero scalar. Show that the new formula gives the same $\hat{\mathbf{y}}$.

32. Let $\{\mathbf{v}_1, \mathbf{v}_2\}$ be an orthogonal set of nonzero vectors, and let c_1, c_2 be any nonzero scalars. Show that $\{c_1\mathbf{v}_1, c_2\mathbf{v}_2\}$ is also an orthogonal set. Since orthogonality of a set is defined in terms of pairs of vectors, this shows that if the vectors in an orthogonal set are normalized, the new set will still be orthogonal.

33. Given $\mathbf{u} \neq \mathbf{0}$ in $\mathbb{R}^n$, let $L = \text{Span}\{\mathbf{u}\}$. Show that the mapping $\mathbf{x} \mapsto \text{proj}_L \mathbf{x}$ is a linear transformation.

34. Given $\mathbf{u} \neq \mathbf{0}$ in $\mathbb{R}^n$, let $L = \text{Span}\{\mathbf{u}\}$. For $\mathbf{y}$ in $\mathbb{R}^n$, the **reflection of $\mathbf{y}$ in L** is the point $\text{refl}_L \mathbf{y}$ defined by

$$\text{refl}_L \mathbf{y} = 2 \cdot \text{proj}_L \mathbf{y} - \mathbf{y}$$

See the figure, which shows that $\text{refl}_L \mathbf{y}$ is the sum of $\hat{\mathbf{y}} = \text{proj}_L \mathbf{y}$ and $\hat{\mathbf{y}} - \mathbf{y}$. Show that the mapping $\mathbf{y} \mapsto \text{refl}_L \mathbf{y}$ is a linear transformation.

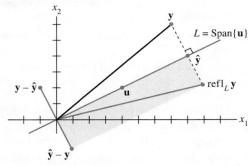

The reflection of $\mathbf{y}$ in a line through the origin.

35. **[M]** Show that the columns of the matrix A are orthogonal by making an appropriate matrix calculation. State the calculation you use.

$$A = \begin{bmatrix} -6 & -3 & 6 & 1 \\ -1 & 2 & 1 & -6 \\ 3 & 6 & 3 & -2 \\ 6 & -3 & 6 & -1 \\ 2 & -1 & 2 & 3 \\ -3 & 6 & 3 & 2 \\ -2 & -1 & 2 & -3 \\ 1 & 2 & 1 & 6 \end{bmatrix}$$

36. **[M]** In parts (a)–(d), let U be the matrix formed by normalizing each column of the matrix A in Exercise 35.

a. Compute $U^T U$ and $U U^T$. How do they differ?

b. Generate a random vector $\mathbf{y}$ in $\mathbb{R}^8$, and compute $\mathbf{p} = U U^T \mathbf{y}$ and $\mathbf{z} = \mathbf{y} - \mathbf{p}$. Explain why $\mathbf{p}$ is in Col A. Verify that $\mathbf{z}$ is orthogonal to $\mathbf{p}$.

c. Verify that $\mathbf{z}$ is orthogonal to each column of U.

d. Notice that $\mathbf{y} = \mathbf{p} + \mathbf{z}$, with $\mathbf{p}$ in Col A. Explain why $\mathbf{z}$ is in $(\text{Col } A)^\perp$. (The significance of this decomposition of $\mathbf{y}$ will be explained in the next section.)

SOLUTIONS TO PRACTICE PROBLEMS

1. The vectors are orthogonal because

$$\mathbf{u}_1 \cdot \mathbf{u}_2 = -2/5 + 2/5 = 0$$

They are unit vectors because

$$\|\mathbf{u}_1\|^2 = (-1/\sqrt{5})^2 + (2/\sqrt{5})^2 = 1/5 + 4/5 = 1$$
$$\|\mathbf{u}_2\|^2 = (2/\sqrt{5})^2 + (1/\sqrt{5})^2 = 4/5 + 1/5 = 1$$

In particular, the set $\{\mathbf{u}_1, \mathbf{u}_2\}$ is linearly independent, and hence is a basis for $\mathbb{R}^2$ since there are two vectors in the set.

2. When $\mathbf{y} = \begin{bmatrix} 7 \\ 6 \end{bmatrix}$ and $\mathbf{u} = \begin{bmatrix} 2 \\ 1 \end{bmatrix}$,

$$\hat{\mathbf{y}} = \frac{\mathbf{y} \cdot \mathbf{u}}{\mathbf{u} \cdot \mathbf{u}} \mathbf{u} = \frac{20}{5} \begin{bmatrix} 2 \\ 1 \end{bmatrix} = 4 \begin{bmatrix} 2 \\ 1 \end{bmatrix} = \begin{bmatrix} 8 \\ 4 \end{bmatrix}$$

This is the same $\hat{\mathbf{y}}$ found in Example 3. The orthogonal projection does not seem to depend on the $\mathbf{u}$ chosen on the line. See Exercise 31.

3. $U\mathbf{y} = \begin{bmatrix} 1/\sqrt{2} & 2/3 \\ 1/\sqrt{2} & -2/3 \\ 0 & 1/3 \end{bmatrix} \begin{bmatrix} -3\sqrt{2} \\ 6 \end{bmatrix} = \begin{bmatrix} 1 \\ -7 \\ 2 \end{bmatrix}$

Also, from Example 6, $\mathbf{x} = \begin{bmatrix} \sqrt{2} \\ 3 \end{bmatrix}$ and $U\mathbf{x} = \begin{bmatrix} 3 \\ -1 \\ 1 \end{bmatrix}$. Hence

$$U\mathbf{x} \cdot U\mathbf{y} = 3 + 7 + 2 = 12, \quad \text{and} \quad \mathbf{x} \cdot \mathbf{y} = -6 + 18 = 12$$

6.3 | ORTHOGONAL PROJECTIONS

The orthogonal projection of a point in $\mathbb{R}^2$ onto a line through the origin has an important analogue in $\mathbb{R}^n$. Given a vector $\mathbf{y}$ and a subspace W in $\mathbb{R}^n$, there is a vector $\hat{\mathbf{y}}$ in W such that (1) $\hat{\mathbf{y}}$ is the unique vector in W for which $\mathbf{y} - \hat{\mathbf{y}}$ is orthogonal to W, and (2) $\hat{\mathbf{y}}$ is the unique vector in W closest to $\mathbf{y}$. See Fig. 1. These two properties of $\hat{\mathbf{y}}$ provide the key to finding least-squares solutions of linear systems, mentioned in the introductory example for this chapter. The full story will be told in Section 6.5.

To prepare for the first theorem, observe that whenever a vector $\mathbf{y}$ is written as a linear combination of vectors $\mathbf{u}_1, \ldots, \mathbf{u}_n$ in $\mathbb{R}^n$, the terms in the sum for $\mathbf{y}$ can be grouped into two parts so that $\mathbf{y}$ can be written as

$$\mathbf{y} = \mathbf{z}_1 + \mathbf{z}_2$$

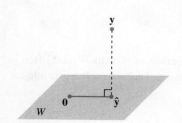

FIGURE 1

where $\mathbf{z}_1$ is a linear combination of some of the $\mathbf{u}_i$ and $\mathbf{z}_2$ is a linear combination of the rest of the $\mathbf{u}_i$. This idea is particularly useful when $\{\mathbf{u}_1, \ldots, \mathbf{u}_n\}$ is an orthogonal basis. Recall from Section 6.1 that $W^\perp$ denotes the set of all vectors orthogonal to a subspace W.

EXAMPLE 1 Let $\{\mathbf{u}_1, \ldots, \mathbf{u}_5\}$ be an orthogonal basis for $\mathbb{R}^5$ and let

$$\mathbf{y} = c_1 \mathbf{u}_1 + \cdots + c_5 \mathbf{u}_5$$

Consider the subspace $W = \text{Span}\{\mathbf{u}_1, \mathbf{u}_2\}$, and write $\mathbf{y}$ as the sum of a vector $\mathbf{z}_1$ in W and a vector $\mathbf{z}_2$ in $W^\perp$.

SOLUTION Write

$$\mathbf{y} = \underbrace{c_1 \mathbf{u}_1 + c_2 \mathbf{u}_2}_{\mathbf{z}_1} + \underbrace{c_3 \mathbf{u}_3 + c_4 \mathbf{u}_4 + c_5 \mathbf{u}_5}_{\mathbf{z}_2}$$

where $\quad\quad\quad \mathbf{z}_1 = c_1 \mathbf{u}_1 + c_2 \mathbf{u}_2 \quad$ is in $\text{Span}\{\mathbf{u}_1, \mathbf{u}_2\}$

and $\quad\quad\quad\quad \mathbf{z}_2 = c_3 \mathbf{u}_3 + c_4 \mathbf{u}_4 + c_5 \mathbf{u}_5 \quad$ is in $\text{Span}\{\mathbf{u}_3, \mathbf{u}_4, \mathbf{u}_5\}$.

To show that $\mathbf{z}_2$ is in $W^\perp$, it suffices to show that $\mathbf{z}_2$ is orthogonal to the vectors in the basis $\{\mathbf{u}_1, \mathbf{u}_2\}$ for W. (See Section 6.1.) Using properties of the inner product, compute

$$\begin{aligned} \mathbf{z}_2 \cdot \mathbf{u}_1 &= (c_3 \mathbf{u}_3 + c_4 \mathbf{u}_4 + c_5 \mathbf{u}_5) \cdot \mathbf{u}_1 \\ &= c_3 \mathbf{u}_3 \cdot \mathbf{u}_1 + c_4 \mathbf{u}_4 \cdot \mathbf{u}_1 + c_5 \mathbf{u}_5 \cdot \mathbf{u}_1 \\ &= 0 \end{aligned}$$

because $\mathbf{u}_1$ is orthogonal to $\mathbf{u}_3$, $\mathbf{u}_4$, and $\mathbf{u}_5$. A similar calculation shows that $\mathbf{z}_2 \cdot \mathbf{u}_2 = 0$. Thus $\mathbf{z}_2$ is in $W^\perp$. ∎

The next theorem shows that the decomposition $\mathbf{y} = \mathbf{z}_1 + \mathbf{z}_2$ in Example 1 can be computed without having an orthogonal basis for $\mathbb{R}^n$. It is enough to have an orthogonal basis only for W.

THEOREM 8

The Orthogonal Decomposition Theorem

Let W be a subspace of $\mathbb{R}^n$. Then each $\mathbf{y}$ in $\mathbb{R}^n$ can be written uniquely in the form

$$\mathbf{y} = \hat{\mathbf{y}} + \mathbf{z} \tag{1}$$

where $\hat{\mathbf{y}}$ is in W and $\mathbf{z}$ is in $W^\perp$. In fact, if $\{\mathbf{u}_1, \ldots, \mathbf{u}_p\}$ is any orthogonal basis of W, then

$$\hat{\mathbf{y}} = \frac{\mathbf{y} \cdot \mathbf{u}_1}{\mathbf{u}_1 \cdot \mathbf{u}_1} \mathbf{u}_1 + \cdots + \frac{\mathbf{y} \cdot \mathbf{u}_p}{\mathbf{u}_p \cdot \mathbf{u}_p} \mathbf{u}_p \tag{2}$$

and $\mathbf{z} = \mathbf{y} - \hat{\mathbf{y}}$.

The vector $\hat{\mathbf{y}}$ in (1) is called the **orthogonal projection of y onto** W and often is written as $\text{proj}_W \mathbf{y}$. See Fig. 2. When W is a one-dimensional subspace, the formula for $\hat{\mathbf{y}}$ matches the formula given in Section 6.2.

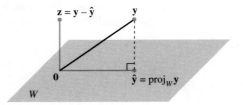

FIGURE 2 The orthogonal projection of $\mathbf{y}$ onto W.

PROOF Let $\{\mathbf{u}_1, \ldots, \mathbf{u}_p\}$ be any orthogonal basis for W, and define $\hat{\mathbf{y}}$ by (2).[1] Then $\hat{\mathbf{y}}$ is in W because $\hat{\mathbf{y}}$ is a linear combination of the basis $\mathbf{u}_1, \ldots, \mathbf{u}_p$. Let $\mathbf{z} = \mathbf{y} - \hat{\mathbf{y}}$. Since $\mathbf{u}_1$ is orthogonal to $\mathbf{u}_2, \ldots, \mathbf{u}_p$, it follows from (2) that

$$\mathbf{z} \cdot \mathbf{u}_1 = (\mathbf{y} - \hat{\mathbf{y}}) \cdot \mathbf{u}_1 = \mathbf{y} \cdot \mathbf{u}_1 - \left(\frac{\mathbf{y} \cdot \mathbf{u}_1}{\mathbf{u}_1 \cdot \mathbf{u}_1} \right) \mathbf{u}_1 \cdot \mathbf{u}_1 - 0 - \cdots - 0$$

$$= \mathbf{y} \cdot \mathbf{u}_1 - \mathbf{y} \cdot \mathbf{u}_1 = 0$$

Thus $\mathbf{z}$ is orthogonal to $\mathbf{u}_1$. Similarly, $\mathbf{z}$ is orthogonal to each $\mathbf{u}_j$ in the basis for W. Hence $\mathbf{z}$ is orthogonal to every vector in W. That is, $\mathbf{z}$ is in $W^\perp$.

To show that the decomposition in (1) is unique, suppose $\mathbf{y}$ can also be written as $\mathbf{y} = \hat{\mathbf{y}}_1 + \mathbf{z}_1$, with $\hat{\mathbf{y}}_1$ in W and $\mathbf{z}_1$ in $W^\perp$. Then $\hat{\mathbf{y}} + \mathbf{z} = \hat{\mathbf{y}}_1 + \mathbf{z}_1$ (since both sides equal $\mathbf{y}$), and so

$$\hat{\mathbf{y}} - \hat{\mathbf{y}}_1 = \mathbf{z}_1 - \mathbf{z}$$

This equality shows that the vector $\mathbf{v} = \hat{\mathbf{y}} - \hat{\mathbf{y}}_1$ is in W and in $W^\perp$ (because $\mathbf{z}_1$ and $\mathbf{z}$ are both in $W^\perp$, and $W^\perp$ is a subspace). Hence $\mathbf{v} \cdot \mathbf{v} = 0$, which shows that $\mathbf{v} = \mathbf{0}$. This proves that $\hat{\mathbf{y}} = \hat{\mathbf{y}}_1$ and also $\mathbf{z}_1 = \mathbf{z}$. ∎

The uniqueness of the decomposition (1) shows that the orthogonal projection $\hat{\mathbf{y}}$ depends only on W and not on the particular basis used in (2).

---

[1] We may assume that W is not the zero subspace, for otherwise $W^\perp = \mathbb{R}^n$ and (1) is simply $\mathbf{y} = \mathbf{0} + \mathbf{y}$. The next section will show that any nonzero subspace of $\mathbb{R}^n$ has an orthogonal basis.

EXAMPLE 2   Let $\mathbf{u}_1 = \begin{bmatrix} 2 \\ 5 \\ -1 \end{bmatrix}$, $\mathbf{u}_2 = \begin{bmatrix} -2 \\ 1 \\ 1 \end{bmatrix}$, and $\mathbf{y} = \begin{bmatrix} 1 \\ 2 \\ 3 \end{bmatrix}$. Observe that $\{\mathbf{u}_1, \mathbf{u}_2\}$ is an orthogonal basis for $W = \text{Span}\{\mathbf{u}_1, \mathbf{u}_2\}$. Write $\mathbf{y}$ as the sum of a vector in W and a vector orthogonal to W.

SOLUTION   The orthogonal projection of $\mathbf{y}$ onto W is

$$\hat{\mathbf{y}} = \frac{\mathbf{y} \cdot \mathbf{u}_1}{\mathbf{u}_1 \cdot \mathbf{u}_1} \mathbf{u}_1 + \frac{\mathbf{y} \cdot \mathbf{u}_2}{\mathbf{u}_2 \cdot \mathbf{u}_2} \mathbf{u}_2$$

$$= \frac{9}{30} \begin{bmatrix} 2 \\ 5 \\ -1 \end{bmatrix} + \frac{3}{6} \begin{bmatrix} -2 \\ 1 \\ 1 \end{bmatrix} = \frac{9}{30} \begin{bmatrix} 2 \\ 5 \\ -1 \end{bmatrix} + \frac{15}{30} \begin{bmatrix} -2 \\ 1 \\ 1 \end{bmatrix} = \begin{bmatrix} -2/5 \\ 2 \\ 1/5 \end{bmatrix}$$

Also

$$\mathbf{y} - \hat{\mathbf{y}} = \begin{bmatrix} 1 \\ 2 \\ 3 \end{bmatrix} - \begin{bmatrix} -2/5 \\ 2 \\ 1/5 \end{bmatrix} = \begin{bmatrix} 7/5 \\ 0 \\ 14/5 \end{bmatrix}$$

Theorem 8 ensures that $\mathbf{y} - \hat{\mathbf{y}}$ is in $W^{\perp}$. To check the calculations, however, it is a good idea to verify that $\mathbf{y} - \hat{\mathbf{y}}$ is orthogonal to both $\mathbf{u}_1$ and $\mathbf{u}_2$ and hence to all of W. The desired decomposition of $\mathbf{y}$ is

$$\mathbf{y} = \begin{bmatrix} 1 \\ 2 \\ 3 \end{bmatrix} = \begin{bmatrix} -2/5 \\ 2 \\ 1/5 \end{bmatrix} + \begin{bmatrix} 7/5 \\ 0 \\ 14/5 \end{bmatrix}$$ ∎

A Geometric Interpretation of the Orthogonal Projection

When W is a one-dimensional subspace, the formula (2) for $\text{proj}_W \mathbf{y}$ contains just one term. Thus, when $\dim W > 1$, each term in (2) is itself an orthogonal projection of $\mathbf{y}$ onto a one-dimensional subspace spanned by one of the $\mathbf{u}$'s in the basis for W. Figure 3 illustrates this when W is a subspace of $\mathbb{R}^3$ spanned by $\mathbf{u}_1$ and $\mathbf{u}_2$. Here $\hat{\mathbf{y}}_1$ and $\hat{\mathbf{y}}_2$ denote the projections of $\mathbf{y}$ onto the lines spanned by $\mathbf{u}_1$ and $\mathbf{u}_2$, respectively. The orthogonal projection $\hat{\mathbf{y}}$ of $\mathbf{y}$ onto W is the sum of the projections of $\mathbf{y}$ onto one-dimensional subspaces that are orthogonal to each other. The vector $\hat{\mathbf{y}}$ in Fig. 3 corresponds to the vector $\mathbf{y}$ in Fig. 4 of Section 6.2, because now it is $\hat{\mathbf{y}}$ that is in W.

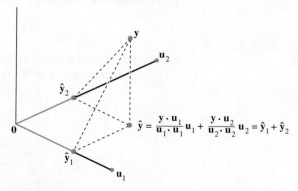

FIGURE 3   The orthogonal projection of $\mathbf{y}$ is the sum of its projections onto one-dimensional subspaces that are mutually orthogonal.

Properties of Orthogonal Projections

If $\{\mathbf{u}_1, \ldots, \mathbf{u}_p\}$ is an orthogonal basis for W and if $\mathbf{y}$ happens to be in W, then the formula for $\text{proj}_W \mathbf{y}$ is exactly the same as the representation of $\mathbf{y}$ given in Theorem 5 in Section 6.2. In this case, $\text{proj}_W \mathbf{y} = \mathbf{y}$.

If $\mathbf{y}$ is in $W = \text{Span}\{\mathbf{u}_1, \ldots, \mathbf{u}_p\}$, then $\text{proj}_W \mathbf{y} = \mathbf{y}$.

This fact also follows from the next theorem.

THEOREM 9

The Best Approximation Theorem

Let W be a subspace of $\mathbb{R}^n$, let $\mathbf{y}$ be any vector in $\mathbb{R}^n$, and let $\hat{\mathbf{y}}$ be the orthogonal projection of $\mathbf{y}$ onto W. Then $\hat{\mathbf{y}}$ is the closest point in W to $\mathbf{y}$, in the sense that

$$\|\mathbf{y} - \hat{\mathbf{y}}\| < \|\mathbf{y} - \mathbf{v}\| \tag{3}$$

for all $\mathbf{v}$ in W distinct from $\hat{\mathbf{y}}$.

The vector $\hat{\mathbf{y}}$ in Theorem 9 is called **the best approximation to y by elements of** W. Later sections in the text will examine problems where a given $\mathbf{y}$ must be replaced, or *approximated*, by a vector $\mathbf{v}$ in some fixed subspace W. The distance from $\mathbf{y}$ to $\mathbf{v}$, given by $\|\mathbf{y} - \mathbf{v}\|$, can be regarded as the "error" of using $\mathbf{v}$ in place of $\mathbf{y}$. Theorem 9 says that this error is minimized when $\mathbf{v} = \hat{\mathbf{y}}$.

Inequality (3) leads to a new proof that $\hat{\mathbf{y}}$ does not depend on the particular orthogonal basis used to compute it. If a different orthogonal basis for W were used to construct an orthogonal projection of $\mathbf{y}$, then this projection would also be the closest point in W to $\mathbf{y}$, namely, $\hat{\mathbf{y}}$.

PROOF Take $\mathbf{v}$ in W distinct from $\hat{\mathbf{y}}$. See Fig. 4. Then $\hat{\mathbf{y}} - \mathbf{v}$ is in W. By the Orthogonal Decomposition Theorem, $\mathbf{y} - \hat{\mathbf{y}}$ is orthogonal to W. In particular, $\mathbf{y} - \hat{\mathbf{y}}$ is orthogonal to $\hat{\mathbf{y}} - \mathbf{v}$ (which is in W). Since

$$\mathbf{y} - \mathbf{v} = (\mathbf{y} - \hat{\mathbf{y}}) + (\hat{\mathbf{y}} - \mathbf{v})$$

the Pythagorean Theorem gives

$$\|\mathbf{y} - \mathbf{v}\|^2 = \|\mathbf{y} - \hat{\mathbf{y}}\|^2 + \|\hat{\mathbf{y}} - \mathbf{v}\|^2$$

(See the colored right triangle in Fig. 4. The length of each side is labeled.) Now $\|\hat{\mathbf{y}} - \mathbf{v}\|^2 > 0$ because $\hat{\mathbf{y}} - \mathbf{v} \neq \mathbf{0}$, and so inequality (3) follows immediately. ∎

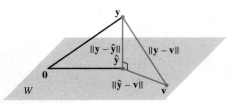

FIGURE 4 The orthogonal projection of $\mathbf{y}$ onto W is the closest point in W to $\mathbf{y}$.

EXAMPLE 3   If $\mathbf{u}_1 = \begin{bmatrix} 2 \\ 5 \\ -1 \end{bmatrix}$, $\mathbf{u}_2 = \begin{bmatrix} -2 \\ 1 \\ 1 \end{bmatrix}$, $\mathbf{y} = \begin{bmatrix} 1 \\ 2 \\ 3 \end{bmatrix}$, and $W = \text{Span}\{\mathbf{u}_1, \mathbf{u}_2\}$, as in Example 2, then the closest point in W to $\mathbf{y}$ is

$$\hat{\mathbf{y}} = \frac{\mathbf{y} \cdot \mathbf{u}_1}{\mathbf{u}_1 \cdot \mathbf{u}_1} \mathbf{u}_1 + \frac{\mathbf{y} \cdot \mathbf{u}_2}{\mathbf{u}_2 \cdot \mathbf{u}_2} \mathbf{u}_2 = \begin{bmatrix} -2/5 \\ 2 \\ 1/5 \end{bmatrix}$$ ∎

EXAMPLE 4   The distance from a point $\mathbf{y}$ in $\mathbb{R}^n$ to a subspace W is defined as the distance from $\mathbf{y}$ to the nearest point in W. Find the distance from $\mathbf{y}$ to $W = \text{Span}\{\mathbf{u}_1, \mathbf{u}_2\}$, where

$$\mathbf{y} = \begin{bmatrix} -1 \\ -5 \\ 10 \end{bmatrix}, \quad \mathbf{u}_1 = \begin{bmatrix} 5 \\ -2 \\ 1 \end{bmatrix}, \quad \mathbf{u}_2 = \begin{bmatrix} 1 \\ 2 \\ -1 \end{bmatrix}$$

SOLUTION   By the Best Approximation Theorem, the distance from $\mathbf{y}$ to W is $\|\mathbf{y} - \hat{\mathbf{y}}\|$, where $\hat{\mathbf{y}} = \text{proj}_W \mathbf{y}$. Since $\{\mathbf{u}_1, \mathbf{u}_2\}$ is an orthogonal basis for W,

$$\hat{\mathbf{y}} = \frac{15}{30}\mathbf{u}_1 + \frac{-21}{6}\mathbf{u}_2 = \frac{1}{2}\begin{bmatrix} 5 \\ -2 \\ 1 \end{bmatrix} - \frac{7}{2}\begin{bmatrix} 1 \\ 2 \\ -1 \end{bmatrix} = \begin{bmatrix} -1 \\ -8 \\ 4 \end{bmatrix}$$

$$\mathbf{y} - \hat{\mathbf{y}} = \begin{bmatrix} -1 \\ -5 \\ 10 \end{bmatrix} - \begin{bmatrix} -1 \\ -8 \\ 4 \end{bmatrix} = \begin{bmatrix} 0 \\ 3 \\ 6 \end{bmatrix}$$

$$\|\mathbf{y} - \hat{\mathbf{y}}\|^2 = 3^2 + 6^2 = 45$$

The distance from $\mathbf{y}$ to W is $\sqrt{45} = 3\sqrt{5}$.   ∎

The final theorem in this section shows how formula (2) for $\text{proj}_W \mathbf{y}$ is simplified when the basis for W is an orthonormal set.

THEOREM 10

If $\{\mathbf{u}_1, \ldots, \mathbf{u}_p\}$ is an orthonormal basis for a subspace W of $\mathbb{R}^n$, then

$$\text{proj}_W \mathbf{y} = (\mathbf{y} \cdot \mathbf{u}_1)\mathbf{u}_1 + (\mathbf{y} \cdot \mathbf{u}_2)\mathbf{u}_2 + \cdots + (\mathbf{y} \cdot \mathbf{u}_p)\mathbf{u}_p \qquad (4)$$

If $U = [\,\mathbf{u}_1 \quad \mathbf{u}_2 \quad \cdots \quad \mathbf{u}_p\,]$, then

$$\text{proj}_W \mathbf{y} = UU^T\mathbf{y} \quad \text{for all } \mathbf{y} \text{ in } \mathbb{R}^n \qquad (5)$$

PROOF   Formula (4) follows immediately from (2) in Theorem 8. Also, (4) shows that $\text{proj}_W \mathbf{y}$ is a linear combination of the columns of U using the weights $\mathbf{y} \cdot \mathbf{u}_1$, $\mathbf{y} \cdot \mathbf{u}_2, \ldots, \mathbf{y} \cdot \mathbf{u}_p$. The weights can be written as $\mathbf{u}_1^T\mathbf{y}, \mathbf{u}_2^T\mathbf{y}, \ldots, \mathbf{u}_p^T\mathbf{y}$, showing that they are the entries in $U^T\mathbf{y}$ and justifying (5).   ∎

WEB   Suppose U is an $n \times p$ matrix with orthonormal columns, and let W be the column space of U. Then

$$U^T U\mathbf{x} = I_p\mathbf{x} = \mathbf{x} \quad \text{for all } \mathbf{x} \text{ in } \mathbb{R}^p \qquad \text{Theorem 6}$$

$$UU^T\mathbf{y} = \text{proj}_W \mathbf{y} \quad \text{for all } \mathbf{y} \text{ in } \mathbb{R}^n \qquad \text{Theorem 10}$$

If U is an $n \times n$ (square) matrix with orthonormal columns, then U is an *orthogonal* matrix, the column space W is all of $\mathbb{R}^n$, and $UU^T\mathbf{y} = I\mathbf{y} = \mathbf{y}$ for all $\mathbf{y}$ in $\mathbb{R}^n$.

Although formula (4) is important for theoretical purposes, in practice it usually involves calculations with square roots of numbers (in the entries of the $\mathbf{u}_i$). Formula (2) is recommended for hand calculations.

> **PRACTICE PROBLEM**
>
> Let $\mathbf{u}_1 = \begin{bmatrix} -7 \\ 1 \\ 4 \end{bmatrix}$, $\mathbf{u}_2 = \begin{bmatrix} -1 \\ 1 \\ -2 \end{bmatrix}$, $\mathbf{y} = \begin{bmatrix} -9 \\ 1 \\ 6 \end{bmatrix}$, and $W = \text{Span}\{\mathbf{u}_1, \mathbf{u}_2\}$. Use the fact that $\mathbf{u}_1$ and $\mathbf{u}_2$ are orthogonal to compute $\text{proj}_W \mathbf{y}$.

6.3 EXERCISES

In Exercises 1 and 2, you may assume that $\{\mathbf{u}_1, \ldots, \mathbf{u}_4\}$ is an orthogonal basis for $\mathbb{R}^4$.

1. $\mathbf{u}_1 = \begin{bmatrix} 0 \\ 1 \\ -4 \\ -1 \end{bmatrix}$, $\mathbf{u}_2 = \begin{bmatrix} 3 \\ 5 \\ 1 \\ 1 \end{bmatrix}$, $\mathbf{u}_3 = \begin{bmatrix} 1 \\ 0 \\ 1 \\ -4 \end{bmatrix}$, $\mathbf{u}_4 = \begin{bmatrix} 5 \\ -3 \\ -1 \\ 1 \end{bmatrix}$,

$\mathbf{x} = \begin{bmatrix} 10 \\ -8 \\ 2 \\ 0 \end{bmatrix}$. Write $\mathbf{x}$ as the sum of two vectors, one in

Span $\{\mathbf{u}_1, \mathbf{u}_2, \mathbf{u}_3\}$ and the other in Span $\{\mathbf{u}_4\}$.

2. $\mathbf{u}_1 = \begin{bmatrix} 1 \\ 2 \\ 1 \\ 1 \end{bmatrix}$, $\mathbf{u}_2 = \begin{bmatrix} -2 \\ 1 \\ -1 \\ 1 \end{bmatrix}$, $\mathbf{u}_3 = \begin{bmatrix} 1 \\ 1 \\ -2 \\ -1 \end{bmatrix}$, $\mathbf{u}_4 = \begin{bmatrix} -1 \\ 1 \\ 1 \\ -2 \end{bmatrix}$,

$\mathbf{v} = \begin{bmatrix} 4 \\ 5 \\ -3 \\ 3 \end{bmatrix}$. Write $\mathbf{v}$ as the sum of two vectors, one in

Span $\{\mathbf{u}_1\}$ and the other in Span $\{\mathbf{u}_2, \mathbf{u}_3, \mathbf{u}_4\}$.

In Exercises 3–6, verify that $\{\mathbf{u}_1, \mathbf{u}_2\}$ is an orthogonal set, and then find the orthogonal projection of $\mathbf{y}$ onto Span $\{\mathbf{u}_1, \mathbf{u}_2\}$.

3. $\mathbf{y} = \begin{bmatrix} -1 \\ 4 \\ 3 \end{bmatrix}$, $\mathbf{u}_1 = \begin{bmatrix} 1 \\ 1 \\ 0 \end{bmatrix}$, $\mathbf{u}_2 = \begin{bmatrix} -1 \\ 1 \\ 0 \end{bmatrix}$

4. $\mathbf{y} = \begin{bmatrix} 6 \\ 3 \\ -2 \end{bmatrix}$, $\mathbf{u}_1 = \begin{bmatrix} 3 \\ 4 \\ 0 \end{bmatrix}$, $\mathbf{u}_2 = \begin{bmatrix} -4 \\ 3 \\ 0 \end{bmatrix}$

5. $\mathbf{y} = \begin{bmatrix} -1 \\ 2 \\ 6 \end{bmatrix}$, $\mathbf{u}_1 = \begin{bmatrix} 3 \\ -1 \\ 2 \end{bmatrix}$, $\mathbf{u}_2 = \begin{bmatrix} 1 \\ -1 \\ -2 \end{bmatrix}$

6. $\mathbf{y} = \begin{bmatrix} 6 \\ 4 \\ 1 \end{bmatrix}$, $\mathbf{u}_1 = \begin{bmatrix} -4 \\ -1 \\ 1 \end{bmatrix}$, $\mathbf{u}_2 = \begin{bmatrix} 0 \\ 1 \\ 1 \end{bmatrix}$

In Exercises 7–10, let W be the subspace spanned by the $\mathbf{u}$'s, and write $\mathbf{y}$ as the sum of a vector in W and a vector orthogonal to W.

7. $\mathbf{y} = \begin{bmatrix} 1 \\ 3 \\ 5 \end{bmatrix}$, $\mathbf{u}_1 = \begin{bmatrix} 1 \\ 3 \\ -2 \end{bmatrix}$, $\mathbf{u}_2 = \begin{bmatrix} 5 \\ 1 \\ 4 \end{bmatrix}$

8. $\mathbf{y} = \begin{bmatrix} -1 \\ 4 \\ 3 \end{bmatrix}$, $\mathbf{u}_1 = \begin{bmatrix} 1 \\ 1 \\ 1 \end{bmatrix}$, $\mathbf{u}_2 = \begin{bmatrix} -1 \\ 3 \\ -2 \end{bmatrix}$

9. $\mathbf{y} = \begin{bmatrix} 4 \\ 3 \\ 3 \\ -1 \end{bmatrix}$, $\mathbf{u}_1 = \begin{bmatrix} 1 \\ 1 \\ 0 \\ 1 \end{bmatrix}$, $\mathbf{u}_2 = \begin{bmatrix} -1 \\ 3 \\ 1 \\ -2 \end{bmatrix}$, $\mathbf{u}_3 = \begin{bmatrix} -1 \\ 0 \\ 1 \\ 1 \end{bmatrix}$

10. $\mathbf{y} = \begin{bmatrix} 3 \\ 4 \\ 5 \\ 6 \end{bmatrix}$, $\mathbf{u}_1 = \begin{bmatrix} 1 \\ 1 \\ 0 \\ -1 \end{bmatrix}$, $\mathbf{u}_2 = \begin{bmatrix} 1 \\ 0 \\ 1 \\ 1 \end{bmatrix}$, $\mathbf{u}_3 = \begin{bmatrix} 0 \\ -1 \\ 1 \\ -1 \end{bmatrix}$

In Exercises 11 and 12, find the closest point to $\mathbf{y}$ in the subspace W spanned by $\mathbf{v}_1$ and $\mathbf{v}_2$.

11. $\mathbf{y} = \begin{bmatrix} 3 \\ 1 \\ 5 \\ 1 \end{bmatrix}$, $\mathbf{v}_1 = \begin{bmatrix} 3 \\ 1 \\ -1 \\ 1 \end{bmatrix}$, $\mathbf{v}_2 = \begin{bmatrix} 1 \\ -1 \\ 1 \\ -1 \end{bmatrix}$

12. $\mathbf{y} = \begin{bmatrix} 3 \\ -1 \\ 1 \\ 13 \end{bmatrix}$, $\mathbf{v}_1 = \begin{bmatrix} 1 \\ -2 \\ -1 \\ 2 \end{bmatrix}$, $\mathbf{v}_2 = \begin{bmatrix} -4 \\ 1 \\ 0 \\ 3 \end{bmatrix}$

In Exercises 13 and 14, find the best approximation to $\mathbf{z}$ by vectors of the form $c_1\mathbf{v}_1 + c_2\mathbf{v}_2$.

13. $\mathbf{z} = \begin{bmatrix} 3 \\ -7 \\ 2 \\ 3 \end{bmatrix}$, $\mathbf{v}_1 = \begin{bmatrix} 2 \\ -1 \\ -3 \\ 1 \end{bmatrix}$, $\mathbf{v}_2 = \begin{bmatrix} 1 \\ 1 \\ 0 \\ -1 \end{bmatrix}$

14. $\mathbf{z} = \begin{bmatrix} 2 \\ 4 \\ 0 \\ -1 \end{bmatrix}$, $\mathbf{v}_1 = \begin{bmatrix} 2 \\ 0 \\ -1 \\ -3 \end{bmatrix}$, $\mathbf{v}_2 = \begin{bmatrix} 5 \\ -2 \\ 4 \\ 2 \end{bmatrix}$

15. Let $\mathbf{y} = \begin{bmatrix} 5 \\ -9 \\ 5 \end{bmatrix}$, $\mathbf{u}_1 = \begin{bmatrix} -3 \\ -5 \\ 1 \end{bmatrix}$, $\mathbf{u}_2 = \begin{bmatrix} -3 \\ 2 \\ 1 \end{bmatrix}$. Find the

distance from $\mathbf{y}$ to the plane in $\mathbb{R}^3$ spanned by $\mathbf{u}_1$ and $\mathbf{u}_2$.

16. Let $\mathbf{y}$, $\mathbf{v}_1$, and $\mathbf{v}_2$ be as in Exercise 12. Find the distance from $\mathbf{y}$ to the subspace of $\mathbb{R}^4$ spanned by $\mathbf{v}_1$ and $\mathbf{v}_2$.

17. Let $\mathbf{y} = \begin{bmatrix} 4 \\ 8 \\ 1 \end{bmatrix}$, $\mathbf{u}_1 = \begin{bmatrix} 2/3 \\ 1/3 \\ 2/3 \end{bmatrix}$, $\mathbf{u}_2 = \begin{bmatrix} -2/3 \\ 2/3 \\ 1/3 \end{bmatrix}$, and

$W = \text{Span}\{\mathbf{u}_1, \mathbf{u}_2\}$.

a. Let $U = [\,\mathbf{u}_1 \quad \mathbf{u}_2\,]$. Compute $U^T U$ and $U U^T$.

b. Compute $\text{proj}_W \mathbf{y}$ and $(U U^T)\mathbf{y}$.

18. Let $\mathbf{y} = \begin{bmatrix} 7 \\ 9 \end{bmatrix}$, $\mathbf{u}_1 = \begin{bmatrix} 1/\sqrt{10} \\ -3/\sqrt{10} \end{bmatrix}$, and $W = \text{Span}\{\mathbf{u}_1\}$.

a. Let U be the 2×1 matrix whose only column is $\mathbf{u}_1$. Compute $U^T U$ and $U U^T$.

b. Compute $\text{proj}_W \mathbf{y}$ and $(U U^T)\mathbf{y}$.

19. Let $\mathbf{u}_1 = \begin{bmatrix} 1 \\ 1 \\ -2 \end{bmatrix}$, $\mathbf{u}_2 = \begin{bmatrix} 5 \\ -1 \\ 2 \end{bmatrix}$, and $\mathbf{u}_3 = \begin{bmatrix} 0 \\ 0 \\ 1 \end{bmatrix}$. Note that $\mathbf{u}_1$ and $\mathbf{u}_2$ are orthogonal but that $\mathbf{u}_3$ is not orthogonal to $\mathbf{u}_1$ or $\mathbf{u}_2$. It can be shown that $\mathbf{u}_3$ is not in the subspace W spanned by $\mathbf{u}_1$ and $\mathbf{u}_2$. Use this fact to construct a nonzero vector $\mathbf{v}$ in $\mathbb{R}^3$ that is orthogonal to $\mathbf{u}_1$ and $\mathbf{u}_2$.

20. Let $\mathbf{u}_1$ and $\mathbf{u}_2$ be as in Exercise 19, and let $\mathbf{u}_4 = \begin{bmatrix} 0 \\ 1 \\ 0 \end{bmatrix}$. It can be shown that $\mathbf{u}_4$ is not in the subspace W spanned by $\mathbf{u}_1$ and $\mathbf{u}_2$. Use this fact to construct a nonzero vector $\mathbf{v}$ in $\mathbb{R}^3$ that is orthogonal to $\mathbf{u}_1$ and $\mathbf{u}_2$.

In Exercises 21 and 22, all vectors and subspaces are in $\mathbb{R}^n$. Mark each statement True or False. Justify each answer.

21. a. If $\mathbf{z}$ is orthogonal to $\mathbf{u}_1$ and to $\mathbf{u}_2$ and if $W = \text{Span}\{\mathbf{u}_1, \mathbf{u}_2\}$, then $\mathbf{z}$ must be in $W^\perp$.

b. For each $\mathbf{y}$ and each subspace W, the vector $\mathbf{y} - \text{proj}_W \mathbf{y}$ is orthogonal to W.

c. The orthogonal projection $\hat{\mathbf{y}}$ of $\mathbf{y}$ onto a subspace W can sometimes depend on the orthogonal basis for W used to compute $\hat{\mathbf{y}}$.

d. If $\mathbf{y}$ is in a subspace W, then the orthogonal projection of $\mathbf{y}$ onto W is $\mathbf{y}$ itself.

e. If the columns of an $n \times p$ matrix U are orthonormal, then $U U^T \mathbf{y}$ is the orthogonal projection of $\mathbf{y}$ onto the column space of U.

22. a. If W is a subspace of $\mathbb{R}^n$ and if $\mathbf{v}$ is in both W and $W^\perp$, then $\mathbf{v}$ must be the zero vector.

b. In the Orthogonal Decomposition Theorem, each term in formula (2) for $\hat{\mathbf{y}}$ is itself an orthogonal projection of $\mathbf{y}$ onto a subspace of W.

c. If $\mathbf{y} = \mathbf{z}_1 + \mathbf{z}_2$, where $\mathbf{z}_1$ is in a subspace W and $\mathbf{z}_2$ is in $W^\perp$, then $\mathbf{z}_1$ must be the orthogonal projection of $\mathbf{y}$ onto W.

d. The best approximation to $\mathbf{y}$ by elements of a subspace W is given by the vector $\mathbf{y} - \text{proj}_W \mathbf{y}$.

e. If an $n \times p$ matrix U has orthonormal columns, then $U U^T \mathbf{x} = \mathbf{x}$ for all $\mathbf{x}$ in $\mathbb{R}^n$.

23. Let A be an $m \times n$ matrix. Prove that every vector $\mathbf{x}$ in $\mathbb{R}^n$ can be written in the form $\mathbf{x} = \mathbf{p} + \mathbf{u}$, where $\mathbf{p}$ is in Row A and $\mathbf{u}$ is in Nul A. Also, show that if the equation $A\mathbf{x} = \mathbf{b}$ is consistent, then there is a unique $\mathbf{p}$ in Row A such that $A\mathbf{p} = \mathbf{b}$.

24. Let W be a subspace of $\mathbb{R}^n$ with an orthogonal basis $\{\mathbf{w}_1, \ldots, \mathbf{w}_p\}$, and let $\{\mathbf{v}_1, \ldots, \mathbf{v}_q\}$ be an orthogonal basis for $W^\perp$.

a. Explain why $\{\mathbf{w}_1, \ldots, \mathbf{w}_p, \mathbf{v}_1, \ldots, \mathbf{v}_q\}$ is an orthogonal set.

b. Explain why the set in part (a) spans $\mathbb{R}^n$.

c. Show that $\dim W + \dim W^\perp = n$.

25. [M] Let U be the 8×4 matrix in Exercise 36 in Section 6.2. Find the closest point to $\mathbf{y} = (1, 1, 1, 1, 1, 1, 1, 1)$ in Col U. Write the keystrokes or commands you use to solve this problem.

26. [M] Let U be the matrix in Exercise 25. Find the distance from $\mathbf{b} = (1, 1, 1, 1, -1, -1, -1, -1)$ to Col U.

SOLUTION TO PRACTICE PROBLEM

Compute

$$\text{proj}_W \mathbf{y} = \frac{\mathbf{y} \cdot \mathbf{u}_1}{\mathbf{u}_1 \cdot \mathbf{u}_1} \mathbf{u}_1 + \frac{\mathbf{y} \cdot \mathbf{u}_2}{\mathbf{u}_2 \cdot \mathbf{u}_2} \mathbf{u}_2 = \frac{88}{66} \mathbf{u}_1 + \frac{-2}{6} \mathbf{u}_2$$

$$= \frac{4}{3} \begin{bmatrix} -7 \\ 1 \\ 4 \end{bmatrix} - \frac{1}{3} \begin{bmatrix} -1 \\ 1 \\ -2 \end{bmatrix} = \begin{bmatrix} -9 \\ 1 \\ 6 \end{bmatrix} = \mathbf{y}$$

In this case, $\mathbf{y}$ happens to be a linear combination of $\mathbf{u}_1$ and $\mathbf{u}_2$, so $\mathbf{y}$ is in W. The closest point in W to $\mathbf{y}$ is $\mathbf{y}$ itself.

6.4 | THE GRAM–SCHMIDT PROCESS

The Gram–Schmidt process is a simple algorithm for producing an orthogonal or orthonormal basis for any nonzero subspace of $\mathbb{R}^n$. The first two examples of the process are aimed at hand calculation.

EXAMPLE 1 Let $W = \text{Span}\{\mathbf{x}_1, \mathbf{x}_2\}$, where $\mathbf{x}_1 = \begin{bmatrix} 3 \\ 6 \\ 0 \end{bmatrix}$ and $\mathbf{x}_2 = \begin{bmatrix} 1 \\ 2 \\ 2 \end{bmatrix}$. Construct an orthogonal basis $\{\mathbf{v}_1, \mathbf{v}_2\}$ for W.

SOLUTION The subspace W is shown in Fig. 1, along with $\mathbf{x}_1$, $\mathbf{x}_2$, and the projection $\mathbf{p}$ of $\mathbf{x}_2$ onto $\mathbf{x}_1$. The component of $\mathbf{x}_2$ orthogonal to $\mathbf{x}_1$ is $\mathbf{x}_2 - \mathbf{p}$, which is in W because it is formed from $\mathbf{x}_2$ and a multiple of $\mathbf{x}_1$. Let $\mathbf{v}_1 = \mathbf{x}_1$ and

$$\mathbf{v}_2 = \mathbf{x}_2 - \mathbf{p} = \mathbf{x}_2 - \frac{\mathbf{x}_2 \cdot \mathbf{x}_1}{\mathbf{x}_1 \cdot \mathbf{x}_1}\mathbf{x}_1 = \begin{bmatrix} 1 \\ 2 \\ 2 \end{bmatrix} - \frac{15}{45}\begin{bmatrix} 3 \\ 6 \\ 0 \end{bmatrix} = \begin{bmatrix} 0 \\ 0 \\ 2 \end{bmatrix}$$

Then $\{\mathbf{v}_1, \mathbf{v}_2\}$ is an orthogonal set of nonzero vectors in W. Since $\dim W = 2$, the set $\{\mathbf{v}_1, \mathbf{v}_2\}$ is a basis for W. ∎

The next example fully illustrates the Gram–Schmidt process. Study it carefully.

EXAMPLE 2 Let $\mathbf{x}_1 = \begin{bmatrix} 1 \\ 1 \\ 1 \\ 1 \end{bmatrix}$, $\mathbf{x}_2 = \begin{bmatrix} 0 \\ 1 \\ 1 \\ 1 \end{bmatrix}$, and $\mathbf{x}_3 = \begin{bmatrix} 0 \\ 0 \\ 1 \\ 1 \end{bmatrix}$. Then $\{\mathbf{x}_1, \mathbf{x}_2, \mathbf{x}_3\}$ is clearly linearly independent and thus is a basis for a subspace W of $\mathbb{R}^4$. Construct an orthogonal basis for W.

SOLUTION

Step 1. Let $\mathbf{v}_1 = \mathbf{x}_1$ and $W_1 = \text{Span}\{\mathbf{x}_1\} = \text{Span}\{\mathbf{v}_1\}$.

Step 2. Let $\mathbf{v}_2$ be the vector produced by subtracting from $\mathbf{x}_2$ its projection onto the subspace W_1. That is, let

$$\mathbf{v}_2 = \mathbf{x}_2 - \text{proj}_{W_1}\mathbf{x}_2$$
$$= \mathbf{x}_2 - \frac{\mathbf{x}_2 \cdot \mathbf{v}_1}{\mathbf{v}_1 \cdot \mathbf{v}_1}\mathbf{v}_1 \qquad \text{Since } \mathbf{v}_1 = \mathbf{x}_1$$
$$= \begin{bmatrix} 0 \\ 1 \\ 1 \\ 1 \end{bmatrix} - \frac{3}{4}\begin{bmatrix} 1 \\ 1 \\ 1 \\ 1 \end{bmatrix} = \begin{bmatrix} -3/4 \\ 1/4 \\ 1/4 \\ 1/4 \end{bmatrix}$$

As in Example 1, $\mathbf{v}_2$ is the component of $\mathbf{x}_2$ orthogonal to $\mathbf{x}_1$, and $\{\mathbf{v}_1, \mathbf{v}_2\}$ is an orthogonal basis for the subspace W_2 spanned by $\mathbf{x}_1$ and $\mathbf{x}_2$.

Step 2′ (optional). If appropriate, scale $\mathbf{v}_2$ to simplify later computations. Since $\mathbf{v}_2$ has fractional entries, it is convenient to scale it by a factor of 4 and replace $\{\mathbf{v}_1, \mathbf{v}_2\}$ by the orthogonal basis

$$\mathbf{v}_1 = \begin{bmatrix} 1 \\ 1 \\ 1 \\ 1 \end{bmatrix}, \quad \mathbf{v}_2' = \begin{bmatrix} -3 \\ 1 \\ 1 \\ 1 \end{bmatrix}$$

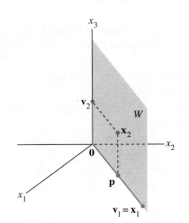

FIGURE 1

Construction of an orthogonal basis $\{\mathbf{v}_1, \mathbf{v}_2\}$.

Step 3. Let $\mathbf{v}_3$ be the vector produced by subtracting from $\mathbf{x}_3$ its projection onto the subspace W_2. Use the orthogonal basis $\{\mathbf{v}_1, \mathbf{v}_2'\}$ to compute this projection onto W_2:

$$\operatorname{proj}_{W_2} \mathbf{x}_3 = \overbrace{\frac{\mathbf{x}_3 \cdot \mathbf{v}_1}{\mathbf{v}_1 \cdot \mathbf{v}_1} \mathbf{v}_1}^{\substack{\text{Projection of} \\ \mathbf{x}_3 \text{ onto } \mathbf{v}_1}} + \overbrace{\frac{\mathbf{x}_3 \cdot \mathbf{v}_2'}{\mathbf{v}_2' \cdot \mathbf{v}_2'} \mathbf{v}_2'}^{\substack{\text{Projection of} \\ \mathbf{x}_3 \text{ onto } \mathbf{v}_2'}} = \frac{2}{4} \begin{bmatrix} 1 \\ 1 \\ 1 \\ 1 \end{bmatrix} + \frac{2}{12} \begin{bmatrix} -3 \\ 1 \\ 1 \\ 1 \end{bmatrix} = \begin{bmatrix} 0 \\ 2/3 \\ 2/3 \\ 2/3 \end{bmatrix}$$

Then $\mathbf{v}_3$ is the component of $\mathbf{x}_3$ orthogonal to W_2, namely,

$$\mathbf{v}_3 = \mathbf{x}_3 - \operatorname{proj}_{W_2} \mathbf{x}_3 = \begin{bmatrix} 0 \\ 0 \\ 1 \\ 1 \end{bmatrix} - \begin{bmatrix} 0 \\ 2/3 \\ 2/3 \\ 2/3 \end{bmatrix} = \begin{bmatrix} 0 \\ -2/3 \\ 1/3 \\ 1/3 \end{bmatrix}$$

See Fig. 2 for a diagram of this construction. Observe that $\mathbf{v}_3$ is in W, because $\mathbf{x}_3$ and $\operatorname{proj}_{W_2} \mathbf{x}_3$ are both in W. Thus $\{\mathbf{v}_1, \mathbf{v}_2', \mathbf{v}_3\}$ is an orthogonal set of nonzero vectors and hence a linearly independent set in W. Note that W is three-dimensional since it was defined by a basis of three vectors. Hence, by the Basis Theorem in Section 4.5, $\{\mathbf{v}_1, \mathbf{v}_2', \mathbf{v}_3\}$ is an orthogonal basis for W. ∎

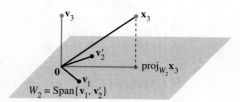

FIGURE 2 The construction of $\mathbf{v}_3$ from $\mathbf{x}_3$ and W_2.

The proof of the next theorem shows that this strategy really works. Scaling of vectors is not mentioned because that is used only to simplify hand calculations.

THEOREM 11

The Gram–Schmidt Process

Given a basis $\{\mathbf{x}_1, \ldots, \mathbf{x}_p\}$ for a nonzero subspace W of $\mathbb{R}^n$, define

$$\mathbf{v}_1 = \mathbf{x}_1$$

$$\mathbf{v}_2 = \mathbf{x}_2 - \frac{\mathbf{x}_2 \cdot \mathbf{v}_1}{\mathbf{v}_1 \cdot \mathbf{v}_1} \mathbf{v}_1$$

$$\mathbf{v}_3 = \mathbf{x}_3 - \frac{\mathbf{x}_3 \cdot \mathbf{v}_1}{\mathbf{v}_1 \cdot \mathbf{v}_1} \mathbf{v}_1 - \frac{\mathbf{x}_3 \cdot \mathbf{v}_2}{\mathbf{v}_2 \cdot \mathbf{v}_2} \mathbf{v}_2$$

$$\vdots$$

$$\mathbf{v}_p = \mathbf{x}_p - \frac{\mathbf{x}_p \cdot \mathbf{v}_1}{\mathbf{v}_1 \cdot \mathbf{v}_1} \mathbf{v}_1 - \frac{\mathbf{x}_p \cdot \mathbf{v}_2}{\mathbf{v}_2 \cdot \mathbf{v}_2} \mathbf{v}_2 - \cdots - \frac{\mathbf{x}_p \cdot \mathbf{v}_{p-1}}{\mathbf{v}_{p-1} \cdot \mathbf{v}_{p-1}} \mathbf{v}_{p-1}$$

Then $\{\mathbf{v}_1, \ldots, \mathbf{v}_p\}$ is an orthogonal basis for W. In addition

$$\operatorname{Span}\{\mathbf{v}_1, \ldots, \mathbf{v}_k\} = \operatorname{Span}\{\mathbf{x}_1, \ldots, \mathbf{x}_k\} \qquad \text{for } 1 \le k \le p \qquad (1)$$

PROOF For $1 \leq k \leq p$, let $W_k = \text{Span}\{x_1, \ldots, x_k\}$. Set $v_1 = x_1$, so that $\text{Span}\{v_1\} = \text{Span}\{x_1\}$. Suppose, for some $k < p$, we have constructed $v_1, \ldots, v_k$ so that $\{v_1, \ldots, v_k\}$ is an orthogonal basis for W_k. Define

$$v_{k+1} = x_{k+1} - \text{proj}_{W_k} x_{k+1} \tag{2}$$

By the Orthogonal Decomposition Theorem, v_{k+1} is orthogonal to W_k. Note that $\text{proj}_{W_k} x_{k+1}$ is in W_k and hence also in W_{k+1}. Since x_{k+1} is in W_{k+1}, so is v_{k+1} (because W_{k+1} is a subspace and is closed under subtraction). Furthermore, $v_{k+1} \neq 0$ because x_{k+1} is not in $W_k = \text{Span}\{x_1, \ldots, x_k\}$. Hence $\{v_1, \ldots, v_{k+1}\}$ is an orthogonal set of nonzero vectors in the $(k+1)$-dimensional space W_{k+1}. By the Basis Theorem in Section 4.5, this set is an orthogonal basis for W_{k+1}. Hence $W_{k+1} = \text{Span}\{v_1, \ldots, v_{k+1}\}$. When $k + 1 = p$, the process stops. ∎

Theorem 11 shows that any nonzero subspace W of $\mathbb{R}^n$ has an orthogonal basis, because an ordinary basis $\{x_1, \ldots, x_p\}$ is always available (by Theorem 11 in Section 4.5), and the Gram–Schmidt process depends only on the existence of orthogonal projections onto subspaces of W that already have orthogonal bases.

Orthonormal Bases

An orthonormal basis is constructed easily from an orthogonal basis $\{v_1, \ldots, v_p\}$: simply normalize (i.e., "scale") all the v_k. When working problems by hand, this is easier than normalizing each v_k as soon as it is found (because it avoids unnecessary writing of square roots).

EXAMPLE 3 Example 1 constructed the orthogonal basis

$$v_1 = \begin{bmatrix} 3 \\ 6 \\ 0 \end{bmatrix}, \quad v_2 = \begin{bmatrix} 0 \\ 0 \\ 2 \end{bmatrix}$$

An orthonormal basis is

$$u_1 = \frac{1}{\|v_1\|} v_1 = \frac{1}{\sqrt{45}} \begin{bmatrix} 3 \\ 6 \\ 0 \end{bmatrix} = \begin{bmatrix} 1/\sqrt{5} \\ 2/\sqrt{5} \\ 0 \end{bmatrix}$$

$$u_2 = \frac{1}{\|v_2\|} v_2 = \begin{bmatrix} 0 \\ 0 \\ 1 \end{bmatrix}$$ ∎

QR Factorization of Matrices

WEB

If an $m \times n$ matrix A has linearly independent columns $x_1, \ldots, x_n$, then applying the Gram–Schmidt process (with normalizations) to $x_1, \ldots, x_n$ amounts to *factoring A*, as described in the next theorem. This factorization is widely used in computer algorithms for various computations, such as solving equations (discussed in Section 6.5) and finding eigenvalues (mentioned in the exercises for Section 5.2).

THEOREM 12

The QR Factorization

If A is an $m \times n$ matrix with linearly independent columns, then A can be factored as $A = QR$, where Q is an $m \times n$ matrix whose columns form an orthonormal basis for Col A and R is an $n \times n$ upper triangular invertible matrix with positive entries on its diagonal.

PROOF The columns of A form a basis $\{\mathbf{x}_1, \ldots, \mathbf{x}_n\}$ for Col A. Construct an orthonormal basis $\{\mathbf{u}_1, \ldots, \mathbf{u}_n\}$ for $W = \text{Col } A$ with property (1) in Theorem 11. This basis may be constructed by the Gram–Schmidt process or some other means. Let

$$Q = [\, \mathbf{u}_1 \quad \mathbf{u}_2 \quad \cdots \quad \mathbf{u}_n \,]$$

For $k = 1, \ldots, n, \mathbf{x}_k$ is in Span $\{\mathbf{x}_1, \ldots, \mathbf{x}_k\} = \text{Span}\{\mathbf{u}_1, \ldots, \mathbf{u}_k\}$. So there are constants, $r_{1k}, \ldots, r_{kk}$, such that

$$\mathbf{x}_k = r_{1k}\mathbf{u}_1 + \cdots + r_{kk}\mathbf{u}_k + 0 \cdot \mathbf{u}_{k+1} + \cdots + 0 \cdot \mathbf{u}_n$$

We may assume that $r_{kk} \geq 0$. (If $r_{kk} < 0$, multiply both r_{kk} and $\mathbf{u}_k$ by -1.) This shows that $\mathbf{x}_k$ is a linear combination of the columns of Q using as weights the entries in the vector

$$\mathbf{r}_k = \begin{bmatrix} r_{1k} \\ \vdots \\ r_{kk} \\ 0 \\ \vdots \\ 0 \end{bmatrix}$$

That is, $\mathbf{x}_k = Q\mathbf{r}_k$ for $k = 1, \ldots, n$. Let $R = [\, \mathbf{r}_1 \quad \cdots \quad \mathbf{r}_n \,]$. Then

$$A = [\, \mathbf{x}_1 \quad \cdots \quad \mathbf{x}_n \,] = [\, Q\mathbf{r}_1 \quad \cdots \quad Q\mathbf{r}_n \,] = QR$$

The fact that R is invertible follows easily from the fact that the columns of A are linearly independent (Exercise 19). Since R is clearly upper triangular, its nonnegative diagonal entries must be positive. ∎

EXAMPLE 4 Find a QR factorization of $A = \begin{bmatrix} 1 & 0 & 0 \\ 1 & 1 & 0 \\ 1 & 1 & 1 \\ 1 & 1 & 1 \end{bmatrix}$.

SOLUTION The columns of A are the vectors $\mathbf{x}_1$, $\mathbf{x}_2$, and $\mathbf{x}_3$ in Example 2. An orthogonal basis for Col $A = \text{Span}\{\mathbf{x}_1, \mathbf{x}_2, \mathbf{x}_3\}$ was found in that example:

$$\mathbf{v}_1 = \begin{bmatrix} 1 \\ 1 \\ 1 \\ 1 \end{bmatrix}, \quad \mathbf{v}_2' = \begin{bmatrix} -3 \\ 1 \\ 1 \\ 1 \end{bmatrix}, \quad \mathbf{v}_3 = \begin{bmatrix} 0 \\ -2/3 \\ 1/3 \\ 1/3 \end{bmatrix}$$

To simplify the arithmetic that follows, scale $\mathbf{v}_3$ by letting $\mathbf{v}_3' = 3\mathbf{v}_3$. Then normalize the three vectors to obtain $\mathbf{u}_1$, $\mathbf{u}_2$, and $\mathbf{u}_3$, and use these vectors as the columns of Q:

$$Q = \begin{bmatrix} 1/2 & -3/\sqrt{12} & 0 \\ 1/2 & 1/\sqrt{12} & -2/\sqrt{6} \\ 1/2 & 1/\sqrt{12} & 1/\sqrt{6} \\ 1/2 & 1/\sqrt{12} & 1/\sqrt{6} \end{bmatrix}$$

By construction, the first k columns of Q are an orthonormal basis of Span $\{x_1, \ldots, x_k\}$. From the proof of Theorem 12, $A = QR$ for some R. To find R, observe that $Q^T Q = I$, because the columns of Q are orthonormal. Hence

$$Q^T A = Q^T(QR) = IR = R$$

and

$$R = \begin{bmatrix} 1/2 & 1/2 & 1/2 & 1/2 \\ -3/\sqrt{12} & 1/\sqrt{12} & 1/\sqrt{12} & 1/\sqrt{12} \\ 0 & -2/\sqrt{6} & 1/\sqrt{6} & 1/\sqrt{6} \end{bmatrix} \begin{bmatrix} 1 & 0 & 0 \\ 1 & 1 & 0 \\ 1 & 1 & 1 \\ 1 & 1 & 1 \end{bmatrix}$$

$$= \begin{bmatrix} 2 & 3/2 & 1 \\ 0 & 3/\sqrt{12} & 2/\sqrt{12} \\ 0 & 0 & 2/\sqrt{6} \end{bmatrix} \qquad \blacksquare$$

NUMERICAL NOTES

1. When the Gram–Schmidt process is run on a computer, roundoff error can build up as the vectors u_k are calculated, one by one. For j and k large but unequal, the inner products $u_j^T u_k$ may not be sufficiently close to zero. This loss of orthogonality can be reduced substantially by rearranging the order of the calculations.[1] However, a different computer-based QR factorization is usually preferred to this modified Gram–Schmidt method because it yields a more accurate orthonormal basis, even though the factorization requires about twice as much arithmetic.

2. To produce a QR factorization of a matrix A, a computer program usually left-multiplies A by a sequence of orthogonal matrices until A is transformed into an upper triangular matrix. This construction is analogous to the left-multiplication by elementary matrices that produces an LU factorization of A.

PRACTICE PROBLEM

Let $W = \text{Span}\{x_1, x_2\}$, where $x_1 = \begin{bmatrix} 1 \\ 1 \\ 1 \end{bmatrix}$ and $x_2 = \begin{bmatrix} 1/3 \\ 1/3 \\ -2/3 \end{bmatrix}$. Construct an orthonormal basis for W.

6.4 EXERCISES

In Exercises 1–6, the given set is a basis for a subspace W. Use the Gram–Schmidt process to produce an orthogonal basis for W.

1. $\begin{bmatrix} 3 \\ 0 \\ -1 \end{bmatrix}, \begin{bmatrix} 8 \\ 5 \\ -6 \end{bmatrix}$

2. $\begin{bmatrix} 0 \\ 4 \\ 2 \end{bmatrix}, \begin{bmatrix} 5 \\ 6 \\ -7 \end{bmatrix}$

3. $\begin{bmatrix} 2 \\ -5 \\ 1 \end{bmatrix}, \begin{bmatrix} 4 \\ -1 \\ 2 \end{bmatrix}$

4. $\begin{bmatrix} 3 \\ -4 \\ 5 \end{bmatrix}, \begin{bmatrix} -3 \\ 14 \\ -7 \end{bmatrix}$

5. $\begin{bmatrix} 1 \\ -4 \\ 0 \\ 1 \end{bmatrix}, \begin{bmatrix} 7 \\ -7 \\ -4 \\ 1 \end{bmatrix}$

6. $\begin{bmatrix} 3 \\ -1 \\ 2 \\ -1 \end{bmatrix}, \begin{bmatrix} -5 \\ 9 \\ -9 \\ 3 \end{bmatrix}$

---

[1] See *Fundamentals of Matrix Computations*, by David S. Watkins (New York: John Wiley & Sons, 1991), pp. 167–180.

7. Find an orthonormal basis of the subspace spanned by the vectors in Exercise 3.

8. Find an orthonormal basis of the subspace spanned by the vectors in Exercise 4.

Find an orthogonal basis for the column space of each matrix in Exercises 9–12.

9. $\begin{bmatrix} 3 & -5 & 1 \\ 1 & 1 & 1 \\ -1 & 5 & -2 \\ 3 & -7 & 8 \end{bmatrix}$

10. $\begin{bmatrix} -1 & 6 & 6 \\ 3 & -8 & 3 \\ 1 & -2 & 6 \\ 1 & -4 & -3 \end{bmatrix}$

11. $\begin{bmatrix} 1 & 2 & 5 \\ -1 & 1 & -4 \\ -1 & 4 & -3 \\ 1 & -4 & 7 \\ 1 & 2 & 1 \end{bmatrix}$

12. $\begin{bmatrix} 1 & 3 & 5 \\ -1 & -3 & 1 \\ 0 & 2 & 3 \\ 1 & 5 & 2 \\ 1 & 5 & 8 \end{bmatrix}$

In Exercises 13 and 14, the columns of Q were obtained by applying the Gram–Schmidt process to the columns of A. Find an upper triangular matrix R such that $A = QR$. Check your work.

13. $A = \begin{bmatrix} 5 & 9 \\ 1 & 7 \\ -3 & -5 \\ 1 & 5 \end{bmatrix}$, $Q = \begin{bmatrix} 5/6 & -1/6 \\ 1/6 & 5/6 \\ -3/6 & 1/6 \\ 1/6 & 3/6 \end{bmatrix}$

14. $A = \begin{bmatrix} -2 & 3 \\ 5 & 7 \\ 2 & -2 \\ 4 & 6 \end{bmatrix}$, $Q = \begin{bmatrix} -2/7 & 5/7 \\ 5/7 & 2/7 \\ 2/7 & -4/7 \\ 4/7 & 2/7 \end{bmatrix}$

15. Find a QR factorization of the matrix in Exercise 11.

16. Find a QR factorization of the matrix in Exercise 12.

In Exercises 17 and 18, all vectors and subspaces are in $\mathbb{R}^n$. Mark each statement True or False. Justify each answer.

17. a. If $\{\mathbf{v}_1, \mathbf{v}_2, \mathbf{v}_3\}$ is an orthogonal basis for W, then multiplying $\mathbf{v}_3$ by a scalar c gives a new orthogonal basis $\{\mathbf{v}_1, \mathbf{v}_2, c\mathbf{v}_3\}$.

b. The Gram–Schmidt process produces from a linearly independent set $\{\mathbf{x}_1, \ldots, \mathbf{x}_p\}$ an orthogonal set $\{\mathbf{v}_1, \ldots, \mathbf{v}_p\}$ with the property that for each k, the vectors $\mathbf{v}_1, \ldots, \mathbf{v}_k$ span the same subspace as that spanned by $\mathbf{x}_1, \ldots, \mathbf{x}_k$.

c. If $A = QR$, where Q has orthonormal columns, then $R = Q^T A$.

18. a. If $W = \text{Span}\{\mathbf{x}_1, \mathbf{x}_2, \mathbf{x}_3\}$ with $\{\mathbf{x}_1, \mathbf{x}_2, \mathbf{x}_3\}$ linearly independent, and if $\{\mathbf{v}_1, \mathbf{v}_2, \mathbf{v}_3\}$ is an orthogonal set in W, then $\{\mathbf{v}_1, \mathbf{v}_2, \mathbf{v}_3\}$ is a basis for W.

b. If $\mathbf{x}$ is not in a subspace W, then $\mathbf{x} - \text{proj}_W \mathbf{x}$ is not zero.

c. In a QR factorization, say $A = QR$ (when A has linearly independent columns), the columns of Q form an orthonormal basis for the column space of A.

19. Suppose $A = QR$, where Q is $m \times n$ and R is $n \times n$. Show that if the columns of A are linearly independent, then R must be invertible. [*Hint:* Study the equation $R\mathbf{x} = \mathbf{0}$ and use the fact that $A = QR$.]

20. Suppose $A = QR$, where R is an invertible matrix. Show that A and Q have the same column space. [*Hint:* Given $\mathbf{y}$ in Col A, show that $\mathbf{y} = Q\mathbf{x}$ for some $\mathbf{x}$. Also, given $\mathbf{y}$ in Col Q, show that $\mathbf{y} = A\mathbf{x}$ for some $\mathbf{x}$.]

21. Given $A = QR$ as in Theorem 12, describe how to find an orthogonal $m \times m$ (square) matrix Q_1 and an invertible $n \times n$ upper triangular matrix R such that

$$A = Q_1 \begin{bmatrix} R \\ 0 \end{bmatrix}$$

The MATLAB `qr` command supplies this "full" QR factorization when rank $A = n$.

22. Let $\mathbf{u}_1, \ldots, \mathbf{u}_p$ be an orthogonal basis for a subspace W of $\mathbb{R}^n$, and let $T : \mathbb{R}^n \to \mathbb{R}^n$ be defined by $T(\mathbf{x}) = \text{proj}_W \mathbf{x}$. Show that T is a linear transformation.

23. Suppose $A = QR$ is a QR factorization of an $m \times n$ matrix A (with linearly independent columns). Partition A as $[A_1 \ A_2]$, where A_1 has p columns. Show how to obtain a QR factorization of A_1, and explain why your factorization has the appropriate properties.

24. [M] Use the Gram–Schmidt process as in Example 2 to produce an orthogonal basis for the column space of

$$A = \begin{bmatrix} -10 & 13 & 7 & -11 \\ 2 & 1 & -5 & 3 \\ -6 & 3 & 13 & -3 \\ 16 & -16 & -2 & 5 \\ 2 & 1 & -5 & -7 \end{bmatrix}$$

25. [M] Use the method in this section to produce a QR factorization of the matrix in Exercise 24.

26. [M] For a matrix program, the Gram–Schmidt process works better with orthonormal vectors. Starting with $\mathbf{x}_1, \ldots, \mathbf{x}_p$ as in Theorem 11, let $A = [\mathbf{x}_1 \ \cdots \ \mathbf{x}_p]$. Suppose Q is an $n \times k$ matrix whose columns form an orthonormal basis for the subspace W_k spanned by the first k columns of A. Then for $\mathbf{x}$ in $\mathbb{R}^n$, $QQ^T\mathbf{x}$ is the orthogonal projection of $\mathbf{x}$ onto W_k (Theorem 10 in Section 6.3). If $\mathbf{x}_{k+1}$ is the next column of A, then equation (2) in the proof of Theorem 11 becomes

$$\mathbf{v}_{k+1} = \mathbf{x}_{k+1} - Q(Q^T\mathbf{x}_{k+1})$$

(The parentheses above reduce the number of arithmetic operations.) Let $\mathbf{u}_{k+1} = \mathbf{v}_{k+1}/\|\mathbf{v}_{k+1}\|$. The new Q for the next step is $[Q \ \mathbf{u}_{k+1}]$. Use this procedure to compute the QR factorization of the matrix in Exercise 24. Write the keystrokes or commands you use.

WEB

SOLUTION TO PRACTICE PROBLEM

Let $\mathbf{v}_1 = \mathbf{x}_1 = \begin{bmatrix} 1 \\ 1 \\ 1 \end{bmatrix}$ and $\mathbf{v}_2 = \mathbf{x}_2 - \dfrac{\mathbf{x}_2 \cdot \mathbf{v}_1}{\mathbf{v}_1 \cdot \mathbf{v}_1} \mathbf{v}_1 = \mathbf{x}_2 - 0\mathbf{v}_1 = \mathbf{x}_2$. So $\{\mathbf{x}_1, \mathbf{x}_2\}$ is already orthogonal. All that is needed is to normalize the vectors. Let

$$\mathbf{u}_1 = \frac{1}{\|\mathbf{v}_1\|} \mathbf{v}_1 = \frac{1}{\sqrt{3}} \begin{bmatrix} 1 \\ 1 \\ 1 \end{bmatrix} = \begin{bmatrix} 1/\sqrt{3} \\ 1/\sqrt{3} \\ 1/\sqrt{3} \end{bmatrix}$$

Instead of normalizing $\mathbf{v}_2$ directly, normalize $\mathbf{v}_2' = 3\mathbf{v}_2$ instead:

$$\mathbf{u}_2 = \frac{1}{\|\mathbf{v}_2'\|} \mathbf{v}_2' = \frac{1}{\sqrt{1^2 + 1^2 + (-2)^2}} \begin{bmatrix} 1 \\ 1 \\ -2 \end{bmatrix} = \begin{bmatrix} 1/\sqrt{6} \\ 1/\sqrt{6} \\ -2/\sqrt{6} \end{bmatrix}$$

Then $\{\mathbf{u}_1, \mathbf{u}_2\}$ is an orthonormal basis for W.

6.5 | LEAST-SQUARES PROBLEMS

The chapter's introductory example described a massive problem $A\mathbf{x} = \mathbf{b}$ that had no solution. Inconsistent systems arise often in applications, though usually not with such an enormous coefficient matrix. When a solution is demanded and none exists, the best one can do is to find an $\mathbf{x}$ that makes $A\mathbf{x}$ as close as possible to $\mathbf{b}$.

Think of $A\mathbf{x}$ as an *approximation* to $\mathbf{b}$. The smaller the distance between $\mathbf{b}$ and $A\mathbf{x}$, given by $\|\mathbf{b} - A\mathbf{x}\|$, the better the approximation. The **general least-squares problem** is to find an $\mathbf{x}$ that makes $\|\mathbf{b} - A\mathbf{x}\|$ as small as possible. The adjective "least-squares" arises from the fact that $\|\mathbf{b} - A\mathbf{x}\|$ is the square root of a sum of squares.

DEFINITION

If A is $m \times n$ and $\mathbf{b}$ is in $\mathbb{R}^m$, a **least-squares solution** of $A\mathbf{x} = \mathbf{b}$ is an $\hat{\mathbf{x}}$ in $\mathbb{R}^n$ such that

$$\|\mathbf{b} - A\hat{\mathbf{x}}\| \leq \|\mathbf{b} - A\mathbf{x}\|$$

for all $\mathbf{x}$ in $\mathbb{R}^n$.

The most important aspect of the least-squares problem is that no matter what $\mathbf{x}$ we select, the vector $A\mathbf{x}$ will necessarily be in the column space, Col A. So we seek an $\mathbf{x}$ that makes $A\mathbf{x}$ the closest point in Col A to $\mathbf{b}$. See Fig. 1. (Of course, if $\mathbf{b}$ happens to be in Col A, then $\mathbf{b}$ *is* $A\mathbf{x}$ for some $\mathbf{x}$, and such an $\mathbf{x}$ is a "least-squares solution.")

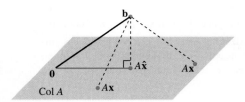

FIGURE 1 The vector $\mathbf{b}$ is closer to $A\hat{\mathbf{x}}$ than to $A\mathbf{x}$ for other $\mathbf{x}$.

Solution of the General Least-Squares Problem

Given A and $\mathbf{b}$ as above, apply the Best Approximation Theorem in Section 6.3 to the subspace Col A. Let

$$\hat{\mathbf{b}} = \text{proj}_{\text{Col } A}\,\mathbf{b}$$

Because $\hat{\mathbf{b}}$ is in the column space of A, the equation $A\mathbf{x} = \hat{\mathbf{b}}$ *is* consistent, and there is an $\hat{\mathbf{x}}$ in $\mathbb{R}^n$ such that

$$A\hat{\mathbf{x}} = \hat{\mathbf{b}} \tag{1}$$

Since $\hat{\mathbf{b}}$ is the closest point in Col A to $\mathbf{b}$, a vector $\hat{\mathbf{x}}$ is a least-squares solution of $A\mathbf{x} = \mathbf{b}$ if and only if $\hat{\mathbf{x}}$ satisfies (1). Such an $\hat{\mathbf{x}}$ in $\mathbb{R}^n$ is a list of weights that will build $\hat{\mathbf{b}}$ out of the columns of A. See Fig. 2. [There are many solutions of (1) if the equation has free variables.]

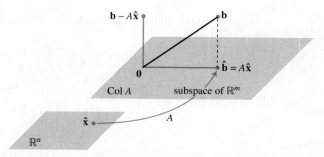

FIGURE 2 The least-squares solution $\hat{\mathbf{x}}$ is in $\mathbb{R}^n$.

Suppose $\hat{\mathbf{x}}$ satisfies $A\hat{\mathbf{x}} = \hat{\mathbf{b}}$. By the Orthogonal Decomposition Theorem in Section 6.3, the projection $\hat{\mathbf{b}}$ has the property that $\mathbf{b} - \hat{\mathbf{b}}$ is orthogonal to Col A, so $\mathbf{b} - A\hat{\mathbf{x}}$ is orthogonal to each column of A. If $\mathbf{a}_j$ is any column of A, then $\mathbf{a}_j \cdot (\mathbf{b} - A\hat{\mathbf{x}}) = 0$, and $\mathbf{a}_j^T (\mathbf{b} - A\hat{\mathbf{x}}) = 0$. Since each $\mathbf{a}_j^T$ is a row of A^T,

$$A^T (\mathbf{b} - A\hat{\mathbf{x}}) = \mathbf{0} \tag{2}$$

(This equation also follows from Theorem 3 in Section 6.1.) Thus

$$A^T\mathbf{b} - A^TA\hat{\mathbf{x}} = \mathbf{0}$$
$$A^TA\hat{\mathbf{x}} = A^T\mathbf{b}$$

These calculations show that each least-squares solution of $A\mathbf{x} = \mathbf{b}$ satisfies the equation

$$A^TA\mathbf{x} = A^T\mathbf{b} \tag{3}$$

The matrix equation (3) represents a system of equations called the **normal equations** for $A\mathbf{x} = \mathbf{b}$. A solution of (3) is often denoted by $\hat{\mathbf{x}}$.

THEOREM 13

> The set of least-squares solutions of $A\mathbf{x} = \mathbf{b}$ coincides with the nonempty set of solutions of the normal equations $A^TA\mathbf{x} = A^T\mathbf{b}$.

PROOF As shown above, the set of least-squares solutions is nonempty and each least-squares solution $\hat{\mathbf{x}}$ satisfies the normal equations. Conversely, suppose $\hat{\mathbf{x}}$ satisfies $A^TA\hat{\mathbf{x}} = A^T\mathbf{b}$. Then $\hat{\mathbf{x}}$ satisfies (2) above, which shows that $\mathbf{b} - A\hat{\mathbf{x}}$ is orthogonal to the rows of A^T and hence is orthogonal to the columns of A. Since the columns of A span Col A, the vector $\mathbf{b} - A\hat{\mathbf{x}}$ is orthogonal to all of Col A. Hence the equation

$$\mathbf{b} = A\hat{\mathbf{x}} + (\mathbf{b} - A\hat{\mathbf{x}})$$

is a decomposition of $\mathbf{b}$ into the sum of a vector in Col A and a vector orthogonal to Col A. By the uniqueness of the orthogonal decomposition, $A\hat{\mathbf{x}}$ must be the orthogonal projection of $\mathbf{b}$ onto Col A. That is, $A\hat{\mathbf{x}} = \hat{\mathbf{b}}$, and $\hat{\mathbf{x}}$ is a least-squares solution. ∎

EXAMPLE 1 Find a least-squares solution of the inconsistent system $A\mathbf{x} = \mathbf{b}$ for

$$A = \begin{bmatrix} 4 & 0 \\ 0 & 2 \\ 1 & 1 \end{bmatrix}, \quad \mathbf{b} = \begin{bmatrix} 2 \\ 0 \\ 11 \end{bmatrix}$$

SOLUTION To use normal equations (3), compute:

$$A^T A = \begin{bmatrix} 4 & 0 & 1 \\ 0 & 2 & 1 \end{bmatrix} \begin{bmatrix} 4 & 0 \\ 0 & 2 \\ 1 & 1 \end{bmatrix} = \begin{bmatrix} 17 & 1 \\ 1 & 5 \end{bmatrix}$$

$$A^T \mathbf{b} = \begin{bmatrix} 4 & 0 & 1 \\ 0 & 2 & 1 \end{bmatrix} \begin{bmatrix} 2 \\ 0 \\ 11 \end{bmatrix} = \begin{bmatrix} 19 \\ 11 \end{bmatrix}$$

Then the equation $A^T A\mathbf{x} = A^T \mathbf{b}$ becomes

$$\begin{bmatrix} 17 & 1 \\ 1 & 5 \end{bmatrix} \begin{bmatrix} x_1 \\ x_2 \end{bmatrix} = \begin{bmatrix} 19 \\ 11 \end{bmatrix}$$

Row operations can be used to solve this system, but since $A^T A$ is invertible and 2×2, it is probably faster to compute

$$(A^T A)^{-1} = \frac{1}{84} \begin{bmatrix} 5 & -1 \\ -1 & 17 \end{bmatrix}$$

and then to solve $A^T A\mathbf{x} = A^T \mathbf{b}$ as

$$\hat{\mathbf{x}} = (A^T A)^{-1} A^T \mathbf{b}$$

$$= \frac{1}{84} \begin{bmatrix} 5 & -1 \\ -1 & 17 \end{bmatrix} \begin{bmatrix} 19 \\ 11 \end{bmatrix} = \frac{1}{84} \begin{bmatrix} 84 \\ 168 \end{bmatrix} = \begin{bmatrix} 1 \\ 2 \end{bmatrix} \qquad ∎$$

In many calculations, $A^T A$ is invertible, but this is not always the case. The next example involves a matrix of the sort that appears in what are called *analysis of variance* problems in statistics.

EXAMPLE 2 Find a least-squares solution of $A\mathbf{x} = \mathbf{b}$ for

$$A = \begin{bmatrix} 1 & 1 & 0 & 0 \\ 1 & 1 & 0 & 0 \\ 1 & 0 & 1 & 0 \\ 1 & 0 & 1 & 0 \\ 1 & 0 & 0 & 1 \\ 1 & 0 & 0 & 1 \end{bmatrix}, \quad \mathbf{b} = \begin{bmatrix} -3 \\ -1 \\ 0 \\ 2 \\ 5 \\ 1 \end{bmatrix}$$

SOLUTION Compute

$$
A^T A = \begin{bmatrix} 1 & 1 & 1 & 1 & 1 & 1 \\ 1 & 1 & 0 & 0 & 0 & 0 \\ 0 & 0 & 1 & 1 & 0 & 0 \\ 0 & 0 & 0 & 0 & 1 & 1 \end{bmatrix} \begin{bmatrix} 1 & 1 & 0 & 0 \\ 1 & 1 & 0 & 0 \\ 1 & 0 & 1 & 0 \\ 1 & 0 & 1 & 0 \\ 1 & 0 & 0 & 1 \\ 1 & 0 & 0 & 1 \end{bmatrix} = \begin{bmatrix} 6 & 2 & 2 & 2 \\ 2 & 2 & 0 & 0 \\ 2 & 0 & 2 & 0 \\ 2 & 0 & 0 & 2 \end{bmatrix}
$$

$$
A^T \mathbf{b} = \begin{bmatrix} 1 & 1 & 1 & 1 & 1 & 1 \\ 1 & 1 & 0 & 0 & 0 & 0 \\ 0 & 0 & 1 & 1 & 0 & 0 \\ 0 & 0 & 0 & 0 & 1 & 1 \end{bmatrix} \begin{bmatrix} -3 \\ -1 \\ 0 \\ 2 \\ 5 \\ 1 \end{bmatrix} = \begin{bmatrix} 4 \\ -4 \\ 2 \\ 6 \end{bmatrix}
$$

The augmented matrix for $A^T A \mathbf{x} = A^T \mathbf{b}$ is

$$
\begin{bmatrix} 6 & 2 & 2 & 2 & 4 \\ 2 & 2 & 0 & 0 & -4 \\ 2 & 0 & 2 & 0 & 2 \\ 2 & 0 & 0 & 2 & 6 \end{bmatrix} \sim \begin{bmatrix} 1 & 0 & 0 & 1 & 3 \\ 0 & 1 & 0 & -1 & -5 \\ 0 & 0 & 1 & -1 & -2 \\ 0 & 0 & 0 & 0 & 0 \end{bmatrix}
$$

The general solution is $x_1 = 3 - x_4$, $x_2 = -5 + x_4$, $x_3 = -2 + x_4$, and x_4 is free. So the general least-squares solution of $A\mathbf{x} = \mathbf{b}$ has the form

$$
\hat{\mathbf{x}} = \begin{bmatrix} 3 \\ -5 \\ -2 \\ 0 \end{bmatrix} + x_4 \begin{bmatrix} -1 \\ 1 \\ 1 \\ 1 \end{bmatrix}
$$

∎

The next theorem gives useful criteria for determining when there is only one least-squares solution of $A\mathbf{x} = \mathbf{b}$. (Of course, the orthogonal projection $\hat{\mathbf{b}}$ is always unique.)

THEOREM 14

Let A be an $m \times n$ matrix. The following statements are logically equivalent:

a. The equation $A\mathbf{x} = \mathbf{b}$ has a unique least-squares solution for each $\mathbf{b}$ in $\mathbb{R}^m$.

b. The columns of A are linearly indpendent.

c. The matrix $A^T A$ is invertible.

When these statements are true, the least-squares solution $\hat{\mathbf{x}}$ is given by

$$
\hat{\mathbf{x}} = (A^T A)^{-1} A^T \mathbf{b} \tag{4}
$$

The main elements of a proof of Theorem 14 are outlined in Exercises 19–21, which also review concepts from Chapter 4. Formula (4) for $\hat{\mathbf{x}}$ is useful mainly for theoretical purposes and for hand calculations when $A^T A$ is a 2×2 invertible matrix.

When a least-squares solution $\hat{\mathbf{x}}$ is used to produce $A\hat{\mathbf{x}}$ as an approximation to $\mathbf{b}$, the distance from $\mathbf{b}$ to $A\hat{\mathbf{x}}$ is called the **least-squares error** of this approximation.

EXAMPLE 3   Given A and $\mathbf{b}$ as in Example 1, determine the least-squares error in the least-squares solution of $A\mathbf{x} = \mathbf{b}$.

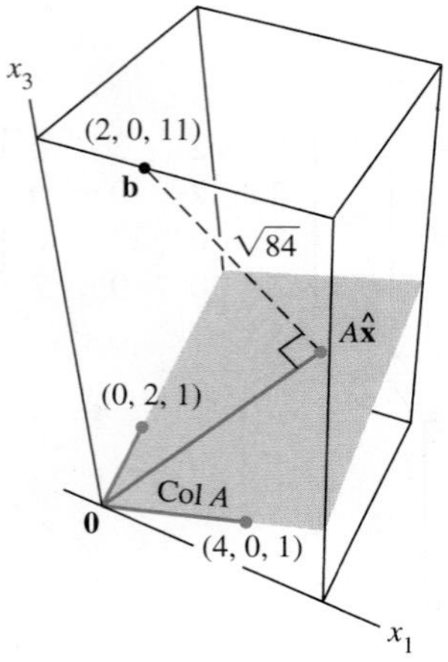

FIGURE 3

SOLUTION From Example 1,

$$\mathbf{b} = \begin{bmatrix} 2 \\ 0 \\ 11 \end{bmatrix} \quad \text{and} \quad A\hat{\mathbf{x}} = \begin{bmatrix} 4 & 0 \\ 0 & 2 \\ 1 & 1 \end{bmatrix} \begin{bmatrix} 1 \\ 2 \end{bmatrix} = \begin{bmatrix} 4 \\ 4 \\ 3 \end{bmatrix}$$

Hence

$$\mathbf{b} - A\hat{\mathbf{x}} = \begin{bmatrix} 2 \\ 0 \\ 11 \end{bmatrix} - \begin{bmatrix} 4 \\ 4 \\ 3 \end{bmatrix} = \begin{bmatrix} -2 \\ -4 \\ 8 \end{bmatrix}$$

and

$$\|\mathbf{b} - A\hat{\mathbf{x}}\| = \sqrt{(-2)^2 + (-4)^2 + 8^2} = \sqrt{84}$$

The least-squares error is $\sqrt{84}$. For any $\mathbf{x}$ in $\mathbb{R}^2$, the distance between $\mathbf{b}$ and the vector $A\mathbf{x}$ is at least $\sqrt{84}$. See Fig. 3. Note that the least-squares solution $\hat{\mathbf{x}}$ itself does not appear in the figure. ■

Alternative Calculations of Least-Squares Solutions

The next example shows how to find a least-squares solution of $A\mathbf{x} = \mathbf{b}$ when the columns of A are orthogonal. Such matrices often appear in linear regression problems, discussed in the next section.

EXAMPLE 4 Find a least-squares solution of $A\mathbf{x} = \mathbf{b}$ for

$$A = \begin{bmatrix} 1 & -6 \\ 1 & -2 \\ 1 & 1 \\ 1 & 7 \end{bmatrix}, \quad \mathbf{b} = \begin{bmatrix} -1 \\ 2 \\ 1 \\ 6 \end{bmatrix}$$

SOLUTION Because the columns $\mathbf{a}_1$ and $\mathbf{a}_2$ of A are orthogonal, the orthogonal projection of $\mathbf{b}$ onto Col A is given by

$$\hat{\mathbf{b}} = \frac{\mathbf{b} \cdot \mathbf{a}_1}{\mathbf{a}_1 \cdot \mathbf{a}_1} \mathbf{a}_1 + \frac{\mathbf{b} \cdot \mathbf{a}_2}{\mathbf{a}_2 \cdot \mathbf{a}_2} \mathbf{a}_2 = \frac{8}{4} \mathbf{a}_1 + \frac{45}{90} \mathbf{a}_2 \tag{5}$$

$$= \begin{bmatrix} 2 \\ 2 \\ 2 \\ 2 \end{bmatrix} + \begin{bmatrix} -3 \\ -1 \\ 1/2 \\ 7/2 \end{bmatrix} = \begin{bmatrix} -1 \\ 1 \\ 5/2 \\ 11/2 \end{bmatrix}$$

Now that $\hat{\mathbf{b}}$ is known, we can solve $A\hat{\mathbf{x}} = \hat{\mathbf{b}}$. But this is trivial, since we already know what weights to place on the columns of A to produce $\hat{\mathbf{b}}$. It is clear from (5) that

$$\hat{\mathbf{x}} = \begin{bmatrix} 8/4 \\ 45/90 \end{bmatrix} = \begin{bmatrix} 2 \\ 1/2 \end{bmatrix} \qquad ■$$

In some cases, the normal equations for a least-squares problem can be *ill-conditioned*; that is, small errors in the calculations of the entries of $A^T A$ can sometimes cause relatively large errors in the solution $\hat{\mathbf{x}}$. If the columns of A are linearly independent, the least-squares solution can often be computed more reliably through a QR factorization of A (described in Section 6.4).[1]

---

[1] The QR method is compared with the standard normal equation method in G. Golub and C. Van Loan, *Matrix Computations*, 3rd ed. (Baltimore: Johns Hopkins Press, 1996), pp. 230–231.

THEOREM 15

Given an $m \times n$ matrix A with linearly independent columns, let $A = QR$ be a QR factorization of A as in Theorem 12. Then, for each **b** in $\mathbb{R}^m$, the equation $A\mathbf{x} = \mathbf{b}$ has a unique least-squares solution, given by

$$\hat{\mathbf{x}} = R^{-1}Q^T\mathbf{b} \qquad (6)$$

PROOF Let $\hat{\mathbf{x}} = R^{-1}Q^T\mathbf{b}$. Then

$$A\hat{\mathbf{x}} = QR\hat{\mathbf{x}} = QRR^{-1}Q^T\mathbf{b} = QQ^T\mathbf{b}$$

By Theorem 12, the columns of Q form an orthonormal basis for Col A. Hence, by Theorem 10, $QQ^T\mathbf{b}$ is the orthogonal projection $\hat{\mathbf{b}}$ of **b** onto Col A. Then $A\hat{\mathbf{x}} = \hat{\mathbf{b}}$, which shows that $\hat{\mathbf{x}}$ is a least-squares solution of $A\mathbf{x} = \mathbf{b}$. The uniqueness of $\hat{\mathbf{x}}$ follows from Theorem 14. ■

NUMERICAL NOTE

Since R in Theorem 15 is upper triangular, $\hat{\mathbf{x}}$ should be calculated as the exact solution of the equation

$$R\mathbf{x} = Q^T\mathbf{b} \qquad (7)$$

It is much faster to solve (7) by back-substitution or row operations than to compute R^{-1} and use (6).

EXAMPLE 5   Find the least-squares solution of $A\mathbf{x} = \mathbf{b}$ for

$$A = \begin{bmatrix} 1 & 3 & 5 \\ 1 & 1 & 0 \\ 1 & 1 & 2 \\ 1 & 3 & 3 \end{bmatrix}, \quad \mathbf{b} = \begin{bmatrix} 3 \\ 5 \\ 7 \\ -3 \end{bmatrix}$$

SOLUTION   The QR factorization of A can be obtained as in Section 6.4.

$$A = QR = \begin{bmatrix} 1/2 & 1/2 & 1/2 \\ 1/2 & -1/2 & -1/2 \\ 1/2 & -1/2 & 1/2 \\ 1/2 & 1/2 & -1/2 \end{bmatrix} \begin{bmatrix} 2 & 4 & 5 \\ 0 & 2 & 3 \\ 0 & 0 & 2 \end{bmatrix}$$

Then

$$Q^T\mathbf{b} = \begin{bmatrix} 1/2 & 1/2 & 1/2 & 1/2 \\ 1/2 & -1/2 & -1/2 & 1/2 \\ 1/2 & -1/2 & 1/2 & -1/2 \end{bmatrix} \begin{bmatrix} 3 \\ 5 \\ 7 \\ -3 \end{bmatrix} = \begin{bmatrix} 6 \\ -6 \\ 4 \end{bmatrix}$$

The least-squares solution $\hat{\mathbf{x}}$ satisfies $R\mathbf{x} = Q^T\mathbf{b}$; that is,

$$\begin{bmatrix} 2 & 4 & 5 \\ 0 & 2 & 3 \\ 0 & 0 & 2 \end{bmatrix} \begin{bmatrix} x_1 \\ x_2 \\ x_3 \end{bmatrix} = \begin{bmatrix} 6 \\ -6 \\ 4 \end{bmatrix}$$

This equation is solved easily and yields $\hat{\mathbf{x}} = \begin{bmatrix} 10 \\ -6 \\ 2 \end{bmatrix}$. ■

| PRACTICE PROBLEMS

1. Let $A = \begin{bmatrix} 1 & -3 & -3 \\ 1 & 5 & 1 \\ 1 & 7 & 2 \end{bmatrix}$ and $\mathbf{b} = \begin{bmatrix} 5 \\ -3 \\ -5 \end{bmatrix}$. Find a least-squares solution of $A\mathbf{x} = \mathbf{b}$, and compute the associated least-squares error.

2. What can you say about the least-squares solution of $A\mathbf{x} = \mathbf{b}$ when $\mathbf{b}$ is orthogonal to the columns of A?

6.5 EXERCISES

In Exercises 1–4, find a least-squares solution of $A\mathbf{x} = \mathbf{b}$ by (a) constructing the normal equations for $\hat{\mathbf{x}}$ and (b) solving for $\hat{\mathbf{x}}$.

1. $A = \begin{bmatrix} -1 & 2 \\ 2 & -3 \\ -1 & 3 \end{bmatrix}$, $\mathbf{b} = \begin{bmatrix} 4 \\ 1 \\ 2 \end{bmatrix}$

2. $A = \begin{bmatrix} 2 & 1 \\ -2 & 0 \\ 2 & 3 \end{bmatrix}$, $\mathbf{b} = \begin{bmatrix} -5 \\ 8 \\ 1 \end{bmatrix}$

3. $A = \begin{bmatrix} 1 & -2 \\ -1 & 2 \\ 0 & 3 \\ 2 & 5 \end{bmatrix}$, $\mathbf{b} = \begin{bmatrix} 3 \\ 1 \\ -4 \\ 2 \end{bmatrix}$

4. $A = \begin{bmatrix} 1 & 3 \\ 1 & -1 \\ 1 & 1 \end{bmatrix}$, $\mathbf{b} = \begin{bmatrix} 5 \\ 1 \\ 0 \end{bmatrix}$

In Exercises 5 and 6, describe all least-squares solutions of the equation $A\mathbf{x} = \mathbf{b}$.

5. $A = \begin{bmatrix} 1 & 1 & 0 \\ 1 & 1 & 0 \\ 1 & 0 & 1 \\ 1 & 0 & 1 \end{bmatrix}$, $\mathbf{b} = \begin{bmatrix} 1 \\ 3 \\ 8 \\ 2 \end{bmatrix}$

6. $A = \begin{bmatrix} 1 & 1 & 0 \\ 1 & 1 & 0 \\ 1 & 1 & 0 \\ 1 & 0 & 1 \\ 1 & 0 & 1 \\ 1 & 0 & 1 \end{bmatrix}$, $\mathbf{b} = \begin{bmatrix} 7 \\ 2 \\ 3 \\ 6 \\ 5 \\ 4 \end{bmatrix}$

7. Compute the least-squares error associated with the least-squares solution found in Exercise 3.

8. Compute the least-squares error associated with the least-squares solution found in Exercise 4.

In Exercises 9–12, find (a) the orthogonal projection of $\mathbf{b}$ onto Col A and (b) a least-squares solution of $A\mathbf{x} = \mathbf{b}$.

9. $A = \begin{bmatrix} 1 & 5 \\ 3 & 1 \\ -2 & 4 \end{bmatrix}$, $\mathbf{b} = \begin{bmatrix} 4 \\ -2 \\ -3 \end{bmatrix}$

10. $A = \begin{bmatrix} 1 & 2 \\ -1 & 4 \\ 1 & 2 \end{bmatrix}$, $\mathbf{b} = \begin{bmatrix} 3 \\ -1 \\ 5 \end{bmatrix}$

11. $A = \begin{bmatrix} 4 & 0 & 1 \\ 1 & -5 & 1 \\ 6 & 1 & 0 \\ 1 & -1 & -5 \end{bmatrix}$, $\mathbf{b} = \begin{bmatrix} 9 \\ 0 \\ 0 \\ 0 \end{bmatrix}$

12. $A = \begin{bmatrix} 1 & 1 & 0 \\ 1 & 0 & -1 \\ 0 & 1 & 1 \\ -1 & 1 & -1 \end{bmatrix}$, $\mathbf{b} = \begin{bmatrix} 2 \\ 5 \\ 6 \\ 6 \end{bmatrix}$

13. Let $A = \begin{bmatrix} 3 & 4 \\ -2 & 1 \\ 3 & 4 \end{bmatrix}$, $\mathbf{b} = \begin{bmatrix} 11 \\ -9 \\ 5 \end{bmatrix}$, $\mathbf{u} = \begin{bmatrix} 5 \\ -1 \end{bmatrix}$, and $\mathbf{v} = \begin{bmatrix} 5 \\ -2 \end{bmatrix}$. Compute $A\mathbf{u}$ and $A\mathbf{v}$, and compare them with $\mathbf{b}$. Could $\mathbf{u}$ possibly be a least-squares solution of $A\mathbf{x} = \mathbf{b}$? (Answer this without computing a least-squares solution.)

14. Let $A = \begin{bmatrix} 2 & 1 \\ -3 & -4 \\ 3 & 2 \end{bmatrix}$, $\mathbf{b} = \begin{bmatrix} 5 \\ 4 \\ 4 \end{bmatrix}$, $\mathbf{u} = \begin{bmatrix} 4 \\ -5 \end{bmatrix}$, and $\mathbf{v} = \begin{bmatrix} 6 \\ -5 \end{bmatrix}$. Compute $A\mathbf{u}$ and $A\mathbf{v}$, and compare them with $\mathbf{b}$. Is it possible that at least one of $\mathbf{u}$ or $\mathbf{v}$ could be a least-squares solution of $A\mathbf{x} = \mathbf{b}$? (Answer this without computing a least-squares solution.)

In Exercises 15 and 16, use the factorization $A = QR$ to find the least-squares solution of $A\mathbf{x} = \mathbf{b}$.

15. $A = \begin{bmatrix} 2 & 3 \\ 2 & 4 \\ 1 & 1 \end{bmatrix} = \begin{bmatrix} 2/3 & -1/3 \\ 2/3 & 2/3 \\ 1/3 & -2/3 \end{bmatrix} \begin{bmatrix} 3 & 5 \\ 0 & 1 \end{bmatrix}$, $\mathbf{b} = \begin{bmatrix} 7 \\ 3 \\ 1 \end{bmatrix}$

16. $A = \begin{bmatrix} 1 & -1 \\ 1 & 4 \\ 1 & -1 \\ 1 & 4 \end{bmatrix} = \begin{bmatrix} 1/2 & -1/2 \\ 1/2 & 1/2 \\ 1/2 & -1/2 \\ 1/2 & 1/2 \end{bmatrix} \begin{bmatrix} 2 & 3 \\ 0 & 5 \end{bmatrix}$, $\mathbf{b} = \begin{bmatrix} -1 \\ 6 \\ 5 \\ 7 \end{bmatrix}$

In Exercises 17 and 18, A is an $m \times n$ matrix and $\mathbf{b}$ is in $\mathbb{R}^m$. Mark each statement True or False. Justify each answer.

17. a. The general least-squares problem is to find an $\mathbf{x}$ that makes $A\mathbf{x}$ as close as possible to $\mathbf{b}$.

b. A least-squares solution of $A\mathbf{x} = \mathbf{b}$ is a vector $\hat{\mathbf{x}}$ that satisfies $A\hat{\mathbf{x}} = \hat{\mathbf{b}}$, where $\hat{\mathbf{b}}$ is the orthogonal projection of $\mathbf{b}$ onto Col A.

c. A least-squares solution of $A\mathbf{x} = \mathbf{b}$ is a vector $\hat{\mathbf{x}}$ such that $\|\mathbf{b} - A\mathbf{x}\| \le \|\mathbf{b} - A\hat{\mathbf{x}}\|$ for all $\mathbf{x}$ in $\mathbb{R}^n$.

d. Any solution of $A^T A\mathbf{x} = A^T\mathbf{b}$ is a least-squares solution of $A\mathbf{x} = \mathbf{b}$.

e. If the columns of A are linearly independent, then the equation $A\mathbf{x} = \mathbf{b}$ has exactly one least-squares solution.

18. a. If $\mathbf{b}$ is in the column space of A, then every solution of $A\mathbf{x} = \mathbf{b}$ is a least-squares solution.

b. The least-squares solution of $A\mathbf{x} = \mathbf{b}$ is the point in the column space of A closest to $\mathbf{b}$.

c. A least-squares solution of $A\mathbf{x} = \mathbf{b}$ is a list of weights that, when applied to the columns of A, produces the orthogonal projection of $\mathbf{b}$ onto Col A.

d. If $\hat{\mathbf{x}}$ is a least-squares solution of $A\mathbf{x} = \mathbf{b}$, then $\hat{\mathbf{x}} = (A^T A)^{-1} A^T\mathbf{b}$.

e. The normal equations always provide a reliable method for computing least-squares solutions.

f. If A has a QR factorization, say $A = QR$, then the best way to find the least-squares solution of $A\mathbf{x} = \mathbf{b}$ is to compute $\hat{\mathbf{x}} = R^{-1} Q^T\mathbf{b}$.

19. Let A be an $m \times n$ matrix. Use the steps below to show that a vector $\mathbf{x}$ in $\mathbb{R}^n$ satisfies $A\mathbf{x} = \mathbf{0}$ if and only if $A^T A\mathbf{x} = \mathbf{0}$. This will show that Nul $A = $ Nul $A^T A$.

a. Show that if $A\mathbf{x} = \mathbf{0}$, then $A^T A\mathbf{x} = \mathbf{0}$.

b. Suppose $A^T A\mathbf{x} = \mathbf{0}$. Explain why $\mathbf{x}^T A^T A\mathbf{x} = \mathbf{0}$, and use this to show that $A\mathbf{x} = \mathbf{0}$.

20. Let A be an $m \times n$ matrix such that $A^T A$ is invertible. Show that the columns of A are linearly independent. [*Careful:* You may not assume that A is invertible; it may not even be square.]

21. Let A be an $m \times n$ matrix whose columns are linearly independent. [*Careful: A* need not be square.]

a. Use Exercise 19 to show that $A^T A$ is an invertible matrix.

b. Explain why A must have at least as many rows as columns.

c. Determine the rank of A.

22. Use Exercise 19 to show that rank $A^T A = $ rank A. [*Hint:* How many columns does $A^T A$ have? How is this connected with the rank of $A^T A$?]

23. Suppose A is $m \times n$ with linearly independent columns and $\mathbf{b}$ is in $\mathbb{R}^m$. Use the normal equations to produce a formula for $\hat{\mathbf{b}}$, the projection of $\mathbf{b}$ onto Col A. [*Hint:* Find $\hat{\mathbf{x}}$ first. The formula does not require an orthogonal basis for Col A.]

24. Find a formula for the least-squares solution of $A\mathbf{x} = \mathbf{b}$ when the columns of A are orthonormal.

25. Describe all least-squares solutions of the system

$$x + y = 2$$
$$x + y = 4$$

26. [M] Example 3 in Section 4.8 displayed a low-pass linear filter that changed a signal $\{y_k\}$ into $\{y_{k+1}\}$ and changed a higher-frequency signal $\{w_k\}$ into the zero signal, where $y_k = \cos(\pi k/4)$ and $w_k = \cos(3\pi k/4)$. The following calculations will design a filter with approximately those properties. The filter equation is

$$a_0 y_{k+2} + a_1 y_{k+1} + a_2 y_k = z_k \qquad \text{for all } k \qquad (8)$$

Because the signals are periodic, with period 8, it suffices to study equation (8) for $k = 0, \dots, 7$. The action on the two signals described above translates into two sets of eight equations, shown below:

$$
\begin{array}{c}
\\ k=0 \\ k=1 \\ \vdots \\ \\ \\ \\ \\ k=7
\end{array}
\begin{bmatrix}
y_{k+2} & y_{k+1} & y_k \\
0 & .7 & 1 \\
-.7 & 0 & .7 \\
-1 & -.7 & 0 \\
-.7 & -1 & -.7 \\
0 & -.7 & -1 \\
.7 & 0 & -.7 \\
1 & .7 & 0 \\
.7 & 1 & .7
\end{bmatrix}
\begin{bmatrix} a_0 \\ a_1 \\ a_2 \end{bmatrix}
=
\begin{bmatrix}
y_{k+1} \\
.7 \\
0 \\
-.7 \\
-1 \\
-.7 \\
0 \\
.7 \\
1
\end{bmatrix}
$$

$$
\begin{array}{c}
\\ k=0 \\ k=1 \\ \vdots \\ \\ \\ \\ \\ k=7
\end{array}
\begin{bmatrix}
w_{k+2} & w_{k+1} & w_k \\
0 & -.7 & 1 \\
.7 & 0 & -.7 \\
-1 & .7 & 0 \\
.7 & -1 & .7 \\
0 & .7 & -1 \\
-.7 & 0 & .7 \\
1 & -.7 & 0 \\
-.7 & 1 & -.7
\end{bmatrix}
\begin{bmatrix} a_0 \\ a_1 \\ a_2 \end{bmatrix}
=
\begin{bmatrix}
0 \\ 0 \\ 0 \\ 0 \\ 0 \\ 0 \\ 0 \\ 0
\end{bmatrix}
$$

Write an equation $A\mathbf{x} = \mathbf{b}$, where A is a 16×3 matrix formed from the two coefficient matrices above and where $\mathbf{b}$ in $\mathbb{R}^{16}$ is formed from the two right sides of the equations. Find a_0, a_1, and a_2 given by the least-squares solution of $A\mathbf{x} = \mathbf{b}$. (The .7 in the data above was used as an approximation for $\sqrt{2}/2$, to illustrate how a typical computation in an applied problem might proceed. If .707 were used instead, the resulting filter coefficients would agree to at least seven decimal places with $\sqrt{2}/4$, $1/2$, and $\sqrt{2}/4$, the values produced by exact arithmetic calculations.)

WEB

> **SOLUTIONS TO PRACTICE PROBLEMS**

1. First, compute

$$A^T A = \begin{bmatrix} 1 & 1 & 1 \\ -3 & 5 & 7 \\ -3 & 1 & 2 \end{bmatrix} \begin{bmatrix} 1 & -3 & -3 \\ 1 & 5 & 1 \\ 1 & 7 & 2 \end{bmatrix} = \begin{bmatrix} 3 & 9 & 0 \\ 9 & 83 & 28 \\ 0 & 28 & 14 \end{bmatrix}$$

$$A^T \mathbf{b} = \begin{bmatrix} 1 & 1 & 1 \\ -3 & 5 & 7 \\ -3 & 1 & 2 \end{bmatrix} \begin{bmatrix} 5 \\ -3 \\ -5 \end{bmatrix} = \begin{bmatrix} -3 \\ -65 \\ -28 \end{bmatrix}$$

Next, row reduce the augmented matrix for the normal equations, $A^T A \mathbf{x} = A^T \mathbf{b}$:

$$\begin{bmatrix} 3 & 9 & 0 & -3 \\ 9 & 83 & 28 & -65 \\ 0 & 28 & 14 & -28 \end{bmatrix} \sim \begin{bmatrix} 1 & 3 & 0 & -1 \\ 0 & 56 & 28 & -56 \\ 0 & 28 & 14 & -28 \end{bmatrix} \sim \cdots \sim \begin{bmatrix} 1 & 0 & -3/2 & 2 \\ 0 & 1 & 1/2 & -1 \\ 0 & 0 & 0 & 0 \end{bmatrix}$$

The general least-squares solution is $x_1 = 2 + \frac{3}{2}x_3$, $x_2 = -1 - \frac{1}{2}x_3$, with x_3 free. For one specific solution, take $x_3 = 0$ (for example), and get

$$\hat{\mathbf{x}} = \begin{bmatrix} 2 \\ -1 \\ 0 \end{bmatrix}$$

To find the least-squares error, compute

$$\hat{\mathbf{b}} = A\hat{\mathbf{x}} = \begin{bmatrix} 1 & -3 & -3 \\ 1 & 5 & 1 \\ 1 & 7 & 2 \end{bmatrix} \begin{bmatrix} 2 \\ -1 \\ 0 \end{bmatrix} = \begin{bmatrix} 5 \\ -3 \\ -5 \end{bmatrix}$$

It turns out that $\hat{\mathbf{b}} = \mathbf{b}$, so $\|\mathbf{b} - \hat{\mathbf{b}}\| = 0$. The least-squares error is zero because $\mathbf{b}$ happens to be in Col A.

2. If $\mathbf{b}$ is orthogonal to the columns of A, then the projection of $\mathbf{b}$ onto the column space of A is $\mathbf{0}$. In this case, a least-squares solution $\hat{\mathbf{x}}$ of $A\mathbf{x} = \mathbf{b}$ satisfies $A\hat{\mathbf{x}} = \mathbf{0}$.

6.6 | APPLICATIONS TO LINEAR MODELS

A common task in science and engineering is to analyze and understand relationships among several quantities that vary. This section describes a variety of situations in which data are used to build or verify a formula that predicts the value of one variable as a function of other variables. In each case, the problem will amount to solving a least-squares problem.

For easy application of the discussion to real problems that you may encounter later in your career, we choose notation that is commonly used in the statistical analysis of scientific and engineering data. Instead of $A\mathbf{x} = \mathbf{b}$, we write $X\boldsymbol{\beta} = \mathbf{y}$ and refer to X as the **design matrix**, $\boldsymbol{\beta}$ as the **parameter vector**, and $\mathbf{y}$ as the **observation vector**.

Least-Squares Lines

The simplest relation between two variables x and y is the linear equation $y = \beta_0 + \beta_1 x$.[1] Experimental data often produce points $(x_1, y_1), \ldots, (x_n, y_n)$ that,

---

[1] This notation is commonly used for least-squares lines instead of $y = mx + b$.

when graphed, seem to lie close to a line. We want to determine the parameters β_0 and β_1 that make the line as "close" to the points as possible.

Suppose β_0 and β_1 are fixed, and consider the line $y = \beta_0 + \beta_1 x$ in Fig. 1. Corresponding to each data point (x_j, y_j) there is a point $(x_j, \beta_0 + \beta_1 x_j)$ on the line with the same x-coordinate. We call y_j the *observed* value of y and $\beta_0 + \beta_1 x_j$ the *predicted* y-value (determined by the line). The difference between an observed y-value and a predicted y-value is called a *residual*.

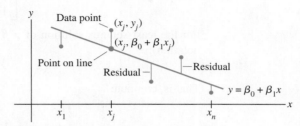

FIGURE 1 Fitting a line to experimental data.

There are several ways to measure how "close" the line is to the data. The usual choice (primarily because the mathematical calculations are simple) is to add the squares of the residuals. The **least-squares line** is the line $y = \beta_0 + \beta_1 x$ that minimizes the sum of the squares of the residuals. This line is also called a **line of regression of y on x**, because any errors in the data are assumed to be only in the y-coordinates. The coefficients β_0, β_1 of the line are called (linear) **regression coefficients**.[2]

If the data points were on the line, the parameters β_0 and β_1 would satisfy the equations

Predicted y-value		Observed y-value
$\beta_0 + \beta_1 x_1$	$=$	y_1
$\beta_0 + \beta_1 x_2$	$=$	y_2
$\vdots$		$\vdots$
$\beta_0 + \beta_1 x_n$	$=$	y_n

We can write this system as

$$X\boldsymbol{\beta} = \mathbf{y}, \quad \text{where } X = \begin{bmatrix} 1 & x_1 \\ 1 & x_2 \\ \vdots & \vdots \\ 1 & x_n \end{bmatrix}, \quad \boldsymbol{\beta} = \begin{bmatrix} \beta_0 \\ \beta_1 \end{bmatrix}, \quad \mathbf{y} = \begin{bmatrix} y_1 \\ y_2 \\ \vdots \\ y_n \end{bmatrix} \tag{1}$$

Of course, if the data points don't lie on a line, then there are no parameters β_0, β_1 for which the predicted y-values in $X\boldsymbol{\beta}$ equal the observed y-values in $\mathbf{y}$, and $X\boldsymbol{\beta} = \mathbf{y}$ has no solution. This is a least-squares problem, $A\mathbf{x} = \mathbf{b}$, with different notation!

The square of the distance between the vectors $X\boldsymbol{\beta}$ and $\mathbf{y}$ is precisely the sum of the squares of the residuals. The $\boldsymbol{\beta}$ that minimizes this sum also minimizes the distance between $X\boldsymbol{\beta}$ and $\mathbf{y}$. *Computing the least-squares solution of $X\boldsymbol{\beta} = \mathbf{y}$ is equivalent to finding the $\boldsymbol{\beta}$ that determines the least-squares line in Fig. 1.*

---

[2]If the measurement errors are in x instead of y, simply interchange the coordinates of the data (x_j, y_j) before plotting the points and computing the regression line. If both coordinates are subject to possible error, then you might choose the line that minimizes the sum of the squares of the *orthogonal* (perpendicular) distances from the points to the line. See the Practice Problems for Section 7.5.

EXAMPLE 1 Find the equation $y = \beta_0 + \beta_1 x$ of the least-squares line that best fits the data points $(2, 1)$, $(5, 2)$, $(7, 3)$, and $(8, 3)$.

SOLUTION Use the x-coordinates of the data to build the design matrix X in (1) and the y-coordinates to build the observation vector $\mathbf{y}$:

$$X = \begin{bmatrix} 1 & 2 \\ 1 & 5 \\ 1 & 7 \\ 1 & 8 \end{bmatrix}, \quad \mathbf{y} = \begin{bmatrix} 1 \\ 2 \\ 3 \\ 3 \end{bmatrix}$$

For the least-squares solution of $X\boldsymbol{\beta} = \mathbf{y}$, obtain the normal equations (with the new notation):

$$X^T X \boldsymbol{\beta} = X^T \mathbf{y}$$

That is, compute

$$X^T X = \begin{bmatrix} 1 & 1 & 1 & 1 \\ 2 & 5 & 7 & 8 \end{bmatrix} \begin{bmatrix} 1 & 2 \\ 1 & 5 \\ 1 & 7 \\ 1 & 8 \end{bmatrix} = \begin{bmatrix} 4 & 22 \\ 22 & 142 \end{bmatrix}$$

$$X^T \mathbf{y} = \begin{bmatrix} 1 & 1 & 1 & 1 \\ 2 & 5 & 7 & 8 \end{bmatrix} \begin{bmatrix} 1 \\ 2 \\ 3 \\ 3 \end{bmatrix} = \begin{bmatrix} 9 \\ 57 \end{bmatrix}$$

The normal equations are

$$\begin{bmatrix} 4 & 22 \\ 22 & 142 \end{bmatrix} \begin{bmatrix} \beta_0 \\ \beta_1 \end{bmatrix} = \begin{bmatrix} 9 \\ 57 \end{bmatrix}$$

Hence

$$\begin{bmatrix} \beta_0 \\ \beta_1 \end{bmatrix} = \begin{bmatrix} 4 & 22 \\ 22 & 142 \end{bmatrix}^{-1} \begin{bmatrix} 9 \\ 57 \end{bmatrix} = \frac{1}{84} \begin{bmatrix} 142 & -22 \\ -22 & 4 \end{bmatrix} \begin{bmatrix} 9 \\ 57 \end{bmatrix} = \frac{1}{84} \begin{bmatrix} 24 \\ 30 \end{bmatrix} = \begin{bmatrix} 2/7 \\ 5/14 \end{bmatrix}$$

Thus the least-squares line has the equation

$$y = \frac{2}{7} + \frac{5}{14} x$$

See Fig. 2. ∎

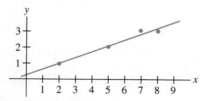

FIGURE 2 The least-squares line $y = \frac{2}{7} + \frac{5}{14} x$.

A common practice before computing a least-squares line is to compute the average $\overline{x}$ of the original x-values and form a new variable $x^* = x - \overline{x}$. The new x-data are said to be in **mean-deviation form**. In this case, the two columns of the design matrix will be orthogonal. Solution of the normal equations is simplified, just as in Example 4 in Section 6.5. See Exercises 17 and 18.

The General Linear Model

In some applications, it is necessary to fit data points with something other than a straight line. In the examples that follow, the matrix equation is still $X\beta = \mathbf{y}$, but the specific form of X changes from one problem to the next. Statisticians usually introduce a **residual vector** ϵ, defined by $\epsilon = \mathbf{y} - X\beta$, and write

$$\mathbf{y} = X\beta + \epsilon$$

Any equation of this form is referred to as a **linear model**. Once X and $\mathbf{y}$ are determined, the goal is to minimize the length of ϵ, which amounts to finding a least-squares solution of $X\beta = \mathbf{y}$. In each case, the least-squares solution $\hat{\beta}$ is a solution of the normal equations

$$X^T X \beta = X^T \mathbf{y}$$

Least-Squares Fitting of Other Curves

When data points $(x_1, y_1), \ldots, (x_n, y_n)$ on a scatter plot do not lie close to any line, it may be appropriate to postulate some other functional relationship between x and y.

The next two examples show how to fit data by curves that have the general form

$$y = \beta_0 f_0(x) + \beta_1 f_1(x) + \cdots + \beta_k f_k(x) \tag{2}$$

where $f_0, \ldots, f_k$ are known functions and $\beta_0, \ldots, \beta_k$ are parameters that must be determined. As we will see, equation (2) describes a linear model because it is linear in the unknown parameters.

For a particular value of x, (2) gives a predicted, or "fitted," value of y. The difference between the observed value and the predicted value is the residual. The parameters $\beta_0, \ldots, \beta_k$ must be determined so as to minimize the sum of the squares of the residuals.

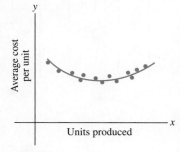

FIGURE 3

Average cost curve.

EXAMPLE 2   Suppose data points $(x_1, y_1), \ldots, (x_n, y_n)$ appear to lie along some sort of parabola instead of a straight line. For instance, if the x-coordinate denotes the production level for a company, and y denotes the average cost per unit of operating at a level of x units per day, then a typical average cost curve looks like a parabola that opens upward (Fig. 3). In ecology, a parabolic curve that opens downward is used to model the net primary production of nutrients in a plant, as a function of the surface area of the foliage (Fig. 4). Suppose we wish to approximate the data by an equation of the form

$$y = \beta_0 + \beta_1 x + \beta_2 x^2 \tag{3}$$

Describe the linear model that produces a "least-squares fit" of the data by equation (3).

SOLUTION   Equation (3) describes the ideal relationship. Suppose the actual values of the parameters are $\beta_0, \beta_1, \beta_2$. Then the coordinates of the first data point (x_1, y_1) satisfy an equation of the form

$$y_1 = \beta_0 + \beta_1 x_1 + \beta_2 x_1^2 + \epsilon_1$$

where ϵ_1 is the residual error between the observed value y_1 and the predicted y-value $\beta_0 + \beta_1 x_1 + \beta_2 x_1^2$. Each data point determines a similar equation:

$$y_1 = \beta_0 + \beta_1 x_1 + \beta_2 x_1^2 + \epsilon_1$$
$$y_2 = \beta_0 + \beta_1 x_2 + \beta_2 x_2^2 + \epsilon_2$$
$$\vdots \qquad \vdots$$
$$y_n = \beta_0 + \beta_1 x_n + \beta_2 x_n^2 + \epsilon_n$$

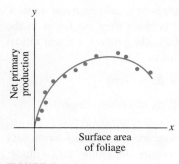

FIGURE 4

Production of nutrients.

It is a simple matter to write this system of equations in the form $\mathbf{y} = X\boldsymbol{\beta} + \boldsymbol{\epsilon}$. To find X, inspect the first few rows of the system and look for the pattern.

$$\begin{bmatrix} y_1 \\ y_2 \\ \vdots \\ y_n \end{bmatrix} = \begin{bmatrix} 1 & x_1 & x_1^2 \\ 1 & x_2 & x_2^2 \\ \vdots & \vdots & \vdots \\ 1 & x_n & x_n^2 \end{bmatrix} \begin{bmatrix} \beta_0 \\ \beta_1 \\ \beta_2 \end{bmatrix} + \begin{bmatrix} \epsilon_1 \\ \epsilon_2 \\ \vdots \\ \epsilon_n \end{bmatrix}$$

$$\mathbf{y} \qquad = \qquad X \qquad\quad \boldsymbol{\beta} \quad + \quad \boldsymbol{\epsilon}$$

■

EXAMPLE 3 If data points tend to follow a pattern such as in Fig. 5, then an appropriate model might be an equation of the form

$$y = \beta_0 + \beta_1 x + \beta_2 x^2 + \beta_3 x^3$$

Such data, for instance, could come from a company's total costs, as a function of the level of production. Describe the linear model that gives a least-squares fit of this type to data $(x_1, y_1), \ldots, (x_n, y_n)$.

SOLUTION By an analysis similar to that in Example 2, we obtain

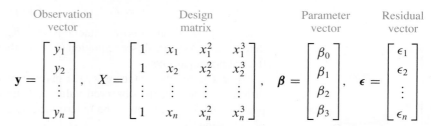

$$\begin{array}{cccc} \text{Observation} & \text{Design} & \text{Parameter} & \text{Residual} \\ \text{vector} & \text{matrix} & \text{vector} & \text{vector} \end{array}$$

$$\mathbf{y} = \begin{bmatrix} y_1 \\ y_2 \\ \vdots \\ y_n \end{bmatrix}, \quad X = \begin{bmatrix} 1 & x_1 & x_1^2 & x_1^3 \\ 1 & x_2 & x_2^2 & x_2^3 \\ \vdots & \vdots & \vdots & \vdots \\ 1 & x_n & x_n^2 & x_n^3 \end{bmatrix}, \quad \boldsymbol{\beta} = \begin{bmatrix} \beta_0 \\ \beta_1 \\ \beta_2 \\ \beta_3 \end{bmatrix}, \quad \boldsymbol{\epsilon} = \begin{bmatrix} \epsilon_1 \\ \epsilon_2 \\ \vdots \\ \epsilon_n \end{bmatrix}$$

■

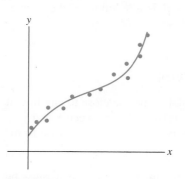

FIGURE 5
Data points along a cubic curve.

Multiple Regression

Suppose an experiment involves two independent variables—say, u and v—and one dependent variable, y. A simple equation for predicting y from u and v has the form

$$y = \beta_0 + \beta_1 u + \beta_2 v \tag{4}$$

A more general prediction equation might have the form

$$y = \beta_0 + \beta_1 u + \beta_2 v + \beta_3 u^2 + \beta_4 uv + \beta_5 v^2 \tag{5}$$

This equation is used in geology, for instance, to model erosion surfaces, glacial cirques, soil pH, and other quantities. In such cases, the least-squares fit is called a *trend surface*.

Equations (4) and (5) both lead to a linear model because they are linear in the unknown parameters (even though u and v are multiplied). In general, a linear model will arise whenever y is to be predicted by an equation of the form

$$y = \beta_0 f_0(u, v) + \beta_1 f_1(u, v) + \cdots + \beta_k f_k(u, v)$$

with $f_0, \ldots, f_k$ any sort of known functions and $\beta_0, \ldots, \beta_k$ unknown weights.

EXAMPLE 4 In geography, local models of terrain are constructed from data $(u_1, v_1, y_1), \ldots, (u_n, v_n, y_n)$, where u_j, v_j, and y_j are latitude, longitude, and altitude, respectively. Describe the linear model based on (4) that gives a least-squares fit to such data. The solution is called the *least-squares plane*. See Fig. 6.

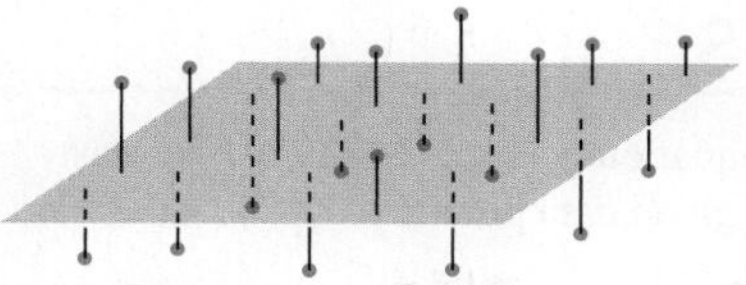

FIGURE 6 A least-squares plane.

SOLUTION We expect the data to satisfy the following equations:

$$y_1 = \beta_0 + \beta_1 u_1 + \beta_2 v_1 + \epsilon_1$$
$$y_2 = \beta_0 + \beta_1 u_2 + \beta_2 v_2 + \epsilon_2$$
$$\vdots \qquad\qquad \vdots$$
$$y_n = \beta_0 + \beta_1 u_n + \beta_2 v_n + \epsilon_n$$

This system has the matrix form $\mathbf{y} = X\boldsymbol{\beta} + \boldsymbol{\epsilon}$, where

Observation vector	Design matrix	Parameter vector	Residual vector

$$\mathbf{y} = \begin{bmatrix} y_1 \\ y_2 \\ \vdots \\ y_n \end{bmatrix}, \quad X = \begin{bmatrix} 1 & u_1 & v_1 \\ 1 & u_2 & v_2 \\ \vdots & \vdots & \vdots \\ 1 & u_n & v_n \end{bmatrix}, \quad \boldsymbol{\beta} = \begin{bmatrix} \beta_0 \\ \beta_1 \\ \beta_2 \end{bmatrix}, \quad \boldsymbol{\epsilon} = \begin{bmatrix} \epsilon_1 \\ \epsilon_2 \\ \vdots \\ \epsilon_n \end{bmatrix}$$

∎

Example 4 shows that the linear model for multiple regression has the same abstract form as the model for the simple regression in the earlier examples. Linear algebra gives us the power to understand the general principle behind all the linear models. Once X is defined properly, the normal equations for $\boldsymbol{\beta}$ have the same matrix form, no matter how many variables are involved. Thus, for any linear model where $X^T X$ is invertible, the least-squares $\hat{\boldsymbol{\beta}}$ is given by $(X^T X)^{-1} X^T \mathbf{y}$.

SG | The Geometry of a Linear Model 6-19

Further Reading

Ferguson, J., *Introduction to Linear Algebra in Geology* (New York: Chapman & Hall, 1994).

Krumbein, W. C., and F. A. Graybill, *An Introduction to Statistical Models in Geology* (New York: McGraw-Hill, 1965).

Legendre, P., and L. Legendre, *Numerical Ecology* (Amsterdam: Elsevier, 1998).

Unwin, David J., *An Introduction to Trend Surface Analysis*, Concepts and Techniques in Modern Geography, No. 5 (Norwich, England: Geo Books, 1975).

> **PRACTICE PROBLEM**

When the monthly sales of a product are subject to seasonal fluctuations, a curve that approximates the sales data might have the form

$$y = \beta_0 + \beta_1 x + \beta_2 \sin(2\pi x/12)$$

where x is the time in months. The term $\beta_0 + \beta_1 x$ gives the basic sales trend, and the sine term reflects the seasonal changes in sales. Give the design matrix and the parameter vector for the linear model that leads to a least-squares fit of the equation above. Assume the data are $(x_1, y_1), \ldots, (x_n, y_n)$.

6.6 EXERCISES

In Exercises 1–4, find the equation $y = \beta_0 + \beta_1 x$ of the least-squares line that best fits the given data points.

1. $(0, 1)$, $(1, 1)$, $(2, 2)$, $(3, 2)$

2. $(1, 0)$, $(2, 1)$, $(4, 2)$, $(5, 3)$

3. $(-1, 0)$, $(0, 1)$, $(1, 2)$, $(2, 4)$

4. $(2, 3)$, $(3, 2)$, $(5, 1)$, $(6, 0)$

5. Let X be the design matrix used to find the least-squares line to fit data $(x_1, y_1), \ldots, (x_n, y_n)$. Use a theorem in Section 6.5 to show that the normal equations have a unique solution if and only if the data include at least two data points with different x-coordinates.

6. Let X be the design matrix in Example 2 corresponding to a least-squares fit of a parabola to data $(x_1, y_1), \ldots, (x_n, y_n)$. Suppose x_1, x_2, and x_3 are distinct. Explain why there is only one parabola that fits the data best, in a least-squares sense. (See Exercise 5.)

7. A certain experiment produces the data $(1, 1.8)$, $(2, 2.7)$, $(3, 3.4)$, $(4, 3.8)$, $(5, 3.9)$. Describe the model that produces a least-squares fit of these points by a function of the form

$$y = \beta_1 x + \beta_2 x^2$$

Such a function might arise, for example, as the revenue from the sale of x units of a product, when the amount offered for sale affects the price to be set for the product.

    a. Give the design matrix, the observation vector, and the unknown parameter vector.

    b. **[M]** Find the associated least-squares curve for the data.

8. A simple curve that often makes a good model for the variable costs of a company, as a function of the sales level x, has the form $y = \beta_1 x + \beta_2 x^2 + \beta_3 x^3$. There is no constant term because fixed costs are not included.

    a. Give the design matrix and the parameter vector for the linear model that leads to a least-squares fit of the equation above, with data $(x_1, y_1), \ldots, (x_n, y_n)$.

    b. **[M]** Find the least-squares curve of the form above to fit the data $(4, 1.58)$, $(6, 2.08)$, $(8, 2.5)$, $(10, 2.8)$, $(12, 3.1)$, $(14, 3.4)$, $(16, 3.8)$, and $(18, 4.32)$, with values in thousands. If possible, produce a graph that shows the data points and the graph of the cubic approximation.

9. A certain experiment produces the data $(1, 7.9)$, $(2, 5.4)$, and $(3, -.9)$. Describe the model that produces a least-squares fit of these points by a function of the form

$$y = A \cos x + B \sin x$$

10. Suppose radioactive substances A and B have decay constants of .02 and .07, respectively. If a mixture of these two substances at time $t = 0$ contains M_A grams of A and M_B grams of B, then a model for the total amount y of the mixture present at time t is

$$y = M_A e^{-.02t} + M_B e^{-.07t} \tag{6}$$

Suppose the initial amounts M_A and M_B are unknown, but a scientist is able to measure the total amounts present at several times and records the following points (t_i, y_i): $(10, 21.34)$, $(11, 20.68)$, $(12, 20.05)$, $(14, 18.87)$, and $(15, 18.30)$.

    a. Describe a linear model that can be used to estimate M_A and M_B.

    b. **[M]** Find the least-squares curve based on (6).

Halley's Comet last appeared in 1986 and will reappear in 2061.

11. **[M]** According to Kepler's first law, a comet should have an elliptic, parabolic, or hyperbolic orbit (with gravitational attractions from the planets ignored). In suitable polar coordinates, the position (r, ϑ) of a comet satisfies an equation of the form

$$r = \beta + e(r \cdot \cos \vartheta)$$

where β is a constant and e is the *eccentricity* of the orbit, with $0 \le e < 1$ for an ellipse, $e = 1$ for a parabola, and $e > 1$ for a hyperbola. Suppose observations of a newly discovered comet provide the data below. Determine the type of orbit, and predict where the comet will be when $\vartheta = 4.6$ (radians).[3]

ϑ	.88	1.10	1.42	1.77	2.14
r	3.00	2.30	1.65	1.25	1.01

12. **[M]** A healthy child's systolic blood pressure p (in millimeters of mercury) and weight w (in pounds) are approximately related by the equation

$$\beta_0 + \beta_1 \ln w = p$$

Use the following experimental data to estimate the systolic blood pressure of a healthy child weighing 100 pounds.

---

[3] The basic idea of least-squares fitting of data is due to K. F. Gauss (and, independently, to A. Legendre), whose initial rise to fame occurred in 1801 when he used the method to determine the path of the asteroid *Ceres*. Forty days after the asteroid was discovered, it disappeared behind the sun. Gauss predicted it would appear ten months later and gave its location. The accuracy of the prediction astonished the European scientific community.

w	44	61	81	113	131
$\ln w$	3.78	4.11	4.39	4.73	4.88
p	91	98	103	110	112

13. [M] To measure the takeoff performance of an airplane, the horizontal position of the plane was measured every second, from $t = 0$ to $t = 12$. The positions (in feet) were: 0, 8.8, 29.9, 62.0, 104.7, 159.1, 222.0, 294.5, 380.4, 471.1, 571.7, 686.8, and 809.2.

    a. Find the least-squares cubic curve $y = \beta_0 + \beta_1 t + \beta_2 t^2 + \beta_3 t^3$ for these data.

    b. Use the result of part (a) to estimate the velocity of the plane when $t = 4.5$ seconds.

14. Let $\bar{x} = \dfrac{1}{n}(x_1 + \cdots + x_n)$ and $\bar{y} = \dfrac{1}{n}(y_1 + \cdots + y_n)$. Show that the least-squares line for the data $(x_1, y_1), \ldots, (x_n, y_n)$ must pass through $(\bar{x}, \bar{y})$. That is, show that $\bar{x}$ and $\bar{y}$ satisfy the linear equation $\bar{y} = \hat{\beta}_0 + \hat{\beta}_1 \bar{x}$. [*Hint:* Derive this equation from the vector equation $\mathbf{y} = X\hat{\beta} + \boldsymbol{\epsilon}$. Denote the first column of X by **1**. Use the fact that the residual vector $\boldsymbol{\epsilon}$ is orthogonal to the column space of X and hence is orthogonal to **1**.]

Given data for a least-squares problem, $(x_1, y_1), \ldots, (x_n, y_n)$, the following abbreviations are helpful:

$$\sum x = \sum_{i=1}^{n} x_i, \quad \sum x^2 = \sum_{i=1}^{n} x_i^2,$$
$$\sum y = \sum_{i=1}^{n} y_i, \quad \sum xy = \sum_{i=1}^{n} x_i y_i$$

The normal equations for a least-squares line $y = \hat{\beta}_0 + \hat{\beta}_1 x$ may be written in the form

$$\begin{aligned} n\hat{\beta}_0 + \hat{\beta}_1 \sum x &= \sum y \\ \hat{\beta}_0 \sum x + \hat{\beta}_1 \sum x^2 &= \sum xy \end{aligned} \quad (7)$$

15. Derive the normal equations (7) from the matrix form given in this section.

16. Use a matrix inverse to solve the system of equations in (7) and thereby obtain formulas for $\hat{\beta}_0$ and $\hat{\beta}_1$ that appear in many statistics texts.

17. a. Rewrite the data in Example 1 with new x-coordinates in mean deviation form. Let X be the associated design matrix. Why are the columns of X orthogonal?

    b. Write the normal equations for the data in part (a), and solve them to find the least-squares line, $y = \beta_0 + \beta_1 x^*$, where $x^* = x - 5.5$.

18. Suppose the x-coordinates of the data $(x_1, y_1), \ldots, (x_n, y_n)$ are in mean deviation form, so that $\sum x_i = 0$. Show that if X is the design matrix for the least-squares line in this case, then $X^T X$ is a diagonal matrix.

Exercises 19 and 20 involve a design matrix X with two or more columns and a least-squares solution $\hat{\beta}$ of $\mathbf{y} = X\beta$. Consider the following numbers.

(i) $\|X\hat{\beta}\|^2$—the sum of the squares of the "regression term." Denote this number by SS(R).

(ii) $\|\mathbf{y} - X\hat{\beta}\|^2$—the sum of the squares for error term. Denote this number by SS(E).

(iii) $\|\mathbf{y}\|^2$—the "total" sum of the squares of the y-values. Denote this number by SS(T).

Every statistics text that discusses regression and the linear model $\mathbf{y} = X\beta + \boldsymbol{\epsilon}$ introduces these numbers, though terminology and notation vary somewhat. To simplify matters, assume that the mean of the y-values is zero. In this case, SS(T) is proportional to what is called the *variance* of the set of y-values.

19. Justify the equation SS(T) = SS(R) + SS(E). [*Hint:* Use a theorem, and explain why the hypotheses of the theorem are satisfied.] This equation is extremely important in statistics, both in regression theory and in the analysis of variance.

20. Show that $\|X\hat{\beta}\|^2 = \hat{\beta}^T X^T \mathbf{y}$. [*Hint:* Rewrite the left side and use the fact that $\hat{\beta}$ satisfies the normal equations.] This formula for SS(R) is used in statistics. From this and from Exercise 19, obtain the standard formula for SS(E):

$$\text{SS(E)} = \mathbf{y}^T \mathbf{y} - \hat{\beta}^T X^T \mathbf{y}$$

SOLUTION TO PRACTICE PROBLEM

Sales trend with seasonal fluctuations.

Construct X and β so that the kth row of $X\beta$ is the predicted y-value that corresponds to the data point (x_k, y_k), namely,

$$\beta_0 + \beta_1 x_k + \beta_2 \sin(2\pi x_k / 12)$$

It should be clear that

$$X = \begin{bmatrix} 1 & x_1 & \sin(2\pi x_1/12) \\ \vdots & \vdots & \vdots \\ 1 & x_n & \sin(2\pi x_n/12) \end{bmatrix}, \quad \beta = \begin{bmatrix} \beta_0 \\ \beta_1 \\ \beta_2 \end{bmatrix}$$

6.7 | INNER PRODUCT SPACES

Notions of length, distance, and orthogonality are often important in applications involving a vector space. For $\mathbb{R}^n$, these concepts were based on the properties of the inner product listed in Theorem 1 of Section 6.1. For other spaces, we need analogues of the inner product with the same properties. The conclusions of Theorem 1 now become *axioms* in the following definition.

DEFINITION

An **inner product** on a vector space V is a function that, to each pair of vectors **u** and **v** in V, associates a real number $\langle \mathbf{u}, \mathbf{v} \rangle$ and satisfies the following axioms, for all **u**, **v**, **w** in V and all scalars c:

1. $\langle \mathbf{u}, \mathbf{v} \rangle = \langle \mathbf{v}, \mathbf{u} \rangle$
2. $\langle \mathbf{u} + \mathbf{v}, \mathbf{w} \rangle = \langle \mathbf{u}, \mathbf{w} \rangle + \langle \mathbf{v}, \mathbf{w} \rangle$
3. $\langle c\mathbf{u}, \mathbf{v} \rangle = c \langle \mathbf{u}, \mathbf{v} \rangle$
4. $\langle \mathbf{u}, \mathbf{u} \rangle \geq 0$ and $\langle \mathbf{u}, \mathbf{u} \rangle = 0$ if and only if $\mathbf{u} = \mathbf{0}$

A vector space with an inner product is called an **inner product space**.

The vector space $\mathbb{R}^n$ with the standard inner product is an inner product space, and nearly everything discussed in this chapter for $\mathbb{R}^n$ carries over to inner product spaces. The examples in this section and the next lay the foundation for a variety of applications treated in courses in engineering, physics, mathematics, and statistics.

EXAMPLE 1 Fix any two positive numbers—say, 4 and 5—and for vectors $\mathbf{u} = (u_1, u_2)$ and $\mathbf{v} = (v_1, v_2)$ in $\mathbb{R}^2$, set

$$\langle \mathbf{u}, \mathbf{v} \rangle = 4u_1v_1 + 5u_2v_2 \tag{1}$$

Show that equation (1) defines an inner product.

SOLUTION Certainly Axiom 1 is satisfied, because $\langle \mathbf{u}, \mathbf{v} \rangle = 4u_1v_1 + 5u_2v_2 = 4v_1u_1 + 5v_2u_2 = \langle \mathbf{v}, \mathbf{u} \rangle$. If $\mathbf{w} = (w_1, w_2)$, then

$$\langle \mathbf{u} + \mathbf{v}, \mathbf{w} \rangle = 4(u_1 + v_1)w_1 + 5(u_2 + v_2)w_2$$
$$= 4u_1w_1 + 5u_2w_2 + 4v_1w_1 + 5v_2w_2$$
$$= \langle \mathbf{u}, \mathbf{w} \rangle + \langle \mathbf{v}, \mathbf{w} \rangle$$

This verifies Axiom 2. For Axiom 3, compute

$$\langle c\mathbf{u}, \mathbf{v} \rangle = 4(cu_1)v_1 + 5(cu_2)v_2 = c(4u_1v_1 + 5u_2v_2) = c \langle \mathbf{u}, \mathbf{v} \rangle$$

For Axiom 4, note that $\langle \mathbf{u}, \mathbf{u} \rangle = 4u_1^2 + 5u_2^2 \geq 0$, and $4u_1^2 + 5u_2^2 = 0$ only if $u_1 = u_2 = 0$, that is, if $\mathbf{u} = \mathbf{0}$. Also, $\langle \mathbf{0}, \mathbf{0} \rangle = 0$. So (1) defines an inner product on $\mathbb{R}^2$. ∎

Inner products similar to (1) can be defined on $\mathbb{R}^n$. They arise naturally in connection with "weighted least-squares" problems, in which weights are assigned to the various entries in the sum for the inner product in such a way that more importance is given to the more reliable measurements.

From now on, when an inner product space involves polynomials or other functions, we will write the functions in the familiar way, rather than use the boldface type for vectors. Nevertheless, it is important to remember that each function *is* a vector when it is treated as an element of a vector space.

EXAMPLE 2   Let $t_0, \ldots, t_n$ be distinct real numbers. For p and q in $\mathbb{P}_n$, define

$$\langle p, q \rangle = p(t_0)q(t_0) + p(t_1)q(t_1) + \cdots + p(t_n)q(t_n) \tag{2}$$

Inner product Axioms 1–3 are readily checked. For Axiom 4, note that

$$\langle p, p \rangle = [p(t_0)]^2 + [p(t_1)]^2 + \cdots + [p(t_n)]^2 \geq 0$$

Also, $\langle \mathbf{0}, \mathbf{0} \rangle = 0$. (The boldface zero here denotes the zero polynomial, the zero vector in $\mathbb{P}_n$.) If $\langle p, p \rangle = 0$, then p must vanish at $n + 1$ points: $t_0, \ldots, t_n$. This is possible only if p is the zero polynomial, because the degree of p is less than $n + 1$. Thus (2) defines an inner product on $\mathbb{P}_n$.   ∎

EXAMPLE 3   Let V be $\mathbb{P}_2$, with the inner product from Example 2, where $t_0 = 0$, $t_1 = \frac{1}{2}$, and $t_2 = 1$. Let $p(t) = 12t^2$ and $q(t) = 2t - 1$. Compute $\langle p, q \rangle$ and $\langle q, q \rangle$.

SOLUTION

$$\begin{aligned}
\langle p, q \rangle &= p(0)q(0) + p\left(\tfrac{1}{2}\right)q\left(\tfrac{1}{2}\right) + p(1)q(1) \\
&= (0)(-1) + (3)(0) + (12)(1) = 12 \\
\langle q, q \rangle &= [q(0)]^2 + \left[q\left(\tfrac{1}{2}\right)\right]^2 + [q(1)]^2 \\
&= (-1)^2 + (0)^2 + (1)^2 = 2
\end{aligned}$$

∎

Lengths, Distances, and Orthogonality

Let V be an inner product space, with the inner product denoted by $\langle \mathbf{u}, \mathbf{v} \rangle$. Just as in $\mathbb{R}^n$, we define the **length**, or **norm**, of a vector $\mathbf{v}$ to be the scalar

$$\|\mathbf{v}\| = \sqrt{\langle \mathbf{v}, \mathbf{v} \rangle}$$

Equivalently, $\|\mathbf{v}\|^2 = \langle \mathbf{v}, \mathbf{v} \rangle$. (This definition makes sense because $\langle \mathbf{v}, \mathbf{v} \rangle \geq 0$, but the definition *does not* say that $\langle \mathbf{v}, \mathbf{v} \rangle$ is a "sum of squares," because $\mathbf{v}$ need not be an element of $\mathbb{R}^n$.)

A **unit vector** is one whose length is 1. The **distance between u and v** is $\|\mathbf{u} - \mathbf{v}\|$. Vectors $\mathbf{u}$ and $\mathbf{v}$ are **orthogonal** if $\langle \mathbf{u}, \mathbf{v} \rangle = 0$.

EXAMPLE 4   Let $\mathbb{P}_2$ have the inner product (2) of Example 3. Compute the lengths of the vectors $p(t) = 12t^2$ and $q(t) = 2t - 1$.

SOLUTION

$$\begin{aligned}
\|p\|^2 = \langle p, p \rangle &= [p(0)]^2 + \left[p\left(\tfrac{1}{2}\right)\right]^2 + [p(1)]^2 \\
&= 0 + [3]^2 + [12]^2 = 153 \\
\|p\| &= \sqrt{153}
\end{aligned}$$

From Example 3, $\langle q, q \rangle = 2$. Hence $\|q\| = \sqrt{2}$.   ∎

The Gram–Schmidt Process

The existence of orthogonal bases for finite-dimensional subspaces of an inner product space can be established by the Gram–Schmidt process, just as in $\mathbb{R}^n$. Certain orthogonal bases that arise frequently in applications can be constructed by this process.

The orthogonal projection of a vector onto a subspace W with an orthogonal basis can be constructed as usual. The projection does not depend on the choice of orthogonal basis, and it has the properties described in the Orthogonal Decomposition Theorem and the Best Approximation Theorem.

EXAMPLE 5 Let V be $\mathbb{P}_4$ with the inner product in Example 2, involving evaluation of polynomials at -2, -1, 0, 1, and 2, and view $\mathbb{P}_2$ as a subspace of V. Produce an orthogonal basis for $\mathbb{P}_2$ by applying the Gram–Schmidt process to the polynomials 1, t, and t^2.

SOLUTION The inner product depends only on the values of a polynomial at $-2, \ldots, 2$, so we list the values of each polynomial as a vector in $\mathbb{R}^5$, underneath the name of the polynomial:[1]

$$
\begin{array}{cccc}
\text{Polynomial:} & 1 & t & t^2 \\[4pt]
\text{Vector of values:} &
\begin{bmatrix} 1 \\ 1 \\ 1 \\ 1 \\ 1 \end{bmatrix}, &
\begin{bmatrix} -2 \\ -1 \\ 0 \\ 1 \\ 2 \end{bmatrix}, &
\begin{bmatrix} 4 \\ 1 \\ 0 \\ 1 \\ 4 \end{bmatrix}
\end{array}
$$

The inner product of two polynomials in V equals the (standard) inner product of their corresponding vectors in $\mathbb{R}^5$. Observe that t is orthogonal to the constant function 1. So take $p_0(t) = 1$ and $p_1(t) = t$. For p_2, use the vectors in $\mathbb{R}^5$ to compute the projection of t^2 onto Span $\{p_0, p_1\}$:

$$\langle t^2, p_0 \rangle = \langle t^2, 1 \rangle = 4 + 1 + 0 + 1 + 4 = 10$$

$$\langle p_0, p_0 \rangle = 5$$

$$\langle t^2, p_1 \rangle = \langle t^2, t \rangle = -8 + (-1) + 0 + 1 + 8 = 0$$

The orthogonal projection of t^2 onto Span $\{1, t\}$ is $\frac{10}{5} p_0 + 0 p_1$. Thus

$$p_2(t) = t^2 - 2p_0(t) = t^2 - 2$$

An orthogonal basis for the subspace $\mathbb{P}_2$ of V is:

$$
\begin{array}{cccc}
\text{Polynomial:} & p_0 & p_1 & p_2 \\[4pt]
\text{Vector of values:} &
\begin{bmatrix} 1 \\ 1 \\ 1 \\ 1 \\ 1 \end{bmatrix}, &
\begin{bmatrix} -2 \\ -1 \\ 0 \\ 1 \\ 2 \end{bmatrix}, &
\begin{bmatrix} 2 \\ -1 \\ -2 \\ -1 \\ 2 \end{bmatrix}
\end{array}
\qquad (3)
$$

■

Best Approximation in Inner Product Spaces

A common problem in applied mathematics involves a vector space V whose elements are functions. The problem is to approximate a function f in V by a function g from a specified subspace W of V. The "closeness" of the approximation of f depends on the way $\| f - g \|$ is defined. We will consider only the case in which the distance between f and g is determined by an inner product. In this case, the *best approximation to f by functions in W* is the orthogonal projection of f onto the subspace W.

EXAMPLE 6 Let V be $\mathbb{P}_4$ with the inner product in Example 5, and let p_0, p_1, and p_2 be the orthogonal basis found in Example 5 for the subspace $\mathbb{P}_2$. Find the best approximation to $p(t) = 5 - \frac{1}{2}t^4$ by polynomials in $\mathbb{P}_2$.

---

[1] Each polynomial in $\mathbb{P}_4$ is uniquely determined by its value at the five numbers $-2, \ldots, 2$. In fact, the correspondence between p and its vector of values is an isomorphism, that is, a one-to-one mapping onto $\mathbb{R}^5$ that preserves linear combinations.

SOLUTION The values of p_0, p_1, and p_2 at the numbers -2, -1, 0, 1, and 2 are listed in $\mathbb{R}^5$ vectors in (3) above. The corresponding values for p are -3, $9/2$, 5, $9/2$, and -3. Compute

$$\langle p, p_0 \rangle = 8, \qquad \langle p, p_1 \rangle = 0, \qquad \langle p, p_2 \rangle = -31$$
$$\langle p_0, p_0 \rangle = 5, \qquad\qquad\qquad\qquad\quad \langle p_2, p_2 \rangle = 14$$

Then the best approximation in V to p by polynomials in $\mathbb{P}_2$ is

$$\hat{p} = \text{proj}_{\mathbb{P}_2} \, p = \frac{\langle p, p_0 \rangle}{\langle p_0, p_0 \rangle} p_0 + \frac{\langle p, p_1 \rangle}{\langle p_1, p_1 \rangle} p_1 + \frac{\langle p, p_2 \rangle}{\langle p_2, p_2 \rangle} p_2$$

$$= \tfrac{8}{5} p_0 + \tfrac{-31}{14} p_2 = \tfrac{8}{5} - \tfrac{31}{14}(t^2 - 2).$$

This polynomial is the closest to p of all polynomials in $\mathbb{P}_2$, when the distance between polynomials is measured only at -2, -1, 0, 1, and 2. See Fig. 1. ∎

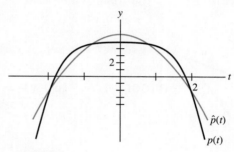

FIGURE 1

The polynomials p_0, p_1, and p_2 in Examples 5 and 6 belong to a class of polynomials that are referred to in statistics as *orthogonal polynomials*.[2] The orthogonality refers to the type of inner product described in Example 2.

Two Inequalities

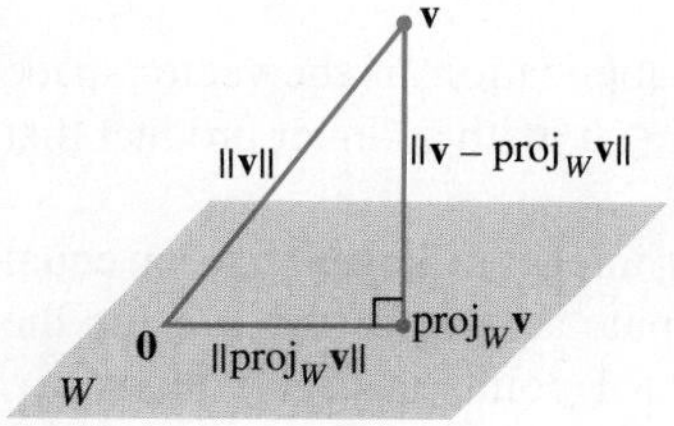

Given a vector $\mathbf{v}$ in an inner product space V and given a finite-dimensional subspace W, we may apply the Pythagorean Theorem to the orthogonal decomposition of $\mathbf{v}$ with respect to W and obtain

$$\|\mathbf{v}\|^2 = \|\text{proj}_W \, \mathbf{v}\|^2 + \|\mathbf{v} - \text{proj}_W \, \mathbf{v}\|^2$$

FIGURE 2
The hypotenuse is the longest side.

See Fig. 2. In particular, this shows that the norm of the projection of $\mathbf{v}$ onto W does not exceed the norm of $\mathbf{v}$ itself. This simple observation leads to the following important inequality.

THEOREM 16

The Cauchy–Schwarz Inequality

For all $\mathbf{u}$, $\mathbf{v}$ in V,

$$|\langle \mathbf{u}, \mathbf{v} \rangle| \leq \|\mathbf{u}\| \, \|\mathbf{v}\| \tag{4}$$

---

[2] See *Statistics and Experimental Design in Engineering and the Physical Sciences*, 2nd ed., by Norman L. Johnson and Fred C. Leone (New York: John Wiley & Sons, 1977). Tables there list "Orthogonal Polynomials," which are simply the values of the polynomial at numbers such as -2, -1, 0, 1, and 2.

PROOF If $\mathbf{u} = \mathbf{0}$, then both sides of (4) are zero, and hence the inequality is true in this case. (See Practice Problem 1.) If $\mathbf{u} \neq \mathbf{0}$, let W be the subspace spanned by $\mathbf{u}$. Recall that $\|c\mathbf{u}\| = |c| \, \|\mathbf{u}\|$ for any scalar c. Thus

$$\| \operatorname{proj}_W \mathbf{v} \| = \left\| \frac{\langle \mathbf{v}, \mathbf{u} \rangle}{\langle \mathbf{u}, \mathbf{u} \rangle} \mathbf{u} \right\| = \frac{|\langle \mathbf{v}, \mathbf{u} \rangle|}{|\langle \mathbf{u}, \mathbf{u} \rangle|} \|\mathbf{u}\| = \frac{|\langle \mathbf{v}, \mathbf{u} \rangle|}{\|\mathbf{u}\|^2} \|\mathbf{u}\| = \frac{|\langle \mathbf{u}, \mathbf{v} \rangle|}{\|\mathbf{u}\|}$$

Since $\| \operatorname{proj}_W \mathbf{v} \| \leq \|\mathbf{v}\|$, we have $\dfrac{|\langle \mathbf{u}, \mathbf{v} \rangle|}{\|\mathbf{u}\|} \leq \|\mathbf{v}\|$, which gives (4). ■

The Cauchy–Schwarz inequality is useful in many branches of mathematics. A few simple applications are presented in the exercises. Our main need for this inequality here is to prove another fundamental inequality involving norms of vectors. See Fig. 3.

THEOREM 17

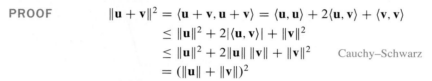

The Triangle Inequality

For all $\mathbf{u}, \mathbf{v}$ in V,

$$\|\mathbf{u} + \mathbf{v}\| \leq \|\mathbf{u}\| + \|\mathbf{v}\|$$

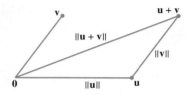

FIGURE 3
The lengths of the sides of a triangle.

PROOF

$$\begin{aligned}
\|\mathbf{u} + \mathbf{v}\|^2 &= \langle \mathbf{u} + \mathbf{v}, \mathbf{u} + \mathbf{v} \rangle = \langle \mathbf{u}, \mathbf{u} \rangle + 2\langle \mathbf{u}, \mathbf{v} \rangle + \langle \mathbf{v}, \mathbf{v} \rangle \\
&\leq \|\mathbf{u}\|^2 + 2|\langle \mathbf{u}, \mathbf{v} \rangle| + \|\mathbf{v}\|^2 \\
&\leq \|\mathbf{u}\|^2 + 2\|\mathbf{u}\| \, \|\mathbf{v}\| + \|\mathbf{v}\|^2 \qquad \text{Cauchy–Schwarz} \\
&= (\|\mathbf{u}\| + \|\mathbf{v}\|)^2
\end{aligned}$$

The triangle inequality follows immediately by taking square roots of both sides. ■

An Inner Product for $C[a, b]$ (Calculus required)

Probably the most widely used inner product space for applications is the vector space $C[a, b]$ of all continuous functions on an interval $a \leq t \leq b$, with an inner product that we will describe.

We begin by considering a polynomial p and any integer n larger than or equal to the degree of p. Then p is in $\mathbb{P}_n$, and we may compute a "length" for p using the inner product of Example 2 involving evaluation at $n + 1$ points in $[a, b]$. However, this length of p captures the behavior at only those $n + 1$ points. Since p is in $\mathbb{P}_n$ for all large n, we could use a much larger n, with many more points for the "evaluation" inner product. See Fig. 4.

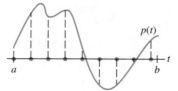
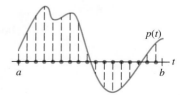

FIGURE 4 Using different numbers of evaluation points in $[a, b]$ to compute $\|p\|^2$.

Let us partition $[a, b]$ into $n + 1$ subintervals of length $\Delta t = (b - a)/(n + 1)$, and let $t_0, \ldots, t_n$ be arbitrary points in these subintervals.

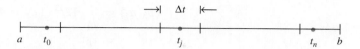

If n is large, the inner product on $\mathbb{P}_n$ determined by $t_0, \ldots, t_n$ will tend to give a large value to $\langle p, p \rangle$, so we scale it down and divide by $n + 1$. Observe that $1/(n + 1) = \Delta t/(b - a)$, and define

$$\langle p, q \rangle = \frac{1}{n + 1} \sum_{j=0}^{n} p(t_j)q(t_j) = \frac{1}{b - a} \left[\sum_{j=0}^{n} p(t_j)q(t_j)\Delta t \right]$$

Now, let n increase without bound. Since polynomials p and q are continuous functions, the expression in brackets is a Riemann sum that approaches a definite integral, and we are led to consider the *average value* of $p(t)q(t)$ on the interval $[a, b]$:

$$\frac{1}{b - a} \int_a^b p(t)q(t) \, dt$$

This quantity is defined for polynomials of any degree (in fact, for all continuous functions), and it has all the properties of an inner product, as the next example shows. The scale factor $1/(b - a)$ is inessential and is often omitted for simplicity.

EXAMPLE 7   For f, g in $C[a, b]$, set

$$\langle f, g \rangle = \int_a^b f(t)g(t) \, dt \tag{5}$$

Show that (5) defines an inner product on $C[a, b]$.

SOLUTION Inner product Axioms 1–3 follow from elementary properties of definite integrals. For Axiom 4, observe that

$$\langle f, f \rangle = \int_a^b [f(t)]^2 \, dt \geq 0$$

The function $[f(t)]^2$ is continuous and nonnegative on $[a, b]$. If the definite integral of $[f(t)]^2$ is zero, then $[f(t)]^2$ must be identically zero on $[a, b]$, by a theorem in advanced calculus, in which case f is the zero function. Thus $\langle f, f \rangle = 0$ implies that f is the zero function on $[a, b]$. So (5) defines an inner product on $C[a, b]$.   ∎

EXAMPLE 8   Let V be the space $C[0, 1]$ with the inner product of Example 7, and let W be the subspace spanned by the polynomials $p_1(t) = 1$, $p_2(t) = 2t - 1$, and $p_3(t) = 12t^2$. Use the Gram–Schmidt process to find an orthogonal basis for W.

SOLUTION Let $q_1 = p_1$, and compute

$$\langle p_2, q_1 \rangle = \int_0^1 (2t - 1)(1) \, dt = (t^2 - t) \Big|_0^1 = 0$$

So p_2 is already orthogonal to q_1, and we can take $q_2 = p_2$. For the projection of p_3 onto $W_2 = \text{Span}\{q_1, q_2\}$, compute

$$\langle p_3, q_1 \rangle = \int_0^1 12t^2 \cdot 1 \, dt = 4t^3 \Big|_0^1 = 4$$

$$\langle q_1, q_1 \rangle = \int_0^1 1 \cdot 1 \, dt = t \Big|_0^1 = 1$$

$$\langle p_3, q_2 \rangle = \int_0^1 12t^2(2t - 1) \, dt = \int_0^1 (24t^3 - 12t^2) \, dt = 2$$

$$\langle q_2, q_2 \rangle = \int_0^1 (2t - 1)^2 \, dt = \frac{1}{6}(2t - 1)^3 \Big|_0^1 = \frac{1}{3}$$

Then

$$\text{proj}_{W_2} \, p_3 = \frac{\langle p_3, q_1 \rangle}{\langle q_1, q_1 \rangle} q_1 + \frac{\langle p_3, q_2 \rangle}{\langle q_2, q_2 \rangle} q_2 = \frac{4}{1} q_1 + \frac{2}{1/3} q_2 = 4q_1 + 6q_2$$

and

$$q_3 = p_3 - \text{proj}_{W_2} \, p_3 = p_3 - 4q_1 - 6q_2$$

As a function, $q_3(t) = 12t^2 - 4 - 6(2t - 1) = 12t^2 - 12t + 2$. The orthogonal basis for the subspace W is $\{q_1, q_2, q_3\}$. ∎

PRACTICE PROBLEMS

Use the inner product axioms to verify the following statements.

1. $\langle \mathbf{v}, \mathbf{0} \rangle = \langle \mathbf{0}, \mathbf{v} \rangle = 0$.
2. $\langle \mathbf{u}, \mathbf{v} + \mathbf{w} \rangle = \langle \mathbf{u}, \mathbf{v} \rangle + \langle \mathbf{u}, \mathbf{w} \rangle$.

6.7 EXERCISES

1. Let $\mathbb{R}^2$ have the inner product of Example 1, and let $\mathbf{x} = (1, 1)$ and $\mathbf{y} = (5, -1)$.
   a. Find $\|\mathbf{x}\|$, $\|\mathbf{y}\|$, and $|\langle \mathbf{x}, \mathbf{y} \rangle|^2$.
   b. Describe all vectors (z_1, z_2) that are orthogonal to $\mathbf{y}$.

2. Let $\mathbb{R}^2$ have the inner product of Example 1. Show that the Cauchy–Schwarz inequality holds for $\mathbf{x} = (3, -2)$ and $\mathbf{y} = (-2, 1)$. [*Suggestion:* Study $|\langle \mathbf{x}, \mathbf{y} \rangle|^2$.]

Exercises 3–8 refer to $\mathbb{P}_2$ with the inner product given by evaluation at $-1, 0$, and 1. (See Example 2.)

3. Compute $\langle p, q \rangle$, where $p(t) = 4 + t$, $q(t) = 5 - 4t^2$.

4. Compute $\langle p, q \rangle$, where $p(t) = 3t - t^2$, $q(t) = 3 + 2t^2$.

5. Compute $\|p\|$ and $\|q\|$, for p and q in Exercise 3.

6. Compute $\|p\|$ and $\|q\|$, for p and q in Exercise 4.

7. Compute the orthogonal projection of q onto the subspace spanned by p, for p and q in Exercise 3.

8. Compute the orthogonal projection of q onto the subspace spanned by p, for p and q in Exercise 4.

9. Let $\mathbb{P}_3$ have the inner product given by evaluation at $-3, -1$, 1, and 3. Let $p_0(t) = 1$, $p_1(t) = t$, and $p_2(t) = t^2$.
   a. Compute the orthogonal projection of p_2 onto the subspace spanned by p_0 and p_1.
   b. Find a polynomial q that is orthogonal to p_0 and p_1, such that $\{p_0, p_1, q\}$ is an orthogonal basis for $\text{Span}\{p_0, p_1, p_2\}$. Scale the polynomial q so that its vector of values at $(-3, -1, 1, 3)$ is $(1, -1, -1, 1)$.

10. Let $\mathbb{P}_3$ have the inner product as in Exercise 9, with p_0, p_1, and q the polynomials described there. Find the best approximation to $p(t) = t^3$ by polynomials in $\text{Span}\{p_0, p_1, q\}$.

11. Let p_0, p_1, and p_2 be the orthogonal polynomials described in Example 5, where the inner product on $\mathbb{P}_4$ is given by evaluation at $-2, -1, 0, 1$, and 2. Find the orthogonal projection of t^3 onto $\text{Span}\{p_0, p_1, p_2\}$.

12. Find a polynomial p_3 such that $\{p_0, p_1, p_2, p_3\}$ (see Exercise 11) is an orthogonal basis for the subspace $\mathbb{P}_3$ of $\mathbb{P}_4$. Scale the polynomial p_3 so that its vector of values is $(-1, 2, 0, -2, 1)$.

13. Let A be any invertible $n \times n$ matrix. Show that for $\mathbf{u}, \mathbf{v}$ in $\mathbb{R}^n$, the formula $\langle \mathbf{u}, \mathbf{v} \rangle = (A\mathbf{u}) \cdot (A\mathbf{v}) = (A\mathbf{u})^T (A\mathbf{v})$ defines an inner product on $\mathbb{R}^n$.

14. Let T be a one-to-one linear transformation from a vector space V into $\mathbb{R}^n$. Show that for $\mathbf{u}, \mathbf{v}$ in V, the formula $\langle \mathbf{u}, \mathbf{v} \rangle = T(\mathbf{u}) \cdot T(\mathbf{v})$ defines an inner product on V.

Use the inner product axioms and other results of this section to verify the statements in Exercises 15–18.

15. $\langle \mathbf{u}, c\mathbf{v} \rangle = c \langle \mathbf{u}, \mathbf{v} \rangle$ for all scalars c.

16. If $\{\mathbf{u}, \mathbf{v}\}$ is an orthonormal set in V, then $\|\mathbf{u} - \mathbf{v}\| = \sqrt{2}$.

17. $\langle \mathbf{u}, \mathbf{v} \rangle = \frac{1}{4}\|\mathbf{u} + \mathbf{v}\|^2 - \frac{1}{4}\|\mathbf{u} - \mathbf{v}\|^2$.

18. $\|\mathbf{u} + \mathbf{v}\|^2 + \|\mathbf{u} - \mathbf{v}\|^2 = 2\|\mathbf{u}\|^2 + 2\|\mathbf{v}\|^2$.

19. Given $a \geq 0$ and $b \geq 0$, let $\mathbf{u} = \begin{bmatrix} \sqrt{a} \\ \sqrt{b} \end{bmatrix}$ and $\mathbf{v} = \begin{bmatrix} \sqrt{b} \\ \sqrt{a} \end{bmatrix}$. Use the Cauchy–Schwarz inequality to compare the geometric mean $\sqrt{ab}$ with the arithmetic mean $(a + b)/2$.

20. Let $\mathbf{u} = \begin{bmatrix} a \\ b \end{bmatrix}$ and $\mathbf{v} = \begin{bmatrix} 1 \\ 1 \end{bmatrix}$. Use the Cauchy–Schwarz inequality to show that

$$\left(\frac{a + b}{2} \right)^2 \leq \frac{a^2 + b^2}{2}$$

Exercises 21–24 refer to $V = C[0, 1]$, with the inner product given by an integral, as in Example 7.

21. Compute $\langle f, g \rangle$, where $f(t) = 1 - 3t^2$ and $g(t) = t - t^3$.

22. Compute $\langle f, g \rangle$, where $f(t) = 5t - 3$ and $g(t) = t^3 - t^2$.

23. Compute $\|f\|$ for f in Exercise 21.

24. Compute $\|g\|$ for g in Exercise 22.

25. Let V be the space $C[-1, 1]$ with the inner product of Example 7. Find an orthogonal basis for the subspace spanned by the polynomials 1, t, and t^2. The polynomials in this basis are called *Legendre polynomials*.

26. Let V be the space $C[-2, 2]$ with the inner product of Example 7. Find an orthogonal basis for the subspace spanned by the polynomials 1, t, and t^2.

27. **[M]** Let $\mathbb{P}_4$ have the inner product as in Example 5, and let p_0, p_1, p_2 be the orthogonal polynomials from that example. Using your matrix program, apply the Gram–Schmidt process to the set $\{p_0, p_1, p_2, t^3, t^4\}$ to create an orthogonal basis for $\mathbb{P}_4$.

28. **[M]** Let V be the space $C[0, 2\pi]$ with the inner product of Example 7. Use the Gram–Schmidt process to create an orthogonal basis for the subspace spanned by $\{1, \cos t, \cos^2 t, \cos^3 t\}$. Use a matrix program or computational program to compute the appropriate definite integrals.

| **SOLUTIONS TO PRACTICE PROBLEMS**

1. By Axiom 1, $\langle \mathbf{v}, \mathbf{0} \rangle = \langle \mathbf{0}, \mathbf{v} \rangle$. Then $\langle \mathbf{0}, \mathbf{v} \rangle = \langle 0\mathbf{v}, \mathbf{v} \rangle = 0\langle \mathbf{v}, \mathbf{v} \rangle$, by Axiom 3, so $\langle \mathbf{0}, \mathbf{v} \rangle = 0$.

2. By Axioms 1, 2, and then 1 again, $\langle \mathbf{u}, \mathbf{v} + \mathbf{w} \rangle = \langle \mathbf{v} + \mathbf{w}, \mathbf{u} \rangle = \langle \mathbf{v}, \mathbf{u} \rangle + \langle \mathbf{w}, \mathbf{u} \rangle = \langle \mathbf{u}, \mathbf{v} \rangle + \langle \mathbf{u}, \mathbf{w} \rangle$.

6.8   APPLICATIONS OF INNER PRODUCT SPACES

The examples in this section suggest how the inner product spaces defined in Section 6.7 arise in practical problems. The first example is connected with the massive least-squares problem of updating the North American Datum, described in the chapter's introductory example.

Weighted Least-Squares

Let $\mathbf{y}$ be a vector of n observations, $y_1, \ldots, y_n$, and suppose we wish to approximate $\mathbf{y}$ by a vector $\hat{\mathbf{y}}$ that belongs to some specified subspace of $\mathbb{R}^n$. (In Section 6.5, $\hat{\mathbf{y}}$ was written as $A\mathbf{x}$ so that $\hat{\mathbf{y}}$ was in the column space of A.) Denote the entries in $\hat{\mathbf{y}}$ by $\hat{y}_1, \ldots, \hat{y}_n$. Then the *sum of the squares for error*, or SS(E), in approximating $\mathbf{y}$ by $\hat{\mathbf{y}}$ is

$$\text{SS(E)} = (y_1 - \hat{y}_1)^2 + \cdots + (y_n - \hat{y}_n)^2 \tag{1}$$

This is simply $\|\mathbf{y} - \hat{\mathbf{y}}\|^2$, using the standard length in $\mathbb{R}^n$.

Now suppose the measurements that produced the entries in **y** are not equally reliable. (This was the case for the North American Datum, since measurements were made over a period of 140 years.) As another example, the entries in **y** might be computed from various samples of measurements, with unequal sample sizes.) Then it becomes appropriate to weight the squared errors in (1) in such a way that more importance is assigned to the more reliable measurements.[1] If the weights are denoted by $w_1^2, \ldots, w_n^2$, then the weighted sum of the squares for error is

$$\text{Weighted SS(E)} = w_1^2(y_1 - \hat{y}_1)^2 + \cdots + w_n^2(y_n - \hat{y}_n)^2 \tag{2}$$

This is the square of the length of $\mathbf{y} - \hat{\mathbf{y}}$, where the length is derived from an inner product analogous to that in Example 1 in Section 6.7, namely,

$$\langle \mathbf{x}, \mathbf{y} \rangle = w_1^2 x_1 y_1 + \cdots + w_n^2 x_n y_n$$

It is sometimes convenient to transform a weighted least-squares problem into an equivalent ordinary least-squares problem. Let W be the diagonal matrix with (positive) $w_1, \ldots, w_n$ on its diagonal, so that

$$
W\mathbf{y} =
\begin{bmatrix}
w_1 & 0 & \cdots & 0 \\
0 & w_2 & & \\
\vdots & & \ddots & \vdots \\
0 & & \cdots & w_n
\end{bmatrix}
\begin{bmatrix}
y_1 \\
y_2 \\
\vdots \\
y_n
\end{bmatrix}
=
\begin{bmatrix}
w_1 y_1 \\
w_2 y_2 \\
\vdots \\
w_n y_n
\end{bmatrix}
$$

with a similar expression for $W\hat{\mathbf{y}}$. Observe that the jth term in (2) can be written as

$$w_j^2(y_j - \hat{y}_j)^2 = (w_j y_j - w_j \hat{y}_j)^2$$

It follows that the weighted SS(E) in (2) is the square of the ordinary length in $\mathbb{R}^n$ of $W\mathbf{y} - W\hat{\mathbf{y}}$, which we write as $\| W\mathbf{y} - W\hat{\mathbf{y}} \|^2$.

Now suppose the approximating vector $\hat{\mathbf{y}}$ is to be constructed from the columns of a matrix A. Then we seek an $\hat{\mathbf{x}}$ that makes $A\hat{\mathbf{x}} = \hat{\mathbf{y}}$ as close to **y** as possible. However, the measure of closeness is the weighted error,

$$\| W\mathbf{y} - W\hat{\mathbf{y}} \|^2 = \| W\mathbf{y} - WA\hat{\mathbf{x}} \|^2$$

Thus $\hat{\mathbf{x}}$ is the (ordinary) least-squares solution of the equation

$$WA\mathbf{x} = W\mathbf{y}$$

The normal equation for the least-squares solution is

$$(WA)^T WA\mathbf{x} = (WA)^T W\mathbf{y}$$

EXAMPLE 1 Find the least-squares line $y = \beta_0 + \beta_1 x$ that best fits the data $(-2, 3), (-1, 5), (0, 5), (1, 4)$, and $(2, 3)$. Suppose the errors in measuring the y-values of the last two data points are greater than for the other points. Weight these data half as much as the rest of the data.

---

[1] Note for readers with a background in statistics: Suppose the errors in measuring the y_i are independent random variables with means equal to zero and variances of $\sigma_1^2, \ldots, \sigma_n^2$. Then the appropriate weights in (2) are $w_i^2 = 1/\sigma_i^2$. The larger the variance of the error, the smaller the weight.

SOLUTION As in Section 6.6, write X for the matrix A and β for the vector $\mathbf{x}$, and obtain

$$X = \begin{bmatrix} 1 & -2 \\ 1 & -1 \\ 1 & 0 \\ 1 & 1 \\ 1 & 2 \end{bmatrix}, \quad \beta = \begin{bmatrix} \beta_0 \\ \beta_1 \end{bmatrix}, \quad \mathbf{y} = \begin{bmatrix} 3 \\ 5 \\ 5 \\ 4 \\ 3 \end{bmatrix}$$

For a weighting matrix, choose W with diagonal entries 2, 2, 2, 1, and 1. Left-multiplication by W scales the rows of X and $\mathbf{y}$:

$$WX = \begin{bmatrix} 2 & -4 \\ 2 & -2 \\ 2 & 0 \\ 1 & 1 \\ 1 & 2 \end{bmatrix}, \quad W\mathbf{y} = \begin{bmatrix} 6 \\ 10 \\ 10 \\ 4 \\ 3 \end{bmatrix}$$

For the normal equation, compute

$$(WX)^T WX = \begin{bmatrix} 14 & -9 \\ -9 & 25 \end{bmatrix} \quad \text{and} \quad (WX)^T W\mathbf{y} = \begin{bmatrix} 59 \\ -34 \end{bmatrix}$$

and solve

$$\begin{bmatrix} 14 & -9 \\ -9 & 25 \end{bmatrix}\begin{bmatrix} \beta_0 \\ \beta_1 \end{bmatrix} = \begin{bmatrix} 59 \\ -34 \end{bmatrix}$$

The solution of the normal equation is (to two significant digits) $\beta_0 = 4.3$ and $\beta_1 = .20$. The desired line is

$$y = 4.3 + .20x$$

In contrast, the ordinary least-squares line for these data is

$$y = 4.0 - .10x$$

Both lines are displayed in Fig. 1. ∎

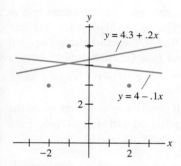

FIGURE 1
Weighted and ordinary least-squares lines.

Trend Analysis of Data

Let f represent an unknown function whose values are known (perhaps only approximately) at $t_0, \ldots, t_n$. If there is a "linear trend" in the data $f(t_0), \ldots, f(t_n)$, then we might expect to approximate the values of f by a function of the form $\beta_0 + \beta_1 t$. If there is a "quadratic trend" to the data, then we would try a function of the form $\beta_0 + \beta_1 t + \beta_2 t^2$. This was discussed in Section 6.6, from a different point of view.

In some statistical problems, it is important to be able to separate the linear trend from the quadratic trend (and possibly cubic or higher-order trends). For instance, suppose engineers are analyzing the performance of a new car, and $f(t)$ represents the distance between the car at time t and some reference point. If the car is traveling at constant velocity, then the graph of $f(t)$ should be a straight line whose slope is the car's velocity. If the gas pedal is suddenly pressed to the floor, the graph of $f(t)$ will change to include a quadratic term and possibly a cubic term (due to the acceleration). To analyze the ability of the car to pass another car, for example, engineers may want to separate the quadratic and cubic components from the linear term.

If the function is approximated by a curve of the form $y = \beta_0 + \beta_1 t + \beta_2 t^2$, the coefficient β_2 may not give the desired information about the quadratic trend in the data, because it may not be "independent" in a statistical sense from the other β_i. To make what is known as a **trend analysis** of the data, we introduce an inner product on the space $\mathbb{P}_n$ analogous to that given in Example 2 in Section 6.7. For p, q in $\mathbb{P}_n$, define

$$\langle p, q \rangle = p(t_0)q(t_0) + \cdots + p(t_n)q(t_n)$$

In practice, statisticians seldom need to consider trends in data of degree higher than cubic or quartic. So let p_0, p_1, p_2, p_3 denote an orthogonal basis of the subspace $\mathbb{P}_3$ of $\mathbb{P}_n$, obtained by applying the Gram–Schmidt process to the polynomials $1, t, t^2,$ and t^3. By Supplementary Exercise 11 in Chapter 2, there is a polynomial g in $\mathbb{P}_n$ whose values at $t_0, \ldots, t_n$ coincide with those of the unknown function f. Let $\hat{g}$ be the orthogonal projection (with respect to the given inner product) of g onto $\mathbb{P}_3$, say,

$$\hat{g} = c_0 p_0 + c_1 p_1 + c_2 p_2 + c_3 p_3$$

Then $\hat{g}$ is called a cubic **trend function**, and $c_0, \ldots, c_3$ are the **trend coefficients** of the data. The coefficient c_1 measures the linear trend, c_2 the quadratic trend, and c_3 the cubic trend. It turns out that if the data have certain properties, these coefficients are statistically independent.

Since $p_0, \ldots, p_3$ are orthogonal, the trend coefficients may be computed one at a time, independently of one another. (Recall that $c_i = \langle g, p_i \rangle / \langle p_i, p_i \rangle$.) We can ignore p_3 and c_3 if we want only the quadratic trend. And if, for example, we needed to determine the quartic trend, we would have to find (via Gram–Schmidt) only a polynomial p_4 in $\mathbb{P}_4$ that is orthogonal to $\mathbb{P}_3$ and compute $\langle g, p_4 \rangle / \langle p_4, p_4 \rangle$.

EXAMPLE 2 The simplest and most common use of trend analysis occurs when the points $t_0, \ldots, t_n$ can be adjusted so that they are evenly spaced and sum to zero. Fit a quadratic trend function to the data $(-2, 3)$, $(-1, 5)$, $(0, 5)$, $(1, 4)$, and $(2, 3)$.

SOLUTION The t-coordinates are suitably scaled to use the orthogonal polynomials found in Example 5 of Section 6.7:

Polynomial:	p_0	p_1	p_2	Data: g
Vector of values:	$\begin{bmatrix} 1 \\ 1 \\ 1 \\ 1 \\ 1 \end{bmatrix}$,	$\begin{bmatrix} -2 \\ -1 \\ 0 \\ 1 \\ 2 \end{bmatrix}$,	$\begin{bmatrix} 2 \\ -1 \\ -2 \\ -1 \\ 2 \end{bmatrix}$,	$\begin{bmatrix} 3 \\ 5 \\ 5 \\ 4 \\ 3 \end{bmatrix}$

The calculations involve only these vectors, not the specific formulas for the orthogonal polynomials. The best approximation to the data by polynomials in $\mathbb{P}_2$ is the orthogonal projection given by

$$\hat{p} = \frac{\langle g, p_0 \rangle}{\langle p_0, p_0 \rangle} p_0 + \frac{\langle g, p_1 \rangle}{\langle p_1, p_1 \rangle} p_1 + \frac{\langle g, p_2 \rangle}{\langle p_2, p_2 \rangle} p_2$$

$$= \frac{20}{5} p_0 - \frac{1}{10} p_1 - \frac{7}{14} p_2$$

and

$$\hat{p}(t) = 4 - .1t - .5(t^2 - 2) \tag{3}$$

Since the coefficient of p_2 is not extremely small, it would be reasonable to conclude that the trend is at least quadratic. This is confirmed by the graph in Fig. 2. ∎

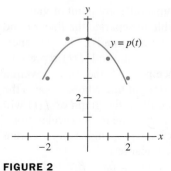

FIGURE 2

Approximation by a quadratic trend function.

Fourier Series (Calculus required)

Continuous functions are often approximated by linear combinations of sine and cosine functions. For instance, a continuous function might represent a sound wave, an electric signal of some type, or the movement of a vibrating mechanical system.

For simplicity, we consider functions on $0 \le t \le 2\pi$. It turns out that any function in $C[0, 2\pi]$ can be approximated as closely as desired by a function of the form

$$\frac{a_0}{2} + a_1 \cos t + \cdots + a_n \cos nt + b_1 \sin t + \cdots + b_n \sin nt \tag{4}$$

for a sufficiently large value of n. The function (4) is called a **trigonometric polynomial**. If a_n and b_n are not both zero, the polynomial is said to be of **order n**. The connection between trigonometric polynomials and other functions in $C[0, 2\pi]$ depends on the fact that for any $n \ge 1$, the set

$$\{1, \cos t, \cos 2t, \ldots, \cos nt, \sin t, \sin 2t, \ldots, \sin nt\} \tag{5}$$

is orthogonal with respect to the inner product

$$\langle f, g \rangle = \int_0^{2\pi} f(t)g(t)\, dt \tag{6}$$

This orthogonality is verified as in the following example and in Exercises 5 and 6.

EXAMPLE 3 Let $C[0, 2\pi]$ have the inner product (6), and let m and n be unequal positive integers. Show that $\cos mt$ and $\cos nt$ are orthogonal.

SOLUTION Use a trigonometric identity. When $m \ne n$,

$$\langle \cos mt, \cos nt \rangle = \int_0^{2\pi} \cos mt \cos nt\, dt$$

$$= \frac{1}{2} \int_0^{2\pi} [\cos(mt + nt) + \cos(mt - nt)]\, dt$$

$$= \frac{1}{2} \left[\frac{\sin(mt + nt)}{m + n} + \frac{\sin(mt - nt)}{m - n} \right]\Big|_0^{2\pi} = 0 \qquad \blacksquare$$

Let W be the subspace of $C[0, 2\pi]$ spanned by the functions in (5). Given f in $C[0, 2\pi]$, the best approximation to f by functions in W is called the **nth-order Fourier approximation** to f on $[0, 2\pi]$. Since the functions in (5) are orthogonal, the best approximation is given by the orthogonal projection onto W. In this case, the coefficients a_k and b_k in (4) are called the **Fourier coefficients** of f. The standard formula for an orthogonal projection shows that

$$a_k = \frac{\langle f, \cos kt \rangle}{\langle \cos kt, \cos kt \rangle}, \quad b_k = \frac{\langle f, \sin kt \rangle}{\langle \sin kt, \sin kt \rangle}, \quad k \ge 1$$

Exercise 7 asks you to show that $\langle \cos kt, \cos kt \rangle = \pi$ and $\langle \sin kt, \sin kt \rangle = \pi$. Thus

$$a_k = \frac{1}{\pi} \int_0^{2\pi} f(t) \cos kt\, dt, \quad b_k = \frac{1}{\pi} \int_0^{2\pi} f(t) \sin kt\, dt \tag{7}$$

The coefficient of the (constant) function 1 in the orthogonal projection is

$$\frac{\langle f, 1 \rangle}{\langle 1, 1 \rangle} = \frac{1}{2\pi} \int_0^{2\pi} f(t) \cdot 1\, dt = \frac{1}{2} \left[\frac{1}{\pi} \int_0^{2\pi} f(t) \cos(0 \cdot t)\, dt \right] = \frac{a_0}{2}$$

where a_0 is defined by (7) for $k = 0$. This explains why the constant term in (4) is written as $a_0/2$.

EXAMPLE 4 Find the nth-order Fourier approximation to the function $f(t) = t$ on the interval $[0, 2\pi]$.

SOLUTION Compute

$$\frac{a_0}{2} = \frac{1}{2} \cdot \frac{1}{\pi} \int_0^{2\pi} t \, dt = \frac{1}{2\pi} \left[\frac{1}{2} t^2 \Big|_0^{2\pi} \right] = \pi$$

and for $k > 0$, using integration by parts,

$$a_k = \frac{1}{\pi} \int_0^{2\pi} t \cos kt \, dt = \frac{1}{\pi} \left[\frac{1}{k^2} \cos kt + \frac{t}{k} \sin kt \right]_0^{2\pi} = 0$$

$$b_k = \frac{1}{\pi} \int_0^{2\pi} t \sin kt \, dt = \frac{1}{\pi} \left[\frac{1}{k^2} \sin kt - \frac{t}{k} \cos kt \right]_0^{2\pi} = -\frac{2}{k}$$

Thus the nth-order Fourier approximation of $f(t) = t$ is

$$\pi - 2 \sin t - \sin 2t - \frac{2}{3} \sin 3t - \cdots - \frac{2}{n} \sin nt$$

Figure 3 shows the third- and fourth-order Fourier approximations of f. ∎

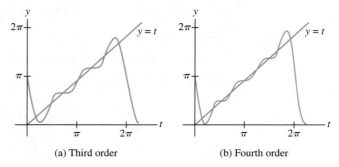

(a) Third order   (b) Fourth order

FIGURE 3 Fourier approximations of the function $f(t) = t$.

The norm of the difference between f and a Fourier approximation is called the **mean square error** in the approximation. (The term *mean* refers to the fact that the norm is determined by an integral.) It can be shown that the mean square error approaches zero as the order of the Fourier approximation increases. For this reason, it is common to write

$$f(t) = \frac{a_0}{2} + \sum_{m=1}^{\infty} (a_m \cos mt + b_m \sin mt)$$

This expression for $f(t)$ is called the **Fourier series** for f on $[0, 2\pi]$. The term $a_m \cos mt$, for example, is the projection of f onto the one-dimensional subspace spanned by $\cos mt$.

PRACTICE PROBLEMS

1. Let $q_1(t) = 1$, $q_2(t) = t$, and $q_3(t) = 3t^2 - 4$. Verify that $\{q_1, q_2, q_3\}$ is an orthogonal set in $C[-2, 2]$ with the inner product of Example 7 in Section 6.7 (integration from -2 to 2).

2. Find the first-order and third-order Fourier approximations to

$$f(t) = 3 - 2 \sin t + 5 \sin 2t - 6 \cos 2t$$

6.8 EXERCISES

1. Find the least-squares line $y = \beta_0 + \beta_1 x$ that best fits the data $(-2,0)$, $(-1,0)$, $(0,2)$, $(1,4)$, and $(2,4)$, assuming that the first and last data points are less reliable. Weight them half as much as the three interior points.

2. Suppose 5 out of 25 data points in a weighted least-squares problem have a y-measurement that is less reliable than the others, and they are to be weighted half as much as the other 20 points. One method is to weight the 20 points by a factor of 1 and the other 5 by a factor of $\frac{1}{2}$. A second method is to weight the 20 points by a factor of 2 and the other 5 by a factor of 1. Do the two methods produce different results? Explain.

3. Fit a cubic trend function to the data in Example 2. The orthogonal cubic polynomial is $p_3(t) = \frac{5}{6}t^3 - \frac{17}{6}t$.

4. To make a trend analysis of six evenly spaced data points, one can use orthogonal polynomials with respect to evaluation at the points $t = -5, -3, -1, 1, 3$, and 5.

   a. Show that the first three orthogonal polynomials are

   $$p_0(t) = 1, \quad p_1(t) = t, \quad \text{and} \quad p_2(t) = \frac{3}{8}t^2 - \frac{35}{8}$$

   (The polynomial p_2 has been scaled so that its values at the evaluation points are small integers.)

   b. Fit a quadratic trend function to the data

   $$(-5, 1), (-3, 1), (-1, 4), (1, 4), (3, 6), (5, 8)$$

In Exercises 5–14, the space is $C[0, 2\pi]$ with the inner product (6).

5. Show that $\sin mt$ and $\sin nt$ are orthogonal when $m \neq n$.

6. Show that $\sin mt$ and $\cos nt$ are orthogonal for all positive integers m and n.

7. Show that $\| \cos kt \|^2 = \pi$ and $\| \sin kt \|^2 = \pi$ for $k > 0$.

8. Find the third-order Fourier approximation to $f(t) = t - 1$.

9. Find the third-order Fourier approximation to $f(t) = 2\pi - t$.

10. Find the third-order Fourier approximation to the *square wave function*, $f(t) = 1$ for $0 \leq t < \pi$ and $f(t) = -1$ for $\pi \leq t < 2\pi$.

11. Find the third-order Fourier approximation to $\sin^2 t$, without performing any integration calculations.

12. Find the third-order Fourier approximation to $\cos^3 t$, without performing any integration calculations.

13. Explain why a Fourier coefficient of the sum of two functions is the sum of the corresponding Fourier coefficients of the two functions.

14. Suppose the first few Fourier coefficients of some function f in $C[0, 2\pi]$ are a_0, a_1, a_2, and b_1, b_2, b_3. Which of the following trigonometric polynomials is closer to f? Defend your answer.

$$g(t) = \frac{a_0}{2} + a_1 \cos t + a_2 \cos 2t + b_1 \sin t$$

$$h(t) = \frac{a_0}{2} + a_1 \cos t + a_2 \cos 2t + b_1 \sin t + b_2 \sin 2t$$

15. [M] Refer to the data in Exercise 13 in Section 6.6, concerning the takeoff performance of an airplane. Suppose the possible measurement errors become greater as the speed of the airplane increases, and let W be the diagonal weighting matrix whose diagonal entries are $1, 1, 1, .9, .9, .8, .7, .6, .5,$ $.4, .3, .2$, and $.1$. Find the cubic curve that fits the data with minimum weighted least-squares error, and use it to estimate the velocity of the plane when $t = 4.5$ seconds.

16. [M] Let f_4 and f_5 be the fourth-order and fifth-order Fourier approximations in $C[0, 2\pi]$ to the square wave function in Exercise 10. Produce separate graphs of f_4 and f_5 on the interval $[0, 2\pi]$, and produce a graph of f_5 on $[-2\pi, 2\pi]$.

SG | The Linearity of an Orthogonal Projection 6-25

SOLUTIONS TO PRACTICE PROBLEMS

1. Compute

$$\langle q_1, q_2 \rangle = \int_{-2}^{2} 1 \cdot t \, dt = \left. \frac{1}{2}t^2 \right|_{-2}^{2} = 0$$

$$\langle q_1, q_3 \rangle = \int_{-2}^{2} 1 \cdot (3t^2 - 4) \, dt = \left. (t^3 - 4t) \right|_{-2}^{2} = 0$$

$$\langle q_2, q_3 \rangle = \int_{-2}^{2} t \cdot (3t^2 - 4) \, dt = \left. \left(\frac{3}{4}t^4 - 2t^2 \right) \right|_{-2}^{2} = 0$$

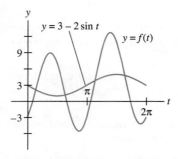

First- and third-order approximations to $f(t)$.

2. The third-order Fourier approximation to f is the best approximation in $C[0, 2\pi]$ to f by functions (vectors) in the subspace spanned by 1, $\cos t$, $\cos 2t$, $\cos 3t$, $\sin t$, $\sin 2t$, and $\sin 3t$. But f is obviously *in* this subspace, so f is its own best approximation:

$$f(t) = 3 - 2\sin t + 5\sin 2t - 6\cos 2t$$

For the first-order approximation, the closest function to f in the subspace $W = \text{Span}\{1, \cos t, \sin t\}$ is $3 - 2\sin t$. The other two terms in the formula for $f(t)$ are orthogonal to the functions in W, so they contribute nothing to the integrals that give the Fourier coefficients for a first-order approximation.

CHAPTER 6 SUPPLEMENTARY EXERCISES

1. The following statements refer to vectors in $\mathbb{R}^n$ (or $\mathbb{R}^m$) with the standard inner product. Mark each statement True or False. Justify each answer.

a. The length of every vector is a positive number.

b. A vector $\mathbf{v}$ and its negative $-\mathbf{v}$ have equal lengths.

c. The distance between $\mathbf{u}$ and $\mathbf{v}$ is $\|\mathbf{u} - \mathbf{v}\|$.

d. If r is any scalar, then $\|r\mathbf{v}\| = r\|\mathbf{v}\|$.

e. If two vectors are orthogonal, they are linearly independent.

f. If $\mathbf{x}$ is orthogonal to both $\mathbf{u}$ and $\mathbf{v}$, then $\mathbf{x}$ must be orthogonal to $\mathbf{u} - \mathbf{v}$.

g. If $\|\mathbf{u} + \mathbf{v}\|^2 = \|\mathbf{u}\|^2 + \|\mathbf{v}\|^2$, then $\mathbf{u}$ and $\mathbf{v}$ are orthogonal.

h. If $\|\mathbf{u} - \mathbf{v}\|^2 = \|\mathbf{u}\|^2 + \|\mathbf{v}\|^2$, then $\mathbf{u}$ and $\mathbf{v}$ are orthogonal.

i. The orthogonal projection of $\mathbf{y}$ onto $\mathbf{u}$ is a scalar multiple of $\mathbf{y}$.

j. If a vector $\mathbf{y}$ coincides with its orthogonal projection onto a subspace W, then $\mathbf{y}$ is in W.

k. The set of all vectors in $\mathbb{R}^n$ orthogonal to one fixed vector is a subspace of $\mathbb{R}^n$.

l. If W is a subspace of $\mathbb{R}^n$, then W and $W^\perp$ have no vectors in common.

m. If $\{\mathbf{v}_1, \mathbf{v}_2, \mathbf{v}_3\}$ is an orthogonal set and if c_1, c_2, and c_3 are scalars, then $\{c_1\mathbf{v}_1, c_2\mathbf{v}_2, c_3\mathbf{v}_3\}$ is an orthogonal set.

n. If a matrix U has orthonormal columns, then $UU^T = I$.

o. A square matrix with orthogonal columns is an orthogonal matrix.

p. If a square matrix has orthonormal columns, then it also has orthonormal rows.

q. If W is a subspace, then $\|\text{proj}_W \mathbf{v}\|^2 + \|\mathbf{v} - \text{proj}_W \mathbf{v}\|^2 = \|\mathbf{v}\|^2$.

r. A least-squares solution of $A\mathbf{x} = \mathbf{b}$ is the vector $A\hat{\mathbf{x}}$ in Col A closest to $\mathbf{b}$, so that $\|\mathbf{b} - A\hat{\mathbf{x}}\| \le \|\mathbf{b} - A\mathbf{x}\|$ for all $\mathbf{x}$.

s. The normal equations for a least-squares solution of $A\mathbf{x} = \mathbf{b}$ are given by $\hat{\mathbf{x}} = (A^TA)^{-1}A^T\mathbf{b}$.

2. Let $\{\mathbf{v}_1, \ldots, \mathbf{v}_p\}$ be an orthonormal set. Verify the following equality by induction, beginning with $p = 2$. If $\mathbf{x} = c_1\mathbf{v}_1 + \cdots + c_p\mathbf{v}_p$, then

$$\|\mathbf{x}\|^2 = |c_1|^2 + \cdots + |c_p|^2$$

3. Let $\{\mathbf{v}_1, \ldots, \mathbf{v}_p\}$ be an orthonormal set in $\mathbb{R}^n$. Verify the following inequality, called *Bessel's inequality*, which is true for each $\mathbf{x}$ in $\mathbb{R}^n$:

$$\|\mathbf{x}\|^2 \ge |\mathbf{x} \cdot \mathbf{v}_1|^2 + |\mathbf{x} \cdot \mathbf{v}_2|^2 + \cdots + |\mathbf{x} \cdot \mathbf{v}_p|^2$$

4. Let U be an $n \times n$ orthogonal matrix. Show that if $\{\mathbf{v}_1, \ldots, \mathbf{v}_n\}$ is an orthonormal basis for $\mathbb{R}^n$, then so is $\{U\mathbf{v}_1, \ldots, U\mathbf{v}_n\}$.

5. Show that if an $n \times n$ matrix U satisfies $(U\mathbf{x}) \cdot (U\mathbf{y}) = \mathbf{x} \cdot \mathbf{y}$ for all $\mathbf{x}$ and $\mathbf{y}$ in $\mathbb{R}^n$, then U is an orthogonal matrix.

6. Show that if U is an orthogonal matrix, then any real eigenvalue of U must be ± 1.

7. A *Householder matrix*, or an *elementary reflector*, has the form $Q = I - 2\mathbf{u}\mathbf{u}^T$ where $\mathbf{u}$ is a unit vector. (See Exercise 13 in the Supplementary Exercises for Chapter 2.) Show that Q is an orthogonal matrix. (Elementary reflectors are often used in computer programs to produce a QR factorization of a matrix A. If A has linearly independent columns, then left-multiplication by a sequence of elementary reflectors can produce an upper triangular matrix.)

8. Let $T : \mathbb{R}^n \to \mathbb{R}^n$ be a linear transformation that preserves lengths; that is, $\|T(\mathbf{x})\| = \|\mathbf{x}\|$ for all $\mathbf{x}$ in $\mathbb{R}^n$.

    a. Show that T also preserves orthogonality; that is, $T(\mathbf{x}) \cdot T(\mathbf{y}) = 0$ whenever $\mathbf{x} \cdot \mathbf{y} = 0$.

    b. Show that the standard matrix of T is an orthogonal matrix.

9. Let $\mathbf{u}$ and $\mathbf{v}$ be linearly independent vectors in $\mathbb{R}^n$ that are *not* orthogonal. Describe how to find the best approximation to $\mathbf{z}$ in $\mathbb{R}^n$ by vectors of the form $x_1\mathbf{u} + x_2\mathbf{v}$ without first constructing an orthogonal basis for Span $\{\mathbf{u}, \mathbf{v}\}$.

10. Suppose the columns of A are linearly independent. Determine what happens to the least-squares solution $\hat{\mathbf{x}}$ of $A\mathbf{x} = \mathbf{b}$ when $\mathbf{b}$ is replaced by $c\mathbf{b}$ for some nonzero scalar c.

11. If a, b, and c are distinct numbers, then the following system is inconsistent because the graphs of the equations are parallel planes. Show that the set of all least-squares solutions of the system is precisely the plane whose equation is $x - 2y + 5z = (a + b + c)/3$.

$$x - 2y + 5z = a$$
$$x - 2y + 5z = b$$
$$x - 2y + 5z = c$$

12. Consider the problem of finding an eigenvalue of an $n \times n$ matrix A when an approximate eigenvector $\mathbf{v}$ is known. Since $\mathbf{v}$ is not exactly correct, the equation

$$A\mathbf{v} = \lambda\mathbf{v} \tag{1}$$

will probably not have a solution. However, λ can be estimated by a least-squares solution when (1) is viewed properly. Think of $\mathbf{v}$ as an $n \times 1$ matrix V, think of λ as a vector in $\mathbb{R}^1$, and denote the vector $A\mathbf{v}$ by the symbol $\mathbf{b}$. Then (1) becomes $\mathbf{b} = \lambda V$, which may also be written as $V\lambda = \mathbf{b}$. Find the least-squares solution of this system of n equations in the one unknown λ, and write this solution using the original symbols. The resulting estimate for λ is called a *Rayleigh quotient*. See Exercises 11 and 12 in Section 5.8.

13. Use the steps below to prove the following relations among the four fundamental subspaces determined by an $m \times n$ matrix A.

$$\text{Row } A = (\text{Nul } A)^\perp, \quad \text{Col } A = (\text{Nul } A^T)^\perp$$

    a. Show that Row A is contained in $(\text{Nul } A)^\perp$. (Show that if $\mathbf{x}$ is in Row A, then $\mathbf{x}$ is orthogonal to every $\mathbf{u}$ in Nul A.)

    b. Suppose rank $A = r$. Find dim Nul A and dim $(\text{Nul } A)^\perp$, and then deduce from part (a) that Row $A = (\text{Nul } A)^\perp$. [*Hint:* Study the exercises for Section 6.3.]

    c. Explain why Col $A = (\text{Nul } A^T)^\perp$.

14. Explain why an equation $A\mathbf{x} = \mathbf{b}$ has a solution if and only if $\mathbf{b}$ is orthogonal to all solutions of the equation $A^T\mathbf{x} = \mathbf{0}$.

Exercises 15 and 16 concern the (real) *Schur factorization* of an $n \times n$ matrix A in the form $A = URU^T$, where U is an orthogonal matrix and R is an $n \times n$ upper triangular matrix.[1]

15. Show that if A admits a (real) Schur factorization, $A = URU^T$, then A has n real eigenvalues, counting multiplicities.

16. Let A be an $n \times n$ matrix with n real eigenvalues, counting multiplicities, denoted by $\lambda_1, \ldots, \lambda_n$. It can be shown that A admits a (real) Schur factorization. Parts (a) and (b) show the key ideas in the proof. The rest of the proof amounts to repeating (a) and (b) for successively smaller matrices, and then piecing together the results.

    a. Let $\mathbf{u}_1$ be a unit eigenvector corresponding to λ_1, let $\mathbf{u}_2, \ldots, \mathbf{u}_n$ be any other vectors such that $\{\mathbf{u}_1, \ldots, \mathbf{u}_n\}$ is an orthonormal basis for $\mathbb{R}^n$, and then let $U = [\,\mathbf{u}_1 \ \ \mathbf{u}_2 \ \ \cdots \ \ \mathbf{u}_n\,]$. Show that the first column of $U^T A U$ is $\lambda_1 \mathbf{e}_1$, where $\mathbf{e}_1$ is the first column of the $n \times n$ identity matrix.

    b. Part (a) implies that $U^T A U$ has the form shown below. Explain why the eigenvalues of A_1 are $\lambda_2, \ldots, \lambda_n$. [*Hint:* See the Supplementary Exercises for Chapter 5.]

$$U^T A U = \begin{bmatrix} \lambda_1 & * & * & * & * \\ 0 & & & & \\ \vdots & & & A_1 & \\ 0 & & & & \end{bmatrix}$$

[**M**] When the right side of an equation $A\mathbf{x} = \mathbf{b}$ is changed slightly—say, to $A\mathbf{x} = \mathbf{b} + \Delta\mathbf{b}$ for some vector $\Delta\mathbf{b}$—the solution changes from $\mathbf{x}$ to $\mathbf{x} + \Delta\mathbf{x}$, where $\Delta\mathbf{x}$ satisfies $A(\Delta\mathbf{x}) = \Delta\mathbf{b}$. The quotient $\|\Delta\mathbf{b}\|/\|\mathbf{b}\|$ is called the **relative change** in $\mathbf{b}$ (or the **relative error** in $\mathbf{b}$ when $\Delta\mathbf{b}$ represents possible error in the entries of $\mathbf{b}$). The relative change in the solution is $\|\Delta\mathbf{x}\|/\|\mathbf{x}\|$. When A is invertible, the **condition number** of A, written as cond(A), produces a bound on how large the relative change in $\mathbf{x}$ can be:

$$\frac{\|\Delta\mathbf{x}\|}{\|\mathbf{x}\|} \leq \text{cond}(A) \cdot \frac{\|\Delta\mathbf{b}\|}{\|\mathbf{b}\|} \tag{2}$$

In Exercises 17–20, solve $A\mathbf{x} = \mathbf{b}$ and $A(\Delta\mathbf{x}) = \Delta\mathbf{b}$, and show that the inequality (2) holds in each case. (See the discussion of *ill-conditioned* matrices in Exercises 41–43 in Section 2.3.)

17. $A = \begin{bmatrix} 4.5 & 3.1 \\ 1.6 & 1.1 \end{bmatrix}$, $\mathbf{b} = \begin{bmatrix} 19.249 \\ 6.843 \end{bmatrix}$, $\Delta\mathbf{b} = \begin{bmatrix} .001 \\ -.003 \end{bmatrix}$

18. $A = \begin{bmatrix} 4.5 & 3.1 \\ 1.6 & 1.1 \end{bmatrix}$, $\mathbf{b} = \begin{bmatrix} .500 \\ -1.407 \end{bmatrix}$, $\Delta\mathbf{b} = \begin{bmatrix} .001 \\ -.003 \end{bmatrix}$

---

[1] If complex numbers are allowed, *every* $n \times n$ matrix A admits a (complex) Schur factorization, $A = URU^{-1}$, where R is upper triangular and U^{-1} is the *conjugate* transpose of U. This very useful fact is discussed in *Matrix Analysis*, by Roger A. Horn and Charles R. Johnson (Cambridge: Cambridge University Press, 1985), pp. 79–100.

19. $A = \begin{bmatrix} 7 & -6 & -4 & 1 \\ -5 & 1 & 0 & -2 \\ 10 & 11 & 7 & -3 \\ 19 & 9 & 7 & 1 \end{bmatrix}$, $\mathbf{b} = \begin{bmatrix} .100 \\ 2.888 \\ -1.404 \\ 1.462 \end{bmatrix}$,

$\Delta\mathbf{b} = 10^{-4} \begin{bmatrix} .49 \\ -1.28 \\ 5.78 \\ 8.04 \end{bmatrix}$

20. $A = \begin{bmatrix} 7 & -6 & -4 & 1 \\ -5 & 1 & 0 & -2 \\ 10 & 11 & 7 & -3 \\ 19 & 9 & 7 & 1 \end{bmatrix}$, $\mathbf{b} = \begin{bmatrix} 4.230 \\ -11.043 \\ 49.991 \\ 69.536 \end{bmatrix}$,

$\Delta\mathbf{b} = 10^{-4} \begin{bmatrix} .27 \\ 7.76 \\ -3.77 \\ 3.93 \end{bmatrix}$

7

Symmetric Matrices and Quadratic Forms

Multichannel Image Processing

Around the world in little more than 80 *minutes*, the two Landsat satellites streak silently across the sky in near polar orbits, recording images of terrain and coastline, in swaths 185 kilometers wide. Every 16 days, each satellite passes over almost every square kilometer of the earth's surface, so any location can be monitored every 8 days.

The Landsat images are useful for many purposes. Developers and urban planners use them to study the rate and direction of urban growth, industrial development, and other changes in land usage. Rural countries can analyze soil moisture, classify the vegetation in remote regions, and locate inland lakes and streams. Governments can detect and assess damage from natural disasters, such as forest fires, lava flows, floods, and hurricanes. Environmental agencies can identify pollution from smokestacks and measure water temperatures in lakes and rivers near power plants.

Sensors aboard the satellite acquire seven simultaneous images of any region on earth to be studied. The sensors record energy from separate wavelength bands — three in the visible light spectrum and four in infrared and thermal bands. Each image is digitized and stored as a rectangular array of numbers, each number indicating the signal intensity at a corresponding small point (or *pixel*)

on the image. Each of the seven images is one channel of a *multichannel* or *multispectral image*.

The seven Landsat images of one fixed region typically contain much redundant information, since some features will appear in several images. Yet other features, because of their color or temperature, may reflect light that is recorded by only one or two sensors. One goal of multichannel image processing is to view the data in a way that extracts information better than studying each image separately.

Principal component analysis is an effective way to suppress redundant information and provide in only one or two composite images most of the information from the initial data. Roughly speaking, the goal is to find a special linear combination of the images, that is, a list of weights that at each pixel combine all seven corresponding image values into one new value. The weights are chosen in a way that makes the range of light intensities — the *scene variance* — in the composite image (called the *first principal component*) greater than that in any of the original images. Additional *component* images can also be constructed, by criteria that will be explained in Section 7.5.

Principal component analysis is illustrated in the photos below, taken over Railroad Valley, Nevada. Images from three Landsat spectral bands are shown in (a)–(c). The total information in the three bands is rearranged in the three principal component images in (d)–(f). The first component (d) displays (or "explains") 93.5% of the scene variance present in the initial data. In this way, the three-channel initial data have been reduced to one-channel data, with a loss in some sense of only 6.5% of the scene variance.

Earth Satellite Corporation of Rockville, Maryland, which kindly supplied the photos shown here, is experimenting with images from 224 separate spectral bands. Principal component analysis, essential for such massive data sets, typically reduces the data to about 15 usable principal components.

WEB

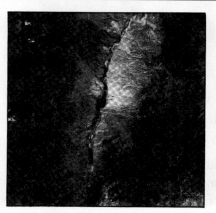

(a) Spectral band 1: Visible blue.

(b) Spectral band 4: Near infrared.

(c) Spectral band 7: Mid-infrared.

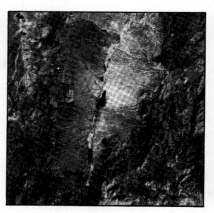

(d) Principal component 1: 93.5%.

(e) Principal component 2: 5.3%.

(f) Principal component 3: 1.2%.

Symmetric matrices arise more often in applications, in one way or another, than any other major class of matrices. The theory is rich and beautiful, depending in an essential way on both diagonalization from Chapter 5 and orthogonality from Chapter 6. The diagonalization of a symmetric matrix, described in Section 7.1, is the foundation for the discussion in Sections 7.2 and 7.3 concerning quadratic forms. Section 7.3, in turn, is needed for the final two sections on the singular value decomposition and on the image processing described in the introductory example. Throughout the chapter, all vectors and matrices have real entries.

7.1 | DIAGONALIZATION OF SYMMETRIC MATRICES

A **symmetric** matrix is a matrix A such that $A^T = A$. Such a matrix is necessarily square. Its main diagonal entries are arbitrary, but its other entries occur in pairs—on opposite sides of the main diagonal.

EXAMPLE 1   Of the following matrices, only the first three are symmetric:

$$\text{Symmetric:} \quad \begin{bmatrix} 1 & 0 \\ 0 & -3 \end{bmatrix}, \quad \begin{bmatrix} 0 & -1 & 0 \\ -1 & 5 & 8 \\ 0 & 8 & -7 \end{bmatrix}, \quad \begin{bmatrix} a & b & c \\ b & d & e \\ c & e & f \end{bmatrix}$$

$$\text{Nonsymmetric:} \quad \begin{bmatrix} 1 & -3 \\ 3 & 0 \end{bmatrix}, \quad \begin{bmatrix} 1 & -4 & 0 \\ -6 & 1 & -4 \\ 0 & -6 & 1 \end{bmatrix}, \quad \begin{bmatrix} 5 & 4 & 3 & 2 \\ 4 & 3 & 2 & 1 \\ 3 & 2 & 1 & 0 \end{bmatrix} \quad \blacksquare$$

To begin the study of symmetric matrices, it is helpful to review the diagonalization process of Section 5.3.

EXAMPLE 2   If possible, diagonalize the matrix $A = \begin{bmatrix} 6 & -2 & -1 \\ -2 & 6 & -1 \\ -1 & -1 & 5 \end{bmatrix}$.

SOLUTION   The characteristic equation of A is

$$0 = -\lambda^3 + 17\lambda^2 - 90\lambda + 144 = -(\lambda - 8)(\lambda - 6)(\lambda - 3)$$

Standard calculations produce a basis for each eigenspace:

$$\lambda = 8: \ \mathbf{v}_1 = \begin{bmatrix} -1 \\ 1 \\ 0 \end{bmatrix}; \quad \lambda = 6: \ \mathbf{v}_2 = \begin{bmatrix} -1 \\ -1 \\ 2 \end{bmatrix}; \quad \lambda = 3: \ \mathbf{v}_3 = \begin{bmatrix} 1 \\ 1 \\ 1 \end{bmatrix}$$

These three vectors form a basis for $\mathbb{R}^3$. In fact, it is easy to check that $\{\mathbf{v}_1, \mathbf{v}_2, \mathbf{v}_3\}$ is an *orthogonal* basis for $\mathbb{R}^3$. Experience from Chapter 6 suggests that an *orthonormal* basis might be useful for calculations, so here are the normalized (unit) eigenvectors.

$$\mathbf{u}_1 = \begin{bmatrix} -1/\sqrt{2} \\ 1/\sqrt{2} \\ 0 \end{bmatrix}, \quad \mathbf{u}_2 = \begin{bmatrix} -1/\sqrt{6} \\ -1/\sqrt{6} \\ 2/\sqrt{6} \end{bmatrix}, \quad \mathbf{u}_3 = \begin{bmatrix} 1/\sqrt{3} \\ 1/\sqrt{3} \\ 1/\sqrt{3} \end{bmatrix}$$

Let

$$P = \begin{bmatrix} -1/\sqrt{2} & -1/\sqrt{6} & 1/\sqrt{3} \\ 1/\sqrt{2} & -1/\sqrt{6} & 1/\sqrt{3} \\ 0 & 2/\sqrt{6} & 1/\sqrt{3} \end{bmatrix}, \quad D = \begin{bmatrix} 8 & 0 & 0 \\ 0 & 6 & 0 \\ 0 & 0 & 3 \end{bmatrix}$$

Then $A = PDP^{-1}$, as usual. But this time, since P is square and has orthonormal columns, P is an *orthogonal* matrix, and P^{-1} is simply P^T. (See Section 6.2.) $\blacksquare$

Theorem 1 explains why the eigenvectors in Example 2 are orthogonal—they correspond to distinct eigenvalues.

THEOREM 1

> If A is symmetric, then any two eigenvectors from different eigenspaces are orthogonal.

PROOF Let $\mathbf{v}_1$ and $\mathbf{v}_2$ be eigenvectors that correspond to distinct eigenvalues, say, λ_1 and λ_2. To show that $\mathbf{v}_1 \cdot \mathbf{v}_2 = 0$, compute

$$\lambda_1 \mathbf{v}_1 \cdot \mathbf{v}_2 = (\lambda_1 \mathbf{v}_1)^T \mathbf{v}_2 = (A\mathbf{v}_1)^T \mathbf{v}_2 \qquad \text{Since } \mathbf{v}_1 \text{ is an eigenvector}$$
$$= (\mathbf{v}_1^T A^T)\mathbf{v}_2 = \mathbf{v}_1^T (A\mathbf{v}_2) \qquad \text{Since } A^T = A$$
$$= \mathbf{v}_1^T (\lambda_2 \mathbf{v}_2) \qquad\qquad \text{Since } \mathbf{v}_2 \text{ is an eigenvector}$$
$$= \lambda_2 \mathbf{v}_1^T \mathbf{v}_2 = \lambda_2 \mathbf{v}_1 \cdot \mathbf{v}_2$$

Hence $(\lambda_1 - \lambda_2)\mathbf{v}_1 \cdot \mathbf{v}_2 = 0$. But $\lambda_1 - \lambda_2 \neq 0$, so $\mathbf{v}_1 \cdot \mathbf{v}_2 = 0$. ∎

The special type of diagonalization in Example 2 is crucial for the theory of symmetric matrices. An $n \times n$ matrix A is said to be **orthogonally diagonalizable** if there are an orthogonal matrix P (with $P^{-1} = P^T$) and a diagonal matrix D such that

$$A = PDP^T = PDP^{-1} \tag{1}$$

Such a diagonalization requires n linearly independent and orthonormal eigenvectors. When is this possible? If A is orthogonally diagonalizable as in (1), then

$$A^T = (PDP^T)^T = P^{TT} D^T P^T = PDP^T = A$$

Thus A is symmetric! Theorem 2 below shows that, conversely, every symmetric matrix is orthogonally diagonalizable. The proof is much harder and is omitted; the main idea for a proof will be given after Theorem 3.

THEOREM 2

An $n \times n$ matrix A is orthogonally diagonalizable if and only if A is a symmetric matrix.

This theorem is rather amazing, because the work in Chapter 5 would suggest that it is usually impossible to tell when a matrix is diagonalizable. But this is not the case for symmetric matrices.

The next example treats a matrix whose eigenvalues are not all distinct.

EXAMPLE 3   Orthogonally diagonalize the matrix $A = \begin{bmatrix} 3 & -2 & 4 \\ -2 & 6 & 2 \\ 4 & 2 & 3 \end{bmatrix}$, whose characteristic equation is

$$0 = -\lambda^3 + 12\lambda^2 - 21\lambda - 98 = -(\lambda - 7)^2(\lambda + 2)$$

SOLUTION The usual calculations produce bases for the eigenspaces:

$$\lambda = 7: \mathbf{v}_1 = \begin{bmatrix} 1 \\ 0 \\ 1 \end{bmatrix}, \mathbf{v}_2 = \begin{bmatrix} -1/2 \\ 1 \\ 0 \end{bmatrix}; \qquad \lambda = -2: \mathbf{v}_3 = \begin{bmatrix} -1 \\ -1/2 \\ 1 \end{bmatrix}$$

Although $\mathbf{v}_1$ and $\mathbf{v}_2$ are linearly independent, they are not orthogonal. Recall from Section 6.2 that the projection of $\mathbf{v}_2$ onto $\mathbf{v}_1$ is $\dfrac{\mathbf{v}_2 \cdot \mathbf{v}_1}{\mathbf{v}_1 \cdot \mathbf{v}_1}\mathbf{v}_1$, and the component of $\mathbf{v}_2$ orthogonal to $\mathbf{v}_1$ is

$$\mathbf{z}_2 = \mathbf{v}_2 - \frac{\mathbf{v}_2 \cdot \mathbf{v}_1}{\mathbf{v}_1 \cdot \mathbf{v}_1}\mathbf{v}_1 = \begin{bmatrix} -1/2 \\ 1 \\ 0 \end{bmatrix} - \frac{-1/2}{2}\begin{bmatrix} 1 \\ 0 \\ 1 \end{bmatrix} = \begin{bmatrix} -1/4 \\ 1 \\ 1/4 \end{bmatrix}$$

Then $\{\mathbf{v}_1, \mathbf{z}_2\}$ is an orthogonal set in the eigenspace for $\lambda = 7$. (Note that $\mathbf{z}_2$ is a linear combination of the eigenvectors $\mathbf{v}_1$ and $\mathbf{v}_2$, so $\mathbf{z}_2$ is in the eigenspace. This construction of $\mathbf{z}_2$ is just the Gram–Schmidt process of Section 6.4.) Since the eigenspace is two-dimensional (with basis $\mathbf{v}_1, \mathbf{v}_2$), the orthogonal set $\{\mathbf{v}_1, \mathbf{z}_2\}$ is an *orthogonal basis* for the eigenspace, by the Basis Theorem. (See Section 2.9 or 4.5.)

Normalize $\mathbf{v}_1$ and $\mathbf{z}_2$ to obtain the following orthonormal basis for the eigenspace for $\lambda = 7$:

$$\mathbf{u}_1 = \begin{bmatrix} 1/\sqrt{2} \\ 0 \\ 1/\sqrt{2} \end{bmatrix}, \quad \mathbf{u}_2 = \begin{bmatrix} -1/\sqrt{18} \\ 4/\sqrt{18} \\ 1/\sqrt{18} \end{bmatrix}$$

An orthonormal basis for the eigenspace for $\lambda = -2$ is

$$\mathbf{u}_3 = \frac{1}{\|2\mathbf{v}_3\|} 2\mathbf{v}_3 = \frac{1}{3} \begin{bmatrix} -2 \\ -1 \\ 2 \end{bmatrix} = \begin{bmatrix} -2/3 \\ -1/3 \\ 2/3 \end{bmatrix}$$

By Theorem 1, $\mathbf{u}_3$ is orthogonal to the other eigenvectors $\mathbf{u}_1$ and $\mathbf{u}_2$. Hence $\{\mathbf{u}_1, \mathbf{u}_2, \mathbf{u}_3\}$ is an orthonormal set. Let

$$P = [\,\mathbf{u}_1 \quad \mathbf{u}_2 \quad \mathbf{u}_3\,] = \begin{bmatrix} 1/\sqrt{2} & -1/\sqrt{18} & -2/3 \\ 0 & 4/\sqrt{18} & -1/3 \\ 1/\sqrt{2} & 1/\sqrt{18} & 2/3 \end{bmatrix}, \quad D = \begin{bmatrix} 7 & 0 & 0 \\ 0 & 7 & 0 \\ 0 & 0 & -2 \end{bmatrix}$$

Then P orthogonally diagonalizes A, and $A = PDP^{-1}$. ■

In Example 3, the eigenvalue 7 has multiplicity two and the eigenspace is two-dimensional. This fact is not accidental, as the next theorem shows.

The Spectral Theorem

The set of eigenvalues of a matrix A is sometimes called the *spectrum* of A, and the following description of the eigenvalues is called a *spectral theorem*.

THEOREM 3

The Spectral Theorem for Symmetric Matrices

An $n \times n$ symmetric matrix A has the following properties:

a. A has n real eigenvalues, counting multiplicities.

b. The dimension of the eigenspace for each eigenvalue λ equals the multiplicity of λ as a root of the characteristic equation.

c. The eigenspaces are mutually orthogonal, in the sense that eigenvectors corresponding to different eigenvalues are orthogonal.

d. A is orthogonally diagonalizable.

Part (a) follows from Exercise 24 in Section 5.5. Part (b) follows easily from part (d). (See Exercise 31.) Part (c) is Theorem 1. Because of (a), a proof of (d) can be given using Exercise 32 and the Schur factorization discussed in Supplementary Exercise 16 in Chapter 6. The details are omitted.

Spectral Decomposition

Suppose $A = PDP^{-1}$, where the columns of P are orthonormal eigenvectors $\mathbf{u}_1, \ldots, \mathbf{u}_n$ of A and the corresponding eigenvalues $\lambda_1, \ldots, \lambda_n$ are in the diagonal matrix D. Then, since $P^{-1} = P^T$,

$$A = PDP^T = \begin{bmatrix} \mathbf{u}_1 & \cdots & \mathbf{u}_n \end{bmatrix} \begin{bmatrix} \lambda_1 & & 0 \\ & \ddots & \\ 0 & & \lambda_n \end{bmatrix} \begin{bmatrix} \mathbf{u}_1^T \\ \vdots \\ \mathbf{u}_n^T \end{bmatrix}$$

$$= \begin{bmatrix} \lambda_1 \mathbf{u}_1 & \cdots & \lambda_n \mathbf{u}_n \end{bmatrix} \begin{bmatrix} \mathbf{u}_1^T \\ \vdots \\ \mathbf{u}_n^T \end{bmatrix}$$

Using the column–row expansion of a product (Theorem 10 in Section 2.4), we can write

$$A = \lambda_1 \mathbf{u}_1 \mathbf{u}_1^T + \lambda_2 \mathbf{u}_2 \mathbf{u}_2^T + \cdots + \lambda_n \mathbf{u}_n \mathbf{u}_n^T \tag{2}$$

This representation of A is called a **spectral decomposition** of A because it breaks up A into pieces determined by the spectrum (eigenvalues) of A. Each term in (2) is an $n \times n$ matrix of rank 1. For example, every column of $\lambda_1 \mathbf{u}_1 \mathbf{u}_1^T$ is a multiple of $\mathbf{u}_1$. Furthermore, each matrix $\mathbf{u}_j \mathbf{u}_j^T$ is a **projection matrix** in the sense that for each $\mathbf{x}$ in $\mathbb{R}^n$, the vector $(\mathbf{u}_j \mathbf{u}_j^T)\mathbf{x}$ is the orthogonal projection of $\mathbf{x}$ onto the subspace spanned by $\mathbf{u}_j$. (See Exercise 35.)

EXAMPLE 4 Construct a spectral decomposition of the matrix A that has the orthogonal diagonalization

$$A = \begin{bmatrix} 7 & 2 \\ 2 & 4 \end{bmatrix} = \begin{bmatrix} 2/\sqrt{5} & -1/\sqrt{5} \\ 1/\sqrt{5} & 2/\sqrt{5} \end{bmatrix} \begin{bmatrix} 8 & 0 \\ 0 & 3 \end{bmatrix} \begin{bmatrix} 2/\sqrt{5} & 1/\sqrt{5} \\ -1/\sqrt{5} & 2/\sqrt{5} \end{bmatrix}$$

SOLUTION Denote the columns of P by $\mathbf{u}_1$ and $\mathbf{u}_2$. Then

$$A = 8\mathbf{u}_1 \mathbf{u}_1^T + 3\mathbf{u}_2 \mathbf{u}_2^T$$

To verify this decomposition of A, compute

$$\mathbf{u}_1 \mathbf{u}_1^T = \begin{bmatrix} 2/\sqrt{5} \\ 1/\sqrt{5} \end{bmatrix} \begin{bmatrix} 2/\sqrt{5} & 1/\sqrt{5} \end{bmatrix} = \begin{bmatrix} 4/5 & 2/5 \\ 2/5 & 1/5 \end{bmatrix}$$

$$\mathbf{u}_2 \mathbf{u}_2^T = \begin{bmatrix} -1/\sqrt{5} \\ 2/\sqrt{5} \end{bmatrix} \begin{bmatrix} -1/\sqrt{5} & 2/\sqrt{5} \end{bmatrix} = \begin{bmatrix} 1/5 & -2/5 \\ -2/5 & 4/5 \end{bmatrix}$$

and

$$8\mathbf{u}_1 \mathbf{u}_1^T + 3\mathbf{u}_2 \mathbf{u}_2^T = \begin{bmatrix} 32/5 & 16/5 \\ 16/5 & 8/5 \end{bmatrix} + \begin{bmatrix} 3/5 & -6/5 \\ -6/5 & 12/5 \end{bmatrix} = \begin{bmatrix} 7 & 2 \\ 2 & 4 \end{bmatrix} = A \qquad \blacksquare$$

NUMERICAL NOTE

When A is symmetric and not too large, modern high-performance computer algorithms calculate eigenvalues and eigenvectors with great precision. They apply a sequence of similarity transformations to A involving orthogonal matrices. The diagonal entries of the transformed matrices converge rapidly to the eigenvalues of A. (See the Numerical Notes in Section 5.2.) Using orthogonal matrices generally prevents numerical errors from accumulating during the process. When A is symmetric, the sequence of orthogonal matrices combines to form an orthogonal matrix whose columns are eigenvectors of A.

A nonsymmetric matrix cannot have a full set of orthogonal eigenvectors, but the algorithm still produces fairly accurate eigenvalues. After that, nonorthogonal techniques are needed to calculate eigenvectors.

PRACTICE PROBLEMS

1. Show that if A is a symmetric matrix, then A^2 is symmetric.
2. Show that if A is orthogonally diagonalizable, then so is A^2.

7.1 EXERCISES

Determine which of the matrices in Exercises 1–6 are symmetric.

1. $\begin{bmatrix} 3 & 5 \\ 5 & -7 \end{bmatrix}$

2. $\begin{bmatrix} -3 & 5 \\ -5 & 3 \end{bmatrix}$

3. $\begin{bmatrix} 2 & 2 \\ 4 & 4 \end{bmatrix}$

4. $\begin{bmatrix} 0 & 8 & 3 \\ 8 & 0 & -2 \\ 3 & -2 & 0 \end{bmatrix}$

5. $\begin{bmatrix} -6 & 2 & 0 \\ 0 & -6 & 2 \\ 0 & 0 & -6 \end{bmatrix}$

6. $\begin{bmatrix} 1 & 2 & 1 & 2 \\ 2 & 1 & 2 & 1 \\ 1 & 2 & 1 & 2 \end{bmatrix}$

Determine which of the matrices in Exercises 7–12 are orthogonal. If orthogonal, find the inverse.

7. $\begin{bmatrix} .6 & .8 \\ .8 & -.6 \end{bmatrix}$

8. $\begin{bmatrix} 1/\sqrt{2} & -1/\sqrt{2} \\ 1/\sqrt{2} & 1/\sqrt{2} \end{bmatrix}$

9. $\begin{bmatrix} -5 & 2 \\ 2 & 5 \end{bmatrix}$

10. $\begin{bmatrix} -1 & 2 & 2 \\ 2 & -1 & 2 \\ 2 & 2 & -1 \end{bmatrix}$

11. $\begin{bmatrix} 2/3 & 2/3 & 1/3 \\ 0 & 1/\sqrt{5} & -2/\sqrt{5} \\ \sqrt{5}/3 & -4/\sqrt{45} & -2/\sqrt{45} \end{bmatrix}$

12. $\begin{bmatrix} .5 & .5 & -.5 & -.5 \\ -.5 & .5 & -.5 & .5 \\ .5 & .5 & .5 & .5 \\ -.5 & .5 & .5 & -.5 \end{bmatrix}$

Orthogonally diagonalize the matrices in Exercises 13–22, giving an orthogonal matrix P and a diagonal matrix D. To save you time, the eigenvalues in Exercises 17–22 are: (17) 5, 2, −2; (18) 25, 3, −50; (19) 7, −2; (20) 13, 7, 1; (21) 9, 5, 1; (22) 2, 0.

13. $\begin{bmatrix} 3 & 1 \\ 1 & 3 \end{bmatrix}$

14. $\begin{bmatrix} 1 & 5 \\ 5 & 1 \end{bmatrix}$

15. $\begin{bmatrix} 16 & -4 \\ -4 & 1 \end{bmatrix}$

16. $\begin{bmatrix} -7 & 24 \\ 24 & 7 \end{bmatrix}$

17. $\begin{bmatrix} 1 & 1 & 3 \\ 1 & 3 & 1 \\ 3 & 1 & 1 \end{bmatrix}$

18. $\begin{bmatrix} -2 & -36 & 0 \\ -36 & -23 & 0 \\ 0 & 0 & 3 \end{bmatrix}$

19. $\begin{bmatrix} 3 & -2 & 4 \\ -2 & 6 & 2 \\ 4 & 2 & 3 \end{bmatrix}$

20. $\begin{bmatrix} 7 & -4 & 4 \\ -4 & 5 & 0 \\ 4 & 0 & 9 \end{bmatrix}$

21. $\begin{bmatrix} 4 & 1 & 3 & 1 \\ 1 & 4 & 1 & 3 \\ 3 & 1 & 4 & 1 \\ 1 & 3 & 1 & 4 \end{bmatrix}$

22. $\begin{bmatrix} 2 & 0 & 0 & 0 \\ 0 & 1 & 0 & 1 \\ 0 & 0 & 2 & 0 \\ 0 & 1 & 0 & 1 \end{bmatrix}$

23. Let $A = \begin{bmatrix} 3 & 1 & 1 \\ 1 & 3 & 1 \\ 1 & 1 & 3 \end{bmatrix}$ and $\mathbf{v} = \begin{bmatrix} 1 \\ 1 \\ 1 \end{bmatrix}$. Verify that 2 is an eigenvalue of A and $\mathbf{v}$ is an eigenvector. Then orthogonally diagonalize A.

24. Let $A = \begin{bmatrix} 5 & -4 & -2 \\ -4 & 5 & 2 \\ -2 & 2 & 2 \end{bmatrix}$, $\mathbf{v}_1 = \begin{bmatrix} -2 \\ 2 \\ 1 \end{bmatrix}$, and $\mathbf{v}_2 = \begin{bmatrix} 1 \\ 1 \\ 0 \end{bmatrix}$. Verify that $\mathbf{v}_1$ and $\mathbf{v}_2$ are eigenvectors of A. Then orthogonally diagonalize A.

In Exercises 25 and 26, mark each statement True or False. Justify each answer.

25. a. An $n \times n$ matrix that is orthogonally diagonalizable must be symmetric.

b. If $A^T = A$ and if vectors $\mathbf{u}$ and $\mathbf{v}$ satisfy $A\mathbf{u} = 3\mathbf{u}$ and $A\mathbf{v} = 4\mathbf{v}$, then $\mathbf{u} \cdot \mathbf{v} = 0$.

c. An $n \times n$ symmetric matrix has n distinct real eigenvalues.

d. For a nonzero $\mathbf{v}$ in $\mathbb{R}^n$, the matrix $\mathbf{v}\mathbf{v}^T$ is called a projection matrix.

26. a. Every symmetric matrix is orthogonally diagonalizable.

b. If $B = PDP^T$, where $P^T = P^{-1}$ and D is a diagonal matrix, then B is a symmetric matrix.

c. An orthogonal matrix is orthogonally diagonalizable.

d. The dimension of an eigenspace of a symmetric matrix equals the multiplicity of the corresponding eigenvalue.

27. Suppose A is a symmetric $n \times n$ matrix and B is any $n \times m$ matrix. Show that $B^T AB$, $B^T B$, and BB^T are symmetric matrices.

28. Show that if A is an $n \times n$ symmetric matrix, then $(A\mathbf{x}) \cdot \mathbf{y} = \mathbf{x} \cdot (A\mathbf{y})$ for all $\mathbf{x}, \mathbf{y}$ in $\mathbb{R}^n$.

29. Suppose A is invertible and orthogonally diagonalizable. Explain why A^{-1} is also orthogonally diagonalizable.

30. Suppose A and B are both orthogonally diagonalizable and $AB = BA$. Explain why AB is also orthogonally diagonalizable.

31. Let $A = PDP^{-1}$, where P is orthogonal and D is diagonal, and let λ be an eigenvalue of A of multiplicity k. Then λ appears k times on the diagonal of D. Explain why the dimension of the eigenspace for λ is k.

32. Suppose $A = PRP^{-1}$, where P is orthogonal and R is upper triangular. Show that if A is symmetric, then R is symmetric and hence is actually a diagonal matrix.

33. Construct a spectral decomposition of A from Example 2.

34. Construct a spectral decomposition of A from Example 3.

35. Let $\mathbf{u}$ be a unit vector in $\mathbb{R}^n$, and let $B = \mathbf{u}\mathbf{u}^T$.

a. Given any $\mathbf{x}$ in $\mathbb{R}^n$, compute $B\mathbf{x}$ and show that $B\mathbf{x}$ is the orthogonal projection of $\mathbf{x}$ onto $\mathbf{u}$, as described in Section 6.2.

b. Show that B is a symmetric matrix and $B^2 = B$.

c. Show that $\mathbf{u}$ is an eigenvector of B. What is the corresponding eigenvalue?

36. Let B be an $n \times n$ symmetric matrix such that $B^2 = B$. Any such matrix is called a **projection matrix** (or an **orthogonal projection matrix**). Given any $\mathbf{y}$ in $\mathbb{R}^n$, let $\hat{\mathbf{y}} = B\mathbf{y}$ and $\mathbf{z} = \mathbf{y} - \hat{\mathbf{y}}$.

a. Show that $\mathbf{z}$ is orthogonal to $\hat{\mathbf{y}}$.

b. Let W be the column space of B. Show that $\mathbf{y}$ is the sum of a vector in W and a vector in $W^\perp$. Why does this prove that $B\mathbf{y}$ is the orthogonal projection of $\mathbf{y}$ onto the column space of B?

[M] Orthogonally diagonalize the matrices in Exercises 37–40. To practice the methods of this section, do not use an eigenvector routine from your matrix program. Instead, use the program to find the eigenvalues, and, for each eigenvalue λ, find an orthonormal basis for $\text{Nul}(A - \lambda I)$, as in Examples 2 and 3.

37. $\begin{bmatrix} 5 & 2 & 9 & -6 \\ 2 & 5 & -6 & 9 \\ 9 & -6 & 5 & 2 \\ -6 & 9 & 2 & 5 \end{bmatrix}$

38. $\begin{bmatrix} .38 & -.18 & -.06 & -.04 \\ -.18 & .59 & -.04 & .12 \\ -.06 & -.04 & .47 & -.12 \\ -.04 & .12 & -.12 & .41 \end{bmatrix}$

39. $\begin{bmatrix} .31 & .58 & .08 & .44 \\ .58 & -.56 & .44 & -.58 \\ .08 & .44 & .19 & -.08 \\ .44 & -.58 & -.08 & .31 \end{bmatrix}$

40. $\begin{bmatrix} 10 & 2 & 2 & -6 & 9 \\ 2 & 10 & 2 & -6 & 9 \\ 2 & 2 & 10 & -6 & 9 \\ -6 & -6 & -6 & 26 & 9 \\ 9 & 9 & 9 & 9 & -19 \end{bmatrix}$

SOLUTIONS TO PRACTICE PROBLEMS

1. $(A^2)^T = (AA)^T = A^T A^T$, by a property of transposes. By hypothesis, $A^T = A$. So $(A^2)^T = AA = A^2$, which shows that A^2 is symmetric.

2. If A is orthogonally diagonalizable, then A is symmetric, by Theorem 2. By Practice Problem 1, A^2 is symmetric and hence is orthogonally diagonalizable (Theorem 2).

7.2 │ QUADRATIC FORMS

Until now, our attention in this text has focused on linear equations, except for the sums of squares encountered in Chapter 6 when computing $\mathbf{x}^T\mathbf{x}$. Such sums and more general expressions, called *quadratic forms*, occur frequently in applications of linear algebra to engineering (in design criteria and optimization) and signal processing (as output noise power). They also arise, for example, in physics (as potential and kinetic energy), differential geometry (as normal curvature of surfaces), economics (as utility functions), and statistics (in confidence ellipsoids). Some of the mathematical background for such applications flows easily from our work on symmetric matrices.

A **quadratic form** on $\mathbb{R}^n$ is a function Q defined on $\mathbb{R}^n$ whose value at a vector $\mathbf{x}$ in $\mathbb{R}^n$ can be computed by an expression of the form $Q(\mathbf{x}) = \mathbf{x}^TA\mathbf{x}$, where A is an $n \times n$ symmetric matrix. The matrix A is called the **matrix of the quadratic form**.

The simplest example of a nonzero quadratic form is $Q(\mathbf{x}) = \mathbf{x}^TI\mathbf{x} = \|\mathbf{x}\|^2$. Examples 1 and 2 show the connection between any symmetric matrix A and the quadratic form $\mathbf{x}^TA\mathbf{x}$.

EXAMPLE 1   Let $\mathbf{x} = \begin{bmatrix} x_1 \\ x_2 \end{bmatrix}$. Compute $\mathbf{x}^TA\mathbf{x}$ for the following matrices:

a. $A = \begin{bmatrix} 4 & 0 \\ 0 & 3 \end{bmatrix}$ b. $A = \begin{bmatrix} 3 & -2 \\ -2 & 7 \end{bmatrix}$

SOLUTION

a. $\mathbf{x}^TA\mathbf{x} = \begin{bmatrix} x_1 & x_2 \end{bmatrix} \begin{bmatrix} 4 & 0 \\ 0 & 3 \end{bmatrix} \begin{bmatrix} x_1 \\ x_2 \end{bmatrix} = \begin{bmatrix} x_1 & x_2 \end{bmatrix} \begin{bmatrix} 4x_1 \\ 3x_2 \end{bmatrix} = 4x_1^2 + 3x_2^2.$

b. There are two -2 entries in A. Watch how they enter the calculations. The $(1, 2)$-entry in A is in boldface type.

$$\mathbf{x}^TA\mathbf{x} = \begin{bmatrix} x_1 & x_2 \end{bmatrix} \begin{bmatrix} 3 & \mathbf{-2} \\ -2 & 7 \end{bmatrix} \begin{bmatrix} x_1 \\ x_2 \end{bmatrix} = \begin{bmatrix} x_1 & x_2 \end{bmatrix} \begin{bmatrix} 3x_1 - \mathbf{2}x_2 \\ -2x_1 + 7x_2 \end{bmatrix}$$

$$= x_1(3x_1 - \mathbf{2}x_2) + x_2(-2x_1 + 7x_2)$$

$$= 3x_1^2 - \mathbf{2}x_1x_2 - 2x_2x_1 + 7x_2^2$$

$$= 3x_1^2 - 4x_1x_2 + 7x_2^2 \qquad \blacksquare$$

The presence of $-4x_1x_2$ in the quadratic form in Example 1(b) is due to the -2 entries off the diagonal in the matrix A. In contrast, the quadratic form associated with the diagonal matrix A in Example 1(a) has no x_1x_2 *cross-product* term.

EXAMPLE 2   For $\mathbf{x}$ in $\mathbb{R}^3$, let $Q(\mathbf{x}) = 5x_1^2 + 3x_2^2 + 2x_3^2 - x_1x_2 + 8x_2x_3$. Write this quadratic form as $\mathbf{x}^TA\mathbf{x}$.

SOLUTION The coefficients of x_1^2, x_2^2, x_3^2 go on the diagonal of A. To make A symmetric, the coefficient of x_ix_j for $i \neq j$ must be split evenly between the (i, j)- and (j, i)-entries in A. The coefficient of x_1x_3 is 0. It is readily checked that

$$Q(\mathbf{x}) = \mathbf{x}^TA\mathbf{x} = \begin{bmatrix} x_1 & x_2 & x_3 \end{bmatrix} \begin{bmatrix} 5 & -1/2 & 0 \\ -1/2 & 3 & 4 \\ 0 & 4 & 2 \end{bmatrix} \begin{bmatrix} x_1 \\ x_2 \\ x_3 \end{bmatrix} \qquad \blacksquare$$

EXAMPLE 3 Let $Q(\mathbf{x}) = x_1^2 - 8x_1x_2 - 5x_2^2$. Compute the value of $Q(\mathbf{x})$ for $\mathbf{x} = \begin{bmatrix} -3 \\ 1 \end{bmatrix}$, $\begin{bmatrix} 2 \\ -2 \end{bmatrix}$, and $\begin{bmatrix} 1 \\ -3 \end{bmatrix}$.

SOLUTION

$$Q(-3, 1) = (-3)^2 - 8(-3)(1) - 5(1)^2 = 28$$
$$Q(2, -2) = (2)^2 - 8(2)(-2) - 5(-2)^2 = 16$$
$$Q(1, -3) = (1)^2 - 8(1)(-3) - 5(-3)^2 = -20$$
∎

In some cases, quadratic forms are easier to use when they have no cross-product terms—that is, when the matrix of the quadratic form is a diagonal matrix. Fortunately, the cross-product term can be eliminated by making a suitable change of variable.

Change of Variable in a Quadratic Form

If $\mathbf{x}$ represents a variable vector in $\mathbb{R}^n$, then a **change of variable** is an equation of the form

$$\mathbf{x} = P\mathbf{y}, \qquad \text{or equivalently,} \qquad \mathbf{y} = P^{-1}\mathbf{x} \tag{1}$$

where P is an invertible matrix and $\mathbf{y}$ is a new variable vector in $\mathbb{R}^n$. Here $\mathbf{y}$ is the coordinate vector of $\mathbf{x}$ relative to the basis of $\mathbb{R}^n$ determined by the columns of P. (See Section 4.4.)

If the change of variable (1) is made in a quadratic form $\mathbf{x}^T A \mathbf{x}$, then

$$\mathbf{x}^T A \mathbf{x} = (P\mathbf{y})^T A (P\mathbf{y}) = \mathbf{y}^T P^T A P \mathbf{y} = \mathbf{y}^T (P^T A P) \mathbf{y} \tag{2}$$

and the new matrix of the quadratic form is $P^T A P$. Since A is symmetric, Theorem 2 guarantees that there is an *orthogonal* matrix P such that $P^T A P$ is a diagonal matrix D, and the quadratic form in (2) becomes $\mathbf{y}^T D \mathbf{y}$. This is the strategy of the next example.

EXAMPLE 4 Make a change of variable that transforms the quadratic form in Example 3 into a quadratic form with no cross-product term.

SOLUTION The matrix of the quadratic form in Example 3 is

$$A = \begin{bmatrix} 1 & -4 \\ -4 & -5 \end{bmatrix}$$

The first step is to orthogonally diagonalize A. Its eigenvalues turn out to be $\lambda = 3$ and $\lambda = -7$. Associated unit eigenvectors are

$$\lambda = 3: \begin{bmatrix} 2/\sqrt{5} \\ -1/\sqrt{5} \end{bmatrix}; \qquad \lambda = -7: \begin{bmatrix} 1/\sqrt{5} \\ 2/\sqrt{5} \end{bmatrix}$$

These vectors are automatically orthogonal (because they correspond to distinct eigenvalues) and so provide an orthonormal basis for $\mathbb{R}^2$. Let

$$P = \begin{bmatrix} 2/\sqrt{5} & 1/\sqrt{5} \\ -1/\sqrt{5} & 2/\sqrt{5} \end{bmatrix}, \qquad D = \begin{bmatrix} 3 & 0 \\ 0 & -7 \end{bmatrix}$$

Then $A = PDP^{-1}$ and $D = P^{-1}AP = P^TAP$, as pointed out earlier. A suitable change of variable is

$$\mathbf{x} = P\mathbf{y}, \qquad \text{where } \mathbf{x} = \begin{bmatrix} x_1 \\ x_2 \end{bmatrix} \quad \text{and} \quad \mathbf{y} = \begin{bmatrix} y_1 \\ y_2 \end{bmatrix}$$

Then

$$x_1^2 - 8x_1x_2 - 5x_2^2 = \mathbf{x}^T A \mathbf{x} = (P\mathbf{y})^T A(P\mathbf{y})$$
$$= \mathbf{y}^T P^T A P \mathbf{y} = \mathbf{y}^T D \mathbf{y}$$
$$= 3y_1^2 - 7y_2^2 \qquad \blacksquare$$

To illustrate the meaning of the equality of quadratic forms in Example 4, we can compute $Q(\mathbf{x})$ for $\mathbf{x} = (2, -2)$ using the new quadratic form. First, since $\mathbf{x} = P\mathbf{y}$,

$$\mathbf{y} = P^{-1}\mathbf{x} = P^T\mathbf{x}$$

so

$$\mathbf{y} = \begin{bmatrix} 2/\sqrt{5} & -1/\sqrt{5} \\ 1/\sqrt{5} & 2/\sqrt{5} \end{bmatrix} \begin{bmatrix} 2 \\ -2 \end{bmatrix} = \begin{bmatrix} 6/\sqrt{5} \\ -2/\sqrt{5} \end{bmatrix}$$

Hence

$$3y_1^2 - 7y_2^2 = 3(6/\sqrt{5})^2 - 7(-2/\sqrt{5})^2 = 3(36/5) - 7(4/5)$$
$$= 80/5 = 16$$

This is the value of $Q(\mathbf{x})$ in Example 3 when $\mathbf{x} = (2, -2)$. See Fig. 1.

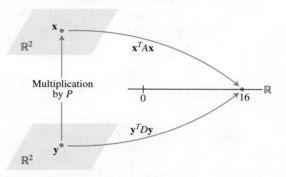

FIGURE 1 Change of variable in $\mathbf{x}^T A \mathbf{x}$.

Example 4 illustrates the following theorem. The proof of the theorem was essentially given before Example 4.

THEOREM 4

The Principal Axes Theorem

Let A be an $n \times n$ symmetric matrix. Then there is an orthogonal change of variable, $\mathbf{x} = P\mathbf{y}$, that transforms the quadratic form $\mathbf{x}^T A \mathbf{x}$ into a quadratic form $\mathbf{y}^T D \mathbf{y}$ with no cross-product term.

The columns of P in the theorem are called the **principal axes** of the quadratic form $\mathbf{x}^T A \mathbf{x}$. The vector $\mathbf{y}$ is the coordinate vector of $\mathbf{x}$ relative to the orthonormal basis of $\mathbb{R}^n$ given by these principal axes.

A Geometric View of Principal Axes

Suppose $Q(\mathbf{x}) = \mathbf{x}^T A \mathbf{x}$, where A is an invertible 2×2 symmetric matrix, and let c be a constant. It can be shown that the set of all $\mathbf{x}$ in $\mathbb{R}^2$ that satisfy

$$\mathbf{x}^T A \mathbf{x} = c \qquad (3)$$

either corresponds to an ellipse (or circle), a hyperbola, two intersecting lines, or a single point, or contains no points at all. If A is a diagonal matrix, the graph is in *standard position*, such as in Fig. 2. If A is not a diagonal matrix, the graph of equation (3) is

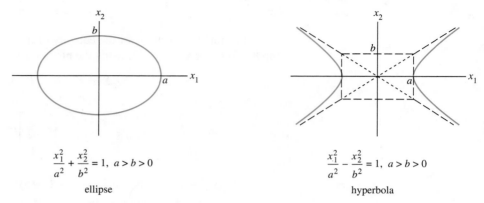

$$\frac{x_1^2}{a^2} + \frac{x_2^2}{b^2} = 1, \ a > b > 0$$

ellipse

$$\frac{x_1^2}{a^2} - \frac{x_2^2}{b^2} = 1, \ a > b > 0$$

hyperbola

FIGURE 2   An ellipse and a hyperbola in standard position.

rotated out of standard position, as in Fig. 3. Finding the *principal axes* (determined by the eigenvectors of A) amounts to finding a new coordinate system with respect to which the graph is in standard position.

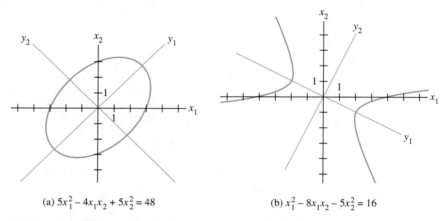

(a) $5x_1^2 - 4x_1x_2 + 5x_2^2 = 48$

(b) $x_1^2 - 8x_1x_2 - 5x_2^2 = 16$

FIGURE 3   An ellipse and a hyperbola *not* in standard position.

The hyperbola in Fig. 3(b) is the graph of the equation $\mathbf{x}^T A \mathbf{x} = 16$, where A is the matrix in Example 4. The positive y_1-axis in Fig. 3(b) is in the direction of the first column of the matrix P in Example 4, and the positive y_2-axis is in the direction of the second column of P.

EXAMPLE 5   The ellipse in Fig. 3(a) is the graph of the equation $5x_1^2 - 4x_1x_2 + 5x_2^2 = 48$. Find a change of variable that removes the cross-product term from the equation.

SOLUTION The matrix of the quadratic form is $A = \begin{bmatrix} 5 & -2 \\ -2 & 5 \end{bmatrix}$. The eigenvalues of A turn out to be 3 and 7, with corresponding unit eigenvectors

$$\mathbf{u}_1 = \begin{bmatrix} 1/\sqrt{2} \\ 1/\sqrt{2} \end{bmatrix}, \quad \mathbf{u}_2 = \begin{bmatrix} -1/\sqrt{2} \\ 1/\sqrt{2} \end{bmatrix}$$

Let $P = [\, \mathbf{u}_1 \quad \mathbf{u}_2 \,] = \begin{bmatrix} 1/\sqrt{2} & -1/\sqrt{2} \\ 1/\sqrt{2} & 1/\sqrt{2} \end{bmatrix}$. Then P orthogonally diagonalizes A, so the change of variable $\mathbf{x} = P\mathbf{y}$ produces the quadratic form $\mathbf{y}^T D\mathbf{y} = 3y_1^2 + 7y_2^2$. The new axes for this change of variable are shown in Fig. 3(a).   ∎

Classifying Quadratic Forms

When A is an $n \times n$ matrix, the quadratic form $Q(\mathbf{x}) = \mathbf{x}^T A\mathbf{x}$ is a real-valued function with domain $\mathbb{R}^n$. Figure 4 displays the graphs of four quadratic forms with domain $\mathbb{R}^2$. For each point $\mathbf{x} = (x_1, x_2)$ in the domain of a quadratic form Q, the graph displays the point (x_1, x_2, z) where $z = Q(\mathbf{x})$. Notice that except at $\mathbf{x} = \mathbf{0}$, the values of $Q(\mathbf{x})$ are all positive in Fig. 4(a) and all negative in Fig. 4(d). The horizontal cross-sections of the graphs are ellipses in Figs. 4(a) and 4(d) and hyperbolas in Fig. 4(c).

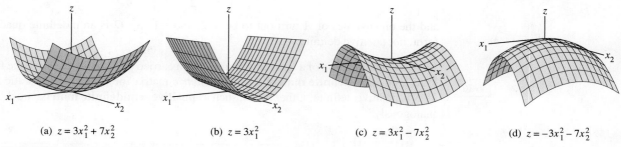

(a) $z = 3x_1^2 + 7x_2^2$ (b) $z = 3x_1^2$ (c) $z = 3x_1^2 - 7x_2^2$ (d) $z = -3x_1^2 - 7x_2^2$

FIGURE 4 Graphs of quadratic forms.

The simple 2×2 examples in Fig. 4 illustrate the following definitions.

DEFINITION

A quadratic form Q is:

a. **positive definite** if $Q(\mathbf{x}) > 0$ for all $\mathbf{x} \neq \mathbf{0}$,

b. **negative definite** if $Q(\mathbf{x}) < 0$ for all $\mathbf{x} \neq \mathbf{0}$,

c. **indefinite** if $Q(\mathbf{x})$ assumes both positive and negative values.

Also, Q is said to be **positive semidefinite** if $Q(\mathbf{x}) \geq 0$ for all $\mathbf{x}$, and to be **negative semidefinite** if $Q(\mathbf{x}) \leq 0$ for all $\mathbf{x}$. The quadratic forms in parts (a) and (b) of Fig. 4 are both positive semidefinite, but the form in (a) is better described as positive definite. Theorem 5 characterizes some quadratic forms in terms of eigenvalues.

THEOREM 5

Quadratic Forms and Eigenvalues

Let A be an $n \times n$ symmetric matrix. Then a quadratic form $\mathbf{x}^T A\mathbf{x}$ is:

a. positive definite if and only if the eigenvalues of A are all positive,

b. negative definite if and only if the eigenvalues of A are all negative, or

c. indefinite if and only if A has both positive and negative eigenvalues.

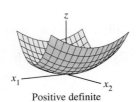

Positive definite

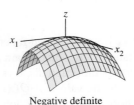

Negative definite

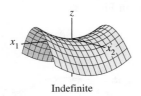

Indefinite

PROOF By the Principal Axes Theorem, there exists an orthogonal change of variable $\mathbf{x} = P\mathbf{y}$ such that

$$Q(\mathbf{x}) = \mathbf{x}^T A \mathbf{x} = \mathbf{y}^T D \mathbf{y} = \lambda_1 y_1^2 + \lambda_2 y_2^2 + \cdots + \lambda_n y_n^2 \qquad (4)$$

where $\lambda_1, \ldots, \lambda_n$ are the eigenvalues of A. Since P is invertible, there is a one-to-one correspondence between all nonzero $\mathbf{x}$ and all nonzero $\mathbf{y}$. Thus the values of $Q(\mathbf{x})$ for $\mathbf{x} \neq \mathbf{0}$ coincide with the values of the expression on the right side of (4), which is obviously controlled by the signs of the eigenvalues $\lambda_1, \ldots, \lambda_n$, in the three ways described in the theorem. ∎

EXAMPLE 6 Is $Q(\mathbf{x}) = 3x_1^2 + 2x_2^2 + x_3^2 + 4x_1x_2 + 4x_2x_3$ positive definite?

SOLUTION Because of all the plus signs, this form "looks" positive definite. But the matrix of the form is

$$A = \begin{bmatrix} 3 & 2 & 0 \\ 2 & 2 & 2 \\ 0 & 2 & 1 \end{bmatrix}$$

and the eigenvalues of A turn out to be 5, 2, and -1. So Q is an indefinite quadratic form, not positive definite. ∎

The classification of a quadratic form is often carried over to the matrix of the form. Thus a **positive definite matrix** A is a *symmetric* matrix for which the quadratic form $\mathbf{x}^T A \mathbf{x}$ is positive definite. Other terms, such as **positive semidefinite matrix**, are defined analogously.

WEB

NUMERICAL NOTE

A fast way to determine whether a symmetric matrix A is positive definite is to attempt to factor A in the form $A = R^T R$, where R is upper triangular with positive diagonal entries. (A slightly modified algorithm for an LU factorization is one approach.) Such a *Cholesky factorization* is possible if and only if A is positive definite. See Supplementary Exercise 7 at the end of Chapter 7.

PRACTICE PROBLEM

Describe a positive semidefinite matrix A in terms of its eigenvalues.

WEB

7.2 EXERCISES

1. Compute the quadratic form $\mathbf{x}^T A \mathbf{x}$, when $A = \begin{bmatrix} 5 & 1/3 \\ 1/3 & 1 \end{bmatrix}$ and

   a. $\mathbf{x} = \begin{bmatrix} x_1 \\ x_2 \end{bmatrix}$ b. $\mathbf{x} = \begin{bmatrix} 6 \\ 1 \end{bmatrix}$ c. $\mathbf{x} = \begin{bmatrix} 1 \\ 3 \end{bmatrix}$

2. Compute the quadratic form $\mathbf{x}^T A \mathbf{x}$, for $A = \begin{bmatrix} 4 & 3 & 0 \\ 3 & 2 & 1 \\ 0 & 1 & 1 \end{bmatrix}$ and

   a. $\mathbf{x} = \begin{bmatrix} x_1 \\ x_2 \\ x_3 \end{bmatrix}$ b. $\mathbf{x} = \begin{bmatrix} 2 \\ -1 \\ 5 \end{bmatrix}$ c. $\mathbf{x} = \begin{bmatrix} 1/\sqrt{3} \\ 1/\sqrt{3} \\ 1/\sqrt{3} \end{bmatrix}$

3. Find the matrix of the quadratic form. Assume $\mathbf{x}$ is in $\mathbb{R}^2$.

   a. $10x_1^2 - 6x_1x_2 - 3x_2^2$ b. $5x_1^2 + 3x_1x_2$

4. Find the matrix of the quadratic form. Assume $\mathbf{x}$ is in $\mathbb{R}^2$.

   a. $20x_1^2 + 15x_1x_2 - 10x_2^2$ b. x_1x_2

5. Find the matrix of the quadratic form. Assume $\mathbf{x}$ is in $\mathbb{R}^3$.
   a. $8x_1^2 + 7x_2^2 - 3x_3^2 - 6x_1x_2 + 4x_1x_3 - 2x_2x_3$
   b. $4x_1x_2 + 6x_1x_3 - 8x_2x_3$

6. Find the matrix of the quadratic form. Assume $\mathbf{x}$ is in $\mathbb{R}^3$.
   a. $5x_1^2 - x_2^2 + 7x_3^2 + 5x_1x_2 - 3x_1x_3$
   b. $x_3^2 - 4x_1x_2 + 4x_2x_3$

7. Make a change of variable, $\mathbf{x} = P\mathbf{y}$, that transforms the quadratic form $x_1^2 + 10x_1x_2 + x_2^2$ into a quadratic form with no cross-product term. Give P and the new quadratic form.

8. Let A be the matrix of the quadratic form
   $$9x_1^2 + 7x_2^2 + 11x_3^2 - 8x_1x_2 + 8x_1x_3$$
   It can be shown that the eigenvalues of A are 3, 9, and 15. Find an orthogonal matrix P such that the change of variable $\mathbf{x} = P\mathbf{y}$ transforms $\mathbf{x}^TA\mathbf{x}$ into a quadratic form with no cross-product term. Give P and the new quadratic form.

Classify the quadratic forms in Exercises 9–18. Then make a change of variable, $\mathbf{x} = P\mathbf{y}$, that transforms the quadratic form into one with no cross-product term. Write the new quadratic form. Construct P using the methods of Section 7.1.

9. $3x_1^2 - 4x_1x_2 + 6x_2^2$ 10. $9x_1^2 - 8x_1x_2 + 3x_2^2$

11. $2x_1^2 + 10x_1x_2 + 2x_2^2$ 12. $-5x_1^2 + 4x_1x_2 - 2x_2^2$

13. $x_1^2 - 6x_1x_2 + 9x_2^2$ 14. $8x_1^2 + 6x_1x_2$

15. **[M]** $-2x_1^2 - 6x_2^2 - 9x_3^2 - 9x_4^2 + 4x_1x_2 + 4x_1x_3 + 4x_1x_4 + 6x_3x_4$

16. **[M]** $4x_1^2 + 4x_2^2 + 4x_3^2 + 4x_4^2 + 3x_1x_2 + 3x_3x_4 - 4x_1x_4 + 4x_2x_3$

17. **[M]** $x_1^2 + x_2^2 + x_3^2 + x_4^2 + 9x_1x_2 - 12x_1x_4 + 12x_2x_3 + 9x_3x_4$

18. **[M]** $11x_1^2 - x_2^2 - 12x_1x_2 - 12x_1x_3 - 12x_1x_4 - 2x_3x_4$

19. What is the largest possible value of the quadratic form $5x_1^2 + 8x_2^2$ if $\mathbf{x} = (x_1, x_2)$ and $\mathbf{x}^T\mathbf{x} = 1$, that is, if $x_1^2 + x_2^2 = 1$? (Try some examples of $\mathbf{x}$.)

20. What is the largest value of the quadratic form $5x_1^2 - 3x_2^2$ if $\mathbf{x}^T\mathbf{x} = 1$?

In Exercises 21 and 22, matrices are $n \times n$ and vectors are in $\mathbb{R}^n$. Mark each statement True or False. Justify each answer.

21. a. The matrix of a quadratic form is a symmetric matrix.
    b. A quadratic form has no cross-product terms if and only if the matrix of the quadratic form is a diagonal matrix.
    c. The principal axes of a quadratic form $\mathbf{x}^TA\mathbf{x}$ are eigenvectors of A.
    d. A positive definite quadratic form Q satisfies $Q(\mathbf{x}) > 0$ for all $\mathbf{x}$ in $\mathbb{R}^n$.

    e. If the eigenvalues of a symmetric matrix A are all positive, then the quadratic form $\mathbf{x}^TA\mathbf{x}$ is positive definite.
    f. A Cholesky factorization of a symmetric matrix A has the form $A = R^TR$, for an upper triangular matrix R with positive diagonal entries.

22. a. The expression $\|\mathbf{x}\|^2$ is a quadratic form.
    b. If A is symmetric and P is an orthogonal matrix, then the change of variable $\mathbf{x} = P\mathbf{y}$ transforms $\mathbf{x}^TA\mathbf{x}$ into a quadratic form with no cross-product term.
    c. If A is a 2×2 symmetric matrix, then the set of $\mathbf{x}$ such that $\mathbf{x}^TA\mathbf{x} = c$ (for a constant c) corresponds to either a circle, an ellipse, or a hyperbola.
    d. An indefinite quadratic form is either positive semidefinite or negative semidefinite.
    e. If A is symmetric and the quadratic form $\mathbf{x}^TA\mathbf{x}$ has only negative values for $\mathbf{x} \neq \mathbf{0}$, then the eigenvalues of A are all negative.

Exercises 23 and 24 show how to classify a quadratic form $Q(\mathbf{x}) = \mathbf{x}^TA\mathbf{x}$, when $A = \begin{bmatrix} a & b \\ b & d \end{bmatrix}$ and $\det A \neq 0$, without finding the eigenvalues of A.

23. If λ_1 and λ_2 are the eigenvalues of A, then the characteristic polynomial of A can be written in two ways: $\det(A - \lambda I)$ and $(\lambda - \lambda_1)(\lambda - \lambda_2)$. Use this fact to show that $\lambda_1 + \lambda_2 = a + d$ (the diagonal entries of A) and $\lambda_1\lambda_2 = \det A$.

24. Verify the following statements.
    a. Q is positive definite if $\det A > 0$ and $a > 0$.
    b. Q is negative definite if $\det A > 0$ and $a < 0$.
    c. Q is indefinite if $\det A < 0$.

25. Show that if B is $m \times n$, then B^TB is positive semidefinite; and if B is $n \times n$ and invertible, then B^TB is positive definite.

26. Show that if an $n \times n$ matrix A is positive definite, then there exists a positive definite matrix B such that $A = B^TB$. [*Hint:* Write $A = PDP^T$, with $P^T = P^{-1}$. Produce a diagonal matrix C such that $D = C^TC$, and let $B = PCP^T$. Show that B works.]

27. Let A and B be symmetric $n \times n$ matrices whose eigenvalues are all positive. Show that the eigenvalues of $A + B$ are all positive. [*Hint:* Consider quadratic forms.]

28. Let A be an $n \times n$ invertible symmetric matrix. Show that if the quadratic form $\mathbf{x}^TA\mathbf{x}$ is positive definite, then so is the quadratic form $\mathbf{x}^TA^{-1}\mathbf{x}$. [*Hint:* Consider eigenvalues.]

| SG | Mastering: Diagonalization and Quadratic Forms 7-7 |

| SOLUTION TO PRACTICE PROBLEM

Make an orthogonal change of variable $\mathbf{x} = P\mathbf{y}$, and write

$$\mathbf{x}^T A \mathbf{x} = \mathbf{y}^T D \mathbf{y} = \lambda_1 y_1^2 + \lambda_2 y_2^2 + \cdots + \lambda_n y_n^2$$

as in equation (4). If an eigenvalue—say, λ_i—were negative, then $\mathbf{x}^T A \mathbf{x}$ would be negative for the $\mathbf{x}$ corresponding to $\mathbf{y} = \mathbf{e}_i$ (the ith column of I_n). So the eigenvalues of a positive semidefinite quadratic form must all be nonnegative. Conversely, if the eigenvalues are nonnegative, the expansion above shows that $\mathbf{x}^T A \mathbf{x}$ must be positive semidefinite.

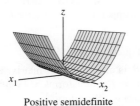

Positive semidefinite

7.3 | CONSTRAINED OPTIMIZATION

Engineers, economists, scientists, and mathematicians often need to find the maximum or minimum value of a quadratic form $Q(\mathbf{x})$ for $\mathbf{x}$ in some specified set. Typically, the problem can be arranged so that $\mathbf{x}$ varies over the set of unit vectors. This *constrained optimization problem* has an interesting and elegant solution. Example 6 below and the discussion in Section 7.5 will illustrate how such problems arise in practice.

The requirement that a vector $\mathbf{x}$ in $\mathbb{R}^n$ be a unit vector can be stated in several equivalent ways:

$$\|\mathbf{x}\| = 1, \qquad \|\mathbf{x}\|^2 = 1, \qquad \mathbf{x}^T \mathbf{x} = 1$$

and

$$x_1^2 + x_2^2 + \cdots + x_n^2 = 1 \tag{1}$$

The expanded version (1) of $\mathbf{x}^T \mathbf{x} = 1$ is commonly used in applications.

When a quadratic form Q has no cross-product terms, it is easy to find the maximum and minimum of $Q(\mathbf{x})$ for $\mathbf{x}^T \mathbf{x} = 1$.

EXAMPLE 1   Find the maximum and minimum values of $Q(\mathbf{x}) = 9x_1^2 + 4x_2^2 + 3x_3^2$ subject to the constraint $\mathbf{x}^T \mathbf{x} = 1$.

SOLUTION   Since x_2^2 and x_3^2 are nonnegative, note that

$$4x_2^2 \leq 9x_2^2 \qquad \text{and} \qquad 3x_3^2 \leq 9x_3^2$$

and hence

$$
\begin{aligned}
Q(\mathbf{x}) &= 9x_1^2 + 4x_2^2 + 3x_3^2 \\
&\leq 9x_1^2 + 9x_2^2 + 9x_3^2 \\
&= 9(x_1^2 + x_2^2 + x_3^2) \\
&= 9
\end{aligned}
$$

whenever $x_1^2 + x_2^2 + x_3^2 = 1$. So the maximum value of $Q(\mathbf{x})$ cannot exceed 9 when $\mathbf{x}$ is a unit vector. Furthermore, $Q(\mathbf{x}) = 9$ when $\mathbf{x} = (1, 0, 0)$. Thus 9 is the maximum value of $Q(\mathbf{x})$ for $\mathbf{x}^T \mathbf{x} = 1$.

To find the minimum value of $Q(\mathbf{x})$, observe that

$$9x_1^2 \geq 3x_1^2, \qquad 4x_2^2 \geq 3x_2^2$$

and hence

$$Q(\mathbf{x}) \geq 3x_1^2 + 3x_2^2 + 3x_3^2 = 3(x_1^2 + x_2^2 + x_3^2) = 3$$

whenever $x_1^2 + x_2^2 + x_3^2 = 1$. Also, $Q(\mathbf{x}) = 3$ when $x_1 = 0$, $x_2 = 0$, and $x_3 = 1$. So 3 is the minimum value of $Q(\mathbf{x})$ when $\mathbf{x}^T \mathbf{x} = 1$.   ∎

It is easy to see in Example 1 that the matrix of the quadratic form Q has eigenvalues 9, 4, and 3 and that the greatest and least eigenvalues equal, respectively, the (constrained) maximum and minimum of $Q(\mathbf{x})$. The same holds true for any quadratic form, as we shall see.

EXAMPLE 2 Let $A = \begin{bmatrix} 3 & 0 \\ 0 & 7 \end{bmatrix}$, and let $Q(\mathbf{x}) = \mathbf{x}^T A \mathbf{x}$ for $\mathbf{x}$ in $\mathbb{R}^2$. Figure 1 displays the graph of Q. Figure 2 shows only the portion of the graph inside a cylinder; the intersection of the cylinder with the surface is the set of points (x_1, x_2, z) such that $z = Q(x_1, x_2)$ and $x_1^2 + x_2^2 = 1$. The "heights" of these points are the constrained values of $Q(\mathbf{x})$. Geometrically, the constrained optimization problem is to locate the highest and lowest points on the intersection curve.

The two highest points on the curve are 7 units above the $x_1 x_2$-plane, occurring where $x_1 = 0$ and $x_2 = \pm 1$. These points correspond to the eigenvalue 7 of A and the eigenvectors $\mathbf{x} = (0, 1)$ and $-\mathbf{x} = (0, -1)$. Similarly, the two lowest points on the curve are 3 units above the $x_1 x_2$-plane. They correspond to the eigenvalue 3 and the eigenvectors $(1, 0)$ and $(-1, 0)$. ■

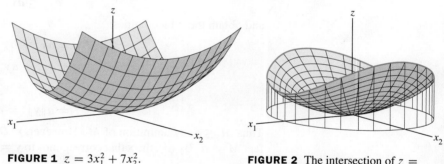

FIGURE 1 $z = 3x_1^2 + 7x_2^2$.

FIGURE 2 The intersection of $z = 3x_1^2 + 7x_2^2$ and the cylinder $x_1^2 + x_2^2 = 1$.

Every point on the intersection curve in Fig. 2 has a z-coordinate between 3 and 7, and for any number t between 3 and 7, there is a unit vector $\mathbf{x}$ such that $Q(\mathbf{x}) = t$. In other words, the set of all possible values of $\mathbf{x}^T A \mathbf{x}$, for $\|\mathbf{x}\| = 1$, is the closed interval $3 \le t \le 7$.

It can be shown that for any symmetric matrix A, the set of all possible values of $\mathbf{x}^T A \mathbf{x}$, for $\|\mathbf{x}\| = 1$, is a closed interval on the real axis. (See Exercise 13.) Denote the left and right endpoints of this interval by m and M, respectively. That is, let

$$m = \min \{\mathbf{x}^T A \mathbf{x} : \|\mathbf{x}\| = 1\}, \quad M = \max \{\mathbf{x}^T A \mathbf{x} : \|\mathbf{x}\| = 1\} \tag{2}$$

Exercise 12 asks you to prove that if λ is an eigenvalue of A, then $m \le \lambda \le M$. The next theorem says that m and M are themselves eigenvalues of A, just as in Example 2.[1]

THEOREM 6

Let A be a symmetric matrix, and define m and M as in (2). Then M is the greatest eigenvalue λ_1 of A and m is the least eigenvalue of A. The value of $\mathbf{x}^T A \mathbf{x}$ is M when $\mathbf{x}$ is a unit eigenvector $\mathbf{u}_1$ corresponding to M. The value of $\mathbf{x}^T A \mathbf{x}$ is m when $\mathbf{x}$ is a unit eigenvector corresponding to m.

---

[1] The use of *minimum* and *maximum* in (2), and *least* and *greatest* in the theorem, refers to the natural ordering of the real numbers, not to magnitudes.

PROOF Orthogonally diagonalize A as PDP^{-1}. We know that

$$\mathbf{x}^T A\mathbf{x} = \mathbf{y}^T D\mathbf{y} \quad \text{when } \mathbf{x} = P\mathbf{y} \tag{3}$$

Also,

$$\|\mathbf{x}\| = \|P\mathbf{y}\| = \|\mathbf{y}\| \quad \text{for all } \mathbf{y}$$

because $P^T P = I$ and $\|P\mathbf{y}\|^2 = (P\mathbf{y})^T (P\mathbf{y}) = \mathbf{y}^T P^T P\mathbf{y} = \mathbf{y}^T\mathbf{y} = \|\mathbf{y}\|^2$. In particular, $\|\mathbf{y}\| = 1$ if and only if $\|\mathbf{x}\| = 1$. Thus $\mathbf{x}^T A\mathbf{x}$ and $\mathbf{y}^T D\mathbf{y}$ assume the same set of values as $\mathbf{x}$ and $\mathbf{y}$ range over the set of all unit vectors.

To simplify notation, suppose that A is a 3×3 matrix with eigenvalues $a \geq b \geq c$. Arrange the (eigenvector) columns of P so that $P = [\,\mathbf{u}_1 \quad \mathbf{u}_2 \quad \mathbf{u}_3\,]$ and

$$D = \begin{bmatrix} a & 0 & 0 \\ 0 & b & 0 \\ 0 & 0 & c \end{bmatrix}$$

Given any unit vector $\mathbf{y}$ in $\mathbb{R}^3$ with coordinates y_1, y_2, y_3, observe that

$$ay_1^2 = ay_1^2$$
$$by_2^2 \leq ay_2^2$$
$$cy_3^2 \leq ay_3^2$$

and obtain these inequalities:

$$\mathbf{y}^T D\mathbf{y} = ay_1^2 + by_2^2 + cy_3^2$$
$$\leq ay_1^2 + ay_2^2 + ay_3^2$$
$$= a(y_1^2 + y_2^2 + y_3^2)$$
$$= a\|\mathbf{y}\|^2 = a$$

Thus $M \leq a$, by definition of M. However, $\mathbf{y}^T D\mathbf{y} = a$ when $\mathbf{y} = \mathbf{e}_1 = (1, 0, 0)$, so in fact $M = a$. By (3), the $\mathbf{x}$ that corresponds to $\mathbf{y} = \mathbf{e}_1$ is the eigenvector $\mathbf{u}_1$ of A, because

$$\mathbf{x} = P\mathbf{e}_1 = \begin{bmatrix} \mathbf{u}_1 & \mathbf{u}_2 & \mathbf{u}_3 \end{bmatrix} \begin{bmatrix} 1 \\ 0 \\ 0 \end{bmatrix} = \mathbf{u}_1$$

Thus $M = a = \mathbf{e}_1^T D\mathbf{e}_1 = \mathbf{u}_1^T A\mathbf{u}_1$, which proves the statement about M. A similar argument shows that m is the least eigenvalue, c, and this value of $\mathbf{x}^T A\mathbf{x}$ is attained when $\mathbf{x} = P\mathbf{e}_3 = \mathbf{u}_3$. ∎

EXAMPLE 3 Let $A = \begin{bmatrix} 3 & 2 & 1 \\ 2 & 3 & 1 \\ 1 & 1 & 4 \end{bmatrix}$. Find the maximum value of the quadratic form $\mathbf{x}^T A\mathbf{x}$ subject to the constraint $\mathbf{x}^T\mathbf{x} = 1$, and find a unit vector at which this maximum value is attained.

SOLUTION By Theorem 6, the desired maximum value is the greatest eigenvalue of A. The characteristic equation turns out to be

$$0 = -\lambda^3 + 10\lambda^2 - 27\lambda + 18 = -(\lambda - 6)(\lambda - 3)(\lambda - 1)$$

The greatest eigenvalue is 6.

The constrained maximum of $\mathbf{x}^T A\mathbf{x}$ is attained when $\mathbf{x}$ is a unit eigenvector for $\lambda = 6$. Solve $(A - 6I)\mathbf{x} = 0$ and find an eigenvector $\begin{bmatrix} 1 \\ 1 \\ 1 \end{bmatrix}$. Set $\mathbf{u}_1 = \begin{bmatrix} 1/\sqrt{3} \\ 1/\sqrt{3} \\ 1/\sqrt{3} \end{bmatrix}$. ∎

In Theorem 7 and in later applications, the values of $x^T Ax$ are computed with additional constraints on the unit vector x.

THEOREM 7

Let A, λ_1, and u_1 be as in Theorem 6. Then the maximum value of $x^T Ax$ subject to the constraints

$$x^T x = 1, \quad x^T u_1 = 0$$

is the second greatest eigenvalue, λ_2, and this maximum is attained when x is an eigenvector u_2 corresponding to λ_2.

Theorem 7 can be proved by an argument similar to the one above in which the theorem is reduced to the case where the matrix of the quadratic form is diagonal. The next example gives an idea of the proof for the case of a diagonal matrix.

EXAMPLE 4   Find the maximum value of $9x_1^2 + 4x_2^2 + 3x_3^2$ subject to the constraints $x^T x = 1$ and $x^T u_1 = 0$, where $u_1 = (1, 0, 0)$. Note that u_1 is a unit eigenvector corresponding to the greatest eigenvalue $\lambda = 9$ of the matrix of the quadratic form.

SOLUTION   If the coordinates of x are x_1, x_2, x_3, then the constraint $x^T u_1 = 0$ means simply that $x_1 = 0$. For such a unit vector, $x_2^2 + x_3^2 = 1$, and

$$\begin{aligned} 9x_1^2 + 4x_2^2 + 3x_3^2 &= 4x_2^2 + 3x_3^2 \\ &\leq 4x_2^2 + 4x_3^2 \\ &= 4(x_2^2 + x_3^2) \\ &= 4 \end{aligned}$$

Thus the constrained maximum of the quadratic form does not exceed 4. And this value is attained for $x = (0, 1, 0)$, which is an eigenvector for the second greatest eigenvalue of the matrix of the quadratic form. ∎

EXAMPLE 5   Let A be the matrix in Example 3 and let u_1 be a unit eigenvector corresponding to the greatest eigenvalue of A. Find the maximum value of $x^T Ax$ subject to the conditions

$$x^T x = 1, \quad x^T u_1 = 0 \tag{4}$$

SOLUTION   From Example 3, the second greatest eigenvalue of A is $\lambda = 3$. Solve $(A - 3I)x = 0$ to find an eigenvector, and normalize it to obtain

$$u_2 = \begin{bmatrix} 1/\sqrt{6} \\ 1/\sqrt{6} \\ -2/\sqrt{6} \end{bmatrix}$$

The vector u_2 is automatically orthogonal to u_1 because the vectors correspond to different eigenvalues. Thus the maximum of $x^T Ax$ subject to the constraints in (4) is 3, attained when $x = u_2$. ∎

The next theorem generalizes Theorem 7 and, together with Theorem 6, gives a useful characterization of *all* the eigenvalues of A. The proof is omitted.

THEOREM 8

> Let A be a symmetric $n \times n$ matrix with an orthogonal diagonalization $A = PDP^{-1}$, where the entries on the diagonal of D are arranged so that $\lambda_1 \geq \lambda_2 \geq \cdots \geq \lambda_n$ and where the columns of P are corresponding unit eigenvectors $\mathbf{u}_1, \ldots, \mathbf{u}_n$. Then for $k = 2, \ldots, n$, the maximum value of $\mathbf{x}^T A \mathbf{x}$ subject to the constraints
>
> $$\mathbf{x}^T \mathbf{x} = 1, \quad \mathbf{x}^T \mathbf{u}_1 = 0, \quad \ldots, \quad \mathbf{x}^T \mathbf{u}_{k-1} = 0$$
>
> is the eigenvalue λ_k, and this maximum is attained at $\mathbf{x} = \mathbf{u}_k$.

Theorem 8 will be helpful in Sections 7.4 and 7.5. The following application requires only Theorem 6.

EXAMPLE 6 During the next year, a county government is planning to repair x hundred miles of public roads and bridges and to improve y hundred acres of parks and recreation areas. The county must decide how to allocate its resources (funds, equipment, labor, etc.) between these two projects. If it is more cost-effective to work simultaneously on both projects rather than on only one, then x and y might satisfy a *constraint* such as

$$4x^2 + 9y^2 \leq 36$$

See Fig. 3. Each point (x, y) in the shaded *feasible set* represents a possible public works schedule for the year. The points on the constraint curve, $4x^2 + 9y^2 = 36$, use the maximum amounts of resources available.

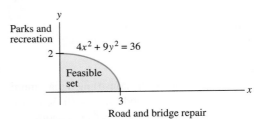

FIGURE 3 Public works schedules.

In choosing its public works schedule, the county wants to consider the opinions of the county residents. To measure the value, or *utility*, that the residents would assign to the various work schedules (x, y), economists sometimes use a function such as

$$q(x, y) = xy$$

The set of points (x, y) at which $q(x, y)$ is a constant is called an *indifference curve*. Three such curves are shown in Fig. 4. Points along an indifference curve correspond to alternatives that county residents as a group would find equally valuable.[2] Find the public works schedule that maximizes the utility function q.

SOLUTION The constraint equation $4x^2 + 9y^2 = 36$ does not describe a set of unit vectors, but a change of variable can fix that problem. Rewrite the constraint in the form

$$\left(\frac{x}{3}\right)^2 + \left(\frac{y}{2}\right)^2 = 1$$

---

[2] Indifference curves are discussed in Michael D. Intriligator, Ronald G. Bodkin, and Cheng Hsiao, *Econometric Models, Techniques, and Applications* (Upper Saddle River, NJ: Prentice-Hall, 1996).

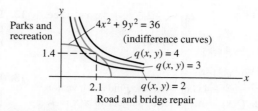

FIGURE 4 The optimum public works schedule is (2.1, 1.4).

and define

$$x_1 = \frac{x}{3}, \quad x_2 = \frac{y}{2}, \quad \text{that is,} \quad x = 3x_1 \quad \text{and} \quad y = 2x_2$$

Then the constraint equation becomes

$$x_1^2 + x_2^2 = 1$$

and the utility function becomes $q(3x_1, 2x_2) = (3x_1)(2x_2) = 6x_1x_2$. Let $\mathbf{x} = \begin{bmatrix} x_1 \\ x_2 \end{bmatrix}$.

Then the problem is to maximize $Q(\mathbf{x}) = 6x_1x_2$ subject to $\mathbf{x}^T\mathbf{x} = 1$. Note that $Q(\mathbf{x}) = \mathbf{x}^T A \mathbf{x}$, where

$$A = \begin{bmatrix} 0 & 3 \\ 3 & 0 \end{bmatrix}$$

The eigenvalues of A are ± 3, with eigenvectors $\begin{bmatrix} 1/\sqrt{2} \\ 1/\sqrt{2} \end{bmatrix}$ for $\lambda = 3$ and $\begin{bmatrix} -1/\sqrt{2} \\ 1/\sqrt{2} \end{bmatrix}$ for $\lambda = -3$. Thus the maximum value of $Q(\mathbf{x}) = q(x_1, x_2)$ is 3, attained when $x_1 = 1/\sqrt{2}$ and $x_2 = 1/\sqrt{2}$.

In terms of the original variables, the optimum public works schedule is $x = 3x_1 = 3/\sqrt{2} \approx 2.1$ hundred miles of roads and bridges and $y = 2x_2 = \sqrt{2} \approx 1.4$ hundred acres of parks and recreational areas. The optimum public works schedule is the point where the constraint curve and the indifference curve $q(x, y) = 3$ just meet. Points (x, y) with a higher utility lie on indifference curves that do not touch the constraint curve. See Fig. 4. ∎

PRACTICE PROBLEMS

1. Let $Q(\mathbf{x}) = 3x_1^2 + 3x_2^2 + 2x_1x_2$. Find a change of variable that transforms Q into a quadratic form with no cross-product term, and give the new quadratic form.

2. With Q as in Problem 1, find the maximum value of $Q(\mathbf{x})$ subject to the constraint $\mathbf{x}^T\mathbf{x} = 1$, and find a unit vector at which the maximum is attained.

7.3 EXERCISES

In Exercises 1 and 2, find the change of variable $\mathbf{x} = P\mathbf{y}$ that transforms the quadratic form $\mathbf{x}^T A \mathbf{x}$ into $\mathbf{y}^T D \mathbf{y}$ as shown.

1. $5x_1^2 + 6x_2^2 + 7x_3^2 + 4x_1x_2 - 4x_2x_3 = 9y_1^2 + 6y_2^2 + 3y_3^2$

2. $3x_1^2 + 2x_2^2 + 2x_3^2 + 2x_1x_2 + 2x_1x_3 + 4x_2x_3 = 5y_1^2 + 2y_2^2$
   [*Hint:* $\mathbf{x}$ and $\mathbf{y}$ must have the same number of coordinates, so the quadratic form shown here must have a coefficient of zero for y_3^2.]

In Exercises 3–6, find (a) the maximum value of $Q(\mathbf{x})$ subject to the constraint $\mathbf{x}^T\mathbf{x} = 1$, (b) a unit vector $\mathbf{u}$ where this maximum is attained, and (c) the maximum of $Q(\mathbf{x})$ subject to the constraints $\mathbf{x}^T\mathbf{x} = 1$ and $\mathbf{x}^T\mathbf{u} = 0$.

3. $Q(\mathbf{x}) = 5x_1^2 + 6x_2^2 + 7x_3^2 + 4x_1x_2 - 4x_2x_3$
   (See Exercise 1.)

4. $Q(\mathbf{x}) = 3x_1^2 + 2x_2^2 + 2x_3^2 + 2x_1x_2 + 2x_1x_3 + 4x_2x_3$ (See Exercise 2.)

5. $Q(\mathbf{x}) = 5x_1^2 + 5x_2^2 - 4x_1x_2$

6. $Q(\mathbf{x}) = 7x_1^2 + 3x_2^2 + 3x_1x_2$

7. Let $Q(\mathbf{x}) = -2x_1^2 - x_2^2 + 4x_1x_2 + 4x_2x_3$. Find a unit vector $\mathbf{x}$ in $\mathbb{R}^3$ at which $Q(\mathbf{x})$ is maximized, subject to $\mathbf{x}^T\mathbf{x} = 1$. [*Hint:* The eigenvalues of the matrix of the quadratic form Q are 2, −1, and −4.]

8. Let $Q(\mathbf{x}) = 7x_1^2 + x_2^2 + 7x_3^2 - 8x_1x_2 - 4x_1x_3 - 8x_2x_3$. Find a unit vector $\mathbf{x}$ in $\mathbb{R}^3$ at which $Q(\mathbf{x})$ is maximized, subject to $\mathbf{x}^T\mathbf{x} = 1$. [*Hint:* The eigenvalues of the matrix of the quadratic form Q are 9 and −3.]

9. Find the maximum value of $Q(\mathbf{x}) = 7x_1^2 + 3x_2^2 - 2x_1x_2$, subject to the constraint $x_1^2 + x_2^2 = 1$. (Do not go on to find a vector where the maximum is attained.)

10. Find the maximum value of $Q(\mathbf{x}) = -3x_1^2 + 5x_2^2 - 2x_1x_2$, subject to the constraint $x_1^2 + x_2^2 = 1$. (Do not go on to find a vector where the maximum is attained.)

11. Suppose $\mathbf{x}$ is a unit eigenvector of a matrix A corresponding to an eigenvalue 3. What is the value of $\mathbf{x}^TA\mathbf{x}$?

12. Let λ be any eigenvalue of a symmetric matrix A. Justify the statement made in this section that $m \le \lambda \le M$, where m and M are defined as in (2). [*Hint:* Find an $\mathbf{x}$ such that $\lambda = \mathbf{x}^TA\mathbf{x}$.]

13. Let A be an $n \times n$ symmetric matrix, let M and m denote the maximum and minimum values of the quadratic form $\mathbf{x}^TA\mathbf{x}$, and denote corresponding unit eigenvectors by $\mathbf{u}_1$ and $\mathbf{u}_n$. The following calculations show that given any number t between M and m, there is a unit vector $\mathbf{x}$ such that $t = \mathbf{x}^TA\mathbf{x}$. Verify that $t = (1 - \alpha)m + \alpha M$ for some number α between 0 and 1. Then let $\mathbf{x} = \sqrt{1 - \alpha}\,\mathbf{u}_n + \sqrt{\alpha}\,\mathbf{u}_1$, and show that $\mathbf{x}^T\mathbf{x} = 1$ and $\mathbf{x}^TA\mathbf{x} = t$.

[**M**] In Exercises 14–17, follow the instructions given for Exercises 3–6.

14. $x_1x_2 + 3x_1x_3 + 30x_1x_4 + 30x_2x_3 + 3x_2x_4 + x_3x_4$

15. $3x_1x_2 + 5x_1x_3 + 7x_1x_4 + 7x_2x_3 + 5x_2x_4 + 3x_3x_4$

16. $4x_1^2 - 6x_1x_2 - 10x_1x_3 - 10x_1x_4 - 6x_2x_3 - 6x_2x_4 - 2x_3x_4$

17. $-6x_1^2 - 10x_2^2 - 13x_3^2 - 13x_4^2 - 4x_1x_2 - 4x_1x_3 - 4x_1x_4 + 6x_3x_4$

SOLUTIONS TO PRACTICE PROBLEMS

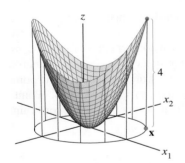

The maximum value of $Q(\mathbf{x})$ subject to $\mathbf{x}^T\mathbf{x} = 1$ is 4.

1. The matrix of the quadratic form is $A = \begin{bmatrix} 3 & 1 \\ 1 & 3 \end{bmatrix}$. It is easy to find the eigenvalues, 4 and 2, and corresponding unit eigenvectors, $\begin{bmatrix} 1/\sqrt{2} \\ 1/\sqrt{2} \end{bmatrix}$ and $\begin{bmatrix} -1/\sqrt{2} \\ 1/\sqrt{2} \end{bmatrix}$. So the desired change of variable is $\mathbf{x} = P\mathbf{y}$, where $P = \begin{bmatrix} 1/\sqrt{2} & -1/\sqrt{2} \\ 1/\sqrt{2} & 1/\sqrt{2} \end{bmatrix}$. (A common error here is to forget to normalize the eigenvectors.) The new quadratic form is $\mathbf{y}^TD\mathbf{y} = 4y_1^2 + 2y_2^2$.

2. The maximum of $Q(\mathbf{x})$ for $\mathbf{x}$ a unit vector is 4, and the maximum is attained at the unit eigenvector $\begin{bmatrix} 1/\sqrt{2} \\ 1/\sqrt{2} \end{bmatrix}$. [A common incorrect answer is $\begin{bmatrix} 1 \\ 0 \end{bmatrix}$. This vector maximizes the quadratic form $\mathbf{y}^TD\mathbf{y}$ instead of $Q(\mathbf{x})$.]

7.4 | THE SINGULAR VALUE DECOMPOSITION

The diagonalization theorems in Sections 5.3 and 7.1 play a part in many interesting applications. Unfortunately, as we know, not all matrices can be factored as $A = PDP^{-1}$ with D diagonal. However, a factorization $A = QDP^{-1}$ *is* possible for *any* $m \times n$ matrix A! A special factorization of this type, called the *singular value decomposition*, is one of the most useful matrix factorizations in applied linear algebra.

The singular value decomposition is based on the following property of the ordinary diagonalization that can be imitated for rectangular matrices: The absolute values of the eigenvalues of a symmetric matrix A measure the amounts that A stretches or shrinks certain vectors (the eigenvectors). If $A\mathbf{x} = \lambda\mathbf{x}$ and $\|\mathbf{x}\| = 1$, then

$$\|A\mathbf{x}\| = \|\lambda\mathbf{x}\| = |\lambda|\,\|\mathbf{x}\| = |\lambda| \tag{1}$$

If λ_1 is the eigenvalue with the greatest magnitude, then a corresponding unit eigenvector $\mathbf{v}_1$ identifies a direction in which the stretching effect of A is greatest. That is, the length of $A\mathbf{x}$ is maximized when $\mathbf{x} = \mathbf{v}_1$, and $\|A\mathbf{v}_1\| = |\lambda_1|$, by (1). This description of $\mathbf{v}_1$ and $|\lambda_1|$ has an analogue for rectangular matrices that will lead to the singular value decomposition.

EXAMPLE 1 If $A = \begin{bmatrix} 4 & 11 & 14 \\ 8 & 7 & -2 \end{bmatrix}$, then the linear transformation $\mathbf{x} \mapsto A\mathbf{x}$ maps the unit sphere $\{\mathbf{x} : \|\mathbf{x}\| = 1\}$ in $\mathbb{R}^3$ onto an ellipse in $\mathbb{R}^2$, shown in Fig. 1. Find a unit vector $\mathbf{x}$ at which the length $\|A\mathbf{x}\|$ is maximized, and compute this maximum length.

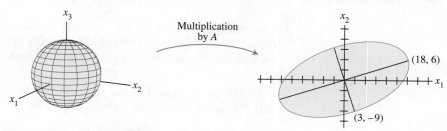

FIGURE 1 A transformation from $\mathbb{R}^3$ to $\mathbb{R}^2$.

SOLUTION The quantity $\|A\mathbf{x}\|^2$ is maximized at the same $\mathbf{x}$ that maximizes $\|A\mathbf{x}\|$, and $\|A\mathbf{x}\|^2$ is easier to study. Observe that

$$\|A\mathbf{x}\|^2 = (A\mathbf{x})^T (A\mathbf{x}) = \mathbf{x}^T A^T A\mathbf{x} = \mathbf{x}^T (A^T A)\mathbf{x}$$

Also, $A^T A$ is a symmetric matrix, since $(A^T A)^T = A^T A^{TT} = A^T A$. So the problem now is to maximize the quadratic form $\mathbf{x}^T (A^T A)\mathbf{x}$ subject to the constraint $\|\mathbf{x}\| = 1$. By Theorem 6 in Section 7.3, the maximum value is the greatest eigenvalue λ_1 of $A^T A$. Also, the maximum value is attained at a unit eigenvector of $A^T A$ corresponding to λ_1.

For the matrix A in this example,

$$A^T A = \begin{bmatrix} 4 & 8 \\ 11 & 7 \\ 14 & -2 \end{bmatrix} \begin{bmatrix} 4 & 11 & 14 \\ 8 & 7 & -2 \end{bmatrix} = \begin{bmatrix} 80 & 100 & 40 \\ 100 & 170 & 140 \\ 40 & 140 & 200 \end{bmatrix}$$

The eigenvalues of $A^T A$ are $\lambda_1 = 360$, $\lambda_2 = 90$, and $\lambda_3 = 0$. Corresponding unit eigenvectors are, respectively,

$$\mathbf{v}_1 = \begin{bmatrix} 1/3 \\ 2/3 \\ 2/3 \end{bmatrix}, \quad \mathbf{v}_2 = \begin{bmatrix} -2/3 \\ -1/3 \\ 2/3 \end{bmatrix}, \quad \mathbf{v}_3 = \begin{bmatrix} 2/3 \\ -2/3 \\ 1/3 \end{bmatrix}$$

The maximum value of $\|A\mathbf{x}\|^2$ is 360, attained when $\mathbf{x}$ is the unit vector $\mathbf{v}_1$. The vector $A\mathbf{v}_1$ is a point on the ellipse in Fig. 1 farthest from the origin, namely,

$$A\mathbf{v}_1 = \begin{bmatrix} 4 & 11 & 14 \\ 8 & 7 & -2 \end{bmatrix} \begin{bmatrix} 1/3 \\ 2/3 \\ 2/3 \end{bmatrix} = \begin{bmatrix} 18 \\ 6 \end{bmatrix}$$

For $\|\mathbf{x}\| = 1$, the maximum value of $\|A\mathbf{x}\|$ is $\|A\mathbf{v}_1\| = \sqrt{360} = 6\sqrt{10}$. ∎

Example 1 suggests that the effect of A on the unit sphere in $\mathbb{R}^3$ is related to the quadratic form $\mathbf{x}^T (A^T A)\mathbf{x}$. In fact, the entire geometric behavior of the transformation $\mathbf{x} \mapsto A\mathbf{x}$ is captured by this quadratic form, as we shall see.

The Singular Values of an $m \times n$ Matrix

Let A be an $m \times n$ matrix. Then A^TA is symmetric and can be orthogonally diagonalized. Let $\{\mathbf{v}_1, \ldots, \mathbf{v}_n\}$ be an orthonormal basis for $\mathbb{R}^n$ consisting of eigenvectors of A^TA, and let $\lambda_1, \ldots, \lambda_n$ be the associated eigenvalues of A^TA. Then, for $1 \leq i \leq n$,

$$\begin{aligned}
\|A\mathbf{v}_i\|^2 = (A\mathbf{v}_i)^TA\mathbf{v}_i &= \mathbf{v}_i^T A^TA\mathbf{v}_i \\
&= \mathbf{v}_i^T(\lambda_i\mathbf{v}_i) \qquad \text{Since } \mathbf{v}_i \text{ is an eigenvector of } A^TA \\
&= \lambda_i \qquad\qquad \text{Since } \mathbf{v}_i \text{ is a unit vector}
\end{aligned} \tag{2}$$

So the eigenvalues of A^TA are all nonnegative. By renumbering, if necessary, we may assume that the eigenvalues are arranged so that

$$\lambda_1 \geq \lambda_2 \geq \cdots \geq \lambda_n \geq 0$$

The **singular values** of A are the square roots of the eigenvalues of A^TA, denoted by $\sigma_1, \ldots, \sigma_n$, and they are arranged in decreasing order. That is, $\sigma_i = \sqrt{\lambda_i}$ for $1 \leq i \leq n$. By equation (2), *the singular values of A are the lengths of the vectors $A\mathbf{v}_1, \ldots, A\mathbf{v}_n$.*

EXAMPLE 2 Let A be the matrix in Example 1. Since the eigenvalues of A^TA are 360, 90, and 0, the singular values of A are

$$\sigma_1 = \sqrt{360} = 6\sqrt{10}, \quad \sigma_2 = \sqrt{90} = 3\sqrt{10}, \quad \sigma_3 = 0$$

From Example 1, the first singular value of A is the maximum of $\|A\mathbf{x}\|$ over all unit vectors, and the maximum is attained at the unit eigenvector $\mathbf{v}_1$. Theorem 7 in Section 7.3 shows that the second singular value of A is the maximum of $\|A\mathbf{x}\|$ over all unit vectors that are *orthogonal to* $\mathbf{v}_1$, and this maximum is attained at the second unit eigenvector, $\mathbf{v}_2$ (Exercise 22). For the $\mathbf{v}_2$ in Example 1,

$$A\mathbf{v}_2 = \begin{bmatrix} 4 & 11 & 14 \\ 8 & 7 & -2 \end{bmatrix} \begin{bmatrix} -2/3 \\ -1/3 \\ 2/3 \end{bmatrix} = \begin{bmatrix} 3 \\ -9 \end{bmatrix}$$

This point is on the minor axis of the ellipse in Fig. 1, just as $A\mathbf{v}_1$ is on the major axis. (See Fig. 2.) The first two singular values of A are the lengths of the major and minor semiaxes of the ellipse. ∎

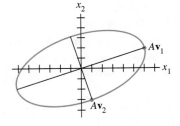

FIGURE 2

The fact that $A\mathbf{v}_1$ and $A\mathbf{v}_2$ are orthogonal in Fig. 2 is no accident, as the next theorem shows.

THEOREM 9

Suppose $\{\mathbf{v}_1, \ldots, \mathbf{v}_n\}$ is an orthonormal basis of $\mathbb{R}^n$ consisting of eigenvectors of A^TA, arranged so that the corresponding eigenvalues of A^TA satisfy $\lambda_1 \geq \cdots \geq \lambda_n$, and suppose A has r nonzero singular values. Then $\{A\mathbf{v}_1, \ldots, A\mathbf{v}_r\}$ is an orthogonal basis for Col A, and rank $A = r$.

PROOF Because $\mathbf{v}_i$ and $\lambda_j\mathbf{v}_j$ are orthogonal for $i \neq j$,

$$(A\mathbf{v}_i)^T(A\mathbf{v}_j) = \mathbf{v}_i^T A^TA\mathbf{v}_j = \mathbf{v}_i^T(\lambda_j\mathbf{v}_j) = 0$$

Thus $\{A\mathbf{v}_1, \ldots, A\mathbf{v}_n\}$ is an orthogonal set. Furthermore, since the lengths of the vectors $A\mathbf{v}_1, \ldots, A\mathbf{v}_n$ are the singular values of A, and since there are r nonzero singular values, $A\mathbf{v}_i \neq \mathbf{0}$ if and only if $1 \leq i \leq r$. So $A\mathbf{v}_1, \ldots, A\mathbf{v}_r$ are linearly independent vectors, and they are in Col A. Finally, for any $\mathbf{y}$ in Col A—say, $\mathbf{y} = A\mathbf{x}$—we can write $\mathbf{x} = c_1\mathbf{v}_1 + \cdots + c_n\mathbf{v}_n$, and

$$\begin{aligned} \mathbf{y} = A\mathbf{x} &= c_1 A\mathbf{v}_1 + \cdots + c_r A\mathbf{v}_r + c_{r+1} A\mathbf{v}_{r+1} + \cdots + c_n A\mathbf{v}_n \\ &= c_1 A\mathbf{v}_1 + \cdots + c_r A\mathbf{v}_r + 0 + \cdots + 0 \end{aligned}$$

Thus $\mathbf{y}$ is in Span $\{A\mathbf{v}_1, \ldots, A\mathbf{v}_r\}$, which shows that $\{A\mathbf{v}_1, \ldots, A\mathbf{v}_r\}$ is an (orthogonal) basis for Col A. Hence rank $A = \dim \text{Col } A = r$. ∎

NUMERICAL NOTE

In some cases, the rank of A may be very sensitive to small changes in the entries of A. The obvious method of counting the number of pivot columns in A does not work well if A is row reduced by a computer. Roundoff error often creates an echelon form with full rank.

In practice, the most reliable way to estimate the rank of a large matrix A is to count the number of nonzero singular values. In this case, extremely small nonzero singular values are assumed to be zero for all practical purposes, and the *effective rank* of the matrix is the number obtained by counting the remaining nonzero singular values.[1]

The Singular Value Decomposition

The decomposition of A involves an $m \times n$ "diagonal" matrix Σ of the form

$$\Sigma = \begin{bmatrix} D & 0 \\ 0 & 0 \end{bmatrix} \begin{matrix} \\ \leftarrow m - r \text{ rows} \end{matrix} \tag{3}$$

$$\underset{n - r \text{ columns}}{\uparrow}$$

where D is an $r \times r$ diagonal matrix for some r not exceeding the smaller of m and n. (If r equals m or n or both, some or all of the zero matrices do not appear.)

THEOREM 10

The Singular Value Decomposition

Let A be an $m \times n$ matrix with rank r. Then there exists an $m \times n$ matrix Σ as in (3) for which the diagonal entries in D are the first r singular values of A, $\sigma_1 \geq \sigma_2 \geq \cdots \geq \sigma_r > 0$, and there exist an $m \times m$ orthogonal matrix U and an $n \times n$ orthogonal matrix V such that

$$A = U\Sigma V^T$$

Any factorization $A = U\Sigma V^T$, with U and V orthogonal, Σ as in (3), and positive diagonal entries in D, is called a **singular value decomposition** (or **SVD**) of A. The matrices U and V are not uniquely determined by A, but the diagonal entries of Σ are necessarily the singular values of A. See Exercise 19. The columns of U in such a decomposition are called **left singular vectors** of A, and the columns of V are called **right singular vectors** of A.

---

[1] In general, rank estimation is not a simple problem. For a discussion of the subtle issues involved, see Philip E. Gill, Walter Murray, and Margaret H. Wright, *Numerical Linear Algebra and Optimization*, vol. 1 (Redwood City, CA: Addison-Wesley, 1991), Sec. 5.8.

PROOF Let λ_i and $\mathbf{v}_i$ be as in Theorem 9, so that $\{A\mathbf{v}_1, \ldots, A\mathbf{v}_r\}$ is an orthogonal basis for Col A. Normalize each $A\mathbf{v}_i$ to obtain an orthonormal basis $\{\mathbf{u}_1, \ldots, \mathbf{u}_r\}$, where

$$\mathbf{u}_i = \frac{1}{\|A\mathbf{v}_i\|}A\mathbf{v}_i = \frac{1}{\sigma_i}A\mathbf{v}_i$$

and

$$A\mathbf{v}_i = \sigma_i \mathbf{u}_i \qquad (1 \leq i \leq r) \tag{4}$$

Now extend $\{\mathbf{u}_1, \ldots, \mathbf{u}_r\}$ to an orthonormal basis $\{\mathbf{u}_1, \ldots, \mathbf{u}_m\}$ of $\mathbb{R}^m$, and let

$$U = [\,\mathbf{u}_1 \quad \mathbf{u}_2 \quad \cdots \quad \mathbf{u}_m\,] \quad \text{and} \quad V = [\,\mathbf{v}_1 \quad \mathbf{v}_2 \quad \cdots \quad \mathbf{v}_n\,]$$

By construction, U and V are orthogonal matrices. Also, from (4),

$$AV = [\,A\mathbf{v}_1 \quad \cdots \quad A\mathbf{v}_r \quad \mathbf{0} \quad \cdots \quad \mathbf{0}\,] = [\,\sigma_1\mathbf{u}_1 \quad \cdots \quad \sigma_r\mathbf{u}_r \quad \mathbf{0} \quad \cdots \quad \mathbf{0}\,]$$

Let D be the diagonal matrix with diagonal entries $\sigma_1, \ldots, \sigma_r$, and let Σ be as in (3) above. Then

$$U\Sigma = [\,\mathbf{u}_1 \quad \mathbf{u}_2 \quad \cdots \quad \mathbf{u}_m\,]\begin{bmatrix} \sigma_1 & & & & 0 & \\ & \sigma_2 & & & & 0 \\ & & \ddots & & & \\ 0 & & & \sigma_r & & \\ & & 0 & & & 0 \end{bmatrix}$$

$$= [\,\sigma_1\mathbf{u}_1 \quad \cdots \quad \sigma_r\mathbf{u}_r \quad \mathbf{0} \quad \cdots \quad \mathbf{0}\,]$$

$$= AV$$

Since V is an orthogonal matrix, $U\Sigma V^T = AVV^T = A$. ∎

The next two examples focus attention on the internal structure of a singular value decomposition. An efficient and numerically stable algorithm for this decomposition would use a different approach. See the Numerical Note at the end of the section.

EXAMPLE 3 Use the results of Examples 1 and 2 to construct a singular value decomposition of $A = \begin{bmatrix} 4 & 11 & 14 \\ 8 & 7 & -2 \end{bmatrix}$.

SOLUTION A construction can be divided into three steps.

SG | Computing an SVD 7-10

Step 1. **Find an orthogonal diagonalization of A^TA.** That is, find the eigenvalues of A^TA and a corresponding orthonormal set of eigenvectors. If A had only two columns, the calculations could be done by hand. Larger matrices usually require a matrix program.[2] However, for the matrix A here, the eigendata for A^TA are provided in Example 1.

Step 2. **Set up V and Σ.** Arrange the eigenvalues of A^TA in decreasing order. In Example 1, the eigenvalues are already listed in decreasing order: 360, 90, and 0. The corresponding unit eigenvectors, $\mathbf{v}_1, \mathbf{v}_2$, and $\mathbf{v}_3$, are the right singular vectors of A. Using Example 1, construct

$$V = [\,\mathbf{v}_1 \quad \mathbf{v}_2 \quad \mathbf{v}_3\,] = \begin{bmatrix} 1/3 & -2/3 & 2/3 \\ 2/3 & -1/3 & -2/3 \\ 2/3 & 2/3 & 1/3 \end{bmatrix}$$

---

[2] See the *Study Guide* for software and graphing calculator commands. MATLAB, for instance, can produce both the eigenvalues and the eigenvectors with one command, `eig`.

The square roots of the eigenvalues are the singular values:

$$\sigma_1 = 6\sqrt{10}, \quad \sigma_2 = 3\sqrt{10}, \quad \sigma_3 = 0$$

The nonzero singular values are the diagonal entries of D. The matrix Σ is the same size as A, with D in its upper left corner and with 0's elsewhere.

$$D = \begin{bmatrix} 6\sqrt{10} & 0 \\ 0 & 3\sqrt{10} \end{bmatrix}, \quad \Sigma = [\, D \;\; 0 \,] = \begin{bmatrix} 6\sqrt{10} & 0 & 0 \\ 0 & 3\sqrt{10} & 0 \end{bmatrix}$$

Step 3. ***Construct U.*** When A has rank r, the first r columns of U are the normalized vectors obtained from $A\mathbf{v}_1, \ldots, A\mathbf{v}_r$. In this example, A has two nonzero singular values, so rank $A = 2$. Recall from equation (2) and the paragraph before Example 2 that $\|A\mathbf{v}_1\| = \sigma_1$ and $\|A\mathbf{v}_2\| = \sigma_2$. Thus

$$\mathbf{u}_1 = \frac{1}{\sigma_1} A\mathbf{v}_1 = \frac{1}{6\sqrt{10}} \begin{bmatrix} 18 \\ 6 \end{bmatrix} = \begin{bmatrix} 3/\sqrt{10} \\ 1/\sqrt{10} \end{bmatrix}$$

$$\mathbf{u}_2 = \frac{1}{\sigma_2} A\mathbf{v}_2 = \frac{1}{3\sqrt{10}} \begin{bmatrix} 3 \\ -9 \end{bmatrix} = \begin{bmatrix} 1/\sqrt{10} \\ -3/\sqrt{10} \end{bmatrix}$$

Note that $\{\mathbf{u}_1, \mathbf{u}_2\}$ is already a basis for $\mathbb{R}^2$. Thus no additional vectors are needed for U, and $U = [\, \mathbf{u}_1 \;\; \mathbf{u}_2 \,]$. The singular value decomposition of A is

$$A = \underset{U}{\begin{bmatrix} 3/\sqrt{10} & 1/\sqrt{10} \\ 1/\sqrt{10} & -3/\sqrt{10} \end{bmatrix}} \underset{\Sigma}{\begin{bmatrix} 6\sqrt{10} & 0 & 0 \\ 0 & 3\sqrt{10} & 0 \end{bmatrix}} \underset{V^T}{\begin{bmatrix} 1/3 & 2/3 & 2/3 \\ -2/3 & -1/3 & 2/3 \\ 2/3 & -2/3 & 1/3 \end{bmatrix}} \quad \blacksquare$$

EXAMPLE 4 Find a singular value decomposition of $A = \begin{bmatrix} 1 & -1 \\ -2 & 2 \\ 2 & -2 \end{bmatrix}$.

SOLUTION First, compute $A^T A = \begin{bmatrix} 9 & -9 \\ -9 & 9 \end{bmatrix}$. The eigenvalues of $A^T A$ are 18 and 0, with corresponding unit eigenvectors

$$\mathbf{v}_1 = \begin{bmatrix} 1/\sqrt{2} \\ -1/\sqrt{2} \end{bmatrix}, \quad \mathbf{v}_2 = \begin{bmatrix} 1/\sqrt{2} \\ 1/\sqrt{2} \end{bmatrix}$$

These unit vectors form the columns of V:

$$V = [\, \mathbf{v}_1 \;\; \mathbf{v}_2 \,] = \begin{bmatrix} 1/\sqrt{2} & 1/\sqrt{2} \\ -1/\sqrt{2} & 1/\sqrt{2} \end{bmatrix}$$

The singular values are $\sigma_1 = \sqrt{18} = 3\sqrt{2}$ and $\sigma_2 = 0$. Since there is only one nonzero singular value, the "matrix" D may be written as a single number. That is, $D = 3\sqrt{2}$. The matrix Σ is the same size as A, with D in its upper left corner:

$$\Sigma = \begin{bmatrix} D & 0 \\ 0 & 0 \\ 0 & 0 \end{bmatrix} = \begin{bmatrix} 3\sqrt{2} & 0 \\ 0 & 0 \\ 0 & 0 \end{bmatrix}$$

To construct U, first construct $A\mathbf{v}_1$ and $A\mathbf{v}_2$:

$$A\mathbf{v}_1 = \begin{bmatrix} 2/\sqrt{2} \\ -4/\sqrt{2} \\ 4/\sqrt{2} \end{bmatrix}, \quad A\mathbf{v}_2 = \begin{bmatrix} 0 \\ 0 \\ 0 \end{bmatrix}$$

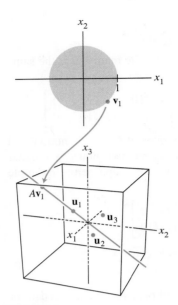

FIGURE 3

As a check on the calculations, verify that $\|A\mathbf{v}_1\| = \sigma_1 = 3\sqrt{2}$. Of course, $A\mathbf{v}_2 = \mathbf{0}$ because $\|A\mathbf{v}_2\| = \sigma_2 = 0$. The only column found for U so far is

$$\mathbf{u}_1 = \frac{1}{3\sqrt{2}} A\mathbf{v}_1 = \begin{bmatrix} 1/3 \\ -2/3 \\ 2/3 \end{bmatrix}$$

The other columns of U are found by extending the set $\{\mathbf{u}_1\}$ to an orthonormal basis for $\mathbb{R}^3$. In this case, we need two orthogonal unit vectors $\mathbf{u}_2$ and $\mathbf{u}_3$ that are orthogonal to $\mathbf{u}_1$. (See Fig. 3.) Each vector must satisfy $\mathbf{u}_1^T \mathbf{x} = 0$, which is equivalent to the equation $x_1 - 2x_2 + 2x_3 = 0$. A basis for the solution set of this equation is

$$\mathbf{w}_1 = \begin{bmatrix} 2 \\ 1 \\ 0 \end{bmatrix}, \quad \mathbf{w}_2 = \begin{bmatrix} -2 \\ 0 \\ 1 \end{bmatrix}$$

(Check that $\mathbf{w}_1$ and $\mathbf{w}_2$ are each orthogonal to $\mathbf{u}_1$.) Apply the Gram–Schmidt process (with normalizations) to $\{\mathbf{w}_1, \mathbf{w}_2\}$, and obtain

$$\mathbf{u}_2 = \begin{bmatrix} 2/\sqrt{5} \\ 1/\sqrt{5} \\ 0 \end{bmatrix}, \quad \mathbf{u}_3 = \begin{bmatrix} -2/\sqrt{45} \\ 4/\sqrt{45} \\ 5/\sqrt{45} \end{bmatrix}$$

Finally, set $U = [\mathbf{u}_1 \ \ \mathbf{u}_2 \ \ \mathbf{u}_3]$, take Σ and V^T from above, and write

$$A = \begin{bmatrix} 1 & -1 \\ -2 & 2 \\ 2 & -2 \end{bmatrix} = \begin{bmatrix} 1/3 & 2/\sqrt{5} & -2/\sqrt{45} \\ -2/3 & 1/\sqrt{5} & 4/\sqrt{45} \\ 2/3 & 0 & 5/\sqrt{45} \end{bmatrix} \begin{bmatrix} 3\sqrt{2} & 0 \\ 0 & 0 \\ 0 & 0 \end{bmatrix} \begin{bmatrix} 1/\sqrt{2} & -1/\sqrt{2} \\ 1/\sqrt{2} & 1/\sqrt{2} \end{bmatrix}$$

∎

Applications of the Singular Value Decomposition

The SVD is often used to estimate the rank of a matrix, as noted above. Several other numerical applications are described briefly below, and an application to image processing is presented in Section 7.5.

EXAMPLE 5 (The Condition Number) Most numerical calculations involving an equation $A\mathbf{x} = \mathbf{b}$ are as reliable as possible when the SVD of A is used. The two orthogonal matrices U and V do not affect lengths of vectors or angles between vectors (Theorem 7 in Section 6.2). Any possible instabilities in numerical calculations are identified in Σ. If the singular values of A are extremely large or small, roundoff errors are almost inevitable, but an error analysis is aided by knowing the entries in Σ and V.

If A is an invertible $n \times n$ matrix, then the ratio σ_1/σ_n of the largest and smallest singular values gives the **condition number** of A. Exercises 41–43 in Section 2.3 showed how the condition number affects the sensitivity of a solution of $A\mathbf{x} = \mathbf{b}$ to changes (or errors) in the entries of A. (Actually, a "condition number" of A can be computed in several ways, but the definition given here is widely used for studying $A\mathbf{x} = \mathbf{b}$.)

∎

EXAMPLE 6 (Bases for Fundamental Subspaces) Given an SVD for an $m \times n$ matrix A, let $\mathbf{u}_1, \ldots, \mathbf{u}_m$ be the left singular vectors, $\mathbf{v}_1, \ldots, \mathbf{v}_n$ the right singular vectors, and $\sigma_1, \ldots, \sigma_n$ the singular values, and let r be the rank of A. By Theorem 9,

$$\{\mathbf{u}_1, \ldots, \mathbf{u}_r\} \tag{5}$$

is an orthonormal basis for Col A.

Recall from Theorem 3 in Section 6.1 that $(\text{Col } A)^{\perp} = \text{Nul } A^T$. Hence

$$\{\mathbf{u}_{r+1}, \ldots, \mathbf{u}_m\} \tag{6}$$

is an orthonormal basis for Nul A^T.

Since $\|A\mathbf{v}_i\| = \sigma_i$ for $1 \leq i \leq n$, and σ_i is 0 if and only if $i > r$, the vectors $\mathbf{v}_{r+1}, \ldots, \mathbf{v}_n$ span a subspace of Nul A of dimension $n - r$. By the Rank Theorem, dim Nul $A = n - \text{rank } A$. It follows that

$$\{\mathbf{v}_{r+1}, \ldots, \mathbf{v}_n\} \tag{7}$$

is an orthonormal basis for Nul A, by the Basis Theorem (in Section 4.5).

From (5) and (6), the orthogonal complement of Nul A^T is Col A. Interchanging A and A^T, note that $(\text{Nul } A)^{\perp} = \text{Col } A^T = \text{Row } A$. Hence, from (7),

$$\{\mathbf{v}_1, \ldots, \mathbf{v}_r\} \tag{8}$$

is an orthonormal basis for Row A.

Figure 4 summarizes (5)–(8), but shows the orthogonal basis $\{\sigma_1 \mathbf{u}_1, \ldots, \sigma_r \mathbf{u}_r\}$ for Col A instead of the normalized basis, to remind you that $A\mathbf{v}_i = \sigma_i \mathbf{u}_i$ for $1 \leq i \leq r$. Explicit orthonormal bases for the four fundamental subspaces determined by A are useful in some calculations, particularly in constrained optimization problems. ∎

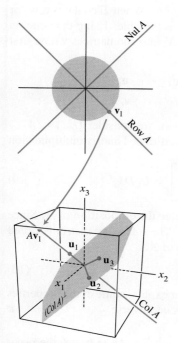

The fundamental subspaces in Example 4.

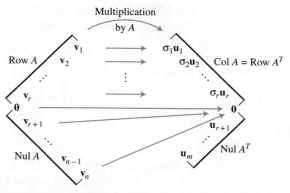

FIGURE 4 The four fundamental subspaces and the action of A.

The four fundamental subspaces and the concept of singular values provide the final statements of the Invertible Matrix Theorem. (Recall that statements about A^T have been omitted from the theorem, to avoid nearly doubling the number of statements.) The other statements were given in Sections 2.3, 2.9, 3.2, 4.6, and 5.2.

THEOREM

The Invertible Matrix Theorem (concluded)

Let A be an $n \times n$ matrix. Then the following statements are each equivalent to the statement that A is an invertible matrix.

u. $(\text{Col } A)^{\perp} = \{\mathbf{0}\}$.

v. $(\text{Nul } A)^{\perp} = \mathbb{R}^n$.

w. Row $A = \mathbb{R}^n$.

x. A has n nonzero singular values.

EXAMPLE 7 (Reduced SVD and the Pseudoinverse of A) When Σ contains rows or columns of zeros, a more compact decomposition of A is possible. Using the notation established above, let $r = \text{rank } A$, and partition U and V into submatrices whose first blocks contain r columns:

$$U = [\, U_r \quad U_{m-r} \,], \quad \text{where } U_r = [\, \mathbf{u}_1 \quad \cdots \quad \mathbf{u}_r \,]$$
$$V = [\, V_r \quad V_{n-r} \,], \quad \text{where } V_r = [\, \mathbf{v}_1 \quad \cdots \quad \mathbf{v}_r \,]$$

Then U_r is $m \times r$ and V_r is $n \times r$. (To simplify notation, we consider U_{m-r} or V_{n-r} even though one of them may have no columns.) Then partitioned matrix multiplication shows that

$$A = [\, U_r \quad U_{m-r} \,] \begin{bmatrix} D & 0 \\ 0 & 0 \end{bmatrix} \begin{bmatrix} V_r^T \\ V_{n-r}^T \end{bmatrix} = U_r D V_r^T \tag{9}$$

This factorization of A is called a **reduced singular value decomposition** of A. Since the diagonal entries in D are nonzero, D is invertible. The following matrix is called the **pseudoinverse** (also, the **Moore–Penrose inverse**) of A:

$$A^+ = V_r D^{-1} U_r^T \tag{10}$$

Supplementary Exercises 12–14 at the end of the chapter explore some of the properties of the reduced singular value decomposition and the pseudoinverse. ∎

EXAMPLE 8 (Least-Squares Solution) Given the equation $A\mathbf{x} = \mathbf{b}$, use the pseudoinverse of A in (10) to define

$$\hat{\mathbf{x}} = A^+ \mathbf{b} = V_r D^{-1} U_r^T \mathbf{b}$$

Then, from the SVD in (9),

$$
\begin{aligned}
A\hat{\mathbf{x}} &= (U_r D V_r^T)(V_r D^{-1} U_r^T \mathbf{b}) \\
&= U_r D D^{-1} U_r^T \mathbf{b} \qquad \text{Because } V_r^T V_r = I_r \\
&= U_r U_r^T \mathbf{b}
\end{aligned}
$$

It follows from (5) that $U_r U_r^T \mathbf{b}$ is the orthogonal projection $\hat{\mathbf{b}}$ of $\mathbf{b}$ onto Col A. (See Theorem 10 in Section 6.3.) Thus $\hat{\mathbf{x}}$ is a least-squares solution of $A\mathbf{x} = \mathbf{b}$. In fact, this $\hat{\mathbf{x}}$ has the smallest length among all least-squares solutions of $A\mathbf{x} = \mathbf{b}$. See Supplementary Exercise 14. ∎

NUMERICAL NOTE

Examples 1–4 and the exercises illustrate the concept of singular values and suggest how to perform calculations by hand. In practice, the computation of $A^T A$ should be avoided, since any errors in the entries of A are squared in the entries of $A^T A$. There exist fast iterative methods that produce the singular values and singular vectors of A accurately to many decimal places.

Further Reading

Horn, Roger A., and Charles R. Johnson, *Matrix Analysis* (Cambridge: Cambridge University Press, 1990).

Long, Cliff, "Visualization of Matrix Singular Value Decomposition." *Mathematics Magazine* **56** (1983), pp. 161–167.

Moler, C. B., and D. Morrison, "Singular Value Analysis of Cryptograms." *Amer. Math. Monthly* **90** (1983), pp. 78–87.

Strang, Gilbert, *Linear Algebra and Its Applications,* 4th ed. (Belmont, CA: Brooks/ Cole, 2005).

Watkins, David S., *Fundamentals of Matrix Computations* (New York: Wiley, 1991), pp. 390–398, 409–421.

| PRACTICE PROBLEM

WEB   Given a singular value decomposition, $A = U \Sigma V^T$, find an SVD of A^T. How are the singular values of A and A^T related?

7.4 EXERCISES

Find the singular values of the matrices in Exercises 1–4.

1. $\begin{bmatrix} 1 & 0 \\ 0 & -3 \end{bmatrix}$ **2.** $\begin{bmatrix} -5 & 0 \\ 0 & 0 \end{bmatrix}$

3. $\begin{bmatrix} \sqrt{6} & 1 \\ 0 & \sqrt{6} \end{bmatrix}$ **4.** $\begin{bmatrix} \sqrt{3} & 2 \\ 0 & \sqrt{3} \end{bmatrix}$

Find an SVD of each matrix in Exercises 5–12. [*Hint:* In Exercise 11, one choice for U is $\begin{bmatrix} -1/3 & 2/3 & 2/3 \\ 2/3 & -1/3 & 2/3 \\ 2/3 & 2/3 & -1/3 \end{bmatrix}$. In

Exercise 12, one column of U can be $\begin{bmatrix} 1/\sqrt{6} \\ -2/\sqrt{6} \\ 1/\sqrt{6} \end{bmatrix}$.]

5. $\begin{bmatrix} -3 & 0 \\ 0 & 0 \end{bmatrix}$ **6.** $\begin{bmatrix} -2 & 0 \\ 0 & -1 \end{bmatrix}$

7. $\begin{bmatrix} 2 & -1 \\ 2 & 2 \end{bmatrix}$ **8.** $\begin{bmatrix} 2 & 3 \\ 0 & 2 \end{bmatrix}$

9. $\begin{bmatrix} 7 & 1 \\ 0 & 0 \\ 5 & 5 \end{bmatrix}$ **10.** $\begin{bmatrix} 4 & -2 \\ 2 & -1 \\ 0 & 0 \end{bmatrix}$

11. $\begin{bmatrix} -3 & 1 \\ 6 & -2 \\ 6 & -2 \end{bmatrix}$ **12.** $\begin{bmatrix} 1 & 1 \\ 0 & 1 \\ -1 & 1 \end{bmatrix}$

13. Find the SVD of $A = \begin{bmatrix} 3 & 2 & 2 \\ 2 & 3 & -2 \end{bmatrix}$ [*Hint:* Work with A^T.]

14. In Exercise 7, find a unit vector **x** at which $A\mathbf{x}$ has maximum length.

15. Suppose the factorization below is an SVD of a matrix A, with the entries in U and V rounded to two decimal places.

$$A = \begin{bmatrix} .40 & -.78 & .47 \\ .37 & -.33 & -.87 \\ -.84 & -.52 & -.16 \end{bmatrix} \begin{bmatrix} 7.10 & 0 & 0 \\ 0 & 3.10 & 0 \\ 0 & 0 & 0 \end{bmatrix}$$
$$\times \begin{bmatrix} .30 & -.51 & -.81 \\ .76 & .64 & -.12 \\ .58 & -.58 & .58 \end{bmatrix}$$

a. What is the rank of A?

b. Use this decomposition of A, with no calculations, to write a basis for Col A and a basis for Nul A. [*Hint:* First write the columns of V.]

16. Repeat Exercise 15 for the following SVD of a 3×4 matrix A:

$$A = \begin{bmatrix} -.86 & -.11 & -.50 \\ .31 & .68 & -.67 \\ .41 & -.73 & -.55 \end{bmatrix} \begin{bmatrix} 12.48 & 0 & 0 & 0 \\ 0 & 6.34 & 0 & 0 \\ 0 & 0 & 0 & 0 \end{bmatrix}$$
$$\times \begin{bmatrix} .66 & -.03 & -.35 & .66 \\ -.13 & -.90 & -.39 & -.13 \\ .65 & .08 & -.16 & -.73 \\ -.34 & .42 & -.84 & -.08 \end{bmatrix}$$

In Exercises 17–24, A is an $m \times n$ matrix with a singular value decomposition $A = U \Sigma V^T$, where U is an $m \times m$ orthogonal matrix, Σ is an $m \times n$ "diagonal" matrix with r positive entries and no negative entries, and V is an $n \times n$ orthogonal matrix. Justify each answer.

17. Suppose A is square and invertible. Find a singular value decomposition of A^{-1}.

18. Show that if A is square, then $|\det A|$ is the product of the singular values of A.

19. Show that the columns of V are eigenvectors of $A^T A$, the columns of U are eigenvectors of AA^T, and the diagonal entries of Σ are the singular values of A. [*Hint:* Use the SVD to compute $A^T A$ and AA^T.]

20. Show that if A is an $n \times n$ positive definite matrix, then an orthogonal diagonalization $A = PDP^T$ is a singular value decomposition of A.

21. Show that if P is an orthogonal $m \times m$ matrix, then PA has the same singular values as A.

22. Justify the statement in Example 2 that the second singular value of a matrix A is the maximum of $\|A\mathbf{x}\|$ as $\mathbf{x}$ varies over all unit vectors orthogonal to $\mathbf{v}_1$, with $\mathbf{v}_1$ a right singular vector corresponding to the first singular value of A. [*Hint:* Use Theorem 7 in Section 7.3.]

23. Let $U = [\,\mathbf{u}_1 \ \cdots \ \mathbf{u}_m\,]$ and $V = [\,\mathbf{v}_1 \ \cdots \ \mathbf{v}_n\,]$, where the $\mathbf{u}_i$ and $\mathbf{v}_i$ are as in Theorem 10. Show that

$$A = \sigma_1 \mathbf{u}_1 \mathbf{v}_1^T + \sigma_2 \mathbf{u}_2 \mathbf{v}_2^T + \cdots + \sigma_r \mathbf{u}_r \mathbf{v}_r^T.$$

24. Using the notation of Exercise 23, show that $A^T \mathbf{u}_j = \sigma_j \mathbf{v}_j$ for $1 \le j \le r = \operatorname{rank} A$.

25. Let $T : \mathbb{R}^n \to \mathbb{R}^m$ be a linear transformation. Describe how to find a basis $\mathcal{B}$ for $\mathbb{R}^n$ and a basis $\mathcal{C}$ for $\mathbb{R}^m$ such that the matrix for T relative to $\mathcal{B}$ and $\mathcal{C}$ is an $m \times n$ "diagonal" matrix.

[M] Compute an SVD of each matrix in Exercises 26 and 27. Report the final matrix entries accurate to two decimal places. Use the method of Examples 3 and 4.

26. $A = \begin{bmatrix} -18 & 13 & -4 & 4 \\ 2 & 19 & -4 & 12 \\ -14 & 11 & -12 & 8 \\ -2 & 21 & 4 & 8 \end{bmatrix}$

27. $A = \begin{bmatrix} 6 & -8 & -4 & 5 & -4 \\ 2 & 7 & -5 & -6 & 4 \\ 0 & -1 & -8 & 2 & 2 \\ -1 & -2 & 4 & 4 & -8 \end{bmatrix}$

28. **[M]** Compute the singular values of the 4×4 matrix in Exercise 9 in Section 2.3, and compute the condition number σ_1/σ_4.

29. **[M]** Compute the singular values of the 5×5 matrix in Exercise 10 in Section 2.3, and compute the condition number σ_1/σ_5.

| SOLUTION TO PRACTICE PROBLEM

If $A = U\Sigma V^T$, where Σ is $m \times n$, then $A^T = (V^T)^T \Sigma^T U^T = V\Sigma^T U^T$. This is an SVD of A^T because V and U are orthogonal matrices and Σ^T is an $n \times m$ "diagonal" matrix. Since Σ and Σ^T have the same nonzero diagonal entries, A and A^T have the same nonzero singular values. [*Note:* If A is $2 \times n$, then AA^T is only 2×2 and its eigenvalues may be easier to compute (by hand) than the eigenvalues of A^TA.]

7.5 | APPLICATIONS TO IMAGE PROCESSING AND STATISTICS

The satellite photographs in this chapter's introduction provide an example of multidimensional, or *multivariate*, data—information organized so that each datum in the data set is identified with a point (vector) in $\mathbb{R}^n$. The main goal of this section is to explain a technique, called *principal component analysis*, used to analyze such multivariate data. The calculations will illustrate the use of orthogonal diagonalization and the singular value decomposition.

Principal component analysis can be applied to any data that consist of lists of measurements made on a collection of objects or individuals. For instance, consider a chemical process that produces a plastic material. To monitor the process, 300 samples are taken of the material produced, and each sample is subjected to a battery of eight tests, such as melting point, density, ductility, tensile strength, and so on. The laboratory report for each sample is a vector in $\mathbb{R}^8$, and the set of such vectors forms an 8×300 matrix, called the **matrix of observations**.

Loosely speaking, we can say that the process control data are eight-dimensional. The next two examples describe data that can be visualized graphically.

EXAMPLE 1 An example of two-dimensional data is given by a set of weights and heights of N college students. Let $\mathbf{X}_j$ denote the **observation vector** in $\mathbb{R}^2$ that lists the weight and height of the jth student. If w denotes weight and h height, then the matrix of observations has the form

$$\begin{bmatrix} w_1 & w_2 & \cdots & w_N \\ h_1 & h_2 & \cdots & h_N \end{bmatrix}$$
$$\uparrow \quad \uparrow \qquad \uparrow$$
$$\mathbf{X}_1 \quad \mathbf{X}_2 \qquad \mathbf{X}_N$$

The set of observation vectors can be visualized as a two-dimensional *scatter plot*. See Fig. 1. ∎

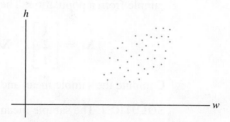

FIGURE 1 A scatter plot of observation vectors $\mathbf{X}_1, \ldots, \mathbf{X}_N$.

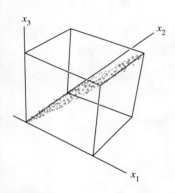

FIGURE 2

A scatter plot of spectral data for a satellite image.

EXAMPLE 2 The first three photographs of Railroad Valley, Nevada, shown in the chapter introduction, can be viewed as *one* image of the region, with *three spectral components*, because simultaneous measurements of the region were made at three separate wavelengths. Each photograph gives different information about the same physical region. For instance, the first pixel in the upper-left corner of each photograph corresponds to the same place on the ground (about 30 meters by 30 meters). To each pixel there corresponds an observation vector in $\mathbb{R}^3$ that lists the signal intensities for that pixel in the three spectral bands.

Typically, the image is 2000×2000 pixels, so there are 4 million pixels in the image. The data for the image form a matrix with 3 rows and 4 million columns (with columns arranged in any convenient order). In this case, the "multidimensional" character of the data refers to the three *spectral* dimensions rather than the two *spatial* dimensions that naturally belong to any photograph. The data can be visualized as a cluster of 4 million points in $\mathbb{R}^3$, perhaps as in Fig. 2. ∎

Mean and Covariance

To prepare for principal component analysis, let $[\,\mathbf{X}_1 \ \cdots \ \mathbf{X}_N\,]$ be a $p \times N$ matrix of observations, such as described above. The **sample mean**, $\mathbf{M}$, of the observation vectors $\mathbf{X}_1, \ldots, \mathbf{X}_N$ is given by

$$\mathbf{M} = \frac{1}{N}(\mathbf{X}_1 + \cdots + \mathbf{X}_N)$$

For the data in Fig. 1, the sample mean is the point in the "center" of the scatter plot. For $k = 1, \ldots, N$, let

$$\hat{\mathbf{X}}_k = \mathbf{X}_k - \mathbf{M}$$

The columns of the $p \times N$ matrix

$$B = [\,\hat{\mathbf{X}}_1 \ \ \hat{\mathbf{X}}_2 \ \ \cdots \ \ \hat{\mathbf{X}}_N\,]$$

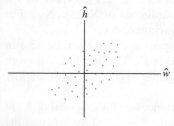

FIGURE 3

Weight–height data in mean-deviation form.

have a zero sample mean, and B is said to be in **mean-deviation form**. When the sample mean is subtracted from the data in Fig. 1, the resulting scatter plot has the form in Fig. 3.

The (**sample**) **covariance matrix** is the $p \times p$ matrix S defined by

$$S = \frac{1}{N-1} BB^T$$

Since any matrix of the form BB^T is positive semidefinite, so is S. (See Exercise 25 in Section 7.2 with B and B^T interchanged.)

EXAMPLE 3   Three measurements are made on each of four individuals in a random sample from a population. The observation vectors are

$$\mathbf{X}_1 = \begin{bmatrix} 1 \\ 2 \\ 1 \end{bmatrix}, \quad \mathbf{X}_2 = \begin{bmatrix} 4 \\ 2 \\ 13 \end{bmatrix}, \quad \mathbf{X}_3 = \begin{bmatrix} 7 \\ 8 \\ 1 \end{bmatrix}, \quad \mathbf{X}_4 = \begin{bmatrix} 8 \\ 4 \\ 5 \end{bmatrix}$$

Compute the sample mean and the covariance matrix.

SOLUTION The sample mean is

$$\mathbf{M} = \frac{1}{4} \left(\begin{bmatrix} 1 \\ 2 \\ 1 \end{bmatrix} + \begin{bmatrix} 4 \\ 2 \\ 13 \end{bmatrix} + \begin{bmatrix} 7 \\ 8 \\ 1 \end{bmatrix} + \begin{bmatrix} 8 \\ 4 \\ 5 \end{bmatrix} \right) = \frac{1}{4} \begin{bmatrix} 20 \\ 16 \\ 20 \end{bmatrix} = \begin{bmatrix} 5 \\ 4 \\ 5 \end{bmatrix}$$

Subtract the sample mean from $\mathbf{X}_1, \dots, \mathbf{X}_4$ to obtain

$$\hat{\mathbf{X}}_1 = \begin{bmatrix} -4 \\ -2 \\ -4 \end{bmatrix}, \quad \hat{\mathbf{X}}_2 = \begin{bmatrix} -1 \\ -2 \\ 8 \end{bmatrix}, \quad \hat{\mathbf{X}}_3 = \begin{bmatrix} 2 \\ 4 \\ -4 \end{bmatrix}, \quad \hat{\mathbf{X}}_4 = \begin{bmatrix} 3 \\ 0 \\ 0 \end{bmatrix}$$

and

$$B = \begin{bmatrix} -4 & -1 & 2 & 3 \\ -2 & -2 & 4 & 0 \\ -4 & 8 & -4 & 0 \end{bmatrix}$$

The sample covariance matrix is

$$S = \frac{1}{3} \begin{bmatrix} -4 & -1 & 2 & 3 \\ -2 & -2 & 4 & 0 \\ -4 & 8 & -4 & 0 \end{bmatrix} \begin{bmatrix} -4 & -2 & -4 \\ -1 & -2 & 8 \\ 2 & 4 & -4 \\ 3 & 0 & 0 \end{bmatrix}$$

$$= \frac{1}{3} \begin{bmatrix} 30 & 18 & 0 \\ 18 & 24 & -24 \\ 0 & -24 & 96 \end{bmatrix} = \begin{bmatrix} 10 & 6 & 0 \\ 6 & 8 & -8 \\ 0 & -8 & 32 \end{bmatrix} \qquad ∎$$

To discuss the entries in $S = [s_{ij}]$, let $\mathbf{X}$ represent a vector that varies over the set of observation vectors and denote the coordinates of $\mathbf{X}$ by $x_1, \dots, x_p$. Then x_1, for example, is a scalar that varies over the set of first coordinates of $\mathbf{X}_1, \dots, \mathbf{X}_N$. For $j = 1, \dots, p$, the diagonal entry s_{jj} in S is called the **variance** of x_j.

The variance of x_j measures the spread of the values of x_j. (See Exercise 13.) In Example 3, the variance of x_1 is 10 and the variance of x_3 is 32. The fact that 32 is more than 10 indicates that the set of third entries in the response vectors contains a wider spread of values than the set of first entries.

The **total variance** of the data is the sum of the variances on the diagonal of S. In general, the sum of the diagonal entries of a square matrix S is called the **trace** of the matrix, written $\text{tr}(S)$. Thus

$$\{\text{total variance}\} = \text{tr}(S)$$

The entry s_{ij} in S for $i \neq j$ is called the **covariance** of x_i and x_j. Observe that in Example 3, the covariance between x_1 and x_3 is 0 because the $(1, 3)$-entry in S is 0. Statisticians say that x_1 and x_3 are **uncorrelated**. Analysis of the multivariate data in $\mathbf{X}_1, \ldots, \mathbf{X}_N$ is greatly simplified when most or all of the variables $x_1, \ldots, x_p$ are uncorrelated, that is, when the covariance matrix of $\mathbf{X}_1, \ldots, \mathbf{X}_N$ is diagonal or nearly diagonal.

Principal Component Analysis

For simplicity, assume that the matrix $[\, \mathbf{X}_1 \;\; \cdots \;\; \mathbf{X}_N \,]$ is already in mean-deviation form. The goal of principal component analysis is to find an orthogonal $p \times p$ matrix $P = [\, \mathbf{u}_1 \;\; \cdots \;\; \mathbf{u}_p \,]$ that determines a change of variable, $\mathbf{X} = P\mathbf{Y}$, or

$$
\begin{bmatrix} x_1 \\ x_2 \\ \vdots \\ x_p \end{bmatrix} = \begin{bmatrix} \mathbf{u}_1 & \mathbf{u}_2 & \cdots & \mathbf{u}_p \end{bmatrix} \begin{bmatrix} y_1 \\ y_2 \\ \vdots \\ y_p \end{bmatrix}
$$

with the property that the new variables $y_1, \ldots, y_p$ are uncorrelated and are arranged in order of decreasing variance.

The orthogonal change of variable $\mathbf{X} = P\mathbf{Y}$ means that each observation vector $\mathbf{X}_k$ receives a "new name," $\mathbf{Y}_k$, such that $\mathbf{X}_k = P\mathbf{Y}_k$. Notice that $\mathbf{Y}_k$ is the coordinate vector of $\mathbf{X}_k$ with respect to the columns of P, and $\mathbf{Y}_k = P^{-1}\mathbf{X}_k = P^T\mathbf{X}_k$ for $k = 1, \ldots, N$.

It is not difficult to verify that for any orthogonal P, the covariance matrix of $\mathbf{Y}_1, \ldots, \mathbf{Y}_N$ is P^TSP (Exercise 11). So the desired orthogonal matrix P is one that makes P^TSP diagonal. Let D be a diagonal matrix with the eigenvalues $\lambda_1, \ldots, \lambda_p$ of S on the diagonal, arranged so that $\lambda_1 \geq \lambda_2 \geq \cdots \geq \lambda_p \geq 0$, and let P be an orthogonal matrix whose columns are the corresponding unit eigenvectors $\mathbf{u}_1, \ldots, \mathbf{u}_p$. Then $S = PDP^T$ and $P^TSP = D$.

The unit eigenvectors $\mathbf{u}_1, \ldots, \mathbf{u}_p$ of the covariance matrix S are called the **principal components** of the data (in the matrix of observations). The **first principal component** is the eigenvector corresponding to the largest eigenvalue of S, the **second principal component** is the eigenvector corresponding to the second largest eigenvalue, and so on.

The first principal component $\mathbf{u}_1$ determines the new variable y_1 in the following way. Let $c_1, \ldots, c_p$ be the entries in $\mathbf{u}_1$. Since $\mathbf{u}_1^T$ is the first row of P^T, the equation $\mathbf{Y} = P^T\mathbf{X}$ shows that

$$
y_1 = \mathbf{u}_1^T\mathbf{X} = c_1x_1 + c_2x_2 + \cdots + c_px_p
$$

Thus y_1 is a linear combination of the original variables $x_1, \ldots, x_p$, using the entries in the eigenvector $\mathbf{u}_1$ as weights. In a similar fashion, $\mathbf{u}_2$ determines the variable y_2, and so on.

EXAMPLE 4 The initial data for the multispectral image of Railroad Valley (Example 2) consisted of 4 million vectors in $\mathbb{R}^3$. The associated covariance matrix is[1]

$$
S = \begin{bmatrix} 2382.78 & 2611.84 & 2136.20 \\ 2611.84 & 3106.47 & 2553.90 \\ 2136.20 & 2553.90 & 2650.71 \end{bmatrix}
$$

---

[1] Data for Example 4 and Exercises 5 and 6 were provided by Earth Satellite Corporation, Rockville, Maryland.

Find the principal components of the data, and list the new variable determined by the first principal component.

SOLUTION The eigenvalues of S and the associated principal components (the unit eigenvectors) are

$$\lambda_1 = 7614.23 \qquad \lambda_2 = 427.63 \qquad \lambda_3 = 98.10$$

$$\mathbf{u}_1 = \begin{bmatrix} .5417 \\ .6295 \\ .5570 \end{bmatrix} \qquad \mathbf{u}_2 = \begin{bmatrix} -.4894 \\ -.3026 \\ .8179 \end{bmatrix} \qquad \mathbf{u}_3 = \begin{bmatrix} .6834 \\ -.7157 \\ .1441 \end{bmatrix}$$

Using two decimal places for simplicity, the variable for the first principal component is

$$y_1 = .54x_1 + .63x_2 + .56x_3$$

This equation was used to create photograph (d) in the chapter introduction. The variables x_1, x_2, x_3 are the signal intensities in the three spectral bands. The values of x_1, converted to a gray scale between black and white, produced photograph (a). Similarly, the values of x_2 and x_3 produced photographs (b) and (c), respectively. At each pixel in photograph (d), the gray scale value is computed from y_1, a weighted linear combination of x_1, x_2, x_3. In this sense, photograph (d) "displays" the first principal component of the data. ∎

In Example 4, the covariance matrix for the transformed data, using variables y_1, y_2, y_3, is

$$D = \begin{bmatrix} 7614.23 & 0 & 0 \\ 0 & 427.63 & 0 \\ 0 & 0 & 98.10 \end{bmatrix}$$

Although D is obviously simpler than the original covariance matrix S, the merit of constructing the new variables is not yet apparent. However, the variances of the variables y_1, y_2, y_3 appear on the diagonal of D, and obviously the first variance in D is much larger than the other two. As we shall see, this fact will permit us to view the data as essentially one-dimensional rather than three-dimensional.

Reducing the Dimension of Multivariate Data

Principal component analysis is potentially valuable for applications in which most of the variation, or dynamic range, in the data is due to variations in *only a few* of the new variables, $y_1, \ldots, y_p$.

It can be shown that an orthogonal change of variables, $\mathbf{X} = P\mathbf{Y}$, does not change the total variance of the data. (Roughly speaking, this is true because left-multiplication by P does not change the lengths of vectors or the angles between them. See Exercise 12.) This means that if $S = PDP^T$, then

$$\begin{Bmatrix} \text{total variance} \\ \text{of } x_1, \ldots, x_p \end{Bmatrix} = \begin{Bmatrix} \text{total variance} \\ \text{of } y_1, \ldots, y_p \end{Bmatrix} = \text{tr}(D) = \lambda_1 + \cdots + \lambda_p$$

The variance of y_j is λ_j, and the quotient $\lambda_j / \text{tr}(S)$ measures the fraction of the total variance that is "explained" or "captured" by y_j.

EXAMPLE 5 Compute the various percentages of variance of the Railroad Valley multispectral data that are displayed in the principal component photographs, (d)–(f), shown in the chapter introduction.

SOLUTION The total variance of the data is

$$\text{tr}(D) = 7614.23 + 427.63 + 98.10 = 8139.96$$

[Verify that this number also equals $\text{tr}(S)$.] The percentages of the total variance explained by the principal components are

First component	Second component	Third component
$\dfrac{7614.23}{8139.96} = 93.5\%$	$\dfrac{427.63}{8139.96} = 5.3\%$	$\dfrac{98.10}{8139.96} = 1.2\%$

In a sense, 93.5% of the information collected by Landsat for the Railroad Valley region is displayed in photograph (d), with 5.3% in (e) and only 1.2% remaining for (f). ∎

The calculations in Example 5 show that the data have practically no variance in the third (new) coordinate. The values of y_3 are all close to zero. Geometrically, the data points lie nearly in the plane $y_3 = 0$, and their locations can be determined fairly accurately by knowing only the values of y_1 and y_2. In fact, y_2 also has relatively small variance, which means that the points lie approximately along a line, and the data are essentially one-dimensional. See Fig. 2, in which the data resemble a popsicle stick.

Characterizations of Principal Component Variables

If $y_1, \ldots, y_p$ arise from a principal component analysis of a $p \times N$ matrix of observations, then the variance of y_1 is as large as possible in the following sense: If $\mathbf{u}$ is any unit vector and if $y = \mathbf{u}^T \mathbf{X}$, then the variance of the values of y as $\mathbf{X}$ varies over the original data $\mathbf{X}_1, \ldots, \mathbf{X}_N$ turns out to be $\mathbf{u}^T S \mathbf{u}$. By Theorem 8 in Section 7.3, the maximum value of $\mathbf{u}^T S \mathbf{u}$, over all unit vectors $\mathbf{u}$, is the largest eigenvalue λ_1 of S, and this variance is attained when $\mathbf{u}$ is the corresponding eigenvector $\mathbf{u}_1$. In the same way, Theorem 8 shows that y_2 has maximum possible variance among all variables $y = \mathbf{u}^T \mathbf{X}$ that are *uncorrelated* with y_1. Likewise, y_3 has maximum possible variance among all variables uncorrelated with both y_1 and y_2, and so on.

NUMERICAL NOTE

The singular value decomposition is the main tool for performing principal component analysis in practical applications. If B is a $p \times N$ matrix of observations in mean-deviation form, and if $A = \left(1/\sqrt{N-1} \right) B^T$, then $A^T A$ is the covariance matrix, S. The squares of the singular values of A are the p eigenvalues of S, and the right singular vectors of A are the principal components of the data.

As mentioned in Section 7.4, iterative calculation of the SVD of A is faster and more accurate than an eigenvalue decomposition of S. This is particularly true, for instance, in the hyperspectral image processing (with $p = 224$) mentioned in the chapter introduction. Principal component analysis is completed in seconds on specialized workstations.

Further Reading

Lillesand, Thomas M., and Ralph W. Kiefer, *Remote Sensing and Image Interpretation*, 4th ed. (New York: John Wiley, 2000).

| PRACTICE PROBLEMS

The following table lists the weights and heights of five boys:

Boy	#1	#2	#3	#4	#5
Weight (lb)	120	125	125	135	145
Height (in.)	61	60	64	68	72

1. Find the covariance matrix for the data.

2. Make a principal component analysis of the data to find a single *size index* that explains most of the variation in the data.

7.5 EXERCISES

In Exercises 1 and 2, convert the matrix of observations to mean-deviation form, and construct the sample covariance matrix.

1. $\begin{bmatrix} 19 & 22 & 6 & 3 & 2 & 20 \\ 12 & 6 & 9 & 15 & 13 & 5 \end{bmatrix}$

2. $\begin{bmatrix} 1 & 5 & 2 & 6 & 7 & 3 \\ 3 & 11 & 6 & 8 & 15 & 11 \end{bmatrix}$

3. Find the principal components of the data for Exercise 1.

4. Find the principal components of the data for Exercise 2.

5. [M] A Landsat image with three spectral components was made of Homestead Air Force Base in Florida (after the base was hit by hurricane Andrew in 1992). The covariance matrix of the data is shown below. Find the first principal component of the data, and compute the percentage of the total variance that is contained in this component.

$$S = \begin{bmatrix} 164.12 & 32.73 & 81.04 \\ 32.73 & 539.44 & 249.13 \\ 81.04 & 249.13 & 189.11 \end{bmatrix}$$

6. [M] The covariance matrix below was obtained from a Landsat image of the Columbia River in Washington, using data from three spectral bands. Let x_1, x_2, x_3 denote the spectral components of each pixel in the image. Find a new variable of the form $y_1 = c_1 x_1 + c_2 x_2 + c_3 x_3$ that has maximum possible variance, subject to the constraint that $c_1^2 + c_2^2 + c_3^2 = 1$. What percentage of the total variance in the data is explained by y_1?

$$S = \begin{bmatrix} 29.64 & 18.38 & 5.00 \\ 18.38 & 20.82 & 14.06 \\ 5.00 & 14.06 & 29.21 \end{bmatrix}$$

7. Let x_1, x_2 denote the variables for the two-dimensional data in Exercise 1. Find a new variable y_1 of the form $y_1 = c_1 x_1 + c_2 x_2$, with $c_1^2 + c_2^2 = 1$, such that y_1 has maximum possible variance over the given data. How much of the variance in the data is explained by y_1?

8. Repeat Exercise 7 for the data in Exercise 2.

9. Suppose three tests are administered to a random sample of college students. Let $\mathbf{X}_1, \ldots, \mathbf{X}_N$ be observation vectors in $\mathbb{R}^3$ that list the three scores of each student, and for $j = 1, 2, 3$, let x_j denote a student's score on the jth exam. Suppose the covariance matrix of the data is

$$S = \begin{bmatrix} 5 & 2 & 0 \\ 2 & 6 & 2 \\ 0 & 2 & 7 \end{bmatrix}$$

Let y be an "index" of student performance, with $y = c_1 x_1 + c_2 x_2 + c_3 x_3$ and $c_1^2 + c_2^2 + c_3^2 = 1$. Choose c_1, c_2, c_3 so that the variance of y over the data set is as large as possible. [*Hint:* The eigenvalues of the sample covariance matrix are $\lambda = 3, 6,$ and 9.]

10. [M] Repeat Exercise 9 with $S = \begin{bmatrix} 5 & 4 & 2 \\ 4 & 11 & 4 \\ 2 & 4 & 5 \end{bmatrix}$.

11. Given multivariate data $\mathbf{X}_1, \ldots, \mathbf{X}_N$ (in $\mathbb{R}^p$) in mean-deviation form, let P be a $p \times p$ matrix, and define $\mathbf{Y}_k = P^T \mathbf{X}_k$ for $k = 1, \ldots, N$.

    a. Show that $\mathbf{Y}_1, \ldots, \mathbf{Y}_N$ are in mean-deviation form. [*Hint:* Let $\mathbf{w}$ be the vector in $\mathbb{R}^N$ with a 1 in each entry. Then $[\mathbf{X}_1 \ \cdots \ \mathbf{X}_N]\mathbf{w} = \mathbf{0}$ (the zero vector in $\mathbb{R}^p$).]

    b. Show that if the covariance matrix of $\mathbf{X}_1, \ldots, \mathbf{X}_N$ is S, then the covariance matrix of $\mathbf{Y}_1, \ldots, \mathbf{Y}_N$ is $P^T S P$.

12. Let $\mathbf{X}$ denote a vector that varies over the columns of a $p \times N$ matrix of observations, and let P be a $p \times p$ orthogonal matrix. Show that the change of variable $\mathbf{X} = P\mathbf{Y}$ does not change the total variance of the data. [*Hint:* By Exercise 11, it suffices to show that tr $(P^T S P) = $ tr (S). Use a property of the trace mentioned in Exercise 25 in Section 5.4.]

13. The sample covariance matrix is a generalization of a formula for the variance of a sample of N scalar measurements, say, $t_1, \ldots, t_N$. If m is the average of $t_1, \ldots, t_N$, then the *sample variance* is given by

$$\frac{1}{N-1} \sum_{k=1}^{n} (t_k - m)^2 \tag{1}$$

Show how the sample covariance matrix, S, defined prior to Example 3, may be written in a form similar to (1). [*Hint:* Use partitioned matrix multiplication to write S as $1/(N-1)$ times the sum of N matrices of size $p \times p$. For $1 \le k \le N$, write $\mathbf{X}_k - \mathbf{M}$ in place of $\hat{\mathbf{X}}_k$.]

SOLUTIONS TO PRACTICE PROBLEMS

1. First arrange the data in mean-deviation form. The sample mean vector is easily seen to be $\mathbf{M} = \begin{bmatrix} 130 \\ 65 \end{bmatrix}$. Subtract $\mathbf{M}$ from the observation vectors (the columns in the table) and obtain

$$B = \begin{bmatrix} -10 & -5 & -5 & 5 & 15 \\ -4 & -5 & -1 & 3 & 7 \end{bmatrix}$$

Then the sample covariance matrix is

$$S = \frac{1}{5-1} \begin{bmatrix} -10 & -5 & -5 & 5 & 15 \\ -4 & -5 & -1 & 3 & 7 \end{bmatrix} \begin{bmatrix} -10 & -4 \\ -5 & -5 \\ -5 & -1 \\ 5 & 3 \\ 15 & 7 \end{bmatrix}$$

$$= \frac{1}{4} \begin{bmatrix} 400 & 190 \\ 190 & 100 \end{bmatrix} = \begin{bmatrix} 100.0 & 47.5 \\ 47.5 & 25.0 \end{bmatrix}$$

2. The eigenvalues of S are (to two decimal places)

$$\lambda_1 = 123.02 \quad \text{and} \quad \lambda_2 = 1.98$$

The unit eigenvector corresponding to λ_1 is $\mathbf{u} = \begin{bmatrix} .900 \\ .436 \end{bmatrix}$. (Since S is 2×2, the computations can be done by hand if a matrix program is not available.) For the *size index*, set

$$y = .900\hat{w} + .436\hat{h}$$

where $\hat{w}$ and $\hat{h}$ are weight and height, respectively, in mean-deviation form. The variance of this index over the data set is 123.02. Because the total variance is $\text{tr}(S) = 100 + 25 = 125$, the size index accounts for practically all (98.4%) of the variance of the data.

The original data for Practice Problem 1 and the line determined by the first principal component $\mathbf{u}$ are shown in Fig. 4. (In parametric vector form, the line is $\mathbf{x} = \mathbf{M} + t\mathbf{u}$.) It can be shown that the line is the best approximation to the data,

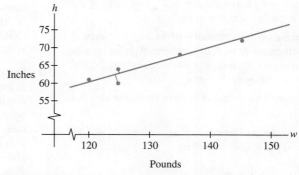

FIGURE 4 An orthogonal regression line determined by the first principal component of the data.

in the sense that the sum of the squares of the *orthogonal* distances to the line is minimized. In fact, principal component analysis is equivalent to what is termed *orthogonal regression*, but that is a story for another day. Perhaps we'll meet again.

CHAPTER 7 SUPPLEMENTARY EXERCISES

1. Mark each statement True or False. Justify each answer. In each part, A represents an $n \times n$ matrix.

   a. If A is orthogonally diagonalizable, then A is symmetric.

   b. If A is an orthogonal matrix, then A is symmetric.

   c. If A is an orthogonal matrix, then $\|A\mathbf{x}\| = \|\mathbf{x}\|$ for all $\mathbf{x}$ in $\mathbb{R}^n$.

   d. The principal axes of a quadratic form $\mathbf{x}^T A \mathbf{x}$ can be the columns of any matrix P that diagonalizes A.

   e. If P is an $n \times n$ matrix with orthogonal columns, then $P^T = P^{-1}$.

   f. If every coefficient in a quadratic form is positive, then the quadratic form is positive definite.

   g. If $\mathbf{x}^T A \mathbf{x} > 0$ for some $\mathbf{x}$, then the quadratic form $\mathbf{x}^T A \mathbf{x}$ is positive definite.

   h. By a suitable change of variable, any quadratic form can be changed into one with no cross-product term.

   i. The largest value of a quadratic form $\mathbf{x}^T A \mathbf{x}$, for $\|\mathbf{x}\| = 1$, is the largest entry on the diagonal of A.

   j. The maximum value of a positive definite quadratic form $\mathbf{x}^T A \mathbf{x}$ is the greatest eigenvalue of A.

   k. A positive definite quadratic form can be changed into a negative definite form by a suitable change of variable $\mathbf{x} = P\mathbf{u}$, for some orthogonal matrix P.

   l. An indefinite quadratic form is one whose eigenvalues are not definite.

   m. If P is an $n \times n$ orthogonal matrix, then the change of variable $\mathbf{x} = P\mathbf{u}$ transforms $\mathbf{x}^T A \mathbf{x}$ into a quadratic form whose matrix is $P^{-1}AP$.

   n. If U is $m \times n$ with orthogonal columns, then $UU^T\mathbf{x}$ is the orthogonal projection of $\mathbf{x}$ onto Col U.

   o. If B is $m \times n$ and $\mathbf{x}$ is a unit vector in $\mathbb{R}^n$, then $\|B\mathbf{x}\| \le \sigma_1$, where σ_1 is the first singular value of B.

   p. A singular value decomposition of an $m \times n$ matrix B can be written as $B = P\Sigma Q$, where P is an $m \times m$ orthogonal matrix, Q is an $n \times n$ orthogonal matrix, and Σ is an $m \times n$ "diagonal" matrix.

   q. If A is $n \times n$, then A and $A^T A$ have the same singular values.

2. Let $\{\mathbf{u}_1, \ldots, \mathbf{u}_n\}$ be an orthonormal basis for $\mathbb{R}^n$, and let $\lambda_1, \ldots, \lambda_n$ be any real scalars. Define

$$A = \lambda_1 \mathbf{u}_1 \mathbf{u}_1^T + \cdots + \lambda_n \mathbf{u}_n \mathbf{u}_n^T$$

   a. Show that A is symmetric.

   b. Show that $\lambda_1, \ldots, \lambda_n$ are the eigenvalues of A.

3. Let A be an $n \times n$ symmetric matrix of rank r. Explain why the spectral decomposition of A represents A as the sum of r rank 1 matrices.

4. Let A be an $n \times n$ symmetric matrix.

   a. Show that $(\text{Col } A)^\perp = \text{Nul } A$. [*Hint:* See Section 6.1.]

   b. Show that each $\mathbf{y}$ in $\mathbb{R}^n$ can be written in the form $\mathbf{y} = \hat{\mathbf{y}} + \mathbf{z}$, with $\hat{\mathbf{y}}$ in Col A and $\mathbf{z}$ in Nul A.

5. Show that if $\mathbf{v}$ is an eigenvector of an $n \times n$ matrix A and $\mathbf{v}$ corresponds to a nonzero eigenvalue of A, then $\mathbf{v}$ is in Col A. [*Hint:* Use the definition of an eigenvector.]

6. Let A be an $n \times n$ symmetric matrix. Use Exercise 5 and an eigenvector basis for $\mathbb{R}^n$ to give a second proof of the decomposition in Exercise 4(b).

7. Prove that an $n \times n$ matrix A is positive definite if and only if A admits a *Cholesky factorization*, namely, $A = R^T R$ for some invertible upper triangular matrix R whose diagonal entries are all positive. [*Hint:* Use a QR factorization and Exercise 26 in Section 7.2.]

8. Use Exercise 7 to show that if A is positive definite, then A has an LU factorization, $A = LU$, where U has positive pivots on its diagonal. (The converse is true, too.)

If A is $m \times n$, then the matrix $G = A^T A$ is called the *Gram matrix of A*. In this case, the entries of G are the inner products of the columns of A. (See Exercises 9 and 10.)

9. Show that the Gram matrix of any matrix A is positive semidefinite, with the same rank as A. (See the Exercises in Section 6.5.)

10. Show that if an $n \times n$ matrix G is positive semidefinite and has rank r, then G is the Gram matrix of some $r \times n$ matrix A. This is called a *rank-revealing factorization* of G. [*Hint:* Consider the spectral decomposition of G, and first write G as BB^T for an $n \times r$ matrix B.]

11. Prove that any $n \times n$ matrix A admits a *polar decomposition* of the form $A = PQ$, where P is an $n \times n$ positive semidefinite matrix with the same rank as A and where Q is an $n \times n$ orthogonal matrix. [*Hint:* Use a singular value decomposition, $A = U\Sigma V^T$, and observe that $A = (U\Sigma U^T)(UV^T)$.] This decomposition is used, for instance, in mechanical engineering to model the deformation of a material. The matrix P describes the stretching or compression of the material (in the directions of the eigenvectors of P), and Q describes the rotation of the material in space.

Exercises 12–14 concern an $m \times n$ matrix A with a reduced singular value decomposition, $A = U_r D V_r^T$, and the pseudoinverse $A^+ = V_r D^{-1} U_r^T$.

12. Verify the properties of A^+:

    a. For each $\mathbf{y}$ in $\mathbb{R}^m$, $AA^+\mathbf{y}$ is the orthogonal projection of $\mathbf{y}$ onto Col A.

    b. For each $\mathbf{x}$ in $\mathbb{R}^n$, $A^+A\mathbf{x}$ is the orthogonal projection of $\mathbf{x}$ onto Row A.

    c. $AA^+A = A$ and $A^+AA^+ = A^+$.

13. Suppose the equation $A\mathbf{x} = \mathbf{b}$ is consistent, and let $\mathbf{x}^+ = A^+\mathbf{b}$. By Exercise 23 in Section 6.3, there is exactly one vector $\mathbf{p}$ in Row A such that $A\mathbf{p} = \mathbf{b}$. The following steps prove that $\mathbf{x}^+ = \mathbf{p}$ and $\mathbf{x}^+$ is the *minimum length solution* of $A\mathbf{x} = \mathbf{b}$.

    a. Show that $\mathbf{x}^+$ is in Row A. [*Hint:* Write $\mathbf{b}$ as $A\mathbf{x}$ for some $\mathbf{x}$, and use Exercise 12.]

    b. Show that $\mathbf{x}^+$ is a solution of $A\mathbf{x} = \mathbf{b}$.

    c. Show that if $\mathbf{u}$ is any solution of $A\mathbf{x} = \mathbf{b}$, then $\|\mathbf{x}^+\| \le \|\mathbf{u}\|$, with equality only if $\mathbf{u} = \mathbf{x}^+$.

14. Given any $\mathbf{b}$ in $\mathbb{R}^m$, adapt Exercise 13 to show that $A^+\mathbf{b}$ is the *least-squares solution of minimum length*. [*Hint:* Consider the equation $A\mathbf{x} = \hat{\mathbf{b}}$, where $\hat{\mathbf{b}}$ is the orthogonal projection of $\mathbf{b}$ onto Col A.]

[**M**] In Exercises 15 and 16, construct the pseudoinverse of A. Begin by using a matrix program to produce the SVD of A, or, if that is not available, begin with an orthogonal diagonalization of A^TA. Use the pseudoinverse to solve $A\mathbf{x} = \mathbf{b}$, for $\mathbf{b} = (6, -1, -4, 6)$, and let $\hat{\mathbf{x}}$ be the solution. Make a calculation to verify that $\hat{\mathbf{x}}$ is in Row A. Find a nonzero vector $\mathbf{u}$ in Nul A, and verify that $\|\hat{\mathbf{x}}\| < \|\hat{\mathbf{x}} + \mathbf{u}\|$, which must be true by Exercise 13(c).

15. $A = \begin{bmatrix} -3 & -3 & -6 & 6 & 1 \\ -1 & -1 & -1 & 1 & -2 \\ 0 & 0 & -1 & 1 & -1 \\ 0 & 0 & -1 & 1 & -1 \end{bmatrix}$

16. $A = \begin{bmatrix} 4 & 0 & -1 & -2 & 0 \\ -5 & 0 & 3 & 5 & 0 \\ 2 & 0 & -1 & -2 & 0 \\ 6 & 0 & -3 & -6 & 0 \end{bmatrix}$

8

The Geometry of Vector Spaces

The Platonic Solids

In the city of Athens in 387 B.C., the Greek philosopher Plato founded an Academy, sometimes referred to as the world's first university. While the curriculum included astronomy, biology, political theory, and philosophy, the subject closest to his heart was geometry. Indeed, inscribed over the doors of his academy were these words: "*Let no one destitute of geometry enter my doors*."

The Greeks were greatly impressed by geometric patterns such as the regular solids. A polyhedron is called regular if its faces are congruent regular polygons and all the angles at the vertices are equal. As early as 150 years before Euclid, the Pythagoreans knew at least three of the regular solids: the tetrahedron (4 triangular faces), the cube (6 square faces), and the octahedron (8 triangular faces). (See Fig. 1.) These shapes occur naturally as crystals of common minerals. There are only five such regular solids, the remaining two being the dodecahedron (12 pentagonal faces) and the icosahedron (20 triangular faces).

Plato discussed the basic theory of these five solids in Book XIII of his *Elements*, and since then they have carried his name: the Platonic solids.

For centuries there was no need to envision geometric objects in more than three dimensions. But nowadays mathematicians regularly deal with objects in vector spaces having four, five, or even hundreds of dimensions. It is not necessarily clear what geometrical properties one might ascribe to these objects in higher dimensions.

For example, what properties do lines have in 2-space and planes have in 3-space that would be useful in higher dimensions? How can one characterize such objects? Sections 8.1 and 8.4 provide some answers. The hyperplanes of Section 8.4 will be important for understanding the multi-dimensional nature of the linear programming problems in Chapter 9.

What would the analogue of a polyhedron "look like" in more than three dimensions? A partial answer is provided by two-dimensional projections of the four-dimensional object, created in a manner analogous to two-dimensional projections of a three-dimensional object. Section 8.5 illustrates this idea for the four-dimensional "cube" and the four-dimensional "simplex."

The study of geometry in higher dimensions not only provides new ways of visualizing abstract algebraic concepts, but also creates tools that may be applied in $\mathbb{R}^3$. For instance, Sections 8.2 and 8.6 include applications to computer graphics, and Section 8.5 outlines a proof (in Exercise 21) that there are only five regular polyhedra in $\mathbb{R}^3$.

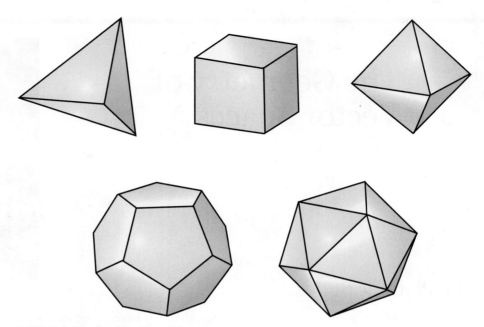

FIGURE 1 The five Platonic solids.

Most applications in earlier chapters involved algebraic calculations with subspaces and linear combinations of vectors. This chapter studies sets of vectors that can be visualized as geometric objects such as line segments, polygons, and solid objects. Individual vectors are viewed as points. The concepts introduced here are used in computer graphics, linear programming (in Chapter 9), and other areas of mathematics.[1]

Throughout the chapter, sets of vectors are described by linear combinations, but with various restrictions on the weights used in the combinations. For instance, in Section 8.1, the sum of the weights is 1, while in Section 8.2, the weights are positive and sum to 1. The visualizations are in $\mathbb{R}^2$ or $\mathbb{R}^3$, of course, but the concepts also apply to $\mathbb{R}^n$ and other vector spaces.

8.1 | AFFINE COMBINATIONS

An affine combination of vectors is a special kind of linear combination. Given vectors (or "points") $\mathbf{v}_1, \mathbf{v}_2, \ldots, \mathbf{v}_p$ in $\mathbb{R}^n$ and scalars $c_1, \ldots, c_p$, an **affine combination** of $\mathbf{v}_1, \mathbf{v}_2, \ldots, \mathbf{v}_p$ is a linear combination

$$c_1\mathbf{v}_1 + \cdots + c_p\mathbf{v}_p$$

such that the weights satisfy $c_1 + \cdots + c_p = 1$.

---

[1] See Foley, van Dam, Feiner, and Hughes, *Computer Graphics—Principles and Practice*, 2nd edition (Boston: Addison-Wesley, 1996), pp. 1083–1112. That material also discusses coordinate-free "affine spaces."

DEFINITION

The set of all affine combinations of points in a set S is called the **affine hull** (or **affine span**) of S, denoted by aff S.

The affine hull of a single point $\mathbf{v}_1$ is just the set $\{\mathbf{v}_1\}$, since it has the form $c_1\mathbf{v}_1$ where $c_1 = 1$. The affine hull of two distinct points is often written in a special way. Suppose $\mathbf{y} = c_1\mathbf{v}_1 + c_2\mathbf{v}_2$ with $c_1 + c_2 = 1$. Write t in place of c_2, so that $c_1 = 1 - c_2 = 1 - t$. Then the affine hull of $\{\mathbf{v}_1, \mathbf{v}_2\}$ is the set

$$\mathbf{y} = (1 - t)\mathbf{v}_1 + t\mathbf{v}_2, \quad \text{with } t \text{ in } \mathbb{R} \tag{1}$$

This set of points includes $\mathbf{v}_1$ (when $t = 0$) and $\mathbf{v}_2$ (when $t = 1$). If $\mathbf{v}_2 = \mathbf{v}_1$, then (1) again describes just one point. Otherwise, (1) describes the *line* through $\mathbf{v}_1$ and $\mathbf{v}_2$. To see this, rewrite (1) in the form

$$\mathbf{y} = \mathbf{v}_1 + t(\mathbf{v}_2 - \mathbf{v}_1) = \mathbf{p} + t\mathbf{u}, \quad \text{with } t \text{ in } \mathbb{R}$$

where $\mathbf{p}$ is $\mathbf{v}_1$ and $\mathbf{u}$ is $\mathbf{v}_2 - \mathbf{v}_1$. The set of all multiples of $\mathbf{u}$ is Span $\{\mathbf{u}\}$, the line through $\mathbf{u}$ and the origin. Adding $\mathbf{p}$ to each point on this line translates Span $\{\mathbf{u}\}$ into the line through $\mathbf{p}$ parallel to the line through $\mathbf{u}$ and the origin. See Fig. 1. (Compare this figure with Fig. 5 in Section 1.5.)

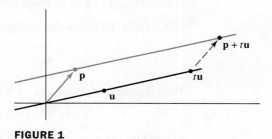

FIGURE 1

Figure 2 uses the original points $\mathbf{v}_1$ and $\mathbf{v}_2$, and displays aff $\{\mathbf{v}_1, \mathbf{v}_2\}$ as the line through $\mathbf{v}_1$ and $\mathbf{v}_2$.

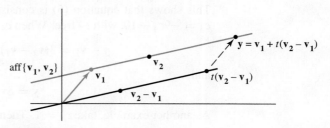

FIGURE 2

Notice that while the point $\mathbf{y}$ in Fig. 2 is an affine combination of $\mathbf{v}_1$ and $\mathbf{v}_2$, the point $\mathbf{y} - \mathbf{v}_1$ equals $t(\mathbf{v}_2 - \mathbf{v}_1)$, which is a linear combination (in fact, a multiple) of $\mathbf{v}_2 - \mathbf{v}_1$. This relation between $\mathbf{y}$ and $\mathbf{y} - \mathbf{v}_1$ holds for any affine combination of points, as the following theorem shows.

THEOREM 1

A point $\mathbf{y}$ in $\mathbb{R}^n$ is an affine combination of $\mathbf{v}_1, \ldots, \mathbf{v}_p$ in $\mathbb{R}^n$ if and only if $\mathbf{y} - \mathbf{v}_1$ is a linear combination of the translated points $\mathbf{v}_2 - \mathbf{v}_1, \ldots, \mathbf{v}_p - \mathbf{v}_1$.

PROOF If $\mathbf{y} - \mathbf{v}_1$ is a linear combination of $\mathbf{v}_2 - \mathbf{v}_1, \ldots, \mathbf{v}_p - \mathbf{v}_1$, there exist weights $c_2, \ldots, c_p$ such that

$$\mathbf{y} - \mathbf{v}_1 = c_2(\mathbf{v}_2 - \mathbf{v}_1) + \cdots + c_p(\mathbf{v}_p - \mathbf{v}_1) \tag{2}$$

Then

$$\mathbf{y} = (1 - c_2 - \cdots - c_p)\mathbf{v}_1 + c_2\mathbf{v}_2 + \cdots + c_p\mathbf{v}_p \tag{3}$$

and the weights in this linear combination sum to 1. So $\mathbf{y}$ is an affine combination of $\mathbf{v}_1, \ldots, \mathbf{v}_p$. Conversely, suppose

$$\mathbf{y} = c_1\mathbf{v}_1 + c_2\mathbf{v}_2 + \cdots + c_p\mathbf{v}_p \tag{4}$$

where $c_1 + \cdots + c_p = 1$. Since $c_1 = 1 - c_2 - \cdots - c_p$, equation (4) may be written as in (3), and this leads to (2), which shows that $\mathbf{y} - \mathbf{v}_1$ is a linear combination of $\mathbf{v}_2 - \mathbf{v}_1, \ldots, \mathbf{v}_p - \mathbf{v}_1$. ∎

In the statement of Theorem 1, the point $\mathbf{v}_1$ could be replaced by any of the other points in the list $\mathbf{v}_1, \ldots, \mathbf{v}_p$. Only the notation in the proof would change.

EXAMPLE 1   Let $\mathbf{v}_1 = \begin{bmatrix} 1 \\ 2 \end{bmatrix}$, $\mathbf{v}_2 = \begin{bmatrix} 2 \\ 5 \end{bmatrix}$, $\mathbf{v}_3 = \begin{bmatrix} 1 \\ 3 \end{bmatrix}$, $\mathbf{v}_4 = \begin{bmatrix} -2 \\ 2 \end{bmatrix}$, and $\mathbf{y} = \begin{bmatrix} 4 \\ 1 \end{bmatrix}$. If possible, write $\mathbf{y}$ as an affine combination of $\mathbf{v}_1, \mathbf{v}_2, \mathbf{v}_3$, and $\mathbf{v}_4$.

SOLUTION Compute the translated points

$$\mathbf{v}_2 - \mathbf{v}_1 = \begin{bmatrix} 1 \\ 3 \end{bmatrix}, \quad \mathbf{v}_3 - \mathbf{v}_1 = \begin{bmatrix} 0 \\ 1 \end{bmatrix}, \quad \mathbf{v}_4 - \mathbf{v}_1 = \begin{bmatrix} -3 \\ 0 \end{bmatrix}, \quad \mathbf{y} - \mathbf{v}_1 = \begin{bmatrix} 3 \\ -1 \end{bmatrix}$$

To find scalars c_2, c_3, and c_4 such that

$$c_2(\mathbf{v}_2 - \mathbf{v}_1) + c_3(\mathbf{v}_3 - \mathbf{v}_1) + c_4(\mathbf{v}_4 - \mathbf{v}_1) = \mathbf{y} - \mathbf{v}_1 \tag{5}$$

row reduce the augmented matrix having these points as columns:

$$\begin{bmatrix} 1 & 0 & -3 & 3 \\ 3 & 1 & 0 & -1 \end{bmatrix} \sim \begin{bmatrix} 1 & 0 & -3 & 3 \\ 0 & 1 & 9 & -10 \end{bmatrix}$$

This shows that equation (5) is consistent, and the general solution is $c_2 = 3c_4 + 3$, $c_3 = -9c_4 - 10$, with c_4 free. When $c_4 = 0$,

$$\mathbf{y} - \mathbf{v}_1 = 3(\mathbf{v}_2 - \mathbf{v}_1) - 10(\mathbf{v}_3 - \mathbf{v}_1) + 0(\mathbf{v}_4 - \mathbf{v}_1)$$

and

$$\mathbf{y} = 8\mathbf{v}_1 + 3\mathbf{v}_2 - 10\mathbf{v}_3$$

As another example, take $c_4 = 1$. Then $c_2 = 6$ and $c_3 = -19$, so

$$\mathbf{y} - \mathbf{v}_1 = 6(\mathbf{v}_2 - \mathbf{v}_1) - 19(\mathbf{v}_3 - \mathbf{v}_1) + 1(\mathbf{v}_4 - \mathbf{v}_1)$$

and

$$\mathbf{y} = 13\mathbf{v}_1 + 6\mathbf{v}_2 - 19\mathbf{v}_3 + \mathbf{v}_4 \quad \blacksquare$$

While the procedure in Example 1 works for arbitrary points $\mathbf{v}_1, \mathbf{v}_2, \ldots, \mathbf{v}_p$ in $\mathbb{R}^n$, the question can be answered more directly if the chosen points $\mathbf{v}_i$ are a basis for $\mathbb{R}^n$. For example, let $\mathcal{B} = \{\mathbf{b}_1, \ldots, \mathbf{b}_n\}$ be such a basis. Then any $\mathbf{y}$ in $\mathbb{R}^n$ is a unique *linear* combination of $\mathbf{b}_1, \ldots, \mathbf{b}_n$. This combination is an affine combination of the $\mathbf{b}$'s if and only if the weights sum to 1. (These weights are just the $\mathcal{B}$-coordinates of $\mathbf{y}$, as in Section 4.4.)

EXAMPLE 2 Let $\mathbf{b}_1 = \begin{bmatrix} 4 \\ 0 \\ 3 \end{bmatrix}$, $\mathbf{b}_2 = \begin{bmatrix} 0 \\ 4 \\ 2 \end{bmatrix}$, $\mathbf{b}_3 = \begin{bmatrix} 5 \\ 2 \\ 4 \end{bmatrix}$, $\mathbf{p}_1 = \begin{bmatrix} 2 \\ 0 \\ 0 \end{bmatrix}$, and $\mathbf{p}_2 = \begin{bmatrix} 1 \\ 2 \\ 2 \end{bmatrix}$.

The set $\mathcal{B} = \{\mathbf{b}_1, \mathbf{b}_2, \mathbf{b}_3\}$ is a basis for $\mathbb{R}^3$. Determine whether the points $\mathbf{p}_1$ and $\mathbf{p}_2$ are affine combinations of the points in $\mathcal{B}$.

SOLUTION Find the $\mathcal{B}$-coordinates of $\mathbf{p}_1$ and $\mathbf{p}_2$. These two calculations can be combined by row reducing the matrix $[\, \mathbf{b}_1 \quad \mathbf{b}_2 \quad \mathbf{b}_3 \quad \mathbf{p}_1 \quad \mathbf{p}_2 \,]$, with two augmented columns:

$$
\begin{bmatrix} 4 & 0 & 5 & 2 & 1 \\ 0 & 4 & 2 & 0 & 2 \\ 3 & 2 & 4 & 0 & 2 \end{bmatrix} \sim \begin{bmatrix} 1 & 0 & 0 & -2 & \frac{2}{3} \\ 0 & 1 & 0 & -1 & \frac{2}{3} \\ 0 & 0 & 1 & 2 & -\frac{1}{3} \end{bmatrix}
$$

Read column 4 to build $\mathbf{p}_1$, and read column 5 to build $\mathbf{p}_2$:

$$\mathbf{p}_1 = -2\mathbf{b}_1 - \mathbf{b}_2 + 2\mathbf{b}_3 \quad \text{and} \quad \mathbf{p}_2 = \tfrac{2}{3}\mathbf{b}_1 + \tfrac{2}{3}\mathbf{b}_2 - \tfrac{1}{3}\mathbf{b}_3$$

The sum of the weights in the linear combination for $\mathbf{p}_1$ is -1, not 1, so $\mathbf{p}_1$ is *not* an affine combination of the $\mathbf{b}$'s. However, $\mathbf{p}_2$ *is* an affine combination of the $\mathbf{b}$'s, because the sum of the weights for $\mathbf{p}_2$ is 1. ∎

DEFINITION

> A set S is **affine** if $\mathbf{p}, \mathbf{q} \in S$ implies that $(1 - t)\mathbf{p} + t\mathbf{q} \in S$ for each real number t.

Geometrically, a set is affine if whenever two points are in the set, the entire line through these points is in the set. (If S contains only one point, $\mathbf{p}$, then the line through $\mathbf{p}$ and $\mathbf{p}$ is just a point, a "degenerate" line.) Algebraically, for a set S to be affine, the definition requires that every affine combination of two points of S belong to S. Remarkably, this is equivalent to requiring that S contain every affine combination of an arbitrary number of points of S.

THEOREM 2

> A set S is affine if and only if every affine combination of points of S lies in S. That is, S is affine if and only if $S = \text{aff } S$.

PROOF Suppose that S is affine and use induction on the number m of points of S occurring in an affine combination. When m is 1 or 2, an affine combination of m points of S lies in S, by the definition of an affine set. Now, assume that every affine combination of k or fewer points of S yields a point in S, and consider a combination of $k + 1$ points. Take $\mathbf{v}_i$ in S for $i = 1, \ldots, k + 1$, and let $\mathbf{y} = c_1\mathbf{v}_1 + \cdots + c_k\mathbf{v}_k + c_{k+1}\mathbf{v}_{k+1}$, where $c_1 + \cdots + c_{k+1} = 1$. Since the c_i's sum to 1, at least one of them must not be equal to 1. By re-indexing the $\mathbf{v}_i$ and c_i, if necessary, we may assume that $c_{k+1} \neq 1$. Let $t = c_1 + \cdots + c_k$. Then $t = 1 - c_{k+1} \neq 0$, and

$$\mathbf{y} = (1 - c_{k+1}) \left(\frac{c_1}{t}\mathbf{v}_1 + \cdots + \frac{c_k}{t}\mathbf{v}_k \right) + c_{k+1}\mathbf{v}_{k+1} \tag{6}$$

By the induction hypothesis, the point $\mathbf{z} = (c_1/t)\mathbf{v}_1 + \cdots + (c_k/t)\mathbf{v}_k$ is in S, since the coefficients sum to 1. Thus (6) displays $\mathbf{y}$ as an affine combination of two points in S, and so $\mathbf{y} \in S$. By the principle of induction, every affine combination of such points lies in S. That is, $\text{aff } S \subset S$. But the reverse inclusion, $S \subset \text{aff } S$, always applies. Thus, when S is affine, $S = \text{aff } S$. Conversely, if $S = \text{aff } S$, then affine combinations of two (or more) points of S lie in S, so S is affine. ∎

The next definition provides terminology for affine sets that emphasizes their close connection with subspaces of $\mathbb{R}^n$.

DEFINITION

A translate of a set S in $\mathbb{R}^n$ by a vector $\mathbf{p}$ is the set $S + \mathbf{p} = \{\mathbf{s} + \mathbf{p} : \mathbf{s} \in S\}$.[2] A **flat** in $\mathbb{R}^n$ is a translate of a subspace of $\mathbb{R}^n$. Two flats are **parallel** if one is a translate of the other. The **dimension of a flat** is the dimension of the corresponding parallel subspace. The **dimension of a set** S, written as dim S, is the dimension of the smallest flat containing S. A **line** in $\mathbb{R}^n$ is a flat of dimension 1. A **hyperplane** in $\mathbb{R}^n$ is a flat of dimension $n - 1$.

In $\mathbb{R}^3$, the proper subspaces[3] consist of the origin $\mathbf{0}$, the set of all lines through $\mathbf{0}$, and the set of all planes through $\mathbf{0}$. Thus the proper flats in $\mathbb{R}^3$ are points (zero-dimensional), lines (one-dimensional), and planes (two-dimensional), which may or may not pass through the origin.

The next theorem shows that these geometric descriptions of lines and planes in $\mathbb{R}^3$ (as translates of subspaces) actually coincide with their earlier algebraic descriptions as sets of all affine combinations of two or three points, respectively.

THEOREM 3

A nonempty set S is affine if and only if it is a flat.

PROOF Suppose that S is affine. Let $\mathbf{p}$ be any fixed point in S and let $W = S + (-\mathbf{p})$, so that $S = W + \mathbf{p}$. To show that S is a flat, it suffices to show that W is a subspace of $\mathbb{R}^n$. Since $\mathbf{p}$ is in S, the zero vector is in W. To show that W is closed under sums and scalar multiples, it suffices to show that if $\mathbf{u}_1$ and $\mathbf{u}_2$ are elements of W, then $\mathbf{u}_1 + t\mathbf{u}_2$ is in W for every real t. Since $\mathbf{u}_1$ and $\mathbf{u}_2$ are in W, there exist $\mathbf{s}_1$ and $\mathbf{s}_2$ in S such that $\mathbf{u}_1 = \mathbf{s}_1 - \mathbf{p}$ and $\mathbf{u}_2 = \mathbf{s}_2 - \mathbf{p}$. So, for each real t,

$$\mathbf{u}_1 + t\mathbf{u}_2 = (\mathbf{s}_1 - \mathbf{p}) + t(\mathbf{s}_2 - \mathbf{p})$$
$$= (1 - t)\mathbf{s}_1 + t(\mathbf{s}_1 + \mathbf{s}_2 - \mathbf{p}) - \mathbf{p}$$

Let $\mathbf{y} = \mathbf{s}_1 + \mathbf{s}_2 - \mathbf{p}$. Then $\mathbf{y}$ is an affine combination of points in S. Since S is affine, $\mathbf{y}$ is in S (by Theorem 2). But then $(1 - t)\mathbf{s}_1 + t\mathbf{y}$ is also in S. So $\mathbf{u}_1 + t\mathbf{u}_2$ is in $-\mathbf{p} + S = W$. This shows that W is a subspace of $\mathbb{R}^n$. Thus S is a flat, because $S = W + \mathbf{p}$.

Conversely, suppose S is a flat. That is, $S = W + \mathbf{p}$ for some $\mathbf{p} \in \mathbb{R}^n$ and some subspace W. To show that S is affine, it suffices to show that for any pair $\mathbf{s}_1$ and $\mathbf{s}_2$ of points in S, the line through $\mathbf{s}_1$ and $\mathbf{s}_2$ lies in S. By definition of W, there exist $\mathbf{u}_1$ and $\mathbf{u}_2$ in W such that $\mathbf{s}_1 = \mathbf{u}_1 + \mathbf{p}$ and $\mathbf{s}_2 = \mathbf{u}_2 + \mathbf{p}$. So, for each real t,

$$(1 - t)\mathbf{s}_1 + t\mathbf{s}_2 = (1 - t)(\mathbf{u}_1 + \mathbf{p}) + t(\mathbf{u}_2 + \mathbf{p})$$
$$= (1 - t)\mathbf{u}_1 + t\mathbf{u}_2 + \mathbf{p}$$

Since W is a subspace, $(1 - t)\mathbf{u}_1 + t\mathbf{u}_2 \in W$ and so $(1 - t)\mathbf{s}_1 + t\mathbf{s}_2 \in W + \mathbf{p} = S$. Thus S is affine. ∎

---

[2] If $\mathbf{p} = \mathbf{0}$, then the translate is just S itself. See Fig. 4 in Section 1.5.

[3] A subset A of a set B is called a **proper** subset of B if $A \neq B$. The same condition applies to proper subspaces and proper flats in $\mathbb{R}^n$: they are not equal to $\mathbb{R}^n$.

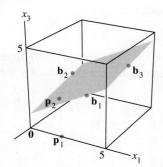

FIGURE 3

Theorem 3 provides a geometric way to view the affine hull of a set: it is the flat that consists of all the affine combinations of points in the set. For instance, Fig. 3 shows the points studied in Example 2. Although the set of all *linear* combinations of $\mathbf{b}_1$, $\mathbf{b}_2$, and $\mathbf{b}_3$ is all of $\mathbb{R}^3$, the set of all *affine* combinations is only the plane through $\mathbf{b}_1$, $\mathbf{b}_2$, and $\mathbf{b}_3$. Note that $\mathbf{p}_2$ (from Example 2) is in the plane through $\mathbf{b}_1$, $\mathbf{b}_2$, and $\mathbf{b}_3$, while $\mathbf{p}_1$ is not in that plane. Also, see Exercise 14.

The next example takes a fresh look at a familiar set—the set of all solutions of a system $A\mathbf{x} = \mathbf{b}$.

EXAMPLE 3 Suppose that the solutions of an equation $A\mathbf{x} = \mathbf{b}$ are all of the form $\mathbf{x} = x_3\mathbf{u} + \mathbf{p}$, where $\mathbf{u} = \begin{bmatrix} 2 \\ -3 \\ 1 \end{bmatrix}$ and $\mathbf{p} = \begin{bmatrix} 4 \\ 0 \\ -3 \end{bmatrix}$. Recall from Section 1.5 that this set is parallel to the solution set of $A\mathbf{x} = \mathbf{0}$, which consists of all points of the form $x_3\mathbf{u}$. Find points $\mathbf{v}_1$ and $\mathbf{v}_2$ such that the solution set of $A\mathbf{x} = \mathbf{b}$ is aff $\{\mathbf{v}_1, \mathbf{v}_2\}$.

SOLUTION The solution set is a line through $\mathbf{p}$ in the direction of $\mathbf{u}$, as in Fig. 1. Since aff $\{\mathbf{v}_1, \mathbf{v}_2\}$ is a line through $\mathbf{v}_1$ and $\mathbf{v}_2$, identify two points on the line $\mathbf{x} = x_3\mathbf{u} + \mathbf{p}$. Two simple choices appear when $x_3 = 0$ and $x_3 = 1$. That is, take $\mathbf{v}_1 = \mathbf{p}$ and $\mathbf{v}_2 = \mathbf{u} + \mathbf{p}$, so that

$$\mathbf{v}_2 = \mathbf{u} + \mathbf{p} = \begin{bmatrix} 2 \\ -3 \\ 1 \end{bmatrix} + \begin{bmatrix} 4 \\ 0 \\ -3 \end{bmatrix} = \begin{bmatrix} 6 \\ -3 \\ -2 \end{bmatrix}.$$

In this case, the solution set is described as the set of all affine combinations of the form

$$\mathbf{x} = (1 - x_3)\begin{bmatrix} 4 \\ 0 \\ -3 \end{bmatrix} + x_3 \begin{bmatrix} 6 \\ -3 \\ -2 \end{bmatrix}. \qquad \blacksquare$$

Earlier, Theorem 1 displayed an important connection between affine combinations and linear combinations. The next theorem provides another view of affine combinations, which for $\mathbb{R}^2$ and $\mathbb{R}^3$ is closely connected to applications in computer graphics, discussed in the next section (and in Section 2.7).

DEFINITION

For $\mathbf{v}$ in $\mathbb{R}^n$, the standard **homogeneous form** of $\mathbf{v}$ is the point $\tilde{\mathbf{v}} = \begin{bmatrix} \mathbf{v} \\ 1 \end{bmatrix}$ in $\mathbb{R}^{n+1}$.

THEOREM 4

A point $\mathbf{y}$ in $\mathbb{R}^n$ is an affine combination of $\mathbf{v}_1, \ldots, \mathbf{v}_p$ in $\mathbb{R}^n$ if and only if the homogeneous form of $\mathbf{y}$ is in Span $\{\tilde{\mathbf{v}}_1, \ldots, \tilde{\mathbf{v}}_p\}$. In fact, $\mathbf{y} = c_1\mathbf{v}_1 + \cdots + c_p\mathbf{v}_p$, with $c_1 + \cdots + c_p = 1$, if and only if $\tilde{\mathbf{y}} = c_1\tilde{\mathbf{v}}_1 + \cdots + c_p\tilde{\mathbf{v}}_p$.

PROOF A point $\mathbf{y}$ is in aff $\{\mathbf{v}_1, \ldots, \mathbf{v}_p\}$ if and only if there exist weights $c_1, \ldots, c_p$ such that

$$\begin{bmatrix} \mathbf{y} \\ 1 \end{bmatrix} = c_1 \begin{bmatrix} \mathbf{v}_1 \\ 1 \end{bmatrix} + c_2 \begin{bmatrix} \mathbf{v}_2 \\ 1 \end{bmatrix} + \cdots + c_p \begin{bmatrix} \mathbf{v}_p \\ 1 \end{bmatrix}$$

This happens if and only if $\tilde{\mathbf{y}}$ is in Span $\{\tilde{\mathbf{v}}_1, \tilde{\mathbf{v}}_2, \ldots, \tilde{\mathbf{v}}_p\}$. $\qquad \blacksquare$

EXAMPLE 4 Let $\mathbf{v}_1 = \begin{bmatrix} 3 \\ 1 \\ 1 \end{bmatrix}$, $\mathbf{v}_2 = \begin{bmatrix} 1 \\ 2 \\ 2 \end{bmatrix}$, $\mathbf{v}_3 = \begin{bmatrix} 1 \\ 7 \\ 1 \end{bmatrix}$, and $\mathbf{p} = \begin{bmatrix} 4 \\ 3 \\ 0 \end{bmatrix}$. Use Theorem 4 to write $\mathbf{p}$ as an affine combination of $\mathbf{v}_1$, $\mathbf{v}_2$, and $\mathbf{v}_3$, if possible.

SOLUTION Row reduce the augmented matrix for the equation

$$x_1 \tilde{\mathbf{v}}_1 + x_2 \tilde{\mathbf{v}}_2 + x_3 \tilde{\mathbf{v}}_3 = \tilde{\mathbf{p}}$$

To simplify the arithmetic, move the fourth row of 1's to the top (equivalent to three row interchanges). After this, the number of arithmetic operations here is basically the same as the number needed for the method using Theorem 1.

$$[\,\tilde{\mathbf{v}}_1 \quad \tilde{\mathbf{v}}_2 \quad \tilde{\mathbf{v}}_3 \quad \tilde{\mathbf{p}}\,] \sim \begin{bmatrix} 1 & 1 & 1 & 1 \\ 3 & 1 & 1 & 4 \\ 1 & 2 & 7 & 3 \\ 1 & 2 & 1 & 0 \end{bmatrix} \sim \begin{bmatrix} 1 & 1 & 1 & 1 \\ 0 & -2 & -2 & 1 \\ 0 & 1 & 6 & 2 \\ 0 & 1 & 0 & -1 \end{bmatrix}$$

$$\sim \cdots \sim \begin{bmatrix} 1 & 0 & 0 & 1.5 \\ 0 & 1 & 0 & -1 \\ 0 & 0 & 1 & .5 \\ 0 & 0 & 0 & 0 \end{bmatrix}$$

By Theorem 4, $1.5\mathbf{v}_1 - \mathbf{v}_2 + .5\mathbf{v}_3 = \mathbf{p}$. See Fig. 4, which shows the plane that contains $\mathbf{v}_1, \mathbf{v}_2, \mathbf{v}_3$, and $\mathbf{p}$ (together with points on the coordinate axes). ∎

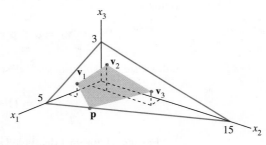

FIGURE 4

PRACTICE PROBLEM

Plot the points $\mathbf{v}_1 = \begin{bmatrix} 1 \\ 0 \end{bmatrix}$, $\mathbf{v}_2 = \begin{bmatrix} -1 \\ 2 \end{bmatrix}$, $\mathbf{v}_3 = \begin{bmatrix} 3 \\ 1 \end{bmatrix}$, and $\mathbf{p} = \begin{bmatrix} 4 \\ 3 \end{bmatrix}$ on graph paper, and explain why $\mathbf{p}$ *must* be an affine combination of $\mathbf{v}_1, \mathbf{v}_2$, and $\mathbf{v}_3$. Then find the affine combination for $\mathbf{p}$. [*Hint:* What is the dimension of aff $\{\mathbf{v}_1, \mathbf{v}_2, \mathbf{v}_3\}$?]

8.1 EXERCISES

In Exercises 1–4, write $\mathbf{y}$ as an affine combination of the other points listed, if possible.

1. $\mathbf{v}_1 = \begin{bmatrix} 1 \\ 2 \end{bmatrix}$, $\mathbf{v}_2 = \begin{bmatrix} -2 \\ 2 \end{bmatrix}$, $\mathbf{v}_3 = \begin{bmatrix} 0 \\ 4 \end{bmatrix}$, $\mathbf{v}_4 = \begin{bmatrix} 3 \\ 7 \end{bmatrix}$, $\mathbf{y} = \begin{bmatrix} 5 \\ 3 \end{bmatrix}$

2. $\mathbf{v}_1 = \begin{bmatrix} 1 \\ 1 \end{bmatrix}$, $\mathbf{v}_2 = \begin{bmatrix} -1 \\ 2 \end{bmatrix}$, $\mathbf{v}_3 = \begin{bmatrix} 3 \\ 2 \end{bmatrix}$, $\mathbf{y} = \begin{bmatrix} 5 \\ 7 \end{bmatrix}$

3. $\mathbf{v}_1 = \begin{bmatrix} -3 \\ 1 \\ 1 \end{bmatrix}$, $\mathbf{v}_2 = \begin{bmatrix} 0 \\ 4 \\ -2 \end{bmatrix}$, $\mathbf{v}_3 = \begin{bmatrix} 4 \\ -2 \\ 6 \end{bmatrix}$, $\mathbf{y} = \begin{bmatrix} 17 \\ 1 \\ 5 \end{bmatrix}$

4. $\mathbf{v}_1 = \begin{bmatrix} 1 \\ 2 \\ 0 \end{bmatrix}$, $\mathbf{v}_2 = \begin{bmatrix} 2 \\ -6 \\ 7 \end{bmatrix}$, $\mathbf{v}_3 = \begin{bmatrix} 4 \\ 3 \\ 1 \end{bmatrix}$, $\mathbf{y} = \begin{bmatrix} -3 \\ 4 \\ -4 \end{bmatrix}$

In Exercises 5 and 6, let $\mathbf{b}_1 = \begin{bmatrix} 2 \\ 1 \\ 1 \end{bmatrix}$, $\mathbf{b}_2 = \begin{bmatrix} 1 \\ 0 \\ -2 \end{bmatrix}$, $\mathbf{b}_3 = \begin{bmatrix} 2 \\ -5 \\ 1 \end{bmatrix}$, and $S = \{\mathbf{b}_1, \mathbf{b}_2, \mathbf{b}_3\}$. Note that S is an orthogonal basis for $\mathbb{R}^3$. Write each of the given points as an affine combination of the points in the set S, if possible. [*Hint:* Use Theorem 5 in Section 6.2 instead of row reduction to find the weights.]

5. a. $\mathbf{p}_1 = \begin{bmatrix} 3 \\ 8 \\ 4 \end{bmatrix}$ b. $\mathbf{p}_2 = \begin{bmatrix} 6 \\ -3 \\ 3 \end{bmatrix}$ c. $\mathbf{p}_3 = \begin{bmatrix} 0 \\ -1 \\ -5 \end{bmatrix}$

6. a. $\mathbf{p}_1 = \begin{bmatrix} 0 \\ -19 \\ -5 \end{bmatrix}$ b. $\mathbf{p}_2 = \begin{bmatrix} 1.5 \\ -1.3 \\ -.5 \end{bmatrix}$ c. $\mathbf{p}_3 = \begin{bmatrix} 5 \\ -4 \\ 0 \end{bmatrix}$

7. Let

$$\mathbf{v}_1 = \begin{bmatrix} 1 \\ 0 \\ 3 \\ 0 \end{bmatrix}, \quad \mathbf{v}_2 = \begin{bmatrix} 2 \\ -1 \\ 0 \\ 4 \end{bmatrix}, \quad \mathbf{v}_3 = \begin{bmatrix} -1 \\ 2 \\ 1 \\ 1 \end{bmatrix},$$

$$\mathbf{p}_1 = \begin{bmatrix} 5 \\ -3 \\ 5 \\ 3 \end{bmatrix}, \quad \mathbf{p}_2 = \begin{bmatrix} -9 \\ 10 \\ 9 \\ -13 \end{bmatrix}, \quad \mathbf{p}_3 = \begin{bmatrix} 4 \\ 2 \\ 8 \\ 5 \end{bmatrix},$$

and $S = \{\mathbf{v}_1, \mathbf{v}_2, \mathbf{v}_3\}$. It can be shown that S is linearly independent.

a. Is $\mathbf{p}_1$ in Span S? Is $\mathbf{p}_1$ in aff S?

b. Is $\mathbf{p}_2$ in Span S? Is $\mathbf{p}_2$ in aff S?

c. Is $\mathbf{p}_3$ in Span S? Is $\mathbf{p}_3$ in aff S?

8. Repeat Exercise 7 when

$$\mathbf{v}_1 = \begin{bmatrix} 1 \\ 0 \\ 3 \\ -2 \end{bmatrix}, \quad \mathbf{v}_2 = \begin{bmatrix} 2 \\ 1 \\ 6 \\ -5 \end{bmatrix}, \quad \mathbf{v}_3 = \begin{bmatrix} 3 \\ 0 \\ 12 \\ -6 \end{bmatrix},$$

$$\mathbf{p}_1 = \begin{bmatrix} 4 \\ -1 \\ 15 \\ -7 \end{bmatrix}, \quad \mathbf{p}_2 = \begin{bmatrix} -5 \\ 3 \\ -8 \\ 6 \end{bmatrix}, \quad \text{and} \quad \mathbf{p}_3 = \begin{bmatrix} 1 \\ 6 \\ -6 \\ -8 \end{bmatrix}.$$

9. Suppose that the solutions of an equation $A\mathbf{x} = \mathbf{b}$ are all of the form $\mathbf{x} = x_3\mathbf{u} + \mathbf{p}$, where $\mathbf{u} = \begin{bmatrix} 4 \\ -2 \end{bmatrix}$ and $\mathbf{p} = \begin{bmatrix} -3 \\ 0 \end{bmatrix}$. Find points $\mathbf{v}_1$ and $\mathbf{v}_2$ such that the solution set of $A\mathbf{x} = \mathbf{b}$ is aff $\{\mathbf{v}_1, \mathbf{v}_2\}$.

10. Suppose that the solutions of an equation $A\mathbf{x} = \mathbf{b}$ are all of the form $\mathbf{x} = x_3\mathbf{u} + \mathbf{p}$, where $\mathbf{u} = \begin{bmatrix} 5 \\ 1 \\ -2 \end{bmatrix}$ and $\mathbf{p} = \begin{bmatrix} 1 \\ -3 \\ 4 \end{bmatrix}$. Find points $\mathbf{v}_1$ and $\mathbf{v}_2$ such that the solution set of $A\mathbf{x} = \mathbf{b}$ is aff $\{\mathbf{v}_1, \mathbf{v}_2\}$.

In Exercises 11 and 12, mark each statement True or False. Justify each answer.

11. a. The set of all affine combinations of points in a set S is called the affine hull of S.

b. If $\{\mathbf{b}_1, \ldots, \mathbf{b}_k\}$ is a linearly independent subset of $\mathbb{R}^n$ and if $\mathbf{p}$ is a linear combination of $\mathbf{b}_1, \ldots, \mathbf{b}_k$, then $\mathbf{p}$ is an affine combination of $\mathbf{b}_1, \ldots, \mathbf{b}_k$.

c. The affine hull of two distinct points is called a line.

d. A flat is a subspace.

e. A plane in $\mathbb{R}^3$ is a hyperplane.

12. a. If $S = \{\mathbf{x}\}$, then aff S is the empty set.

b. A set is affine if and only if it contains its affine hull.

c. A flat of dimension 1 is called a line.

d. A flat of dimension 2 is called a hyperplane.

e. A flat through the origin is a subspace.

13. Suppose $\{\mathbf{v}_1, \mathbf{v}_2, \mathbf{v}_3\}$ is a basis for $\mathbb{R}^3$. Show that Span $\{\mathbf{v}_2 - \mathbf{v}_1, \mathbf{v}_3 - \mathbf{v}_1\}$ is a plane in $\mathbb{R}^3$. [*Hint:* What can you say about $\mathbf{u}$ and $\mathbf{v}$ when Span $\{\mathbf{u}, \mathbf{v}\}$ is a plane?]

14. Show that if $\{\mathbf{v}_1, \mathbf{v}_2, \mathbf{v}_3\}$ is a basis for $\mathbb{R}^3$, then aff $\{\mathbf{v}_1, \mathbf{v}_2, \mathbf{v}_3\}$ is the plane through $\mathbf{v}_1$, $\mathbf{v}_2$, and $\mathbf{v}_3$.

15. Let A be an $m \times n$ matrix and, given $\mathbf{b}$ in $\mathbb{R}^m$, show that the set S of all solutions of $A\mathbf{x} = \mathbf{b}$ is an affine subset of $\mathbb{R}^n$.

16. Let $\mathbf{v} \in \mathbb{R}^n$ and let $k \in \mathbb{R}$. Prove that $S = \{\mathbf{x} \in \mathbb{R}^n : \mathbf{x} \cdot \mathbf{v} = k\}$ is an affine subset of $\mathbb{R}^n$.

17. Choose a set S of three points such that aff S is the plane in $\mathbb{R}^3$ whose equation is $x_3 = 5$. Justify your work.

18. Choose a set S of four distinct points in $\mathbb{R}^3$ such that aff S is the plane $2x_1 + x_2 - 3x_3 = 12$. Justify your work.

19. Let S be an affine subset of $\mathbb{R}^n$, suppose $f : \mathbb{R}^n \to \mathbb{R}^m$ is a linear transformation, and let $f(S)$ denote the set of images $\{f(\mathbf{x}) : \mathbf{x} \in S\}$. Prove that $f(S)$ is an affine subset of $\mathbb{R}^m$.

20. Let $f : \mathbb{R}^n \to \mathbb{R}^m$ be a linear transformation, let T be an affine subset of $\mathbb{R}^m$, and let $S = \{\mathbf{x} \in \mathbb{R}^n : f(\mathbf{x}) \in T\}$. Show that S is an affine subset of $\mathbb{R}^n$.

In Exercises 21–26, prove the given statement about subsets A and B of $\mathbb{R}^n$, or provide the required example in $\mathbb{R}^2$. A proof for an exercise may use results from earlier exercises (as well as theorems already available in the text).

21. If $A \subset B$ and B is affine, then aff $A \subset B$.

22. If $A \subset B$, then aff $A \subset$ aff B.

23. $[(\text{aff } A) \cup (\text{aff } B)] \subset \text{aff } (A \cup B)$. [*Hint:* To show that $D \cup E \subset F$, show that $D \subset F$ and $E \subset F$.]

24. Find an example in $\mathbb{R}^2$ to show that equality need not hold in the statement of Exercise 23. [*Hint:* Consider sets A and B, each of which contains only one or two points.]

25. aff $(A \cap B) \subset (\text{aff } A \cap \text{aff } B)$.

26. Find an example in $\mathbb{R}^2$ to show that equality need not hold in the statement of Exercise 25.

SOLUTION TO PRACTICE PROBLEM

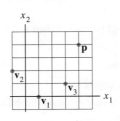

Since the points $\mathbf{v}_1$, $\mathbf{v}_2$, and $\mathbf{v}_3$ are not collinear (that is, not on a single line), aff $\{\mathbf{v}_1, \mathbf{v}_2, \mathbf{v}_3\}$ cannot be one-dimensional. Thus, aff $\{\mathbf{v}_1, \mathbf{v}_2, \mathbf{v}_3\}$ must equal $\mathbb{R}^2$. To find the actual weights used to express $\mathbf{p}$ as an affine combination of $\mathbf{v}_1$, $\mathbf{v}_2$, and $\mathbf{v}_3$, first compute

$$\mathbf{v}_2 - \mathbf{v}_1 = \begin{bmatrix} -2 \\ 2 \end{bmatrix}, \quad \mathbf{v}_3 - \mathbf{v}_1 = \begin{bmatrix} 2 \\ 1 \end{bmatrix}, \quad \text{and} \quad \mathbf{p} - \mathbf{v}_1 = \begin{bmatrix} 3 \\ 3 \end{bmatrix}$$

To write $\mathbf{p} - \mathbf{v}_1$ as a linear combination of $\mathbf{v}_2 - \mathbf{v}_1$ and $\mathbf{v}_3 - \mathbf{v}_1$, row reduce the matrix having these points as columns:

$$\begin{bmatrix} -2 & 2 & 3 \\ 2 & 1 & 3 \end{bmatrix} \sim \begin{bmatrix} 1 & 0 & \frac{1}{2} \\ 0 & 1 & 2 \end{bmatrix}$$

Thus $\mathbf{p} - \mathbf{v}_1 = \frac{1}{2}(\mathbf{v}_2 - \mathbf{v}_1) + 2(\mathbf{v}_3 - \mathbf{v}_1)$, which shows that

$$\mathbf{p} = \left(1 - \tfrac{1}{2} - 2\right)\mathbf{v}_1 + \tfrac{1}{2}\mathbf{v}_2 + 2\mathbf{v}_3 = -\tfrac{3}{2}\mathbf{v}_1 + \tfrac{1}{2}\mathbf{v}_2 + 2\mathbf{v}_3$$

This expresses $\mathbf{p}$ as an affine combination of $\mathbf{v}_1$, $\mathbf{v}_2$, and $\mathbf{v}_3$, because the coefficients sum to 1.

Alternatively, use the method of Example 3 and row reduce:

$$\begin{bmatrix} \mathbf{v}_1 & \mathbf{v}_2 & \mathbf{v}_3 & \mathbf{p} \\ 1 & 1 & 1 & 1 \end{bmatrix} \sim \begin{bmatrix} 1 & 1 & 1 & 1 \\ 1 & -1 & 3 & 4 \\ 0 & 2 & 1 & 3 \end{bmatrix} \sim \begin{bmatrix} 1 & 0 & 0 & -\frac{3}{2} \\ 0 & 1 & 0 & \frac{1}{2} \\ 0 & 0 & 1 & 2 \end{bmatrix}$$

This shows that $\mathbf{p} = -\tfrac{3}{2}\mathbf{v}_1 + \tfrac{1}{2}\mathbf{v}_2 + 2\mathbf{v}_3$.

8.2 | AFFINE INDEPENDENCE

This section continues to explore the relation between linear concepts and affine concepts. Consider first a set of three vectors in $\mathbb{R}^3$, say $S = \{\mathbf{v}_1, \mathbf{v}_2, \mathbf{v}_3\}$. If S is linearly dependent, then one of the vectors is a linear combination of the other two vectors. What happens when one of the vectors is an *affine* combination of the others? For instance, suppose that

$$\mathbf{v}_3 = (1 - t)\mathbf{v}_1 + t\mathbf{v}_2, \quad \text{for some } t \text{ in } \mathbb{R}.$$

Then

$$(1 - t)\mathbf{v}_1 + t\mathbf{v}_2 - \mathbf{v}_3 = \mathbf{0}.$$

This is a linear dependence relation because not all the weights are zero. But more is true—the weights in the dependence relation sum to 0:

$$(1 - t) + t + (-1) = 0.$$

This is the additional property needed to define *affine dependence*.

DEFINITION

An indexed set of points $\{\mathbf{v}_1, \ldots, \mathbf{v}_p\}$ in $\mathbb{R}^n$ is **affinely dependent** if there exist real numbers $c_1, \ldots, c_p$, not all zero, such that

$$c_1 + \cdots + c_p = 0 \quad \text{and} \quad c_1\mathbf{v}_1 + \cdots + c_p\mathbf{v}_p = \mathbf{0} \tag{1}$$

Otherwise, the set is **affinely independent**.

An affine combination is a special type of linear combination, and affine dependence is a restricted type of linear dependence. Thus, each affinely dependent set is automatically linearly dependent.

A set $\{\mathbf{v}_1\}$ of only one point (even the zero vector) must be affinely independent because the required properties of the coefficients c_i cannot be satisfied when there is only one coefficient. For $\{\mathbf{v}_1\}$, the first equation in (1) is just $c_1 = 0$, and yet at least one (the only one) coefficient must be nonzero.

Exercise 13 asks you to show that an indexed set $\{\mathbf{v}_1, \mathbf{v}_2\}$ is affinely dependent if and only if $\mathbf{v}_1 = \mathbf{v}_2$. The following theorem handles the general case and shows how the concept of affine dependence is analogous to that of linear dependence. Parts (c) and (d) give useful methods for determining whether a set is affinely dependent. Recall from Section 8.1 that if $\mathbf{v}$ is in $\mathbb{R}^n$, then the vector $\tilde{\mathbf{v}}$ in $\mathbb{R}^{n+1}$ denotes the homogeneous form of $\mathbf{v}$.

THEOREM 5

Given an indexed set $S = \{\mathbf{v}_1, \ldots, \mathbf{v}_p\}$ in $\mathbb{R}^n$, with $p \geq 2$, the following statements are logically equivalent. That is, either they are all true statements or they are all false.

a. S is affinely dependent.
b. One of the points in S is an affine combination of the other points in S.
c. The set $\{\mathbf{v}_2 - \mathbf{v}_1, \ldots, \mathbf{v}_p - \mathbf{v}_1\}$ in $\mathbb{R}^n$ is linearly dependent.
d. The set $\{\tilde{\mathbf{v}}_1, \ldots, \tilde{\mathbf{v}}_p\}$ of homogeneous forms in $\mathbb{R}^{n+1}$ is linearly dependent.

PROOF Suppose statement (a) is true, and let $c_1, \ldots, c_p$ satisfy (1). By renaming the points if necessary, one may assume that $c_1 \neq 0$ and divide both equations in (1) by c_1, so that $1 + (c_2/c_1) + \cdots + (c_p/c_1) = 0$ and

$$\mathbf{v}_1 = (-c_2/c_1)\mathbf{v}_2 + \cdots + (-c_p/c_1)\mathbf{v}_p \tag{2}$$

Note that the coefficients on the right side of (2) sum to 1. Thus (a) implies (b). Now, suppose that (b) is true. By renaming the points if necessary, one may assume that $\mathbf{v}_1 = c_2\mathbf{v}_2 + \cdots + c_p\mathbf{v}_p$, where $c_2 + \cdots + c_p = 1$. Then

$$(c_2 + \cdots + c_p)\mathbf{v}_1 = c_2\mathbf{v}_2 + \cdots + c_p\mathbf{v}_p \tag{3}$$

and

$$c_2(\mathbf{v}_2 - \mathbf{v}_1) + \cdots + c_p(\mathbf{v}_p - \mathbf{v}_1) = \mathbf{0} \tag{4}$$

Not all of $c_2, \ldots, c_p$ can be zero because they sum to 1. So (b) implies (c).

Next, if (c) is true, then there exist weights $c_2, \ldots, c_p$, not all zero, such that (4) holds. Rewrite (4) as (3) and set $c_1 = -(c_2 + \cdots + c_p)$. Then $c_1 + \cdots + c_p = 0$. Thus (3) shows that (1) is true. So (c) implies (a), which proves that (a), (b), and (c) are logically equivalent. Finally, (d) is equivalent to (a) because the two equations in (1) are equivalent to the following equation involving the homogeneous forms of the points in S:

$$c_1 \begin{bmatrix} \mathbf{v}_1 \\ 1 \end{bmatrix} + \cdots + c_p \begin{bmatrix} \mathbf{v}_p \\ 1 \end{bmatrix} = \begin{bmatrix} \mathbf{0} \\ 0 \end{bmatrix} \qquad \blacksquare$$

In statement (c) of Theorem 5, $\mathbf{v}_1$ could be replaced by any of the other points in the list $\mathbf{v}_1, \ldots, \mathbf{v}_p$. Only the notation in the proof would change. So, to test whether a set is affinely dependent, subtract one point in the set from the other points, and check whether the translated set of $p - 1$ points is linearly dependent.

EXAMPLE 1   The affine hull of two distinct points $\mathbf{p}$ and $\mathbf{q}$ is a line. If a third point $\mathbf{r}$ is on the line, then $\{\mathbf{p}, \mathbf{q}, \mathbf{r}\}$ is an affinely dependent set. If a point $\mathbf{s}$ is not on the line through $\mathbf{p}$ and $\mathbf{q}$, then these three points are not collinear and $\{\mathbf{p}, \mathbf{q}, \mathbf{s}\}$ is an affinely independent set. See Fig. 1. ∎

FIGURE 1 $\{\mathbf{p}, \mathbf{q}, \mathbf{r}\}$ is affinely dependent.

EXAMPLE 2   Let $\mathbf{v}_1 = \begin{bmatrix} 1 \\ 3 \\ 7 \end{bmatrix}$, $\mathbf{v}_2 = \begin{bmatrix} 2 \\ 7 \\ 6.5 \end{bmatrix}$, $\mathbf{v}_3 = \begin{bmatrix} 0 \\ 4 \\ 7 \end{bmatrix}$, and $S = \{\mathbf{v}_1, \mathbf{v}_2, \mathbf{v}_3\}$. Determine whether S is affinely independent.

SOLUTION   Compute $\mathbf{v}_2 - \mathbf{v}_1 = \begin{bmatrix} 1 \\ 4 \\ -.5 \end{bmatrix}$ and $\mathbf{v}_3 - \mathbf{v}_1 = \begin{bmatrix} -1 \\ 1 \\ 0 \end{bmatrix}$. These two points are not multiples and hence form a linearly independent set, S'. So all statements in Theorem 5 are false, and S is affinely independent. Figure 2 shows S and the translated set S'. Notice that Span S' is a plane through the origin and aff S is a parallel plane through $\mathbf{v}_1, \mathbf{v}_2$, and $\mathbf{v}_3$. (Only a portion of each plane is shown here, of course.) ∎

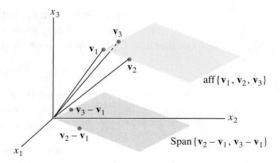

FIGURE 2   An affinely independent set $\{\mathbf{v}_1, \mathbf{v}_2, \mathbf{v}_3\}$.

EXAMPLE 3   Let $\mathbf{v}_1 = \begin{bmatrix} 1 \\ 3 \\ 7 \end{bmatrix}$, $\mathbf{v}_2 = \begin{bmatrix} 2 \\ 7 \\ 6.5 \end{bmatrix}$, $\mathbf{v}_3 = \begin{bmatrix} 0 \\ 4 \\ 7 \end{bmatrix}$, and $\mathbf{v}_4 = \begin{bmatrix} 0 \\ 14 \\ 6 \end{bmatrix}$, and let $S = \{\mathbf{v}_1, \ldots, \mathbf{v}_4\}$. Is S affinely dependent?

SOLUTION   Compute $\mathbf{v}_2 - \mathbf{v}_1 = \begin{bmatrix} 1 \\ 4 \\ -.5 \end{bmatrix}$, $\mathbf{v}_3 - \mathbf{v}_1 = \begin{bmatrix} -1 \\ 1 \\ 0 \end{bmatrix}$, and $\mathbf{v}_4 - \mathbf{v}_1 = \begin{bmatrix} -1 \\ 11 \\ -1 \end{bmatrix}$, and row reduce the matrix:

$$\begin{bmatrix} 1 & -1 & -1 \\ 4 & 1 & 11 \\ -.5 & 0 & -1 \end{bmatrix} \sim \begin{bmatrix} 1 & -1 & -1 \\ 0 & 5 & 15 \\ 0 & -.5 & -1.5 \end{bmatrix} \sim \begin{bmatrix} 1 & -1 & -1 \\ 0 & 5 & 15 \\ 0 & 0 & 0 \end{bmatrix}$$

Recall from Section 4.6 (or Section 2.8) that the columns are linearly dependent because not every column is a pivot column; so $\mathbf{v}_2 - \mathbf{v}_1, \mathbf{v}_3 - \mathbf{v}_1$, and $\mathbf{v}_4 - \mathbf{v}_1$ are linearly dependent. By statement (c) in Theorem 5, $\{\mathbf{v}_1, \mathbf{v}_2, \mathbf{v}_3, \mathbf{v}_4\}$ is affinely dependent. This dependence can also be established using (d) in Theorem 5 instead of (c). ∎

The calculations in Example 3 show that $\mathbf{v}_4 - \mathbf{v}_1$ is a linear combination of $\mathbf{v}_2 - \mathbf{v}_1$ and $\mathbf{v}_3 - \mathbf{v}_1$, which means that $\mathbf{v}_4 - \mathbf{v}_1$ is in Span $\{\mathbf{v}_2 - \mathbf{v}_1, \mathbf{v}_3 - \mathbf{v}_1\}$. By Theorem 1 in Section 8.1, $\mathbf{v}_4$ is in aff $\{\mathbf{v}_1, \mathbf{v}_2, \mathbf{v}_3\}$. In fact, complete row reduction of the matrix in Example 3 would show that

$$\mathbf{v}_4 - \mathbf{v}_1 = 2(\mathbf{v}_2 - \mathbf{v}_1) + 3(\mathbf{v}_3 - \mathbf{v}_1) \tag{5}$$

$$\mathbf{v}_4 = -4\mathbf{v}_1 + 2\mathbf{v}_2 + 3\mathbf{v}_3 \tag{6}$$

See Fig. 3.

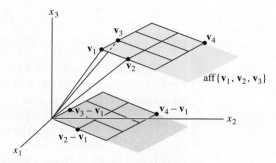

FIGURE 3 $\mathbf{v}_4$ is in the plane aff $\{\mathbf{v}_1, \mathbf{v}_2, \mathbf{v}_3\}$.

Figure 3 shows grids on both Span$\{\mathbf{v}_2 - \mathbf{v}_1, \mathbf{v}_3 - \mathbf{v}_1\}$ and aff $\{\mathbf{v}_1, \mathbf{v}_2, \mathbf{v}_3\}$. The grid on aff $\{\mathbf{v}_1, \mathbf{v}_2, \mathbf{v}_3\}$ is based on (5). Another "coordinate system" can be based on (6), in which the coefficients -4, 2, and 3 are called *affine* or *barycentric* coordinates of $\mathbf{v}_4$.

Barycentric Coordinates

The definition of barycentric coordinates depends on the following affine version of the Unique Representation Theorem in Section 4.4. See Exercise 17 in this section for the proof.

THEOREM 6

Let $S = \{\mathbf{v}_1, \ldots, \mathbf{v}_k\}$ be an affinely independent set in $\mathbb{R}^n$. Then each $\mathbf{p}$ in aff S has a unique representation as an affine combination of $\mathbf{v}_1, \ldots, \mathbf{v}_k$. That is, for each $\mathbf{p}$ there exists a unique set of scalars $c_1, \ldots, c_k$ such that

$$\mathbf{p} = c_1 \mathbf{v}_1 + \cdots + c_k \mathbf{v}_k \quad \text{and} \quad c_1 + \cdots + c_k = 1 \tag{7}$$

DEFINITION

Let $S = \{\mathbf{v}_1, \ldots, \mathbf{v}_k\}$ be an affinely independent set. Then for each point $\mathbf{p}$ in aff S, the coefficients $c_1, \ldots, c_p$ in the unique representation (7) of $\mathbf{p}$ are called the **barycentric** (or, sometimes, **affine**) **coordinates** of $\mathbf{p}$.

Observe that (7) is equivalent to the single equation

$$\begin{bmatrix} \mathbf{p} \\ 1 \end{bmatrix} = c_1 \begin{bmatrix} \mathbf{v}_1 \\ 1 \end{bmatrix} + \cdots + c_k \begin{bmatrix} \mathbf{v}_k \\ 1 \end{bmatrix} \tag{8}$$

involving the homogeneous forms of the points. Row reduction of the augmented matrix $\begin{bmatrix} \tilde{\mathbf{v}}_1 & \cdots & \tilde{\mathbf{v}}_k & \tilde{\mathbf{p}} \end{bmatrix}$ for (8) produces the barycentric coordinates of $\mathbf{p}$.

EXAMPLE 4 Let $\mathbf{a} = \begin{bmatrix} 1 \\ 7 \end{bmatrix}$, $\mathbf{b} = \begin{bmatrix} 3 \\ 0 \end{bmatrix}$, $\mathbf{c} = \begin{bmatrix} 9 \\ 3 \end{bmatrix}$, and $\mathbf{p} = \begin{bmatrix} 5 \\ 3 \end{bmatrix}$. Find the barycentric coordinates of $\mathbf{p}$ determined by the affinely independent set $\{\mathbf{a}, \mathbf{b}, \mathbf{c}\}$.

SOLUTION Row reduce the augmented matrix of points in homogeneous form, moving the last row of ones to the top to simplify the arithmetic:

$$\begin{bmatrix} \tilde{\mathbf{a}} & \tilde{\mathbf{b}} & \tilde{\mathbf{c}} & \tilde{\mathbf{p}} \end{bmatrix} = \begin{bmatrix} 1 & 3 & 9 & 5 \\ 7 & 0 & 3 & 3 \\ 1 & 1 & 1 & 1 \end{bmatrix} \sim \begin{bmatrix} 1 & 1 & 1 & 1 \\ 1 & 3 & 9 & 5 \\ 7 & 0 & 3 & 3 \end{bmatrix}$$

$$\sim \begin{bmatrix} 1 & 0 & 0 & \frac{1}{4} \\ 0 & 1 & 0 & \frac{1}{3} \\ 0 & 0 & 1 & \frac{5}{12} \end{bmatrix}$$

The coordinates are $\frac{1}{4}$, $\frac{1}{3}$, and $\frac{5}{12}$, so $\mathbf{p} = \frac{1}{4}\mathbf{a} + \frac{1}{3}\mathbf{b} + \frac{5}{12}\mathbf{c}$. ∎

Barycentric coordinates have both physical and geometric interpretations. They were originally defined by A. F. Moebius in 1827 for a point $\mathbf{p}$ inside a triangular region with vertices $\mathbf{a}$, $\mathbf{b}$, and $\mathbf{c}$. He wrote that the barycentric coordinates of $\mathbf{p}$ are three nonnegative numbers $m_{\mathbf{a}}$, $m_{\mathbf{b}}$, and $m_{\mathbf{c}}$ such that $\mathbf{p}$ is the center of mass of a system consisting of the triangle (with no mass) and masses $m_{\mathbf{a}}$, $m_{\mathbf{b}}$, and $m_{\mathbf{c}}$ at the corresponding vertices. The masses are uniquely determined by requiring that their sum be 1. This view is still useful in physics today.[1]

Figure 4 gives a geometric interpretation to the barycentric coordinates in Example 4, showing the triangle $\triangle\mathbf{abc}$ and three small triangles $\triangle\mathbf{pbc}$, $\triangle\mathbf{apc}$, and $\triangle\mathbf{abp}$. The areas of the small triangles are proportional to the barycentric coordinates of $\mathbf{p}$. In fact,

$$\text{area}(\triangle\mathbf{pbc}) = \frac{1}{4} \cdot \text{area}(\triangle\mathbf{abc})$$

$$\text{area}(\triangle\mathbf{apc}) = \frac{1}{3} \cdot \text{area}(\triangle\mathbf{abc}) \qquad (9)$$

$$\text{area}(\triangle\mathbf{abp}) = \frac{5}{12} \cdot \text{area}(\triangle\mathbf{abc})$$

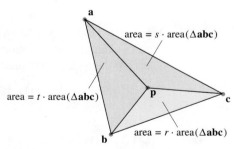

FIGURE 4 $\mathbf{p} = r\mathbf{a} + s\mathbf{b} + t\mathbf{c}$. Here, $r = \frac{1}{4}$, $s = \frac{1}{3}$, $t = \frac{5}{12}$.

The formulas in Fig. 4 are verified in Exercises 21–23. Analogous equalities for volumes of tetrahedrons hold for the case when $\mathbf{p}$ is a point inside a tetrahedron in $\mathbb{R}^3$, with vertices $\mathbf{a}$, $\mathbf{b}$, $\mathbf{c}$, and $\mathbf{d}$.

---

[1] See Exercise 29 in Section 1.3. In astronomy, however, "barycentric coordinates" usually refer to ordinary $\mathbb{R}^3$ coordinates of points in what is now called the *International Celestial Reference System*, a Cartesian coordinate system for outer space, with the origin at the center of mass (the barycenter) of the solar system.

When a point is not inside the triangle (or tetrahedron), some or all of the barycentric coordinates will be negative. The case of a triangle is illustrated in Fig. 5, for vertices $\mathbf{a}$, $\mathbf{b}$, $\mathbf{c}$, and coordinate values r, s, t, as above. The points on the line through $\mathbf{b}$ and $\mathbf{c}$, for instance, have $r = 0$ because they are affine combinations of only $\mathbf{b}$ and $\mathbf{c}$. The parallel line through $\mathbf{a}$ identifies points with $r = 1$.

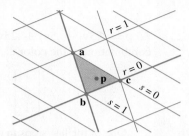

FIGURE 5 Barycentric coordinates for points in aff $\{\mathbf{a}, \mathbf{b}, \mathbf{c}\}$.

Barycentric Coordinates in Computer Graphics

When working with geometric objects in a computer graphics program, a designer may use a "wire-frame" approximation to an object at certain key points in the process of creating a realistic final image.[2] For instance, if the surface of part of an object consists of small flat triangular surfaces, then a graphics program can easily add color, lighting, and shading to each small surface when that information is known only at the vertices. Barycentric coordinates provide the tool for smoothly interpolating the vertex information over the interior of a triangle. The interpolation at a point is simply the linear combination of the vertex values using the barycentric coordinates as weights.

Colors on a computer screen are often described by RGB coordinates. A triple (r, g, b) indicates the amount of each color—red, green, and blue—with the parameters varying from 0 to 1. For example, pure red is $(1, 0, 0)$, white is $(1, 1, 1)$, and black is $(0, 0, 0)$.

EXAMPLE 5 Let $\mathbf{v}_1 = \begin{bmatrix} 3 \\ 1 \\ 5 \end{bmatrix}$, $\mathbf{v}_2 = \begin{bmatrix} 4 \\ 3 \\ 4 \end{bmatrix}$, $\mathbf{v}_3 = \begin{bmatrix} 1 \\ 5 \\ 1 \end{bmatrix}$, and $\mathbf{p} = \begin{bmatrix} 3 \\ 3 \\ 3.5 \end{bmatrix}$. The colors at the vertices $\mathbf{v}_1$, $\mathbf{v}_2$, and $\mathbf{v}_3$ of a triangle are magenta $(1, 0, 1)$, light magenta $(1, .4, 1)$, and purple $(.6, 0, 1)$, respectively. Find the interpolated color at $\mathbf{p}$. See Fig. 6.

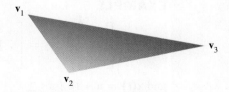

FIGURE 6 Interpolated colors.

---

[2] The Introductory Example for Chapter 2 shows a wire-frame model of a Boeing 777 airplane, used to visualize the flow of air over the surface of the plane.

SOLUTION First, find the barycentric coordinates of **p**. Here is the calculation using homogeneous forms of the points, with the first step moving row 4 to row 1:

$$\begin{bmatrix} \tilde{\mathbf{v}}_1 & \tilde{\mathbf{v}}_2 & \tilde{\mathbf{v}}_3 & \tilde{\mathbf{p}} \end{bmatrix} \sim \begin{bmatrix} 1 & 1 & 1 & 1 \\ 3 & 4 & 1 & 3 \\ 1 & 3 & 5 & 3 \\ 5 & 4 & 1 & 3.5 \end{bmatrix} \sim \begin{bmatrix} 1 & 0 & 0 & .25 \\ 0 & 1 & 0 & .50 \\ 0 & 0 & 1 & .25 \\ 0 & 0 & 0 & 0 \end{bmatrix}$$

So $\mathbf{p} = .25\mathbf{v}_1 + .5\mathbf{v}_2 + .25\mathbf{v}_3$. Use the barycentric coordinates of **p** to make a linear combination of the color data. The RGB values for **p** are

$$.25\begin{bmatrix} 1 \\ 0 \\ 1 \end{bmatrix} + .50\begin{bmatrix} 1 \\ .4 \\ 1 \end{bmatrix} + .25\begin{bmatrix} .6 \\ 0 \\ 1 \end{bmatrix} = \begin{bmatrix} .9 \\ .2 \\ 1 \end{bmatrix} \begin{matrix} \text{red} \\ \text{green} \\ \text{blue} \end{matrix}$$ ∎

One of the last steps in preparing a graphics scene for display on a computer screen is to remove "hidden surfaces" that should not be visible on the screen. Imagine the viewing screen as consisting of, say, a million pixels, and consider a ray or "line of sight" from the viewer's eye through a pixel and into the collection of objects that make up the 3D display. The color and other information displayed in the pixel on the screen should come from the object that the ray first intersects. See Fig. 7. When the objects in the graphics scene are approximated by wire frames with triangular patches, the hidden surface problem can be solved using barycentric coordinates.

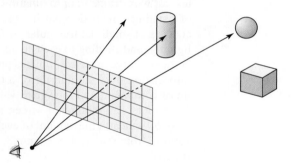

FIGURE 7 A ray from the eye through the screen to the nearest object.

The mathematics for finding the ray-triangle intersections can also be used to perform extremely realistic shading of objects. Currently, this *ray-tracing* method is too slow for real-time rendering, but recent advances in hardware implementation may change that in the future.[3]

EXAMPLE 6 Let

$$\mathbf{v}_1 = \begin{bmatrix} 1 \\ 1 \\ -6 \end{bmatrix}, \quad \mathbf{v}_2 = \begin{bmatrix} 8 \\ 1 \\ -4 \end{bmatrix}, \quad \mathbf{v}_3 = \begin{bmatrix} 5 \\ 11 \\ -2 \end{bmatrix}, \quad \mathbf{a} = \begin{bmatrix} 0 \\ 0 \\ 10 \end{bmatrix}, \quad \mathbf{b} = \begin{bmatrix} .7 \\ .4 \\ -3 \end{bmatrix},$$

and $\mathbf{x}(t) = \mathbf{a} + t\mathbf{b}$ for $t \geq 0$. Find the point where the ray $\mathbf{x}(t)$ intersects the plane that contains the triangle with vertices $\mathbf{v}_1$, $\mathbf{v}_2$, and $\mathbf{v}_3$. Is this point inside the triangle?

---

[3] See Joshua Fender and Jonathan Rose, "A High-Speed Ray Tracing Engine Built on a Field-Programmable System," in *Proc. Int. Conf on Field-Programmable Technology*, IEEE (2003). (A single processor can calculate 600 million ray-triangle intersections per second.)

SOLUTION The plane is aff $\{v_1, v_2, v_3\}$. A typical point in this plane may be written as $(1 - c_2 - c_3)v_1 + c_2v_2 + c_3v_3$ for some c_2 and c_3. (The weights in this combination sum to 1.) The ray $x(t)$ intersects the plane when c_2, c_3, and t satisfy

$$(1 - c_2 - c_3)v_1 + c_2v_2 + c_3v_3 = a + tb$$

Rearrange this as $c_2(v_2 - v_1) + c_3(v_3 - v_1) + t(-b) = a - v_1$. In matrix form,

$$\begin{bmatrix} v_2 - v_1 & v_3 - v_1 & -b \end{bmatrix} \begin{bmatrix} c_2 \\ c_3 \\ t \end{bmatrix} = a - v_1$$

For the specific points given here,

$$v_2 - v_1 = \begin{bmatrix} 7 \\ 0 \\ 2 \end{bmatrix}, \quad v_3 - v_1 = \begin{bmatrix} 4 \\ 10 \\ 4 \end{bmatrix}, \quad a - v_1 = \begin{bmatrix} -1 \\ -1 \\ 16 \end{bmatrix}$$

Row reduction of the augmented matrix above produces

$$\begin{bmatrix} 7 & 4 & -.7 & -1 \\ 0 & 10 & -.4 & -1 \\ 2 & 4 & 3 & 16 \end{bmatrix} \sim \begin{bmatrix} 1 & 0 & 0 & .3 \\ 0 & 1 & 0 & .1 \\ 0 & 0 & 1 & 5 \end{bmatrix}$$

Thus $c_2 = .3$, $c_3 = .1$, and $t = 5$. Therefore, the intersection point is

$$x(5) = a + 5b = \begin{bmatrix} 0 \\ 0 \\ 10 \end{bmatrix} + 5 \begin{bmatrix} .7 \\ .4 \\ -3 \end{bmatrix} = \begin{bmatrix} 3.5 \\ 2.0 \\ -5.0 \end{bmatrix}$$

Also,

$$x(5) = (1 - .3 - .1)v_1 + .3v_2 + .1v_3$$

$$= .6 \begin{bmatrix} 1 \\ 1 \\ -6 \end{bmatrix} + .3 \begin{bmatrix} 8 \\ 1 \\ -4 \end{bmatrix} + .1 \begin{bmatrix} 5 \\ 11 \\ -2 \end{bmatrix} = \begin{bmatrix} 3.5 \\ 2.0 \\ -5.0 \end{bmatrix}$$

The intersection point is inside the triangle because the barycentric weights for $x(5)$ are all positive. ∎

PRACTICE PROBLEMS

1. Describe a fast way to determine when three points are collinear.

2. The points $v_1 = \begin{bmatrix} 4 \\ 1 \end{bmatrix}$, $v_2 = \begin{bmatrix} 1 \\ 0 \end{bmatrix}$, $v_3 = \begin{bmatrix} 5 \\ 4 \end{bmatrix}$, and $v_4 = \begin{bmatrix} 1 \\ 2 \end{bmatrix}$ form an affinely dependent set. Find weights $c_1, \dots, c_4$ that produce an **affine dependence relation** $c_1v_1 + \cdots + c_4v_4 = 0$, where $c_1 + \cdots + c_4 = 0$ and not all c_i are zero. [*Hint:* See the end of the proof of Theorem 5.]

8.2 EXERCISES

In Exercises 1–6, determine if the set of points is affinely dependent. (See Practice Problem 2.) If so, construct an affine dependence relation for the points.

1. $\begin{bmatrix} 3 \\ -3 \end{bmatrix}, \begin{bmatrix} 0 \\ 6 \end{bmatrix}, \begin{bmatrix} 2 \\ 0 \end{bmatrix}$
   2. $\begin{bmatrix} 2 \\ 1 \end{bmatrix}, \begin{bmatrix} 5 \\ 4 \end{bmatrix}, \begin{bmatrix} -3 \\ -2 \end{bmatrix}$

3. $\begin{bmatrix} 1 \\ 2 \\ -1 \end{bmatrix}, \begin{bmatrix} -2 \\ -4 \\ 8 \end{bmatrix}, \begin{bmatrix} 2 \\ -1 \\ 11 \end{bmatrix}, \begin{bmatrix} 0 \\ 15 \\ -9 \end{bmatrix}$

4. $\begin{bmatrix} -2 \\ 5 \\ 3 \end{bmatrix}, \begin{bmatrix} 0 \\ -3 \\ 7 \end{bmatrix}, \begin{bmatrix} 1 \\ -2 \\ -6 \end{bmatrix}, \begin{bmatrix} -2 \\ 7 \\ -3 \end{bmatrix}$

5. $\begin{bmatrix} 1 \\ 0 \\ -2 \end{bmatrix}, \begin{bmatrix} 0 \\ 1 \\ 1 \end{bmatrix}, \begin{bmatrix} -1 \\ 5 \\ 1 \end{bmatrix}, \begin{bmatrix} 0 \\ 5 \\ -3 \end{bmatrix}$

6. $\begin{bmatrix} 1 \\ 3 \\ 1 \end{bmatrix}, \begin{bmatrix} 0 \\ -1 \\ -2 \end{bmatrix}, \begin{bmatrix} 2 \\ 5 \\ 2 \end{bmatrix}, \begin{bmatrix} 3 \\ 5 \\ 0 \end{bmatrix}$

In Exercises 7 and 8, find the barycentric coordinates of **p** with respect to the affinely independent set of points that precedes it.

7. $\begin{bmatrix} 1 \\ -1 \\ 2 \\ 1 \end{bmatrix}, \begin{bmatrix} 2 \\ 1 \\ 0 \\ 1 \end{bmatrix}, \begin{bmatrix} 1 \\ 2 \\ -2 \\ 0 \end{bmatrix}, \mathbf{p} = \begin{bmatrix} 5 \\ 4 \\ -2 \\ 2 \end{bmatrix}$

8. $\begin{bmatrix} 0 \\ 1 \\ -2 \\ 1 \end{bmatrix}, \begin{bmatrix} 1 \\ 1 \\ 0 \\ 2 \end{bmatrix}, \begin{bmatrix} 1 \\ 4 \\ -6 \\ 5 \end{bmatrix}, \mathbf{p} = \begin{bmatrix} -1 \\ 1 \\ -4 \\ 0 \end{bmatrix}$

In Exercises 9 and 10, mark each statement True or False. Justify each answer.

9. a. If $\mathbf{v}_1, \dots, \mathbf{v}_p$ are in $\mathbb{R}^n$ and if the set $\{\mathbf{v}_1 - \mathbf{v}_2, \mathbf{v}_3 - \mathbf{v}_2, \dots, \mathbf{v}_p - \mathbf{v}_2\}$ is linearly dependent, then $\{\mathbf{v}_1, \dots, \mathbf{v}_p\}$ is affinely dependent. (Read this carefully.)

   b. If $\mathbf{v}_1, \dots \mathbf{v}_p$ are in $\mathbb{R}^n$ and if the set of homogeneous forms $\{\tilde{\mathbf{v}}_1, \dots, \tilde{\mathbf{v}}_p\}$ in $\mathbb{R}^{n+1}$ is linearly independent, then $\{\mathbf{v}_1, \dots, \mathbf{v}_p\}$ is affinely dependent.

   c. A finite set of points $\{\mathbf{v}_1, \dots, \mathbf{v}_k\}$ is affinely dependent if there exist real numbers $c_1, \dots, c_k$, not all zero, such that $c_1 + \cdots + c_k = 1$ and $c_1\mathbf{v}_1 + \cdots + c_k\mathbf{v}_k = \mathbf{0}$.

   d. If $S = \{\mathbf{v}_1, \dots, \mathbf{v}_p\}$ is an affinely independent set in $\mathbb{R}^n$ and if **p** in $\mathbb{R}^n$ has a negative barycentric coordinate determined by S, then **p** is not in aff S.

   e. If $\mathbf{v}_1, \mathbf{v}_2, \mathbf{v}_3, \mathbf{a}$, and **b** are in $\mathbb{R}^3$ and if a ray $\mathbf{a} + t\mathbf{b}$ for $t \geq 0$ intersects the triangle with vertices $\mathbf{v}_1, \mathbf{v}_2$, and $\mathbf{v}_3$, then the barycentric coordinates of the intersection point are all nonnegative.

10. a. If $\{\mathbf{v}_1, \dots \mathbf{v}_p\}$ is an affinely dependent set in $\mathbb{R}^n$, then the set $\{\tilde{\mathbf{v}}_1, \dots, \tilde{\mathbf{v}}_p\}$ in $\mathbb{R}^{n+1}$ of homogeneous forms may be linearly independent.

   b. If $\mathbf{v}_1, \mathbf{v}_2, \mathbf{v}_3$, and $\mathbf{v}_4$ are in $\mathbb{R}^3$ and if the set $\{\mathbf{v}_2 - \mathbf{v}_1, \mathbf{v}_3 - \mathbf{v}_1, \mathbf{v}_4 - \mathbf{v}_1\}$ is linearly independent, then $\{\mathbf{v}_1, \dots, \mathbf{v}_4\}$ is affinely independent.

   c. Given $S = \{\mathbf{b}_1, \dots, \mathbf{b}_k\}$ in $\mathbb{R}^n$, each **p** in aff S has a unique representation as an affine combination of $\mathbf{b}_1, \dots, \mathbf{b}_k$.

   d. When color information is specified at each vertex $\mathbf{v}_1, \mathbf{v}_2, \mathbf{v}_3$ of a triangle in $\mathbb{R}^3$, then the color may be interpolated at a point **p** in aff $\{\mathbf{v}_1, \mathbf{v}_2, \mathbf{v}_3\}$ using the barycentric coordinates of **p**.

   e. If T is a triangle in $\mathbb{R}^2$ and if a point **p** is on an edge of the triangle, then the barycentric coordinates of **p** (for this triangle) are not all positive.

11. Explain why any set of five or more points in $\mathbb{R}^3$ must be affinely dependent.

12. Show that a set $\{\mathbf{v}_1, \dots, \mathbf{v}_p\}$ in $\mathbb{R}^n$ is affinely dependent when $p \geq n + 2$.

13. Use only the definition of affine dependence to show that an indexed set $\{\mathbf{v}_1, \mathbf{v}_2\}$ in $\mathbb{R}^n$ is affinely dependent if and only if $\mathbf{v}_1 = \mathbf{v}_2$.

14. The conditions for affine dependence are stronger than those for linear dependence, so an affinely dependent set is automatically linearly dependent. Also, a linearly independent set cannot be affinely dependent and therefore must be affinely independent. Construct two linearly dependent indexed sets S_1 and S_2 in $\mathbb{R}^2$ such that S_1 is affinely dependent and S_2 is affinely independent. In each case, the set should contain either one, two, or three nonzero points.

15. Let $\mathbf{v}_1 = \begin{bmatrix} -1 \\ 2 \end{bmatrix}$, $\mathbf{v}_2 = \begin{bmatrix} 0 \\ 4 \end{bmatrix}$, $\mathbf{v}_3 = \begin{bmatrix} 2 \\ 0 \end{bmatrix}$, and let $S = \{\mathbf{v}_1, \mathbf{v}_2, \mathbf{v}_3\}$.

   a. Show that the set S is affinely independent.

   b. Find the barycentric coordinates of $\mathbf{p}_1 = \begin{bmatrix} 2 \\ 3 \end{bmatrix}$, $\mathbf{p}_2 = \begin{bmatrix} 1 \\ 2 \end{bmatrix}$, $\mathbf{p}_3 = \begin{bmatrix} -2 \\ 1 \end{bmatrix}$, $\mathbf{p}_4 = \begin{bmatrix} 1 \\ -1 \end{bmatrix}$, and $\mathbf{p}_5 = \begin{bmatrix} 1 \\ 1 \end{bmatrix}$, with respect to S.

   c. Let T be the triangle with vertices $\mathbf{v}_1, \mathbf{v}_2$, and $\mathbf{v}_3$. When the sides of T are extended, the lines divide $\mathbb{R}^2$ into seven regions. See Fig. 8. Note the signs of the barycentric coordinates of the points in each region. For example, $\mathbf{p}_5$ is inside the triangle T and all its barycentric coordinates are positive. Point $\mathbf{p}_1$ has coordinates $(-, +, +)$. Its third coordinate is positive because $\mathbf{p}_1$ is on the $\mathbf{v}_3$ side of the line through $\mathbf{v}_1$ and $\mathbf{v}_2$. Its first coordinate is negative because $\mathbf{p}_1$ is opposite the $\mathbf{v}_1$ side of the line through $\mathbf{v}_2$ and $\mathbf{v}_3$. Point $\mathbf{p}_2$ is on the $\mathbf{v}_2\mathbf{v}_3$ edge of T. Its coordinates are $(0, +, +)$. Without calculating the actual values, determine the signs of the barycentric coordinates of points $\mathbf{p}_6, \mathbf{p}_7$, and $\mathbf{p}_8$ as shown in Fig. 8.

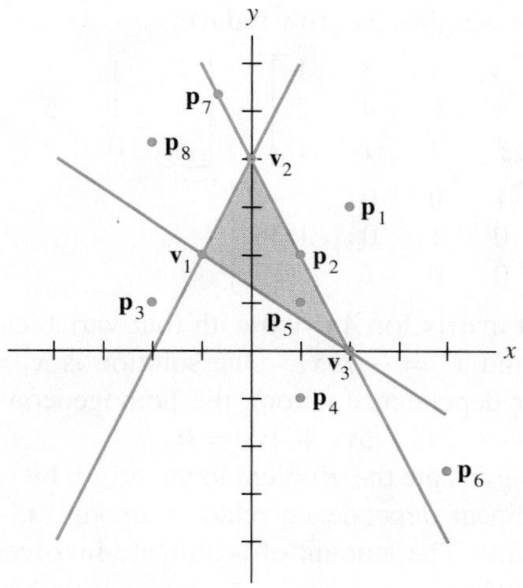

FIGURE 8

16. Let $\mathbf{v}_1 = \begin{bmatrix} 0 \\ 1 \end{bmatrix}$, $\mathbf{v}_2 = \begin{bmatrix} 1 \\ 5 \end{bmatrix}$, $\mathbf{v}_3 = \begin{bmatrix} 4 \\ 3 \end{bmatrix}$, $\mathbf{p}_1 = \begin{bmatrix} 3 \\ 5 \end{bmatrix}$,

$\mathbf{p}_2 = \begin{bmatrix} 5 \\ 1 \end{bmatrix}$, $\mathbf{p}_3 = \begin{bmatrix} 2 \\ 3 \end{bmatrix}$, $\mathbf{p}_4 = \begin{bmatrix} -1 \\ 0 \end{bmatrix}$, $\mathbf{p}_5 = \begin{bmatrix} 0 \\ 4 \end{bmatrix}$,

$\mathbf{p}_6 = \begin{bmatrix} 1 \\ 2 \end{bmatrix}$, $\mathbf{p}_7 = \begin{bmatrix} 6 \\ 4 \end{bmatrix}$, and $S = \{\mathbf{v}_1, \mathbf{v}_2, \mathbf{v}_3\}$.

a. Show that the set S is affinely independent.

b. Find the barycentric coordinates of $\mathbf{p}_1$, $\mathbf{p}_2$, and $\mathbf{p}_3$ with respect to S.

c. On graph paper, sketch the triangle T with vertices $\mathbf{v}_1$, $\mathbf{v}_2$, and $\mathbf{v}_3$, extend the sides as in Fig. 5, and plot the points $\mathbf{p}_4$, $\mathbf{p}_5$, $\mathbf{p}_6$, and $\mathbf{p}_7$. Without calculating the actual values, determine the signs of the barycentric coordinates of points $\mathbf{p}_4$, $\mathbf{p}_5$, $\mathbf{p}_6$, and $\mathbf{p}_7$.

17. Prove Theorem 6 for an affinely independent set $S = \{\mathbf{v}_1, \ldots, \mathbf{v}_k\}$ in $\mathbb{R}^n$. [*Hint:* One method is to mimic the proof of Theorem 7 in Section 4.4.]

18. Let T be a tetrahedron in "standard" position, with three edges along the three positive coordinate axes in $\mathbb{R}^3$, and suppose the vertices are $a\mathbf{e}_1$, $b\mathbf{e}_2$, $c\mathbf{e}_3$, and $\mathbf{0}$, where $[\mathbf{e}_1 \ \mathbf{e}_2 \ \mathbf{e}_3] = I_3$. Find formulas for the barycentric coordinates of an arbitrary point $\mathbf{p}$ in $\mathbb{R}^3$.

19. Let $\{\mathbf{p}_1, \mathbf{p}_2, \mathbf{p}_3\}$ be an affinely dependent set of points in $\mathbb{R}^n$ and let $f : \mathbb{R}^n \to \mathbb{R}^m$ be a linear transformation. Show that $\{f(\mathbf{p}_1), f(\mathbf{p}_2), f(\mathbf{p}_3)\}$ is affinely dependent in $\mathbb{R}^m$.

20. Suppose that $\{\mathbf{p}_1, \mathbf{p}_2, \mathbf{p}_3\}$ is an affinely independent set in $\mathbb{R}^n$ and $\mathbf{q}$ is an arbitrary point in $\mathbb{R}^n$. Show that the translated set $\{\mathbf{p}_1 + \mathbf{q}, \mathbf{p}_2 + \mathbf{q}, \mathbf{p}_3 + \mathbf{q}\}$ is also affinely independent.

In Exercises 21–24, **a**, **b**, and **c** are noncollinear points in $\mathbb{R}^2$ and **p** is any other point in $\mathbb{R}^2$. Let $\triangle\mathbf{abc}$ denote the closed triangular region determined by **a**, **b**, and **c**, and let $\triangle\mathbf{pbc}$ be the region determined by **p**, **b**, and **c**. For convenience, assume that **a**, **b**, and **c** are arranged so that $\det[\tilde{\mathbf{a}} \ \tilde{\mathbf{b}} \ \tilde{\mathbf{c}}]$ is positive, where $\tilde{\mathbf{a}}$, $\tilde{\mathbf{b}}$, and $\tilde{\mathbf{c}}$ are the standard homogeneous forms for the points.

21. Show that the area of $\triangle\mathbf{abc}$ is $\det[\tilde{\mathbf{a}} \ \tilde{\mathbf{b}} \ \tilde{\mathbf{c}}]/2$. [*Hint:* Consult Sections 3.2 and 3.3, including the Exercises.]

22. Let **p** be a point on the line through **a** and **b**. Show that $\det[\tilde{\mathbf{a}} \ \tilde{\mathbf{b}} \ \tilde{\mathbf{p}}] = 0$.

23. Let **p** be any point in the interior of $\triangle\mathbf{abc}$, with barycentric coordinates (r, s, t), so that

$$[\tilde{\mathbf{a}} \quad \tilde{\mathbf{b}} \quad \tilde{\mathbf{c}}] \begin{bmatrix} r \\ s \\ t \end{bmatrix} = \tilde{\mathbf{p}}$$

Use Exercise 19 and a fact about determinants (Chapter 3) to show that

$r = (\text{area of } \triangle\mathbf{pbc})/(\text{area of } \triangle\mathbf{abc})$

$s = (\text{area of } \triangle\mathbf{apc})/(\text{area of } \triangle\mathbf{abc})$

$t = (\text{area of } \triangle\mathbf{abp})/(\text{area of } \triangle\mathbf{abc})$

24. Take **q** on the line segment from **b** to **c** and consider the line through **q** and **a**, which may be written as $\mathbf{p} = (1 - x)\mathbf{q} + x\mathbf{a}$ for all real x. Show that, for each x, $\det[\tilde{\mathbf{p}} \ \tilde{\mathbf{b}} \ \tilde{\mathbf{c}}] = x \cdot \det[\tilde{\mathbf{a}} \ \tilde{\mathbf{b}} \ \tilde{\mathbf{c}}]$. From this and earlier work, conclude that the parameter x is the first barycentric coordinate of **p**. However, by construction, the parameter x also determines the relative distance between **p** and **q** along the segment from **q** to **a**. (When $x = 1$, $\mathbf{p} = \mathbf{a}$.) When this fact is applied to Example 5, it shows that the colors at vertex **a** and the point **q** are smoothly interpolated as **p** moves along the line between **a** and **q**.

SOLUTIONS TO PRACTICE PROBLEMS

1. From Example 1, the problem is to determine if the points are affinely dependent. Use the method of Example 2 and subtract one point from the other two. If one of these two new points is a multiple of the other, the original three points lie on a line.

2. The proof of Theorem 5 essentially points out that an affine dependence relation among points corresponds to a linear dependence relation among the homogeneous forms of the points, using the *same* weights. So, row reduce:

$$\begin{bmatrix} \tilde{\mathbf{v}}_1 & \tilde{\mathbf{v}}_2 & \tilde{\mathbf{v}}_3 & \tilde{\mathbf{v}}_4 \end{bmatrix} = \begin{bmatrix} 4 & 1 & 5 & 1 \\ 1 & 0 & 4 & 2 \\ 1 & 1 & 1 & 1 \end{bmatrix} \sim \begin{bmatrix} 1 & 1 & 1 & 1 \\ 4 & 1 & 5 & 1 \\ 1 & 0 & 4 & 2 \end{bmatrix}$$

$$\sim \begin{bmatrix} 1 & 0 & 0 & -1 \\ 0 & 1 & 0 & 1.25 \\ 0 & 0 & 1 & .75 \end{bmatrix}$$

View this matrix as the coefficient matrix for $A\mathbf{x} = \mathbf{0}$ with four variables. Then x_4 is free, $x_1 = x_4$, $x_2 = -1.25x_4$, and $x_3 = -.75x_4$. One solution is $x_1 = x_4 = 4$, $x_2 = -5$, and $x_3 = -3$. A linear dependence among the homogeneous forms is $4\tilde{\mathbf{v}}_1 - 5\tilde{\mathbf{v}}_2 - 3\tilde{\mathbf{v}}_3 + 4\tilde{\mathbf{v}}_4 = \mathbf{0}$. So $4\mathbf{v}_1 - 5\mathbf{v}_2 - 3\mathbf{v}_3 + 4\mathbf{v}_4 = \mathbf{0}$.

Another solution method is to translate the problem to the origin by subtracting $\mathbf{v}_1$ from the other points, find a linear dependence relation among the translated points, and then rearrange the terms. The amount of arithmetic involved is about the same as in the approach shown above.

8.3 │ CONVEX COMBINATIONS

Section 8.1 considered special linear combinations of the form

$$c_1\mathbf{v}_1 + c_2\mathbf{v}_2 + \cdots + c_k\mathbf{v}_k, \quad \text{where } c_1 + c_2 + \cdots + c_k = 1$$

This section further restricts the weights to be nonnegative.

DEFINITION

A **convex combination** of points $\mathbf{v}_1, \mathbf{v}_2, \ldots, \mathbf{v}_k$ in $\mathbb{R}^n$ is a linear combination of the form

$$c_1\mathbf{v}_1 + c_2\mathbf{v}_2 + \cdots + c_k\mathbf{v}_k$$

such that $c_1 + c_2 + \cdots + c_k = 1$ and $c_i \geq 0$ for all i. The set of all convex combinations of points in a set S is called the **convex hull** of S, denoted by conv S.

The convex hull of a single point $\mathbf{v}_1$ is just the set $\{\mathbf{v}_1\}$, the same as the affine hull. In other cases, the convex hull is properly contained in the affine hull. Recall that the affine hull of distinct points $\mathbf{v}_1$ and $\mathbf{v}_2$ is the line

$$\mathbf{y} = (1 - t)\mathbf{v}_1 + t\mathbf{v}_2, \quad \text{with } t \text{ in } \mathbb{R}$$

Because the weights in a convex combination are nonnegative, the points in conv $\{\mathbf{v}_1, \mathbf{v}_2\}$ may be written as

$$\mathbf{y} = (1 - t)\mathbf{v}_1 + t\mathbf{v}_2, \quad \text{with } 0 \leq t \leq 1$$

which is the **line segment** between $\mathbf{v}_1$ and $\mathbf{v}_2$, hereafter denoted by $\overline{\mathbf{v}_1\mathbf{v}_2}$.

If a set S is affinely independent and if $\mathbf{p} \in \text{aff } S$, then $\mathbf{p} \in \text{conv } S$ if and only if the barycentric coordinates of $\mathbf{p}$ are nonnegative. Example 1 shows a special situation in which S is much more than just affinely independent.

EXAMPLE 1 Let

$$\mathbf{v}_1 = \begin{bmatrix} 3 \\ 0 \\ 6 \\ -3 \end{bmatrix}, \quad \mathbf{v}_2 = \begin{bmatrix} -6 \\ 3 \\ 3 \\ 0 \end{bmatrix}, \quad \mathbf{v}_3 = \begin{bmatrix} 3 \\ 6 \\ 0 \\ 3 \end{bmatrix}, \quad \mathbf{p}_1 = \begin{bmatrix} 0 \\ 3 \\ 3 \\ 0 \end{bmatrix}, \quad \mathbf{p}_2 = \begin{bmatrix} -10 \\ 5 \\ 11 \\ -4 \end{bmatrix},$$

and $S = \{\mathbf{v}_1, \mathbf{v}_2, \mathbf{v}_3\}$. Note that S is an orthogonal set. Determine whether $\mathbf{p}_1$ is in Span S, aff S, and conv S. Then do the same for $\mathbf{p}_2$.

SOLUTION If $\mathbf{p}_1$ is at least a *linear* combination of the points in S, then the weights are easily found, because S is an orthogonal set. Let W be the subspace spanned by S. A calculation as in Section 6.3 shows that the orthogonal projection of $\mathbf{p}_1$ onto W is $\mathbf{p}_1$ itself:

$$\mathrm{proj}_W \, \mathbf{p}_1 = \frac{\mathbf{p}_1 \cdot \mathbf{v}_1}{\mathbf{v}_1 \cdot \mathbf{v}_1} \mathbf{v}_1 + \frac{\mathbf{p}_1 \cdot \mathbf{v}_2}{\mathbf{v}_2 \cdot \mathbf{v}_2} \mathbf{v}_2 + \frac{\mathbf{p}_1 \cdot \mathbf{v}_3}{\mathbf{v}_3 \cdot \mathbf{v}_3} \mathbf{v}_3$$

$$= \frac{18}{54}\mathbf{v}_1 + \frac{18}{54}\mathbf{v}_2 + \frac{18}{54}\mathbf{v}_3$$

$$= \frac{1}{3}\begin{bmatrix} 3 \\ 0 \\ 6 \\ -3 \end{bmatrix} + \frac{1}{3}\begin{bmatrix} -6 \\ 3 \\ 3 \\ 0 \end{bmatrix} + \frac{1}{3}\begin{bmatrix} 3 \\ 6 \\ 0 \\ 3 \end{bmatrix} = \begin{bmatrix} 0 \\ 3 \\ 3 \\ 0 \end{bmatrix} = \mathbf{p}_1$$

This shows that $\mathbf{p}_1$ is *in* Span S. Also, since the coefficients sum to 1, $\mathbf{p}_1$ is in aff S. In fact, $\mathbf{p}_1$ is in conv S, because the coefficients are also nonnegative.

For $\mathbf{p}_2$, a similar calculation shows that $\mathrm{proj}_W \, \mathbf{p}_2 \neq \mathbf{p}_2$. Since $\mathrm{proj}_W \, \mathbf{p}_2$ is the closest point in Span S to $\mathbf{p}_2$, the point $\mathbf{p}_2$ is not in Span S. In particular, $\mathbf{p}_2$ cannot be in aff S or conv S. ∎

Recall that a set S is affine if it contains all lines determined by pairs of points in S. When attention is restricted to convex combinations, the appropriate condition involves line segments rather than lines.

DEFINITION   A set S is **convex** if for each $\mathbf{p}, \mathbf{q} \in S$, the line segment $\overline{\mathbf{p}\mathbf{q}}$ is contained in S.

Intuitively, a set S is convex if every two points in the set can "see" each other without the line of sight leaving the set. Figure 1 illustrates this idea.

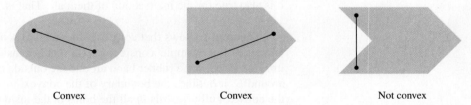

Convex   Convex   Not convex

FIGURE 1

The next result is analogous to Theorem 2 for affine sets.

THEOREM 7   A set S is convex if and only if every convex combination of points of S lies in S. That is, S is convex if and only if $S = \mathrm{conv}\, S$.

PROOF The argument is similar to the proof of Theorem 2. The only difference is in the induction step. When taking a convex combination of $k + 1$ points, consider $\mathbf{y} = c_1\mathbf{v}_1 + \cdots + c_k\mathbf{v}_k + c_{k+1}\mathbf{v}_{k+1}$, where $c_1 + \cdots + c_{k+1} = 1$ and $0 \leq c_i \leq 1$ for all i. If $c_{k+1} = 1$, then $\mathbf{y} = \mathbf{v}_{k+1}$, which belongs to S, and there is nothing further to prove. If $c_{k+1} < 1$, let $t = c_1 + \cdots + c_k$. Then $t = 1 - c_{k+1} > 0$ and

$$\mathbf{y} = (1 - c_{k+1})\left(\frac{c_1}{t}\mathbf{v}_1 + \cdots + \frac{c_k}{t}\mathbf{v}_k\right) + c_{k+1}\mathbf{v}_{k+1} \qquad (1)$$

By the induction hypothesis, the point $\mathbf{z} = (c_1/t)\mathbf{v}_1 + \cdots + (c_k/t)\mathbf{v}_k$ is in S, since the nonnegative coefficients sum to 1. Thus equation (1) displays $\mathbf{y}$ as a convex combination of two points in S. By the principle of induction, every convex combination of such points lies in S. ■

Theorem 9 below provides a more geometric characterization of the convex hull of a set. It requires a preliminary result on intersections of sets. Recall from Section 4.1 (Exercise 32) that the intersection of two subspaces is itself a subspace. In fact, the intersection of any collection of subspaces is itself a subspace. A similar result holds for affine sets and convex sets.

THEOREM 8

Let $\{S_\alpha : \alpha \in \mathcal{A}\}$ be any collection of convex sets. Then $\cap_{\alpha \in \mathcal{A}} S_\alpha$ is convex. If $\{T_\beta : \beta \in \mathcal{B}\}$ is any collection of affine sets, then $\cap_{\beta \in \mathcal{B}} T_\beta$ is affine.

PROOF If $\mathbf{p}$ and $\mathbf{q}$ are in $\cap S_\alpha$, then $\mathbf{p}$ and $\mathbf{q}$ are in each S_α. Since each S_α is convex, the line segment between $\mathbf{p}$ and $\mathbf{q}$ is in S_α for all α and hence that segment is contained in $\cap S_\alpha$. The proof of the affine case is similar. ■

THEOREM 9

For any set S, the convex hull of S is the intersection of all the convex sets that contain S.

PROOF Let T denote the intersection of all the convex sets containing S. Since conv S is a convex set containing S, it follows that $T \subset$ conv S. On the other hand, let C be any convex set containing S. Then C contains every convex combination of points of C (Theorem 7), and hence also contains every convex combination of points of the subset S. That is, conv $S \subset C$. Since this is true for every convex set C containing S, it is also true for the intersection of them all. That is, conv $S \subset T$. ■

Theorem 9 shows that conv S is in a natural sense the "smallest" convex set containing S. For example, consider a set S that lies inside some large rectangle in $\mathbb{R}^2$, and imagine stretching a rubber band around the outside of S. As the rubber band contracts around S, it outlines the boundary of the convex hull of S. Or to use another analogy, the convex hull of S fills *in* all the holes in the inside of S and fills *out* all the dents in the boundary of S.

EXAMPLE 2

a. The convex hulls of sets S and T in $\mathbb{R}^2$ are shown below.

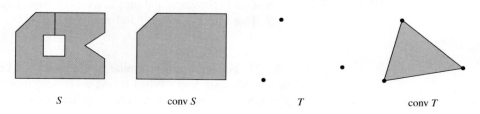

| S | conv S | T | conv T |

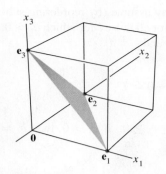

FIGURE 2

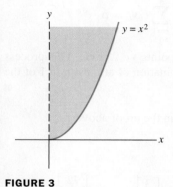

FIGURE 3

b. Let S be the set consisting of the standard basis for $\mathbb{R}^3$, $S = \{\mathbf{e}_1, \mathbf{e}_2, \mathbf{e}_3\}$. Then conv S is a triangular surface in $\mathbb{R}^3$, with vertices $\mathbf{e}_1$, $\mathbf{e}_2$, and $\mathbf{e}_3$. See Fig. 2. ∎

EXAMPLE 3 Let $S = \left\{ \begin{bmatrix} x \\ y \end{bmatrix} : x \geq 0 \text{ and } y = x^2 \right\}$. Show that the convex hull of S is the union of the origin and $\left\{ \begin{bmatrix} x \\ y \end{bmatrix} : x > 0 \text{ and } y \geq x^2 \right\}$. See Fig. 3.

SOLUTION Every point in conv S must lie on a line segment that connects two points of S. The dashed line in Fig. 3 indicates that, except for the origin, the positive y-axis is not in conv S, because the origin is the only point of S on the y-axis. It may seem reasonable that Fig. 3 does show conv S, but how can you be sure that the point $(10^{-2}, 10^4)$, for example, is on a line segment from the origin to a point on the curve in S?

Consider any point $\mathbf{p}$ in the shaded region of Fig. 3, say

$$\mathbf{p} = \begin{bmatrix} a \\ b \end{bmatrix}, \quad \text{with } a > 0 \text{ and } b \geq a^2$$

The line through $\mathbf{0}$ and $\mathbf{p}$ has the equation $y = (b/a)t$ for t real. That line intersects S where t satisfies $(b/a)t = t^2$, that is, when $t = b/a$. Thus, $\mathbf{p}$ is on the line segment from $\mathbf{0}$ to $\begin{bmatrix} b/a \\ b^2/a^2 \end{bmatrix}$, which shows that Fig. 3 is correct. ∎

The following theorem is basic in the study of convex sets. It was first proved by Constantin Caratheodory in 1907. If $\mathbf{p}$ is in the convex hull of S, then, by definition, $\mathbf{p}$ must be a convex combination of points of S. But the definition makes no stipulation as to how many points of S are required to make the combination. Caratheodory's remarkable theorem says that in an n-dimensional space, the number of points of S in the convex combination never has to be more than $n + 1$.

THEOREM 10 **(Caratheodory)** If S is a nonempty subset of $\mathbb{R}^n$, then every point in conv S can be expressed as a convex combination of $n + 1$ or fewer points of S.

PROOF Given $\mathbf{p}$ in conv S, one may write $\mathbf{p} = c_1\mathbf{v}_1 + \cdots + c_k\mathbf{v}_k$, where $\mathbf{v}_i \in S$, $c_1 + \cdots + c_k = 1$, and $c_i \geq 0$, for some k and $i = 1, \ldots, k$. The goal is to show that such an expression exists for $\mathbf{p}$ with $k \leq n + 1$.

If $k > n + 1$, then $\{\mathbf{v}_1, \ldots, \mathbf{v}_k\}$ is affinely dependent, by Exercise 12 in Section 8.2. Thus there exist scalars $d_1, \ldots, d_k$, not all zero, such that

$$\sum_{i=1}^{k} d_i\mathbf{v}_i = \mathbf{0} \quad \text{and} \quad \sum_{i=1}^{k} d_i = 0$$

Consider the two equations

$$c_1\mathbf{v}_1 + c_2\mathbf{v}_2 + \cdots + c_k\mathbf{v}_k = \mathbf{p}$$

and

$$d_1\mathbf{v}_1 + d_2\mathbf{v}_2 + \cdots + d_k\mathbf{v}_k = \mathbf{0}$$

By subtracting an appropriate multiple of the second equation from the first, we now eliminate one of the $\mathbf{v}_i$ terms and obtain a convex combination of fewer than k elements of S that is equal to $\mathbf{p}$.

Since not all of the d_i coefficients are zero, we may assume (by reordering subscripts if necessary) that $d_k > 0$ and that $c_k/d_k \leq c_i/d_i$ for all those i for which $d_i > 0$. For $i = 1, \ldots, k$, let $b_i = c_i - (c_k/d_k)d_i$. Then $b_k = 0$ and

$$\sum_{i=1}^{k} b_i = \sum_{i=1}^{k} c_i - \frac{c_k}{d_k} \sum_{i=1}^{k} d_i = 1 - 0 = 1$$

Furthermore, each $b_i \geq 0$. Indeed, if $d_i \leq 0$, then $b_i \geq c_i \geq 0$. If $d_i > 0$, then $b_i = d_i(c_i/d_i - c_k/d_k) \geq 0$. By construction,

$$\sum_{i=1}^{k-1} b_i \mathbf{v}_i = \sum_{i=1}^{k} b_i \mathbf{v}_i = \sum_{i=1}^{k} \left(c_i - \frac{c_k}{d_k} d_i \right) \mathbf{v}_i$$

$$= \sum_{i=1}^{k} c_i \mathbf{v}_i - \frac{c_k}{d_k} \sum_{i=1}^{k} d_i \mathbf{v}_i = \sum_{i=1}^{k} c_i \mathbf{v}_i = \mathbf{p}$$

Thus $\mathbf{p}$ is now a convex combination of $k - 1$ of the points $\mathbf{v}_1, \ldots, \mathbf{v}_k$. This process may be repeated until $\mathbf{p}$ is expressed as a convex combination of at most $n + 1$ of the points of S. ∎

The following example illustrates the calculations in the proof above.

EXAMPLE 4 Let

$$\mathbf{v}_1 = \begin{bmatrix} 1 \\ 0 \end{bmatrix}, \quad \mathbf{v}_2 = \begin{bmatrix} 2 \\ 3 \end{bmatrix}, \quad \mathbf{v}_3 = \begin{bmatrix} 5 \\ 4 \end{bmatrix}, \quad \mathbf{v}_4 = \begin{bmatrix} 3 \\ 0 \end{bmatrix}, \quad \mathbf{p} = \begin{bmatrix} \frac{10}{3} \\ \frac{5}{2} \end{bmatrix},$$

and $S = \{\mathbf{v}_1, \mathbf{v}_2, \mathbf{v}_3, \mathbf{v}_4\}$. Then

$$\tfrac{1}{4}\mathbf{v}_1 + \tfrac{1}{6}\mathbf{v}_2 + \tfrac{1}{2}\mathbf{v}_3 + \tfrac{1}{12}\mathbf{v}_4 = \mathbf{p} \tag{2}$$

Use the procedure in the proof of Caratheodory's Theorem to express $\mathbf{p}$ as a convex combination of three points of S.

SOLUTION The set S is affinely dependent. Use the technique of Section 8.2 to obtain an affine dependence relation

$$-5\mathbf{v}_1 + 4\mathbf{v}_2 - 3\mathbf{v}_3 + 4\mathbf{v}_4 = \mathbf{0} \tag{3}$$

Next, choose the points $\mathbf{v}_2$ and $\mathbf{v}_4$ in (3), whose coefficients are positive. For each point, compute the ratio of the quotients in equations (2) and (3). The ratio for $\mathbf{v}_2$ is $\tfrac{1}{6} \div 4 = \tfrac{1}{24}$, and that for $\mathbf{v}_4$ is $\tfrac{1}{12} \div 4 = \tfrac{1}{48}$. The ratio for $\mathbf{v}_4$ is smaller, so subtract $\tfrac{1}{48}$ times equation (3) from equation (2) to eliminate $\mathbf{v}_4$:

$$\left(\tfrac{1}{4} + \tfrac{5}{48}\right)\mathbf{v}_1 + \left(\tfrac{1}{6} - \tfrac{4}{48}\right)\mathbf{v}_2 + \left(\tfrac{1}{2} + \tfrac{3}{48}\right)\mathbf{v}_3 + \left(\tfrac{1}{12} - \tfrac{4}{48}\right)\mathbf{v}_4 = \mathbf{p}$$

$$\tfrac{17}{48}\mathbf{v}_1 + \tfrac{4}{48}\mathbf{v}_2 + \tfrac{27}{48}\mathbf{v}_3 = \mathbf{p}$$ ∎

This result cannot, in general, be improved by decreasing the required number of points. Indeed, given any three non-collinear points in $\mathbb{R}^2$, the centroid of the triangle formed by them is in the convex hull of all three, but is not in the convex hull of any two.

PRACTICE PROBLEMS

1. Let $\mathbf{v}_1 = \begin{bmatrix} 6 \\ 2 \\ 2 \end{bmatrix}$, $\mathbf{v}_2 = \begin{bmatrix} 7 \\ 1 \\ 5 \end{bmatrix}$, $\mathbf{v}_3 = \begin{bmatrix} -2 \\ 4 \\ -1 \end{bmatrix}$, $\mathbf{p}_1 = \begin{bmatrix} 1 \\ 3 \\ 1 \end{bmatrix}$, and $\mathbf{p}_2 = \begin{bmatrix} 3 \\ 2 \\ 1 \end{bmatrix}$, and let $S = \{\mathbf{v}_1, \mathbf{v}_2, \mathbf{v}_3\}$. Determine whether $\mathbf{p}_1$ and $\mathbf{p}_2$ are in conv S.

2. Let S be the set of points on the curve $y = 1/x$ for $x > 0$. Explain geometrically why conv S consists of all points on and above the curve S.

8.3 EXERCISES

1. In $\mathbb{R}^2$, let $S = \left\{ \begin{bmatrix} 0 \\ y \end{bmatrix} : 0 \le y < 1 \right\} \cup \left\{ \begin{bmatrix} 2 \\ 0 \end{bmatrix} \right\}$. Describe (or sketch) the convex hull of S.

2. Describe the convex hull of the set S of points $\begin{bmatrix} x \\ y \end{bmatrix}$ in $\mathbb{R}^2$ that satisfy the given conditions. Justify your answers. (Show that an arbitrary point $\mathbf{p}$ in S belongs to conv S.)
   a. $y = 1/x$ and $x \ge 1/2$
   b. $y = \sin x$
   c. $y = x^{1/2}$ and $x \ge 0$

3. Consider the points in Exercise 5 in Section 8.1. Which of $\mathbf{p}_1$, $\mathbf{p}_2$, and $\mathbf{p}_3$ are in conv S?

4. Consider the points in Exercise 6 in Section 8.1. Which of $\mathbf{p}_1$, $\mathbf{p}_2$, and $\mathbf{p}_3$ are in conv S?

5. Let
$$\mathbf{v}_1 = \begin{bmatrix} -1 \\ -3 \\ 4 \end{bmatrix}, \mathbf{v}_2 = \begin{bmatrix} 0 \\ -3 \\ 1 \end{bmatrix}, \mathbf{v}_3 = \begin{bmatrix} 1 \\ -1 \\ 4 \end{bmatrix}, \mathbf{v}_4 = \begin{bmatrix} 1 \\ 1 \\ -2 \end{bmatrix},$$
$$\mathbf{p}_1 = \begin{bmatrix} 1 \\ -1 \\ 2 \end{bmatrix}, \mathbf{p}_2 = \begin{bmatrix} 0 \\ -2 \\ 2 \end{bmatrix},$$
and $S = \{\mathbf{v}_1, \mathbf{v}_2, \mathbf{v}_3, \mathbf{v}_4\}$. Determine whether $\mathbf{p}_1$ and $\mathbf{p}_2$ are in conv S.

6. Let $\mathbf{v}_1 = \begin{bmatrix} 2 \\ 0 \\ -1 \\ 2 \end{bmatrix}$, $\mathbf{v}_2 = \begin{bmatrix} 0 \\ -2 \\ 2 \\ 1 \end{bmatrix}$, $\mathbf{v}_3 = \begin{bmatrix} -2 \\ 1 \\ 0 \\ 2 \end{bmatrix}$, $\mathbf{p}_1 = \begin{bmatrix} -1 \\ 2 \\ -\frac{3}{2} \\ \frac{5}{2} \end{bmatrix}$,

$\mathbf{p}_2 = \begin{bmatrix} -\frac{1}{2} \\ 0 \\ \frac{1}{4} \\ \frac{7}{4} \end{bmatrix}$, $\mathbf{p}_3 = \begin{bmatrix} 6 \\ -4 \\ 1 \\ -1 \end{bmatrix}$, and $\mathbf{p}_4 = \begin{bmatrix} -1 \\ -2 \\ 0 \\ 4 \end{bmatrix}$, and let S be

the orthogonal set $\{\mathbf{v}_1, \mathbf{v}_2, \mathbf{v}_3\}$. Determine whether each $\mathbf{p}_i$ is in Span S, aff S, or conv S.
   a. $\mathbf{p}_1$ b. $\mathbf{p}_2$ c. $\mathbf{p}_3$ d. $\mathbf{p}_4$

Exercises 7–10 use the terminology from Section 8.2.

7. a. Let $T = \left\{ \begin{bmatrix} -1 \\ 0 \end{bmatrix}, \begin{bmatrix} 2 \\ 3 \end{bmatrix}, \begin{bmatrix} 4 \\ 1 \end{bmatrix} \right\}$, and let
   $$\mathbf{p}_1 = \begin{bmatrix} 2 \\ 1 \end{bmatrix}, \mathbf{p}_2 = \begin{bmatrix} 3 \\ 2 \end{bmatrix}, \mathbf{p}_3 = \begin{bmatrix} 2 \\ 0 \end{bmatrix}, \text{ and } \mathbf{p}_4 = \begin{bmatrix} 0 \\ 2 \end{bmatrix}.$$
   Find the barycentric coordinates of $\mathbf{p}_1$, $\mathbf{p}_2$, $\mathbf{p}_3$, and $\mathbf{p}_4$ with respect to T.

   b. Use your answers in part (a) to determine whether each of $\mathbf{p}_1, \ldots, \mathbf{p}_4$ in part (a) is inside, outside, or on the edge of conv T, a triangular region.

8. Repeat Exercise 7 for $T = \left\{ \begin{bmatrix} 2 \\ 0 \end{bmatrix}, \begin{bmatrix} 0 \\ 5 \end{bmatrix}, \begin{bmatrix} -1 \\ 1 \end{bmatrix} \right\}$ and
   $$\mathbf{p}_1 = \begin{bmatrix} 2 \\ 1 \end{bmatrix}, \mathbf{p}_2 = \begin{bmatrix} 1 \\ 1 \end{bmatrix}, \mathbf{p}_3 = \begin{bmatrix} 1 \\ \frac{1}{3} \end{bmatrix}, \text{ and } \mathbf{p}_4 = \begin{bmatrix} 1 \\ 0 \end{bmatrix}.$$

9. Let $S = \{\mathbf{v}_1, \mathbf{v}_2, \mathbf{v}_3, \mathbf{v}_4\}$ be an affinely independent set. Consider the points $\mathbf{p}_1, \ldots, \mathbf{p}_5$ whose barycentric coordinates with respect to S are given by $(2, 0, 0, -1)$, $(0, \frac{1}{2}, \frac{1}{4}, \frac{1}{4})$, $(\frac{1}{2}, 0, \frac{3}{2}, -1)$, $(\frac{1}{3}, \frac{1}{4}, \frac{1}{4}, \frac{1}{6})$, and $(\frac{1}{3}, 0, \frac{2}{3}, 0)$, respectively. Determine whether each of $\mathbf{p}_1, \ldots, \mathbf{p}_5$ is inside, outside, or on the surface of conv S, a tetrahedron. Are any of these points on an edge of conv S?

10. Repeat Exercise 9 for the points $\mathbf{q}_1, \ldots, \mathbf{q}_5$ whose barycentric coordinates with respect to S are given by $(\frac{1}{8}, \frac{1}{4}, \frac{1}{8}, \frac{1}{2})$, $(\frac{3}{4}, -\frac{1}{4}, 0, \frac{1}{2})$, $(0, \frac{3}{4}, \frac{1}{4}, 0)$, $(0, -2, 0, 3)$, and $(\frac{1}{3}, \frac{1}{3}, \frac{1}{3}, 0)$, respectively.

In Exercises 11 and 12, mark each statement True or False. Justify each answer.

11. a. If $\mathbf{y} = c_1\mathbf{v}_1 + c_2\mathbf{v}_2 + c_3\mathbf{v}_3$ and $c_1 + c_2 + c_3 = 1$, then $\mathbf{y}$ is a convex combination of $\mathbf{v}_1$, $\mathbf{v}_2$, and $\mathbf{v}_3$.

    b. If S is a nonempty set, then conv S contains some points that are not in S.

    c. If S and T are convex sets, then $S \cup T$ is also convex.

12. a. A set is convex if $\mathbf{x}, \mathbf{y} \in S$ implies that the line segment between $\mathbf{x}$ and $\mathbf{y}$ is contained in S.

    b. If S and T are convex sets, then $S \cap T$ is also convex.

c. If S is a nonempty subset of $\mathbb{R}^5$ and $\mathbf{y} \in \text{conv } S$, then there exist distinct points $\mathbf{v}_1, \ldots, \mathbf{v}_6$ in S such that $\mathbf{y}$ is a convex combination of $\mathbf{v}_1, \ldots, \mathbf{v}_6$.

13. Let S be a convex subset of $\mathbb{R}^n$ and suppose that $f : \mathbb{R}^n \to \mathbb{R}^m$ is a linear transformation. Prove that the set $f(S) = \{ f(\mathbf{x}) : \mathbf{x} \in S \}$ is a convex subset of $\mathbb{R}^m$.

14. Let $f : \mathbb{R}^n \to \mathbb{R}^m$ be a linear transformation and let T be a convex subset of $\mathbb{R}^m$. Prove that the set $S = \{ \mathbf{x} \in \mathbb{R}^n : f(\mathbf{x}) \in T \}$ is a convex subset of $\mathbb{R}^n$.

15. Let $\mathbf{v}_1 = \begin{bmatrix} 1 \\ 0 \end{bmatrix}$, $\mathbf{v}_2 = \begin{bmatrix} 1 \\ 2 \end{bmatrix}$, $\mathbf{v}_3 = \begin{bmatrix} 4 \\ 2 \end{bmatrix}$, $\mathbf{v}_4 = \begin{bmatrix} 4 \\ 0 \end{bmatrix}$, and $\mathbf{p} = \begin{bmatrix} 2 \\ 1 \end{bmatrix}$. Confirm that

$$\mathbf{p} = \tfrac{1}{3}\mathbf{v}_1 + \tfrac{1}{3}\mathbf{v}_2 + \tfrac{1}{6}\mathbf{v}_3 + \tfrac{1}{6}\mathbf{v}_4 \quad \text{and} \quad \mathbf{v}_1 - \mathbf{v}_2 + \mathbf{v}_3 - \mathbf{v}_4 = \mathbf{0}.$$

Use the procedure in the proof of Caratheodory's Theorem to express $\mathbf{p}$ as a convex combination of three of the $\mathbf{v}_i$'s. Do this in *two* ways.

16. Repeat Exercise 9 for points $\mathbf{v}_1 = \begin{bmatrix} -1 \\ 0 \end{bmatrix}$, $\mathbf{v}_2 = \begin{bmatrix} 0 \\ 3 \end{bmatrix}$, $\mathbf{v}_3 = \begin{bmatrix} 3 \\ 1 \end{bmatrix}$, $\mathbf{v}_4 = \begin{bmatrix} 1 \\ -1 \end{bmatrix}$, and $\mathbf{p} = \begin{bmatrix} 1 \\ 2 \end{bmatrix}$, given that

$$\mathbf{p} = \tfrac{1}{121}\mathbf{v}_1 + \tfrac{72}{121}\mathbf{v}_2 + \tfrac{37}{121}\mathbf{v}_3 + \tfrac{1}{11}\mathbf{v}_4$$

and

$$10\mathbf{v}_1 - 6\mathbf{v}_2 + 7\mathbf{v}_3 - 11\mathbf{v}_4 = \mathbf{0}.$$

In Exercises 17–20, prove the given statement about subsets A and B of $\mathbb{R}^n$. A proof for an exercise may use results of earlier exercises.

17. If $A \subset B$ and B is convex, then conv $A \subset B$.

18. If $A \subset B$, then conv $A \subset$ conv B.

19. a. $[(\text{conv } A) \cup (\text{conv } B)] \subset \text{conv}\,(A \cup B)$

b. Find an example in $\mathbb{R}^2$ to show that equality need not hold in part (a).

20. a. conv $(A \cap B) \subset [(\text{conv } A) \cap (\text{conv } B)]$

b. Find an example in $\mathbb{R}^2$ to show that equality need not hold in part (a).

21. Let $\mathbf{p}_0$, $\mathbf{p}_1$, and $\mathbf{p}_2$ be points in $\mathbb{R}^n$, and define $\mathbf{f}_0(t) = (1-t)\mathbf{p}_0 + t\mathbf{p}_1$, $\mathbf{f}_1(t) = (1-t)\mathbf{p}_1 + t\mathbf{p}_2$, and $\mathbf{g}(t) = (1-t)\mathbf{f}_0(t) + t\mathbf{f}_1(t)$ for $0 \le t \le 1$. For the points as shown below, draw a picture that shows $\mathbf{f}_0\left(\tfrac{1}{2}\right), \mathbf{f}_1\left(\tfrac{1}{2}\right)$, and $\mathbf{g}\left(\tfrac{1}{2}\right)$.

22. Repeat Exercise 21 for $\mathbf{f}_0\left(\tfrac{3}{4}\right), \mathbf{f}_1\left(\tfrac{3}{4}\right)$, and $\mathbf{g}\left(\tfrac{3}{4}\right)$.

23. Let $\mathbf{g}(t)$ be defined as in Exercise 21. Its graph is called a *quadratic Bézier curve*, and it is used in some computer graphics designs. The points $\mathbf{p}_0$, $\mathbf{p}_1$, and $\mathbf{p}_2$ are called the *control points* for the curve. Compute a formula for $\mathbf{g}(t)$ that involves only $\mathbf{p}_0$, $\mathbf{p}_1$, and $\mathbf{p}_2$. Then show that $\mathbf{g}(t)$ is in conv $\{\mathbf{p}_0, \mathbf{p}_1, \mathbf{p}_2\}$ for $0 \le t \le 1$.

24. Given control points $\mathbf{p}_0$, $\mathbf{p}_1$, $\mathbf{p}_2$, and $\mathbf{p}_3$ in $\mathbb{R}^n$, let $\mathbf{g}_1(t)$ for $0 \le t \le 1$ be the quadratic Bézier curve from Exercise 23 determined by $\mathbf{p}_0$, $\mathbf{p}_1$, and $\mathbf{p}_2$, and let $\mathbf{g}_2(t)$ be defined similarly for $\mathbf{p}_1$, $\mathbf{p}_2$, and $\mathbf{p}_3$. For $0 \le t \le 1$, define $\mathbf{h}(t) = (1-t)\mathbf{g}_1(t) + t\mathbf{g}_2(t)$. Show that the graph of $\mathbf{h}(t)$ lies in the convex hull of the four control points. This curve is called a *cubic Bézier curve*, and its definition here is one step in an algorithm for constructing Bézier curves (discussed later in Section 8.6). A Bézier curve of degree k is determined by $k + 1$ control points, and its graph lies in the convex hull of these control points.

SOLUTIONS TO PRACTICE PROBLEMS

1. The points $\mathbf{v}_1$, $\mathbf{v}_2$, and $\mathbf{v}_3$ are not orthogonal, so compute

$$\mathbf{v}_2 - \mathbf{v}_1 = \begin{bmatrix} 1 \\ -1 \\ 3 \end{bmatrix}, \; \mathbf{v}_3 - \mathbf{v}_1 = \begin{bmatrix} -8 \\ 2 \\ -3 \end{bmatrix}, \; \mathbf{p}_1 - \mathbf{v}_1 = \begin{bmatrix} -5 \\ 1 \\ -1 \end{bmatrix}, \; \text{and } \mathbf{p}_2 - \mathbf{v}_1 = \begin{bmatrix} -3 \\ 0 \\ -1 \end{bmatrix}$$

Augment the matrix $[\, \mathbf{v}_2 - \mathbf{v}_1 \quad \mathbf{v}_3 - \mathbf{v}_1 \,]$ with both $\mathbf{p}_1 - \mathbf{v}_1$ and $\mathbf{p}_2 - \mathbf{v}_1$, and row reduce:

$$\begin{bmatrix} 1 & -8 & -5 & -3 \\ -1 & 2 & 1 & 0 \\ 3 & -3 & -1 & -1 \end{bmatrix} \sim \begin{bmatrix} 1 & 0 & \tfrac{1}{3} & 1 \\ 0 & 1 & \tfrac{2}{3} & \tfrac{1}{2} \\ 0 & 0 & 0 & -\tfrac{5}{2} \end{bmatrix}$$

The third column shows that $\mathbf{p}_1 - \mathbf{v}_1 = \tfrac{1}{3}(\mathbf{v}_2 - \mathbf{v}_1) + \tfrac{2}{3}(\mathbf{v}_3 - \mathbf{v}_1)$, which leads to $\mathbf{p}_1 = 0\mathbf{v}_1 + \tfrac{1}{3}\mathbf{v}_2 + \tfrac{2}{3}\mathbf{v}_3$. Thus $\mathbf{p}_1$ is in conv S. In fact, $\mathbf{p}_1$ is in conv $\{\mathbf{v}_2, \mathbf{v}_3\}$.

The last column of the matrix shows that $\mathbf{p}_2 - \mathbf{v}_1$ is not a linear combination of $\mathbf{v}_2 - \mathbf{v}_1$ and $\mathbf{v}_3 - \mathbf{v}_1$. Thus $\mathbf{p}_2$ is not an affine combination of $\mathbf{v}_1$, $\mathbf{v}_2$, and $\mathbf{v}_3$, so $\mathbf{p}_2$ cannot possibly be in conv S.

An alternative method of solution is to row reduce the augmented matrix of homogeneous forms:

$$\begin{bmatrix} \tilde{\mathbf{v}}_1 & \tilde{\mathbf{v}}_2 & \tilde{\mathbf{v}}_3 & \tilde{\mathbf{p}}_1 & \tilde{\mathbf{p}}_2 \end{bmatrix} \sim \begin{bmatrix} 1 & 0 & 0 & 0 & 0 \\ 0 & 1 & 0 & \frac{1}{3} & 0 \\ 0 & 0 & 1 & \frac{2}{3} & 0 \\ 0 & 0 & 0 & 0 & 1 \end{bmatrix}$$

2. If $\mathbf{p}$ is a point above S, then the line through $\mathbf{p}$ with slope -1 will intersect S at two points before it reaches the positive x- and y-axes.

8.4 | HYPERPLANES

Hyperplanes play a special role in the geometry of $\mathbb{R}^n$ because they divide the space into two disjoint pieces, just as a plane separates $\mathbb{R}^3$ into two parts and a line cuts through $\mathbb{R}^2$. The key to working with hyperplanes is to use simple *implicit* descriptions, rather than the *explicit* or parametric representations of lines and planes used in the earlier work with affine sets.[1]

An implicit equation of a line in $\mathbb{R}^2$ has the form $ax + by = d$. An implicit equation of a plane in $\mathbb{R}^3$ has the form $ax + by + cz = d$. Both equations describe the line or plane as the set of all points at which a linear expression (also called a *linear functional*) has a fixed value, d.

DEFINITION

A **linear functional** on $\mathbb{R}^n$ is a linear transformation f from $\mathbb{R}^n$ into $\mathbb{R}$. For each scalar d in $\mathbb{R}$, the symbol $[f:d]$ denotes the set of all $\mathbf{x}$ in $\mathbb{R}^n$ at which the value of f is d. That is,

$$[f:d] \quad \text{is the set} \quad \{\mathbf{x} \in \mathbb{R}^n : f(\mathbf{x}) = d\}$$

The **zero** *functional* is the transformation such that $f(\mathbf{x}) = 0$ for all $\mathbf{x}$ in $\mathbb{R}^n$. All other linear functionals on $\mathbb{R}^n$ are said to be **nonzero**.

EXAMPLE 1   In $\mathbb{R}^2$, the line $x - 4y = 13$ is a hyperplane in $\mathbb{R}^2$, and it is the set of points at which the linear functional $f(x, y) = x - 4y$ has the value 13. That is, the line is the set $[f:13]$. ∎

EXAMPLE 2   In $\mathbb{R}^3$, the plane $5x - 2y + 3z = 21$ is a hyperplane, the set of points at which the linear functional $g(x, y, z) = 5x - 2y + 3z$ has the value 21. This hyperplane is the set $[g:21]$. ∎

If f is a linear functional on $\mathbb{R}^n$, then the standard matrix of this linear transformation f is a $1 \times n$ matrix A, say $A = [a_1 \quad a_2 \quad \cdots \quad a_n]$. So

$$[f:0] \quad \text{is the same as} \quad \{\mathbf{x} \in \mathbb{R}^n : A\mathbf{x} = 0\} = \text{Nul } A \tag{1}$$

---

[1] Parametric representations were introduced in Section 1.5.

If f is a nonzero functional, then rank $A = 1$, and dim Nul $A = n - 1$, by the Rank Theorem.[2] Thus, the subspace $[f : 0]$ has dimension $n - 1$ and so is a hyperplane. Also, if d is any number in $\mathbb{R}$, then

$$[f : d] \quad \text{is the same as} \quad \{\mathbf{x} \in \mathbb{R}^n : A\mathbf{x} = d\} \tag{2}$$

Recall from Theorem 6 in Section 1.5 that the set of solutions of $A\mathbf{x} = \mathbf{b}$ is obtained by translating the solution set of $A\mathbf{x} = \mathbf{0}$, using any particular solution $\mathbf{p}$ of $A\mathbf{x} = \mathbf{b}$. When A is the standard matrix of the transformation f, this theorem says that

$$[f : d] = [f : 0] + \mathbf{p} \quad \text{for any } \mathbf{p} \text{ in } [f : d] \tag{3}$$

Thus the sets $[f : d]$ are hyperplanes parallel to $[f : 0]$. See Fig. 1.

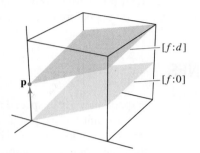

FIGURE 1 Parallel hyperplanes, with $f(\mathbf{p}) = d$.

When A is a $1 \times n$ matrix, the equation $A\mathbf{x} = d$ may be written with an inner product $\mathbf{n} \cdot \mathbf{x}$, using $\mathbf{n}$ in $\mathbb{R}^n$ with the same entries as A. Thus, from (2),

$$[f : d] \quad \text{is the same as} \quad \{\mathbf{x} \in \mathbb{R}^n : \mathbf{n} \cdot \mathbf{x} = d\} \tag{4}$$

Then $[f : 0] = \{\mathbf{x} \in \mathbb{R}^n : \mathbf{n} \cdot \mathbf{x} = 0\}$, which shows that $[f : 0]$ is the orthogonal complement of the subspace spanned by $\mathbf{n}$. In the terminology of calculus and geometry for $\mathbb{R}^3$, $\mathbf{n}$ is called a **normal** vector to $[f : 0]$. (A "normal" vector in this sense need not have unit length.) Also, $\mathbf{n}$ is said to be **normal** to each parallel hyperplane $[f : d]$, even though $\mathbf{n} \cdot \mathbf{x}$ is not zero when $d \neq 0$.

Another name for $[f : d]$ is a *level set* of f, and $\mathbf{n}$ is sometimes called the *gradient* of f when $f(\mathbf{x}) = \mathbf{n} \cdot \mathbf{x}$ for each $\mathbf{x}$.

EXAMPLE 3 Let $\mathbf{n} = \begin{bmatrix} 3 \\ 4 \end{bmatrix}$ and $\mathbf{v} = \begin{bmatrix} 1 \\ -6 \end{bmatrix}$, and let $H = \{\mathbf{x} : \mathbf{n} \cdot \mathbf{x} = 12\}$, so $H = [f : 12]$, where $f(x, y) = 3x + 4y$. Thus H is the line $3x + 4y = 12$. Find an implicit description of the parallel hyperplane (line) $H_1 = H + \mathbf{v}$.

SOLUTION First, find a point $\mathbf{p}$ in H_1. To do this, find a point in H and add $\mathbf{v}$ to it. For instance, $\begin{bmatrix} 0 \\ 3 \end{bmatrix}$ is in H, so $\mathbf{p} = \begin{bmatrix} 1 \\ -6 \end{bmatrix} + \begin{bmatrix} 0 \\ 3 \end{bmatrix} = \begin{bmatrix} 1 \\ -3 \end{bmatrix}$ is in H_1. Now, compute $\mathbf{n} \cdot \mathbf{p} = -9$. This shows that $H_1 = [f : -9]$. See Fig. 2, which also shows the subspace $H_0 = \{\mathbf{x} : \mathbf{n} \cdot \mathbf{x} = 0\}$. ∎

The next three examples show connections between implicit and explicit descriptions of hyperplanes. Example 4 begins with an implicit form.

---

[2] See Theorem 14 in Section 2.9 or Theorem 14 in Section 4.6.

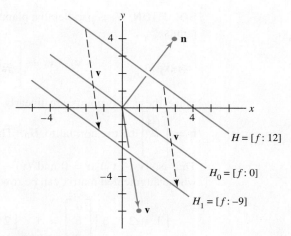

FIGURE 2

EXAMPLE 4 In $\mathbb{R}^2$, give an explicit description of the line $x - 4y = 13$ in parametric vector form.

SOLUTION This amounts to solving a nonhomogeneous equation $A\mathbf{x} = \mathbf{b}$, where $A = [1 \ -4]$ and $\mathbf{b}$ is the number 13 in $\mathbb{R}$. Write $x = 13 + 4y$, where y is a free variable. In parametric form, the solution is

$$\mathbf{x} = \begin{bmatrix} x \\ y \end{bmatrix} = \begin{bmatrix} 13 + 4y \\ y \end{bmatrix} = \begin{bmatrix} 13 \\ 0 \end{bmatrix} + y \begin{bmatrix} 4 \\ 1 \end{bmatrix} = \mathbf{p} + y\mathbf{q}, \quad y \in \mathbb{R} \quad \blacksquare$$

Converting an explicit description of a line into implicit form is more involved. The basic idea is to construct $[f:0]$ and then find d for $[f:d]$.

EXAMPLE 5 Let $\mathbf{v}_1 = \begin{bmatrix} 1 \\ 2 \end{bmatrix}$ and $\mathbf{v}_2 = \begin{bmatrix} 6 \\ 0 \end{bmatrix}$, and let L_1 be the line through $\mathbf{v}_1$ and $\mathbf{v}_2$. Find a linear functional f and a constant d such that $L_1 = [f:d]$.

SOLUTION The line L_1 is parallel to the translated line L_0 through $\mathbf{v}_2 - \mathbf{v}_1$ and the origin. The defining equation for L_0 has the form

$$[a \ \ b]\begin{bmatrix} x \\ y \end{bmatrix} = 0 \quad \text{or} \quad \mathbf{n} \cdot \mathbf{x} = 0, \quad \text{where} \quad \mathbf{n} = \begin{bmatrix} a \\ b \end{bmatrix} \tag{5}$$

Since $\mathbf{n}$ is orthogonal to the subspace L_0, which contains $\mathbf{v}_2 - \mathbf{v}_1$, compute

$$\mathbf{v}_2 - \mathbf{v}_1 = \begin{bmatrix} 6 \\ 0 \end{bmatrix} - \begin{bmatrix} 1 \\ 2 \end{bmatrix} = \begin{bmatrix} 5 \\ -2 \end{bmatrix}$$

and solve

$$[a \ \ b]\begin{bmatrix} 5 \\ -2 \end{bmatrix} = 0$$

By inspection, a solution is $[a \ \ b] = [2 \ \ 5]$. Let $f(x, y) = 2x + 5y$. From (5), $L_0 = [f:0]$, and $L_1 = [f:d]$ for some d. Since $\mathbf{v}_1$ is on line L_1, $d = f(\mathbf{v}_1) = 2(1) + 5(2) = 12$. Thus, the equation for L_1 is $2x + 5y = 12$. As a check, note that $f(\mathbf{v}_2) = f(6, 0) = 2(6) + 5(0) = 12$, so $\mathbf{v}_2$ is on L_1, too. $\blacksquare$

EXAMPLE 6 Let $\mathbf{v}_1 = \begin{bmatrix} 1 \\ 1 \\ 1 \end{bmatrix}$, $\mathbf{v}_2 = \begin{bmatrix} 2 \\ -1 \\ 4 \end{bmatrix}$, and $\mathbf{v}_3 = \begin{bmatrix} 3 \\ 1 \\ 2 \end{bmatrix}$. Find an implicit description $[f:d]$ of the plane H_1 that passes through $\mathbf{v}_1$, $\mathbf{v}_2$, and $\mathbf{v}_3$.

SOLUTION H_1 is parallel to a plane H_0 through the origin that contains the translated points

$$\mathbf{v}_2 - \mathbf{v}_1 = \begin{bmatrix} 1 \\ -2 \\ 3 \end{bmatrix} \quad \text{and} \quad \mathbf{v}_3 - \mathbf{v}_1 = \begin{bmatrix} 2 \\ 0 \\ 1 \end{bmatrix}$$

Since these two points are linearly independent, $H_0 = \text{Span} \{\mathbf{v}_2 - \mathbf{v}_1, \mathbf{v}_3 - \mathbf{v}_1\}$. Let $\mathbf{n} = \begin{bmatrix} a \\ b \\ c \end{bmatrix}$ be the normal to H_0. Then $\mathbf{v}_2 - \mathbf{v}_1$ and $\mathbf{v}_3 - \mathbf{v}_1$ are each orthogonal to $\mathbf{n}$.
That is, $(\mathbf{v}_2 - \mathbf{v}_1) \cdot \mathbf{n} = 0$ and $(\mathbf{v}_3 - \mathbf{v}_1) \cdot \mathbf{n} = 0$. These two equations form a system whose augmented matrix can be row reduced:

$$\begin{bmatrix} 1 & -2 & 3 \end{bmatrix} \begin{bmatrix} a \\ b \\ c \end{bmatrix} = 0, \quad \begin{bmatrix} 2 & 0 & 1 \end{bmatrix} \begin{bmatrix} a \\ b \\ c \end{bmatrix} = 0, \quad \begin{bmatrix} 1 & -2 & 3 & 0 \\ 2 & 0 & 1 & 0 \end{bmatrix}$$

Row operations yield $a = (-\frac{2}{4})c$, $b = (\frac{5}{4})c$, with c free. Set $c = 4$, for instance. Then $\mathbf{n} = \begin{bmatrix} -2 \\ 5 \\ 4 \end{bmatrix}$ and $H_0 = [f : 0]$, where $f(\mathbf{x}) = -2x_1 + 5x_2 + 4x_3$.

The parallel hyperplane H_1 is $[f : d]$. To find d, use the fact that $\mathbf{v}_1$ is in H_1, and compute $d = f(\mathbf{v}_1) = f(1, 1, 1) = -2(1) + 5(1) + 4(1) = 7$. As a check, compute $f(\mathbf{v}_2) = f(2, -1, 4) = -2(2) + 5(-1) + 4(4) = 16 - 9 = 7$. ∎

The procedure in Example 6 generalizes to higher dimensions. However, for the special case of $\mathbb{R}^3$, one can also use the **cross-product** formula to compute $\mathbf{n}$, using a symbolic determinant as a mnemonic device:

$$\mathbf{n} = (\mathbf{v}_2 - \mathbf{v}_1) \times (\mathbf{v}_3 - \mathbf{v}_1)$$

$$= \begin{vmatrix} 1 & 2 & \mathbf{i} \\ -2 & 0 & \mathbf{j} \\ 3 & 1 & \mathbf{k} \end{vmatrix} = \begin{vmatrix} -2 & 0 \\ 3 & 1 \end{vmatrix} \mathbf{i} - \begin{vmatrix} 1 & 2 \\ 3 & 1 \end{vmatrix} \mathbf{j} + \begin{vmatrix} 1 & 2 \\ -2 & 0 \end{vmatrix} \mathbf{k}$$

$$= -2\mathbf{i} + 5\mathbf{j} + 4\mathbf{k} = \begin{bmatrix} -2 \\ 5 \\ 4 \end{bmatrix}$$

If only the formula for f is needed, the cross-product calculation may be written as an ordinary determinant:

$$f(x_1, x_2, x_3) = \begin{vmatrix} 1 & 2 & x_1 \\ -2 & 0 & x_2 \\ 3 & 1 & x_3 \end{vmatrix} = \begin{vmatrix} -2 & 0 \\ 3 & 1 \end{vmatrix} x_1 - \begin{vmatrix} 1 & 2 \\ 3 & 1 \end{vmatrix} x_2 + \begin{vmatrix} 1 & 2 \\ -2 & 0 \end{vmatrix} x_3$$

$$= -2x_1 + 5x_2 + 4x_3$$

So far, every hyperplane examined has been described as $[f : d]$ for some linear functional f and some d in $\mathbb{R}$, or equivalently as $\{\mathbf{x} \in \mathbb{R}^n : \mathbf{n} \cdot \mathbf{x} = d\}$ for some $\mathbf{n}$ in $\mathbb{R}^n$. The following theorem shows that *every* hyperplane has these equivalent descriptions.

THEOREM 11

A subset H of $\mathbb{R}^n$ is a hyperplane if and only if $H = [f : d]$ for some nonzero linear functional f and some scalar d in $\mathbb{R}$. Thus, if H is a hyperplane, there exist a nonzero vector $\mathbf{n}$ and a real number d such that $H = \{\mathbf{x} : \mathbf{n} \cdot \mathbf{x} = d\}$.

PROOF Suppose that H is a hyperplane, take $\mathbf{p} \in H$, and let $H_0 = H - \mathbf{p}$. Then H_0 is an $(n-1)$-dimensional subspace. Next, take any point $\mathbf{y}$ that is not in H_0. By the Orthogonal Decomposition Theorem in Section 6.3,

$$\mathbf{y} = \mathbf{y}_1 + \mathbf{n}$$

where $\mathbf{y}_1$ is a vector in H_0 and $\mathbf{n}$ is orthogonal to every vector in H_0. The function f defined by

$$f(\mathbf{x}) = \mathbf{n} \cdot \mathbf{x} \quad \text{for } \mathbf{x} \in \mathbb{R}^n$$

is a linear functional, by properties of the inner product. Now, $[f:0]$ is a hyperplane that contains H_0, by construction of $\mathbf{n}$. It follows that

$$H_0 = [f:0]$$

[Argument: H_0 contains a basis S of $n-1$ vectors, and since S is in the $(n-1)$-dimensional subspace $[f:0]$, S must also be a basis for $[f:0]$, by the Basis Theorem.] Finally, let $d = f(\mathbf{p}) = \mathbf{n} \cdot \mathbf{p}$. Then, as in (3) shown earlier,

$$[f:d] = [f:0] + \mathbf{p} = H_0 + \mathbf{p} = H$$

The converse statement that $[f:d]$ is a hyperplane follows from (1) and (3) above. ∎

Many important applications of hyperplanes depend on the possibility of "separating" two sets by a hyperplane. Intuitively, this means that one of the sets is on one side of the hyperplane and the other set is on the other side. The following terminology and notation will help to make this idea more precise.

TOPOLOGY IN $\mathbb{R}^n$: TERMS AND FACTS

For any point $\mathbf{p}$ in $\mathbb{R}^n$ and any real $\delta > 0$, the **open ball** $B(\mathbf{p}, \delta)$ with center $\mathbf{p}$ and radius δ is given by

$$B(\mathbf{p}, \delta) = \{\mathbf{x} : \|\mathbf{x} - \mathbf{p}\| < \delta\}$$

Given a set S in $\mathbb{R}^n$, a point $\mathbf{p}$ is an **interior point** of S if there exists a $\delta > 0$ such that $B(\mathbf{p}, \delta) \subset S$. If every open ball centered at $\mathbf{p}$ intersects both S and the complement of S, then $\mathbf{p}$ is called a **boundary point** of S. A set is **open** if it contains none of its boundary points. (This is equivalent to saying that all of its points are interior points.) A set is **closed** if it contains all of its boundary points. (If S contains some but not all of its boundary points, then S is neither open nor closed.) A set S is **bounded** if there exists a $\delta > 0$ such that $S \subset B(\mathbf{0}, \delta)$. A set in $\mathbb{R}^n$ is **compact** if it is closed and bounded.

Theorem: The convex hull of an open set is open, and the convex hull of a compact set is compact. (The convex hull of a closed set need not be closed. See Exercise 27.)

EXAMPLE 7 Let

$$S = \text{conv}\left\{\begin{bmatrix} -2 \\ 2 \end{bmatrix}, \begin{bmatrix} -2 \\ -2 \end{bmatrix}, \begin{bmatrix} 2 \\ -2 \end{bmatrix}, \begin{bmatrix} 2 \\ 2 \end{bmatrix}\right\}, \quad \mathbf{p}_1 = \begin{bmatrix} -1 \\ 0 \end{bmatrix}, \quad \text{and} \quad \mathbf{p}_2 = \begin{bmatrix} 2 \\ 1 \end{bmatrix},$$

as shown in Fig. 3. Then $\mathbf{p}_1$ is an interior point since $B\left(\mathbf{p}, \frac{3}{4}\right) \subset S$. The point $\mathbf{p}_2$ is a boundary point since every open ball centered at $\mathbf{p}_2$ intersects both S and the complement of S. The set S is closed since it contains all its boundary points. The set S is bounded since $S \subset B(\mathbf{0}, 3)$. Thus S is also compact. ∎

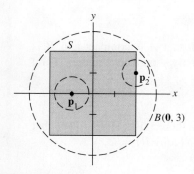

FIGURE 3

The set S is closed and bounded.

Notation: If f is a linear functional, then $f(A) \le d$ means $f(\mathbf{x}) \le d$ for each $\mathbf{x} \in A$. Corresponding notations will be used when the inequalities are reversed or when they are strict.

DEFINITION

The hyperplane $H = [f:d]$ **separates** two sets A and B if one of the following holds:

(i) $f(A) \le d$ and $f(B) \ge d$,   or

(ii) $f(A) \ge d$ and $f(B) \le d$.

If in the conditions above all the weak inequalities are replaced by strict inequalities, then H is said to **strictly separate** A and B.

Notice that strict separation requires that the two sets be disjoint, while mere separation does not. Indeed, if two circles in the plane are externally tangent, then their common tangent line separates them (but does not separate them strictly).

Although it is necessary that two sets be disjoint in order to strictly separate them, this condition is not sufficient, even for closed convex sets. For example, let

$$A = \left\{ \begin{bmatrix} x \\ y \end{bmatrix} : x \ge \frac{1}{2} \text{ and } \frac{1}{x} \le y \le 2 \right\} \quad \text{and} \quad B = \left\{ \begin{bmatrix} x \\ y \end{bmatrix} : x \ge 0 \text{ and } y = 0 \right\}$$

Then A and B are disjoint closed convex sets, but they cannot be strictly separated by a hyperplane (line in $\mathbb{R}^2$). See Fig. 4. Thus the problem of separating (or strictly separating) two sets by a hyperplane is more complex than it might at first appear.

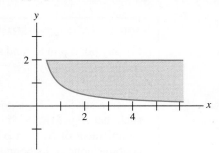

FIGURE 4 Disjoint closed convex sets.

There are many interesting conditions on the sets A and B that imply the existence of a separating hyperplane, but the following two theorems are sufficient for this section. The proof of the first theorem requires quite a bit of preliminary material,[3] but the second theorem follows easily from the first.

THEOREM 12

Suppose A and B are nonempty convex sets such that A is compact and B is closed. Then there exists a hyperplane H that strictly separates A and B if and only if $A \cap B = \varnothing$.

THEOREM 13

Suppose A and B are nonempty compact sets. Then there exists a hyperplane that strictly separates A and B if and only if $(\text{conv } A) \cap (\text{conv } B) = \varnothing$.

[3] A proof of Theorem 12 is given in Steven R. Lay, *Convex Sets and Their Applications* (New York: John Wiley & Sons, 1982; Mineola, NY: Dover Publications, 2007), pp. 34–39.

PROOF Suppose that $(\text{conv } A) \cap (\text{conv } B) = \varnothing$. Since the convex hull of a compact set is compact, Theorem 12 ensures that there is a hyperplane H that strictly separates conv A and conv B. Clearly, H also strictly separates the smaller sets A and B.

Conversely, suppose the hyperplane $H = [f:d]$ strictly separates A and B. Without loss of generality, assume that $f(A) < d$ and $f(B) > d$. Let $\mathbf{x} = c_1 \mathbf{x}_1 + \cdots + c_k \mathbf{x}_k$ be any convex combination of elements of A. Then

$$f(\mathbf{x}) = c_1 f(\mathbf{x}_1) + \cdots + c_k f(\mathbf{x}_k) < c_1 d + \cdots + c_k d = d$$

since $c_1 + \cdots + c_k = 1$. Thus $f(\text{conv } A) < d$. Likewise, $f(\text{conv } B) > d$, so $H = [f:d]$ strictly separates conv A and conv B. By Theorem 12, conv A and conv B must be disjoint. ∎

EXAMPLE 8 Let

$$\mathbf{a}_1 = \begin{bmatrix} 2 \\ 1 \\ 1 \end{bmatrix}, \quad \mathbf{a}_2 = \begin{bmatrix} -3 \\ 2 \\ 1 \end{bmatrix}, \quad \mathbf{a}_3 = \begin{bmatrix} 3 \\ 4 \\ 0 \end{bmatrix}, \quad \mathbf{b}_1 = \begin{bmatrix} 1 \\ 0 \\ 2 \end{bmatrix}, \quad \text{and} \quad \mathbf{b}_2 = \begin{bmatrix} 2 \\ -1 \\ 5 \end{bmatrix},$$

and let $A = \{\mathbf{a}_1, \mathbf{a}_2, \mathbf{a}_3\}$ and $B = \{\mathbf{b}_1, \mathbf{b}_2\}$. Show that the hyperplane $H = [f:5]$, where $f(x_1, x_2, x_3) = 2x_1 - 3x_2 + x_3$, does not separate A and B. Is there a hyperplane parallel to H that does separate A and B? Do the convex hulls of A and B intersect?

SOLUTION Evaluate the linear functional f at each of the points in A and B:

$$f(\mathbf{a}_1) = 2, \quad f(\mathbf{a}_2) = -11, \quad f(\mathbf{a}_3) = -6, \quad f(\mathbf{b}_1) = 4, \quad \text{and} \quad f(\mathbf{b}_2) = 12$$

Since $f(\mathbf{b}_1) = 4$ is less than 5 and $f(\mathbf{b}_2) = 12$ is greater than 5, points of B lie on both sides of $H = [f:5]$ and so H does not separate A and B.

Since $f(A) < 3$ and $f(B) > 3$, the parallel hyperplane $[f:3]$ strictly separates A and B. By Theorem 13, $(\text{conv } A) \cap (\text{conv } B) = \varnothing$.

Caution: If there were no hyperplane parallel to H that strictly separated A and B, this would *not* necessarily imply that their convex hulls intersect. It might be that some other hyperplane not parallel to H would strictly separate them. ∎

PRACTICE PROBLEM

Let $\mathbf{p}_1 = \begin{bmatrix} 1 \\ 0 \\ 2 \end{bmatrix}$, $\mathbf{p}_2 = \begin{bmatrix} -1 \\ 2 \\ 1 \end{bmatrix}$, $\mathbf{n}_1 = \begin{bmatrix} 1 \\ 1 \\ -2 \end{bmatrix}$, $\mathbf{n}_2 = \begin{bmatrix} -2 \\ 1 \\ 3 \end{bmatrix}$, let H_1 be the hyperplane (plane) in $\mathbb{R}^3$ passing through the point $\mathbf{p}_1$ and having normal vector $\mathbf{n}_1$, and let H_2 be the hyperplane passing through the point $\mathbf{p}_2$ and having normal vector $\mathbf{n}_2$. Give an explicit description of $H_1 \cap H_2$ by a formula that shows how to generate all points in $H_1 \cap H_2$.

8.4 EXERCISES

1. Let L be the line in $\mathbb{R}^2$ through the points $\begin{bmatrix} -1 \\ 4 \end{bmatrix}$ and $\begin{bmatrix} 3 \\ 1 \end{bmatrix}$. Find a linear functional f and a real number d such that $L = [f:d]$.

2. Let L be the line in $\mathbb{R}^2$ through the points $\begin{bmatrix} 1 \\ 4 \end{bmatrix}$ and $\begin{bmatrix} -2 \\ -1 \end{bmatrix}$. Find a linear functional f and a real number d such that $L = [f:d]$.

In Exercises 3 and 4, determine whether each set is open or closed or neither open nor closed.

3. a. $\{(x, y) : y > 0\}$
   b. $\{(x, y) : x = 2 \text{ and } 1 \le y \le 3\}$
   c. $\{(x, y) : x = 2 \text{ and } 1 < y < 3\}$
   d. $\{(x, y) : xy = 1 \text{ and } x > 0\}$
   e. $\{(x, y) : xy \ge 1 \text{ and } x > 0\}$

4. a. $\{(x, y) : x^2 + y^2 = 1\}$
   b. $\{(x, y) : x^2 + y^2 > 1\}$
   c. $\{(x, y) : x^2 + y^2 \le 1 \text{ and } y > 0\}$
   d. $\{(x, y) : y \ge x^2\}$
   e. $\{(x, y) : y < x^2\}$

In Exercises 5 and 6, determine whether or not each set is compact and whether or not it is convex.

5. Use the sets from Exercise 3.

6. Use the sets from Exercise 4.

In Exercises 7–10, let H be the hyperplane through the listed points. (a) Find a vector $\mathbf{n}$ that is normal to the hyperplane. (b) Find a linear functional f and a real number d such that $H = [f : d]$.

7. $\begin{bmatrix} 1 \\ 1 \\ 3 \end{bmatrix}, \begin{bmatrix} 2 \\ 4 \\ 1 \end{bmatrix}, \begin{bmatrix} -1 \\ -2 \\ 5 \end{bmatrix}$
8. $\begin{bmatrix} 1 \\ -2 \\ 1 \end{bmatrix}, \begin{bmatrix} 4 \\ -2 \\ 3 \end{bmatrix}, \begin{bmatrix} 7 \\ -4 \\ 4 \end{bmatrix}$

9. $\begin{bmatrix} 1 \\ 0 \\ 1 \\ 0 \end{bmatrix}, \begin{bmatrix} 2 \\ 3 \\ 1 \\ 0 \end{bmatrix}, \begin{bmatrix} 1 \\ 2 \\ 2 \\ 0 \end{bmatrix}, \begin{bmatrix} 1 \\ 1 \\ 1 \\ 1 \end{bmatrix}$

10. $\begin{bmatrix} 1 \\ 2 \\ 0 \\ 0 \end{bmatrix}, \begin{bmatrix} 2 \\ 2 \\ -1 \\ -3 \end{bmatrix}, \begin{bmatrix} 1 \\ 3 \\ 2 \\ 7 \end{bmatrix}, \begin{bmatrix} 3 \\ 2 \\ -1 \\ -1 \end{bmatrix}$

11. Let $\mathbf{p} = \begin{bmatrix} 1 \\ -3 \\ 1 \\ 2 \end{bmatrix}$, $\mathbf{n} = \begin{bmatrix} 2 \\ 1 \\ 5 \\ -1 \end{bmatrix}$, $\mathbf{v}_1 = \begin{bmatrix} 0 \\ 1 \\ 1 \\ 1 \end{bmatrix}$, $\mathbf{v}_2 = \begin{bmatrix} -2 \\ 0 \\ 1 \\ 3 \end{bmatrix}$,

and $\mathbf{v}_3 = \begin{bmatrix} 1 \\ 4 \\ 0 \\ 4 \end{bmatrix}$, and let H be the hyperplane in $\mathbb{R}^4$ with normal $\mathbf{n}$ and passing through $\mathbf{p}$. Which of the points $\mathbf{v}_1$, $\mathbf{v}_2$, and $\mathbf{v}_3$ are on the same side of H as the origin, and which are not?

12. Let $\mathbf{a}_1 = \begin{bmatrix} 2 \\ -1 \\ 5 \end{bmatrix}$, $\mathbf{a}_2 = \begin{bmatrix} 3 \\ 1 \\ 3 \end{bmatrix}$, $\mathbf{a}_3 = \begin{bmatrix} -1 \\ 6 \\ 0 \end{bmatrix}$, $\mathbf{b}_1 = \begin{bmatrix} 0 \\ 5 \\ -1 \end{bmatrix}$,

$\mathbf{b}_2 = \begin{bmatrix} 1 \\ -3 \\ -2 \end{bmatrix}$, $\mathbf{b}_3 = \begin{bmatrix} 2 \\ 2 \\ 1 \end{bmatrix}$, and $\mathbf{n} = \begin{bmatrix} 3 \\ 1 \\ -2 \end{bmatrix}$, and let $A = \{\mathbf{a}_1, \mathbf{a}_2, \mathbf{a}_3\}$ and $B = \{\mathbf{b}_1, \mathbf{b}_2, \mathbf{b}_3\}$. Find a hyperplane H

with normal $\mathbf{n}$ that separates A and B. Is there a hyperplane parallel to H that strictly separates A and B?

13. Let $\mathbf{p}_1 = \begin{bmatrix} 2 \\ -3 \\ 1 \\ 2 \end{bmatrix}$, $\mathbf{p}_2 = \begin{bmatrix} 1 \\ 2 \\ -1 \\ 3 \end{bmatrix}$, $\mathbf{n}_1 = \begin{bmatrix} 1 \\ 2 \\ 4 \\ 2 \end{bmatrix}$, and

$\mathbf{n}_2 = \begin{bmatrix} 2 \\ 3 \\ 1 \\ 5 \end{bmatrix}$, let H_1 be the hyperplane in $\mathbb{R}^4$ through $\mathbf{p}_1$ with

normal $\mathbf{n}_1$, and let H_2 be the hyperplane through $\mathbf{p}_2$ with normal $\mathbf{n}_2$. Give an explicit description of $H_1 \cap H_2$. [*Hint:* Find a point $\mathbf{p}$ in $H_1 \cap H_2$ and two linearly independent vectors $\mathbf{v}_1$ and $\mathbf{v}_2$ that span a subspace parallel to the 2-dimensional flat $H_1 \cap H_2$.]

14. Let F_1 and F_2 be 4-dimensional flats in $\mathbb{R}^6$, and suppose that $F_1 \cap F_2 \ne \varnothing$. What are the possible dimensions of $F_1 \cap F_2$?

In Exercises 15–20, write a formula for a linear functional f and specify a number d, so that $[f : d]$ is the hyperplane H described in the exercise.

15. Let A be the 1×4 matrix $\begin{bmatrix} 1 & -3 & 4 & -2 \end{bmatrix}$ and let $b = 5$. Let $H = \{\mathbf{x} \text{ in } \mathbb{R}^4 : A\mathbf{x} = b\}$.

16. Let A be the 1×5 matrix $\begin{bmatrix} 2 & 5 & -3 & 0 & 6 \end{bmatrix}$. Note that Nul A is in $\mathbb{R}^5$. Let $H = \text{Nul } A$.

17. Let H be the plane in $\mathbb{R}^3$ spanned by the rows of $B = \begin{bmatrix} 1 & 3 & 5 \\ 0 & 2 & 4 \end{bmatrix}$. That is, $H = \text{Row } B$. [*Hint:* How is H related to Nul B? See Section 6.1.]

18. Let H be the plane in $\mathbb{R}^3$ spanned by the rows of $B = \begin{bmatrix} 1 & 4 & -5 \\ 0 & -2 & 8 \end{bmatrix}$. That is, $H = \text{Row } B$.

19. Let H be the column space of the matrix $B = \begin{bmatrix} 1 & 0 \\ 4 & 2 \\ -7 & -6 \end{bmatrix}$. That is, $H = \text{Col } B$. [*Hint:* How is Col B related to Nul B^T? See Section 6.1.]

20. Let H be the column space of the matrix $B = \begin{bmatrix} 1 & 0 \\ 5 & 2 \\ -4 & -4 \end{bmatrix}$. That is, $H = \text{Col } B$.

In Exercises 21 and 22, mark each statement True or False. Justify each answer.

21. a. A linear transformation from $\mathbb{R}$ to $\mathbb{R}^n$ is called a linear functional.
    b. If f is a linear functional defined on $\mathbb{R}^n$, then there exists a real number k such that $f(\mathbf{x}) = k\mathbf{x}$ for all $\mathbf{x}$ in $\mathbb{R}^n$.
    c. If a hyperplane strictly separates sets A and B, then $A \cap B = \varnothing$.
    d. If A and B are closed convex sets and $A \cap B = \varnothing$, then there exists a hyperplane that strictly separates A and B.

22. a. If d is a real number and f is a nonzero linear functional defined on $\mathbb{R}^n$, then $[f : d]$ is a hyperplane in $\mathbb{R}^n$.

b. Given any vector $\mathbf{n}$ and any real number d, the set $\{\mathbf{x} : \mathbf{n} \cdot \mathbf{x} = d\}$ is a hyperplane.

c. If A and B are nonempty disjoint sets such that A is compact and B is closed, then there exists a hyperplane that strictly separates A and B.

d. If there exists a hyperplane H such that H does not strictly separate two sets A and B, then $(\text{conv } A) \cap (\text{conv } B) \neq \varnothing$.

23. Let $\mathbf{v}_1 = \begin{bmatrix} 1 \\ 1 \end{bmatrix}$, $\mathbf{v}_2 = \begin{bmatrix} 3 \\ 0 \end{bmatrix}$, $\mathbf{v}_3 = \begin{bmatrix} 5 \\ 3 \end{bmatrix}$, and $\mathbf{p} = \begin{bmatrix} 4 \\ 1 \end{bmatrix}$. Find a hyperplane $[f : d]$ (in this case, a line) that strictly separates $\mathbf{p}$ from conv $\{\mathbf{v}_1, \mathbf{v}_2, \mathbf{v}_3\}$.

24. Repeat Exercise 23 for $\mathbf{v}_1 = \begin{bmatrix} 1 \\ 2 \end{bmatrix}$, $\mathbf{v}_2 = \begin{bmatrix} 5 \\ 1 \end{bmatrix}$, $\mathbf{v}_3 = \begin{bmatrix} 4 \\ 4 \end{bmatrix}$, and $\mathbf{p} = \begin{bmatrix} 2 \\ 3 \end{bmatrix}$.

25. Let $\mathbf{p} = \begin{bmatrix} 4 \\ 1 \end{bmatrix}$. Find a hyperplane $[f : d]$ that strictly separates $B(\mathbf{0}, 3)$ and $B(\mathbf{p}, 1)$. [*Hint:* After finding f, show that the point $\mathbf{v} = (1 - .75)\mathbf{0} + .75\mathbf{p}$ is neither in $B(\mathbf{0}, 3)$ nor in $B(\mathbf{p}, 1)$.]

26. Let $\mathbf{q} = \begin{bmatrix} 2 \\ 3 \end{bmatrix}$ and $\mathbf{p} = \begin{bmatrix} 6 \\ 1 \end{bmatrix}$. Find a hyperplane $[f : d]$ that strictly separates $B(\mathbf{q}, 3)$ and $B(\mathbf{p}, 1)$.

27. Give an example of a closed subset S of $\mathbb{R}^2$ such that conv S is not closed.

28. Give an example of a compact set A and a closed set B in $\mathbb{R}^2$ such that $(\text{conv } A) \cap (\text{conv } B) = \varnothing$ but A and B cannot be strictly separated by a hyperplane.

29. Prove that the open ball $B(\mathbf{p}, \delta) = \{\mathbf{x} : \|\mathbf{x} - \mathbf{p}\| < \delta\}$ is a convex set. [*Hint:* Use the Triangle Inequality.]

30. Prove that the convex hull of a bounded set is bounded.

SOLUTION TO PRACTICE PROBLEM

First, compute $\mathbf{n}_1 \cdot \mathbf{p}_1 = -3$ and $\mathbf{n}_2 \cdot \mathbf{p}_2 = 7$. The hyperplane H_1 is the solution set of the equation $x_1 + x_2 - 2x_3 = -3$, and H_2 is the solution set of the equation $-2x_1 + x_2 + 3x_3 = 7$. Then

$$H_1 \cap H_2 = \{\mathbf{x} : x_1 + x_2 - 2x_3 = -3 \text{ and } -2x_1 + x_2 + 3x_3 = 7\}$$

This is an implicit description of $H_1 \cap H_2$. To find an explicit description, solve the system of equations by row reduction:

$$\begin{bmatrix} 1 & 1 & -2 & -3 \\ -2 & 1 & 3 & 7 \end{bmatrix} \sim \begin{bmatrix} 1 & 0 & -\frac{5}{3} & -\frac{10}{3} \\ 0 & 1 & -\frac{1}{3} & \frac{1}{3} \end{bmatrix}$$

Thus $x_1 = -\frac{10}{3} + \frac{5}{3}x_3$, $x_2 = \frac{1}{3} + \frac{1}{3}x_3$, $x_3 = x_3$. Let $\mathbf{p} = \begin{bmatrix} -\frac{10}{3} \\ \frac{1}{3} \\ 0 \end{bmatrix}$ and $\mathbf{v} = \begin{bmatrix} \frac{5}{3} \\ \frac{1}{3} \\ 1 \end{bmatrix}$. The general solution can be written as $\mathbf{x} = \mathbf{p} + x_3\mathbf{v}$. Thus $H_1 \cap H_2$ is the line through $\mathbf{p}$ in the direction of $\mathbf{v}$. Note that $\mathbf{v}$ is orthogonal to both $\mathbf{n}_1$ and $\mathbf{n}_2$.

8.5 | POLYTOPES

This section studies geometric properties of an important class of compact convex sets called polytopes. These sets arise in all sorts of applications, including game theory (Section 9.1), linear programming (Sections 9.2 to 9.4), and more general optimization problems, such as the design of feedback controls for engineering systems.

A **polytope** in $\mathbb{R}^n$ is the convex hull of a finite set of points. In $\mathbb{R}^2$, a polytope is simply a polygon. In $\mathbb{R}^3$, a polytope is called a polyhedron. Important features of a polyhedron are its faces, edges, and vertices. For example, the cube has 6 square faces, 12 edges, and 8 vertices. The following definitions provide terminology for higher dimensions as well as $\mathbb{R}^2$ and $\mathbb{R}^3$. Recall that the dimension of a set in $\mathbb{R}^n$ is the dimension of the smallest flat that contains it. Also, note that a polytope is a special type of compact convex set, because a finite set in $\mathbb{R}^n$ is compact and the convex hull of this set is compact, by the theorem in the topology terms and facts box in Section 8.4.

DEFINITION

Let S be a compact convex subset of $\mathbb{R}^n$. A nonempty subset F of S is called a (proper) **face** of S if $F \neq S$ and there exists a hyperplane $H = [f:d]$ such that $F = S \cap H$ and either $f(S) \leq d$ or $f(S) \geq d$. The hyperplane H is called a **supporting hyperplane** to S. If the dimension of F is k, then F is called a **k-face** of S.

If P is a polytope of dimension k, then P is called a **k-polytope**. A 0-face of P is called a **vertex** (plural: **vertices**), a 1-face is an **edge**, and a $(k-1)$-dimensional face is a **facet** of S.

EXAMPLE 1 Suppose S is a cube in $\mathbb{R}^3$. When a plane H is translated through $\mathbb{R}^3$ until it just touches (supports) the cube but does not cut through the interior of the cube, there are three possibilities for $H \cap S$, depending on the orientation of H. (See Figure 1.)

$H \cap S$ may be a 2-dimensional square face (facet) of the cube.

$H \cap S$ may be a 1-dimensional edge of the cube.

$H \cap S$ may be a 0-dimensional vertex of the cube. ∎

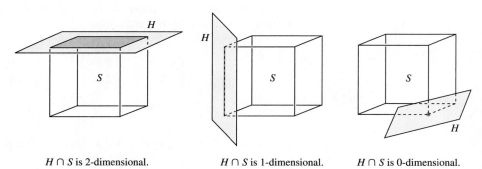

$H \cap S$ is 2-dimensional. $H \cap S$ is 1-dimensional. $H \cap S$ is 0-dimensional.

FIGURE 1

Most applications of polytopes involve the vertices in some way, because they have a special property that is identified in the following definition.

DEFINITION

Let S be a convex set. A point $\mathbf{p}$ in S is called an **extreme point** of S if $\mathbf{p}$ is not in the interior of any line segment that lies in S. More precisely, if $\mathbf{x}, \mathbf{y} \in S$ and $\mathbf{p} \in \overline{xy}$, then $\mathbf{p} = \mathbf{x}$ or $\mathbf{p} = \mathbf{y}$. The set of all extreme points of S is called the **profile** of S.

A vertex of any compact convex set S is automatically an extreme point of S. This fact is proved during the proof of Theorem 14, below. In working with a polytope, say $P = \text{conv}\{\mathbf{v}_1, \ldots, \mathbf{v}_k\}$ for $\mathbf{v}_1, \ldots, \mathbf{v}_k$ in $\mathbb{R}^n$, it is usually helpful to know that $\mathbf{v}_1, \ldots, \mathbf{v}_k$ are the extreme points of P. However, such a list might contain extraneous points. For example, some vector $\mathbf{v}_i$ could be the midpoint of an edge of the polytope. Of course, in this case $\mathbf{v}_i$ is not really needed to generate the convex hull. The following definition describes the property of the vertices that will make them all extreme points.

DEFINITION

> The set $\{\mathbf{v}_1, \ldots, \mathbf{v}_k\}$ is a **minimal representation** of the polytope P if $P = \text{conv}\{\mathbf{v}_1, \ldots, \mathbf{v}_k\}$ and for each $i = 1, \ldots, k, \mathbf{v}_i \notin \text{conv}\{\mathbf{v}_j : j \neq i\}$.

Every polytope has a minimal representation. For if $P = \text{conv}\{\mathbf{v}_1, \ldots, \mathbf{v}_k\}$ and if some $\mathbf{v}_i$ is a convex combination of the other points, then $\mathbf{v}_i$ may be deleted from the set of points without changing the convex hull. This process may be repeated until the minimal representation is left. It can be shown that the minimal representation is unique.

THEOREM 14

> Suppose $M = \{\mathbf{v}_1, \ldots, \mathbf{v}_k\}$ is the minimal representation of the polytope P. Then the following three statements are equivalent:
>
> a. $\mathbf{p} \in M$.
>
> b. $\mathbf{p}$ is a vertex of P.
>
> c. $\mathbf{p}$ is an extreme point of P.

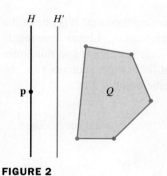

H H'

$\mathbf{p}$

Q

FIGURE 2

PROOF (a) $\Rightarrow$ (b) Suppose $\mathbf{p} \in M$ and let $Q = \text{conv}\{\mathbf{v} : \mathbf{v} \in M \text{ and } \mathbf{v} \neq \mathbf{p}\}$. It follows from the definition of M that $\mathbf{p} \notin Q$, and since Q is compact, Theorem 13 implies the existence of a hyperplane H' that strictly separates $\{\mathbf{p}\}$ and Q. Let H be the hyperplane through $\mathbf{p}$ parallel to H'. See Fig. 2.

Then Q lies in one of the closed half-spaces H^+ bounded by H and so $P \subseteq H^+$. Thus H supports P at $\mathbf{p}$. Furthermore, $\mathbf{p}$ is the only point of P that can lie on H, so $H \cap P = \{\mathbf{p}\}$ and $\mathbf{p}$ is a vertex of P.

(b) $\Rightarrow$ (c) Let $\mathbf{p}$ be a vertex of P. Then there exists a hyperplane $H = [f:d]$ such that $H \cap P = \{\mathbf{p}\}$ and $f(P) \geq d$. If $\mathbf{p}$ were not an extreme point, then there would exist points $\mathbf{x}$ and $\mathbf{y}$ in P such that $\mathbf{p} = (1 - c)\mathbf{x} + c\mathbf{y}$ with $0 < c < 1$. That is,

$$c\mathbf{y} = \mathbf{p} - (1 - c)\mathbf{x} \quad \text{and} \quad \mathbf{y} = \left(\frac{1}{c}\right)(\mathbf{p}) - \left(\frac{1}{c} - 1\right)(\mathbf{x})$$

It follows that $f(\mathbf{y}) = \dfrac{1}{c}f(\mathbf{p}) - \left(\dfrac{1}{c} - 1\right)f(\mathbf{x})$. But $f(\mathbf{p}) = d$ and $f(\mathbf{x}) \geq d$, so

$$f(\mathbf{y}) \leq \left(\frac{1}{c}\right)(d) - \left(\frac{1}{c} - 1\right)(d) = d$$

On the other hand, $\mathbf{y} \in P$, so $f(\mathbf{y}) \geq d$. It follows that $f(\mathbf{y}) = d$ and that $\mathbf{y} \in H \cap P$. This contradicts the fact that $\mathbf{p}$ is a vertex. So $\mathbf{p}$ must be an extreme point. (Note that this part of the proof does not depend on P being a polytope. It holds for any compact convex set.)

(c) $\Rightarrow$ (a) It is clear that any extreme point of P must be a member of M. ∎

EXAMPLE 2 Recall that the profile of a set S is the set of extreme points of S. Theorem 14 shows that the profile of a polygon in $\mathbb{R}^2$ is the set of vertices. (See Fig. 3.) The profile of a closed ball is its boundary. An open set has no extreme points, so its profile is empty. A closed half-space has no extreme points, so its profile is empty. ∎

FIGURE 3

Exercise 18 asks you to show that a point **p** in a convex set S is an extreme point of S if and only if, when **p** is removed from S, the remaining points still form a convex set. It follows that if S^* is any subset of S such that conv S^* is equal to S, then S^* must contain the profile of S. The sets in Example 2 show that in general S^* may have to be larger than the profile of S. It is true, however, that when S is compact we may actually take S^* to be the profile of S, as Theorem 15 will show. Thus every nonempty compact set S has an extreme point, and the set of all extreme points is the smallest subset of S whose convex hull is equal to S.

THEOREM 15

> Let S be a nonempty compact convex set. Then S is the convex hull of its profile (the set of extreme points of S).

PROOF The proof is by induction on the dimension of the set S.[1] ∎

One important application of Theorem 15 is the following theorem. It is one of the key theoretical results in the development of linear programming. Linear functionals are continuous, and continuous functions always attain their maximum and minimum on a compact set. The significance of Theorem 16 is that for compact convex sets, the maximum (and minimum) is actually attained at an extreme point of S.

THEOREM 16

> Let f be a linear functional defined on a nonempty compact convex set S. Then there exist extreme points $\hat{\mathbf{v}}$ and $\hat{\mathbf{w}}$ of S such that
> $$f(\hat{\mathbf{v}}) = \max_{\mathbf{v} \in S} f(\mathbf{v}) \quad \text{and} \quad f(\hat{\mathbf{w}}) = \min_{\mathbf{v} \in S} f(\mathbf{v})$$

PROOF Assume that f attains its maximum m on S at some point $\mathbf{v}'$ in S. That is, $f(\mathbf{v}') = m$. We wish to show that there exists an extreme point in S with the same property. By Theorem 15, $\mathbf{v}'$ is a convex combination of the extreme points of S. That is, there exist extreme points $\mathbf{v}_1, \ldots, \mathbf{v}_k$ of S and nonnegative $c_1, \ldots, c_k$ such that

$$\mathbf{v}' = c_1 \mathbf{v}_1 + \cdots + c_k \mathbf{v}_k \quad \text{with } c_1 + \cdots + c_k = 1$$

If none of the extreme points of S satisfies $f(\mathbf{v}) = m$, then

$$f(\mathbf{v}_i) < m \quad \text{for } i = 1, \ldots, k$$

---

[1] The details may be found in Steven R. Lay, *Convex Sets and Their Applications* (New York: John Wiley & Sons, 1982; Mineola, NY: Dover Publications, 2007, p. 43.

since m is the maximum of f on S. But then, because f is linear,

$$
\begin{aligned}
m = f(\mathbf{v}') &= f(c_1\mathbf{v}_1 + \cdots + c_k\mathbf{v}_k) \\
&= c_1 f(\mathbf{v}_1) + \cdots + c_k f(\mathbf{v}_k) \\
&< c_1 m + \cdots + c_k m = m(c_1 + \cdots + c_k) = m
\end{aligned}
$$

This contradiction implies that some extreme point $\hat{\mathbf{v}}$ of S must satisfy $f(\hat{\mathbf{v}}) = m$. The proof for $\hat{\mathbf{w}}$ is similar. ∎

EXAMPLE 3 Given points $\mathbf{p}_1 = \begin{bmatrix} -1 \\ 0 \end{bmatrix}$, $\mathbf{p}_2 = \begin{bmatrix} 3 \\ 1 \end{bmatrix}$, and $\mathbf{p}_3 = \begin{bmatrix} 1 \\ 2 \end{bmatrix}$ in $\mathbb{R}^2$, let $S = \text{conv}\{\mathbf{p}_1, \mathbf{p}_2, \mathbf{p}_3\}$. For each linear functional f, find the maximum value m of f on the set S, and find all points $\mathbf{x}$ in S at which $f(\mathbf{x}) = m$.

a. $f_1(x_1, x_2) = x_1 + x_2$ b. $f_2(x_1, x_2) = -3x_1 + x_2$ c. $f_3(x_1, x_2) = x_1 + 2x_2$

SOLUTION By Theorem 16, the maximum value is attained at one of the extreme points of S. So to find m, evaluate f at each extreme point and select the largest value.

a. $f_1(\mathbf{p}_1) = -1$, $f_1(\mathbf{p}_2) = 4$, and $f_1(\mathbf{p}_3) = 3$, so $m_1 = 4$. Graph the line $f_1(x_1, x_2) = m_1$, that is, $x_1 + x_2 = 4$, and note that $\mathbf{x} = \mathbf{p}_2$ is the only point in S at which $f_1(\mathbf{x}) = 4$. See Fig. 4(a).

b. $f_2(\mathbf{p}_1) = 3$, $f_2(\mathbf{p}_2) = -8$, and $f_2(\mathbf{p}_3) = -1$, so $m_1 = 3$. Graph the line $f_2(x_1, x_2) = m_2$, that is, $-3x_1 + x_2 = 3$, and note that $\mathbf{x} = \mathbf{p}_1$ is the only point in S at which $f_2(\mathbf{x}) = 3$. See Fig. 4(b).

c. $f_3(\mathbf{p}_1) = -1$, $f_3(\mathbf{p}_2) = 5$, and $f_3(\mathbf{p}_3) = 5$, so $m_1 = 5$. Graph the line $f_3(x_1, x_2) = m_3$, that is, $x_1 + 2x_2 = 5$. Here, f_3 attains its maximum value at $\mathbf{p}_2$, at $\mathbf{p}_3$, and at every point in the convex hull of $\mathbf{p}_2$ and $\mathbf{p}_3$. See Fig. 4(c). ∎

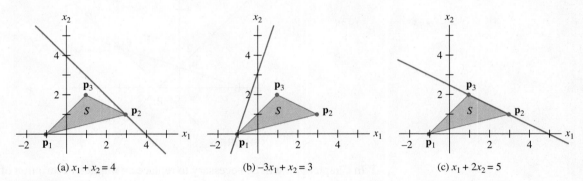

(a) $x_1 + x_2 = 4$ (b) $-3x_1 + x_2 = 3$ (c) $x_1 + 2x_2 = 5$

FIGURE 4

The situation illustrated in Example 3 for $\mathbb{R}^2$ also applies in higher dimensions. The maximum value of a linear functional f on a polytope P occurs at the intersection of a supporting hyperplane and P. This intersection is either a single extreme point of P, or the convex hull of 2 or more extreme points of P. In either case, the intersection is a polytope, and its extreme points form a subset of the extreme points of P.

By definition, a polytope is the convex hull of a finite set of points. This is an explicit representation of the polytope since it identifies points in the set. A polytope may also be represented implicitly as the intersection of a finite number of closed half-spaces. Example 4 illustrates this in $\mathbb{R}^2$.

EXAMPLE 4   Let

$$\mathbf{p}_1 = \begin{bmatrix} 0 \\ 1 \end{bmatrix}, \quad \mathbf{p}_2 = \begin{bmatrix} 1 \\ 0 \end{bmatrix}, \quad \text{and} \quad \mathbf{p}_3 = \begin{bmatrix} 3 \\ 2 \end{bmatrix}$$

in $\mathbb{R}^2$, and let $S = \text{conv}\,\{\mathbf{p}_1, \mathbf{p}_2, \mathbf{p}_3\}$. Simple algebra shows that the line through $\mathbf{p}_1$ and $\mathbf{p}_2$ is given by $x_1 + x_2 = 1$, and S is on the side of this line where

$$x_1 + x_2 \geq 1 \quad \text{or, equivalently,} \quad -x_1 - x_2 \leq -1.$$

Similarly, the line through $\mathbf{p}_2$ and $\mathbf{p}_3$ is $x_1 - x_2 = 1$, and S is on the side where

$$x_1 - x_2 \leq 1$$

Also, the line through $\mathbf{p}_3$ and $\mathbf{p}_1$ is $-x_1 + 3x_2 = 3$, and S is on the side where

$$-x_1 + 3x_2 \leq 3.$$

See Figure 5. It follows that S can be described as the solution set of the system of linear inequalities

$$-x_1 - x_2 \leq -1$$
$$x_1 - x_2 \leq 1$$
$$-x_1 + 3x_2 \leq 3$$

This system may be written as $A\mathbf{x} \leq \mathbf{b}$, where

$$A = \begin{bmatrix} -1 & -1 \\ 1 & -1 \\ -1 & 3 \end{bmatrix}, \quad \mathbf{x} = \begin{bmatrix} x_1 \\ x_2 \end{bmatrix}, \quad \text{and} \quad \mathbf{b} = \begin{bmatrix} -1 \\ 1 \\ 3 \end{bmatrix}.$$

Note that an inequality between two vectors, such as $A\mathbf{x}$ and $\mathbf{b}$, applies to each of the corresponding coordinates in those vectors.   ■

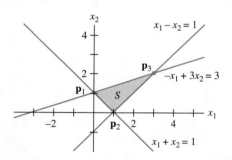

FIGURE 5

In Chapter 9, it will be necessary to replace an implicit description of a polytope by a minimal representation of the polytope, listing all the extreme points of the polytope. In simple cases, a graphical solution is feasible. The following example shows how to handle the situation when several points of interest are too close to identify easily on a graph.

EXAMPLE 5   Let P be the set of points in $\mathbb{R}^2$ that satisfy $A\mathbf{x} \leq \mathbf{b}$, where

$$A = \begin{bmatrix} 1 & 3 \\ 1 & 1 \\ 3 & 2 \end{bmatrix} \quad \text{and} \quad \mathbf{b} = \begin{bmatrix} 18 \\ 8 \\ 21 \end{bmatrix}$$

and $\mathbf{x} \geq \mathbf{0}$. Find the minimal representation of P.

SOLUTION The condition $\mathbf{x} \geq \mathbf{0}$ places P in the first quadrant of $\mathbb{R}^2$, a typical condition in linear programming problems. The three inequalities in $A\mathbf{x} \leq \mathbf{b}$ involve three boundary lines:

$$(1) \ \ x_1 + 3x_2 = 18 \quad (2) \ \ x_1 + x_2 = 8 \quad (3) \ \ 3x_1 + 2x_2 = 21$$

All three lines have negative slopes, so a general idea of the shape of P is easy to visualize. Even a rough sketch of the graphs of these lines will reveal that $(0, 0)$, $(7, 0)$, and $(0, 6)$ are vertices of the polytope P.

What about the intersections of the lines (1), (2), and (3)? Sometimes it is clear from the graph which intersections to include. But if not, then the following algebraic procedure will work well:

When an intersection point is found that corresponds to two inequalities, test it in the other inequalities to see whether the point is in the polytope.

The intersection of (1) and (2) is $\mathbf{p}_{12} = (3, 5)$. Both coordinates are nonnegative, so $\mathbf{p}_{12}$ satisfies all inequalities except possibly the third inequality. Test this:

$$3(3) + 2(5) = 19 < 21$$

This intersection point satisfies the inequality for (3), so $\mathbf{p}_{12}$ is in the polytope.

The intersection of (2) and (3) is $\mathbf{p}_{23} = (5, 3)$. This satisfies all inequalities except possibly the inequality for (1). Test this:

$$1(5) + 3(3) = 14 < 18$$

This shows that $\mathbf{p}_{23}$ is in the polytope.

Finally, the intersection of (1) and (3) is $\mathbf{p}_{13} = \left(\frac{27}{7}, \frac{33}{7}\right)$. Test this in the inequality for (2):

$$1\left(\frac{27}{7}\right) + 1\left(\frac{33}{7}\right) = \frac{60}{7} \approx 8.6 > 8$$

Thus $\mathbf{p}_{13}$ does **not** satisfy the second inequality, which shows that $\mathbf{p}_{13}$ is **not** in P. In conclusion, the minimal representation of the polytope P is

$$\left\{ \begin{bmatrix} 0 \\ 0 \end{bmatrix}, \begin{bmatrix} 7 \\ 0 \end{bmatrix}, \begin{bmatrix} 3 \\ 5 \end{bmatrix}, \begin{bmatrix} 5 \\ 3 \end{bmatrix}, \begin{bmatrix} 0 \\ 6 \end{bmatrix} \right\}. \qquad \blacksquare$$

The remainder of this section discusses the construction of two basic polytopes in $\mathbb{R}^3$ (and higher dimensions). The first appears in linear programming problems, the subject of Chapter 9. Both polytopes provide opportunities to visualize $\mathbb{R}^4$ in a remarkable way.

Simplex

A **simplex** is the convex hull of an affinely independent finite set of vectors. To construct a k-dimensional simplex (or k-simplex), proceed as follows:

0-simplex S^0: a single point $\{\mathbf{v}_1\}$

1-simplex S^1: $\text{conv}(S^0 \cup \{\mathbf{v}_2\})$, with $\mathbf{v}_2$ not in aff S^0

2-simplex S^2: $\text{conv}(S^1 \cup \{\mathbf{v}_3\})$, with $\mathbf{v}_3$ not in aff S^1

$$\vdots$$

k-simplex S^k: $\text{conv}(S^{k-1} \cup \{\mathbf{v}_{k+1}\})$, with $\mathbf{v}_{k+1}$ not in aff S^{k-1}

The simplex S^1 is a line segment. The triangle S^2 comes from choosing a point $\mathbf{v}_3$ that is not in the line containing S_1 and then forming the convex hull with S_2. (See

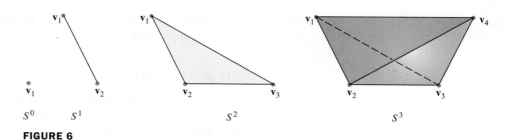

FIGURE 6

Fig. 6.) The tetrahedron S_3 is produced by choosing a point $\mathbf{v}_4$ not in the plane of S^2 and then forming the convex hull with S^2.

Before continuing, consider some of the patterns that are appearing. The triangle S^2 has three edges. Each of these edges is a line segment like S^1. Where do these three line segments come from? One of them is S^1. One of them comes by joining the endpoint $\mathbf{v}_2$ to the new point $\mathbf{v}_3$. The third comes from joining the other endpoint $\mathbf{v}_1$ to $\mathbf{v}_3$. You might say that each endpoint in S^1 is stretched out into a line segment in S^2.

The tetrahedron S^3 in Fig. 6 has four triangular faces. One of these is the original triangle S^2, and the other three come from stretching the edges of S^2 out to the new point $\mathbf{v}_4$. Notice too that the vertices of S^2 get stretched out into edges in S^3. The other edges in S^3 come from the edges in S^2. This suggests how to "visualize" the four-dimensional S^4.

The construction of S^4, called a pentatope, involves forming the convex hull of S^3 with a point $\mathbf{v}_5$ not in the 3-space of S^3. A complete picture is impossible, of course, but Fig. 7 is suggestive: S^4 has five vertices, and any four of the vertices determine a facet in the shape of a tetrahedron. For example, the figure emphasizes the facet with vertices $\mathbf{v}_1$, $\mathbf{v}_2$, $\mathbf{v}_4$, and $\mathbf{v}_5$ and the facet with vertices $\mathbf{v}_2$, $\mathbf{v}_3$, $\mathbf{v}_4$, and $\mathbf{v}_5$. There are five

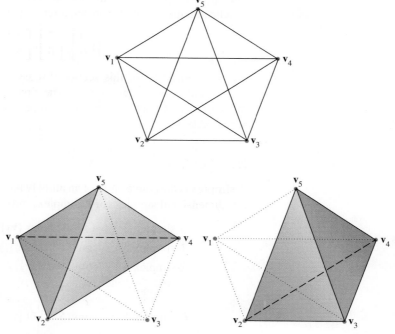

FIGURE 7 The 4-dimensional simplex S^4 projected onto $\mathbb{R}^2$, with two tetrahedral facets emphasized.

such facets. Figure 7 identifies all ten edges of S^4, and these can be used to visualize the ten triangular faces.

Figure 8 shows another representation of the 4-dimensional simplex S^4. This time the fifth vertex appears "inside" the tetrahedron S^3. The highlighted tetrahedral facets also appear to be "inside" S^3.

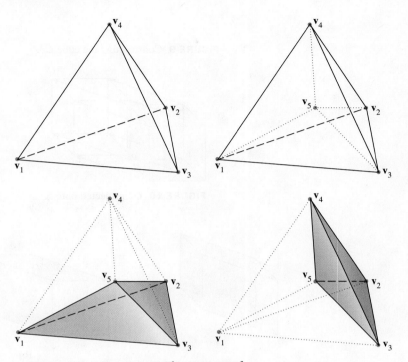

FIGURE 8 The fifth vertex of S^4 is "inside" S^3.

Hypercube

Let $I_i = \overline{\mathbf{0}\mathbf{e}_i}$ be the line segment from the origin $\mathbf{0}$ to the standard basis vector $\mathbf{e}_i$ in $\mathbb{R}^n$. Then for k such that $1 \le k \le n$, the vector sum[2]

$$C^k = I_1 + I_2 + \cdots + I_k$$

is called a k-dimensional **hypercube**.

To visualize the construction of C^k, start with the simple cases. The hypercube C^1 is the line segment I_1. If C^1 is translated by $\mathbf{e}_2$, the convex hull of its initial and final positions describes a square C^2. (See Fig. 9 on page 478.) Translating C^2 by $\mathbf{e}_3$ creates the cube C^3. A similar translation of C^3 by the vector $\mathbf{e}_4$ yields the 4-dimensional hypercube C^4.

Again, this is hard to visualize, but Fig. 10 shows a 2-dimensional projection of C^4. Each of the edges of C^3 is stretched into a square face of C^4. And each of the square faces of C^3 is stretched into a cubic face of C^4. Figure 11 shows three facets of C^4. Part (a) highlights the cube that comes from the left square face of C^3. Part (b) shows the cube that comes from the front square face of C^3. And part (c) emphasizes the cube that comes from the top square face of C^3.

---

[2] The vector sum of two sets A and B is defined by $A + B = \{\mathbf{c} : \mathbf{c} = \mathbf{a} + \mathbf{b}$ for some $\mathbf{a} \in A$ and $\mathbf{b} \in B\}$.

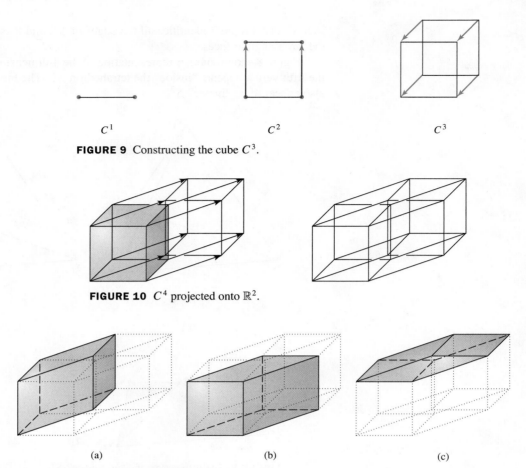

FIGURE 9 Constructing the cube C^3.

FIGURE 10 C^4 projected onto $\mathbb{R}^2$.

(a) (b) (c)

FIGURE 11 Three of the cubic facets of C^4.

Figure 12 shows another representation of C^4 in which the translated cube is placed "inside" C^3. This makes it easier to visualize the cubic facets of C^4, since there is less distortion.

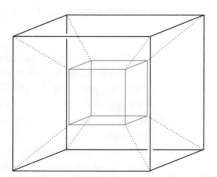

FIGURE 12 The translated image of C^3 is placed "inside" C^3 to obtain C^4.

Altogether, the 4-dimensional cube C^4 has eight cubic faces. Two come from the original and translated images of C^3, and six come from the square faces of C^3 that are stretched into cubes. The square 2-dimensional faces of C^4 come from the square faces of C^3 and its translate, and the edges of C^3 that are stretched into squares. Thus there are $2 \times 6 + 12 = 24$ square faces. To count the edges, take 2 times the number of edges in C^3 and add the number of vertices in C^3. This makes $2 \times 12 + 8 = 32$ edges in C^4. The vertices in C^4 all come from C^3 and its translate, so there are $2 \times 8 = 16$ vertices.

One of the truly remarkable results in the study of polytopes is the following formula, first proved by Leonard Euler (1707–1783). It establishes a simple relationship between the number of faces of different dimensions in a polytope. To simplify the statement of the formula, let $f_k(P)$ denote the number of k-dimensional faces of an n-dimensional polytope P.[3]

$$\text{Euler's formula:} \qquad \sum_{k=0}^{n-1} (-1)^k f_k(P) = 1 + (-1)^{n-1}$$

In particular, when $n = 3$, $v - e + f = 2$, where v, e, and f denote the number of vertices, edges, and facets (respectively) of P.

PRACTICE PROBLEM

1. Find the minimal representation of the polytope P defined by the inequalities $A\mathbf{x} \le \mathbf{b}$ and $\mathbf{x} \ge \mathbf{0}$, when $A = \begin{bmatrix} 1 & 3 \\ 1 & 2 \\ 2 & 1 \end{bmatrix}$ and $\mathbf{b} = \begin{bmatrix} 12 \\ 9 \\ 12 \end{bmatrix}$.

8.5 EXERCISES

1. Given points $\mathbf{p}_1 = \begin{bmatrix} 1 \\ 0 \end{bmatrix}$, $\mathbf{p}_2 = \begin{bmatrix} 2 \\ 3 \end{bmatrix}$, and $\mathbf{p}_3 = \begin{bmatrix} -1 \\ 2 \end{bmatrix}$ in $\mathbb{R}^2$, let $S = \text{conv}\{\mathbf{p}_1, \mathbf{p}_2, \mathbf{p}_3\}$. For each linear functional f, find the maximum value m of f on the set S, and find all points $\mathbf{x}$ in S at which $f(\mathbf{x}) = m$.

   a. $f(x_1, x_2) = x_1 - x_2$ b. $f(x_1, x_2) = x_1 + x_2$

   c. $f(x_1, x_2) = -3x_1 + x_2$

2. Given points $\mathbf{p}_1 = \begin{bmatrix} 0 \\ -1 \end{bmatrix}$, $\mathbf{p}_2 = \begin{bmatrix} 2 \\ 1 \end{bmatrix}$, and $\mathbf{p}_3 = \begin{bmatrix} 1 \\ 2 \end{bmatrix}$ in $\mathbb{R}^2$, let $S = \text{conv}\{\mathbf{p}_1, \mathbf{p}_2, \mathbf{p}_3\}$. For each linear functional f, find the maximum value m of f on the set S, and find all points $\mathbf{x}$ in S at which $f(\mathbf{x}) = m$.

   a. $f(x_1, x_2) = x_1 + x_2$ b. $f(x_1, x_2) = x_1 - x_2$

   c. $f(x_1, x_2) = -2x_1 + x_2$

3. Repeat Exercise 1 where m is the *minimum* value of f on S instead of the maximum value.

4. Repeat Exercise 2 where m is the *minimum* value of f on S instead of the maximum value.

In Exercises 5–8, find the minimal representation of the polytope defined by the inequalities $A\mathbf{x} \le \mathbf{b}$ and $\mathbf{x} \ge \mathbf{0}$.

5. $A = \begin{bmatrix} 1 & 2 \\ 3 & 1 \end{bmatrix}$, $\mathbf{b} = \begin{bmatrix} 10 \\ 15 \end{bmatrix}$

6. $A = \begin{bmatrix} 2 & 3 \\ 4 & 1 \end{bmatrix}$, $\mathbf{b} = \begin{bmatrix} 18 \\ 16 \end{bmatrix}$

7. $A = \begin{bmatrix} 1 & 3 \\ 1 & 1 \\ 4 & 1 \end{bmatrix}$, $\mathbf{b} = \begin{bmatrix} 18 \\ 10 \\ 28 \end{bmatrix}$

8. $A = \begin{bmatrix} 2 & 1 \\ 1 & 1 \\ 1 & 2 \end{bmatrix}$, $\mathbf{b} = \begin{bmatrix} 8 \\ 6 \\ 7 \end{bmatrix}$

9. Let $S = \{(x, y) : x^2 + (y - 1)^2 \le 1\} \cup \{(3, 0)\}$. Is the origin an extreme point of conv S? Is the origin a vertex of conv S?

10. Find an example of a closed convex set S in $\mathbb{R}^2$ such that its profile P is nonempty but conv $P \ne S$.

11. Find an example of a bounded convex set S in $\mathbb{R}^2$ such that its profile P is nonempty but conv $P \ne S$.

12. a. Determine the number of k-faces of the 5-dimensional simplex S^5 for $k = 0, 1, \ldots, 4$. Verify that your answer satisfies Euler's formula.

    b. Make a chart of the values of $f_k(S^n)$ for $n = 1, \ldots, 5$ and $k = 0, 1, \ldots, 4$. Can you see a pattern? Guess a general formula for $f_k(S^n)$.

---

[3] A proof is presented in Steven R. Lay, *Convex Sets and Their Applications* (New York: John Wiley & Sons, 1982; Mineola, NY: Dover Publications, 2007), p. 131.

13. a. Determine the number of k-faces of the 5-dimensional hypercube C^5 for $k = 0, 1, \ldots, 4$. Verify that your answer satisfies Euler's formula.

b. Make a chart of the values of $f_k(C^n)$ for $n = 1, \ldots, 5$ and $k = 0, 1, \ldots, 4$. Can you see a pattern? Guess a general formula for $f_k(C^n)$.

14. Suppose $v_1, \ldots, v_k$ are linearly independent vectors in $\mathbb{R}^n$ ($1 \le k \le n$). Then the set $X^k = \text{conv} \{\pm v_1, \ldots, \pm v_k\}$ is called a **k-crosspolytope**.

a. Sketch X^1 and X^2.

b. Determine the number of k-faces of the 3-dimensional crosspolytope X^3 for $k = 0, 1, 2$. What is another name for X^3?

c. Determine the number of k-faces of the 4-dimensional crosspolytope X^4 for $k = 0, 1, 2, 3$. Verify that your answer satisfies Euler's formula.

d. Find a formula for $f_k(X^n)$, the number of k-faces of X^n, for $0 \le k \le n - 1$.

15. A **k-pyramid** P^k is the convex hull of a $(k - 1)$-polytope Q and a point $x \notin \text{aff } Q$. Find a formula for each of the following in terms of $f_j(Q)$, $j = 0, \ldots, n - 1$.

a. The number of vertices of P^n: $f_0(P^n)$.

b. The number of k-faces of P^n: $f_k(P^n)$, for $1 \le k \le n - 2$.

c. The number of $(n - 1)$-dimensional facets of P^n: $f_{n-1}(P^n)$.

In Exercises 16 and 17, mark each statement True or False. Justify each answer.

16. a. A polytope is the convex hull of a finite set of points.

b. Let p be an extreme point of a convex set S. If $u, v \in S$, $p \in \overline{uv}$, and $p \neq u$, then $p = v$.

c. If S is a nonempty convex subset of $\mathbb{R}^n$, then S is the convex hull of its profile.

d. The 4-dimensional simplex S^4 has exactly five facets, each of which is a 3-dimensional tetrahedron.

17. a. A cube in $\mathbb{R}^3$ has five facets.

b. A point p is an extreme point of a polytope P if and only if p is a vertex of P.

c. If S is a nonempty compact convex set and a linear functional attains its maximum at a point p, then p is an extreme point of S.

d. A 2-dimensional polytope always has the same number of vertices and edges.

18. Let v be an element of the convex set S. Prove that v is an extreme point of S if and only if the set $\{x \in S : x \neq v\}$ is convex.

19. If $c \in \mathbb{R}$ and S is a set, define $cS = \{cx : x \in S\}$. Let S be a convex set and suppose $c > 0$ and $d > 0$. Prove that $cS + dS = (c + d)S$.

20. Find an example to show that the convexity of S is necessary in Exercise 19.

21. If A and B are convex sets, prove that $A + B$ is convex.

22. A polyhedron (3-polytope) is called **regular** if all its facets are congruent regular polygons and all the angles at the vertices are equal. Supply the details in the following proof that there are only five regular polyhedra.

a. Suppose that a regular polyhedron has r facets, each of which is a k-sided regular polygon, and that s edges meet at each vertex. Letting v and e denote the numbers of vertices and edges in the polyhedron, explain why $kr = 2e$ and $sv = 2e$.

b. Use Euler's formula to show that $\dfrac{1}{s} + \dfrac{1}{k} = \dfrac{1}{2} + \dfrac{1}{e}$.

c. Find all the integral solutions of the equation in part (b) that satisfy the geometric constraints of the problem. (How small can k and s be?)

For your information, the five regular polyhedra are the tetrahedron (4, 6, 4), the cube (8, 12, 6), the octahedron (6, 12, 8), the dodecahedron (20, 30, 12), and the icosahedron (12, 30, 20). (The numbers in parentheses indicate the numbers of vertices, edges, and faces, respectively.)

SOLUTION TO PRACTICE PROBLEM

1. The matrix inequality $Ax \le b$ yields the following system of inequalities:

$$\text{(a) } x_1 + 3x_2 \le 12$$
$$\text{(b) } x_1 + 2x_2 \le 9$$
$$\text{(c) } 2x_1 + x_2 \le 12$$

The condition $x \ge 0$, places the polytope in the first quadrant of the plane. One vertex is $(0, 0)$. The x_1-intercepts of the three lines (when $x_2 = 0$) are 12, 9, and 6, so $(6, 0)$ is a vertex. The x_2-intercepts of the three lines (when $x_1 = 0$) are 4, 4.5, and 12, so $(0, 4)$ is a vertex.

How do the three boundary lines intersect for positive values of x_1 and x_2? The intersection of (a) and (b) is at $\mathbf{p}_{ab} = (3, 3)$. Testing $\mathbf{p}_{ab}$ in (c) gives $2(3) + 1(3) = 9 < 12$, so $\mathbf{p}_{ab}$ is in P. The intersection of (b) and (c) is at $\mathbf{p}_{bc} = (5, 2)$. Testing $\mathbf{p}_{bc}$ in (a) gives $1(5) + 3(2) = 11 < 12$, so $\mathbf{p}_{bc}$ is in P. The intersection of (a) and (c) is at $\mathbf{p}_{ac} = (4.8, 2.4)$. Testing $\mathbf{p}_{ac}$ in (b) gives $1(4.8) + 2(2.4) = 9.6 > 9$. So $\mathbf{p}_{ac}$ is not in P.

Finally, the three vertices (extreme points) of the polytopes are $(0, 0)$, $(6, 0)$, $(5, 2)$ $(3, 3)$, and $(0, 4)$. These points form the minimal representation of P. This is displayed graphically in Fig. 13.

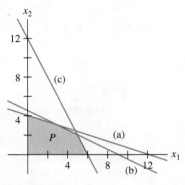

FIGURE 13

8.6 | CURVES AND SURFACES

For thousands of years, builders used long thin strips of wood to create the hull of a boat. In more recent times, designers used long, flexible metal strips to lay out the surfaces of cars and airplanes. Weights and pegs shaped the strips into smooth curves called *natural cubic splines*. The curve between two successive control points (pegs or weights) has a parametric representation using cubic polynomials. Unfortunately, such curves have the property that moving one control point affects the shape of the entire curve, because of physical forces that the pegs and weights exert on the strip. Design engineers had long wanted local control of the curve—in which movement of one control point would affect only a small portion of the curve. In 1962, a French automotive engineer, Pierre Bézier, solved this problem by adding extra control points and using a class of curves now called by his name.

Bézier Curves

The curves described below play an important role in computer graphics as well as engineering. For example, they are used in Adobe Illustrator and Macromedia Freehand, and in application programming languages such as OpenGL. These curves permit a program to store exact information about curved segments and surfaces in a relatively small number of control points. All graphics commands for the segments and surfaces have only to be computed for the control points. The special structure of these curves also speeds up other calculations in the "graphics pipeline" that creates the final display on the viewing screen.

Exercises in Section 8.3 introduced quadratic Bézier curves and showed one method for constructing Bézier curves of higher degree. The discussion here focuses on quadratic and cubic Bézier curves, which are determined by three or four control points, denoted by $\mathbf{p}_0$, $\mathbf{p}_1$, $\mathbf{p}_2$, and $\mathbf{p}_3$. These points can be in $\mathbb{R}^2$ or $\mathbb{R}^3$, or they can be represented by homogeneous forms in $\mathbb{R}^3$ or $\mathbb{R}^4$. The standard parametric descriptions of these curves, for $0 \le t \le 1$, are

$$\mathbf{w}(t) = (1-t)^2\mathbf{p}_0 + 2t(1-t)\mathbf{p}_1 + t^2\mathbf{p}_2 \tag{1}$$

$$\mathbf{x}(t) = (1-t)^3\mathbf{p}_0 + 3t(1-t)^2\mathbf{p}_1 + 3t^2(1-t)\mathbf{p}_2 + t^3\mathbf{p}_3 \tag{2}$$

Figure 1 shows two typical curves. Usually, the curves pass through only the initial and terminal control points, but a Bézier curve is always in the convex hull of its control points. (See Exercises 21–24 in Section 8.3.)

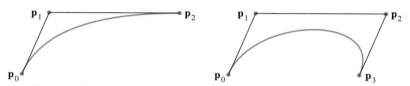

FIGURE 1 Quadratic and cubic Bézier curves.

Bézier curves are useful in computer graphics because their essential properties are preserved under the action of linear transformations and translations. For instance, if A is a matrix of appropriate size, then from the linearity of matrix multiplication, for $0 \le t \le 1$,

$$A\mathbf{x}(t) = A[(1-t)^3\mathbf{p}_0 + 3t(1-t)^2\mathbf{p}_1 + 3t^2(1-t)\mathbf{p}_2 + t^3\mathbf{p}_3]$$
$$= (1-t)^3 A\mathbf{p}_0 + 3t(1-t)^2 A\mathbf{p}_1 + 3t^2(1-t)A\mathbf{p}_2 + t^3 A\mathbf{p}_3$$

The new control points are $A\mathbf{p}_0, \dots, A\mathbf{p}_3$. Translations of Bézier curves are considered in Exercise 1.

The curves in Fig. 1 suggest that the control points determine the tangent lines to the curves at the initial and terminal control points. Recall from calculus that for any parametric curve, say $\mathbf{y}(t)$, the direction of the tangent line to the curve at a point $\mathbf{y}(t)$ is given by the derivative $\mathbf{y}'(t)$, called the **tangent vector** of the curve. (This derivative is computed entry by entry.)

EXAMPLE 1 Determine how the tangent vector of the quadratic Bézier curve $\mathbf{w}(t)$ is related to the control points of the curve, at $t = 0$ and $t = 1$.

SOLUTION Write the weights in equation (1) as simple polynomials

$$\mathbf{w}(t) = (1 - 2t + t^2)\mathbf{p}_0 + (2t - 2t^2)\mathbf{p}_1 + t^2\mathbf{p}_2$$

Then, because differentiation is a linear transformation on functions,

$$\mathbf{w}'(t) = (-2 + 2t)\mathbf{p}_0 + (2 - 4t)\mathbf{p}_1 + 2t\mathbf{p}_2$$

So

$$\mathbf{w}'(0) = -2\mathbf{p}_0 + 2\mathbf{p}_1 = 2(\mathbf{p}_1 - \mathbf{p}_0)$$
$$\mathbf{w}'(1) = -2\mathbf{p}_1 + 2\mathbf{p}_2 = 2(\mathbf{p}_2 - \mathbf{p}_1)$$

The tangent vector at $\mathbf{p}_0$, for instance, points from $\mathbf{p}_0$ to $\mathbf{p}_1$, but it is twice as long as the segment from $\mathbf{p}_0$ to $\mathbf{p}_1$. Notice that $\mathbf{w}'(0) = \mathbf{0}$ when $\mathbf{p}_1 = \mathbf{p}_0$. In this case, $\mathbf{w}(t) = (1 - t^2)\mathbf{p}_1 + t^2\mathbf{p}_2$, and the graph of $\mathbf{w}(t)$ is the line segment from $\mathbf{p}_1$ to $\mathbf{p}_2$. ∎

Connecting Two Bézier Curves

Two basic Bézier curves can be joined end to end, with the terminal point of the first curve $\mathbf{x}(t)$ being the initial point $\mathbf{p}_2$ of the second curve $\mathbf{y}(t)$. The combined curve is said to have G^0 *geometric continuity* (at $\mathbf{p}_2$) because the two segments join at $\mathbf{p}_2$. If the tangent line to curve 1 at $\mathbf{p}_2$ has a different direction than the tangent line to curve 2, then a "corner," or abrupt change of direction, may be apparent at $\mathbf{p}_2$. See Fig. 2.

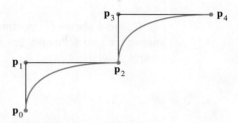

FIGURE 2 G^0 continuity at $\mathbf{p}_2$.

To avoid a sharp bend, it usually suffices to adjust the curves to have what is called G^1 *geometric continuity*, where both tangent vectors at $\mathbf{p}_2$ point in the same direction. That is, the derivatives $\mathbf{x}'(1)$ and $\mathbf{y}'(0)$ point in the same direction, even though their magnitudes may be different. When the tangent vectors are actually equal at $\mathbf{p}_2$, the tangent vector is continuous at $\mathbf{p}_2$, and the combined curve is said to have C^1 continuity, or C^1 *parametric continuity*. Figure 3 shows G^1 continuity in (a) and C^1 continuity in (b).

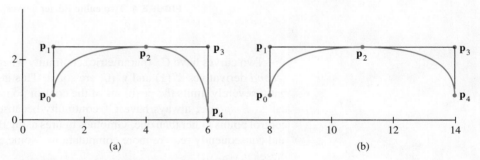

(a)   (b)

FIGURE 3 (a) G^1 continuity and (b) C^1 continuity.

EXAMPLE 2 Let $\mathbf{x}(t)$ and $\mathbf{y}(t)$ determine two quadratic Bézier curves, with control points $\{\mathbf{p}_0, \mathbf{p}_1, \mathbf{p}_2\}$ and $\{\mathbf{p}_2, \mathbf{p}_3, \mathbf{p}_4\}$, respectively. The curves are joined at $\mathbf{p}_2 = \mathbf{x}(1) = \mathbf{y}(0)$.

a. Suppose the combined curve has G^1 continuity (at $\mathbf{p}_2$). What algebraic restriction does this condition impose on the control points? Express this restriction in geometric language.

b. Repeat part (a) for C^1 continuity.

SOLUTION

a. From Example 1, $\mathbf{x}'(1) = 2(\mathbf{p}_2 - \mathbf{p}_1)$. Also, using the control points for $\mathbf{y}(t)$ in place of $\mathbf{w}(t)$, Example 1 shows that $\mathbf{y}'(0) = 2(\mathbf{p}_3 - \mathbf{p}_2)$. G^1 continuity means that $\mathbf{y}'(0) = k\mathbf{x}'(1)$ for some positive constant k. Equivalently,

$$\mathbf{p}_3 - \mathbf{p}_2 = k(\mathbf{p}_2 - \mathbf{p}_1), \quad \text{with } k > 0 \tag{3}$$

Geometrically, (3) implies that $\mathbf{p}_2$ lies on the line segment from $\mathbf{p}_1$ to $\mathbf{p}_3$. To prove this, let $t = (k+1)^{-1}$, and note that $0 < t < 1$. Solve for k to obtain $k = (1-t)/t$. When this expression is used for k in (3), a rearrangement shows that $\mathbf{p}_2 = (1-t)\mathbf{p}_1 + t\mathbf{p}_3$, which verifies the assertion about $\mathbf{p}_2$.

b. C^1 continuity means that $\mathbf{y}'(0) = \mathbf{x}'(1)$. Thus $2(\mathbf{p}_3 - \mathbf{p}_2) = 2(\mathbf{p}_2 - \mathbf{p}_1)$, so $\mathbf{p}_3 - \mathbf{p}_2 = \mathbf{p}_2 - \mathbf{p}_1$, and $\mathbf{p}_2 = (\mathbf{p}_1 + \mathbf{p}_3)/2$. Geometrically, $\mathbf{p}_2$ is the midpoint of the line segment from $\mathbf{p}_1$ to $\mathbf{p}_3$. See Fig. 3. ∎

Figure 4 shows C^1 continuity for two cubic Bézier curves. Notice how the point joining the two segments lies in the middle of the line segment between the adjacent control points.

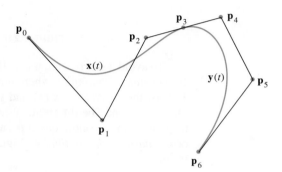

FIGURE 4 Two cubic Bézier curves.

Two curves have C^2 (parametric) continuity when they have C^1 continuity and the *second* derivatives $\mathbf{x}''(1)$ and $\mathbf{y}''(0)$ are equal. This is possible for cubic Bézier curves, but it severely limits the positions of the control points. Another class of cubic curves, called *B-splines*, always have C^2 continuity because each pair of curves share three control points rather than one. Graphics figures using B-splines have more control points and consequently require more computations. Some exercises for this section examine these curves.

Surprisingly, if $\mathbf{x}(t)$ and $\mathbf{y}(t)$ join at $\mathbf{p}_3$, the apparent smoothness of the curve at $\mathbf{p}_3$ is usually the same for both G^1 continuity and C^1 continuity. This is because the magnitude of $\mathbf{x}'(t)$ is not related to the physical shape of the curve. The magnitude reflects only the mathematical parameterization of the curve. For instance, if a new vector function $\mathbf{z}(t)$ equals $\mathbf{x}(2t)$, then the point $\mathbf{z}(t)$ traverses the curve from $\mathbf{p}_0$ to $\mathbf{p}_3$ twice as fast as the original version, because $2t$ reaches 1 when t is .5. But, by the chain rule of calculus, $\mathbf{z}'(t) = 2 \cdot \mathbf{x}'(2t)$, so the tangent vector to $\mathbf{z}(t)$ at $\mathbf{p}_3$ is twice the tangent vector to $\mathbf{x}(t)$ at $\mathbf{p}_3$.

In practice, many simple Bézier curves are joined to create graphics objects. Type-setting programs provide one important application, because many letters in a type font involve curved segments. Each letter in a PostScript® font, for example, is stored as a set of control points, along with information on how to construct the "outline" of the letter using line segments and Bézier curves. Enlarging such a letter basically requires multiplying the coordinates of each control point by one constant scale factor. Once the outline of the letter has been computed, the appropriate solid parts of the letter are filled in. Figure 5 illustrates this for a character in a PostScript font. Note the control points.

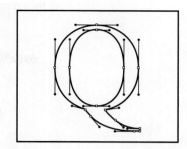

FIGURE 5 A PostScript character.

Matrix Equations for Bézier Curves

Since a Bézier curve is a linear combination of control points using polynomials as weights, the formula for $\mathbf{x}(t)$ may be written as

$$\mathbf{x}(t) = \begin{bmatrix} \mathbf{p}_0 & \mathbf{p}_1 & \mathbf{p}_2 & \mathbf{p}_3 \end{bmatrix} \begin{bmatrix} (1-t)^3 \\ 3t(1-t)^2 \\ 3t^2(1-t) \\ t^3 \end{bmatrix}$$

$$= \begin{bmatrix} \mathbf{p}_0 & \mathbf{p}_1 & \mathbf{p}_2 & \mathbf{p}_3 \end{bmatrix} \begin{bmatrix} 1 - 3t + 3t^2 - t^3 \\ 3t - 6t^2 + 3t^3 \\ 3t^2 - 3t^3 \\ t^3 \end{bmatrix}$$

$$= \begin{bmatrix} \mathbf{p}_0 & \mathbf{p}_1 & \mathbf{p}_2 & \mathbf{p}_3 \end{bmatrix} \begin{bmatrix} 1 & -3 & 3 & -1 \\ 0 & 3 & -6 & 3 \\ 0 & 0 & 3 & -3 \\ 0 & 0 & 0 & 1 \end{bmatrix} \begin{bmatrix} 1 \\ t \\ t^2 \\ t^3 \end{bmatrix}$$

The matrix whose columns are the four control points is called a **geometry matrix**, G. The 4×4 matrix of polynomial coefficients is the **Bézier basis matrix**, M_B. If $\mathbf{u}(t)$ is the column vector of powers of t, then the Bézier curve is given by

$$\mathbf{x}(t) = GM_B\mathbf{u}(t) \tag{4}$$

Other parametric cubic curves in computer graphics are written in this form, too. For instance, if the entries in the matrix M_B are changed appropriately, the resulting curves are B-splines. They are "smoother" than Bézier curves, but they do not pass through any of the control points. A **Hermite** cubic curve arises when the matrix M_B is replaced by a Hermite basis matrix. In this case, the columns of the geometry matrix consist of the starting and ending points of the curves and the tangent vectors to the curves at those points.[1]

The Bézier curve in equation (4) can also be "factored" in another way, to be used in the discussion of Bézier surfaces. For convenience later, the parameter t is replaced

---

[1] The term *basis matrix* comes from the rows of the matrix that list the coefficients of the *blending* polynomials used to define the curve. For a cubic Bézier curve, the four polynomials are $(1-t)^3$, $3t(1-t)^2$, $3t^2(1-t)$, and t^3. They form a basis for the space $\mathbb{P}_3$ of polynomials of degree 3 or less. Each entry in the vector $\mathbf{x}(t)$ is a linear combination of these polynomials. The weights come from the rows of the geometry matrix G in (4).

by a parameter s:

$$\mathbf{x}(s) = \mathbf{u}(s)^T M_B^T \begin{bmatrix} \mathbf{p}_0 \\ \mathbf{p}_1 \\ \mathbf{p}_2 \\ \mathbf{p}_3 \end{bmatrix} = \begin{bmatrix} 1 & s & s^2 & s^3 \end{bmatrix} \begin{bmatrix} 1 & 0 & 0 & 0 \\ -3 & 3 & 0 & 0 \\ 3 & -6 & 3 & 0 \\ -1 & 3 & -3 & 1 \end{bmatrix} \begin{bmatrix} \mathbf{p}_0 \\ \mathbf{p}_1 \\ \mathbf{p}_2 \\ \mathbf{p}_3 \end{bmatrix}$$

$$= \begin{bmatrix} (1-s)^3 & 3s(1-s)^2 & 3s^2(1-s) & s^3 \end{bmatrix} \begin{bmatrix} \mathbf{p}_0 \\ \mathbf{p}_1 \\ \mathbf{p}_2 \\ \mathbf{p}_3 \end{bmatrix} \tag{5}$$

This formula is not quite the same as the transpose of the product on the right of (4), because $\mathbf{x}(s)$ and the control points appear in (5) without transpose symbols. The matrix of control points in (5) is called a **geometry vector**. This should be viewed as a 4×1 block (partitioned) matrix whose entries are column vectors. The matrix to the left of the geometry vector, in the second part of (5), can be viewed as a block matrix, too, with a scalar in each block. The partitioned matrix multiplication makes sense, because each (vector) entry in the geometry vector can be left-multiplied by a scalar as well as by a matrix. Thus, the column vector $\mathbf{x}(s)$ is represented by (5).

Bézier Surfaces

A 3D bicubic surface patch can be constructed from a set of four Bézier curves. Consider the four geometry matrices

$$\begin{bmatrix} \mathbf{p}_{11} & \mathbf{p}_{12} & \mathbf{p}_{13} & \mathbf{p}_{14} \end{bmatrix}$$
$$\begin{bmatrix} \mathbf{p}_{21} & \mathbf{p}_{22} & \mathbf{p}_{23} & \mathbf{p}_{24} \end{bmatrix}$$
$$\begin{bmatrix} \mathbf{p}_{31} & \mathbf{p}_{32} & \mathbf{p}_{33} & \mathbf{p}_{34} \end{bmatrix}$$
$$\begin{bmatrix} \mathbf{p}_{41} & \mathbf{p}_{42} & \mathbf{p}_{43} & \mathbf{p}_{44} \end{bmatrix}$$

and recall from equation (4) that a Bézier curve is produced when any one of these matrices is multiplied on the right by the following vector of weights:

$$M_B \mathbf{u}(t) = \begin{bmatrix} (1-t)^3 \\ 3t(1-t)^2 \\ 3t^2(1-t) \\ t^3 \end{bmatrix}$$

Let G be the block (partitioned) 4×4 matrix whose entries are the control points $\mathbf{p}_{ij}$ displayed above. Then the following product is a block 4×1 matrix, and each entry is a Bézier curve:

$$GM_B \mathbf{u}(t) = \begin{bmatrix} \mathbf{p}_{11} & \mathbf{p}_{12} & \mathbf{p}_{13} & \mathbf{p}_{14} \\ \mathbf{p}_{21} & \mathbf{p}_{22} & \mathbf{p}_{23} & \mathbf{p}_{24} \\ \mathbf{p}_{31} & \mathbf{p}_{32} & \mathbf{p}_{33} & \mathbf{p}_{34} \\ \mathbf{p}_{41} & \mathbf{p}_{42} & \mathbf{p}_{43} & \mathbf{p}_{44} \end{bmatrix} \begin{bmatrix} (1-t)^3 \\ 3t(1-t)^2 \\ 3t^2(1-t) \\ t^3 \end{bmatrix}$$

In fact,

$$GM_B \mathbf{u}(t) = \begin{bmatrix} (1-t)^3 \mathbf{p}_{11} + 3t(1-t)^2 \mathbf{p}_{12} + 3t^2(1-t)\mathbf{p}_{13} + t^3 \mathbf{p}_{14} \\ (1-t)^3 \mathbf{p}_{21} + 3t(1-t)^2 \mathbf{p}_{22} + 3t^2(1-t)\mathbf{p}_{23} + t^3 \mathbf{p}_{24} \\ (1-t)^3 \mathbf{p}_{31} + 3t(1-t)^2 \mathbf{p}_{32} + 3t^2(1-t)\mathbf{p}_{33} + t^3 \mathbf{p}_{34} \\ (1-t)^3 \mathbf{p}_{41} + 3t(1-t)^2 \mathbf{p}_{42} + 3t^2(1-t)\mathbf{p}_{43} + t^3 \mathbf{p}_{44} \end{bmatrix}$$

Now fix t. Then $GM_B\mathbf{u}(t)$ is a column vector that can be used as a geometry vector in equation (5) for a Bézier curve in another variable s. This observation produces the **Bézier bicubic surface**:

$$\mathbf{x}(s,t) = \mathbf{u}(s)^T M_B^T GM_B\mathbf{u}(t), \quad \text{where } 0 \le s, t \le 1 \tag{6}$$

The formula for $\mathbf{x}(s,t)$ is a linear combination of the sixteen control points. If one imagines that these control points are arranged in a fairly uniform rectangular array, as in Fig. 6, then the Bézier surface is controlled by a web of eight Bézier curves, four in the "s-direction" and four in the "t-direction." The surface actually passes through the four control points at its "corners." When it is in the middle of a larger surface, the sixteen-point surface shares its twelve boundary control points with its neighbors.

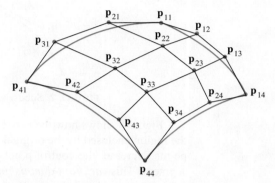

FIGURE 6 Sixteen control points for a Bézier bicubic surface patch.

Approximations to Curves and Surfaces

In CAD programs and in programs used to create realistic computer games, the designer often works at a graphics workstation to compose a "scene" involving various geometric structures. This process requires interaction between the designer and the geometric objects. Each slight repositioning of an object requires new mathematical computations by the graphics program. Bézier curves and surfaces can be useful in this process because they involve fewer control points than objects approximated by many polygons. This dramatically reduces the computation time and speeds up the designer's work.

After the scene composition, however, the final image preparation has different computational demands that are more easily met by objects consisting of flat surfaces and straight edges, such as polyhedra. The designer needs to *render* the scene, by introducing light sources, adding color and texture to surfaces, and simulating reflections from the surfaces.

Computing the direction of a reflected light at a point $\mathbf{p}$ on a surface, for instance, requires knowing the directions of both the incoming light and the *surface normal*—the vector perpendicular to the tangent plane at $\mathbf{p}$. Computing such normal vectors is much easier on a surface composed of, say, tiny flat polygons than on a curved surface whose normal vector changes continuously as $\mathbf{p}$ moves. If $\mathbf{p}_1$, $\mathbf{p}_2$, and $\mathbf{p}_3$ are adjacent vertices of a flat polygon, then the surface normal is just plus or minus the cross product $(\mathbf{p}_2 - \mathbf{p}_1) \times (\mathbf{p}_2 - \mathbf{p}_3)$. When the polygon is small, only one normal vector is needed for rendering the entire polygon. Also, two widely used shading routines, Gouraud shading and Phong shading, both require a surface to be defined by polygons.

As a result of these needs for flat surfaces, the Bézier curves and surfaces from the scene composition stage now are usually approximated by straight line segments and polyhedral surfaces. The basic idea for approximating a Bézier curve or surface is to divide the curve or surface into smaller pieces, with more and more control points.

Recursive Subdivision of Bézier Curves and Surfaces

Figure 7 shows the four control points $\mathbf{p}_0, \ldots, \mathbf{p}_3$ for a Bézier curve, along with control points for two new curves, each coinciding with half of the original curve. The "left" curve begins at $\mathbf{q}_0 = \mathbf{p}_0$ and ends at $\mathbf{q}_3$, at the midpoint of the original curve. The "right" curve begins at $\mathbf{r}_0 = \mathbf{q}_3$ and ends at $\mathbf{r}_3 = \mathbf{p}_3$.

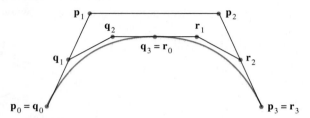

FIGURE 7 Subdivision of a Bézier curve.

Figure 8 shows how the new control points enclose regions that are "thinner" than the region enclosed by the original control points. As the distances between the control points decrease, the control points of each curve segment also move closer to a line segment. This *variation-diminishing property* of Bézier curves depends on the fact that a Bézier curve always lies in the convex hull of the control points.

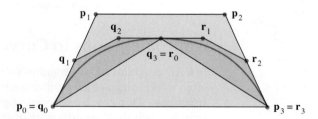

FIGURE 8 Convex hulls of the control points.

The new control points are related to the original control points by simple formulas. Of course, $\mathbf{q}_0 = \mathbf{p}_0$ and $\mathbf{r}_3 = \mathbf{p}_3$. The midpoint of the original curve $\mathbf{x}(t)$ occurs at $\mathbf{x}(.5)$ when $\mathbf{x}(t)$ has the standard parameterization,

$$\mathbf{x}(t) = (1 - 3t + 3t^2 - t^3)\mathbf{p}_0 + (3t - 6t^2 + 3t^3)\mathbf{p}_1 + (3t^2 - 3t^3)\mathbf{p}_2 + t^3\mathbf{p}_3 \quad (7)$$

for $0 \le t \le 1$. Thus, the new control points $\mathbf{q}_3$ and $\mathbf{r}_0$ are given by

$$\mathbf{q}_3 = \mathbf{r}_0 = \mathbf{x}(.5) = \tfrac{1}{8}(\mathbf{p}_0 + 3\mathbf{p}_1 + 3\mathbf{p}_2 + \mathbf{p}_3) \quad (8)$$

The formulas for the remaining "interior" control points are also simple, but the derivation of the formulas requires some work involving the tangent vectors of the curves. By definition, the tangent vector to a parameterized curve $\mathbf{x}(t)$ is the derivative $\mathbf{x}'(t)$. This vector shows the direction of the line tangent to the curve at $\mathbf{x}(t)$. For the Bézier curve in (7),

$$\mathbf{x}'(t) = (-3 + 6t - 3t^2)\mathbf{p}_0 + (3 - 12t + 9t^2)\mathbf{p}_1 + (6t - 9t^2)\mathbf{p}_2 + 3t^2\mathbf{p}_3$$

for $0 \le t \le 1$. In particular,

$$\mathbf{x}'(0) = 3(\mathbf{p}_1 - \mathbf{p}_0) \quad \text{and} \quad \mathbf{x}'(1) = 3(\mathbf{p}_3 - \mathbf{p}_2) \quad (9)$$

Geometrically, $\mathbf{p}_1$ is on the line tangent to the curve at $\mathbf{p}_0$, and $\mathbf{p}_2$ is on the line tangent to the curve at $\mathbf{p}_3$. See Fig. 8. Also, from $\mathbf{x}'(t)$, compute

$$\mathbf{x}'(.5) = \tfrac{3}{4}(-\mathbf{p}_0 - \mathbf{p}_1 + \mathbf{p}_2 + \mathbf{p}_3) \tag{10}$$

Let $\mathbf{y}(t)$ be the Bézier curve determined by $\mathbf{q}_0, \dots, \mathbf{q}_3$, and let $\mathbf{z}(t)$ be the Bézier curve determined by $\mathbf{r}_0, \dots, \mathbf{r}_3$. Since $\mathbf{y}(t)$ traverses the same path as $\mathbf{x}(t)$ but only gets to $\mathbf{x}(.5)$ as t goes from 0 to 1, $\mathbf{y}(t) = \mathbf{x}(.5t)$ for $0 \le t \le 1$. Similarly, since $\mathbf{z}(t)$ starts at $\mathbf{x}(.5)$ when $t = 0$, $\mathbf{z}(t) = \mathbf{x}(.5 + .5t)$ for $0 \le t \le 1$. By the chain rule for derivatives,

$$\mathbf{y}'(t) = .5\mathbf{x}'(.5t) \quad \text{and} \quad \mathbf{z}'(t) = .5\mathbf{x}'(.5 + .5t) \quad \text{for } 0 \le t \le 1 \tag{11}$$

From (9) with $\mathbf{y}'(0)$ in place of $\mathbf{x}'(0)$, from (11) with $t = 0$, and from (9), the control points for $\mathbf{y}(t)$ satisfy

$$3(\mathbf{q}_1 - \mathbf{q}_0) = \mathbf{y}'(0) = .5\mathbf{x}'(0) = \tfrac{3}{2}(\mathbf{p}_1 - \mathbf{p}_0) \tag{12}$$

From (9) with $\mathbf{y}'(1)$ in place of $\mathbf{x}'(1)$, from (11) with $t = 1$, and from (10),

$$3(\mathbf{q}_3 - \mathbf{q}_2) = \mathbf{y}'(1) = .5\mathbf{x}'(.5) = \tfrac{3}{8}(-\mathbf{p}_0 - \mathbf{p}_1 + \mathbf{p}_2 + \mathbf{p}_3) \tag{13}$$

Equations (8), (9), (10), (12), and (13) can be solved to produce the formulas for $\mathbf{q}_0, \dots,$ $\mathbf{q}_3$ shown in Exercise 13. Geometrically, the formulas are displayed in Fig. 9. The interior control points $\mathbf{q}_1$ and $\mathbf{r}_2$ are the midpoints, respectively, of the segment from $\mathbf{p}_0$ to $\mathbf{p}_1$ and the segment from $\mathbf{p}_2$ to $\mathbf{p}_3$. When the midpoint of the segment from $\mathbf{p}_1$ to $\mathbf{p}_2$ is connected to $\mathbf{q}_1$, the resulting line segment has $\mathbf{q}_2$ in the middle!

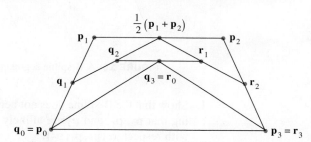

FIGURE 9 Geometric structure of new control points.

This completes one step of the subdivision process. The "recursion" begins, and both new curves are subdivided. The recursion continues to a depth at which all curves are sufficiently straight. Alternatively, at each step the recursion can be "adaptive" and not subdivide one of the two new curves if that curve is sufficiently straight. Once the subdivision completely stops, the endpoints of each curve are joined by line segments, and the scene is ready for the next step in the final image preparation.

A Bézier bicubic surface has the same variation-diminishing property as the Bézier curves that make up each cross-section of the surface, so the process described above can be applied in each cross-section. With the details omitted, here is the basic strategy. Consider the four "parallel" Bézier curves whose parameter is s, and apply the subdivision process to each of them. This produces four sets of eight control points; each set determines a curve as s varies from 0 to 1. As t varies, however, there are eight curves, each with four control points. Apply the subdivision process to each of these sets of four points, creating a total of 64 control points. Adaptive recursion is possible in this setting, too, but there are some subtleties involved.[2]

---

[2] See Foley, van Dam, Feiner, and Hughes, *Computer Graphics—Principles and Practice*, 2nd Ed. (Boston: Addison-Wesley, 1996), pp. 527–528.

A *spline* usually refers to a curve that passes through specified points. A B-spline, however, usually does not pass through its control points. A single segment has the parametric form

$$\mathbf{x}(t) = \tfrac{1}{6}\big[(1-t)^3\mathbf{p}_0 + (3t^3 - 6t^2 + 4)\mathbf{p}_1 \\ + (-3t^3 + 3t^2 + 3t + 1)\mathbf{p}_2 + t^3\mathbf{p}_3\big] \tag{14}$$

for $0 \le t \le 1$, where $\mathbf{p}_0$, $\mathbf{p}_1$, $\mathbf{p}_2$, and $\mathbf{p}_3$ are the control points. When t varies from 0 to 1, $\mathbf{x}(t)$ creates a short curve that lies close to $\overline{\mathbf{p}_1\mathbf{p}_2}$. Basic algebra shows that the B-spline formula can also be written as

$$\mathbf{x}(t) = \tfrac{1}{6}\big[(1-t)^3\mathbf{p}_0 + (3t(1-t)^2 - 3t + 4)\mathbf{p}_1 \\ + (3t^2(1-t) + 3t + 1)\mathbf{p}_2 + t^3\mathbf{p}_3\big] \tag{15}$$

This shows the similarity with the Bézier curve. Except for the $1/6$ factor at the front, the $\mathbf{p}_0$ and $\mathbf{p}_3$ terms are the same. The $\mathbf{p}_1$ component has been increased by $-3t + 4$ and the $\mathbf{p}_2$ component has been increased by $3t + 1$. These components move the curve closer to $\overline{\mathbf{p}_1\mathbf{p}_2}$ than the Bézier curve. The $1/6$ factor is necessary to keep the sum of the coefficients equal to 1. Figure 10 compares a B-spline with a Bézier curve that has the same control points.

FIGURE 10 A B-spline segment and a Bézier curve.

1. Show that the B-spline does not begin at $\mathbf{p}_0$, but $\mathbf{x}(0)$ is in conv $\{\mathbf{p}_0, \mathbf{p}_1, \mathbf{p}_2\}$. Assuming that $\mathbf{p}_0$, $\mathbf{p}_1$, and $\mathbf{p}_2$ are affinely independent, find the affine coordinates of $\mathbf{x}(0)$ with respect to $\{\mathbf{p}_0, \mathbf{p}_1, \mathbf{p}_2\}$.

2. Show that the B-spline does not end at $\mathbf{p}_3$, but $\mathbf{x}(1)$ is in conv $\{\mathbf{p}_1, \mathbf{p}_2, \mathbf{p}_3\}$. Assuming that $\mathbf{p}_1$, $\mathbf{p}_2$, and $\mathbf{p}_3$ are affinely independent, find the affine coordinates of $\mathbf{x}(1)$ with respect to $\{\mathbf{p}_1, \mathbf{p}_2, \mathbf{p}_3\}$.

8.6 EXERCISES

1. Suppose a Bézier curve is translated to $\mathbf{x}(t) + \mathbf{b}$. That is, for $0 \le t \le 1$, the new curve is

$$\mathbf{x}(t) = (1-t)^3\mathbf{p}_0 + 3t(1-t)^2\mathbf{p}_1 \\ + 3t^2(1-t)\mathbf{p}_2 + t^3\mathbf{p}_3 + \mathbf{b}$$

Show that this new curve is again a Bézier curve. [*Hint:* Where are the new control points?]

2. The parametric vector form of a B-spline curve was defined in the Practice Problems as

$$\mathbf{x}(t) = \tfrac{1}{6}\big[(1-t)^3\mathbf{p}_0 + (3t(1-t) - 3t + 4)\mathbf{p}_1 \\ + (3t^2(1-t) + 3t + 1)\mathbf{p}_2 + t^3\mathbf{p}_3\big] \quad \text{for } 0 \le t \le 1,$$

where $\mathbf{p}_0$, $\mathbf{p}_1$, $\mathbf{p}_2$, and $\mathbf{p}_3$ are the control points.

a. Show that for $0 \le t \le 1$, $\mathbf{x}(t)$ is in the convex hull of the control points.

b. Suppose that a B-spline curve $\mathbf{x}(t)$ is translated to $\mathbf{x}(t) + \mathbf{b}$ (as in Exercise 1). Show that this new curve is again a B-spline.

3. Let $\mathbf{x}(t)$ be a cubic Bézier curve determined by points $\mathbf{p}_0$, $\mathbf{p}_1$, $\mathbf{p}_2$, and $\mathbf{p}_3$.

a. Compute the *tangent* vector $\mathbf{x}'(t)$. Determine how $\mathbf{x}'(0)$ and $\mathbf{x}'(1)$ are related to the control points, and give geometric descriptions of the *directions* of these tangent vectors. Is it possible to have $\mathbf{x}'(1) = \mathbf{0}$?

b. Compute the second derivative $\mathbf{x}''(t)$ and determine how $\mathbf{x}''(0)$ and $\mathbf{x}''(1)$ are related to the control points. Draw a figure based on Fig. 10, and construct a line segment that points in the direction of $\mathbf{x}''(0)$. [*Hint:* Use $\mathbf{p}_1$ as the origin of the coordinate system.]

4. Let $\mathbf{x}(t)$ be the B-spline in Exercise 2, with control points $\mathbf{p}_0$, $\mathbf{p}_1, \mathbf{p}_2$, and $\mathbf{p}_3$.

   a. Compute the tangent vector $\mathbf{x}'(t)$ and determine how the derivatives $\mathbf{x}'(0)$ and $\mathbf{x}'(1)$ are related to the control points. Give geometric descriptions of the *directions* of these tangent vectors. Explore what happens when both $\mathbf{x}'(0)$ and $\mathbf{x}'(1)$ equal $\mathbf{0}$. Justify your assertions.

   b. Compute the second derivative $\mathbf{x}''(t)$ and determine how $\mathbf{x}''(0)$ and $\mathbf{x}''(1)$ are related to the control points. Draw a figure based on Fig. 10, and construct a line segment that points in the direction of $\mathbf{x}''(1)$. [*Hint:* Use $\mathbf{p}_2$ as the origin of the coordinate system.]

5. Let $\mathbf{x}(t)$ and $\mathbf{y}(t)$ be cubic Bézier curves with control points $\{\mathbf{p}_0, \mathbf{p}_1, \mathbf{p}_2, \mathbf{p}_3\}$ and $\{\mathbf{p}_3, \mathbf{p}_4, \mathbf{p}_5, \mathbf{p}_6\}$, respectively, so that $\mathbf{x}(t)$ and $\mathbf{y}(t)$ are joined at $\mathbf{p}_3$. The following questions refer to the curve consisting of $\mathbf{x}(t)$ followed by $\mathbf{y}(t)$. For simplicity, assume that the curve is in $\mathbb{R}^2$.

   a. What condition on the control points will guarantee that the curve has C^1 continuity at $\mathbf{p}_3$? Justify your answer.

   b. What happens when $\mathbf{x}'(1)$ and $\mathbf{y}'(0)$ are both the zero vector?

6. A B-spline is built out of B-spline segments, described in Exercise 2. Let $\mathbf{p}_0, \ldots, \mathbf{p}_4$ be control points. For $0 \le t \le 1$, let $\mathbf{x}(t)$ and $\mathbf{y}(t)$ be determined by the geometry matrices $[\mathbf{p}_0 \ \mathbf{p}_1 \ \mathbf{p}_2 \ \mathbf{p}_3]$ and $[\mathbf{p}_1 \ \mathbf{p}_2 \ \mathbf{p}_3 \ \mathbf{p}_4]$, respectively. Notice how the two segments share three control points. The two segments do not overlap, however—they join at a common endpoint, close to $\mathbf{p}_2$.

   a. Show that the combined curve has G^0 continuity—that is, $\mathbf{x}(1) = \mathbf{y}(0)$.

   b. Show that the curve has C^1 continuity at the join point, $\mathbf{x}(1)$. That is, show that $\mathbf{x}'(1) = \mathbf{y}'(0)$.

7. Let $\mathbf{x}(t)$ and $\mathbf{y}(t)$ be Bézier curves from Exercise 5, and suppose the combined curve has C^2 continuity (which includes C^1 continuity) at $\mathbf{p}_3$. Set $\mathbf{x}''(1) = \mathbf{y}''(0)$ and show that $\mathbf{p}_5$ is completely determined by $\mathbf{p}_1, \mathbf{p}_2$, and $\mathbf{p}_3$. Thus, the points $\mathbf{p}_0, \ldots, \mathbf{p}_3$ and the C^2 condition determine all but one of the control points for $\mathbf{y}(t)$.

8. Let $\mathbf{x}(t)$ and $\mathbf{y}(t)$ be segments of a B-spline as in Exercise 6. Show that the curve has C^2 continuity (as well as C^1 continuity) at $\mathbf{x}(1)$. That is, show that $\mathbf{x}''(1) = \mathbf{y}''(0)$. This higher-order continuity is desirable in CAD applications such as automotive body design, since the curves and surfaces appear much smoother. However, B-splines require three times the computation of Bézier curves, for curves of comparable length. For surfaces, B-splines require nine times the computation of Bézier surfaces. Programmers often choose Bézier surfaces for applications (such as an airplane cockpit simulator) that require real-time rendering.

9. A quartic Bézier curve is determined by five control points, $\mathbf{p}_0, \mathbf{p}_1, \mathbf{p}_2, \mathbf{p}_3$, and $\mathbf{p}_4$:

$$\mathbf{x}(t) = (1-t)^4\mathbf{p}_0 + 4t(1-t)^3\mathbf{p}_1 + 6t^2(1-t)^2\mathbf{p}_2$$
$$+ 4t^3(1-t)\mathbf{p}_3 + t^4\mathbf{p}_4 \quad \text{for } 0 \le t \le 1$$

Construct the quartic basis matrix M_B for $\mathbf{x}(t)$.

10. The "B" in B-spline refers to the fact that a segment $\mathbf{x}(t)$ may be written in terms of a basis matrix, M_S, in a form similar to a Bézier curve. That is,

$$\mathbf{x}(t) = GM_S\mathbf{u}(t) \quad \text{for } 0 \le t \le 1$$

where G is the geometry matrix $[\mathbf{p}_0 \ \mathbf{p}_1 \ \mathbf{p}_2 \ \mathbf{p}_3]$ and $\mathbf{u}(t)$ is the column vector $(1, t, t^2, t^3)$. In a *uniform* B-spline, each segment uses the same basis matrix, but the geometry matrix changes. Construct the basis matrix M_S for $\mathbf{x}(t)$.

In Exercises 11 and 12, mark each statement True or False. Justify each answer.

11. a. The cubic Bézier curve is based on four control points.

   b. Given a quadratic Bézier curve $\mathbf{x}(t)$ with control points $\mathbf{p}_0, \mathbf{p}_1$, and $\mathbf{p}_2$, the directed line segment $\mathbf{p}_1 - \mathbf{p}_0$ (from $\mathbf{p}_0$ to $\mathbf{p}_1$) is the tangent vector to the curve at $\mathbf{p}_0$.

   c. When two quadratic Bézier curves with control points $\{\mathbf{p}_0, \mathbf{p}_1, \mathbf{p}_2\}$ and $\{\mathbf{p}_2, \mathbf{p}_3, \mathbf{p}_4\}$ are joined at $\mathbf{p}_2$, the combined Bézier curve will have C^1 continuity at $\mathbf{p}_2$ if $\mathbf{p}_2$ is the midpoint of the line segment between $\mathbf{p}_1$ and $\mathbf{p}_3$.

12. a. The essential properties of Bézier curves are preserved under the action of linear transformations, but not translations.

   b. When two Bézier curves $\mathbf{x}(t)$ and $\mathbf{y}(t)$ are joined at the point where $\mathbf{x}(1) = \mathbf{y}(0)$, the combined curve has G^0 continuity at that point.

   c. The Bézier basis matrix is a matrix whose columns are the control points of the curve.

Exercises 13–15 concern the subdivision of a Bézier curve shown in Fig. 7. Let $\mathbf{x}(t)$ be the Bézier curve, with control points $\mathbf{p}_0, \ldots, \mathbf{p}_3$, and let $\mathbf{y}(t)$ and $\mathbf{z}(t)$ be the subdividing Bézier curves as in the text, with control points $\mathbf{q}_0, \ldots, \mathbf{q}_3$ and $\mathbf{r}_0, \ldots, \mathbf{r}_3$, respectively.

13. a. Use equation (12) to show that $\mathbf{q}_1$ is the midpoint of the segment from $\mathbf{p}_0$ to $\mathbf{p}_1$.

   b. Use equation (13) to show that

$$8\mathbf{q}_2 = 8\mathbf{q}_3 + \mathbf{p}_0 + \mathbf{p}_1 - \mathbf{p}_2 - \mathbf{p}_3.$$

   c. Use part (b), equation (8), and part (a) to show that $\mathbf{q}_2$ is the midpoint of the segment from $\mathbf{q}_1$ to the midpoint of the segment from $\mathbf{p}_1$ to $\mathbf{p}_2$. That is, $\mathbf{q}_2 = \frac{1}{2}[\mathbf{q}_1 + \frac{1}{2}(\mathbf{p}_1 + \mathbf{p}_2)]$.

14. a. Justify each equals sign:

$$3(\mathbf{r}_3 - \mathbf{r}_2) = \mathbf{z}'(1) = .5\mathbf{x}'(1) = \tfrac{3}{2}(\mathbf{p}_3 - \mathbf{p}_2).$$

b. Show that $\mathbf{r}_2$ is the midpoint of the segment from $\mathbf{p}_2$ to $\mathbf{p}_3$.

c. Justify each equals sign: $3(\mathbf{r}_1 - \mathbf{r}_0) = \mathbf{z}'(0) = .5\mathbf{x}'(.5)$.

d. Use part (c) to show that $8\mathbf{r}_1 = -\mathbf{p}_0 - \mathbf{p}_1 + \mathbf{p}_2 + \mathbf{p}_3 + 8\mathbf{r}_0$.

e. Use part (d), equation (8), and part (a) to show that $\mathbf{r}_1$ is the midpoint of the segment from $\mathbf{r}_2$ to the midpoint of the segment from $\mathbf{p}_1$ to $\mathbf{p}_2$. That is, $\mathbf{r}_1 = \frac{1}{2}[\mathbf{r}_2 + \frac{1}{2}(\mathbf{p}_1 + \mathbf{p}_2)]$.

15. Sometimes only one half of a Bézier curve needs further subdividing. For example, subdivision of the "left" side is accomplished with parts (a) and (c) of Exercise 13 and equation (8). When both halves of the curve $\mathbf{x}(t)$ are divided, it is possible to organize calculations efficiently to calculate both left and right control points concurrently, without using equation (8) directly.

a. Show that the tangent vectors $\mathbf{y}'(1)$ and $\mathbf{z}'(0)$ are equal.

b. Use part (a) to show that $\mathbf{q}_3$ (which equals $\mathbf{r}_0$) is the midpoint of the segment from $\mathbf{q}_2$ to $\mathbf{r}_1$.

c. Using part (b) and the results of Exercises 13 and 14, write an algorithm that computes the control points for both $\mathbf{y}(t)$ and $\mathbf{z}(t)$ in an efficient manner. The only operations needed are sums and division by 2.

16. Explain why a cubic Bézier curve is completely determined by $\mathbf{x}(0)$, $\mathbf{x}'(0)$, $\mathbf{x}(1)$, and $\mathbf{x}'(1)$.

17. TrueType® fonts, created by Apple Computer and Adobe Systems, use quadratic Bézier curves, while PostScript® fonts, created by Microsoft, use cubic Bézier curves. The cubic curves provide more flexibility for typeface design, but it is important to Microsoft that every typeface using quadratic curves can be transformed into one that uses cubic curves. Suppose that $\mathbf{w}(t)$ is a quadratic curve, with control points $\mathbf{p}_0$, $\mathbf{p}_1$, and $\mathbf{p}_2$.

a. Find control points $\mathbf{r}_0$, $\mathbf{r}_1$, $\mathbf{r}_2$, and $\mathbf{r}_3$ such that the cubic Bézier curve $\mathbf{x}(t)$ with these control points has the property that $\mathbf{x}(t)$ and $\mathbf{w}(t)$ have the same initial and terminal points and the same tangent vectors at $t = 0$ and $t = 1$. (See Exercise 16.)

b. Show that if $\mathbf{x}(t)$ is constructed as in part (a), then $\mathbf{x}(t) = \mathbf{w}(t)$ for $0 \le t \le 1$.

18. Use partitioned matrix multiplication to compute the following matrix product, which appears in the alternative formula (5) for a Bézier curve:

$$\begin{bmatrix} 1 & 0 & 0 & 0 \\ -3 & 3 & 0 & 0 \\ 3 & -6 & 3 & 0 \\ -1 & 3 & -3 & 1 \end{bmatrix} \begin{bmatrix} \mathbf{p}_0 \\ \mathbf{p}_1 \\ \mathbf{p}_2 \\ \mathbf{p}_3 \end{bmatrix}$$

SOLUTIONS TO PRACTICE PROBLEMS

1. From equation (14) with $t = 0$, $\mathbf{x}(0) \ne \mathbf{p}_0$ because

$$\mathbf{x}(0) = \tfrac{1}{6}[\mathbf{p}_0 + 4\mathbf{p}_1 + \mathbf{p}_2] = \tfrac{1}{6}\mathbf{p}_0 + \tfrac{2}{3}\mathbf{p}_1 + \tfrac{1}{6}\mathbf{p}_2.$$

The coefficients are nonnegative and sum to 1, so $\mathbf{x}(0)$ is in conv $\{\mathbf{p}_0, \mathbf{p}_1, \mathbf{p}_2\}$, and the affine coordinates with respect to $\{\mathbf{p}_0, \mathbf{p}_1, \mathbf{p}_2\}$ are $\left(\tfrac{1}{6}, \tfrac{2}{3}, \tfrac{1}{6}\right)$.

2. From equation (14) with $t = 1$, $\mathbf{x}(1) \ne \mathbf{p}_3$ because

$$\mathbf{x}(1) = \tfrac{1}{6}[\mathbf{p}_1 + 4\mathbf{p}_2 + \mathbf{p}_3] = \tfrac{1}{6}\mathbf{p}_1 + \tfrac{2}{3}\mathbf{p}_2 + \tfrac{1}{6}\mathbf{p}_3.$$

The coefficients are nonnegative and sum to 1, so $\mathbf{x}(1)$ is in conv $\{\mathbf{p}_1, \mathbf{p}_2, \mathbf{p}_3\}$, and the affine coordinates with respect to $\{\mathbf{p}_1, \mathbf{p}_2, \mathbf{p}_3\}$ are $\left(\tfrac{1}{6}, \tfrac{2}{3}, \tfrac{1}{6}\right)$.

Uniqueness of the Reduced Echelon Form

THEOREM

Uniqueness of the Reduced Echelon Form

Each $m \times n$ matrix A is row equivalent to a unique reduced echelon matrix U.

PROOF The proof uses the idea from Section 4.3 that the columns of row-equivalent matrices have exactly the same linear dependence relations.

The row reduction algorithm shows that there exists at least one such matrix U. Suppose that A is row equivalent to matrices U and V in reduced echelon form. The leftmost nonzero entry in a row of U is a "leading 1." Call the location of such a leading 1 a pivot position, and call the column that contains it a pivot column. (This definition uses only the echelon nature of U and V and does not assume the uniqueness of the reduced echelon form.)

The pivot columns of U and V are precisely the nonzero columns that are *not* linearly dependent on the columns to their left. (This condition is satisfied automatically by a *first* column if it is nonzero.) Since U and V are row equivalent (both being row equivalent to A), their columns have the same linear dependence relations. Hence, the pivot columns of U and V appear in the same locations. If there are r such columns, then since U and V are in reduced echelon form, their pivot columns are the first r columns of the $m \times m$ identity matrix. Thus, *corresponding pivot columns of U and V are equal.*

Finally, consider any nonpivot column of U, say column j. This column is either zero or a linear combination of the pivot columns to its left (because those pivot columns are a basis for the space spanned by the columns to the left of column j). Either case can be expressed by writing $U\mathbf{x} = \mathbf{0}$ for some $\mathbf{x}$ whose jth entry is 1. Then $V\mathbf{x} = \mathbf{0}$, too, which says that column j of V is either zero or the *same* linear combination of the pivot columns of V to *its* left. Since corresponding pivot columns of U and V are equal, columns j of U and V are also equal. This holds for all nonpivot columns, so $V = U$, which proves that U is unique.

APPENDIX

B

Complex Numbers

A **complex number** is a number written in the form

$$z = a + bi$$

where a and b are real numbers and i is a formal symbol satisfying the relation $i^2 = -1$. The number a is the **real part** of z, denoted by Re z, and b is the **imaginary part** of z, denoted by Im z. Two complex numbers are considered equal if and only if their real and imaginary parts are equal. For example, if $z = 5 + (-2)i$, then Re $z = 5$ and Im $z = -2$. For simplicity, we write $z = 5 - 2i$.

A real number a is considered as a special type of complex number, by identifying a with $a + 0i$. Furthermore, arithmetic operations on real numbers can be extended to the set of complex numbers.

The **complex number system**, denoted by $\mathbb{C}$, is the set of all complex numbers, together with the following operations of addition and multiplication:

$$(a + bi) + (c + di) = (a + c) + (b + d)i \tag{1}$$

$$(a + bi)(c + di) = (ac - bd) + (ad + bc)i \tag{2}$$

These rules reduce to ordinary addition and multiplication of real numbers when b and d are zero in (1) and (2). It is readily checked that the usual laws of arithmetic for $\mathbb{R}$ also hold for $\mathbb{C}$. For this reason, multiplication is usually computed by algebraic expansion, as in the following example.

EXAMPLE 1
$$\begin{aligned}
(5 - 2i)(3 + 4i) &= 15 + 20i - 6i - 8i^2 \\
&= 15 + 14i - 8(-1) \\
&= 23 + 14i
\end{aligned}$$

That is, multiply each term of $5 - 2i$ by each term of $3 + 4i$, use $i^2 = -1$, and write the result in the form $a + bi$. ∎

Subtraction of complex numbers z_1 and z_2 is defined by

$$z_1 - z_2 = z_1 + (-1)z_2$$

In particular, we write $-z$ in place of $(-1)z$.

The **conjugate** of $z = a + bi$ is the complex number $\overline{z}$ (read as "z bar"), defined by

$$\overline{z} = a - bi$$

Obtain $\overline{z}$ from z by reversing the sign of the imaginary part.

EXAMPLE 2   The conjugate of $-3 + 4i$ is $-3 - 4i$; write $\overline{-3 + 4i} = -3 - 4i$. ∎

Observe that if $z = a + bi$, then

$$z\overline{z} = (a + bi)(a - bi) = a^2 - abi + bai - b^2i^2 = a^2 + b^2 \tag{3}$$

Since $z\overline{z}$ is real and nonnegative, it has a square root. The **absolute value** (or **modulus**) of z is the real number $|z|$ defined by

$$|z| = \sqrt{z\overline{z}} = \sqrt{a^2 + b^2}$$

If z is a real number, then $z = a + 0i$, and $|z| = \sqrt{a^2}$, which equals the ordinary absolute value of a.

Some useful properties of conjugates and absolute value are listed below; w and z denote complex numbers.

1. $\overline{z} = z$ if and only if z is a real number.
2. $\overline{w + z} = \overline{w} + \overline{z}$.
3. $\overline{wz} = \overline{w}\,\overline{z}$; in particular, $\overline{rz} = r\overline{z}$ if r is a real number.
4. $z\overline{z} = |z|^2 \geq 0$.
5. $|wz| = |w||z|$.
6. $|w + z| \leq |w| + |z|$.

If $z \neq 0$, then $|z| > 0$ and z has a multiplicative inverse, denoted by $1/z$ or z^{-1} and given by

$$\frac{1}{z} = z^{-1} = \frac{\overline{z}}{|z|^2}$$

Of course, a quotient w/z simply means $w \cdot (1/z)$.

EXAMPLE 3   Let $w = 3 + 4i$ and $z = 5 - 2i$. Compute $z\overline{z}$, $|z|$, and w/z.

SOLUTION   From equation (3),

$$z\overline{z} = 5^2 + (-2)^2 = 25 + 4 = 29$$

For the absolute value, $|z| = \sqrt{z\overline{z}} = \sqrt{29}$. To compute w/z, first multiply both the numerator and the denominator by $\overline{z}$, the conjugate of the denominator. Because of (3), this eliminates the i in the denominator:

$$\frac{w}{z} = \frac{3+4i}{5-2i}$$

$$= \frac{3+4i}{5-2i} \cdot \frac{5+2i}{5+2i}$$

$$= \frac{15+6i+20i-8}{5^2+(-2)^2}$$

$$= \frac{7+26i}{29}$$

$$= \frac{7}{29} + \frac{26}{29}i \qquad \blacksquare$$

Geometric Interpretation

Each complex number $z = a + bi$ corresponds to a point (a, b) in the plane $\mathbb{R}^2$, as in Fig. 1. The horizontal axis is called the **real axis** because the points $(a, 0)$ on it correspond to the real numbers. The vertical axis is the **imaginary axis** because the points $(0, b)$ on it correspond to the **pure imaginary numbers** of the form $0 + bi$, or simply bi. The conjugate of z is the mirror image of z in the real axis. The absolute value of z is the distance from (a, b) to the origin.

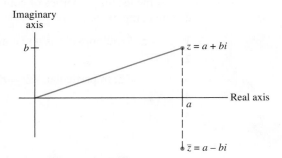

FIGURE 1 The complex conjugate is a mirror image.

Addition of complex numbers $z = a + bi$ and $w = c + di$ corresponds to vector addition of (a, b) and (c, d) in $\mathbb{R}^2$, as in Fig. 2.

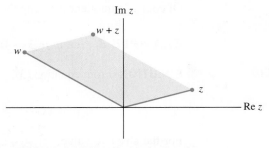

FIGURE 2 Addition of complex numbers.

To give a graphical representation of complex multiplication, we use **polar coordinates** in $\mathbb{R}^2$. Given a nonzero complex number $z = a + bi$, let φ be the angle between the positive real axis and the point (a, b), as in Fig. 3 where $-\pi < \varphi \leq \pi$. The angle φ is called the **argument** of z; we write $\varphi = \arg z$. From trigonometry,

$$a = |z| \cos \varphi, \qquad b = |z| \sin \varphi$$

and so

$$z = a + bi = |z|(\cos \varphi + i \sin \varphi)$$

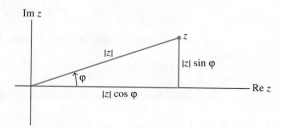

FIGURE 3 Polar coordinates of z.

If w is another nonzero complex number, say,

$$w = |w| (\cos \vartheta + i \sin \vartheta)$$

then, using standard trigonometric identities for the sine and cosine of the sum of two angles, one can verify that

$$wz = |w|\,|z|\,[\cos(\vartheta + \varphi) + i \sin(\vartheta + \varphi)] \qquad (4)$$

See Fig. 4. A similar formula may be written for quotients in polar form. The formulas for products and quotients can be stated in words as follows.

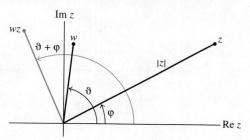

FIGURE 4 Multiplication with polar coordinates.

The product of two nonzero complex numbers is given in polar form by the product of their absolute values and the sum of their arguments. The quotient of two nonzero complex numbers is given by the quotient of their absolute values and the difference of their arguments.

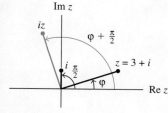

Multiplication by i.

EXAMPLE 4

a. If w has absolute value 1, then $w = \cos \vartheta + i \sin \vartheta$, where ϑ is the argument of w. Multiplication of any nonzero number z by w simply rotates z through the angle ϑ.

b. The argument of i itself is $\pi/2$ radians, so multiplication of z by i rotates z through an angle of $\pi/2$ radians. For example, $3 + i$ is rotated into $(3 + i)i = -1 + 3i$. ∎

Powers of a Complex Number

Formula (4) applies when $z = w = r(\cos \varphi + i \sin \varphi)$. In this case

$$z^2 = r^2(\cos 2\varphi + i \sin 2\varphi)$$

and

$$\begin{aligned} z^3 &= z \cdot z^2 \\ &= r(\cos \varphi + i \sin \varphi) \cdot r^2(\cos 2\varphi + i \sin 2\varphi) \\ &= r^3(\cos 3\varphi + i \sin 3\varphi) \end{aligned}$$

In general, for any positive integer k,

$$z^k = r^k(\cos k\varphi + i \sin k\varphi)$$

This fact is known as *De Moivre's Theorem*.

Complex Numbers and $\mathbb{R}^2$

Although the elements of $\mathbb{R}^2$ and $\mathbb{C}$ are in one-to-one correspondence, and the operations of addition are essentially the same, there is a logical distinction between $\mathbb{R}^2$ and $\mathbb{C}$. In $\mathbb{R}^2$ we can only multiply a vector by a real scalar, whereas in $\mathbb{C}$ we can multiply any two complex numbers to obtain a third complex number. (The dot product in $\mathbb{R}^2$ doesn't count, because it produces a scalar, not an element of $\mathbb{R}^2$.) We use scalar notation for elements in $\mathbb{C}$ to emphasize this distinction.

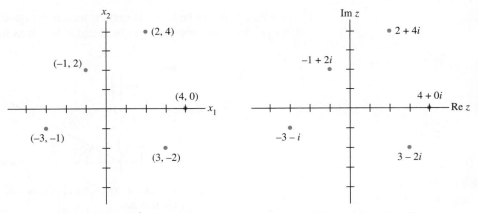

The real plane $\mathbb{R}^2$.   The complex plane $\mathbb{C}$.

Glossary

A

adjugate (or **classical adjoint**): The matrix adj A formed from a square matrix A by replacing the (i, j)-entry of A by the (i, j)-cofactor, for all i and j, and then transposing the resulting matrix.

affine combination: A linear combination of vectors (points in $\mathbb{R}^n$) in which the sum of the weights involved is 1.

affine dependence relation: An equation of the form $c_1\mathbf{v}_1 + \cdots + c_p\mathbf{v}_p = \mathbf{0}$, where the weights $c_1, \ldots, c_p$ are not all zero, and $c_1 + \cdots + c_p = 0$.

affine hull (or **affine span**) of a set S: The set of all affine combinations of points in S, denoted by aff S.

affinely dependent set: A set $\{\mathbf{v}_1, \ldots, \mathbf{v}_p\}$ in $\mathbb{R}^n$ such that there are real numbers $c_1, \ldots, c_p$, not all zero, such that $c_1 + \cdots + c_p = 0$ and $c_1\mathbf{v}_1 + \cdots + c_p\mathbf{v}_p = \mathbf{0}$

affinely independent set: A set $\{\mathbf{v}_1, \ldots, \mathbf{v}_p\}$ in $\mathbb{R}^n$ that is not affinely dependent.

affine set (or **affine subset**): A set S of points such that if $\mathbf{p}$ and $\mathbf{q}$ are in S, then $(1 - t)\mathbf{p} + t\mathbf{q} \in S$ for each real number t.

affine transformation: A mapping $T: \mathbb{R}^n \to \mathbb{R}^m$ of the form $T(\mathbf{x}) = A\mathbf{x} + \mathbf{b}$, with A an $m \times n$ matrix and $\mathbf{b}$ in $\mathbb{R}^m$.

algebraic multiplicity: The multiplicity of an eigenvalue as a root of the characteristic equation.

angle (between nonzero vectors $\mathbf{u}$ and $\mathbf{v}$ in $\mathbb{R}^2$ or $\mathbb{R}^3$): The angle ϑ between the two directed line segments from the origin to the points $\mathbf{u}$ and $\mathbf{v}$. Related to the scalar product by

$$\mathbf{u} \cdot \mathbf{v} = \|\mathbf{u}\| \, \|\mathbf{v}\| \cos \vartheta$$

associative law of multiplication: $A(BC) = (AB)C$, for all A, B, C.

attractor (of a dynamical system in $\mathbb{R}^2$): The origin when all trajectories tend toward $\mathbf{0}$.

augmented matrix: A matrix made up of a coefficient matrix for a linear system and one or more columns to the right. Each extra column contains the constants from the right side of a system with the given coefficient matrix.

auxiliary equation: A polynomial equation in a variable r, created from the coefficients of a homogeneous difference equation.

B

back-substitution (with matrix notation): The backward phase of row reduction of an augmented matrix that transforms an echelon matrix into a reduced echelon matrix; used to find the solution(s) of a system of linear equations.

backward phase (of row reduction): The last part of the algorithm that reduces a matrix in echelon form to a reduced echelon form.

band matrix: A matrix whose nonzero entries lie within a band along the main diagonal.

barycentric coordinates (of a point $\mathbf{p}$ with respect to an affinely independent set $S = \{\mathbf{v}_1 \ldots, \mathbf{v}_k\}$): The (unique) set of weights $c_1, \ldots, c_k$ such that $\mathbf{p} = c_1\mathbf{v}_1 + \cdots + c_k\mathbf{v}_k$ and $c_1 + \cdots + c_k = 1$. (Sometimes also called the **affine coordinates** of $\mathbf{p}$ with respect to S.)

basic variable: A variable in a linear system that corresponds to a pivot column in the coefficient matrix.

basis (for a nontrivial subspace H of a vector space V): An indexed set $\mathcal{B} = \{\mathbf{v}_1, \ldots, \mathbf{v}_p\}$ in V such that: (i) $\mathcal{B}$ is a linearly independent set and (ii) the subspace spanned by $\mathcal{B}$ coincides with H, that is, $H = \text{Span}\{\mathbf{v}_1, \ldots, \mathbf{v}_p\}$.

$\mathcal{B}$-coordinates of x: *See* coordinates of $\mathbf{x}$ relative to the basis $\mathcal{B}$.

best approximation: The closest point in a given subspace to a given vector.

bidiagonal matrix: A matrix whose nonzero entries lie on the main diagonal and on one diagonal adjacent to the main diagonal.

block diagonal (matrix): A partitioned matrix $A = [A_{ij}]$ such that each block A_{ij} is a zero matrix for $i \neq j$.

block matrix: *See* partitioned matrix.

block matrix multiplication: The row–column multiplication of partitioned matrices as if the block entries were scalars.

block upper triangular (matrix): A partitioned matrix $A = [A_{ij}]$ such that each block A_{ij} is a zero matrix for $i > j$.

boundary point of a set S in $\mathbb{R}^n$: A point $\mathbf{p}$ such that every open ball in $\mathbb{R}^n$ centered at $\mathbf{p}$ intersects both S and the complement of S.

bounded set in $\mathbb{R}^n$: A set that is contained in an open ball $B(\mathbf{0}, \delta)$ for some $\delta > 0$.

$\mathcal{B}$-matrix (for T): A matrix $[T]_{\mathcal{B}}$ for a linear transformation $T: V \to V$ relative to a basis $\mathcal{B}$ for V, with the property that $[T(\mathbf{x})]_{\mathcal{B}} = [T]_{\mathcal{B}}[\mathbf{x}]_{\mathcal{B}}$ for all $\mathbf{x}$ in V.

C

Cauchy–Schwarz inequality: $|\langle \mathbf{u}, \mathbf{v} \rangle| \leq \|u\| \cdot \|v\|$ for all $\mathbf{u}, \mathbf{v}$.

change of basis: *See* change-of-coordinates matrix.

change-of-coordinates matrix (from a basis $\mathcal{B}$ to a basis $\mathcal{C}$): A matrix $\underset{\mathcal{C} \leftarrow \mathcal{B}}{P}$ that transforms $\mathcal{B}$-coordinate vectors into $\mathcal{C}$-coordinate vectors: $[\mathbf{x}]_{\mathcal{C}} = \underset{\mathcal{C} \leftarrow \mathcal{B}}{P}[\mathbf{x}]_{\mathcal{B}}$. If $\mathcal{C}$ is the standard basis for $\mathbb{R}^n$, then $\underset{\mathcal{C} \leftarrow \mathcal{B}}{P}$ is sometimes written as $P_{\mathcal{B}}$.

characteristic equation (of A): $\det(A - \lambda I) = 0$.

characteristic polynomial (of A): $\det(A - \lambda I)$ or, in some texts, $\det(\lambda I - A)$.

Cholesky factorization: A factorization $A = R^T R$, where R is an invertible upper triangular matrix whose diagonal entries are all positive.

closed ball (in $\mathbb{R}^n$): A set $\{\mathbf{x} : \|\mathbf{x} - \mathbf{p}\| < \delta\}$ in $\mathbb{R}^n$, where $\mathbf{p}$ is in $\mathbb{R}^n$ and $\delta > 0$.

closed set (in $\mathbb{R}^n$): A set that contains all of its boundary points.

codomain (of a transformation $T : \mathbb{R}^n \to \mathbb{R}^m$): The set $\mathbb{R}^m$ that contains the range of T. In general, if T maps a vector space V into a vector space W, then W is called the codomain of T.

coefficient matrix: A matrix whose entries are the coefficients of a system of linear equations.

cofactor: A number $C_{ij} = (-1)^{i+j} \det A_{ij}$, called the (i, j)-cofactor of A, where A_{ij} is the submatrix formed by deleting the ith row and the jth column of A.

cofactor expansion: A formula for $\det A$ using cofactors associated with one row or one column, such as for row 1:
$$\det A = a_{11} C_{11} + \cdots + a_{1n} C_{1n}$$

column–row expansion: The expression of a product AB as a sum of outer products: $\operatorname{col}_1(A) \operatorname{row}_1(B) + \cdots + \operatorname{col}_n(A) \operatorname{row}_n(B)$, where n is the number of columns of A.

column space (of an $m \times n$ matrix A): The set Col A of all linear combinations of the columns of A. If $A = [\mathbf{a}_1 \cdots \mathbf{a}_n]$, then Col $A = \operatorname{Span} \{\mathbf{a}_1, \ldots, \mathbf{a}_n\}$. Equivalently,
$$\operatorname{Col} A = \{\mathbf{y} : \mathbf{y} = A\mathbf{x} \text{ for some } \mathbf{x} \text{ in } \mathbb{R}^n\}$$

column sum: The sum of the entries in a column of a matrix.

column vector: A matrix with only one column, or a single column of a matrix that has several columns.

commuting matrices: Two matrices A and B such that $AB = BA$.

compact set (in $\mathbb{R}^n$): A set in $\mathbb{R}^n$ that is both closed and bounded.

companion matrix: A special form of matrix whose characteristic polynomial is $(-1)^n p(\lambda)$ when $p(\lambda)$ is a specified polynomial whose leading term is λ^n.

complex eigenvalue: A nonreal root of the characteristic equation of an $n \times n$ matrix.

complex eigenvector: A nonzero vector $\mathbf{x}$ in $\mathbb{C}^n$ such that $A\mathbf{x} = \lambda \mathbf{x}$, where A is an $n \times n$ matrix and λ is a complex eigenvalue.

component of y orthogonal to u (for $\mathbf{u} \neq \mathbf{0}$): The vector $\mathbf{y} - \dfrac{\mathbf{y} \cdot \mathbf{u}}{\mathbf{u} \cdot \mathbf{u}} \mathbf{u}$.

composition of linear transformations: A mapping produced by applying two or more linear transformations in succession. If the transformations are matrix transformations, say left-multiplication by B followed by left-multiplication by A, then the composition is the mapping $\mathbf{x} \mapsto A(B\mathbf{x})$.

condition number (of A): The quotient σ_1 / σ_n, where σ_1 is the largest singular value of A and σ_n is the smallest singular value. The condition number is $+\infty$ when σ_n is zero.

conformable for block multiplication: Two partitioned matrices A and B such that the block product AB is defined: The column partition of A must match the row partition of B.

consistent linear system: A linear system with at least one solution.

constrained optimization: The problem of maximizing a quantity such as $\mathbf{x}^T A \mathbf{x}$ or $\|A\mathbf{x}\|$ when $\mathbf{x}$ is subject to one or more constraints, such as $\mathbf{x}^T \mathbf{x} = 1$ or $\mathbf{x}^T \mathbf{v} = 0$.

consumption matrix: A matrix in the Leontief input–output model whose columns are the unit consumption vectors for the various sectors of an economy.

contraction: A mapping $\mathbf{x} \mapsto r\mathbf{x}$ for some scalar r, with $0 \leq r \leq 1$.

controllable (pair of matrices): A matrix pair (A, B) where A is $n \times n$, B has n rows, and
$$\operatorname{rank} [\, B \quad AB \quad A^2 B \quad \cdots \quad A^{n-1} B \,] = n$$
Related to a state-space model of a control system and the difference equation $\mathbf{x}_{k+1} = A\mathbf{x}_k + B\mathbf{u}_k$ ($k = 0, 1, \ldots$).

convergent (sequence of vectors): A sequence $\{\mathbf{x}_k\}$ such that the entries in $\mathbf{x}_k$ can be made as close as desired to the entries in some fixed vector for all k sufficiently large.

convex combination (of points $\mathbf{v}_1, \ldots, \mathbf{v}_k$ in $\mathbb{R}^n$): A linear combination of vectors (points) in which the weights in the combination are nonnegative and the sum of the weights is 1.

convex hull (of a set S): The set of all convex combinations of points in S, denoted by: conv S.

convex set: A set S with the property that for each $\mathbf{p}$ and $\mathbf{q}$ in S, the line segment $\overline{\mathbf{p}\mathbf{q}}$ is contained in S.

coordinate mapping (determined by an ordered basis $\mathcal{B}$ in a vector space V): A mapping that associates to each $\mathbf{x}$ in V its coordinate vector $[\mathbf{x}]_{\mathcal{B}}$.

coordinates of x relative to the basis $\mathcal{B} = \{\mathbf{b}_1, \ldots, \mathbf{b}_n\}$: The weights $c_1, \ldots, c_n$ in the equation $\mathbf{x} = c_1 \mathbf{b}_1 + \cdots + c_n \mathbf{b}_n$.

coordinate vector of x relative to $\mathcal{B}$: The vector $[\mathbf{x}]_{\mathcal{B}}$ whose entries are the coordinates of $\mathbf{x}$ relative to the basis $\mathcal{B}$.

covariance (of variables x_i and x_j, for $i \neq j$): The entry s_{ij} in the covariance matrix S for a matrix of observations, where x_i and x_j vary over the ith and jth coordinates, respectively, of the observation vectors.

covariance matrix (or **sample covariance matrix**): The $p \times p$ matrix S defined by $S = (N - 1)^{-1} BB^T$, where B is a $p \times N$ matrix of observations in mean-deviation form.

Cramer's rule: A formula for each entry in the solution $\mathbf{x}$ of the equation $A\mathbf{x} = \mathbf{b}$ when A is an invertible matrix.

cross-product term: A term $cx_i x_j$ in a quadratic form, with $i \neq j$.

cube: A three-dimensional solid object bounded by six square faces, with three faces metting at each vertex.

D

decoupled system: A difference equation $\mathbf{y}_{k+1} = A\mathbf{y}_k$, or a differential equation $\mathbf{y}'(t) = A\mathbf{y}(t)$, in which A is a diagonal matrix. The discrete evolution of each entry in $\mathbf{y}_k$ (as a function of k), or the continuous evolution of each entry in the vector-valued function $\mathbf{y}(t)$, is unaffected by what happens to the other entries as $k \to \infty$ or $t \to \infty$.

design matrix: The matrix X in the linear model $\mathbf{y} = X\boldsymbol{\beta} + \boldsymbol{\epsilon}$, where the columns of X are determined in some way by the observed values of some independent variables.

determinant (of a square matrix A): The number det A defined inductively by a cofactor expansion along the first row of A. Also, $(-1)^r$ times the product of the diagonal entries in any echelon form U obtained from A by row replacements and r row interchanges (but no scaling operations).

diagonal entries (in a matrix): Entries having equal row and column indices.

diagonalizable (matrix): A matrix that can be written in factored form as PDP^{-1}, where D is a diagonal matrix and P is an invertible matrix.

diagonal matrix: A square matrix whose entries *not* on the main diagonal are all zero.

difference equation (or **linear recurrence relation**): An equation of the form $\mathbf{x}_{k+1} = A\mathbf{x}_k$ $(k = 0, 1, 2, \ldots)$ whose solution is a sequence of vectors, $\mathbf{x}_0, \mathbf{x}_1, \ldots$.

dilation: A mapping $\mathbf{x} \mapsto r\mathbf{x}$ for some scalar r, with $1 < r$.

dimension:
of a flat S: The dimension of the corresponding parallel subspace.
of a set S: The dimension of the smallest flat containing S.
of a subspace S: The number of vectors in a basis for S, written as dim S.
of a vector space V: The number of vectors in a basis for V, written as dim V. The dimension of the zero space is 0.

discrete linear dynamical system: A difference equation of the form $\mathbf{x}_{k+1} = A\mathbf{x}_k$ that describes the changes in a system (usually a physical system) as time passes. The physical system is measured at discrete times, when $k = 0, 1, 2, \ldots$, and the **state** of the system at time k is a vector $\mathbf{x}_k$ whose entries provide certain facts of interest about the system.

distance between u and v: The length of the vector $\mathbf{u} - \mathbf{v}$, denoted by dist $(\mathbf{u}, \mathbf{v})$.

distance to a subspace: The distance from a given point (vector) $\mathbf{v}$ to the nearest point in the subspace.

distributive laws: (left) $A(B + C) = AB + AC$, and (right) $(B + C)A = BA + CA$, for all A, B, C.

domain (of a transformation T): The set of all vectors $\mathbf{x}$ for which $T(\mathbf{x})$ is defined.

dot product: *See* inner product.

dynamical system: *See* discrete linear dynamical system.

E

echelon form (or **row echelon form**, of a matrix): An echelon matrix that is row equivalent to the given matrix.

echelon matrix (or **row echelon matrix**): A rectangular matrix that has three properties: (1) All nonzero rows are above any row of all zeros. (2) Each leading entry of a row is in a column to the right of the leading entry of the row above it. (3) All entries in a column below a leading entry are zero.

eigenfunctions (of a differential equation $\mathbf{x}'(t) = A\mathbf{x}(t)$): A function $\mathbf{x}(t) = \mathbf{v}e^{\lambda t}$, where $\mathbf{v}$ is an eigenvector of A and λ is the corresponding eigenvalue.

eigenspace (of A corresponding to λ): The set of *all* solutions of $A\mathbf{x} = \lambda\mathbf{x}$, where λ is an eigenvalue of A. Consists of the zero vector and all eigenvectors corresponding to λ.

eigenvalue (of A): A scalar λ such that the equation $A\mathbf{x} = \lambda\mathbf{x}$ has a solution for some nonzero vector $\mathbf{x}$.

eigenvector (of A): A *nonzero* vector $\mathbf{x}$ such that $A\mathbf{x} = \lambda\mathbf{x}$ for some scalar λ.

eigenvector basis: A basis consisting entirely of eigenvectors of a given matrix.

eigenvector decomposition (of $\mathbf{x}$): An equation, $\mathbf{x} = c_1\mathbf{v}_1 + \cdots + c_n\mathbf{v}_n$, expressing $\mathbf{x}$ as a linear combination of eigenvectors of a matrix.

elementary matrix: An invertible matrix that results by performing one elementary row operation on an identity matrix.

elementary row operations: (1) (Replacement) Replace one row by the sum of itself and a multiple of another row. (2) Interchange two rows. (3) (Scaling) Multiply all entries in a row by a nonzero constant.

equal vectors: Vectors in $\mathbb{R}^n$ whose corresponding entries are the same.

equilibrium prices: A set of prices for the total output of the various sectors in an economy, such that the income of each sector exactly balances its expenses.

equilibrium vector: *See* steady-state vector.

equivalent (linear) systems: Linear systems with the same solution set.

exchange model: *See* Leontief exchange model.

existence question: Asks, "Does a solution to the system exist?" That is, "Is the system consistent?" Also, "Does a solution of $A\mathbf{x} = \mathbf{b}$ exist for *all* possible $\mathbf{b}$?"

expansion by cofactors: *See* cofactor expansion.

explicit description (of a subspace W of $\mathbb{R}^n$): A parametric representation of W as the set of all linear combinations of a set of specified vectors.

extreme point (of a convex set S): A point $\mathbf{p}$ in S such that $\mathbf{p}$ is not in the interior of any line segment that lies in S. (That is, if **x**, **y** are in S and **p** is on the line segment $\overline{\mathbf{xy}}$, then $\mathbf{p} = \mathbf{x}$ or $\mathbf{p} = \mathbf{y}$.)

F

factorization (of A): An equation that expresses A as a product of two or more matrices.

final demand vector (or **bill of final demands**): The vector **d** in the Leontief input–output model that lists the dollar values of the goods and services demanded from the various sectors by the nonproductive part of the economy. The vector **d** can represent consumer demand, government consumption, surplus production, exports, or other external demand.

finite-dimensional (vector space): A vector space that is spanned by a finite set of vectors.

flat (in $\mathbb{R}^n$): A translate of a subspace of $\mathbb{R}^n$.

flexibility matrix: A matrix whose jth column gives the deflections of an elastic beam at specified points when a unit force is applied at the jth point on the beam.

floating point arithmetic: Arithmetic with numbers represented as decimals $\pm .d_1 \cdots d_p \times 10^r$, where r is an integer and the number p of digits to the right of the decimal point is usually between 8 and 16.

flop: One arithmetic operation $(+, -, *, /)$ on two real floating point numbers.

forward phase (of row reduction): The first part of the algorithm that reduces a matrix to echelon form.

Fourier approximation (of order n): The closest point in the subspace of nth-order trigonometric polynomials to a given function in $C[0, 2\pi]$.

Fourier coefficients: The weights used to make a trigonometric polynomial as a Fourier approximation to a function.

Fourier series: An infinite series that converges to a function in the inner product space $C[0, 2\pi]$, with the inner product given by a definite integral.

free variable: Any variable in a linear system that is not a basic variable.

full rank (matrix): An $m \times n$ matrix whose rank is the smaller of m and n.

fundamental set of solutions: A basis for the set of all solutions of a homogeneous linear difference or differential equation.

fundamental subspaces (determined by A): The null space and column space of A, and the null space and column space of A^T, with Col A^T commonly called the row space of A.

G

Gaussian elimination: *See* row reduction algorithm.

general least-squares problem: Given an $m \times n$ matrix A and a vector **b** in $\mathbb{R}^m$, find $\hat{\mathbf{x}}$ in $\mathbb{R}^n$ such that $\|\mathbf{b} - A\hat{\mathbf{x}}\| \le \|\mathbf{b} - A\mathbf{x}\|$ for all **x** in $\mathbb{R}^n$.

general solution (of a linear system): A parametric description of a solution set that expresses the basic variables in terms of the free variables (the parameters), if any. After Section 1.5, the parametric description is written in vector form.

Givens rotation: A linear transformation from $\mathbb{R}^n$ to $\mathbb{R}^n$ used in computer programs to create zero entries in a vector (usually a column of a matrix).

Gram matrix (of A): The matrix $A^T A$.

Gram–Schmidt process: An algorithm for producing an orthogonal or orthonormal basis for a subspace that is spanned by a given set of vectors.

H

homogeneous coordinates: In $\mathbb{R}^3$, the representation of (x, y, z) as (X, Y, Z, H) for any $H \ne 0$, where $x = X/H$, $y = Y/H$, and $z = Z/H$. In $\mathbb{R}^2$, H is usually taken as 1, and the homogeneous coordinates of (x, y) are written as $(x, y, 1)$.

homogeneous equation: An equation of the form $A\mathbf{x} = \mathbf{0}$, possibly written as a vector equation or as a system of linear equations.

homogeneous form of (a vector) **v** in $\mathbb{R}^n$: The point $\tilde{\mathbf{v}} = \begin{bmatrix} \mathbf{v} \\ 1 \end{bmatrix}$ in $\mathbb{R}^{n+1}$.

Householder reflection: A transformation $\mathbf{x} \mapsto Q\mathbf{x}$, where $Q = I - 2\mathbf{uu}^T$ and **u** is a unit vector ($\mathbf{u}^T\mathbf{u} = 1$).

hyperplane (in $\mathbb{R}^n$): A flat in $\mathbb{R}^n$ of dimension $n - 1$. Also: a translate of a subspace of dimension $n - 1$.

I

identity matrix (denoted by I or I_n): A square matrix with ones on the diagonal and zeros elsewhere.

ill-conditioned matrix: A square matrix with a large (or possibly infinite) condition number; a matrix that is singular or can become singular if some of its entries are changed ever so slightly.

image (of a vector **x** under a transformation T): The vector $T(\mathbf{x})$ assigned to **x** by T.

implicit description (of a subspace W of $\mathbb{R}^n$): A set of one or more homogeneous equations that characterize the points of W.

Im x: The vector in $\mathbb{R}^n$ formed from the imaginary parts of the entries of a vector **x** in $\mathbb{C}^n$.

inconsistent linear system: A linear system with no solution.

indefinite matrix: A symmetric matrix A such that $\mathbf{x}^T A\mathbf{x}$ assumes both positive and negative values.

indefinite quadratic form: A quadratic form Q such that $Q(\mathbf{x})$ assumes both positive and negative values.

infinite-dimensional (vector space): A nonzero vector space V that has no finite basis.

inner product: The scalar $\mathbf{u}^T\mathbf{v}$, usually written as $\mathbf{u} \cdot \mathbf{v}$, where **u** and **v** are vectors in $\mathbb{R}^n$ viewed as $n \times 1$ matrices. Also called the **dot product** of **u** and **v**. In general, a function on a vector space that assigns to each pair of vectors **u** and **v** a number $\langle \mathbf{u}, \mathbf{v} \rangle$, subject to certain axioms. See Section 6.7.

inner product space: A vector space on which is defined an inner product.

input–output matrix: *See* consumption matrix.

input–output model: *See* Leontief input–output model.

interior point (of a set S in $\mathbb{R}^n$): A point **p** in S such that for some $\delta > 0$, the open ball $\mathbf{B}(\mathbf{p}, \delta)$ centered at **p** is contained in S.

intermediate demands: Demands for goods or services that will be consumed in the process of producing other goods and services for consumers. If **x** is the production level and C is the consumption matrix, then $C\mathbf{x}$ lists the intermediate demands.

interpolating polynomial: A polynomial whose graph passes through every point in a set of data points in $\mathbb{R}^2$.

invariant subspace (for A): A subspace H such that $A\mathbf{x}$ is in H whenever **x** is in H.

inverse (of an $n \times n$ matrix A): An $n \times n$ matrix A^{-1} such that $AA^{-1} = A^{-1}A = I_n$.

inverse power method: An algorithm for estimating an eigenvalue λ of a square matrix, when a good initial estimate of λ is available.

invertible linear transformation: A linear transformation $T : \mathbb{R}^n \to \mathbb{R}^n$ such that there exists a function $S : \mathbb{R}^n \to \mathbb{R}^n$ satisfying both $T(S(\mathbf{x})) = \mathbf{x}$ and $S(T(\mathbf{x})) = \mathbf{x}$ for all **x** in $\mathbb{R}^n$.

invertible matrix: A square matrix that possesses an inverse.

isomorphic vector spaces: Two vector spaces V and W for which there is a one-to-one linear transformation T that maps V onto W.

isomorphism: A one-to-one linear mapping from one vector space onto another.

K

kernel (of a linear transformation $T : V \to W$): The set of **x** in V such that $T(\mathbf{x}) = \mathbf{0}$.

Kirchhoff's laws: (1) (**voltage law**) The algebraic sum of the RI voltage drops in one direction around a loop equals the algebraic sum of the voltage sources in the same direction around the loop. (2) (**current law**) The current in a branch is the algebraic sum of the loop currents flowing through that branch.

L

ladder network: An electrical network assembled by connecting in series two or more electrical circuits.

leading entry: The leftmost nonzero entry in a row of a matrix.

least-squares error: The distance $\|\mathbf{b} - A\hat{\mathbf{x}}\|$ from **b** to $A\hat{\mathbf{x}}$, when $\hat{\mathbf{x}}$ is a least-squares solution of $A\mathbf{x} = \mathbf{b}$.

least-squares line: The line $y = \hat{\beta}_0 + \hat{\beta}_1 x$ that minimizes the least-squares error in the equation $\mathbf{y} = X\beta + \epsilon$.

least-squares solution (of $A\mathbf{x} = \mathbf{b}$): A vector $\hat{\mathbf{x}}$ such that $\|\mathbf{b} - A\hat{\mathbf{x}}\| \le \|\mathbf{b} - A\mathbf{x}\|$ for all **x** in $\mathbb{R}^n$.

left inverse (of A): Any rectangular matrix C such that $CA = I$.

left-multiplication (by A): Multiplication of a vector or matrix on the left by A.

left singular vectors (of A): The columns of U in the singular value decomposition $A = U\Sigma V^T$.

length (or **norm**, of **v**): The scalar $\|\mathbf{v}\| = \sqrt{\mathbf{v} \cdot \mathbf{v}} = \sqrt{\langle \mathbf{v}, \mathbf{v} \rangle}$.

Leontief exchange (or **closed**) **model**: A model of an economy where inputs and outputs are fixed, and where a set of prices for the outputs of the sectors is sought such that the income of each sector equals its expenditures. This "equilibrium" condition is expressed as a system of linear equations, with the prices as the unknowns.

Leontief input–output model (or **Leontief production equation**): The equation $\mathbf{x} = C\mathbf{x} + \mathbf{d}$, where **x** is production, **d** is final demand, and C is the consumption (or input–output) matrix. The jth column of C lists the inputs that sector j consumes per unit of output.

level set (or **gradient**) of a linear functional f on $\mathbb{R}^n$: A set $[f : d] = \{\mathbf{x} \in \mathbb{R}^n : f(\mathbf{x}) = d\}$

linear combination: A sum of scalar multiples of vectors. The scalars are called the *weights*.

linear dependence relation: A homogeneous vector equation where the weights are all specified and at least one weight is nonzero.

linear equation (in the variables $x_1, \ldots, x_n$): An equation that can be written in the form $a_1x_1 + a_2x_2 + \cdots + a_nx_n = b$, where b and the coefficients $a_1, \ldots, a_n$ are real or complex numbers.

linear filter: A linear difference equation used to transform discrete-time signals.

linear functional (on $\mathbb{R}^n$): A linear transformation f from $\mathbb{R}^n$ into $\mathbb{R}$.

linearly dependent (vectors): An indexed set $\{\mathbf{v}_1, \ldots, \mathbf{v}_p\}$ with the property that there exist weights $c_1, \ldots, c_p$, not all zero, such that $c_1\mathbf{v}_1 + \cdots + c_p\mathbf{v}_p = \mathbf{0}$. That is, the vector equation $c_1\mathbf{v}_1 + c_2\mathbf{v}_2 + \cdots + c_p\mathbf{v}_p = \mathbf{0}$ has a *nontrivial* solution.

linearly independent (vectors): An indexed set $\{\mathbf{v}_1, \ldots, \mathbf{v}_p\}$ with the property that the vector equation $c_1\mathbf{v}_1 + c_2\mathbf{v}_2 + \cdots + c_p\mathbf{v}_p = \mathbf{0}$ has *only* the trivial solution, $c_1 = \cdots = c_p = 0$.

linear model (in statistics): Any equation of the form $\mathbf{y} = X\beta + \epsilon$, where X and **y** are known and β is to be chosen to minimize the length of the **residual vector**, ϵ.

linear system: A collection of one or more linear equations involving the same variables, say, $x_1, \ldots, x_n$.

linear transformation T (from a vector space V into a vector space W): A rule T that assigns to each vector **x** in V a unique vector $T(\mathbf{x})$ in W, such that (i) $T(\mathbf{u} + \mathbf{v}) = T(\mathbf{u}) + T(\mathbf{v})$ for all **u**, **v** in V, and (ii) $T(c\mathbf{u}) = cT(\mathbf{u})$ for all **u** in V and all scalars c. Notation:

$T: V \rightarrow W$; also, $\mathbf{x} \mapsto A\mathbf{x}$ when $T: \mathbb{R}^n \rightarrow \mathbb{R}^m$ and A is the standard matrix for T.

line through p parallel to v: The set $\{\mathbf{p} + t\mathbf{v} : t \text{ in } \mathbb{R}\}$.

loop current: The amount of electric current flowing through a loop that makes the algebraic sum of the RI voltage drops around the loop equal to the algebraic sum of the voltage sources in the loop.

lower triangular matrix: A matrix with zeros above the main diagonal.

lower triangular part (of A): A lower triangular matrix whose entries on the main diagonal and below agree with those in A.

LU factorization: The representation of a matrix A in the form $A = LU$ where L is a square lower triangular matrix with ones on the diagonal (a unit lower triangular matrix) and U is an echelon form of A.

M

magnitude (of a vector): *See* norm.

main diagonal (of a matrix): The entries with equal row and column indices.

mapping: *See* transformation.

Markov chain: A sequence of probability vectors $\mathbf{x}_0$, $\mathbf{x}_1$, $\mathbf{x}_2, \ldots$, together with a stochastic matrix P such that $\mathbf{x}_{k+1} = P\mathbf{x}_k$ for $k = 0, 1, 2, \ldots$.

matrix: A rectangular array of numbers.

matrix equation: An equation that involves at least one matrix; for instance, $A\mathbf{x} = \mathbf{b}$.

matrix for T relative to bases $\mathcal{B}$ and $\mathcal{C}$: A matrix M for a linear transformation $T: V \rightarrow W$ with the property that $[T(\mathbf{x})]_{\mathcal{C}} = M[\mathbf{x}]_{\mathcal{B}}$ for all $\mathbf{x}$ in V, where $\mathcal{B}$ is a basis for V and $\mathcal{C}$ is a basis for W. When $W = V$ and $\mathcal{C} = \mathcal{B}$, the matrix M is called the $\mathcal{B}$-matrix for T and is denoted by $[T]_{\mathcal{B}}$.

matrix of observations: A $p \times N$ matrix whose columns are observation vectors, each column listing p measurements made on an individual or object in a specified population or set.

matrix transformation: A mapping $\mathbf{x} \mapsto A\mathbf{x}$, where A is an $m \times n$ matrix and $\mathbf{x}$ represents any vector in $\mathbb{R}^n$.

maximal linearly independent set (in V): A linearly independent set $\mathcal{B}$ in V such that if a vector $\mathbf{v}$ in V but not in $\mathcal{B}$ is added to $\mathcal{B}$, then the new set is linearly dependent.

mean-deviation form (of a matrix of observations): A matrix whose row vectors are in mean-deviation form. For each row, the entries sum to zero.

mean-deviation form (of a vector): A vector whose entries sum to zero.

mean square error: The error of an approximation in an inner product space, where the inner product is defined by a definite integral.

migration matrix: A matrix that gives the percentage movement between different locations, from one period to the next.

minimal spanning set (for a subspace H): A set $\mathcal{B}$ that spans H and has the property that if one of the elements of $\mathcal{B}$ is removed from $\mathcal{B}$, then the new set does not span H.

$m \times n$ matrix: A matrix with m rows and n columns.

Moore–Penrose inverse: *See* pseudoinverse.

multiple regression: A linear model involving several independent variables and one dependent variable.

N

nearly singular matrix: An ill-conditioned matrix.

negative definite matrix: A symmetric matrix A such that $\mathbf{x}^T A \mathbf{x} < 0$ for all $\mathbf{x} \neq \mathbf{0}$.

negative definite quadratic form: A quadratic form Q such that $Q(\mathbf{x}) < 0$ for all $\mathbf{x} \neq \mathbf{0}$.

negative semidefinite matrix: A symmetric matrix A such that $\mathbf{x}^T A \mathbf{x} \leq 0$ for all $\mathbf{x}$.

negative semidefinite quadratic form: A quadratic form Q such that $Q(\mathbf{x}) \leq 0$ for all $\mathbf{x}$.

nonhomogeneous equation: An equation of the form $A\mathbf{x} = \mathbf{b}$ with $\mathbf{b} \neq \mathbf{0}$, possibly written as a vector equation or as a system of linear equations.

nonsingular (matrix): An invertible matrix.

nontrivial solution: A nonzero solution of a homogeneous equation or system of homogeneous equations.

nonzero (matrix or vector): A matrix (with possibly only one row or column) that contains at least one nonzero entry.

norm (or **length**, of $\mathbf{v}$): The scalar $\|\mathbf{v}\| = \sqrt{\mathbf{v} \cdot \mathbf{v}} = \sqrt{\langle \mathbf{v}, \mathbf{v} \rangle}$.

normal equations: The system of equations represented by $A^T A \mathbf{x} = A^T \mathbf{b}$, whose solution yields all least-squares solutions of $A\mathbf{x} = \mathbf{b}$. In statistics, a common notation is $X^T X \boldsymbol{\beta} = X^T \mathbf{y}$.

normalizing (a nonzero vector $\mathbf{v}$): The process of creating a unit vector $\mathbf{u}$ that is a positive multiple of $\mathbf{v}$.

normal vector (to a subspace V of $\mathbb{R}^n$): A vector $\mathbf{n}$ in $\mathbb{R}^n$ such that $\mathbf{n} \cdot \mathbf{x} = 0$ for all $\mathbf{x}$ in V.

null space (of an $m \times n$ matrix A): The set Nul A of all solutions to the homogeneous equation $A\mathbf{x} = \mathbf{0}$. Nul $A = \{\mathbf{x} : \mathbf{x} \text{ is in } \mathbb{R}^n \text{ and } A\mathbf{x} = \mathbf{0}\}$.

O

observation vector: The vector $\mathbf{y}$ in the linear model $\mathbf{y} = X\boldsymbol{\beta} + \boldsymbol{\epsilon}$, where the entries in $\mathbf{y}$ are the observed values of a dependent variable.

one-to-one (mapping): A mapping $T: \mathbb{R}^n \rightarrow \mathbb{R}^m$ such that each $\mathbf{b}$ in R^m is the image of *at most* one $\mathbf{x}$ in $\mathbb{R}^n$.

onto (mapping): A mapping $T: \mathbb{R}^n \rightarrow \mathbb{R}^m$ such that each $\mathbf{b}$ in R^m is the image of *at least* one $\mathbf{x}$ in $\mathbb{R}^n$.

open ball B(p, δ) in $\mathbb{R}^n$: The set $\{\mathbf{x} : \|\mathbf{x} - \mathbf{p}\| < \delta\}$ in $\mathbb{R}^n$, where $\delta > 0$.

open set S in $\mathbb{R}^n$: A set that contains none of its boundary points. (Equivalently, S is open if every point of S is an interior point.)

origin: The zero vector.

orthogonal basis: A basis that is also an orthogonal set.

orthogonal complement (of W): The set $W^\perp$ of all vectors orthogonal to W.

orthogonal decomposition: The representation of a vector $\mathbf{y}$ as the sum of two vectors, one in a specified subspace W and the other in $W^\perp$. In general, a decomposition $\mathbf{y} = c_1\mathbf{u}_1 + \cdots + c_p\mathbf{u}_p$, where $\{\mathbf{u}_1, \ldots, \mathbf{u}_p\}$ is an orthogonal basis for a subspace that contains $\mathbf{y}$.

orthogonally diagonalizable (matrix): A matrix A that admits a factorization, $A = PDP^{-1}$, with P an orthogonal matrix ($P^{-1} = P^T$) and D diagonal.

orthogonal matrix: A square invertible matrix U such that $U^{-1} = U^T$.

orthogonal projection of $\mathbf{y}$ onto $\mathbf{u}$ (or onto the line through $\mathbf{u}$ and the origin, for $\mathbf{u} \neq \mathbf{0}$): The vector $\hat{\mathbf{y}}$ defined by $\hat{\mathbf{y}} = \dfrac{\mathbf{y} \cdot \mathbf{u}}{\mathbf{u} \cdot \mathbf{u}}\mathbf{u}$.

orthogonal projection of $\mathbf{y}$ onto W: The unique vector $\hat{\mathbf{y}}$ in W such that $\mathbf{y} - \hat{\mathbf{y}}$ is orthogonal to W. Notation: $\hat{\mathbf{y}} = \text{proj}_W \mathbf{y}$.

orthogonal set: A set S of vectors such that $\mathbf{u} \cdot \mathbf{v} = 0$ for each distinct pair $\mathbf{u}, \mathbf{v}$ in S.

orthogonal to W: Orthogonal to every vector in W.

orthonormal basis: A basis that is an orthogonal set of unit vectors.

orthonormal set: An orthogonal set of unit vectors.

outer product: A matrix product $\mathbf{u}\mathbf{v}^T$ where $\mathbf{u}$ and $\mathbf{v}$ are vectors in $\mathbb{R}^n$ viewed as $n \times 1$ matrices. (The transpose symbol is on the "outside" of the symbols $\mathbf{u}$ and $\mathbf{v}$.)

overdetermined system: A system of equations with more equations than unknowns.

P

parallel flats: Two or more flats such that each flat is a translate of the other flats.

parallelogram rule for addition: A geometric interpretation of the sum of two vectors $\mathbf{u}, \mathbf{v}$ as the diagonal of the parallelogram determined by $\mathbf{u}, \mathbf{v}$, and $\mathbf{0}$.

parameter vector: The unknown vector $\boldsymbol{\beta}$ in the linear model $\mathbf{y} = X\boldsymbol{\beta} + \boldsymbol{\epsilon}$.

parametric equation of a line: An equation of the form $\mathbf{x} = \mathbf{p} + t\mathbf{v}$ (t in $\mathbb{R}$).

parametric equation of a plane: An equation of the form $\mathbf{x} = \mathbf{p} + s\mathbf{u} + t\mathbf{v}$ (s, t in $\mathbb{R}$), with $\mathbf{u}$ and $\mathbf{v}$ linearly independent.

partitioned matrix (or **block matrix**): A matrix whose entries are themselves matrices of appropriate sizes.

permuted lower triangular matrix: A matrix such that a permutation of its rows will form a lower triangular matrix.

permuted LU factorization: The representation of a matrix A in the form $A = LU$ where L is a square matrix such that a permutation of its rows will form a unit lower triangular matrix, and U is an echelon form of A.

pivot: A nonzero number that either is used in a pivot position to create zeros through row operations or is changed into a leading 1, which in turn is used to create zeros.

pivot column: A column that contains a pivot position.

pivot position: A position in a matrix A that corresponds to a leading entry in an echelon form of A.

plane through u, v, and the origin: A set whose parametric equation is $\mathbf{x} = s\mathbf{u} + t\mathbf{v}$ (s, t in $\mathbb{R}$), with $\mathbf{u}$ and $\mathbf{v}$ linearly independent.

polar decomposition (of A): A factorization $A = PQ$, where P is an $n \times n$ positive semidefinite matrix with the same rank as A, and Q is an $n \times n$ orthogonal matrix.

polygon: A polytope in $\mathbb{R}^2$.

polyhedron: A polytope in $\mathbb{R}^3$.

polytope: The convex hull of a finite set of points in $\mathbb{R}^n$ (a special type of compact convex set).

positive combination (of points $\mathbf{v}_1, \ldots, \mathbf{v}_m$ in $\mathbb{R}^n$): A linear combination $c_1\mathbf{v}_1 + \cdots + c_m\mathbf{v}_m$, where all $c_i \geq 0$.

positive definite matrix: A symmetric matrix A such that $\mathbf{x}^T A \mathbf{x} > 0$ for all $\mathbf{x} \neq \mathbf{0}$.

positive definite quadratic form: A quadratic form Q such that $Q(\mathbf{x}) > 0$ for all $\mathbf{x} \neq \mathbf{0}$.

positive hull (of a set S): The set of all positive combinations of points in S, denoted by pos S.

positive semidefinite matrix: A symmetric matrix A such that $\mathbf{x}^T A \mathbf{x} \geq 0$ for all $\mathbf{x}$.

positive semidefinite quadratic form: A quadratic form Q such that $Q(\mathbf{x}) \geq 0$ for all $\mathbf{x}$.

power method: An algorithm for estimating a strictly dominant eigenvalue of a square matrix.

principal axes (of a quadratic form $\mathbf{x}^T A \mathbf{x}$): The orthonormal columns of an orthogonal matrix P such that $P^{-1}AP$ is diagonal. (These columns are unit eigenvectors of A.) Usually the columns of P are ordered in such a way that the corresponding eigenvalues of A are arranged in decreasing order of magnitude.

principal components (of the data in a matrix B of observations): The unit eigenvectors of a sample covariance matrix S for B, with the eigenvectors arranged so that the corresponding eigenvalues of S decrease in magnitude. If B is in mean-deviation form, then the principal components are the right singular vectors in a singular value decomposition of B^T.

probability vector: A vector in $\mathbb{R}^n$ whose entries are nonnegative and sum to one.

product Ax: The linear combination of the columns of A using the corresponding entries in **x** as weights.

production vector: The vector in the Leontief input–output model that lists the amounts that are to be produced by the various sectors of an economy.

profile (of a set S in $\mathbb{R}^n$): The set of extreme points of S.

projection matrix (or **orthogonal projection matrix**): A symmetric matrix B such that $B^2 = B$. A simple example is $B = \mathbf{vv}^T$, where **v** is a unit vector.

proper subset of a set S: A subset of S that does not equal S itself.

proper subspace: Any subspace of a vector space V other than V itself.

pseudoinverse (of A): The matrix $VD^{-1}U^T$, when UDV^T is a reduced singular value decomposition of A.

Q

QR factorization: A factorization of an $m \times n$ matrix A with linearly independent columns, $A = QR$, where Q is an $m \times n$ matrix whose columns form an orthonormal basis for Col A, and R is an $n \times n$ upper triangular invertible matrix with positive entries on its diagonal.

quadratice Bézier curve: A curve whose description may be written in the form $\mathbf{g}(t) = (1-t)\mathbf{f}_0(t) + t\mathbf{f}_1(t)$ for $0 \le t \le 1$, where $\mathbf{f}_0(t) = (1-t)\mathbf{p}_0 + t\mathbf{p}_1$ and $\mathbf{f}_1(t) = (1-t)\mathbf{p}_1 + t\mathbf{p}_2$. The points $\mathbf{p}_0, \mathbf{p}_1, \mathbf{p}_2$ are called the *control points* for the curve.

quadratic form: A function Q defined for **x** in $\mathbb{R}^n$ by $Q(\mathbf{x}) = \mathbf{x}^T A\mathbf{x}$, where A is an $n \times n$ symmetric matrix (called the **matrix of the quadratic form**).

R

range (of a linear transformation T): The set of all vectors of the form $T(\mathbf{x})$ for some **x** in the domain of T.

rank (of a matrix A): The dimension of the column space of A, denoted by rank A.

Rayleigh quotient: $R(\mathbf{x}) = (\mathbf{x}^T A\mathbf{x})/(\mathbf{x}^T\mathbf{x})$. An estimate of an eigenvalue of A (usually a symmetric matrix).

recurrence relation: *See* difference equation.

reduced echelon form (or **reduced row echelon form**): A reduced echelon matrix that is row equivalent to a given matrix.

reduced echelon matrix: A rectangular matrix in echelon form that has these additional properties: The leading entry in each nonzero row is 1, and each leading 1 is the only nonzero entry in its column.

reduced singular value decomposition: A factorization $A = UDV^T$, for an $m \times n$ matrix A of rank r, where U is $m \times r$ with orthonormal columns, D is an $r \times r$ diagonal matrix with the r nonzero singular values of A on its diagonal, and V is $n \times r$ with orthonormal columns.

regression coefficients: The coefficients β_0 and β_1 in the least-squares line $y = \beta_0 + \beta_1 x$.

regular solid: One of the five possible regular polyhedrons in $\mathbb{R}^3$: the tetrahedron (4 equal triangular faces), the cube (6 square faces), the octahedron (8 equal triangular faces), the dodecahedron (12 equal pentagonal faces), and the icosahedron (20 equal triangular faces).

regular stochastic matrix: A stochastic matrix P such that some matrix power P^k contains only strictly positive entries.

relative change or **relative error** (in **b**): The quantity $\|\Delta\mathbf{b}\| / \|\mathbf{b}\|$ when **b** is changed to $\mathbf{b} + \Delta\mathbf{b}$.

repellor (of a dynamical system in $\mathbb{R}^2$): The origin when all trajectories except the constant zero sequence or function tend away from **0**.

residual vector: The quantity $\boldsymbol{\epsilon}$ that appears in the general linear model: $\mathbf{y} = X\boldsymbol{\beta} + \boldsymbol{\epsilon}$; that is, $\boldsymbol{\epsilon} = \mathbf{y} - X\boldsymbol{\beta}$, the difference between the observed values and the predicted values (of y).

Re x: The vector in $\mathbb{R}^n$ formed from the real parts of the entries of a vector **x** in $\mathbb{C}^n$.

right inverse (of A): Any rectangular matrix C such that $AC = I$.

right-multiplication (by A): Multiplication of a matrix on the right by A.

right singular vectors (of A): The columns of V in the singular value decomposition $A = U\Sigma V^T$.

roundoff error: Error in floating point arithmetic caused when the result of a calculation is rounded (or truncated) to the number of floating point digits stored. Also, the error that results when the decimal representation of a number such as 1/3 is approximated by a floating point number with a finite number of digits.

row–column rule: The rule for computing a product AB in which the (i, j)-entry of AB is the sum of the products of corresponding entries from row i of A and column j of B.

row equivalent (matrices): Two matrices for which there exists a (finite) sequence of row operations that transforms one matrix into the other.

row reduction algorithm: A systematic method using elementary row operations that reduces a matrix to echelon form or reduced echelon form.

row replacement: An elementary row operation that replaces one row of a matrix by the sum of the row and a multiple of another row.

row space (of a matrix A): The set Row A of all linear combinations of the vectors formed from the rows of A; also denoted by Col A^T.

row sum: The sum of the entries in a row of a matrix.

row vector: A matrix with only one row, or a single row of a matrix that has several rows.

row–vector rule for computing Ax: The rule for computing a product Ax in which the ith entry of Ax is the sum of the products of corresponding entries from row i of A and from the vector **x**.

S

saddle point (of a dynamical system in $\mathbb{R}^2$): The origin when some trajectories are attracted to **0** and other trajectories are repelled from **0**.

same direction (as a vector **v**): A vector that is a positive multiple of **v**.

sample mean: The average M of a set of vectors, $\mathbf{X}_1, \ldots, \mathbf{X}_N$, given by $M = (1/N)(\mathbf{X}_1 + \cdots + \mathbf{X}_N)$.

scalar: A (real) number used to multiply either a vector or a matrix.

scalar multiple of u by c: The vector $c\mathbf{u}$ obtained by multiplying each entry in **u** by c.

scale (a vector): Multiply a vector (or a row or column of a matrix) by a nonzero scalar.

Schur complement: A certain matrix formed from the blocks of a 2×2 partitioned matrix $A = [A_{ij}]$. If A_{11} is invertible, its Schur complement is given by $A_{22} - A_{21}A_{11}^{-1}A_{12}$. If A_{22} is invertible, its Schur complement is given by $A_{11} - A_{12}A_{22}^{-1}A_{21}$.

Schur factorization (of A, for real scalars): A factorization $A = URU^T$ of an $n \times n$ matrix A having n real eigenvalues, where U is an $n \times n$ orthogonal matrix and R is an upper triangular matrix.

set spanned by $\{\mathbf{v}_1, \ldots, \mathbf{v}_p\}$: The set Span $\{\mathbf{v}_1, \ldots, \mathbf{v}_p\}$.

signal (or **discrete-time signal**): A doubly infinite sequence of numbers, $\{y_k\}$; a function defined on the integers; belongs to the vector space $\mathbb{S}$.

similar (matrices): Matrices A and B such that $P^{-1}AP = B$, or equivalently, $A = PBP^{-1}$, for some invertible matrix P.

similarity transformation: A transformation that changes A into $P^{-1}AP$.

simplex: The convex hull of an affinely independent finite set of vectors in $\mathbb{R}^n$.

singular (matrix): A square matrix that has no inverse.

singular value decomposition (of an $m \times n$ matrix A): $A = U\Sigma V^T$, where U is an $m \times m$ orthogonal matrix, V is an $n \times n$ orthogonal matrix, and Σ is an $m \times n$ matrix with nonnegative entries on the main diagonal (arranged in decreasing order of magnitude) and zeros elsewhere. If rank $A = r$, then Σ has exactly r positive entries (the nonzero singular values of A) on the diagonal.

singular values (of A): The (positive) square roots of the eigenvalues of A^TA, arranged in decreasing order of magnitude.

size (of a matrix): Two numbers, written in the form $m \times n$, that specify the number of rows (m) and columns (n) in the matrix.

solution (of a linear system involving variables $x_1, \ldots, x_n$): A list $(s_1, s_2, \ldots, s_n)$ of numbers that makes each equation in the system a true statement when the values $s_1, \ldots, s_n$ are substituted for $x_1, \ldots, x_n$, respectively.

solution set: The set of all possible solutions of a linear system. The solution set is empty when the linear system is inconsistent.

Span $\{\mathbf{v}_1, \ldots, \mathbf{v}_p\}$: The set of all linear combinations of $\mathbf{v}_1, \ldots, \mathbf{v}_p$. Also, the *subspace spanned* (or *generated*) by $\mathbf{v}_1, \ldots, \mathbf{v}_p$.

spanning set (for a subspace H): Any set $\{\mathbf{v}_1, \ldots, \mathbf{v}_p\}$ in H such that $H = \text{Span}\{\mathbf{v}_1, \ldots, \mathbf{v}_p\}$.

spectral decomposition (of A): A representation

$$A = \lambda_1 \mathbf{u}_1 \mathbf{u}_1^T + \cdots + \lambda_n \mathbf{u}_n \mathbf{u}_n^T$$

where $\{\mathbf{u}_1, \ldots, \mathbf{u}_n\}$ is an orthonormal basis of eigenvectors of A, and $\lambda_1, \ldots, \lambda_n$ are the corresponding eigenvalues of A.

spiral point (of a dynamical system in $\mathbb{R}^2$): The origin when the trajectories spiral about **0**.

stage-matrix model: A difference equation $\mathbf{x}_{k+1} = A\mathbf{x}_k$ where $\mathbf{x}_k$ lists the number of females in a population at time k, with the females classified by various stages of development (such as juvenile, subadult, and adult).

standard basis: The basis $\mathcal{E} = \{\mathbf{e}_1, \ldots, \mathbf{e}_n\}$ for $\mathbb{R}^n$ consisting of the columns of the $n \times n$ identity matrix, or the basis $\{1, t, \ldots, t^n\}$ for $\mathbb{P}_n$.

standard matrix (for a linear transformation T): The matrix A such that $T(\mathbf{x}) = A\mathbf{x}$ for all **x** in the domain of T.

standard position: The position of the graph of an equation $\mathbf{x}^TA\mathbf{x} = c$, when A is a diagonal matrix.

state vector: A probability vector. In general, a vector that describes the "state" of a physical system, often in connection with a difference equation $\mathbf{x}_{k+1} = A\mathbf{x}_k$.

steady-state vector (for a stochastic matrix P): A probability vector **q** such that $P\mathbf{q} = \mathbf{q}$.

stiffness matrix: The inverse of a flexibility matrix. The jth column of a stiffness matrix gives the loads that must be applied at specified points on an elastic beam in order to produce a unit deflection at the jth point on the beam.

stochastic matrix: A square matrix whose columns are probability vectors.

strictly dominant eigenvalue: An eigenvalue λ_1 of a matrix A with the property that $|\lambda_1| > |\lambda_k|$ for all other eigenvalues λ_k of A.

submatrix (of A): Any matrix obtained by deleting some rows and/or columns of A; also, A itself.

subspace: A subset H of some vector space V such that H has these properties: (1) the zero vector of V is in H; (2) H is closed under vector addition; and (3) H is closed under multiplication by scalars.

supporting hyperplane (to a compact convex set S in $\mathbb{R}^n$): A hyperplane $H = [f : d]$ such that $H \cap S \neq \varnothing$ and either $f(x) \leq d$ for all x in S or $f(x) \geq d$ for all x in S.

symmetric matrix: A matrix A such that $A^T = A$.

system of linear equations (or a **linear system**): A collection of one or more linear equations involving the same set of variables, say, $x_1, \ldots, x_n$.

T

tetrahedron: A three-dimensional solid object bounded by four equal triangular faces, with three faces meeting at each vertex.

total variance: The trace of the covariance matrix S of a matrix of observations.

trace (of a square matrix A): The sum of the diagonal entries in A, denoted by tr A.

trajectory: The graph of a solution $\{\mathbf{x}_0, \mathbf{x}_1, \mathbf{x}_2, \ldots\}$ of a dynamical system $\mathbf{x}_{k+1} = A\mathbf{x}_k$, often connected by a thin curve to make the trajectory easier to see. Also, the graph of $\mathbf{x}(t)$ for $t \geq 0$, when $\mathbf{x}(t)$ is a solution of a differential equation $\mathbf{x}'(t) = A\mathbf{x}(t)$.

transfer matrix: A matrix A associated with an electrical circuit having input and output terminals, such that the output vector is A times the input vector.

transformation (or **function**, or **mapping**) T from $\mathbb{R}^n$ to $\mathbb{R}^m$: A rule that assigns to each vector $\mathbf{x}$ in $\mathbb{R}^n$ a unique vector $T(\mathbf{x})$ in $\mathbb{R}^m$. Notation: $T: \mathbb{R}^n \to \mathbb{R}^m$. Also, $T: V \to W$ denotes a rule that assigns to each $\mathbf{x}$ in V a unique vector $T(\mathbf{x})$ in W.

translation (by a vector $\mathbf{p}$): The operation of adding $\mathbf{p}$ to a vector or to each vector in a given set.

transpose (of A): An $n \times m$ matrix A^T whose columns are the corresponding rows of the $m \times n$ matrix A.

trend analysis: The use of orthogonal polynomials to fit data, with the inner product given by evaluation at a finite set of points.

triangle inequality: $\|\mathbf{u} + \mathbf{v}\| \leq \|\mathbf{u}\| + \|\mathbf{v}\|$ for all $\mathbf{u}, \mathbf{v}$.

triangular matrix: A matrix A with either zeros above or zeros below the diagonal entries.

trigonometric polynomial: A linear combination of the constant function 1 and sine and cosine functions such as $\cos nt$ and $\sin nt$.

trivial solution: The solution $\mathbf{x} = \mathbf{0}$ of a homogeneous equation $A\mathbf{x} = \mathbf{0}$.

U

uncorrelated variables: Any two variables x_i and x_j (with $i \neq j$) that range over the ith and jth coordinates of the observation vectors in an observation matrix, such that the covariance s_{ij} is zero.

underdetermined system: A system of equations with fewer equations than unknowns.

uniqueness question: Asks, "If a solution of a system exists, is it unique—that is, is it the only one?"

unit consumption vector: A column vector in the Leontief input–output model that lists the inputs a sector needs for each unit of its output; a column of the consumption matrix.

unit lower triangular matrix: A square lower triangular matrix with ones on the main diagonal.

unit vector: A vector $\mathbf{v}$ such that $\|\mathbf{v}\| = 1$.

upper triangular matrix: A matrix U (not necessarily square) with zeros below the diagonal entries $u_{11}, u_{22}, \ldots$.

V

Vandermonde matrix: An $n \times n$ matrix V or its transpose, when V has the form

$$
V = \begin{bmatrix} 1 & x_1 & x_1^2 & \cdots & x_1^{n-1} \\ 1 & x_2 & x_2^2 & \cdots & x_2^{n-1} \\ \vdots & \vdots & \vdots & & \vdots \\ 1 & x_n & x_n^2 & \cdots & x_n^{n-1} \end{bmatrix}
$$

variance (of a variable x_j): The diagonal entry s_{jj} in the covariance matrix S for a matrix of observations, where x_j varies over the jth coordinates of the observation vectors.

vector: A list of numbers; a matrix with only one column. In general, any element of a vector space.

vector addition: Adding vectors by adding corresponding entries.

vector equation: An equation involving a linear combination of vectors with undetermined weights.

vector space: A set of objects, called vectors, on which two operations are defined, called addition and multiplication by scalars. Ten axioms must be satisfied. See the first definition in Section 4.1.

vector subtraction: Computing $\mathbf{u} + (-1)\mathbf{v}$ and writing the result as $\mathbf{u} - \mathbf{v}$.

W

weighted least squares: Least-squares problems with a weighted inner product such as

$$
\langle \mathbf{x}, \mathbf{y} \rangle = w_1^2 x_1 y_1 + \cdots + w_n^2 x_n y_n.
$$

weights: The scalars used in a linear combination.

Z

zero subspace: The subspace $\{\mathbf{0}\}$ consisting of only the zero vector.

zero vector: The unique vector, denoted by $\mathbf{0}$, such that $\mathbf{u} + \mathbf{0} = \mathbf{u}$ for all $\mathbf{u}$. In $\mathbb{R}^n$, $\mathbf{0}$ is the vector whose entries are all zeros.

Answers to Odd-Numbered Exercises

Chapter 1

Section 1.1, page 10

1. The solution is $(x_1, x_2) = (-8, 3)$, or simply $(-8, 3)$.

3. $(2, 1)$

5. Replace Row 2 by its sum with -4 times Row 3, and then replace Row 1 by its sum with 3 times Row 3.

7. The solution set is empty.

9. $(16, 21, 14, 4)$ **11.** Inconsistent

13. $(5, 3, -1)$ **15.** Inconsistent

17. Calculations show that the system is inconsistent, so the three lines have no point in common.

19. $h \neq 2$ **21.** All h

23. Mark a statement True only if the statement is *always* true. Giving you the answers here would defeat the purpose of the true–false questions, which is to help you learn to read the text carefully. The *Study Guide* will tell you where to look for the answers, but you should not consult it until you have made an honest attempt to find the answers yourself.

25. $k - 2g + h = 0$

27. The row reduction of
$$\begin{bmatrix} a & b & f \\ c & d & g \end{bmatrix} \text{ to } \begin{bmatrix} a & b & f \\ 0 & d - b(\frac{c}{a}) & g - f(\frac{c}{a}) \end{bmatrix}$$
shows that $d - b(\frac{c}{a})$ must be nonzero, since f and g are arbitrary. Otherwise, for some choices of f and g the second row could correspond to an equation of the form $0 = q$, where q is nonzero. Thus $ad \neq bc$.

29. Swap Row 1 and Row 3; swap Row 1 and Row 3.

31. Replace Row 3 by Row 3 + (−4)Row 1; replace Row 3 by Row 3 + (4)Row 1.

33. Review Practice Problem 1 and then *write* a solution. The *Study Guide* has a solution.

Section 1.2, page 21

1. Reduced echelon form: a and b. Echelon form: d. Not in echelon form: c.

3. $\begin{bmatrix} 1 & 2 & 0 & -8 \\ 0 & 0 & 1 & 4 \\ 0 & 0 & 0 & 0 \end{bmatrix}$.

Pivot cols 1 and 3: $\begin{bmatrix} 1 & 2 & 4 & 8 \\ 2 & 4 & 6 & 8 \\ 3 & 6 & 9 & 12 \end{bmatrix}$.

5. $\begin{bmatrix} \blacksquare & * \\ 0 & \blacksquare \end{bmatrix}, \begin{bmatrix} \blacksquare & * \\ 0 & 0 \end{bmatrix}, \begin{bmatrix} 0 & \blacksquare \\ 0 & 0 \end{bmatrix}$

7. $\begin{cases} x_1 = -5 - 3x_2 \\ x_2 \text{ is free.} \\ x_3 = 3 \end{cases}$

9. $\begin{cases} x_1 = 3 + 2x_3 \\ x_2 = 3 + 2x_3 \\ x_3 \text{ is free.} \end{cases}$

11. $\begin{cases} x_1 = \frac{2}{3}x_2 - \frac{4}{3}x_3 \\ x_2 \text{ is free.} \\ x_3 \text{ is free.} \end{cases}$

13. $\begin{cases} x_1 = 5 + 3x_5 \\ x_2 = 1 + 4x_5 \\ x_3 \text{ is free.} \\ x_4 = 4 - 9x_5 \\ x_5 \text{ is free.} \end{cases}$

Note: The *Study Guide* discusses the common mistake $x_3 = 0$.

15. **a.** Consistent, with many solutions
  b. Consistent, with many solutions

17. All h

19. **a.** Inconsistent when $h = 2$ and $k \neq 8$
  b. Unique solution when $h \neq 2$
  c. Many solutions when $h = 2$ and $k = 8$

21. Read the text carefully, and write your answers before you consult the *Study Guide*. Remember, a statement is true only if it is true in all cases.

23. Since there are four pivots (one in each column of the coefficient matrix), the augmented matrix must reduce to the form
$$\begin{bmatrix} 1 & 0 & 0 & 0 & a \\ 0 & 1 & 0 & 0 & b \\ 0 & 0 & 1 & 0 & c \\ 0 & 0 & 0 & 1 & d \end{bmatrix}$$
and so
$$\begin{aligned} x_1 &&&= a \\ &x_2 &&= b \\ &&x_3 &= c \\ &&&x_4 = d \end{aligned}$$

No matter what the values of a, b, c and d, the solution exists and is unique.

25. If the coefficient matrix has a pivot position in every row, then there is a pivot position in the bottom row, and there is no room for a pivot in the augmented column. So, the system is consistent, by Theorem 2.

27. If a linear system is consistent, then the solution is unique if and only if *every column in the coefficient matrix is a pivot column; otherwise, there are infinitely many solutions.*

29. An underdetermined system always has more variables than equations. There cannot be more basic variables than there are equations, so there must be at least one free variable. Such a variable may be assigned infinitely many different values. If the system is consistent, each different value of a free variable will produce a different solution.

31. Yes, a system of linear equations with more equations than unknowns can be consistent. The following system has a solution ($x_1 = x_2 = 1$):

$$\begin{aligned} x_1 + x_2 &= 2 \\ x_1 - x_2 &= 0 \\ 3x_1 + 2x_2 &= 5 \end{aligned}$$

33. $p(t) = 1 + 3t + 2t^2$

Section 1.3, page 32

1. $\begin{bmatrix} -4 \\ 1 \end{bmatrix}, \begin{bmatrix} 5 \\ 4 \end{bmatrix}$

3.

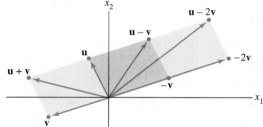

5. $x_1 \begin{bmatrix} 3 \\ -2 \\ 8 \end{bmatrix} + x_2 \begin{bmatrix} 5 \\ 0 \\ -9 \end{bmatrix} = \begin{bmatrix} 2 \\ -3 \\ 8 \end{bmatrix}$,

$\begin{bmatrix} 3x_1 \\ -2x_1 \\ 8x_1 \end{bmatrix} + \begin{bmatrix} 5x_2 \\ 0 \\ -9x_2 \end{bmatrix} = \begin{bmatrix} 2 \\ -3 \\ 8 \end{bmatrix}$,

$\begin{bmatrix} 3x_1 + 5x_2 \\ -2x_1 \\ 8x_1 - 9x_2 \end{bmatrix} = \begin{bmatrix} 2 \\ -3 \\ 8 \end{bmatrix}$

$\begin{aligned} 3x_1 + 5x_2 &= 2 \\ -2x_1 &= -3 \\ 8x_1 - 9x_2 &= 8 \end{aligned}$

Usually the intermediate steps are not displayed.

7. $\mathbf{a} = \mathbf{u} - 2\mathbf{v}$, $\mathbf{b} = 2\mathbf{u} - 2\mathbf{v}$, $\mathbf{c} = 2\mathbf{u} - 3.5\mathbf{v}$, $\mathbf{d} = 3\mathbf{u} - 4\mathbf{v}$
Yes, every vector in $\mathbb{R}^2$ is a linear combination of $\mathbf{u}$ and $\mathbf{v}$.

9. $x_1 \begin{bmatrix} 0 \\ 4 \\ -1 \end{bmatrix} + x_2 \begin{bmatrix} 1 \\ 6 \\ 3 \end{bmatrix} + x_3 \begin{bmatrix} 5 \\ -1 \\ -8 \end{bmatrix} = \begin{bmatrix} 0 \\ 0 \\ 0 \end{bmatrix}$

11. No, $\mathbf{b}$ is *not* a linear combination of $\mathbf{a}_1$, $\mathbf{a}_2$, and $\mathbf{a}_3$.

13. No, $\mathbf{b}$ is *not* a linear combination of the columns of A.

15. $h = 3$

17. Noninteger weights are acceptable, of course, but some simple choices are $0 \cdot \mathbf{v}_1 + 0 \cdot \mathbf{v}_2 = \mathbf{0}$, and

$$1 \cdot \mathbf{v}_1 + 0 \cdot \mathbf{v}_2 = \begin{bmatrix} 3 \\ 1 \\ 2 \end{bmatrix}, \qquad 0 \cdot \mathbf{v}_1 + 1 \cdot \mathbf{v}_2 = \begin{bmatrix} -4 \\ 0 \\ 1 \end{bmatrix}$$

$$1 \cdot \mathbf{v}_1 + 1 \cdot \mathbf{v}_2 = \begin{bmatrix} -1 \\ 1 \\ 3 \end{bmatrix}, \qquad 1 \cdot \mathbf{v}_1 - 1 \cdot \mathbf{v}_2 = \begin{bmatrix} 7 \\ 1 \\ 1 \end{bmatrix}$$

19. Span $\{\mathbf{v}_1, \mathbf{v}_2\}$ is the set of points on the line through $\mathbf{v}_1$ and $\mathbf{0}$, because $\mathbf{v}_2$ is a multiple of $\mathbf{v}_1$.

21. *Hint:* Show that $\begin{bmatrix} 2 & 2 & h \\ -1 & 1 & k \end{bmatrix}$ is consistent for all h and k. Explain what this calculation shows about Span $\{\mathbf{u}, \mathbf{v}\}$.

23. Before you consult your *Study Guide*, read the entire section carefully. Pay special attention to definitions and theorem statements, and note any remarks that precede or follow them.

25. **a.** No, three   **b.** Yes, infinitely many
    c. $\mathbf{a}_1 = 1 \cdot \mathbf{a}_1 + 0 \cdot \mathbf{a}_2 + 0 \cdot \mathbf{a}_3$

27. **a.** $5\mathbf{v}_1$ is the output of 5 days of operation of mine #1.
    b. The total output is $x_1\mathbf{v}_1 + x_2\mathbf{v}_2$, so x_1 and x_2 should satisfy $x_1\mathbf{v}_1 + x_2\mathbf{v}_2 = \begin{bmatrix} 240 \\ 2824 \end{bmatrix}$.
    c. [M] 1.73 days for mine #1 and 4.70 days for mine #2

29. $(17/14, -34/14, 16/14) = (17/14, -17/7, 8/7)$

31. **a.** $\begin{bmatrix} 10/3 \\ 2 \end{bmatrix}$
    b. Add 3.5 g at $(0, 1)$, add 0.5 g at $(8, 1)$, and add 2 g at $(2, 4)$.

33. Review Practice Problem 1 and then *write* a solution. The *Study Guide* has a solution.

Section 1.4, page 40

1. The product is not defined because the number of columns (2) in the 3×2 matrix does not match the number of entries (3) in the vector.

3. **a.** $A\mathbf{x} = \begin{bmatrix} 1 & 2 \\ -3 & 1 \\ 1 & 6 \end{bmatrix} \begin{bmatrix} -2 \\ 3 \end{bmatrix} = -2 \cdot \begin{bmatrix} 1 \\ -3 \\ 1 \end{bmatrix} + 3 \cdot \begin{bmatrix} 2 \\ 1 \\ 6 \end{bmatrix}$

$= \begin{bmatrix} -2 \\ 6 \\ -2 \end{bmatrix} + \begin{bmatrix} 6 \\ 3 \\ 18 \end{bmatrix} = \begin{bmatrix} 4 \\ 9 \\ 16 \end{bmatrix}$

b. $A\mathbf{x} = \begin{bmatrix} 1 & 2 \\ -3 & 1 \\ 1 & 6 \end{bmatrix} \begin{bmatrix} -2 \\ 3 \end{bmatrix} = \begin{bmatrix} 1\cdot(-2) + 2\cdot(3) \\ (-3)\cdot(-2) + 1\cdot(3) \\ 1\cdot(-2) + 6\cdot(3) \end{bmatrix}$

$= \begin{bmatrix} 4 \\ 9 \\ 16 \end{bmatrix}.$

Show your work here and for Exercises 4–6, but thereafter perform the calculations mentally.

5. $2\cdot\begin{bmatrix} 1 \\ -2 \end{bmatrix} - 1\cdot\begin{bmatrix} 2 \\ -3 \end{bmatrix} + 1\cdot\begin{bmatrix} -3 \\ 1 \end{bmatrix} - 1\cdot\begin{bmatrix} 1 \\ -1 \end{bmatrix} = \begin{bmatrix} -4 \\ 1 \end{bmatrix}$

7. $\begin{bmatrix} 4 & -5 & 7 \\ -1 & 3 & -8 \\ 7 & -5 & 0 \\ -4 & 1 & 2 \end{bmatrix} \begin{bmatrix} x_1 \\ x_2 \\ x_3 \end{bmatrix} = \begin{bmatrix} 6 \\ -8 \\ 0 \\ -7 \end{bmatrix}$

9. $x_1\begin{bmatrix} 5 \\ 0 \end{bmatrix} + x_2\begin{bmatrix} 1 \\ 2 \end{bmatrix} + x_3\begin{bmatrix} -3 \\ 4 \end{bmatrix} = \begin{bmatrix} 8 \\ 0 \end{bmatrix}$ and

$\begin{bmatrix} 5 & 1 & -3 \\ 0 & 2 & 4 \end{bmatrix} \begin{bmatrix} x_1 \\ x_2 \\ x_3 \end{bmatrix} = \begin{bmatrix} 8 \\ 0 \end{bmatrix}$

11. $\begin{bmatrix} 1 & 3 & -4 & -2 \\ 1 & 5 & 2 & 4 \\ -3 & -7 & 6 & 12 \end{bmatrix},\ \mathbf{x} = \begin{bmatrix} x_1 \\ x_2 \\ x_3 \end{bmatrix} = \begin{bmatrix} -11 \\ 3 \\ 0 \end{bmatrix}$

13. Yes. (Justify your answer.)

15. The equation $A\mathbf{x} = \mathbf{b}$ is not consistent when $3b_1 + b_2$ is nonzero. (Show your work.) The set of $\mathbf{b}$ for which the equation *is* consistent is a line through the origin—the set of all points (b_1, b_2) satisfying $b_2 = -3b_1$.

17. Only three rows contain a pivot position. The equation $A\mathbf{x} = \mathbf{b}$ does *not* have a solution for each $\mathbf{b}$ in $\mathbb{R}^4$, by Theorem 4.

19. The work in Exercise 17 shows that statement (d) in Theorem 4 is false. So all four statements in Theorem 4 are false. Thus, not all vectors in $\mathbb{R}^4$ can be written as a linear combination of the columns of A. Also, the columns of A do *not* span $\mathbb{R}^4$.

21. The matrix $[\mathbf{v}_1\ \ \mathbf{v}_2\ \ \mathbf{v}_3]$ does not have a pivot in each row, so the columns of the matrix do not span $\mathbb{R}^4$, by Theorem 4. That is, $\{\mathbf{v}_1, \mathbf{v}_2, \mathbf{v}_3\}$ does not span $\mathbb{R}^4$.

23. Read the text carefully and try to mark each exercise statement True or False before you consult the *Study Guide*. Several parts of Exercises 29 and 30 are *implications* of the form

"If ⟨statement 1⟩, then ⟨statement 2⟩"

or equivalently,

"⟨statement 2⟩, if ⟨statement 1⟩"

Mark such an implication as True if ⟨statement 2⟩ is true in all cases when ⟨statement 1⟩ is true.

25. $c_1 = -3,\ c_2 = -1,\ c_3 = 2$

27. The matrix equation can be written as $c_1\mathbf{v}_1 + c_2\mathbf{v}_2 + c_3\mathbf{v}_3 + c_4\mathbf{v}_4 + c_5\mathbf{v}_5 = \mathbf{v}_6$, where $c_1 = -3$,

$c_2 = 1,\ c_3 = 2,\ c_4 = -1,\ c_5 = 2$, and

$\mathbf{v}_1 = \begin{bmatrix} -3 \\ 5 \end{bmatrix},\quad \mathbf{v}_2 = \begin{bmatrix} 5 \\ 8 \end{bmatrix},\quad \mathbf{v}_3 = \begin{bmatrix} -4 \\ 1 \end{bmatrix},$

$\mathbf{v}_4 = \begin{bmatrix} 9 \\ -2 \end{bmatrix},\quad \mathbf{v}_5 = \begin{bmatrix} 7 \\ -4 \end{bmatrix},\quad \mathbf{v}_6 = \begin{bmatrix} 11 \\ -11 \end{bmatrix}.$

29. *Hint:* Start with any 3×3 matrix B in echelon form that has three pivot positions.

31. *Write* your solution before you check the *Study Guide*.

33. *Hint:* How many pivot columns does A have? Why?

35. Suppose $\mathbf{y}$ and $\mathbf{z}$ satisfy $A\mathbf{y} = \mathbf{z}$. Then $5\mathbf{z} = 5A\mathbf{y}$. By Theorem 5(b), $5A\mathbf{y} = A(5\mathbf{y})$. So $5\mathbf{z} = A(5\mathbf{y})$, which shows that $5\mathbf{y}$ is a solution of $A\mathbf{x} = 5\mathbf{z}$. Thus the equation $A\mathbf{x} = 5\mathbf{z}$ is consistent.

37. [M] The columns do not span $\mathbb{R}^4$.

39. [M] The columns span $\mathbb{R}^4$.

41. [M] Delete column 4 of the matrix in Exercise 39. It is also possible to delete column 3 instead of column 4.

Section 1.5, page 47

1. The system has a nontrivial solution because there is a free variable, x_3.

3. The system has a nontrivial solution because there is a free variable, x_3.

5. $\mathbf{x} = \begin{bmatrix} x_1 \\ x_2 \\ x_3 \end{bmatrix} = x_3\begin{bmatrix} -1 \\ -1 \\ 1 \end{bmatrix}$

7. $\mathbf{x} = \begin{bmatrix} x_1 \\ x_2 \\ x_3 \\ x_4 \end{bmatrix} = x_3\begin{bmatrix} -9 \\ 4 \\ 1 \\ 0 \end{bmatrix} + x_4\begin{bmatrix} 8 \\ -5 \\ 0 \\ 1 \end{bmatrix}$

9. $\mathbf{x} = x_2\begin{bmatrix} 2 \\ 1 \\ 0 \end{bmatrix}$

11. *Hint:* The system derived from the *reduced* echelon form is

$\begin{aligned} x_1 - 4x_2 \qquad\quad + 5x_6 &= 0 \\ x_3 \quad - x_6 &= 0 \\ x_5 - 4x_6 &= 0 \\ 0 &= 0 \end{aligned}$

The basic variables are x_1, x_3, and x_5. The remaining variables are free. The *Study Guide* discusses two mistakes that are often made on this type of problem.

13. $\mathbf{x} = \begin{bmatrix} 5 \\ -2 \\ 0 \end{bmatrix} + x_3\begin{bmatrix} 4 \\ -7 \\ 1 \end{bmatrix} = \mathbf{p} + x_3\mathbf{q}$. Geometrically, the

solution set is the line through $\begin{bmatrix} 5 \\ -2 \\ 0 \end{bmatrix}$ parallel to $\begin{bmatrix} 4 \\ -7 \\ 1 \end{bmatrix}$.

15. Let $\mathbf{u} = \begin{bmatrix} -5 \\ 1 \\ 0 \end{bmatrix}$, $\mathbf{v} = \begin{bmatrix} 3 \\ 0 \\ 1 \end{bmatrix}$, $\mathbf{p} = \begin{bmatrix} -2 \\ 0 \\ 0 \end{bmatrix}$. The solution of
the homogeneous equation is $\mathbf{x} = x_2\mathbf{u} + x_3\mathbf{v}$, the plane
through the origin spanned by $\mathbf{u}$ and $\mathbf{v}$. The solution set of
the nonhomogeneous system is $\mathbf{x} = \mathbf{p} + x_2\mathbf{u} + x_3\mathbf{v}$, the
plane through $\mathbf{p}$ parallel to the solution set of the
homogeneous equation.

17. $\mathbf{x} = \begin{bmatrix} x_1 \\ x_2 \\ x_3 \end{bmatrix} = \begin{bmatrix} 8 \\ -4 \\ 0 \end{bmatrix} + x_3 \begin{bmatrix} -1 \\ -1 \\ 1 \end{bmatrix}$. The solution set is the
line through $\begin{bmatrix} 8 \\ -4 \\ 0 \end{bmatrix}$, parallel to the line that is the solution
set of the homogeneous system in Exercise 5.

19. $\mathbf{x} = \mathbf{a} + t\mathbf{b}$, where t represents a parameter, or
$\mathbf{x} = \begin{bmatrix} x_1 \\ x_2 \end{bmatrix} = \begin{bmatrix} -2 \\ 0 \end{bmatrix} + t \begin{bmatrix} -5 \\ 3 \end{bmatrix}$, or $\begin{cases} x_1 = -2 - 5t \\ x_2 = 3t \end{cases}$

21. $\mathbf{x} = \mathbf{p} + t(\mathbf{q} - \mathbf{p}) = \begin{bmatrix} 3 \\ -3 \end{bmatrix} + t \begin{bmatrix} 1 \\ 4 \end{bmatrix}$

23. It is important to read the text carefully and write your
answers. After that, check the *Study Guide*, if necessary.

25. a. $A\mathbf{w} = A(\mathbf{p} + \mathbf{v}_h) = A\mathbf{p} + A\mathbf{v}_h = \mathbf{b} + \mathbf{0} = \mathbf{b}$
  b. $A\mathbf{v}_h = A(\mathbf{w} - \mathbf{p}) = A\mathbf{w} - A\mathbf{p} = \mathbf{b} - \mathbf{b} = \mathbf{0}$

27. (*Geometric argument using Theorem 6*) Since the equation
$A\mathbf{x} = \mathbf{b}$ is consistent, its solution set is obtained by
translating the solution set of $A\mathbf{x} = \mathbf{0}$, by Theorem 6. So
the solution set of $A\mathbf{x} = \mathbf{b}$ is a single vector if and only if
the solution set of $A\mathbf{x} = \mathbf{0}$ is a single vector, and that
happens if and only if $A\mathbf{x} = \mathbf{0}$ has only the trivial solution.

(*Proof using free variables*) If $A\mathbf{x} = \mathbf{b}$ has a solution, then
the solution is unique if and only if there are no free
variables in the corresponding system of equations, that is,
if and only if every column of A is a pivot column. This
happens if and only if the equation $A\mathbf{x} = \mathbf{0}$ has only the
trivial solution.

29. a. When A is a 4×4 matrix with three pivot positions, the
equation $A\mathbf{x} = \mathbf{0}$ has a free variable and hence has
nontrivial solutions.
  b. With three pivot positions, A does not have a pivot
position in each of its four rows. By Theorem 4 in
Section 1.4, the equation $A\mathbf{x} = \mathbf{b}$ does not have a
solution for every possible $\mathbf{b}$. The word "possible" in
the exercise means that the only vectors considered in
this case are those in $\mathbb{R}^4$, because A has four rows.

31. a. When A is a 3×2 matrix with two pivot positions, each
column is a pivot column. So the equation $A\mathbf{x} = \mathbf{0}$ has
no free variables and hence no nontrivial solution.
  b. With two pivot positions and three rows, A cannot have
a pivot in every row. So the equation $A\mathbf{x} = \mathbf{b}$ cannot
have a solution for every possible $\mathbf{b}$ (in $\mathbb{R}^3$), by
Theorem 4 in Section 1.4.

33. Your example should have the property that the sum of the
entries in each row is zero. Why?

35. One answer: $\mathbf{x} = \begin{bmatrix} 3 \\ -1 \end{bmatrix}$

37. One answer is $A = \begin{bmatrix} 1 & -4 \\ 1 & -4 \end{bmatrix}$. The *Study Guide* shows
how to analyze the problem in order to construct A. If $\mathbf{b}$ is
any vector *not* a multiple of the first column of A, then the
solution set of $A\mathbf{x} = \mathbf{b}$ is empty and thus cannot be formed
by translating the solution set of $A\mathbf{x} = \mathbf{0}$. This does not
contradict Theorem 6, because that theorem applies when
the equation $A\mathbf{x} = \mathbf{b}$ has a nonempty solution set.

39. Suppose $A\mathbf{v} = \mathbf{0}$ and $A\mathbf{w} = \mathbf{0}$. Then, since
$A(\mathbf{v} + \mathbf{w}) = A\mathbf{v} + A\mathbf{w}$ by Theorem 5(a) in Section 1.4,
$A(\mathbf{v} + \mathbf{w}) = A\mathbf{v} + A\mathbf{w} = \mathbf{0} + \mathbf{0} = \mathbf{0}$. Now, let c and d be
scalars. Using both parts of Theorem 5, $A(c\mathbf{v} + d\mathbf{w}) =$
$A(c\mathbf{v}) + A(d\mathbf{w}) = cA\mathbf{v} + dA\mathbf{w} = c\mathbf{0} + d\mathbf{0} = \mathbf{0}$.

Section 1.6, page 54

1. The general solution is $p_G = .875p_S$, with p_S free. One
equilibrium solution is $p_S = 1000$ and $p_G = 875$. Using
fractions, the general solution could be written
$p_G = (7/8)p_S$, and a natural choice of prices might be
$p_S = 80$ and $p_G = 70$. Only the *ratio* of the prices is
important. The economic equilibrium is unaffected by a
proportional change in prices.

3. a.

	Distribution of Output from:				
	F&P	Man.	Ser.		
Output	↓	↓	↓	Input	Purchased by:
	.1	.1	.2	→	F&P
	.8	.1	.4	→	Man.
	.1	.8	.4	→	Ser.

  b. $\begin{bmatrix} .9 & -.1 & -.2 & 0 \\ -.8 & .9 & -.4 & 0 \\ -.1 & -.8 & .6 & 0 \end{bmatrix}$

  c. [M] $p_{F\&P} \approx 30$, $p_M \approx 71$, $p_S = 100$.

5. a.

	Distribution of Output from:					
	Ag.	Man.	Ser.	Transp.		
Output	↓	↓	↓	↓	Input	Purchased by:
	.20	.35	.10	.20	→	Ag.
	.20	.10	.20	.30	→	Man.
	.30	.35	.50	.20	→	Ser.
	.30	.20	.20	.30	→	Transp.

  b. One solution is $p_A = 7.99$, $p_M = 8.36$, $p_S = 14.65$,
and $p_T = 10.00$.

c.   Distribution of Output from:
            Ag. Man. Ser. Transp.

Output	$\downarrow$	$\downarrow$	$\downarrow$	$\downarrow$	Input	Purchased by:
	.20	.35	.10	.20	$\rightarrow$	Ag.
	.10	.10	.20	.30	$\rightarrow$	Man.
	.40	.35	.50	.20	$\rightarrow$	Ser.
	.30	.20	.20	.30	$\rightarrow$	Transp.

d. One solution is $p_A = 7.81$, $p_M = 7.67$, $p_S = 15.62$, and $p_T = 10.00$.

The campaign has benefited Services the most.

7. $3NaHCO_3 + H_3C_6H_5O_7 \rightarrow Na_3C_6H_5O_7 + 3H_2O + 3CO_2$

9. $B_2S_3 + 6H_2O \rightarrow 2H_3BO_3 + 3H_2S$

11. [M] $16MnS + 13As_2Cr_{10}O_{35} + 374H_2SO_4$
$$\rightarrow 16HMnO_4 + 26AsH_3 + 130CrS_3O_{12} + 327H_2O$$

13. a. $\begin{cases} x_1 = x_3 - 40 \\ x_2 = x_3 + 10 \\ x_3 \text{ is free} \\ x_4 = x_6 + 50 \\ x_5 = x_6 + 60 \\ x_6 \text{ is free} \end{cases}$ **b.** $\begin{cases} x_2 = 50 \\ x_3 = 40 \\ x_4 = 50 \\ x_5 = 60 \end{cases}$

15. $\begin{cases} x_1 = 60 + x_6 \\ x_2 = -10 + x_6 \\ x_3 = 90 + x_6 \\ x_4 = x_6 \\ x_5 = 80 + x_6 \\ x_6 \text{ is free} \end{cases}$

In order for the flow to be nonnegative, $x_6 \geq 10$

Section 1.7, page 60

Justify your answers to Exercises 1–22.

1. Lin. indep.   **3.** Lin. depen.

5. Lin. indep.   **7.** Lin. depen.

9. a. No h **b.** All h

11. $h = -4$ **13.** All h

15. Lin. depen. **17.** Lin. depen. **19.** Lin. indep.

21. If you consult your *Study Guide* before you make a good effort to answer the true-false questions, you will destroy most of their value.

23. $\begin{bmatrix} \blacksquare & * \\ 0 & 0 \end{bmatrix}, \begin{bmatrix} 0 & \blacksquare \\ 0 & 0 \end{bmatrix}, \begin{bmatrix} 0 & 0 \\ 0 & 0 \end{bmatrix}$

25. $\begin{bmatrix} \blacksquare & * \\ 0 & \blacksquare \\ 0 & 0 \\ 0 & 0 \end{bmatrix}$ and $\begin{bmatrix} 0 & \blacksquare \\ 0 & 0 \\ 0 & 0 \\ 0 & 0 \end{bmatrix}$

27. All four columns of the 6×4 matrix A must be pivot columns. Otherwise, the equation $A\mathbf{x} = \mathbf{0}$ would have a free variable, in which case the columns of A would be linearly dependent.

29. A: Any 3×2 matrix with the second column a multiple of the first will have the desired property.

B: Any 3×2 matrix with two nonzero columns such that neither column is a multiple of the other will work. In this case, the columns form a linearly independent set, and so the equation $B\mathbf{x} = \mathbf{0}$ has only the trivial solution.

31. $\mathbf{x} = \begin{bmatrix} 1 \\ 1 \\ -1 \end{bmatrix}$

33. True, by Theorem 7. (The *Study Guide* adds another justification.)

35. True, by Theorem 9.

37. True. A linear dependence relation among $\mathbf{v}_1$, $\mathbf{v}_2$, and $\mathbf{v}_3$ may be extended to a linear dependence relation among $\mathbf{v}_1$, $\mathbf{v}_2$, $\mathbf{v}_3$, and $\mathbf{v}_4$ by placing a zero weight on $\mathbf{v}_4$.

39. You should be able to work this important problem without help. *Write* your solution before you consult the *Study Guide*.

41. [M] Using the pivot columns of A,
$$B = \begin{bmatrix} 3 & -4 & 7 \\ -5 & -3 & -11 \\ 4 & 3 & 2 \\ 8 & -7 & 4 \end{bmatrix}$$

Other choices are possible.

43. [M] Each column of A that is not a column of B is in the set spanned by the columns of B.

Section 1.8, page 68

1. $\begin{bmatrix} 2 \\ -6 \end{bmatrix}, \begin{bmatrix} 2a \\ 2b \end{bmatrix}$ **3.** $\mathbf{x} = \begin{bmatrix} 7 \\ 6 \\ 3 \end{bmatrix}$, unique solution

5. $\mathbf{x} = \begin{bmatrix} 3 \\ 1 \\ 0 \end{bmatrix}$, not unique   **7.** $a = 5, b = 6$

9. $\mathbf{x} = x_3 \begin{bmatrix} 4 \\ 3 \\ 1 \\ 0 \end{bmatrix}$

11. Yes, because the system represented by $[\, A \quad \mathbf{b}\,]$ is consistent.

13.

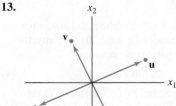

A reflection through the origin

15.

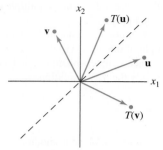

A reflection through the line $x_1 = x_2$.

17. $\begin{bmatrix} 8 \\ 2 \end{bmatrix}, \begin{bmatrix} -3 \\ 9 \end{bmatrix}, \begin{bmatrix} 5 \\ 11 \end{bmatrix}$ **19.** $\begin{bmatrix} 13 \\ 7 \end{bmatrix}, \begin{bmatrix} 2x_1 - x_2 \\ 5x_1 + 6x_2 \end{bmatrix}$

21. Read the text carefully and write your answers before you check the *Study Guide*. Notice that Exercise 21(e) is a sentence of the form

"⟨statement 1⟩ if and only if ⟨statement 2⟩"

Mark such a sentence as True if ⟨statement 1⟩ is true whenever ⟨statement 2⟩ is true *and* also ⟨statement 2⟩ is true whenever ⟨statement 1⟩ is true.

23. a. When $b = 0$, $f(x) = mx$. In this case, for all x, y in $\mathbb{R}$ and all scalars c and d,

$$f(cx + dy) = m(cx + dy) = mcx + mdy$$
$$= c(mx) + d(my)$$
$$= c \cdot f(x) + d \cdot f(y)$$

This shows that f is linear.

b. When $f(x) = mx + b$, with b nonzero, $f(0) = m(0) + b = b \neq 0$.

c. In calculus, f is called a "linear function" because the graph of f is a line.

25. *Hint*: Show that the image of a line (that is, the set of images of all points on a line) can be represented by the parametric equation of a line.

27. Any point $\mathbf{x}$ on the plane P satisfies the parametric equation $\mathbf{x} = s\mathbf{u} + t\mathbf{v}$ for some values of s and t. By linearity, the image $T(\mathbf{x})$ satisfies the parametric equation

$$T(\mathbf{x}) = sT(\mathbf{u}) + tT(\mathbf{v}) \quad (s, t \text{ in } \mathbb{R}) \tag{*}$$

The set of images is just Span $\{T(\mathbf{u}), T(\mathbf{v})\}$. If $T(\mathbf{u})$ and $T(\mathbf{v})$ are linearly independent, Span $\{T(\mathbf{u}), T(\mathbf{v})\}$ is a plane through $T(\mathbf{u})$, $T(\mathbf{v})$, and $\mathbf{0}$. If $T(\mathbf{u})$ and $T(\mathbf{v})$ are linearly dependent and not both zero, then Span $\{T(\mathbf{u}), T(\mathbf{v})\}$ is a line through $\mathbf{0}$. If $T(\mathbf{u}) = T(\mathbf{v}) = \mathbf{0}$, then Span $\{T(\mathbf{u}), T(\mathbf{v})\}$ is $\{\mathbf{0}\}$.

29.

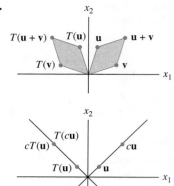

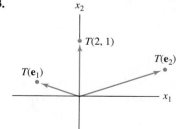

31. *Hint*: Since $\{\mathbf{v}_1, \mathbf{v}_2, \mathbf{v}_3\}$ is linearly dependent, you can write a certain equation and work with it.

33. One possibility is to show that T does not map the zero vector into the zero vector, something that every linear transformation *does* do: $T(0, 0) = (0, -3, 0)$.

35. Take $\mathbf{u}$ and $\mathbf{v}$ in $\mathbb{R}^3$ and let c and d be scalars. Then

$$c\mathbf{u} + d\mathbf{v} = (cu_1 + dv_1, cu_2 + dv_2, cu_3 + dv_3)$$

The transformation T is linear because

$$T(c\mathbf{u} + d\mathbf{v}) = (cu_1 + dv_1, 0, cu_3 + dv_3)$$
$$= (cu_1, 0, cu_3) + (dv_1, 0, dv_3)$$
$$= c(u_1, 0, u_3) + d(v_1, 0, v_3)$$
$$= cT(\mathbf{u}) + dT(\mathbf{v})$$

37. [M] All multiples of $(-1, -1, 1, 0)$

39. [M] Yes. One choice for $\mathbf{x}$ is $(1, 2, 0, 0)$.

Section 1.9, page 78

1. $\begin{bmatrix} 3 & -5 \\ 1 & 2 \\ 3 & 0 \\ 1 & 0 \end{bmatrix}$ **3.** $\begin{bmatrix} 1 & 0 \\ -3 & 1 \end{bmatrix}$ **5.** $\begin{bmatrix} 0 & -1 \\ 1 & 0 \end{bmatrix}$

7. $\begin{bmatrix} -1/\sqrt{2} & 1/\sqrt{2} \\ 1/\sqrt{2} & 1/\sqrt{2} \end{bmatrix}$ **9.** $\begin{bmatrix} 0 & -1 \\ -1 & 0 \end{bmatrix}$

11. The described transformation T maps $\mathbf{e}_1$ into $-\mathbf{e}_1$ and maps $\mathbf{e}_2$ into $-\mathbf{e}_2$. A rotation through π radians also maps $\mathbf{e}_1$ into $-\mathbf{e}_1$ and maps $\mathbf{e}_2$ into $-\mathbf{e}_2$. Since a linear transformation is completely determined by what it does to the columns of the identity matrix, the rotation transformation has the same effect as T on every vector in $\mathbb{R}^2$.

13.

15. $\begin{bmatrix} 2 & -4 & 0 \\ 1 & 0 & -1 \\ 0 & -1 & 3 \end{bmatrix}$

17. $\begin{bmatrix} 1 & 2 & 0 & 0 \\ 0 & 0 & 0 & 0 \\ 0 & 2 & 0 & 1 \\ 0 & 1 & 0 & -1 \end{bmatrix}$

19. $\begin{bmatrix} 1 & -5 & 4 \\ 0 & 1 & -6 \end{bmatrix}$

21. $\mathbf{x} = \begin{bmatrix} 7 \\ -4 \end{bmatrix}$

23. Answer the questions before checking the *Study Guide*.

Justify your answers to Exercises 25–28.

25. Not one-to-one and does not map $\mathbb{R}^4$ onto $\mathbb{R}^4$

27. Not one-to-one but maps $\mathbb{R}^3$ onto $\mathbb{R}^2$

29. $\begin{bmatrix} \blacksquare & * & * \\ 0 & \blacksquare & * \\ 0 & 0 & \blacksquare \\ 0 & 0 & 0 \end{bmatrix}$

31. n. (Explain why, and then check the *Study Guide*.)

33. *Hint:* If $\mathbf{e}_j$ is the jth column of I_n, then $B\mathbf{e}_j$ is the jth column of B.

35. *Hint:* Is it possible that $m > n$? What about $m < n$?

37. [M] No. (Explain why.)

39. [M] No. (Explain why.)

Section 1.10, page 86

1. a. $x_1 \begin{bmatrix} 110 \\ 4 \\ 20 \\ 2 \end{bmatrix} + x_2 \begin{bmatrix} 130 \\ 3 \\ 18 \\ 5 \end{bmatrix} = \begin{bmatrix} 295 \\ 9 \\ 48 \\ 8 \end{bmatrix}$, where x_1 is the number of servings of Cheerios and x_2 is the number of servings of 100% Natural Cereal.

b. $\begin{bmatrix} 110 & 130 \\ 4 & 3 \\ 20 & 18 \\ 2 & 5 \end{bmatrix} \begin{bmatrix} x_1 \\ x_2 \end{bmatrix} = \begin{bmatrix} 295 \\ 9 \\ 48 \\ 8 \end{bmatrix}$. Mix 1.5 servings of Cheerios together with 1 serving of 100% Natural Cereal.

3. a. She should mix .99 serving of Mac and Cheese, 1.54 servings of broccoli, and .79 serving of chicken to get her desired nutritional content.

b. She should mix 1.09 servings of shells and white cheddar, .88 serving of broccoli, and 1.03 servings of chicken to get her desired nutritional content. Notice that this mix contains significantly less broccoli, so she should like it better.

5. $R\mathbf{i} = \mathbf{v}$, $\begin{bmatrix} 11 & -5 & 0 & 0 \\ -5 & 10 & -1 & 0 \\ 0 & -1 & 9 & -2 \\ 0 & 0 & -2 & 10 \end{bmatrix} \begin{bmatrix} I_1 \\ I_2 \\ I_3 \\ I_4 \end{bmatrix} = \begin{bmatrix} 50 \\ -40 \\ 30 \\ -30 \end{bmatrix}$

[M]: $\mathbf{i} = \begin{bmatrix} I_1 \\ I_2 \\ I_3 \\ I_4 \end{bmatrix} = \begin{bmatrix} 3.68 \\ -1.90 \\ 2.57 \\ -2.49 \end{bmatrix}$

7. $R\mathbf{i} = \mathbf{v}$, $\begin{bmatrix} 12 & -7 & 0 & -4 \\ -7 & 15 & -6 & 0 \\ 0 & -6 & 14 & -5 \\ -4 & 0 & -5 & 13 \end{bmatrix} \begin{bmatrix} I_1 \\ I_2 \\ I_3 \\ I_4 \end{bmatrix} = \begin{bmatrix} 40 \\ 30 \\ 20 \\ -10 \end{bmatrix}$

[M]: $\mathbf{i} = \begin{bmatrix} I_1 \\ I_2 \\ I_3 \\ I_4 \end{bmatrix} = \begin{bmatrix} 11.43 \\ 10.55 \\ 8.04 \\ 5.84 \end{bmatrix}$

9. $\mathbf{x}_{k+1} = M\mathbf{x}_k$ for $k = 0, 1, 2, \ldots$, where
$$M = \begin{bmatrix} .93 & .05 \\ .07 & .95 \end{bmatrix} \quad \text{and} \quad \mathbf{x}_0 = \begin{bmatrix} 800{,}000 \\ 500{,}000 \end{bmatrix}.$$
The population in 2012 (for $k = 2$) is $\mathbf{x}_2 = \begin{bmatrix} 741{,}720 \\ 558{,}280 \end{bmatrix}$.

11. a. $M = \begin{bmatrix} .98363 & .00167 \\ .01637 & .99833 \end{bmatrix}$

b. [M] $\mathbf{x}_6 = \begin{bmatrix} 30{,}754{,}500 \\ 229{,}449{,}000 \end{bmatrix}$

13. [M]

a. The population of the city decreases. After 7 years, the populations are about equal, but the city population continues to decline. After 20 years, there are only 417,000 persons in the city (417,456 rounded off). However, the changes in population seem to grow smaller each year.

b. The city population is increasing slowly, and the suburban population is decreasing. After 20 years, the city population has grown from 350,000 to about 370,000.

Chapter 1 Supplementary Exercises, page 88

1. a. F **b.** F **c.** T **d.** F **e.** T **f.** T
g. F **h.** F **i.** T **j.** F **k.** T **l.** F
m. T **n.** T **o.** T **p.** T **q.** F **r.** T
s. F **t.** T **u.** F **v.** F **w.** F **x.** T
y. T **z.** F

3. a. Any consistent linear system whose echelon form is

$\begin{bmatrix} \blacksquare & * & * & * \\ 0 & \blacksquare & * & * \\ 0 & 0 & 0 & 0 \end{bmatrix}$ or $\begin{bmatrix} \blacksquare & * & * & * \\ 0 & 0 & \blacksquare & * \\ 0 & 0 & 0 & 0 \end{bmatrix}$

or $\begin{bmatrix} 0 & \blacksquare & * & * \\ 0 & 0 & \blacksquare & * \\ 0 & 0 & 0 & 0 \end{bmatrix}$

b. Any consistent linear system whose reduced echelon form is I_3.

c. Any inconsistent linear system of three equations in three variables.

5. a. The solution set: (i) is empty if $h = 12$ and $k \neq 2$; (ii) contains a unique soltution if $h \neq 12$; (iii) contains infinitely many solutions if $h = 12$ and $k = 2$.

b. The solution set is empty if $k + 3h = 0$; otherwise, the solution set contains a unique solution.

7. a. Set $\mathbf{v}_1 = \begin{bmatrix} 2 \\ -5 \\ 7 \end{bmatrix}$, $\mathbf{v}_2 = \begin{bmatrix} -4 \\ 1 \\ -5 \end{bmatrix}$, $\mathbf{v}_3 = \begin{bmatrix} -2 \\ 1 \\ -3 \end{bmatrix}$, and

$\mathbf{b} = \begin{bmatrix} b_1 \\ b_2 \\ b_3 \end{bmatrix}$. "Determine if $\mathbf{v}_1, \mathbf{v}_2, \mathbf{v}_3$ span $\mathbb{R}^3$."

Solution: No.

b. Set $A = \begin{bmatrix} 2 & -4 & -2 \\ -5 & 1 & 1 \\ 7 & -5 & -3 \end{bmatrix}$. "Determine if the columns of A span $\mathbb{R}^3$."

c. Define $T(\mathbf{x}) = A\mathbf{x}$. "Determine if T maps $\mathbb{R}^3$ onto $\mathbb{R}^3$."

9. $\begin{bmatrix} 5 \\ 6 \end{bmatrix} = \frac{4}{3}\begin{bmatrix} 2 \\ 1 \end{bmatrix} + \frac{7}{3}\begin{bmatrix} 1 \\ 2 \end{bmatrix}$ or $\begin{bmatrix} 5 \\ 6 \end{bmatrix} = \begin{bmatrix} 8/3 \\ 4/3 \end{bmatrix} + \begin{bmatrix} 7/3 \\ 14/3 \end{bmatrix}$

10. *Hint:* Construct a "grid" on the $x_1 x_2$-plane determined by $\mathbf{a}_1$ and $\mathbf{a}_2$.

11. A solution set is a line when the system has one free variable. If the coefficient matrix is 2×3, then two of the columns should be pivot columns. For instance, take $\begin{bmatrix} 1 & 2 & * \\ 0 & 3 & * \end{bmatrix}$. Put anything in column 3. The resulting matrix will be in echelon form. Make one row replacement operation on the second row to create a matrix *not* in echelon form, such as $\begin{bmatrix} 1 & 2 & 1 \\ 0 & 3 & 1 \end{bmatrix} \sim \begin{bmatrix} 1 & 2 & 1 \\ 1 & 5 & 2 \end{bmatrix}$.

12. *Hint:* How many free variables are in the equation $A\mathbf{x} = \mathbf{0}$?

13. $E = \begin{bmatrix} 1 & 0 & -3 \\ 0 & 1 & 2 \\ 0 & 0 & 0 \end{bmatrix}$

15. a. If the three vectors are linearly independent, then a, c, and f must all be nonzero.

b. The numbers $a, \ldots, f$ can have any values.

16. *Hint:* List the columns from right to left as $\mathbf{v}_1, \ldots, \mathbf{v}_4$.

17. *Hint*: Use Theorem 7.

19. Let M be the line through the origin that is parallel to the line through $\mathbf{v}_1, \mathbf{v}_2,$ and $\mathbf{v}_3$. Then $\mathbf{v}_2 - \mathbf{v}_1$ and $\mathbf{v}_3 - \mathbf{v}_1$ are both on M. So one of these two vectors is a multiple of the other, say $\mathbf{v}_2 - \mathbf{v}_1 = k(\mathbf{v}_3 - \mathbf{v}_1)$. This equation produces a linear dependence relation: $(k-1)\mathbf{v}_1 + \mathbf{v}_2 - k\mathbf{v}_3 = \mathbf{0}$.

A second solution: A parametric equation of the line is $\mathbf{x} = \mathbf{v}_1 + t(\mathbf{v}_2 - \mathbf{v}_1)$. Since $\mathbf{v}_3$ is on the line, there is some t_0 such that $\mathbf{v}_3 = \mathbf{v}_1 + t_0(\mathbf{v}_2 - \mathbf{v}_1) = (1 - t_0)\mathbf{v}_1 + t_0\mathbf{v}_2$. So $\mathbf{v}_3$ is a linear combination of $\mathbf{v}_1$ and $\mathbf{v}_2$, and $\{\mathbf{v}_1, \mathbf{v}_2, \mathbf{v}_3\}$ is linearly dependent.

21. $\begin{bmatrix} 1 & 0 & 0 \\ 0 & -1 & 0 \\ 0 & 0 & 1 \end{bmatrix}$

23. $a = 4/5, b = -3/5$

25. a. The vector lists the numbers of three-, two-, and one-bedroom apartments provided when x_1 floors of plan A are constructed.

b. $x_1\begin{bmatrix} 3 \\ 7 \\ 8 \end{bmatrix} + x_2\begin{bmatrix} 4 \\ 4 \\ 8 \end{bmatrix} + x_3\begin{bmatrix} 5 \\ 3 \\ 9 \end{bmatrix}$

c. **[M]** Use 2 floors of plan A and 15 floors of plan B. Or, use 6 floors of plan A, 2 floors of plan B, and 8 floors of plan C. These are the only feasible solutions. There are other mathematical solutions, but they require a negative number (or a fractional number) of floors of one or two of the plans, which makes no physical sense.

Chapter 2

Section 2.1, page 100

1. $\begin{bmatrix} -4 & 0 & 2 \\ -8 & 10 & -4 \end{bmatrix}, \begin{bmatrix} 3 & -5 & 3 \\ -7 & 6 & -7 \end{bmatrix}$, not defined, $\begin{bmatrix} 1 & 13 \\ -7 & -6 \end{bmatrix}$

3. $\begin{bmatrix} 1 & 5 \\ -3 & 5 \end{bmatrix}, \begin{bmatrix} 6 & -15 \\ 9 & -6 \end{bmatrix}$

5. a. $A\mathbf{b}_1 = \begin{bmatrix} -10 \\ 0 \\ 26 \end{bmatrix}$, $A\mathbf{b}_2 = \begin{bmatrix} 11 \\ 8 \\ -19 \end{bmatrix}$,

$AB = \begin{bmatrix} -10 & 11 \\ 0 & 8 \\ 26 & -19 \end{bmatrix}$

b. $AB = \begin{bmatrix} -1 \cdot 4 + 3(-2) & -1(-2) + 3 \cdot 3 \\ 2 \cdot 4 + 4(-2) & 2(-2) + 4 \cdot 3 \\ 5 \cdot 4 - 3(-2) & 5(-2) - 3 \cdot 3 \end{bmatrix}$

$= \begin{bmatrix} -10 & 11 \\ 0 & 8 \\ 26 & -19 \end{bmatrix}$

7. 3×7 **9.** $k = -2$

11. $AD = \begin{bmatrix} 5 & 6 & 6 \\ 10 & 12 & 10 \\ 15 & 15 & 12 \end{bmatrix}$, $DA = \begin{bmatrix} 5 & 10 & 15 \\ 6 & 12 & 15 \\ 6 & 10 & 12 \end{bmatrix}$

Right-multiplication (that is, multiplication on the right) by D multiplies each *column* of A by the corresponding diagonal entry of D. Left-multiplication by D multiplies each *row* of A by the corresponding diagonal entry of D. The *Study Guide* tells how to make $AB = BA$, but you should try this yourself before looking there.

13. *Hint*: One of the two matrices is Q.

15. Answer the questions before looking in the *Study Guide*.

17. $\mathbf{b}_1 = \begin{bmatrix} 3 \\ 2 \end{bmatrix}, \mathbf{b}_2 = \begin{bmatrix} 1 \\ 4 \end{bmatrix}$

19. The third column of AB is the sum of the first two columns of AB. Here's why. Write $B = [\,\mathbf{b}_1 \quad \mathbf{b}_2 \quad \mathbf{b}_3\,]$. By definition, the third column of AB is $A\mathbf{b}_3$. If $\mathbf{b}_3 = \mathbf{b}_1 + \mathbf{b}_2$, then $A\mathbf{b}_3 = A(\mathbf{b}_1 + \mathbf{b}_2) = A\mathbf{b}_1 + A\mathbf{b}_2$, by a property of matrix-vector multiplication.

21. The columns of A are linearly dependent. Why?

23. *Hint:* Suppose $\mathbf{x}$ satisfies $A\mathbf{x} = \mathbf{0}$, and show that $\mathbf{x}$ must be $\mathbf{0}$.

25. *Hint*: Use the results of Exercises 23 and 24, and apply the associative law of multiplication to the product CAD.

27. $\mathbf{u}^T\mathbf{v} = \mathbf{v}^T\mathbf{u} = -3a + 2b - 5c$,

$$\mathbf{uv}^T = \begin{bmatrix} -3a & -3b & -3c \\ 2a & 2b & 2c \\ -5a & -5b & -5c \end{bmatrix},$$

$$\mathbf{vu}^T = \begin{bmatrix} -3a & 2a & -5a \\ -3b & 2b & -5b \\ -3c & 2c & -5c \end{bmatrix}$$

29. *Hint*: For Theorem 2(b), show that the (i, j)-entry of $A(B + C)$ equals the (i, j)-entry of $AB + AC$.

31. *Hint*: Use the definition of the product $I_m A$ and the fact that $I_m\mathbf{x} = \mathbf{x}$ for $\mathbf{x}$ in $\mathbb{R}^m$.

33. *Hint*: First write the (i, j)-entry of $(AB)^T$, which is the (j, i)-entry of AB. Then, to compute the (i, j)-entry in B^TA^T, use the facts that the entries in row i of B^T are $b_{1i}, \ldots, b_{ni}$, because they come from column i of B, and the entries in column j of A^T are $a_{j1}, \ldots, a_{jn}$, because they come from row j of A.

35. [M] The answer here depends on the choice of matrix program. For MATLAB, use the `help` command to read about `zeros`, `ones`, `eye`, and `diag`. For the TI-86, study the `dim`, `fill`, and `iden` instructions. The TI-86 does not have a "diagonal" command.

37. [M] Display your results and report your conclusions.

39. [M] Display your results and report your conclusions.

41. The matrices appear to approach the matrix
$$\begin{bmatrix} 1/3 & 1/3 & 1/3 \\ 1/3 & 1/3 & 1/3 \\ 1/3 & 1/3 & 1/3 \end{bmatrix}.$$

Section 2.2, page 109

1. $\begin{bmatrix} 2 & -3 \\ -5/2 & 4 \end{bmatrix}$ **3.** $-\dfrac{1}{3}\begin{bmatrix} -3 & -3 \\ 6 & 7 \end{bmatrix}$ or $\begin{bmatrix} 1 & 1 \\ -2 & -7/3 \end{bmatrix}$

5. $x_1 = 7$ and $x_2 = -9$

7. $\mathbf{a}$ and $\mathbf{b}$: $\begin{bmatrix} -9 \\ 4 \end{bmatrix}, \begin{bmatrix} 11 \\ -5 \end{bmatrix}, \begin{bmatrix} 6 \\ -2 \end{bmatrix}$, and $\begin{bmatrix} 13 \\ -5 \end{bmatrix}$

9. Write out your answers before checking the *Study Guide*.

11. The proof can be modeled after the proof of Theorem 5.

13. $AB = AC \Rightarrow A^{-1}AB = A^{-1}AC \Rightarrow IB = IC \Rightarrow B = C$. No, in general, B and C can be different when A is not invertible. See Exercise 10 in Section 2.1.

15. *Hint*: Apply the elementary matrices used to row reduce A to I, to the matrix $[A,\ B]$.

17. $D = C^{-1}B^{-1}A^{-1}$. Show that D works.

19. After you find $X = CB - A$, show that X is a solution.

21. *Hint*: Consider the equation $A\mathbf{x} = \mathbf{0}$.

23. *Hint*: If $A\mathbf{x} = \mathbf{0}$ has only the trivial solution, then there are no free variables in the equation $A\mathbf{x} = \mathbf{0}$, and each column of A is a pivot column.

25. *Hint*: Consider the case $a = b = 0$. Then consider the vector $\begin{bmatrix} -b \\ a \end{bmatrix}$, and use the fact that $ad - bc = 0$.

27. *Hint*: For part (a), interchange A and B in the box following Example 6 in Section 2.1, and then replace B by the identity matrix. For parts (b) and (c), begin by writing
$$A = \begin{bmatrix} \text{row}_1(A) \\ \text{row}_2(A) \\ \text{row}_3(A) \end{bmatrix}$$

29. $\dfrac{1}{3}\begin{bmatrix} -9 & 3 \\ -4 & 1 \end{bmatrix}$ **31.** $\begin{bmatrix} 8 & 3 & 1 \\ 10 & 4 & 1 \\ 7/2 & 3/2 & 1/2 \end{bmatrix}$

33. The general form of A^{-1} is
$$A^{-1} = B = \begin{bmatrix} 1 & 0 & 0 & \cdots & 0 \\ -1 & 1 & 0 & & 0 \\ 0 & -1 & 1 & & \\ \vdots & & & \ddots & \vdots \\ 0 & 0 & \cdots & -1 & 1 \end{bmatrix}.$$

Hint: For $j = 1, \ldots, n$, let $\mathbf{a}_j$, $\mathbf{b}_j$, and $\mathbf{e}_j$ denote the jth columns of A, B, and I, respectively. Use the facts that $\mathbf{a}_j - \mathbf{a}_{j+1} = \mathbf{e}_j$ and $\mathbf{b}_j = \mathbf{e}_j - \mathbf{e}_{j+1}$ for $j = 1, \ldots, n-1$, and $\mathbf{a}_n = \mathbf{b}_n = \mathbf{e}_n$.

35. $\begin{bmatrix} 3 \\ 0 \\ -1 \end{bmatrix}$. Find this by row reducing $[A\ \ \mathbf{e}_3]$.

37. $C = \begin{bmatrix} 1 & 1 & -1 \\ -1 & 1 & 0 \end{bmatrix}$

39. [M] The deflections are .62, .66, and .52 inches, respectively.

41. [M] .95, 6.19, 11.43, and 3.81 newtons, respectively

Section 2.3, page 115
The abbreviation IMT (here and in the *Study Guide*) denotes the Invertible Matrix Theorem (Theorem 8).

1. Invertible, by the IMT. Neither column of the matrix is a multiple of the other column, so they are linearly independent. Also, the matrix is invertible by Theorem 4 in Section 2.2 because the determinant is nonzero.

3. Notice that A^T has a pivot in every column, so by IMT, A^T is invertible. Hence by IMT, A is also invertible.

5. Not invertible, by the IMT. Since this matrix has a column of zeros, its columns form a linearly dependent set and hence the matrix is not invertible.

7. Invertible, by the IMT. The matrix row reduces to

$$\begin{bmatrix} -1 & -3 & 0 & 1 \\ 0 & -4 & 8 & 0 \\ 0 & 0 & 3 & 0 \\ 0 & 0 & 0 & 1 \end{bmatrix}$$

and has four pivot positions.

9. [M] The 4×4 matrix has four pivot positions, so it is invertible by the IMT.

11. The *Study Guide* will help, but first try to answer the questions based on your careful reading of the text.

13. A square upper triangular matrix is invertible if and only if all the entries on the diagonal are nonzero. Why?

Note: The answers below for Exercises 15–29 mention the IMT. In many cases, part or all of an acceptable answer could also be based on results that were used to establish the IMT.

15. No, because statement (h) of the IMT is then false. A 4×4 matrix cannot be invertible when its columns do not span $\mathbb{R}^4$.

17. If A has two identical columns, then its columns are linearly dependent. Part (e) of the IMT shows that A cannot be invertible.

19. By statement (e) of the IMT, D is invertible. Thus the equation $D\mathbf{x} = \mathbf{b}$ has a solution for each $\mathbf{b}$ in $\mathbb{R}^7$, by statement (g) of the IMT. Can you say more?

21. The matrix C cannot be invertible, by Theorem 5 in Section 2.2 or by the paragraph following the IMT. So statement (g) of the IMT is false and so is (h). The columns of C do not span $\mathbb{R}^n$.

23. Statement (g) of the IMT is false for F, so statement (d) is false, too. That is, the equation $F\mathbf{x} = \mathbf{0}$ has a nontrivial solution.

25. *Hint:* Use the IMT first.

27. Let W be the inverse of AB. Then $ABW = I$ and $A(BW) = I$. Unfortunately, this equation by itself does not prove that A is invertible. Why not? Finish the proof before you check the *Study Guide*.

29. Since the transformation $\mathbf{x} \mapsto A\mathbf{x}$ is one-to-one, statement (f) of the IMT is true. Then statement (i) is also true and the transformation $\mathbf{x} \mapsto A\mathbf{x}$ maps $\mathbb{R}^n$ onto $\mathbb{R}^n$. Also, A is invertible, which implies that the transformation $\mathbf{x} \mapsto A\mathbf{x}$ is invertible, by Theorem 9.

31. *Hint:* If the equation $A\mathbf{x} = \mathbf{b}$ has a solution for each $\mathbf{b}$, then A has a pivot in each row (Theorem 4 in Section 1.4). Could there be free variables in an equation $A\mathbf{x} = \mathbf{b}$?

33. *Hint:* First show that the standard matrix of T is invertible. Then use a theorem or theorems to show that $T^{-1}(\mathbf{x}) = B\mathbf{x}$, where $B = \begin{bmatrix} 7 & 9 \\ 4 & 5 \end{bmatrix}$.

35. *Hint:* To show that T is one-to-one, suppose that $T(\mathbf{u}) = T(\mathbf{v})$ for some vectors $\mathbf{u}$ and $\mathbf{v}$ in $\mathbb{R}^n$. Deduce that $\mathbf{u} = \mathbf{v}$. To show that T is onto, suppose $\mathbf{y}$ represents an arbitrary vector in $\mathbb{R}^n$ and use the inverse S to produce an $\mathbf{x}$ such that $T(\mathbf{x}) = \mathbf{y}$. A second proof can be given using Theorem 9 together with a theorem from Section 1.9.

37. *Hint:* Consider the standard matrices of T and U.

39. If T maps $\mathbb{R}^n$ onto $\mathbb{R}^n$, then the columns of its standard matrix A span $\mathbb{R}^n$, by Theorem 12 in Section 1.9. By the IMT, A is invertible. Hence, by Theorem 9, T is invertible, and A^{-1} is the standard matrix of T^{-1}. Since A^{-1} is also invertible, by the IMT, its columns are linearly independent and span $\mathbb{R}^n$. Applying Theorem 12 in Section 1.9 to the transformation T^{-1} shows that T^{-1} is a one-to-one mapping of $\mathbb{R}^n$ onto $\mathbb{R}^n$.

41. [M]
   a. The exact solution of (3) is $x_1 = 3.94$ and $x_2 = .49$. The exact solution of review (4) is $x_1 = 2.90$ and $x_2 = 2.00$.
   b. When the solution of (4) is used as an approximation for the solution in (3), the error in using the value of 2.90 for x_1 is about 26%, and the error in using 2.0 for x_2 is about 308%.
   c. The condition number of the coefficient matrix is 3363. The percentage change in the solution from (3) to (4) is about 7700 times the percentage change in the right side of the equation. This is the same order of magnitude as the condition number. The condition number gives a rough measure of how sensitive the solution of $A\mathbf{x} = \mathbf{b}$ can be to changes in $\mathbf{b}$. Further information about the condition number is given at the end of Chapter 6 and in Chapter 7.

43. [M] cond$(A) \approx 69{,}000$, which is between 10^4 and 10^5. So about 4 or 5 digits of accuracy may be lost. Several experiments with MATLAB should verify that $\mathbf{x}$ and $\mathbf{x}_1$ agree to 11 or 12 digits.

45. [M] Some versions of MATLAB issue a warning when asked to invert a Hilbert matrix of order about 12 or larger using floating-point arithmetic. The product AA^{-1} should have several off-diagonal entries that are far from being zero. If not, try a larger matrix.

Section 2.4, page 121

1. $\begin{bmatrix} A & B \\ EA + C & EB + D \end{bmatrix}$ **3.** $\begin{bmatrix} C & D \\ A & B \end{bmatrix}$

5. $Y = B^{-1}$ (explain why), $X = -B^{-1}A$, $Z = C$

7. $X = A^{-1}$ (why?), $Y = -BA^{-1}$, $Z = 0$ (why?)

9. $A_{21} = -B_{21}B_{11}^{-1}$, $A_{31} = -B_{31}B_{11}^{-1}$, $C_{22} = B_{22} - B_{21}B_{11}^{-1}B_{12}$

11. You can check your answers in the *Study Guide*.

13. *Hint:* Suppose A is invertible, and let $A^{-1} = \begin{bmatrix} D & E \\ F & G \end{bmatrix}$. Show that $BD = I$ and $CG = I$. This implies that B and

C are invertible. (Explain why!) Conversely, suppose B and C are invertible. To prove that A is invertible, guess what A^{-1} must be and check that it works.

15. $G_{k+1} = \begin{bmatrix} X_k & \mathbf{x}_{k+1} \end{bmatrix} \begin{bmatrix} X_k^T \\ \mathbf{x}_{k+1}^T \end{bmatrix} = X_k X_k^T + \mathbf{x}_{k+1}\mathbf{x}_{k+1}^T$
$= G_k + \mathbf{x}_{k+1}\mathbf{x}_{k+1}^T$
Only the outer product matrix $\mathbf{x}_{k+1}\mathbf{x}_{k+1}^T$ needs to be computed (and then added to G_k).

17. The inverse of $\begin{bmatrix} I & 0 \\ X & I \end{bmatrix}$ is $\begin{bmatrix} I & 0 \\ -X & I \end{bmatrix}$. Similarly, $\begin{bmatrix} I & Y \\ 0 & I \end{bmatrix}$ has an inverse. From equation (7), one obtains

$\begin{bmatrix} I & 0 \\ -X & I \end{bmatrix} \begin{bmatrix} A_{11} & A_{12} \\ A_{21} & A_{22} \end{bmatrix} \begin{bmatrix} I & -Y \\ 0 & I \end{bmatrix} = \begin{bmatrix} A_{11} & 0 \\ 0 & S \end{bmatrix}$ (*)

If A is invertible, then the matrix on the right side of (*) is a product of invertible matrices and hence is invertible. By Exercise 13, A_{11} and S must be invertible.

19. $W(s) = I_m - C(A - sI_n)^{-1}B$. This is the Schur complement of $A - sI_n$ in the system matrix.

21. a. $A^2 = \begin{bmatrix} 1 & 0 \\ 2 & -1 \end{bmatrix} \begin{bmatrix} 1 & 0 \\ 2 & -1 \end{bmatrix}$

$= \begin{bmatrix} 1+0 & 0+0 \\ 2-2 & 0+(-1)^2 \end{bmatrix} = \begin{bmatrix} 1 & 0 \\ 0 & 1 \end{bmatrix}$

b. $M^2 = \begin{bmatrix} A & 0 \\ I & -A \end{bmatrix} \begin{bmatrix} A & 0 \\ I & -A \end{bmatrix}$

$= \begin{bmatrix} A^2+0 & 0+0 \\ A-A & 0+(-A)^2 \end{bmatrix} = \begin{bmatrix} I & 0 \\ 0 & I \end{bmatrix}$

23. If A_1 and B_1 are $(k+1) \times (k+1)$ and lower triangular, then write $A_1 = \begin{bmatrix} a & \mathbf{0}^T \\ \mathbf{v} & A \end{bmatrix}$ and $B_1 = \begin{bmatrix} b & \mathbf{0}^T \\ \mathbf{w} & B \end{bmatrix}$, where A and B are $k \times k$ and lower triangular, $\mathbf{v}$ and $\mathbf{w}$ are in $\mathbb{R}^k$, and a and b are suitable scalars. Assume that the product of $k \times k$ lower triangular matrices is lower triangular, and compute the product $A_1 B_1$. What do you conclude?

25. Use Exercise 13 to find the inverse of a matrix of the form $B = \begin{bmatrix} B_{11} & 0 \\ 0 & B_{22} \end{bmatrix}$, where B_{11} is $p \times p$, B_{22} is $q \times q$, and B is invertible. Partition the matrix A, and apply your result twice to find that

$A^{-1} = \begin{bmatrix} -5 & 2 & 0 & 0 & 0 \\ 3 & -1 & 0 & 0 & 0 \\ 0 & 0 & 1/2 & 0 & 0 \\ 0 & 0 & 0 & 3 & -4 \\ 0 & 0 & 0 & -5/2 & 7/2 \end{bmatrix}$

27. a., b. The commands to be used in these exercises will depend on the matrix program.

c. The algebra needed comes from the block matrix equation

$\begin{bmatrix} A_{11} & 0 \\ A_{21} & A_{22} \end{bmatrix} \begin{bmatrix} \mathbf{x}_1 \\ \mathbf{x}_2 \end{bmatrix} = \begin{bmatrix} \mathbf{b}_1 \\ \mathbf{b}_2 \end{bmatrix}$

where $\mathbf{x}_1$ and $\mathbf{b}_1$ are in $\mathbb{R}^{20}$ and $\mathbf{x}_2$ and $\mathbf{b}_2$ are in $\mathbb{R}^{30}$. Then $A_{11}\mathbf{x}_1 = \mathbf{b}_1$, which can be solved to produce $\mathbf{x}_1$.

The equation $A_{21}\mathbf{x}_1 + A_{22}\mathbf{x}_2 = \mathbf{b}_2$ yields $A_{22}\mathbf{x}_2 = \mathbf{b}_2 - A_{21}\mathbf{x}_1$, which can be solved for $\mathbf{x}_2$ by row reducing the matrix $[A_{22} \quad \mathbf{c}]$, where $\mathbf{c} = \mathbf{b}_2 - A_{21}\mathbf{x}_1$.

Section 2.5, page 129

1. $L\mathbf{y} = \mathbf{b} \Rightarrow \mathbf{y} = \begin{bmatrix} -7 \\ -2 \\ 6 \end{bmatrix}$, $U\mathbf{x} = \mathbf{y} \Rightarrow \mathbf{x} = \begin{bmatrix} 3 \\ 4 \\ -6 \end{bmatrix}$

3. $\mathbf{y} = \begin{bmatrix} 6 \\ 12 \\ 0 \end{bmatrix}$, $\mathbf{x} = \begin{bmatrix} -5 \\ -4 \\ 0 \end{bmatrix}$ **5.** $\mathbf{y} = \begin{bmatrix} 1 \\ 3 \\ 1 \\ -4 \end{bmatrix}$, $\mathbf{x} = \begin{bmatrix} 38 \\ 16 \\ 8.5 \\ -4 \end{bmatrix}$

7. $LU = \begin{bmatrix} 1 & 0 \\ -3/2 & 1 \end{bmatrix} \begin{bmatrix} 2 & 5 \\ 0 & 7/2 \end{bmatrix}$

9. $\begin{bmatrix} 1 & 0 & 0 \\ -3 & 1 & 0 \\ 3 & 2 & 1 \end{bmatrix} \begin{bmatrix} 3 & 1 & 2 \\ 0 & 3 & 2 \\ 0 & 0 & 4 \end{bmatrix}$

11. $\begin{bmatrix} 1 & 0 & 0 \\ 2 & 1 & 0 \\ -1 & 1 & 1 \end{bmatrix} \begin{bmatrix} 3 & 7 & 2 \\ 0 & 5 & 0 \\ 0 & 0 & 5 \end{bmatrix}$

13. $\begin{bmatrix} 1 & 0 & 0 & 0 \\ -1 & 1 & 0 & 0 \\ 4 & 5 & 1 & 0 \\ -2 & -1 & 0 & 1 \end{bmatrix} \begin{bmatrix} 1 & 3 & -5 & -3 \\ 0 & -2 & 3 & 1 \\ 0 & 0 & 0 & 0 \\ 0 & 0 & 0 & 0 \end{bmatrix}$

15. $\begin{bmatrix} 1 & 0 & 0 \\ -3 & 1 & 0 \\ 2 & 3 & 1 \end{bmatrix} \begin{bmatrix} 2 & 0 & 5 & 2 \\ 0 & 3 & 2 & 3 \\ 0 & 0 & 0 & 4 \end{bmatrix}$

17. $U^{-1} = \begin{bmatrix} 1/2 & -3/4 & -2 \\ 0 & -1/4 & -1 \\ 0 & 0 & -1/2 \end{bmatrix}$,

$L^{-1} = \begin{bmatrix} 1 & 0 & 0 \\ 2 & 1 & 0 \\ -2 & -1 & 1 \end{bmatrix}$,

$A^{-1} = \begin{bmatrix} 3 & 5/4 & -2 \\ 3/2 & 3/4 & -1 \\ 1 & 1/2 & -1/2 \end{bmatrix}$

19. *Hint:* Think about row reducing $[\, A \quad I \,]$.

21. *Hint:* Represent the row operations by a sequence of elementary matrices.

23. a. Denote the rows of D as transposes of column vectors. Then partitioned matrix multiplication yields

$A = CD = \begin{bmatrix} \mathbf{c}_1 & \cdots & \mathbf{c}_4 \end{bmatrix} \begin{bmatrix} \mathbf{v}_1^T \\ \vdots \\ \mathbf{v}_4^T \end{bmatrix}$

$= \mathbf{c}_1\mathbf{v}_1^T + \cdots + \mathbf{c}_4\mathbf{v}_4^T$

b. A has 40,000 entries. Since C has 1600 entries and D has 400 entries, together they occupy only 5% of the memory needed to store A.

25. Explain why U, D, and V^T are invertible. Then use a theorem on the inverse of a product of invertible matrices.

27. a.

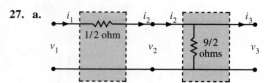

b.

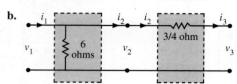

29. a. $\begin{bmatrix} 1 + R_3/R_2 & -R_1 - R_3 - (R_1R_3)/R_2 \\ -1/R_2 & 1 + R_1/R_2 \end{bmatrix}$

b. $A = \begin{bmatrix} 3 & -12 \\ -1/3 & 5/3 \end{bmatrix}$

$= \begin{bmatrix} 1 & -6 \\ 0 & 1 \end{bmatrix} \begin{bmatrix} 1 & 0 \\ -1/3 & 1 \end{bmatrix} \begin{bmatrix} 1 & -2 \\ 0 & 1 \end{bmatrix}$

Set $R_1 = 2$ ohms, $R_2 = 3$ ohms, and $R_3 = 6$ ohms

31. [M]

a. $L = \begin{bmatrix} 1 & 0 & 0 & 0 & 0 & 0 & 0 & 0 \\ -.25 & 1 & 0 & 0 & 0 & 0 & 0 & 0 \\ -.25 & -.0667 & 1 & 0 & 0 & 0 & 0 & 0 \\ 0 & -.2667 & -.2857 & 1 & 0 & 0 & 0 & 0 \\ 0 & 0 & -.2679 & -.0833 & 1 & 0 & 0 & 0 \\ 0 & 0 & 0 & -.2917 & -.2921 & 1 & 0 & 0 \\ 0 & 0 & 0 & 0 & -.2697 & -.0861 & 1 & 0 \\ 0 & 0 & 0 & 0 & 0 & -.2948 & -.2931 & 1 \end{bmatrix}$

$U = \begin{bmatrix} 4 & -1 & -1 & 0 & 0 & 0 & 0 & 0 \\ 0 & 3.75 & -.25 & -1 & 0 & 0 & 0 & 0 \\ 0 & 0 & 3.7333 & -1.0667 & -1 & 0 & 0 & 0 \\ 0 & 0 & 0 & 3.4286 & -.2857 & -1 & 0 & 0 \\ 0 & 0 & 0 & 0 & 3.7083 & -1.0833 & -1 & 0 \\ 0 & 0 & 0 & 0 & 0 & 3.3919 & -.2921 & -1 \\ 0 & 0 & 0 & 0 & 0 & 0 & 3.7052 & -1.0861 \\ 0 & 0 & 0 & 0 & 0 & 0 & 0 & 3.3868 \end{bmatrix}$

b. $\mathbf{x} =$
$(3.9569, 6.5885, 4.2392, 7.3971, 5.6029, 8.7608, 9.4115, 12.0431)$

c. $A^{-1} = \begin{bmatrix} .2953 & .0866 & .0945 & .0509 & .0318 & .0227 & .0100 & .0082 \\ .0866 & .2953 & .0509 & .0945 & .0227 & .0318 & .0082 & .0100 \\ .0945 & .0509 & .3271 & .1093 & .1045 & .0591 & .0318 & .0227 \\ .0509 & .0945 & .1093 & .3271 & .0591 & .1045 & .0227 & .0318 \\ .0318 & .0227 & .1045 & .0591 & .3271 & .1093 & .0945 & .0509 \\ .0227 & .0318 & .0591 & .1045 & .1093 & .3271 & .0509 & .0945 \\ .0100 & .0082 & .0318 & .0227 & .0945 & .0509 & .2953 & .0866 \\ .0082 & .0100 & .0227 & .0318 & .0509 & .0945 & .0866 & .2953 \end{bmatrix}$

Obtain A^{-1} directly and then compute $A^{-1} - U^{-1}L^{-1}$ to compare the two methods for inverting a matrix.

Section 2.6, page 136

1. $C = \begin{bmatrix} .10 & .60 & .60 \\ .30 & .20 & 0 \\ .30 & .10 & .10 \end{bmatrix}$, $\left\{ \begin{matrix} \text{intermediate} \\ \text{demand} \end{matrix} \right\} = \begin{bmatrix} 60 \\ 20 \\ 10 \end{bmatrix}$

3. $\mathbf{x} = \begin{bmatrix} 44.44 \\ 16.67 \\ 16.67 \end{bmatrix}$ **5.** $\mathbf{x} = \begin{bmatrix} 110 \\ 120 \end{bmatrix}$

7. a. $\begin{bmatrix} 1.6 \\ 1.2 \end{bmatrix}$ **b.** $\begin{bmatrix} 111.6 \\ 121.2 \end{bmatrix}$

c. *Hint:* Find a formula involving $(I - C)^{-1}$. See the *Study Guide*.

9. $\mathbf{x} = \begin{bmatrix} 82.8 \\ 131.0 \\ 110.3 \end{bmatrix}$

11. *Hint:* Use properties of transposes to obtain $\mathbf{p}^T = \mathbf{p}^T C + \mathbf{v}^T$, so that $\mathbf{p}^T\mathbf{x} = (\mathbf{p}^T C + \mathbf{v}^T)\mathbf{x} = \mathbf{p}^T C\mathbf{x} + \mathbf{v}^T\mathbf{x}$. Now compute $\mathbf{p}^T\mathbf{x}$ from the production equation.

13. [M] $\mathbf{x} =$
$(99576, 97703, 51231, 131570, 49488, 329554, 13835)$.
The entries in $\mathbf{x}$ suggest more precision in the answer than is warranted by the entries in $\mathbf{d}$, which appear to be accurate only to perhaps the nearest thousand. So a more realistic answer for $\mathbf{x}$ might be
$\mathbf{x} = 1000 \times (100, 98, 51, 132, 49, 330, 14)$.

15. [M] $\mathbf{x}^{(12)}$ is the first vector whose entries are accurate to the nearest thousand. The calculation of $\mathbf{x}^{(12)}$ takes about 1260 flops, while row reduction of $[(I - C)\ \mathbf{d}]$ takes only about 550 flops. If C is larger than 20×20, then fewer flops are needed to compute $\mathbf{x}^{(12)}$ by iteration than to compute the equilibrium vector $\mathbf{x}$ by row reduction. As the size of C grows, the advantage of the iterative method increases. Also, because C becomes more sparse for larger models of the economy, fewer iterations are needed for reasonable accuracy.

Section 2.7, page 144

1. $\begin{bmatrix} 1 & .25 & 0 \\ 0 & 1 & 0 \\ 0 & 0 & 1 \end{bmatrix}$ **3.** $\begin{bmatrix} 0 & -1 & -1 \\ 1 & 0 & 2 \\ 0 & 0 & 1 \end{bmatrix}$

5. $\begin{bmatrix} 1/\sqrt{2} & 1/\sqrt{2} & 0 \\ 1/\sqrt{2} & -1/\sqrt{2} & 0 \\ 0 & 0 & 1 \end{bmatrix}$

7. $\begin{bmatrix} 1/2 & -\sqrt{3}/2 & 3 + 4\sqrt{3} \\ \sqrt{3}/2 & 1/2 & 4 - 3\sqrt{3} \\ 0 & 0 & 1 \end{bmatrix}$
See the Practice Problem.

9. $A(BD)$ requires 800 multiplications. $(AB)D$ requires 408 multiplications. The first method uses about twice as many multiplications. If D had 10,000 columns, the counts would be 80,000 and 40,008, respectively.

11. Use the fact that
$$\sec \varphi - \tan \varphi \sin \varphi = \frac{1}{\cos \varphi} - \frac{\sin^2 \varphi}{\cos \varphi} = \cos \varphi$$

13. $\begin{bmatrix} A & \mathbf{p} \\ \mathbf{0}^T & 1 \end{bmatrix} = \begin{bmatrix} I & \mathbf{p} \\ \mathbf{0}^T & 1 \end{bmatrix} \begin{bmatrix} A & 0 \\ \mathbf{0}^T & 1 \end{bmatrix}$. First apply the linear transformation A, and then translate by $\mathbf{p}$.

15. $(12, -6, -3)$ **17.** $\begin{bmatrix} 1 & 0 & 0 & 0 \\ 0 & 1/2 & -\sqrt{3}/2 & 0 \\ 0 & \sqrt{3}/2 & 1/2 & 0 \\ 0 & 0 & 0 & 1 \end{bmatrix}$

19. The triangle with vertices at $(7, 2, 0)$, $(7.5, 5, 0)$, and $(5, 5, 0)$

21. [M] $\begin{bmatrix} 2.2586 & -1.0395 & -.3473 \\ -1.3495 & 2.3441 & .0696 \\ .0910 & -.3046 & 1.2777 \end{bmatrix} \begin{bmatrix} X \\ Y \\ Z \end{bmatrix} = \begin{bmatrix} R \\ G \\ B \end{bmatrix}$

Section 2.8, page 151

1. The set is closed under sums but not under multiplication by a negative scalar. (Sketch an example.)

3. The set is not closed under sums or scalar multiples.

5. Yes. The system corresponding to $\begin{bmatrix} \mathbf{v}_1 & \mathbf{v}_2 & \mathbf{w} \end{bmatrix}$ is consistent.

7. a. The three vectors $\mathbf{v}_1$, $\mathbf{v}_2$, and $\mathbf{v}_3$
   b. Infinitely many vectors
   c. Yes, because $A\mathbf{x} = \mathbf{p}$ has a solution.

9. No, because $A\mathbf{p} \neq \mathbf{0}$.

11. $p = 4$ and $q = 3$. Nul A is a subspace of $\mathbb{R}^4$ because solutions of $A\mathbf{x} = \mathbf{0}$ must have four entries, to match the columns of A. Col A is a subspace of $\mathbb{R}^3$ because each column vector has three entries.

13. For Nul A, choose $(1, -2, 1, 0)$ or $(-1, 4, 0, 1)$, for example. For Col A, select any column of A.

15. Let A be the matrix whose columns are the vectors given. Then A is invertible because its determinant is nonzero, and so its columns form a basis for $\mathbb{R}^2$, by the IMT (or by Example 5). (Other reasons for the invertibility of A could be given.)

17. Let A be the matrix whose columns are the vectors given. Row reduction shows three pivots, so A is invertible. By the IMT, the columns of A form a basis for $\mathbb{R}^3$.

19. Let A be the 3×2 matrix whose columns are the vectors given. The columns of A cannot possibly span $\mathbb{R}^3$ because A cannot have a pivot in every row. So the columns are not a basis for $\mathbb{R}^3$. (They are a basis for a plane in $\mathbb{R}^3$.)

21. Read the section carefully, and write your answers before checking the *Study Guide*. This section has terms and key concepts that you must learn now before going on.

23. Basis for Col A: $\begin{bmatrix} 4 \\ 6 \\ 3 \end{bmatrix}, \begin{bmatrix} 5 \\ 5 \\ 4 \end{bmatrix}$

Basis for Nul A: $\begin{bmatrix} 4 \\ -5 \\ 1 \\ 0 \end{bmatrix}, \begin{bmatrix} -7 \\ 6 \\ 0 \\ 1 \end{bmatrix}$

25. Basis for Col A: $\begin{bmatrix} 1 \\ -1 \\ -2 \\ 3 \end{bmatrix}, \begin{bmatrix} 4 \\ 2 \\ 2 \\ 6 \end{bmatrix}, \begin{bmatrix} -3 \\ 3 \\ 5 \\ -5 \end{bmatrix}$

Basis for Nul A: $\begin{bmatrix} 2 \\ -2.5 \\ 1 \\ 0 \\ 0 \end{bmatrix}, \begin{bmatrix} -7 \\ .5 \\ 0 \\ -4 \\ 1 \end{bmatrix}$

27. Construct a nonzero 3×3 matrix A, and construct $\mathbf{b}$ to be almost any convenient linear combination of the columns of A.

29. *Hint:* You need a nonzero matrix whose columns are linearly dependent.

31. If Col $F \neq \mathbb{R}^5$, then the columns of F do not span $\mathbb{R}^5$. Since F is square, the IMT shows that F is not invertible and the equation $F\mathbf{x} = \mathbf{0}$ has a nontrivial solution. That is, Nul F contains a nonzero vector. Another way to describe this is to write Nul $F \neq \{\mathbf{0}\}$.

33. If Nul $C = \{\mathbf{0}\}$, then the equation $C\mathbf{x} = \mathbf{0}$ has only the trivial solution. Since C is square, the IMT shows that C is invertible and the equation $C\mathbf{x} = \mathbf{b}$ has a solution for each $\mathbf{b}$ in $\mathbb{R}^6$. Also, each solution is unique, by Theorem 5 in Section 2.2.

35. If Nul B contains nonzero vectors, then the equation $B\mathbf{x} = \mathbf{0}$ has nontrivial solutions. Since B is square, the IMT shows that B is not invertible and the columns of B do not span $\mathbb{R}^5$. So Col B is a subspace of $\mathbb{R}^5$, but Col $B \neq \mathbb{R}^5$.

37. [M] Display the reduced echelon form of A, and select the pivot columns of A as a basis for Col A. For Nul A, write the solution of $A\mathbf{x} = \mathbf{0}$ in parametric vector form.

Basis for Col A: $\begin{bmatrix} 3 \\ -7 \\ -5 \\ 3 \end{bmatrix}, \begin{bmatrix} -5 \\ 9 \\ 7 \\ -7 \end{bmatrix}$

Basis for Nul A: $\begin{bmatrix} -2.5 \\ -1.5 \\ 1 \\ 0 \\ 0 \end{bmatrix}, \begin{bmatrix} 4.5 \\ 2.5 \\ 0 \\ 1 \\ 0 \end{bmatrix}, \begin{bmatrix} -3.5 \\ -1.5 \\ 0 \\ 0 \\ 1 \end{bmatrix}$

Section 2.9, page 157

1. $\mathbf{x} = 3\mathbf{b}_1 + 2\mathbf{b}_2 = 3\begin{bmatrix} 1 \\ 1 \end{bmatrix} + 2\begin{bmatrix} 2 \\ -1 \end{bmatrix} = \begin{bmatrix} 7 \\ 1 \end{bmatrix}$

3. $\begin{bmatrix} 1 \\ 2 \end{bmatrix}$ **5.** $\begin{bmatrix} 4 \\ 1 \end{bmatrix}$ **7.** $[\mathbf{w}]_{\mathcal{B}} = \begin{bmatrix} 2 \\ -1 \end{bmatrix}$, $[\mathbf{x}]_{\mathcal{B}} = \begin{bmatrix} 1.5 \\ .5 \end{bmatrix}$

9. Basis for Col A: $\begin{bmatrix} 1 \\ 3 \\ 2 \\ 5 \end{bmatrix}, \begin{bmatrix} 2 \\ 1 \\ -1 \\ 0 \end{bmatrix}, \begin{bmatrix} -6 \\ 5 \\ 9 \\ 14 \end{bmatrix}$; dim Col $A = 3$

Basis for Nul A: $\begin{bmatrix} -3 \\ 1 \\ 0 \\ 0 \end{bmatrix}$; $\dim \text{Nul } A = 1$

11. Basis for Col A: $\begin{bmatrix} 2 \\ 3 \\ 0 \\ -3 \end{bmatrix}, \begin{bmatrix} -5 \\ -8 \\ 9 \\ -7 \end{bmatrix}$; $\dim \text{Col } A = 2$

Basis for Nul A: $\begin{bmatrix} -2 \\ 1 \\ 0 \\ 0 \\ 0 \end{bmatrix}, \begin{bmatrix} -1 \\ 0 \\ 0 \\ 1 \\ 0 \end{bmatrix}, \begin{bmatrix} -1 \\ 0 \\ -1 \\ 0 \\ 1 \end{bmatrix}$; $\dim \text{Nul } A = 3$

13. The vectors $\mathbf{v}_1$, $\mathbf{v}_3$, and $\mathbf{v}_4$ form a basis for the given subspace, H. So, dimension $H = 3$.

15. Col $A = \mathbb{R}^4$, because A has a pivot in each row, and so the columns of A span $\mathbb{R}^4$. Nul A *cannot* equal $\mathbb{R}^2$, because Nul A is a subspace of $\mathbb{R}^6$. It is true, however, that Nul A is two-dimensional. Reason: The equation $A\mathbf{x} = \mathbf{0}$ has two free variables, because A has six columns and only four of them are pivot columns.

17. See the *Study Guide* after you write your justifications.

19. The fact that the solution space of $A\mathbf{x} = \mathbf{0}$ has a basis of three vectors means that $\dim \text{Nul } A = 3$. Since a 5×7 matrix A has seven columns, the Rank Theorem shows that rank $A = 7 - \dim \text{Nul } A = 4$. See the *Study Guide* for a justification that does not explicitly mention the Rank Theorem.

21. A 9×8 matrix has eight columns. By the Rank Theorem, $\dim \text{Nul } A = 8 - \text{rank } A$. Since the rank is 7, $\dim \text{Nul } A = 1$. That is, the dimension of the solution space of $A\mathbf{x} = \mathbf{0}$ is 1.

23. Create a 3×5 matrix A with two pivot columns. The remaining three columns will correspond to free variables in the equation $A\mathbf{x} = \mathbf{0}$. So the desired construction is possible.

25. The p columns of A span Col A by definition. If $\dim \text{Col } A = p$, then the spanning set of p columns is automatically a basis for Col A, by the Basis Theorem. In particular, the columns are linearly independent.

27. **a.** *Hint:* The columns of B span W, and each vector $\mathbf{a}_j$ is in W. The vector $\mathbf{c}_j$ is in $\mathbb{R}^p$ because B has p columns.
  b. *Hint:* What is the size of C?
  c. *Hint:* How are B and C related to A?

29. **[M]** Your calculations should show that the matrix $[\,\mathbf{v}_1 \ \ \mathbf{v}_2 \ \ \mathbf{x}\,]$ corresponds to a consistent system. The $\mathcal{B}$-coordinate vector of $\mathbf{x}$ is $(2, -1)$.

Chapter 2 Supplementary Exercises, page 160

1. a. T **b.** F **c.** T **d.** F **e.** F **f.** F
  g. T **h.** T **i.** T **j.** F **k.** T **l.** F

m. F **n.** T **o.** F **p.** T

3. I

5. $A^2 = 2A - I$. Multiply by A: $A^3 = 2A^2 - A$.
Substitute $A^2 = 2A - I$:
$A^3 = 2(2A - I) - A = 3A - 2I$.
Multiply by A again: $A^4 = A(3A - 2I) = 3A^2 - 2A$.
Substitute the identity $A^2 = 2A - I$ again:
$A^4 = 3(2A - I) - 2A = 4A - 3I$.

7. $\begin{bmatrix} 10 & -1 \\ 9 & 10 \\ -5 & -3 \end{bmatrix}$ **9.** $\begin{bmatrix} -3 & 13 \\ -8 & 27 \end{bmatrix}$

11. a. $p(x_i) = c_0 + c_1 x_i + \cdots + c_{n-1} x_i^{n-1}$
$= \text{row}_i (V) \cdot \begin{bmatrix} c_0 \\ \vdots \\ c_{n-1} \end{bmatrix} = \text{row}_i (V\mathbf{c}) = y_i$

  b. Suppose $x_1, \ldots, x_n$ are distinct, and suppose $V\mathbf{c} = \mathbf{0}$ for some vector $\mathbf{c}$. Then the entries in $\mathbf{c}$ are the coefficients of a polynomial whose value is zero at the distinct points $x_1, \ldots, x_n$. However, a nonzero polynomial of degree $n - 1$ cannot have n zeros, so the polynomial must be identically zero. That is, the entries in $\mathbf{c}$ must all be zero. This shows that the columns of V are linearly independent.

  c. *Hint:* When $x_1, \ldots, x_n$ are distinct, there is a vector $\mathbf{c}$ such that $V\mathbf{c} = \mathbf{y}$. Why?

13. a. $P^2 = (\mathbf{u}\mathbf{u}^T)(\mathbf{u}\mathbf{u}^T) = \mathbf{u}(\mathbf{u}^T\mathbf{u})\mathbf{u}^T = \mathbf{u}(1)\mathbf{u}^T = P$
  b. $P^T = (\mathbf{u}\mathbf{u}^T)^T = \mathbf{u}^{TT}\mathbf{u}^T = \mathbf{u}\mathbf{u}^T = P$
  c. $Q^2 = (I - 2P)(I - 2P)$
$= I - I(2P) - 2PI + 2P(2P)$
$= I - 4P + 4P^2 = I$, because of part (a).

15. Left-multiplication by an elementary matrix produces an elementary row operation:
$B \sim E_1 B \sim E_2 E_1 B \sim E_3 E_2 E_1 B = C$
So B is row equivalent to C. Since row operations are reversible, C is row equivalent to B. (Alternatively, show C being changed into B by row operations using the inverses of the E_i.)

17. Since B is 4×6 (with more columns than rows), its six columns are linearly dependent and there is a nonzero $\mathbf{x}$ such that $B\mathbf{x} = \mathbf{0}$. Thus $AB\mathbf{x} = A\mathbf{0} = \mathbf{0}$, which shows that the matrix AB is not invertible, by the Invertible Matrix Theorem.

19. **[M]** To four decimal places, as k increases,
$A^k \to \begin{bmatrix} .2857 & .2857 & .2857 \\ .4286 & .4286 & .4286 \\ .2857 & .2857 & .2857 \end{bmatrix}$ and
$B^k \to \begin{bmatrix} .2022 & .2022 & .2022 \\ .3708 & .3708 & .3708 \\ .4270 & .4270 & .4270 \end{bmatrix}$

or, in rational format,

$$A^k \to \begin{bmatrix} 2/7 & 2/7 & 2/7 \\ 3/7 & 3/7 & 3/7 \\ 2/7 & 2/7 & 2/7 \end{bmatrix} \quad \text{and}$$

$$B^k \to \begin{bmatrix} 18/89 & 18/89 & 18/89 \\ 33/89 & 33/89 & 33/89 \\ 38/89 & 38/89 & 38/89 \end{bmatrix}$$

Chapter 3

Section 3.1, page 167

1. 1   **3.** −5   **5.** −23   **7.** 4

9. 10. Start with row 3.

11. −12. Start with column 1 or row 4.

13. 6. Start with row 2 or column 2.

15. 1   **17.** −5

19. $ad - bc, cb - da$. Interchanging two rows changes the sign of the determinant.

21. $-2, (18 + 12k) - (20 + 12k) = -2$. A row replacement does not change the value of a determinant.

23. $-5, k(4) - k(2) + k(-7) = -5k$. Scaling a row by a constant k multiplies the determinant by k.

25. 1   **27.** k **29.** −1

31. 1. The matrix is upper or lower triangular, with only 1's on the diagonal. The determinant is 1, the product of the diagonal entries.

33. $\det EA = \det \begin{bmatrix} c & d \\ a & b \end{bmatrix} = cb - ad = (-1)(ad - bc)$
$= (\det E)(\det A)$

35. $\det EA = \det \begin{bmatrix} a + kc & b + kd \\ c & d \end{bmatrix}$
$= (a + kc)d - (b + kd)c$
$= ad + kcd - bc - kdc = (+1)(ad - bc)$
$= (\det E)(\det A)$

37. $5A = \begin{bmatrix} 15 & 5 \\ 20 & 10 \end{bmatrix}$; no

39. Hints are in the *Study Guide*.

41. The area of the parallelogram and the determinant of $[\mathbf{u} \quad \mathbf{v}]$ both equal 6. If $\mathbf{v} = \begin{bmatrix} x \\ 2 \end{bmatrix}$ for any x, the area is still 6. In each case the base of the parallelogram is unchanged, and the altitude remains 2 because the second coordinate of $\mathbf{v}$ is always 2.

43. [M] In general, $\det(A + B)$ is not equal to $\det A + \det B$.

45. [M] You can check your conjectures when you get to Section 3.2.

Section 3.2, page 175

1. Interchanging two rows reverses the sign of the determinant.

3. A row replacement operation does not change the determinant.

5. 3   **7.** 0   **9.** 3   **11.** 120

13. 6   **15.** 35   **17.** −7   **19.** 14

21. Invertible   **23.** Not invertible

25. Linearly independent   **27.** See the *Study Guide*.

29. −32

31. *Hint:* Show that $(\det A)(\det A^{-1}) = 1$.

33. *Hint:* Use Theorem 6.

35. *Hint:* Use Theorem 6 and another theorem.

37. $\det AB = \det \begin{bmatrix} 6 & 0 \\ 17 & 4 \end{bmatrix} = 24$;
$(\det A)(\det B) = 3 \cdot 8 = 24$

39. a. −12   **b.** 500   **c.** −3   **d.** $\frac{1}{4}$ **e.** 64

41. $\det A = (a + e)d - (b + f)c = ad + ed - bc - fc$
$= (ad - bc) + (ed - fc) = \det B + \det C$

43. *Hint:* Compute $\det A$ by a cofactor expansion down column 3.

45. [M] See the *Study Guide* after you have made a conjecture about $A^T A$ and AA^T.

Section 3.3, page 184

1. $\begin{bmatrix} 5/6 \\ -1/6 \end{bmatrix}$ **3.** $\begin{bmatrix} 4 \\ 5/2 \end{bmatrix}$ **5.** $\begin{bmatrix} 3/2 \\ 4 \\ -7/2 \end{bmatrix}$

7. $s \neq \pm\sqrt{3}$; $x_1 = \dfrac{5s + 4}{6(s^2 - 3)}, x_2 = \dfrac{-4s - 15}{4(s^2 - 3)}$

9. $s \neq 0, -1$; $x_1 = \dfrac{1}{3(s + 1)}, x_2 = \dfrac{4s + 3}{6s(s + 1)}$

11. $\text{adj } A = \begin{bmatrix} 0 & 1 & 0 \\ -3 & -1 & -3 \\ 3 & 2 & 6 \end{bmatrix}, A^{-1} = \dfrac{1}{3} \begin{bmatrix} 0 & 1 & 0 \\ -3 & -1 & -3 \\ 3 & 2 & 6 \end{bmatrix}$

13. $\text{adj } A = \begin{bmatrix} -1 & -1 & 5 \\ 1 & -5 & 1 \\ 1 & 7 & -5 \end{bmatrix}, A^{-1} = \dfrac{1}{6} \begin{bmatrix} -1 & -1 & 5 \\ 1 & -5 & 1 \\ 1 & 7 & -5 \end{bmatrix}$

15. $\text{adj } A = \begin{bmatrix} 2 & 0 & 0 \\ 2 & 6 & 0 \\ -1 & -9 & 3 \end{bmatrix}, A^{-1} = \dfrac{1}{6} \begin{bmatrix} 2 & 0 & 0 \\ 2 & 6 & 0 \\ -1 & -9 & 3 \end{bmatrix}$

17. If $A = \begin{bmatrix} a & b \\ c & d \end{bmatrix}$, then $C_{11} = d, C_{12} = -c, C_{21} = -b$, $C_{22} = a$. The adjugate matrix is the transpose of cofactors:

$\text{adj } A = \begin{bmatrix} d & -b \\ -c & a \end{bmatrix}$

Following Theorem 8, we divide by $\det A$; this produces the formula from Section 2.2.

19. 8   **21.** 14   **23.** 22

25. A 3×3 matrix A is not invertible if and only if its columns are linearly dependent (by the Invertible Matrix Theorem). This happens if and only if one of the columns is in the plane spanned by the other two columns, which is equivalent to the condition that the parallelepiped determined by these columns has zero volume, which in turn is equivalent to the condition that $\det A = 0$.

27. 24   **29.** $\frac{1}{2}|\det[\,\mathbf{v}_1 \quad \mathbf{v}_2\,]|$

31. a. See Example 5.   **b.** $4\pi abc/3$

33. [M] In MATLAB, the entries in $B - \text{inv}(A)$ are approximately 10^{-15} or smaller. See the *Study Guide* for suggestions that may save you keystrokes as you work.

35. [M] MATLAB Student Version 4.0 uses 57,771 flops for $\text{inv}(A)$, and 14,269,045 flops for the inverse formula. The `inv(A)` command requires only about 0.4% of the operations for the inverse formula. The *Study Guide* shows how to use the `flops` command.

Chapter 3 Supplementary Exercises, page 185

1. a. T   **b.** T   **c.** F   **d.** F   **e.** F   **f.** F
g. T   **h.** T   **i.** F   **j.** F   **k.** T   **l.** F
m. F   **n.** T   **o.** F   **p.** T

The solution for Exercise 3 is based on the fact that if a matrix contains two rows (or two columns) that are multiples of each other, then the determinant of the matrix is zero, by Theorem 4, because the matrix cannot be invertible.

3. Make two row replacement operations, and then factor out a common multiple in row 2 and a common multiple in row 3.

$$\begin{vmatrix} 1 & a & b+c \\ 1 & b & a+c \\ 1 & c & a+b \end{vmatrix} = \begin{vmatrix} 1 & a & b+c \\ 0 & b-a & a-b \\ 0 & c-a & a-c \end{vmatrix}$$

$$= (b-a)(c-a)\begin{vmatrix} 1 & a & b+c \\ 0 & 1 & -1 \\ 0 & 1 & -1 \end{vmatrix}$$

$$= 0$$

5. -12

7. When the determinant is expanded by cofactors of the first row, the equation has the form $ax + by + c = 0$, where at least one of a and b is not zero. This is the equation of a line. It is clear that (x_1, y_1) and (x_2, y_2) are on the line, because when the coordinates of one of the points are substituted for x and y, two rows of the matrix are equal and so the determinant is zero.

9. $T \sim \begin{bmatrix} 1 & a & a^2 \\ 0 & b-a & b^2-a^2 \\ 0 & c-a & c^2-a^2 \end{bmatrix}$. Thus, by Theorem 3,

$$\det T = (b-a)(c-a)\det\begin{bmatrix} 1 & a & a^2 \\ 0 & 1 & b+a \\ 0 & 1 & c+a \end{bmatrix}$$

$$= (b-a)(c-a)\det\begin{bmatrix} 1 & a & a^2 \\ 0 & 1 & b+a \\ 0 & 0 & c-b \end{bmatrix}$$

$$= (b-a)(c-a)(c-b)$$

11. Area $= 12$. If one vertex is subtracted from all four vertices, and if the new vertices are $\mathbf{0}, \mathbf{v}_1, \mathbf{v}_2$, and $\mathbf{v}_3$, then the translated figure (and hence the original figure) will be a parallelogram if and only if one of $\mathbf{v}_1, \mathbf{v}_2$, and $\mathbf{v}_3$ is the sum of the other two vectors.

13. By the Inverse Formula, $(\text{adj }A) \cdot \dfrac{1}{\det A} A = A^{-1}A = I$. By the Invertible Matrix Theorem, $\text{adj }A$ is invertible and $(\text{adj }A)^{-1} = \dfrac{1}{\det A} A$.

15. a. $X = CA^{-1}, Y = D - CA^{-1}B$. Now use Exercise 14(c).

   b. From part (a), and the multiplicative property of determinants,

$$\det\begin{bmatrix} A & B \\ C & D \end{bmatrix} = \det[\,A(D - CA^{-1}B)\,]$$

$$= \det[\,AD - ACA^{-1}B\,]$$

$$= \det[\,AD - CAA^{-1}B\,]$$

$$= \det[\,AD - CB\,]$$

   where the equality $AC = CA$ was used in the third step.

17. First consider the case $n = 2$, and prove that the result holds by directly computing the determinants of B and C. Now assume that the formula holds for all $(k-1) \times (k-1)$ matrices, and let A, B, and C be $k \times k$ matrices. Use a cofactor expansion along the first column and the inductive hypothesis to find $\det B$. Use row replacement operations on C to create zeros below the first pivot and produce a triangular matrix. Find the determinant of this matrix and add it to $\det B$ to get the result.

19. [M] Compute:

$$\begin{vmatrix} 1 & 1 & 1 \\ 1 & 2 & 2 \\ 1 & 2 & 3 \end{vmatrix} = 1, \qquad \begin{vmatrix} 1 & 1 & 1 & 1 \\ 1 & 2 & 2 & 2 \\ 1 & 2 & 3 & 3 \\ 1 & 2 & 3 & 4 \end{vmatrix} = 1,$$

$$\begin{vmatrix} 1 & 1 & 1 & 1 & 1 \\ 1 & 2 & 2 & 2 & 2 \\ 1 & 2 & 3 & 3 & 3 \\ 1 & 2 & 3 & 4 & 4 \\ 1 & 2 & 3 & 4 & 5 \end{vmatrix} = 1$$

Conjecture:

$$\begin{vmatrix} 1 & 1 & 1 & \cdots & 1 \\ 1 & 2 & 2 & & 2 \\ 1 & 2 & 3 & & 3 \\ \vdots & & & \ddots & \vdots \\ 1 & 2 & 3 & \cdots & n \end{vmatrix} = 1$$

To confirm the conjecture, use row replacement operations to create zeros below the first pivot, then the second pivot, and so on. The resulting matrix is

$$\begin{vmatrix} 1 & 1 & 1 & \cdots & 1 \\ 0 & 1 & 1 & & 1 \\ 0 & 0 & 1 & & 1 \\ \vdots & & & \ddots & \vdots \\ 0 & 0 & 0 & \cdots & 1 \end{vmatrix}$$

which is an upper triangular matrix with determinant 1.

Chapter 4

Section 4.1, page 195

1. **a.** $\mathbf{u} + \mathbf{v}$ is in V because its entries will both be nonnegative.

   b. *Example:* If $\mathbf{u} = \begin{bmatrix} 2 \\ 2 \end{bmatrix}$ and $c = -1$, then $\mathbf{u}$ is in V, but $c\mathbf{u}$ is not in V.

3. *Example:* If $\mathbf{u} = \begin{bmatrix} .5 \\ .5 \end{bmatrix}$ and $c = 4$, then $\mathbf{u}$ is in H, but $c\mathbf{u}$ is not in H.

5. Yes, by Theorem 1, because the set is Span $\{t^2\}$.

7. No, the set is not closed under multiplication by scalars that are not integers.

9. $H = $ Span $\{\mathbf{v}\}$, where $\mathbf{v} = \begin{bmatrix} -2 \\ 5 \\ 3 \end{bmatrix}$. By Theorem 1, H is a subspace of $\mathbb{R}^3$.

11. $W = $ Span $\{\mathbf{u}, \mathbf{v}\}$, where $\mathbf{u} = \begin{bmatrix} 2 \\ -1 \\ 0 \end{bmatrix}$, $\mathbf{v} = \begin{bmatrix} 3 \\ 0 \\ 2 \end{bmatrix}$. By Theorem 1, W is a subspace of $\mathbb{R}^3$.

13. **a.** There are only three vectors in $\{\mathbf{v}_1, \mathbf{v}_2, \mathbf{v}_3\}$, and $\mathbf{w}$ is not one of them.

   b. There are infinitely many vectors in Span $\{\mathbf{v}_1, \mathbf{v}_2, \mathbf{v}_3\}$.

   c. $\mathbf{w}$ is in Span $\{\mathbf{v}_1, \mathbf{v}_2, \mathbf{v}_3\}$ because $\mathbf{w} = \mathbf{v}_1 + \mathbf{v}_2$.

15. W is not a vector space because the zero vector is not in W.

17. $S = \left\{ \begin{bmatrix} 2 \\ 0 \\ -1 \\ 0 \end{bmatrix}, \begin{bmatrix} -1 \\ 3 \\ 0 \\ 3 \end{bmatrix}, \begin{bmatrix} 0 \\ -1 \\ 3 \\ 0 \end{bmatrix} \right\}$

19. *Hint:* Use Theorem 1.

Warning: Although the *Study Guide* has complete solutions for *every* odd-numbered exercise whose answer here is only a "Hint," you *must* really try to work the solution yourself. Otherwise, you will not benefit from the exercise.

21. Yes. The conditions for a subspace are obviously satisfied: The zero matrix is in H, the sum of two upper triangular matrices is upper triangular, and any scalar multiple of an upper triangular matrix is again upper triangular.

23. See the *Study Guide* after you have written your answers.

25. 4   27. **a.** 8   **b.** 3   **c.** 5   **d.** 4

29. $\mathbf{u} + (-1)\mathbf{u} = 1\mathbf{u} + (-1)\mathbf{u}$ Axiom 10
    $= [1 + (-1)]\mathbf{u}$ Axiom 8
    $= 0\mathbf{u} = \mathbf{0}$ Exercise 27
   From Exercise 26, it follows that $(-1)\mathbf{u} = -\mathbf{u}$.

31. Any subspace H that contains $\mathbf{u}$ and $\mathbf{v}$ must also contain all scalar multiples of $\mathbf{u}$ and $\mathbf{v}$ and hence must contain all sums of scalar multiples of $\mathbf{u}$ and $\mathbf{v}$. Thus H must contain Span $\{\mathbf{u}, \mathbf{v}\}$.

33. *Hint:* For part of the solution, consider $\mathbf{w}_1$ and $\mathbf{w}_2$ in $H + K$, and write $\mathbf{w}_1$ and $\mathbf{w}_2$ in the form $\mathbf{w}_1 = \mathbf{u}_1 + \mathbf{v}_1$ and $\mathbf{w}_2 = \mathbf{u}_2 + \mathbf{v}_2$, where $\mathbf{u}_1$ and $\mathbf{u}_2$ are in H, and $\mathbf{v}_1$ and $\mathbf{v}_2$ are in K.

35. **[M]** The reduced echelon form of $[\mathbf{v}_1, \mathbf{v}_2, \mathbf{v}_3, \mathbf{w}]$ shows that $\mathbf{w} = \mathbf{v}_1 - 2\mathbf{v}_2 + \mathbf{v}_3$. Hence $\mathbf{w}$ is in the subspace spanned by $\mathbf{v}_1, \mathbf{v}_2,$ and $\mathbf{v}_3$.

37. **[M]** The functions are $\cos 4t$ and $\cos 6t$. See Exercise 34 in Section 4.5.

Section 4.2, page 205

1. $\begin{bmatrix} 3 & -5 & -3 \\ 6 & -2 & 0 \\ -8 & 4 & 1 \end{bmatrix} \begin{bmatrix} 1 \\ 3 \\ -4 \end{bmatrix} = \begin{bmatrix} 0 \\ 0 \\ 0 \end{bmatrix}$,
   so $\mathbf{w}$ is in Nul A.

3. $\begin{bmatrix} 2 \\ -3 \\ 1 \\ 0 \end{bmatrix}, \begin{bmatrix} -4 \\ 2 \\ 0 \\ 1 \end{bmatrix}$ 5. $\begin{bmatrix} 4 \\ 1 \\ 0 \\ 0 \\ 0 \end{bmatrix}, \begin{bmatrix} -2 \\ 0 \\ 5 \\ 1 \\ 0 \end{bmatrix}$

7. W is not a subspace of $\mathbb{R}^3$ because the zero vector $(0, 0, 0)$ is not in W.

9. W is a subspace of $\mathbb{R}^4$ because W is the set of solutions of the system

   $p - 3q - 4s \qquad = 0$
   $2p \qquad - s - 5r = 0$

11. W is not a subspace because $\mathbf{0}$ is not in W. *Justification:* If a typical element $(s - 2t, 3 + 3s, 3s + t, 2s)$ were zero, then $3 + 3s = 0$ and $2s = 0$, which is impossible.

13. $W = \text{Col } A$ for $A = \begin{bmatrix} 1 & -6 \\ 0 & 1 \\ 1 & 0 \end{bmatrix}$, so W is a vector space by Theorem 3.

15. $\begin{bmatrix} 0 & 2 & 1 \\ 1 & -1 & 2 \\ 3 & 1 & 0 \\ 2 & -1 & -1 \end{bmatrix}$

17. a. 2   **b.** 4   **19. a.** 5   **b.** 2

21. The vector $\begin{bmatrix} 2 \\ 3 \end{bmatrix}$ is in Nul A and the vector $\begin{bmatrix} 6 \\ -3 \\ -9 \\ 9 \end{bmatrix}$ is in Col A. Other answers are possible.

23. $\mathbf{w}$ is in both Nul A and Col A.

25. See the *Study Guide*. By now you should know how to use it properly.

27. Let $\mathbf{x} = \begin{bmatrix} 3 \\ 2 \\ -1 \end{bmatrix}$ and $A = \begin{bmatrix} 1 & -3 & -3 \\ -2 & 4 & 2 \\ -1 & 5 & 7 \end{bmatrix}$. Then $\mathbf{x}$ is in Nul A. Since Nul A is a subspace of $\mathbb{R}^3$, $10\mathbf{x}$ is in Nul A.

29. a. $A0 = 0$, so the zero vector is in Col A.

b. By a property of matrix multiplication, $A\mathbf{x} + A\mathbf{w} = A(\mathbf{x} + \mathbf{w})$, which shows that $A\mathbf{x} + A\mathbf{w}$ is a linear combination of the columns of A and hence is in Col A.

c. $c(A\mathbf{x}) = A(c\mathbf{x})$, which shows that $c(A\mathbf{x})$ is in Col A for all scalars c.

31. a. For arbitrary polynomials $\mathbf{p}, \mathbf{q}$ in $\mathbb{P}_2$ and any scalar c,

$$T(\mathbf{p} + \mathbf{q}) = \begin{bmatrix} (\mathbf{p}+\mathbf{q})(0) \\ (\mathbf{p}+\mathbf{q})(1) \end{bmatrix} = \begin{bmatrix} \mathbf{p}(0) + \mathbf{q}(0) \\ \mathbf{p}(1) + \mathbf{q}(1) \end{bmatrix}$$

$$= \begin{bmatrix} \mathbf{p}(0) \\ \mathbf{p}(1) \end{bmatrix} + \begin{bmatrix} \mathbf{q}(0) \\ \mathbf{q}(1) \end{bmatrix} = T(\mathbf{p}) + T(\mathbf{q})$$

$$T(c\mathbf{p}) = \begin{bmatrix} c\mathbf{p}(0) \\ c\mathbf{p}(1) \end{bmatrix} = c\begin{bmatrix} \mathbf{p}(0) \\ \mathbf{p}(1) \end{bmatrix} = cT(\mathbf{p})$$

So T is a linear transformation from $\mathbb{P}_2$ into $\mathbb{P}_2$.

b. Any quadratic polynomial that vanishes at 0 and 1 must be a multiple of $\mathbf{p}(t) = t(t-1)$. The range of T is $\mathbb{R}^2$.

33. a. For A, B in $M_{2\times2}$ and any scalar c,

$$T(A + B) = (A + B) + (A + B)^T$$
$$= A + B + A^T + B^T \quad \text{Transpose property}$$
$$= (A + A^T) + (B + B^T) = T(A) + T(B)$$
$$T(cA) = (cA) + (cA)^T = cA + cA^T$$
$$= c(A + A^T) = cT(A)$$

So T is a linear transformation from $M_{2\times2}$ into $M_{2\times2}$.

b. If B is any element in $M_{2\times2}$ with the property that $B^T = B$, and if $A = \frac{1}{2}B$, then

$$T(A) = \frac{1}{2}B + \left(\frac{1}{2}B\right)^T = \frac{1}{2}B + \frac{1}{2}B = B$$

c. Part (b) showed that the range of T contains all B such that $B^T = B$. So it suffices to show that any B in the range of T has this property. If $B = T(A)$, then by properties of transposes,

$$B^T = (A + A^T)^T = A^T + A^{TT} = A^T + A = B$$

d. The kernel of T is $\left\{ \begin{bmatrix} 0 & b \\ -b & 0 \end{bmatrix} : b \text{ real} \right\}$.

35. *Hint:* Check the three conditions for a subspace. Typical elements of $T(U)$ have the form $T(\mathbf{u}_1)$ and $T(\mathbf{u}_2)$, where $\mathbf{u}_1, \mathbf{u}_2$ are in U.

37. [M] $\mathbf{w}$ is in Col A but not in Nul A. (Explain why.)

39. [M] The reduced echelon form of A is

$$\begin{bmatrix} 1 & 0 & 1/3 & 0 & 10/3 \\ 0 & 1 & 1/3 & 0 & -26/3 \\ 0 & 0 & 0 & 1 & -4 \\ 0 & 0 & 0 & 0 & 0 \end{bmatrix}$$

Section 4.3, page 213

1. The 3×3 matrix $A = \begin{bmatrix} 1 & 1 & 1 \\ 0 & 1 & 1 \\ 0 & 0 & 1 \end{bmatrix}$ has three pivot positions. By the Invertible Matrix Theorem, A is invertible and its columns form a basis for $\mathbb{R}^3$. (See Example 3.)

3. This set does not form a basis for $\mathbb{R}^3$. The set is linearly dependent and does not span $\mathbb{R}^3$.

5. This set does not form a basis for $\mathbb{R}^3$. The set is linearly dependent because the zero vector is in the set. However,

$$\begin{bmatrix} 3 & -3 & 0 & 0 \\ -3 & 7 & 0 & -3 \\ 0 & 0 & 0 & 5 \end{bmatrix} \sim \begin{bmatrix} 1 & -1 & 0 & 0 \\ 0 & 4 & 0 & -3 \\ 0 & 0 & 0 & 5 \end{bmatrix}$$

The matrix has a pivot in each row and hence its columns span $\mathbb{R}^3$.

7. This set does not form a basis for $\mathbb{R}^3$. The set is linearly independent because one vector is not a multiple of the other. However, the vectors do not span $\mathbb{R}^3$. The matrix $\begin{bmatrix} -2 & 6 \\ 3 & -1 \\ 0 & 5 \end{bmatrix}$ can have at most two pivots since it has only two columns. So there will not be a pivot in each row.

9. $\begin{bmatrix} 2 \\ -1 \\ 1 \\ 0 \end{bmatrix}$ **11.** $\begin{bmatrix} 3 \\ 1 \\ 0 \end{bmatrix}, \begin{bmatrix} -2 \\ 0 \\ 1 \end{bmatrix}$

13. Basis for Nul A: $\begin{bmatrix} -6 \\ -5/2 \\ 1 \\ 0 \end{bmatrix}, \begin{bmatrix} -5 \\ -3/2 \\ 0 \\ 1 \end{bmatrix}$

Basis for Col A: $\begin{bmatrix} -2 \\ 2 \\ -3 \end{bmatrix}, \begin{bmatrix} 4 \\ -6 \\ 8 \end{bmatrix}$

15. $\{\mathbf{v}_1, \mathbf{v}_2, \mathbf{v}_4, \mathbf{v}_5\}$ **17.** [M] $\{\mathbf{v}_1, \mathbf{v}_2, \mathbf{v}_3, \mathbf{v}_5\}$

19. The three simplest answers are $\{\mathbf{v}_1, \mathbf{v}_2\}$, $\{\mathbf{v}_1, \mathbf{v}_3\}$, and $\{\mathbf{v}_2, \mathbf{v}_3\}$. Other answers are possible.

21. See the *Study Guide* for hints.

23. *Hint:* Use the Invertible Matrix Theorem.

25. No. (Why is the set not a basis for H?)

27. $\{\cos \omega t, \sin \omega t\}$

29. Let A be the $n \times k$ matrix $[\mathbf{v}_1 \cdots \mathbf{v}_k]$. Since A has fewer columns than rows, there cannot be a pivot position in each row of A. By Theorem 4 in Section 1.4, the columns of A do not span $\mathbb{R}^n$ and hence are not a basis for $\mathbb{R}^n$.

31. *Hint:* If $\{\mathbf{v}_1, \dots, \mathbf{v}_p\}$ is linearly dependent, then there exist $c_1, \dots, c_p$, not all zero, such that $c_1\mathbf{v}_1 + \cdots + c_p\mathbf{v}_p = \mathbf{0}$. Use this equation.

33. Neither polynomial is a multiple of the other polynomial, so $\{\mathbf{p}_1, \mathbf{p}_2\}$ is a linearly independent set in $\mathbb{P}_3$.

35. Let $\{\mathbf{v}_1, \mathbf{v}_3\}$ be any linearly independent set in the vector space V, and let $\mathbf{v}_2$ and $\mathbf{v}_4$ be linear combinations of $\mathbf{v}_1$ and $\mathbf{v}_3$. Then $\{\mathbf{v}_1, \mathbf{v}_3\}$ is a basis for Span$\{\mathbf{v}_1, \mathbf{v}_2, \mathbf{v}_3, \mathbf{v}_4\}$.

37. [M] You could be clever and find special values of t that produce several zeros in (5), and thereby create a system of equations that can be solved easily by hand. Or, you could use values of t such as $t = 0, .1, .2, \dots$ to create a system of equations that you can solve with a matrix program.

Section 4.4, page 222

1. $\begin{bmatrix} 3 \\ -7 \end{bmatrix}$ **3.** $\begin{bmatrix} -7 \\ 4 \\ 3 \end{bmatrix}$ **5.** $\begin{bmatrix} 2 \\ -1 \end{bmatrix}$ **7.** $\begin{bmatrix} -1 \\ -1 \\ 3 \end{bmatrix}$

9. $\begin{bmatrix} 1 & 2 \\ -3 & -5 \end{bmatrix}$ **11.** $\begin{bmatrix} 5 \\ 1 \end{bmatrix}$ **13.** $\begin{bmatrix} 2 \\ 6 \\ -1 \end{bmatrix}$

15. The *Study Guide* has hints.

17. $\begin{bmatrix} 1 \\ 1 \end{bmatrix} = 5\mathbf{v}_1 - 2\mathbf{v}_2 = 10\mathbf{v}_1 - 3\mathbf{v}_2 + \mathbf{v}_3 = -\mathbf{v}_2 - \mathbf{v}_3$
(infinitely many answers)

19. *Hint:* By hypothesis, the zero vector has a unique representation as a linear combination of elements of S.

21. $\begin{bmatrix} 9 & 2 \\ 4 & 1 \end{bmatrix}$

23. *Hint:* Suppose that $[\mathbf{u}]_{\mathcal{B}} = [\mathbf{w}]_{\mathcal{B}}$ for some $\mathbf{u}$ and $\mathbf{w}$ in V, and denote the entries in $[\mathbf{u}]_{\mathcal{B}}$ by $c_1, \dots, c_n$. Use the definition of $[\mathbf{u}]_{\mathcal{B}}$.

25. One possible approach: First, show that if $\mathbf{u}_1, \dots, \mathbf{u}_p$ are linearly *dependent*, then $[\mathbf{u}_1]_{\mathcal{B}}, \dots, [\mathbf{u}_p]_{\mathcal{B}}$ are linearly dependent. Second, show that if $[\mathbf{u}_1]_{\mathcal{B}}, \dots, [\mathbf{u}_p]_{\mathcal{B}}$ are linearly dependent, then $\mathbf{u}_1, \dots, \mathbf{u}_p$ are linearly *dependent*. Use the two equations displayed in the exercise. A slightly different proof is given in the *Study Guide*.

27. Linearly independent. (Justify answers to Exercises 27–34.)

29. Linearly dependent.

31. a. The coordinate vectors $\begin{bmatrix} 1 \\ -3 \\ 5 \end{bmatrix}$, $\begin{bmatrix} -3 \\ 5 \\ -7 \end{bmatrix}$, $\begin{bmatrix} -4 \\ 5 \\ -6 \end{bmatrix}$, $\begin{bmatrix} 1 \\ 0 \\ -1 \end{bmatrix}$ do not span $\mathbb{R}^3$. Because of the isomorphism between $\mathbb{R}^3$ and $\mathbb{P}_2$, the corresponding polynomials do not span $\mathbb{P}_2$.

b. The coordinate vectors $\begin{bmatrix} 0 \\ 5 \\ 1 \end{bmatrix}$, $\begin{bmatrix} 1 \\ -8 \\ -2 \end{bmatrix}$, $\begin{bmatrix} -3 \\ 4 \\ 2 \end{bmatrix}$, $\begin{bmatrix} 2 \\ -3 \\ 0 \end{bmatrix}$ span $\mathbb{R}^3$. Because of the isomorphism between $\mathbb{R}^3$ and $\mathbb{P}_2$, the corresponding polynomials span $\mathbb{P}_2$.

33. [M] The coordinate vectors $\begin{bmatrix} 3 \\ 7 \\ 0 \\ 0 \end{bmatrix}$, $\begin{bmatrix} 5 \\ 1 \\ 0 \\ -2 \end{bmatrix}$, $\begin{bmatrix} 0 \\ 1 \\ -2 \\ 0 \end{bmatrix}$, $\begin{bmatrix} 1 \\ 16 \\ -6 \\ 2 \end{bmatrix}$ are a linearly dependent subset of $\mathbb{R}^4$. Because of the isomorphism between $\mathbb{R}^4$ and $\mathbb{P}_3$, the corresponding polynomials form a linearly dependent subset of $\mathbb{P}_3$, and thus cannot be a basis for $\mathbb{P}_3$.

35. [M] $[\mathbf{x}]_{\mathcal{B}} = \begin{bmatrix} -5/3 \\ 8/3 \end{bmatrix}$ **37.** [M] $\begin{bmatrix} 1.3 \\ 0 \\ 0.8 \end{bmatrix}$

Section 4.5, page 229

1. $\begin{bmatrix} 1 \\ 1 \\ 0 \end{bmatrix}$, $\begin{bmatrix} -2 \\ 1 \\ 3 \end{bmatrix}$; dim is 2

3. $\begin{bmatrix} 0 \\ 1 \\ 0 \\ 1 \end{bmatrix}$, $\begin{bmatrix} 0 \\ -1 \\ 1 \\ 2 \end{bmatrix}$, $\begin{bmatrix} 2 \\ 0 \\ -3 \\ 0 \end{bmatrix}$; dim is 3

5. $\begin{bmatrix} 1 \\ 2 \\ 0 \\ -3 \end{bmatrix}$, $\begin{bmatrix} -2 \\ 0 \\ -2 \\ 0 \end{bmatrix}$, $\begin{bmatrix} 0 \\ 5 \\ 2 \\ 6 \end{bmatrix}$; dim is 3

7. No basis; dim is 0   **9.** 2   **11.** 3   **13.** 2, 3

15. 2, 3   **17.** 0, 3

19. See the *Study Guide*.

21. *Hint:* You need only show that the first four Hermite polynomials are linearly independent. Why?

23. $[\mathbf{p}]_{\mathcal{B}} = (3, 6, 2, 1)$

25. *Hint:* Suppose S does span V, and use the Spanning Set Theorem. This leads to a contradiction, which shows that the spanning hypothesis is false.

27. *Hint:* Use the fact that each $\mathbb{P}_n$ is a subspace of $\mathbb{P}$.

29. Justify each answer.
   a. True **b.** True **c.** True

31. *Hint:* Since H is a nonzero subspace of a finite-dimensional space, H is finite-dimensional and has a basis, say, $\mathbf{v}_1, \ldots, \mathbf{v}_p$. First show that $\{T(\mathbf{v}_1), \ldots, T(\mathbf{v}_p)\}$ spans $T(H)$.

33. [M] **a.** One basis is $\{\mathbf{v}_1, \mathbf{v}_2, \mathbf{v}_3, \mathbf{e}_2, \mathbf{e}_3\}$. In fact, any two of the vectors $\mathbf{e}_2, \ldots, \mathbf{e}_5$ will extend $\{\mathbf{v}_1, \mathbf{v}_2, \mathbf{v}_3\}$ to a basis of $\mathbb{R}^5$.

Section 4.6, page 236

1. rank $A = 2$; dim Nul $A = 2$;

Basis for Col A: $\begin{bmatrix} 1 \\ -1 \\ 5 \end{bmatrix}, \begin{bmatrix} -4 \\ 2 \\ -6 \end{bmatrix}$

Basis for Row A: $(1, 0, -1, 5), (0, -2, 5, -6)$

Basis for Nul A: $\begin{bmatrix} 1 \\ 5/2 \\ 1 \\ 0 \end{bmatrix}, \begin{bmatrix} -5 \\ -3 \\ 0 \\ 1 \end{bmatrix}$

3. rank $A = 3$; dim Nul $A = 3$;

Basis for Col A: $\begin{bmatrix} 2 \\ -2 \\ 4 \\ -2 \end{bmatrix}, \begin{bmatrix} 6 \\ -3 \\ 9 \\ 3 \end{bmatrix}, \begin{bmatrix} 3 \\ 0 \\ 3 \\ 3 \end{bmatrix}$

Basis for Row A: $(2, 6, -6, 6, 3, 6), (0, 3, 0, 3, 3, 0),$
$(0, 0, 0, 0, 3, 0)$

Basis for Nul A: $\begin{bmatrix} 3 \\ 0 \\ 1 \\ 0 \\ 0 \\ 0 \end{bmatrix}, \begin{bmatrix} 0 \\ -1 \\ 0 \\ 1 \\ 0 \\ 0 \end{bmatrix}, \begin{bmatrix} -3 \\ 0 \\ 0 \\ 0 \\ 0 \\ 1 \end{bmatrix}$

5. 4, 3, 3

7. Yes; no. Since Col A is a four-dimensional subspace of $\mathbb{R}^4$, it coincides with $\mathbb{R}^4$. The null space cannot be $\mathbb{R}^3$, because the vectors in Nul A have 7 entries. Nul A is a three-dimensional subspace of $\mathbb{R}^7$, by the Rank Theorem.

9. 3, no. Notice that the columns of a 4×6 matrix are in $\mathbb{R}^4$, rather than $\mathbb{R}^3$. Col A is a three-dimensional subspace of $\mathbb{R}^4$.

11. 2

13. 5, 5. In both cases, the number of pivots cannot exceed the number of columns or the number of rows.

15. 4 **17.** See the *Study Guide*.

19. Yes. Try to write an explanation before you consult the *Study Guide*.

21. No. Explain why.

23. Yes. Only six homogeneous linear equations are necessary.

25. No. Explain why.

27. Row A and Nul A are in $\mathbb{R}^n$; Col A and Nul A^T are in $\mathbb{R}^m$. There are only four distinct subspaces because Row $A^T = $ Col A and Col $A^T = $ Row A.

29. Recall that dim Col $A = m$ precisely when Col $A = \mathbb{R}^m$, or equivalently, when the equation $A\mathbf{x} = \mathbf{b}$ is consistent for all $\mathbf{b}$. By Exercise 28(b), dim Col $A = m$ precisely when dim Nul $A^T = 0$, or equivalently, when the equation $A^T\mathbf{x} = \mathbf{0}$ has only the trivial solution.

31. $\mathbf{u}\mathbf{v}^T = \begin{bmatrix} 2a & 2b & 2c \\ -3a & -3b & -3c \\ 5a & 5b & 5c \end{bmatrix}$. The columns are all multiples of $\mathbf{u}$, so Col $\mathbf{u}\mathbf{v}^T$ is one-dimensional, unless $a = b = c = 0$.

33. *Hint:* Let $A = [\,\mathbf{u} \quad \mathbf{u}_2 \quad \mathbf{u}_3\,]$. If $\mathbf{u} \neq \mathbf{0}$, then $\mathbf{u}$ is a basis for Col A. Why?

35. [M] *Hint:* See Exercise 28 and the remarks before Example 4.

37. [M] The matrices C and R given for Exercise 35 work here, and $A = CR$.

Section 4.7, page 242

1. a. $\begin{bmatrix} 6 & 9 \\ -2 & -4 \end{bmatrix}$ **b.** $\begin{bmatrix} 0 \\ -2 \end{bmatrix}$

3. (ii)

5. a. $\begin{bmatrix} 4 & -1 & 0 \\ -1 & 1 & 1 \\ 0 & 1 & -2 \end{bmatrix}$ **b.** $\begin{bmatrix} 8 \\ 2 \\ 2 \end{bmatrix}$

7. $\underset{C \leftarrow B}{P} = \begin{bmatrix} -3 & 1 \\ -5 & 2 \end{bmatrix}$, $\underset{B \leftarrow C}{P} = \begin{bmatrix} -2 & 1 \\ -5 & 3 \end{bmatrix}$

9. $\underset{C \leftarrow B}{P} = \begin{bmatrix} 2 & 3 \\ 0 & -1 \end{bmatrix}$, $\underset{B \leftarrow C}{P} = \frac{1}{2}\begin{bmatrix} 1 & 3 \\ 0 & -2 \end{bmatrix}$

11. See the *Study Guide*.

13. $\underset{C \leftarrow B}{P} = \begin{bmatrix} 1 & 3 & 0 \\ -2 & -5 & 2 \\ 1 & 4 & 3 \end{bmatrix}$, $[-1 + 2t]_B = \begin{bmatrix} 5 \\ -2 \\ 1 \end{bmatrix}$

15. a. $\mathcal{B}$ is a basis for V.
   b. The coordinate mapping is a linear transformation.
   c. The product of a matrix and a vector
   d. The coordinate vector of $\mathbf{v}$ relative to $\mathcal{B}$

17. a. [M]

$$P^{-1} = \frac{1}{32}\begin{bmatrix} 32 & 0 & 16 & 0 & 12 & 0 & 10 \\ & 32 & 0 & 24 & 0 & 20 & 0 \\ & & 16 & 0 & 16 & 0 & 15 \\ & & & 8 & 0 & 10 & 0 \\ & & & & 4 & 0 & 6 \\ & & & & & 2 & 0 \\ & & & & & & 1 \end{bmatrix}$$

b. $\cos^2 t = (1/2)[1 + \cos 2t]$
$\cos^3 t = (1/4)[3 \cos t + \cos 3t]$
$\cos^4 t = (1/8)[3 + 4 \cos 2t + \cos 4t]$
$\cos^5 t = (1/16)[10 \cos t + 5 \cos 3t + \cos 5t]$
$\cos^6 t = (1/32)[10 + 15 \cos 2t + 6 \cos 4t + \cos 6t]$

19. [M] *Hint:* Let $\mathcal{C}$ be the basis $\{\mathbf{v}_1, \mathbf{v}_2, \mathbf{v}_3\}$. Then the columns of P are $[\mathbf{u}_1]_\mathcal{C}$, $[\mathbf{u}_2]_\mathcal{C}$, and $[\mathbf{u}_3]_\mathcal{C}$. Use the definition of $\mathcal{C}$-coordinate vectors and matrix algebra to compute $\mathbf{u}_1$, $\mathbf{u}_2$, $\mathbf{u}_3$. The solution method is discussed in the *Study Guide*. Here are the numerical answers:

a. $\mathbf{u}_1 = \begin{bmatrix} -6 \\ -5 \\ 21 \end{bmatrix}$, $\mathbf{u}_2 = \begin{bmatrix} -6 \\ -9 \\ 32 \end{bmatrix}$, $\mathbf{u}_3 = \begin{bmatrix} -5 \\ 0 \\ 3 \end{bmatrix}$

b. $\mathbf{w}_1 = \begin{bmatrix} 28 \\ -9 \\ -3 \end{bmatrix}$, $\mathbf{w}_2 = \begin{bmatrix} 38 \\ -13 \\ 2 \end{bmatrix}$, $\mathbf{w}_3 = \begin{bmatrix} 21 \\ -7 \\ 3 \end{bmatrix}$

Section 4.8, page 251

1. If $y_k = 2^k$, then $y_{k+1} = 2^{k+1}$ and $y_{k+2} = 2^{k+2}$. Substituting these formulas into the left side of the equation gives

$$y_{k+2} + 2y_{k+1} - 8y_k = 2^{k+2} + 2 \cdot 2^{k+1} - 8 \cdot 2^k$$
$$= 2^k(2^2 + 2 \cdot 2 - 8)$$
$$= 2^k(0) = 0 \quad \text{for all } k$$

Since the difference equation holds for all k, 2^k is a solution. A similar calculation works for $y_k = (-4)^k$.

3. The signals 2^k and $(-4)^k$ are linearly independent because neither is a multiple of the other. For instance, there is no scalar c such that $2^k = c(-4)^k$ *for all k*. By Theorem 17, the solution set H of the difference equation in Exercise 1 is two-dimensional. By the Basis Theorem in Section 4.5, the two linearly independent signals 2^k and $(-4)^k$ form a basis for H.

5. If $y_k = (-2)^k$, then

$$y_{k+2} + 4y_{k+1} + 4y_k = (-2)^{k+2} + 4(-2)^{k+1} + 4(-2)^k$$
$$= (-2)^k[(-2)^2 + 4(-2) + 4]$$
$$= (-2)^k(0) = 0 \quad \text{for all } k$$

Similarly, if $y_k = k(-2)^k$, then

$$y_{k+2} + 4y_{k+1} + 4y_k$$
$$= (k + 2)(-2)^{k+2} + 4(k + 1)(-2)^{k+1} + 4k(-2)^k$$
$$= (-2)^k[(k + 2)(-2)^2 + 4(k + 1)(-2) + 4k]$$
$$= (-2)^k[4k + 8 - 8k - 8 + 4k]$$
$$= (-2)^k(0) \quad \text{for all } k$$

Thus both $(-2)^k$ and $k(-2)^k$ are in the solution space H of the difference equation. Also, there is no scalar c such that $k(-2)^k = c(-2)^k$ *for all k*, because c must be chosen independently of k. Likewise, there is no scalar c such that $(-2)^k = ck(-2)^k$ *for all k*. So the two signals are linearly

independent. Since $\dim H = 2$, the signals form a basis for H, by the Basis Theorem.

7. Yes   **9.** Yes

11. No, two signals cannot span the three-dimensional solution space.

13. $\left(\frac{1}{3}\right)^k, \left(\frac{2}{3}\right)^k$ **15.** $\left(\frac{1}{2}\right)^k, \left(-\frac{2}{3}\right)^k$

17. $Y_k = c_1(.8)^k + c_2(.5)^k + 10 \rightarrow 10 \quad$ as $k \rightarrow \infty$

19. $y_k = c_1(-2 + \sqrt{3})^k + c_2(-2 - \sqrt{3})^k$

21. $7, 5, 4, 3, 4, 5, 6, 6, 7, 8, 9, 8, 7$; see figure:

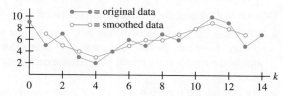

23. a. $y_{k+1} - 1.01y_k = -450$, $y_0 = 10{,}000$

25. $k^2 + c_1 \cdot (-4)^k + c_2$ **27.** $-2 + k + c_1 \cdot 2^k + c_2 \cdot (-2)^k$

29. $\mathbf{x}_{k+1} = A\mathbf{x}_k$, where

$$A = \begin{bmatrix} 0 & 1 & 0 & 0 \\ 0 & 0 & 1 & 0 \\ 0 & 0 & 0 & 1 \\ 2 & -6 & 8 & -3 \end{bmatrix}, \quad \mathbf{x} = \begin{bmatrix} y_k \\ y_{k+1} \\ y_{k+2} \\ y_{k+3} \end{bmatrix}$$

31. The equation holds for all k, so it holds with k replaced by $k - 1$, which transforms the equation into

$$y_{k+2} + 5y_{k+1} + 6y_k = 0 \quad \text{for all } k$$

The equation is of order 2.

33. For all k, the Casorati matrix $C(k)$ is not invertible. In this case, the Casorati matrix gives no information about the linear independence/dependence of the set of signals. In fact, neither signal is a multiple of the other, so they are linearly independent.

35. *Hint:* Verify the two properties that define a linear transformation. For $\{y_k\}$ and $\{z_k\}$ in $\mathbb{S}$, study $T(\{y_k\} + \{z_k\})$. Note that if r is any scalar, then the kth term of $r\{y_k\}$ is ry_k; so $T(r\{y_k\})$ is the sequence $\{w_k\}$ given by

$$w_k = ry_{k+2} + a(ry_{k+1}) + b(ry_k)$$

37. *Hint:* Find $TD(y_0, y_1, y_2, \ldots)$ and $DT(y_0, y_1, y_2, \ldots)$.

Section 4.9, page 260

1. a. From:

	N	M	To:
	.7	.6	News
	.3	.4	Music

b. $\begin{bmatrix} 1 \\ 0 \end{bmatrix}$ **c.** 33%

3. a. From:

	H	I	To:
	.95	.45	Healthy
	.05	.55	Ill

b. 15%, 12.5%

c. .925; use $\mathbf{x}_0 = \begin{bmatrix} 1 \\ 0 \end{bmatrix}$.

5. $\begin{bmatrix} 5/14 \\ 9/14 \end{bmatrix}$ **7.** $\begin{bmatrix} 1/4 \\ 1/2 \\ 1/4 \end{bmatrix}$

9. Yes, because P^2 has all positive entries.

11. a. $\begin{bmatrix} 2/3 \\ 1/3 \end{bmatrix}$ **b.** 2/3

13. a. $\begin{bmatrix} .9 \\ .1 \end{bmatrix}$ **b.** .10, no

15. [M] About 13.9% of the United States population

17. a. The entries in a column of P sum to 1. A column in the matrix $P - I$ has the same entries as in P except that one of the entries is decreased by 1. Hence each column sum is 0.

b. By part (a), the bottom row of $P - I$ is the negative of the sum of the other rows.

c. By part (b) and the Spanning Set Theorem, the bottom row of $P - I$ can be removed and the remaining $(n - 1)$ rows will still span the row space. Alternatively, use part (a) and the fact that row operations do not change the row space. Let A be the matrix obtained from $P - I$ by adding to the bottom row all the other rows. By part (a), the row space is spanned by the first $(n - 1)$ rows of A.

d. By the Rank Theorem and part (c), the dimension of the column space of $P - I$ is less than n, and hence the null space is nontrivial. Instead of the Rank Theorem, you may use the Invertible Matrix Theorem, since $P - I$ is a square matrix.

19. a. The product $S\mathbf{x}$ equals the sum of the entries in $\mathbf{x}$. For a probability vector, this sum must be 1.

b. $P = [\, \mathbf{p}_1 \quad \mathbf{p}_2 \quad \cdots \quad \mathbf{p}_n \,]$, where the $\mathbf{p}_i$ are probability vectors. By matrix multiplication and part (a),

$$SP = \begin{bmatrix} S\mathbf{p}_1 & S\mathbf{p}_2 & \cdots & S\mathbf{p}_n \end{bmatrix} = \begin{bmatrix} 1 & 1 & \cdots & 1 \end{bmatrix} = S$$

c. By part (b), $S(P\mathbf{x}) = (SP)\mathbf{x} = S\mathbf{x} = 1$. Also, the entries in $P\mathbf{x}$ are nonnegative (because P and $\mathbf{x}$ have nonnegative entries). Hence, by (a), $P\mathbf{x}$ is a probability vector.

Chapter 4 Supplementary Exercises, page 262

1. a. T **b.** T **c.** F **d.** F **e.** T **f.** T
 g. F **h.** F **i.** T **j.** F **k.** F **l.** F
 m. T **n.** F **o.** T **p.** T **q.** F **r.** T
 s. T **t.** F

3. The set of all (b_1, b_2, b_3) satisfying $b_1 + 2b_2 + b_3 = 0$.

5. The vector $\mathbf{p}_1$ is not zero and $\mathbf{p}_2$ is not a multiple of $\mathbf{p}_1$, so keep both of these vectors. Since $\mathbf{p}_3 = 2\mathbf{p}_1 + 2\mathbf{p}_2$, discard $\mathbf{p}_3$. Since $\mathbf{p}_4$ has a t^2 term, it cannot be a linear combination of $\mathbf{p}_1$ and $\mathbf{p}_2$, so keep $\mathbf{p}_4$. Finally, $\mathbf{p}_5 = \mathbf{p}_1 + \mathbf{p}_4$, so discard $\mathbf{p}_5$. The resulting basis is $\{\mathbf{p}_1, \mathbf{p}_2, \mathbf{p}_4\}$.

7. You would have to know that the solution set of the homogeneous system is spanned by two solutions. In this case, the null space of the 18×20 coefficient matrix A is at most two-dimensional. By the Rank Theorem, $\dim \text{Col } A \geq 20 - 2 = 18$, which means that $\text{Col } A = \mathbb{R}^{18}$, because A has 18 rows, and every equation $A\mathbf{x} = \mathbf{b}$ is consistent.

9. Let A be the standard $m \times n$ matrix of the transformation T.

a. If T is one-to-one, then the columns of A are linearly independent (Theorem 12 in Section 1.9), so $\dim \text{Nul } A = 0$. By the Rank Theorem, $\dim \text{Col } A = \text{rank } A = n$. Since the range of T is Col A, the dimension of the range of T is n.

b. If T is onto, then the columns of A span $\mathbb{R}^m$ (Theorem 12 in Section 1.9), so $\dim \text{Col } A = m$. By the Rank Theorem, $\dim \text{Nul } A = n - \dim \text{Col } A = n - m$. Since the kernel of T is Nul A, the dimension of the kernel of T is $n - m$.

11. If S is a finite spanning set for V, then a subset of S—say S'—is a basis for V. Since S' must span V, S' cannot be a proper subset of S because of the minimality of S. Thus $S' = S$, which proves that S is a basis for V.

12. a. *Hint:* Any $\mathbf{y}$ in Col AB has the form $\mathbf{y} = AB\mathbf{x}$ for some $\mathbf{x}$.

13. By Exercise 9, rank $PA \leq$ rank A, and rank $A =$ rank $P^{-1}PA \leq$ rank PA. Thus rank $PA =$ rank A.

15. The equation $AB = 0$ shows that each column of B is in Nul A. Since Nul A is a subspace, all linear combinations of the columns of B are in Nul A, so Col B is a subspace of Nul A. By Theorem 11 in Section 4.5, $\dim \text{Col } B \leq \dim \text{Nul } A$. Applying the Rank Theorem, we find that

$$n = \text{rank } A + \dim \text{Nul } A \geq \text{rank } A + \text{rank } B$$

17. a. Let A_1 consist of the r pivot columns in A. The columns of A_1 are linearly independent. So A_1 is an $m \times r$ submatrix with rank r.

b. By the Rank Theorem applied to A_1, the dimension of Row A is r, so A_1 has r linearly independent rows. Use them to form A_2. Then A_2 is $r \times r$ with linearly independent rows. By the Invertible Matrix Theorem, A_2 is invertible.

19. $\begin{bmatrix} B & AB & A^2B \end{bmatrix} = \begin{bmatrix} 0 & 1 & 0 \\ 1 & -.9 & .81 \\ 1 & .5 & .25 \end{bmatrix}$

$\sim \begin{bmatrix} 1 & -.9 & .81 \\ 0 & 1 & 0 \\ 0 & 0 & -.56 \end{bmatrix}$

This matrix has rank 3, so the pair (A, B) is controllable.

21. [M] rank $\begin{bmatrix} B & AB & A^2B & A^3B \end{bmatrix} = 3$. The pair (A, B) is not controllable.

Chapter 5

Section 5.1, page 271

1. Yes   **3.** Yes, $\lambda = -2$ **5.** Yes, $\lambda = -5$

7. Yes, $\begin{bmatrix} 1 \\ 1 \\ -1 \end{bmatrix}$

9. $\lambda = 1$: $\begin{bmatrix} 0 \\ 1 \end{bmatrix}$; $\lambda = 3$: $\begin{bmatrix} 1 \\ 1 \end{bmatrix}$

11. $\lambda = -1$: $\begin{bmatrix} 3 \\ 2 \end{bmatrix}$; $\lambda = 7$: $\begin{bmatrix} -1 \\ 2 \end{bmatrix}$

13. $\lambda = 1$: $\begin{bmatrix} 0 \\ 1 \\ 0 \end{bmatrix}$; $\lambda = 2$: $\begin{bmatrix} -1 \\ 2 \\ 2 \end{bmatrix}$; $\lambda = 3$: $\begin{bmatrix} -1 \\ 1 \\ 1 \end{bmatrix}$

15. $\begin{bmatrix} -1 \\ 1 \\ 0 \end{bmatrix}, \begin{bmatrix} -1 \\ 0 \\ 1 \end{bmatrix}$ **17.** $0, 3, -2$

19. 0. Justify your answer.

21. See the *Study Guide*, after you have written your answers.

23. *Hint:* Use Theorem 2.

25. *Hint:* Use the equation $A\mathbf{x} = \lambda\mathbf{x}$ to find an equation involving A^{-1}.

27. *Hint:* For any λ, $(A - \lambda I)^T = A^T - \lambda I$. By a theorem (which one?), $A^T - \lambda I$ is invertible if and only if $A - \lambda I$ is invertible.

29. Let $\mathbf{v}$ be the vector in $\mathbb{R}^n$ whose entries are all 1's. Then $A\mathbf{v} = s\mathbf{v}$.

31. *Hint:* If A is the standard matrix of T, look for a nonzero vector $\mathbf{v}$ (a point in the plane) such that $A\mathbf{v} = \mathbf{v}$.

33. a. $\mathbf{x}_{k+1} = c_1 \lambda^{k+1} \mathbf{u} + c_2 \mu^{k+1} \mathbf{v}$

b. $A\mathbf{x}_k = A(c_1 \lambda^k \mathbf{u} + c_2 \mu^k \mathbf{v})$
$= c_1 \lambda^k A\mathbf{u} + c_2 \mu^k A\mathbf{v}$ Linearity
$= c_1 \lambda^k \lambda \mathbf{u} + c_2 \mu^k \mu \mathbf{v}$ **u** and **v** are eigenvectors.
$= \mathbf{x}_{k+1}$

35.

37. **[M]** $\lambda = 5$: $\begin{bmatrix} -1 \\ -1 \\ 2 \end{bmatrix}$; $\lambda = 10$: $\begin{bmatrix} -3 \\ 2 \\ 1 \end{bmatrix}$; $\lambda = 15$: $\begin{bmatrix} 2 \\ 2 \\ 1 \end{bmatrix}$

39. **[M]** $\lambda = -4$: $\begin{bmatrix} 0 \\ 1 \\ 2 \\ 0 \\ 1 \end{bmatrix}$; $\lambda = -8$: $\begin{bmatrix} 6 \\ 3 \\ 3 \\ 2 \\ 0 \end{bmatrix}, \begin{bmatrix} 0 \\ 0 \\ -1 \\ 0 \\ 1 \end{bmatrix}$;

$\lambda = 12$: $\begin{bmatrix} 0 \\ 0 \\ -1 \\ 1 \\ 0 \end{bmatrix}, \begin{bmatrix} 2 \\ 1 \\ 2 \\ 0 \\ 1 \end{bmatrix}$

Section 5.2, page 279

1. $\lambda^2 - 4\lambda - 45$; $9, -5$ **3.** $\lambda^2 - 3\lambda - 40$; $-5, 8$

5. $\lambda^2 - 16\lambda + 48$; $4, 12$

7. $\lambda^2 - 9\lambda + 32$; no real eigenvalues

9. $-\lambda^3 + 10\lambda^2 - 33\lambda + 36$ **11.** $-\lambda^3 + 8\lambda^2 - 19\lambda + 12$

13. $-\lambda^3 + 18\lambda^2 - 95\lambda + 150$ **15.** $2, 3, 5, 5$

17. $3, 3, 1, 1, 0$

19. *Hint:* The equation given holds for all λ.

21. The *Study Guide* has hints.

23. *Hint:* Find an invertible matrix P such that $RQ = P^{-1}AP$.

25. a. $\{\mathbf{v}_1, \mathbf{v}_2\}$, where $\mathbf{v}_2 = \begin{bmatrix} -1 \\ 1 \end{bmatrix}$ is an eigenvector for $\lambda = .3$

b. $\mathbf{x}_0 = \mathbf{v}_1 - \frac{1}{14}\mathbf{v}_2$

c. $\mathbf{x}_1 = \mathbf{v}_1 - \frac{1}{14}(.3)\mathbf{v}_2$, $\mathbf{x}_2 = \mathbf{v}_1 - \frac{1}{14}(.3)^2\mathbf{v}_2$, and $\mathbf{x}_k = \mathbf{v}_1 - \frac{1}{14}(.3)^k\mathbf{v}_2$. As $k \to \infty$, $(.3)^k \to 0$ and $\mathbf{x}_k \to \mathbf{v}_1$.

27. a. $A\mathbf{v}_1 = \mathbf{v}_1$, $A\mathbf{v}_2 = .5\mathbf{v}_2$, $A\mathbf{v}_3 = .2\mathbf{v}_3$. (This also shows that the eigenvalues of A are 1, .5, and .2.)

b. $\{\mathbf{v}_1, \mathbf{v}_2, \mathbf{v}_3\}$ is linearly independent because the eigenvectors correspond to distinct eigenvalues (Theorem 2). Since there are 3 vectors in the set, the set is a basis for $\mathbb{R}^3$. So there exist (unique) constants such that

$$\mathbf{x}_0 = c_1 \mathbf{v}_1 + c_2 \mathbf{v}_2 + c_3 \mathbf{v}_3$$

Then

$$\mathbf{w}\mathbf{x}_0 = c_1 \mathbf{w}^T\mathbf{v}_1 + c_2 \mathbf{w}^T\mathbf{v}_2 + c_3 \mathbf{w}^T\mathbf{v}_3 \qquad (*)$$

Since $\mathbf{x}_0$ and $\mathbf{v}_1$ are probability vectors and since the entries in $\mathbf{v}_2$ and in $\mathbf{v}_3$ each sum to 0, $(*)$ shows that $1 = c_1$.

c. By part (b),

$$\mathbf{x}_0 = \mathbf{v}_1 + c_2 \mathbf{v}_2 + c_3 \mathbf{v}_3$$

Using part (a),

$$\mathbf{x}_k = A^k\mathbf{x}_0 = A^k\mathbf{v}_1 + c_2 A^k\mathbf{v}_2 + c_3 A^k\mathbf{v}_3$$
$$= \mathbf{v}_1 + c_2(.5)^k\mathbf{v}_2 + c_3(.2)^k\mathbf{v}_3$$
$$\to \mathbf{v}_1 \text{ as } k \to \infty$$

29. [M] Report your results and conclusions. You can avoid tedious calculations if you use the program `gauss` discussed in the *Study Guide*.

Section 5.3, page 286

1. $\begin{bmatrix} 226 & -525 \\ 90 & -209 \end{bmatrix}$ **3.** $\begin{bmatrix} a^k & 0 \\ 2(a^k - b^k) & b^k \end{bmatrix}$

5. $\lambda = 2: \begin{bmatrix} -1 \\ 1 \\ -1 \end{bmatrix}$; $\lambda = 3: \begin{bmatrix} 1 \\ -1 \\ 0 \end{bmatrix}, \begin{bmatrix} 0 \\ -1 \\ 1 \end{bmatrix}$

When an answer involves a diagonalization, $A = PDP^{-1}$, the factors P and D are not unique, so your answer may differ from that given here.

7. $P = \begin{bmatrix} 1 & 0 \\ 3 & 1 \end{bmatrix}, D = \begin{bmatrix} 1 & 0 \\ 0 & -1 \end{bmatrix}$

9. Not diagonalizable

11. $P = \begin{bmatrix} 1 & -1 & -1 \\ 2 & 1 & 0 \\ 3 & 0 & 1 \end{bmatrix}, D = \begin{bmatrix} 5 & 0 & 0 \\ 0 & -1 & 0 \\ 0 & 0 & -1 \end{bmatrix}$

13. $P = \begin{bmatrix} -1 & 2 & 1 \\ -1 & -1 & 0 \\ 1 & 0 & 1 \end{bmatrix}, D = \begin{bmatrix} 5 & 0 & 0 \\ 0 & 1 & 0 \\ 0 & 0 & 1 \end{bmatrix}$

15. $P = \begin{bmatrix} -1 & -1 & -1 \\ 1 & 1 & 0 \\ -1 & 0 & 1 \end{bmatrix}, D = \begin{bmatrix} 0 & 0 & 0 \\ 0 & 1 & 0 \\ 0 & 0 & 1 \end{bmatrix}$

17. Not diagonalizable

19. $P = \begin{bmatrix} 1 & 3 & -1 & -1 \\ 0 & 2 & -1 & 2 \\ 0 & 0 & 1 & 0 \\ 0 & 0 & 0 & 1 \end{bmatrix}, D = \begin{bmatrix} 5 & 0 & 0 & 0 \\ 0 & 3 & 0 & 0 \\ 0 & 0 & 2 & 0 \\ 0 & 0 & 0 & 2 \end{bmatrix}$

21. See the *Study Guide*.   **23.** Yes. (Explain why.)

25. No, A must be diagonalizable. (Explain why.)

27. *Hint:* Write $A = PDP^{-1}$. Since A is invertible, 0 is not an eigenvalue of A, so D has nonzero entries on its diagonal.

29. One answer is $P_1 = \begin{bmatrix} 1 & 1 \\ -2 & -1 \end{bmatrix}$, whose columns are eigenvectors corresponding to the eigenvalues in D_1.

31. *Hint:* Construct a suitable 2×2 triangular matrix.

33. [M] $P = \begin{bmatrix} 2 & 0 & -1 & 1 \\ 7 & -1 & 0 & -4 \\ 7 & 0 & 2 & 0 \\ 0 & 1 & 0 & 3 \end{bmatrix}$,

$D = \begin{bmatrix} -12 & 0 & 0 & 0 \\ 0 & -12 & 0 & 0 \\ 0 & 0 & 13 & 0 \\ 0 & 0 & 0 & 13 \end{bmatrix}$

35. [M] $P = \begin{bmatrix} 2 & 1 & -3 & 1 & 0 \\ 1 & 0 & 0 & 1 & 0 \\ 0 & 1 & 0 & 0 & -1 \\ 0 & 1 & 0 & 1 & 0 \\ 0 & 0 & 2 & 0 & 1 \end{bmatrix}$,

$D = \begin{bmatrix} 7 & 0 & 0 & 0 & 0 \\ 0 & 7 & 0 & 0 & 0 \\ 0 & 0 & 7 & 0 & 0 \\ 0 & 0 & 0 & -14 & 0 \\ 0 & 0 & 0 & 0 & -14 \end{bmatrix}$

Section 5.4, page 293

1. $\begin{bmatrix} 3 & -1 & 0 \\ -5 & 6 & 4 \end{bmatrix}$

3. a. $T(\mathbf{e}_1) = \mathbf{b}_3, T(\mathbf{e}_2) = -\mathbf{b}_1 - 2\mathbf{b}_2, T(\mathbf{e}_3) = 2\mathbf{b}_1 + 3\mathbf{b}_3$

b. $[T(\mathbf{e}_1)]_{\mathcal{B}} = \begin{bmatrix} 0 \\ 0 \\ 1 \end{bmatrix}, [T(\mathbf{e}_2)]_{\mathcal{B}} = \begin{bmatrix} -1 \\ -2 \\ 0 \end{bmatrix}$,

$[T(\mathbf{e}_3)]_{\mathcal{B}} = \begin{bmatrix} 2 \\ 0 \\ 3 \end{bmatrix}$

c. $\begin{bmatrix} 0 & -1 & 2 \\ 0 & -2 & 0 \\ 1 & 0 & 3 \end{bmatrix}$

5. a. $9 - 3t + t^2 + t^3$

b. For any $\mathbf{p}, \mathbf{q}$ in $\mathbb{P}_2$ and any scalar c,

$$T[\mathbf{p}(t) + \mathbf{q}(t)] = (t + 3)[\mathbf{p}(t) + \mathbf{q}(t)]$$
$$= (t + 3)\mathbf{p}(t) + (t + 3)\mathbf{q}(t)$$
$$= T[\mathbf{p}(t)] + T[\mathbf{q}(t)]$$
$$T[c \cdot \mathbf{p}(t)] = (t + 3)[c \cdot \mathbf{p}(t)] = c \cdot (t + 3)\mathbf{p}(t)$$
$$= c \cdot T[\mathbf{p}(t)]$$

c. $\begin{bmatrix} 3 & 0 & 0 \\ 1 & 3 & 0 \\ 0 & 1 & 3 \\ 0 & 0 & 1 \end{bmatrix}$

7. $\begin{bmatrix} 3 & 0 & 0 \\ 5 & -2 & 0 \\ 0 & 4 & 1 \end{bmatrix}$

9. a. $\begin{bmatrix} 2 \\ 5 \\ 8 \end{bmatrix}$

b. *Hint:* Compute $T(\mathbf{p} + \mathbf{q})$ and $T(c \cdot \mathbf{p})$ for arbitrary $\mathbf{p}, \mathbf{q}$ in $\mathbb{P}_2$ and an arbitrary scalar c.

c. $\begin{bmatrix} 1 & -1 & 1 \\ 1 & 0 & 0 \\ 1 & 1 & 1 \end{bmatrix}$

11. $\begin{bmatrix} -2 & -2 \\ 0 & -1 \end{bmatrix}$ **13.** $\mathbf{b}_1 = \begin{bmatrix} 1 \\ 1 \end{bmatrix}, \mathbf{b}_2 = \begin{bmatrix} 1 \\ 3 \end{bmatrix}$

15. $\mathbf{b}_1 = \begin{bmatrix} 2 \\ 1 \end{bmatrix}, \mathbf{b}_2 = \begin{bmatrix} -1 \\ 3 \end{bmatrix}$

17. **a.** $A\mathbf{b}_1 = 3\mathbf{b}_1$, so $\mathbf{b}_1$ is an eigenvector of A. However, A has only one eigenvalue, $\lambda = 3$, and the eigenspace is only one-dimensional, so A is not diagonalizable.

b. $\begin{bmatrix} 3 & 1 \\ 0 & 3 \end{bmatrix}$

19. By definition, if A is similar to B, there exists an invertible matrix P such that $P^{-1}AP = B$. (See Section 5.2.) Then B is invertible because it is the product of invertible matrices. To show that A^{-1} is similar to B^{-1}, use the equation $P^{-1}AP = B$. See the *Study Guide*.

21. *Hint:* Review Practice Problem 2.

23. *Hint:* Compute $B(P^{-1}\mathbf{x})$.

25. *Hint:* Write $A = PBP^{-1} = (PB)P^{-1}$, and use the trace property.

27. For each j, $I(\mathbf{b}_j) = \mathbf{b}_j$. Since the standard coordinate vector of any vector in $\mathbb{R}^n$ is just the vector itself, $[I(\mathbf{b}_j)]_{\mathcal{E}} = \mathbf{b}_j$. Thus the matrix for I relative to $\mathcal{B}$ and the standard basis $\mathcal{E}$ is simply $[\,\mathbf{b}_1 \quad \mathbf{b}_2 \quad \cdots \quad \mathbf{b}_n\,]$. This matrix is precisely the *change-of-coordinates* matrix $P_{\mathcal{B}}$ defined in Section 4.4.

29. The $\mathcal{B}$-matrix for the identity transformation is I_n, because the $\mathcal{B}$-coordinate vector of the jth basis vector $\mathbf{b}_j$ is the jth column of I_n.

31. **[M]** $\begin{bmatrix} -7 & -2 & -6 \\ 0 & -4 & -6 \\ 0 & 0 & -1 \end{bmatrix}$

Section 5.5, page 300

1. $\lambda = 2 + i, \begin{bmatrix} -1+i \\ 1 \end{bmatrix}; \quad \lambda = 2 - i, \begin{bmatrix} -1-i \\ 1 \end{bmatrix}$

3. $\lambda = 3 + 2i, \begin{bmatrix} -1-i \\ 4 \end{bmatrix}; \quad \lambda = 3 - 2i, \begin{bmatrix} -1+i \\ 4 \end{bmatrix}$

5. $\lambda = 4 + i, \begin{bmatrix} 1-i \\ 2 \end{bmatrix}; \quad \lambda = 4 - i, \begin{bmatrix} 1+i \\ 2 \end{bmatrix}$

7. $\lambda = \sqrt{3} \pm i, \varphi = \pi/6$ radian, $r = 2$

9. $\lambda = \pm 2i, \varphi = -\pi/2$ radians, $r = 2$

11. $\lambda = -\sqrt{3} \pm i, \varphi = -5\pi/6$ radian, $r = 2$

In Exercises 13–20, other answers are possible. Any P that makes $P^{-1}AP$ equal to the given C or to C^T is a satisfactory answer. First find P; then compute $P^{-1}AP$.

13. $P = \begin{bmatrix} -1 & -1 \\ 1 & 0 \end{bmatrix}, C = \begin{bmatrix} 2 & -1 \\ 1 & 2 \end{bmatrix}$

15. $P = \begin{bmatrix} 3 & 1 \\ 0 & 2 \end{bmatrix}, C = \begin{bmatrix} 1 & 3 \\ -3 & 1 \end{bmatrix}$

17. $P = \begin{bmatrix} 1 & 2 \\ 0 & -5 \end{bmatrix}, C = \begin{bmatrix} -3 & 4 \\ -4 & -3 \end{bmatrix}$

19. $P = \begin{bmatrix} 2 & -1 \\ 2 & 0 \end{bmatrix}, C = \begin{bmatrix} .96 & -.28 \\ .28 & .96 \end{bmatrix}$

21. $\mathbf{y} = \begin{bmatrix} 2 \\ -1+2i \end{bmatrix} = \dfrac{-1+2i}{5} \begin{bmatrix} -2-4i \\ 5 \end{bmatrix}$

23. **a.** Properties of conjugates and the fact that $\overline{\mathbf{x}}^T = \overline{\mathbf{x}^T}$;

b. $\overline{A\mathbf{x}} = A\overline{\mathbf{x}}$ and A is real; (c) because $\mathbf{x}^T A\overline{\mathbf{x}}$ is a scalar and hence may be viewed as a 1×1 matrix; (d) properties of transposes; (e) $A^T = A$, definition of q

25. *Hint:* First write $\mathbf{x} = \operatorname{Re}\mathbf{x} + i(\operatorname{Im}\mathbf{x})$.

27. **[M]** $P = \begin{bmatrix} -1 & 1 & -1 & -1 \\ 0 & -1 & 0 & 2 \\ 1 & 0 & 0 & -2 \\ 0 & 0 & 2 & 0 \end{bmatrix}$,

$C = \begin{bmatrix} -2 & 5 & 0 & 0 \\ -5 & -2 & 0 & 0 \\ 0 & 0 & -4 & 10 \\ 0 & 0 & -10 & -4 \end{bmatrix}$

Other choices are possible, but C must equal $P^{-1}AP$.

Section 5.6, page 309

1. **a.** *Hint:* Find c_1, c_2 such that $\mathbf{x}_0 = c_1\mathbf{v}_1 + c_2\mathbf{v}_2$. Use this representation and the fact that $\mathbf{v}_1$ and $\mathbf{v}_2$ are eigenvectors of A to compute $\mathbf{x}_1 = \begin{bmatrix} 49/3 \\ 41/3 \end{bmatrix}$.

b. In general, $\mathbf{x}_k = 5(3)^k \mathbf{v}_1 - 4(\frac{1}{3})^k \mathbf{v}_2$ for $k \ge 0$.

3. When $p = .2$, the eigenvalues of A are .9 and .7, and

$$\mathbf{x}_k = c_1(.9)^k \begin{bmatrix} 1 \\ 1 \end{bmatrix} + c_2(.7)^k \begin{bmatrix} 2 \\ 1 \end{bmatrix} \to \mathbf{0} \quad \text{as } k \to \infty$$

The higher predation rate cuts down the owls' food supply, and eventually both predator and prey populations perish.

5. If $p = .325$, the eigenvalues are 1.05 and .55. Since $1.05 > 1$, both populations will grow at 5% per year. An eigenvector for 1.05 is $(6, 13)$, so eventually there will be approximately 6 spotted owls to every 13 (thousand) flying squirrels.

7. **a.** The origin is a saddle point because A has one eigenvalue larger than 1 and one smaller than 1 (in absolute value).

b. The direction of greatest attraction is given by the eigenvector corresponding to the eigenvalue $1/3$, namely, $\mathbf{v}_2$. All vectors that are multiples of $\mathbf{v}_2$ are attracted to the origin. The direction of greatest repulsion is given by the eigenvector $\mathbf{v}_1$. All multiples of $\mathbf{v}_1$ are repelled.

c. See the *Study Guide*.

9. Saddle point; eigenvalues: 2, .5; direction of greatest repulsion: the line through $(0, 0)$ and $(-1, 1)$; direction of greatest attraction: the line through $(0, 0)$ and $(1, 4)$

11. Attractor; eigenvalues: .9, .8; greatest attraction: line through $(0, 0)$ and $(5, 4)$

13. Repeller; eigenvalues: 1.2, 1.1; greatest repulsion: line through $(0, 0)$ and $(3, 4)$

15. $\mathbf{x}_k = \mathbf{v}_1 + .1(.5)^k \begin{bmatrix} 2 \\ -3 \\ 1 \end{bmatrix} + .3(.2)^k \begin{bmatrix} -1 \\ 0 \\ 1 \end{bmatrix} \rightarrow \mathbf{v}_1$ as $k \rightarrow \infty$

17. a. $A = \begin{bmatrix} 0 & 1.6 \\ .3 & .8 \end{bmatrix}$

b. The population is growing because the largest eigenvalue of A is 1.2, which is larger than 1 in magnitude. The eventual growth rate is 1.2, which is 20% per year. The eigenvector $(4, 3)$ for $\lambda_1 = 1.2$ shows that there will be 4 juveniles for every 3 adults.

c. **[M]** The juvenile–adult ratio seems to stabilize after about 5 or 6 years. The *Study Guide* describes how to construct a matrix program to generate a data matrix whose columns list the numbers of juveniles and adults each year. Graphing the data is also discussed.

Section 5.7, page 317

1. $\mathbf{x}(t) = \dfrac{5}{2}\begin{bmatrix} -3 \\ 1 \end{bmatrix}e^{4t} - \dfrac{3}{2}\begin{bmatrix} -1 \\ 1 \end{bmatrix}e^{2t}$

3. $-\dfrac{5}{2}\begin{bmatrix} -3 \\ 1 \end{bmatrix}e^{t} + \dfrac{9}{2}\begin{bmatrix} -1 \\ 1 \end{bmatrix}e^{-t}$. The origin is a saddle point. The direction of greatest attraction is the line through $(-1, 1)$ and the origin. The direction of greatest repulsion is the line through $(-3, 1)$ and the origin.

5. $-\dfrac{1}{2}\begin{bmatrix} 1 \\ 3 \end{bmatrix}e^{4t} + \dfrac{7}{2}\begin{bmatrix} 1 \\ 1 \end{bmatrix}e^{6t}$. The origin is a repeller. The direction of greatest repulsion is the line through $(1, 1)$ and the origin.

7. Set $P = \begin{bmatrix} 1 & 1 \\ 3 & 1 \end{bmatrix}$ and $D = \begin{bmatrix} 4 & 0 \\ 0 & 6 \end{bmatrix}$. Then $A = PDP^{-1}$. Substituting $\mathbf{x} = P\mathbf{y}$ into $\mathbf{x}' = A\mathbf{x}$, we have

$$\frac{d}{dt}(P\mathbf{y}) = A(P\mathbf{y})$$
$$P\mathbf{y}' = PDP^{-1}(P\mathbf{y}) = PD\mathbf{y}$$

Left-multiplying by P^{-1} gives

$$\mathbf{y}' = D\mathbf{y}, \quad \text{or} \quad \begin{bmatrix} y_1'(t) \\ y_2'(t) \end{bmatrix} = \begin{bmatrix} 4 & 0 \\ 0 & 6 \end{bmatrix}\begin{bmatrix} y_1(t) \\ y_2(t) \end{bmatrix}$$

9. (complex solution):

$c_1\begin{bmatrix} 1-i \\ 1 \end{bmatrix}e^{(-2+i)t} + c_2\begin{bmatrix} 1+i \\ 1 \end{bmatrix}e^{(-2-i)t}$

(real solution):

$c_1\begin{bmatrix} \cos t + \sin t \\ \cos t \end{bmatrix}e^{-2t} + c_2\begin{bmatrix} \sin t - \cos t \\ \sin t \end{bmatrix}e^{-2t}$

The trajectories spiral in toward the origin.

11. (complex): $c_1\begin{bmatrix} -3+3i \\ 2 \end{bmatrix}e^{3it} + c_2\begin{bmatrix} -3-3i \\ 2 \end{bmatrix}e^{-3it}$

(real):

$c_1\begin{bmatrix} -3\cos 3t - 3\sin 3t \\ 2\cos 3t \end{bmatrix} + c_2\begin{bmatrix} -3\sin 3t + 3\cos 3t \\ 2\sin 3t \end{bmatrix}$

The trajectories are ellipses about the origin.

13. (complex): $c_1\begin{bmatrix} 1+i \\ 2 \end{bmatrix}e^{(1+3i)t} + c_2\begin{bmatrix} 1-i \\ 2 \end{bmatrix}e^{(1-3i)t}$

(real): $c_1\begin{bmatrix} \cos 3t - \sin 3t \\ 2\cos 3t \end{bmatrix}e^t + c_2\begin{bmatrix} \sin 3t + \cos 3t \\ 2\sin 3t \end{bmatrix}e^t$

The trajectories spiral out, away from the origin.

15. **[M]** $\mathbf{x}(t) = c_1\begin{bmatrix} -1 \\ 0 \\ 1 \end{bmatrix}e^{-2t} + c_2\begin{bmatrix} -6 \\ 1 \\ 5 \end{bmatrix}e^{-t} + c_3\begin{bmatrix} -4 \\ 1 \\ 4 \end{bmatrix}e^{t}$

The origin is a saddle point. A solution with $c_3 = 0$ is attracted to the origin. A solution with $c_1 = c_2 = 0$ is repelled.

17. **[M]** (complex):

$c_1\begin{bmatrix} -3 \\ 1 \\ 1 \end{bmatrix}e^{t} + c_2\begin{bmatrix} 23-34i \\ -9+14i \\ 3 \end{bmatrix}e^{(5+2i)t} +$

$c_3\begin{bmatrix} 23+34i \\ -9-14i \\ 3 \end{bmatrix}e^{(5-2i)t}$

(real): $c_1\begin{bmatrix} -3 \\ 1 \\ 1 \end{bmatrix}e^{t} + c_2\begin{bmatrix} 23\cos 2t + 34\sin 2t \\ -9\cos 2t - 14\sin 2t \\ 3\cos 2t \end{bmatrix}e^{5t} +$

$c_3\begin{bmatrix} 23\sin 2t - 34\cos 2t \\ -9\sin 2t + 14\cos 2t \\ 3\sin 2t \end{bmatrix}e^{5t}$

The origin is a repeller. The trajectories spiral outward, away from the origin.

19. **[M]** $A = \begin{bmatrix} -2 & 3/4 \\ 1 & -1 \end{bmatrix}$,

$\begin{bmatrix} v_1(t) \\ v_2(t) \end{bmatrix} = \dfrac{5}{2}\begin{bmatrix} 1 \\ 2 \end{bmatrix}e^{-.5t} - \dfrac{1}{2}\begin{bmatrix} -3 \\ 2 \end{bmatrix}e^{-2.5t}$

21. **[M]** $A = \begin{bmatrix} -1 & -8 \\ 5 & -5 \end{bmatrix}$,

$\begin{bmatrix} i_L(t) \\ v_C(t) \end{bmatrix} = \begin{bmatrix} -20\sin 6t \\ 15\cos 6t - 5\sin 6t \end{bmatrix}e^{-3t}$

Section 5.8, page 324

1. Eigenvector: $\mathbf{x}_4 = \begin{bmatrix} 1 \\ .3326 \end{bmatrix}$, or $A\mathbf{x}_4 = \begin{bmatrix} 4.9978 \\ 1.6652 \end{bmatrix}$; $\lambda \approx 4.9978$

3. Eigenvector: $\mathbf{x}_4 = \begin{bmatrix} .5188 \\ 1 \end{bmatrix}$, or $A\mathbf{x}_4 = \begin{bmatrix} .4594 \\ .9075 \end{bmatrix}$; $\lambda \approx .9075$

5. $\mathbf{x} = \begin{bmatrix} -.7999 \\ 1 \end{bmatrix}$, $A\mathbf{x} = \begin{bmatrix} 4.0015 \\ -5.0020 \end{bmatrix}$; estimated $\lambda = -5.0020$

7. **[M]**

$\mathbf{x}_k: \begin{bmatrix} .75 \\ 1 \end{bmatrix}, \begin{bmatrix} 1 \\ .9565 \end{bmatrix}, \begin{bmatrix} .9932 \\ 1 \end{bmatrix}, \begin{bmatrix} 1 \\ .9990 \end{bmatrix}, \begin{bmatrix} .9998 \\ 1 \end{bmatrix}$

$\mu_k:$ 11.5, 12.78, 12.96, 12.9948, 12.9990

9. **[M]** $\mu_5 = 8.4233$, $\mu_6 = 8.4246$; actual value: 8.42443 (accurate to 5 places)

11. μ_k: 5.8000, 5.9655, 5.9942, 5.9990 $(k = 1, 2, 3, 4)$;
$R(\mathbf{x}_k)$: 5.9655, 5.9990, 5.99997, 5.9999993

13. Yes, but the sequences may converge very slowly.

15. *Hint:* Write $A\mathbf{x} - \alpha\mathbf{x} = (A - \alpha I)\mathbf{x}$, and use the fact that $(A - \alpha I)$ is invertible when α is *not* an eigenvalue of A.

17. [M] $\nu_0 = 3.3384$, $\nu_1 = 3.32119$ (accurate to 4 places with rounding), $\nu_2 = 3.3212209$. Actual value: 3.3212201 (accurate to 7 places)

19. [M] **a.** $\mu_6 = 30.2887 = \mu_7$ to four decimal places. To six places, the largest eigenvalue is 30.288685, with eigenvector (.957629, .688937, 1, .943782).

  b. The inverse power method (with $\alpha = 0$) produces $\mu_1^{-1} = .010141$, $\mu_2^{-1} = .010150$. To seven places, the smallest eigenvalue is .0101500, with eigenvector $(-.603972, 1, -.251135, .148953)$. The reason for the rapid convergence is that the next-to-smallest eigenvalue is near .85.

21. **a.** If the eigenvalues of A are all less than 1 in magnitude, and if $\mathbf{x} \neq \mathbf{0}$, then $A^k\mathbf{x}$ is approximately an eigenvector for large k.

  b. If the strictly dominant eigenvalue is 1, and if $\mathbf{x}$ has a component in the direction of the corresponding eigenvector, then $\{A^k\mathbf{x}\}$ will converge to a multiple of that eigenvector.

  c. If the eigenvalues of A are all greater than 1 in magnitude, and if $\mathbf{x}$ is not an eigenvector, then the distance from $A^k\mathbf{x}$ to the nearest eigenvector will *increase* as $k \to \infty$.

Chapter 5 Supplementary Exercises, page 326

1. a. T   **b.** F   **c.** T   **d.** F   **e.** T   **f.** T
  g. F   **h.** T   **i.** F   **j.** T   **k.** F   **l.** F
  m. F   **n.** T   **o.** F   **p.** T   **q.** F   **r.** T
  s. F   **t.** T   **u.** T   **v.** T   **w.** F   **x.** T

3. a. Suppose $A\mathbf{x} = \lambda\mathbf{x}$, with $\mathbf{x} \neq \mathbf{0}$. Then

$$(5I - A)\mathbf{x} = 5\mathbf{x} - A\mathbf{x} = 5\mathbf{x} - \lambda\mathbf{x} = (5 - \lambda)\mathbf{x}.$$

  The eigenvalue is $5 - \lambda$.

  b. $(5I - 3A + A^2)\mathbf{x} = 5\mathbf{x} - 3A\mathbf{x} + A(A\mathbf{x})$
$$= 5\mathbf{x} - 3\lambda\mathbf{x} + \lambda^2\mathbf{x}$$
$$= (5 - 3\lambda + \lambda^2)\mathbf{x}.$$

  The eigenvalue is $5 - 3\lambda + \lambda^2$.

5. Suppose $A\mathbf{x} = \lambda\mathbf{x}$, with $\mathbf{x} \neq \mathbf{0}$. Then

$$p(A)\mathbf{x} = (c_0 I + c_1 A + c_2 A^2 + \cdots + c_n A^n)\mathbf{x}$$
$$= c_0\mathbf{x} + c_1 A\mathbf{x} + c_2 A^2\mathbf{x} + \cdots + c_n A^n\mathbf{x}$$
$$= c_0\mathbf{x} + c_1\lambda\mathbf{x} + c_2\lambda^2\mathbf{x} + \cdots + c_n\lambda^n\mathbf{x} = p(\lambda)\mathbf{x}$$

So $p(\lambda)$ is an eigenvalue of the matrix $p(A)$.

7. If $A = PDP^{-1}$, then $p(A) = Pp(D)P^{-1}$, as shown in Exercise 6. If the (j, j) entry in D is λ, then the (j, j) entry in D^k is λ^k, and so the (j, j) entry in $p(D)$ is $p(\lambda)$. If p is the characteristic polynomial of A, then $p(\lambda) = 0$

for each diagonal entry of D, because these entries in D are the eigenvalues of A. Thus $p(D)$ is the zero matrix. Thus $p(A) = P \cdot 0 \cdot P^{-1} = 0$.

9. If $I - A$ were not invertible, then the equation $(I - A)\mathbf{x} = \mathbf{0}$ would have a nontrivial solution $\mathbf{x}$. Then $\mathbf{x} - A\mathbf{x} = \mathbf{0}$ and $A\mathbf{x} = 1 \cdot \mathbf{x}$, which shows that A would have 1 as an eigenvalue. This cannot happen if all the eigenvalues are less than 1 in magnitude. So $I - A$ must be invertible.

11. a. Take $\mathbf{x}$ in H. Then $\mathbf{x} = c\mathbf{u}$ for some scalar c. So $A\mathbf{x} = A(c\mathbf{u}) = c(A\mathbf{u}) = c(\lambda\mathbf{u}) = (c\lambda)\mathbf{u}$, which shows that $A\mathbf{x}$ is in H.

  b. Let $\mathbf{x}$ be a nonzero vector in K. Since K is one-dimensional, K must be the set of all scalar multiples of $\mathbf{x}$. If K is invariant under A, then $A\mathbf{x}$ is in K and hence $A\mathbf{x}$ is a multiple of $\mathbf{x}$. Thus $\mathbf{x}$ is an eigenvector of A.

13. 1, 3, 7

15. Replace a by $a - \lambda$ in the determinant formula from Exercise 16 in Chapter 3 Supplementary Exercises:

$$\det(A - \lambda I) = (a - b - \lambda)^{n-1}[a - \lambda + (n - 1)b]$$

This determinant is zero only if $a - b - \lambda = 0$ or $a - \lambda + (n - 1)b = 0$. Thus λ is an eigenvalue of A if and only if $\lambda = a - b$ or $\lambda = a + (n - 1)b$. From the formula for $\det(A - \lambda I)$ above, the algebraic multiplicity is $n - 1$ for $a - b$ and 1 for $a + (n - 1)b$.

17. $\det(A - \lambda I) = (a_{11} - \lambda)(a_{22} - \lambda) - a_{12}a_{21} =$
$\lambda^2 - (a_{11} + a_{22})\lambda + (a_{11}a_{22} - a_{12}a_{21}) =$
$\lambda^2 - (\text{tr } A)\lambda + \det A$. Use the quadratic formula to solve the characteristic equation:

$$\lambda = \frac{\text{tr } A \pm \sqrt{(\text{tr } A)^2 - 4\det A}}{2}$$

The eigenvalues are both real if and only if the discriminant is nonnegative, that is, $(\text{tr } A)^2 - 4\det A \geq 0$. This inequality simplifies to $(\text{tr } A)^2 \geq 4\det A$ and $\left(\dfrac{\text{tr } A}{2}\right)^2 \geq \det A$.

19. $C_p = \begin{bmatrix} 0 & 1 \\ -6 & 5 \end{bmatrix}$; $\det(C_p - \lambda I) = 6 - 5\lambda + \lambda^2 = p(\lambda)$

21. If p is a polynomial of order 2, then a calculation such as in Exercise 19 shows that the characteristic polynomial of C_p is $p(\lambda) = (-1)^2 p(\lambda)$, so the result is true for $n = 2$. Suppose the result is true for $n = k$ for some $k \geq 2$, and consider a polynomial p of degree $k + 1$. Then, expanding $\det(C_p - \lambda I)$ by cofactors down the first column, the determinant of $C_p - \lambda I$ equals

$$(-\lambda) \det \begin{bmatrix} -\lambda & 1 & \cdots & 0 \\ \vdots & & & \vdots \\ 0 & & & 1 \\ -a_1 & -a_2 & \cdots & -a_k - \lambda \end{bmatrix} + (-1)^{k+1} a_0$$

The $k \times k$ matrix shown is $C_q - \lambda I$, where $q(t) = a_1 + a_2 t + \cdots + a_k t^{k-1} + t^k$. By the induction assumption, the determinant of $C_q - \lambda I$ is $(-1)^k q(\lambda)$. Thus

$$
\begin{aligned}
\det(C_p - \lambda I) &= (-1)^{k+1} a_0 + (-\lambda)(-1)^k q(\lambda) \\
&= (-1)^{k+1}[a_0 + \lambda(a_1 + \cdots + a_k \lambda^{k-1} + \lambda^k)] \\
&= (-1)^{k+1} p(\lambda)
\end{aligned}
$$

So the formula holds for $n = k + 1$ when it holds for $n = k$. By the principle of induction, the formula for $\det(C_p - \lambda I)$ is true for all $n \geq 2$.

23. From Exercise 22, the columns of the Vandermonde matrix V are eigenvectors of C_p, corresponding to the eigenvalues $\lambda_1, \lambda_2, \lambda_3$ (the roots of the polynomial p). Since these eigenvalues are distinct, the eigenvectors form a linearly independent set, by Theorem 2 in Section 5.1. Thus V has linearly independent columns and hence is invertible, by the Invertible Matrix Theorem. Finally, since the columns of V are eigenvectors of C_p, the Diagonalization Theorem (Theorem 5 in Section 5.3) shows that $V^{-1} C_p V$ is diagonal.

25. **[M]** If your matrix program computes eigenvalues and eigenvectors by iterative methods rather than symbolic calculations, you may have some difficulties. You should find that $AP - PD$ has extremely small entries and PDP^{-1} is close to A. (This was true just a few years ago, but the situation could change as matrix programs continue to improve.) If you constructed P from the program's eigenvectors, check the condition number of P. This may indicate that you do not really have three linearly independent eigenvectors.

Chapter 6

Section 6.1, page 336

1. $5, 8, \frac{8}{5}$ 3. $\begin{bmatrix} 3/35 \\ -1/35 \\ -1/7 \end{bmatrix}$ 5. $\begin{bmatrix} 8/13 \\ 12/13 \end{bmatrix}$

7. $\sqrt{35}$ 9. $\begin{bmatrix} -.6 \\ .8 \end{bmatrix}$ 11. $\begin{bmatrix} 7/\sqrt{69} \\ 2/\sqrt{69} \\ 4/\sqrt{69} \end{bmatrix}$

13. $5\sqrt{5}$ 15. Not orthogonal 17. Orthogonal

19. Refer to the *Study Guide* after you have written your answers.

21. *Hint:* Use Theorems 3 and 2 from Section 2.1.

23. $\mathbf{u} \cdot \mathbf{v} = 0$, $\|\mathbf{u}\|^2 = 30$, $\|\mathbf{v}\|^2 = 101$, $\|\mathbf{u} + \mathbf{v}\|^2 = (-5)^2 + (-9)^2 + 5^2 = 131 = 30 + 101$

25. The set of all multiples of $\begin{bmatrix} -b \\ a \end{bmatrix}$ (when $\mathbf{v} \neq \mathbf{0}$)

27. *Hint:* Use the definition of orthogonality.

29. *Hint:* Consider a typical vector $\mathbf{w} = c_1 \mathbf{v}_1 + \cdots + c_p \mathbf{v}_p$ in W.

31. *Hint:* If $\mathbf{x}$ is in $W^\perp$, then $\mathbf{x}$ is orthogonal to every vector in W.

33. **[M]** State your conjecture and verify it algebraically.

Section 6.2, page 344

1. Not orthogonal 3. Not orthogonal 5. Orthogonal

7. Show $\mathbf{u}_1 \cdot \mathbf{u}_2 = 0$, mention Theorem 4, and observe that two linearly independent vectors in $\mathbb{R}^2$ form a basis. Then obtain

$$
\mathbf{x} = \frac{39}{13} \begin{bmatrix} 2 \\ -3 \end{bmatrix} + \frac{26}{52} \begin{bmatrix} 6 \\ 4 \end{bmatrix} = 3 \begin{bmatrix} 2 \\ -3 \end{bmatrix} + \frac{1}{2} \begin{bmatrix} 6 \\ 4 \end{bmatrix}
$$

9. Show $\mathbf{u}_1 \cdot \mathbf{u}_2 = 0$, $\mathbf{u}_1 \cdot \mathbf{u}_3 = 0$, and $\mathbf{u}_2 \cdot \mathbf{u}_3 = 0$. Mention Theorem 4, and observe that three linearly independent vectors in $\mathbb{R}^3$ form a basis. Then obtain

$$
\mathbf{x} = \frac{5}{2} \mathbf{u}_1 - \frac{27}{18} \mathbf{u}_2 + \frac{18}{9} \mathbf{u}_3 = \frac{5}{2} \mathbf{u}_1 - \frac{3}{2} \mathbf{u}_2 + 2\mathbf{u}_3
$$

11. $\begin{bmatrix} -2 \\ 1 \end{bmatrix}$ 13. $\mathbf{y} = \begin{bmatrix} -4/5 \\ 7/5 \end{bmatrix} + \begin{bmatrix} 14/5 \\ 8/5 \end{bmatrix}$

15. $\mathbf{y} - \hat{\mathbf{y}} = \begin{bmatrix} .6 \\ -.8 \end{bmatrix}$, distance is 1

17. $\begin{bmatrix} 1/\sqrt{3} \\ 1/\sqrt{3} \\ 1/\sqrt{3} \end{bmatrix}$, $\begin{bmatrix} -1/\sqrt{2} \\ 0 \\ 1/\sqrt{2} \end{bmatrix}$

19. Orthonormal 21. Orthonormal

23. See the *Study Guide*.

25. *Hint:* $\|U\mathbf{x}\|^2 = (U\mathbf{x})^T (U\mathbf{x})$. Also, parts (a) and (c) follow from (b).

27. *Hint:* You need two theorems, one of which applies only to *square* matrices.

29. *Hint:* If you have a candidate for an inverse, you can check to see whether the candidate works.

31. Suppose $\hat{\mathbf{y}} = \dfrac{\mathbf{y} \cdot \mathbf{u}}{\mathbf{u} \cdot \mathbf{u}} \mathbf{u}$. Replace $\mathbf{u}$ by $c\mathbf{u}$ with $c \neq 0$; then

$$
\frac{\mathbf{y} \cdot (c\mathbf{u})}{(c\mathbf{u}) \cdot (c\mathbf{u})} (c\mathbf{u}) = \frac{c(\mathbf{y} \cdot \mathbf{u})}{c^2 \mathbf{u} \cdot \mathbf{u}} (c)\mathbf{u} = \hat{\mathbf{y}}
$$

33. Let $L = \text{Span}\{\mathbf{u}\}$, where $\mathbf{u}$ is nonzero, and let $T(\mathbf{x}) = \text{proj}_L \mathbf{x}$. By definition,

$$
T(\mathbf{x}) = \frac{\mathbf{x} \cdot \mathbf{u}}{\mathbf{u} \cdot \mathbf{u}} \mathbf{u} = (\mathbf{x} \cdot \mathbf{u})(\mathbf{u} \cdot \mathbf{u})^{-1} \mathbf{u}
$$

For $\mathbf{x}$ and $\mathbf{y}$ in $\mathbb{R}^n$ and any scalars c and d, properties of the inner product (Theorem 1) show that

$$
\begin{aligned}
T(c\mathbf{x} + d\mathbf{y}) &= [(c\mathbf{x} + d\mathbf{y}) \cdot \mathbf{u}](\mathbf{u} \cdot \mathbf{u})^{-1} \mathbf{u} \\
&= [c(\mathbf{x} \cdot \mathbf{u}) + d(\mathbf{y} \cdot \mathbf{u})](\mathbf{u} \cdot \mathbf{u})^{-1} \mathbf{u} \\
&= c(\mathbf{x} \cdot \mathbf{u})(\mathbf{u} \cdot \mathbf{u})^{-1} \mathbf{u} + d(\mathbf{y} \cdot \mathbf{u})(\mathbf{u} \cdot \mathbf{u})^{-1} \mathbf{u} \\
&= cT(\mathbf{x}) + dT(\mathbf{y})
\end{aligned}
$$

Thus T is linear.

Section 6.3, page 352

1. $\mathbf{x} = -\frac{8}{9}\mathbf{u}_1 - \frac{2}{9}\mathbf{u}_2 + \frac{2}{3}\mathbf{u}_3 + 2\mathbf{u}_4;$ $\mathbf{x} = \begin{bmatrix} 0 \\ -2 \\ 4 \\ -2 \end{bmatrix} + \begin{bmatrix} 10 \\ -6 \\ -2 \\ 2 \end{bmatrix}$

3. $\begin{bmatrix} -1 \\ 4 \\ 0 \end{bmatrix}$ **5.** $\begin{bmatrix} -1 \\ 2 \\ 6 \end{bmatrix} = \mathbf{y}$

7. $\mathbf{y} = \begin{bmatrix} 10/3 \\ 2/3 \\ 8/3 \end{bmatrix} + \begin{bmatrix} -7/3 \\ 7/3 \\ 7/3 \end{bmatrix}$ **9.** $\mathbf{y} = \begin{bmatrix} 2 \\ 4 \\ 0 \\ 0 \end{bmatrix} + \begin{bmatrix} 2 \\ -1 \\ 3 \\ -1 \end{bmatrix}$

11. $\begin{bmatrix} 3 \\ -1 \\ 1 \\ -1 \end{bmatrix}$ **13.** $\begin{bmatrix} -1 \\ -3 \\ -2 \\ 3 \end{bmatrix}$ **15.** $\sqrt{40}$

17. a. $U^TU = \begin{bmatrix} 1 & 0 \\ 0 & 1 \end{bmatrix}, UU^T = \begin{bmatrix} 8/9 & -2/9 & 2/9 \\ -2/9 & 5/9 & 4/9 \\ 2/9 & 4/9 & 5/9 \end{bmatrix}$

b. $\text{proj}_W\, \mathbf{y} = 6\mathbf{u}_1 + 3\mathbf{u}_2 = \begin{bmatrix} 2 \\ 4 \\ 5 \end{bmatrix}, (UU^T)\mathbf{y} = \begin{bmatrix} 2 \\ 4 \\ 5 \end{bmatrix}$

19. Any multiple of $\begin{bmatrix} 0 \\ 2/5 \\ 1/5 \end{bmatrix}$, such as $\begin{bmatrix} 0 \\ 2 \\ 1 \end{bmatrix}$

21. Write your answers before checking the *Study Guide*.

23. *Hint:* Use Theorem 3 and the Orthogonal Decomposition Theorem. For the uniqueness, suppose $A\mathbf{p} = \mathbf{b}$ and $A\mathbf{p}_1 = \mathbf{b}$, and consider the equations $\mathbf{p} = \mathbf{p}_1 + (\mathbf{p} - \mathbf{p}_1)$ and $\mathbf{p} = \mathbf{p} + \mathbf{0}$.

Section 6.4, page 358

1. $\begin{bmatrix} 3 \\ 0 \\ -1 \end{bmatrix}, \begin{bmatrix} -1 \\ 5 \\ -3 \end{bmatrix}$ **3.** $\begin{bmatrix} 2 \\ -5 \\ 1 \end{bmatrix}, \begin{bmatrix} 3 \\ 3/2 \\ 3/2 \end{bmatrix}$

5. $\begin{bmatrix} 1 \\ -4 \\ 0 \\ 1 \end{bmatrix}, \begin{bmatrix} 5 \\ 1 \\ -4 \\ -1 \end{bmatrix}$ **7.** $\begin{bmatrix} 2/\sqrt{30} \\ -5/\sqrt{30} \\ 1/\sqrt{30} \end{bmatrix}, \begin{bmatrix} 2/\sqrt{6} \\ 1/\sqrt{6} \\ 1/\sqrt{6} \end{bmatrix}$

9. $\begin{bmatrix} 3 \\ 1 \\ -1 \\ 3 \end{bmatrix}, \begin{bmatrix} 1 \\ 3 \\ 3 \\ -1 \end{bmatrix}, \begin{bmatrix} -3 \\ 1 \\ 1 \\ 3 \end{bmatrix}$ **11.** $\begin{bmatrix} 1 \\ -1 \\ -1 \\ 1 \\ 1 \end{bmatrix}, \begin{bmatrix} 3 \\ 0 \\ 3 \\ -3 \\ 3 \end{bmatrix}, \begin{bmatrix} 2 \\ 0 \\ 2 \\ 2 \\ -2 \end{bmatrix}$

13. $R = \begin{bmatrix} 6 & 12 \\ 0 & 6 \end{bmatrix}$

15. $Q = \begin{bmatrix} 1/\sqrt{5} & 1/2 & 1/2 \\ -1/\sqrt{5} & 0 & 0 \\ -1/\sqrt{5} & 1/2 & 1/2 \\ 1/\sqrt{5} & -1/2 & 1/2 \\ 1/\sqrt{5} & 1/2 & -1/2 \end{bmatrix}$,

$R = \begin{bmatrix} \sqrt{5} & -\sqrt{5} & 4\sqrt{5} \\ 0 & 6 & -2 \\ 0 & 0 & 4 \end{bmatrix}$

17. See the *Study Guide*.

19. Suppose $\mathbf{x}$ satisfies $R\mathbf{x} = \mathbf{0}$; then $QR\mathbf{x} = Q\mathbf{0} = \mathbf{0}$, and $A\mathbf{x} = \mathbf{0}$. Since the columns of A are linearly independent, $\mathbf{x}$ must be zero. This fact, in turn, shows that the columns of R are linearly independent. Since R is square, it is invertible, by the Invertible Matrix Theorem.

21. Denote the columns of Q by $\mathbf{q}_1, \ldots, \mathbf{q}_n$. Note that $n \le m$, because A is $m \times n$ and has linearly independent columns. Use the fact that the columns of Q can be extended to an orthonormal basis for $\mathbb{R}^m$, say, $\{\mathbf{q}_1, \ldots, \mathbf{q}_m\}$. (The *Study Guide* describes one method.) Let $Q_0 = [\,\mathbf{q}_{n+1} \;\cdots\; \mathbf{q}_m\,]$ and $Q_1 = [\, Q \quad Q_0\,]$. Then, using partitioned matrix multiplication, $Q_1 \begin{bmatrix} R \\ 0 \end{bmatrix} = QR = A$.

23. *Hint:* Partition R as a 2×2 block matrix.

25. [M] The diagonal entries of R are 20, 6, 10.3923, and 7.0711, to four decimal places.

Section 6.5, page 366

1. a. $\begin{bmatrix} 6 & -11 \\ -11 & 22 \end{bmatrix}\begin{bmatrix} x_1 \\ x_2 \end{bmatrix} = \begin{bmatrix} -4 \\ 11 \end{bmatrix}$ **b.** $\hat{\mathbf{x}} = \begin{bmatrix} 3 \\ 2 \end{bmatrix}$

3. a. $\begin{bmatrix} 6 & 6 \\ 6 & 42 \end{bmatrix}\begin{bmatrix} x_1 \\ x_2 \end{bmatrix} = \begin{bmatrix} 6 \\ -6 \end{bmatrix}$ **b.** $\hat{\mathbf{x}} = \begin{bmatrix} 4/3 \\ -1/3 \end{bmatrix}$

5. $\hat{\mathbf{x}} = \begin{bmatrix} 5 \\ -3 \\ 0 \end{bmatrix} + x_3\begin{bmatrix} -1 \\ 1 \\ 1 \end{bmatrix}$ **7.** $2\sqrt{5}$

9. a. $\hat{\mathbf{b}} = \begin{bmatrix} 1 \\ 1 \\ 0 \end{bmatrix}$ **b.** $\hat{\mathbf{x}} = \begin{bmatrix} 2/7 \\ 1/7 \end{bmatrix}$

11. a. $\hat{\mathbf{b}} = \begin{bmatrix} 3 \\ 1 \\ 4 \\ -1 \end{bmatrix}$ **b.** $\hat{\mathbf{x}} = \begin{bmatrix} 2/3 \\ 0 \\ 1/3 \end{bmatrix}$

13. $A\mathbf{u} = \begin{bmatrix} 11 \\ -11 \\ 11 \end{bmatrix}$, $A\mathbf{v} = \begin{bmatrix} 7 \\ -12 \\ 7 \end{bmatrix}$,

$\mathbf{b} - A\mathbf{u} = \begin{bmatrix} 0 \\ 2 \\ -6 \end{bmatrix}$, $\mathbf{b} - A\mathbf{v} = \begin{bmatrix} 4 \\ 3 \\ -2 \end{bmatrix}$. No, $\mathbf{u}$ could not possibly be a least-squares solution of $A\mathbf{x} = \mathbf{b}$. Why?

15. $\hat{\mathbf{x}} = \begin{bmatrix} 4 \\ -1 \end{bmatrix}$ **17.** See the *Study Guide*.

19. a. If $A\mathbf{x} = \mathbf{0}$, then $A^T A\mathbf{x} = A^T \mathbf{0} = \mathbf{0}$. This shows that Nul A is contained in Nul $A^T A$.

b. If $A^T A\mathbf{x} = \mathbf{0}$, then $\mathbf{x}^T A^T A\mathbf{x} = \mathbf{x}^T \mathbf{0} = 0$. So $(A\mathbf{x})^T (A\mathbf{x}) = 0$ (which means that $\|A\mathbf{x}\|^2 = 0$), and hence $A\mathbf{x} = \mathbf{0}$. This shows that Nul $A^T A$ is contained in Nul A.

21. *Hint:* For part (a), use an important theorem from Chapter 2.

23. By Theorem 14, $\hat{\mathbf{b}} = A\hat{\mathbf{x}} = A(A^T A)^{-1} A^T \mathbf{b}$. The matrix $A(A^T A)^{-1} A^T$ occurs frequently in statistics, where it is sometimes called the *hat-matrix*.

25. The normal equations are $\begin{bmatrix} 2 & 2 \\ 2 & 2 \end{bmatrix} \begin{bmatrix} x \\ y \end{bmatrix} = \begin{bmatrix} 6 \\ 6 \end{bmatrix}$, whose solution is the set of (x, y) such that $x + y = 3$. The solutions correspond to points on the line midway between the lines $x + y = 2$ and $x + y = 4$.

Section 6.6, page 374

1. $y = .9 + .4x$ **3.** $y = 1.1 + 1.3x$

5. If two data points have different x-coordinates, then the two columns of the design matrix X cannot be multiples of each other and hence are linearly independent. By Theorem 14 in Section 6.5, the normal equations have a unique solution.

7. a. $\mathbf{y} = X\boldsymbol{\beta} + \boldsymbol{\epsilon}$, where $\mathbf{y} = \begin{bmatrix} 1.8 \\ 2.7 \\ 3.4 \\ 3.8 \\ 3.9 \end{bmatrix}$, $X = \begin{bmatrix} 1 & 1 \\ 2 & 4 \\ 3 & 9 \\ 4 & 16 \\ 5 & 25 \end{bmatrix}$,

$\boldsymbol{\beta} = \begin{bmatrix} \beta_1 \\ \beta_2 \end{bmatrix}$, $\boldsymbol{\epsilon} = \begin{bmatrix} \epsilon_1 \\ \epsilon_2 \\ \epsilon_3 \\ \epsilon_4 \\ \epsilon_5 \end{bmatrix}$

b. [M] $y = 1.76x - .20x^2$

9. $\mathbf{y} = X\boldsymbol{\beta} + \boldsymbol{\epsilon}$, where $\mathbf{y} = \begin{bmatrix} 7.9 \\ 5.4 \\ -.9 \end{bmatrix}$, $X = \begin{bmatrix} \cos 1 & \sin 1 \\ \cos 2 & \sin 2 \\ \cos 3 & \sin 3 \end{bmatrix}$,

$\boldsymbol{\beta} = \begin{bmatrix} A \\ B \end{bmatrix}$, $\boldsymbol{\epsilon} = \begin{bmatrix} \epsilon_1 \\ \epsilon_2 \\ \epsilon_3 \end{bmatrix}$

11. [M] $\beta = 1.45$ and $e = .811$; the orbit is an ellipse. The equation $r = \beta/(1 - e \cdot \cos \vartheta)$ produces $r = 1.33$ when $\vartheta = 4.6$.

13. [M] **a.** $y = -.8558 + 4.7025t + 5.5554t^2 - .0274t^3$

b. The velocity function is $v(t) = 4.7025 + 11.1108t - .0822t^2$, and $v(4.5) = 53.0$ ft/sec.

15. *Hint:* Write X and $\mathbf{y}$ as in equation (1), and compute $X^T X$ and $X^T \mathbf{y}$.

17. a. The mean of the x-data is $\bar{x} = 5.5$. The data in mean-deviation form are $(-3.5, 1)$, $(-.5, 2)$, $(1.5, 3)$,

and $(2.5, 3)$. The columns of X are orthogonal because the entries in the second column sum to 0.

b. $\begin{bmatrix} 4 & 0 \\ 0 & 21 \end{bmatrix} \begin{bmatrix} \beta_0 \\ \beta_1 \end{bmatrix} = \begin{bmatrix} 9 \\ 7.5 \end{bmatrix}$,

$y = \frac{9}{4} + \frac{5}{14}x^* = \frac{9}{4} + \frac{5}{14}(x - 5.5)$

19. *Hint:* The equation has a nice geometric interpretation.

Section 6.7, page 382

1. a. $3, \sqrt{105}, 225$ **b.** All multiples of $\begin{bmatrix} 1 \\ 4 \end{bmatrix}$

3. 28   **5.** $5\sqrt{2}, 3\sqrt{3}$ **7.** $\frac{56}{25} + \frac{14}{25}t$

9. a. Constant polynomial, $p(t) = 5$

b. $t^2 - 5$ is orthogonal to p_0 and p_1; values: $(4, -4, -4, 4)$; answer: $q(t) = \frac{1}{4}(t^2 - 5)$

11. $\frac{17}{5}t$

13. Verify each of the four axioms. For instance:

1. $\langle \mathbf{u}, \mathbf{v} \rangle = (A\mathbf{u}) \cdot (A\mathbf{v})$ Definition
$= (A\mathbf{v}) \cdot (A\mathbf{u})$ Property of the dot product
$= \langle \mathbf{v}, \mathbf{u} \rangle$ Definition

15. $\langle \mathbf{u}, c\mathbf{v} \rangle = \langle c\mathbf{v}, \mathbf{u} \rangle$ Axiom 1
$= c\langle \mathbf{v}, \mathbf{u} \rangle$ Axiom 3
$= c\langle \mathbf{u}, \mathbf{v} \rangle$ Axiom 1

17. *Hint:* Compute 4 times the right-hand side.

19. $\langle \mathbf{u}, \mathbf{v} \rangle = \sqrt{a}\sqrt{b} + \sqrt{b}\sqrt{a} = 2\sqrt{ab}$, $\|\mathbf{u}\|^2 = (\sqrt{a})^2 + (\sqrt{b})^2 = a + b$. Since a and b are nonnegative, $\|\mathbf{u}\| = \sqrt{a + b}$. Similarly, $\|\mathbf{v}\| = \sqrt{b + a}$. By Cauchy–Schwarz, $2\sqrt{ab} \leq \sqrt{a + b}\sqrt{b + a} = a + b$. Hence, $\sqrt{ab} \leq \dfrac{a + b}{2}$.

21. 0   **23.** $2/\sqrt{5}$ **25.** $1, t, 3t^2 - 1$

27. [M] The new orthogonal polynomials are multiples of $-17t + 5t^3$ and $72 - 155t^2 + 35t^4$. Scale these polynomials so their values at $-2, -1, 0, 1$, and 2 are small integers.

Section 6.8, page 389

1. $y = 2 + \frac{3}{2}t$

3. $p(t) = 4p_0 - .1p_1 - .5p_2 + .2p_3$
$= 4 - .1t - .5(t^2 - 2) + .2\left(\frac{5}{6}t^3 - \frac{17}{6}t\right)$
(This polynomial happens to fit the data exactly.)

5. Use the identity
$$\sin mt \sin nt = \frac{1}{2}[\cos(mt - nt) - \cos(mt + nt)]$$

7. Use the identity $\cos^2 kt = \dfrac{1 + \cos 2kt}{2}$.

9. $\pi + 2\sin t + \sin 2t + \frac{2}{3}\sin 3t$ [*Hint:* Save time by using the results from Example 4.]

11. $\frac{1}{2} - \frac{1}{2}\cos 2t$ (Why?)

13. *Hint:* Take functions f and g in $C[0, 2\pi]$, and fix an integer $m \geq 0$. Write the Fourier coefficient of $f + g$ that involves $\cos mt$, and write the Fourier coefficient that involves $\sin mt \, (m > 0)$.

15. [M] The cubic curve is the graph of $g(t) = -.2685 + 3.6095t + 5.8576t^2 - .0477t^3$. The velocity at $t = 4.5$ seconds is $g'(4.5) = 53.4$ ft/sec. This is about .7% faster than the estimate obtained in Exercise 13 in Section 6.6.

Chapter 6 Supplementary Exercises, page 390

1. a. F **b.** T **c.** T **d.** F **e.** F **f.** T
  g. T **h.** T **i.** F **j.** T **k.** T **l.** F
  m. T **n.** F **o.** F **p.** T **q.** T **r.** F
  s. F

2. *Hint:* If $\{\mathbf{v}_1, \mathbf{v}_2\}$ is an orthonormal set and $\mathbf{x} = c_1\mathbf{v}_1 + c_2\mathbf{v}_2$, then the vectors $c_1\mathbf{v}_1$ and $c_2\mathbf{v}_2$ are orthogonal, and

$$\|\mathbf{x}\|^2 = \|c_1\mathbf{v}_1 + c_2\mathbf{v}_2\|^2 = \|c_1\mathbf{v}_1\|^2 + \|c_2\mathbf{v}_2\|^2$$
$$= (|c_1| \|\mathbf{v}_1\|)^2 + (|c_2| \|\mathbf{v}_2\|)^2 = |c_1|^2 + |c_2|^2$$

(Explain why.) So the stated equality holds for $p = 2$. Suppose that the equality holds for $p = k$, with $k \geq 2$, let $\{\mathbf{v}_1, \ldots, \mathbf{v}_{k+1}\}$ be an orthonormal set, and consider $\mathbf{x} = c_1\mathbf{v}_1 + \cdots + c_k\mathbf{v}_k + c_{k+1}\mathbf{v}_{k+1} = \mathbf{u}_k + c_{k+1}\mathbf{v}_{k+1}$, where $\mathbf{u}_k = c_1\mathbf{v}_1 + \cdots + c_k\mathbf{v}_k$.

3. Given $\mathbf{x}$ and an orthonormal set $\{\mathbf{v}_1, \ldots, \mathbf{v}_p\}$ in $\mathbb{R}^n$, let $\hat{\mathbf{x}}$ be the orthogonal projection of $\mathbf{x}$ onto the subspace spanned by $\mathbf{v}_1, \ldots, \mathbf{v}_p$. By Theorem 10 in Section 6.3,

$$\hat{\mathbf{x}} = (\mathbf{x} \cdot \mathbf{v}_1)\mathbf{v}_1 + \cdots + (\mathbf{x} \cdot \mathbf{v}_p)\mathbf{v}_p$$

By Exercise 2, $\|\hat{\mathbf{x}}\|^2 = |\mathbf{x} \cdot \mathbf{v}_1|^2 + \cdots + |\mathbf{x} \cdot \mathbf{v}_p|^2$. Bessel's inequality follows from the fact that $\|\hat{\mathbf{x}}\|^2 \leq \|\mathbf{x}\|^2$, noted before the statement of the Cauchy–Schwarz inequality, in Section 6.7.

5. Suppose $(U\mathbf{x}) \cdot (U\mathbf{y}) = \mathbf{x} \cdot \mathbf{y}$ for all $\mathbf{x}, \mathbf{y}$ in $\mathbb{R}^n$, and let $\mathbf{e}_1, \ldots, \mathbf{e}_n$ be the standard basis for $\mathbb{R}^n$. For $j = 1, \ldots, n$, $U\mathbf{e}_j$ is the jth column of U. Since $\|U\mathbf{e}_j\|^2 = (U\mathbf{e}_j) \cdot (U\mathbf{e}_j) = \mathbf{e}_j \cdot \mathbf{e}_j = 1$, the columns of U are unit vectors; since $(U\mathbf{e}_j) \cdot (U\mathbf{e}_k) = \mathbf{e}_j \cdot \mathbf{e}_k = 0$ for $j \neq k$, the columns are pairwise orthogonal.

7. *Hint:* Compute $Q^T Q$, using the fact that $(\mathbf{u}\mathbf{u}^T)^T = \mathbf{u}^{TT}\mathbf{u}^T = \mathbf{u}\mathbf{u}^T$.

9. Let $W = \text{Span}\{\mathbf{u}, \mathbf{v}\}$. Given $\mathbf{z}$ in $\mathbb{R}^n$, let $\hat{\mathbf{z}} = \text{proj}_W \mathbf{z}$. Then $\hat{\mathbf{z}}$ is in Col A, where $A = [\mathbf{u} \quad \mathbf{v}]$, say, $\hat{\mathbf{z}} = A\hat{\mathbf{x}}$ for some $\hat{\mathbf{x}}$ in $\mathbb{R}^2$. So $\hat{\mathbf{x}}$ is a least-squares solution of $A\mathbf{x} = \mathbf{z}$. The normal equations can be solved to produce $\hat{\mathbf{x}}$, and then $\hat{\mathbf{z}}$ is found by computing $A\hat{\mathbf{x}}$.

11. *Hint:* Let $\mathbf{x} = \begin{bmatrix} x \\ y \\ z \end{bmatrix}$, $\mathbf{b} = \begin{bmatrix} a \\ b \\ c \end{bmatrix}$, $\mathbf{v} = \begin{bmatrix} 1 \\ -2 \\ 5 \end{bmatrix}$, and

$A = \begin{bmatrix} \mathbf{v}^T \\ \mathbf{v}^T \\ \mathbf{v}^T \end{bmatrix} = \begin{bmatrix} 1 & -2 & 5 \\ 1 & -2 & 5 \\ 1 & -2 & 5 \end{bmatrix}$. The given set of

equations is $A\mathbf{x} = \mathbf{b}$, and the set of all least-squares solutions coincides with the set of solutions of $A^TA\mathbf{x} = A^T\mathbf{b}$ (Theorem 13 in Section 6.5). Study this equation, and use the fact that $(\mathbf{v}\mathbf{v}^T)\mathbf{x} = \mathbf{v}(\mathbf{v}^T\mathbf{x}) = (\mathbf{v}^T\mathbf{x})\mathbf{v}$, because $\mathbf{v}^T\mathbf{x}$ is a scalar.

13. a. The row–column calculation of $A\mathbf{u}$ shows that each row of A is orthogonal to every $\mathbf{u}$ in Nul A. So each row of A is in $(\text{Nul } A)^\perp$. Since $(\text{Nul } A)^\perp$ is a subspace, it must contain all linear combinations of the rows of A; hence $(\text{Nul } A)^\perp$ contains Row A.

  b. If rank $A = r$, then dim Nul $A = n - r$, by the Rank Theorem. By Exercise 24(c) in Section 6.3,

  $$\dim \text{Nul } A + \dim(\text{Nul } A)^\perp = n$$

  So dim$(\text{Nul } A)^\perp$ must be r. But Row A is an r-dimensional subspace of $(\text{Nul } A)^\perp$, by the Rank Theorem and part (a). Therefore, Row A must coincide with $(\text{Nul } A)^\perp$.

  c. Replace A by A^T in part (b) and conclude that Row A^T coincides with $(\text{Nul } A^T)^\perp$. Since Row $A^T = \text{Col } A$, this proves (c).

15. If $A = URU^T$ with U orthogonal, then A is similar to R (because U is invertible and $U^T = U^{-1}$) and so A has the same eigenvalues as R (by Theorem 4 in Section 5.2), namely, the n real numbers on the diagonal of R.

17. [M] $\dfrac{\|\Delta\mathbf{x}\|}{\|\mathbf{x}\|} = .4618$,

$\text{cond}(A) \times \dfrac{\|\Delta\mathbf{b}\|}{\|\mathbf{b}\|} = 3363 \times (1.548 \times 10^{-4}) = .5206$.
Observe that $\|\Delta\mathbf{x}\|/\|\mathbf{x}\|$ almost equals $\text{cond}(A)$ times $\|\Delta\mathbf{b}\|/\|\mathbf{b}\|$.

19. [M] $\dfrac{\|\Delta\mathbf{x}\|}{\|\mathbf{x}\|} = 7.178 \times 10^{-8}$, $\dfrac{\|\Delta\mathbf{b}\|}{\|\mathbf{b}\|} = 2.832 \times 10^{-4}$.
Observe that the relative change in $\mathbf{x}$ is *much* smaller than the relative change in $\mathbf{b}$. In fact, since

$$\text{cond}(A) \times \dfrac{\|\Delta\mathbf{b}\|}{\|\mathbf{b}\|} = 23{,}683 \times (2.832 \times 10^{-4}) = 6.707$$

the theoretical bound on the relative change in $\mathbf{x}$ is 6.707 (to four significant figures). This exercise shows that even when a condition number is large, the relative error in a solution need not be as large as you might expect.

Chapter 7

Section 7.1, page 399

1. Symmetric   **3.** Not symmetric   **5.** Not symmetric

7. Orthogonal, $\begin{bmatrix} .6 & .8 \\ .8 & -.6 \end{bmatrix}$ **9.** Not orthogonal

11. Orthogonal, $\begin{bmatrix} 2/3 & 0 & \sqrt{5}/3 \\ 2/3 & 1/\sqrt{5} & -4/\sqrt{45} \\ 1/3 & -2/\sqrt{5} & -2/\sqrt{45} \end{bmatrix}$

13. $P = \begin{bmatrix} 1/\sqrt{2} & -1/\sqrt{2} \\ 1/\sqrt{2} & 1/\sqrt{2} \end{bmatrix}$, $D = \begin{bmatrix} 4 & 0 \\ 0 & 2 \end{bmatrix}$

15. $P = \begin{bmatrix} -4/\sqrt{17} & 1/\sqrt{17} \\ 1/\sqrt{17} & 4/\sqrt{17} \end{bmatrix}$, $D = \begin{bmatrix} 17 & 0 \\ 0 & 0 \end{bmatrix}$

17. $P = \begin{bmatrix} 1/\sqrt{3} & 1/\sqrt{6} & -1/\sqrt{2} \\ 1/\sqrt{3} & -2/\sqrt{6} & 0 \\ 1/\sqrt{3} & 1/\sqrt{6} & 1/\sqrt{2} \end{bmatrix}$,

$D = \begin{bmatrix} 5 & 0 & 0 \\ 0 & 2 & 0 \\ 0 & 0 & -2 \end{bmatrix}$

19. $P = \begin{bmatrix} -1/\sqrt{5} & 4/\sqrt{45} & -2/3 \\ 2/\sqrt{5} & 2/\sqrt{45} & -1/3 \\ 0 & 5/\sqrt{45} & 2/3 \end{bmatrix}$,

$D = \begin{bmatrix} 7 & 0 & 0 \\ 0 & 7 & 0 \\ 0 & 0 & -2 \end{bmatrix}$

21. $P = \begin{bmatrix} .5 & -.5 & -1/\sqrt{2} & 0 \\ .5 & .5 & 0 & -1/\sqrt{2} \\ .5 & -.5 & 1/\sqrt{2} & 0 \\ .5 & .5 & 0 & 1/\sqrt{2} \end{bmatrix}$,

$D = \begin{bmatrix} 9 & 0 & 0 & 0 \\ 0 & 5 & 0 & 0 \\ 0 & 0 & 1 & 0 \\ 0 & 0 & 0 & 1 \end{bmatrix}$

23. $P = \begin{bmatrix} 1/\sqrt{3} & -1/\sqrt{2} & -1/\sqrt{6} \\ 1/\sqrt{3} & 1/\sqrt{2} & -1/\sqrt{6} \\ 1/\sqrt{3} & 0 & 2/\sqrt{6} \end{bmatrix}$,

$D = \begin{bmatrix} 5 & 0 & 0 \\ 0 & 2 & 0 \\ 0 & 0 & 2 \end{bmatrix}$

25. See the *Study Guide*.

27. $(B^T A B)^T = B^T A^T B^{TT}$ Product of transposes in reverse order
$= B^T A B$ Because A is symmetric

The result about $B^T B$ is a special case when $A = I$.
$(BB^T)^T = B^{TT} B^T = BB^T$, so BB^T is symmetric.

29. *Hint:* Use an orthogonal diagonalization of A, or appeal to Theorem 2.

31. The Diagonalization Theorem in Section 5.3 says that the columns of P are (linearly independent) eigenvectors corresponding to the eigenvalues of A listed on the diagonal of D. So P has exactly k columns of eigenvectors corresponding to λ. These k columns form a basis for the eigenspace.

33. $A = 8\mathbf{u}_1\mathbf{u}_1^T + 6\mathbf{u}_2\mathbf{u}_2^T + 3\mathbf{u}_3\mathbf{u}_3^T$

$= 8\begin{bmatrix} 1/2 & -1/2 & 0 \\ -1/2 & 1/2 & 0 \\ 0 & 0 & 0 \end{bmatrix}$

$+ 6\begin{bmatrix} 1/6 & 1/6 & -2/6 \\ 1/6 & 1/6 & -2/6 \\ -2/6 & -2/6 & 4/6 \end{bmatrix}$

$+ 3\begin{bmatrix} 1/3 & 1/3 & 1/3 \\ 1/3 & 1/3 & 1/3 \\ 1/3 & 1/3 & 1/3 \end{bmatrix}$

35. *Hint:* $(\mathbf{uu}^T)\mathbf{x} = \mathbf{u}(\mathbf{u}^T\mathbf{x}) = (\mathbf{u}^T\mathbf{x})\mathbf{u}$, because $\mathbf{u}^T\mathbf{x}$ is a scalar.

Section 7.2, page 406

1. a. $5x_1^2 + \frac{2}{3}x_1x_2 + x_2^2$ **b.** 185   **c.** 16

3. a. $\begin{bmatrix} 10 & -3 \\ -3 & -3 \end{bmatrix}$ **b.** $\begin{bmatrix} 5 & 3/2 \\ 3/2 & 0 \end{bmatrix}$

5. a. $\begin{bmatrix} 8 & -3 & 2 \\ -3 & 7 & -1 \\ 2 & -1 & -3 \end{bmatrix}$ **b.** $\begin{bmatrix} 0 & 2 & 3 \\ 2 & 0 & -4 \\ 3 & -4 & 0 \end{bmatrix}$

7. $\mathbf{x} = P\mathbf{y}$, where $P = \dfrac{1}{\sqrt{2}}\begin{bmatrix} 1 & -1 \\ 1 & 1 \end{bmatrix}$, $\mathbf{y}^T D\mathbf{y} = 6y_1^2 - 4y_2^2$

In Exercises 9–14, other answers (change of variables and new quadratic form) are possible.

9. Positive definite; eigenvalues are 7 and 2
Change of variable: $\mathbf{x} = P\mathbf{y}$, with $P = \dfrac{1}{\sqrt{5}}\begin{bmatrix} -1 & 2 \\ 2 & 1 \end{bmatrix}$
New quadratic form: $7y_1^2 + 2y_2^2$

11. Indefinite; eigenvalues are 7 and -3
Change of variable: $\mathbf{x} = P\mathbf{y}$, with $P = \dfrac{1}{\sqrt{2}}\begin{bmatrix} 1 & -1 \\ 1 & 1 \end{bmatrix}$
New quadratic form: $7y_1^2 - 3y_2^2$

13. Positive semidefinite; eigenvalues are 10 and 0
Change of variable: $\mathbf{x} = P\mathbf{y}$, with $P = \dfrac{1}{\sqrt{10}}\begin{bmatrix} 1 & 3 \\ -3 & 1 \end{bmatrix}$
New quadratic form: $10y_1^2$

15. [M] Negative semidefinite; eigenvalues are $0, -6, -8, -12$
Change of variable: $\mathbf{x} = P\mathbf{y}$;

$P = \begin{bmatrix} 3/\sqrt{12} & 0 & -1/2 & 0 \\ 1/\sqrt{12} & -2/\sqrt{6} & 1/2 & 0 \\ 1/\sqrt{12} & 1/\sqrt{6} & 1/2 & -1/\sqrt{2} \\ 1/\sqrt{12} & 1/\sqrt{6} & 1/2 & 1/\sqrt{2} \end{bmatrix}$

New quadratic form: $-6y_2^2 - 8y_3^2 - 12y_4^2$

17. [M] Indefinite; eigenvalues are 8.5 and -6.5
Change of variable: $\mathbf{x} = P\mathbf{y}$;

$P = \dfrac{1}{\sqrt{50}}\begin{bmatrix} 3 & -4 & 3 & 4 \\ 5 & 0 & -5 & 0 \\ 4 & 3 & 4 & -3 \\ 0 & 5 & 0 & 5 \end{bmatrix}$

New quadratic form: $8.5y_1^2 + 8.5y_2^2 - 6.5y_3^2 - 6.5y_4^2$

19. 8   **21.** See the *Study Guide*.

23. Write the characteristic polynomial in two ways:

$$\det(A - \lambda I) = \det\begin{bmatrix} a - \lambda & b \\ b & d - \lambda \end{bmatrix}$$
$$= \lambda^2 - (a + d)\lambda + ad - b^2$$

and

$$(\lambda - \lambda_1)(\lambda - \lambda_2) = \lambda^2 - (\lambda_1 + \lambda_2)\lambda + \lambda_1\lambda_2$$

Equate coefficients to obtain $\lambda_1 + \lambda_2 = a + d$ and $\lambda_1\lambda_2 = ad - b^2 = \det A$.

25. Exercise 27 in Section 7.1 showed that $B^T B$ is symmetric. Also, $\mathbf{x}^T B^T B\mathbf{x} = (B\mathbf{x})^T B\mathbf{x} = \|B\mathbf{x}\|^2 \geq 0$, so the quadratic form is positive semidefinite, and we say that the matrix $B^T B$ is positive semidefinite. *Hint:* To show that $B^T B$ is positive definite when B is square and invertible, suppose that $\mathbf{x}^T B^T B\mathbf{x} = 0$ and deduce that $\mathbf{x} = \mathbf{0}$.

27. *Hint:* Show that $A + B$ is symmetric and the quadratic form $\mathbf{x}^T(A + B)\mathbf{x}$ is positive definite.

Section 7.3, page 413

1. $\mathbf{x} = P\mathbf{y}$, where $P = \begin{bmatrix} 1/3 & 2/3 & -2/3 \\ 2/3 & 1/3 & 2/3 \\ -2/3 & 2/3 & 1/3 \end{bmatrix}$

3. a. 9   **b.** $\pm\begin{bmatrix} 1/3 \\ 2/3 \\ -2/3 \end{bmatrix}$ **c.** 6

5. a. 7   **b.** $\pm\begin{bmatrix} -1/\sqrt{2} \\ 1/\sqrt{2} \end{bmatrix}$ **c.** 3

7. $\pm\begin{bmatrix} 1/3 \\ 2/3 \\ 2/3 \end{bmatrix}$ **9.** $5 + \sqrt{5}$ **11.** 3

13. *Hint:* If $m = M$, take $\alpha = 0$ in the formula for $\mathbf{x}$. That is, let $\mathbf{x} = \mathbf{u}_n$, and verify that $\mathbf{x}^T A\mathbf{x} = m$. If $m < M$ and if t is a number between m and M, then $0 \leq t - m \leq M - m$ and $0 \leq (t - m)/(M - m) \leq 1$. So let $\alpha = (t - m)/(M - m)$. Solve the expression for α to see that $t = (1 - \alpha)m + \alpha M$. As α goes from 0 to 1, t goes from m to M. Construct $\mathbf{x}$ as in the statement of the exercise, and verify its properties.

15. [M] **a.** 7.5   **b.** $\begin{bmatrix} .5 \\ .5 \\ .5 \\ .5 \end{bmatrix}$ **c.** $-.5$

17. [M] **a.** -4 **b.** $\begin{bmatrix} -3/\sqrt{12} \\ 1/\sqrt{12} \\ 1/\sqrt{12} \\ 1/\sqrt{12} \end{bmatrix}$ **c.** -10

Section 7.4, page 423

1. 3, 1   **3.** 3, 2

The answers in Exercises 5–13 are not the only possibilities.

5. $\begin{bmatrix} -3 & 0 \\ 0 & 0 \end{bmatrix} = \begin{bmatrix} -1 & 0 \\ 0 & 1 \end{bmatrix}\begin{bmatrix} 3 & 0 \\ 0 & 0 \end{bmatrix}\begin{bmatrix} 1 & 0 \\ 0 & 1 \end{bmatrix}$

7. $\begin{bmatrix} 1/\sqrt{5} & -2/\sqrt{5} \\ 2/\sqrt{5} & 1/\sqrt{5} \end{bmatrix}\begin{bmatrix} 3 & 0 \\ 0 & 2 \end{bmatrix}\begin{bmatrix} 2/\sqrt{5} & 1/\sqrt{5} \\ -1/\sqrt{5} & 2/\sqrt{5} \end{bmatrix}$

9. $\begin{bmatrix} 1/\sqrt{2} & -1/\sqrt{2} & 0 \\ 0 & 0 & 1 \\ 1/\sqrt{2} & 1/\sqrt{2} & 0 \end{bmatrix}\begin{bmatrix} 3\sqrt{10} & 0 \\ 0 & \sqrt{10} \\ 0 & 0 \end{bmatrix}$
$\times \begin{bmatrix} 2/\sqrt{5} & 1/\sqrt{5} \\ -1/\sqrt{5} & 2/\sqrt{5} \end{bmatrix}$

11. $\begin{bmatrix} -1/3 & 2/3 & 2/3 \\ 2/3 & -1/3 & 2/3 \\ 2/3 & 2/3 & -1/3 \end{bmatrix}\begin{bmatrix} 3\sqrt{10} & 0 \\ 0 & 0 \\ 0 & 0 \end{bmatrix}$
$\times \begin{bmatrix} 3/\sqrt{10} & -1/\sqrt{10} \\ 1/\sqrt{10} & 3/\sqrt{10} \end{bmatrix}$

13. $\begin{bmatrix} 3 & 2 & 2 \\ 2 & 3 & -2 \end{bmatrix}$
$= \begin{bmatrix} 1/\sqrt{2} & -1/\sqrt{2} \\ 1/\sqrt{2} & 1/\sqrt{2} \end{bmatrix}\begin{bmatrix} 5 & 0 & 0 \\ 0 & 3 & 0 \end{bmatrix}$
$\times \begin{bmatrix} 1/\sqrt{2} & 1/\sqrt{2} & 0 \\ -1/\sqrt{18} & 1/\sqrt{18} & -4/\sqrt{18} \\ -2/3 & 2/3 & 1/3 \end{bmatrix}$

15. a. rank $A = 2$

b. Basis for Col A: $\begin{bmatrix} .40 \\ .37 \\ -.84 \end{bmatrix}$, $\begin{bmatrix} -.78 \\ -.33 \\ -.52 \end{bmatrix}$

Basis for Nul A: $\begin{bmatrix} .58 \\ -.58 \\ .58 \end{bmatrix}$

(Remember that V^T appears in the SVD.)

17. Let $A = U\Sigma V^T = U\Sigma V^{-1}$. Since A is square and invertible, rank $A = n$, and all the entries on the diagonal of Σ must be nonzero. So $A^{-1} = (U\Sigma V^{-1})^{-1} = V\Sigma^{-1}U^{-1} = V\Sigma^{-1}U^T$.

19. *Hint:* Since U and V are orthogonal,

$$A^T A = (U\Sigma V^T)^T U\Sigma V^T = V\Sigma^T U^T U\Sigma V^T$$
$$= V(\Sigma^T\Sigma)V^{-1}$$

Thus V diagonalizes $A^T A$. What does this tell you about V?

21. Let $A = U\Sigma V^T$. The matrix PU is orthogonal, because P and U are both orthogonal. (See Exercise 29 in Section 6.2.) So the equation $PA = (PU)\Sigma V^T$ has the form required for a singular value decomposition. By Exercise 19, the diagonal entries in Σ are the singular values of PA.

23. *Hint:* Use a column–row expansion of $(U\Sigma)V^T$.

25. *Hint:* Consider the SVD for the standard matrix of T—say, $A = U\Sigma V^T = U\Sigma V^{-1}$. Let $\mathcal{B} = \{\mathbf{v}_1, \ldots, \mathbf{v}_n\}$ and $\mathcal{C} = \{\mathbf{u}_1, \ldots, \mathbf{u}_m\}$ be bases constructed from the columns of V and U, respectively. Compute the matrix for T relative to $\mathcal{B}$ and $\mathcal{C}$, as in Section 5.4. To do this, you must show that $V^{-1}\mathbf{v}_j = \mathbf{e}_j$, the jth column of I_n.

27. **[M]**
$$
\begin{bmatrix}
-.57 & -.65 & -.42 & .27 \\
.63 & -.24 & -.68 & -.29 \\
.07 & -.63 & .53 & -.56 \\
-.51 & .34 & -.29 & -.73
\end{bmatrix}
$$
$$
\times
\begin{bmatrix}
16.46 & 0 & 0 & 0 & 0 \\
0 & 12.16 & 0 & 0 & 0 \\
0 & 0 & 4.87 & 0 & 0 \\
0 & 0 & 0 & 4.31 & 0
\end{bmatrix}
$$
$$
\times
\begin{bmatrix}
-.10 & .61 & -.21 & -.52 & .55 \\
-.39 & .29 & .84 & -.14 & -.19 \\
-.74 & -.27 & -.07 & .38 & .49 \\
.41 & -.50 & .45 & -.23 & .58 \\
-.36 & -.48 & -.19 & -.72 & -.29
\end{bmatrix}
$$

29. **[M]** 25.9343, 16.7554, 11.2917, 1.0785, .00037793; $\sigma_1/\sigma_5 = 68{,}622$

Section 7.5, page 430

1. $M = \begin{bmatrix} 12 \\ 10 \end{bmatrix}$; $B = \begin{bmatrix} 7 & 10 & -6 & -9 & -10 & 8 \\ 2 & -4 & -1 & 5 & 3 & -5 \end{bmatrix}$;

$S = \begin{bmatrix} 86 & -27 \\ -27 & 16 \end{bmatrix}$

3. $\begin{bmatrix} .95 \\ -.32 \end{bmatrix}$ for $\lambda = 95.2$, $\begin{bmatrix} .32 \\ .95 \end{bmatrix}$ for $\lambda = 6.8$

5. **[M]** (.130, .874, .468), 75.9% of the variance

7. $y_1 = .95x_1 - .32x_2$; y_1 explains 93.3% of the variance.

9. $c_1 = 1/3$, $c_2 = 2/3$, $c_3 = 2/3$; the variance of y is 9.

11. a. If $\mathbf{w}$ is the vector in $\mathbb{R}^N$ with a 1 in each position, then

$$\begin{bmatrix} \mathbf{X}_1 & \cdots & \mathbf{X}_N \end{bmatrix} \mathbf{w} = \mathbf{X}_1 + \cdots + \mathbf{X}_N = \mathbf{0}$$

because the $\mathbf{X}_k$ are in mean-deviation form. Then

$$
\begin{aligned}
&\begin{bmatrix} \mathbf{Y}_1 & \cdots & \mathbf{Y}_N \end{bmatrix} \mathbf{w} \\
&= \begin{bmatrix} P^T\mathbf{X}_1 & \cdots & P^T\mathbf{X}_N \end{bmatrix} \mathbf{w} \qquad \text{By definition} \\
&= P^T \begin{bmatrix} \mathbf{X}_1 & \cdots & \mathbf{X}_N \end{bmatrix} \mathbf{w} = P^T\mathbf{0} = \mathbf{0}
\end{aligned}
$$

That is, $\mathbf{Y}_1 + \cdots + \mathbf{Y}_N = \mathbf{0}$, so the $\mathbf{Y}_k$ are in mean-deviation form.

b. *Hint:* Because the $\mathbf{X}_j$ are in mean-deviation form, the covariance matrix of the $\mathbf{X}_j$ is

$$1/(N-1)\begin{bmatrix} \mathbf{X}_1 & \cdots & \mathbf{X}_N \end{bmatrix}\begin{bmatrix} \mathbf{X}_1 & \cdots & \mathbf{X}_N \end{bmatrix}^T$$

Compute the covariance matrix of the $\mathbf{Y}_j$, using part (a).

13. If $B = \begin{bmatrix} \hat{\mathbf{X}}_1 & \cdots & \hat{\mathbf{X}}_N \end{bmatrix}$, then

$$
\begin{aligned}
S &= \frac{1}{N-1}BB^T = \frac{1}{N-1}\begin{bmatrix} \hat{\mathbf{X}}_1 & \cdots & \hat{\mathbf{X}}_n \end{bmatrix}\begin{bmatrix} \hat{\mathbf{X}}_1^T \\ \vdots \\ \hat{\mathbf{X}}_N^T \end{bmatrix} \\
&= \frac{1}{N-1}\sum_1^N \hat{\mathbf{X}}_k\hat{\mathbf{X}}_k^T = \frac{1}{N-1}\sum_1^N (\mathbf{X}_k - \mathbf{M})(\mathbf{X}_k - \mathbf{M})^T
\end{aligned}
$$

Chapter 7 Supplementary Exercises, page 432

1. a. T **b.** F **c.** T **d.** F **e.** F **f.** F
g. F **h.** T **i.** F **j.** F **k.** F **l.** F
m. T **n.** F **o.** T **p.** T **q.** F

3. If rank $A = r$, then dim Nul $A = n - r$, by the Rank Theorem. So 0 is an eigenvalue of multiplicity $n - r$. Hence, of the n terms in the spectral decomposition of A, exactly $n - r$ are zero. The remaining r terms (corresponding to the nonzero eigenvalues) are all rank 1 matrices, as mentioned in the discussion of the spectral decomposition.

5. If $A\mathbf{v} = \lambda\mathbf{v}$ for some nonzero λ, then $\mathbf{v} = \lambda^{-1}A\mathbf{v} = A(\lambda^{-1}\mathbf{v})$, which shows that $\mathbf{v}$ is a linear combination of the columns of A.

7. *Hint:* If $A = R^TR$, where R is invertible, then A is positive definite, by Exercise 25 in Section 7.2. Conversely, suppose that A is positive definite. Then by Exercise 26 in Section 7.2, $A = B^TB$ for some positive definite matrix B. Explain why B admits a QR factorization, and use it to create the Cholesky factorization of A.

9. If A is $m \times n$ and $\mathbf{x}$ is in $\mathbb{R}^n$, then $\mathbf{x}^TA^TA\mathbf{x} = (A\mathbf{x})^T(A\mathbf{x}) = \|A\mathbf{x}\|^2 \geq 0$. Thus A^TA is positive semidefinite. By Exercise 22 in Section 6.5, rank $A^TA = $ rank A.

11. *Hint:* Write an SVD of A in the form $A = U\Sigma V^T = PQ$, where $P = U\Sigma U^T$ and $Q = UV^T$. Show that P is symmetric and has the same eigenvalues as Σ. Explain why Q is an orthogonal matrix.

13. a. If $\mathbf{b} = A\mathbf{x}$, then $\mathbf{x}^+ = A^+\mathbf{b} = A^+A\mathbf{x}$. By Exercise 12(a), $\mathbf{x}^+$ is the orthogonal projection of $\mathbf{x}$ onto Row A.

b. From (a) and then Exercise 12(c), $A\mathbf{x}^+ = A(A^+A\mathbf{x}) = (AA^+A)\mathbf{x} = A\mathbf{x} = \mathbf{b}$.

c. Since $\mathbf{x}^+$ is the orthogonal projection onto Row A, the Pythagorean Theorem shows that $\|\mathbf{u}\|^2 = \|\mathbf{x}^+\|^2 + \|\mathbf{u} - \mathbf{x}^+\|^2$. Part (c) follows immediately.

15. **[M]** $A^+ = \frac{1}{40} \cdot \begin{bmatrix} -2 & -14 & 13 & 13 \\ -2 & -14 & 13 & 13 \\ -2 & 6 & -7 & -7 \\ 2 & -6 & 7 & 7 \\ 4 & -12 & -6 & -6 \end{bmatrix}$, $\hat{\mathbf{x}} = \begin{bmatrix} .7 \\ .7 \\ -.8 \\ .8 \\ .6 \end{bmatrix}$

The reduced echelon form of $\begin{bmatrix} A \\ \mathbf{x}^T \end{bmatrix}$ is the same as the reduced echelon form of A, except for an extra row of zeros. So adding scalar multiples of the rows of A to $\mathbf{x}^T$ can produce the zero vector, which shows that $\mathbf{x}^T$ is in Row A.

$$\text{Basis for Nul } A: \begin{bmatrix} -1 \\ 1 \\ 0 \\ 0 \\ 0 \end{bmatrix}, \begin{bmatrix} 0 \\ 0 \\ 1 \\ 1 \\ 0 \end{bmatrix}$$

Chapter 8

Section 8.1, page 442

1. Some possible answers: $\mathbf{y} = 2\mathbf{v}_1 - 1.5\mathbf{v}_2 + .5\mathbf{v}_3$, $\mathbf{y} = 2\mathbf{v}_1 - 2\mathbf{v}_3 + \mathbf{v}_4$, $\mathbf{y} = 2\mathbf{v}_1 + 3\mathbf{v}_2 - 7\mathbf{v}_3 + 3\mathbf{v}_4$

3. $\mathbf{y} = -3\mathbf{v}_1 + 2\mathbf{v}_2 + 2\mathbf{v}_3$. The weights sum to 1, so this is an affine sum.

5. a. $\mathbf{p}_1 = 3\mathbf{b}_1 - \mathbf{b}_2 - \mathbf{b}_3 \in \text{aff } S$ since the coefficients sum to 1.
   b. $\mathbf{p}_2 = 2\mathbf{b}_1 + 0\mathbf{b}_2 + \mathbf{b}_3 \notin \text{aff } S$ since the coefficients do not sum to 1.
   c. $\mathbf{p}_3 = -\mathbf{b}_1 + 2\mathbf{b}_2 + 0\mathbf{b}_3 \in \text{aff } S$ since the coefficients sum to 1.

7. a. $\mathbf{p}_1 \in \text{Span } S$, but $\mathbf{p}_1 \notin \text{aff } S$
   b. $\mathbf{p}_2 \in \text{Span } S$, and $\mathbf{p}_2 \in \text{aff } S$
   c. $\mathbf{p}_3 \notin \text{Span } S$, so $\mathbf{p}_3 \notin \text{aff } S$

9. $\mathbf{v}_1 = \begin{bmatrix} -3 \\ 0 \end{bmatrix}$ and $\mathbf{v}_2 = \begin{bmatrix} 1 \\ -2 \end{bmatrix}$. Other answers are possible.

11. See the *Study Guide*.

13. Span $\{\mathbf{v}_2 - \mathbf{v}_1, \mathbf{v}_3 - \mathbf{v}_1\}$ is a plane if and only if $\{\mathbf{v}_2 - \mathbf{v}_1, \mathbf{v}_3 - \mathbf{v}_1\}$ is linearly independent. Suppose c_2 and c_3 satisfy $c_2(\mathbf{v}_2 - \mathbf{v}_1) + c_3(\mathbf{v}_3 - \mathbf{v}_1) = \mathbf{0}$. Show that this implies $c_2 = c_3 = 0$.

15. Let $S = \{\mathbf{x} : A\mathbf{x} = \mathbf{b}\}$. To show that S is affine, it suffices to show that S is a flat, by Theorem 3. Let $W = \{\mathbf{x} : A\mathbf{x} = \mathbf{0}\}$. Then W is a subspace of $\mathbb{R}^n$, by Theorem 2 in Section 4.2 (or Theorem 12 in Section 2.8). Since $S = W + \mathbf{p}$, where $\mathbf{p}$ satisfies $A\mathbf{p} = \mathbf{b}$, by Theorem 6 in Section 1.5, S is a translate of W, and hence S is a flat.

17. A suitable set consists of any three vectors that are not collinear and have 5 as their third entry. If 5 is their third entry, they lie in the plane $z = 5$. If the vectors are not collinear, their affine hull cannot be a line, so it must be the plane.

19. If $\mathbf{p}, \mathbf{q} \in f(S)$, then there exist $\mathbf{r}, \mathbf{s} \in S$ such that $f(\mathbf{r}) = \mathbf{p}$ and $f(\mathbf{s}) = \mathbf{q}$. Given any $t \in \mathbb{R}$, we must show that $\mathbf{z} = (1 - t)\mathbf{p} + t\mathbf{q}$ is in $f(S)$. Now use definitions of $\mathbf{p}$ and $\mathbf{q}$, and the fact that f is linear. The complete proof is presented in the *Study Guide*.

21. Since B is affine, Theorem 1 implies that B contains all affine combinations of points of B. Hence B contains all affine combinations of points of A. That is, aff $A \subset B$.

23. Since $A \subset (A \cup B)$, it follows from Exercise 22 that aff $A \subset \text{aff } (A \cup B)$. Similarly, aff $B \subset \text{aff } (A \cup B)$, so $[\text{aff } A \cup \text{aff } B] \subset \text{aff } (A \cup B)$.

25. To show that $D \subset E \cap F$, show that $D \subset E$ and $D \subset F$. The complete proof is presented in the *Study Guide*.

Section 8.2, page 452

1. Affinely dependent and $2\mathbf{v}_1 + \mathbf{v}_2 - 3\mathbf{v}_3 = \mathbf{0}$

3. The set is affinely independent. If the points are called $\mathbf{v}_1$, $\mathbf{v}_2$, $\mathbf{v}_3$, and $\mathbf{v}_4$, then $\{\mathbf{v}_1, \mathbf{v}_2, \mathbf{v}_3\}$ is a basis for $\mathbb{R}^3$ and $\mathbf{v}_4 = 16\mathbf{v}_1 + 5\mathbf{v}_2 - 3\mathbf{v}_3$, but the weights in the linear combination do not sum to 1.

5. $-4\mathbf{v}_1 + 5\mathbf{v}_2 - 4\mathbf{v}_3 + 3\mathbf{v}_4 = \mathbf{0}$

7. The barycentric coordinates are $(-2, 4, -1)$.

9. See the *Study Guide*.

11. When a set of five points is translated by subtracting, say, the first point, the new set of four points must be linearly dependent, by Theorem 8 in Section 1.7, because the four points are in $\mathbb{R}^3$. By Theorem 5, the original set of five points is affinely dependent.

13. If $\{\mathbf{v}_1, \mathbf{v}_2\}$ is affinely dependent, then there exist c_1 and c_2, not both zero, such that $c_1 + c_2 = 0$ and $c_1\mathbf{v}_1 + c_2\mathbf{v}_2 = \mathbf{0}$. Show that this implies $\mathbf{v}_1 = \mathbf{v}_2$. For the converse, suppose $\mathbf{v}_1 = \mathbf{v}_2$ and select specific c_1 and c_2 that show their affine dependence. The details are in the *Study Guide*.

15. a. The vectors $\mathbf{v}_2 - \mathbf{v}_1 = \begin{bmatrix} 1 \\ 2 \end{bmatrix}$ and $\mathbf{v}_3 - \mathbf{v}_1 = \begin{bmatrix} 3 \\ -2 \end{bmatrix}$ are not multiples and hence are linearly independent. By Theorem 5, S is affinely independent.
   b. $\mathbf{p}_1 \leftrightarrow \left(-\frac{6}{8}, \frac{9}{8}, \frac{5}{8}\right)$, $\mathbf{p}_2 \leftrightarrow \left(0, \frac{1}{2}, \frac{1}{2}\right)$, $\mathbf{p}_3 \leftrightarrow \left(\frac{14}{8}, -\frac{5}{8}, -\frac{1}{8}\right)$, $\mathbf{p}_4 \leftrightarrow \left(\frac{6}{8}, -\frac{5}{8}, \frac{7}{8}\right)$, $\mathbf{p}_5 \leftrightarrow \left(\frac{1}{4}, \frac{1}{8}, \frac{5}{8}\right)$
   c. $\mathbf{p}_6$ is $(-, -, +)$, $\mathbf{p}_7$ is $(0, +, -)$, and $\mathbf{p}_8$ is $(+, +, -)$.

17. Suppose $S = \{\mathbf{b}_1, \ldots, \mathbf{b}_k\}$ is an affinely independent set. Then equation (7) has a solution, because $\mathbf{p}$ is in aff S. Hence equation (8) has a solution. By Theorem 5, the homogeneous forms of the points in S are linearly independent. Thus (8) has a unique solution. Then (7) also has a unique solution, because (8) encodes both equations that appear in (7).

The following argument mimics the proof of Theorem 7 in Section 4.4. If $S = \{\mathbf{b}_1, \ldots, \mathbf{b}_k\}$ is an affinely independent set, then scalars $c_1, \ldots, c_k$ exist that satisfy (7), by definition of aff S. Suppose $\mathbf{x}$ also has the representation

$$\mathbf{x} = d_1\mathbf{b}_1 + \cdots + d_k\mathbf{b}_k \quad \text{and} \quad d_1 + \cdots + d_k = 1 \quad (7a)$$

for scalars $d_1, \ldots, d_k$. Then subtraction produces the equation

$$\mathbf{0} = \mathbf{x} - \mathbf{x} = (c_1 - d_1)\mathbf{b}_1 + \cdots + (c_k - d_k)\mathbf{b}_k \quad (7b)$$

The weights in (7b) sum to 0 because the c's and the d's separately sum to 1. This is impossible, unless each weight in (8) is 0, because S is an affinely independent set. This proves that $c_i = d_i$ for $i = 1, \ldots, k$.

19. If $\{\mathbf{p}_1, \mathbf{p}_2, \mathbf{p}_3\}$ is an affinely dependent set, then there exist scalars c_1, c_2, and c_3, not all zero, such that $c_1\mathbf{p}_1 + c_2\mathbf{p}_2 + c_3\mathbf{p}_3 = \mathbf{0}$ and $c_1 + c_2 + c_3 = 0$. Now use the linearity of f.

21. Let $\mathbf{a} = \begin{bmatrix} a_1 \\ a_2 \end{bmatrix}$, $\mathbf{b} = \begin{bmatrix} b_1 \\ b_2 \end{bmatrix}$, and $\mathbf{c} = \begin{bmatrix} c_1 \\ c_2 \end{bmatrix}$. Then

$$\det[\,\tilde{\mathbf{a}} \quad \tilde{\mathbf{b}} \quad \tilde{\mathbf{c}}\,] = \det \begin{bmatrix} a_1 & b_1 & c_1 \\ a_2 & b_2 & c_2 \\ 1 & 1 & 1 \end{bmatrix} =$$

$\det \begin{bmatrix} a_1 & a_2 & 1 \\ b_1 & b_2 & 1 \\ c_1 & c_2 & 1 \end{bmatrix}$, by the transpose property of the

determinant (Theorem 5 in Section 3.2). By Exercise 30 in Section 3.3, this determinant equals 2 times the area of the triangle with vertices at $\mathbf{a}$, $\mathbf{b}$, and $\mathbf{c}$.

23. If $[\,\tilde{\mathbf{a}} \quad \tilde{\mathbf{b}} \quad \tilde{\mathbf{c}}\,] \begin{bmatrix} r \\ s \\ t \end{bmatrix} = \tilde{\mathbf{p}}$, then Cramer's rule gives

$r = \det[\,\tilde{\mathbf{p}} \quad \tilde{\mathbf{b}} \quad \tilde{\mathbf{c}}\,] / \det[\,\tilde{\mathbf{a}} \quad \tilde{\mathbf{b}} \quad \tilde{\mathbf{c}}\,]$. By Exercise 21, the numerator of this quotient is twice the area of $\triangle\mathbf{pbc}$, and the denominator is twice the area of $\triangle\mathbf{abc}$. This proves the formula for r. The other formulas are proved using Cramer's rule for s and t.

Section 8.3, page 459

1. See the *Study Guide*.

3. None are in conv S.

5. $\mathbf{p}_1 = -\frac{1}{6}\mathbf{v}_1 + \frac{1}{3}\mathbf{v}_2 + \frac{2}{3}\mathbf{v}_3 + \frac{1}{6}\mathbf{v}_4$, so $\mathbf{p}_1 \notin$ conv S.
$\mathbf{p}_2 = \frac{1}{3}\mathbf{v}_1 + \frac{1}{3}\mathbf{v}_2 + \frac{1}{6}\mathbf{v}_3 + \frac{1}{6}\mathbf{v}_4$, so $\mathbf{p}_2 \in$ conv S.

7. a. The barycentric coordinates of $\mathbf{p}_1$, $\mathbf{p}_2$, $\mathbf{p}_3$, and $\mathbf{p}_4$ are, respectively, $\left(\frac{1}{3}, \frac{1}{6}, \frac{1}{2}\right)$, $\left(0, \frac{1}{2}, \frac{1}{2}\right)$, $\left(\frac{1}{2}, -\frac{1}{4}, \frac{3}{4}\right)$, and $\left(\frac{1}{2}, \frac{3}{4}, -\frac{1}{4}\right)$.
b. $\mathbf{p}_3$ and $\mathbf{p}_4$ are outside conv T. $\mathbf{p}_1$ is inside conv T. $\mathbf{p}_2$ is on the edge $\overline{\mathbf{v}_2\mathbf{v}_3}$ of conv T.

9. $\mathbf{p}_1$ and $\mathbf{p}_3$ are outside the tetrahedron conv S. $\mathbf{p}_2$ is on the face containing the vertices $\mathbf{v}_2$, $\mathbf{v}_3$, and $\mathbf{v}_4$. $\mathbf{p}_4$ is inside conv S. $\mathbf{p}_5$ is on the edge between $\mathbf{v}_1$ and $\mathbf{v}_3$.

11. See the *Study Guide*.

13. If $\mathbf{p}, \mathbf{q} \in f(S)$, then there exist $\mathbf{r}, \mathbf{s} \in S$ such that $f(\mathbf{r}) = \mathbf{p}$ and $f(\mathbf{s}) = \mathbf{q}$. The goal is to show that the line segment $\mathbf{y} = (1 - t)\mathbf{p} + t\mathbf{q}$, for $0 \le t \le 1$, is in $f(S)$. Use the linearity of f and the convexity of S to show that $\mathbf{y} = f(\mathbf{w})$ for some $\mathbf{w}$ in S. This will show that $\mathbf{y}$ is in $f(S)$ and that $f(S)$ is convex.

15. $\mathbf{p} = \frac{1}{6}\mathbf{v}_1 + \frac{1}{2}\mathbf{v}_2 + \frac{1}{3}\mathbf{v}_4$ and $\mathbf{p} = \frac{1}{2}\mathbf{v}_1 + \frac{1}{6}\mathbf{v}_2 + \frac{1}{3}\mathbf{v}_3$.

17. Suppose $A \subset B$, where B is convex. Then, since B is convex, Theorem 7 implies that B contains all convex combinations of points of B. Hence B contains all convex combinations of points of A. That is, conv $A \subset B$.

19. a. Use Exercise 18 to show that conv A and conv B are both subsets of conv $(A \cup B)$. This will imply that their union is also a subset of conv $(A \cup B)$.

b. One possibility is to let A be two adjacent corners of a square and let B be the other two corners. Then what is (conv A) $\cup$ (conv B), and what is conv $(A \cup B)$?

21.

23. $\mathbf{g}(t) = (1 - t)\mathbf{f}_0(t) + t\mathbf{f}_1(t)$
$= (1 - t)[(1 - t)\mathbf{p}_0 + t\mathbf{p}_1] + t[(1 - t)\mathbf{p}_1 + t\mathbf{p}_2]$
$= (1 - t)^2\mathbf{p}_0 + 2t(1 - t)\mathbf{p}_1 + t^2\mathbf{p}_2$.
The sum of the weights in the linear combination for $\mathbf{g}$ is $(1 - t)^2 + 2t(1 - t) + t^2$, which equals $(1 - 2t + t^2) + (2t - 2t^2) + t^2 = 1$. The weights are each between 0 and 1 when $0 \le t \le 1$, so $\mathbf{g}(t)$ is in conv $\{\mathbf{p}_0, \mathbf{p}_1, \mathbf{p}_2\}$.

Section 8.4, page 467

1. $f(x_1, x_2) = 3x_1 + 4x_2$ and $d = 13$

3. a. Open   **b.** Closed   **c.** Neither
    d. Closed   **e.** Closed

5. a. Not compact, convex
    b. Compact, convex
    c. Not compact, convex
    d. Not compact, not convex
    e. Not compact, convex

7. a. $\mathbf{n} = \begin{bmatrix} 0 \\ 2 \\ 3 \end{bmatrix}$ or a multiple
b. $f(\mathbf{x}) = 2x_2 + 3x_3$, $d = 11$

9. a. $\mathbf{n} = \begin{bmatrix} 3 \\ -1 \\ 2 \\ 1 \end{bmatrix}$ or a multiple
b. $f(\mathbf{x}) = 3x_1 - x_2 + 2x_3 + x_4$, $d = 5$

11. $\mathbf{v}_2$ is on the same side as $\mathbf{0}$, $\mathbf{v}_1$ is on the other side, and $\mathbf{v}_3$ is in H.

13. One possibility is $\mathbf{p} = \begin{bmatrix} 32 \\ -14 \\ 0 \\ 0 \end{bmatrix}$, $\mathbf{v}_1 = \begin{bmatrix} 10 \\ -7 \\ 1 \\ 0 \end{bmatrix}$,

$\mathbf{v}_2 = \begin{bmatrix} -4 \\ 1 \\ 0 \\ 1 \end{bmatrix}$.

15. $f(x_1, x_2, x_3) = x_1 - 3x_2 + 4x_3 - 2x_4$, and $d = 5$

17. $f(x_1, x_2, x_3) = x_1 - 2x_2 + x_3$, and $d = 0$

19. $f(x_1, x_2, x_3) = -5x_1 + 3x_2 + x_3$, and $d = 0$

21. See the *Study Guide*.

23. $f(x_1, x_2) = 3x_1 - 2x_2$ with d satisfying $9 < d < 10$ is one possibility.

25. $f(x, y) = 4x + 1$. A natural choice for d is 12.75, which equals $f(3, .75)$. The point $(3, .75)$ is three-fourths of the distance between the center of $B(\mathbf{0}, 3)$ and the center of $B(\mathbf{p}, 1)$.

27. Exercise 2(a) in Section 8.3 gives one possibility. Or let $S = \{(x, y) : x^2 y^2 = 1 \text{ and } y > 0\}$. Then conv S is the upper (open) half-plane.

29. Let $\mathbf{x}, \mathbf{y} \in B(\mathbf{p}, \delta)$ and suppose $\mathbf{z} = (1 - t)\mathbf{x} + t\mathbf{y}$, where $0 \le t \le 1$. Then show that

$$\|\mathbf{z} - \mathbf{p}\| = \|[(1 - t)\mathbf{x} + t\mathbf{y}] - \mathbf{p}\|$$
$$= \|(1 - t)(\mathbf{x} - \mathbf{p}) + t(\mathbf{y} - \mathbf{p})\| < \delta.$$

Section 8.5, page 479

1. a. $m = 1$ at the point $\mathbf{p}_1$ **b.** $m = 5$ at the point $\mathbf{p}_2$
c. $m = 5$ at the point $\mathbf{p}_3$

3. a. $m = -3$ at the point $\mathbf{p}_3$
b. $m = 1$ on the set conv $\{\mathbf{p}_1, \mathbf{p}_3\}$
c. $m = -3$ on the set conv $\{\mathbf{p}_1, \mathbf{p}_2\}$

5. $\left\{ \begin{bmatrix} 0 \\ 0 \end{bmatrix}, \begin{bmatrix} 5 \\ 0 \end{bmatrix}, \begin{bmatrix} 4 \\ 3 \end{bmatrix}, \begin{bmatrix} 0 \\ 5 \end{bmatrix} \right\}$

7. $\left\{ \begin{bmatrix} 0 \\ 0 \end{bmatrix}, \begin{bmatrix} 7 \\ 0 \end{bmatrix}, \begin{bmatrix} 6 \\ 4 \end{bmatrix}, \begin{bmatrix} 0 \\ 6 \end{bmatrix} \right\}$

9. The origin is an extreme point, but it is not a vertex. Explain why.

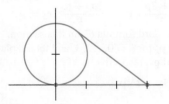

11. One possibility is to let S be a square that includes part of the boundary but not all of it. For example, include just two adjacent edges. The convex hull of the profile P is a triangular region.

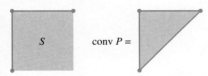

S conv $P =$

13. a. $f_0(C^5) = 32$, $f_1(C^5) = 80$, $f_2(C^5) = 80$,
$f_3(C^5) = 40$, $f_4(C^5) = 10$, and
$32 - 80 + 80 - 40 + 10 = 2$.

b.

	f_0	f_1	f_2	f_3	f_4
S^1	2				
S^2	4	4			
S^3	8	12	6		
S^4	16	32	24	8	
S^5	32	80	80	40	10

For a general formula, see the *Study Guide*.

15. a. $f_0(P^n) = f_0(Q) + 1$
b. $f_k(P^n) = f_k(Q) + f_{k-1}(Q)$
c. $f_{n-1}(P^n) = f_{n-2}(Q) + 1$

17. See the *Study Guide*.

19. Let S be convex and let $\mathbf{x} \in cS + dS$, where $c > 0$ and $d > 0$. Then there exist $\mathbf{s}_1$ and $\mathbf{s}_2$ in S such that $\mathbf{x} = c\mathbf{s}_1 + d\mathbf{s}_2$. But then

$$\mathbf{x} = c\mathbf{s}_1 + d\mathbf{s}_2 = (c + d)\left(\frac{c}{c + d}\mathbf{s}_1 + \frac{d}{c + d}\mathbf{s}_2 \right).$$

Now show that the expression on the right side is a member of $(c + d)S$.

For the converse, pick a typical point in $(c + d)S$ and show it is in $cS + dS$.

21. *Hint:* Suppose A and B are convex. Let $\mathbf{x}, \mathbf{y} \in A + B$. Then there exist $\mathbf{a}, \mathbf{c} \in A$ and $\mathbf{b}, \mathbf{d} \in B$ such that $\mathbf{x} = \mathbf{a} + \mathbf{b}$ and $\mathbf{y} = \mathbf{c} + \mathbf{d}$. For any t such that $0 \le t \le 1$, show that

$$\mathbf{w} = (1 - t)\mathbf{x} + t\mathbf{y} = (1 - t)(\mathbf{a} + \mathbf{b}) + t(\mathbf{c} + \mathbf{d})$$

represents a point in $A + B$.

Section 8.6, page 490

1. The control points for $\mathbf{x}(t) + \mathbf{b}$ should be $\mathbf{p}_0 + \mathbf{b}$, $\mathbf{p}_1 + \mathbf{b}$, and $\mathbf{p}_3 + \mathbf{b}$. Write the Bézier curve through these points, and show algebraically that this curve is $\mathbf{x}(t) + \mathbf{b}$. See the *Study Guide*.

3. a. $\mathbf{x}'(t) = (-3 + 6t - 3t_2)\mathbf{p}_0 + (3 - 12t + 9t^2)\mathbf{p}_1 + (6t - 9t^2)\mathbf{p}_2 + 3t^2\mathbf{p}_3$, so
$\mathbf{x}'(0) = -3\mathbf{p}_0 + 3\mathbf{p}_1 = 3(\mathbf{p}_1 - \mathbf{p}_0)$, and
$\mathbf{x}'(1) = -3\mathbf{p}_2 + 3\mathbf{p}_3 = 3(\mathbf{p}_3 - \mathbf{p}_2)$. This shows that the tangent vector $\mathbf{x}'(0)$ points in the direction from $\mathbf{p}_0$ to $\mathbf{p}_1$ and is three times the length of $\mathbf{p}_1 - \mathbf{p}_0$. Likewise, $\mathbf{x}'(1)$ points in the direction from $\mathbf{p}_2$ to $\mathbf{p}_3$ and is three times the length of $\mathbf{p}_3 - \mathbf{p}_2$. In particular, $\mathbf{x}'(1) = \mathbf{0}$ if and only if $\mathbf{p}_3 = \mathbf{p}_2$.

b. $\mathbf{x}''(t) = (6 - 6t)\mathbf{p}_0 + (-12 + 18t)\mathbf{p}_1$
$\qquad + (6 - 18t)\mathbf{p}_2 + 6t\mathbf{p}_3$, so that
$\mathbf{x}''(0) = 6\mathbf{p}_0 - 12\mathbf{p}_1 + 6\mathbf{p}_2 = 6(\mathbf{p}_0 - \mathbf{p}_1) + 6(\mathbf{p}_2 - \mathbf{p}_1)$
and
$\mathbf{x}''(1) = 6\mathbf{p}_1 - 12\mathbf{p}_2 + 6\mathbf{p}_3 = 6(\mathbf{p}_1 - \mathbf{p}_2) + 6(\mathbf{p}_3 - \mathbf{p}_2)$
For a picture of $\mathbf{x}''(0)$, construct a coordinate system with the origin at $\mathbf{p}_1$, temporarily, label $\mathbf{p}_0$ as $\mathbf{p}_0 - \mathbf{p}_1$, and label $\mathbf{p}_2$ as $\mathbf{p}_2 - \mathbf{p}_1$. Finally, construct a line from

this new origin through the sum of $\mathbf{p}_0 - \mathbf{p}_1$ and $\mathbf{p}_2 - \mathbf{p}_1$, extended out a bit. That line points in the direction of $\mathbf{x}''(0)$.

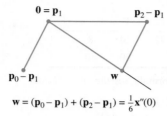

$$\mathbf{w} = (\mathbf{p}_0 - \mathbf{p}_1) + (\mathbf{p}_2 - \mathbf{p}_1) = \tfrac{1}{6}\mathbf{x}''(0)$$

5. a. From Exercise 3(a) or equation (9) in the text,

$$\mathbf{x}'(1) = 3(\mathbf{p}_3 - \mathbf{p}_2)$$

Use the formula for $\mathbf{x}'(0)$, with the control points from $\mathbf{y}(t)$, and obtain

$$\mathbf{y}'(0) = 3\mathbf{p}_3 + 3\mathbf{p}_4 = 3(\mathbf{p}_4 - \mathbf{p}_3)$$

For C^1 continuity, $3(\mathbf{p}_3 - \mathbf{p}_2) = 3(\mathbf{p}_4 - \mathbf{p}_3)$, so $\mathbf{p}_3 = (\mathbf{p}_4 + \mathbf{p}_2)/2$, and $\mathbf{p}_3$ is the midpoint of the line segment from $\mathbf{p}_2$ to $\mathbf{p}_4$.

b. If $\mathbf{x}'(1) = \mathbf{y}'(0) = \mathbf{0}$, then $\mathbf{p}_2 = \mathbf{p}_3$ and $\mathbf{p}_3 = \mathbf{p}_4$. Thus, the "line segment" from $\mathbf{p}_2$ to $\mathbf{p}_4$ is just the point $\mathbf{p}_3$. [*Note:* In this case, the combined curve is still C^1 continuous, by definition. However, some choices of the other "control" points, $\mathbf{p}_0, \mathbf{p}_1, \mathbf{p}_5$, and $\mathbf{p}_6$, can produce a curve with a visible corner at $\mathbf{p}_3$, in which case the curve is not G^1 continuous at $\mathbf{p}_3$.]

7. *Hint:* Use $\mathbf{x}''(t)$ from Exercise 3 and adapt this for the second curve to see that

$$\mathbf{y}''(t) = 6(1-t)\mathbf{p}_3 + 6(-2+3t)\mathbf{p}_4 + 6(1-3t)\mathbf{p}_5 + 6t\mathbf{p}_6$$

Then set $\mathbf{x}''(1) = \mathbf{y}''(0)$. Since the curve is C^1 continuous at $\mathbf{p}_3$, Exercise 5(a) says that the point $\mathbf{p}_3$ is the midpoint of the segment from $\mathbf{p}_2$ to $\mathbf{p}_4$. This implies that $\mathbf{p}_4 - \mathbf{p}_3 = \mathbf{p}_3 - \mathbf{p}_2$. Use this substitution to show that $\mathbf{p}_4$ and $\mathbf{p}_5$ are uniquely determined by $\mathbf{p}_1, \mathbf{p}_2$, and $\mathbf{p}_3$. Only $\mathbf{p}_6$ can be chosen arbitrarily.

9. Write a vector of the polynomial weights for $\mathbf{x}(t)$, expand the polynomial weights, and factor the vector as $M_B \mathbf{u}(t)$:

$$\begin{bmatrix} 1 - 4t + 6t^2 - 4t^3 + t^4 \\ 4t - 12t^2 + 12t^3 - 4t^4 \\ 6t^2 - 12t^3 + 6t^4 \\ 4t^3 - 4t^4 \\ t^4 \end{bmatrix}$$

$$= \begin{bmatrix} 1 & -4 & 6 & -4 & 1 \\ 0 & 4 & -12 & 12 & -4 \\ 0 & 0 & 6 & -12 & 6 \\ 0 & 0 & 0 & 4 & -4 \\ 0 & 0 & 0 & 0 & 1 \end{bmatrix} \begin{bmatrix} 1 \\ t \\ t^2 \\ t^3 \\ t^4 \end{bmatrix},$$

$$M_B = \begin{bmatrix} 1 & -4 & 6 & -4 & 1 \\ 0 & 4 & -12 & 12 & -4 \\ 0 & 0 & 6 & -12 & 6 \\ 0 & 0 & 0 & 4 & -4 \\ 0 & 0 & 0 & 0 & 1 \end{bmatrix}$$

11. See the *Study Guide*.

13. a. *Hint:* Use the fact that $\mathbf{q}_0 = \mathbf{p}_0$.

   b. Multiply the first and last parts of equation (13) by $\tfrac{8}{3}$ and solve for $8\mathbf{q}_2$.

   c. Use equation (8) to substitute for $8\mathbf{q}_3$ and then apply part (a).

15. a. From equation (11), $\mathbf{y}'(1) = .5\mathbf{x}'(.5) = \mathbf{z}'(0)$.

   b. Observe that $\mathbf{y}'(1) = 3(\mathbf{q}_3 - \mathbf{q}_2)$. This follows from equation (9), with $\mathbf{y}(t)$ and its control points in place of $\mathbf{x}(t)$ and its control points. Similarly, for $\mathbf{z}(t)$ and its control points, $\mathbf{z}'(0) = 3(\mathbf{r}_1 - \mathbf{r}_0)$. By part (a), $3(\mathbf{q}_3 - \mathbf{q}_2) = 3(\mathbf{r}_1 - \mathbf{r}_0)$. Replace $\mathbf{r}_0$ by $\mathbf{q}_3$, and obtain $\mathbf{q}_3 - \mathbf{q}_2 = \mathbf{r}_1 - \mathbf{q}_3$, and hence $\mathbf{q}_3 = (\mathbf{q}_2 + \mathbf{r}_1)/2$.

   c. Set $\mathbf{q}_0 = \mathbf{p}_0$ and $\mathbf{r}_3 = \mathbf{p}_3$. Compute $\mathbf{q}_1 = (\mathbf{p}_0 + \mathbf{p}_1)/2$ and $\mathbf{r}_2 = (\mathbf{p}_2 + \mathbf{p}_3)/2$. Compute $\mathbf{m} = (\mathbf{p}_1 + \mathbf{p}_2)/2$. Compute $\mathbf{q}_2 = (\mathbf{q}_1 + \mathbf{m})/2$ and $\mathbf{r}_1 = (\mathbf{m} + \mathbf{r}_2)/2$. Compute $\mathbf{q}_3 = (\mathbf{q}_2 + \mathbf{r}_1)/2$ and set $\mathbf{r}_0 = \mathbf{q}_3$.

17. a. $\mathbf{r}_0 = \mathbf{p}_0,\ \mathbf{r}_1 = \dfrac{\mathbf{p}_0 + 2\mathbf{p}_1}{3},\ \mathbf{r}_2 = \dfrac{2\mathbf{p}_1 + \mathbf{p}_2}{3},\ \mathbf{r}_3 = \mathbf{p}_2$

   b. *Hint:* Write the standard formula (7) in this section, with $\mathbf{r}_i$ in place of $\mathbf{p}_i$ for $i = 0, \ldots, 3$, and then replace $\mathbf{r}_0$ and $\mathbf{r}_3$ by $\mathbf{p}_0$ and $\mathbf{p}_2$, respectively:

$$\mathbf{x}(t) = (1 - 3t + 3t^2 - t^3)\mathbf{p}_0 \\ + (3t - 6t^2 + 3t^3)\mathbf{r}_1 \qquad (iii) \\ + (3t^2 - 3t^3)\mathbf{r}_2 + t^3\mathbf{p}_2$$

   Use the formulas for $\mathbf{r}_1$ and $\mathbf{r}_2$ from part (a) to examine the second and third terms in this expression for $\mathbf{x}(t)$.

Answers to Even-Numbered Exercises

Chapter 1

Section 1.1, page 10

2. The solution is $(x_1, x_2) = (9, -5)$, or simply $(9, -5)$.

4. The point of intersection is $(-3, -5)$

6. Replace Row 4 by its sum with -4 times Row 3. After that, scale Row 4 by $-1/7$.

8. $(0, 0, 0, 0)$ **10.** $(-47, 12, 2, -2)$

12. Inconsistent   **14.** $(2, -1, 2)$ **16.** Consistent

18. The three planes have one point in common.

20. $h \neq -4$ **22.** $h = 6$

23. a. True. See the remarks following the box titled "Elementary Row Operations."

    b. False. A 5×6 matrix has five rows.

    c. False. The description applies to a single solution. The solution *set* consists of all possible solutions. Only in special cases does the solution set consist of exactly one solution. A statement should be marked True only if the statement is *always* true.

    d. True. See the box before Example 2.

24. a. False. The definition of *row equivalent* requires that there exist a sequence of row operations that transforms one matrix into the other.

    b. True. Elementary row operations do not change the solution set.

    c. False. The definition of *equivalent systems* is in the second paragraph after equation (2).

    d. True. By definition, a consistent system has *at least one* solution.

26. $d \neq 2c$

28. Answers may vary. The systems corresponding to the following matrices each have the solution set $x_1 = 3$, $x_2 = -2$, $x_3 = -1$. (The tildes represent row equivalence.)

$$\begin{bmatrix} 1 & 0 & 0 & 3 \\ 0 & 1 & 0 & -2 \\ 0 & 0 & 1 & -1 \end{bmatrix} \sim \begin{bmatrix} 1 & 0 & 0 & 3 \\ 2 & 1 & 0 & 4 \\ 0 & 0 & 1 & -1 \end{bmatrix}$$

$$\sim \begin{bmatrix} 1 & 0 & 0 & 3 \\ 2 & 1 & 0 & 4 \\ 2 & 0 & 1 & 5 \end{bmatrix}$$

30. Scale Row 3 by $-1/5$; scale Row 3 by -5.

32. Replace Row 3 by Row 3 + (-4)Row 2; replace Row 3 by Row 3 + (4)Row 2.

34. $(20, 27.5, 30, 22.5)$

Section 1.2, page 21

2. Reduced echelon form: a. Echelon form: b and d. Not in echelon form: c.

4. $\begin{bmatrix} 1 & 0 & 0 & 1 \\ 0 & 1 & 0 & -2 \\ 0 & 0 & 1 & 2 \end{bmatrix}$

Pivot cols 1, 2, and 3: $\begin{bmatrix} 1 & 2 & 4 & 5 \\ 2 & 4 & 5 & 4 \\ 4 & 5 & 4 & 2 \end{bmatrix}$.

6. $\begin{bmatrix} \blacksquare & * \\ 0 & \blacksquare \\ 0 & 0 \end{bmatrix}, \begin{bmatrix} \blacksquare & * \\ 0 & 0 \\ 0 & 0 \end{bmatrix}, \begin{bmatrix} 0 & \blacksquare \\ 0 & 0 \\ 0 & 0 \end{bmatrix}$

8. $\begin{cases} x_1 = 4 \\ x_2 = 3 \\ x_3 \text{ is free.} \end{cases}$ **10.** $\begin{cases} x_1 = 2 + 2x_2 \\ x_2 \text{ is free.} \\ x_3 = -2 \end{cases}$

12. No solutions   **14.** $\begin{cases} x_1 = 3 + 5x_3 \\ x_2 = 6 - 4x_3 + x_4 \\ x_3 \text{ is free.} \\ x_4 \text{ is free.} \\ x_5 = 0 \end{cases}$

16. a. Inconsistent   **b.** Consistent, with many solutions

18. All h.

20. a. Inconsistent when $h = -6$ and $k \neq 2$

    b. Unique solution when $h \neq -6$

    c. Many solutions when $h = -6$ and $k = 2$

21. a. False. See Theorem 1.

    b. False. See the second paragraph in this section.

    c. True. Basic variables are defined after equation (4).

    d. True. A similar statement appears at the beginning of "Parametric Descriptions of Solution Sets."

    e. False. The row shown corresponds to the equation $5x_4 = 0$, which does not by itself lead to a

contradiction. So the system might be consistent or it
might be inconsistent.

22. a. True. See Theorem 1.

    b. False. See Theorem 2.

    c. False. See the beginning of the subsection "Pivot
Positions." The pivot positions in a matrix are
determined completely by the positions of the leading
entries in the nonzero rows of any echelon form
obtained from the matrix.

    d. True. See the paragraph just before Example 4.

    e. False. The existence of at least one solution is not
related to the presence or absence of free variables. If
the system is inconsistent, the solution set is empty. See
the solution of Practice Problem 2.

24. The system is consistent because there is no pivot in the last
column of the augmented matrix.

26. Yes. The system is consistent because with three pivots,
there must be a pivot in every row of the coefficient matrix.
The reduced echelon form of the augmented matrix cannot
contain a row of the form $\begin{bmatrix} 0 & 0 & 0 & 0 & 0 & 1 \end{bmatrix}$

28. Every column in the augmented matrix *except the rightmost
column* is a pivot column, and the rightmost column is *not* a
pivot column.

30. Example: $\begin{aligned} x_1 + x_2 + x_3 &= 4 \\ 2x_1 + 2x_2 + 2x_3 &= 5 \end{aligned}$

32. When $n = 20$, the backward phase requires about 7% of the
total flops. When $n = 200$, the backward phase requires
about 0.7% of the total flops. The *Instructor's Solution
Manual* has the details.

34. [M] $p(t) = 1.7125t - 1.1948t^2 + .6615t^3 - .0701t^4 +$
$.0026t^5$, and $p(7.5) = 64.6$ hundred lb.
[*Note:* $p(7.5) = 64.8$ when the coefficients of $p(t)$ are
retained as originally computed.] If a polynomial of lower
degree is used, the resulting system of equations is
overdetermined. The augmented matrix for such a system is
the same as the one used to find p, except that at least
column 6 is missing. When the augmented matrix is row
reduced, the sixth row of the augmented matrix will be all
zeros except for a nonzero entry in the augmented column,
indicating that no solution exists.

Section 1.3, page 32

2. $\begin{bmatrix} 1 \\ 3 \end{bmatrix}, \begin{bmatrix} 7 \\ -9 \end{bmatrix}$

4.

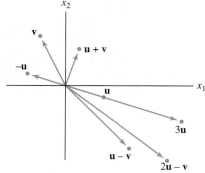

6. $\begin{aligned} 3x_1 + 7x_2 - 2x_3 &= 0 \\ -2x_1 + 3x_2 + x_3 &= 0 \end{aligned}$

8. $\mathbf{w} = 2\mathbf{v} - \mathbf{u}$, $\mathbf{x} = -2\mathbf{u} + 2\mathbf{v}$, $\mathbf{y} = -2\mathbf{u} + 3.5\mathbf{v}$,
$\mathbf{z} = -3\mathbf{u} + 4\mathbf{v}$ Yes, every vector in $\mathbb{R}^2$ is a linear
combination of $\mathbf{u}$ and $\mathbf{v}$.

10. $x_1 \begin{bmatrix} 3 \\ -2 \\ 5 \end{bmatrix} + x_2 \begin{bmatrix} -2 \\ -7 \\ 4 \end{bmatrix} + x_3 \begin{bmatrix} 4 \\ 5 \\ -3 \end{bmatrix} = \begin{bmatrix} 3 \\ 1 \\ 2 \end{bmatrix}$

12. Yes, $\mathbf{b}$ is a linear combination of $\mathbf{a}_1$, $\mathbf{a}_2$, and $\mathbf{a}_3$.

14. Yes, $\mathbf{b}$ is a linear combination of the columns of A.

16. $h = -2$

18. Noninteger weights are acceptable, of course, but some
simple choices are $0 \cdot \mathbf{v}_1 + 0 \cdot \mathbf{v}_2 = \mathbf{0}$, and

$1 \cdot \mathbf{v}_1 + 0 \cdot \mathbf{v}_2 = \begin{bmatrix} 1 \\ 1 \\ -2 \end{bmatrix}$, $0 \cdot \mathbf{v}_1 + 1 \cdot \mathbf{v}_2 = \begin{bmatrix} -2 \\ 3 \\ 0 \end{bmatrix}$

$1 \cdot \mathbf{v}_1 + 1 \cdot \mathbf{v}_2 = \begin{bmatrix} -1 \\ 4 \\ -2 \end{bmatrix}$, $1 \cdot \mathbf{v}_1 - 1 \cdot \mathbf{v}_2 = \begin{bmatrix} 3 \\ -2 \\ -2 \end{bmatrix}$

20. Span $\{\mathbf{v}_1, \mathbf{v}_2\}$ is a plane in $\mathbb{R}^3$.

22. Construct any 3×4 matrix in echelon form that
corresponds to an inconsistent system. Perform sufficient
row operations on the matrix to eliminate all zero entries in
the first three columns.

23. a. False. The alternative notation is $(-4, 3)$, using
parentheses and a comma.

    b. False. Plot the points to verify this. Or, see the
statement preceding Example 3. If $\begin{bmatrix} -5 \\ 2 \end{bmatrix}$ were on the
line through $\begin{bmatrix} -2 \\ 5 \end{bmatrix}$ and the origin, then $\begin{bmatrix} -5 \\ 2 \end{bmatrix}$ would
have to be a multiple of $\begin{bmatrix} -2 \\ 5 \end{bmatrix}$, which is not the case.

    c. True. See the line displayed just before Example 4.

    d. True. See the box that discusses the matrix in (5).

    e. False. The statement is often true, but Span $\{\mathbf{u}, \mathbf{v}\}$ is not
a plane when $\mathbf{v}$ is a multiple of $\mathbf{u}$, or when $\mathbf{u}$ is the zero
vector.

24. a. False. Span $\{\mathbf{u}, \mathbf{v}\}$ can be a plane.

b. True. See the beginning of the subsection "Vectors in $\mathbb{R}^n$."

c. True. See the comment following the definition of the Span of a set of vectors.

d. False. $\mathbf{u} - \mathbf{v} + \mathbf{v} = \mathbf{u}$.

e. False. Setting all the coefficients equal to zero is a legitimate linear combination of a set of vectors.

26. a. No, $\mathbf{b}$ is *not* a linear combination of the columns of A; that is, $\mathbf{b}$ is *not* in W.

b. The second column of A is in W because $\mathbf{a}_2 = 0 \cdot \mathbf{a}_1 + 1 \cdot \mathbf{a}_2 + 0 \cdot \mathbf{a}_3$.

28. a. The amount of heat produced when the steam plant burns x_1 tons of anthracite and x_2 tons of bituminous coal is $27.6x_1 + 30.2x_2$ million Btu.

b. The vector $x_1 \begin{bmatrix} 27.6 \\ 3100 \\ 250 \end{bmatrix} + x_2 \begin{bmatrix} 30.2 \\ 6400 \\ 360 \end{bmatrix}$ gives the total output produced by x_1 tons of anthracite and x_2 tons of bituminous coal.

c. [M] Solve $x_1 \begin{bmatrix} 27.6 \\ 3100 \\ 250 \end{bmatrix} + x_2 \begin{bmatrix} 30.2 \\ 6400 \\ 360 \end{bmatrix} = \begin{bmatrix} 162 \\ 23{,}610 \\ 1{,}623 \end{bmatrix}$.

The steam plant burned 3.9 tons of anthracite coal and 1.8 tons of bituminous coal.

30. Let m be the total mass of the system. By definition,
$$\mathbf{v} = \frac{1}{m}(m_1\mathbf{v}_1 + \cdots + m_k\mathbf{v}_k) = \frac{m_1}{m}\mathbf{v}_1 + \cdots + \frac{m_k}{m}\mathbf{v}_k$$

The second expression displays $\mathbf{v}$ as a linear combination of $\mathbf{v}_1, \ldots, \mathbf{v}_k$, which shows that $\mathbf{v}$ is in Span$\{\mathbf{v}_1, \ldots, \mathbf{v}_k\}$.

32. On the figure, draw a parallelogram to show that the equation $x_1\mathbf{v}_1 + x_2\mathbf{v}_2 + 0 \cdot \mathbf{v}_3 = \mathbf{b}$ has a solution. Draw another parallelogram to show that the equation $x_1\mathbf{v}_1 + 0 \cdot \mathbf{v}_2 + x_3\mathbf{v}_3 = \mathbf{b}$ also has a solution. (See the *Instructor's Solution Manual.*) Thus the equation $x_1\mathbf{v}_1 + x_2\mathbf{v}_2 + x_3\mathbf{v}_3 = \mathbf{b}$ has at least two solutions, not just one unique solution.

34. a. For $j = 1, \ldots, n$, $u_j + (-1)u_j = (-1)u_j + u_j = 0$, by properties of $\mathbb{R}$. By vector equality,
$$\mathbf{u} + (-\mathbf{u}) = \mathbf{u} + (-1)\mathbf{u} = (-1)\mathbf{u} + \mathbf{u} = (-\mathbf{u}) + \mathbf{u} = \mathbf{0}$$

b. For scalars c and d, the jth entries of $c(d\mathbf{u})$ and $(cd)\mathbf{u}$ are $c(du_j)$ and $(cd)u_j$, respectively. These entries in $\mathbb{R}$ are equal, so the vectors $c(d\mathbf{u})$ and $(cd)\mathbf{u}$ are equal.

Section 1.4, page 40

2. The product is not defined because the number of columns (1) in the 3×1 matrix does not match the number of entries (2) in the vector.

4. a. $A\mathbf{x} = \begin{bmatrix} 1 & 3 & -4 \\ 3 & 2 & 1 \end{bmatrix} \begin{bmatrix} 1 \\ 2 \\ 1 \end{bmatrix}$

$= 1 \cdot \begin{bmatrix} 1 \\ 3 \end{bmatrix} + 2 \cdot \begin{bmatrix} 3 \\ 2 \end{bmatrix} + 1 \cdot \begin{bmatrix} -4 \\ 1 \end{bmatrix} = \begin{bmatrix} 3 \\ 8 \end{bmatrix}$,

b. $A\mathbf{x} = \begin{bmatrix} 1 & 3 & -4 \\ 3 & 2 & 1 \end{bmatrix} \begin{bmatrix} 1 \\ 2 \\ 1 \end{bmatrix}$

$= \begin{bmatrix} 1 \cdot 1 + 3 \cdot 2 + (-4) \cdot 1 \\ 3 \cdot 1 + 2 \cdot 2 + 1 \cdot 1 \end{bmatrix} = \begin{bmatrix} 3 \\ 8 \end{bmatrix}$

6. $-3 \cdot \begin{bmatrix} 2 \\ 3 \\ 8 \\ -2 \end{bmatrix} + 5 \cdot \begin{bmatrix} -3 \\ 2 \\ -5 \\ 1 \end{bmatrix} = \begin{bmatrix} -21 \\ 1 \\ -49 \\ 11 \end{bmatrix}$

8. $\begin{bmatrix} 2 & -1 & -4 & 0 \\ -4 & 5 & 3 & 2 \end{bmatrix} \begin{bmatrix} z_1 \\ z_2 \\ z_3 \\ z_4 \end{bmatrix} = \begin{bmatrix} 5 \\ 12 \end{bmatrix}$

10. $x_1 \begin{bmatrix} 4 \\ 5 \\ 3 \end{bmatrix} + x_2 \begin{bmatrix} -1 \\ 3 \\ -1 \end{bmatrix} = \begin{bmatrix} 8 \\ 2 \\ 1 \end{bmatrix}$ and

$\begin{bmatrix} 4 & -1 \\ 5 & 3 \\ 3 & -1 \end{bmatrix} \begin{bmatrix} x_1 \\ x_2 \end{bmatrix} = \begin{bmatrix} 8 \\ 2 \\ 1 \end{bmatrix}$

12. $\begin{bmatrix} 1 & 2 & -1 & 1 \\ -3 & -4 & 2 & 2 \\ 5 & 2 & 3 & -3 \end{bmatrix}, \mathbf{x} = \begin{bmatrix} x_1 \\ x_2 \\ x_3 \end{bmatrix} = \begin{bmatrix} -4 \\ 4 \\ 3 \end{bmatrix}$

14. No. The equation $A\mathbf{x} = \mathbf{u}$ has no solution.

16. The equation $A\mathbf{x} = \mathbf{b}$ is consistent if and only if $6b_1 + 7b_2 + 2b_3 = 0$. The set of such $\mathbf{b}$ is a plane through the origin in $\mathbb{R}^3$.

18. Only three rows of B contain pivot positions, hence not every vector in $\mathbb{R}^4$ can be written as a linear combination of the columns of B. Since the vectors in the columns of B are not in $\mathbb{R}^3$, they cannot span $\mathbb{R}^3$.

20. Only three rows of B contain pivot positions. The columns of B do not span $\mathbb{R}^4$ and the equation $B\mathbf{x} = \mathbf{y}$ does *not* have a solution for every $\mathbf{y}$ in $\mathbb{R}^4$, by Theorem 4.

22. The matrix $[\mathbf{v}_1 \quad \mathbf{v}_2 \quad \mathbf{v}_3]$ has a pivot in each row, so the columns of the matrix span $\mathbb{R}^3$, by Theorem 4. That is, $\{\mathbf{v}_1, \mathbf{v}_2, \mathbf{v}_3\}$ spans $\mathbb{R}^3$.

23. a. False. See the paragraph following equation (3). The text calls $A\mathbf{x} = \mathbf{b}$ a *matrix equation*.

b. True. See the box before Example 3.

c. False. See the warning following Theorem 4.

d. True. See Example 4.

e. True. See parts (c) and (a) in Theorem 4.

f. True. In Theorem 4, statement (a) is false if and only if statement (d) is also false.

24. a. True. This is part of Theorem 3.

b. True. See the box following Theorem 3.

c. True, by the definition of $A\mathbf{x}$.

d. False. In Theorem 4, statement (d) is true if and only if (a) is true.

e. True, by Theorem 3.

f. False. In Theorem 4, statement (c) is false if and only if statement (a) is also false.

26. $2\mathbf{u} - 3\mathbf{v} - \mathbf{w} = \mathbf{0}$ can be rewritten as $2\mathbf{u} - 3\mathbf{v} = \mathbf{w}$. So, a solution is $x_1 = 2$, $x_2 = -3$.

28. $Q\mathbf{x} = \mathbf{v}$, where $Q = \begin{bmatrix} \mathbf{q}_1 & \mathbf{q}_2 & \mathbf{q}_3 \end{bmatrix}$ and $\mathbf{x} = \begin{bmatrix} x_1 \\ x_2 \\ x_3 \end{bmatrix}$

Note: If your answer is the equation $A\mathbf{x} = \mathbf{b}$, you must specify what A and $\mathbf{b}$ are.

30. Start with any nonzero 3×3 matrix B in echelon form that has fewer than three pivot positions. Perform a row operation that creates a matrix A that is *not* in echelon form. Then A has the desired property. Since A does not have a pivot position in every row, the columns of A do not span $\mathbb{R}^3$, by Theorem 4.

32. A set of three vectors in $\mathbb{R}^4$ cannot span $\mathbb{R}^4$. Reason: The matrix A whose columns are these three vectors has four rows. To have a pivot in each row, A would have to have at least four columns (one for each pivot), which is not the case. Since A does not have a pivot in every row, its columns do not span $\mathbb{R}^4$, by Theorem 4. In general, a set of n vectors in $\mathbb{R}^m$ cannot span $\mathbb{R}^m$ when n is less than m.

34. Given $A\mathbf{u}_1 = \mathbf{v}_1$ and $A\mathbf{u}_2 = \mathbf{v}_2$, you are asked to show that the equation $A\mathbf{x} = \mathbf{w}$ has a solution, where $\mathbf{w} = \mathbf{v}_1 + \mathbf{v}_2$. Observe that $\mathbf{w} = A\mathbf{u}_1 + A\mathbf{u}_2$ and use Theorem 5(a). That is, $\mathbf{w} = A\mathbf{u}_1 + A\mathbf{u}_2 = A(\mathbf{u}_1 + \mathbf{u}_2)$. So the vector $\mathbf{x} = \mathbf{u}_1 + \mathbf{u}_2$ is a solution of $\mathbf{w} = A\mathbf{x}$.

36. If the equation $A\mathbf{x} = \mathbf{b}$ has a unique solution, then the associated system of equations does not have any free variables. If every variable is a basic variable, then each column of A is a pivot column. So the reduced echelon form of A must be $\begin{bmatrix} 1 & 0 & 0 & 0 \\ 0 & 1 & 0 & 0 \\ 0 & 0 & 1 & 0 \\ 0 & 0 & 0 & 1 \end{bmatrix}$. Now it is clear that A has a pivot position in each *row*. By Theorem 4, the columns of A span $\mathbb{R}^4$.

38. [M] The matrix has pivots in only three rows. So, the columns of the original matrix do not span $\mathbb{R}^4$, by Theorem 4.

40. [M] The matrix has a pivot in every row, so its columns span $\mathbb{R}^4$, by Theorem 4.

42. [M] Delete column 3 of the matrix in Exercise 40. (It is also possible to delete column 2 instead of column 3.) No, you cannot delete more than one column.

Section 1.5, page 47

2. There is no free variable; the system has only the trivial solution.

4. The variable x_3 is free; the system has nontrivial solutions.

6. $\mathbf{x} = x_3 \begin{bmatrix} 1 \\ 1 \\ 1 \end{bmatrix}$

8. $\mathbf{x} = x_3 \begin{bmatrix} 2 \\ -2 \\ 1 \\ 0 \end{bmatrix} + x_4 \begin{bmatrix} 7 \\ 4 \\ 0 \\ 1 \end{bmatrix}$

10. $\mathbf{x} = x_3 \begin{bmatrix} 0 \\ 0 \\ 1 \\ 0 \end{bmatrix} + x_4 \begin{bmatrix} -4 \\ 0 \\ 0 \\ 1 \end{bmatrix}$

12. $\mathbf{x} = x_2 \begin{bmatrix} 2 \\ 1 \\ 0 \\ 0 \\ 0 \\ 0 \end{bmatrix} + x_3 \begin{bmatrix} -3 \\ 0 \\ 1 \\ 0 \\ 0 \\ 0 \end{bmatrix} + x_5 \begin{bmatrix} -29 \\ 0 \\ 0 \\ -4 \\ 1 \\ 0 \end{bmatrix}$

14. $\mathbf{x} = \begin{bmatrix} 0 \\ 3 \\ 2 \\ 0 \end{bmatrix} + x_4 \begin{bmatrix} 5 \\ -2 \\ 5 \\ 1 \end{bmatrix} = \mathbf{p} + x_4\mathbf{q}$. The solution set is the line through $\mathbf{p}$ parallel to $\mathbf{q}$.

16. Let $\mathbf{u} = \begin{bmatrix} 2 \\ 1 \\ 0 \end{bmatrix}$, $\mathbf{v} = \begin{bmatrix} -3 \\ 0 \\ 1 \end{bmatrix}$, $\mathbf{p} = \begin{bmatrix} 4 \\ 0 \\ 0 \end{bmatrix}$. The solution of the homogeneous equation is $\mathbf{x} = x_2\mathbf{u} + x_3\mathbf{v}$, the plane through the origin spanned by $\mathbf{u}$ and $\mathbf{v}$. The solution of the nonhomogeneous system is $\mathbf{x} = \mathbf{p} + x_2\mathbf{u} + x_3\mathbf{v}$, the plane through $\mathbf{p}$ parallel to the solution of the homogeneous equation.

18. $\mathbf{x} = \begin{bmatrix} x_1 \\ x_2 \\ x_3 \end{bmatrix} = \begin{bmatrix} 7 \\ -1 \\ 0 \end{bmatrix} + x_3 \begin{bmatrix} 1 \\ 1 \\ 1 \end{bmatrix}$. The solution set is the line through $\begin{bmatrix} -7 \\ 1 \\ 0 \end{bmatrix}$, parallel to the line that is the solution set of the homogeneous system in Exercise 6.

20. $\mathbf{x} = \mathbf{a} + t\mathbf{b}$, where t represents a parameter, or $\mathbf{x} = \begin{bmatrix} x_1 \\ x_2 \end{bmatrix} = \begin{bmatrix} 3 \\ -2 \end{bmatrix} + t \begin{bmatrix} -7 \\ 6 \end{bmatrix}$, or $\begin{cases} x_1 = 3 - 7t \\ x_2 = -2 + 6t \end{cases}$

22. $\mathbf{x} = \mathbf{p} + t(\mathbf{q} - \mathbf{p}) = \begin{bmatrix} -3 \\ 2 \end{bmatrix} + t \begin{bmatrix} 3 \\ -5 \end{bmatrix}$

23. a. True. See the first paragraph of the subsection titled "Homogeneous Linear Systems."

b. False. The equation $A\mathbf{x} = \mathbf{0}$ gives an *implicit* description of its solution set. See the subsection titled "Parametric Vector Form."

c. False. The equation $A\mathbf{x} = \mathbf{0}$ *always* has the trivial solution. The box before Example 1 uses the word "nontrivial" instead of "trivial."

d. False. The line goes through $\mathbf{p}$ parallel to $\mathbf{v}$. See the paragraph that precedes Fig. 5.

e. False. The solution set could be *empty*! The statement (from Theorem 6) is true only when there exists a vector $\mathbf{p}$ such that $A\mathbf{p} = \mathbf{b}$.

24. a. False. The trivial solution is always a solution of a homogeneous system of equations.

   b. False. A nontrivial solution of $A\mathbf{x} = \mathbf{0}$ is any nonzero $\mathbf{x}$ that satisfies the equation. See the sentence before Example 2.

   c. True. See Example 3 and the paragraph following it.

   d. True. If the zero vector is a solution, then $\mathbf{b} = A\mathbf{x} = A\mathbf{0} = \mathbf{0}$.

   e. False. The equation $A\mathbf{x} = \mathbf{b}$ may not have any solution.

26. When A is the 3×3 zero matrix, *every* $\mathbf{x}$ in $\mathbb{R}^3$ satisfies $A\mathbf{x} = \mathbf{0}$. So the solution set is all vectors in $\mathbb{R}^3$.

28. a. No   **b.** Yes   **30. a.** Yes   **b.** Yes

32. No. If the solution set of $A\mathbf{x} = \mathbf{b}$ contained the origin, then $\mathbf{0}$ would satisfy $A\mathbf{0} = \mathbf{b}$, which is not true since $\mathbf{b}$ is not the zero vector.

34. Construct any 3×3 matrix such that the sum of twice the first column and the third column equals the second column.

36. One answer: $\mathbf{x} = \begin{bmatrix} 2 \\ 3 \end{bmatrix}$

38. If c is a scalar, then $A(c\mathbf{w}) = cA\mathbf{w}$, by Theorem 5(b) in Section 1.4. If $\mathbf{w}$ satisfies $A\mathbf{x} = \mathbf{0}$, then $A\mathbf{w} = \mathbf{0}$, $cA\mathbf{w} = c \cdot \mathbf{0} = \mathbf{0}$, and so $A(c\mathbf{w}) = \mathbf{0}$.

40. No. If $A\mathbf{x} = \mathbf{b}$ has no solution, then A cannot have a pivot in each row. Since A is 3×3, it has at most two pivot positions. So the equation $A\mathbf{x} = \mathbf{y}$ for any $\mathbf{y}$ has at most two basic variables and at least one free variable. Thus the solution set for $A\mathbf{x} = \mathbf{y}$ is either empty or has infinitely many elements.

Section 1.6, page 54

2. Take some other value for p_S, say, 200 million dollars. The other equilibrium prices are then $p_C = 188$ million, $p_E = 170$ million. Any constant nonnegative multiple of these prices forms a set of equilibrium prices, because the solution set of the system of equations consists of all multiples of one vector. Changing the unit of measurement to, say, Japanese yen has the same effect as multiplying all equilibrium prices by a constant. The *ratios* of the prices remain the same, no matter what currency is used.

4. a.

Output	Distribution of Output from:				Purchased by:
	Mining	Lumber	Energy	Transp.	Input
	.30	.15	.20	.20	→ Mining
	.10	.15	.15	.10	→ Lumber
	.60	.50	.45	.50	→ Energy
	0	.20	.20	.20	→ Transp.

   b. One solution is $p_M = 137$, $p_L = 84$, $p_E = 316$, and $p_T = 100$.

6. $2Al_2O_3 + 3C \rightarrow 4Al + 3CO_2$

8. $2H_3O + CaCO_3 \rightarrow 3H_2O + Ca + CO_2$

10. $15PbN_6 + 44CrMn_2O_8 \rightarrow$
$5Pb_3O_4 + 22Cr_2O_3 + 88MnO_2 + 90NO$

12. $\begin{cases} x_1 = 20 \\ x_2 = 20 + x_4 \\ x_3 = -80 + x_4 \\ x_4 \text{ is free} \end{cases}$

For the network flows to be nonnegative, $x_4 \geq 80$.

14. a. $\begin{cases} x_1 = 80 - x_5 \\ x_2 = -180 + x_4 + x_5 \\ x_3 = -90 + x_4 + x_5 \\ x_4 \text{ is free} \\ x_5 \text{ is free} \end{cases}$ **b.** $\begin{cases} x_1 = 80 \\ x_2 = -180 + x_4 \\ x_3 = -90 + x_4 \\ x_4 \text{ is free} \end{cases}$

   c. The minimum value of x_4 when $x_5 = 0$ is 180.

Section 1.7, page 60

Answers to Exercises 1–20 are justified in the *Instructor's Solution Manual*.

2. Lin. indep.   **4.** Lin. indep.   **6.** Lin. indep.

8. Lin. depen.   **10. a.** No h **b.** All h

12. $h = -18$ **14.** $h = -38$ **16.** Lin. depen.

18. Lin. depen.   **20.** Lin. depen.

21. a. False. A homogenous system *always* has the trivial solution. See the box before Example 2.

   b. False. See the warning after Theorem 7.

   c. True. See Fig. 3, after Theorem 8.

   d. True. See the remark following Example 4.

22. a. True. See Theorem 7.

   b. True. See Example 4.

   c. False. For instance, the set consisting of $\begin{bmatrix} 1 \\ -2 \\ 3 \end{bmatrix}$ and $\begin{bmatrix} 2 \\ -4 \\ 6 \end{bmatrix}$ is linearly dependent. See the warning after Theorem 8.

   d. False. For instance, the set consisting of $\begin{bmatrix} 1 \\ -2 \\ 3 \end{bmatrix}$ and $\begin{bmatrix} 2 \\ -4 \\ 6 \end{bmatrix}$ consists of two linearly dependent vectors in $\mathbb{R}^3$.

24. $\begin{bmatrix} \blacksquare & * & * \\ 0 & \blacksquare & * \\ 0 & 0 & \blacksquare \end{bmatrix}$ **26.** $\begin{bmatrix} \blacksquare & * & * \\ 0 & \blacksquare & * \\ 0 & 0 & \blacksquare \\ 0 & 0 & 0 \end{bmatrix}$

28. It must have 4 pivot columns. (By Theorem 4 in Section 1.4, A has a pivot in each of its four rows. Since each pivot position is in a different column, A has four pivot columns.)

30. a. n

b. The columns of A are linearly independent if and only if the equation $A\mathbf{x} = \mathbf{0}$ has only the trivial solution. This happens if and only if $A\mathbf{x} = \mathbf{0}$ has no free variables, which in turn happens if and only if every variable is a basic variable, that is, if and only if every column of A is a pivot column.

32. $\mathbf{x} = \begin{bmatrix} 1 \\ -3 \\ -1 \end{bmatrix}$ **34.** True, by Example 3.

36. False. Counterexample: Take $\mathbf{v}_2$, to be a multiple of $\mathbf{v}_1$. Take $\mathbf{v}_3$ to be *not* a multiple of $\mathbf{v}_1$.

38. True. If the equation $x_1\mathbf{v}_1 + x_2\mathbf{v}_2 + x_3\mathbf{v}_3 = \mathbf{0}$ had a nontrivial solution (with at least one of x_1, x_2, x_3 nonzero), then so would the equation $x_1\mathbf{v}_1 + x_2\mathbf{v}_2 + x_3\mathbf{v}_3 + 0 \cdot \mathbf{v}_4 = \mathbf{0}$. But that can't happen because $\{\mathbf{v}_1, \mathbf{v}_2, \mathbf{v}_3, \mathbf{v}_4\}$ is linearly independent. So $\{\mathbf{v}_1, \mathbf{v}_2, \mathbf{v}_3\}$ must be linearly independent. This problem can also be solved using Exercise 37.

40. An $m \times n$ matrix with n pivot columns has a pivot in each column. So the equation $A\mathbf{x} = \mathbf{b}$ has no free variables. If there is a solution, it must be unique.

42. [M] Using the pivot columns of A,
$$B = \begin{bmatrix} 12 & 10 & -6 & 4 \\ -7 & -6 & 4 & -7 \\ 9 & 9 & -9 & 9 \\ -4 & -3 & -1 & -8 \\ 8 & 7 & -5 & 1 \end{bmatrix}$$
Other choices are possible.

44. [M] Each column of A that is not a column of B is in the set spanned by the columns of B.

Section 1.8, page 68

2. $T(\mathbf{u}) = \begin{bmatrix} 1 \\ 2 \\ -3 \end{bmatrix}$, $T(\mathbf{v}) = \begin{bmatrix} a/3 \\ b/3 \\ c/3 \end{bmatrix}$

4. $\mathbf{x} = \begin{bmatrix} -17 \\ -7 \\ -1 \end{bmatrix}$, unique solution

6. $\mathbf{x} = \begin{bmatrix} 10 \\ 3 \\ 0 \end{bmatrix}$, not unique

8. 7 rows and 5 columns

10. $\mathbf{x} = x_3 \begin{bmatrix} -2 \\ -2 \\ 1 \\ 0 \end{bmatrix} + x_4 \begin{bmatrix} 4 \\ -3 \\ 0 \\ 1 \end{bmatrix}$

12. No, because the system represented by $[A \quad \mathbf{b}]$ is inconsistent.

14.

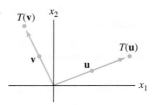

A scaling by a factor of 2.

16.

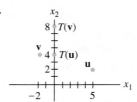

A projection onto the x_2 axis and a scaling by 2.

18.

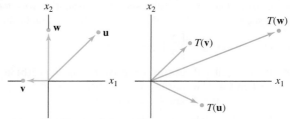

20. $\begin{bmatrix} -3 & 7 \\ 5 & -2 \end{bmatrix}$

21. a. True. Functions from $\mathbb{R}^n$ to $\mathbb{R}^m$ are defined before Fig. 2. A linear transformation is a function with certain properties.

b. False. The domain is $\mathbb{R}^5$. See the paragraph before Example 1.

c. False. The range is the set of all linear combinations of the columns of A. See the paragraph before Example 1.

d. False. See the paragraph after the definition of *linear transformation*.

e. True. See the paragraph following the box that contains equation (4).

22. a. True. See the subsection on Matrix Transformations.

b. True. See the subsection on Matrix Transformations.

c. False. The question is an existence question. See the remark about Example 1(d), following the solution of Example 1.

d. True. See the discussion following the definition of *linear transformation*.

e. True. $T(\mathbf{0}) = \mathbf{0}$. See the box after the definition of *linear transformation*.

24. Let $T(\mathbf{x}) = A\mathbf{x} + \mathbf{b}$ for $\mathbf{x}$ in $\mathbb{R}^n$. If $\mathbf{b}$ is not zero, $T(\mathbf{0}) = A\mathbf{0} + \mathbf{b} = \mathbf{b} \neq \mathbf{0}$. (Actually, one can show that T fails both properties of a linear transformation.)

26. a. The line through $\mathbf{p}$ and $\mathbf{q}$ is parallel to $\mathbf{q} - \mathbf{p}$. (See the figure that accompanies Exercises 21 and 22 in Section

1.5.) Since **p** is on the line, the equation of the line is $\mathbf{x} = \mathbf{p} + t(\mathbf{q} - \mathbf{p})$. Rewrite this as $\mathbf{x} = \mathbf{p} - t\mathbf{p} + t\mathbf{q}$ and $\mathbf{x} = (1 - t)\mathbf{p} + t\mathbf{q}$.

b. Consider $\mathbf{x} = (1 - t)\mathbf{p} + t\mathbf{q}$ for t such that $0 \leq t \leq 1$. Then, by the linearity of T, for $0 \leq t \leq 1$

$$T(\mathbf{x}) = T((1 - t)\mathbf{p} + t\mathbf{q})$$
$$= (1 - t)T(\mathbf{p}) + tT(\mathbf{q}) \qquad (1)$$

If $T(\mathbf{p})$ and $T(\mathbf{q})$ are distinct, then (1) is the equation for the line segment between $T(\mathbf{p})$ and $T(\mathbf{q})$, as shown in part (a). Otherwise, the set of images is just the single point $T(\mathbf{p})$, because $(1 - t)T(\mathbf{p}) + tT(\mathbf{q}) = (1 - t)T(\mathbf{p}) + tT(\mathbf{p}) = T(\mathbf{p})$.

28. Consider a point **x** in the parallelogram determined by **u** and **v**, say $\mathbf{x} = a\mathbf{u} + b\mathbf{v}$ for $0 \leq a \leq 1, 0 \leq b \leq 1$. By linearity of T, the image of **x** is

$$T(\mathbf{x}) = T(a\mathbf{u} + b\mathbf{v})$$
$$= aT(\mathbf{u}) + bT(\mathbf{v}), \quad \text{for } 0 \leq a \leq 1, 0 \leq b \leq 1$$

This image point lies in the parallelogram determined by $T(\mathbf{u})$ and $T(\mathbf{v})$. Special "degenerate" cases arise when $T(\mathbf{u})$ and $T(\mathbf{v})$ are linearly dependent. See the *Instructor's Solution Manual*.

30. Given any **x** in $\mathbb{R}^n$, there are constants $c_1, \ldots, c_p$ such that $\mathbf{x} = c_1\mathbf{v}_1 + \cdots + c_p\mathbf{v}_p$, because $\mathbf{v}_1, \ldots, \mathbf{v}_p$ span $\mathbb{R}^n$. Then, from property (5) of a linear transformation,

$$T(\mathbf{x}) = c_1T(\mathbf{v}_1) + \cdots + c_pT(\mathbf{v}_p) = c_1\mathbf{0} + \cdots + c_p\mathbf{0} = \mathbf{0}$$

32. Take any vector (x_1, x_2) with $x_2 \neq 0$, and use a negative scalar. For instance, $T(0, 1) = (-2, -4)$, but $T(-1 \cdot (0, 1)) = T(0, -1) = (-2, 4) \neq (-1) \cdot T(0, 1)$.

34. Take **u** and **v** in $\mathbb{R}^3$ and let c and d be scalars. Then

$$c\mathbf{u} + d\mathbf{v} = (cu_1 + dv_1, cu_2 + dv_2, cu_3 + dv_3)$$

The transformation T is linear because

$$T(c\mathbf{u} + d\mathbf{v}) = (cu_1 + dv_1, cu_2 + dv_2, -(cu_3 + dv_3))$$
$$= (cu_1 + dv_1, cu_2 + du_2, -cu_3 - dv_3)$$
$$= (cu_1, cu_2, -cu_3) + (dv_1, dv_3, -dv_3)$$
$$= c(u_1, u_2, -u_3) + d(v_1, v_2, -v_3)$$
$$= cT(\mathbf{u}) + dT(\mathbf{v})$$

36. Suppose $\{\mathbf{u}, \mathbf{v}\}$ is a linearly independent set in $\mathbb{R}^n$, and yet $T(\mathbf{u})$ and $T(\mathbf{v})$ are linearly dependent. Then there exist weights c_1, c_2, not both zero, such that

$$c_1T(\mathbf{u}) + c_2T(\mathbf{v}) = \mathbf{0}$$

Because T is linear, $T(c_1\mathbf{u} + c_2\mathbf{v}) = \mathbf{0}$. That is, the vector $\mathbf{x} = c_1\mathbf{u} + c_2\mathbf{v}$ satisfies $T(\mathbf{x}) = \mathbf{0}$. Furthermore, **x** cannot be the zero vector, since that would mean that a nontrivial linear combination of **u** and **v** is zero, which is impossible because **u** and **v** are linearly independent. Thus, the equation $T(\mathbf{x}) = \mathbf{0}$ has a nontrivial solution.

38. [M] All multiples of $(-1, -1, -1, 1)$

40. [M] Yes. One choice for **x** is $(1, 2, 1, 0)$.

Section 1.9, page 78

2. $\begin{bmatrix} 1 & -2 & 3 \\ 4 & 9 & -8 \end{bmatrix}$ **4.** $\begin{bmatrix} 1 & 2 \\ 0 & 1 \end{bmatrix}$

6. $\begin{bmatrix} 0 & -1 \\ 1 & 0 \end{bmatrix}$ **8.** $\begin{bmatrix} 0 & -1 \\ -1 & -2 \end{bmatrix}$ **10.** $\begin{bmatrix} 0 & -1 \\ 1 & 0 \end{bmatrix}$

12. The transformation T in Exercise 10 maps $\mathbf{e}_1$ into $\mathbf{e}_2$ and maps $\mathbf{e}_2$ into $-\mathbf{e}_1$. A counterclockwise rotation about the origin through $\pi/2$ radians also maps $\mathbf{e}_1$ into $\mathbf{e}_2$ and maps $\mathbf{e}_2$ into $-\mathbf{e}_1$. Since a linear transformation is completely determined by what it does to the columns of the identity matrix, the rotation transformation has the same effect as T on every vector in $\mathbb{R}^2$.

14.

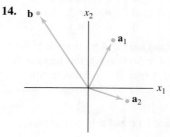

$$T\left(\begin{bmatrix} 1 \\ -2 \end{bmatrix}\right) = \mathbf{a}_1 - 2\mathbf{a}_2 = \mathbf{b}.$$

16. $\begin{bmatrix} 3 & -2 \\ 1 & 4 \\ 0 & 1 \end{bmatrix}$ **18.** $\begin{bmatrix} 1 & 4 \\ 0 & 0 \\ 1 & -3 \\ 1 & 0 \end{bmatrix}$

20. $\begin{bmatrix} 3 & 0 & 4 & -2 \end{bmatrix}$ **22.** $\mathbf{x} = (1, 2)$

23. a. True. See Theorem 10.
   b. True. See Example 3.
   c. False. See the paragraph before Table 1.
   d. False. See the definition of *onto*. *Any* function from $\mathbb{R}^n$ to $\mathbb{R}^m$ maps each vector onto another vector.
   e. False. See Example 5.

24. a. False. See Theorem 12.
   b. True. See Theorem 10.
   c. True. See Theorem 10.
   d. False. See the definition of *one-to-one*. Any *function* from $\mathbb{R}^n$ to $\mathbb{R}^m$ maps a vector onto a single (unique) vector.
   e. False. See Table 3.

26. The transformation in Exercise 2 is not one-to-one, by Theorem 12, because the standard matrix is 2×3 and so has linearly dependent columns. However, the matrix has a pivot in each row and so the columns span $\mathbb{R}^2$. By Theorem 12, the transformation maps $\mathbb{R}^3$ onto $\mathbb{R}^2$.

28. The standard matrix A of the transformation T in Exercise 14 has linearly independent columns, because the figure shows that $\mathbf{a}_1$ and $\mathbf{a}_2$ are not multiples of each other. So T is one-to-one, by Theorem 12. Also, A must have a pivot in each column because the equation $A\mathbf{x} = \mathbf{0}$ has no free variables. Thus, the echelon form of A is $\begin{bmatrix} \blacksquare & * \\ 0 & \blacksquare \end{bmatrix}$. Since A has a pivot in each row, the columns of A span $\mathbb{R}^2$. So T maps $\mathbb{R}^2$ onto $\mathbb{R}^2$. An alternative argument for the second part is to observe directly from the figure that $\mathbf{a}_1$ and $\mathbf{a}_2$ span $\mathbb{R}^2$. This is more or less evident, based on experience with grids such as those in Fig. 8 (and the figure with Exercises 7 and 8) in Section 1.3.

30. $\begin{bmatrix} \blacksquare & * & * & * \\ 0 & \blacksquare & * & * \\ 0 & 0 & \blacksquare & * \end{bmatrix}$, $\begin{bmatrix} \blacksquare & * & * & * \\ 0 & \blacksquare & * & * \\ 0 & 0 & 0 & \blacksquare \end{bmatrix}$,

$\begin{bmatrix} \blacksquare & * & * & * \\ 0 & 0 & \blacksquare & * \\ 0 & 0 & 0 & \blacksquare \end{bmatrix}$, $\begin{bmatrix} 0 & \blacksquare & * & * \\ 0 & 0 & \blacksquare & * \\ 0 & 0 & 0 & \blacksquare \end{bmatrix}$

32. A has m pivot columns if and only if A has a pivot position in each row. By Theorem 4 in Section 1.4, this happens if and only if the columns of A span $\mathbb{R}^m$, and this in turn happens (by Theorem 12) if and only if T maps $\mathbb{R}^n$ onto $\mathbb{R}^m$.

34. Take any $\mathbf{u}$ and $\mathbf{v}$ in $\mathbb{R}^p$, and let c and d be any scalars. Then

$$T(S(c\mathbf{u} + d\mathbf{v})) = T(c \cdot S(\mathbf{u}) + d \cdot S(\mathbf{v}))$$

because S is linear

$$= c \cdot T(S(\mathbf{u})) + d \cdot T(S(\mathbf{v}))$$

because T is linear

This calculation shows that the mapping $\mathbf{x} \mapsto T(S(\mathbf{x}))$ is linear. See equation (4) in Section 1.8.

36. The transformation T maps $\mathbb{R}^n$ *onto* $\mathbb{R}^m$ if and only if for each $\mathbf{y}$ in $\mathbb{R}^m$ *there exists* an $\mathbf{x}$ in $\mathbb{R}^n$ such that $\mathbf{y} = T(\mathbf{x})$.

38. [M] Yes. There is a pivot in every column of the standard matrix A, so the equation $A\mathbf{x} = \mathbf{0}$ has only the trivial solution. By Theorem 11, the transformation T is one-to-one.

40. [M] Yes. There is a pivot in every row, so the columns of the standard matrix do span $\mathbb{R}^5$. By Theorem 12, the transformation T does map $\mathbb{R}^5$ onto $\mathbb{R}^5$.

Section 1.10, page 86

2. a. $B = \begin{bmatrix} SW & Cpx \end{bmatrix} = \begin{bmatrix} 160 & 110 \\ 5 & 2 \\ 6 & .1 \\ 1 & .4 \end{bmatrix}$, $\mathbf{u} = \begin{bmatrix} 3 \\ 2 \end{bmatrix}$

b. [M] Mix .4 serving of Shredded Wheat with .6 serving of Crispix.

4. a. $\begin{bmatrix} 36 & 51 & 13 & 80 \\ 52 & 34 & 74 & 0 \\ 0 & 7 & 1.1 & 3.4 \\ 1.26 & .19 & .8 & .18 \end{bmatrix} \begin{bmatrix} x_1 \\ x_2 \\ x_3 \\ x_4 \end{bmatrix} = \begin{bmatrix} 33 \\ 45 \\ 3 \\ .8 \end{bmatrix}$,

where $x_1, \ldots, x_4$ represent the numbers of units (100 g) of nonfat milk, soy flour, whey, and isolated soy protein, respectively, to be used in the mixture

b. [M] The "solution" is $x_1 = .64$, $x_2 = .54$, $x_3 = -.09$, and $x_4 = -.21$. This solution is not feasible, because the mixture cannot include negative amounts of whey and isolated soy protein.

6. $\begin{bmatrix} 6 & -1 & 0 & 0 \\ -1 & 9 & -4 & 0 \\ 0 & -4 & 7 & -2 \\ 0 & 0 & -2 & 7 \end{bmatrix} \begin{bmatrix} I_1 \\ I_2 \\ I_3 \\ I_4 \end{bmatrix} = \begin{bmatrix} 30 \\ 20 \\ 40 \\ 10 \end{bmatrix}$

[M]: $\begin{bmatrix} I_1 \\ I_2 \\ I_3 \\ I_4 \end{bmatrix} = \begin{bmatrix} 6.36 \\ 8.14 \\ 11.73 \\ 4.78 \end{bmatrix}$

8. $\begin{bmatrix} 9 & -1 & 0 & -1 & -4 \\ -1 & 7 & -2 & 0 & -3 \\ 0 & -2 & 10 & -3 & -3 \\ -1 & 0 & -3 & 7 & -2 \\ -4 & -3 & -3 & -2 & 12 \end{bmatrix} \begin{bmatrix} I_1 \\ I_2 \\ I_3 \\ I_4 \\ I_5 \end{bmatrix} = \begin{bmatrix} 50 \\ -30 \\ 20 \\ -40 \\ 0 \end{bmatrix}$

[M]: $\begin{bmatrix} I_1 \\ I_2 \\ I_3 \\ I_4 \\ I_5 \end{bmatrix} = \begin{bmatrix} 4.00 \\ -4.38 \\ -0.90 \\ -5.80 \\ -0.96 \end{bmatrix}$

10. $\mathbf{x}_{k+1} = M\mathbf{x}_k$ for $k = 0, 1, 2, \ldots$, where $M = \begin{bmatrix} .94 & .04 \\ .06 & .96 \end{bmatrix}$ and $\mathbf{x}_0 = \begin{bmatrix} 10{,}000{,}000 \\ 800{,}000 \end{bmatrix}$. The population in 2012 (for $k = 2$) is $\mathbf{x}_2 = \begin{bmatrix} 8{,}920{,}800 \\ 1{,}879{,}200 \end{bmatrix}$.

12. $\mathbf{x}_0 = \begin{bmatrix} 295 \\ 55 \\ 150 \end{bmatrix}$, $\mathbf{x}_2 \approx \begin{bmatrix} 312 \\ 58 \\ 130 \end{bmatrix}$

14. [M] Here are the solution temperatures (in degrees) for the two figures shown:

   (*A*) (10, 10, 10, 10)

   (*B*) (10, 17.5, 20, 12.5)

a. At each interior point in the figure for Exercises 33 and 34 in Section 1.1, the temperature is the sum of the temperatures at the corresponding interior points in plates A and B here.

b. If the boundary temperatures in plate A are changed by a factor of 3, the interior temperatures should also change by a factor of 3.

c. The correspondence from the list of eight boundary temperatures to the list of four interior temperatures is a linear transformation. A verification of this statement is not expected.

Chapter 1 Supplementary Exercises, page 88

1. a. False. (The word "reduced" is missing.) Counterexample:

$$A = \begin{bmatrix} 1 & 2 \\ 3 & 4 \end{bmatrix}, \quad B = \begin{bmatrix} 1 & 2 \\ 0 & 2 \end{bmatrix}, \quad C = \begin{bmatrix} 1 & 2 \\ 0 & 1 \end{bmatrix}$$

The matrix A is row equivalent to matrices B and C, both in echelon form.

b. False. Counterexample: Let A be any $n \times n$ matrix with fewer than n pivot columns. Then the equation $A\mathbf{x} = \mathbf{0}$ has infinitely many solutions.

c. True. If a linear system has more than one solution, it is a consistent system and has a free variable. By the Existence and Uniqueness Theorem in Section 1.2, the system has infinitely many solutions.

d. False. Counterexample: The following system has no free variables and no solution:

$$x_1 + x_2 = 1$$
$$x_2 = 5$$
$$x_1 + x_2 = 2$$

e. True. See the box after the definition of elementary row operations, in Section 1.1. If $[A \quad \mathbf{b}]$ is transformed into $[C \quad \mathbf{d}]$ by elementary row operations, then the two augmented matrices are row equivalent.

f. True. Theorem 6 in Section 1.5 essentially says that when $A\mathbf{x} = \mathbf{b}$ is consistent, the solution sets of the nonhomogeneous equation and the homogeneous equation are translates of each other. In this case, the two equations have the same number of solutions.

g. False. For the columns of A to span $\mathbb{R}^m$, the equation $A\mathbf{x} = \mathbf{b}$ must be consistent for *all* $\mathbf{b}$ in $\mathbb{R}^m$, not for just one vector $\mathbf{b}$ in $\mathbb{R}^m$.

h. False. *Any* matrix can be transformed by elementary row operations into reduced echelon form.

i. True. If A is row equivalent to B, then A can be transformed by elementary row operations first into B and then further transformed into the reduced echelon form U of B. Since the reduced echelon form of A is unique, it must be U.

j. False. Every equation $A\mathbf{x} = \mathbf{0}$ has the trivial solution whether or not some variables are free.

k. True, by Theorem 4 in Section 1.4. If the equation $A\mathbf{x} = \mathbf{b}$ is consistent for every $\mathbf{b}$ in $\mathbb{R}^m$, then A must have a pivot position in every one of its m rows. If A has m pivot positions, then A has m pivot columns, each containing one pivot position.

l. False. The word "unique" should be deleted. Let A be any $m \times n$ matrix with m pivot columns but more than m columns altogether. Then the equation $A\mathbf{x} = \mathbf{b}$ is consistent and has m basic variables and at least one free variable. Thus the equation does not have a unique solution.

m. True. If A has n pivot positions, it has a pivot in each of its n columns and in each of its n rows. The reduced echelon form has a 1 in each pivot position, so the reduced echelon form is the $n \times n$ identity marix.

n. True. Both matrices A and B can be row reduced to the 3×3 identity matrix, as discussed in the previous question. Since the row operations that transform B into

I_3 are reversible, A can be transformed first into I_3 and then into B.

o. True. The reason is essentially the same as that given for question f.

p. True. If the columns of A span $\mathbb{R}^m$, then the reduced echelon form of A is a matrix U with a pivot in each row, by Theorem 4 in Section 1.4. Since B is row equivalent to A, B can be transformed by row operations first into A and then further transformed into U. Since U has a pivot in each row, so does B. By Theorem 4, the columns of B span $\mathbb{R}^m$.

q. False. See Example 5 in Section 1.7.

r. True. Any set of three vectors in $\mathbb{R}^2$ would have to be linearly dependent, by Theorem 8 in Section 1.7.

s. False. If a set $\{\mathbf{v}_1, \mathbf{v}_2, \mathbf{v}_3, \mathbf{v}_4\}$ were to span $\mathbb{R}^5$, then the matrix $A = [\mathbf{v}_1 \quad \mathbf{v}_2 \quad \mathbf{v}_3 \quad \mathbf{v}_4]$ would have a pivot position in each of its five rows, which is impossible since A has only four columns.

t. True. The vector $-\mathbf{u}$ is a linear combination of $\mathbf{u}$ and $\mathbf{v}$, namely, $-\mathbf{u} = (-1)\mathbf{u} + 0\mathbf{v}$.

u. False. If nonzero $\mathbf{u}$ and $\mathbf{v}$ are multiples, then Span$\{\mathbf{u}, \mathbf{v}\}$ is a line, and $\mathbf{w}$ need not be on that line.

v. False. Let $\mathbf{u}$ and $\mathbf{v}$ be any linearly independent pair of vectors and let $\mathbf{w} = 2\mathbf{v}$. Then $\mathbf{w} = 0\mathbf{u} + 2\mathbf{v}$, so $\mathbf{w}$ is a linear combination of $\mathbf{u}$ and $\mathbf{v}$. However, $\mathbf{u}$ cannot be a linear combination of $\mathbf{v}$ and $\mathbf{w}$ because if it were, $\mathbf{u}$ would be a multiple of $\mathbf{v}$. That is not possible since $\{\mathbf{u}, \mathbf{v}\}$ is linearly independent.

w. False. The statement would be true if the condition $\mathbf{v}_1$ is not zero were present. See Theorem 7 in Section 1.7. However, if $\mathbf{v}_1 = \mathbf{0}$, then $\{\mathbf{v}_1, \mathbf{v}_2, \mathbf{v}_3\}$ is linearly dependent, no matter what else might be true about $\mathbf{v}_2$ and $\mathbf{v}_3$.

x. True. "Function" is another word used for "transformation" (as mentioned in the definition of "transformation" in Section 1.8), and a linear transformation is a special type of transformation.

y. True. For the transformation $\mathbf{x} \mapsto A\mathbf{x}$ to map $\mathbb{R}^5$ onto $\mathbb{R}^6$, the matrix A would have to have a pivot in every row and hence have six pivot columns. This is impossible because A has only five columns.

z. False. For the transformation $\mathbf{x} \mapsto A\mathbf{x}$ to be one-to-one, A must have a pivot in each column. Since A has n columns and m pivots, m might be less than n.

2. If $a \neq 0$, then $x = b/a$; the solution is unique. If $a = 0$, and $b \neq 0$, the solution set is empty, because $0x = 0 \neq b$. If $a = 0$ and $b = 0$, the equation $0x = 0$ has infinitely many solutions.

4. Since there are three pivots (one in each column of A), the *augmented* matrix must reduce to the form

$$\begin{bmatrix} \blacksquare & * & * & * \\ 0 & \blacksquare & * & * \\ 0 & 0 & \blacksquare & * \end{bmatrix}.$$ A solution of $A\mathbf{x} = \mathbf{b}$ exists for all

b because there is a pivot in each row of A. Each solution is unique because there are no free variables.

6. a. Set $\mathbf{v}_1 = \begin{bmatrix} 4 \\ 8 \end{bmatrix}$, $\mathbf{v}_2 = \begin{bmatrix} -2 \\ -3 \end{bmatrix}$, $\mathbf{v}_3 = \begin{bmatrix} 7 \\ 10 \end{bmatrix}$, and

$\mathbf{b} = \begin{bmatrix} -5 \\ -3 \end{bmatrix}$. "Determine if **b** is a linear combination of $\mathbf{v}_1, \mathbf{v}_2, \mathbf{v}_3$." Or, "Determine if **b** is in Span $\{\mathbf{v}_1, \mathbf{v}_2, \mathbf{v}_3\}$." To do this, compute

$$\begin{bmatrix} 4 & -2 & 7 & -5 \\ 8 & -3 & 10 & -3 \end{bmatrix} \sim \begin{bmatrix} 4 & -2 & 7 & -5 \\ 0 & 1 & -4 & 7 \end{bmatrix}$$

The system is consistent, so **b** *is* in Span$\{\mathbf{v}_1, \mathbf{v}_2, \mathbf{v}_3\}$.

b. Set $A = \begin{bmatrix} 4 & -2 & 7 \\ 8 & -3 & 10 \end{bmatrix}$, $\mathbf{b} = \begin{bmatrix} -5 \\ -3 \end{bmatrix}$. "Determine if **b** is a linear combination of the columns of A."

c. Define $T(\mathbf{x}) = A\mathbf{x}$. "Determine if **b** is in the range of T."

8. a. $\begin{bmatrix} \blacksquare & * & * \\ 0 & \blacksquare & * \end{bmatrix}$, $\begin{bmatrix} \blacksquare & * & * \\ 0 & 0 & \blacksquare \end{bmatrix}$, $\begin{bmatrix} 0 & \blacksquare & * \\ 0 & 0 & \blacksquare \end{bmatrix}$

b. $\begin{bmatrix} \blacksquare & * & * \\ 0 & \blacksquare & * \\ 0 & 0 & \blacksquare \end{bmatrix}$

10. The line through $\mathbf{a}_1$ and the origin and the line through $\mathbf{a}_2$ and the origin determine a "grid" on the $x_1 x_2$-plane, as shown below. Every point in $\mathbb{R}^2$ can be described uniquely in terms of this grid. Thus **b** can be reached from the origin by traveling a certain number of units in the $\mathbf{a}_1$-direction and a certain number of units in the $\mathbf{a}_2$-direction. So the equation $A\mathbf{x} = \mathbf{b}$ has a solution, and that solution is unique.

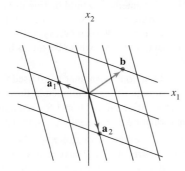

12. A solution set is a plane where there are two free variables. If the coefficient matrix is 2×3, then only one column can be a pivot column. The echelon form will have all zeros in the second row. Use a row replacement to create a matrix not in echelon form, for instance: $\begin{bmatrix} 1 & 2 & 3 \\ 1 & 2 & 3 \end{bmatrix}$.

14. The vectors are linearly independent for all a except $a = 2$ and $a = -1$.

16. Denote the columns from right to left by $\mathbf{v}_1, \ldots, \mathbf{v}_4$. The "first" vector $\mathbf{v}_1$ is nonzero, $\mathbf{v}_2$ is not a multiple of $\mathbf{v}_1$ (because the third entry of $\mathbf{v}_2$ is nonzero), and $\mathbf{v}_3$ is not a linear combination of $\mathbf{v}_1$ and $\mathbf{v}_2$ (because the second entry of $\mathbf{v}_3$ is nonzero). Finally, by looking at first entries in the

vectors, $\mathbf{v}_4$ cannot be a linear combination of $\mathbf{v}_1, \mathbf{v}_2$, and $\mathbf{v}_3$. By Theorem 7 in Section 1.7, the columns are linearly independent.

18. Suppose c_1 and c_2 are constants such that

$$c_1\mathbf{v}_1 + c_2(\mathbf{v}_1 + \mathbf{v}_2) = \mathbf{0} \tag{$*$}$$

Then $(c_1 + c_2)\mathbf{v}_1 + c_2\mathbf{v}_2 = \mathbf{0}$. Since $\mathbf{v}_1$ and $\mathbf{v}_2$ are linearly independent, both $c_1 + c_2 = 0$ and $c_2 = 0$. It follows that both c_1 and c_2 in ($*$) must be zero, which shows that $\{\mathbf{v}_1, \mathbf{v}_1 + \mathbf{v}_2\}$ is linearly independent.

20. If $T(\mathbf{u}) = \mathbf{v}$, then, since T is linear, $T(-\mathbf{u}) = T((-1)\mathbf{u}) = (-1)T(\mathbf{u}) = -\mathbf{v}$.

22. By Theorem 12 in Section 1.9, the columns of A span $\mathbb{R}^3$. By Theorem 4 in Section 1.4, A has a pivot in each of its three rows. Since A has three columns, each column must be a pivot column. So the equation $A\mathbf{x} = \mathbf{0}$ has no free variables, and the columns of A are linearly independent. By Theorem 12 in Section 1.9, the transformation $\mathbf{x} \mapsto A\mathbf{x}$ is one-to-one.

24. $a = 1/\sqrt{5}$, $b = -2/\sqrt{5}$

Chapter 2

Section 2.1, page 100

2. $A + 3B = \begin{bmatrix} 23 & -15 & 2 \\ 7 & -17 & -7 \end{bmatrix}$, $2C - 3E$ is not defined,

$DB = \begin{bmatrix} 26 & -35 & -12 \\ -3 & -11 & -13 \end{bmatrix}$, EC is not defined.

4. $A - 5I_3 = \begin{bmatrix} 0 & -1 & 3 \\ -4 & -2 & -6 \\ -3 & 1 & -3 \end{bmatrix}$,

$(5I_3)A = \begin{bmatrix} 25 & -5 & 15 \\ -20 & 15 & -30 \\ -15 & 5 & 10 \end{bmatrix}$

6. a. $A\mathbf{b}_1 = \begin{bmatrix} -5 \\ 12 \\ 3 \end{bmatrix}$, $A\mathbf{b}_2 = \begin{bmatrix} 22 \\ -22 \\ -2 \end{bmatrix}$, $AB = \begin{bmatrix} -5 & 22 \\ 12 & -22 \\ 3 & -2 \end{bmatrix}$

b. $AB = \begin{bmatrix} 4 \cdot 1 - 3 \cdot 3 & 4 \cdot 4 - 3 \cdot (-2) \\ -3 \cdot 1 + 5 \cdot 3 & -3 \cdot 4 + 5 \cdot (-2) \\ 0 \cdot 1 + 1 \cdot 3 & 0 \cdot 4 + 1 \cdot (-2) \end{bmatrix}$

$= \begin{bmatrix} -5 & 22 \\ 12 & -22 \\ 3 & -2 \end{bmatrix}$

8. B has 5 rows.

10. $AB = AC = \begin{bmatrix} -21 & -21 \\ 7 & 7 \end{bmatrix}$

12. By inspection of A, a suitable column for B is any multiple of $(2, 1)$. For example: $B = \begin{bmatrix} 2 & 6 \\ 1 & 3 \end{bmatrix}$.

14. By definition, $UQ = [\, U\mathbf{q}_1 \;\; \cdots \;\; U\mathbf{q}_4 \,]$. From Example 6 in Section 1.8, the first column of UQ lists the total costs for materials, labor, and overhead used to manufacture products B and C during the first quarter of the year. Columns 2, 3, and 4 of UQ list the total amounts spent to manufacture B and C during the 2nd, 3rd, and 4th quarters, respectively.

15. a. False. See the definition of AB.

  b. False. The roles of A and B should be reversed in the second half of the statement. See the box after Example 3.

  c. True. See Theorem 2(b), read right to left.

  d. True. See Theorem 3(b), read right to left.

  e. False. The phase "in the same order" should be "in the reverse order." See the box after Theorem 3.

16. a. True. See the box after Example 4.

  b. False. AB must be a 3×3 matrix, but the formula given here for AB implies that it is 3×1. The plus signs should be just spaces (between columns). This is a common mistake.

  c. True. Apply Theorem 3(d) to $A^2 = AA$.

  d. False. The left-to-right order of $(ABC)^T$, is $C^T B^T A^T$. The order cannot be changed, in general.

  e. True. This general statement follows from Theorem 3(b).

18. The third column of AB is also all zeros because $A\mathbf{b}_3 = A\mathbf{0} = \mathbf{0}$.

20. The first two columns of AB are $A\mathbf{b}_1$ and $A\mathbf{b}_2$. They are equal because $\mathbf{b}_1$ and $\mathbf{b}_2$ are equal.

22. If the columns of B are linearly dependent, then there exists a nonzero vector $\mathbf{x}$ such that $B\mathbf{x} = \mathbf{0}$. From this, $A(B\mathbf{x}) = A\mathbf{0} = \mathbf{0}$ and $(AB)\mathbf{x} = \mathbf{0}$ (by associativity). Since $\mathbf{x}$ is nonzero, the columns of AB must be linearly dependent.

24. Write $I_3 = [\mathbf{e}_1 \ \ \mathbf{e}_2 \ \ \mathbf{e}_3]$ and $D = [\mathbf{d}_1 \ \ \mathbf{d}_2 \ \ \mathbf{d}_3]$. By definition of AD, the equation $AD = I_3$ is equivalent to the three equations $A\mathbf{d}_1 = \mathbf{e}_1$, $A\mathbf{d}_2 = \mathbf{e}_2$, and $A\mathbf{d}_3 = \mathbf{e}_3$. Each of these equations has at least one solution because the columns of A span $\mathbb{R}^3$. (See Theorem 4 in Section 1.4.) Select one solution of each equation, and use them for the columns of D. Then $AD = I_3$.

26. Take any $\mathbf{b}$ in $\mathbb{R}^m$. By hypothesis, $AD\mathbf{b} = I_m\mathbf{b} = \mathbf{b}$. Rewrite this equation as $A(D\mathbf{b}) = \mathbf{b}$. Thus, the vector $\mathbf{x} = D\mathbf{b}$ satisfies $A\mathbf{x} = \mathbf{b}$. This proves that the equation $A\mathbf{x} = \mathbf{b}$ has a solution for each $\mathbf{b}$ in $\mathbb{R}^m$. By Theorem 4 in Section 1.4, A has a pivot position in each row. Since each pivot is in a different column, A must have at least as many columns as rows.

28. Since the inner product $\mathbf{u}^T \mathbf{v}$ is a real number, it equals its transpose. That is, $\mathbf{u}^T \mathbf{v} = (\mathbf{u}^T \mathbf{v})^T = \mathbf{v}^T (\mathbf{u}^T)^T = \mathbf{v}^T \mathbf{u}$, by Theorem 3(d) regarding the transpose of a product of matrices and by Theorem 3(a). The outer product $\mathbf{u}\mathbf{v}^T$ is an $n \times n$ matrix. By Theorem 3, $(\mathbf{u}\mathbf{v}^T)^T = (\mathbf{v}^T)^T\mathbf{u}^T = \mathbf{v}\mathbf{u}^T$.

30. The (i, j)-entries of $r(AB)$, $(rA)B$, and $A(rB)$ are all equal, because

$$r\sum_{k=1}^{n} a_{ik}b_{kj} = \sum_{k=1}^{n} (ra_{ik})b_{kj} = \sum_{k=1}^{n} a_{ik}(rb_{kj})$$

32. Let $\mathbf{e}_j$ and $\mathbf{a}_j$ denote the jth columns of I_n and A, respectively. By definition, the jth column of AI_n is $A\mathbf{e}_j$, which is simply $\mathbf{a}_j$ because $\mathbf{e}_j$ has 1 in the jth position and 0's elsewhere. Thus corresponding columns of AI_n and A are equal. Hence $AI_n = A$.

34. By Theorem 3(d), $(AB\mathbf{x})^T = \mathbf{x}^T(AB)^T = \mathbf{x}^T B^T A^T$.

36. [M] The answer will depend on the choice of matrix program. In MATLAB, the command `rand(5,6)` creates a 5×6 matrix with random entries uniformly distributed between 0 and 1. The command

`round(19*(rand(4,4) - .5))`

creates a random 4×4 matrix with integer entries between -9 and 9. The same result is produced by the command `randomint(4,4)` in the Laydata Toolbox on the text web site. On the TI-86 calculator, the corresponding command is `randM(4,4)`.

38. The equation $(A + I)(A - I) = A^2 - I$ is correct, however $(A + B)(A - B) \neq A^2 - B^2$ most of the time and hence will usually fail for random matrices.

40. The ones move out a super-diagonal for each consecutive power. Note $S^5 = S^6 = 0$.

Section 2.2, page 109

2. $\begin{bmatrix} -5 & 2 \\ 8 & -3 \end{bmatrix}$

4. $\frac{1}{4}\begin{bmatrix} -6 & 4 \\ -4 & 2 \end{bmatrix}$ or $\begin{bmatrix} -3/2 & 1 \\ -1 & 1/2 \end{bmatrix}$

6. $x_1 = -5$ and $x_2 = 26/3$

8. Left-multiply each side of $A = PBP^{-1}$ by P^{-1}:

$P^{-1}A = P^{-1}PBP^{-1}$, $P^{-1}A = IBP^{-1}$,

$P^{-1}A = BP^{-1}$

Then right-multiply each side of the result by P:

$P^{-1}AP = BP^{-1}P$, $P^{-1}AP = BI$,

$P^{-1}AP = B$

9. a. True, by the definition of *invertible*.

  b. False. See Theorem 6(b).

  c. False. If $A = \begin{bmatrix} 1 & 1 \\ 0 & 0 \end{bmatrix}$, then $ab - cd = 1 - 0 \neq 0$, but Theorem 4 shows that this matrix is not invertible, because $ad - bc = 0$.

  d. True. This follows from Theorem 5, which also says that the solution of $A\mathbf{x} = \mathbf{b}$ is unique for each $\mathbf{b}$.

  e. True, by the box just before Example 6.

10. a. False. The last part of Theorem 7 is misstated here.

b. True, by Theorem 6(a).

c. False. The product matrix is invertible, but the product of the inverses should be in the reverse order. See Theorem 6(b).

d. True. See the subsection "Another View of Matrix Inversion".

e. True, by Theorem 7.

12. $AD = I \Rightarrow A^{-1}AD = A^{-1}I \Rightarrow ID = A^{-1} \Rightarrow D = A^{-1}$. Parentheses are routinely suppressed because of the associative property of matrix multiplication.

14. Right-multiply each side of the equation $(B - C)D = 0$ by D^{-1}, and obtain

$$(B - C)DD^{-1} = 0D^{-1}, \quad (B - C)I = 0$$

Thus $B - C = 0$, and $B = C$.

16. Let $C = AB$. Since B is invertible, use B^{-1} to solve for A:

$$CB^{-1} = ABB^{-1}, \quad CB^{-1} = AI = A$$

This shows that A is the product of invertible matrices and hence is invertible, by Theorem 6.

18. $A = BCB^{-1}$

20. **a.** Left-multiply both sides of $(A - AX)^{-1} = X^{-1}B$ by X to see that B is invertible because it is the product of invertible matrices.

b. $X = (A + B^{-1})^{-1}A$. A careful proof should justify $A - AX = B^{-1}X$ and show that $A + B^{-1}$ is invertible.

22. Suppose A is invertible. By Theorem 5, the equation $A\mathbf{x} = \mathbf{b}$ has a solution (in fact, a unique solution) for each **b**. By Theorem 4 in Section 1.4, the columns of A span $\mathbb{R}^n$.

24. If the equation $A\mathbf{x} = \mathbf{b}$ has a solution for each **b** in $\mathbb{R}^n$, then A has a pivot position in each row, by Theorem 4 in Section 1.4. Since A is square, the pivots must be on the diagonal of A. It follows that A is row equivalent to I_n. By Theorem 7, A is invertible.

26. $\begin{bmatrix} d & -b \\ -c & a \end{bmatrix} \begin{bmatrix} a & b \\ c & d \end{bmatrix} = \begin{bmatrix} da - bc & 0 \\ 0 & -cb + ad \end{bmatrix}$

$\begin{bmatrix} a & b \\ c & d \end{bmatrix} \begin{bmatrix} d & -b \\ -c & a \end{bmatrix} = \begin{bmatrix} ad - bc & 0 \\ 0 & -cb + da \end{bmatrix}$

Divide both sides of each equation by $ad - bc$ to get $A^{-1}A = I$ and $AA^{-1} = I$.

28. When row 2 of A is replaced by $\text{row}_2(A) - 3 \cdot \text{row}_1(A)$, the

result may be written as

$$\begin{bmatrix} \text{row}_1(A) \\ \text{row}_2(A) - 3 \cdot \text{row}_1(A) \\ \text{row}_3(A) \end{bmatrix}$$

$$= \begin{bmatrix} \text{row}_1(I) \cdot A \\ \text{row}_2(I) \cdot A - 3 \cdot \text{row}_1(I) \cdot A \\ \text{row}_3(I) \cdot A \end{bmatrix}$$

$$= \begin{bmatrix} \text{row}_1(I) \cdot A \\ [\text{row}_2(I) - 3 \cdot \text{row}_1(I)] \cdot A \\ \text{row}_3(I) \cdot A \end{bmatrix}$$

$$= \begin{bmatrix} \text{row}_1(I) \\ \text{row}_2(I) - 3 \cdot \text{row}_1(I) \\ \text{row}_3(I) \end{bmatrix} A = EA$$

Here E is obtained by replacing $\text{row}_2(I)$ by $\text{row}_2(I) - 3 \cdot \text{row}_1(I)$.

30. $\frac{-1}{3}\begin{bmatrix} 7 & -6 \\ -4 & 3 \end{bmatrix}$ **32.** Not invertible

34. $\begin{bmatrix} 1 & 0 & 0 & \cdots & 0 \\ -1 & 1/2 & 0 & & \\ 0 & -1/2 & 1/3 & & \\ \vdots & & & \ddots & \vdots \\ 0 & 0 & & -1/(n-1) & 1/n \end{bmatrix}$

The *Instructor's Solutions Manual* has a proof that this matrix is the desired inverse.

36. [M] Write $B = [\,A \quad F\,]$, where F consists of the last two columns of I_3, and row reduce. The last two columns of A^{-1} are

$$\begin{bmatrix} 0.1126 & -0.1559 \\ -0.5611 & 1.0077 \\ 0.0828 & -0.1915 \end{bmatrix}$$

38. $D = \begin{bmatrix} 1 & 0 \\ 1 & 1 \\ 1 & 1 \\ 0 & 1 \end{bmatrix}$

There is *no* 4×2 matrix C such that $CA = I_4$. If this were true, then $CA\mathbf{x}$ would equal **x** for all **x** in $\mathbb{R}^4$. This cannot happen because the columns of A are linearly dependent and so $A\mathbf{x} = \mathbf{0}$ for some nonzero vector **x**. For such an **x**, $CA\mathbf{x} = C(\mathbf{0}) = \mathbf{0}$. Or, see Exercise 23 or 25 in Section 2.1.

40. [M] $D^{-1} = \frac{100}{3}\begin{bmatrix} 3 & -1 & 0 \\ -1 & 4 & -1 \\ 0 & -1 & 3 \end{bmatrix}$, $\mathbf{f} = \begin{bmatrix} 0 \\ -4/3 \\ 4 \end{bmatrix}$ pounds

42. [M] The forces at the four points are -10.476, 31.429, -10.476, and 0 newtons, respectively.

Section 2.3, page 115

The abbreviation IMT (here and in the *Study Guide*) denotes the Invertible Matrix Theorem (Theorem 8).

2. Not invertible, by Theorem 4 in Section 2.2, because the determinant is zero. Less obvious is the fact that the columns are linearly dependent—the second column is $-1/2$ times the first column. From this and the IMT, it follows that the matrix is singular.

4. The matrix contains a row of zeros and hence cannot be row reduced to I. Hence, by IMT, the matrix is not invertible.

6. Invertible, by the IMT. The matrix row reduces to
$$\begin{bmatrix} 1 & -3 & 6 \\ 0 & 4 & 3 \\ 0 & 0 & 1 \end{bmatrix}$$ and is row equivalent to I_3.

8. The 4×4 matrix is invertible, by the IMT, because it is already in echelon form and has four pivot columns.

10. [M] The 5×5 matrix is invertible because it has five pivot positions, by the IMT.

11. a. True, by the IMT. If statement (d) of the IMT is true, then so is statement (b).

b. True. If statement (h) of the IMT is true, then so is statement (e).

c. False. Statement (g) of the IMT is true only for invertible matrices.

d. True, by the IMT. If the equation $A\mathbf{x} = \mathbf{0}$ has a nontrivial solution, then statement (d) of the IMT is false. In this case, all the lettered statements in the IMT are false, including statement (c), which means that A must have fewer than n pivot positions.

e. True, by the IMT. If A^T is not invertible, then statement (l) of the IMT is false, and hence statement (a) must also be false.

12. a. True. If statement (k) of the IMT is true, then so is statement (j). By the uniqueness of inverses, $D = A^{-1}$, so $DA = I$.

b. False. Notice that (i) of the IMT uses the word *onto* rather than the word *into*.

c. True. Since (e) of IMT is true, so is (h).

d. False. Since (g) of IMT is true, so is (f).

e. False, by the IMT. The fact that there is a $\mathbf{b}$ in $\mathbb{R}^n$ such that the equation $A\mathbf{x} = \mathbf{b}$ is consistent, does not imply that statement (g) of the IMT is true, and hence there could be more than one solution.

Note: The answers below for Exercises 14–30 refer mostly to the IMT. In many cases, part or all of an acceptable answer could also be based on various results that were used to establish the IMT.

14. If A is lower triangular with nonzero entries on the diagonal, then these n diagonal entries can be used as pivots to produce zeros below the diagonal. Thus A has n pivots and so is invertible, by the IMT. If one of the diagonal entries in A is zero, A will have fewer than n pivots and hence will be singular.

16. If A is invertible, so is A^T, by statement (l) of IMT. Hence statement (e) is true for A^T.

18. No. If A has two identical rows, then A^T has two identical columns and hence fails statement (e) of the IMT. But then A^T is not invertible, so statement (l) fails for A.

20. By statement (g) of the IMT, A is invertible. Hence each equation $A\mathbf{x} = \mathbf{b}$ has a unique solution, by Theorem 5 in Section 2.2.

22. By the box following the IMT, E and F are invertible and are inverses of each other. So $FE = I = EF$. Thus E and F commute.

24. Statement (b) of the IMT is false for G, so statements (e) and (h) are also false. That is, the columns of G are linearly *dependent* and the columns do *not* span $\mathbb{R}^n$.

26. If the columns of A are linearly independent, then A is invertible, by the IMT. So A^2, which is the product of invertible matrices, is invertible. By the IMT, the columns of A^2 span $\mathbb{R}^n$.

28. Let W be the inverse of AB. Then $WAB = I$ and $(WA)B = I$. By statement (j) of the IMT, applied to B in place of A, the matrix B is invertible, since it is square.

30. Since the transformation $\mathbf{x} \mapsto A\mathbf{x}$ is not one-to-one, statement (f) of the IMT is false. Then statement (i) is also false and the transformation $\mathbf{x} \mapsto A\mathbf{x}$ does not map $\mathbb{R}^n$ onto $\mathbb{R}^n$. Also, A is not invertible, which implies that the transformation $\mathbf{x} \mapsto A\mathbf{x}$ is not invertible, by Theorem 9.

32. If $A\mathbf{x} = \mathbf{0}$ has only the trivial solution, then A must have a pivot in each of its n columns. Since A is square, there must be a pivot in each *row* of A. By Theorem 4 in Section 1.4, the equation $A\mathbf{x} = \mathbf{b}$ has a solution for each $\mathbf{b}$ in $\mathbb{R}^n$.

34. The standard matrix of T is
$$A = \begin{bmatrix} 2 & -8 \\ -2 & 7 \end{bmatrix},$$
which is invertible because $\det A = -2 \neq 0$. By Theorem 9, T is invertible, and $T^{-1}(\mathbf{x}) = B\mathbf{x}$, where
$$B = A^{-1} = -\frac{1}{2}\begin{bmatrix} 7 & 8 \\ 2 & 2 \end{bmatrix}.$$

36. Let A be the standard matrix of T. By hypothesis, T is not a one-to-one mapping. So, by Theorem 12 in Section 1.9, the standard matrix A of T has linearly dependent columns. Since A is square, the columns of A do not span $\mathbb{R}^n$, by the IMT. By Theorem 12, again, T cannot map $\mathbb{R}^n$ onto $\mathbb{R}^n$.

38. Given any $\mathbf{v}$ in $\mathbb{R}^n$, write $\mathbf{v} = T(\mathbf{x})$ for some $\mathbf{x}$, because T is an onto mapping. Then, the assumed properties of S and U show that $S(\mathbf{v}) = S(T(\mathbf{x})) = \mathbf{x}$ and $U(\mathbf{v}) = U(T(\mathbf{x})) = \mathbf{x}$. So $S(\mathbf{v})$ and $U(\mathbf{v})$ are equal for each $\mathbf{v}$. That is, S and U are the same function from $\mathbb{R}^n$ into $\mathbb{R}^n$.

40. Given $\mathbf{u}$, $\mathbf{v}$ in $\mathbb{R}^n$, let $\mathbf{x} = S(\mathbf{u})$ and $\mathbf{y} = S(\mathbf{v})$. Then $T(\mathbf{x}) = T(S(\mathbf{u})) = \mathbf{u}$ and $T(\mathbf{y}) = T(S(\mathbf{v})) = \mathbf{v}$, by equation (2). Hence

$$S(\mathbf{u} + \mathbf{v}) = S(T(\mathbf{x}) + T(\mathbf{y}))$$
$$= S(T(\mathbf{x} + \mathbf{y})) \qquad \text{because } T \text{ is linear}$$
$$= \mathbf{x} + \mathbf{y} \qquad \text{by equation (1)}$$
$$= S(\mathbf{u}) + S(\mathbf{v})$$

So S preserves sums. For any scalar r,

$$S(r\mathbf{u}) = S(rT(\mathbf{x}))$$
$$= S(T(r\mathbf{x})) \qquad T \text{ is linear equation (1)}$$
$$= r\mathbf{x}$$
$$= rS(\mathbf{u})$$

So S preserves scalar multiples. Thus S is a linear transformation.

42. [M] cond$(A) \approx 10$, which is approximately 10^1. If you make several trials with MATLAB, which records 16 digits accurately, you should find that $\mathbf{x}$ and $\mathbf{x}_1$ agree to at least 14 or 15 significant digits. So about 1 significant digit is lost.

44. [M] Solve $A\mathbf{x} = (0, 0, 0, 0, 1)$. MATLAB shows that cond$(A) \approx 4.8 \times 10^5$. With MATLAB, the entries in the computed value of $\mathbf{x}$ should be accurate to at least 11 digits.

Section 2.4, page 121

2. $\begin{bmatrix} EP & EQ \\ FR & FS \end{bmatrix}$ **4.** $\begin{bmatrix} W & X \\ -EW + Y & -EX + Z \end{bmatrix}$

6. Assume that X, Z, A, and C are square. Then $X = A^{-1}$, and $Z = C^{-1}$, by the IMT, and $Y = -C^{-1}BA^{-1}$.

8. Assume that A and X are square. Then $X = A^{-1}$, by the IMT, $Y = 0$, and $Z = -A^{-1}B$.

10. $Q = -B + DA$, $P = -A$, $R = -D$

11. a. True. See definition (1) in the paragraph preceding Example 4.

b. False. See Example 3. The number of columns of A_{11} and A_{12} must match the number of rows of B_1 and B_2, respectively.

12. a. False. Both BA and AB are defined, although they have different dimensions. In fact,
$$AB = \begin{bmatrix} A_1 B_1 & A_1 B_2 \\ A_2 B_1 & A_2 B_2 \end{bmatrix}, \text{ which is the block analogue}$$
of an outer product. See Example 4.

b. False. Q^T and R^T also need to be switched.

14. The calculations in Example 5 showed that if A is invertible, then both A_{11} and A_{22} are invertible. Conversely, suppose A_{11} and A_{22} are both invertible, and define B to be the matrix that Example 5 says should be the inverse of A. A routine calculation shows that $AB = I$. Since A is square, the IMT implies that A is invertible. (Alternatively, one could also show that $BA = I$.)

16. $\begin{bmatrix} A_{11} & A_{12} \\ A_{21} & A_{22} \end{bmatrix} =$
$$\begin{bmatrix} I & 0 \\ A_{21}A_{11}^{-1} & I \end{bmatrix} \begin{bmatrix} A_{11} & 0 \\ 0 & S \end{bmatrix} \begin{bmatrix} I & A_{11}^{-1}A_{12} \\ 0 & I \end{bmatrix}$$
with $S = A_{22} - A_{21}A_{11}^{-1}A_{12}$.

18. The Schur complement of $X^T X$ is
$$\mathbf{x}_0^T \mathbf{x}_0 - (\mathbf{x}_0^T X)(X^T X)^{-1}(X^T \mathbf{x}_0)$$
$$= \mathbf{x}_0^T (I_m - X(X^T X)^{-1} X^T)\mathbf{x}_0 = \mathbf{x}_0^T M \mathbf{x}_0.$$

20. The Schur complement of $A - BC - sI_n$ is
$I_m + C(A - BC - sI_n)^{-1}B$. *Note:* The proof that this function actually is the inverse of $W(s)$ in Exercise 19 involves only matrix algebra, but it is a little tricky. The following algebraic identity is needed:
$$CU^{-1}B - CV^{-1}B = C(U^{-1} - V^{-1})B$$
$$= CU^{-1}(V - U)V^{-1}B$$
for any invertible $n \times n$ matrices U and V and any B and C such that the multiplication is well defined.

22. There are many possible solutions, however $C = I$,
$$A = B = \begin{bmatrix} 1 & O \\ 2 & -1 \end{bmatrix}, \text{ and } D = -A \text{ is one possibility.}$$

24. Let

$$A_n = \begin{bmatrix} 1 & 0 & 0 & \cdots & 0 \\ 1 & 1 & 0 & & 0 \\ 1 & 1 & 1 & & 0 \\ \vdots & & & \ddots & \vdots \\ 1 & 1 & 1 & \cdots & 1 \end{bmatrix},$$

$$B_n = \begin{bmatrix} 1 & 0 & 0 & \cdots & 0 \\ -1 & 1 & 0 & & 0 \\ 0 & -1 & 1 & & 0 \\ \vdots & & & \ddots & \vdots \\ 0 & 0 & \cdots & -1 & 1 \end{bmatrix}$$

By direct computation, $A_2 B_2 = I_2$. Assume that for $n = k$, the matrix $A_k B_k = I_k$, and write
$$A_{k+1} = \begin{bmatrix} 1 & \mathbf{0}^T \\ \mathbf{v} & A_k \end{bmatrix} \quad \text{and} \quad B_{k+1} = \begin{bmatrix} 1 & \mathbf{0}^T \\ \mathbf{w} & B_k \end{bmatrix}$$
where $\mathbf{v}$ and $\mathbf{w}$ are in $\mathbb{R}^k$, $\mathbf{v}^T = [1 \; 1 \; \cdots \; 1]$, and $\mathbf{w}^T = [-1 \; 0 \; \cdots \; 0]$. Then

$$A_{k+1}B_{k+1} = \begin{bmatrix} 1 & \mathbf{0}^T \\ \mathbf{v} & A_k \end{bmatrix} \begin{bmatrix} 1 & \mathbf{0}^T \\ \mathbf{w} & B_k \end{bmatrix}$$
$$= \begin{bmatrix} 1 + \mathbf{0}^T\mathbf{w} & \mathbf{0}^T + \mathbf{0}^T B_k \\ \mathbf{v} + A_k\mathbf{w} & \mathbf{v0}^T + A_k B_k \end{bmatrix}$$
$$= \begin{bmatrix} 1 & \mathbf{0}^T \\ \mathbf{0} & I_k \end{bmatrix} = I_{k+1}$$

The $(2, 1)$ entry is $\mathbf{0}$ because $\mathbf{v}$ equals the first column in A_k, and $A_k\mathbf{w}$ is -1 times the first column of A_k.

26. [M] The commands to be used in these exercises will depend on the matrix program. See the *Instructor's Solution Manual*.

Section 2.5, page 129

2. $L\mathbf{y} = \mathbf{b} \Rightarrow \mathbf{y} = \begin{bmatrix} 2 \\ 0 \\ 6 \end{bmatrix}$, $U\mathbf{x} = \mathbf{y} \Rightarrow \mathbf{x} = \begin{bmatrix} -11 \\ -6 \\ -3 \end{bmatrix}$

4. $\mathbf{y} = \begin{bmatrix} 0 \\ -5 \\ -18 \end{bmatrix}$, $\mathbf{x} = \begin{bmatrix} -5 \\ 1 \\ 3 \end{bmatrix}$

6. $\mathbf{y} = \begin{bmatrix} 1 \\ 0 \\ -4 \\ 3 \end{bmatrix}$, $\mathbf{x} = \begin{bmatrix} 33 \\ -12 \\ 2 \\ 3 \end{bmatrix}$

8. $LU = \begin{bmatrix} 1 & 0 \\ 2 & 1 \end{bmatrix} \begin{bmatrix} 6 & 4 \\ 0 & -3 \end{bmatrix}$

10. $\begin{bmatrix} 1 & 0 & 0 \\ -2 & 1 & 0 \\ -2 & 5 & 1 \end{bmatrix} \begin{bmatrix} -5 & 0 & 4 \\ 0 & 2 & 3 \\ 0 & 0 & 9 \end{bmatrix}$

12. $\begin{bmatrix} 1 & 0 & 0 \\ 2 & 1 & 0 \\ -3 & 2 & 1 \end{bmatrix} \begin{bmatrix} 2 & 3 & 2 \\ 0 & 7 & 5 \\ 0 & 0 & 0 \end{bmatrix}$

14. $\begin{bmatrix} 1 & 0 & 0 & 0 \\ 5 & 1 & 0 & 0 \\ -2 & 1 & 1 & 0 \\ -1 & 2 & 0 & 1 \end{bmatrix} \begin{bmatrix} 1 & 3 & 1 & 5 \\ 0 & 5 & 1 & 6 \\ 0 & 0 & 0 & 0 \\ 0 & 0 & 0 & 0 \end{bmatrix}$

16. $\begin{bmatrix} 1 & 0 & 0 & 0 & 0 \\ -2 & 1 & 0 & 0 & 0 \\ 3 & 2 & 1 & 0 & 0 \\ -3 & 0 & 0 & 1 & 0 \\ 4 & 3 & 0 & 0 & 1 \end{bmatrix} \begin{bmatrix} 2 & -3 & 4 \\ 0 & 2 & 1 \\ 0 & 0 & 0 \\ 0 & 0 & 0 \\ 0 & 0 & 0 \end{bmatrix}$

18. $L = \begin{bmatrix} 1 & 0 & 0 \\ -2 & 1 & 0 \\ 3 & -1 & 1 \end{bmatrix}$, $U = \begin{bmatrix} 2 & -4 & 2 \\ 0 & -3 & 6 \\ 0 & 0 & 1 \end{bmatrix}$,

$L^{-1} = \begin{bmatrix} 1 & 0 & 0 \\ 2 & 1 & 0 \\ -1 & 1 & 1 \end{bmatrix}$, $U^{-1} = \frac{1}{6}\begin{bmatrix} 3 & -4 & 18 \\ 0 & -2 & 12 \\ 0 & 0 & 6 \end{bmatrix}$, and

$A^{-1} = U^{-1}L^{-1} = \frac{1}{6}\begin{bmatrix} -23 & 14 & 18 \\ -16 & 10 & 12 \\ -6 & 6 & 6 \end{bmatrix}$

20. Since L is unit lower triangular, it is invertible and may be row reduced to I by adding suitable multiples of a row to the rows below it, beginning with the top row. If elementary matrices $E_1, \ldots, E_p$ implement these row operations, then

$$E_p \cdots E_1 A = (E_p \cdots E_1 L)U = IU = U$$

This shows that A may be row reduced to U using only row replacement operations.

22. $B = \begin{bmatrix} 1 & 0 & 0 \\ 3 & 1 & 0 \\ 1 & -1 & 1 \\ 2 & 2 & -1 \\ -3 & -3 & 2 \end{bmatrix}$, $C = \begin{bmatrix} 2 & -4 & -2 & 3 \\ 0 & 3 & 1 & -1 \\ 0 & 0 & 0 & 5 \end{bmatrix}$;

if $A = LU$, with only three nonzero rows in U, use the first three columns of L for B and the top three rows of U for C.

24. Since Q is square and $Q^T Q = I$, Q is invertible and $Q^{-1} = Q^T$, by the Invertible Matrix Theorem. Thus A is the product of invertible matrices and hence is invertible. By Theorem 5, the equation $A\mathbf{x} = \mathbf{b}$ has a unique solution for all $\mathbf{b}$. From $A\mathbf{x} = \mathbf{b}$, we have $QR\mathbf{x} = \mathbf{b}$, $Q^T QR\mathbf{x} = Q^T\mathbf{b}$, $R\mathbf{x} = Q^T\mathbf{b}$, and $\mathbf{x} = R^{-1}Q^T\mathbf{b}$. A good algorithm for finding $\mathbf{x}$ is to compute $Q^T\mathbf{b}$ and then row reduce $[R\ Q^T\mathbf{b}]$. (See Exercise 15 in Section 2.2.) The reduction is fast because R is triangular.

26. In general, $A^k = PD^k P^{-1}$, where

$$D^k = \begin{bmatrix} 2^k & 0 & 0 \\ 0 & 3^k & 0 \\ 0 & 0 & 1^k \end{bmatrix}.$$

28. $\begin{bmatrix} 1 & 0 \\ -1/R_3 & 1 \end{bmatrix} \begin{bmatrix} 1 & 0 \\ -1/R_2 & 1 \end{bmatrix} \begin{bmatrix} 1 & 0 \\ -1/R_1 & 1 \end{bmatrix} = \begin{bmatrix} 1 & 0 \\ -(1/R_1 + 1/R_2 + 1/R_3) & 1 \end{bmatrix}.$

The single shunt resistance is $\frac{R_1 R_2 R_3}{R_1 R_2 + R_1 R_3 + R_2 R_3}$.

30. There are many possible factorizations of A. For example,

$$A = \begin{bmatrix} 3 & -12 \\ -1/3 & 5/3 \end{bmatrix}$$

$$= \begin{bmatrix} 1 & -6 \\ 0 & 1 \end{bmatrix} \begin{bmatrix} 1 & 0 \\ -1/6 & 1 \end{bmatrix} \begin{bmatrix} 1 & 0 \\ -1/6 & 1 \end{bmatrix} \begin{bmatrix} 1 & -2 \\ 0 & 1 \end{bmatrix},$$

corresponds to a ladder with a circuit followed by two shunts and another circuit with $R_1 = 2$ ohms, and $R_2 = R_3 = R_4 = 6$ ohms.

32. **[M]**

a. $L = \begin{bmatrix} 1 & & & \\ -1/3 & 1 & & \\ & -3/8 & 1 & \\ & & -8/21 & 1 \end{bmatrix}$

$U = \begin{bmatrix} 3 & -1 & & \\ & 8/3 & -1 & \\ & & 21/8 & -1 \\ & & & 55/21 \end{bmatrix}$

b. Let $\mathbf{s}_k$ satisfy $L\mathbf{s}_k = \mathbf{t}_{k-1}$. Then $\mathbf{t}_k$ satisfies $U\mathbf{t}_k = \mathbf{s}_k$.

$\mathbf{t}_0 = \begin{bmatrix} 10 \\ 15 \\ 15 \\ 10 \end{bmatrix}$; Then

$\mathbf{s}_1 = \begin{bmatrix} 10.0000 \\ 18.3333 \\ 21.8750 \\ 18.3333 \end{bmatrix}$, $\mathbf{t}_1 = \begin{bmatrix} 7.0000 \\ 11.0000 \\ 11.0000 \\ 7.0000 \end{bmatrix}$,

$\mathbf{s}_2 = \begin{bmatrix} 7.0000 \\ 13.3333 \\ 16.0000 \\ 13.0952 \end{bmatrix}$, $\mathbf{t}_2 = \begin{bmatrix} 5.0000 \\ 8.0000 \\ 8.0000 \\ 5.0000 \end{bmatrix}$,

$$s_3 = \begin{bmatrix} 5.0000 \\ 9.6667 \\ 11.6250 \\ 9.4286 \end{bmatrix}, \quad t_3 = \begin{bmatrix} 3.6000 \\ 5.8000 \\ 5.8000 \\ 3.6000 \end{bmatrix},$$

$$s_4 = \begin{bmatrix} 3.6000 \\ 7.0000 \\ 8.4250 \\ 6.8095 \end{bmatrix}, \quad t_4 = \begin{bmatrix} 2.6000 \\ 4.2000 \\ 4.2000 \\ 2.6000 \end{bmatrix}$$

Section 2.6, page 136
Note: Exercises 2, 3, and 4 could be used for students to discover the linearity of the Leontief model.

2. $\begin{bmatrix} 37.03 \\ 38.89 \\ 16.67 \end{bmatrix}$ **4.** $\begin{bmatrix} 81.48 \\ 55.56 \\ 33.33 \end{bmatrix}$ **6.** $\begin{bmatrix} 48.57 \\ 45.71 \end{bmatrix}$

8. a. Since $\mathbf{x}$ satisfies $(I - C)\mathbf{x} = \mathbf{d}$ and $\Delta\mathbf{x}$ satisfies $(I - C)\Delta\mathbf{x} = \Delta\mathbf{d}$, linearity of matrix multiplication shows that

$$(I - C)(\mathbf{x} + \Delta\mathbf{x}) = (I - C)\mathbf{x} + (I - C)\Delta\mathbf{x}$$
$$= \mathbf{d} + \Delta\mathbf{d}$$

which means that $\mathbf{x} + \Delta\mathbf{x}$ satisfies the production equation for a demand of $\mathbf{d} + \Delta\mathbf{d}$.

b. If $\Delta\mathbf{x}$ satisfies $(I - C)\Delta\mathbf{x} = \Delta\mathbf{d}$, then $\Delta\mathbf{x} = (I - C)^{-1}\Delta\mathbf{d}$, which is the first column of $(I - C)^{-1}$ in the case when $\Delta\mathbf{d}$ is the first column of I.

10. By the argument in Exercise 8, the effect of raising the demand for the output of one sector of the economy is given by the entries in the corresponding column of $(I - C)^{-1}$. When these entries are all positive, every sector must increase its output by some positive (though possibly small) quantity. So an increase in demand for *any* sector will increase the demand for *every* sector.

12. $D_{m+1} = I + CD_m$

14. [M] $x = (134034, 131687, 69472, 176912, 66596, 443773, 18431)$. In view of the remarks for Exercise 13, a realistic answer might be
$\mathbf{x} = 1000 \times (134, 132, 69, 177, 67, 444, 18)$.

Section 2.7, page 144

2. $\begin{bmatrix} -1 & 0 \\ 0 & 1 \end{bmatrix}\begin{bmatrix} 4 & 2 & 5 \\ 0 & 2 & 3 \end{bmatrix} = \begin{bmatrix} -4 & -2 & -5 \\ 0 & 2 & 3 \end{bmatrix}$

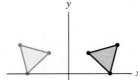

4. $\begin{bmatrix} .5 & 0 & -.5 \\ 0 & 1.5 & 6 \\ 0 & 0 & 1 \end{bmatrix}$

6. $\begin{bmatrix} 1/\sqrt{2} & -1/\sqrt{2} & 0 \\ -1/\sqrt{2} & -1/\sqrt{2} & 0 \\ 0 & 0 & 1 \end{bmatrix}$

8. $\begin{bmatrix} \sqrt{2}/2 & -\sqrt{2}/2 & 3 + 2\sqrt{2} \\ \sqrt{2}/2 & \sqrt{2}/2 & 7 - 5\sqrt{2} \\ 0 & 0 & 1 \end{bmatrix}$

10. D commutes with R but not with T; R does not commute with T.

12. Two identities:
$\tan\varphi/2 = (1 - \cos\varphi)/(\sin\varphi) = (\sin\varphi)/(1 + \cos\varphi)$. The first identity shows that $1 - (\tan\varphi/2)(\sin\varphi) = \cos\varphi$, and hence

$$\begin{bmatrix} 1 & -\tan\varphi/2 & 0 \\ 0 & 1 & 0 \\ 0 & 0 & 1 \end{bmatrix}\begin{bmatrix} 1 & 0 & 0 \\ \sin\varphi & 1 & 0 \\ 0 & 0 & 1 \end{bmatrix}$$
$$= \begin{bmatrix} \cos\varphi & -\tan\varphi/2 & 0 \\ \sin\varphi & 1 & 0 \\ 0 & 0 & 1 \end{bmatrix}$$

The second identity shows that
$$(\cos\varphi)(-\tan\varphi/2) - \tan\varphi/2 = -(\cos\varphi + 1)(\tan\varphi/2)$$
$$= -\sin\varphi$$

Hence

$$\begin{bmatrix} \cos\varphi & -\tan\varphi/2 & 0 \\ \sin\varphi & 1 & 0 \\ 0 & 0 & 1 \end{bmatrix}\begin{bmatrix} 1 & -\tan\varphi/2 & 0 \\ 0 & 1 & 0 \\ 0 & 0 & 1 \end{bmatrix}$$
$$= \begin{bmatrix} \cos\varphi & -\sin\varphi & 0 \\ \sin\varphi & \cos\varphi & 0 \\ 0 & 0 & 1 \end{bmatrix}$$

14. The matrix from Exercise 7 may be written as

$$\begin{bmatrix} 1/2 & -\sqrt{3}/2 & 3 + 4\sqrt{3} \\ \sqrt{3}/2 & 1/2 & 4 - 3\sqrt{3} \\ 0 & 0 & 1 \end{bmatrix}$$
$$= \begin{bmatrix} 1 & 0 & 3 + 4\sqrt{3} \\ 0 & 1 & 4 - 3\sqrt{3} \\ 0 & 0 & 1 \end{bmatrix}\begin{bmatrix} 1/2 & -\sqrt{3}/2 & 0 \\ \sqrt{3}/2 & 1/2 & 0 \\ 0 & 0 & 1 \end{bmatrix}$$

This is a rotation through $60°$ about the origin, followed by a translation by $(3 + 4\sqrt{3}, 4 - 3\sqrt{3})$.

16. Both $(1, -2, -3, 4)$ and $(10, -20, -30, 40)$ are homogeneous coordinates for $(1/4, -1/2, -3/4)$ because of the formulas $x = X/H$, $y = Y/H$, and $z = Z/H$.

18. $\begin{bmatrix} \sqrt{3}/2 & 1/2 & 0 & 5 \\ -1/2 & \sqrt{3}/2 & 0 & -2 \\ 0 & 0 & 1 & 1 \\ 0 & 0 & 0 & 1 \end{bmatrix}$

20. The triangle with vertices $(4.67, 2, 0)$, $(15, 10, 0)$, and $(1.11, 2.22, 0)$.

22. [M] $\begin{bmatrix} 1.0031 & .9548 & .6179 \\ .9968 & -.2707 & -.6448 \\ 1.0085 & -1.1105 & 1.6996 \end{bmatrix} \begin{bmatrix} Y \\ I \\ Q \end{bmatrix} = \begin{bmatrix} R \\ G \\ B \end{bmatrix}$

Section 2.8, page 151

2. The set is closed under scalar multiplication, but not under sums. For instance, the sum of $(-1, 0)$ and $(0, -1)$ is not in the set.

4. The set is closed under sums, but not under multiplication by a negative scalar.

6. No.   **8.** Yes   **10.** Yes, $A\mathbf{u} = \mathbf{0}$.

12. $p = 3, q = 5$. Nul A is a subspace of $\mathbb{R}^3$ because solutions of $A\mathbf{x} = \mathbf{0}$ must have three entries, to match the columns of A. Col A is a subspace of $\mathbb{R}^5$ because each column vector has five entries.

14. Nul A: $\begin{bmatrix} -1 \\ 5 \\ -3 \end{bmatrix}$, or any nonzero multiple of this vector

   Col A: any column of A

16. One vector is a multiple of the other, so they are linearly dependent and hence cannot be a basis for any subspace.

18. Let A be the matrix whose columns are the vectors given. A row reduces to a matrix with only 2 pivots, so A is not invertible by the IMT and its columns do not form a basis for $\mathbb{R}^3$.

20. The vectors are linearly dependent because there are more vectors in the set than entries in each vector (Theorem 8 in Section 1.7). So the vectors cannot be a basis for any subspace.

21. a. False. See the definition at the beginning of the section. The critical phrase "for each" is missing in (ii) and (iii). (This is a common student error!)
   b. True. See the paragraph before Example 4.
   c. False. See Theorem 12. The null space is a subspace of $\mathbb{R}^n$, not $\mathbb{R}^m$.
   d. True. See Example 5.
   e. True. See the first part of the solution of Example 8.

22. a. False. See the definition at the beginning of the section. The condition about the zero vector is only one of the conditions for a subspace.
   b. False. See the Warning that follows Theorem 13.
   c. True. See Example 3.
   d. False. Since y need not be in H, it is not guaranteed by the definition of a subspace that $x + y$ will be in H.
   e. True. See the paragraph after Example 4.

24. Basis for Col A: $\begin{bmatrix} 3 \\ 2 \\ 3 \end{bmatrix}, \begin{bmatrix} 9 \\ 7 \\ 6 \end{bmatrix}$

   Basis for Nul A: $\begin{bmatrix} 2 \\ 1 \\ 0 \\ 0 \end{bmatrix}, \begin{bmatrix} 6 \\ 0 \\ -2 \\ 1 \end{bmatrix}$

26. Basis for Col A: $\begin{bmatrix} 3 \\ 3 \\ 0 \\ 6 \end{bmatrix}, \begin{bmatrix} -1 \\ 1 \\ 3 \\ 3 \end{bmatrix}, \begin{bmatrix} -1 \\ 0 \\ -1 \\ -2 \end{bmatrix}$

   Basis for Nul A: $\begin{bmatrix} 0 \\ -3 \\ 1 \\ 0 \\ 0 \end{bmatrix}, \begin{bmatrix} -4/3 \\ 2 \\ 0 \\ 2 \\ 1 \end{bmatrix}$

28. The easiest construction is to write a 3×3 matrix in echelon form that has only two pivots, and let $\mathbf{b}$ be any vector in $\mathbb{R}^3$ whose third entry is nonzero.

30. Since Col A is the set of all linear combinations of $\mathbf{a}_1, \ldots, \mathbf{a}_p$, the set $\{\mathbf{a}_1, \ldots, \mathbf{a}_p\}$ spans Col A. Because $\{\mathbf{a}_1, \ldots, \mathbf{a}_p\}$ is also linearly independent, it is a basis for Col A. (There is no need to discuss pivot columns and Theorem 13, though a proof could be given using this information.)

32. If Col $B = \mathbb{R}^7$, then the columns of B span $\mathbb{R}^7$. Since B is square, the IMT shows that B is invertible and the equation $B\mathbf{x} = \mathbf{b}$ has a solution for each $\mathbf{b}$ in $\mathbb{R}^7$. Also, each solution is unique, by Theorem 5 in Section 2.2.

34. If the columns of A form a basis, they are linearly independent. This means that A cannot have more columns than rows. Since the columns also span $\mathbb{R}^m$, A must have a pivot in each row, which means that A cannot have more rows than columns. As a result, A must be a square matrix.

36. If the columns of C are linearly independent, then the equation $C\mathbf{x} = \mathbf{0}$ has only the trivial (zero) solution. That is, Nul $C = \{\mathbf{0}\}$.

38. [M] Display the reduced echelon form of A, and select the pivot columns of A as a basis for Col A. For Nul A, write the solution of $A\mathbf{x} = \mathbf{0}$ in parametric vector form.

   Basis for Col A: $\begin{bmatrix} 5 \\ 4 \\ 5 \\ -7 \end{bmatrix}, \begin{bmatrix} 3 \\ 1 \\ 1 \\ -5 \end{bmatrix}, \begin{bmatrix} -6 \\ -8 \\ 5 \\ 8 \end{bmatrix}, \begin{bmatrix} -8 \\ -7 \\ 19 \\ 5 \end{bmatrix}.$

   Basis for Nul A: $\begin{bmatrix} -1 \\ 1 \\ 1 \\ 0 \\ 0 \end{bmatrix}$

Section 2.9, page 157

2. $\mathbf{x} = (-1)\mathbf{b}_1 + 2\mathbf{b}_2 = (-1)\begin{bmatrix} -3 \\ 1 \end{bmatrix} + 2\begin{bmatrix} 3 \\ 2 \end{bmatrix} = \begin{bmatrix} 9 \\ 3 \end{bmatrix}$

4. $\begin{bmatrix} -3 \\ -2 \end{bmatrix}$ **6.** $\begin{bmatrix} 3 \\ 2 \end{bmatrix}$

8. $[\mathbf{x}]_{\mathcal{B}} = \begin{bmatrix} 2 \\ -1 \end{bmatrix}$, $[\mathbf{y}]_{\mathcal{B}} = \begin{bmatrix} 1.5 \\ 1.0 \end{bmatrix}$, $[\mathbf{z}]_{\mathcal{B}} = \begin{bmatrix} -1 \\ -.5 \end{bmatrix}$

10. Basis for Col A: $\begin{bmatrix} 1 \\ 2 \\ -2 \\ 3 \end{bmatrix}, \begin{bmatrix} -2 \\ -1 \\ 0 \\ 1 \end{bmatrix}, \begin{bmatrix} 5 \\ 5 \\ 1 \\ 1 \end{bmatrix}, \begin{bmatrix} 4 \\ 6 \\ -6 \\ 5 \end{bmatrix}$;

dim Col $A = 4$

Basis for Nul A: $\begin{bmatrix} -1 \\ -1 \\ 1 \\ 0 \\ 0 \end{bmatrix}$; dim Nul $A = 1$

12. Basis of Col A: $\begin{bmatrix} 1 \\ 5 \\ 4 \\ 3 \end{bmatrix}, \begin{bmatrix} 2 \\ 1 \\ 6 \\ 4 \end{bmatrix}, \begin{bmatrix} -4 \\ -9 \\ -9 \\ -5 \end{bmatrix}$;

dim Col $A = 3$

Basis for Nul A: $\begin{bmatrix} 0 \\ -2 \\ 0 \\ 1 \\ 0 \end{bmatrix}, \begin{bmatrix} 0 \\ -1 \\ 1 \\ 0 \\ 1 \end{bmatrix}$; dim Nul $A = 2$

14. Let A be the matrix whose columns are given, and let $H = \text{Col } A$. Columns 1, 2, and 4 of A form a basis for H, so dim $H = 3$.

16. Col A *cannot* equal $\mathbb{R}^3$, because the columns of A have 4 entries. (In fact, Col A is a three-dimensional subspace of $\mathbb{R}^4$, because the 3 pivot columns of A form a basis for Col A.) Since A has 7 columns and 3 pivot columns, the equation $A\mathbf{x} = \mathbf{0}$ has 4 free variables. So, dim Nul $A = 4$.

17. a. True. This is the definition of a B-coordinate vector.

b. False. A line must be through the origin in $\mathbb{R}^n$ to be a subspace of $\mathbb{R}^n$.

c. True. The sentence before Example 3 concludes that the number of pivot columns of A is the rank of A, which is the dimension of Col A by definition.

d. True. This is equivalent to the Rank Theorem because rank A *is* the dimension of Col A.

e. True, by the Basis Theorem. In this case, the spanning set is automatically a linearly independent set.

18. a. True. This fact is justified in the second paragraph of this section.

b. False. The dimension of Nul A is the number of *free* variables in the equation $A\mathbf{x} = \mathbf{0}$. See Example 2.

c. True, by the definition of *rank*.

d. True. See the second paragraph after Fig. 1.

e. True, by the Basis Theorem. In this case, the linearly independent set is automatically a spanning set.

20. A 6×8 matrix A has eight columns. By the Rank Theorem, rank $A = 8 - \dim \text{Nul } A$. Since the null space is three-dimensional, rank $A = 5$.

22. Let H be a four-dimensional subspace spanned by a set S of five vectors. If S were linearly independent, it would be a

basis for H. This is impossible, by the statement just before the definition of *dimension* in Section 2.9, which essentially says that *every* basis of a p-dimensional subspace consists of p vectors. Thus, S must be linearly dependent.

24. A rank 1 matrix has a one-dimensional column space. Every column is a multiple of some fixed vector. To construct a 3×4 matrix, choose any nonzero vector in $\mathbb{R}^3$, and use it for one column. Choose any multiples of the vector for the other three columns.

26. If columns $\mathbf{a}_1, \mathbf{a}_3, \mathbf{a}_4, \mathbf{a}_5$, and $\mathbf{a}_7$ of A are linearly independent and if dim Col $A = 5$, then $\{\mathbf{a}_1, \mathbf{a}_3, \mathbf{a}_4, \mathbf{a}_5, \mathbf{a}_7\}$ is a linearly independent set in a five-dimensional column space. By the Basis Theorem, this set of five vectors is a basis for the column space.

28. If $\mathcal{A}$ contained more vectors than $\mathcal{B}$, then $\mathcal{A}$ would be linearly dependent, by Exercise 27, because $\mathcal{B}$ spans W. Repeat the argument with $\mathcal{B}$ and $\mathcal{A}$ interchanged to conclude that $\mathcal{B}$ cannot contain more vectors than $\mathcal{A}$.

30. [M] The first three columns of $[\,\mathbf{v}_1 \quad \mathbf{v}_2 \quad \mathbf{v}_3 \quad \mathbf{x}\,]$ are pivot columns and so form a basis for H. The fourth column of the reduced echelon form of this matrix shows that the $\mathcal{B}$-coordinate vector of $\mathbf{x}$ is $(-2, 1, 1)$.

Chapter 2 Supplementary Exercises, page 160

1. a. True. If A and B are $m \times n$, then B^T has as many rows as A has columns, so AB^T is defined. Also, $A^T B$ is defined because A^T has m columns and B has m rows.

b. False. B must have two columns. A has as many columns as B has rows.

c. True. The ith row of A has the form $(0, \ldots, d_i, \ldots, 0)$. So the ith row of AB is $(0, \ldots, d_i, \ldots, 0)B$, which is d_i times the ith row of B.

d. False. Take the zero matrix for B. Or, construct a matrix B such that the equation $B\mathbf{x} = \mathbf{0}$ has nontrivial solutions, and construct C and D so that $C \neq D$ and the columns of $C - D$ satisfy the equation $B\mathbf{x} = \mathbf{0}$. Then $B(C - D) = 0$ and $BC = BD$.

e. False. Counterexample: $A = \begin{bmatrix} 1 & 0 \\ 0 & 0 \end{bmatrix}$ and $C = \begin{bmatrix} 0 & 0 \\ 0 & 1 \end{bmatrix}$.

f. False. $(A + B)(A - B) = A^2 - AB + BA - B^2$. This equals $A^2 - B^2$ if and only if A commutes with B.

g. True. An $n \times n$ replacement matrix has $n + 1$ nonzero entries. The $n \times n$ scale and interchange matrices have n nonzero entries.

h. True. The transpose of an elementary matrix is an elementary matrix of the same type.

i. True. An $n \times n$ elementary matrix is obtained by a row operation on I_n.

j. False. Elementary matrices are invertible, so a product of such matrices is invertible. But not every square matrix is invertible.

k. True. If A is 3×3 with three pivot positions, then A is row equivalent to I_3.

l. False. A must be square in order to conclude from the equation $AB = I$ that A is invertible.

m. False. AB is invertible, but $(AB)^{-1} = B^{-1}A^{-1}$, and this product is not always equal to $A^{-1}B^{-1}$.

n. True. Given $AB = BA$, left-multiply by A^{-1} to get $B = A^{-1}BA$, and then right-multiply by A^{-1} to obtain $BA^{-1} = A^{-1}B$.

o. False. The correct equation is $(rA)^{-1} = r^{-1}A^{-1}$, because $(rA)(r^{-1}A^{-1}) = (rr^{-1})(AA^{-1}) = 1 \cdot I = I$.

p. True. If the equation $A\mathbf{x} = \begin{bmatrix} 1 \\ 0 \\ 0 \end{bmatrix}$ has a unique solution, then there are no free variables in this equation, which means that A must have three pivot positions (since A is 3×3). By the Invertible Matrix Theorem, A is invertible.

2. $C = (C^{-1})^{-1} = \dfrac{1}{-2}\begin{bmatrix} 7 & -5 \\ -6 & 4 \end{bmatrix} = \begin{bmatrix} -7/2 & 5/2 \\ 3 & -2 \end{bmatrix}$

4. $I + A + A^2 + \cdots + A^{n-1}$

6. By computation, $A^2 = I$, $B^2 = I$, and
$$AB = \begin{bmatrix} 0 & 1 \\ -1 & 0 \end{bmatrix} = -BA$$

7. See Exercise 12 in Section 2.2.

8. By definition of matrix multiplication, the matrix A satisfies
$$A\begin{bmatrix} 1 & 2 \\ 3 & 7 \end{bmatrix} = \begin{bmatrix} 1 & 3 \\ 1 & 1 \end{bmatrix}$$

Right-multiply both sides by the inverse of $\begin{bmatrix} 1 & 2 \\ 3 & 7 \end{bmatrix}$:
$$A = \begin{bmatrix} 1 & 3 \\ 1 & 1 \end{bmatrix}\begin{bmatrix} 7 & -2 \\ -3 & 1 \end{bmatrix} = \begin{bmatrix} -2 & 1 \\ 4 & -1 \end{bmatrix}$$

10. Since A is invertible, so is A^T, by the Invertible Matrix Theorem. Then A^TA is the product of invertible matrices and so is invertible. Thus, the formula $(A^TA)^{-1}A^T$ makes sense. By Theorem 6 in Section 2.2,
$$(A^TA)^{-1} \cdot A^T = A^{-1}(A^T)^{-1}A^T = A^{-1}I = A^{-1}$$

An alternative calculation:
$(A^TA)^{-1}A^T \cdot A = (A^TA)^{-1}(A^TA) = I$. Since A is invertible, this equation shows that its inverse is $(A^TA)^{-1}A^T$.

11. c. When $x_1, \ldots, x_n$ are distinct, the columns of V are linearly independent, by part (b). By the Invertible Matrix Theorem, V is invertible and its columns span $\mathbb{R}^n$. So, for every $\mathbf{y} = (y_1, \ldots, y_n)$ in $\mathbb{R}^n$, there is a vector $\mathbf{c}$ such that $V\mathbf{c} = \mathbf{y}$. Let p be the polynomial whose coefficients are listed in $\mathbf{c}$. Then, by part (a), p is an interpolating polynomial for $(x_1, y_1), \ldots, (x_n, y_n)$.

12. If $A = LU$, then $\text{col}_1(A) = L \cdot \text{col}_1(U)$. Since $\text{col}_1(U)$ has a zero in every entry except possibly the first, $L \cdot \text{col}_1(U)$ is a linear combination of the columns of L in which all

weights except possibly the first are zero. So $\text{col}_1(A)$ is a multiple of $\text{col}_1(L)$.

Similarly, $\text{col}_2(A) = L \cdot \text{col}_2(U)$, which is a linear combination of the columns of L using the first two entries in $\text{col}_2(U)$ as weights, because the other entries in $\text{col}_2(U)$ are zero. Thus $\text{col}_2(A)$ is a linear combination of the first two columns of L.

14. $P\mathbf{x} = \begin{bmatrix} 0 \\ 0 \\ 3 \end{bmatrix}$, $Q\mathbf{x} = \begin{bmatrix} 1 \\ 5 \\ -3 \end{bmatrix}$

16. Since A is not invertible, there is a nonzero vector $\mathbf{v}$ in $\mathbb{R}^n$ such that $A\mathbf{v} = \mathbf{0}$. Place n copies of $\mathbf{v}$ into an $n \times n$ matrix B. Then $AB = A[\mathbf{v} \ \cdots \ \mathbf{v}] = [A\mathbf{v} \ \cdots \ A\mathbf{v}] = 0$.

18. Suppose $\mathbf{x}$ satisfies $A\mathbf{x} = \mathbf{b}$. Then $CA\mathbf{x} = C\mathbf{b}$. Since $CA = I$, $\mathbf{x}$ must be $C\mathbf{b}$. This shows that $C\mathbf{b}$ is the only solution of $A\mathbf{x} = \mathbf{b}$.

20. [M] If J denotes the $n \times n$ matrix of 1's, then
$$A_n = J - I_n \quad \text{and} \quad A_n^{-1} = \frac{1}{n-1} \cdot J - I_n$$
Proof: Observe that $J^2 = nJ$ and $A_nJ = (J - I)J = J^2 - J = (n-1)J$. Now compute $A_n((n-1)^{-1}J - I) = (n-1)^{-1}A_nJ - A_n = J - (J - I) = I$. Since A_n is square, A_n is invertible and its inverse is $(n-1)^{-1}J - I$.

Chapter 3

Section 3.1, page 167

2. 2   **4.** 20   **6.** 1   **8.** −11

10. −6. Start with row 2.

12. 36. Start with row 1 or column 4.

14. 9. Start with row 4 or column 5.

16. 2   **18.** 20

20. $ad - bc$, $a(kd) - b(kc) = k(ad - bc)$. Scaling a row by k multiplies the determinant by k.

22. $ad - bc$, $(ad + kcd) - (bc + kdc) = ad - bc$. Row replacement does not change a determinant.

24. $2a - 6b + 3c$, $-2a + 6b - 3c$. Interchanging two rows reverses the sign of the determinant.

26. 1   **28.** k **30.** −1

32. k. A scaling matrix is diagonal, with k on the diagonal and with 1's as the other diagonal entries. The determinant is the product of the diagonal entries.

34. $\det EA = \begin{vmatrix} a & b \\ kc & kd \end{vmatrix} = akd - bkc = k(ad - bc)$
$\qquad = (\det E)(\det A)$

36. $\det EA = \begin{vmatrix} a & b \\ ka + c & kb + d \end{vmatrix} = a(kb + d) - b(ka + c)$
$\qquad = akb + ad - bka - bc = (+1)(ad - bc)$
$\qquad = (\det E)(\det A)$

38. $\det kA = k^2 \cdot \det A$

39. a. True. See the paragraph preceding the definition of $\det A$.

b. False. See the definition of cofactor, preceding Theorem 1.

40. a. False. See Theorem 1.

b. False. See Theorem 2.

42. The area of the parallelogram and the determinant of $[\,\mathbf{v}\quad\mathbf{u}\,]$ are both bc. The determinant of $[\,\mathbf{u}\quad\mathbf{v}\,]$ is $-bc$. Both matrices determine the same parallelogram, with base of length c and height b.

44. [M] Theorem 6 in Section 3.2 will show that $\det AB = (\det A)(\det B)$.

46. [M] If A is invertible, then $\det A \neq 0$, by Theorem 4 in Section 3.2. Students will be asked in Exercise 31 of Section 3.2 to prove that $\det A^{-1} = 1/(\det A)$.

Section 3.2, page 175

2. A constant may be factored out of one row.

4. A row replacement operation does not change the determinant.

6. -18 **8.** 0 **10.** 24 **12.** 114

14. 0 **16.** 21 **18.** 7 **20.** 7

22. Not invertible **24.** Linearly independent

26. Linearly dependent

27. a. True. Theorem 3(a).

b. False. If scaling operations are used to produce U, then the formula described may not give $\det A$. See the paragraph following Example 2.

c. True. See the remark following Theorem 4.

d. False. See the warning after Example 5.

28. a. True. By Theorem 3(b), the first interchange changes only the sign of the determinant, so the second interchange restores the original sign of the determinant.

b. False. True when A is triangular (Theorem 2 in Section 3.1).

c. False. The conditions described provide only some cases when $\det A$ is zero. See the paragraph after Theorem 4.

d. False. See Theorem 5.

30. If two rows are equal, interchange them. This doesn't change the matrix, but the sign of the determinant is reversed. This is possible only if the determinant is zero. The result about columns can be explained the same way, or one can remark that if A has two equal columns, then A^T has two equal rows. In this case, $\det A^T = 0$. So $\det A = 0$, too, by Theorem 5.

32. $\det(rA) = r^n \cdot \det A$

34. $\det(PAP^{-1}) = (\det P)(\det A)(\det P^{-1})$ By Theorem 6
$\qquad\qquad\quad\ = (\det P)(\det A)(\det P)^{-1}$ By Exercise 31
$\qquad\qquad\quad\ = \det A$

36. $0 = \det A^4 = (\det A)^4$, by Theorem 6. So $\det A = 0$, which implies that A is not invertible, by Theorem 4.

38. $\det AB = \det \begin{bmatrix} 6 & 0 \\ -2 & 0 \end{bmatrix} = 0$

$(\det A)(\det B) = (-6 + 6)(-4 + 2) = 0$

40. a. -2 **b.** 32 **c.** -16 **d.** 1 **e.** -1

42. $\det(A + B) = \det \begin{bmatrix} 1 + a & b \\ c & 1 + d \end{bmatrix}$
$\qquad\qquad\quad\ = 1 + a + d + ad - bc.$
Also $\det A + \det B = 1 + (ad - bc)$. Since $\det(A + B) - (\det A + \det B) = a + d$, we have $\det(A + B) = \det A + \det B$ if and only if $a + d = 0$.

44. $\det AE = \det(AE)^T$ Theorem 5
$\qquad\quad\ = \det E^T A^T$ Section 2.1
$\qquad\quad\ = (\det E^T)(\det A^T)$ Theorem 6
$\qquad\quad\ = (\det E)(\det A)$ Theorem 5 used twice

46. [M] For A as in Exercise 9 in Section 2.3, $\det A = 1$ and cond $A = 23683$. Although A is nearly singular, it has an inverse:

$$A^{-1} = \begin{bmatrix} -19 & -14 & 0 & 7 \\ -549 & -401 & -2 & 196 \\ 267 & 195 & 1 & -95 \\ -278 & -203 & -1 & 99 \end{bmatrix}$$

The determinant is sensitive to scaling, but the condition number does not change:

$\det(10A) = 10^4(-1), \det(0.1A) = 10^{-4}(-1)$ but

$\text{cond}(10A) = \text{cond}(0.1A) = \text{cond } A$

The same things happen when $A = I_4$.

Section 3.3, page 184

2. $\begin{bmatrix} 5/3 \\ -2/3 \end{bmatrix}$ **4.** $\begin{bmatrix} -3/2 \\ 1/2 \end{bmatrix}$ **6.** $\begin{bmatrix} -4 \\ 13 \\ -1 \end{bmatrix}$

8. All real s; $x_1 = \dfrac{3s + 2}{3(s^2 + 3)}$, $x_2 = \dfrac{2s - 9}{5(s^2 + 3)}$

10. $s \neq 0, 1/4$; $x_1 = \dfrac{6s - 2}{3s(4s - 1)}$, $x_2 = \dfrac{1}{3(4s - 1)}$

12. adj $A = \begin{bmatrix} -1 & 3 & 7 \\ 0 & 0 & 5 \\ 2 & -1 & -4 \end{bmatrix}$, $A^{-1} = \dfrac{1}{5}\begin{bmatrix} -1 & 3 & 7 \\ 0 & 0 & 5 \\ 2 & -1 & -4 \end{bmatrix}$

14. adj $A = \begin{bmatrix} 5 & -3 & -8 \\ 2 & -2 & -3 \\ -4 & 3 & 6 \end{bmatrix}$, $A^{-1} = (-1)\begin{bmatrix} 5 & -3 & -8 \\ 2 & -2 & -3 \\ -4 & 3 & 6 \end{bmatrix}$

16. adj $A = \begin{bmatrix} -9 & -6 & 14 \\ 0 & 3 & -1 \\ 0 & 0 & -3 \end{bmatrix}$, $A^{-1} = -\dfrac{1}{9}\begin{bmatrix} -9 & -6 & 14 \\ 0 & 3 & -1 \\ 0 & 0 & -3 \end{bmatrix}$

18. Each cofactor in A is an integer because it is just a sum of products of entries of A. Hence all the entries in adj A are integers. Since det $A = 1$, the inverse formula in Theorem 8 shows that all the entries in A^{-1} are integers.

20. 7　　　**22.** 21　　　**24.** 15

26. By definition, $\mathbf{p} + S$ is the set of all vectors of the form $\mathbf{p} + \mathbf{v}$, where $\mathbf{v}$ is in S. Applying T to a typical vector in $\mathbf{p} + S$, we have $T(\mathbf{p} + \mathbf{v}) = T(\mathbf{p}) + T(\mathbf{v})$. This vector is in the set denoted by $T(\mathbf{p}) + T(S)$. This proves that T maps the set $\mathbf{p} + S$ into the set $T(\mathbf{p}) + T(S)$.

　Conversely, any vector in $T(\mathbf{p}) + T(S)$ has the form $T(\mathbf{p}) + T(\mathbf{v})$ for some $\mathbf{v}$ in S. This vector may be written as $T(\mathbf{p} + \mathbf{v})$. This shows that every vector in $T(\mathbf{p}) + T(S)$ is the image under T of some point in $\mathbf{p} + S$.

28. Use Theorem 10. Or, compute the vectors that determine the image, namely, the columns of

$$A[\,\mathbf{b}_1 \quad \mathbf{b}_2\,] = \begin{bmatrix} 14 & 2 \\ -3 & 1 \end{bmatrix}$$

The determinant of this matrix is 20.

30. Let $\mathbf{p} = (x_3, y_3)$ and let $R' = R - \mathbf{p}$. The vertices of R' are $\mathbf{v} = (x_1 - x_3, y_1 - y_3)$, $\mathbf{v}_2 = (x_2 - x_3, y_2 - y_3)$, and the origin. Then

$$\{\text{area of } R\} = \{\text{area of } R'\}$$

$$= \frac{1}{2} \left\{ \begin{array}{l} \text{area of parallelogram} \\ \text{determined by } \mathbf{v}_1 \text{ and } \mathbf{v}_2 \end{array} \right\}$$

$$= \frac{1}{2} \left| \det \begin{bmatrix} x_1 - x_3 & x_2 - x_3 \\ y_1 - y_3 & y_2 - y_3 \end{bmatrix} \right| \quad (1)$$

Also, using row operations, we get

$$\det \begin{bmatrix} x_1 & y_1 & 1 \\ x_2 & y_2 & 1 \\ x_3 & y_3 & 1 \end{bmatrix} = \det \begin{bmatrix} x_1 - x_3 & y_1 - y_3 & 0 \\ x_2 - x_3 & y_2 - y_3 & 0 \\ x_3 & y_3 & 1 \end{bmatrix}$$

$$= \det \begin{bmatrix} x_1 - x_3 & y_1 - y_3 \\ x_2 - x_3 & y_2 - y_3 \end{bmatrix}$$

$$= \det \begin{bmatrix} x_1 - x_3 & x_2 - x_3 \\ y_1 - y_3 & y_2 - y_3 \end{bmatrix}$$

This calculation and equation (1) give the desired result.

32. From the formula in the exercise,

$$\{\text{volume of } S\} = \tfrac{1}{3}\{\text{area of base}\} \cdot \{\text{height}\} = \tfrac{1}{6}$$

because the vectors $\mathbf{e}_1$, $\mathbf{e}_2$, and $\mathbf{e}_3$ have unit length. The tetrahedron S' with vertices at $\mathbf{0}$, $\mathbf{v}_1$, $\mathbf{v}_2$, and $\mathbf{v}_3$ is the image of S under the linear transformation T such that $T(\mathbf{e}_1) = \mathbf{v}_1$, $T(\mathbf{e}_2) = \mathbf{v}_2$, and $T(\mathbf{e}_3) = \mathbf{v}_3$. The standard matrix for T is $A = [\,\mathbf{v}_1 \quad \mathbf{v}_2 \quad \mathbf{v}_3\,]$. By Theorem 10,

$$\{\text{volume of } S'\} = |\det A| \cdot \tfrac{1}{6} = \tfrac{1}{6}|\det[\,\mathbf{v}_1 \quad \mathbf{v}_2 \quad \mathbf{v}_3\,]|$$

34. [M] MATLAB:

```
x2 = det([A(:,1)  b  A(:,3:4)])/det(A)
```

Chapter 3 Supplementary Exercises, page 185

1. a. True. The columns of A are linearly dependent.

b. True. See Exercise 30 in Section 3.2.

c. False. See Theorem 3(c); in this case det $5A = 5^3$ det A.

d. False. Consider $A = \begin{bmatrix} 2 & 0 \\ 0 & 1 \end{bmatrix}$, $B = \begin{bmatrix} 1 & 0 \\ 0 & 3 \end{bmatrix}$, and $A + B = \begin{bmatrix} 3 & 0 \\ 0 & 4 \end{bmatrix}$.

e. False. By Theorem 6, det $A^3 = 2^3$.

f. False. See Theorem 3(b).

g. True. See Theorem 3(c).

h. True. See Theorem 3(a).

i. False. See Theorem 5.

j. False. See Theorem 3(c); this statement is false for $n \times n$ invertible matrices with n an even integer.

k. True. See Theorems 6 and 5; det $A^T A = (\det A)^2$.

l. False. The coefficient matrix must be invertible.

m. False. The area of the triangle is 5.

n. True. See Theorem 6; det $A^3 = (\det A)^3$.

o. False. See Exercise 31 in Section 3.2.

p. True. See Theorem 6.

The solutions for Exercises 2 and 4 are based on the fact that if a matrix contains two rows (or two columns) that are multiples of each other, then the determinant of the matrix is zero, by Theorem 4, because the matrix cannot be invertible.

2. $\begin{vmatrix} 12 & 13 & 14 \\ 15 & 16 & 17 \\ 18 & 19 & 20 \end{vmatrix} = \begin{vmatrix} 12 & 13 & 14 \\ 3 & 3 & 3 \\ 6 & 6 & 6 \end{vmatrix} = 0$

4. $\begin{vmatrix} a & b & c \\ a+x & b+x & c+x \\ a+y & b+y & c+y \end{vmatrix} = \begin{vmatrix} a & b & c \\ x & x & x \\ y & y & y \end{vmatrix}$

$= xy \begin{vmatrix} a & b & c \\ 1 & 1 & 1 \\ 1 & 1 & 1 \end{vmatrix} = 0$

6. 12

8. $\det \begin{bmatrix} 1 & x & y \\ 1 & x_1 & y_1 \\ 0 & 1 & m \end{bmatrix} = 0$. When the determinant is expanded by cofactors of the first row, the equation has the form $1 \cdot (mx_1 - y_1) - x(m) + y \cdot 1 = 0$, which can be written as $y - y_1 = m(x - x_1)$.

10. An expansion of the determinant along the top row of V will show that $f(t)$ has the form $f(t) = c_0 + c_1 t + c_2 t^2 + c_3 t^3$ where, by Exercise 9,

$$c_3 = \det \begin{bmatrix} 1 & x_1 & x_1^2 \\ 1 & x_2 & x_2^2 \\ 1 & x_3 & x_3^2 \end{bmatrix}$$

$$= (x_2 - x_1)(x_3 - x_1)(x_3 - x_2) \neq 0$$

So $f(t)$ is a cubic polynomial in t. The points $(x_1, 0)$, $(x_2, 0)$, and $(x_3, 0)$ are on the graph of f, because when x_1, x_2, or x_3 are substituted for t in V, the matrix V will have two rows the same and hence have a zero determinant. That is, $f(x_1) = 0$ for $i = 1, 2, 3$.

12. A 2×2 matrix A is invertible if and only if the parallelogram determined by the columns of A has nonzero area.

14. a. An expansion by cofactors along the last row shows that for $1 \leq k \leq n$,

$$\det \begin{bmatrix} A & O \\ O & I_k \end{bmatrix}$$

$$= 0 + \cdots + 0 + (-1)^{(n+k)+(n+k)} \cdot 1 \cdot \det \begin{bmatrix} A & O \\ O & I_{k-1} \end{bmatrix}$$

When $k = 1$, we interpret I_0 as having *no* rows or columns. Chaining these equalities together gives

$$\det \begin{bmatrix} A & O \\ O & I_k \end{bmatrix} = \cdots = \det \begin{bmatrix} A & O \\ O & I_2 \end{bmatrix}$$

$$= \det \begin{bmatrix} A & O \\ O & 1 \end{bmatrix} = \det A$$

b. An expansion by cofactors along the first row shows that for $1 \leq k \leq n$,

$$\det \begin{bmatrix} I_k & O \\ C_k & D \end{bmatrix} = 1 \cdot \det \begin{bmatrix} I_{k-1} & O \\ C_{k-1} & D \end{bmatrix}$$

where $C_n = C$ and C_{k-1} is formed by deleting the first column in C_k. Chaining these equalities together as in part (a) produces the desired equation.

c. Observe that

$$\begin{bmatrix} A & O \\ C & D \end{bmatrix} = \begin{bmatrix} A & O \\ O & I \end{bmatrix} \begin{bmatrix} I & O \\ C & D \end{bmatrix}$$

From the multiplicative property of determinants and parts (a) and (b),

$$\det \begin{bmatrix} A & O \\ C & D \end{bmatrix} = \det \begin{bmatrix} A & O \\ O & I \end{bmatrix} \cdot \det \begin{bmatrix} I & O \\ C & D \end{bmatrix}$$

$$= (\det A)(\det D)$$

We have proved that *the determinant of a block lower triangular matrix is the product of the determinants of its diagonal entries* (assuming square diagonal entries). The second part of (c) follows from the first part and the fact that the determinant of a matrix equals the determinant of its transpose:

$$\det \begin{bmatrix} A & B \\ O & D \end{bmatrix} = \det \begin{bmatrix} A & B \\ O & D \end{bmatrix}^T$$

$$= \det \begin{bmatrix} A^T & O \\ B^T & D^T \end{bmatrix}$$

$$= (\det A^T)(\det D^T)$$

$$= (\det A)(\det D)$$

16. a. Row replacement operations do not change the determinant of A. The resulting matrix is

$$\begin{bmatrix} a-b & -a+b & 0 & \cdots & 0 \\ 0 & a-b & -a+b & & 0 \\ 0 & 0 & a-b & & 0 \\ \vdots & & & \ddots & \vdots \\ b & b & b & \cdots & a \end{bmatrix}$$

b. Since column replacement operations are equivalent to row operations on A^T and $\det A^T = \det A$, column replacement operations do not change the determinant of the matrix. The resulting matrix is

$$\begin{bmatrix} a-b & 0 & 0 & \cdots & 0 \\ 0 & a-b & 0 & & 0 \\ 0 & 0 & a-b & & 0 \\ \vdots & & & \ddots & \vdots \\ b & 2b & 3b & \cdots & a+(n-1)b \end{bmatrix}$$

c. Since the preceding matrix is a lower triangular matrix with the same determinant as A,

$$\det A = (a-b)^{n-1}(a+(n-1)b)$$

18. [M]

a. $(3-8)^3[3+(3)8] = -3375$

b. $(8-3)^4[8+(4)3] = 12{,}500$

20. [M] Compute:

$$\begin{vmatrix} 1 & 1 & 1 \\ 1 & 3 & 3 \\ 1 & 3 & 6 \end{vmatrix} = 6, \qquad \begin{vmatrix} 1 & 1 & 1 & 1 \\ 1 & 3 & 3 & 3 \\ 1 & 3 & 6 & 6 \\ 1 & 3 & 6 & 9 \end{vmatrix} = 18,$$

$$\begin{vmatrix} 1 & 1 & 1 & 1 & 1 \\ 1 & 3 & 3 & 3 & 3 \\ 1 & 3 & 6 & 6 & 6 \\ 1 & 3 & 6 & 9 & 9 \\ 1 & 3 & 6 & 9 & 12 \end{vmatrix} = 54 = 18 \cdot 3$$

Conjecture:

$$\begin{vmatrix} 1 & 1 & 1 & \cdots & 1 \\ 1 & 3 & 3 & & 3 \\ 1 & 3 & 6 & & 6 \\ \vdots & & & \ddots & \vdots \\ 1 & 3 & 6 & \cdots & 3(n-1) \end{vmatrix} = 2 \cdot 3^{n-2}$$

To confirm the conjecture, use row replacement operations to create zeros below the first pivot and then below the second pivot. The resulting matrix is

$$\begin{vmatrix} 1 & 1 & 1 & 1 & 1 & 1 & \cdots & 1 \\ 0 & 2 & 2 & 2 & 2 & 2 & & 2 \\ 0 & 0 & 3 & 3 & 3 & 3 & & 3 \\ 0 & 0 & 3 & 6 & 6 & 6 & & 6 \\ 0 & 0 & 3 & 6 & 9 & 9 & & 9 \\ \vdots & & & & & & \ddots & \vdots \\ 0 & 0 & 3 & 6 & 9 & 12 & \cdots & 3(n-2) \end{vmatrix}$$

This matrix has the same determinant as the original matrix, and is recognizable as a block matrix of the form

$$\begin{bmatrix} A & B \\ O & D \end{bmatrix}$$

where

$$A = \begin{bmatrix} 1 & 1 \\ 0 & 2 \end{bmatrix},$$

$$D = \begin{bmatrix} 3 & 3 & 3 & 3 & \cdots & 3 \\ 3 & 6 & 6 & 6 & & 6 \\ 3 & 6 & 9 & 9 & & 9 \\ \vdots & & & & \ddots & \vdots \\ 3 & 6 & 9 & 12 & \cdots & 3(n-2) \end{bmatrix}$$

$$= 3 \begin{bmatrix} 1 & 1 & 1 & 1 & \cdots & 1 \\ 1 & 2 & 2 & 2 & & 2 \\ 1 & 2 & 3 & 3 & & 3 \\ \vdots & & & & \ddots & \vdots \\ 1 & 2 & 3 & 4 & \cdots & n-2 \end{bmatrix}$$

Use Exercise 14(c) to find that the determinant of the matrix $\begin{bmatrix} A & B \\ O & D \end{bmatrix}$ is $(\det A)(\det D) = 2 \det D$, and then use Exercise 32 in Section 3.2 and Exercise 19 above to show that $\det D = 3^{n-2}$.

Chapter 4

Section 4.1, page 195

2. a. Given $\begin{bmatrix} x \\ y \end{bmatrix}$ in W and any scalar c, the vector

$c \begin{bmatrix} x \\ y \end{bmatrix} = \begin{bmatrix} cx \\ cy \end{bmatrix}$ is in W because
$(cx)(cy) = c^2(xy) \geq 0$, since $xy \geq 0$.

b. *Example:* If $\mathbf{u} = \begin{bmatrix} -1 \\ -7 \end{bmatrix}$ and $\mathbf{v} = \begin{bmatrix} 2 \\ 3 \end{bmatrix}$, then $\mathbf{u}$ and $\mathbf{v}$ are in W, but $\mathbf{u} + \mathbf{v}$ is not in W.

4.

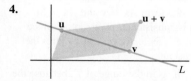

$\mathbf{u}$ and $\mathbf{v}$ are on the line, but $\mathbf{u} + \mathbf{v}$ is not.

6. No, the zero polynomial is not in the set.

8. Yes. The zero vector is in the set, H. If $\mathbf{p}$ and $\mathbf{q}$ are in H, then $(\mathbf{p} + \mathbf{q})(0) = \mathbf{p}(0) + \mathbf{q}(0) = 0$, so $\mathbf{p} + \mathbf{q}$ is in H. Also, for any scalar c, $(c\mathbf{p})(0) = c \cdot \mathbf{p}(0) = c \cdot 0 = 0$, so $c\mathbf{p}$ is in H.

10. $H = \text{Span}\{\mathbf{v}\}$, where $\mathbf{v} = \begin{bmatrix} 3 \\ 0 \\ -7 \end{bmatrix}$. By Theorem 1, H is a subspace of $\mathbb{R}^3$.

12. $W = \text{Span}\{\mathbf{u}, \mathbf{v}\}$, where $\mathbf{u} = \begin{bmatrix} 2 \\ 2 \\ 2 \\ 0 \end{bmatrix}$, $\mathbf{v} = \begin{bmatrix} 4 \\ 0 \\ -3 \\ 5 \end{bmatrix}$. By Theorem 1, W is a subspace of $\mathbb{R}^4$.

14. Yes, because the equation $c_1\mathbf{v}_1 + c_2\mathbf{v}_2 + c_3\mathbf{v}_3 = \mathbf{w}$ has a solution, as revealed by an echelon form of the augmented matrix for this equation.

16. W is not a vector space because the zero vector is not in W

18. $S = \left\{ \begin{bmatrix} 4 \\ 0 \\ 1 \\ 0 \end{bmatrix}, \begin{bmatrix} 3 \\ 0 \\ 3 \\ 3 \end{bmatrix}, \begin{bmatrix} 0 \\ 0 \\ 1 \\ -2 \end{bmatrix} \right\}$

20. a. The constant function $\mathbf{f}(t) = 0$ is continuous. The sum of two continuous functions is continuous. A constant multiple of a continuous function is continuous.

b. Let $H = \{\mathbf{f} \text{ in } C[a, b] : \mathbf{f}(a) = \mathbf{f}(b)\}$. Take $\mathbf{f}$ and $\mathbf{g}$ in H and let c be a real number. Then, the function $\mathbf{f} + \mathbf{g}$ is in $C(a, b)$, because the sum of two continuous functions is continuous. Also,

$$(\mathbf{f} + \mathbf{g})(a) = \mathbf{f}(a) + \mathbf{g}(a) = \mathbf{f}(b) + \mathbf{g}(b) = (\mathbf{f} + \mathbf{g})(b),$$

which shows that $\mathbf{f} + \mathbf{g}$ is in H. Next, using the definition of $c\mathbf{f}$,

$$(c\mathbf{f})(a) = c(\mathbf{f}(a)) = c(\mathbf{f}(b)),$$

because $\mathbf{f}$ is in H. Also, $c(\mathbf{f}(b)) = c\mathbf{f}(b) = (c\mathbf{f})(b)$. This shows that the scalar multiple, $c\mathbf{f}$, is in H. Thus H is closed under sums and scalar multiples, so H is a subspace.

22. Yes. See the proof of Theorem 12 in Section 2.8 for a proof that is similar to the one needed here.

23. a. False. The zero vector in V is the function $\mathbf{f}$ whose values $\mathbf{f}(t)$ are zero *for all t* in $\mathbb{R}$. See Example 5.

b. False. See the definition of a vector. An arrow in three-dimensional space is an example of a vector, but not every vector is such an arrow.

c. False. Exercises 1, 2, and 3 each provide an example of a subset that contains the zero vector but is not a subspace.

d. True. See the paragraph before Example 6.

e. False. Digital signals are used. See Example 3.

24. a. True. See the definition of a vector space.

b. True. See statement (3) in the box before Example 1.

c. True. See the paragraph before Example 6.

d. False. See Example 8.

e. False. The second and third parts of the conditions are stated incorrectly. In part (ii) here, for example, there is no statement that $\mathbf{u}$ and $\mathbf{v}$ represent all possible elements of H.

26. a. 3 **b.** 5 **c.** 4

28. a. 4 **b.** 7 **c.** 3 **d.** 5 **e.** 4

30. $\mathbf{u} = 1 \cdot \mathbf{u}$ Axiom 10
$= c^{-1}c \cdot \mathbf{u} = c^{-1}(c\mathbf{u})$ Axiom 9
$= c^{-1}\mathbf{0} = \mathbf{0}$ Property (2)

32. Both H and K contain the zero vector of V because they are subspaces of V. Hence $\mathbf{0}$ is in $H \cap K$. Take $\mathbf{u}$ and $\mathbf{v}$ in $H \cap K$. Then $\mathbf{u}$ and $\mathbf{v}$ are in both H and K. Since H is a subspace, $\mathbf{u} + \mathbf{v}$ is in H. Likewise, $\mathbf{u} + \mathbf{v}$ is in K. Hence $\mathbf{u} + \mathbf{v}$ is in $H \cap K$. For any scalar c, the vector $c\mathbf{u}$ is in both H and K because they are subspaces. Hence $c\mathbf{u}$ is in $H \cap K$. Thus $H \cap K$ is a subspace.

    The union of two subspaces is not, in general, a subspace. In $\mathbb{R}^2$, let H be the x-axis and K the y-axis. The sum of a nonzero vector in H and a nonzero vector in K is not on either the x-axis or the y-axis. So $H \cup K$ is not closed under vector addition, and $H \cup K$ is not a subspace of $\mathbb{R}^2$.

34. A proof that $H + K = \text{Span}\{\mathbf{u}_1, \ldots, \mathbf{u}_p, \mathbf{v}_1, \ldots, \mathbf{v}_q\}$ has two parts. First, one must show that $H + K$ is a subset of $\text{Span}\{\mathbf{u}_1, \ldots, \mathbf{u}_p, \mathbf{v}_1, \ldots, \mathbf{v}_q\}$. Second, one must show that $\text{Span}\{\mathbf{u}_1, \ldots, \mathbf{u}_p, \mathbf{v}_1, \ldots, \mathbf{v}_q\}$ is a subset of $H + K$.

  a. A typical vector H has the form $c_1\mathbf{u}_1 + \cdots + c_p\mathbf{u}_p$ and a typical vector in K has the form $d_1\mathbf{v}_1 + \cdots + d_q\mathbf{v}_q$. The sum of these two vectors is a linear combination of $\mathbf{u}_1, \ldots, \mathbf{u}_p, \mathbf{v}_1, \ldots, \mathbf{v}_q$ and so belongs to $\text{Span}\{\mathbf{u}_1, \ldots, \mathbf{u}_p, \mathbf{v}_1, \ldots, \mathbf{v}_q\}$. Thus $H + K$ is a subset of $\text{Span}\{\mathbf{u}_1, \ldots, \mathbf{u}_p, \mathbf{v}_1, \ldots, \mathbf{v}_q\}$.

  b. Each of the vectors $\mathbf{u}_1, \ldots, \mathbf{u}_p, \mathbf{v}_1, \ldots, \mathbf{v}_q$ belongs to $H + K$, by Exercise 33(b), and so any linear combination of these vectors belongs to $H + K$, since $H + K$ is a subspace, by Exercise 33(a). Thus, $\text{Span}\{\mathbf{u}_1, \ldots, \mathbf{u}_p, \mathbf{v}_1, \ldots, \mathbf{v}_q\}$ is a subset of $H + K$.

36. [M] Write A as $[\, \mathbf{a}_1 \;\; \mathbf{a}_2 \;\; \mathbf{a}_3 \,]$. The reduced echelon form of $[\, \mathbf{a}_1 \;\; \mathbf{a}_2 \;\; \mathbf{a}_3 \;\; \mathbf{y} \,]$ shows that $\mathbf{y} = -0.2\mathbf{a}_1 - 0.4\mathbf{a}_2 + 0.6\mathbf{a}_3$. Hence $\mathbf{y}$ is in the subspace spanned by $\mathbf{a}_1, \mathbf{a}_2$, and $\mathbf{a}_3$.

38. [M] The functions are $\sin 3t$, $\cos 4t$, and $\sin 5t$.

Section 4.2, page 205

2. $\begin{bmatrix} 2 & 6 & 4 \\ -3 & 2 & 5 \\ -5 & -4 & 1 \end{bmatrix} \begin{bmatrix} 1 \\ -1 \\ 1 \end{bmatrix} = \begin{bmatrix} 0 \\ 0 \\ 0 \end{bmatrix}$, so $\mathbf{w}$ is in Nul A.

4. $\begin{bmatrix} 3 \\ 1 \\ 0 \\ 0 \end{bmatrix}, \begin{bmatrix} 0 \\ 0 \\ 0 \\ 1 \end{bmatrix}$

6. $\begin{bmatrix} -5 \\ 3 \\ 1 \\ 0 \\ 0 \end{bmatrix}, \begin{bmatrix} 6 \\ -1 \\ 0 \\ 1 \\ 0 \end{bmatrix}, \begin{bmatrix} -1 \\ 0 \\ 0 \\ 0 \\ 1 \end{bmatrix}$

8. W is not a subspace because $\mathbf{0}$ is not in W. The vector $(0, 0, 0)$ does not satisfy the condition $3r - 2 = 3s + t$.

10. W is a subspace of $\mathbb{R}^4$ by Theorem 2, because W is the set of solutions of the homogeneous system

$$3a + b - c = 0$$
$$a + b + 2c - 2d = 0$$

12. If $(3p - 5q, 4q, p, q + 1)$ were the zero vector, then $4q = 0$ and $q + 1 = 0$, which is impossible. So $\mathbf{0}$ is not in W, and W is not a subspace.

14. $W = \text{Col } A$ for $A = \begin{bmatrix} -1 & 3 \\ 1 & -2 \\ 5 & -1 \end{bmatrix}$, so W is a vector space by Theorem 3.

16. $\begin{bmatrix} 1 & -1 & 0 \\ 2 & 0 & 3 \\ 1 & 3 & -3 \\ 0 & 1 & 1 \end{bmatrix}$

18. a. 3 **b.** 4   **20. a.** 5 **b.** 1

22. $\begin{bmatrix} 1 \\ 1 \\ -1 \end{bmatrix}$ is in Nul A, $\begin{bmatrix} 5 \\ -1 \\ 0 \\ -5 \end{bmatrix}$ is in Col A. Other answers are possible.

24. The vector $\mathbf{w}$ is in both Nul A and Col A, since $A\mathbf{w} = \mathbf{0}$ and $\mathbf{w} = \mathbf{a}_1 + \mathbf{a}_2$.

25. a. True, by the definition before Example 1.
  b. False. See Theorem 2.
  c. True. See the remark just before Example 4.
  d. False. The equation $A\mathbf{x} = \mathbf{b}$ must be consistent *for every* $\mathbf{b}$. See #7 in the table on p. 232.
  e. True. See Fig. 2. (A subspace is itself a vector space.)
  f. True. See the remark after Theorem 3.

26. a. True. See Theorem 2. (A subspace is itself a vector space.)
  b. True. See Theorem 3.
  c. False. See the box after Theorem 3.
  d. True. See the paragraph after the definition of linear transformation.
  e. True. See Fig. 2. (A subspace is itself a vector space.)
  f. True. See the paragraph before Example 8.

28. The two systems have the form $A\mathbf{x} = \mathbf{v}$ and $A\mathbf{x} = 5\mathbf{v}$. Since the first system is consistent, $\mathbf{v}$ is in Col A. Since Col A is a subspace of $\mathbb{R}^3$, $5\mathbf{v}$ is also in Col A. Thus the second system is consistent.

30. The zero vector $\mathbf{0}_W$ of W is in the range of T, because the linear transformation maps the zero vector of V to $\mathbf{0}_W$. Typical vectors in the range of T are $T(\mathbf{x})$ and $T(\mathbf{w})$, where $\mathbf{x}, \mathbf{w}$ are in V. Since T is a linear transformation,

$$T(\mathbf{x}) + T(\mathbf{w}) = T(\mathbf{x} + \mathbf{w}) \quad \text{In the range of } T$$

Thus the range of T is closed under vector addition. Also, for any scalar c, $c \cdot T(\mathbf{x}) = T(c\mathbf{x})$, since T is a linear transformation. Thus $c \cdot T(\mathbf{x})$ is in the range of T, so the range is closed under scalar multiplication. Hence the range of T is a subspace of W.

32. $\mathbf{p}_1(t) = t$, $\mathbf{p}_2(t) = t^2$. The range of T is $\left\{ \begin{bmatrix} a \\ a \end{bmatrix} : a \text{ real} \right\}$.

34. The kernel of T is $\{\mathbf{0}\}$.

36. Since Z is a subspace of W, the zero vector $\mathbf{0}_W$ of W is in Z. Because T is linear, T maps the zero vector $\mathbf{0}_V$ of V to $\mathbf{0}_W$. Thus $\mathbf{0}_V$ is in $U = \{\mathbf{x} : T(\mathbf{x}) \text{ is in } Z\}$. Now take $\mathbf{u}_1$, $\mathbf{u}_2$ in U. Since T is linear,

$$T(\mathbf{u}_1 + \mathbf{u}_2) = T(\mathbf{u}_1) + T(\mathbf{u}_2) \tag{$*$}$$

By definition of U, $T(\mathbf{u}_1)$ and $T(\mathbf{u}_2)$ are in Z, and so the sum on the right side of equation $(*)$ is in Z because Z is a subspace. This proves that $\mathbf{u}_1 + \mathbf{u}_2$ is in U, so U is closed under vector addition. For any scalar c, $c \cdot T(\mathbf{u}_1)$ is in Z because Z is a subspace. Since T is linear, $T(c\mathbf{u}_1)$ is in Z. Hence $c\mathbf{u}_1$ is in U. Thus U is a subspace of V.

37. [M] $\mathbf{w}$ is in Col A. In fact, $\mathbf{w} = A\mathbf{x}$ for

$$\mathbf{x} = (1/95, -20/19, -172/95, 0)$$

$\mathbf{w}$ is not in Nul A because $A\mathbf{w} = (14, 0, 0, 0)$.

38. [M] $\mathbf{w}$ is in Col A and in Nul A because $\mathbf{w} = A\mathbf{x}$ for $\mathbf{x} = (-2, 3, 0, 1)$, and $A\mathbf{w} = (0, 0, 0, 0)$.

39. [M] The reduced echelon form of A is

$$\begin{bmatrix} 1 & 0 & 1/3 & 0 & 10/3 \\ 0 & 1 & 1/3 & 0 & -26/3 \\ 0 & 0 & 0 & 1 & -4 \\ 0 & 0 & 0 & 0 & 0 \end{bmatrix}$$

a. Most students will row reduce $[\, B \quad \mathbf{a}_3 \,]$ and $[\, B \quad \mathbf{a}_5 \,]$ to show that the equations $B\mathbf{x} = \mathbf{a}_3$ and $B\mathbf{x} = \mathbf{a}_5$ are consistent. You can use a discussion of this part to lead into Examples 8 and 9 in Section 4.3.

b. The method in Example 3 produces $(-1/3, -1/3, 1, 0, 0)$ and $(-10/3, 26/3, 0, 4, 1)$.

c. This part reviews Section 1.9. An echelon form of A shows that the columns of A are linearly dependent and do not span $\mathbb{R}^4$. By Theorem 12 in Section 1.9, T is not one-to-one and T does not map $\mathbf{R}^5$ onto $\mathbf{R}^4$.

40. [M] Row reduction of $[\, \mathbf{v}_1 \quad \mathbf{v}_2 \quad -\mathbf{v}_3 \quad -\mathbf{v}_4 \quad \mathbf{0} \,]$ yields

$$\begin{bmatrix} 1 & 0 & 0 & -10/3 & 0 \\ 0 & 1 & 0 & 26/3 & 0 \\ 0 & 0 & 1 & -4 & 0 \end{bmatrix}$$

The general solution is a multiple of $(10, -26, 12, 3)$. One choice for $\mathbf{w}$ is $10\mathbf{v}_1 - 26\mathbf{v}_2 \ (= 12\mathbf{v}_3 + 3\mathbf{v}_4)$, which is $(24, -48, -24)$. Another choice is $\mathbf{w} = (1, -2, -1)$.

Section 4.3, page 213

2. This set does not form a basis for $\mathbb{R}^3$. The set is linearly dependent because the zero vector is in the set. The columns of $\begin{bmatrix} 1 & 0 & 0 \\ 1 & 0 & 1 \\ 0 & 0 & 1 \end{bmatrix}$ do not span $\mathbb{R}^3$, by the Invertible Matrix Theorem.

4. These vectors form a basis for $\mathbb{R}^3$. See Example 5 for an example of a justification.

6. This set does not form a basis for $\mathbb{R}^3$. Notice $\begin{bmatrix} 1 & -4 \\ 2 & 3 \\ -4 & 6 \end{bmatrix} \sim \begin{bmatrix} 1 & -4 \\ 0 & 1 \\ 0 & 0 \end{bmatrix}$. Thus the matrix does not have a pivot in each row, so its columns do not span $\mathbb{R}^3$ and hence do not form a basis for $\mathbb{R}^3$. Since there is a pivot in every column of the echelon matrix, the columns form a linearly independent set.

8. This set does not form a basis for $\mathbb{R}^3$. The set is linearly dependent because there are more vectors than entries in each vector. However, the vectors do span $\mathbb{R}^3$.

10. $\begin{bmatrix} -2 \\ 1 \\ 0 \\ 1 \\ 0 \end{bmatrix}, \begin{bmatrix} -3 \\ 2 \\ 2 \\ 0 \\ 1 \end{bmatrix}$ **12.** $\begin{bmatrix} 1 \\ -3 \end{bmatrix}$

14. Basis for Nul A: $\begin{bmatrix} -2 \\ 1 \\ 0 \\ 0 \\ 0 \end{bmatrix}, \begin{bmatrix} -2 \\ 0 \\ 2 \\ 1 \\ 0 \end{bmatrix}$

Basis for Col A: $\begin{bmatrix} 1 \\ 1 \\ 2 \\ 3 \end{bmatrix}, \begin{bmatrix} 3 \\ 0 \\ -3 \\ 0 \end{bmatrix}, \begin{bmatrix} 8 \\ 8 \\ 9 \\ 9 \end{bmatrix}$

16. $\{\mathbf{v}_1, \mathbf{v}_2, \mathbf{v}_3\}$ **18.** [M] $\{\mathbf{v}_1, \mathbf{v}_2, \mathbf{v}_3, \mathbf{v}_5\}$

20. The three simplest answers are $\{\mathbf{v}_1, \mathbf{v}_2\}$, $\{\mathbf{v}_1, \mathbf{v}_3\}$, and $\{\mathbf{v}_2, \mathbf{v}_3\}$. Other answers are possible.

21. a. False. The zero vector by itself is linearly dependent. See the paragraph preceding Theorem 4.

b. False. The set $\{\mathbf{b}_1, \ldots, \mathbf{b}_p\}$ must also be linearly independent. See the definition of a basis.

c. True. See Example 3.

d. False. See the subsection "Two Views of a Basis."

e. False. See the shaded area before Example 9.

22. a. False. The subspace spanned by the set must also coincide with H. See the definition of basis.

b. True, by the Spanning Set Theorem, applied to V instead of H. (V is nonzero because the spanning set uses nonzero vectors.)

c. True. See the subsection "Two Views of a Basis."

d. False. See the paragraph before Example 8.

e. False. See the warning after Theorem 6.

24. Let $A = [\, \mathbf{v}_1 \quad \cdots \quad \mathbf{v}_n \,]$. Since A is square and its columns are linearly independent, its columns also span $\mathbb{R}^n$, by the Invertible Matrix Theorem. So $\{\mathbf{v}_1, \ldots, \mathbf{v}_n\}$ is a basis for $\mathbb{R}^n$.

26. A basis is $\{\sin t, \sin 2t\}$ because this set is linearly independent (by inspection), and $\sin t \cos t = \frac{1}{2} \sin 2t$, as pointed out in Example 2.

28. $\{e^{-bt}, te^{-bt}\}$. The set is linearly independent because neither function is a *scalar* multiple of the other, and the set spans H.

30. There are more vectors than there are entries in each vector. By Theorem 8 in Section 1.7, the set is linearly dependent and therefore cannot be a basis for $\mathbb{R}^n$.

32. Suppose that $\{T(\mathbf{v}_1), \ldots, T(\mathbf{v}_p)\}$ is linearly dependent. Then there exist $c_1, \ldots, c_p$, not all zero, such that

$$c_1 T(\mathbf{v}_1) + \cdots + c_p T(\mathbf{v}_p) = \mathbf{0}$$

Since T is linear and $\mathbf{0} = T(\mathbf{0})$,

$$T(c_1 \mathbf{v}_1 + \cdots + c_p \mathbf{v}_p) = T(\mathbf{0})$$

By hypothesis, T is one-to-one, so this equation implies that $c_1 \mathbf{v}_1 + \cdots + c_p \mathbf{v}_p = \mathbf{0}$, which shows that $\{\mathbf{v}_1, \ldots, \mathbf{v}_p\}$ is linearly dependent.

34. By inspection, $\mathbf{p}_3 = \mathbf{p}_1 + \mathbf{p}_2$, or $\mathbf{p}_1 + \mathbf{p}_2 - \mathbf{p}_3 = \mathbf{0}$. By the Spanning Set Theorem, Span $\{\mathbf{p}_1, \mathbf{p}_2, \mathbf{p}_3\}$ = Span $\{\mathbf{p}_1, \mathbf{p}_2\}$. Since neither $\mathbf{p}_1$ nor $\mathbf{p}_2$ is a multiple of the other, they are linearly independent and hence $\{\mathbf{p}_1, \mathbf{p}_2\}$ is a basis for Span $\{\mathbf{p}_1, \mathbf{p}_2, \mathbf{p}_3\}$.

36. **[M]** Row reducing $[\, \mathbf{u}_1 \quad \mathbf{u}_2 \quad \mathbf{u}_3 \,]$ shows that $\mathbf{u}_1$ and $\mathbf{u}_2$ are the pivot columns of this matrix. Thus $\{\mathbf{u}_1, \mathbf{u}_2\}$ is a basis for H.

Row reducing $[\, \mathbf{v}_1 \quad \mathbf{v}_2 \quad \mathbf{v}_3 \,]$ shows that $\mathbf{v}_1, \mathbf{v}_2$, and $\mathbf{v}_3$ are the pivot columns of this matrix. Thus $\{\mathbf{v}_1, \mathbf{v}_2, \mathbf{v}_3\}$ is a basis for K.

Row reducing $[\, \mathbf{u}_1 \quad \mathbf{u}_2 \quad \mathbf{u}_3 \quad \mathbf{v}_1 \quad \mathbf{v}_2 \quad \mathbf{v}_3 \,]$ shows that $\mathbf{u}_1, \mathbf{u}_2, \mathbf{v}_2$, and $\mathbf{v}_3$ are the pivot columns of this matrix. Thus $\{\mathbf{u}_1, \mathbf{u}_2, \mathbf{v}_2, \mathbf{v}_3\}$ is a basis for $H + K$.

38. **[M]** For example, writing

$$c_1 \cdot 1 + c_2 \cdot \cos t + c_3 \cdot \cos^2 t + c_4 \cdot \cos^3 t +$$
$$c_5 \cdot \cos^4 t + c_6 \cdot \cos^5 t + c_7 \cdot \cos^6 t = 0$$

with $t = 0, .1, .2, .3, .4, .5, .6$ gives a 7×7 coefficient matrix A for the homogeneous system $A\mathbf{c} = \mathbf{0}$. The matrix A is invertible, so the system $A\mathbf{c} = \mathbf{0}$ has only the trivial solution and $\{1, \cos t, \cos^2 t, \cos^3 t, \cos^4 t, \cos^5 t, \cos^6 t\}$ is a linearly independent set of functions.

Section 4.4, page 222

2. $\begin{bmatrix} -26 \\ 1 \end{bmatrix}$ **4.** $\begin{bmatrix} 8 \\ -5 \\ 1 \end{bmatrix}$ **6.** $\begin{bmatrix} 3 \\ -2 \end{bmatrix}$ **8.** $\begin{bmatrix} 1 \\ -1 \\ 1 \end{bmatrix}$

10. $\begin{bmatrix} 3 & 2 & 1 \\ 0 & 2 & -2 \\ 6 & -4 & 3 \end{bmatrix}$ **12.** $\begin{bmatrix} -8 \\ 5 \end{bmatrix}$ **14.** $\begin{bmatrix} 3 \\ 2 \\ -1 \end{bmatrix}$

15. a. True, by the definition of the $\mathcal{B}$-coordinate vector.
  b. False. See equation (4).
  c. False. $\mathbb{P}_3$ is isomorphic to $\mathbb{R}^4$. See Example 5.

16. a. True. See Example 2.

  b. False. By definition, the coordinate mapping goes in the reverse direction.
  c. True, when the plane passes through the origin, as in Example 7.

18. Since $\mathbf{b}_1 = 1 \cdot \mathbf{b}_1 + 0 \cdot \mathbf{b}_2 + \cdots + 0 \cdot \mathbf{b}_n$, the $\mathcal{B}$-coordinate vector of $\mathbf{b}_1$ is

$$[\, \mathbf{b}_1 \,]_\mathcal{B} = \begin{bmatrix} 1 \\ 0 \\ \vdots \\ 0 \end{bmatrix} = \mathbf{e}_1$$

For each k, $\mathbf{b}_k = 0 \cdot \mathbf{b}_1 + \cdots + 1 \cdot \mathbf{b}_k + \cdots + 0 \cdot \mathbf{b}_n$, so $[\, \mathbf{b}_k \,]_\mathcal{B} = (0, \ldots, 1, \ldots, 0) = \mathbf{e}_k$.

20. For $\mathbf{w}$ in V, there exist scalars $k_1, \ldots, k_4$ such that

$$\mathbf{w} = k_1 \mathbf{v}_1 + \cdots + k_4 \mathbf{v}_4 \qquad (1)$$

because $\{\mathbf{v}_1, \ldots, \mathbf{v}_4\}$ spans V. Also, because the set is linearly dependent, there exist scalars $c_1, \ldots, c_4$, not all zero, such that

$$\mathbf{0} = c_1 \mathbf{v}_1 + \cdots + c_4 \mathbf{v}_4$$

Adding gives

$$\mathbf{w} = \mathbf{w} + \mathbf{0} = (k_1 + c_1)\mathbf{v}_1 + \cdots + (k_4 + c_4)\mathbf{v}_4$$

At least one of the weights here differs from the corresponding weight in equation (1) because at least one of the c_i is nonzero. So $\mathbf{w}$ is expressed in more than one way as a linear combination of $\mathbf{v}_1, \ldots, \mathbf{v}_4$.

22. Let $P_\mathcal{B} = [\, \mathbf{b}_1 \quad \cdots \quad \mathbf{b}_n \,]$. Then $P_\mathcal{B}[\mathbf{x}]_\mathcal{B} = \mathbf{x}$ and $[\mathbf{x}]_\mathcal{B} = P_\mathcal{B}^{-1}\mathbf{x}$. As mentioned in the text, the correspondence $\mathbf{x} \mapsto P_\mathcal{B}^{-1}\mathbf{x}$ is the coordinate mapping, so the desired matrix is $A = P_\mathcal{B}^{-1}$.

24. Given $\mathbf{y} = (y_1, \ldots, y_n)$ in $\mathbb{R}^n$, let $\mathbf{u} = y_1 \mathbf{b}_1 + \cdots + y_n \mathbf{b}_n$. Then, by definition, $[\mathbf{u}]_\mathcal{B} = \mathbf{y}$. So the coordinate mapping transforms $\mathbf{u}$ into $\mathbf{y}$. Since $\mathbf{y}$ was arbitrary, the coordinate mapping is onto.

26. $\mathbf{w}$ is a linear combination of $\mathbf{u}_1, \ldots, \mathbf{u}_p$ if and only if there exist scalars $c_1, \ldots, c_p$ such that

$$\mathbf{w} = c_1 \mathbf{u}_1 + \cdots + c_p \mathbf{u}_p \qquad (2)$$

Since the coordinate mapping is linear,

$$[\mathbf{w}]_\mathcal{B} = c_1 [\mathbf{u}_1]_\mathcal{B} + \cdots + c_p [\mathbf{u}_p]_\mathcal{B} \qquad (3)$$

Conversely, (3) implies (2) because the coordinate mapping is one-to-one. Thus $\mathbf{w}$ is a linear combination of $\mathbf{u}_1, \ldots, \mathbf{u}_p$ if and only if (3) holds for some $c_1, \ldots, c_p$, which is equivalent to saying that $[\mathbf{w}]_\mathcal{B}$ is a linear combination of $[\mathbf{u}_1]_\mathcal{B}, \ldots, [\mathbf{u}_p]_\mathcal{B}$.

Note: Students need to be urged to *write*, not just to compute, in Exercises 27–34. The language in the *Study Guide* solution of Exercise 31 provides a model for the students. In Exercise 32, students may have difficulty distinguishing between the two isomorphic vector spaces, sometimes giving a vector in $\mathbb{R}^3$ as the answer for part (b).

28. Linearly independent because the coordinate vectors

$$\begin{bmatrix} 1 \\ 0 \\ -2 \\ -1 \end{bmatrix}, \begin{bmatrix} 0 \\ 1 \\ 0 \\ 2 \end{bmatrix}, \begin{bmatrix} 1 \\ 1 \\ -2 \\ 0 \end{bmatrix}$$ are linearly independent.

30. Linearly dependent because the coordinate vectors

$$\begin{bmatrix} 8 \\ -12 \\ 6 \\ -1 \end{bmatrix}, \begin{bmatrix} 9 \\ -6 \\ 1 \\ 0 \end{bmatrix}, \begin{bmatrix} 1 \\ 6 \\ -5 \\ 1 \end{bmatrix}$$ are linearly dependent.

32. a. The coordinate vectors $\begin{bmatrix} 1 \\ 0 \\ 1 \end{bmatrix}, \begin{bmatrix} 0 \\ 1 \\ -3 \end{bmatrix}, \begin{bmatrix} 1 \\ 1 \\ -3 \end{bmatrix}$ span

$\mathbb{R}^3$. Thus these three vectors form a basis for $\mathbb{R}^3$ by the Invertible Matrix Theorem. Because of the isomorphism between $\mathbb{R}^3$ and $\mathbb{P}_2$, the corresponding polynomials form a basis for $\mathbb{P}_2$.

b. Since $[\mathbf{q}]_\mathcal{B} = (-1, 1, 2)$, one may compute

$$\begin{bmatrix} 1 & 0 & 1 \\ 0 & 1 & 1 \\ 1 & -3 & -3 \end{bmatrix} \begin{bmatrix} -1 \\ 1 \\ 2 \end{bmatrix} = \begin{bmatrix} 1 \\ 3 \\ -10 \end{bmatrix}$$

and $\mathbf{q} = 1 + 3t - 11t^2$.

34. [M] The coordinate vectors $\begin{bmatrix} 5 \\ -3 \\ 4 \\ 2 \end{bmatrix}, \begin{bmatrix} 9 \\ 1 \\ 8 \\ -6 \end{bmatrix}, \begin{bmatrix} 6 \\ -2 \\ 5 \\ 0 \end{bmatrix},$

$\begin{bmatrix} 0 \\ 0 \\ 0 \\ 1 \end{bmatrix}$ are linearly dependent. Because of the isomorphism

between $\mathbb{R}^4$ and $\mathbb{P}_3$, the corresponding polynomials are linearly dependent and therefore cannot form a basis for $\mathbb{P}_3$.

36. [M] Row reduction of $[\mathbf{v}_1 \quad \mathbf{v}_2 \quad \mathbf{v}_3]$ shows that there is a pivot in each column, so the columns are linearly independent and hence form a basis for the subspace H which they span.

$$[\mathbf{x}]_\mathcal{B} = \begin{bmatrix} 3 \\ 5 \\ 2 \end{bmatrix}$$

38. [M] $\begin{bmatrix} 1.30 \\ .75 \\ 1.60 \end{bmatrix}$

Section 4.5, page 229

2. $\begin{bmatrix} 2 \\ 0 \\ -2 \end{bmatrix}, \begin{bmatrix} 0 \\ -4 \\ 0 \end{bmatrix}$; dim is 2

4. $\begin{bmatrix} 1 \\ -1 \\ 3 \\ 1 \end{bmatrix}, \begin{bmatrix} 2 \\ 0 \\ -1 \\ 1 \end{bmatrix}$; dim is 2

6. $\begin{bmatrix} 3 \\ 0 \\ -7 \\ -3 \end{bmatrix}, \begin{bmatrix} 0 \\ -1 \\ 6 \\ 0 \end{bmatrix}, \begin{bmatrix} -1 \\ -3 \\ 5 \\ 1 \end{bmatrix}$; dim is 3

8. $\begin{bmatrix} 3 \\ 1 \\ 0 \\ 0 \end{bmatrix}, \begin{bmatrix} -1 \\ 0 \\ 1 \\ 0 \end{bmatrix}, \begin{bmatrix} 0 \\ 0 \\ 0 \\ 1 \end{bmatrix}$; dim is 3

10. 1   **12.** 3   **14.** 3, 4   **16.** 0, 2   **18.** 1, 2

19. a. True. See the box before Example 5.
  b. False, unless the plane is through the origin. Read Example 4 carefully.
  c. False. The dimension is 5. See Example 1.
  d. False. S must have exactly n elements to be a basis for V. See Theorem 10.
  e. True. See Practice Problem 2.

20. a. False. The only subspaces of $\mathbb{R}^3$ are listed in Example 4. $\mathbb{R}^2$ is not even a subset of $\mathbb{R}^3$, because vectors in $\mathbb{R}^3$ have three coordinates. Review Example 8 in Section 4.1.
  b. False. The number of *free* variables equals the dimension of Nul A. See the box before Example 5.
  c. False. Read carefully the definition before Example 1. *Not* being spanned by a finite set is not the same as being spanned by an infinite set. The space $\mathbb{R}^2$ is finite-dimensional, yet it is spanned by the infinite set S of all vectors of the form (x, y), where x and y are integers. (Of course, the two vectors $(1, 0)$ and $(0, 1)$ in S by themselves span $\mathbb{R}^2$.)
  d. False. S must have exactly n elements to be a basis of V. See the Basis Theorem.
  e. True. See Example 4.

22. Obviously, none of the Laguerre polynomials is a linear combination of the Laguerre polynomials of lower degree. By Theorem 4 (Section 4.3), the set of polynomials is linearly independent. Since this set contains four vectors, and $\mathbb{P}_3$ is four-dimensional, the set is a basis of $\mathbb{P}_3$, by the Basis Theorem.

24. $[\mathbf{p}]_\mathcal{B} = (6, 3, -2)$

26. If dim $V = 0$, the statement is obvious. Otherwise, H contains a basis, consisting of n linearly independent vectors. By the Basis Theorem applied to V, the vectors form a basis for V.

28. The space $C(\mathbb{R})$ contains the space $\mathbb{P}$ as a subspace. If $C(\mathbb{R})$ were finite-dimensional, $\mathbb{P}$ would be finite-dimensional, too, by Theorem 11. This is not true, by Exercise 27, so $C(\mathbb{R})$ is infinite-dimensional.

30. a. False. This is *not* Theorem 9. If $\mathbf{x}$ in V is nonzero, the set $\{\mathbf{0}, \mathbf{x}, 2\mathbf{x}, \ldots, (p-1)\mathbf{x}\}$ is linearly dependent, no matter what the dimension of V.
  b. True. If dim V were less than or equal to p, V would have a basis of not more than p elements. Such a set

would span V. Since this is not the case, dim V must be greater than p.

c. False. Counterexample: Take any nonzero vector $\mathbf{v}$, and consider the set $\{\mathbf{v}, 2\mathbf{v}, 3\mathbf{v}, \ldots, (p-1)\mathbf{v}\}$.

32. Let $\{\mathbf{u}_1, \ldots, \mathbf{u}_p\}$ be a basis for H. Then $\{T(\mathbf{u}_1), \ldots, T(\mathbf{u}_p)\}$ spans $T(H)$, as is easily seen. Further, since T is one-to-one, Exercise 32 in Section 4.3 shows that $\{T(\mathbf{u}_1), \ldots, T(\mathbf{u}_p)\}$ is linearly independent. So this set of images is a basis for $T(H)$. So dim $H = p$ and dim $T(H) = p$.

33. [M] a. $\{\mathbf{v}_1, \mathbf{v}_2, \mathbf{v}_3, \mathbf{e}_2, \mathbf{e}_3\}$

b. The first k columns of A are pivot columns because, by assumption, the original k vectors are linearly independent. Col $A = \mathbb{R}^n$, because the columns of A include all the columns of the identity matrix.

34. [M] The $\mathcal{B}$-coordinate vectors of the vectors in $\mathcal{C}$ are the columns of the matrix

$$P = \begin{bmatrix} 1 & 0 & -1 & 0 & 1 & 0 & -1 \\ & 1 & 0 & -3 & 0 & 5 & 0 \\ & & 2 & 0 & -8 & 0 & 18 \\ & & & 4 & 0 & -20 & 0 \\ & & & & 8 & 0 & -48 \\ & & & & & 16 & 0 \\ & & & & & & 32 \end{bmatrix}$$

a. This problem is an [M] exercise because it involves a large matrix. However, one should always think about a problem before rushing to use a matrix program. Actually, neither part of this exercise requires a matrix program. Simply observe that the matrix P is invertible because it is triangular with nonzero entries on the diagonal. So the columns of P are linearly independent. Because the coordinate mapping is an isomorphism, the vectors in $\mathcal{C}$ are linearly independent.

b. dim $H = 7$, because $\mathcal{B}$ is a basis for H with 7 elements. Since $\mathcal{C}$ is linearly independent, and the vectors in $\mathcal{C}$ lie in H (because of the trig identities), $\mathcal{C}$ is a basis for H, by the Basis Theorem. (Another argument is to use the fact that the $\mathcal{B}$-coordinate vectors of the vectors in $\mathcal{C}$ span $\mathbb{R}^7$, so the vectors in $\mathcal{C}$ span H. But you must distinguish between vectors in $\mathbb{R}^7$ and vectors in H.)

Section 4.6, page 236

2. rank $A = 3$; dim Nul $A = 2$;

Basis for Col A: $\begin{bmatrix} 1 \\ 2 \\ 3 \\ 3 \end{bmatrix}, \begin{bmatrix} 4 \\ 6 \\ 3 \\ 0 \end{bmatrix}, \begin{bmatrix} 2 \\ -3 \\ -3 \\ 0 \end{bmatrix}$

Basis for Row A: $(1, 3, 4, -1, 2), (0, 0, 1, -1, 1),$
$(0, 0, 0, 0, -5)$

Basis for Nul A: $\begin{bmatrix} -3 \\ 1 \\ 0 \\ 0 \\ 0 \end{bmatrix}, \begin{bmatrix} -3 \\ 0 \\ 1 \\ 1 \\ 0 \end{bmatrix}$

4. rank $A = 5$; dim Nul $A = 1$;

Basis for Col A: $\begin{bmatrix} 1 \\ 1 \\ 1 \\ 1 \\ 1 \end{bmatrix}, \begin{bmatrix} 1 \\ 2 \\ -1 \\ -2 \\ -2 \end{bmatrix}, \begin{bmatrix} -2 \\ -3 \\ 0 \\ 2 \\ 1 \end{bmatrix}, \begin{bmatrix} 1 \\ -2 \\ 1 \\ -3 \\ 2 \end{bmatrix}, \begin{bmatrix} -2 \\ -3 \\ 6 \\ 0 \\ -1 \end{bmatrix}$

Basis for Row A: $(1, 1, -2, 0, 1, -2), (0, 1, -1, 0, -3, -1),$
$(0, 0, 1, 1, -13, -1), (0, 0, 0, 0, 1, -1), (0, 0, 0, 0, 0, 1)$

Basis for Nul A: $\begin{bmatrix} -1 \\ -1 \\ -1 \\ 1 \\ 0 \\ 0 \end{bmatrix}$

6. $3, 2, 2$

8. 4. It is impossible for Col A to be $\mathbb{R}^4$ because the vectors in Col A have 6 entries. Col A is a four-dimensional subspace of $\mathbb{R}^6$.

10. 2   12. 2

14. 4, 4. If A is 5×4, its rows are in $\mathbb{R}^4$ and there can be at most four linearly independent vectors in such a set. If A is 4×5, it cannot have more than four linearly independent rows because there are only four rows.

16. 0

17. a. True. The row vectors in A are identified with the columns of A^T. See the paragraph before Example 1.

b. False. See the warning after Example 2.

c. True. See the Rank Theorem.

d. False. See the Rank Theorem. The sum of the two dimensions equals the number of *columns* in A.

e. True. See the Numerical Note before the Practice Problems.

18. a. False. Review the warning after the proof of Theorem 6 in Section 4.3.

b. False. See the warning after Example 2. For instance, a row interchange usually changes dependence relations among the rows.

c. True. See the remark in the proof of the Rank Theorem.

d. True. This fact was noted in the paragraph before Example 4. It also follows from the fact that the rows of a matrix—say, A^T—are the columns of its transpose, and $A^{TT} = A$.

e. True. See Theorem 13.

20. No. The presence of two free variables indicates that the null space of the coefficient matrix A is two-dimensional. Since there are eight unknowns, A has eight columns and

therefore must have rank 6, by the Rank Theorem. Since there are only six equations, A has six rows, and Col A is a subspace of $\mathbb{R}^6$. Since rank $A = 6$, we conclude that Col $A = \mathbb{R}^6$, which means that the equation $A\mathbf{x} = \mathbf{b}$ is consistent for all $\mathbf{b}$.

22. No. The coefficient matrix A is 10×12 and hence has rank at most 10. By the Rank Theorem, dim Nul A will be at least 2, so Nul A cannot be spanned by one vector.

24. The coefficient matrix A in this case is 7×6. It is possible that for some $\mathbf{b}$ in $\mathbb{R}^7$, the equation $A\mathbf{x} = \mathbf{b}$ has a unique solution. In this case, there are no free variables, so the rank of A must equal the number of columns, by the Rank Theorem. However, in any case, the rank of A cannot exceed 6, and so Col A must be a proper subspace of $\mathbb{R}^7$. Thus there exist vectors in $\mathbb{R}^7$ that are not in Col A. For such right-hand sides, the equation $A\mathbf{x} = \mathbf{b}$ will have no solution.

26. When an $m \times n$ matrix A has more rows than columns, A can have at most n pivot columns. So A has *full* rank when all n columns are pivot columns. This happens if and only if the equation $A\mathbf{x} = \mathbf{0}$ has only the trivial solution, that is, if and only if the columns of A are linearly independent.

28. a. dim Row $A = $ dim Col $A = $ rank A, by the Rank Theorem. So part (a) follows from the second part of that theorem.

    b. Apply part (a) with A replaced by A^T and use the fact that Row A^T is just Col A.

30. The equation $A\mathbf{x} = \mathbf{b}$ is consistent if and only if

    rank $[\, A \quad \mathbf{b}\,] = $ rank A

    because the two ranks are equal if and only if $\mathbf{b}$ is not a pivot column of $[\, A \quad \mathbf{b}\,]$. The result follows now from Theorem 2 in Section 1.2.

32. $\mathbf{v} = (1, -3, 4) = \begin{bmatrix} 1 \\ -3 \\ 4 \end{bmatrix}$

34. Since A can be reduced to an echelon form U by row operations, there exist invertible $m \times m$ elementary matrices $E_1, \ldots, E_p$, such that $(E_p \cdots E_1)A = U$, and $A = (E_p \cdots E_1)^{-1}U$, since the product of invertible matrices is invertible. Let $E = (E_p \cdots E_1)^{-1}$. Then $A = EU$. Denote the columns of E by $\mathbf{c}_1, \ldots, \mathbf{c}_m$. Since rank $A = r$, its echelon form U has r nonzero rows, which we can denote by $\mathbf{d}_1^T, \ldots, \mathbf{d}_r^T$. By the column–row expansion of EU (Theorem 10 in Section 2.4),

$$A = EU = [\, \mathbf{c}_1 \; \cdots \; \mathbf{c}_m \,] \begin{bmatrix} \mathbf{d}_1^T \\ \vdots \\ \mathbf{d}_r^T \\ 0 \\ \vdots \\ 0 \end{bmatrix} = \mathbf{c}_1\mathbf{d}_1^T + \cdots + \mathbf{c}_r\mathbf{d}_r^T$$

35. [M] a. Many answers are possible. Here are the "canonical" choices, for $A = [\, \mathbf{a}_1 \quad \mathbf{a}_2 \quad \cdots \quad \mathbf{a}_7\,]$:

$$C = [\, \mathbf{a}_1 \quad \mathbf{a}_2 \quad \mathbf{a}_4 \quad \mathbf{a}_6 \,], \quad N = \begin{bmatrix} -13/2 & -5 & 3 \\ -11/2 & -1/2 & -2 \\ 1 & 0 & 0 \\ 0 & 11/2 & -7 \\ 0 & 1 & 0 \\ 0 & 0 & -1 \\ 0 & 0 & 1 \end{bmatrix}$$

$$R = \begin{bmatrix} 1 & 0 & 13/2 & 0 & 5 & 0 & -3 \\ 0 & 1 & 11/2 & 0 & 1/2 & 0 & 2 \\ 0 & 0 & 0 & 1 & -11/2 & 0 & 7 \\ 0 & 0 & 0 & 0 & 0 & 1 & 1 \end{bmatrix}$$

b. $M = [\, 2 \quad 41 \quad 0 \quad -28 \quad 11\,]^T$. The matrix $[\, R^T \quad N\,]$ is 7×7 because the columns of R^T and N are in $\mathbb{R}^7$, and dim Row $A + $ dim Nul $A = 7$. The matrix $[\, C \quad M\,]$ is 5×5 because the columns of C and M are in $\mathbb{R}^5$ and dim Col $A + $ dim Nul $A^T = 5$, by Exercise 28(b). The invertibility of these matrices follows from the fact that their columns are linearly independent, which can be proved from Theorem 3 in Section 6.1.

36. [M] In most cases, C will be 6×4, constructed from the first four columns of A, R will be 4×7, N will be 7×3, and M will be 6×2.

38. [M] In general, if A is nonzero, then $A = CR$ because

$$CR = C[\, \mathbf{r}_1 \quad \mathbf{r}_2 \quad \cdots \quad \mathbf{r}_n\,] = [\, C\mathbf{r}_1 \quad C\mathbf{r}_2 \quad \cdots \quad C\mathbf{r}_n\,]$$

To explain why the matrix on the right is A itself, consider the pivot columns of A (i.e., the columns of C) and then consider the nonpivot columns of A.

The ith pivot column of R is $\mathbf{e}_i$ (the ith column of the identity matrix). So $C\mathbf{e}_i$ is the ith pivot column of A. Since A and R have pivot columns in the same location, when C multiplies a pivot column of R, the result is a pivot column of A, in the correct location.

A nonpivot column of R—say, $\mathbf{r}_j$—contains the weights needed to construct column j of A from the pivot columns in A, as discussed in Example 9 in Section 4.3 and the paragraph preceding that example. Thus $\mathbf{r}_j$ contains the weights needed to construct column j of A from the columns of C, so $C\mathbf{r}_j = \mathbf{a}_j$.

Section 4.7, page 242

2. a. $\begin{bmatrix} -2 & 3 \\ 4 & -6 \end{bmatrix}$ b. $\begin{bmatrix} 5 \\ -10 \end{bmatrix}$ 4. (i)

6. a. $\begin{bmatrix} 2 & 0 & -3 \\ -1 & 3 & 0 \\ 1 & 1 & 2 \end{bmatrix}$ b. $\begin{bmatrix} -4 \\ -7 \\ 3 \end{bmatrix}$

8. $\underset{C \leftarrow B}{P} = \begin{bmatrix} 9 & -8 \\ -10 & 9 \end{bmatrix}$, $\underset{B \leftarrow C}{P} = \begin{bmatrix} 9 & 8 \\ 10 & 9 \end{bmatrix}$

10. $_{C}\underset{B}{\overset{P}{\leftarrow}} = \begin{bmatrix} 3 & 1 \\ -2 & 0 \end{bmatrix}$, $_{B}\underset{C}{\overset{P}{\leftarrow}} = \frac{1}{2}\begin{bmatrix} 0 & -1 \\ 2 & 3 \end{bmatrix}$

11. a. False. See Theorem 15.

b. True. See the first paragraph in the subsection "Change of Basis in $\mathbb{R}^n$."

12. a. True. The columns of $_{C}\underset{B}{\overset{P}{\leftarrow}}$ are coordinate vectors of the linearly independent set $\mathcal{B}$. See the second paragraph after Theorem 15.

b. False. The row reduction is discussed after Example 2. The matrix P obtained there satisfies $[\mathbf{x}]_C = P[\mathbf{x}]_{\mathcal{B}}$.

14. a. $_{C}\underset{B}{\overset{P}{\leftarrow}} = \begin{bmatrix} 1 & 2 & 1 \\ 0 & 1 & 2 \\ -3 & -5 & 0 \end{bmatrix}$

b. Solve $_{C}\underset{B}{\overset{P}{\leftarrow}} \begin{bmatrix} x_1 \\ x_2 \\ x_3 \end{bmatrix} = \begin{bmatrix} 0 \\ 0 \\ 1 \end{bmatrix}$, and obtain

$t^2 = 3(1 - 3t^2) - 2(2 + t - 5t^2) + (1 + 2t)$.

16. a. $[\mathbf{b}_1]_C = Q[\mathbf{b}_1]_{\mathcal{B}} = Q\begin{bmatrix} 1 \\ 0 \\ \vdots \\ 0 \end{bmatrix} = Q\mathbf{e}_1$

b. $[\mathbf{b}_k]_C$

c. $[\mathbf{b}_k]_C = Q[\mathbf{b}_k]_{\mathcal{B}} = Q\mathbf{e}_k$

17. a. [M] $P^{-1} = \dfrac{1}{32}\begin{bmatrix} 32 & 0 & 16 & 0 & 12 & 0 & 10 \\ & 32 & 0 & 24 & 0 & 20 & 0 \\ & & 16 & 0 & 16 & 0 & 15 \\ & & & 8 & 0 & 10 & 0 \\ & & & & 4 & 0 & 6 \\ & & & & & 2 & 0 \\ & & & & & & 1 \end{bmatrix}$

b. $\cos^2 t = (1/2)[1 + \cos 2t]$

$\cos^3 t = (1/4)[3\cos t + \cos 3t]$

$\cos^4 t = (1/8)[3 + 4\cos 2t + \cos 4t]$

$\cos^5 t = (1/16)[10\cos t + 5\cos 3t + \cos 5t]$

$\cos^6 t = (1/32)[10 + 15\cos 2t + 6\cos 4t + \cos 6t]$

18. [M] Let $C = \{\mathbf{y}_0, \ldots, \mathbf{y}_6\}$, where $\mathbf{y}_k$ is the function $\cos kt$. Then the C-coordinate vector of $5\cos^3 t - 6\cos^4 t + 5\cos^5 t - 12\cos^6 t$ is $(0, 0, 0, 5, -6, 5, -12)$. Left-multiplication by the inverse of matrix P in Exercise 17 changes this C-coordinate vector into the $\mathcal{B}$-coordinate vector $(-6, 55/8, -69/8, 45/16, -3, 5/16, -3/8)$. So the integral (10) in this exercise equals

$\displaystyle\int \Big[-6 + \frac{55}{8}\cos t - \frac{69}{8}\cos 2t + \frac{45}{16}\cos 3t - 3\cos 4t$

$+ \frac{5}{16}\cos 5t - \frac{3}{8}\cos 6t \Big] dt$

From calculus, the integral equals

$-6t + \dfrac{55}{8}\sin t - \dfrac{69}{16}\sin 2t + \dfrac{15}{16}\sin 3t - \dfrac{3}{4}\sin 4t$

$+ \dfrac{1}{16}\sin 5t - \dfrac{1}{16}\sin 6t + C$

20. a. $_{D}\underset{B}{\overset{P}{\leftarrow}} = _{D}\underset{C}{\overset{P}{\leftarrow}} \cdot _{C}\underset{B}{\overset{P}{\leftarrow}}$. Reason: $[\mathbf{b}_j]_D = _{D}\underset{C}{\overset{P}{\leftarrow}} [\mathbf{b}_j]_C$. So, by Theorem 15 and the definition of matrix multiplication:

$_{D}\underset{B}{\overset{P}{\leftarrow}} = \begin{bmatrix} [\mathbf{b}_1]_D & [\mathbf{b}_2]_D \end{bmatrix}$

$= \begin{bmatrix} _{D}\underset{C}{\overset{P}{\leftarrow}}[\mathbf{b}_1]_C & _{D}\underset{C}{\overset{P}{\leftarrow}}[\mathbf{b}_2]_C \end{bmatrix}$

$= _{D}\underset{C}{\overset{P}{\leftarrow}}\begin{bmatrix} [\mathbf{b}_1]_C & [\mathbf{b}_2]_C \end{bmatrix}$

$= _{D}\underset{C}{\overset{P}{\leftarrow}} \cdot _{C}\underset{B}{\overset{P}{\leftarrow}}$

b. Answers will vary.

Section 4.8, page 251

2. If $y_k = 5^k$, then $y_{k+1} = 5^{k+1}$ and $y_{k+2} = 5^{k+2}$. Substituting these formulas into the left side of the equation gives

$y_{k+2} - 25y_k = 5^{k+2} - 25 \cdot 5^k$

$= 5^k(5^2 - 25) = 5^k(0) = 0$ for all k

Since the difference equation holds for all k, 5^k is a solution. A similar calculation works for $y_k = (-5)^k$:

$y_{k+2} - 25y_k = (-5)^{k+2} - 25(-5)^k$

$= (-5)^k[(-5)^2 - 25]$

$= (-5)^k(0) = 0$ for all k

4. The signals 5^k and $(-5)^k$ are linearly independent because neither is a multiple of the other. If H is the solution space of the difference equation in Exercise 2, then dim $H = 2$, by Theorem 17. So the two linearly independent solutions form a basis for H, by the Basis Theorem in Section 4.5.

6. If $y_k = 4^k \cos\left(\frac{k\pi}{2}\right)$, then

$y_{k+2} + 16y_k = 4^{k+2}\cos\left(\frac{(k+2)\pi}{2}\right) + 16 \cdot 4^k \cos\left(\frac{k\pi}{2}\right)$

$= 4^{k+2}\left[\cos\left(\frac{k\pi}{2} + \pi\right) + \cos\left(\frac{k\pi}{2}\right)\right]$

$= 0$ for all k

because $\cos(t + \pi) = -\cos t$ for all t. A similar calculation holds for $z_k = 4^k \sin\left(\frac{k\pi}{2}\right)$, using the trigonometric identity $\sin(t + \pi) = -\sin t$. Thus y_k and z_k are both solutions of the difference equation $y_{k+2} + 16y_k = 0$. These solutions are obviously linearly independent because neither is a multiple of the other. Since the solution space H is two-dimensional, y_k and z_k form a basis for H, by the Basis Theorem.

8. Yes   **10.** Yes

12. No, two signals cannot span a four-dimensional solution space.

14. $2^k, 3^k$ **16.** $5^k, (-5)^k$

18. The auxiliary equation is $r^2 - 1.35r + .45 = 0$, with roots .75 and .6. The constant solution of the nonhomogeneous equation is found by solving $T - 1.35T + .45T = 1$, to obtain $T = 10$. The general solution of the nonhomogeneous equation is

$$Y_k = c_1(.75)^k + c_2(.6)^k + 10$$

20. Let $a = -2 + \sqrt{3}$ and $b = -2 - \sqrt{3}$. Then c_1 and c_2 must satisfy

$$\begin{bmatrix} a & b \\ a^N & b^N \end{bmatrix} \begin{bmatrix} c_1 \\ c_2 \end{bmatrix} = \begin{bmatrix} 5000 \\ 0 \end{bmatrix}$$

Solving (by row operations or Cramer's rule), we obtain

$$y_k = c_1 a^k + c_2 b^k = \frac{5000}{ab^N - ba^N}(a^k b^N - a^N b^k)$$

22. $1.4, 0, -1.4, -2, -1.4, 0, 1.4, 2, 1.4$

This signal is 2 times the signal output by the filter when the input (in Example 3) was $\cos(\pi t/4)$. This is what is to be expected since the filter is *linear*. The output should be 2 times the output from $\cos(\pi t/4)$ plus 1 times the (zero) output from $\cos(3\pi t/4)$.

23. b. [M] MATLAB code:

```
pay = 450, y = 10000, m = 0
table = [0 ; y]
while y > 450
    y = 1.01*y - pay
    m = m + 1
    table = [table [m ; y] ]
                %append new column
end
m, y
```

c. [M] At month 26, the last payment is $114.88. The total paid by the borrower is $11,364.88.

24. a. $y_{k+1} - 1.005y_k = 200$, $y_0 = 1,000$

b. [M] MATLAB code:

```
pay = 200, y = 1000, table = [0 ; y]
for m = 1:60
    y = 1.005*y + pay
    table = [table [m ; y] ]
end
interest = y - 60*pay - 1000
```

c. [M] The total is $6213.55 at $k = 24$, $12,090.06 at $k = 48$, and $15,302.86 at $k = 60$. When $k = 60$, the interest earned is $2302.86.

25. To show that the sequence $y_k = k^2$ is a solution of $y_{k+2} + 3y_{k+1} - 4y_k = 10k + 7$, compute

$$
\begin{array}{rcl}
y_{k+2} = (k+2)^2 & = & k^2 + 4k + 4 \\
3y_{k+1} = 3(k+1)^2 = 3(k^2 + 2k + 1) = & & 3k^2 + 6k + 3 \\
-4y_k & = & -4k^2 \\
\hline
& = & 10k + 7
\end{array}
$$

This shows that the sequence $y_k = k^2$ is a solution of the given equation. To find the general solution of the difference equation, study the homogeneous equation, $y_{k+2} + 3y_{k+1} - 4y_k = 0$. Substitute r^k for y_k in this equation, factor the left side, and set the result equal to 0 for all k (the right side of the equation):

$$
\begin{aligned}
r^{k+2} + 3r^{k+1} - 4r^k &= r^k(r^2 + 3r - 4) \\
&= r^k(r - 1)(r + 4) \\
&= 0 \quad \text{(for all } k)
\end{aligned}
$$

This shows that the sequence $y_k = r^k$ satisfies the homogeneous equation for $r = 0, 1$, and -4. The general solution of the nonhomogeneous equation is the sum of a particular solution $y_k = k^2$ and the general solution of the associated homogeneous equation. The solution of the original difference equation is

$$y_k = k^2 + c_1(-4)^k + c_2 1^k$$

or simply $y_k = k^2 + c_1(-4)^k + c_2$

26. $1 + k + c_1 \cdot (1)^k + c_2 \cdot (5)^k$

28. $y_k = 1 + 2k + 5^k + (-5)^k$

30. $\mathbf{x}_{k+1} = A\mathbf{x}_k$, where

$$A = \begin{bmatrix} 0 & 1 & 0 \\ 0 & 0 & 1 \\ -8 & 0 & 5 \end{bmatrix}, \quad \mathbf{x} = \begin{bmatrix} y_k \\ y_{k+1} \\ y_{k+2} \end{bmatrix}$$

32. If $a_3 \neq 0$, the order is 3; if $a_3 = 0$ and $a_2 \neq 0$, the order is 2; if $a_3 = a_2 = 0$ and $a_1 \neq 0$, the order is 1; otherwise, the order is 0 (with only the zero signal for a solution).

34. No, the signals could be linearly dependent. *Example:* The following functions are linearly independent when considered as functions on the real line, because they have different periods and no one of the functions is a linear combination of the other two.

$$f(t) = \sin \pi t, \quad g(t) = \sin 2\pi t, \quad h(t) = \sin 3\pi t$$

Since f, g, and h are zero at every integer, the signals are linearly dependent as vectors in $\mathbb{S}$.

36. Given $\mathbf{z}$ in V, suppose that $\mathbf{x}_p$ in V satisfies $T(\mathbf{x}_p) = \mathbf{z}$. Also if $\mathbf{u}$ is in the kernel of T, then $T(\mathbf{u}) = \mathbf{0}$. Since T is linear, $T(\mathbf{u} + \mathbf{x}_p) = T(\mathbf{u}) + T(\mathbf{x}_p) = \mathbf{z}$. So the vector $\mathbf{x} = \mathbf{u} + \mathbf{x}_p$ satisfies the nonhomogeneous equation $T(\mathbf{x}) = \mathbf{z}$.

37. The linear transformations T and D are defined by

$$T(y_0, y_1, y_2, \dots) = (y_1, y_2, y_3, \dots),$$

often called the "left-shift" operator on the sequences, and

$$D(y_0, y_1, y_2, \dots) = (0, y_0, y_1, y_2, \dots),$$

often called the "right-shift" operator. Using an arbitrary sequence, compute

$$TD(y_0, y_1, y_2, \ldots) = T(0, y_0, y_1, y_2, \ldots)$$
$$= (y_0, y_1, y_2, \ldots),$$

so $TD = I$. However,

$$DT(y_0, y_1, y_2, \ldots) = D(y_1, y_2, y_3, \ldots) = (0, y_1, y_2, \ldots),$$

so $DT \neq I$.

Section 4.9, page 260

2. a.

From:

	1	2	3	To:
	.6	.2	.2	1
	.2	.6	.2	2
	.2	.2	.6	3

b. .28

4. a.

From:

	G	I	B	To:
	.4	.5	.3	G
	.3	.2	.4	I
	.3	.3	.3	B

b. 30%   **c.** 40%

6. $\begin{bmatrix} 4/7 \\ 3/7 \end{bmatrix}$ **8.** $\begin{bmatrix} 5/9 \\ 2/27 \\ 10/27 \end{bmatrix}$

10. No, because P^k has a zero in the lower left corner for all k.

12. Each food will be preferred equally, because $\begin{bmatrix} 1/3 \\ 1/3 \\ 1/3 \end{bmatrix}$ is the steady-state vector.

14. There is a 40% chance of good weather because $\begin{bmatrix} .4 \\ .3 \\ .3 \end{bmatrix}$ is the steady-state vector.

16. [**M**] The steady-state vector is approximately $(.435, .091, .474)$. Of the 2000 cars, about 182 will be rented or available from the downtown location.

18. If $\alpha = \beta = 0$, then $\begin{bmatrix} 1 \\ 0 \end{bmatrix}$ and $\begin{bmatrix} 0 \\ 1 \end{bmatrix}$ are the steady-state vectors. Otherwise, $\frac{1}{\alpha + \beta}\begin{bmatrix} \beta \\ \alpha \end{bmatrix}$ is the only steady-state vector.

20. Let $P = [\mathbf{p}_1 \quad \mathbf{p}_2 \quad \cdots \quad \mathbf{p}_n]$, so that

$$P^2 = [P\mathbf{p}_1 \quad P\mathbf{p}_2 \quad \cdots \quad P\mathbf{p}_n]$$

By Exercise 19(c), the columns of P^2 are probability vectors, so P^2 is a stochastic matrix. Alternatively, $SP = S$, by Exercise 19(b), since P is a stochastic matrix. Right-multiplication by P yields $SP^2 = SP$. The right side is just S, so that $SP^2 = S$. Since the entries in P^2 are obviously nonnegative (they are sums of products of the nonnegative entries in P), this shows that P^2 is also a stochastic matrix.

21. [**M**]
a. To four decimal places,

$$P^4 = P^5 = \begin{bmatrix} .2816 & .2816 & .2816 & .2816 \\ .3355 & .3355 & .3355 & .3355 \\ .1819 & .1819 & .1819 & .1819 \\ .2009 & .2009 & .2009 & .2009 \end{bmatrix},$$

$$\mathbf{q} = \begin{bmatrix} .2816 \\ .3355 \\ .1819 \\ .2009 \end{bmatrix}$$

Note that, due to round-off, the column sums are not 1.

b. To four decimal places,

$$Q^{80} = \begin{bmatrix} .7354 & .7348 & .7351 \\ .0881 & .0887 & .0884 \\ .1764 & .1766 & .1765 \end{bmatrix},$$

$$Q^{116} = Q^{117} = \begin{bmatrix} .7353 & .7353 & .7353 \\ .0882 & .0882 & .0882 \\ .1765 & .1765 & .1765 \end{bmatrix},$$

$$\mathbf{q} = \begin{bmatrix} .7353 \\ .0882 \\ .1765 \end{bmatrix}$$

c. Let P be an $n \times n$ regular stochastic matrix, $\mathbf{q}$ the steady-state vector of P, and $\mathbf{e}_1$ the first column of the identity matrix. Then $P^k\mathbf{e}_1$ is the first column of P^k. By Theorem 18, $P^k\mathbf{e}_1 \to \mathbf{q}$ as $k \to \infty$. Replacing $\mathbf{e}_1$ by the other columns of the identity matrix, we conclude that each column of P^k converges to $\mathbf{q}$ as $k \to \infty$. Thus $P^k \to [\mathbf{q} \quad \mathbf{q} \quad \cdots \quad \mathbf{q}]$.

22. [**M**] (Discussion based on MATLAB Student Version 4.0, running on a 100-MHz 486 laptop computer with 32 Mb of memory) Let A be a random 32×32 stochastic matrix.

Method (1): The following command line will construct A and $\mathbf{q}$ but not display them, and it will announce the elapsed computer processing time and the number of flops used (type all the commands on one line so they can be recalled and rerun several times):

```
A = randomstoc(32); flops(0);
tic, x = nulbasis(A - eye(32));
q = x/sum(x); toc, flops
```

The time ranged from 1.04 to 1.21 seconds, with 35,463 flops.

Method (2):

```
A = randomstoc(32); flops(0);
tic, B = A^100; q = B(:,1); toc, flops
```

The time ranged from 1.37 to 1.48 seconds, with 6,488,082 flops. If only A^{70} is computed, the time is about .94 second, which is faster than method (1), even though it uses about 4,522,000 flops.

Chapter 4 Supplementary Exercises, page 262

1. a. True. Span $\{\mathbf{v}_1, \ldots, \mathbf{v}_p\}$ is a subspace of V, and every subspace is itself a vector space.

b. True. Any linear combination of $\mathbf{v}_1, \ldots, \mathbf{v}_{p-1}$ is also a linear combination of $\mathbf{v}_1, \ldots, \mathbf{v}_{p-1}, \mathbf{v}_p$, using a weight of zero on $\mathbf{v}_p$.

c. False. Take $\mathbf{v}_p = 2\mathbf{v}_1$.

d. False. Let $\{\mathbf{e}_1, \mathbf{e}_2, \mathbf{e}_3\}$ be the standard basis for $\mathbb{R}^3$. Then $\{\mathbf{e}_1, \mathbf{e}_2\}$ is a linearly independent set but is not a basis for $\mathbb{R}^3$.

e. True. See the Spanning Set Theorem (Section 4.3).

f. True. By the Basis Theorem, S must be a basis for V because S contains exactly p vectors, and so must be linearly independent.

g. False. The plane must go through the origin to be a subspace.

h. False. Consider $\begin{bmatrix} 2 & 5 & -2 & 0 \\ 0 & 0 & 7 & 3 \\ 0 & 0 & 0 & 0 \end{bmatrix}$.

i. True. This concept is presented before Theorem 13 in Section 4.6.

j. False. Row operations on A do not change the solutions of $A\mathbf{x} = \mathbf{0}$.

k. False. Consider $\begin{bmatrix} 1 & 2 \\ 3 & 6 \end{bmatrix}$.

l. False. If U has k nonzero rows, then rank $A = k$ and dim Nul $A = n - k$ by the Rank Theorem.

m. True. Row equivalent matrices have the same number of pivot columns.

n. False. The nonzero rows of A span Row A, but they may not be linearly independent.

o. True. The nonzero rows of the reduced row echelon form E form a basis for the row space of each matrix that is row equivalent to E.

p. True. If H is the zero subspace, let A be the 3×3 zero matrix. If dim $H = 1$, let $\{\mathbf{v}\}$ be a basis for H, and set $A = [\,\mathbf{v} \quad \mathbf{v} \quad \mathbf{v}\,]$. If dim $H = 2$, let $\{\mathbf{u}, \mathbf{v}\}$ be a basis for H, and set $A = [\,\mathbf{u} \quad \mathbf{v} \quad \mathbf{v}\,]$, for example. If dim $H = 3$, then any invertible matrix A will work. Or, let $\{\mathbf{u}, \mathbf{v}, \mathbf{w}\}$ be a basis for H, and set $A = [\,\mathbf{u} \quad \mathbf{v} \quad \mathbf{w}\,]$.

q. False. Consider $\begin{bmatrix} 1 & 0 & 0 \\ 0 & 1 & 0 \end{bmatrix}$. If rank $A = n$ (the number of columns in A), then the transformation $\mathbf{x} \mapsto A\mathbf{x}$ is one-to-one.

r. True. If $\mathbf{x} \mapsto A\mathbf{x}$ is onto, then Col $A = \mathbb{R}^m$ and rank $A = m$.

s. True. See the second paragraph after Theorem 15 in Section 4.7.

t. False. The jth column of the change-of-coordinates matrix $\underset{C \leftarrow B}{P}$ is $[\,\mathbf{b}_j\,]_C$.

2. Any two of these three:

$$\left\{ \begin{bmatrix} 1 \\ 2 \\ -1 \\ 3 \end{bmatrix}, \begin{bmatrix} -2 \\ 5 \\ -4 \\ 1 \end{bmatrix}, \begin{bmatrix} 5 \\ -8 \\ 7 \\ 1 \end{bmatrix} \right\}$$

4. The vectors $\mathbf{f}$ and $\mathbf{g}$ are not *scalar* multiples of each other, so $\{\mathbf{f}, \mathbf{g}\}$ is linearly independent.

6. Choose any two polynomials that are not multiples. Since they are linearly independent and belong to a two-dimensional space, they will be a basis for H.

8. The case $n = 0$ is trivial. If $n > 0$, then a basis for H consists of n linearly independent vectors, say, $\mathbf{u}_1, \ldots, \mathbf{u}_n$. These vectors remain linearly independent when considered as elements of V. But any n linearly independent vectors in the n-dimensional space V must form a basis for V, by the Basis Theorem in Section 4.5. So $\mathbf{u}_1, \ldots, \mathbf{u}_n$ span V. Thus $H = \text{Span}\{\mathbf{u}_1, \ldots, \mathbf{u}_n\} = V$.

10. Let $S = \{\mathbf{v}_1, \ldots, \mathbf{v}_p\}$. If S were linearly independent and not a basis for V, then S would not span V. In this case, there would be a vector $\mathbf{v}_{p+1}$ in V that is not in $\text{Span}\{\mathbf{v}_1, \ldots, \mathbf{v}_p\}$. Let $S' = \{\mathbf{v}_1, \ldots, \mathbf{v}_p, \mathbf{v}_{p+1}\}$. Then S' is linearly independent because none of the vectors in S' is a linear combination of vectors that precede it. Since S' is larger than S, this would contradict the maximality of S. Hence S must be a basis for V.

12. a. Any $\mathbf{y}$ in Col AB has the form $AB\mathbf{x}$ for some $\mathbf{x}$. Then $\mathbf{y} = A(B\mathbf{x})$, which shows that $\mathbf{y}$ is a linear combination of the columns of A. Thus Col AB is a subset of Col A; that is, Col AB is a *subspace* of Col A. By Theorem 11 in Section 4.5, dim Col $AB \leq$ dim Col A; that is, rank $AB \leq$ rank A.

b. By the Rank Theorem and part (a):

$$\text{rank } AB = \text{rank}(AB)^T = \text{rank } B^T A^T$$
$$\leq \text{rank } B^T = \text{rank } B$$

14. Note that $(AQ)^T = Q^T A^T$. Since Q^T is invertible, we can use Exercise 13 to conclude that

$$\text{rank}(AQ)^T = \text{rank } Q^T A^T = \text{rank } A^T$$

Since the ranks of a matrix and its transpose are equal (by the Rank Theorem), rank $AQ = $ rank A.

16. Suppose rank $A = r_1$ and rank $B = r_2$. Let the rank factorizations of A and B be $A = C_1 R_1$ and $B = C_2 R_2$. Create an $m \times (r_1 + r_2)$ matrix $C = [\,C_1 \quad C_2\,]$ and an $(r_1 + r_2) \times n$ matrix $R = \begin{bmatrix} R_1 \\ R_2 \end{bmatrix}$. Then

$$A + B = C_1 R_1 + C_2 R_2 = CR$$

By Exercise 12, the rank of $A + B$ cannot exceed the rank of C. Since C has $r_1 + r_2$ columns, rank $C \leq r_1 + r_2$. Thus the rank of $A + B$ cannot exceed $r_1 + r_2 = $ rank $A + $ rank B.

18. a. Using the equation $\mathbf{x}_{k+1} = A\mathbf{x}_k + B\mathbf{u}_k$ for $k = 0, 1, 2, 3, 4$, and letting $\mathbf{x}_0 = \mathbf{0}$, we have

$$\mathbf{x}_1 = A\mathbf{x}_0 + B\mathbf{u}_0 = B\mathbf{u}_0$$
$$\mathbf{x}_2 = A\mathbf{x}_1 + B\mathbf{u}_1 = AB\mathbf{u}_0 + B\mathbf{u}_1$$
$$\mathbf{x}_3 = A\mathbf{x}_2 + B\mathbf{u}_2$$
$$= A(AB\mathbf{u}_0 + B\mathbf{u}_1) + B\mathbf{u}_2$$
$$= A^2 B\mathbf{u}_0 + AB\mathbf{u}_1 + B\mathbf{u}_2$$
$$\mathbf{x}_4 = A\mathbf{x}_3 + B\mathbf{u}_3$$
$$= A(A^2 B\mathbf{u}_0 + AB\mathbf{u}_1 + B\mathbf{u}_2) + B\mathbf{u}_3$$
$$= A^3 B\mathbf{u}_0 + A^2 B\mathbf{u}_1 + AB\mathbf{u}_2 + B\mathbf{u}_3$$
$$= \begin{bmatrix} B & AB & A^2B & A^3B \end{bmatrix} \begin{bmatrix} \mathbf{u}_3 \\ \mathbf{u}_2 \\ \mathbf{u}_1 \\ \mathbf{u}_0 \end{bmatrix}$$
$$= M\mathbf{u}, \text{ where } \mathbf{u} \text{ is in } \mathbb{R}^8.$$

b. If (A, B) is controllable, then the controllability matrix M has rank 4, with a pivot in each row, and the columns of M span $\mathbb{R}^4$. Therefore, for any vector in $\mathbf{v}$ in $\mathbb{R}^4$, there is a vector $\mathbf{u}$ in $\mathbb{R}^8$ such that $\mathbf{v} = M\mathbf{u}$. However, from part (a) we know that $\mathbf{x}_4 = M\mathbf{u}$ when $\mathbf{u}$ is partitioned into a control sequence $\mathbf{u}_0, \ldots, \mathbf{u}_3$. This particular control sequence makes $\mathbf{x}_4 = \mathbf{v}$.

20. $\begin{bmatrix} B & AB & A^2B \end{bmatrix} = \begin{bmatrix} 1 & .5 & .19 \\ 1 & .7 & .45 \\ 0 & 0 & 0 \end{bmatrix}$. This matrix has rank less than 3, so the pair (A, B) is not controllable.

22. [M] rank $\begin{bmatrix} B & AB & A^2B & A^3B \end{bmatrix} = 4$.
The pair (A, B) is controllable.

Chapter 5

Section 5.1, page 271

2. Yes   **4.** Yes, $\lambda = 3$ **6.** No

8. Yes, $\begin{bmatrix} 0 \\ 3 \\ 2 \end{bmatrix}$ **10.** $\begin{bmatrix} -2 \\ 1 \end{bmatrix}$

12. $\lambda = 3$: $\begin{bmatrix} -1 \\ 1 \end{bmatrix}$; $\lambda = 7$: $\begin{bmatrix} 1 \\ 3 \end{bmatrix}$ **14.** $\begin{bmatrix} 1 \\ 2 \\ 1 \end{bmatrix}$

16. $\begin{bmatrix} 1 \\ 1 \\ 1 \\ 0 \end{bmatrix}, \begin{bmatrix} 0 \\ 0 \\ 0 \\ 1 \end{bmatrix}$ **18.** 5, 0, 3

20. $\lambda = 0$. Eigenvectors for $\lambda = 0$ have entries that produce linear dependence relations among the columns of A. Any nonzero vector (in $\mathbb{R}^3$) whose entries sum to 0 will work. Find any two such vectors that are not multiples of each other; for example, $\begin{bmatrix} 1 \\ 1 \\ -2 \end{bmatrix}$ and $\begin{bmatrix} 1 \\ -1 \\ 0 \end{bmatrix}$.

21. a. False. The equation $A\mathbf{x} = \lambda\mathbf{x}$ must have a *nontrivial* solution.

b. True. See the paragraph after Example 5.
c. True. See the discussion of equation (3).
d. True. See Example 2 and the paragraph preceding it. Also, see the Numerical Note.
e. False. See the warning after Example 3.

22. a. False. The vector $\mathbf{x}$ in $A\mathbf{x} = \lambda\mathbf{x}$ must be *nonzero*.
b. False. See Example 4 for a two-dimensional eigenspace, which contains two linearly independent eigenvectors corresponding to the same eigenvalue. The statement given is not at all the same as Theorem 2. In fact, it is the converse of Theorem 2 (for the case $r = 2$).
c. True. See the paragraph after Example 1.
d. False. Theorem 1 concerns a *triangular* matrix. See Examples 3 and 4 for counterexamples.
e. True. See the paragraph following Example 3. The eigenspace of A corresponding to λ is the null space of the matrix $A - \lambda I$.

24. Any triangular matrix with the same number in both diagonal entries, such as $\begin{bmatrix} 4 & 5 \\ 0 & 4 \end{bmatrix}$.

26. If $A\mathbf{x} = \lambda\mathbf{x}$ for some $\mathbf{x} \neq \mathbf{0}$, then
$$A^2\mathbf{x} = A(A\mathbf{x}) = A(\lambda\mathbf{x}) = \lambda A\mathbf{x} = \lambda^2\mathbf{x}.$$
However, $A^2\mathbf{x} = \mathbf{0}$ because $A^2 = 0$. Therefore, $\mathbf{0} = \lambda^2\mathbf{x}$. Since $\mathbf{x} \neq \mathbf{0}$, we conclude that λ must be zero. Thus each eigenvalue of A is zero.

28. If A is lower triangular, then A^T is upper triangular and has the same diagonal entries as A. Hence, by the part of Theorem 1 already proved in the text, these diagonal entries are eigenvalues of A^T. By Exercise 27, they are also eigenvalues of A.

30. By Exercise 29 applied to A^T in place of A, we conclude that s is an eigenvalue of A^T. By Exercise 27, s is an eigenvalue of A.

32. Suppose T rotates points about some line L that passes through the origin in $\mathbb{R}^3$. That line consists of all multiples of some nonzero vector $\mathbf{v}$. The points on this line do not move under the action of T. So $T(\mathbf{v}) = \mathbf{v}$. If A is the standard matrix of T, then $A\mathbf{v} = \mathbf{v}$. Thus $\mathbf{v}$ is an eigenvector of A corresponding to the eigenvalue 1. The eigenspace is Span $\{\mathbf{v}\}$.

If the rotation happens to be half of a full rotation, that is, through an angle of 180 degrees, let P be a plane through the origin that is perpendicular to the line L. Each point $\mathbf{p}$ in this plane rotates into $-\mathbf{p}$. That is, each point in P is an eigenvector of A corresponding to the eigenvalue -1.

34. You could try to write $\mathbf{x}_0$ as a linear combination of eigenvectors, $\mathbf{v}_1, \ldots, \mathbf{v}_p$, of A. If $\lambda_1, \ldots, \lambda_p$ are corresponding eigenvalues, and if $\mathbf{x}_0 = c_1\mathbf{v}_1 + \cdots + c_p\mathbf{v}_p$, then you could *define*
$$\mathbf{x}_k = c_1\lambda_1^k\mathbf{v}_1 + \cdots + c_p\lambda_p^k\mathbf{v}_p$$
In this case, for $k = 0, 1, 2, \ldots,$

$$Ax_k = A(c_1\lambda_1^k v_1 + \cdots + c_p\lambda_p^k v_p)$$
$$= c_1\lambda_1^k A v_1 + \cdots + c_p\lambda_p^k A v_p \qquad \text{Linearity}$$
$$= c_1\lambda_1^{k+1} v_1 + \cdots + c_p\lambda_p^{k+1} v_p \qquad \text{The } v_i \text{ are}$$
$$\text{eigenvectors.}$$
$$= x_{k+1}$$

36.

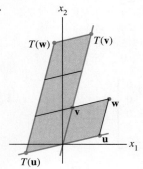

38. [M] $\lambda = -1$: $\begin{bmatrix} 1 \\ 1 \\ 0 \\ 1 \end{bmatrix}$; $\lambda = 1$: $\begin{bmatrix} 2 \\ 2 \\ 0 \\ 1 \end{bmatrix}$;

$\lambda = -2$: $\begin{bmatrix} 0 \\ 1 \\ 1 \\ 0 \end{bmatrix}$; $\lambda = 2$: $\begin{bmatrix} 2 \\ 2 \\ 1 \\ 1 \end{bmatrix}$;

40. [M] $\lambda = -14$: $\begin{bmatrix} -1 \\ 0 \\ 1 \\ 0 \\ 0 \end{bmatrix}$, $\begin{bmatrix} 10 \\ 6 \\ 0 \\ 5 \\ 3 \end{bmatrix}$;

$\lambda = 42$: $\begin{bmatrix} 0 \\ 1 \\ -2 \\ 5 \\ 0 \end{bmatrix}$, $\begin{bmatrix} -5 \\ -1 \\ -3 \\ 0 \\ 5 \end{bmatrix}$

Section 5.2, page 279

2. $\lambda^2 + 3\lambda + 2; -2, -1$

4. $\lambda^2 - 11\lambda + 18; 2, 9$

6. $\lambda^2 - 14\lambda + 49; 7, 7$

8. $\lambda^2 + 3\lambda - 10; -5, 2$

10. $-\lambda^3 + 15\lambda^2 - 73\lambda + 115$

12. $-\lambda^3 + 2\lambda^2 + 1\lambda + 4$

14. $-\lambda^3 + 7\lambda^2 - 4\lambda - 30$

16. $3, 2, 6, -5$

18. $h = 3$

20. $\det(A^T - \lambda I) = \det(A^T - \lambda I^T)$
$$= \det(A - \lambda I)^T \qquad \text{Transpose property}$$
$$= \det(A - \lambda I) \qquad \text{Theorem 3(c)}$$

21. a. False (although true for a triangular matrix). See Example 1 for a matrix whose determinant is not the product of its diagonal entries.

  b. False. However, a row replacement operation does not change the determinant. See Theorem 3(e).

  c. True. See Theorem 3(b).

d. False. See the solution of Example 4. The monomial $\lambda + 5$ is a factor of the characteristic polynomial if and only if -5 is an eigenvalue of A; it may also happen that 5 is an eigenvalue.

22. a. False. The absolute value of det A equals the volume. See the paragraph before Theorem 3.

  b. False. A and A^T have the same determinant. See Theorem 3(c).

  c. True. See the paragraph before Example 4.

  d. False. See the warning after Theorem 4.

24. First observe that if P is invertible, then Theorem 3(b) shows that $1 = \det(I) = \det(PP^{-1}) = (\det P)(\det P^{-1})$. Then, if $A = PBP^{-1}$, Theorem 3(b) again shows that

$$\det A = \det(PBP^{-1})$$
$$= (\det P)(\det B)(\det P^{-1}) = \det B$$

26. If $a \neq 0$, then
$$A = \begin{bmatrix} a & b \\ c & d \end{bmatrix} \sim \begin{bmatrix} a & b \\ 0 & d - ca^{-1}b \end{bmatrix} = U, \text{ and}$$
$\det A = (a)(d - ca^{-1}b) = ad - bc$. If $a = 0$, then
$$A = \begin{bmatrix} 0 & b \\ c & d \end{bmatrix} \sim \begin{bmatrix} c & d \\ 0 & b \end{bmatrix} = U \text{ (with one interchange)},$$
so $\det A = (-1)^1(cb) = 0 - bc$.

28. [M] In general, the eigenvectors of A are not the same as the eigenvectors of A^T, unless, of course, $A^T = A$.

30. [M] $a = 32$: $\lambda = 1, 1, 2$;
$a = 31.9$: $\lambda = .2958, 1, 2.7042$;
$a = 31.8$: $\lambda = -.1279, 1, 3.1279$;
$a = 32.1$: $\lambda = 1, 1.5 \pm .9747i$;
$a = 32.2$: $\lambda = 1, 1.5 \pm 1.4663i$

Section 5.3, page 286

2. $\begin{bmatrix} 321 & -160 \\ 480 & -239 \end{bmatrix}$

4. $\begin{bmatrix} -3 \cdot (-3)^k + 4 \cdot (-2)^k & 6 \cdot (-3)^k - 6 \cdot (-2)^k \\ -2 \cdot (-3)^k + 2 \cdot (-2)^k & 4 \cdot (-3)^k - 3 \cdot (-2)^k \end{bmatrix}$

6. $\lambda = 3$: $\begin{bmatrix} 3 \\ 0 \\ 1 \end{bmatrix}$, $\begin{bmatrix} -1 \\ -3 \\ 0 \end{bmatrix}$; $\lambda = 4$: $\begin{bmatrix} 0 \\ 1 \\ 0 \end{bmatrix}$

When an answer involves a diagonalization, $A = PDP^{-1}$, the factors P and D are not unique, so your answer may differ from that given here.

8. Not diagonalizable. The eigenvalue 3 has multiplicity two, but the associated eigenspace is only one-dimensional.

10. $P = \begin{bmatrix} -1 & 3 \\ 1 & 4 \end{bmatrix}$, $D = \begin{bmatrix} -2 & 0 \\ 0 & 5 \end{bmatrix}$

12. $P = \begin{bmatrix} -1 & -1 & 1 \\ 1 & 0 & 1 \\ 0 & 1 & 1 \end{bmatrix}$, $D = \begin{bmatrix} 2 & 0 & 0 \\ 0 & 2 & 0 \\ 0 & 0 & 5 \end{bmatrix}$

14. $P = \begin{bmatrix} -1 & 0 & -2 \\ 1 & 1 & 0 \\ 0 & 0 & 1 \end{bmatrix}, D = \begin{bmatrix} 2 & 0 & 0 \\ 0 & 3 & 0 \\ 0 & 0 & 3 \end{bmatrix}$

16. Not diagonalizable. The only real eigenvalue is 0 and its eigenspace is only one-dimensional.

18. $P = \begin{bmatrix} 1 & 2 & 1 \\ 1 & 1 & 1 \\ 1 & 2 & 0 \end{bmatrix}, D = \begin{bmatrix} -2 & 0 & 0 \\ 0 & -1 & 0 \\ 0 & 0 & 0 \end{bmatrix}$

20. Not diagonalizable. Notice that the eigenvalue $\lambda = 3$ has multiplicity 2, but the eigenspace is only one-dimensional.

21. a. False. The symbol D does not automatically denote a diagonal matrix.

b. True. See the remark after the statement of the Diagonalization Theorem.

c. False. The 3×3 matrix in Example 4 has 3 eigenvalues, counting multiplicities, but it is not diagonalizable.

d. False. Invertibility depends on 0 not being an eigenvalue. (See the Invertible Matrix Theorem.) A diagonalizable matrix may or may not have 0 as an eigenvalue. See Examples 3 and 5 for both possibilities.

22. a. False. The n eigenvectors must be linearly independent. See the Diagonalization Theorem.

b. False. The matrix in Example 3 is diagonalizable, but it has only 2 distinct eigenvalues. (The statement given is the *converse* of Theorem 6.)

c. True. This follows from $AP = PD$ and formulas (1) and (2) in the proof of the Diagonalization Theorem.

d. False. See Example 4. The matrix there is invertible because 0 is not an eigenvalue, but the matrix is not diagonalizable.

24. No, by Theorem 7(b). Here is an explanation that does not appeal to Theorem 7: Let v_1 and v_2 be eigenvectors that span the two one-dimensional eigenspaces. If v is any other eigenvector, then it belongs to one of the eigenspaces and hence is a multiple of either v_1 or v_2. So there cannot exist three linearly independent eigenvectors. By the Diagonalization Theorem, A cannot be diagonalizable.

26. Yes, if the third eigenspace is only one-dimensional. In this case, the sum of the dimensions of the eigenspaces will be six, whereas the matrix is 7×7. See Theorem 7(b). An argument similar to that for Exercise 24 can also be given.

28. If A has n linearly independent eigenvectors, then by the Diagonalization Theorem, $A = PDP^{-1}$ for some invertible P and diagonal D. Using properties of transposes,

$A^T = (PDP^{-1})^T = (P^{-1})^T D^T P^T$
$= (P^T)^{-1} D P^T = QDQ^{-1}$

where $Q = (P^T)^{-1}$. Thus A^T is diagonalizable. By the Diagonalization Theorem, the columns of Q are n linearly independent eigenvectors of A^T.

30. A nonzero multiple of an eigenvector is another eigenvector. To produce P_2, simply multiply one or both columns of P by a nonzero scalar unequal to 1.

32. Construct a 2×2 matrix with two distinct eigenvalues, one of which is zero. Simple examples for a and b nonzero:

$\begin{bmatrix} a & b \\ 0 & 0 \end{bmatrix}$, $\begin{bmatrix} 0 & 0 \\ a & b \end{bmatrix}$, $\begin{bmatrix} 0 & a \\ 0 & b \end{bmatrix}$

34. [M] $P = \begin{bmatrix} 2 & -1 & 2 & -2 & 3 \\ -1 & 1 & 0 & 7 & 7 \\ 1 & 0 & 0 & -5 & -5 \\ 0 & 1 & 0 & 5 & 0 \\ 0 & 0 & 1 & 0 & 5 \end{bmatrix}$,

$D = \begin{bmatrix} 5 & 0 & 0 & 0 & 0 \\ 0 & 5 & 0 & 0 & 0 \\ 0 & 0 & 5 & 0 & 0 \\ 0 & 0 & 0 & -2 & 0 \\ 0 & 0 & 0 & 0 & -2 \end{bmatrix}$

36. [M] $P = \begin{bmatrix} 1 & 1 & -1 & 0 & 0 \\ 1 & 0 & 3 & 1 & 0 \\ 0 & 3 & 0 & 0 & -1 \\ 1 & 1 & 0 & 1 & 0 \\ 0 & 0 & 3 & 0 & 1 \end{bmatrix}$,

$D = \begin{bmatrix} 24 & 0 & 0 & 0 & 0 \\ 0 & 36 & 0 & 0 & 0 \\ 0 & 0 & 36 & 0 & 0 \\ 0 & 0 & 0 & -48 & 0 \\ 0 & 0 & 0 & 0 & -48 \end{bmatrix}$

Section 5.4, page 293

2. $\begin{bmatrix} 3 & -2 \\ -3 & 5 \end{bmatrix}$ **4.** $\begin{bmatrix} 2 & -3 & 1 \\ -2 & 0 & 5 \end{bmatrix}$

6. a. $3 - 2t + 7t^2 - 4t^3 + 2t^4$

b. For any p, q in $\mathbb{P}_2$ and any scalar c,

$$T(p + q) = [p(t) + q(t)] + 2t^2[p(t) + q(t)]$$
$$= [p(t) + 2t^2 p(t)] + [q(t) + 2t^2 q(t)]$$
$$= T(p) + T(q)$$

$$T(cp) = [c \cdot p(t)] + 2t^2[c \cdot p(t)]$$
$$= c \cdot [p(t) + 2t^2 p(t)]$$
$$= c \cdot T(p)$$

c. $\begin{bmatrix} 1 & 0 & 0 \\ 0 & 1 & 0 \\ 2 & 0 & 1 \\ 0 & 2 & 0 \\ 0 & 0 & 2 \end{bmatrix}$

8. $5b_2 - 5b_3$

10. a. For any p, q in $\mathbb{P}_3$ and any scalar c,

$$T(\mathbf{p} + \mathbf{q}) = \begin{bmatrix} (\mathbf{p} + \mathbf{q})(-2) \\ (\mathbf{p} + \mathbf{q})(3) \\ (\mathbf{p} + \mathbf{q})(1) \\ (\mathbf{p} + \mathbf{q})(0) \end{bmatrix}$$

$$= \begin{bmatrix} \mathbf{p}(-2) \\ \mathbf{p}(3) \\ \mathbf{p}(1) \\ \mathbf{p}(0) \end{bmatrix} + \begin{bmatrix} \mathbf{q}(-2) \\ \mathbf{q}(3) \\ \mathbf{q}(1) \\ \mathbf{q}(0) \end{bmatrix}$$

$$= T(\mathbf{p}) + T(\mathbf{q})$$

$$T(c \cdot \mathbf{p}) = \begin{bmatrix} (c \cdot \mathbf{p})(-2) \\ (c \cdot \mathbf{p})(3) \\ (c \cdot \mathbf{p})(1) \\ (c \cdot \mathbf{p})(0) \end{bmatrix} = c \cdot \begin{bmatrix} \mathbf{p}(-2) \\ \mathbf{p}(3) \\ \mathbf{p}(1) \\ \mathbf{p}(0) \end{bmatrix}$$

$$= c \cdot T(\mathbf{p})$$

b. $\begin{bmatrix} 1 & -2 & 4 & -8 \\ 1 & 3 & 9 & 27 \\ 1 & 1 & 1 & 1 \\ 1 & 0 & 0 & 0 \end{bmatrix}$

12. $\begin{bmatrix} -4 & 0 \\ 2 & -2 \end{bmatrix}$ 14. $\mathbf{b}_1 = \begin{bmatrix} 1 \\ 1 \end{bmatrix}, \mathbf{b}_2 = \begin{bmatrix} -1 \\ 1 \end{bmatrix}$

16. $\mathbf{b}_1 = \begin{bmatrix} 2 \\ 1 \end{bmatrix}, \mathbf{b}_2 = \begin{bmatrix} -1 \\ 1 \end{bmatrix}$

18. If there is a basis $\mathcal{B}$ such that $[T]_{\mathcal{B}}$ is diagonal, then A is similar to a diagonal matrix, by the second paragraph following Example 3. If A has a set of three eigenvectors that is linearly independent, then there will be a choice of $\mathcal{B}$ such that $[T]_{\mathcal{B}}$ is diagonal. However, since A has only two distinct eigenvalues, it may be the case that a set of linearly independent eigenvectors contains at most two vectors.

20. If $A = PBP^{-1}$, then $A^2 = (PBP^{-1})(PBP^{-1}) = PB(P^{-1}P)BP^{-1} = PB \cdot I \cdot BP^{-1} = PB^2 P^{-1}$. So A^2 is similar to B^2.

22. If A is diagonalizable, then $A = PDP^{-1}$ for some P. Also, if B is similar to A, then $B = QAQ^{-1}$ for some Q. Then

$$B = Q(PDP^{-1})Q^{-1} = (QP)D(P^{-1}Q^{-1})$$
$$= (QP)D(QP)^{-1}$$

So B is diagonalizable.

24. If $A = PBP^{-1}$, then rank $A = $ rank $P(BP^{-1}) = $ rank BP^{-1}, by Supplementary Exercise 13 in Chapter 4. Also, rank $BP^{-1} = $ rank B, by Supplementary Exercise 14 in Chapter 4, since P^{-1} is invertible. Thus rank $A = $ rank B.

26. If $A = PDP^{-1}$ for some P, then the general trace property from Exercise 25 shows that
tr $A = $ tr$[(PD)P^{-1}] = $ tr$[P^{-1}PD] = $ tr D. (Or, one can use the result of Exercise 25 that since A is similar to D, tr $A = $ tr D.) Since the eigenvalues of A are on the main diagonal of D, tr D is the sum of the eigenvalues of A.

28. For each j, $I(\mathbf{b}_j) = \mathbf{b}_j$, and $[I(\mathbf{b}_j)]_{\mathcal{C}} = [\mathbf{b}_j]_{\mathcal{C}}$. By formula (4), the matrix for I relative to the bases $\mathcal{B}$ and $\mathcal{C}$ is

$$M = \begin{bmatrix} [\mathbf{b}_1]_{\mathcal{C}} & [\mathbf{b}_2]_{\mathcal{C}} & \cdots & [\mathbf{b}_n]_{\mathcal{C}} \end{bmatrix}$$

In Theorem 15 in Section 4.7, this matrix was denoted by $\underset{\mathcal{C} \leftarrow \mathcal{B}}{P}$ and was called the *change-of-coordinates matrix from $\mathcal{B}$ to $\mathcal{C}$*.

30. **[M]** $P^{-1}AP = \begin{bmatrix} 2 & -1 & 0 \\ 0 & 3 & 0 \\ 0 & 1 & 4 \end{bmatrix}$

32. **[M]** $\lambda = -2$: $\mathbf{b}_1 = \begin{bmatrix} 1 \\ 1 \\ 1 \\ 0 \end{bmatrix}, \mathbf{b}_2 = \begin{bmatrix} 6 \\ -3 \\ 0 \\ 4 \end{bmatrix};$

$\lambda = 1$: $\mathbf{b}_3 = \begin{bmatrix} 2 \\ -1 \\ -7 \\ 2 \end{bmatrix};$

$\lambda = 5$: $\mathbf{b}_4 = \begin{bmatrix} 2 \\ 1 \\ -1 \\ 2 \end{bmatrix};$

basis: $\mathcal{B} = \{\mathbf{b}_1, \mathbf{b}_2, \mathbf{b}_3, \mathbf{b}_4\}$

Section 5.5, page 300

2. $\lambda = 3 + 3i, \begin{bmatrix} i \\ 1 \end{bmatrix}; \quad \lambda = 3 - 3i, \begin{bmatrix} -i \\ 1 \end{bmatrix}$

4. $\lambda = 2 + i, \begin{bmatrix} -1 + i \\ 1 \end{bmatrix}; \quad \lambda = 2 - i, \begin{bmatrix} -1 - i \\ 1 \end{bmatrix}$

6. $\lambda = 5 + i, \begin{bmatrix} 2 + i \\ 1 \end{bmatrix}; \quad \lambda = 5 - i, \begin{bmatrix} 2 - i \\ 1 \end{bmatrix}$

8. $\lambda = 3 \pm 3\sqrt{3}i, \varphi = -\pi/3$ radians, $r = 6$

10. $\lambda = \pm .5i, \varphi = -\pi/2$ radians, $r = .5$

12. $\lambda = 3 \pm i\sqrt{3}, \varphi = \pi/6$ radians, $r = 2\sqrt{3}$

In Exercises 13–20, other answers are possible. Any P that makes $P^{-1}AP$ equal to the given C or to C^T is a satisfactory answer. First find P; then compute $P^{-1}AP$.

14. $P = \begin{bmatrix} \sqrt{2} & 1 \\ 0 & 1 \end{bmatrix}, C = \begin{bmatrix} 2 & -\sqrt{2} \\ \sqrt{2} & 2 \end{bmatrix}$

16. $P = \begin{bmatrix} 1 & 1 \\ 0 & -1 \end{bmatrix}, C = \begin{bmatrix} 5 & 1 \\ -1 & 5 \end{bmatrix}$

18. $P = \begin{bmatrix} 3 & 1 \\ 0 & -2 \end{bmatrix}, C = \begin{bmatrix} 4 & 3 \\ -3 & 4 \end{bmatrix}$

20. $P = \begin{bmatrix} 1 & -1 \\ 0 & 1 \end{bmatrix}, C = \begin{bmatrix} 1 & -4 \\ 4 & 1 \end{bmatrix}$

22. $A(\mu \mathbf{x}) = \mu(A\mathbf{x}) = \mu(\lambda \mathbf{x}) = \lambda(\mu \mathbf{x})$

24. $\overline{\mathbf{x}}^T A\mathbf{x} = \overline{\mathbf{x}}^T(\lambda \mathbf{x}) = \lambda \cdot \overline{\mathbf{x}}^T \mathbf{x}$ because $\mathbf{x}$ is an eigenvector. It is easy to see that $\overline{\mathbf{x}}^T \mathbf{x}$ is real (and positive) because $\overline{z}z$ is nonnegative for every complex number z. Since $\overline{\mathbf{x}}^T A\mathbf{x}$ is real, by Exercise 23, so is λ. Next, write $\mathbf{x} = \mathbf{u} + i\mathbf{v}$, where $\mathbf{u}$ and $\mathbf{v}$ are real vectors. Then

$$A\mathbf{x} = A(\mathbf{u} + i\mathbf{v}) = A\mathbf{u} + iA\mathbf{v} \quad \text{and} \quad \lambda \mathbf{x} = \lambda \mathbf{u} + i\lambda \mathbf{v}$$

The real part of $A\mathbf{x}$ is $A\mathbf{u}$ because the entries in A, $\mathbf{u}$, and $\mathbf{v}$ are all real. The real part of $\lambda\mathbf{x}$ is $\lambda\mathbf{u}$ because λ and the entries in $\mathbf{u}$ and $\mathbf{v}$ are real. Since $A\mathbf{x}$ and $\lambda\mathbf{x}$ are equal, their real parts are equal, too. (Apply the corresponding statement about complex numbers to each entry of $A\mathbf{x}$.) Thus $A\mathbf{u} = \lambda\mathbf{u}$, which shows that the real part of $\mathbf{x}$ is an eigenvector of A.

26. a. If $\lambda = a - bi$, then

$$A\mathbf{v} = \lambda\mathbf{v} = (a - bi)(\operatorname{Re}\mathbf{v} + i\operatorname{Im}\mathbf{v})$$
$$= \underbrace{(a\operatorname{Re}\mathbf{v} + b\operatorname{Im}\mathbf{v})}_{\operatorname{Re} A\mathbf{v}} + i\underbrace{(a\operatorname{Im}\mathbf{v} - b\operatorname{Re}\mathbf{v})}_{\operatorname{Im} A\mathbf{v}}$$

By Exercise 25,

$$A(\operatorname{Re}\mathbf{v}) = \operatorname{Re} A\mathbf{v} = a\operatorname{Re}\mathbf{v} + b\operatorname{Im}\mathbf{v}$$
$$A(\operatorname{Im}\mathbf{v}) = \operatorname{Im} A\mathbf{v} = -b\operatorname{Re}\mathbf{v} + a\operatorname{Im}\mathbf{v}$$

b. Let $P = [\operatorname{Re}\mathbf{v} \quad \operatorname{Im}\mathbf{v}]$. By part (a),

$$A(\operatorname{Re}\mathbf{v}) = P\begin{bmatrix} a \\ b \end{bmatrix}, \quad A(\operatorname{Im}\mathbf{v}) = P\begin{bmatrix} -b \\ a \end{bmatrix}$$

So

$$AP = [A(\operatorname{Re}\mathbf{v}) \quad A(\operatorname{Im}\mathbf{v})]$$
$$= \begin{bmatrix} P\begin{bmatrix} a \\ b \end{bmatrix} & P\begin{bmatrix} -b \\ a \end{bmatrix} \end{bmatrix} = P\begin{bmatrix} a & -b \\ b & a \end{bmatrix} = PC$$

28. [M] $P = \begin{bmatrix} 1 & 1 & -2 & 0 \\ -4 & 0 & 0 & -2 \\ 0 & 0 & -3 & 1 \\ 2 & 0 & 4 & 0 \end{bmatrix}$,

$$C = \begin{bmatrix} 2 & -5 & 0 & 0 \\ 5 & 2 & 0 & 0 \\ 0 & 0 & 3 & 1 \\ 0 & 0 & -1 & 3 \end{bmatrix}$$

Other choices are possible, but C must equal $P^{-1}AP$.

Section 5.6, page 309

2. $\mathbf{x}_k = 2 \cdot 3^k \begin{bmatrix} 1 \\ 0 \\ -3 \end{bmatrix} + 1 \cdot \left(\dfrac{4}{5}\right)^k \begin{bmatrix} 2 \\ 1 \\ -5 \end{bmatrix} + 2 \cdot \left(\dfrac{3}{5}\right)^k \begin{bmatrix} -3 \\ -3 \\ 7 \end{bmatrix}$

So $\mathbf{x}_k \approx 2 \cdot 3^k \begin{bmatrix} 1 \\ 0 \\ -3 \end{bmatrix}$ for all k sufficiently large.

4. $\mathbf{x}_k = c_1 \begin{bmatrix} 4 \\ 5 \end{bmatrix} + c_2(.6)^k \begin{bmatrix} 4 \\ 1 \end{bmatrix} \to c_1 \begin{bmatrix} 4 \\ 5 \end{bmatrix}$, as $k \to \infty$.

Provided that $c_1 > 0$, the owl and wood rat populations each stabilize in size, and eventually the populations are in the ratio of 4 owls to 5 thousand rats. If some aspect of the model were to change slightly, the characteristic equation would change slightly and the perturbed matrix A might not have 1 as an eigenvalue. If the eigenvalue becomes slightly larger than 1, the two populations will grow; if the eigenvalue becomes slightly less than 1, both populations will decline.

6. When $p = .5$, the eigenvalues of A are .9 and .7, both less than 1 in magnitude. The origin is an attractor for the dynamical system, and each trajectory tends toward $\mathbf{0}$. So populations of both owls and squirrels eventually disappear. For any p, the characteristic equation of A is $\lambda^2 - 1.6\lambda + (.48 + .3p) = 0$. The matrix A has an eigenvalue of 1 when $p = .4$. In this case, both owl and squirrel populations tend toward constant levels, with 1 spotted owl for every 2 (thousand) flying squirrels.

8. Saddle point (because one or more eigenvalues are greater than 1, and one or more eigenvalues are less than 1, in magnitude); direction of greatest repulsion: the line through $(0, 0, 0)$ and $(1, 0, -3)$; direction of greatest attraction: the line through $(0, 0, 0)$ and $(-3, -3, 7)$

10. Attractor; eigenvalues: .9, .5; direction of greatest attraction: the line through $(0, 0)$ and $(2, 1)$

12. Saddle point; eigenvalues: 1.1, .8; greatest repulsion: line through $(0, 0)$ and $(1, 1)$; greatest attraction: line through $(0, 0)$ and $(2, 1)$

14. Repeller; eigenvalues: 1.3, 1.1; greatest repulsion: line through $(0, 0)$ and $(-3, 2)$

16. [M] $\mathbf{v}_k = c_1 \begin{bmatrix} .435 \\ .091 \\ .474 \end{bmatrix} + c_2(.89)^k \begin{bmatrix} -1 \\ 1 \\ 0 \end{bmatrix} + c_3(.81)^k \begin{bmatrix} -1 \\ 0 \\ 1 \end{bmatrix}$

Note: The exact value of the steady-state vector is $\mathbf{q} = (91/99, 19/99, 99/99) \approx (.9192, .1919, 1.0)$.

18. a. $A = \begin{bmatrix} 0 & 0 & .42 \\ .6 & 0 & 0 \\ 0 & .75 & .95 \end{bmatrix}$

b. [M] The long-term growth rate is $\lambda_1 = 1.105$. A corresponding eigenvector is approximately $(38, 21, 100)$. For each 100 adults, there will be approximately 38 calves and 21 yearlings.

Section 5.7, page 317

2. $\mathbf{x}(t) = \dfrac{1}{2}\begin{bmatrix} -1 \\ 1 \end{bmatrix}e^{-3t} + \dfrac{5}{2}\begin{bmatrix} 1 \\ 1 \end{bmatrix}e^{-t}$

4. $\dfrac{13}{4}\begin{bmatrix} -1 \\ 1 \end{bmatrix}e^{3t} - \dfrac{5}{4}\begin{bmatrix} -5 \\ 1 \end{bmatrix}e^{-t}$. The origin is a saddle point. The direction of greatest attraction is the line through $(-5, 1)$ and the origin. The direction of greatest repulsion is the line through $(-1, 1)$ and the origin.

6. $5\begin{bmatrix} 1 \\ 1 \end{bmatrix}e^{-t} - \begin{bmatrix} 2 \\ 3 \end{bmatrix}e^{-2t}$. The origin is an attractor. The direction of greatest attraction is the line through $(2, 3)$ and the origin.

8. Set $P = \begin{bmatrix} 2 & 1 \\ 3 & 1 \end{bmatrix}$ and $D = \begin{bmatrix} -2 & 0 \\ 0 & -1 \end{bmatrix}$. Then $A = PDP^{-1}$. Substituting $\mathbf{x} = P\mathbf{y}$ into $\mathbf{x}' = A\mathbf{x}$, we have

$$\dfrac{d}{dt}(P\mathbf{y}) = A(P\mathbf{y})$$
$$P\mathbf{y}' = PDP^{-1}(P\mathbf{y}) = PD\mathbf{y}$$

Left-multiplying by P^{-1} gives

$$\mathbf{y}' = D\mathbf{y}, \quad \text{or} \quad \begin{bmatrix} y_1'(t) \\ y_2'(t) \end{bmatrix} = \begin{bmatrix} -2 & 0 \\ 0 & -1 \end{bmatrix} \begin{bmatrix} y_1(t) \\ y_2(t) \end{bmatrix}$$

10. (complex): $\quad c_1 \begin{bmatrix} 1+i \\ -2 \end{bmatrix} e^{(2+i)t} + c_2 \begin{bmatrix} 1-i \\ -2 \end{bmatrix} e^{(2-i)t}$

(real): $\quad c_1 \begin{bmatrix} \cos t - \sin t \\ -2\cos t \end{bmatrix} e^{2t} + c_2 \begin{bmatrix} \sin t + \cos t \\ -2\sin t \end{bmatrix} e^{2t}$

The trajectories spiral out, away from the origin.

12. (complex): $\quad c_1 \begin{bmatrix} 3-i \\ 2 \end{bmatrix} e^{(-1+2i)t} + c_2 \begin{bmatrix} 3+i \\ 2 \end{bmatrix} e^{(-1-2i)t}$

(real):

$c_1 \begin{bmatrix} 3\cos 2t + \sin 2t \\ 2\cos 2t \end{bmatrix} e^{-t} + c_2 \begin{bmatrix} 3\sin 2t - \cos 2t \\ 2\sin 2t \end{bmatrix} e^{-t}$

The trajectories spiral in, toward the origin.

14. (complex): $\quad c_1 \begin{bmatrix} 1-i \\ 4 \end{bmatrix} e^{2it} + c_2 \begin{bmatrix} 1+i \\ 4 \end{bmatrix} e^{-2it}$

(real): $\quad c_1 \begin{bmatrix} \cos 2t + \sin 2t \\ 4\cos 2t \end{bmatrix} + c_2 \begin{bmatrix} \sin 2t - \cos 2t \\ 4\sin 2t \end{bmatrix}$

The trajectories are ellipses about the origin.

16. [M] $\mathbf{x}(t) = c_1 \begin{bmatrix} 7 \\ -2 \\ 3 \end{bmatrix} e^{4t} + c_2 \begin{bmatrix} 3 \\ -1 \\ 1 \end{bmatrix} e^{3t} + c_3 \begin{bmatrix} 2 \\ 0 \\ 1 \end{bmatrix} e^{2t}$

The origin is a repeller. All trajectories curve away from the origin.

18. [M] (complex): $\quad c_1 \begin{bmatrix} 1 \\ 2 \\ 0 \end{bmatrix} e^{-7t} + c_2 \begin{bmatrix} 6+2i \\ 9+3i \\ 10 \end{bmatrix} e^{(5+i)t} +$

$c_3 \begin{bmatrix} 6-2i \\ 9-3i \\ 10 \end{bmatrix} e^{(5-i)t}$

(real) : $\quad c_1 \begin{bmatrix} 1 \\ 2 \\ 0 \end{bmatrix} e^{-7t} + c_2 \begin{bmatrix} 6\cos t - 2\sin t \\ 9\cos t - 3\sin t \\ 10\cos t \end{bmatrix} e^{5t} +$

$c_3 \begin{bmatrix} 6\sin t + 2\cos t \\ 9\sin t + 3\cos t \\ 10\sin t \end{bmatrix} e^{5t}$

When $c_2 = c_3 = 0$, the trajectories tend straight toward **0**. In other cases, the trajectories spiral outward.

20. [M] $A = \begin{bmatrix} -2 & 1/3 \\ 3/2 & -3/2 \end{bmatrix}$,

$\begin{bmatrix} v_1(t) \\ v_2(t) \end{bmatrix} = \frac{5}{3}\begin{bmatrix} 1 \\ 3 \end{bmatrix} e^{-t} - \frac{2}{3}\begin{bmatrix} -2 \\ 3 \end{bmatrix} e^{-2.5t}$

22. [M] $A = \begin{bmatrix} 0 & 2 \\ -.4 & .8 \end{bmatrix}$,

$\begin{bmatrix} i_L(t) \\ v_C(t) \end{bmatrix} = \begin{bmatrix} 30\sin .8t \\ 12\cos .8t - 6\sin .8t \end{bmatrix} e^{-.4t}$

Section 5.8, page 324

2. Eigenvector: $\mathbf{x}_4 = \begin{bmatrix} -.2520 \\ 1 \end{bmatrix}$, or $A\mathbf{x}_4 = \begin{bmatrix} -1.2536 \\ 5.0064 \end{bmatrix}$; $\lambda \approx 5.0064$

4. Eigenvector: $\mathbf{x}_4 = \begin{bmatrix} 1 \\ .7502 \end{bmatrix}$, or $A\mathbf{x}_4 = \begin{bmatrix} -.4012 \\ -.3009 \end{bmatrix}$; $\lambda \approx -.4012$

6. $\mathbf{x} = \begin{bmatrix} -.4996 \\ 1 \end{bmatrix}$, $A\mathbf{x} = \begin{bmatrix} -2.0008 \\ 4.0024 \end{bmatrix}$; estimated $\lambda = 4.0024$

8. [M]

$\mathbf{x}_k: \begin{bmatrix} .5 \\ 1 \end{bmatrix}, \begin{bmatrix} .2857 \\ 1 \end{bmatrix}, \begin{bmatrix} .2558 \\ 1 \end{bmatrix}, \begin{bmatrix} .2510 \\ 1 \end{bmatrix}, \begin{bmatrix} .2502 \\ 1 \end{bmatrix}$

$\mu_k: \quad 7, \qquad 6.14, \qquad 6.02, \qquad 6.0039, \qquad 6.0006$

10. [M] $\mu_5 = 9.9319$, $\mu_6 = 9.9872$; actual value: 10
Note: Starting with $\mathbf{x}_0 = (0, 0, 1)$ produces $\mu_4 = 9.9993$, $\mu_5 = 9.9999$.

12. $\quad \mu_k: -4.3333, -3.9231, -4.0196, -3.9951$
$R(\mathbf{x}_k): -3.9231, -3.9951, -3.9997, -3.99998$

14. Use the inverse power method, with $\alpha = 4$.

16. $\lambda = \alpha + 1/\mu$

18. [M] $\nu_0 = -1.375$, $\nu_1 = -1.42623$, $\nu_2 = -1.42432$, $\nu_3 = -1.42444$. Actual: -1.424429 (accurate to six places)

20. [M] **a.** $\mu_8 = 19.1820 = \mu_9$ to four decimal places. To six places, the largest eigenvalue is 19.182037, with eigenvector $(.184416, 1, .179615, .407110)$.
b. $\mu_1^{-1} = .012235$, $\mu_2^{-1} = .012205$. To six places, the smallest eigenvalue is .012206, with eigenvector $(1, .222610, -.917993, .660483)$. The other eigenvalues are -2.453128 and -1.741114, to six places.

Chapter 5 Supplementary Exercises, page 326

1. a. True. If A is invertible and if $A\mathbf{x} = 1 \cdot \mathbf{x}$ for some nonzero $\mathbf{x}$, then left-multiply by A^{-1} to obtain $\mathbf{x} = A^{-1}\mathbf{x}$, which may be rewritten as $A^{-1}\mathbf{x} = 1 \cdot \mathbf{x}$. Since $\mathbf{x}$ is nonzero, this shows that 1 is an eigenvalue of A^{-1}.

b. False. If A is row equivalent to the identity matrix, then A is invertible. The matrix in Example 4 in Section 5.3 shows that an invertible matrix need not be diagonalizable. Also, see Exercise 31 in Section 5.3.

c. True. If A contains a row or column of zeros, then A is not row equivalent to the identity matrix and thus is not invertible. By the Invertible Matrix Theorem (as stated in Section 5.2), 0 is an eigenvalue of A.

d. False. Consider a diagonal matrix D whose eigenvalues are 1 and 3; that is, its diagonal entries are 1 and 3. Then D^2 is a diagonal matrix whose eigenvalues (diagonal entries) are 1 and 9. In general, the eigenvalues of A^2 are the *squares* of the eigenvalues of A.

e. True. Suppose a nonzero vector $\mathbf{x}$ satisfies $A\mathbf{x} = \lambda\mathbf{x}$, then

$$A^2\mathbf{x} = A(A\mathbf{x}) = A(\lambda\mathbf{x}) = \lambda A\mathbf{x} = \lambda^2\mathbf{x}$$

This shows that $\mathbf{x}$ is also an eigenvector of A^2.

f. True. Suppose a nonzero vector $\mathbf{x}$ satisfies $A\mathbf{x} = \lambda\mathbf{x}$, then left-multiply by A^{-1} to obtain $\mathbf{x} = A^{-1}(\lambda\mathbf{x}) = \lambda A^{-1}\mathbf{x}$. Since A is invertible, the eigenvalue λ is not zero. So $\lambda^{-1}\mathbf{x} = A^{-1}\mathbf{x}$, which shows that $\mathbf{x}$ is also an eigenvector of A^{-1}.

g. False. Zero is an eigenvalue of each singular square matrix.

h. True. By definition, an eigenvector must be nonzero.

i. False. If the dimension of the eigenspace is at least 2, then there are at least two linearly independent eigenvectors in the same subspace.

j. True. This follows from Theorem 4 in Section 5.2.

k. False. Let A be the 3×3 matrix in Example 3 in Section 5.3. Then A is similar to a diagonal matrix D. The eigenvectors of D are the columns of I_3, but the eigenvectors of A are entirely different.

l. False. Let $A = \begin{bmatrix} 2 & 0 \\ 0 & 3 \end{bmatrix}$. Then $\mathbf{e}_1 = \begin{bmatrix} 1 \\ 0 \end{bmatrix}$ and $\mathbf{e}_2 = \begin{bmatrix} 0 \\ 1 \end{bmatrix}$ are eigenvectors of A, but $\mathbf{e}_1 + \mathbf{e}_2$ is not. (Actually, it can be shown that if two eigenvectors of A correspond to distinct eigenvalues, then their sum cannot be an eigenvector.)

m. False. *All* the diagonal entries of an upper triangular matrix are the eigenvalues of the matrix (Theorem 1 in Section 5.1). A diagonal entry may be zero.

n. True. Matrices A and A^T have the same characteristic polynomial, because $\det(A^T - \lambda I) = \det(A - \lambda I)^T = \det(A - \lambda I)$, by the determinant transpose property.

o. False. Counterexample: Let A be the 5×5 identity matrix.

p. True. For example, let A be the matrix that rotates vectors through $\pi/2$ radians about the origin. Then $A\mathbf{x}$ is not a multiple of $\mathbf{x}$ when $\mathbf{x}$ is nonzero.

q. False. If A is a diagonal matrix with a zero on the diagonal, then the columns of A are not linearly independent.

r. True. If $A\mathbf{x} = \lambda_1\mathbf{x}$ and $A\mathbf{x} = \lambda_2\mathbf{x}$, then $\lambda_1\mathbf{x} = \lambda_2\mathbf{x}$ and $(\lambda_1 - \lambda_2)\mathbf{x} = \mathbf{0}$. If $\mathbf{x} \neq \mathbf{0}$, then λ_1 must equal λ_2.

s. False. Let A be a singular matrix that is diagonalizable. (For instance, let A be a diagonal matrix with a zero on the diagonal.) Then, by Theorem 8 in Section 5.4, the transformation $\mathbf{x} \mapsto A\mathbf{x}$ is represented by a diagonal matrix relative to a coordinate system determined by eigenvectors of A.

t. True. By definition of matrix multiplication,

$$A = AI = A[\,\mathbf{e}_1 \quad \mathbf{e}_2 \quad \cdots \quad \mathbf{e}_n\,] = [\,A\mathbf{e}_1 \quad A\mathbf{e}_2 \quad \cdots \quad A\mathbf{e}_n\,]$$

If $A\mathbf{e}_j = d_j\mathbf{e}_j$ for $j = 1, \ldots, n$, then A is a diagonal matrix with diagonal entries $d_1, \ldots, d_n$.

u. True. If $B = PDP^{-1}$, where D is a diagonal matrix, and if $A = QBQ^{-1}$, then $A = Q(PDP^{-1})Q^{-1} = (QP)D(QP)^{-1}$, which shows that A is diagonalizable.

v. True. Since B is invertible, AB is similar to $B(AB)B^{-1}$, which equals BA.

w. False. Having n linearly independent eigenvectors makes an $n \times n$ matrix diagonalizable (by the Diagonalization Theorem in Section 5.3), but not necessarily invertible. One of the eigenvalues of the matrix could be zero.

x. True. If A is diagonalizable, then by the Diagonalization Theorem, A has n linearly independent eigenvectors $\mathbf{v}_1, \ldots, \mathbf{v}_n$ in $\mathbb{R}^n$. By the Basis Theorem, $\{\mathbf{v}_1, \ldots, \mathbf{v}_n\}$ spans $\mathbb{R}^n$. This means that each vector in $\mathbb{R}^n$ can be written as a linear combination of $\mathbf{v}_1, \ldots, \mathbf{v}_n$.

2. Suppose $B\mathbf{x} \neq \mathbf{0}$ and $AB\mathbf{x} = \lambda\mathbf{x}$ for some λ. Then $A(B\mathbf{x}) = \lambda\mathbf{x}$. Left-multiply each side by B, and obtain $BA(B\mathbf{x}) = B(\lambda\mathbf{x}) = \lambda(B\mathbf{x})$. This equation says that $B\mathbf{x}$ is an eigenvector of BA, because $B\mathbf{x} \neq \mathbf{0}$.

4. Assume that $A\mathbf{x} = \lambda\mathbf{x}$ for some nonzero vector $\mathbf{x}$. The desired statement is true for $m = 1$, by the assumption about λ. Suppose the statement holds when $m = k$, for some $k \geq 1$. That is, suppose that $A^k\mathbf{x} = \lambda^k\mathbf{x}$. Then, by the induction hypothesis,

$$A^{k+1}\mathbf{x} = A(A^k\mathbf{x}) = A(\lambda^k\mathbf{x})$$

Continuing, $A^{k+1}\mathbf{x} = \lambda^k A\mathbf{x} = \lambda^{k+1}\mathbf{x}$, because $\mathbf{x}$ is an eigenvector of A corresponding to λ. Since $\mathbf{x}$ is nonzero, this equation shows that λ^{k+1} is an eigenvalue of A^{k+1}, with corresponding eigenvector $\mathbf{x}$. Thus the desired statement is true when $m = k + 1$. By the principal of induction, the statement is true for each positive integer m.

6. a. If $A = PDP^{-1}$, then $A^k = PD^kP^{-1}$, and

$$\begin{aligned} B &= 5I - 3A + A^2 \\ &= 5PIP^{-1} - 3PDP^{-1} + PD^2P^{-1} \\ &= P(5I - 3D + D^2)P^{-1} \end{aligned}$$

Since D is diagonal, so is $5I - 3D + D^2$. Thus B is similar to a diagonal matrix.

b. $\begin{aligned} p(A) &= c_0 I + c_1 PDP^{-1} + c_2 PD^2P^{-1} \\ &\quad + \cdots + c_n PD^nP^{-1} \\ &= P(c_0 I + c_1 D + c_2 D^2 + \cdots + c_n D^n)P^{-1} \\ &= Pp(D)P^{-1} \end{aligned}$

This shows that $p(A)$ is diagonalizable, because $p(D)$ is a linear combination of diagonal matrices and hence is diagonal. In fact, because D is diagonal, it is easy to see that

$$p(D) = \begin{bmatrix} p(2) & 0 \\ 0 & p(7) \end{bmatrix}$$

8. a. If λ is an eigenvalue of an $n \times n$ diagonalizable matrix A, then $A = PDP^{-1}$ for an invertible matrix P and an $n \times n$ diagonal matrix D whose diagonal entries are the eigenvalues of A. If the multiplicity of λ is n, then λ must appear in every diagonal entry of D. That is, $D = \lambda I$. In this case, $A = P(\lambda I)P^{-1} = \lambda PIP^{-1} = \lambda PP^{-1} = \lambda I$.

   b. Since the matrix $A = \begin{bmatrix} 3 & 1 \\ 0 & 3 \end{bmatrix}$ is triangular, its eigenvalues are on the diagonal. Thus 3 is an eigenvalue with multiplicity 2. If the 2×2 matrix A were diagonalizable, then A would be $3I$, by part (a). This is not the case, so A is not diagonalizable.

10. To show that A^k tends to the zero matrix, it suffices to show that each column of A^k can be made as close to the zero vector as desired by taking k sufficiently large. The jth column of A is $A\mathbf{e}_j$, where $\mathbf{e}_j$ is the jth column of the identity matrix. Since A is diagonalizable, there is a basis for $\mathbb{R}^n$ consisting of eigenvectors $\mathbf{v}_1, \ldots, \mathbf{v}_n$, corresponding to eigenvalues $\lambda_1, \ldots, \lambda_n$. So there exist scalars $c_1, \ldots, c_n$, such that

$$\mathbf{e}_j = c_1 \mathbf{v}_1 + \cdots + c_n \mathbf{v}_n \quad \text{(an eigenvector decomposition of } \mathbf{e}_j)$$

Then, for $k = 1, 2, \ldots,$

$$A^k \mathbf{e}_j = c_1 (\lambda_1)^k \mathbf{v}_1 + \cdots + c_n (\lambda_n)^k \mathbf{v}_n \qquad (*)$$

If the eigenvalues are all less than 1 in absolute value, then their kth powers all tend to zero. So $(*)$ shows that $A^k \mathbf{e}_j$ tends to the zero matrix, as desired.

12. Let U and V be echelon forms of A and B, obtained with r and s row interchanges, respectively, and no scaling. Then

$$\det A = (-1)^r \det U \quad \text{and} \quad \det B = (-1)^s \det V$$

Using first the row operations that reduce A to U, we can reduce G to a matrix of the form $G' = \begin{bmatrix} U & Y \\ 0 & B \end{bmatrix}$. Then, using the row operations that reduce B to V, we can further reduce G' to $G'' = \begin{bmatrix} U & Y \\ 0 & V \end{bmatrix}$. There will be $r + s$ row interchanges, and so

$$\det G = \det \begin{bmatrix} A & X \\ 0 & B \end{bmatrix} = (-1)^{r+s} \det \begin{bmatrix} U & Y \\ 0 & V \end{bmatrix}$$

Since $\begin{bmatrix} U & Y \\ 0 & V \end{bmatrix}$ is upper triangular, its determinant equals the product of the diagonal entries, and since U and V are upper triangular, this product also equals $(\det U)(\det V)$. Thus

$$\det G = (-1)^{r+s}(\det U)(\det V) = (\det A)(\det B)$$

For any scalar λ, the matrix $G - \lambda I$ has the same partitioned form as G, with $A - \lambda I$ and $B - \lambda I$ as its diagonal blocks. (Here I represents various identity matrices of appropriate sizes.) Hence the result about $\det G$ shows that

$$\det(G - \lambda I) = \det(A - \lambda I) \cdot \det(B - \lambda I)$$

14. $6, -1, -1, -5$

16. The 3×3 matrix has eigenvalues $1 - 2$ and $1 + (2)(2)$, that is, -1 and 5. The eigenvalues of the 5×5 matrix are $7 - 3$ and $7 + (4)(3)$, that is, 4 and 19.

18. The eigenvalues of A are 1 and .6. Use this to factor A and A^k.

$$A = \begin{bmatrix} -1 & -3 \\ 2 & 2 \end{bmatrix} \begin{bmatrix} 1 & 0 \\ 0 & .6 \end{bmatrix} \frac{1}{4} \begin{bmatrix} 2 & 3 \\ -2 & -1 \end{bmatrix}$$

$$A^k = \begin{bmatrix} -1 & -3 \\ 2 & 2 \end{bmatrix} \begin{bmatrix} 1^k & 0 \\ 0 & .6^k \end{bmatrix} \cdot \frac{1}{4} \begin{bmatrix} 2 & 3 \\ -2 & -1 \end{bmatrix}$$

$$= \frac{1}{4} \begin{bmatrix} -1 & -3 \\ 2 & 2 \end{bmatrix} \begin{bmatrix} 2 & 3 \\ -2 \cdot (.6)^k & -(.6)^k \end{bmatrix}$$

$$= \frac{1}{4} \begin{bmatrix} -2 + 6(.6)^k & -3 + 3(.6)^k \\ 4 - 4(.6)^k & 6 - 2(.6)^k \end{bmatrix}$$

$$\rightarrow \frac{1}{4} \begin{bmatrix} -2 & -3 \\ 4 & 6 \end{bmatrix} \quad \text{as } k \to \infty$$

20. $C_p = \begin{bmatrix} 0 & 1 & 0 \\ 0 & 0 & 1 \\ 24 & -26 & 9 \end{bmatrix}$;

$$\det(C_p - \lambda I) = 24 - 26\lambda + 9\lambda^2 - \lambda^3$$

22. a. $C_p = \begin{bmatrix} 0 & 1 & 0 \\ 0 & 0 & 1 \\ -a_0 & -a_1 & -a_2 \end{bmatrix}$

   b. Since λ is a zero of p, $a_0 + a_1\lambda + a_2\lambda^2 + \lambda^3 = 0$ and $-a_0 - a_1\lambda - a_2\lambda^2 = \lambda^3$. Thus

$$C_p \begin{bmatrix} 1 \\ \lambda \\ \lambda^2 \end{bmatrix} = \begin{bmatrix} \lambda \\ \lambda^2 \\ -a_0 - a_1\lambda - a_2\lambda^2 \end{bmatrix} = \begin{bmatrix} \lambda \\ \lambda^2 \\ \lambda^3 \end{bmatrix}$$

That is, $C_p(1, \lambda, \lambda^2) = \lambda(1, \lambda, \lambda^2)$, which shows that $(1, \lambda, \lambda^2)$ is an eigenvector of C_p corresponding to the eigenvalue λ.

24. [M] The MATLAB command `roots(p)` requires as input a row vector p whose entries are the coefficients of a polynomial, with the highest order coefficient listed first. MATLAB constructs a companion matrix C_p whose characteristic polynomial is p, so the roots of p are the eigenvalues of C_p. The numerical values of the eigenvalues (roots) are found by the same QR algorithm used by the command `eig(A)`.

25. [M] The MATLAB command `[P D] = eig(A)` produces a matrix P, whose condition number is 1.6×10^8, and a diagonal matrix D, whose entries are *almost* 2, 2, 1. However, the exact eigenvalues of A are 2, 2, 1, and A is not diagonalizable.

26. [M] This matrix may cause the same sort of trouble as the matrix in Exercise 25. A matrix program that computes eigenvalues by an interative process may indicate that A

has four distinct eigenvalues, all close to zero. However, the only eigenvalue is 0, with multiplicity 4, because $A^4 = 0$.

Chapter 6

Section 6.1, page 336

2. $35, 5, \frac{1}{7}$ **4.** $\begin{bmatrix} -1/5 \\ 2/5 \end{bmatrix}$ **6.** $\begin{bmatrix} 30/49 \\ -10/49 \\ 15/49 \end{bmatrix}$

8. 7 **10.** $\begin{bmatrix} -6/\sqrt{61} \\ 4/\sqrt{61} \\ -3/\sqrt{61} \end{bmatrix}$ **12.** $\begin{bmatrix} .8 \\ .6 \end{bmatrix}$

14. $2\sqrt{17}$ **16.** Orthogonal **18.** Not orthogonal

19. a. True. See the definition of $\|\mathbf{v}\|$.

  b. True. See Theorem 1(c).

  c. True. See the discussion of Fig. 5.

  d. False. Counterexample: $\begin{bmatrix} 1 & 1 \\ 0 & 0 \end{bmatrix}$.

  e. True. See the box following Example 6.

20. a. True. See Example 1 and Theorem 1(a).

  b. False. The absolute value is missing. See the box before Example 2.

  c. True, by definition of the orthogonal complement.

  d. True, by the Pythagorean Theorem.

  e. True, by Theorem 3.

22. $\mathbf{u} \cdot \mathbf{u} \geq 0$ because $\mathbf{u} \cdot \mathbf{u}$ is a sum of squares of the entries in $\mathbf{u}$. The sum of squares of numbers is zero if and only if all the numbers are themselves zero.

24. $\|\mathbf{u} + \mathbf{v}\|^2 = (\mathbf{u} + \mathbf{v}) \cdot (\mathbf{u} + \mathbf{v}) = \mathbf{u} \cdot \mathbf{u} + 2\mathbf{u} \cdot \mathbf{v} + \mathbf{v} \cdot \mathbf{v}$
$$= \|\mathbf{u}\|^2 + 2\mathbf{u} \cdot \mathbf{v} + \|\mathbf{v}\|^2$$
$\|\mathbf{u} - \mathbf{v}\|^2 = (\mathbf{u} - \mathbf{v}) \cdot (\mathbf{u} - \mathbf{v})$
$$= \mathbf{u} \cdot \mathbf{u} + \mathbf{u} \cdot (-\mathbf{v}) - \mathbf{v} \cdot \mathbf{u} + \mathbf{v} \cdot \mathbf{v}$$
$$= \|\mathbf{u}\|^2 - 2\mathbf{u} \cdot \mathbf{v} + \|\mathbf{v}\|^2$$

When $\|\mathbf{u} + \mathbf{v}\|^2$ and $\|\mathbf{u} - \mathbf{v}\|^2$ are added, the $\mathbf{u} \cdot \mathbf{v}$ terms cancel, and the result is $2\|\mathbf{u}\|^2 + 2\|\mathbf{v}\|^2$.

26. Theorem 2 in Chapter 4, because W is the null space of the $1 \times n$ matrix $\mathbf{u}^T$. W is a plane through the origin of $\mathbb{R}^3$.

28. An arbitrary $\mathbf{w}$ in Span $\{\mathbf{u}, \mathbf{v}\}$ has the form $\mathbf{w} = c_1\mathbf{u} + c_2\mathbf{v}$. If $\mathbf{y}$ is orthogonal to $\mathbf{u}$ and $\mathbf{v}$, then $\mathbf{u} \cdot \mathbf{y} = 0$ and $\mathbf{v} \cdot \mathbf{y} = 0$. By linearity of the inner product [Theorem 1(b) and 1(c)],
$$\mathbf{w} \cdot \mathbf{y} = (c_1\mathbf{u} + c_2\mathbf{v}) \cdot \mathbf{y} = c_1\mathbf{u} \cdot \mathbf{y} + c_2\mathbf{v} \cdot \mathbf{y} = c_10 + c_20 = 0$$

30. a. If $\mathbf{z}$ is in $W^\perp$, $\mathbf{u}$ is in W, and c is any scalar, then $(c\mathbf{z}) \cdot \mathbf{u} = c(\mathbf{z} \cdot \mathbf{u}) = c0 = 0$. Since $\mathbf{u}$ is any element of W, $c\mathbf{z}$ is in $W^\perp$.

  b. Take any $\mathbf{z}_1, \mathbf{z}_2$ in $W^\perp$. Then, for any $\mathbf{u}$ in W, $(\mathbf{z}_1 + \mathbf{z}_2) \cdot \mathbf{u} = \mathbf{z}_1 \cdot \mathbf{u} + \mathbf{z}_2 \cdot \mathbf{u} = 0 + 0 = 0$, which shows that $\mathbf{z}_1 + \mathbf{z}_2$ is in $W^\perp$.

  c. Obviously $\mathbf{0}$ is in $W^\perp$, because $\mathbf{0}$ is orthogonal to every vector. This fact, together with (a) and (b), shows that $W^\perp$ is a subspace.

32. [M] This exercise anticipates Theorem 7 in Section 6.2. The matrix A has orthonormal columns.

33. [M] The mapping $\mathbf{x} \mapsto T(\mathbf{x}) = \left(\dfrac{\mathbf{x} \cdot \mathbf{v}}{\mathbf{v} \cdot \mathbf{v}}\right) \mathbf{v}$ is a linear transformation. In Section 6.2, the mapping will be called the orthogonal projection of $\mathbf{x}$ onto Span $\{\mathbf{v}\}$. To verify the linearity, take any $\mathbf{x}$ and $\mathbf{y}$ in $\mathbb{R}^4$ (or $\mathbb{R}^n$) and any scalar c. Then properties of the inner product (Theorem 1) show that

$$T(\mathbf{x} + \mathbf{y}) = \frac{(\mathbf{x} + \mathbf{y}) \cdot \mathbf{v}}{\mathbf{v} \cdot \mathbf{v}}\mathbf{v} = \frac{\mathbf{x} \cdot \mathbf{v} + \mathbf{y} \cdot \mathbf{v}}{\mathbf{v} \cdot \mathbf{v}}\mathbf{v}$$
$$= \left(\frac{\mathbf{x} \cdot \mathbf{v}}{\mathbf{v} \cdot \mathbf{v}} + \frac{\mathbf{y} \cdot \mathbf{v}}{\mathbf{v} \cdot \mathbf{v}}\right)\mathbf{v} = \left(\frac{\mathbf{x} \cdot \mathbf{v}}{\mathbf{v} \cdot \mathbf{v}}\right)\mathbf{v} + \left(\frac{\mathbf{y} \cdot \mathbf{v}}{\mathbf{v} \cdot \mathbf{v}}\right)\mathbf{v}$$
$$= T(\mathbf{x}) + T(\mathbf{y})$$
$$T(c\mathbf{x}) = \frac{(c\mathbf{x}) \cdot \mathbf{v}}{\mathbf{v} \cdot \mathbf{v}}\mathbf{v} = \frac{c(\mathbf{x} \cdot \mathbf{v})}{\mathbf{v} \cdot \mathbf{v}}\mathbf{v} = c\left(\frac{\mathbf{x} \cdot \mathbf{v}}{\mathbf{v} \cdot \mathbf{v}}\right)\mathbf{v} = cT(\mathbf{x})$$

Another argument is to view T as the composition of three linear mappings: $\mathbf{x} \mapsto a = \mathbf{x} \cdot \mathbf{v}, a \mapsto b = a/(\mathbf{v} \cdot \mathbf{v})$, and $b \mapsto b\mathbf{v}$.

34. [M] $N = \begin{bmatrix} -5 & 1 \\ -1 & 4 \\ 1 & 0 \\ 0 & -1 \\ 0 & 3 \end{bmatrix}$,

$R = \begin{bmatrix} 1 & 0 & 5 & 0 & -1/3 \\ 0 & 1 & 1 & 0 & -4/3 \\ 0 & 0 & 0 & 1 & 1/3 \end{bmatrix}$, $RN = \begin{bmatrix} 0 & 0 \\ 0 & 0 \\ 0 & 0 \end{bmatrix}$

The row–column rule for computing RN produces a 3×2 matrix of zeros, which shows that the rows of R are orthogonal to the columns of N. This is to be expected from Theorem 3, because each row of R is in Row A and each column of N is in Nul A.

Section 6.2, page 344

2. Orthogonal **4.** Orthogonal **6.** Not orthogonal

8. Show $\mathbf{u}_1 \cdot \mathbf{u}_2 = 0$, mention Theorem 4, and observe that two linearly independent vectors in $\mathbb{R}^2$ form a basis. Then obtain
$$\mathbf{x} = -\frac{15}{10}\begin{bmatrix} 3 \\ 1 \end{bmatrix} + \frac{30}{40}\begin{bmatrix} -2 \\ 6 \end{bmatrix} = -\frac{3}{2}\begin{bmatrix} 3 \\ 1 \end{bmatrix} + \frac{3}{4}\begin{bmatrix} -2 \\ 6 \end{bmatrix}$$

10. Show $\mathbf{u}_1 \cdot \mathbf{u}_2 = 0$, $\mathbf{u}_1 \cdot \mathbf{u}_3 = 0$, and $\mathbf{u}_2 \cdot \mathbf{u}_3 = 0$. Mention Theorem 4, and observe that three linearly independent vectors in $\mathbb{R}^3$ form a basis. Then obtain
$$\mathbf{x} = \frac{24}{18}\mathbf{u}_1 + \frac{3}{9}\mathbf{u}_2 + \frac{6}{18}\mathbf{u}_3 = \frac{4}{3}\mathbf{u}_1 + \frac{1}{3}\mathbf{u}_2 + \frac{1}{3}\mathbf{u}_3$$

12. $\begin{bmatrix} .4 \\ -1.2 \end{bmatrix}$ **14.** $\mathbf{y} = \begin{bmatrix} 14/5 \\ 2/5 \end{bmatrix} + \begin{bmatrix} -4/5 \\ 28/5 \end{bmatrix}$

16. $\mathbf{y} - \hat{\mathbf{y}} = \begin{bmatrix} -6 \\ 3 \end{bmatrix}$, distance is $\sqrt{45} = 3\sqrt{5}$

18. Not orthogonal   **20.** $\begin{bmatrix} -2/3 \\ 1/3 \\ 2/3 \end{bmatrix}, \begin{bmatrix} 1/\sqrt{5} \\ 2/\sqrt{5} \\ 0 \end{bmatrix}$

22. Orthonormal

23. a. True. For example, the vectors **u** and **y** in Example 3 are linearly independent but not orthogonal.

 b. True. The formulas for the weights are given in Theorem 5.

 c. False. See the paragraph following Example 5.

 d. False. The matrix must also be square. See the paragraph before Example 7.

 e. False. See Example 4. The distance is $\|\mathbf{y} - \hat{\mathbf{y}}\|$.

24. a. True. But every orthogonal set of *nonzero vectors* is linearly independent. See Theorem 4.

 b. False. To be orthonormal, the vectors in S must be unit vectors as well as being orthogonal to each other.

 c. True. See Theorem 7(a).

 d. True. See the paragraph before Example 3.

 e. True. See the paragraph before Example 7.

26. If $\mathbf{v}_1, \ldots, \mathbf{v}_n$ are nonzero and orthogonal, then they are linearly independent, by Theorem 4. By the Invertible Matrix Theorem, $\{\mathbf{v}_1, \ldots, \mathbf{v}_n\}$ is a basis for $\mathbb{R}^n$. If $W = \text{Span}\{\mathbf{v}_1, \ldots, \mathbf{v}_n\}$, then W must be $\mathbb{R}^n$.

28. If U is an $n \times n$ orthogonal matrix, then $I = UU^{-1} = UU^T$. Since U is the transpose of U^T, Theorem 6 applied to U^T says that U^T has orthonormal columns. In particular, the columns of U^T are linearly independent and hence form a basis for $\mathbb{R}^n$, by the Invertible Matrix Theorem (see Section 4.6). That is, the rows of U form a basis (in fact, an orthonormal basis) for $\mathbb{R}^n$.

30. If U is an orthogonal matrix, its columns are orthonormal. Interchanging the columns does not change their orthonormality, so the new matrix—say, V—still has orthonormal columns. By Theorem 6, $V^TV = I$. Since V is square, $V^T = V^{-1}$ by the Invertible Matrix Theorem.

32. If $\mathbf{v}_1 \cdot \mathbf{v}_2 = 0$, then by Theorem 1(c) in Section 6.1, $(c_1\mathbf{v}_1) \cdot (c_2\mathbf{v}_2) = c_1[\mathbf{v}_1 \cdot (c_2\mathbf{v}_2)] = c_1c_2(\mathbf{v}_1 \cdot \mathbf{v}_2) = c_1c_20 = 0$.

34. Let $L = \text{Span}\{\mathbf{u}\}$, where **u** is nonzero, and let $T(\mathbf{y}) = \text{refl}_L\mathbf{y} = 2 \cdot \text{proj}_L \mathbf{y} - \mathbf{y}$. By Exercise 33, the mapping $\mathbf{y} \mapsto \text{proj}_L \mathbf{y}$ is linear. Thus, for **y** and **z** in $\mathbb{R}^n$ and any scalars c and d,

$$T(c\mathbf{y} + d\mathbf{z}) = 2 \cdot \text{proj}_L(c\mathbf{y} + d\mathbf{z}) - (c\mathbf{y} + d\mathbf{z})$$
$$= 2(c \cdot \text{proj}_L \mathbf{y} + d \cdot \text{proj}_L \mathbf{z}) - c\mathbf{y} - d\mathbf{z}$$
$$= 2c \cdot \text{proj}_L \mathbf{y} - c\mathbf{y} + 2d \cdot \text{proj}_L \mathbf{z} - d\mathbf{z}$$
$$= cT(\mathbf{y}) + dT(\mathbf{z})$$

Thus T is linear.

35. [M] The proof of Theorem 6 shows that the inner products to be checked are actually entries in the matrix product A^TA. A calculation shows that $A^TA = 100I_4$. Since the off-diagonal entries in A^TA are zero, the columns of A are orthogonal.

36. [M] **a.** $U^TU = I_4$, but UU^T is an 8×8 matrix which is nothing like I_8. In fact

$$UU^T =$$
$$(.01)\begin{bmatrix} 82 & 0 & -20 & 8 & 6 & 20 & 24 & 0 \\ 0 & 42 & 24 & 0 & -20 & 6 & 20 & -32 \\ -20 & 24 & 58 & 20 & 0 & 32 & 0 & 6 \\ 8 & 0 & 20 & 82 & 24 & -20 & 6 & 0 \\ 6 & -20 & 0 & 24 & 18 & 0 & -8 & 20 \\ 20 & 6 & 32 & -20 & 0 & 58 & 0 & 24 \\ 24 & 20 & 0 & 6 & -8 & 0 & 18 & -20 \\ 0 & -32 & 6 & 0 & 20 & 24 & -20 & 42 \end{bmatrix}$$

 b. The vector $\mathbf{p} = UU^T\mathbf{y}$ is in Col U because $\mathbf{p} = U(U^T\mathbf{y})$. Since the columns of U are simply scaled versions of the columns of A, Col U = Col A. Thus **p** is in Col A.

 d. From part (c), **z** is orthogonal to each column of A. By Exercise 29 in Section 6.1, **z** must be orthogonal to every vector in Col A; that is, **z** is in $(\text{Col } A)^\perp$.

Section 6.3, page 352

2. $\mathbf{v} = 2\mathbf{u}_1 + \frac{3}{7}\mathbf{u}_2 + \frac{12}{7}\mathbf{u}_3 - \frac{8}{7}\mathbf{u}_4$; $\mathbf{v} = \begin{bmatrix} 2 \\ 4 \\ 2 \\ 2 \end{bmatrix} + \begin{bmatrix} 2 \\ 1 \\ -5 \\ 1 \end{bmatrix}$

4. $\begin{bmatrix} 6 \\ 3 \\ 0 \end{bmatrix}$ **6.** $\begin{bmatrix} 6 \\ 4 \\ 1 \end{bmatrix} = \mathbf{y}$ **8.** $\mathbf{y} = \begin{bmatrix} 3/2 \\ 7/2 \\ 1 \end{bmatrix} + \begin{bmatrix} -5/2 \\ 1/2 \\ 2 \end{bmatrix}$

10. $\mathbf{y} = \begin{bmatrix} 5 \\ 2 \\ 3 \\ 6 \end{bmatrix} + \begin{bmatrix} -2 \\ 2 \\ 2 \\ 0 \end{bmatrix}$ **12.** $\begin{bmatrix} -1 \\ -5 \\ -3 \\ 9 \end{bmatrix}$

14. $\begin{bmatrix} 1 \\ 0 \\ -1/2 \\ -3/2 \end{bmatrix}$ **16.** 8

18. a. $U^TU = [1] = 1$, $UU^T = \begin{bmatrix} .1 & -.3 \\ -.3 & .9 \end{bmatrix}$

 b. $\text{proj}_W \mathbf{y} = \frac{-20}{\sqrt{10}}\mathbf{u}_1 = \begin{bmatrix} -2 \\ 6 \end{bmatrix}$,

 $(UU^T)\mathbf{y} = \begin{bmatrix} .7 - 2.7 \\ -2.1 + 8.1 \end{bmatrix} = \begin{bmatrix} -2 \\ 6 \end{bmatrix}$

20. Any multiple of $\begin{bmatrix} 0 \\ 4/5 \\ 2/5 \end{bmatrix}$, such as $\begin{bmatrix} 0 \\ 2 \\ 1 \end{bmatrix}$

21. a. True. See the calculations for $\mathbf{z}_2$ in Example 1 or the box after Example 6 in Section 6.1.

 b. True, by the Orthogonal Decomposition Theorem.

c. False. See the last paragraph in the proof of Theorem 8, or see the second paragraph after the statement of Theorem 9.

d. True. See the box before The Best Approximation Theorem.

e. True. Theorem 10 applies to the column space W of U because the columns of U are linearly independent and hence form a basis for W.

22. a. True. See the proof of the Orthogonal Decomposition Theorem.

b. True. See the subsection "A Geometric Interpretation of the Orthogonal Projection."

c. True, by the uniqueness of the orthogonal decomposition in Theorem 8.

d. False. The Best Approximation Theorem says that the best approximation to $\mathbf{y}$ is $\text{proj}_W \mathbf{y}$.

e. False, unless $n = p$, because $UU^T\mathbf{x}$ is only the orthogonal projection of $\mathbf{x}$ onto the column space of U. See the paragraph following the proof of Theorem 10.

24. a. By hypothesis, the vectors $\mathbf{w}_1, \ldots, \mathbf{w}_p$ are pairwise orthogonal, and the vectors $\mathbf{v}_1, \ldots, \mathbf{v}_q$ are pairwise orthogonal. Also, $\mathbf{w}_i \cdot \mathbf{v}_j = 0$ for any i and j because the $\mathbf{v}$'s are in the orthogonal complement of W.

b. For any $\mathbf{y}$ in $\mathbb{R}^n$, write $\mathbf{y} = \hat{\mathbf{y}} + \mathbf{z}$ as in the Orthogonal Decomposition Theorem, with $\hat{\mathbf{y}}$ in W and $\mathbf{z}$ in $W^\perp$. Then there exist scalars $c_1, \ldots, c_p$ and $d_1, \ldots, d_q$ such that

$$\mathbf{y} = \hat{\mathbf{y}} + \mathbf{z} = c_1\mathbf{w}_1 + \cdots + c_p\mathbf{w}_p + d_1\mathbf{v}_1 + \cdots + d_q\mathbf{v}_q$$

Thus $\{\mathbf{w}_1, \ldots, \mathbf{w}_p, \mathbf{v}_1, \ldots, \mathbf{v}_q\}$ spans $\mathbb{R}^n$.

c. The set $\{\mathbf{w}_1, \ldots, \mathbf{w}_p, \mathbf{v}_1, \ldots, \mathbf{v}_q\}$ is linearly independent by part (a), spans $\mathbb{R}^n$ by part (b), and thus is a basis for $\mathbb{R}^n$. Hence

$$\dim W + \dim W^\perp = p + q = \dim \mathbb{R}^n = n$$

25. [M] U has orthonormal columns, by Theorem 6 in Section 6.2, because $U^TU = I_4$. The closest point to $\mathbf{y}$ in $\text{Col } U$ is the orthogonal projection $\hat{\mathbf{y}}$ of $\mathbf{y}$ onto $\text{Col } U$. From Theorem 10,

$$\hat{\mathbf{y}} = UU^T\mathbf{y} = (1.2, .4, 1.2, 1.2, .4, 1.2, .4, .4)$$

26. [M] To two decimal places,
$\hat{\mathbf{b}} = UU^T\mathbf{b} = (.20, .92, .44, 1.00, -.20, -.44, .60, -.92)$.
The distance from $\mathbf{b}$ to $\text{Col } U$ is $\|\mathbf{b} - \hat{\mathbf{b}}\| = 2.1166$, to four decimal places.

Section 6.4, page 358

2. $\begin{bmatrix} 0 \\ 4 \\ 2 \end{bmatrix}, \begin{bmatrix} 5 \\ 4 \\ -8 \end{bmatrix}$ **4.** $\begin{bmatrix} 3 \\ -4 \\ 5 \end{bmatrix}, \begin{bmatrix} 3 \\ 6 \\ 3 \end{bmatrix}$

6. $\begin{bmatrix} 3 \\ -1 \\ 2 \\ -1 \end{bmatrix}, \begin{bmatrix} 4 \\ 6 \\ -3 \\ 0 \end{bmatrix}$ **8.** $\begin{bmatrix} 3/\sqrt{50} \\ -4/\sqrt{50} \\ 5/\sqrt{50} \end{bmatrix}, \begin{bmatrix} 1/\sqrt{6} \\ 2/\sqrt{6} \\ 1/\sqrt{6} \end{bmatrix}$

10. $\begin{bmatrix} -1 \\ 3 \\ 1 \\ 1 \end{bmatrix}, \begin{bmatrix} 3 \\ 1 \\ 1 \\ -1 \end{bmatrix}, \begin{bmatrix} 1 \\ 1 \\ -3 \\ 1 \end{bmatrix}$ **12.** $\begin{bmatrix} 1 \\ -1 \\ 0 \\ 1 \\ 1 \end{bmatrix}, \begin{bmatrix} -1 \\ 1 \\ 2 \\ 1 \\ 1 \end{bmatrix}, \begin{bmatrix} 1 \\ 1 \\ 0 \\ -1 \\ 1 \end{bmatrix}$

14. $R = \begin{bmatrix} 7 & 7 \\ 0 & 7 \end{bmatrix}$

16. $Q = \begin{bmatrix} 1/2 & -1/\sqrt{8} & 1/2 \\ -1/2 & 1/\sqrt{8} & 1/2 \\ 0 & 2/\sqrt{8} & 0 \\ 1/2 & 1/\sqrt{8} & -1/2 \\ 1/2 & 1/\sqrt{8} & 1/2 \end{bmatrix}$,

$R = \begin{bmatrix} 2 & 8 & 7 \\ 0 & \sqrt{8} & 12/\sqrt{8} \\ 0 & 0 & 6 \end{bmatrix}$

17. a. False. Scaling was used in Example 2, but the scale factor was nonzero.

b. True. See property (1) in the statement of Theorem 11.

c. True. See the solution of Example 4.

18. a. False. The three orthogonal vectors must be *nonzero* to be a basis for a three-dimensional subspace. (This was the case in Step 3 in the solution of Example 2.)

b. True. If $\mathbf{x}$ is not in a subspace W, then $\mathbf{x}$ cannot equal $\text{proj}_W \mathbf{x}$, because $\text{proj}_W \mathbf{x}$ is in W. This idea was used for $\mathbf{x} = \mathbf{v}_{k+1}$ in the proof of Theorem 11.

c. True, by Theorem 12.

20. If $\mathbf{y}$ is in $\text{Col } A$, then $\mathbf{y} = A\mathbf{x}$ for some $\mathbf{x}$. Then $\mathbf{y} = QR\mathbf{x} = Q(R\mathbf{x})$, which shows that $\mathbf{y}$ is a linear combination of the columns of Q using the entries in $R\mathbf{x}$ as weights. Conversely, suppose $\mathbf{y} = Q\mathbf{x}$ for some $\mathbf{x}$. Since R is invertible, the equation $A = QR$ implies that $Q = AR^{-1}$. So $\mathbf{y} = AR^{-1}\mathbf{x} = A(R^{-1}\mathbf{x})$, which shows that $\mathbf{y}$ is in $\text{Col } A$.

22. We may assume that $\{\mathbf{u}_1, \ldots, \mathbf{u}_p\}$ is an orthonormal basis for W, by normalizing the vectors in the original basis given for W, if necessary. Let U be the matrix whose columns are $\mathbf{u}_1, \ldots, \mathbf{u}_p$. Then, by Theorem 10 in Section 6.3, $T(\mathbf{x}) = \text{proj}_W \mathbf{x} = (UU^T)\mathbf{x}$ for $\mathbf{x}$ in $\mathbb{R}^n$. Thus T is a matrix transformation and hence is a linear transformation, as was shown in Section 1.8.

24. [M] $\begin{bmatrix} -10 \\ 2 \\ -6 \\ 16 \\ 2 \end{bmatrix}, \begin{bmatrix} 3 \\ 3 \\ -3 \\ 0 \\ 3 \end{bmatrix}, \begin{bmatrix} 6 \\ 0 \\ 6 \\ 6 \\ 0 \end{bmatrix}, \begin{bmatrix} 0 \\ 5 \\ 0 \\ 0 \\ -5 \end{bmatrix}$

25. [M] $Q = \begin{bmatrix} -.5 & .5 & .5774 & 0 \\ .1 & .5 & 0 & .7071 \\ -.3 & -.5 & .5774 & 0 \\ .8 & 0 & .5774 & 0 \\ .1 & .5 & 0 & -.7071 \end{bmatrix}$

$$R = \begin{bmatrix} 20 & -20 & -10 & 10 \\ 0 & 6 & -8 & -6 \\ 0 & 0 & 10.3923 & -5.1962 \\ 0 & 0 & 0 & 7.0711 \end{bmatrix}$$

26. [M] In MATLAB, when A has n columns, suitable commands are

```
Q = A(:,1)/norm(A(:,1))
        % The first column of Q
        for j = 2:n
            v = A(:,j) - Q*(Q'*A(:,j))
            Q(:,j) = v/norm(v)
            % Add a new column to Q
        end
```

Section 6.5, page 366

2. a. $\begin{bmatrix} 12 & 8 \\ 8 & 10 \end{bmatrix}\begin{bmatrix} x_1 \\ x_2 \end{bmatrix} = \begin{bmatrix} -24 \\ -2 \end{bmatrix}$ **b.** $\hat{\mathbf{x}} = \begin{bmatrix} -4 \\ 3 \end{bmatrix}$

4. a. $\begin{bmatrix} 3 & 3 \\ 3 & 11 \end{bmatrix}\begin{bmatrix} x_1 \\ x_2 \end{bmatrix} = \begin{bmatrix} 6 \\ 14 \end{bmatrix}$ **b.** $\hat{\mathbf{x}} = \begin{bmatrix} 1 \\ 1 \end{bmatrix}$

6. $\hat{\mathbf{x}} = \begin{bmatrix} 5 \\ -1 \\ 0 \end{bmatrix} + x_3 \begin{bmatrix} -1 \\ 1 \\ 1 \end{bmatrix}$ **8.** $\sqrt{6}$

10. a. $\hat{\mathbf{b}} = \begin{bmatrix} 4 \\ -1 \\ 4 \end{bmatrix}$ **b.** $\hat{\mathbf{x}} = \begin{bmatrix} 3 \\ 1/2 \end{bmatrix}$

12. a. $\hat{\mathbf{b}} = \begin{bmatrix} 5 \\ 2 \\ 3 \\ 6 \end{bmatrix}$ **b.** $\hat{\mathbf{x}} = \begin{bmatrix} 1/3 \\ 14/3 \\ -5/3 \end{bmatrix}$

14. $A\mathbf{u} = \begin{bmatrix} 3 \\ 8 \\ 2 \end{bmatrix}$, $A\mathbf{v} = \begin{bmatrix} 7 \\ 2 \\ 8 \end{bmatrix}$, $\mathbf{b} - A\mathbf{u} = \begin{bmatrix} 2 \\ -4 \\ 2 \end{bmatrix}$,

$\mathbf{b} - A\mathbf{v} = \begin{bmatrix} -2 \\ 2 \\ -4 \end{bmatrix}$. Note that

$\|\mathbf{b} - A\mathbf{u}\| = \|\mathbf{b} - A\mathbf{v}\| = \sqrt{24}$, so $A\mathbf{u}$ and $A\mathbf{v}$ are equally close to $\mathbf{b}$. The orthogonal projection is the *unique* closest point in Col A to $\mathbf{b}$, so neither $A\mathbf{u}$ nor $A\mathbf{v}$ can be $\hat{\mathbf{b}}$. That is, neither $\mathbf{u}$ nor $\mathbf{v}$ can be a least-squares solution of $A\mathbf{x} = \mathbf{b}$.

16. $\hat{\mathbf{x}} = \begin{bmatrix} 2.9 \\ .9 \end{bmatrix}$

17. a. True. See the beginning of the section. The distance from $A\mathbf{x}$ to $\mathbf{b}$ is $\|A\mathbf{x} - \mathbf{b}\|$.

b. True. See the comments about equation (1).

c. False. The inequality points in the wrong direction. See the definition of a least-squares solution.

d. True. See Theorem 13.

e. True. See Theorem 14.

18. a. True. See the paragraph following the definition of a least-squares solution.

b. False. If $\hat{\mathbf{x}}$ is the least-squares solution, then $A\hat{\mathbf{x}}$ is the point in the column space of A closest to $\mathbf{b}$. See Fig. 1 and the paragraph preceding it.

c. True. See the discussion following equation (1).

d. False. The formula applies only when the columns of A are linearly independent. See Theorem 14.

e. False. See the comments after Example 4.

f. False. See the Numerical Note.

20. Suppose that $A\mathbf{x} = \mathbf{0}$. Then $A^T A\mathbf{x} = A^T \mathbf{0} = \mathbf{0}$. Since $A^T A$ is invertible, by hypothesis, $\mathbf{x}$ must be zero. Hence the columns of A are linearly independent.

22. $A^T A$ has n columns because A does. Then

rank $A^T A = n - \dim \text{Nul } A^T A$	The Rank Theorem
$= n - \dim \text{Nul } A$	Exercise 19
$= \text{rank } A$	The Rank Theorem

24. $\hat{\mathbf{x}} = A^T \mathbf{b}$, from the normal equations, because $A^T A = I$.

26. [M] $a_0 = a_2 = .3535$, $a_1 = .5$ (With .707 in place of .7, $a_0 = a_2 \approx .35355339$, $a_1 = .5$.)

Section 6.6, page 374

2. $y = -.6 + .7x$ **4.** $y = 4.3 - .7x$

6. If the columns of X were linearly dependent, then the same dependence relation would hold for the vectors in $\mathbb{R}^3$ formed from the top three entries of the column. In this case, the *Vandermonde* matrix

$$\begin{bmatrix} 1 & x_1 & x_1^2 \\ 1 & x_2 & x_2^2 \\ 1 & x_3 & x_3^2 \end{bmatrix}$$

would be noninvertible. However, it can be shown that since x_1, x_2, and x_3 are distinct, this matrix *is* invertible, which means that the columns of X are, in fact, linearly independent. As in Exercise 5, Theorem 14 implies that there is only one least-squares solution of $\mathbf{y} = X\boldsymbol{\beta}$.

One way to show that the 3×3 matrix above is invertible is to show that its determinant is $(x_2 - x_1)(x_3 - x_1)(x_3 - x_2)$. Another way is to appeal to Supplementary Exercise 11(b) in Chapter 2.

8. a. $X = \begin{bmatrix} x_1 & x_1^2 & x_1^3 \\ \vdots & \vdots & \vdots \\ x_n & x_n^2 & x_n^3 \end{bmatrix}$, $\boldsymbol{\beta} = \begin{bmatrix} \beta_1 \\ \beta_2 \\ \beta_3 \end{bmatrix}$

b. [M] $y = .5132x - .03348x^2 + .001016x^3$, using four significant figures in the coefficients. *Note:* If you use .001 as the coefficient of x^3, your graph will fall somewhat below the last three or four data points.

10. a. $\mathbf{y} = X\boldsymbol{\beta} + \boldsymbol{\epsilon}$, where $\mathbf{y} = \begin{bmatrix} 21.34 \\ 20.68 \\ 20.05 \\ 18.87 \\ 18.30 \end{bmatrix}$,

$$X = \begin{bmatrix} e^{-.02(10)} & e^{-.07(10)} \\ e^{-.02(11)} & e^{-.07(11)} \\ e^{-.02(12)} & e^{-.07(12)} \\ e^{-.02(14)} & e^{-.07(14)} \\ e^{-.02(15)} & e^{-.07(15)} \end{bmatrix}, \; \boldsymbol{\beta} = \begin{bmatrix} M_A \\ M_B \end{bmatrix}, \; \boldsymbol{\epsilon} = \begin{bmatrix} \epsilon_1 \\ \epsilon_2 \\ \epsilon_3 \\ \epsilon_4 \\ \epsilon_5 \end{bmatrix}$$

b. [M] $y = 19.94e^{-.02t} + 10.10e^{-.07t}$, $M_A = 19.94$,
$M_B = 10.10$

12. [M] $p = 18.56 + 19.24 \ln w$ (using text values for $\ln w$).
When w is 100, $p \approx 107$.

14. Write the design matrix as $X = [\,\mathbf{1} \quad \mathbf{x}\,]$. Since the residual
vector, $\boldsymbol{\epsilon} = \mathbf{y} - X\hat{\boldsymbol{\beta}}$, is orthogonal to Col X, we have (using
the notation shown just after Exercise 14)

$$0 = \mathbf{1} \cdot \boldsymbol{\epsilon} = \mathbf{1} \cdot (\mathbf{y} - X\hat{\boldsymbol{\beta}}) = \mathbf{1}^T \mathbf{y} - (\mathbf{1}^T X)\hat{\boldsymbol{\beta}}$$

$$= (y_1 + \cdots + y_n) - [\, n \quad \Sigma x \,] \begin{bmatrix} \hat{\beta}_0 \\ \hat{\beta}_1 \end{bmatrix}$$

$$= \Sigma y - n\hat{\beta}_0 - \hat{\beta}_1 \Sigma x$$

Divide by $-n$, move the first term to the left side of the
equation, and obtain $\bar{y} = \hat{\beta}_0 + \hat{\beta}_1 \bar{x}$.

16. The determinant of the coefficient matrix of the equations
in (7) is $n\Sigma x^2 - (\Sigma x)^2$. Using the 2×2 formula for the
inverse of the coefficient matrix, we have

$$\begin{bmatrix} \hat{\beta}_0 \\ \hat{\beta}_1 \end{bmatrix} = \frac{1}{n\Sigma x^2 - (\Sigma x)^2} \begin{bmatrix} \Sigma x^2 & -\Sigma x \\ -\Sigma x & n \end{bmatrix} \begin{bmatrix} \Sigma y \\ \Sigma xy \end{bmatrix}$$

Hence

$$\hat{\beta}_0 = \frac{(\Sigma x^2)(\Sigma y) - (\Sigma x)(\Sigma xy)}{n\,\Sigma x^2 - (\Sigma x)^2},$$

$$\hat{\beta}_1 = \frac{n\,\Sigma xy - (\Sigma x)(\Sigma y)}{n\,\Sigma x^2 - (\Sigma x)^2}$$

Note: A simple algebraic calculation shows that
$\Sigma y - (\Sigma x)\hat{\beta}_1 = n\hat{\beta}_0$, which provides a simple formula for
$\hat{\beta}_0$, once $\hat{\beta}_1$ is known.

18. $X^T X = \begin{bmatrix} 1 & \cdots & 1 \\ x_1 & \cdots & x_n \end{bmatrix} \begin{bmatrix} 1 & x_1 \\ \vdots & \vdots \\ 1 & x_n \end{bmatrix} = \begin{bmatrix} n & \Sigma x \\ \Sigma x & (\Sigma x)^2 \end{bmatrix}$

This matrix is a diagonal matrix when $\Sigma x = 0$.

20. $\|X\hat{\boldsymbol{\beta}}\|^2 = (X\hat{\boldsymbol{\beta}})^T (X\hat{\boldsymbol{\beta}}) = \hat{\boldsymbol{\beta}}^T X^T X \hat{\boldsymbol{\beta}} = \hat{\boldsymbol{\beta}}^T X^T \mathbf{y}$, because $\hat{\boldsymbol{\beta}}^T$
satisfies the normal equations: $X^T X \boldsymbol{\beta} = X^T \mathbf{y}$. Since
$\|X\hat{\boldsymbol{\beta}}\|^2 = \text{SS(R)}$ and $\mathbf{y}^T \mathbf{y} = \|\mathbf{y}\|^2 = \text{SS(T)}$, Exercise 19
shows that

$$\text{SS(E)} = \text{SS(T)} - \text{SS(R)} = \mathbf{y}^T \mathbf{y} - \hat{\boldsymbol{\beta}}^T X^T \mathbf{y}$$

Section 6.7, page 382

2. $\|\mathbf{x}\|^2 = \langle \mathbf{x}, \mathbf{x} \rangle = 4(3)(3) + 5(-2)(-2) = 56$
$\|\mathbf{y}\|^2 = \langle \mathbf{y}, \mathbf{y} \rangle = 4(-2)(-2) + 5(1)(1) = 21$
$\|\mathbf{x}\|^2 \|\mathbf{y}\|^2 = 56(21) = 1176$
$\langle \mathbf{x}, \mathbf{y} \rangle = 4(3)(-2) + 5(-2)(1) = -34$
$|\langle \mathbf{x}, \mathbf{y} \rangle|^2 = 1156 < 1176 = \|\mathbf{x}\|^2 \|\mathbf{y}\|^2$

4. Polynomials: $3t - t^2$ $3 + 2t^2$

Values: $\begin{bmatrix} -4 \\ 0 \\ 2 \end{bmatrix}$ $\begin{bmatrix} 5 \\ 3 \\ 5 \end{bmatrix}$

$\langle p, q \rangle = -20 + 0 + 10 = -10$

6. $\|p\| = 2\sqrt{5}$, $\|q\| = \sqrt{59}$

8. $\dfrac{\langle q, p \rangle}{\langle p, p \rangle} p(t) = -\dfrac{10}{20} p(t) = -\dfrac{1}{2}(3t - t^2) = -\dfrac{3}{2}t + \dfrac{1}{2}t^2$

10. Polynomials: p_0 p_1 q $p(t) = t^3$

Values: $\begin{bmatrix} 1 \\ 1 \\ 1 \\ 1 \end{bmatrix}$ $\begin{bmatrix} -3 \\ -1 \\ 1 \\ 3 \end{bmatrix}$ $\begin{bmatrix} 1 \\ -1 \\ -1 \\ 1 \end{bmatrix}$ $\begin{bmatrix} -27 \\ -1 \\ 1 \\ 27 \end{bmatrix}$

$\hat{p}(t) = \dfrac{0}{4} p_0 + \dfrac{164}{20} p_1 + \dfrac{0}{4} q = \dfrac{41}{5} t$

12. Use Exercise 11 to get $t^3 - \dfrac{17}{5}t$. Then $5t^3 - 17t$ is also
orthogonal to p_0, p_1, p_2, but its vector of values is
$(-6, 12, 0, -12, 6)$. Answer: $p_3(t) = \dfrac{1}{6}(5t^3 - 17t)$.

14. 1. $\langle \mathbf{u}, \mathbf{v} \rangle = T(\mathbf{u}) \cdot T(\mathbf{v})$ Definition
    $= T(\mathbf{v}) \cdot T(\mathbf{u})$ Property of dot product
    $= \langle \mathbf{v}, \mathbf{u} \rangle$ Definition

2.
$\langle \mathbf{u} + \mathbf{v}, \mathbf{w} \rangle = T(\mathbf{u} + \mathbf{v}) \cdot T(\mathbf{w})$ Definition
    $= [T(\mathbf{u}) + T(\mathbf{v})] \cdot T(\mathbf{w})$ Linearity of T
    $= T(\mathbf{u}) \cdot T(\mathbf{w}) + T(\mathbf{v}) \cdot T(\mathbf{w})$ Property of $\cdot$
    $= \langle \mathbf{u}, \mathbf{w} \rangle + \langle \mathbf{v}, \mathbf{w} \rangle$ Definition

3. $\langle c\mathbf{u}, \mathbf{v} \rangle = T(c\mathbf{u}) \cdot T(\mathbf{v})$ Definition
    $= cT(\mathbf{u}) \cdot T(\mathbf{v})$ Linearity of T
    $= c\langle \mathbf{u}, \mathbf{v} \rangle$ Definition

4. $\langle \mathbf{u}, \mathbf{u} \rangle = T(\mathbf{u}) \cdot T(\mathbf{u}) \geq 0$ Property of dot product

If $\mathbf{u} = \mathbf{0}$, then $T(\mathbf{u}) = \mathbf{0}$, because T is linear, and
$\langle \mathbf{u}, \mathbf{u} \rangle = 0$. Conversely, if $\langle \mathbf{u}, \mathbf{u} \rangle = 0$, then
$T(\mathbf{u}) \cdot T(\mathbf{u}) = 0$, and hence $T(\mathbf{u}) = \mathbf{0}$ by a property of the
dot product. Since T is one-to-one, $\mathbf{u} = \mathbf{0}$.

16. $\|\mathbf{u} - \mathbf{v}\|^2$
    $= \langle \mathbf{u} - \mathbf{v}, \mathbf{u} - \mathbf{v} \rangle$
    $= \langle \mathbf{u}, \mathbf{u} - \mathbf{v} \rangle - \langle \mathbf{v}, \mathbf{u} - \mathbf{v} \rangle$ Axioms 2 and 3
    $= \langle \mathbf{u}, \mathbf{u} \rangle - \langle \mathbf{u}, \mathbf{v} \rangle - \langle \mathbf{v}, \mathbf{u} \rangle + \langle \mathbf{v}, \mathbf{v} \rangle$ Axioms 1–3
    $= \langle \mathbf{u}, \mathbf{u} \rangle - 2\langle \mathbf{u}, \mathbf{v} \rangle + \langle \mathbf{v}, \mathbf{v} \rangle$ Axiom 1
    $= \|\mathbf{u}\|^2 - 2\langle \mathbf{u}, \mathbf{v} \rangle + \|\mathbf{v}\|^2$

If $\{\mathbf{u}, \mathbf{v}\}$ is orthonormal, then $\|\mathbf{u}\|^2 = \|\mathbf{v}\|^2 = 1$ and
$\langle \mathbf{u}, \mathbf{v} \rangle = 0$. So $\|\mathbf{u} - \mathbf{v}\|^2 = 2$.

18. The calculation in Exercise 16 shows that

$$\|\mathbf{u} - \mathbf{v}\|^2 = \|\mathbf{u}\|^2 - 2\langle \mathbf{u}, \mathbf{v} \rangle + \|\mathbf{v}\|^2$$

Similarly,

$$\|\mathbf{u} + \mathbf{v}\|^2 = \|\mathbf{u}\|^2 + 2\langle \mathbf{u}, \mathbf{v} \rangle + \|\mathbf{v}\|^2$$

Adding gives $\|\mathbf{u} + \mathbf{v}\|^2 + \|\mathbf{u} - \mathbf{v}\|^2 = 2\|\mathbf{u}\|^2 + 2\|\mathbf{v}\|^2$.

20. If $\mathbf{u} = (a, b)$ and $\mathbf{v} = (1, 1)$, then $\|\mathbf{u}\|^2 = a^2 + b^2$, $\|\mathbf{v}\|^2 = 2$, and $|\langle \mathbf{u}, \mathbf{v} \rangle| = |a + b|$. The desired inequality follows when the Cauchy–Schwarz inequality is rewritten as

$$\left(\frac{\langle \mathbf{u}, \mathbf{v} \rangle}{2} \right)^2 \le \frac{\|\mathbf{u}\|^2 \|\mathbf{v}\|^2}{4}$$

22. $\int_0^1 (5t - 3)(t^3 - t^2) \, dt = \int_0^1 (5t^4 - 8t^3 + 3t^2) \, dt = 0$

24. $\int_0^1 (t^3 - t^2)^2 \, dt = \int_0^1 (t^6 - 2t^5 + t^4) \, dt = 1/105$,
$\|g\| = 1/\sqrt{105}$

26. $1, t, 3t^2 - 4$

27. **[M]** $p_0(t) = 1$, $p_1(t) = t$, $p_2(t) = -2 + t^2$,
$p_3(t) = (-17t + 5t^3)/6$, $p_4(t) = (72 - 155t^2 + 35t^4)/12$

The columns of the following matrix list the values of the respective polynomials at $-2, -1, 0, 1,$ and 2:

$$A = \begin{bmatrix} 1 & -2 & 2 & -1 & 1 \\ 1 & -1 & -1 & 2 & -4 \\ 1 & 0 & -2 & 0 & 6 \\ 1 & 1 & -1 & -2 & -4 \\ 1 & 2 & 2 & 1 & 1 \end{bmatrix}$$

28. **[M]** The orthogonal basis is $f_0(t) = 1$, $f_1(t) = \cos t$, $f_2(t) = \cos^2 t - \frac{1}{2}$, and $f_3(t) = \cos^3 t - \frac{3}{4} \cos t$. Note that $2f_2(t) = \cos 2t$ and $4f_3(t) = \cos 3t$.

Section 6.8, page 389

2. Let X be the original design matrix, and let $\mathbf{y}$ be the original observation vector. Let W be the weighting matrix for the first method. Then the weighting matrix for the second method is $2W$. The weighted least squares by the first method is equivalent to the ordinary least squares for an equation whose normal equation is

$$(WX)^T WX \hat{\boldsymbol{\beta}} = (WX)^T W\mathbf{y} \qquad (1)$$

while the second method is equivalent to the ordinary least squares for an equation whose normal equation is

$$(2WX)^T (2W)X \hat{\boldsymbol{\beta}} = (2WX)^T (2W)\mathbf{y} \qquad (2)$$

Since equation (2) can be written as $4(WX)^T WX \hat{\boldsymbol{\beta}} = 4(WX)^T W\mathbf{y}$, it has the same solutions as equation (1).

4. a. The vectors of polynomial values are

$p_0 \leftrightarrow (1, 1, 1, 1, 1, 1)$, $\quad p_1 \leftrightarrow (-5, -3, -1, 1, 3, 5)$,
$p_2 \leftrightarrow (5, -1, -4, -4, -1, 5)$

Verify that these vectors in $\mathbb{R}^6$ are mutually orthogonal.
b. $4p_0 + \frac{5}{7} p_1 + \frac{1}{14} p_2$

6. Use the identity

$\sin mt \cos nt = \frac{1}{2}[\sin(mt + nt) + \sin(mt - nt)]$

8. $-1 + \pi - 2 \sin t - \sin 2t - \frac{2}{3} \sin 3t$

10. $\frac{4}{\pi} \sin t + \frac{4}{3\pi} \sin 3t$

12. The trigonometric identity $\cos 3t = 4 \cos^3 t - 3 \cos t$ shows that

$$\cos^3 t = \tfrac{3}{4} \cos t + \tfrac{1}{4} \cos 3t$$

The expression on the right is in the subspace spanned by the trigonometric polynomials of order 3 or less, so this expression *is* the third-order Fourier approximation to $\cos^3 t$.

14. g and h are both in the subspace H spanned by the trigonometric polynomials of order 2 or less. Since h is the second-order Fourier approximation to f, it is closer to f than any other function in the subspace H.

16. **[M]** $f_4(t) = \frac{4}{\pi} \sin t + \frac{4}{3\pi} \sin 3t$,
$f_5(t) = f_4(t) + \frac{4}{5\pi} \sin 5t$

Chapter 6 Supplementary Exercises, page 390

1. a. False. The length of the zero vector is 0.

b. True. By the displayed equation before Example 2 in Section 6.1, with $c = -1$, $\| - \mathbf{x} \| = \|(-1)\mathbf{x}\| = |-1|\|\mathbf{x}\| = \|\mathbf{x}\|$.

c. True. This is the definition of distance.

d. False. The equation would be true if $r \|\mathbf{v}\|$ were replaced by $|r| \|\mathbf{v}\|$.

e. False. Orthogonal *nonzero* vectors are linearly independent.

f. True. If $\mathbf{x} \cdot \mathbf{u} = 0$ and $\mathbf{x} \cdot \mathbf{v} = 0$, then $\mathbf{x} \cdot (\mathbf{u} - \mathbf{v}) = \mathbf{x} \cdot \mathbf{u} - \mathbf{x} \cdot \mathbf{v} = 0$.

g. True. This is the "only if" part of the Pythagorean Theorem in Section 6.1.

h. True. This is the "only if" part of the Pythagorean Theorem in Section 6.1 when $\mathbf{v}$ is replaced by $-\mathbf{v}$, because $\| - \mathbf{v} \|^2$ is the same as $\|\mathbf{v}\|^2$.

i. False. The orthogonal projection of $\mathbf{y}$ onto $\mathbf{u}$ is a scalar multiple of $\mathbf{u}$, not $\mathbf{y}$ (except when $\mathbf{y}$ itself is already a multiple of $\mathbf{u}$).

j. True. The orthogonal projection of any vector $\mathbf{y}$ onto W is always a vector in W.

k. True. This is a special case of the statement in the box following Example 6 in Section 6.1 (and proved in Exercise 30 in Section 6.1).

l. False. The zero vector is in both W and $W^\perp$.

m. True. (See Exercise 32 in Section 6.2.) If $\mathbf{v}_i \cdot \mathbf{v}_j = 0$, then $(c_i \mathbf{v}_i) \cdot (c_j \mathbf{v}_j) = c_i c_j (\mathbf{v}_i \cdot \mathbf{v}_j) = c_i c_j (0) = 0$.

n. False. The statement is true only for a *square* matrix. See Theorem 10 in Section 6.3.

o. False. An orthogonal matrix is square and has ortho*normal* columns.

p. True. See Exercises 27 and 28 in Section 6.2. If U has orthonormal columns, then $U^T U = I$. If U is also square, then the Invertible Matrix Theorem shows that U is invertible and U^T is U^{-1}. In this case, $U^T U = UU^{-1} = I$, which shows that the columns of U^T are orthonormal; that is, the rows of U are orthonormal.

q. True. By the Orthogonal Decomposition Theorem, the vectors $\text{proj}_W \mathbf{v}$ and $\mathbf{v} - \text{proj}_W \mathbf{v}$ are orthogonal, so the stated equality follows from the Pythagorean Theorem.

r. False. A least-squares solution is a vector $\hat{\mathbf{x}}$ (not $A\hat{\mathbf{x}}$) such that the vector $A\hat{\mathbf{x}}$ is the closest point to $\mathbf{b}$ in Col A.

s. False. The equation $\hat{\mathbf{x}} = (A^T A)^{-1} A^T \mathbf{b}$ describes the *solution* of the normal equations, not the matrix form of the normal equations. Furthermore, this equation makes sense only when $A^T A$ is invertible.

2. If $\{\mathbf{v}_1, \mathbf{v}_2\}$ is an orthonormal set and $\mathbf{x} = c_1 \mathbf{v}_1 + c_2 \mathbf{v}_2$, then the vectors $c_1 \mathbf{v}_1$ and $c_2 \mathbf{v}_2$ are orthogonal (Exercise 32 in Section 6.2). By the Pythagorean Theorem and properties of the norm

$$\|\mathbf{x}\|^2 = \|c_1 \mathbf{v}_1 + c_2 \mathbf{v}_2\|^2 = \|c_1 \mathbf{v}_1\|^2 + \|c_2 \mathbf{v}_2\|^2$$
$$= (|c_1| \|\mathbf{v}_1\|)^2 + (|c_1| \|\mathbf{v}_2\|)^2 = |c_1|^2 + |c_2|^2$$

So the stated equality holds for $p = 2$. Now suppose the equality holds for $p = k$, with $k \geq 2$. Let $\{\mathbf{v}_1, \ldots, \mathbf{v}_{k+1}\}$ be an orthonormal set, and consider

$$\mathbf{x} = c_1 \mathbf{v}_1 + \cdots + c_k \mathbf{v}_k + c_{k+1} \mathbf{v}_{k+1} = \mathbf{u}_k + c_{k+1} \mathbf{v}_{k+1}$$

where $\mathbf{u}_k = c_1 \mathbf{v}_1 + \cdots + c_k \mathbf{v}_k$. Observe that $\mathbf{u}_k$ and $c_{k+1} \mathbf{v}_{k+1}$ are orthogonal, because $\mathbf{v}_j \cdot \mathbf{v}_{k+1} = 0$ for $j = 1, \ldots, k$. By the Pythagorean Theorem and the assumption that the stated equality holds for k, and because $\|c_{k+1} \mathbf{v}_{k+1}\|^2 = |c_{k+1}|^2 \|\mathbf{v}_{k+1}\|^2 = |c_{k+1}|^2$,

$$\|\mathbf{x}\|^2 = \|\mathbf{u}_k\|^2 + \|c_{k+1} \mathbf{v}_{k+1}\|^2 = $$
$$|c_1|^2 + \cdots + |c_k|^2 + |c_{k+1}|^2$$

Thus the truth of the equality for $p = k$ implies its truth for $p = k + 1$. By the principle of induction, the equality is true for all integers $p \geq 2$.

4. By parts (a) and (c) of Theorem 7 in Section 6.2, $\{U\mathbf{v}_1, \ldots, U\mathbf{v}_n\}$ is an orthonormal set in $\mathbb{R}^n$. Since there are n vectors in this linearly independent set, the set is a basis for $\mathbb{R}^n$.

6. If $U\mathbf{x} = \lambda \mathbf{x}$ for some $\mathbf{x} \neq \mathbf{0}$, then by Theorem 7(a) in Section 6.2 and by a property of the norm, $\|\mathbf{x}\| = \|U\mathbf{x}\| = \|\lambda \mathbf{x}\| = |\lambda| \|\mathbf{x}\|$, which shows that $|\lambda| = 1$ (because $\|\mathbf{x}\| \neq 0$).

8. a. Suppose $\mathbf{x} \cdot \mathbf{y} = 0$. By the Pythagorean Theorem,

$$\|\mathbf{x}\|^2 + \|\mathbf{y}\|^2 = \|\mathbf{x} + \mathbf{y}\|^2$$

Since T preserves lengths and is linear,

$$\|T(\mathbf{x})\|^2 + \|T(\mathbf{y})\|^2 = \|T(\mathbf{x} + \mathbf{y})\|^2 = \|T(\mathbf{x}) + T(\mathbf{y})\|^2$$

This equation shows that $T(\mathbf{x})$ and $T(\mathbf{y})$ are orthogonal, because of the Pythagorean Theorem. Thus T preserves orthogonality.

b. The standard matrix of T is $[\, T(\mathbf{e}_1) \quad \cdots \quad T(\mathbf{e}_n) \,]$, where $\mathbf{e}_1, \ldots, \mathbf{e}_n$ are the columns of the identity matrix. Then $\{T(\mathbf{e}_1), \ldots, T(\mathbf{e}_n)\}$ is an orthonormal set because T preserves both orthogonality and lengths (and because the columns of the identity matrix form an orthonormal set). Finally, a square matrix with orthonormal columns is an orthogonal matrix, as was observed in Section 6.2.

10. Use Theorem 14 in Section 6.5. If $c \neq 0$, the least-squares solution of $A\mathbf{x} = c\mathbf{b}$ is given by $(A^T A)^{-1} A^T (c\mathbf{b})$, which equals $c(A^T A)^{-1} A^T \mathbf{b}$, by linearity of matrix multiplication. This solution is c times the least-squares solution of $A\mathbf{x} = \mathbf{b}$.

12. Equation (1) in the exercise has been written as $V\lambda = \mathbf{b}$, where V is a single nonzero column vector $\mathbf{v}$, and $\mathbf{b} = A\mathbf{v}$. The least-squares solution $\hat{\lambda}$ of $V\lambda = \mathbf{b}$ is the exact solution of the normal equations $V^T V\lambda = V^T \mathbf{b}$. In the original notation, this equation is $\mathbf{v}^T \mathbf{v}\lambda = \mathbf{v}^T A\mathbf{v}$. Since $\mathbf{v}^T \mathbf{v}$ is nonzero, the least-squares solution $\hat{\lambda}$ is $\mathbf{v}^T A\mathbf{v}/(\mathbf{v}^T \mathbf{v})$. This expression is the Rayleigh quotient discussed in the exercises for Section 5.8.

14. The equation $A\mathbf{x} = \mathbf{b}$ has a solution if and only if $\mathbf{b}$ is in Col A. By Exercise 13(c), $A\mathbf{x} = \mathbf{b}$ has a solution if and only if $\mathbf{b}$ is orthogonal to Nul A^T. This happens if and only if $\mathbf{b}$ is orthogonal to all solutions of $A^T \mathbf{x} = \mathbf{0}$.

16. a. If $U = [\, \mathbf{u}_1 \quad \mathbf{u}_2 \quad \cdots \quad \mathbf{u}_n \,]$, then $AU = [\, \lambda_1 \mathbf{u}_1 \quad A\mathbf{u}_2 \quad \cdots \quad A\mathbf{u}_n \,]$. Since $\mathbf{u}_1$ is a unit vector and $\mathbf{u}_2, \ldots, \mathbf{u}_n$ are orthogonal to $\mathbf{u}_1$, the first column of $U^T AU$ is $U^T (\lambda_1 \mathbf{u}_1) = \lambda_1 U^T \mathbf{u}_1 = \lambda_1 \mathbf{e}_1$.

b. From part (a),

$$U^T AU = \begin{bmatrix} \lambda_1 & * & * & * & * \\ 0 & & & & \\ \vdots & & & A_1 & \\ 0 & & & & \end{bmatrix}$$

View $U^T AU$ as a 2×2 block upper-triangular matrix, with A_1 as the $(2, 2)$-block. Then, from Supplementary Exercise 12 in Chapter 5,

$$\det(U^T AU - \lambda I_n) = \det((\lambda_1 - \lambda)I_1) \cdot \det(A_1 - \lambda I_{n-1})$$
$$= (\lambda_1 - \lambda) \cdot \det(A_1 - \lambda I_{n-1})$$

This shows that the eigenvalues of $U^T AU$, namely, $\lambda_1, \ldots, \lambda_n$, consist of λ_1 and the eigenvalues of A_1. So the eigenvalues of A_1 are $\lambda_2, \ldots, \lambda_n$.

18. [M] $\dfrac{\|\Delta\mathbf{x}\|}{\|\mathbf{x}\|} = .00212$, cond$(A) \times \dfrac{\|\Delta\mathbf{b}\|}{\|\mathbf{b}\|} = $ $3363 \times (.00212) \approx 7.1$. In this case, $\|\Delta\mathbf{x}\|/\|\mathbf{x}\|$ is almost the same as $\|\Delta\mathbf{b}\|/\|\mathbf{b}\|$, even though the large condition number suggests that $\|\Delta\mathbf{x}\|/\|\mathbf{x}\|$ could be much larger.

20. [M]

$$\text{cond}(A) \times \frac{\|\Delta\mathbf{b}\|}{\|\mathbf{b}\|} = 23{,}683 \times (1.097 \times 10^{-5}) = .2598.$$

This calculation shows that the relative change in $\mathbf{x}$, for this particular $\mathbf{b}$ and $\Delta\mathbf{b}$, should not exceed .2598. As it turns out, $\|\Delta\mathbf{x}\|/\|\mathbf{x}\| = .2597$. So the theoretical maximum change is almost achieved.

Chapter 7

Section 7.1, page 399

2. Not symmetric   **4.** Symmetric   **6.** Not symmetric

8. Orthogonal, $\begin{bmatrix} 1/\sqrt{2} & 1/\sqrt{2} \\ -1/\sqrt{2} & 1/\sqrt{2} \end{bmatrix}$ **10.** Not orthogonal

12. Orthogonal, $\begin{bmatrix} .5 & -.5 & .5 & -.5 \\ .5 & .5 & .5 & .5 \\ -.5 & -.5 & .5 & .5 \\ -.5 & .5 & .5 & -.5 \end{bmatrix}$

14. $P = \begin{bmatrix} 1/\sqrt{2} & -1/\sqrt{2} \\ 1/\sqrt{2} & 1/\sqrt{2} \end{bmatrix}$, $D = \begin{bmatrix} 6 & 0 \\ 0 & -4 \end{bmatrix}$

16. $P = \begin{bmatrix} 3/5 & -4/5 \\ 4/5 & 3/5 \end{bmatrix}$, $D = \begin{bmatrix} 25 & 0 \\ 0 & -25 \end{bmatrix}$

18. $P = \begin{bmatrix} -4/5 & 0 & 3/5 \\ 3/5 & 0 & 4/5 \\ 0 & 1 & 0 \end{bmatrix}$, $D = \begin{bmatrix} 25 & 0 & 0 \\ 0 & 3 & 0 \\ 0 & 0 & -50 \end{bmatrix}$

20. $P = \begin{bmatrix} 2/3 & -1/3 & 2/3 \\ -1/3 & 2/3 & 2/3 \\ 2/3 & 2/3 & -1/3 \end{bmatrix}$, $D = \begin{bmatrix} 13 & 0 & 0 \\ 0 & 7 & 0 \\ 0 & 0 & 1 \end{bmatrix}$

22. $P = \begin{bmatrix} 1 & 0 & 0 & 0 \\ 0 & 1/\sqrt{2} & 0 & -1/\sqrt{2} \\ 0 & 0 & 1 & 0 \\ 0 & 1/\sqrt{2} & 0 & 1/\sqrt{2} \end{bmatrix}$,

$D = \begin{bmatrix} 2 & 0 & 0 & 0 \\ 0 & 2 & 0 & 0 \\ 0 & 0 & 2 & 0 \\ 0 & 0 & 0 & 0 \end{bmatrix}$

24. $P = \begin{bmatrix} -2/3 & 1/\sqrt{2} & 1/\sqrt{18} \\ 2/3 & 1/\sqrt{2} & -1/\sqrt{18} \\ 1/3 & 0 & 4/\sqrt{18} \end{bmatrix}$,

$D = \begin{bmatrix} 10 & 0 & 0 \\ 0 & 1 & 0 \\ 0 & 0 & 1 \end{bmatrix}$

25. a. True. See Theorem 2 and the paragraph preceding the theorem.

b. True. This is a particular case of the statement in Theorem 1, when $\mathbf{u}$ and $\mathbf{v}$ are nonzero.

c. False. There are n real eigenvalues (Theorem 3), but they need not be distinct (Example 3).

d. False. See the paragraph following formula (2), in which each $\mathbf{u}$ is a unit vector.

26. a. True, by Theorem 2.

b. True. See the displayed equation in the paragraph before Theorem 2.

c. False. An orthogonal matrix can be symmetric (and hence orthogonally diagonalizable), but not every orthogonal matrix is symmetric. The matrix P in Example 2 is an orthogonal matrix, but it is not symmetric.

d. True, by Theorem 3(b).

28. $(A\mathbf{x}) \cdot \mathbf{y} = (A\mathbf{x})^T\mathbf{y} = \mathbf{x}^TA^T\mathbf{y} = \mathbf{x}^TA\mathbf{y} = \mathbf{x} \cdot (A\mathbf{y})$, because $A^T = A$.

30. If A and B are orthogonally diagonalizable, then A and B are symmetric, by Theorem 2. If $AB = BA$, then $(AB)^T = (BA)^T = A^TB^T = AB$. So AB is symmetric and hence is orthogonally diagonalizable, by Theorem 2.

32. If $A = PRP^{-1}$, then $P^{-1}AP = R$. Since P is orthogonal, $R = P^TAP$. Hence $R^T = (P^TAP)^T = P^TA^TP^{TT} = P^TAP = R$, which shows that R is symmetric. Since R is also upper triangular, its entries above the diagonal must be zeros, to match the zeros below the diagonal. Thus R is a diagonal matrix.

34. $A = 7\mathbf{u}_1\mathbf{u}_1^T + 7\mathbf{u}_2\mathbf{u}_2^T - 2\mathbf{u}_3\mathbf{u}_3^T$, where

$\mathbf{u}_1\mathbf{u}_1^T = \begin{bmatrix} 1/2 & 0 & 1/2 \\ 0 & 0 & 0 \\ 1/2 & 0 & 1/2 \end{bmatrix}$,

$\mathbf{u}_2\mathbf{u}_2^T = \begin{bmatrix} 1/18 & -4/18 & -1/18 \\ -4/18 & 16/18 & 4/18 \\ -1/18 & 4/18 & 1/18 \end{bmatrix}$, and

$\mathbf{u}_3\mathbf{u}_3^T = \begin{bmatrix} 4/9 & 2/9 & -4/9 \\ 2/9 & 1/9 & -2/9 \\ -4/9 & -2/9 & 4/9 \end{bmatrix}$.

36. Given any $\mathbf{y}$ in $\mathbb{R}^n$, let $\hat{\mathbf{y}} = B\mathbf{y}$ and $\mathbf{z} = \mathbf{y} - \hat{\mathbf{y}}$. Suppose $B^T = B$ and $B^2 = B$. Then $B^TB = BB = B$.

a. $\mathbf{z} \cdot \hat{\mathbf{y}} = (\mathbf{y} - B\mathbf{y}) \cdot (B\mathbf{y}) = \mathbf{y} \cdot (B\mathbf{y}) - (B\mathbf{y}) \cdot (B\mathbf{y})$
$= \mathbf{y}^TB\mathbf{y} - (B\mathbf{y})^TB\mathbf{y} = \mathbf{y}^TB\mathbf{y} - \mathbf{y}^TB^TB\mathbf{y} = 0$

So $\mathbf{z}$ is orthogonal to $\hat{\mathbf{y}}$.

b. Any vector in $W = \text{Col } B$ has the form $B\mathbf{u}$ for some $\mathbf{u}$. To show that $\mathbf{y} - \hat{\mathbf{y}}$ is orthogonal to $B\mathbf{u}$, use Exercise 28 since B is symmetric:

$$(\mathbf{y} - \hat{\mathbf{y}}) \cdot B\mathbf{u} = [B(\mathbf{y} - \hat{\mathbf{y}})] \cdot \mathbf{u} = [B\mathbf{y} - BB\mathbf{y}] \cdot \mathbf{u} = 0$$

because $B^2 = B$. So $\mathbf{y} - \hat{\mathbf{y}}$ is in $W^\perp$, and the decomposition $\mathbf{y} = \hat{\mathbf{y}} + (\mathbf{y} - \hat{\mathbf{y}})$ expresses $\mathbf{y}$ as the sum of a vector in W and a vector in $W^\perp$. By the Orthogonal Decomposition Theorem in Section 6.3, this decomposition is unique, and so $\hat{\mathbf{y}}$ must be $\text{proj}_W \mathbf{y}$.

37. [M] $P = \dfrac{1}{2}\begin{bmatrix} -1 & 1 & 1 & 1 \\ 1 & 1 & 1 & -1 \\ -1 & 1 & -1 & -1 \\ 1 & 1 & -1 & 1 \end{bmatrix}$,

$D = \begin{bmatrix} 18 & 0 & 0 & 0 \\ 0 & 10 & 0 & 0 \\ 0 & 0 & 4 & 0 \\ 0 & 0 & 0 & -12 \end{bmatrix}$

38. [M] $P = \begin{bmatrix} .8 & -.2 & .4 & -.4 \\ .4 & -.4 & -.2 & .8 \\ .4 & .4 & -.8 & -.2 \\ .2 & .8 & .4 & .4 \end{bmatrix}$,

$D = \begin{bmatrix} .25 & 0 & 0 & 0 \\ 0 & .30 & 0 & 0 \\ 0 & 0 & .55 & 0 \\ 0 & 0 & 0 & .75 \end{bmatrix}$

39. [M] $P = \begin{bmatrix} .7071 & .4243 & -.4 & -.4 \\ 0 & .5657 & -.2 & .8 \\ 0 & .5657 & .8 & -.2 \\ .7071 & -.4243 & .4 & .4 \end{bmatrix}$,

$D = \begin{bmatrix} .75 & 0 & 0 & 0 \\ 0 & .75 & 0 & 0 \\ 0 & 0 & 0 & 0 \\ 0 & 0 & 0 & -1.25 \end{bmatrix}$.

Note: $.4243 \approx 3/\sqrt{50}$ and $.5657 \approx 4/\sqrt{50}$.

40. [M] $P =$

$\begin{bmatrix} 1/\sqrt{2} & 1/\sqrt{6} & 1/\sqrt{12} & 1/\sqrt{20} & 1/\sqrt{5} \\ -1/\sqrt{2} & 1/\sqrt{6} & 1/\sqrt{12} & 1/\sqrt{20} & 1/\sqrt{5} \\ 0 & -2/\sqrt{6} & 1/\sqrt{12} & 1/\sqrt{20} & 1/\sqrt{5} \\ 0 & 0 & -3/\sqrt{12} & 1/\sqrt{20} & 1/\sqrt{5} \\ 0 & 0 & 0 & -4/\sqrt{20} & 1/\sqrt{5} \end{bmatrix}$,

$D = \begin{bmatrix} 8 & 0 & 0 & 0 & 0 \\ 0 & 8 & 0 & 0 & 0 \\ 0 & 0 & 32 & 0 & 0 \\ 0 & 0 & 0 & -28 & 0 \\ 0 & 0 & 0 & 0 & 17 \end{bmatrix}$

Section 7.2, page 406

2. a. $4x_1^2 + 2x_2^2 + x_3^2 + 6x_1x_2 + 2x_2x_3$ **b.** 21   **c.** 5

4. a. $\begin{bmatrix} 20 & 7.5 \\ 7.5 & -10 \end{bmatrix}$ **b.** $\begin{bmatrix} 0 & .5 \\ .5 & 0 \end{bmatrix}$

6. a. $\begin{bmatrix} 5 & 5/2 & -3/2 \\ 5/2 & -1 & 0 \\ -3/2 & 0 & 7 \end{bmatrix}$ **b.** $\begin{bmatrix} 0 & -2 & 0 \\ -2 & 0 & 2 \\ 0 & 2 & 1 \end{bmatrix}$

8. $P = \frac{1}{3}\begin{bmatrix} 2 & -1 & 2 \\ -1 & 2 & 2 \\ 2 & 2 & -1 \end{bmatrix}$, $\mathbf{y}^T D \mathbf{y} = 15y_1^2 + 9y_2^2 + 3y_3^2$

In Exercises 10–14, other answers (change of variables and new quadratic form) are possible.

10. Positive definite; eigenvalues are 11 and 1
Change of variable: $\mathbf{x} = P\mathbf{y}$, with $P = \dfrac{1}{\sqrt{5}}\begin{bmatrix} 2 & 1 \\ -1 & 2 \end{bmatrix}$
New quadratic form: $11y_1^2 + y_2^2$

12. Negative definite; eigenvalues are -1 and -6
Change of variable: $\mathbf{x} = P\mathbf{y}$, with $P = \dfrac{1}{\sqrt{5}}\begin{bmatrix} 1 & -2 \\ 2 & 1 \end{bmatrix}$
New quadratic form: $-y_1^2 - 6y_2^2$

14. Indefinite; eigenvalues are 9 and -1
Change of variable: $\mathbf{x} = P\mathbf{y}$ where $P = \dfrac{1}{\sqrt{10}}\begin{bmatrix} 3 & -1 \\ 1 & 3 \end{bmatrix}$
New quadratic form: $9y_1^2 - y_2^2$

16. [M] Positive definite; eigenvalues are 6.5 and 1.5
Change of variable: $\mathbf{x} = P\mathbf{y}$;
$P = \dfrac{1}{\sqrt{50}}\begin{bmatrix} 3 & -4 & 3 & 4 \\ 5 & 0 & -5 & 0 \\ 4 & 3 & 4 & -3 \\ 0 & 5 & 0 & 5 \end{bmatrix}$
New quadratic form: $6.5y_1^2 + 6.5y_2^2 + 1.5y_3^2 + 1.5y_4^2$

18. [M] Indefinite; eigenvalues are $17, 1, -1, -7$
Change of variable: $\mathbf{x} = P\mathbf{y}$;
$P = \begin{bmatrix} -3/\sqrt{12} & 0 & 0 & 1/2 \\ 1/\sqrt{12} & 0 & -2/\sqrt{6} & 1/2 \\ 1/\sqrt{12} & -1/\sqrt{2} & 1/\sqrt{6} & 1/2 \\ 1/\sqrt{12} & 1/\sqrt{2} & 1/\sqrt{6} & 1/2 \end{bmatrix}$
New quadratic form: $17y_1^2 + y_2^2 - y_3^2 - 7y_4^2$

20. 5

21. a. True, by the definition before Example 1, even though a nonsymmetric matrix could be used to compute values of a quadratic form.
   b. True. See the paragraph following Example 3.
   c. True, because the columns of P in Theorem 4 are eigenvectors of A. Review the Diagonalization Theorem (Theorem 5) in Section 5.3.
   d. False. $Q(\mathbf{x}) = 0$ when $\mathbf{x} = \mathbf{0}$.
   e. True. Theorem 5(a).
   f. True. See the Numerical Note after Example 6.

22. a. True. See the paragraph before Example 1.
   b. False. The matrix P must be orthogonal and make $P^T A P$ diagonal. See the paragraph before Example 4.
   c. False. There are also "degenerate" cases: a single point, two intersecting lines, or no points at all. See the subsection "A Geometric View of Principal Axes."
   d. False. See the definition before Theorem 5.
   e. True, by Theorem 5(b). If $\mathbf{x}^T A \mathbf{x}$ has only negative values for $\mathbf{x} \ne \mathbf{0}$, then $\mathbf{x}^T A \mathbf{x}$ is negative definite.

24. If $\det A > 0$, then by Exercise 23, $\lambda_1 \lambda_2 > 0$, so that λ_1 and λ_2 have the same sign; also, $ad = \det A + b^2 > 0$.
   a. If $\det A > 0$ and $a > 0$, then $d > 0$, too (because $ad > 0$). By Exercise 23, $\lambda_1 + \lambda_2 = a + d > 0$. Since λ_1 and λ_2 have the same sign, they are both positive. So Q is positive definite, by Theorem 5.

b. If $\det A > 0$ and $a < 0$, then $d < 0$, too. As in (a), we conclude that λ_1 and λ_2 are both negative and that Q is negative definite.

c. If $\det A < 0$, then by Exercise 23, $\lambda_1\lambda_2 < 0$, which shows that λ_1 and λ_2 have opposite signs. By Theorem 5, Q is indefinite.

26. We may assume $A = PDP^T$, with $P^T = P^{-1}$. The eigenvalues of A are all positive; denote them by $\lambda_1, \ldots, \lambda_n$. Let C be the diagonal matrix with $\sqrt{\lambda_1}, \ldots, \sqrt{\lambda_n}$ on the diagonal. Then $D = C^2 = C^TC$. If $B = PCP^T$, then B is positive definite because its eigenvalues are the positive numbers on the diagonal of C. Also,

$$B^TB = (PCP^T)^T(PCP^T) = (P^{TT}C^TP^T)(PCP)$$
$$= PC^TCP \qquad \text{Because } P^TP = I$$
$$= PDP = A$$

28. The eigenvalues of A are all positive, by Theorem 5. Since the eigenvalues of A^{-1} are the reciprocals of the eigenvalues of A (see Exercise 25 in Section 5.1), the eigenvalues of A^{-1} are positive. (Note that A^{-1} is symmetric.) By Theorem 5, the quadratic form $\mathbf{x}^TA^{-1}\mathbf{x}$ is positive definite.

Section 7.3, page 413

2. $\mathbf{x} = P\mathbf{y}$, where $P = \begin{bmatrix} 1/\sqrt{3} & -2/\sqrt{6} & 0 \\ 1/\sqrt{3} & 1/\sqrt{6} & -1/\sqrt{2} \\ 1/\sqrt{3} & 1/\sqrt{6} & 1/\sqrt{2} \end{bmatrix}$

4. a. 5   b. $\pm\begin{bmatrix} 1/\sqrt{3} \\ 1/\sqrt{3} \\ 1/\sqrt{3} \end{bmatrix}$ c. 2

6. a. $\frac{15}{2}$ b. $\pm\begin{bmatrix} 3/\sqrt{10} \\ 1/\sqrt{10} \end{bmatrix}$ c. $\frac{5}{2}$

8. Any unit vector that is a linear combination of $\begin{bmatrix} -2/\sqrt{5} \\ 1/\sqrt{5} \\ 0 \end{bmatrix}$ and $\begin{bmatrix} -1/\sqrt{2} \\ 0 \\ 1/\sqrt{2} \end{bmatrix}$. Equivalently, any unit vector that is orthogonal to $\begin{bmatrix} 1 \\ 2 \\ 1 \end{bmatrix}$.

10. $1 + \sqrt{17}$

12. Let $\mathbf{x}$ be a unit eigenvector for the eigenvalue λ. Then $\mathbf{x}^TA\mathbf{x} = \mathbf{x}^T(\lambda\mathbf{x}) = \lambda$, because $\mathbf{x}^T\mathbf{x} = 1$. So λ must satisfy $m \le \lambda \le M$.

14. [M] a. 17   b. $\begin{bmatrix} .5 \\ .5 \\ .5 \\ .5 \end{bmatrix}$ c. 13

16. [M] a. 9   b. $\begin{bmatrix} -2/\sqrt{6} \\ 0 \\ 1/\sqrt{6} \\ 1/\sqrt{6} \end{bmatrix}$ c. 3

Section 7.4, page 423

2. 5, 0   4. 3, 1

6. $\begin{bmatrix} -2 & 0 \\ 0 & -1 \end{bmatrix} = \begin{bmatrix} -1 & 0 \\ 0 & -1 \end{bmatrix}\begin{bmatrix} 2 & 0 \\ 0 & 1 \end{bmatrix}\begin{bmatrix} 1 & 0 \\ 0 & 1 \end{bmatrix}$

8. $\begin{bmatrix} 2/\sqrt{5} & -1/\sqrt{5} \\ 1/\sqrt{5} & 2/\sqrt{5} \end{bmatrix}\begin{bmatrix} 4 & 0 \\ 0 & 1 \end{bmatrix}\begin{bmatrix} 1/\sqrt{5} & 2/\sqrt{5} \\ -2/\sqrt{5} & 1/\sqrt{5} \end{bmatrix}$

10. $\begin{bmatrix} 2/\sqrt{5} & 1/\sqrt{5} & 0 \\ 1/\sqrt{5} & -2/\sqrt{5} & 0 \\ 0 & 0 & 1 \end{bmatrix}\begin{bmatrix} 5 & 0 \\ 0 & 0 \\ 0 & 0 \end{bmatrix}\begin{bmatrix} 2/\sqrt{5} & -1/\sqrt{5} \\ 1/\sqrt{5} & 2/\sqrt{5} \end{bmatrix}$

12. $\begin{bmatrix} 1/\sqrt{3} & 1/\sqrt{2} & 1/\sqrt{6} \\ 1/\sqrt{3} & 0 & -2/\sqrt{6} \\ 1/\sqrt{3} & -1/\sqrt{2} & 1/\sqrt{6} \end{bmatrix}\begin{bmatrix} \sqrt{3} & 0 \\ 0 & \sqrt{2} \\ 0 & 0 \end{bmatrix}\begin{bmatrix} 0 & 1 \\ 1 & 0 \end{bmatrix}$

14. From Exercise 7, $A = U\Sigma V^T$ with $V = \begin{bmatrix} 2/\sqrt{5} & -1/\sqrt{5} \\ 1/\sqrt{5} & 2/\sqrt{5} \end{bmatrix}$. The first column of V is a unit vector at which $\|A\mathbf{x}\|$ is maximized.

16. a. rank $A = 2$

b. Basis for Col A: $\begin{bmatrix} -.86 \\ .31 \\ .41 \end{bmatrix}, \begin{bmatrix} -.11 \\ .68 \\ -.73 \end{bmatrix}$

Basis for Nul A: $\begin{bmatrix} .65 \\ .08 \\ -.16 \\ -.73 \end{bmatrix}, \begin{bmatrix} -.34 \\ .42 \\ -.84 \\ -.08 \end{bmatrix}$

18. The determinant of an orthogonal matrix U is ± 1, because
$$1 = \det I = \det U^TU = (\det U^T)(\det U) = (\det U)^2$$
Suppose A is square and $A = U\Sigma V^T$. Then Σ is square, and
$$\det A = (\det U)(\det \Sigma)(\det V^T)$$
$$= \pm \det \Sigma = \pm\sigma_1\cdots\sigma_n$$

20. If A is positive definite, then $A = PDP^T$, where P is an orthogonal matrix and D is a diagonal matrix. The diagonal entries of D are positive, because they are the eigenvalues of a positive definite matrix. Also, the matrix P^T is an orthogonal matrix because it is invertible, and the inverse and transpose of P^T coincide, since $(P^T)^{-1} = (P^{-1})^{-1} = P = (P^T)^T$. Thus, the factorization $A = PDP^T$ has the properties that make it a singular value decomposition.

22. The right singular vector $\mathbf{v}_1$ is an eigenvector for the largest eigenvalue λ_1 of A^TA. By Theorem 7 in Section 7.3, the second largest eigenvalue, λ_2, is the maximum of $\mathbf{x}^T(A^TA)\mathbf{x}$ over all unit vectors orthogonal to $\mathbf{v}_1$. Since

$\mathbf{x}^T(A^TA)\mathbf{x} = \|A\mathbf{x}\|^2$, the square root of λ_2, which is the second singular value of A, is the maximum of $\|A\mathbf{x}\|$ over all unit vectors orthogonal to $\mathbf{v}_1$.

24. From Exercise 23, $A^T = \sigma_1\mathbf{v}_1\mathbf{u}_1^T + \cdots + \sigma_r\mathbf{v}_r\mathbf{u}_r^T$. Then

$$\begin{aligned} A^T\mathbf{u}_j &= (\sigma_j\mathbf{v}_j\mathbf{u}_j^T)\mathbf{u}_j \quad \text{Because } \mathbf{u}_i^T\mathbf{u}_j = 0 \text{ for } i \neq j \\ &= \sigma_j\mathbf{v}_j \quad\quad\quad \text{Because } \mathbf{u}_j^T\mathbf{u}_j = 1 \end{aligned}$$

26. **[M]**
$$\begin{bmatrix} .5 & -.5 & -.5 & -.5 \\ .5 & .5 & .5 & -.5 \\ .5 & -.5 & .5 & .5 \\ .5 & .5 & -.5 & .5 \end{bmatrix} \begin{bmatrix} 40 & 0 & 0 & 0 \\ 0 & 20 & 0 & 0 \\ 0 & 0 & 10 & 0 \\ 0 & 0 & 0 & 0 \end{bmatrix}$$
$$\times \begin{bmatrix} -.4 & .8 & -.2 & .4 \\ .8 & .4 & .4 & .2 \\ .4 & -.2 & -.8 & .4 \\ -.2 & -.4 & .4 & .8 \end{bmatrix}$$

The entries in this exercise are simple, to allow students to check their work mentally or by hand. The *Study Guide* contains a sequence of MATLAB commands that produce this SVD.

28. **[M]** 25.9609, 14.4566, 4.6700, 2.2868; $\sigma_1/\sigma_4 = 11.3525$

Section 7.5, page 430

2. $M = \begin{bmatrix} 4 \\ 9 \end{bmatrix}$; $B = \begin{bmatrix} -3 & 1 & -2 & 2 & 3 & -1 \\ -6 & 2 & -3 & -1 & 6 & 2 \end{bmatrix}$,
$S = \begin{bmatrix} 5.6 & 8 \\ 8 & 18 \end{bmatrix}$

4. $\begin{bmatrix} .44 \\ .90 \end{bmatrix}$ for $\lambda = 21.9$, $\begin{bmatrix} -.90 \\ .44 \end{bmatrix}$ for $\lambda = 1.7$

6. **[M]** $y_1 = .62x_1 + .60x_2 + .51x_3$, which explains 64.9% of the total variance.

8. $y_1 = .44x_1 + .90x_2$; y_1 explains 92.9% of the variance.

10. **[M]** $c_1 = .41$, $c_2 = .82$, $c_3 = .41$ to two decimal places, or $c_1 = 1/\sqrt{6}$, $c_2 = 2/\sqrt{6}$, $c_3 = 1/\sqrt{6}$. The variance of y is 15.

12. By Exercise 11, the change of variable $\mathbf{X} = P\mathbf{Y}$ changes the covariance matrix S of $\mathbf{X}$ into the covariance matrix P^TSP of $\mathbf{Y}$. The total variance of the data, as described by $\mathbf{Y}$, is tr(P^TSP). However, since P^TSP is similar to S, they have the same trace (Exercise 25 in Section 5.4). Thus the total variance of the data is unchanged by this change of variable.

Chapter 7 Supplementary Exercises, page 432

1. a. True. This is part of Theorem 2 in Section 7.1. The proof of this fact appears just before the statement of Theorem 2.

b. False. Counterexample: $A = \begin{bmatrix} 0 & -1 \\ 1 & 0 \end{bmatrix}$.

c. True. This is proved in the first part of the proof of Theorem 6 in Section 7.3. It is also a consequence of Theorem 7 in Section 6.2.

d. False. The principal axes of $\mathbf{x}^TA\mathbf{x}$ are the columns of any *orthogonal* matrix P that diagonalizes A. [*Note:* When A has an eigenvalue whose eigenspace has dimension greater than 1 (for example, when $A = I$), the principal axes are not uniquely determined.]

e. False. Counterexample: $P = \begin{bmatrix} 1 & -1 \\ 1 & 1 \end{bmatrix}$. The columns here are orthogonal but not orthonormal. If P is a square matrix with ortho*normal* columns, $P^T = P^{-1}$.

f. False. See Example 6 in Section 7.2.

g. False. Counterexample: $A = \begin{bmatrix} 2 & 0 \\ 0 & -3 \end{bmatrix}$ and $\mathbf{x} = \begin{bmatrix} 1 \\ 0 \end{bmatrix}$, then $\mathbf{x}^TA\mathbf{x} = 2 > 0$, but $\mathbf{x}^TA\mathbf{x}$ is an indefinite quadratic form.

h. True. This is basically the Principal Axes Theorem (Section 7.2). Any quadratic form can be written as $\mathbf{x}^TA\mathbf{x}$ for some symetric matrix A.

i. False. See Example 3 in Section 7.3.

j. False. The maximum value must be computed over the set of *unit* vectors. Without a restriction on the norm of $\mathbf{x}$, the values of $\mathbf{x}^TA\mathbf{x}$ can be made as large as desired.

k. False. Any orthogonal change of variable $\mathbf{x} = P\mathbf{y}$ changes a positive definite form into another positive definite form. Proof: By Theorem 5 in Section 7.2, the classification of a quadratic form is determined by the eigenvalues of the matrix of the form. Given a form $\mathbf{x}^TA\mathbf{x}$, the matrix of the new quadratic form is $P^{-1}AP$, which is similar to A and therefore has the same eigenvalues as A.

l. False. The term "definite eigenvalue" is undefined and therefore meaningless.

m. True. If $\mathbf{x} = P\mathbf{y}$, then $\mathbf{x}^TA\mathbf{x} = (P\mathbf{y})^TA(P\mathbf{y}) = \mathbf{y}^TP^TAP\mathbf{y} = \mathbf{y}^T(P^{-1}AP)\mathbf{y}$.

n. False. Counterexample: Let $U = \begin{bmatrix} 1 & -1 \\ 1 & -1 \end{bmatrix}$. The columns of U must be ortho*normal* to make $UU^T\mathbf{x}$ the orthogonal projection of $\mathbf{x}$ onto Col U.

o. True. This follows from the discussion in Example 2 of Section 7.4, which refers to a proof given in Example 1.

p. True. Theorem 10 in Section 7.4 writes the decomposition in the form $U\Sigma V^T$, where U and V are orthogonal matrices. In this case, V^T is also an orthogonal matrix. [Proof: Because V is orthogonal, V is invertible and $V^{-1} = V^T$. Then $(V^{-1})^T = (V^T)^T$ and $(V^T)^{-1} = (V^T)^T$. Since V^T is square and invertible, the second equality shows that V^T is an orthogonal matrix.]

q. False. Counterexample: The singular values of $A = \begin{bmatrix} 2 & 0 \\ 0 & 1 \end{bmatrix}$ are 2 and 1, but the singular values of A^TA are 4 and 1.

2. a. Each term in the expansion of A is symmetric, by Exercise 35 in Section 7.1. The fact that $(B + C)^T = B^T + C^T$ implies that any sum of symmetric matrices is symmetric. So A is symmetric. A direct calculation also shows that $A^T = A$.

b. $A\mathbf{u}_1 = (\lambda_1 \mathbf{u}_1 \mathbf{u}_1^T)\mathbf{u}_1 + \cdots + (\lambda_n \mathbf{u}_n \mathbf{u}_n^T)\mathbf{u}_1$

$\qquad = \lambda_1 \mathbf{u}_1$

because $\mathbf{u}_1^T \mathbf{u}_1 = 1$ and $\mathbf{u}_j^T \mathbf{u}_1 = 0$ for $j \neq 1$. Since $\mathbf{u}_1 \neq \mathbf{0}$, λ_1 is an eigenvalue of A. A similar argument shows that for $j = 2, \ldots, n$, λ_j is an eigenvalue of A.

4. a. By Theorem 3 in Section 6.1, $(\text{Col } A)^\perp = \text{Nul } A^T = \text{Nul } A$, because $A^T = A$.

b. Take $\mathbf{y}$ in $\mathbb{R}^n$. By the Orthogonal Decomposition Theorem (Section 6.3), $\mathbf{y} = \hat{\mathbf{y}} + \mathbf{z}$, with $\hat{\mathbf{y}}$ in $\text{Col } A$ and $\mathbf{z}$ in $(\text{Col } A)^\perp$. By part (a), $\mathbf{z}$ is in $\text{Nul } A$, which concludes the proof.

6. Because A is symmetric, there is an orthonormal eigenvector basis $\{\mathbf{u}_1, \ldots, \mathbf{u}_n\}$ for $\mathbb{R}^n$. Let $r = \text{rank } A$. If $r = 0$, then $A = 0$, and the decomposition of Exercise 4(b) is $\mathbf{y} = \mathbf{0} + \mathbf{y}$ for each $\mathbf{y}$ in $\mathbb{R}^n$; if $r = n$, then the decomposition is $\mathbf{y} = \mathbf{y} + \mathbf{0}$ for each $\mathbf{y}$.

So, assume $0 < r < n$. Then $\dim \text{Nul } A = n - r$, by the Rank Theorem, and so 0 is an eigenvalue with multiplicity $n - r$. Hence, there are r nonzero eigenvalues, counted according to their multiplicities. Renumber the eigenvector basis, if necessary, so that $\mathbf{u}_1, \ldots, \mathbf{u}_r$ are the eigenvectors corresponding to the nonzero eigenvalues.

By Exercise 5, $\mathbf{u}_1, \ldots, \mathbf{u}_r$ are in $\text{Col } A$. Also, $\mathbf{u}_{r+1}, \ldots, \mathbf{u}_n$ are in $\text{Nul } A$, because these vectors are eigenvectors for $\lambda = 0$. For $\mathbf{y}$ in $\mathbb{R}^n$, there are scalars $c_1, \ldots, c_n$ such that

$$\mathbf{y} = \underbrace{c_1 \mathbf{u}_1 + \cdots + c_r \mathbf{u}_r}_{\hat{\mathbf{y}}} + \underbrace{c_{r+1}\mathbf{u}_{r+1} + \cdots + c_n \mathbf{u}_n}_{\mathbf{z}}$$

This provides the decomposition in Exercise 4(b).

8. Suppose A is positive definite, and consider a Cholesky factorization $A = R^T R$, with R upper triangular and having positive entries on its diagonal. Let D be the diagonal matrix whose diagonal entries are the entries on the diagonal of R. Since right-multiplication by a diagonal matrix scales the columns of the matrix on its left, the matrix $L = R^T D^{-1}$ is lower triangular with 1's on its diagonal. If $U = DR$, then $A = R^T D^{-1} DR = LU$.

10. If rank $G = r$, then $\dim \text{Nul } G = n - r$, by the Rank Theorem. Hence 0 is an eigenvalue of multiplicity $n - r$, and the spectral decomposition of G is

$$G = \lambda_1 \mathbf{u}_1 \mathbf{u}_1^T + \cdots + \lambda_r \mathbf{u}_r \mathbf{u}_r^T$$

Also, $\lambda_1, \ldots, \lambda_r$ are positive because G is positive semidefinite. Thus

$$G = \left(\sqrt{\lambda_1}\mathbf{u}_1\right)\left(\sqrt{\lambda_1}\mathbf{u}_1\right)^T + \cdots + \left(\sqrt{\lambda_r}\mathbf{u}_r\right)\left(\sqrt{\lambda_r}\mathbf{u}_r\right)^T$$

By the column–row expansion of a matrix product, $G = BB^T$, where B is the $n \times r$ matrix:

$$B = \begin{bmatrix} \sqrt{\lambda_1}\mathbf{u}_1 & \cdots & \sqrt{\lambda_r}\mathbf{u}_r \end{bmatrix}$$

Finally, $G = A^T A$ for $A = B^T$.

12. a. Because the columns of V_r are orthonormal, $AA^+\mathbf{y} = (U_r D V_r^T)(V_r D^{-1} U_r^T)\mathbf{y} = (U_r DD^{-1} U_r^T)\mathbf{y} = U_r U_r^T \mathbf{y}$. Since $U_r U_r^T \mathbf{y}$ is the orthogonal projection of $\mathbf{y}$ onto $\text{Col } U_r$ (by Theorem 10 in Section 6.3), and since $\text{Col } U_r = \text{Col } A$ by (5) in Example 6 of Section 7.4, $AA^+\mathbf{y}$ is the orthogonal projection of $\mathbf{y}$ onto $\text{Col } A$.

b. $A^+ A\mathbf{x} = (V_r D^{-1} U_r^T)(U_r D V_r^T)\mathbf{x} = (V_r D^{-1} DV_r^T)\mathbf{x} = V_r V_r^T \mathbf{x}$. Since $V_r V_r^T \mathbf{x}$ is the orthogonal projection of $\mathbf{x}$ onto $\text{Col } V_r$, and since $\text{Col } V_r = \text{Row } A$ by (8) in Example 6 of Section 7.4, $A^+ A\mathbf{x}$ is the orthogonal projection of $\mathbf{x}$ onto $\text{Row } A$.

c. Use the reduced singular value decomposition of A, the definition of A^+, and associativity of matrix multiplication:

$$\begin{aligned} AA^+A &= (U_r D V_r^T)(V_r D^{-1} U_r^T)(U_r D V_r^T) \\ &= (U_r DD^{-1} U_r^T)(U_r D V_r^T) \\ &= (U_r DD^{-1} DV_r^T) \\ &= U_r D V_r^T = A \end{aligned}$$

$$\begin{aligned} A^+ AA^+ &= (V_r D^{-1} U_r^T)(U_r D V_r^T)(V_r D^{-1} U_r^T) \\ &= (V_r D^{-1} DV_r^T)(V_r D^{-1} U_r^T) \\ &= (V_r D^{-1} DD^{-1} U_r^T) \\ &= V_r D^{-1} U_r^T = A^+ \end{aligned}$$

14. The least-squares solutions of $A\mathbf{x} = \mathbf{b}$ are precisely the solutions of $A\mathbf{x} = \hat{\mathbf{b}}$ where $\hat{\mathbf{b}}$ is the orthogonal projection of $\mathbf{b}$ onto $\text{Col } A$. From Exercise 13, the minimum length solution of $A\mathbf{x} = \hat{\mathbf{b}}$ is $A^+\hat{\mathbf{b}}$, so $A^+\hat{\mathbf{b}}$ is the minimum length least-squares solution of $A\mathbf{x} = \mathbf{b}$. However, $\hat{\mathbf{b}} = AA^+\mathbf{b}$, by Exercise 12(a), and hence $A^+\hat{\mathbf{b}} = A^+ AA^+\mathbf{b} = A^+\mathbf{b}$, by Exercise 12(c). Thus $A^+\mathbf{b}$ is the minimum length least-squares solution of $A\mathbf{x} = \mathbf{b}$.

16. $[M]A^+ = \begin{bmatrix} .5 & 0 & -.05 & -.15 \\ 0 & 0 & 0 & 0 \\ 0 & 2 & .50 & 1.50 \\ .5 & -1 & -.35 & -1.05 \\ 0 & 0 & 0 & 0 \end{bmatrix}, \hat{\mathbf{x}} = \begin{bmatrix} 2.3 \\ 0 \\ 5.0 \\ -.9 \\ 0 \end{bmatrix}$

Basis for Nul A: $\begin{bmatrix} 0 \\ 1 \\ 0 \\ 0 \\ 0 \end{bmatrix}, \begin{bmatrix} 0 \\ 0 \\ 0 \\ 0 \\ 1 \end{bmatrix}$. Adding any nonzero vector

$\mathbf{u}$ in Nul A to $\hat{\mathbf{x}}$ changes a zero entry to a nonzero entry; in this case the inequality $\|\hat{\mathbf{x}}\| < \|\hat{\mathbf{x}} + \mathbf{u}\|$ is evident.

Chapter 8

Section 8.1, page 442

2. $\mathbf{y} = -5\mathbf{v}_1 + 2\mathbf{v}_2 + 4\mathbf{v}_3$. The weights sum to 1, so this is an affine sum.

Solution: $\mathbf{v}_1 = \begin{bmatrix} 1 \\ 1 \end{bmatrix}$, $\mathbf{v}_2 = \begin{bmatrix} -1 \\ 2 \end{bmatrix}$, $\mathbf{v}_3 = \begin{bmatrix} 3 \\ 2 \end{bmatrix}$, $\mathbf{y} = \begin{bmatrix} 5 \\ 7 \end{bmatrix}$,

so

$$\mathbf{v}_2 - \mathbf{v}_1 = \begin{bmatrix} -2 \\ 1 \end{bmatrix}, \ \mathbf{v}_3 - \mathbf{v}_1 = \begin{bmatrix} 2 \\ 1 \end{bmatrix}, \text{ and } \mathbf{y} - \mathbf{v}_1 = \begin{bmatrix} 4 \\ 6 \end{bmatrix}$$

Solve $c_2(\mathbf{v}_2 - \mathbf{v}_1) + c_3(\mathbf{v}_3 - \mathbf{v}_1) = \mathbf{y} - \mathbf{v}_1$ by row reducing the augmented matrix: $\begin{bmatrix} -2 & 2 & 4 \\ 1 & 1 & 6 \end{bmatrix}$ The general solution is $c_2 = 2$ and $c_3 = 4$, so $\mathbf{y} - \mathbf{v}_1 = 2(\mathbf{v}_2 - \mathbf{v}_1) + 4(\mathbf{v}_3 - \mathbf{v}_1)$ and $\mathbf{y} = -5\mathbf{v}_1 + 2\mathbf{v}_2 + 4\mathbf{v}_3$. The weights sum to 1, so this is an affine sum.

4. $\mathbf{y} = 2.6\mathbf{v}_1 - .4\mathbf{v}_2 - 1.2\mathbf{v}_3$. The weights sum to 1, so this is an affine sum.

6. **a.** $\mathbf{p}_1 = -4\mathbf{b}_1 + 2\mathbf{b}_2 + 3\mathbf{b}_3 \in$ aff S since the coefficients sum to 1.

   b. $\mathbf{p}_2 = .2\mathbf{b}_1 + .5\mathbf{b}_2 + .3\mathbf{b}_3 \in$ aff S since the coefficients sum to 1.

   c. $\mathbf{p}_3 = \mathbf{b}_1 + \mathbf{b}_2 + \mathbf{b}_3 \notin$ aff S since the coefficients do not sum to 1.

8. The matrix $[\, \mathbf{v}_1 \ \ \mathbf{v}_2 \ \ \mathbf{v}_3 \ \ \mathbf{p}_1 \ \ \mathbf{p}_2 \ \ \mathbf{p}_3 \,]$ reduces to

$$\begin{bmatrix} 1 & 0 & 0 & 3 & 0 & -2 \\ 0 & 1 & 0 & -1 & 0 & 6 \\ 0 & 0 & 1 & 1 & 0 & -3 \\ 0 & 0 & 0 & 0 & 1 & 0 \end{bmatrix}.$$

Parts (a), (b), and (c) use columns 4, 5, and 6, respectively, as the "augmented" column.

   a. $\mathbf{p}_1 = 3\mathbf{v}_1 - \mathbf{v}_2 + \mathbf{v}_3$, so $\mathbf{p}_1$ is in Span S. The weights do not sum to 1, so $\mathbf{p}_1 \notin$ aff S.

   b. $\mathbf{p}_2 \notin$ Span S because $0 \neq 1$ (column 5 is the augmented column), so $\mathbf{p}_2$ cannot possibly be in aff S.

   c. $\mathbf{p}_3 = -2\mathbf{v}_1 + 6\mathbf{v}_2 - 3\mathbf{v}_3$, so $\mathbf{p}_3$ is in Span S. The weights sum to 1, so $\mathbf{p}_3 \in$ aff S.

10. $\mathbf{v}_1 = \begin{bmatrix} 1 \\ -3 \\ 4 \end{bmatrix}$ and $\mathbf{v}_2 = \begin{bmatrix} 6 \\ -2 \\ 2 \end{bmatrix}$. Other answers are possible.

11. **a.** True. See the definition at the beginning of this section.

   b. False. The weights in the linear combination must sum to 1. See the definition.

   c. True. See equation (1).

   d. False. A flat is a translate of a subspace. See the definition prior to Theorem 3.

   e. True. A hyperplane in $\mathbb{R}^3$ has dimension 2, so it is a plane. See the definition prior to Theorem 3.

12. **a.** False. If $S = \{\mathbf{x}\}$, then aff $S = \{\mathbf{x}\}$. See the definition at the beginning of this section.

   b. True. See Theorem 1.

   c. True. See the definition prior to Theorem 3.

   d. False. A flat of dimension 2 is called a hyperplane only if the flat is considered a subset of $\mathbb{R}^3$. In general, a hyperplane is a flat of dimension $n - 1$. See the definition prior to Theorem 3.

   e. True. A flat through the origin is a subspace translated by the $\mathbf{0}$ vector.

14. Since $\{\mathbf{v}_1, \mathbf{v}_2, \mathbf{v}_3\}$ is a basis for $\mathbb{R}^3$, the set $W = \text{Span}\,\{\mathbf{v}_2 - \mathbf{v}_1, \mathbf{v}_3 - \mathbf{v}_1\}$ is a plane in $\mathbb{R}^3$, by Exercise 13. Thus, $W + \mathbf{v}_1$ is a plane parallel to W that contains $\mathbf{v}_1$. Since $\mathbf{v}_2 = (\mathbf{v}_2 - \mathbf{v}_1) + \mathbf{v}_1$, $W + \mathbf{v}_1$ contains $\mathbf{v}_2$. Similarly, $W + \mathbf{v}_1$ contains $\mathbf{v}_3$. Finally, Theorem 1 shows that aff $\{\mathbf{v}_1, \mathbf{v}_2, \mathbf{v}_3\}$ is the plane $W + \mathbf{v}_1$ that contains $\mathbf{v}_1, \mathbf{v}_2$, and $\mathbf{v}_3$.

16. Suppose $\mathbf{p}, \mathbf{q} \in S$ and $t \in \mathbb{R}$. Then, by properties of the dot product (Theorem 1 in Section 6.1),

$$[(1 - t)\mathbf{p} + t\mathbf{q}] \cdot \mathbf{v} = (1 - t)(\mathbf{p} \cdot \mathbf{v}) + t(\mathbf{q} \cdot \mathbf{v})$$
$$= (1 - t)k + tk = k$$

Thus, $[(1 - t)\mathbf{p} + t\mathbf{q}] \in S$, by definition of S. This shows that S is an affine set.

18. A suitable set consists of any four vectors that lie in the plane $2x_1 + x_2 - 3x_3 = 12$ and are not collinear. If the vectors are not collinear, their affine hull cannot be a line, so it must be the plane.

20. Given an affine set T, let $S = \{\mathbf{x} \in \mathbb{R}^n : f(\mathbf{x}) \in T\}$. Consider $\mathbf{x}, \mathbf{y} \in S$ and $t \in \mathbb{R}$. Then

$$f((1 - t)\mathbf{x} + t\mathbf{y}) = (1 - t)f(\mathbf{x}) + tf(\mathbf{y})$$

But $f(\mathbf{x}) \in T$ and $f(\mathbf{y}) \in T$, so $(1 - t)f(\mathbf{x}) + tf(\mathbf{y}) \in T$ because T is an affine set. It follows that $[(1 - t)\mathbf{x} + t\mathbf{y}] \in S$. This is true for all $\mathbf{x}, \mathbf{y} \in S$ and $t \in \mathbb{R}$, so S is an affine set.

22. Since $B \subset$ aff B, we have $A \subset B \subset$ aff B. But aff B is an affine set, so Exercise 21 implies aff $A \subset$ aff B.

24. One possibility is to let $A = \{(0, 1)\}$ and $B = \{(1, 0)\}$. Then (aff A) $\cup$ (aff B) consists of the two coordinate axes, but aff$(A \cup B) = \mathbb{R}^2$.

26. One possibility is to let $A = \{(0, 1)\}$ and $B = \{(0, 2)\}$. Then both aff A and aff B are equal to the x-axis. But $A \cap B = \varnothing$, so aff $(A \cap B) = \varnothing$.

Section 8.2, page 452

2. $\mathbf{v}_1 = \begin{bmatrix} 2 \\ 1 \end{bmatrix}, \mathbf{v}_2 = \begin{bmatrix} 5 \\ 4 \end{bmatrix}, \mathbf{v}_3 = \begin{bmatrix} -3 \\ -2 \end{bmatrix}, \mathbf{v}_2 - \mathbf{v}_1 = \begin{bmatrix} 3 \\ 3 \end{bmatrix},$
$\mathbf{v}_3 - \mathbf{v}_1 = \begin{bmatrix} -5 \\ -3 \end{bmatrix}.$ Since $\mathbf{v}_3 - \mathbf{v}_1$ and $\mathbf{v}_2 - \mathbf{v}_1$ are not multiples, they are linearly independent. By Theorem 5, $\{\mathbf{v}_1, \mathbf{v}_2, \mathbf{v}_3\}$ is affinely independent.

4. $-6\mathbf{v}_1 + 3\mathbf{v}_2 - 2\mathbf{v}_3 + 5\mathbf{v}_4 = \mathbf{0}$
   Solution: Name the points $\mathbf{v}_1, \mathbf{v}_2, \mathbf{v}_3$, and $\mathbf{v}_4$ and use Theorem 5. Compute $\mathbf{v}_2 - \mathbf{v}_1 = \begin{bmatrix} 2 \\ -8 \\ 4 \end{bmatrix},$
   $\mathbf{v}_3 - \mathbf{v}_1 = \begin{bmatrix} 3 \\ -7 \\ -9 \end{bmatrix},$ and $\mathbf{v}_4 - \mathbf{v}_1 = \begin{bmatrix} 0 \\ 2 \\ -6 \end{bmatrix}.$ To study linear independence of these new points, row reduce the augmented matrix for $A\mathbf{x} = \mathbf{0}$:

$$\begin{bmatrix} 2 & 3 & 0 & 0 \\ -8 & -7 & 2 & 0 \\ 4 & -9 & -6 & 0 \end{bmatrix} \sim \begin{bmatrix} 2 & 3 & 0 & 0 \\ 0 & 5 & 2 & 0 \\ 0 & -15 & -6 & 0 \end{bmatrix} \sim$$

$$\begin{bmatrix} 2 & 3 & 0 & 0 \\ 0 & 5 & 2 & 0 \\ 0 & 0 & 0 & 0 \end{bmatrix} \sim \begin{bmatrix} 1 & 0 & -.6 & 0 \\ 0 & 1 & .4 & 0 \\ 0 & 0 & 0 & 0 \end{bmatrix}.$$

The first three columns are linearly dependent, so $\{\mathbf{v}_1, \mathbf{v}_2, \mathbf{v}_3, \mathbf{v}_4\}$ is affinely dependent, by Theorem 5. To find the affine dependence relation, write the general solution of this system: $x_1 = .6x_3$, $x_2 = -.4x_3$, with x_3 free. Set $x_3 = 5$, for instance. Then $x_1 = 3$, $x_2 = -2$, and $x_3 = 5$. Thus, $3(\mathbf{v}_2 - \mathbf{v}_1) - 2(\mathbf{v}_3 - \mathbf{v}_1) + 5(\mathbf{v}_4 - \mathbf{v}_1) = \mathbf{0}$. Rearrange to obtain $-6\mathbf{v}_1 + 3\mathbf{v}_2 - 2\mathbf{v}_3 + 5\mathbf{v}_4 = \mathbf{0}$.

Alternative Solution: Name the points $\mathbf{v}_1$, $\mathbf{v}_2$, $\mathbf{v}_3$, and $\mathbf{v}_4$. Use Theorem 5(d) and study the homogeneous forms of the points. The first step is to move the bottom row of 1's (in the augmented matrix) to the top to simplify the arithmetic:

$$[\,\tilde{\mathbf{v}}_1 \quad \tilde{\mathbf{v}}_2 \quad \tilde{\mathbf{v}}_3 \quad \tilde{\mathbf{v}}_4\,] \sim \begin{bmatrix} 1 & 1 & 1 & 1 \\ -2 & 0 & 1 & -2 \\ 5 & -3 & -2 & 7 \\ 3 & 7 & -6 & -3 \end{bmatrix}$$

$$\sim \begin{bmatrix} 1 & 0 & 0 & 1.2 \\ 0 & 1 & 0 & -.6 \\ 0 & 0 & 1 & .4 \\ 0 & 0 & 0 & 0 \end{bmatrix}$$

Thus, $x_1 + 1.2x_4 = 0$, $x_2 - .6x_4 = 0$, and $x_3 + .4x_4 = 0$, with x_4 free. Take $x_4 = 5$, for example, and get $x_1 = -6$, $x_2 = 3$, and $x_3 = -x_2$. An affine dependence relation is $-6\mathbf{v}_1 + 3\mathbf{v}_2 - 2\mathbf{v}_3 + 5\mathbf{v}_4 = \mathbf{0}$.

6. The set is affinely independent, as the following calculation with homogeneous forms shows:

$$[\,\tilde{\mathbf{v}}_1 \quad \tilde{\mathbf{v}}_2 \quad \tilde{\mathbf{v}}_3 \quad \tilde{\mathbf{v}}_4\,] \sim \begin{bmatrix} 1 & 1 & 1 & 1 \\ 1 & 0 & 2 & 3 \\ 3 & -1 & 5 & 5 \\ 1 & -2 & 2 & 0 \end{bmatrix}$$

$$\sim \begin{bmatrix} 1 & 0 & 0 & 0 \\ 0 & 1 & 0 & 0 \\ 0 & 0 & 1 & 0 \\ 0 & 0 & 0 & 1 \end{bmatrix}$$

Row reduction of $[\,\mathbf{v}_1 \quad \mathbf{v}_2 \quad \mathbf{v}_3 \quad \mathbf{v}_4\,]$ shows that $\{\mathbf{v}_1, \mathbf{v}_2, \mathbf{v}_3\}$ is a basis for $\mathbb{R}^3$ and $\mathbf{v}_4 = -2\mathbf{v}_1 + 1.5\mathbf{v}_2 + 2.5\mathbf{v}_3$, but the weights in the linear combination do not sum to 1. *Instructor*: (Possible exam question.) If the last entry of $\mathbf{v}_4$ is changed from 0 to 1, then row reduction of $[\,\mathbf{v}_1 \quad \mathbf{v}_2 \quad \mathbf{v}_3 \quad \mathbf{v}_4\,]$ shows that $\{\mathbf{v}_1, \mathbf{v}_2, \mathbf{v}_3\}$ is a basis for $\mathbb{R}^3$ and $\mathbf{v}_4 = -3\mathbf{v}_1 + \mathbf{v}_2 + 3\mathbf{v}_3$.

8. The barycentric coordinates are $(2, -1, 0)$.
Solution: Denote the given points as $\mathbf{v}_1$, $\mathbf{v}_2$, $\mathbf{v}_3$, and $\mathbf{p}$. Row reduce the augmented matrix for the equation $x_1\tilde{\mathbf{v}}_1 + x_2\tilde{\mathbf{v}}_2 + x_3\tilde{\mathbf{v}}_3 = \tilde{\mathbf{p}}$.

$$[\,\tilde{\mathbf{v}}_1 \quad \tilde{\mathbf{v}}_2 \quad \tilde{\mathbf{v}}_3 \quad \tilde{\mathbf{p}}\,] \sim \begin{bmatrix} 1 & 1 & 1 & 1 \\ 0 & 1 & 1 & -1 \\ 1 & 1 & 4 & 1 \\ -2 & 0 & -6 & -4 \\ 1 & 2 & 5 & 0 \end{bmatrix}$$

$$\sim \begin{bmatrix} 1 & 0 & 0 & 2 \\ 0 & 1 & 0 & -1 \\ 0 & 0 & 1 & 0 \\ 0 & 0 & 0 & 0 \\ 0 & 0 & 0 & 0 \end{bmatrix}$$

Thus, $\tilde{\mathbf{p}} = 2\tilde{\mathbf{v}}_1 - \tilde{\mathbf{v}}_2 + 0\tilde{\mathbf{v}}_3$, so $\mathbf{p} = 2\mathbf{v}_1 - \mathbf{v}_2$. The barycentric coordinates are $(2, -1, 0)$.
Instructor note: $\mathbf{v}_3 = 3\mathbf{v}_1 + \mathbf{v}_2$

9. a. True. Theorem 5 uses the point $\mathbf{v}_1$ for the translation, but the paragraph after the theorem points out that any one of the points in the set can be used for the translation.
   b. False, by statement (d) of Theorem 5.
   c. False. The weights in the linear combination must sum to 0, not 1. See the definition at the beginning of this section.
   d. False. The only points that have barycentric coordinates determined by S belong to aff S. See the definition after Theorem 6.
   e. True. The barycentric coordinates are zero on the edges of the triangle and positive for interior points. See Example 6.

10. a. False. By Theorem 5, the set of homogeneous forms must be linearly dependent, too.
   b. True. If one statement in Theorem 5 is false, the other statements are false, too.
   c. False. Theorem 6 applies only when S is affinely independent.
   d. False. The color interpolation applies only to points whose barycentric coordinates are nonnegative, since the colors are formed by nonnegative combinations of red, green, and blue. See Example 5.
   e. True. See the discussion of Fig. 5.

12. Suppose $\mathbf{v}_1 \ldots, \mathbf{v}_p$ are in $\mathbb{R}^n$ and $p \geq n + 2$. Since $p - 1 \geq n + 1$, the points $\mathbf{v}_2 - \mathbf{v}_1, \mathbf{v}_3 - \mathbf{v}_1, \ldots, \mathbf{v}_p - \mathbf{v}_1$ are linearly dependent, by Theorem 8 in Section 1.7. By Theorem 5, $\{\mathbf{v}_1, \ldots, \mathbf{v}_p\}$ is affinely dependent.

14. Let S_1 consist of three (distinct) points on a line through the origin. The set is affinely dependent because the third point is on the line determined by the first two points. Let S_2 consist of two (distinct) points on a line through the origin. By Exercise 13, the set is affinely independent because the two points are distinct. (A correct solution should include a justification for the sets presented.)

16. a. The vectors $\mathbf{v}_2 - \mathbf{v}_1 = \begin{bmatrix} 1 \\ 4 \end{bmatrix}$ and $\mathbf{v}_3 - \mathbf{v}_1 = \begin{bmatrix} 4 \\ 2 \end{bmatrix}$ are not multiples and hence are linearly independent. By Theorem 5, S is affinely independent.

b. $\mathbf{p}_1 \leftrightarrow \left(-\frac{2}{7}, \frac{5}{7}, \frac{4}{7}\right)$, $\mathbf{p}_2 \leftrightarrow \left(\frac{2}{7}, -\frac{5}{7}, \frac{10}{7}\right)$, $\mathbf{p}_3 \leftrightarrow \left(\frac{2}{7}, \frac{2}{7}, \frac{3}{7}\right)$

c. $\mathbf{p}_4 \leftrightarrow (+, -, -)$, $\mathbf{p}_5 \leftrightarrow (+, +, -)$, $\mathbf{p}_6 \leftrightarrow (+, +, +)$, $\mathbf{p}_7 \leftrightarrow (-, 0, +)$. See the figure below. Actually, $\mathbf{p}_4 \leftrightarrow \left(\frac{19}{14}, -\frac{2}{14}, -\frac{3}{14}\right)$, $\mathbf{p}_5 \leftrightarrow \left(\frac{5}{14}, \frac{12}{14}, -\frac{3}{14}\right)$, $\mathbf{p}_6 \leftrightarrow \left(\frac{9}{14}, \frac{2}{14}, \frac{3}{14}\right)$, $\mathbf{p}_7 \leftrightarrow \left(-\frac{1}{2}, 0, \frac{3}{2}\right)$.

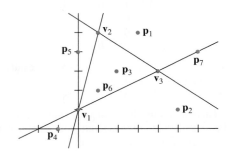

18. Let $\mathbf{p} = \begin{bmatrix} x \\ y \\ z \end{bmatrix}$. Then $\begin{bmatrix} x \\ y \\ z \end{bmatrix} =$

$$\frac{x}{a}\begin{bmatrix} a \\ 0 \\ 0 \end{bmatrix} + \frac{y}{b}\begin{bmatrix} 0 \\ b \\ 0 \end{bmatrix} + \frac{z}{c}\begin{bmatrix} 0 \\ 0 \\ c \end{bmatrix} + \left(1 - \frac{x}{a} - \frac{y}{b} - \frac{z}{c}\right)\begin{bmatrix} 0 \\ 0 \\ 0 \end{bmatrix}.$$

So the barycentric coordinates are x/a, y/b, z/c, and $1 - x/a - y/b - z/c$. This holds for any nonzero choices of a, b, and c.

20. If the translated set $\{\mathbf{p}_1 + \mathbf{q}, \mathbf{p}_2 + \mathbf{q}, \mathbf{p}_3 + \mathbf{q}\}$ were affinely dependent, then there would exist real numbers c_1, c_2, and c_3, not all zero and with $c_1 + c_2 + c_3 = 0$, such that $c_1(\mathbf{p}_1 + \mathbf{q}) + c_2(\mathbf{p}_2 + \mathbf{q}) + c_3(\mathbf{p}_3 + \mathbf{q}) = \mathbf{0}$. But then, $c_1\mathbf{p}_1 + c_2\mathbf{p}_2 + c_3\mathbf{p}_3 + (c_1 + c_2 + c_3)\mathbf{q} = \mathbf{0}$. Since $c_1 + c_2 + c_3 = 0$, this implies $c_1\mathbf{p}_1 + c_2\mathbf{p}_2 + c_3\mathbf{p}_3 = \mathbf{0}$, which would make $\{\mathbf{p}_1, \mathbf{p}_2, \mathbf{p}_3\}$ affinely dependent. But $\{\mathbf{p}_1, \mathbf{p}_2, \mathbf{p}_3\}$ is affinely independent, so the translated set must in fact be affinely independent, too.

22. If $\mathbf{p}$ is on the line through $\mathbf{a}$ and $\mathbf{b}$, then $\mathbf{p}$ is an affine combination of $\mathbf{a}$ and $\mathbf{b}$, so $\tilde{\mathbf{p}}$ is a linear combination of $\tilde{\mathbf{a}}$ and $\tilde{\mathbf{b}}$. Thus the columns of $[\,\tilde{\mathbf{a}} \quad \tilde{\mathbf{b}} \quad \tilde{\mathbf{p}}\,]$ are linearly dependent. So the determinant of this matrix is zero.

24. Let $\mathbf{p} = (1 - x)\mathbf{q} + x\mathbf{a}$, where $\mathbf{q}$ is on the line segment from $\mathbf{b}$ to $\mathbf{c}$. Then, because the determinant is a linear function of the first column when the other columns are fixed (Section 3.2),

$$\det[\,\tilde{\mathbf{p}} \quad \tilde{\mathbf{b}} \quad \tilde{\mathbf{c}}\,] = \det[\,(1 - x)\tilde{\mathbf{q}} + x\tilde{\mathbf{a}} \quad \tilde{\mathbf{b}} \quad \tilde{\mathbf{c}}\,]$$
$$= (1 - x) \cdot \det[\,\tilde{\mathbf{q}} \quad \tilde{\mathbf{b}} \quad \tilde{\mathbf{c}}\,]$$
$$+ x \cdot \det[\,\tilde{\mathbf{a}} \quad \tilde{\mathbf{b}} \quad \tilde{\mathbf{c}}\,]$$

Now, $[\,\tilde{\mathbf{q}} \quad \tilde{\mathbf{b}} \quad \tilde{\mathbf{c}}\,]$ is a singular matrix because $\tilde{\mathbf{q}}$ is a linear combination of $\tilde{\mathbf{b}}$ and $\tilde{\mathbf{c}}$. So $\det[\,\tilde{\mathbf{q}} \quad \tilde{\mathbf{b}} \quad \tilde{\mathbf{c}}\,] = 0$ and $\det[\,\tilde{\mathbf{p}} \quad \tilde{\mathbf{b}} \quad \tilde{\mathbf{c}}\,] = x \cdot \det[\,\tilde{\mathbf{a}} \quad \tilde{\mathbf{b}} \quad \tilde{\mathbf{c}}\,]$.

Section 8.3, page 459

2. a.

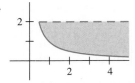

b.

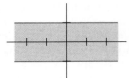

c.

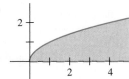

4. $\mathbf{p}_2 \in \text{conv } S$

Solution: From Exercise 6 in Section 8.1, $\mathbf{p}_3$ is not in aff S, so it certainly is not in conv S. Since $\mathbf{p}_1 = -4\mathbf{b}_1 + 2\mathbf{b}_2 + 3\mathbf{b}_3$ and $\mathbf{p}_2 = 0.2\mathbf{b}_1 + 0.5\mathbf{b}_2 + .3\mathbf{b}_3$, and in each case the weights sum to 1, both $\mathbf{p}_1$ and $\mathbf{p}_2$ are in aff S. However, S is affinely independent (because S is linearly independent), so the weights in these combinations are barycentric coordinates. Thus, $\mathbf{p}_2$ is in conv S, because its barycentric coordinates are nonnegative. This is not the case for $\mathbf{p}_1$, so $\mathbf{p}_1 \notin \text{conv } S$.

6. a. $\mathbf{p}_1 = \frac{1}{2}\mathbf{v}_1 - \frac{1}{2}\mathbf{v}_2 + \mathbf{v}_3 \in \text{aff } S$
  b. $\mathbf{p}_2 = \frac{1}{4}\mathbf{v}_1 + \frac{1}{4}\mathbf{v}_2 + \frac{1}{2}\mathbf{v}_3 \in \text{conv } S$
  c. $\mathbf{p}_3 = \mathbf{v}_1 + \mathbf{v}_2 - 2\mathbf{v}_3 \in \text{Span } S$
  d. $\text{proj}_{\text{Span } S}\,\mathbf{p}_4 = \left(-\frac{4}{9}, -\frac{8}{9}, \frac{10}{9}, 4\right) \neq \mathbf{p}_4$, so $\mathbf{p}_4 \notin \text{Span } S$
  Solution: Let W be the subspace spanned by the orthogonal set $S = \{\mathbf{v}_1, \mathbf{v}_2, \mathbf{v}_3\}$. As in Example 1, the barycentric coordinates of the points $\mathbf{p}_1, \ldots, \mathbf{p}_4$ with respect to S are easy to compute, and they determine whether or not a point is in Span S, aff S, or conv S.

  a. $\text{proj}_W\,\mathbf{p}_1 = \dfrac{\mathbf{p}_1 \cdot \mathbf{v}_1}{\mathbf{v}_1 \cdot \mathbf{v}_1}\mathbf{v}_1 + \dfrac{\mathbf{p}_1 \cdot \mathbf{v}_2}{\mathbf{v}_2 \cdot \mathbf{v}_2}\mathbf{v}_2 + \dfrac{\mathbf{p}_1 \cdot \mathbf{v}_3}{\mathbf{v}_3 \cdot \mathbf{v}_3}\mathbf{v}_3$

$$= \frac{1}{2}\begin{bmatrix} 2 \\ 0 \\ -1 \\ 2 \end{bmatrix} - \frac{1}{2}\begin{bmatrix} 0 \\ -2 \\ 2 \\ 1 \end{bmatrix} + \begin{bmatrix} -2 \\ 1 \\ 0 \\ 2 \end{bmatrix}$$

$$= \begin{bmatrix} -1 \\ 2 \\ -\frac{3}{2} \\ \frac{5}{2} \end{bmatrix} = \mathbf{p}_1$$

This shows that $\mathbf{p}_1$ is in $W = \text{Span } S$. Also, since the coefficients sum to 1, $\mathbf{p}_1$ is in aff S. However, $\mathbf{p}_1$ is not in conv S, because the coefficients are not all nonnegative.

b. Similarly, $\text{proj}_W \mathbf{p}_2 = \frac{\frac{9}{4}}{9}\mathbf{v}_1 + \frac{\frac{9}{4}}{9}\mathbf{v}_2 + \frac{\frac{9}{2}}{9}\mathbf{v}_3 =$
$\frac{1}{4}\mathbf{v}_1 + \frac{1}{4}\mathbf{v}_2 + \frac{1}{2}\mathbf{v}_3 = \mathbf{p}_2$. This shows that $\mathbf{p}_2$ lies in Span S. Also, since the coefficients sum to 1, $\mathbf{p}_2$ is in aff S. In fact, $\mathbf{p}_2$ is in conv S, because the coefficients are also nonnegative.

c. $\text{proj}_W \mathbf{p}_3 = \frac{9}{9}\mathbf{v}_1 + \frac{9}{9}\mathbf{v}_2 - \frac{18}{9}\mathbf{v}_3 = \mathbf{v}_1 + \mathbf{v}_2 - 2\mathbf{v}_3 = \mathbf{p}_3$. Thus $\mathbf{p}_3$ is in Span S. However, since the coefficients do not sum to 1, $\mathbf{p}_3$ is not in aff S and certainly not in conv S.

d. $\text{proj}_W \mathbf{p}_4 = \frac{6}{9}\mathbf{v}_1 + \frac{8}{9}\mathbf{v}_2 - \frac{8}{9}\mathbf{v}_3 \neq \mathbf{p}_4$. Since $\text{proj}_W \mathbf{p}_4$ is the closest point in Span S to $\mathbf{p}_4$, the point $\mathbf{p}_4$ is not in Span S. In particular, $\mathbf{p}_4$ cannot be in aff S or conv S.

8. a. The barycentric coordinates of $\mathbf{p}_1, \mathbf{p}_2, \mathbf{p}_3,$ and $\mathbf{p}_4$ are, respectively, $\left(\frac{12}{13}, \frac{3}{13}, -\frac{2}{13}\right)$, $\left(\frac{8}{13}, \frac{2}{13}, \frac{3}{13}\right)$, $\left(\frac{2}{3}, 0, \frac{1}{3}\right)$, and $\left(\frac{9}{13}, -\frac{1}{13}, \frac{5}{13}\right)$.

b. $\mathbf{p}_1$ and $\mathbf{p}_4$ are outside conv T. $\mathbf{p}_2$ is inside conv T. $\mathbf{p}_3$ is on the edge $\overline{\mathbf{v}_1\mathbf{v}_3}$ of conv T.

10. $\mathbf{q}_1$ is inside conv S. $\mathbf{q}_2$ and $\mathbf{q}_4$ are outside the tetrahedron conv S. $\mathbf{q}_3$ is on the edge between $\mathbf{v}_2$ and $\mathbf{v}_3$. $\mathbf{q}_5$ is on the face containing the vertices $\mathbf{v}_1, \mathbf{v}_2,$ and $\mathbf{v}_3$.
Solution: $\mathbf{q}_1$ is inside conv S because the barycentric coordinates are all positive. $\mathbf{q}_2$ is outside conv S because it has one negative barycentric coordinate. $\mathbf{q}_4$ is outside conv S for the same reason. $\mathbf{q}_3$ is on the edge between $\mathbf{v}_2$ and $\mathbf{v}_3$ because $\left(0, \frac{3}{4}, \frac{1}{4}, 0\right)$ shows that $\mathbf{q}_3$ is a convex combination of $\mathbf{v}_2$ and $\mathbf{v}_3$. $\mathbf{q}_5$ is on the face containing the vertices $\mathbf{v}_1, \mathbf{v}_2,$ and $\mathbf{v}_3$ because $\left(\frac{1}{3}, \frac{1}{3}, \frac{1}{3}, 0\right)$ shows that $\mathbf{q}_5$ is a convex combination of those vertices.

11. a. False. In order for $\mathbf{y}$ to be a convex combination, the c's must also all be nonnegative. See the definition at the beginning of this section.

b. False. If S is convex, then conv S is equal to S. See Theorem 7.

c. False. For example, the union of two distinct points is not convex, but the individual points are.

12. a. True. See the definition prior to Theorem 7.

b. True. See Theorem 9.

c. False. The points do not have to be distinct. For example, S might consist of two points in $\mathbb{R}^5$. A point in conv S would be a convex combination of these two points. Caratheodory's Theorem requires $n + 1$ **or fewer** points.

14. Suppose $\mathbf{r}, \mathbf{s} \in S$ and $0 \leq t \leq 1$. Then, since f is a linear transformation,

$$f[(1 - t)\mathbf{r} + t\mathbf{s}] = (1 - t)f(\mathbf{r}) + tf(\mathbf{s})$$

But $f(\mathbf{r}) \in T$ and $f(\mathbf{s}) \in T$, so $(1 - t)f(\mathbf{r}) + tf(\mathbf{s}) \in T$ since T is a convex set. It follows that $(1 - t)\mathbf{r} + t\mathbf{s} \in S$, because S consists of all points that f maps into T. This shows that S is convex.

16. $\mathbf{p} = \frac{3}{5}\mathbf{v}_2 + \frac{3}{10}\mathbf{v}_3 + \frac{1}{10}\mathbf{v}_4$ and $\mathbf{p} = \frac{1}{11}\mathbf{v}_1 + \frac{6}{11}\mathbf{v}_2 + \frac{4}{11}\mathbf{v}_3$.

Solution: $\mathbf{v}_1 = \begin{bmatrix} -1 \\ 0 \end{bmatrix}, \mathbf{v}_2 = \begin{bmatrix} 0 \\ 3 \end{bmatrix}, \mathbf{v}_3 = \begin{bmatrix} 3 \\ 1 \end{bmatrix}$, and $\mathbf{p} = \begin{bmatrix} 1 \\ 2 \end{bmatrix}$. It is straightforward to confirm the equations in the problem:

(1) $\frac{1}{121}\mathbf{v}_1 + \frac{72}{121}\mathbf{v}_2 + \frac{37}{121}\mathbf{v}_3 + \frac{1}{11}\mathbf{v}_4 = \mathbf{p}$

and

(2) $10\mathbf{v}_1 - 6\mathbf{v}_2 + 7\mathbf{v}_3 - 11\mathbf{v}_4 = \mathbf{0}$.

Notice that the coefficients of $\mathbf{v}_1$ and $\mathbf{v}_3$ in equation (2) are positive. With the notation of the proof of Caratheodory's Theorem, $d_1 = 10$ and $d_3 = 7$. The corresponding coefficients in equation (1) are $c_1 = \frac{1}{121}$ and $c_3 = \frac{37}{121}$. The ratios of these coefficients are $c_1/d_1 = \frac{1}{121} \div 10 = \frac{1}{1210}$ and $c_3/d_3 = \frac{37}{121} \div 7 = \frac{37}{847}$. Use the smaller ratio to eliminate $\mathbf{v}_1$ from equation (1). That is, add $-\frac{1}{1210}$ times equation (2) to equation (1):

$\mathbf{p} = \left(\frac{1}{121} - \frac{10}{1210}\right)\mathbf{v}_1 + \left(\frac{72}{121} + \frac{6}{1210}\right)\mathbf{v}_2$
$\qquad + \left(\frac{37}{121} - \frac{7}{1210}\right)\mathbf{v}_3 + \left(\frac{1}{11} + \frac{11}{1210}\right)\mathbf{v}_4$
$= \frac{3}{5}\mathbf{v}_2 + \frac{3}{10}\mathbf{v}_3 + \frac{1}{10}\mathbf{v}_4$

To obtain the second combination, multiply equation (2) by -1 to reverse the signs so that d_2 and d_4 become positive. Repeating the analysis with these terms eliminates the $\mathbf{v}_4$ term.

18. Suppose $A \subset B$. Then $A \subset B \subset$ conv B. Since conv B is convex, Exercise 17 shows that conv $A \subset$ conv B.

20. a. Since $(A \cap B) \subset A$, Exercise 18 shows that conv $(A \cap B) \subset$ conv A. Similarly, conv $(A \cap B) \subset$ conv B. Thus, conv $(A \cap B) \subset [($conv $A) \cap ($conv $B)]$.

b. One possibility is to let A be a pair of opposite vertices of a square and let B be the other pair of opposite vertices. Then conv A and conv B are intersecting diagonals of the square. $A \cap B$ is the empty set, so conv $(A \cap B)$ must be empty, too. But conv $A \cap$ conv B contains the single point where the diagonals intersect. So conv $(A \cap B)$ is a proper subset of conv $A \cap$ conv B.

22.

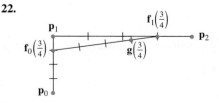

24. $\mathbf{h}(t) = (1 - t)\mathbf{g}_1(t) + t\mathbf{g}_2(t)$. Use the representation for $\mathbf{g}_1(t)$ from Exercise 23, and the analogous representation for $\mathbf{g}_2(t)$, based on the control points $\mathbf{p}_1, \mathbf{p}_2,$ and $\mathbf{p}_3$, and obtain

$$\mathbf{h}(t) = (1-t)[(1-t)^2\mathbf{p}_0 + 2t(1-t)\mathbf{p}_1 + t^2\mathbf{p}_2]$$
$$+ t[(1-t)^2\mathbf{p}_1 + 2t(1-t)\mathbf{p}_2 + t^2\mathbf{p}_3]$$
$$= (1-t)^3\mathbf{p}_0 + 2t(1-2t+t^2)\mathbf{p}_1 + (t^2-t^3)\mathbf{p}_2$$
$$+ t(1-2t+t^2)\mathbf{p}_1 + 2t^2(1-t)\mathbf{p}_2 + t^3\mathbf{p}_3$$
$$= (1-3t+3t^2-t^3)\mathbf{p}_0 + (2t-4t^2+2t^3)\mathbf{p}_1$$
$$+ (t^2-t^3)\mathbf{p}_2 + (t-2t^2+t^3)\mathbf{p}_1 + (2t^2-2t^3)\mathbf{p}_2$$
$$+ t^3\mathbf{p}_3$$
$$= (1-3t+3t^2-t^3)\mathbf{p}_0 + (3t-6t^2+3t^3)\mathbf{p}_1$$
$$+ (3t^2-3t^3)\mathbf{p}_2 + t^3\mathbf{p}_3$$

By inspection, the sum of the weights in this linear combination is 1, for all t. To show that the weights are nonnegative for $0 \le t \le 1$, factor the coefficients and write

$$\mathbf{h}(t) = (1-t)^3\mathbf{p}_0 + 3t(1-t)^2\mathbf{p}_1$$
$$+ 3t^2(1-t)\mathbf{p}_2 + t^3\mathbf{p}_3 \quad \text{for } 0 \le t \le 1$$

Thus, $\mathbf{h}(t)$ is in the convex hull of the control points $\mathbf{p}_0$, $\mathbf{p}_1$, $\mathbf{p}_2$, and $\mathbf{p}_3$.

Section 8.4, page 467

2. $f(x_1, x_2) = 5x_1 - 3x_2$ and $d = -7$

4. a. Closed   **b.** Open   **c.** Neither
   d. Closed   **e.** Open

6. a. Compact, not convex
   b. Not compact, not convex
   c. Not compact, convex
   d. Not compact, convex
   e. Not compact, not convex

8. a. $\mathbf{n} = \begin{bmatrix} 4 \\ 3 \\ -6 \end{bmatrix}$ or a multiple

   b. $f(\mathbf{x}) = 4x_1 + 3x_2 - 6x_3, \ d = -8$

10. a. $\mathbf{n} = \begin{bmatrix} -2 \\ 3 \\ -5 \\ 1 \end{bmatrix}$ or a multiple

   b. $f(\mathbf{x}) = -2x_1 + 3x_2 - 5x_3 + x_4, \ d = 4$

Solution: a. $\mathbf{v}_2 - \mathbf{v}_1 = \begin{bmatrix} 1 \\ 0 \\ -1 \\ -3 \end{bmatrix}$, $\mathbf{v}_3 - \mathbf{v}_1 = \begin{bmatrix} 0 \\ 1 \\ 2 \\ 7 \end{bmatrix}$,

$\mathbf{v}_4 - \mathbf{v}_1 = \begin{bmatrix} 2 \\ 0 \\ -1 \\ -1 \end{bmatrix}$. Solve the equations $(\mathbf{v}_2 - \mathbf{v}_1)\cdot\mathbf{n} = 0$,

$(\mathbf{v}_3 - \mathbf{v}_1)\cdot\mathbf{n} = 0$, and $(\mathbf{v}_4 - \mathbf{v}_1)\cdot\mathbf{n} = 0$. The augmented matrix is

$$\begin{bmatrix} 1 & 0 & -1 & -3 & 0 \\ 0 & 1 & 2 & 7 & 0 \\ 2 & 0 & -1 & -1 & 0 \end{bmatrix} \sim \begin{bmatrix} 1 & 0 & 0 & 0 & 2 \\ 0 & 1 & 0 & 0 & -3 \\ 0 & 0 & 1 & 0 & 5 \end{bmatrix}$$

Thus, $x_1 = -2x_4$, $x_2 = 3x_4$, $x_3 = -5x_4$, with x_4 free. Take

$x_4 = 1$, for example, to get $\mathbf{n} = \begin{bmatrix} -2 \\ 3 \\ -5 \\ 1 \end{bmatrix}$.

   b. Let $f(x_1, x_2, x_3, x_4) = -2x_1 + 3x_2 - 5x_3 + x_4$. Let $d = f(\mathbf{v}_1) = -2(1) + 3(2) + 0 + 0 = 4$.

12. Let $H = [f:d]$, where $f(x_1, x_2, x_3) = 3x_1 + x_2 - 2x_3$ and $d = 4$. There is no hyperplane parallel to H that strictly separates A and B because both sets have a point at which f takes the value of 4. There may be (and in fact is) a hyperplane that is *not* parallel to H that strictly separates A and B.

14. 2, 3, or 4

16. $f(x_1, x_2, x_3, x_4, x_5) = 2x_1 + 5x_2 - 3x_3 + 6x_5$, and $d = 0$

18. $f(x_1, x_2, x_3) = -11x_1 + 4x_2 + x_3$, and $d = 0$

20. $f(x_1, x_2, x_3) = -6x_1 + 2x_2 + x_3$, and $d = 0$

21. a. False. A linear functional goes from $\mathbb{R}^n$ to $\mathbb{R}$. See the definition at the beginning of this section.
   b. False. See the discussion of (1) and (4). There is a $1 \times n$ matrix A such that $f(\mathbf{x}) = A\mathbf{x}$ for all $\mathbf{x}$ in $\mathbb{R}^n$. Equivalently, there is a point $\mathbf{n}$ in $\mathbb{R}^n$ such that $f(\mathbf{x}) = \mathbf{n}\cdot\mathbf{x}$ for all $\mathbf{x}$ in $\mathbb{R}^n$.
   c. True. See the comments after the definition of *strictly separate*.
   d. False. See the sets in Figure 4.

22. a. True. See the statement after (3).
   b. False. The vector $\mathbf{n}$ must be nonzero. If $\mathbf{n} = \mathbf{0}$, then the given set is empty if $d \ne 0$ and the set is all of $\mathbb{R}^n$ if $d = 0$.
   c. False. Theorem 12 requires that the sets A and B be convex. For example, A could be the boundary of a circle and B could be the center of the circle.
   d. False. Some other hyperplane might strictly separate them. See the caution at the end of Example 8.

24. $f(x_1, x_2) = -2x_1 + 3x_2$ with d satisfying $4 < d < 5$ is one possibility.

26. $f(x, y) = 4x - 2y$. A natural choice for d is $f(5, 1.5) = 17$.
Solution: The normal to the separating hyperplane has the direction of the line segment between $\mathbf{p}$ and $\mathbf{q}$. So, let $\mathbf{n} = \mathbf{p} - \mathbf{q} = \begin{bmatrix} 4 \\ -2 \end{bmatrix}$. The distance between $\mathbf{p}$ and $\mathbf{q}$ is $\sqrt{20}$, which is more than the sum of the radii of the two balls. The large ball has center $\mathbf{q}$. A point three-fourths of the distance from $\mathbf{q}$ to $\mathbf{p}$ will be greater than 3 units from $\mathbf{q}$ and greater than 1 unit from $\mathbf{p}$. This point is

$$\mathbf{x} = .75\mathbf{p} + .25\mathbf{q} = .75\begin{bmatrix} 6 \\ 1 \end{bmatrix} + .25\begin{bmatrix} 2 \\ 3 \end{bmatrix} = \begin{bmatrix} 5.0 \\ 1.5 \end{bmatrix}$$

Compute $\mathbf{n}\cdot\mathbf{x} = 17$. The desired hyperplane is

$$\left\{ \begin{bmatrix} x \\ y \end{bmatrix} : 4x - 2y = 17 \right\}.$$

28. One possibility is $A = \{(x, y) : x^2 y^2 = 1 \text{ and } y > 0\}$ and $B = \{(x, y) : |x| \leq 1 \text{ and } y = 0\}$.

30. Let S be a bounded set. Then there exists a $\delta > 0$ such that $S \subset B(\mathbf{0}, \delta)$. But $B(\mathbf{0}, \delta)$ is convex by Exercise 29, so Theorem 9 in Section 8.3 (or Exercise 17 in Section 8.3) implies that conv $S \subset B(\mathbf{p}, \delta)$ and conv S is bounded.

Section 8.5, page 479

2. a. $m = 3$ on the set conv $\{\mathbf{p}_2, \mathbf{p}_3\}$

b. $m = 1$ on the set conv $\{\mathbf{p}_1, \mathbf{p}_2\}$

c. $m = 0$ is at the point $\mathbf{p}_3$

4. a. $m = -1$ at the point $\mathbf{p}_1$

b. $m = -1$ at the point $\mathbf{p}_3$

c. $m = -3$ at the point $\mathbf{p}_2$

6. $\left\{ \begin{bmatrix} 0 \\ 0 \end{bmatrix}, \begin{bmatrix} 4 \\ 0 \end{bmatrix}, \begin{bmatrix} 3 \\ 4 \end{bmatrix}, \begin{bmatrix} 0 \\ 6 \end{bmatrix} \right\}$

8. $\left\{ \begin{bmatrix} 0 \\ 0 \end{bmatrix}, \begin{bmatrix} 4 \\ 0 \end{bmatrix}, \begin{bmatrix} 3 \\ 2 \end{bmatrix}, \begin{bmatrix} 0 \\ 3.5 \end{bmatrix} \right\}$

10. One possibility is a ray. It has an extreme point at one end.

12. a. $f_0(S^5) = 6$, $f_1(S^5) = 15$, $f_2(S^5) = 20$, $f_3(S^5) = 15$, $f_4(S^5) = 6$, and $6 - 15 + 20 - 15 + 6 = 2$.

b.

	f_0	f_1	f_2	f_3	f_4
S^1	2				
S^2	3	3			
S^3	4	6	4		
S^4	5	10	10	5	
S^5	6	15	20	15	6

$f_k(S^n) = \binom{n+1}{k+1}$, where $\binom{a}{b} = \dfrac{a!}{b!\,(a-b)!}$ is the binomial coefficient.

14. a. X^1 is a line segment

X^2 is a parallelogram

b. $f_0(X^3) = 6$, $f_1(X^3) = 12$, $f_2(X^3) = 8$. X^3 is an octahedron.

c. $f_0(X^4) = 8$, $f_1(X^4) = 24$, $f_2(X^4) = 32$, $f_3(X^4) = 16$, $8 - 24 + 32 - 16 = 0$

d. $f_k(X^n) = 2^{k+1} \binom{n}{k+1}$, $0 \leq k \leq n-1$, where $\binom{a}{b} = \dfrac{a!}{b!\,(a-b)!}$ is the binomial coefficient.

16. a. True. See the definition in the second paragraph of this section.

b. True. See the definition after Example 1.

c. False. S must also be compact. See Theorem 15.

d. True. See the paragraph that discusses Figure 7.

17. a. False. It has six facets (faces).

b. True. See Theorem 14.

c. False. The maximum is always attained at some extreme point, but there may be other points that are not extreme points at which the maximum is attained. See Theorem 16.

d. True. Follows from Euler's formula with $n = 2$.

18. Let $\mathbf{x}$ be an extreme point of the convex set S and let $T = \{\mathbf{y} \in S : \mathbf{y} \neq \mathbf{x}\}$. If $\mathbf{y}$ and $\mathbf{z}$ are in T, then $\overline{\mathbf{yz}} \subseteq S$ since S is convex. But since $\mathbf{x}$ is an extreme point of S, $\mathbf{x} \notin \overline{\mathbf{yz}}$, so $\overline{\mathbf{yz}} \subset T$. Thus T is convex.

Conversely, suppose $\mathbf{x} \in S$, but $\mathbf{x}$ is not an extreme point of S. Then there exist $\mathbf{y}$ and $\mathbf{z}$ in S such that $\mathbf{x} \in \overline{\mathbf{yz}}$, with $\mathbf{x} \neq \mathbf{y}$ and $\mathbf{x} \neq \mathbf{z}$. It follows that $\mathbf{y}$ and $\mathbf{z}$ are in T, but $\overline{\mathbf{yz}} \not\subset T$. Hence T is not convex.

20. For example, let $S = \{1, 2\}$ in $\mathbb{R}^1$. Then $2S = \{2, 4\}$, $3S = \{3, 6\}$ and $(2 + 3)S = \{5, 10\}$. However, $2S + 3S = \{2, 4\} + \{3, 6\} = \{2 + 3, 4 + 3, 2 + 6, 4 + 6\} = \{5, 7, 8, 10\} \neq (2 + 3)S$.

21. Suppose A and B are convex. Let $\mathbf{x}, \mathbf{y} \in A + B$. Then there exist $\mathbf{a}, \mathbf{c} \in A$ and $\mathbf{b}, \mathbf{d} \in B$ such that $\mathbf{x} = \mathbf{a} + \mathbf{b}$ and $\mathbf{y} = \mathbf{c} + \mathbf{d}$. For any t such that $0 \leq t \leq 1$, we have

$$\begin{aligned} \mathbf{w} &= (1 - t)\mathbf{x} + t\mathbf{y} \\ &= (1 - t)(\mathbf{a} + \mathbf{b}) + t(\mathbf{c} + \mathbf{d}) \\ &= [(1 - t)\mathbf{a} + t\mathbf{c}] + [(1 - t)\mathbf{b} + t\mathbf{d}] \end{aligned}$$

But $(1 - t)\mathbf{a} + t\mathbf{c} \in A$ since A is convex, and $(1 - t)\mathbf{b} + t\mathbf{d} \in B$ since B is convex. Thus $\mathbf{w}$ is in $A + B$, which shows that $A + B$ is convex.

22. a. Since each edge belongs to two facets, kr is twice the number of edges: $kr = 2e$. Since each edge has two vertices, $sv = 2e$.

b. $v - e + r = 2$, so
$$\frac{2e}{s} - e + \frac{2e}{k} = 2 \Rightarrow \frac{1}{s} + \frac{1}{k} = \frac{1}{2} + \frac{1}{e}$$

c. A polygon must have at least three sides, so $k \geq 3$. At least three edges meet at each vertex, so $s \geq 3$. But both k and s cannot both be greater than 3, for then the left side of the equation in (b) could not exceed $1/2$.

When $k = 3$, we get $\dfrac{1}{s} - \dfrac{1}{6} = \dfrac{1}{e}$ so $s = 3, 4$, or 5. For these values, we get $e = 6, 12$, or 30, corresponding to the tetrahedron, the octahedron, and the icosahedron, respectively.

When $s = 3$, we get $\dfrac{1}{k} - \dfrac{1}{6} = \dfrac{1}{e}$ so $k = 2, 3$, or 5 and $e = 6, 12$, or 30, respectively. These values correspond to the tetrahedron, the cube, and the dodecahedron.

Section 8.6, page 490

2. a. Equation (15) reveals that each polynomial weight is nonnegative for $0 \le t \le 1$, since $4 - 3t > 0$. For the sum of the coefficients, use (15) with the first term expanded: $1 - 3t + 3t^2 - t^3$. The 1 here plus the 4 and 1 in the coefficients of $\mathbf{p}_1$ and $\mathbf{p}_2$, respectively, sum to 6, while the other terms sum to 0. This explains the 1/6 in the formula for $\mathbf{x}(t)$, which makes the coefficients sum to 1. Thus, $\mathbf{x}(t)$ is a convex combination of the control points for $0 \le t \le 1$.

b. Since the coefficients inside the brackets in equation (14) sum to 6, it follows that

$$\mathbf{b} = \tfrac{1}{6}[6\mathbf{b}]$$
$$= \tfrac{1}{6}\big[(1-t)^3\mathbf{b} + (3t^3 - 6t^2 + 4)\mathbf{b}$$
$$\quad + (-3t^3 + 3t^2 + 3t + 1)\mathbf{b} + t^3\mathbf{b}\big]$$

and hence $\mathbf{x}(t) + \mathbf{b}$ may be written in a similar form, with $\mathbf{p}_i$ replaced by $\mathbf{p}_i + \mathbf{b}$ for each i. This shows that $\mathbf{x}(t) + \mathbf{b}$ is a cubic B-spline with control points $\mathbf{p}_i + \mathbf{b}$ for $i = 0, \ldots, 3$.

4. a. $\mathbf{x}'(t) = \tfrac{1}{6}\big[(-3t^2 + 6t - 3)\,\mathbf{p}_0 + (9t^2 - 12t)\,\mathbf{p}_1$
$\qquad\qquad + (-9t^2 + 6t + 3)\,\mathbf{p}_2 + 3t^2\mathbf{p}_3\big]$

$\mathbf{x}'(0) = \tfrac{1}{2}(\mathbf{p}_2 - \mathbf{p}_0)$ and $\mathbf{x}'(1) = \tfrac{1}{2}(\mathbf{p}_3 - \mathbf{p}_1)$

(Verify that, in the first part of Fig. 10, a line drawn through $\mathbf{p}_0$ and $\mathbf{p}_2$ is parallel to the tangent line at the beginning of the B-spline.)

When $\mathbf{x}'(0)$ and $\mathbf{x}'(1)$ are both zero, the figure collapses and the convex hull of the set of control points is the line segment between $\mathbf{p}_0$ and $\mathbf{p}_3$, in which case $\mathbf{x}(t)$ is a straight line. Where does $\mathbf{x}(t)$ start? In this case,

$\mathbf{x}(t) = \tfrac{1}{6}\big[(-4t^3 + 6t^2 + 2)\mathbf{p}_0 + (4t^3 - 6t^2 + 4)\mathbf{p}_3\big]$
$\mathbf{x}(0) = \tfrac{1}{3}\mathbf{p}_0 + \tfrac{2}{3}\mathbf{p}_3$ and $\mathbf{x}(1) = \tfrac{2}{3}\mathbf{p}_0 + \tfrac{1}{3}\mathbf{p}_3$

The curve begins closer to $\mathbf{p}_3$ and finishes closer to $\mathbf{p}_0$. Could it turn around during its travel? Since $\mathbf{x}'(t) = 2t(1 - t)(\mathbf{p}_0 - \mathbf{p}_3)$, the curve travels in the direction $\mathbf{p}_0 - \mathbf{p}_3$, so when $\mathbf{x}'(0) = \mathbf{x}'(1) = 0$, the curve always moves away from $\mathbf{p}_3$ toward $\mathbf{p}_0$ for $0 \le t \le 1$.

b. $\mathbf{x}''(t) = (1 - t)\mathbf{p}_0 + (-2 + 3t)\mathbf{p}_1 + (1 - 3t)\mathbf{p}_2 + t\mathbf{p}_3$
$\mathbf{x}''(0) = \mathbf{p}_0 - 2\mathbf{p}_1 + \mathbf{p}_2 = (\mathbf{p}_0 - \mathbf{p}_1) + (\mathbf{p}_2 - \mathbf{p}_1)$
and
$\mathbf{x}''(1) = \mathbf{p}_1 - 2\mathbf{p}_2 + \mathbf{p}_3 = (\mathbf{p}_1 - \mathbf{p}_2) + (\mathbf{p}_3 - \mathbf{p}_2)$

For a picture of $\mathbf{x}''(0)$, construct a coordinate system with the origin at $\mathbf{p}_1$, temporarily, label $\mathbf{p}_0$ as $\mathbf{p}_0 - \mathbf{p}_1$, and label $\mathbf{p}_2$ as $\mathbf{p}_2 - \mathbf{p}_1$. Finally, construct a line from this new origin to the sum of $\mathbf{p}_0 - \mathbf{p}_1$ and $\mathbf{p}_2 - \mathbf{p}_1$. That segment represents $\mathbf{x}''(0)$.

For a picture of $\mathbf{x}''(1)$, construct a coordinate system with the origin at $\mathbf{p}_2$, temporarily, label $\mathbf{p}_1$ as $\mathbf{p}_1 - \mathbf{p}_2$, and label $\mathbf{p}_3$ as $\mathbf{p}_3 - \mathbf{p}_2$. Finally, construct a line from this new origin to the sum of $\mathbf{p}_1 - \mathbf{p}_2$ and $\mathbf{p}_3 - \mathbf{p}_2$. That segment represents $\mathbf{x}''(1)$.

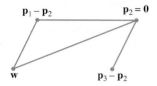

$\mathbf{w} = (\mathbf{p}_1 - \mathbf{p}_2) + (\mathbf{p}_3 - \mathbf{p}_2) = \mathbf{x}''(1)$

6. a. With $\mathbf{x}(t)$ as in Exercise 2,

$$\mathbf{x}(0) = (\mathbf{p}_0 + 4\mathbf{p}_1 + \mathbf{p}_2)/6$$

and

$$\mathbf{x}(1) = (\mathbf{p}_1 + 4\mathbf{p}_2 + \mathbf{p}_3)/6$$

Use the formula for $\mathbf{x}(0)$, but with the shifted control points for $\mathbf{y}(t)$, and obtain

$$\mathbf{y}(0) = (\mathbf{p}_1 + 4\mathbf{p}_2 + \mathbf{p}_3)/6$$

This equals $\mathbf{x}(1)$, so the B-spline is G^0 continuous at the join point.

b. From Exercise 4(a),

$$\mathbf{x}'(1) = (\mathbf{p}_3 - \mathbf{p}_1)/2 \quad \text{and} \quad \mathbf{x}'(0) = (\mathbf{p}_2 - \mathbf{p}_0)/2$$

Use the formula for $\mathbf{x}'(0)$ with the control points for $\mathbf{y}(t)$, and obtain

$$\mathbf{y}'(0) = (\mathbf{p}_3 - \mathbf{p}_1)/2 = \mathbf{x}'(1)$$

Thus the B-spline is C^1 continuous at the join point.

8. From Exercise 4(b), $\mathbf{x}''(0) = \mathbf{p}_0 - 2\mathbf{p}_1 + \mathbf{p}_2$ and $\mathbf{x}''(1) = \mathbf{p}_1 - 2\mathbf{p}_2 + \mathbf{p}_3$. Use the formula for $\mathbf{x}''(0)$, with the shifted control points for $\mathbf{y}(t)$, to get

$$\mathbf{y}''(0) = \mathbf{p}_1 - 2\mathbf{p}_2 + 2\mathbf{p}_3 = \mathbf{x}''(1)$$

Thus the curve has C^2 continuity at $\mathbf{x}(1)$.

10. Write a vector of the polynomial weights for $\mathbf{x}(t)$, expand the polynomial weights, taking care to write the terms in ascending powers of t, and factor the vector as $M_S\mathbf{u}(t)$:

$$\frac{1}{6}\begin{bmatrix} 1 - 3t + 3t^2 - t^3 \\ 4 - 6t^2 + 3t^3 \\ 1 + 3t + 3t^2 - 3t^3 \\ t^3 \end{bmatrix}$$

$$= \frac{1}{6}\begin{bmatrix} 1 & -3 & 3 & -1 \\ 4 & 0 & -6 & 3 \\ 1 & 3 & 3 & -3 \\ 0 & 0 & 0 & 1 \end{bmatrix}\begin{bmatrix} 1 \\ t \\ t^2 \\ t^3 \end{bmatrix} = M_S\mathbf{u}(t),$$

$$M_S = \frac{1}{6}\begin{bmatrix} 1 & -3 & 3 & -1 \\ 4 & 0 & -6 & 3 \\ 1 & 3 & 3 & -3 \\ 0 & 0 & 0 & 1 \end{bmatrix}$$

11. a. True. See equation (2).

b. False. Example 1 shows that the tangent vector $\mathbf{x}'(t)$ at $\mathbf{p}_0$ is <u>two</u> times the directed line segment from $\mathbf{p}_0$ to $\mathbf{p}_1$.

c. True. See Example 2.

12. a. False. The essential properties are preserved under translations as well as linear transformations. See the comment after Figure 1.

b. True. This is the definition of G^0 continuity at a point.

c. False. The Bézier basis matrix is a matrix of the polynomial coefficients of the control points. See the definition before equation (4).

14. a. $3(\mathbf{r}_3 - \mathbf{r}_2) = \mathbf{z}'(1)$, by equation (9) with $\mathbf{z}'(1)$ and $\mathbf{r}_i$ in place of $\mathbf{x}'(1)$ and $\mathbf{p}_i$.
$\mathbf{z}'(1) = .5\mathbf{x}'(1)$, by equation (11) with $t = 1$.
$.5\mathbf{x}'(1) = (.5)3(\mathbf{p}_3 - \mathbf{p}_2)$, by equation (9).

b. From part (a), $6(\mathbf{r}_3 - \mathbf{r}_2) = 3(\mathbf{p}_3 - \mathbf{p}_2)$,
$\mathbf{r}_3 - \mathbf{r}_2 = \frac{1}{2}\mathbf{p}_3 - \frac{1}{2}\mathbf{p}_2$, and $\mathbf{r}_3 - \frac{1}{2}\mathbf{p}_3 + \frac{1}{2}\mathbf{p}_2 = \mathbf{r}_2$. Since $\mathbf{r}_3 = \mathbf{p}_3$, this equation becomes $\mathbf{r}_2 = \frac{1}{2}(\mathbf{p}_3 + \mathbf{p}_2)$.

c. $3(\mathbf{r}_1 - \mathbf{r}_0) = \mathbf{z}'(0)$, by equation (9) with $\mathbf{z}'(0)$ and $\mathbf{r}_i$ in place of $\mathbf{x}'(0)$ and $\mathbf{p}_j$.
$\mathbf{z}'(0) = .5\mathbf{x}'(.5)$, by equation (11) with $t = .5$.

d. Part (c) and equation (10) show that
$3(\mathbf{r}_1 - \mathbf{r}_0) = \frac{3}{8}(-\mathbf{p}_0 - \mathbf{p}_1 + \mathbf{p}_2 + \mathbf{p}_3)$.
Multiply by $\frac{8}{3}$ and rearrange to obtain
$8\mathbf{r}_1 = -\mathbf{p}_0 - \mathbf{p}_1 + \mathbf{p}_2 + \mathbf{p}_3 + 8\mathbf{r}_0$.

e. From equation (8), $8\mathbf{r}_0 = \mathbf{p}_0 + 3\mathbf{p}_1 + 3\mathbf{p}_2 + \mathbf{p}_3$.
Substitute into the equation from part (d), and obtain
$8\mathbf{r}_1 = 2\mathbf{p}_1 + 4\mathbf{p}_2 + 2\mathbf{p}_3$.
Divide by 8 and use part (b) to obtain

$\mathbf{r}_1 = \frac{1}{4}\mathbf{p}_1 + \frac{1}{2}\mathbf{p}_2 + \frac{1}{4}\mathbf{p}_3$
$= (\frac{1}{4}\mathbf{p}_1 + \frac{1}{4}\mathbf{p}_2) + \frac{1}{4}(\mathbf{p}_2 + \mathbf{p}_3)$
$= \frac{1}{2} \cdot \frac{1}{2}(\mathbf{p}_1 + \mathbf{p}_2) + \frac{1}{2}\mathbf{r}_2$

Interchange the terms on the right, and obtain
$\mathbf{r}_1 = \frac{1}{2}\left[\mathbf{r}_2 + \frac{1}{2}(\mathbf{p}_1 + \mathbf{p}_2)\right]$.

16. A Bézier curve is completely determined by its four control points. Two are given directly: $\mathbf{p}_0 = \mathbf{x}(0)$ and $\mathbf{p}_3 = \mathbf{x}(1)$. From equation (9), $\mathbf{x}'(0) = 3(\mathbf{p}_1 - \mathbf{p}_0)$ and $\mathbf{x}'(1) = 3(\mathbf{p}_3 - \mathbf{p}_2)$. Solving gives

$\mathbf{p}_1 = \mathbf{p}_0 + \frac{1}{3}\mathbf{x}'(0)$ and $\mathbf{p}_2 = \mathbf{p}_3 - \frac{1}{3}\mathbf{x}'(1)$

17. a. The quadratic curve is
$\mathbf{w}(t) = (1 - t)^2\mathbf{p}_0 + 2t(1 - t)\mathbf{p}_1 + t^2\mathbf{p}_2$. From Example 1, the tangent vectors at the endpoints are $\mathbf{w}'(0) = 2\mathbf{p}_1 - 2\mathbf{p}_0$ and $\mathbf{w}'(1) = 2\mathbf{p}_2 - 2\mathbf{p}_1$. Denote the control points of $\mathbf{x}(t)$ by $\mathbf{r}_0, \mathbf{r}_1, \mathbf{r}_2$, and $\mathbf{r}_3$. Then

$\mathbf{r}_0 = \mathbf{x}(0) = \mathbf{w}(0) = \mathbf{p}_0$

and

$\mathbf{r}_3 = \mathbf{x}(1) = \mathbf{w}(1) = \mathbf{p}_2$

From equation (9) or Exercise 3(a) (using $\mathbf{r}_i$ in place of $\mathbf{p}_i$) and Example 1, and assuming $\mathbf{x}'(0) = \mathbf{w}'(0)$,

$-3\mathbf{r}_0 + 3\mathbf{r}_1 = \mathbf{x}'(0) = \mathbf{w}'(0) = 2\mathbf{p}_1 - 2\mathbf{p}_0$

so $-\mathbf{p}_0 + \mathbf{r}_1 = -\mathbf{r}_0 + \mathbf{r}_1 = \dfrac{2\mathbf{p}_1 - 2\mathbf{p}_0}{3}$

and

$$\mathbf{r}_1 = \frac{\mathbf{p}_0 + 2\mathbf{p}_1}{3} \tag{i}$$

Similarly, using the tangent data at $t = 1$, along with equation (9) and Example 1, yields

$-3\mathbf{r}_2 + 3\mathbf{r}_3 = \mathbf{x}'(1) = \mathbf{w}'(1) = 2\mathbf{p}_2 - 2\mathbf{p}_1$,

$-\mathbf{r}_2 + \mathbf{p}_2 = \dfrac{2\mathbf{p}_2 - 2\mathbf{p}_1}{3}$, $\mathbf{r}_2 = \mathbf{p}_2 - \dfrac{2\mathbf{p}_2 - 2\mathbf{p}_1}{3}$,

and

$$\mathbf{r}_2 = \frac{2\mathbf{p}_1 + \mathbf{p}_2}{3} \tag{ii}$$

b. Write the standard formula (7) in this section, with $\mathbf{r}_i$ in place of $\mathbf{p}_i$ for $i = 0, \ldots, 3$, and then replace $\mathbf{r}_0$ and $\mathbf{r}_3$ by $\mathbf{p}_0$ and $\mathbf{p}_2$, respectively:

$$\begin{aligned}\mathbf{x}(t) = &(1 - 3t + 3t^2 - t^3)\mathbf{p}_0 \\ &+ (3t - 6t^2 + 3t^3)\mathbf{r}_1 \\ &+ (3t^2 - 3t^3)\mathbf{r}_2 + t^3\mathbf{p}_2\end{aligned} \tag{iii}$$

Use formulas (i) and (ii) for $\mathbf{r}_1$ and $\mathbf{r}_2$ to examine the second and third terms in formula (iii):

$\begin{aligned}(3t - 6t^2 + 3t^3)\mathbf{r}_1 &= \frac{1}{3}(3t - 6t^2 + 3t^3)\mathbf{p}_0 \\ &\quad + \frac{2}{3}(3t - 6t^2 + 3t^3)\mathbf{p}_1 \\ &= (t - 2t^2 + t^3)\mathbf{p}_0 \\ &\quad + (2t - 4t^2 + 2t^3)\mathbf{p}_1\end{aligned}$

$\begin{aligned}(3t^2 - 3t^3)\mathbf{r}_2 &= \frac{2}{3}(3t^2 - 3t^3)\mathbf{p}_1 + \frac{1}{3}(3t^2 - 3t^3)\mathbf{p}_2 \\ &= (2t^2 - 2t^3)\mathbf{p}_1 + (t^2 - t^3)\mathbf{p}_2\end{aligned}$

When these two results are substituted in formula (iii), the coefficient of $\mathbf{p}_0$ is

$(1 - 3t + 3t^2 - t^3) + (t - 2t^2 + t^3)$
$= 1 - 2t + t^2 = (1 - t)^2$

The coefficient of $\mathbf{p}_1$ is

$(2t - 4t^2 + 2t^3) + (2t^2 - 2t^3)$
$= 2t - 2t^2 = 2t(1 - t)$

The coefficient of $\mathbf{p}_2$ is $(t^2 - t^3) + t^3 = t^2$. So $\mathbf{x}(t) = (1 - t)^2\mathbf{p}_0 + 2t(1 - t)\mathbf{p}_1 + t^2\mathbf{p}_2$, which shows that $\mathbf{x}(t)$ is the quadratic Bézier curve $\mathbf{w}(t)$.

18. $\begin{bmatrix} \mathbf{p}_0 \\ -3\mathbf{p}_0 + 3\mathbf{p}_1 \\ 3\mathbf{p}_0 - 6\mathbf{p}_1 + 3\mathbf{p}_2 \\ -\mathbf{p}_0 + 3\mathbf{p}_1 - 3\mathbf{p}_2 + \mathbf{p}_3 \end{bmatrix}$

Index

Photo Credits